Singular Intersection Homology

Intersection homology is a version of homology theory that extends Poincaré duality and its applications to stratified spaces, such as singular varieties. This is the first comprehensive expository book-length introduction to intersection homology from the viewpoint of singular and piecewise linear chains. Recent breakthroughs have made this approach viable by providing intersection homology and cohomology versions of all the standard tools in the homology toolbox, making the subject readily accessible to graduate students and researchers in topology as well as researchers from other fields.

This text includes both new research material and new proofs of previously known results in intersection homology, as well as treatments of many classical topics in algebraic and manifold topology. Written in a detailed but expository style, this book is suitable as an introduction to intersection homology or as a thorough reference.

GREG FRIEDMAN is Professor of Mathematics at Texas Christian University. Professor Friedman's primary research is in geometric and algebraic topology, with particular emphases on stratified spaces and high-dimensional knot theory. He has given introductory lecture series on intersection homology at the University of Lille and the Fields Institute for Research in Mathematical Sciences. He has received grants from the National Science Foundation and the Simons Foundation.

NEW MATHEMATICAL MONOGRAPHS

All the titles listed below can be obtained from good booksellers or from Cambridge University Press. For a complete series listing visit www.cambridge.org/mathematics.

1. M. Cabanes and M. Enguehard *Representation Theory of Finite Reductive Groups*
2. J. B. Garnett and D. E. Marshall *Harmonic Measure*
3. P. Cohn *Free Ideal Rings and Localization in General Rings*
4. E. Bombieri and W. Gubler *Heights in Diophantine Geometry*
5. Y. J. Ionin and M. S. Shrikhande *Combinatorics of Symmetric Designs*
6. S. Berhanu, P. D. Cordaro and J. Hounie *An Introduction to Involutive Structures*
7. A. Shlapentokh *Hilbert's Tenth Problem*
8. G. Michler *Theory of Finite Simple Groups I*
9. A. Baker and G. Wüstholz *Logarithmic Forms and Diophantine Geometry*
10. P. Kronheimer and T. Mrowka *Monopoles and Three-Manifolds*
11. B. Bekka, P. de la Harpe and A. Valette *Kazhdan's Property (T)*
12. J. Neisendorfer *Algebraic Methods in Unstable Homotopy Theory*
13. M. Grandis *Directed Algebraic Topology*
14. G. Michler *Theory of Finite Simple Groups II*
15. R. Schertz *Complex Multiplication*
16. S. Bloch *Lectures on Algebraic Cycles (2nd Edition)*
17. B. Conrad, O. Gabber and G. Prasad *Pseudo-reductive Groups*
18. T. Downarowicz *Entropy in Dynamical Systems*
19. C. Simpson *Homotopy Theory of Higher Categories*
20. E. Fricain and J. Mashreghi *The Theory of H(b) Spaces I*
21. E. Fricain and J. Mashreghi *The Theory of H(b) Spaces II*
22. J. Goubault-Larrecq *Non-Hausdorff Topology and Domain Theory*
23. J. Śniatycki *Differential Geometry of Singular Spaces and Reduction of Symmetry*
24. E. Riehl *Categorical Homotopy Theory*
25. B. A. Munson and I. Volić *Cubical Homotopy Theory*
26. B. Conrad, O. Gabber and G. Prasad *Pseudo-reductive Groups (2nd Edition)*
27. J. Heinonen, P. Koskela, N. Shanmugalingam and J. T. Tyson *Sobolev Spaces on Metric Measure Spaces*
28. Y.-G. Oh *Symplectic Topology and Floer Homology I*
29. Y.-G. Oh *Symplectic Topology and Floer Homology II*
30. A. Bobrowski *Convergence of One-Parameter Operator Semigroups*
31. K. Costello and O. Gwilliam *Factorization Algebras in Quantum Field Theory I*
32. J.-H. Evertse and K. Győry *Discriminant Equations in Diophantine Number Theory*
33. G. Friedman *Singular Intersection Homology*
34. S. Schwede *Global Homotopy Theory*
35. M. Dickmann, N. Schwartz and M. Tressl *Spectral Spaces*
36. A. Baernstein II *Symmetrization in Analysis*
37. A. Defant, D. Garcıa, M. Maestre and P. Sevilla-Peris *Dirichlet Series and Holomorphic Functions in High Dimensions*
38. N. Th. Varopoulos *Potential Theory and Geometry on Lie Groups*
39. D. Arnal and B. Currey *Representations of Solvable Lie Groups*

Singular Intersection Homology

GREG FRIEDMAN
Texas Christian University

CAMBRIDGE
UNIVERSITY PRESS

CAMBRIDGE
UNIVERSITY PRESS

University Printing House, Cambridge CB2 8BS, United Kingdom

One Liberty Plaza, 20th Floor, New York, NY 10006, USA

477 Williamstown Road, Port Melbourne, VIC 3207, Australia

314–321, 3rd Floor, Plot 3, Splendor Forum, Jasola District Centre, New Delhi – 110025, India

79 Anson Road, #06–04/06, Singapore 079906

Cambridge University Press is part of the University of Cambridge.

It furthers the University's mission by disseminating knowledge in the pursuit of education, learning, and research at the highest international levels of excellence.

www.cambridge.org
Information on this title: www.cambridge.org/9781107150744
DOI: 10.1017/9781316584446

First published 2020

A catalogue record for this publication is available from the British Library.

ISBN 978-1-107-15074-4 Hardback

To Angie

Contents

Preface

This book arose thanks to a short course the author was asked to give in Lille in 2013 as an introduction to intersection homology theory. Originally conceived as a set of written lecture notes, the project quickly grew into the more comprehensive volume that follows. The goal has been to provide a single coherent exposition of the basic PL (piecewise linear) and singular chain intersection homology theory as it has come to exist today. Older results have been given more detailed treatments than previously existed in the literature, and several newer, though likely not unexpected, topics have been newly developed here, such as intersection homology Poincaré duality and products over Dedekind rings, including $\mathbb{Z}$.

To say a word about our primary topic, though a more extensive introduction will be provided in Chapter 1, intersection homology was first developed by Mark Goresky and Robert MacPherson in the late 1970s and early 1980s in order to generalize to spaces with singularities some of the most significant tools of manifold theory, including Poincaré duality and signatures. Although originally introduced in the language of PL chain complexes, it was soon reformulated in terms of sheaf theory, and it was in this form that it quickly found much success, particularly in applications to algebraic geometry and representation theory. Early highlights in these directions include a key role in the proof of the Kazhdan–Lusztig conjecture, a singular variety version of the Weil conjectures, and generalizations to singular complex projective varieties of the "Kähler package" for smooth complex projective varieties, including a Lefschetz hyperplane theorem, a hard Lefschetz theorem, and Hodge decomposition and signature theorems[1]. In the time since, intersection homology has exploded. As of 2017, Mathematical Reviews records 700 entries that mention intersection homology or intersection cohomology, and this jumps to over 1100 when including the closely related perverse sheaves, which developed out of intersection homology. Viewpoints have also proliferated. In addition to definitions via PL and singular chains and through sheaf theory, an analytic L^2-cohomology formulation initially due to Jeff Cheeger developed concurrently to the work of Goresky and MacPherson, and another approach via what we might call perverse differential forms is the setting for some of the

[1] The book [140] by Kirwan and Woolf provides an excellent introduction to these applications of intersection homology.

most exciting current work in the field, providing a means to explore an intersection homology version of the rational homotopy type of a singular space. Each of these perspectives has its merits, and, as is often the case in mathematics, sometimes the most powerful results come by considering the interplay among different perspectives.

The intent of this book is to introduce the reader to the PL and singular chain perspectives on intersection homology. By this choice we do not mean at all to undervalue the other approaches. Rather, by sticking to the chain-theoretic context we hope to provide an introduction that will be readily accessible to the student or researcher familiar with the basics of algebraic topology without the need for the additional prerequisites of the sheaf-theoretic or more analytic formulations. This may then motivate the reader on to further study requiring more background; to facilitate this, we provide in Chapter 10 a collection of suggested references for the reader who wishes to pursue these other vantage points and their applications, including references for several excellent introductory textbooks and expositions. We also feel that the time is ripe for such a chain-based text given recent developments that allow for a thorough treatment of intersection homology duality via cup and cap products that completely parallels the modern approach to duality on manifolds as presented, for example, in Hatcher [125]. We provide such a textbook treatment for the first time here.

Prerequisites This book is intended to be as self-contained as possible, with the main prerequisite being a course in algebraic topology, particularly homology and cohomology through Poincaré duality. Some additional background in homological algebra may be useful throughout, and some familiarity with manifold theory and characteristic classes will serve as good motivation in the later chapters. In fact, we hope that this material might make for a good reading course for second- or third-year graduate students, as much of our development parallels and reinforces that of the standard tools of homology theory, though often the proofs need some modifications. The book also includes a number of sections, including the two appendices at the end, that provide some of the less standard background results in detail, as well as some expository sections regarding further directions and applications that there was not space to pursue here. When it is necessary to use facts from further afield, such as some occasional elementary sheaf theory or more advanced algebraic or geometric topology, we have attempted to provide copious references, with a preference for textbooks when at all possible. Our favored sources include topology texts by Hatcher [125], Munkres [181, 182], Dold [71], Spanier [220], Bredon [38], and Davis and Kirk [67]; books on PL topology by Hudson [130] and Rourke and Sanderson [198]; algebra books by Lang [147], Lam [146], and Bourbaki [30]; homological algebra books by Hilton and Stammbach [126], Weibel [238], and Rotman [197]; and introductions to sheaf theory by Bredon [37] and Swan [230].

Acknowledgments This work would likely not have been conceived without the kind invitation from David Chataur, Martin Saralegi, and Daniel Tanré to visit and lecture at Université Lille 1 and Université Artois. I thank those universities for their support, and I thank David,

Martin, and Daniel for the wonderful opportunity to visit and talk intersection homology with them and their students. I would also like to thank my colleagues at TCU who suffered through endless questions about background material and occasional lectures as I sorted things out. In particular, thanks to Scott Nollet, Loren Spice, Efton Park, Ken Richardson, and Igor Prokhorenkov. The book further benefited from conversations with Markus Banagl, Laurențiu Maxim, Jörg Schürmann, and Jonathan Woolf. My perpetual thanks go to my Ph.D. advisor, Sylvain Cappell, for first suggesting that intersection homology would be something I would find interesting to think about and for his continued support throughout my career. Most of all, I would like to thank my collaborator Jim McClure, without whom much of the work on intersection homology I have participated in over the past several years would never have occurred. In particular, the intersection (co)homology cup and cap products presented in this book owe their existence to Jim's deep insights and instincts. More specific thanks also to Jim for reading over various draft sections of the manuscript, for helping with a number of technical issues, and for suggesting additional results to be included.

Many of the diagrams in this book were typeset using the TEX commutative diagrams package by Paul Taylor.

Support During the writing of this book, in addition to primary support from my home institution, Texas Christian University, I received support from a grant from the Simons Foundation (#209127 to Greg Friedman), a grant from the National Science Foundation (DMS-1308306), and funding from Université Lille 1 and Université Artois for my visits. I am grateful to all.

Notations and Conventions

We now describe some conventions, notational and otherwise, we attempt to use throughout the book, making no claim to complete consistency.

Spaces

1. Manifolds and ∂-manifolds are usually denoted M or N. A manifold is a Hausdorff space that is locally homeomorphic to Euclidean space; we do not assume manifolds must be paracompact or second-countable. A ∂-manifold[2] is a Hausdorff space that is locally homeomorphic to Euclidean space or Euclidean half-space $\{(x_1, \ldots, x_k) \in \mathbb{R}^k \mid x_1 \geq 0\}$; in other words, a ∂-manifold is what is often called a "manifold with boundary." The boundary of a ∂-manifold may be empty. "Manifold" will always mean a ∂-manifold with empty boundary. There is also an empty manifold of every dimension.
2. Arbitrary spaces have letters from the end of the alphabet such as Z, though sometimes also other letters. The space with one point is occasionally denoted pt.
3. Open subsets get letters such as U, V, W.
4. Subsets will be denoted $A \subset X$, rather than $A \subseteq X$; in other words $A \subset X$ includes the possibility that $A = X$. The interior of A is denoted $\mathring{A}$ and the closure is denoted $\bar{A}$.
5. Simplicial complexes will be given letters such as K, L. Subdivisions will generally be denoted by a prime, such as K'. We will often abuse notation and use K to represent both the simplicial complex (a space with a combinatorial structure as a union of simplices) and its underlying space as a topological space disregarding the extra structure. When we wish to emphasize the difference, for example in Appendix B, we will use $|K|$ to denote the underlying space.

[2] We mostly avoid the phrase "manifold with boundary," which sounds as though it specifies some particular class of manifolds but which is really a generalization of the concept of "manifold." Furthermore, we take the view that a "manifold with boundary" that has a non-empty boundary is not a manifold! This is because points on the boundary fail to satisfy the property that they should have Euclidean neighborhoods, which we take as part of the definition of being a manifold. The other problem is that "manifold with boundary" implies that there is a boundary and it is tempting to think then that the boundary cannot be empty. As an alternative, some authors have taken to using the notation "∂-manifold" as a replacement for "manifold with boundary." This seems to avoid these issues as well as eliminate some clunky phrasing.

6. When working with product spaces, we may write elements of $X \times Y$ as either (x, y) or $x \times y$. Product maps are usually written $f \times g$.
7. Generic maps between spaces will be denoted by letters such as f or g. The letter i, or variants such as $\mathfrak{i}$, generally denotes an inclusion. The map $\mathbf{d}$ is the diagonal map $\mathbf{d} : Z \to Z \times Z$, $\mathbf{d}(z) = (z, z)$.
8. While we will attempt to parenthesize fairly thoroughly, we will occasionally rely on a few simplifying conventions. In particular, expressions of the form $A - B$ should be understood as $(A) - (B)$. So, for example, $X \times Y - A \times B$ means $(X \times Y) - (A \times B)$ and not $X \times (Y - A) \times B$, and $X - K \cup L$ means $X - (K \cup L)$.
9. For a compact space Z, the space cZ is the open cone $cZ = [0, 1) \times Z/\sim$, where $\sim$ is the relation $(0, w) \sim (0, z)$ for all $w, z \in Z$. We typically denote the vertex of a cone by v. Similarly, the closed cone is $\bar{c}Z = [0, 1] \times Z/\sim$. More generally, for $r > 0$, we let $c_rZ = [0, r) \times Z/\sim$ and $\bar{c}_rZ = [0, r] \times Z/\sim$; in particular, $cZ = c_1Z$. Then $c_rZ \subset \bar{c}_rZ \subset c_sZ \subset \bar{c}_sZ$ whenever $r < s$.
10. For a compact space Z, the (unreduced) suspension is $SZ = [-1, 1] \times X/\sim$, where the relation $\sim$ is such that $(-1, w) \sim (-1, z)$ and $(1, w) \sim (1, z)$ for any $w, z \in Z$. So $SZ = \bar{c}Z \cup_Z \bar{c}Z$.
11. When taking the product of a space with a Euclidean space, interval, or sphere, we usually put the Euclidean space, interval, or sphere on the left, e.g. $\mathbb{R} \times Z$ instead of $Z \times \mathbb{R}$. This has some ramifications for signs. For example, if ξ is a singular cycle in Z and $\bar{c}\xi$ denotes the singular cone on ξ in $\bar{c}Z$ (see Example 3.4.7), this is the convention that is consistent with adding the cone vertex as the first vertex and so gives us $\partial(\bar{c}\xi) = \xi$.
12. We use $\amalg$ to denote disjoint union.
13. Filtered spaces (our main object of study) are generally denoted by capital letters near the end of the alphabet, in particular X (or Y when we talk about multiple filtered spaces at the same time); the filtrations are usually left implicit in the sense that we say "the filtered space X." When we need to refer to the filtration explicitly, we let X^i denote the ith *skeleton* of the filtration, and we let $X_i = X^i - X^{i-1}$; see Section 2.2. The connected components of each $X^i - X^{i-1}$ are called *strata*. The *formal dimension* of a filtered space is generically denoted n (or m for a second filtered space). When we wish to emphasize the formal dimension of X, we write $X = X^n$. The *codimension* of X^i in X^n is $\text{codim}(X^i) = n - i$. If S is a stratum in $X^i - X^{i-1}$, then $\text{codim}(S) = \text{codim}(X^i)$. Subspaces of filtered spaces, which inherit filtrations by intersection with the X^i, have letters like A or B, so we tend to have filtered pairs (X, A) or (Y, B).
14. If we wish to consider the underlying topological space of a filtered space X, i.e. we wish to explicitly disregard the filtration, we may write $|X|$.
15. The *singular locus* of a filtered space $X = X^n$ is defined to be X^{n-1} and can also be written Σ_X, or simply Σ if the space is clear. Strata contained in the singular locus are called *singular strata*.
16. Generic *strata* (see Section 2.2) of a filtered space have letters such as S and T. *Regular strata* are sometimes denoted R.

17. The *links* occurring in locally cone-like spaces (see Section 2.3), in particular CS sets or stratified pseudomanifolds, are denoted L or, occasionally, ℓ. We let $\mathrm{Lk}(x)$ denote the *polyhedral link* of a point in a piecewise linear space, i.e. if x is contained in the piecewise linear space X, then $\mathrm{Lk}(x)$ is the unique PL space such that x has a neighborhood piecewise linearly homeomorphic to $c\mathrm{Lk}(x)$; see [198, Section 1.1].
18. If X is a piecewise linear space, we let $\mathfrak{X}$ denote the filtered space with the underlying space of X but with its intrinsic PL filtration; see Section 2.10. Similarly, if X is a CS set, $\mathfrak{X}$ will denote the underlying space of X with its intrinsic filtration as a CS set.

Algebra

1. G will always be an abelian group, R a commutative ring with unity. In some contexts, R will be assumed to be a Dedekind domain, though this will be established at the relevant time.
2. Subgroups (or submodules) will be denoted $H \subset G$, rather than $H \subseteq G$; in other words $H \subset G$ includes the possibility that $H = G$.
3. We use the standard notations for standard algebraic objects: $\mathbb{Z}$ for integers, $\mathbb{Q}$ for rational numbers, $\mathbb{R}$ for real numbers (which also notates the *space* of real numbers, i.e. one-dimensional Euclidean space).
4. When working with R-modules in the context of a fixed ring R, we write $\mathrm{Hom}(A, B)$ and $A \otimes B$ rather than $\mathrm{Hom}_R(A, B)$ and $A \otimes_R B$.
5. Dedekind domains have cohomological dimension ≤ 1 (this follows from [197, Proposition 8.1] using that Dedekind domains are hereditary by definition [197, page 161]). Therefore, if R is a Dedekind domain, $\mathrm{Ext}^n_R(A, B) = 0$ for $n > 1$ and for any R-modules A, B. Hence, we simply write $\mathrm{Ext}(A, B)$ instead of $\mathrm{Ext}^1_R(A, B)$. Similarly, $\mathrm{Tor}^n_R(A, B) = 0$ for $n > 1$ and for any R-modules A, B; rather than $\mathrm{Tor}^1_R(A, B)$ we write $A * B$.
6. Generic purely algebraic chain complexes are denoted C_*, D_*, etc. Cohomologically graded complexes can be denoted C^*, D^*, etc.
7. For almost[3] all chain complexes, the boundary maps are all denoted ∂. For cohomologically graded complexes, we use d for the coboundary maps. If we wish to emphasize that ∂ is the boundary map of the chain complex C_*, we can write ∂_{C_*}, and analogously for coboundary maps of cochain complexes.
8. Elements of geometric chain complexes are typically denoted by lower-case Greek letters such as ξ, ζ, η, though we sometimes also use x, y, z. Note that we generally abuse notation by using the same symbol to refer to both a homology class and a chain representing it. For example, $\xi \in H_i(C_*)$ means that ξ is a homology class that we also think of as being represented by a cycle in C_i that we also denote ξ. In most contexts, this should not cause much confusion, though in those instances where confusion might reasonably occur, we use ξ just to denote the chain and $[\xi]$ to specify the homology class. We will indicate this notation specifically when it occurs. More generally, $[\cdot]$ indicates some

[3] We will see an exception in Section 6.2 for $I^{\bar{p}}S_*(X)$.

sort of equivalence class, so, depending on context, $[\xi]$ might reference a singular chain $\xi \in S_*(X)$ representing an element $[\xi] \in S_*(X,A)$ or an element $[\xi] \in H_*(X)$ or $[\xi] \in H_*(X,A)$. Similarly, if ξ is a simplicial chain, $[\xi]$ might denote the class in the PL chain complex $\mathfrak{C}_*(X)$ represented by ξ. See "Algebraic topology" notation item 5 below.

9. Elements of cochain complexes are denoted by lower-case Greek letters such as α, β, γ. Again, we typically abuse notation by using the same symbol to refer to both a cohomology class and a cochain representing it. For example, $\alpha \in H^i(C^*)$ means that α is a cohomology class that we also think of as being represented by a cocycle in C^i that we also denote α. In most contexts, this should not cause much confusion, though in those instances where confusion might reasonably occur, we use α just to denote the cochain and $[\alpha]$ to specify the cohomology class. We will indicate this notation specifically when it occurs. More generally, $[\cdot]$ indicates an equivalence class.
10. The connecting morphisms in long exact homology sequences are denoted ∂_*. The connecting morphisms in long exact cohomology sequences are denoted d^*.
11. Augmentation maps of chain complexes are denoted **a**; for example, we might have $\mathbf{a} : S_*(X) \to \mathbb{Z}$.
12. If x is an element of a chain or cochain complex, then we use $|x|$ to indicate the degree x. For example, if $x \in C_i$ or $X \in C^i$, then[4] $|x| = i$.

Algebraic topology

1. Δ^i denotes the standard geometric i-dimensional simplex. For definiteness, we can suppose that Δ^i is embedded in $\mathbb{R}^i$ with vertices

$$(0,\ldots,0),(1,0,\ldots,0),\ldots,(0,\ldots,0,1).$$

By an "open simplex" or an "open face," we mean the interior of a simplex, e.g. the complement in Δ^i of the union of its faces of dimension $< i$.
2. Lower-case Greek letters such as σ, τ, and often others can denote either simplices in a simplicial complex or singular simplices, depending on context. The symbol $\mathring{\sigma}$ denotes an open simplex.
3. Lower-case Greek letters such as ξ and η will typically be used for chains and α and β will typically be used for cochains.
4. If ξ is a chain, then we use $|\xi|$ to indicate its support. If ξ is a simplicial chain, this is the union of the simplices appearing in ξ, while if ξ is singular it is the union of the images of the singular simplices of ξ. If σ is an oriented simplex in a simplicial complex, then we will typically write σ instead of $|\sigma|$ unless we really need to emphasize the notion of σ as a space. Note that $|\xi|$ might also indicate the degree of ξ, depending on context.

[4] Technically, this is not quite the right thing to do as the standard equivalence between homological and cohomological gradings tells us that the notation C^i should be equivalent to the notation C_{-i}. However, matters of degree will arise only when working with signs, and so $|x|$ will really only have significance mod 2. Therefore, we will live with this inconsistency.

5. Simplicial chain complexes are denoted $C_*(X)$, singular chain complexes are denoted $S_*(X)$, PL chain complexes are denoted $\mathfrak{C}_*(X)$. When there are subspaces or coefficients involved, the notations look like $C_*(X, A; G)$ for a subspace A and a coefficient group G. We use the same notation $H_*(X)$ for both homology groups $H_*(C_*(X))$ or $H_*(S_*(X))$, letting context determine which is meant. Since simplicial and PL chains often occur in the same context, we use $\mathfrak{H}_*(X)$ for $H_*(\mathfrak{C}_*(X))$.
6. If $f : X \to Y$ is a map of spaces, we abuse notation by letting f also denote both the induced chain maps of chain complexes defined on the spaces and the induced maps on homology, e.g. we write $f : S_*(X) \to S_*(Y)$ and $f : H_*(X) \to H_*(Y)$. The dualized maps of cochain complexes and cohomology groups are denoted f^*, e.g. $f^* : S^*(Y) \to S^*(X)$ and $f^* : H^*(Y) \to H^*(X)$. Similarly, if $f : C_* \to D_*$ is a purely algebraic map of chain complexes of R-modules, we also write $f : H_*(C_*) \to H_*(D_*)$ for the induced homology map and $f^* : H^*(\text{Hom}(D_*, R)) \to H^*(\text{Hom}(C_*, R))$ for the induced cohomology map.
7. For Mayer–Vietoris sequences, the map $H_*(U) \oplus H_*(V) \to H_*(U \cup V)$ will take (ξ, η) to $\xi + \eta$. Therefore, the map $H_*(U \cap V) \to H_*(U) \oplus H_*(V)$ will take ξ to $(\xi, -\xi)$.
8. The cross product chain map $S_*(X) \otimes S_*(Y) \to S_*(X \times Y)$ (and its variants) can be written as either ε or $\times$. For example, we tend to write $\varepsilon : S_*(X) \otimes S_*(Y) \to S_*(X \times Y)$, but given two specific chains x, y, we may write $x \times y$. Unfortunately, it is common in algebraic topology to use the symbol $\times$ for both chain cross products and cochain cross products. We perpetuate this ambiguity, though context should make clear which is meant.
9. We use $\smile$ for cup products and $\frown$ for cap products. This distinguishes them from $\cup$ and $\cap$ for unions and intersections.
10. Fundamental classes are denoted Γ, with a decoration such as Γ_X if it is necessary to keep track of the space X.
11. The Poincaré duality map, consisting of a signed cap product with a fundamental class, is denoted $\mathcal{D}$.
12. We use $1 \in S^0(X)$ to denote the cocycle that evaluates to 1 on every 0-simplex. This is sometimes called the augmentation cocycle.

Intersection homology and cohomology

1. Perversities (see Section 3.1) are denoted $\bar{p}, \bar{q}, \bar{r}$, etc. In general, perversities will always have bars, with the exception[5] of the special perversities Q that occur in the discussion of the Künneth theorem; see Section 6.4.
2. $\bar{0}$ denotes the perversity that always evaluates to 0. We write $\bar{t}$ for the top perversity: $\bar{t}(S) = \text{codim}(S) - 2$. By $\bar{m}$ and $\bar{n}$ we denote respectively the lower middle perversity

[5] This special case is partly historical, because there is little risk of confusion since Q is not used for anything else, and partly idiosyncratic. Probably we should use $\bar{Q}$.

and upper middle perversity, i.e.

$$\bar{m}(S) = \left\lfloor \frac{\text{codim}(S) - 2}{2} \right\rfloor \qquad \text{(round down)},$$

$$\bar{n}(S) = \left\lceil \frac{\text{codim}(S) - 2}{2} \right\rceil \qquad \text{(round up)}.$$

3. For a perversity $\bar{p}$, we let $D\bar{p}$ be the *dual* or *complementary* perversity with $D\bar{p}(S) = \bar{t}(S) - \bar{p}(S)$ for all singular strata S; see Definition 3.1.7.
4. Throughout the first part of the book, simplicial, PL, and singular perversity $\bar{p}$ intersection chain complexes are written $I^{\bar{p}}C_*^{\text{GM}}(X)$, $I^{\bar{p}}\mathfrak{C}_*^{\text{GM}}(X)$, $I^{\bar{p}}S_*^{\text{GM}}(X)$, with corresponding homology groups $I^{\bar{p}}H_*^{\text{GM}}(X)$, $I^{\bar{p}}\mathfrak{H}_*^{\text{GM}}(X)$, $I^{\bar{p}}H_*^{\text{GM}}(X)$. The GM here stands for "Goresky–MacPherson." In Chapter 6, we introduce the variant "non-GM" intersection homology and the notation becomes simply $I^{\bar{p}}C_*(X)$, $I^{\bar{p}}\mathfrak{C}_*(X)$, and $I^{\bar{p}}S_*(X)$ with corresponding homology groups $I^{\bar{p}}H_*(X)$, $I^{\bar{p}}\mathfrak{H}_*(X)$, and $I^{\bar{p}}H_*(X)$.
5. When we introduce non-GM intersection homology, the definition will use a modified boundary map that we write as $\hat{\partial}$; see Section 6.2.1.
6. For intersection cohomology, we raise the index and lower the perversity marking, e.g. $I_{\bar{p}}S^*(X)$ and $I_{\bar{p}}H^*(X)$. Lowering the perversity symbol has no intrinsic meaning; it is meant as a further distinguishing aid between homology and cohomology.
7. We write the intersection product, which appears primarily in Section 8.5, with the symbol $\pitchfork$. Note that this differs from the use of this symbol in the early intersection homology literature, such as [105], where $A \pitchfork B$ typically means A and B are in (stratified) general position. In [105], the intersection product is written with $\cap$, but for us this risks confusion with the cap product. In other sources the intersection product of chains is sometimes written $\xi \bullet \eta$ or $\xi \cdot \eta$. We prefer to utilize $\pitchfork$ as the intersection pairing and to state transversality in words.

Miscellaneous conventions

Signs

1. We utilize throughout the Koszul sign conventions, so that interchange of elements of degrees i and j usually results in a sign $(-1)^{ij}$. See the Appendix A.1 for details.
2. The standard exception to the Koszul rule, necessary for evaluation to be a chain map, is that the sign occurring in the coboundary map of the chain complex $E^* = \text{Hom}^*(C_*, D_*)$ has the form

$$(d_E^* f)(c) = \partial_{D_*}(f(c)) - (-1)^{|f|} f(\partial_{C_*}(c))$$

for $c \in C_*$ and $f \in \text{Hom}^*(C_*, D_*)$. In particular, if $\alpha \in \text{Hom}^i(C_*, R) = \text{Hom}(C_i, R)$, then $df = (-1)^{i+1} f\partial$.
3. The connecting morphisms of long exact homology sequences have degree -1 and so can generate signs upon interchanges.

Identity map

The expression id is used for the identity function. It can be either a topological or algebraic identity. Context will usually make clear which identity function is meant, though we can make it precise with subscripts such as $\mathrm{id}_X : X \to X$ or $\mathrm{id}_{C_*} : C_* \to C_*$.

Parentheses

1. When a function f acts on an element x of a set, group, etc., we generally write $f(x)$. The standard exception will be boundary maps ∂ acting on a chain ξ, which we will usually write as $\partial\xi$.
2. To avoid the ambiguity inherent in writing expressions such as $\partial\xi \otimes \eta$, we will write either $\partial(\xi \otimes \eta)$ or $(\partial\xi) \otimes \eta$, as appropriate. We also use $\xi \otimes \partial\eta$, as there is no ambiguity here.
3. When parentheses are omitted, expressions compile from the right. For example, if $f : X \to Y$ and $g : Y \to Z$, then, as usual, $gf(x)$ means $g(f(x))$. As a more complex example, $\Phi(\mathrm{id} \otimes \beta)\partial(\xi \otimes \eta)$ means $\Phi((\mathrm{id} \otimes \beta)(\partial(\xi \otimes \eta)))$.
4. We will use an obnoxious number of parentheses to describe spaces as clearly as possible. As noted in "Spaces" notation, item 8 above, one place where we will sometimes avoid this is when considering complements.

1
Introduction

Let us begin with some motivation, followed by some general remarks about the structure of this book and what can be found (and not found!) in it.

1.1 What Is Intersection Homology?

Perhaps the most significant result about the topology of manifolds is the Poincaré Duality Theorem: If M is a closed connected oriented n-dimensional manifold and $\Gamma \in H_n(M) \cong \mathbb{Z}$ is a generator, then the cap product $\frown \Gamma : H^i(M) \to H_{n-i}(M)$ is an isomorphism for all i. There are more general versions with more bells and whistles, but, in any form, Poincaré duality, and related invariants such as signatures and L-classes, is a fundamental tool in the study and classification of manifolds.

Unfortunately, Poincaré duality fails in general for spaces that are not manifolds. In fact, it is enough for a space to have just one point that is not locally Euclidean. For example, let $S^n \vee S^n$ be the one-point union of two n-dimensional spheres, $n > 0$. Then $H_0(S^n \vee S^n) \cong \mathbb{Z}$ but $H^n(S^n \vee S^n) \cong \mathbb{Z} \oplus \mathbb{Z}$. Or, as a slightly more substantive example, one where we cannot simply pull the two pieces apart, consider the suspended torus ST^2 (Figure 1.1). This three-dimensional space has two "singular points," each of which has a neighborhood homeomorphic to the cone on the torus cT^2, and the cone point of cT^2 does not have a neighborhood homeomorphic to $\mathbb{R}^3$. Perhaps the easiest way to show this also illustrates the power of algebraic topology: If we let v be the cone point of cT^2, then, as cones are contractible, the long exact sequence of the pair and homotopy invariance of homology give us

$$H_2(cT^2, cT^2 - \{v\}) \cong H_1(cT^2 - \{v\}) \cong H_1(T^2) \cong \mathbb{Z} \oplus \mathbb{Z}.$$

But if v has a neighborhood homeomorphic to $\mathbb{R}^3$, then by excision we would have

$$H_2(cT^2, cT^2 - \{v\}) \cong H_2(\mathbb{R}^3, \mathbb{R}^3 - \{v\}) \cong H_1(\mathbb{R}^3 - \{v\}) \cong H_1(S^2) \cong \mathbb{Z}.$$

So ST^2 is not a manifold, and routine computations show that

$$\begin{aligned} H_3(ST^2) &= \mathbb{Z}, & H^3(ST^2) &= \mathbb{Z}, \\ H_2(ST^2) &= \mathbb{Z} \oplus \mathbb{Z}, & H^2(ST^2) &= \mathbb{Z} \oplus \mathbb{Z}, \\ H_1(ST^2) &= 0, & H^1(ST^2) &= 0, \\ H_0(ST^2) &= \mathbb{Z}, & H^0(ST^2) &= \mathbb{Z}. \end{aligned}$$

So, for example, $H_2(ST^2) \not\cong H^1(ST^2)$. Poincaré duality fails.

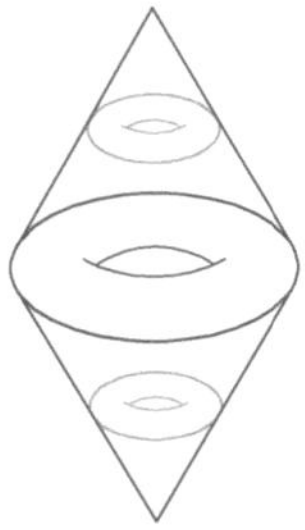

Figure 1.1 The suspended torus ST^2.

But spaces with *singularities*, points that do not have Euclidean neighborhoods, are both important and not always all that pathological. Many of them, such as our suspension example, possess dense open subsets that are manifolds. For example, if we remove the two suspension points from ST^2 we have $(0,1) \times T^2$, a manifold. Much more significant classes of examples come by considering algebraic varieties and orbit spaces of manifolds and varieties by group actions. In general such spaces may have singularities, and they will not necessarily just be isolated points. But with some reasonable assumptions (for example, assuming the group actions are nice enough or that the varieties are complex irreducible – see Section 2.8), such spaces will contain dense open manifold subsets, and, in fact, they will be filtered by closed subsets

$$X = X^n \supset X^{n-1} \supset \cdots \supset X^0 \supset X^{-1} = \emptyset$$

in such a way that each $X^k - X^{k-1}$ will be a manifold or empty. Such filtrations of spaces may be in some way intrinsic to the space (Figure 1.2), or they may be imposed by some other consideration such as the desire to study a manifold together with embedded subspaces (Figure 1.3).

The connected components of the $X^k - X^{k-1}$ are called *strata*. When $k < n$, we say they are *singular strata*, even though, depending on the choice of stratification, they may contain points with Euclidean neighborhoods. The subspace X^{n-1}, which is the union of the singular strata, is also called the *singular locus* or *singular set* and denoted Σ. The components of $X^n - X^{n-1} = X - \Sigma$ are called *regular strata*. It is usually too much to ask

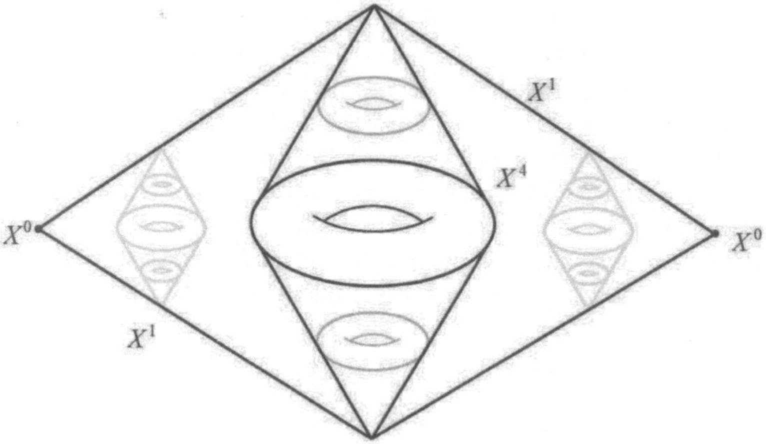

Figure 1.2 The twice suspended torus $X = S(ST^2)$. This space has a natural filtration in which X^0 comprises the suspension points of the second suspension, $X^1 = X^2 = X^3$ is the suspension of the suspension points of the first suspension, and $X^4 = X$. Note that X^0 is a 0-manifold, $X^1 - X^0$ is two open intervals, $X^2 - X^1 = X^3 - X^2 = \emptyset$, and $X^4 - X^3 \cong (-1, 1) \times (-1, 1) \times T^2$.

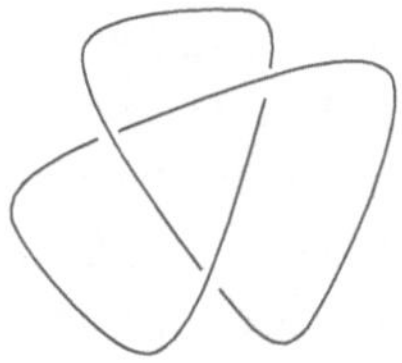

Figure 1.3 A manifold embedded in the ambient manifold S^3 (not shown).

for something like a tubular neighborhood around a singular stratum, i.e. a neighborhood homeomorphic to a fiber bundle, but, perhaps again with some additional conditions, the "normal behavior" along singular strata will be locally uniform. A typical condition is that a point $x \in X^k - X^{k-1}$ should have a neighborhood U of the form $U \cong \mathbb{R}^k \times cL$, where L is a compact filtered space and such that the homeomorphism takes $\mathbb{R}^k \times \{v\}$, again letting v be the cone point, to a neighborhood of x in $X^k - X^{k-1}$. For the remainder of this introductory discussion, we will limit ourselves to discussing the class of stratified spaces called *(stratified) pseudomanifolds*, defined formally in Section 2.4, which possess all of these nice local properties and which is a broad enough class to encompass all irreducible complex analytic varieties and all connected orbit spaces of smooth actions of compact Lie groups on manifolds. For simplicity of discussion, we also assume through this introduction that all spaces are compact, connected, and oriented.

Given all the manifold structure present and the other good behaviors of such spaces, it is reasonable to ask whether there might be some way to recover some version of Poincaré duality after all. This is precisely what Mark Goresky and Robert MacPherson did in [105] by introducing *intersection homology*. Intersection homology is defined by modifying the definition of the homology groups $H_*(X)$ so that only chains satisfying certain extra geometric conditions are allowed. These geometric conditions are governed by a *perversity parameter* $\bar{p}$, which assigns an integer to each singular stratum of the space. The result is the perversity $\bar{p}$ intersection chains $I^{\bar{p}}C_*(X)$ and their homology groups $I^{\bar{p}}H_*(X)$. Furthermore, to each perversity $\bar{p}$ there is a complementary *dual perversity* $D\bar{p}$, and Goresky and MacPherson showed that, given certain assumptions on X and $\bar{p}$, there are *intersection pairings*

$$I^{\bar{p}}H_i(X) \otimes I^{D\bar{p}}H_{n-i}(X) \to \mathbb{Z}$$

that become nonsingular over the rationals, i.e. after tensoring everything with $\mathbb{Q}$.

Let us provide a rough sketch of the basic idea of how and why this all works. We will be very loose about the specific details here, but more about this material and the original construction of the Goresky–MacPherson intersection pairing can be found in Section 8.5 and, of course, in [105].

To get at the idea, we must first ask what it is that makes manifolds so special. One consequence of their locally Euclidean nature is that it is possible to take advantage of general position: If M^n is a smooth manifold and P^p and Q^q are two smooth submanifolds, then it is possible to perturb one of P or Q so that the intersection $P \cap Q$ will be a manifold of dimension $p + q - n$. In particular, we can find a Euclidean neighborhood U_x of any point $x \in P \cap Q$ so that the triple $(U, P \cap U, Q \cap U)$ is homeomorphic to the triple $(\mathbb{R}^n, \mathbb{R}^p \times \{0\}, \{0\} \times \mathbb{R}^q)$, with the intersection of the two subspaces having dimension $p+q-n$ and providing a Euclidean neighborhood of x in $P \cap Q$; see e.g. [38, Section II.15]. Furthermore, if M, P, and Q are all oriented, it is possible to orient $P \cap Q$ by a construction involving bases for these local vector spaces [38, Section VI.11.12][1]. These ideas can be extended so that, if ξ and η are two chains in M (simplicial, piecewise linear, or singular) that satisfy an appropriate notion of general position, then there is defined an intersection $\xi \pitchfork \eta$ of degree $\deg(\xi) + \deg(\eta) - n$. This notion yields a partially-defined product on chains $\pitchfork\colon C_i(M) \otimes C_j(M) \to C_{i+j-n}(M)$. It is not fully defined because we cannot meaningfully intersect chains that are not in general position, just as the intersection of two submanifolds not in general position will not generally be a manifold. However, this intersection pairing is well defined as a map $\pitchfork\colon H_i(M) \otimes H_j(M) \to H_{i+j-n}(M)$ because any two cycles can be pushed into general position without changing their homology classes, and the homology class of the resulting intersection does not depend on the choices. Of particular note are the products $\pitchfork\colon H_i(M) \otimes H_{n-i}(M) \to H_0(M)$ because composing with the augmentation map $\mathbf{a}$ then yields a bilinear pairing $\pitchfork\colon H_i(M) \otimes H_{n-i}(M) \to H_0(M) \xrightarrow{\mathbf{a}} \mathbb{Z}$. As any homomorphism to $\mathbb{Z}$ must take any element of finite order to 0, this intersection pairing

[1] Technically, what we have described here is *transversality*, while *general position* is simply the requirement in an n-manifold that a p-manifold and a q-manifold meet in a subspace of dimension $\leq p + q - n$.

descends to a map $H_i(M)/T_i(M) \otimes H_{n-i}(M)/T_{n-i}(M) \to \mathbb{Z}$, where we let $T_*(M)$ denote the torsion subgroup of $H_*(M)$.

What does this have to do with Poincaré duality? If M is a closed oriented n-manifold, then Poincaré duality and the Universal Coefficient Theorem together yield isomorphisms

$$H_i(M) \cong H^{n-i}(M) \cong \operatorname{Hom}(H_{n-i}(M),\mathbb{Z}) \oplus \operatorname{Ext}(H_{n-i-1}(M),\mathbb{Z}).$$

Some elementary homological algebra then allows us to derive from this an isomorphism

$$H_i(M)/T_i(M) \cong \operatorname{Hom}(H_{n-i}(M)/T_{n-i}(M),\mathbb{Z}).$$

Some slightly more elaborate homological algebra also leads to an isomorphism

$$T_i(M) \cong \operatorname{Hom}(T_{n-i-1}(M),\mathbb{Q}/\mathbb{Z}).$$

Applying the adjunction relation, these two isomorphisms can be interpreted as nonsingular bilinear pairings

$$H_i(M)/T_i(M) \otimes H_{n-i}(M)/T_{n-i}(M) \to \mathbb{Z},$$
$$T_i(M) \otimes T_{n-i-1}(M) \to \mathbb{Q}/\mathbb{Z}.$$

The first of these turns out to be precisely the intersection pairing! And the second is the closely related torsion linking pairing. If $\xi \in C_i(M)$ is a cycle with $k\xi = \partial\zeta$ for some $k \in \mathbb{Z}$, $k \neq 0$, then the linking pairing of $\xi \in T_i(M)$ with $\eta \in T_{n-i-1}(M)$ can be computed as $1/k$ times the intersection number of ζ with η, assuming the chains are all in general position. This number is well defined in $\mathbb{Q}/\mathbb{Z}$.

Prior to the invention of the modern version of cohomology, Poincaré duality was formulated in these terms. These days, most readers will be more familiar with the nonsingular cup product pairing $H^i(M)/T^i(M) \otimes H^{n-i}(M)/T^{n-i}(M) \xrightarrow{\smile} \mathbb{Z}$, which turns out to be isomorphic to the intersection pairing via the Poincaré duality isomorphisms. In general, cup products are simpler to define than intersection products; they are defined at the cochain level $C^i(M) \otimes C^j(M) \xrightarrow{\smile} C^{i+j}(M)$, and, perhaps most importantly, the cup product can be defined on any space, though in general we do not obtain a nonsingular pairing. The only downside to the cup product is that it obfuscates this beautiful geometric interpretation of Poincaré duality, an interpretation that will allow us to see clearly what goes wrong for spaces that are not manifolds.

So, let us return to spaces with singularities. As a simple example, consider $X = M_1 \vee M_2$, the wedge of two n-manifolds, $n > 2$. In a manifold of dimension $n > 2$, any two curves can be perturbed to be disjoint as $1 + 1 - n < 0$. But in $X = M_1 \vee M_2$, any two curves that pass through the wedge point v cannot be separated (unless one only intersects $\{v\}$ at an endpoint). Furthermore, even if $n = 2$ and ξ and η are two 1-chains that have an isolated intersection at v, the lack of a local Euclidean neighborhood makes it unclear how to orient the intersection point, which is a necessary step in defining an intersection product (Figure 1.4). So we see that singularities are not compatible with having well-defined intersection products.

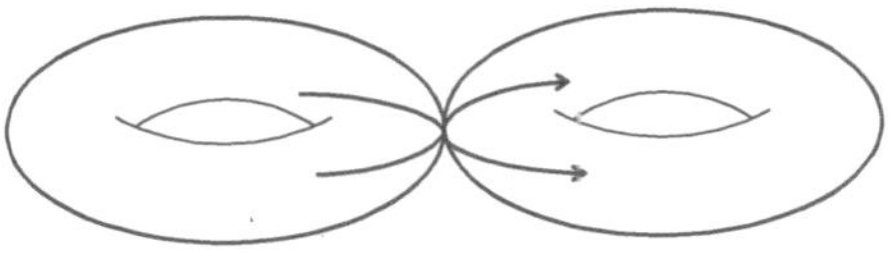

Figure 1.4 A failure of general position. What's the intersection number of the two curves depicted?

Or are they? The fundamental insight of Goresky and MacPherson was that if chains don't intersect well at singularities, perhaps they shouldn't be allowed to interact with the singularities too much. In fact, roughly stated, the allowability condition that a chain ξ must satisfy to be a perversity $\bar{p}$ intersection chain says that, if S is any singular stratum of an n-dimensional space X and if S has dimension k and ξ is an i-chain with support $|\xi|$, then

$$\dim(|\xi| \cap S) \leq i + k - n + \bar{p}(S), \tag{1.1}$$

and a similar condition must hold for $\partial\xi$. There is a way to make this precise with singular chains, but for now the reader will be safe imagining simplicial chains to make better sense of these dimension requirements. Without the $\bar{p}(S)$ summand, inequality (1.1) would be precisely the requirement that $|\xi|$ and S be in general position if X were a manifold. The $\bar{p}(S)$ term allows for some deviation from the strict general position formula; hence *perversity*. The complex of chains satisfying these conditions is the *perversity $\bar{p}$ intersection chain complex $I^{\bar{p}}C_*(X)$*, and the resulting homology groups $I^{\bar{p}}H_*(X)$ are the *perversity $\bar{p}$ intersection homology groups*.

Now suppose ξ is a $\bar{p}$-allowable i-chain, i.e. $\xi \in I^{\bar{p}}C_i(X)$, and that η is a $\bar{q}$-allowable j-chain. We will also suppose that there is a perversity $\bar{r}$ such that for each singular stratum S we have $\bar{p}(S) + \bar{q}(S) \leq \bar{r}(S) \leq \bar{t}(S)$, where $\bar{t}$ is the *top perversity* defined by $\bar{t}(S) = \operatorname{codim}(S) - 2 = \dim(X) - \dim(S) - 2$. Lastly, we suppose that our space X is a *stratified pseudomanifold* and that ξ and η are in *stratified general position*, which means that they should satisfy the general position inequality within each singular stratum:

$$\dim(S \cap |\xi| \cap |\eta|) \leq \dim(S \cap |\xi|) + \dim(S \cap |\eta|) - \dim(S).$$

With these assumptions, it is possible to define an intersection $\xi \pitchfork \eta$ that is an $\bar{r}$-allowable $i + j - n$ chain! Furthermore, work of Clint McCrory [170, 171] shows that it is possible to push any $\bar{p}$-allowable cycle ξ and $\bar{q}$-allowable cycle η into stratified general position and in such a way that the resulting homologies between cycles also satisfy the respective allowability conditions. We therefore arrive at a map

$$\pitchfork : I^{\bar{p}}H_i(X) \otimes I^{\bar{q}}H_j(X) \to I^{\bar{r}}H_{i+j-n}(X),$$

generalizing the intersection product for manifolds. If $\bar{q}$ is the *complementary perversity* $D\bar{p}$, which is defined so that $\bar{p}(S) + D\bar{p}(S) = \bar{t}(S) = \operatorname{codim}(S) - 2$, then by composing with an augmentation map we get a pairing

$$I^{\bar{p}}H_i(X) \otimes I^{D\bar{p}}H_{n-i}(X) \xrightarrow{\pitchfork} I^{\bar{t}}H_0(X) \xrightarrow{\mathbf{a}} \mathbb{Z}.$$

The intersection homology Poincaré Duality Theorem of [105] says that this pairing becomes nonsingular when tensored with $\mathbb{Q}$. If M is a manifold (unstratified), the perversity conditions become vacuous, and this pairing reduces to the intersection pairing over $\mathbb{Z}$, which is nonsingular when tensored with $\mathbb{Q}$.

To give an idea about why this pairing works after having argued that intersection pairings are not so compatible with singularities, notice that if $\xi \pitchfork \eta$ is a $\bar{t}$-allowable 0-chain then its intersection with the singular stratum S must satisfy

$$\dim(|\xi \pitchfork \eta| \cap S) \leq 0 - \operatorname{codim}(S) + \bar{t}(S) = -2.$$

So, in other words, $|\xi \pitchfork \eta|$ must be contained in the dense manifold part of X. In fact, with a bit more work, the allowability and stratified general position conditions imply that $|\xi| \cap |\eta| \subset X - \Sigma$, the dense submanifold of X. So the bad behavior discussed previously cannot happen because the intersection of chains of complementary dimension and complementary perversity is forced to happen in the nice manifold portion of the space, not at the singularities. If ξ and η do not have complementary dimensions, it is possible that $|\xi| \cap |\eta|$ might have a nontrivial intersection with Σ, but the $\bar{r}$-allowability of $\xi \pitchfork \eta$ shows that such intersections within the singular locus are carefully controlled by the perversity data.

Here is another important motivating example that provides some idea of why intersection homology Poincaré duality might work out. Let M be a compact oriented n-dimensional manifold with boundary $\partial M \neq \emptyset$. Let $X = M/\partial M$. So we can think of X as M with its boundary collapsed to a point or, up to homeomorphism, it is M with the closed cone $\bar{c}(\partial M)$ adjoined, $X \cong M \cup_{\partial M} \bar{c}(\partial M)$. If we let v be the cone point, then v will not in general have a Euclidean neighborhood unless, for example, $\partial M \cong S^{n-1}$. So it is natural to stratify X by $\{v\} \subset X$, and any perversity on X is determined by the single value $p := \bar{p}(\{v\})$. Without working carefully through the details here, the basic idea is that if i is small compared to a value depending on p, then the allowability condition (1.1) will prevent i-chains in $I^{\bar{p}}C_i(X)$ from intersecting v. So the low-dimensional chains behave as though the cone point is not there, and we get $I^{\bar{p}}H_i(X) \cong H_i(X - \{v\}) \cong H_i(M)$. On the other hand, if i is large enough then the allowability condition will be satisfied for any i-chain, noting that $\dim(|\xi| \cap \{v\}) \leq 0$ because v is a point, and so all i-chains can be utilized. Therefore, $I^{\bar{p}}H_i(X) \cong H_i(X)$, and so $I^{\bar{p}}H_i(X) \cong H_i(M, \partial M)$ if $i > 0$. It turns out that there is only one middle dimension in which there is a transition between these behaviors, and in that dimension we get $I^{\bar{p}}H_i(X) \cong \operatorname{im}(H_i(M) \to H_i(M, \partial M))$. Altogether, the precise statement works out as follows, assuming $p < n - 1$:

$$I^{\bar{p}}H_i(X) \cong \begin{cases} H_i(M, \partial M), & i > n - p - 1, \\ \operatorname{im}(H_i(M) \to H_i(M, \partial M)), & i = n - p - 1, \\ H_i(M), & i < n - p - 1. \end{cases}$$

But now recall that the Lefschetz Duality Theorem for manifolds with boundary provides a duality isomorphism $\frown \Gamma : H^i(M) \to H_{n-i}(M, \partial M)$, and, modulo torsion, this can also be partially interpreted in terms of a nonsingular intersection pairing $H_i(M) \otimes H_{n-i}(M, \partial M) \to H_0(M) \to \mathbb{Z}$, with the geometric intersections occurring in the interior of M. Lefschetz duality also implies a nondegenerate intersection pairing among the groups $\operatorname{im}(H_i(M) \to H_i(M, \partial M))$; see Section 8.4.5 for more details. As we vary the perversity, intersection homology of X provides all of these groups! And the duality between the perversity $\bar{p}$ and its dual $D\bar{p}$ positions the behavioral transitions in complementary dimensions: Notice that the dual $D\bar{p}$ takes the value $D\bar{p}(\{v\}) = n-2-p$, so indeed $(n-p-1)+(n-(n-2-p)-1) = n$. So the intersection homology pairings generate the Lefschetz duality pairings as special cases!

One seeming deficiency in the intersection homology groups is that the intersection pairing $I^{\bar{p}}H_i(X) \otimes I^{D\bar{p}}H_{n-i}(X) \to \mathbb{Z}$ is not just between complementary dimensions but between complementary perversities. So even when $n = 2k$, we do not necessarily have a middle-dimensional pairing of a group with itself. In manifold theory, if $n = 2k$ then such self-pairings $H_k(M) \otimes H_k(M) \to \mathbb{Z}$ are symmetric for k even and antisymmetric for k odd, and such pairings possess their own algebraic invariants, such as the signature for k even, that play a key role in manifold classification. Given a version of Poincaré duality for stratified spaces, such invariants are the desired consequence. In general, however, there is no self-complementary perversity such that $\bar{p} = D\bar{p}$. However, there are two dual perversities, $\bar{m}$ and $\bar{n} = D\bar{m}$, called the lower and upper middle perversities, and these are as close as possible. If the pseudomanifold X satisfies certain local intersection homology vanishing conditions, then $I^{\bar{m}}H_*(X)$ and $I^{\bar{n}}H_*(X)$ will be isomorphic and we do get a self-pairing. Already in [105], Goresky and MacPherson observed that this is the case for spaces stratifiable by strata only of even codimension, and this includes complex varieties. Important broader classes of such spaces were introduced later, including Witt spaces by Paul Siegel [218] and IP spaces(intersection homology Poincaré spaces) by William Pardon [187]. As is the signature for manifolds, the intersection homology signature (and, in fact, a more refined invariant – the class of the intersection pairing in the Witt group) is a bordism invariant of such spaces, and this has ramifications toward the geometric representation of certain generalized homology theories, including ko-homology and $\mathbb{L}$-homology, by bordisms of stratified spaces. This fact can also be used to construct for such spaces a version of the characteristic L-classes in ordinary homology. We provide an exploration of these topics in our culminating chapter, Chapter 9.

Another seeming shortcoming of intersection homology duality is that the intersection pairing is in general only nonsingular after tensoring with $\mathbb{Q}$. Over $\mathbb{Z}$, the map $I^{\bar{p}}H_i(X) \to \operatorname{Hom}(I^{D\bar{p}}H_{n-i}(X), \mathbb{Z})$ adjoint to the intersection pairing is injective, making the pairing nondegenerate, but it is not necessarily an isomorphism and so the pairing is not necessarily nonsingular. But, in fact, this must be the most we can hope for in general, as the intersection pairing on the groups $\operatorname{im}(H_i(M) \to H_i(M, \partial M))$ for a manifold only need be nondegenerate, not necessarily nonsingular, and we have already seen that this occurs as a special case of intersection homology duality[2]. Yet there are local "torsion-free" conditions due to Goresky and Siegel [111] that can be imposed on a space to imply nonsingularity of the pairing over $\mathbb{Z}$, as well as the existence of nonsingular torsion linking pairings analogous to those for manifolds. More recent work on such spaces has developed intersection cohomology and cup and cap products, so that now intersection Poincaré duality can also be expressed as an isomorphism of the form $\frown \Gamma : I_{\bar{p}}H^i(X) \to I^{D\bar{p}}H_{n-i}(X)$. This formulation was introduced with field coefficients in [100], for which the torsion-free conditions are automatic, and is developed here in Chapters 7 and 8 over more general rings, including $\mathbb{Z}$.

1.2 Simplicial vs. PL vs. Singular

As the reader should be aware from an introductory algebraic topology course, there are several ways to define homology groups on a space, and, assuming the space is nice enough, those definitions that the space admits will yield isomorphic homology groups. Each such definition has its own advantages and disadvantages: homology via CW complexes is difficult to set up technically but then often allows for the simplest computations; simplicial homology is defined combinatorially and very amenable to computations by computer but enforces a somewhat rigid structure on spaces that can make it difficult to prove theorems or work with subspaces; singular homology is defined on arbitrary spaces and is often the best setting to prove theorems but it is usually hopeless for direct computation from the definitions. We encounter the same trade-offs in intersection homology. CW homology is not really available at all, and so we have simplicial and singular homology, each of which will be treated in this book.

There is yet another species of homology we will utilize that occupies something of a middle ground between simplicial and singular homology: piecewise linear (or PL) homology. The basic idea is that PL chains are linear combinations of geometric simplices, just like in simplicial homology, but the simplices are not required to all come from the same triangulation. Technically, a PL chain lives in the direct limit of simplicial chain complexes, with the limit being taken over all suitably compatible triangulations of the space and with the maps in the direct system being induced by geometric subdivision of triangulations. We will see in Theorem 3.3.20 that for any PL filtered space the PL intersection homology groups (and hence ordinary PL homology groups on any PL space) are isomorphic to the simplicial groups with respect to any triangulation satisfying a mild hypothesis (that the triangulation be full). With some other mild assumptions, we will show that these simplicial and PL intersection homology groups are isomorphic to the singular intersection homology groups in Theorem 5.4.2. As we do not have an acyclic carrier theorem available in intersection homology, it would be much more difficult to establish an isomorphism between simplicial and singular intersection homology without using the PL theory as an intermediary.

[2] It's not a bug, it's a feature!

One of the advantages that PL homology has over simplicial homology is that it behaves much better with respect to the consideration of open subsets. An open subset of a triangulated space needs to be given its own triangulation in order to speak of its simplicial chains; but as PL homology already considers all triangulations, a PL chain in X that is supported in an open subset U is already a PL chain in U without worrying about the specific triangulation. Consequently, we obtain excision and Mayer–Vietoris theorems for PL homology that mirror the singular homology theorems and so are more general than what one sees for simplicial homology. Another technical advantage, which we shall only touch upon lightly in Section 8.5, is that PL chains provide a good setting for defining intersection pairings, which Goresky and MacPherson used to demonstrate Poincaré duality for intersection homology when they introduced it in [105].

As we progress to the later stages of the book, however, the technical advantages of the PL approach will begin to lessen as the technical difficulties begin to escalate. For example, as the cup and cap products in intersection homology cannot be defined using an Alexander–Whitney-type formula (as far as we know), the simplicial approach does not provide any utility toward computing these products. At the same time, the direct limit that arises in the definition of PL chains dualizes to an inverse limit for PL cochains, and these can be difficult to work with. Consequently, when we reach intersection cohomology, we will discuss briefly the PL intersection cohomology groups, but we will limit our discussion of products and duality to the singular chain setting.

1.3 A Note about Sheaves and Their Scarcity

As documented by Steven Kleiman in his somewhat controversial historical survey of the early development of intersection homology [141], after first developing PL intersection homology Goresky and MacPherson soon discovered that their work dovetailed with research in algebraic geometry that Pierre Deligne was undertaking from the point of view of sheaf theory[3]. Goresky and MacPherson quickly recognized the power of the tools available working in the derived category of complexes of sheaves on a space, especially a sheaf-theoretic duality theorem called Verdier duality, and intersection homology was reformulated in these terms in [106]. Using Verdier duality, they provided a proof of intersection homology duality on topological pseudomanifolds, extending their duality results beyond the piecewise linear pseudomanifolds of [105]. Furthermore, sheaf theory provides a good axiomatic framework, which enabled them in [106] to prove that, with certain restrictions on $\bar{p}$, the groups $I^{\bar{p}}H_*(X)$ are topological invariants; in other words they do not depend on the choice of stratification.

From here, the sheaf-theoretic perspective on intersection homology largely took over, and it has been the venue of many of the most significant applications of intersection

[3] At the same time, Jeff Cheeger was developing a similar theory from the analytic point of view using L^2-cohomology [59, 61].

homology. For a good introductory survey of what has been accomplished in this vein, including many extensions of significant results concerning the homology and cohomology of nonsingular algebraic varieties to results on the intersection homology and intersection cohomology of singular algebraic varieties, we recommend [140]; for a sheaf-theoretic introduction directed more toward applications to characteristic classes, we suggest [11].

Nonetheless, the working title of this book was "An Introduction to Intersection Homology Without Sheaves," and we have several motivations for providing a text without sheaf theory. One is that there are already several excellent introductions to intersection homology and related topics primarily from that point of view, including [11, 28, 34, 70, 140][4]. Another is that, while common in algebraic geometry, sheaf theory is not always a standard item in the toolbox of the topology student, while singular homology most often is. A singular chain version of intersection homology was introduced by Henry King in [139], and it has developed along its own path, though one often intertwined with the sheaf theory. Each approach offers its own advantages. Historically, sheaf theory has provided more powerful machinery and close connections to the techniques of modern algebraic geometry, and several major results have yet to be formulated any other way. However, sheaf theory, especially in the derived category, can also be very abstract. By contrast, simplicial/PL/singular chains offer a more geometric picture and are more amenable to homotopy-theoretic arguments. And these two sides of the coin can work well in combination, for example in the author's proof of Poincaré duality on homotopically stratified spaces [86], which involves homotopy arguments to provide needed axiomatic properties for sheaves of chains. But, perhaps most importantly, it is only recently that the development of cup and cap products for intersection homology and cohomology has made it possible to provide a chain-theoretic treatment that completely mimics a standard introductory text on algebraic topology. The time seems ripe to do so in a comprehensive way, with the hopes that the reader will both find the chain-theoretic approach to intersection homology useful on its own and also then be better motivated and equipped to go on to learn the sheaf perspective from other sources. Toward this latter end, we have provided some suggestions for further reading at the end of the book in Chapter 10.

1.4 GM vs. Non-GM Intersection Homology and an Important Note about Notation

When Goresky and MacPherson first introduced intersection homology in [105], they required their perversity parameters to satisfy some very specific rules. Over time, it has become apparent that these restrictions can be loosened, and today it is possible to define and prove useful theorems about intersection homology with essentially no restrictions on

[4] Brasselet's work [34] also contains an extensive overview of simplicial intersection homology.

the perversities at all. Unfortunately, however, using certain "high" perversities requires altering the definition of intersection homology a bit in order to still get the best theorems.

More precisely, one of the Goresky–MacPherson requirements is that $\bar{p}(S) \leq \operatorname{codim}(S) - 2$ for all singular strata S. If $\bar{p}$ is a perversity that does not meet this requirement, it is still possible to define simplicial and singular intersection homology exactly as done in [105, 139], but its local properties do not quite behave as one would expect from the sheaf-theoretic formulation. This results, for example, in the failure of the intersection homology duality theorem for these perversities in this version of intersection homology[5]. While it is possible to modify the sheaf theory to reflect the chain theory for these high perversities (see Habegger and Saper [120]), it turns out to be better for our purposes to modify the chain theory to better reflect the expectations of the sheaf theory. Such modifications were developed by the author in [85] and by Martin Saralegi in [205]. With these modifications, we obtain duality theorems for all perversities. See [91] for a more detailed exposition of this discussion.

Over the course of this historical development, various notations have been used. In several papers, the author used the notation $I^{\bar{p}}H(X; G_0)$ to refer to the modified chain-theoretic version of intersection homology, where G is a coefficient group and G_0 refers to a notion of a "stratified coefficient system"; see [85]. However, this notation leaves a bit to be desired as, for a constant coefficient group, it is really the definition of the intersection chain complex that has been modified and so not the group itself that merits a decoration. Given that it is the modern, adapted version of intersection homology that admits the most general duality theorems and that best mirrors the groups obtained from sheaf theory, we have been so bold in [100] as to rechristen these groups $I^{\bar{p}}H_*(X; G)$ (when the coefficient group is G) and to call them the *intersection homology groups*. When $\bar{p}(S) \leq \operatorname{codim}(S) - 2$ for all singular strata S, these are identical to the groups $I^{\bar{p}}H_*(X; G)$ as classically defined by Goresky and MacPherson [105], King [139], or via sheaf theory [28, 106]. However, these modified groups are a bit more involved, and so we prefer in the first few chapters of the book to stick more closely to the original Goresky–MacPherson/King definitions. When doing so, we have chosen to use the notation $I^{\bar{p}}H^{\mathrm{GM}}_*(X; G)$, with the "GM" standing for "Goresky–MacPherson." It is in this context that we will develop many of the basic properties of intersection homology, and then when we introduce $I^{\bar{p}}H_*(X; G)$, we will note which results carry over directly and provide any additional needed arguments at that time. When necessary, we will refer to $I^{\bar{p}}H^{\mathrm{GM}}_*(X; G)$ and $I^{\bar{p}}H_*(X; G)$ respectively as "GM intersection homology" and "non-GM intersection homology," with an unqualified "intersection homology" meaning the latter. While this is perhaps not a perfect notational solution for compatibility with the existing literature, we do note again that so long as $\bar{p}(S) \leq \operatorname{codim}(S) - 2$ for all S, which historically has been a common assumption, we do have $I^{\bar{p}}H^{\mathrm{GM}}_*(X; G) \cong I^{\bar{p}}H_*(X; G)$. This should mitigate some of the possible confusion.

[5] See Section 6.1 for an example.

The technical details concerning non-GM intersection homology and why it is necessary can be found at the beginning of Chapter 6.

1.5 Outline

We now provide a brief summary of what can be found in each chapter, hoping to provide something of an overview of the material contained in the book. In order to avoid getting bogged down in too many technical details here, we will remain somewhat sketchy. For example, a result stated here about "stratified spaces" will often have a more specific class of space in mind. We refer the reader to the precise statements later in the text.

Chapter 2 is an introduction to stratified spaces. We begin with the quite general notion of a filtered space and move progressively through more and more constrained classes, including locally cone-like spaces, manifold stratified spaces, the CS sets of Siebenmann, recursive CS sets (our terminology), and, ultimately, topological and piecewise linear (PL) pseudomanifolds. To facilitate this last definition, we provide some background on PL topology, both at this point and in somewhat more detail in Appendix B. Within the main text, we take the point of view that a PL space is one that has been endowed with a family of compatible triangulations; in the appendix we connect this definition with the version in terms of PL coordinate charts as in Hudson [130]. In the later sections of the chapter, we turn to some more specialized topics, including normalization of pseudomanifolds, which will play a limited role in the book, pseudomanifolds with boundary, and certain other more specialized types of spaces, such as Whitney stratified spaces, Thom–Mather stratified spaces, and homotopically stratified spaces. While treating these in Section 2.8, we pause to observe through the citation of outside results that the class of pseudomanifolds includes many spaces found in nature, such as singular varieties and orbit spaces of group actions.

In Section 2.9, we discuss stratified maps between stratified spaces. And we close Chapter 2 with two sections on even further specialized topics: intrinsic filtrations, and products and joins of stratified spaces. These sections will be utilized later but can be safely skipped on a first reading. In fact, we hope that the reader will tread lightly through Chapter 2 in general, absorbing just enough of the ideas about our spaces of interest to proceed on to intersection homology itself.

Chapter 3 introduces intersection homology, beginning with a discussion of perversity parameters. The treatment of (GM) intersection chains begins with the simplicial version, followed by PL (piecewise linear) intersection chains, and then singular intersection chains. As the reader may be less familiar with the PL category than the others, we provide the relevant background. Simplicial and PL chains are closely related, but for technical reasons to be seen, we will generally focus on the latter.

Having established the basic definitions, in Chapter 4 we develop the basic properties of PL and singular intersection homology. We consider the behavior of the intersection homology groups under appropriately stratified maps and homotopies, demonstrating invariance of the groups under stratified homotopy equivalences. Later in the chapter we introduce relative intersection homology, the long exact sequence of a pair, Mayer–Vietoris sequences, and excision.

We note especially Section 4.2, in which we present the formula for the intersection homology of a cone. In addition, to providing a good basic example of an intersection homology computation, the cone formula plays an essential role throughout the theory: All points in pseudomanifolds have neighborhoods that are stratified homotopy equivalent to cones, and so these computations contain all the local homological data about such a space. Just as it is the local Euclidean nature of manifolds that ultimately leads to deep homological theorems about their topology, so too do the deep theorems about pseudomanifolds follow from these cone computations.

A powerful set of tools for assembling these "local to global" phenomena is known as the "Mayer–Vietoris arguments." In the sheaf cohomology theory, there is a general principle that local cohomology isomorphisms lead to global (hyper)cohomology isomorphisms. When working without sheaves, the Mayer–Vietoris arguments play the same role. They will be the basic device undergirding our most powerful theorems. We begin Chapter 5 by establishing this toolbox and then get to work on more advanced properties of intersection homology.

Section 5.2 contains a detailed discussion of the cross product of chains and the intersection homology Künneth Theorem for the product of a stratified space with a manifold. More background material for this section, including a detailed introduction to the Eilenberg–Zilber shuffle product, can be found in Appendix B. Explicit treatment of this product is often omitted from modern algebraic topology texts in favor of acyclic model arguments, but these are not available for working with intersection homology.

In Section 5.3, we introduce intersection homology with coefficients in groups other than $\mathbb{Z}$. This is a more delicate issue than with ordinary homology, and the Universal Coefficient Theorem does not always hold for intersection homology. It will hold, however, if we assume the vanishing of the torsion of certain local intersection homology groups, and this leads to the Goresky–Siegel "locally torsion-free" condition, which is also necessary for many of our later theorems.

In Section 5.4, we show that the PL and singular intersection homology groups are isomorphic on PL stratified spaces, and Section 5.5 contains the proof that intersection homology is stratification-independent when using certain perversities, including the original ones of Goresky and MacPherson. We close Chapter 5 with a proof that the intersection homology of compact pseudomanifolds is finitely generated.

In Chapter 6, we turn at last to the non-GM version of intersection homology, providing detailed motivation, definitions, and basic properties. The chapter culminates in a Künneth Theorem for the product of two stratified spaces. This Künneth Theorem plays a starring role in Chapter 7, which contains our treatment of intersection cohomology and its cup, cap, and cross products, including full details of the properties of these products. Much of this material has not been worked out previously, and Section 7.3.9 consists of a summary of all of the properties, their conditions, and where they can be found in the text.

Chapter 8 concerns Poincaré duality. We show that oriented pseudomanifolds possess fundamental intersection homology classes in Section 8.1, and we prove the cap product form of the Poincaré Duality Theorem in Section 8.2. This is followed by Lefschetz duality for pseudomanifolds with boundary and derivation of the nonsingular cup product and torsion pairings. Section 8.5 provides an expositional survey of intersection pairings and the original approach of Goresky and MacPherson to intersection homology duality.

We introduce Witt and IP spaces in Chapter 9; these are the spaces on which middle-dimensional self-pairings are possible, and hence signatures. Using the signature, we then provide in detail the Goresky–MacPherson construction of the characteristic L-classes for Witt spaces, which is modeled upon the classical construction for PL manifolds. To conclude, we provide a survey of bordism theories of pseudomanifolds in Section 9.5.

Chapter 10 provides an afterward with suggested further reading in a variety of directions. We also include two appendices: one concerning algebra background of various sorts and the other about piecewise linear (PL) topology.

2
Stratified Spaces

In this chapter, we introduce a succession of classes of space, beginning quite generally and then introducing more and more rigorous requirements. For each class of spaces we provide some examples and examine basic properties. As this chapter is somewhat lengthy, we hope the reader will take it as something of a reference section, to be read only lightly at first on the way toward intersection homology and then returned to as necessary later.

We start in Section 2.1 with a few easy examples to get the reader acclimated to the basic idea of stratified spaces. Then in Section 2.2 we define *filtered spaces*, which are simply spaces with a sequence of closed subspaces

$$X = X^n \supset X^{n-1} \supset \cdots \supset X^0 \supset X^{-1} = \emptyset.$$

This is followed in Section 2.2.2 by the *stratified spaces*, which impose some additional reasonable point-set conditions, essentially that the closure of any stratum (a connected component of $X^k - X^{k-1}$) is a union of strata. *Manifold stratified spaces* then add the condition that each stratum be a manifold.

Locally cone-like spaces and the *CS sets* of Siebenmann are introduced in Section 2.3. These satisfy the added condition that each point should have a neighborhood that looks like a bundle over a neighborhood of the point in its stratum with the fiber being a cone on a compact space, the *link*. The CS sets are spaces that are locally cone-like and manifold stratified.

Section 2.4 introduces *pseudomanifolds*, which are our most important spaces, possessing versions of Poincaré duality. These are CS sets with yet further structure, with each link itself being a pseudomanifold of lower dimension. Pseudomanifolds also possess dense manifold subspaces. Examples include all irreducible complex analytic varieties and the orbit spaces of smooth compact Lie group actions on manifolds.

In Section 2.5, we begin with a review of piecewise linear (PL) topology from a point of view that will be of most use to us. Further details on simplicial complexes and PL topology can be found in Appendix B. We then discuss simplicial and PL pseudomanifolds, which are pseudomanifolds with compatible PL structures.

Section 2.6 contains a brief treatment of *normalization*, a process by which a pseudomanifold may be replaced by something like a resolution whose links are connected. It is often easier to prove results about intersection homology by first proving them for normal pseudomanifolds and then passing the results to more general pseudomanifolds. However, we have managed to avoid the need for this throughout the book, and, as the details of normalization require a certain amount of point-set technology, we have forgone the details in favor of a survey approach.

Just as it is useful to study manifolds with boundary, so too do we want to consider pseudomanifolds that may have boundaries. We refer to these as ∂-pseudomanifolds and study them in Section 2.7. Throughout the text, a "pseudomanifold" will always mean a ∂-pseudomanifold with empty boundary, and we will always use "∂-pseudomanifold" if a boundary may occur.

Section 2.8 concerns some other types of more specialized stratified spaces that one may encounter. This includes the *Whitney stratified spaces* that arise in analytic/algebraic geometry and the *Thom–Mather stratified spaces* that also arise in more analytic settings. We cite some results about these spaces to verify that irreducible complex varieties and orbit spaces of smooth manifolds really are pseudomanifolds. We also mention a more general class of stratified space, the *homotopically stratified spaces*, which were introduced by Quinn in [192] to provide "a setting for the study of purely topological stratified phenomena, particularly group actions on manifolds." This class accommodates orbit spaces of more general group actions on more general manifolds.

In Section 2.9, we finish off the main body of the chapter with a short section on *stratified maps*, i.e. maps that are compatible with stratification data in some way.

The last two sections of the chapter contain more specialized material that we will need later but that can even more certainly be skipped at a first pass. Section 2.10 concerns intrinsic filtrations. While a topological space might be able to carry any number of stratification structures that make it a space of one of the types discussed to this point, sometimes spaces carry a unique natural such structure, hence "intrinsic." In particular, all PL spaces carry intrinsic PL stratifications as PL CS sets, as we see in Section 2.10.1. Section 2.11 is then about products and joins of stratified spaces.

A note about dimension. For pseudomanifolds, "dimension" has a fairly typical meaning: If X is an n-dimensional pseudomanifold, then X contains a dense n-dimensional topological manifold and each point $x \in X$ has a neighborhood of topological dimension n. On the other hand, for the most general filtered spaces, we will want our filtration subsets X^i to be indexed with some set of integers $\{i\}$, which will play a role in our intersection homology computations, but we do not necessarily want this index to correspond to any topological notion of dimension. In these cases, we refer to i as the *formal dimension*. Unfortunately, this occasionally results in some awkwardness. For example, the strata of manifold stratified spaces will be manifolds, which certainly do have a topological dimension, but we will not always want the topological dimension and the formal dimension to correspond. Such situations arise by necessity when working with stratified spaces that are not the closure of a union of strata of a fixed dimension. More specifically, for example, suppose we have a stratified space X whose topological dimension is n, and we also assign the formal dimension to be n, which certainly seems like the sensible thing to do. But now suppose that X has an open subset U that has topological dimension $m < n$; this occurs for example if $X = S^1 \vee S^2$ and U is an open subset of $X - S^2$. For the purposes of intersection homology computations, we will typically want the formal dimensions of the induced strata of subsets, especially open subsets, to be the same as their formal dimensions as strata of X. This is necessary for intersection homology to behave well under inclusion maps. So the formal dimension of U must remain n, rather than the more natural m. There are perhaps other ways to formulate the inclusion of subsets, but they would require an unfortunate amount of index bookkeeping. So while the notion of formal dimension is not always completely natural, we adopt it as the lesser evil. Unless made clear otherwise, the dimensions and codimensions of strata for the purposes of intersection homology computations will always be these formal dimensions. Though one happy consequence of the dimensional homogeneity of pseudomanifolds is that every open set of an n-dimensional pseudomanifold will have topological dimension n, and so for pseudomanifolds we can and will assume that the topological and formal dimensions always agree[1].

2.1 First Examples of Stratified Spaces

We have already encountered some simple examples of stratified spaces in Section 1.1, such as the wedge of spheres $S^2 \vee S^2$, the suspended torus ST^2 (Figure 1.1), and the twice suspended torus (Figure 1.2). Even in these basic examples we can see several interesting features:

1. Even though none of these spaces is a manifold, each is "mostly" a manifold, meaning that each possesses a dense subset that is a manifold. For $S^2 \vee S^2$, if we remove the wedge point we have two open disks. For ST^2, if we remove the "north and south poles," we have left $(-1, 1) \times T^2$. For $S(ST^2)$, if we remove the suspension of the suspension points of ST^2 we have $(-1, 1) \times (-1, 1) \times T^2$.
2. As computed in Section 1.1 for ST^2, just these few "bad points" are enough to ruin Poincaré duality.
3. In each of these examples, the *singularities*, i.e. the "bad points," are themselves not too bad. In fact, the *singular sets* of $S^2 \vee S^2$ and ST^2 are each a discrete set of points, and so a zero-dimensional manifold. The singular set of $X = S(ST^2)$ is homeomorphic

[1] This has the potential to cause a bit of confusion when working with non-open subsets of pseudomanifolds that are themselves pseudomanifolds, for example when we define the L-classes for Witt spaces in Section 9.4. In such settings we will have to be careful about when we are treating such subspaces with their inherited formal dimensions as opposed to their intrinsic dimensions. In context, there should not be too much confusion.

to S^1, although with the filtration we have given X this circle carries its own nontrivial filtration with X^0 consisting of two points and $X^1 - X^0$ consisting of two open intervals.

These features will be typical of the spaces we intend to study: they are assembled from manifold pieces of various dimensions, including a dense top-dimensional piece, but this is not generally enough for the space to possess Poincaré duality in the usual sense. Instead we will need intersection homology to obtain duality results.

It is not difficult to construct other, more elaborate, non-manifold spaces with similar features. For example, we can construct more spaces with zero-dimensional singularities, sometimes called *isolated singularities*, by taking a compact manifold with boundary and attaching a cone on the boundary or by starting with a manifold with multiple boundary components and coning each off separately. However, it is also not difficult to form spaces with singular sets of higher dimension. For example, take our suspension ST^2 and a manifold M and form the product space $ST^2 \times M$. Then $(-1, 1) \times T^2 \times M$ is our dense manifold, while the singular set of non-manifold points will be homeomorphic to two disjoint copies of M. Once again the singular set is a manifold, but this shows that it may be a manifold of any dimension.

How about a space in which the singular set is not itself a manifold but can be similarly disassembled into manifold pieces? Let's start with the suspended torus ST^2, cross it with a circle to get $ST^2 \times S^1$, and then suspend this whole space to arrive at $X = S(ST^2 \times S^1)$. Keeping track of the singular points, $ST^2 \times S^1$ has a singular set consisting of two copies of the circle, and X has a singular set consisting of the suspension of the two circles, which is a set Σ homeomorphic to two copies of the sphere S^2 with their north poles attached to each other and their south poles attached to each other. Notice that $X - \Sigma$ is a manifold homeomorphic to $(-1, 1) \times (-1, 1) \times T^2 \times S^1$. Meanwhile Σ is not a manifold, but it is a suspension of a manifold, so if we let $\{n, s\}$ be the north and south poles of the last suspension, $\Sigma - \{n, s\} \cong (-1, 1) \times (S^1 \amalg S^1)$, which is a two-dimensional manifold. So we begin to see spaces whose singular sets are also "nearly" manifolds, except for their own singular subsets!

The reader is invited to think through more examples obtained by repeated applications of suspension or crossing with a manifold (or even crossing with a singular space such as SM!). What is the dense manifold? What is the singular set? Are there singularities that prevent the singular set from being a manifold? And so on.

Some less artificial examples come from algebraic geometry. For example, if X is an irreducible complex algebraic variety, then the singular set Σ of X is itself an algebraic variety described by a larger set of polynomial equations. And Σ itself will be a union of irreducible complex varieties (possibly of different dimensions), each of which will itself consist of a dense manifold set and lower-dimensional singular sets, which will be unions of irreducible varieties, and so on. Eventually this decomposition process bottoms out, and we see that the singular set of X is naturally layered, or *stratified*, in a way similar to our somewhat artificial constructions involving suspensions and products.

Example 2.1.1 Consider the subspace of $\mathbb{R}^3$ determined by the polynomial equation $xyz = 0$. This is the union of the three coordinate planes. It is a manifold except along the three axes, whose union is described by the system of equations $\{xyz = 0, xy+yz+xz = 0\}$. This is in turn a manifold except at the origin, which we can describe by the system $\{xyz = 0, xy + yz + xz = 0, x + y + z = 0\}$. △

Although our work will be completely topological, we will briefly discuss stratifications of algebraic spaces further in Section 2.8. The study of algebraic varieties through stratified space methods has consistently been a leading motivation for development in the field of study and remains an active area of research.

2.2 Filtered and Stratified Spaces

So far, we have seen spaces with various layers of singularities such that each layer is the closure of a manifold. We wish to make this concept more precise through a series of definitions. We will begin with very general notions and then define more and more specific types of spaces. The reader should be aware that the definitions in the literature are not always consistent and so some care should be exercised.

2.2.1 Filtered Spaces

We assume all spaces are Hausdorff, sometimes without further mention. It is common to find additional point-set topological assumptions, such as paracompactness or second-countability, in the definitions of the various types of spaces we will be considering. One of the benefits of our approach is that these additional assumptions do not seem to be necessary for any of the results we will encounter. In particular, we do not need to assume that manifolds are paracompact or second-countable.

Our most general level of definition is that of a filtered space.

Definition 2.2.1 A *filtered space* is a Hausdorff topological space X together with a sequence of closed subspaces[2]

$$\emptyset = X^{-1} \subset X^0 \subset X^1 \subset X^2 \subset \cdots \subset X^{n-1} \subset X^n = X$$

for some integer $n \geq -1$. The smallest index is always -1 and X^{-1} is always empty; we will typically not mention X^{-1} explicitly. We will generally refer to "the filtered space X," leaving the filtration tacit. If we wish explicitly to consider the filtered space X devoid of its filtration information we will write $|X|$.

The space X^i is called the *i-skeleton*. The index i is called the *formal dimension of the*

[2] For us, the symbols $X \subset Y$ always includes the possibility that $X = Y$.

skeleton[3], and we say that X has formal dimension n. This notion of formal dimension does not necessarily have anything to do with other concepts of dimension, though for many of the spaces we consider below, particularly pseudomanifolds, the skeleton dimension will be the same as the topological dimension. Notice that it is possible to have $X^i = X^{i-1}$. The intuition in this case is similar to that for CW complexes, where we would say that the k-skeleton of X contains all cells of dimension $\leq k$, but the k- and $(k+1)$-skeleta are equal if there are no $(k+1)$-cells.

We will use throughout the notation $X_i = X^i - X^{i-1}$ for $i \geq 0$. The connected components[4] of X_i are called the *strata* of X of formal dimension i. If X has formal dimension n and $S \subset X_i = X^i - X^{i-1}$ is a stratum of X, we say that S is a stratum of formal dimension i and formal *codimension* $n - i$. Note that the strata of all codimensions partition X.

Example 2.2.2 Any finite-dimensional simplicial or CW complex is a filtered space, filtered by its simplicial or cellular skeleta[5]. The strata are, respectively, the open simplices (i.e. the interiors of the simplices) of X or the interiors of cells of X. CW complexes may satisfy the condition that $X^i = X^{i-1}$. For example, if we think of the n-sphere $X = S^n$ as being composed of one 0-cell and one n-cell, attached in the unique way to the 0-cell, then $X^0 = X^1 = \cdots = X^{n-1}$. △

Example 2.2.3 Let X be a finite simplicial complex filtered by its simplicial skeleta, and let $p : E \to X$ be a fibration. Then we can filter E with the filtration $E^i = p^{-1}(X^i)$. △

When we use filtrations for which $X^i = X^{i-1}$ for some i, it is inconvenient to have to list all the skeleta, so we often employ an abbreviated notation. For example, suppose X is a filtered space for which $X^i = \emptyset$ for $i < 3$, $X^3 = X^4$, and $X = X^5$. Then we will write the filtration of the space simply as $X^3 \subset X^5$. Note also that the statement $X = X^5$ is meant to imply that X has formal dimension 5; we will continue to use this convention below.

Example 2.2.4 Let $X = X^5$ be a five-dimensional simplicial complex, and let X^2 be its simplicial 2-skeleton. Then $X^2 \subset X^5$ is a filtration of X.

One particular point of care (and possibly of confusion) with the shortened notation is that we could just as well have defined X^2 to be the simplicial 3-skeleton of X, since the definition of a filtered space does not require that the formal dimension of a skeleton necessarily has any connection with the topological dimension of the space. Then the notation $X^2 \subset X^5$ has some ambiguity: does X^2 refer to the simplicial 2-skeleton or to the formal 2-skeleton of the filtration? In most situations below in which there is a topological notion of dimension available, the topological and formal definitions of a skeleton will coincide, and so the

3 Note that we use the phrase "formal dimension" differently from e.g. Siebenmann [217, page 127], who defines the formal dimension to be $\max\{i \mid X^i - X^{i-1} \neq \emptyset\}$. This will be necessary when dealing with subsets; see Remark 2.2.15 and Section 4.3 for more details. We will often omit the word "formal" when no confusion can arise, especially when formal dimension agrees with topological dimension.

4 In some of the literature, the word *stratum* is reserved for the entire set $X_i = X^i - X^{i-1}$ and not just its connected components.

5 See Appendix B for a review of simplicial complexes.

reader is free to trust his or her instincts. When the formal and topological dimensions do not coincide, we will be explicit. △

Example 2.2.5 If M^m is a smooth manifold and N^n is a closed smooth submanifold of M, we have the filtered space $N^n \subset M^m$. As in the previous example, and in lieu of statements to the contrary, this notation is taken to mean that M has formal dimension m (corresponding to its topological dimension), N is taken to be the n-skeleton (as well as the k-skeleton for all $n \leq k < m$), and the skeleta of formal dimension less than n are empty.

The n-dimensional strata in this example are the connected components of N, and the m-dimensional strata are the connected components of $M - N$. All other strata are empty. △

Example 2.2.6 Let $X = X^n$ be a finite-dimensional simplicial complex filtered by its simplicial skeleta X^i. We can form a filtration on the path space PX, which is the space of maps $\gamma : [0,1] \to X$ with the compact-open topology. We can define a filtration by letting $(PX)^i = \{\gamma \in PX \mid \gamma(1) \in X^i\}$. Notice that in this case the skeleton dimension i of $(PX)^i$ will usually not correspond to the dimension in any geometric sense since path spaces are usually infinite-dimensional. A stratum E of PX is a set of paths such that for each $\gamma_1, \gamma_2 \in E$, $\gamma_1(1)$ and $\gamma_2(1)$ lie in the same simplex interior of X. In fact, given two such paths, it is not difficult to verify that they are homotopic through paths all of which lie in the same stratum, and so, with these filtrations, there is a bijection between strata of X and strata of PX. △

Example 2.2.7 Here is another example demonstrating that we must be careful not to trust intuition too much when working with formal dimensions. For any natural number n, we have the filtered space $X = X^n$ such that $X^i = \emptyset$ for all i, $-1 \leq i \leq n$. △

Example 2.2.8 Let $X = X^2$ be the union of the open upper half-plane $\{(x, y) \in \mathbb{R}^2 \mid y > 0\}$ with the y-axis $\{(x, y) \in \mathbb{R}^2 \mid x = 0\}$. If we let X^1 be the y-axis Y, we can filter X as $Y \subset X$. Here the strata are Y and the two components of $X - Y$. △

Example 2.2.9 (Subspace filtrations) If X is a filtered space and $Z \subset X$ is a subspace, we can define the *subspace filtration* on Z by $Z^i = Z \cap X^i$. In this case, we give Z the same formal dimension as X. △

Example 2.2.10 (Product filtrations) If X, Y are filtered spaces of respective formal dimensions n, m, then $X \times Y$ has a natural filtration of formal dimension $m + n$ such that $(X \times Y)^i = \bigcup_{j+k=i} X^j \times Y^k$. In this case, the strata have the form $S \times T$ where $S \subset X$ and $T \subset Y$ are strata of X and Y, respectively. △

Example 2.2.11 (Cones) An extremely important way to create new filtered spaces from old ones is by taking cones. If X is a compact filtered space of formal dimension $n - 1$, there is a natural filtration of formal dimension n on the open cone $cX = [0,1) \times X/\sim$, where $\sim$ is the relation $(0, x) \sim (0, y)$ for all $x, y \in X$. We simply define the filtration on cX so that $(cX)^i = c(X^{i-1})$ for all i with $0 \leq i \leq n$. We utilize here the common convention that the cone on the empty set is a point, in this case the cone point $v \in cX$ represented in the quotient by any point $(0, x)$. With this definition cX has formal dimension n, and the strata of cX are the cone point and the products of the strata of X with the open interval. Notice that we always have $(cX)^0 = \{v\}$, with the possibility that $(cX)^i = \emptyset$ for some $i > 0$ if $X^{i-1} = \emptyset$. By Definition 2.2.1, the skeleton $(cX)^{-1}$ is empty.

The suspension of a compact filtered space can be filtered analogously. △

As cones will play such an important role for us, we also pause here to establish the following notation:

Definition 2.2.12 For a compact space Z, the space cZ is the *open cone* $cZ = [0, 1) \times Z/\sim$, where $\sim$ is the relation $(0, w) \sim (0, z)$ for all $w, z \in Z$. We typically denote the vertex of a cone by v. Similarly, the closed cone is $\bar{c}Z = [0, 1] \times Z/\sim$. More generally, for $r > 0$, we let $c_r Z = [0, r) \times Z/\sim$ and $\bar{c}_r Z = [0, r] \times Z/\sim$; in particular, $cZ = c_1 Z$. Then $c_r Z \subset \bar{c}_r Z \subset c_s Z \subset \bar{c}_s Z$ whenever $r < s$.

In the study of filtered spaces, the strata of the highest possible dimension play an important role; for example, if X is an n-dimensional stratified pseudomanifold (see Definition 2.4.1, below), then X is the closure of the union of its n-dimensional strata. Hence the following definition is useful:

Definition 2.2.13 If X is a filtered space of (formal) dimension n, the components of $X_n = X^n - X^{n-1}$ are called the *regular strata* of X and all other strata are called *singular strata*. We sometimes let Σ_X denote the union of the singular strata; the set Σ_X is called the *singular set* or the *singular locus* of X. Of course $\Sigma_X = X^{n-1}$, but the notation Σ_X is a convenient way to refer to the singular locus without explicitly referencing the formal dimension of X.

Example 2.2.14 Let X be an n-dimensional simplicial complex, filtered by its simplicial skeleta as in Example 2.2.2. Then the interiors of the n-simplices are the regular strata, and the interiors of all other faces are the singular strata. △

Remark 2.2.15 As we have defined filtered spaces, it is possible for there to be no regular strata. For example, it is allowable within the definitions to have X be homeomorphic to the circle S^1 and to have $X^1 = S^1$ but to consider X as having formal dimension 2 so that $X = X^2 = X^1$. In this case $X^2 - X^1 = \emptyset$, and S^1 is a singular stratum of formal dimension 1. While we do not generally intend to study spaces with no regular strata, unfortunately they become unavoidable (and something of a nuisance) in arguments that require us to consider subsets of more well-behaved spaces.

This phenomenon is related to the issue mentioned in the introduction to this chapter concerning possible discrepancies between topological and formal dimensions, especially when restricting to subsets. For example, let us consider again the example $X = S^1 \vee S^2$. A fairly reasonable filtration here would be to let $X^1 = S^1$ and $X^2 = X$ so that X has formal dimension 2. The problem is that when we want to think of $X^1 = S^1$ as a subspace of X, for reasons that will become clear below, we will want to continue to think of X^1 as having codimension one, and so our subspace X^1 will have to be a subspace of formal dimension 2 corresponding precisely to the example of the preceding paragraph. This is also consistent with taking the subspace filtration on S^1 as defined in Example 2.2.9.

Since there's a lot of room for confusion here, we encourage the reader for now to treat all spaces as though they have regular strata (in fact it would not be too problematic on a first pass through the book to imagine that the regular strata are always dense in X). We will assume that all spaces have this property unless stated explicitly otherwise or required when working with subspaces.

See Section 4.3 below for more details concerning these issues, especially Section 4.3.1.

2.2.2 Stratified Spaces

We next introduce the *Frontier Condition*, which eliminates some of the possible pathologies in how the strata of a filtered space can fit together.

Definition 2.2.16 If T is a stratum of the filtered space X, let $\bar{T}$ denote the closure of T. The filtered set X satisfies the *Frontier Condition* if for any two strata S,T of X such that $S \cap \bar{T} \neq \emptyset$ then $S \subset \bar{T}$.

This condition does not hold for Example 2.2.8, which had the form $Y \subset X$ with Y the y-axis in the plane and X the union of Y with the open upper half-plane (Figure 2.1). Here Y certainly intersects the closure of $X - Y$ in X (which is the union of the upper half plane and the origin), but it is not contained in that closure. The Frontier Condition does not always hold for CW complexes with their natural filtrations either: Let X be the CW complex obtained by starting with an interval (with its standard CW structure with two 0-cells and one 1-cell) and attaching a 2-cell by gluing its boundary to the midpoint of the interval. Then the interiors of the 1- and 2-cells are strata and the 1-cell intersects the closure of the 2-cell but is not contained in that closure.

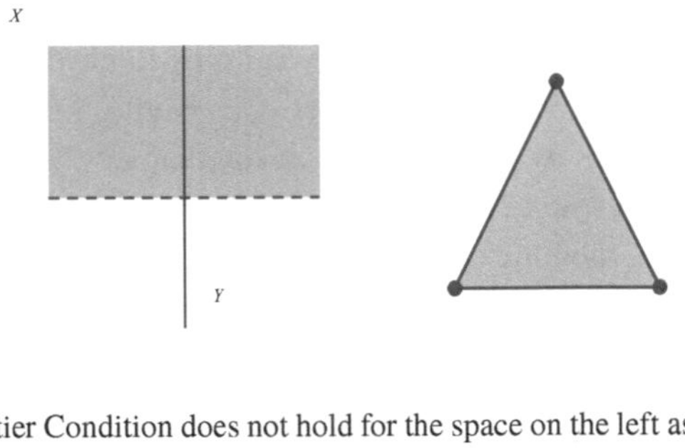

Figure 2.1 The Frontier Condition does not hold for the space on the left as the closure of $X - Y$ does not contain Y. However, the 2-simplex on the right with its simplicial filtration does satisfy the Frontier Condition as the closure of each face interior contains all of the faces it intersects.

By contrast, the Frontier Condition does hold for our other examples:

1. An embedded k-dimensional smooth submanifold of a smooth n-manifold is certainly contained in the closure of its complement if $k < n$.
2. Each open simplex (i.e. simplex interior) of a finite-dimensional simplicial complex intersects only the closures of the interiors of the simplices of which it is a face, and it is then contained in those simplices of which it is a face (Figure 2.1).
3. For our path space example, Example 2.2.6, let us sketch an argument that the Frontier Condition is satisfied. Suppose γ is in a stratum S of PX consisting of paths with endpoint in the interior of the i-simplex σ of X and that γ is also in the closure of another stratum S' of PX. Let σ' be the j-simplex such that $\gamma'(1)$ is in the interior of σ' if $\gamma' \in S'$. Then there are paths in S' arbitrarily close to γ in the compact-open topology of PX. But then we must have that σ is a face of σ', since $\gamma(1)$ must be a limit point of the $\gamma'(1)$. So now to show that $S \subset \bar{S}'$, we need only observe that if now η is *any* path in X with $\eta(1)$ in the interior of σ, then there are arbitrarily nearby paths η' with $\eta'(1)$ in the interior of σ'; for example we can just extend η a bit into the interior of σ'. Thus $S \subset \bar{S}'$.

Definition 2.2.17 We will say that a filtered space X is a *stratified space* if it satisfies the Frontier Condition.

Remark 2.2.18 This is not a standard definition in all sources. Other conditions are sometimes required, for example that the collection of strata be locally finite (i.e. that every point of the space has a neighborhood that intersects only finitely many strata). Alternatively, stratified spaces are sometimes defined without reference to a filtration by simply declaring a space to be a disjoint union of a locally finite collection of subsets called strata such that the Frontier Condition holds and the strata are locally closed[6]; see, for example, [133]. We will not require the local finiteness or locally closed conditions for arbitrary stratified spaces, but these properties will follow from the definitions for other types of stratified spaces we will consider; see Lemma 2.3.8. It will be necessary when discussing intersection homology to assume that all spaces are filtered so that codimension of strata is a well-defined concept.

Remark 2.2.19 In settings where we are working explicitly with stratified spaces, we will tend to use the words "filtration" and "stratification" interchangeably.

The benefit of working with stratified spaces, rather than general filtered spaces, is that the set of strata possesses some nice structure. For example, we get a partial order $\prec$ defined by $S \prec T$ if $S \subset \bar{T}$.

Proposition 2.2.20 *If X is a stratified space, then $\prec$ is a partial order. Furthermore, the closure of any stratum is a union of strata of lower dimension; in fact $\bar{T} = \bigcup_{S \prec T} S$.*

Proof Reflexivity of the relation is evident, and transitivity follows from basic topological properties of closure. To demonstrate antisymmetry, we need to see that if $S \subset \bar{T}$ and $T \subset \bar{S}$, then $S = T$. We may assume that S and T are not empty as, if either is empty, then so

6 Recall that a subset $Z \subset X$ is *locally closed* if it is the intersection of an open set in X and a closed set in X.

must be the other and the conclusion follows. Now, suppose that $S \subset \bar{T}$ and $T \subset \bar{S}$ and that $S \subset X_i = X^i - X^{i-1}$ and $T \subset X_j$. As X^i is closed in X, we have $\bar{S} \subset X^i$ and $\bar{T} \subset X^j$, and it follows that $T \subset \bar{S} \subset X^i$ so that $j \leq i$. But similarly from $S \subset \bar{T}$, we have $i \leq j$, so $i = j$. Thus S and T are each connected components of $X_i = X_j$, and each is thus closed in X_i [181, p. 160]. So, there is a closed set C in X such that $X_i \cap C = S$, and we must have $\bar{S} \subset C$. But then $T \subset \bar{S} \subset C$ and $T \subset X_i$, implying $T \subset X_i \cap C = S$. We thus have $T \subset S$ and, by a symmetric argument, $S \subset T$. Therefore, $S = T$ as desired.

Next, it is clear from the definitions that $\bigcup_{S \prec T} S \subset \bar{T}$. Now suppose $x \in \bar{T}$. Then since the strata partition X, the point x is in some stratum S and by the Frontier Condition, $S \subset \bar{T}$ and so $S \prec T$. Therefore, $x \in \bigcup_{S \prec T} S \subset \bar{T}$. It follows that $\bar{T} = \bigcup_{S \prec T} S$. □

Example 2.2.21 Let X be a finite-dimensional simplicial complex filtered by its simplicial skeleta. Then the strata are open simplices (i.e. the interiors of simplices), and $S \prec T$ if and only if S is an open face of the closed simplex $\bar{T}$, so X is a stratified space. In fact, the closure of an open simplex is the disjoint union of all its open faces. See again Figure 2.1. △

(a) Manifold Stratified Spaces

Our preferred stratified spaces will be those all of whose strata are manifolds.

Definition 2.2.22 A *manifold stratified space* is a stratified space all of whose i-dimensional strata are i-dimensional manifolds[7].

Example 2.2.23 A finite-dimensional simplicial complex filtered as in Example 2.2.2 is a manifold stratified space. Its strata are the open faces of its simplices, each of which is homeomorphic to some Euclidean space. △

Example 2.2.24 An m-dimensional (topological) manifold M with the trivial filtration (i.e. the filtration whose only strata are the components of M in dimension m) is naturally an m-dimensional manifold stratified space. If M^m is a smooth manifold with smooth closed submanifold $N^n \subset M^m$, $n < m$, then M with the filtration $N \subset M$ is naturally an m-dimensional manifold stratified space whose n-dimensional strata are the components of N and whose m-dimensional strata are the components of $M - N$. △

Example 2.2.25 (Product filtrations) If X is a manifold stratified space and M is an m-dimensional manifold, then $M \times X$ has a natural structure as a manifold stratified space with skeleta $(M \times X)^i = M \times X^{i-m}$. Each stratum of $M \times X$ has the form $M \times S$ for some stratum $S \subset X$ and so is also a manifold, and it is easy to verify the Frontier Condition for $M \times X$ using that the Frontier Condition holds on X. △

[7] Recall from our established notation that by our definitions a "manifold" may not possess a boundary. What are often called "manifolds with boundary" will be called ∂-manifolds. We also recall that there is an empty manifold of every dimension, so the empty set with any formal dimension is a manifold stratified space.

Example 2.2.26 (Cones) If X is a manifold stratified space, there is a natural manifold stratified space structure on the open cone cX. We define the filtration on cX as in Example 2.2.11 so that $(cX)^0$ is the cone point and so that for $i > 0$ we have $(cX)^i = (0, 1) \times X^{i-1}$. The strata are then the cone point and the products of the strata of X with the interval, each of which is a manifold if X is manifold stratified. Note that the Frontier Condition holds on cX as a consequence of it holding for X and that the cone point is in the closure of every non-empty stratum.

Similarly, the suspension of a manifold stratified space is a manifold stratified space with filtration $(SX)^i = S(X^{i-1})$, the suspension of the empty set consisting of two points. △

Remark 2.2.27 For a manifold stratified space, the formal dimension of a stratum will always agree with its topological dimension as a manifold, but the formal dimension of the manifold stratified space itself might be larger. For example, we can have the manifold stratified space X with filtration $X^0 \subset X^1 \subset X^2 = X$ with $X^0 = \emptyset$ and $X^1 = X^2 = S^1$. Here X has formal dimension 2, but the stratum $S^1 = X_1 = X^1 - X^0$ has formal dimension 1. Once again, such situations will not be typical, but they will occur. Unless stated otherwise or when working with subspaces, we will assume that if X is an n-dimensional manifold stratified space then $X^n - X^{n-1}$ is non-empty so that X possesses regular strata. Unfortunately, as mentioned in Remark 2.2.15, it will be necessary within some arguments to utilize manifold stratified spaces without regular strata. See Section 4.3, especially Section 4.3.1, for more details.

2.2.3 Depth

For running induction arguments, it is helpful to have the notion of *depth*, which is a measure of how many layers of strata there are in a stratified space. Before giving the definition, let us give some motivating examples.

Example 2.2.28 Suppose X is a stratified space of formal dimension n such that $X^i = \emptyset$ for all $i < n$. In this case, we often say that X is *unfiltered* or has the *trivial filtration*, and we define its depth to be 0. Similarly, if we have a space as in Remark 2.2.15 such X has formal dimension n and is such that $X^i = \emptyset$ for all $i < m$ and $X^m = X^{m+1} = \cdots = X^n$, then again there is only one dimension possessing non-empty strata (namely dimension m), and so we say that X has depth 0.

If we have an n-dimensional stratified space $X^k \subset X^n$ such that $X^i = \emptyset$ for $i < k$, $X^i = X^k$ for $k \leq i < n$, and $X^{n-1} \neq X^n$, then we say that X has depth 1. The stratified space in Examples 2.2.24 has depth 1. △

Building on these examples, we present the formal definition:

Definition 2.2.29 Let S and $\{S_i\}$ be strata of the stratified space X such that $S = S_d < S_{d-1} < S_{d-2} < \cdots < S_0$, where $<$ is the partial ordering of the strata, with $S_i \neq S_j$ for $i \neq j$ and so that this is the (not necessarily unique) longest such chain of strata containing S as its minimal element. Then we call d the *depth* of S.

The *depth* of a stratified space X is defined to be the maximum of the depths of its strata. This is well defined, as all filtered spaces are assumed to have finite formal dimension.

Example 2.2.30 If X is an n-dimensional simplex filtered by its simplicial skeleta, then each open i-face has depth $n - i$, and X has depth n.

If X is filtered as $N \subset M = X$, where N is a smooth non-empty n-dimensional submanifold of a smooth m-manifold M, $n < m$, then the stratum N has depth 1, the stratum $M - N$ has depth 0, and X has depth 1.

The suspended torus of Figure 1.1 has depth 1, and the twice suspended torus of Figure 1.2 has depth 2.

If X is the disjoint union of spheres $X = S^2 \amalg S^3$, filtered by $S^2 \subset X$, each stratum has depth 0 and X has depth 0.

Suppose X is the simplicial complex $\Delta^2 \vee \Delta^1$, with the two simplices attached at a common vertex, filtered by its simplicial skeleta. Each vertex of Δ^2 has depth 2, while the vertex of Δ^1 that is not shared with Δ^2 has depth 1. The stratified space X has depth 2.

Notice that the formal dimension of X is irrelevant for considerations of depth. △

Remark 2.2.31 There are other reasonable definitions of depth. This one will work well for us. In [217], Siebenmann defines depth of a stratified space X to be

$$\sup\{m - n \mid X^m - X^{m-l} \neq \emptyset \neq X^n - X^{n-1}\}.$$

It is also sometimes useful to define depth to be one less than the number of distinct non-negative formal dimensions such that X has a non-empty stratum of that dimension.

2.3 Locally Cone-like Spaces and CS Sets

Manifold stratified spaces decompose into partially ordered sets of strata, each of which is a manifold. But it turns out that even this is too general a setting to expect nice results. For example, a wild embedding of a manifold into a higher-dimensional manifold would give us a manifold stratified space, but in such cases the relation between strata can be very complicated. To avoid pathological cases in manifold theory, one typically imposes conditions such as *local flatness* by requiring that each point in the embedded manifold $K^k \subset M^m$ have a neighborhood pair $(N, N \cap K)$ that is homeomorphic to the standard Euclidean $(\mathbb{R}^m, \mathbb{R}^k)$. The following definition does something analogous for stratified spaces by imposing local topological structure at each point.

Definition 2.3.1 A filtered space X of formal dimension n is *locally cone-like* if for all i, $0 \leq i \leq n$, and for each $x \in X_i$ there is an open neighborhood U of x in X_i, a neighborhood N of x in X, a compact filtered space L (which may be empty), and a homeomorphism $h : U \times cL \to N$ such that $h(U \times c(L^k)) = X^{i+k+1} \cap N$. In this case L is called a *link* of x

and N is called a *distinguished neighborhood* of x; see Figure 2.2. For a given x, the space L is not necessarily uniquely determined[8]; see Example 2.3.12.

Locally cone-like filtered spaces whose i-dimensional strata are i-dimensional manifolds are called *CS sets*; see [139, 217][9]. This is equivalent to every point in an i-dimensional stratum having a distinguished neighborhood homeomorphic to $\mathbb{R}^i \times cL$, so we will generally restrict attention to distinguished neighborhoods of this form.

We also consider the empty set with any formal dimension to be a CS set.

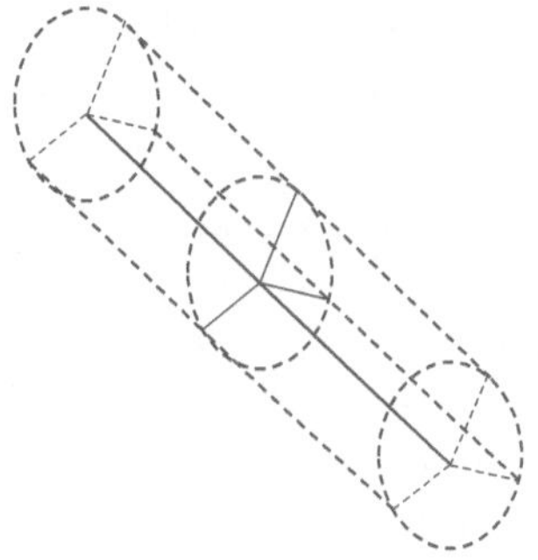

Figure 2.2 An example of a distinguished neighborhood homeomorphic to $\mathbb{R}^1 \times cL$. Here the link L is the one-dimensional circle filtered with a 0-skeleton consisting of three points. The neighborhood depicted has seven strata, including the stratum $\mathbb{R}^1 \times \{v\}$ which contains the points of which this is a distinguished neighborhood.

Notice that for a CS set the condition $h(U \times c(L^k)) = X^{i+k+1} \cap N$ is consistent with the product and cone filtrations introduced in Examples 2.2.25 and 2.2.11: Since U is an i-dimensional stratum, we expect $U \times c(L^k)$ to have dimension $i + k + 1$. Notice also that when $k = -1$, then L^k is empty, so $U \times c(L^k) = U \times \{v\} \cong U$, and $h(U \times \{v\}) = X^i \cap N$.

The type of homeomorphism that comes up in the definition of locally cone-like spaces is sufficiently useful that we pause to provide a definition, although we will study more general maps of filtered spaces in Section 2.9.

Definition 2.3.2 Suppose that X, Y are filtered spaces of the same formal dimension and that $f : X \to Y$ is a homeomorphism that takes the i-skeleton of X onto the i-skeleton of Y for all i. Then we say that the homeomorphism f is *filtration-preserving*. We also say that it is a *filtered homeomorphism* and that the two spaces are *filtered homeomorphic*.

It will often be convenient to make the filtered homeomorphisms that arise in the definition of distinguished neighborhoods for locally cone-like spaces implicit and so to simply treat a distinguished neighborhood as having the form $U \times cL$.

8 There are settings in which the links *are* unique; in particular this will be the case for piecewise linear (PL) CS sets; see Lemma 2.5.18.

9 In these sources CS sets are assumed to be metrizable, but we will not need this here.

The locally cone-like condition functions as a sort of homogeneity condition for filtered spaces. Unlike manifolds, in which all points have arbitrarily small Euclidean neighborhoods, two points of a stratified space generally will not have the same local topology, especially if they are contained in different strata. However, if x and y are two points in a CS set that are sufficiently close together to be contained in the same distinguished neighborhood, then they will have homeomorphic neighborhoods of all scales, as we can see by shrinking Euclidean neighborhoods and cones along their cone rays. Thus such points will have the same local topology.

Example 2.3.3 The n-dimensional manifold M is a CS set with the trivial filtration $\emptyset \subset M^n$. Each point $x \in M$ has a neighborhood filtered homeomorphic to $\mathbb{R}^n \times c\emptyset$, with $\emptyset$ having formal dimension -1. △

Example 2.3.4 Let M be a compact connected manifold of dimension $n - 1$. Suppose we filter the suspension $X = SM$ so that $X^0 = \{n, s\}$ consists of the cone points of the suspension, and the full filtration is $\{n, s\} \subset SM$. This is a CS set: $X - \{n, s\}$ is the manifold $(-1, 1) \times M$, and so each point in this stratum has a distinguished neighborhood of the form $\mathbb{R}^n \times c\emptyset = \mathbb{R}^n$, with $\emptyset$ filtered to have formal dimension -1. The strata $\{n\}$ and $\{s\}$ are each 0-manifolds and each has a distinguished neighborhood homeomorphic to $\mathbb{R}^0 \times cM \cong cM$ with M trivially filtered. So the points n and s each have links homeomorphic to M.

If we consider the space $Y = S^1 \times SM$, filtered by the product filtration as in Example 2.2.25, then again we obtain a CS set with manifold strata $Y - S^1 \times \{n, s\} \cong S^1 \times (-1, 1) \times M$, $S^1 \times \{n\}$ and $S^1 \times \{s\}$. Points in $Y - S^1 \times \{n, s\}$ have distinguished neighborhoods filtered homeomorphic to $\mathbb{R}^{n+1} \times c\emptyset$, while points in the other strata have distinguished neighborhoods filtered homeomorphic to $\mathbb{R}^1 \times cM$. △

Example 2.3.5 Suppose that X is a CS set and that $x \in X$ has a distinguished neighborhood filtered homeomorphic to $\mathbb{R}^k \times cL$. Then, if we form $(0, 1) \times X$ using the product filtration (see Example 2.2.25), each point $(t, x) \in (0, 1) \times X$ will have a neighborhood filtered homeomorphic to $[(0, 1) \times \mathbb{R}^k] \times cL \cong \mathbb{R}^{k+1} \times cL$. In particular, if L is a link of x in X, then L will also be a link of (t, x) in $(0, 1) \times X$.

Using this, we see that if X is a compact CS set then cX is a CS set: the cone point has a neighborhood $\mathbb{R}^0 \times cX$, while each point of $cX - \{v\}$ has a distinguished neighborhood by the preceding paragraph.

△

Example 2.3.6 Here is an example to demonstrate why it is not possible in a CS set to refer to "the" link of a point: it is possible for two non-homeomorphic spaces to have homeomorphic cones. This example comes from the famous Double Suspension Theorem [41] which states that a double suspension of a homology sphere is homeomorphic to a sphere. Let Σ^{n-2} be a homology sphere; then $X = S(S\Sigma) \cong S^n$. Now, filter $X = S(S\Sigma)$ by $\{n, s\} \subset X$, where $\{n, s\}$ are the vertices of the second suspension. Then we certainly have a CS set: $X - \{n, s\}$ is an n-manifold, homeomorphic to S^n with two points removed. Points in this stratum have distinguished neighborhoods filtered homeomorphic to $\mathbb{R}^n \times c\emptyset$. The vertices n and s each have neighborhoods homeomorphic to $\mathbb{R}^0 \times c(S\Sigma) = c(S\Sigma)$, but they also have Euclidean neighborhoods that are homeomorphic to $\mathbb{R}^0 \times c(S^{n-1}) \cong c(S^{n-1})$. The links $S\Sigma$ and S^{n-1} are the same only in the event that $S\Sigma$ itself happens to be a sphere, but in that case we can stratify $S\Sigma$ itself as a CS set whose suspension vertices have neighborhoods both of the form $\mathbb{R}^0 \times c\Sigma$ and $\mathbb{R}^0 \times c(S^{n-2})$, and now certainly the links are not homeomorphic unless Σ is itself a sphere. Since there exist homology spheres that are not spheres, there is non-uniqueness of links in this example (even if we can't tell which pair of links are non-homeomorphic!).

However, this lack of uniqueness is not as problematic as it may appear as it turns out that all possible links of points in the same stratum of a CS set will have the same intersection homology groups, as we'll prove below in Corollary 5.3.14. △

In first defining CS sets in [217], Siebenmann did not explicitly require the Frontier Condition, but the following lemma shows that it is satisfied automatically.

Lemma 2.3.7 *A CS set satisfies the Frontier Condition.*

Proof Suppose S, T are strata of X with S a component of X_i and that $S \cap \bar{T} \neq \emptyset$. We must show that $S \subset \bar{T}$; we can assume that $S \neq T$ since in that case it is immediate that $S \subset \bar{T}$. Since $\bar{T}$ is a closed set in X, the set $S \cap \bar{T}$ must be closed in S in its subspace topology. It suffices then to show that $S \cap \bar{T}$ is open in S, so that it must then be the entire connected component S of X_i. So suppose $x \in S \cap \bar{T}$. Then there is a neighborhood $U \cong \mathbb{R}^i$ of x in S and a neighborhood N of x in X such that $N \cong U \times cL$ by some filtered homeomorphism for some compact filtered link L. As will become our common practice, we leave the homeomorphism tacit and identify N with the structure $U \times cL$. As x must be in the closure of a stratum $T \neq S$, the link L cannot be empty. The strata of N beside U all have the form $(U \times cs) - U$ for some stratum s of L, though some of these strata may be contained in the same stratum of X; so $T \cap N$ is the union of strata of N of this form. In particular, $\bar{T} \cap N$ is the closure of the union of some strata $(U \times cs) - U$ of N, and this closure contains all of U. Thus $U \subset S \cap \bar{T}$, and U is open in S. Since each point of $S \cap \bar{T}$ has such an open neighborhood in S, this shows that $S \cap \bar{T}$ is open in S. □

CS sets also satisfy some other nice properties one typically wants in a stratified space:

Lemma 2.3.8 *Let X be a CS set.*

1. *The strata of X are locally closed, i.e. each is the intersection of a closed set of X with an open set of X.*
2. *The stratification of X is locally finite, i.e. every point has a neighborhood intersecting only finitely many strata of X.*
3. *If X is a compact CS set then it possesses a finite number of strata.*

Proof For the first statement, let S be a stratum with $S \subset X_i$ and, for each $x \in S$, let W_x be a distinguished neighborhood of x. By definition, $W_x \cap X^i \subset S$. Let $W = \bigcup_x W_x$. Then $S = W \cap X^i$, with W open and X^i closed in X.

The third statement follows easily from the second: as every point of X has a neighborhood that intersects only finitely many strata and as X is covered by a finite number of such neighborhoods, X has only finitely many strata.

To prove the second statement, we will argue by contradiction. Suppose $x \in X$ is a point such that every neighborhood of x intersects infinitely many strata. Let N be a distinguished neighborhood of x so that N is filtered homeomorphic to $\mathbb{R}^i \times cL$ for some compact filtered space L (we tacitly identify N with $\mathbb{R}^i \times cL$ via this homeomorphism). Aside from the stratum $\mathbb{R}^i \times \{v\}$, where v is the cone vertex, all the strata of $\mathbb{R}^i \times cL$ have the form $\mathbb{R}^i \times (cS - \{v\})$, where S is a stratum of L. Each stratum of X must intersect N in a union[10] of such strata of N, and as N must intersect infinitely many strata of X, it follows that L must have an infinite number of strata.

Let us identify L with the subspace $\{0\} \times \{1/2\} \times L \subset \mathbb{R}^i \times cL$, where (abusing notation) $\{1/2\} \times L$ denotes the image of $\{1/2\} \times L \subset [0,1) \times L$ under the quotient map to cL. Let $\{S_\alpha\}$ be the strata of X that intersect N, excluding the stratum containing x. Let $\{y_\alpha\}$ be an infinite set of points such that $y_\alpha \in L \cap S_\alpha$. Then the compactness of L implies that $\{y_\alpha\}$ must have a limit point, say $y \in L$. So every neighborhood of y contains infinitely many of the y_α and so intersects infinitely many strata of X. But from the structure of distinguished neighborhoods, y must be contained in a stratum of X of higher dimension than the stratum containing x.

So we have shown that if $x \in X$ is a point such that every neighborhood of x intersects infinitely many strata and if x is contained in the stratum S, then there is a point $y \in X$ contained in a higher-dimensional stratum T such that $S \prec T$ and such that y also has this property. But since we assume all of our spaces are finite-dimensional, this eventually leads to a contradiction. If we use this procedure to construct a sequence of points with this property, each contained in a stratum of higher dimension, eventually we wind up with a point z in a stratum M of maximal dimension, meaning that there is no stratum R of X such that $M \prec R$. In this case, a sufficiently small neighborhood of z intersects only the stratum M, a contradiction. □

Example 2.3.9 Here is an example of a compact manifold stratified set that is not locally cone-like and possesses an infinite number of strata. Let X be $\{x \in \mathbb{R} \mid x = 1/n, n \in \mathbb{Z}\} \cup \{0\}$ endowed with the trivial filtration with a single skeleton X^0. The connected components of X are the points of X, which are each embedded 0-manifolds, and the Frontier Condition holds, but no neighborhood of the point $\{0\}$ is homeomorphic to a cone. △

A common strengthening of the definition of a CS set requires that each link also be a CS set whose own links are CS sets, and so on. We will call such CS sets *recursive*:

Definition 2.3.10 A *recursive CS set* is a CS set such that every point has a link that is either empty or itself a recursive CS set.

10 By analogy, think of the stratification of S^1 as $\{x_0\} \subset S^1$ for any $x_0 \in S^1$; then $S^1 - \{x_0\}$ is a single stratum that intersects any distinguished neighborhood of x_0 in two components.

This definition at first appears circular, but by definition every link of a CS set X must have formal dimension less than that of X itself. So the definition is really inductive with the zero-dimensional recursive CS sets being disjoint unions of points.

Siebenmann intentionally did not assume that the links of a CS set be themselves CS sets; see the second remark on page 128 of [217]. However, it is common in many sources (e.g. [56, 133]) to include such a condition, which can be useful for making inductive arguments. Typically authors still call these "CS sets," and we introduce the "recursive" nomenclature here. Unless stated explicitly, we do not assume that CS sets are necessarily recursive.

Example 2.3.11 The CS sets of Examples 2.3.4 and 2.3.6 were both recursive CS sets, since all links were manifolds or suspensions of manifolds. △

Example 2.3.12 In the introduction to [217], Siebenmann provides an example of a CS set that is not evidently a recursive CS set. This is a compact non-manifold X such that $X \times \mathbb{R} \cong S^3 \times \mathbb{R}$. It follows that the cone on X is a CS set with two nontrivial skeleta – the cone point and $X \times \mathbb{R}$. But the obvious link of the cone vertex is X, which we know is not a manifold. Siebenmann notes that conjecturally it might be possible to find a manifold link that provides cX the structure of a recursive CS set, but the question is not settled there. △

The following lemma contains the useful fact that an open subset of a CS set is a CS set, and similarly for recursive CS sets.

Lemma 2.3.13 *If X is a (recursive) CS set and $V \subset X$ is an open subspace filtered by the subspace filtration $V^i = V \cap X^i$, then V is a (recursive) CS set.*

Proof It is only necessary to show that V is locally cone-like. Since any open subset of a manifold is a manifold, it will then follow from Lemma 2.3.7 that V satisfies the Frontier Condition.

Suppose $x \in V_i$. By assumption, x has a neighborhood N filtered homeomorphic to $\mathbb{R}^i \times cL$ in X. Let $U = N \cap V \cap X_i$. Then U is a neighborhood of x in X_i, and since U can be identified with an open subset of $\mathbb{R}^i$, we can choose a neighborhood D^i of x in X_i such that $\bar{D}^i \subset U$, $D^i \cong \mathbb{R}^i$, and $\bar{D}^i$ is compact. So consider $D^i \times cL \subset \mathbb{R}^i \times cL$. The idea now is that if $cL = [0,1) \times L/\sim$ then there is some $t \in (0,1)$ such that the subcone $c_tL = [0,t) \times L/\sim \subset cL$ satisfies $D^i \times c_tL \subset N \cap V$, where again we abuse notation by identifying N identically with $\mathbb{R}^i \times cL$. Since $N \cap V$ is an open neighborhood of $\bar{D}^i$, by the Tube Lemma [181, Lemma 26.8] $\bar{D}^i$ has a neighborhood in $N \cap V$ of the form $\bar{D}^i \times W$, where W is an open neighborhood of the vertex of cL. But now again applying the Tube Lemma to $[0,1) \times L$ and using the definition of the quotient topology for cL, we see that we can find a c_tL such that $c_tL \subset W$. Then $D^i \times c_tL$ is a neighborhood of x in $N \cap V$.

Since X and V use the same links, if X is recursive, so is V. □

Remark 2.3.14 Notice that in our applications of the Tube Lemma in the proof of the preceding lemma we have made critical use of the assumption that links are compact.

We close this section with some further observations about the point-set topology of CS sets that will be needed below.

Lemma 2.3.15 *CS sets are locally compact.*

Proof If X is a CS set, then, by definition, every point $x \in X$ has a neighborhood homeomorphic to $\mathbb{R}^i \times cL$, where L is a compact space and the image of x under the homeomorphism has the form (z, v) with $z \in \mathbb{R}^i$ and v the cone point of cL. Let $\bar{D}$ be the closed disk of unit radius about z in $\mathbb{R}^i$, and let $\bar{c}_r Z$ be as in Definition 2.2.12 for $0 < r < 1$. Then $\bar{D} \times \bar{c}_r Z$ is a compact neighborhood of (z, v) in $\mathbb{R}^i \times cL$, and it follows that x has a compact neighborhood in X. □

Corollary 2.3.16 *CS sets are completely regular*[11]*. In particular, they are regular.*

Proof It is a general fact of point-set topology that locally compact Hausdorff spaces are completely regular, and so regular. See [247, Theorem 19.3 and Definition 14.8]. □

Corollary 2.3.17 *If X is a CS set, $Z_1 \subset X$ is compact, $Z_2 \subset X$ is closed, and $Z_1 \cap Z_2 = \emptyset$, then there are disjoint open subspaces $U_1, U_2 \subset X$ such that $Z_1 \subset U_1$ and $Z_2 \subset U_2$.*

Proof Suppose $x \in Z_1$, and notice that Z_1 is contained in the open subset $X - Z_2$. As X is locally compact Hausdorff, there is a neighborhood V_x of x in $X - Z_2$ such that $\bar{V}_x \subset X - Z_2$ [181, Theorem 29.2]. As x ranges over the elements of Z_1, the V_x provide an open cover of Z_1; as Z_1 is compact, there is a finite subcover $\{V_{x_i}\}_{i=1}^m$. Then $U_1 = \bigcup_i V_{x_i}$ is an open subset of X containing Z_1, while $U_2 = \bigcap_i (X - \bar{V}_{x_i})$ is an open subset of X containing Z_2 and $U_1 \cap U_2 = \emptyset$. □

Corollary 2.3.18 *If X is a CS set and $K \subset W \subset X$ with K compact and W open, then there is a neighborhood V of K in X with $\bar{V} \subset W$.*

Proof This follows from the preceding corollary by letting $Z_1 = K$, $Z_2 = X - W$, and $V = \bar{U}_1$. □

2.4 Pseudomanifolds

We now arrive at the definition of a stratified pseudomanifold; these are the spaces that we shall eventually show possess an intersection homology version of Poincaré duality. Stratified pseudomanifolds are a special kind of recursive CS set. The additional idea is that a pseudomanifold should have a sort of dimensional homogeneity. To illustrate the idea, suppose M^m and N^n are compact manifolds of different dimensions, and consider the cone on the disjoint union $X = c(M \amalg N)$. It is easy to verify that this is a manifold

[11] A space is completely regular if and only if whenever $A \subset X$ is a closed subset and $x \notin A$ there is a continuous function $f : X \to I$ such that $f(x) = 0$ and $f(A) = 1$. As $f^{-1}([0, 1/2))$ and $f^{-1}((1/2, 1])$ are open sets separating x and A, it follows that a completely regular space is regular.

stratified space, and in fact a recursive CS set. However, X is essentially made up of two pieces of different dimensions: there are points whose neighborhoods are homeomorphic to $\mathbb{R}^{m+1}$ and others whose neighborhoods are homeomorphic to $\mathbb{R}^{n+1}$. What dimension could a fundamental class be? The definition of a stratified pseudomanifold is designed to avoid this sort of problem.

Definition 2.4.1 An n-dimensional recursive CS set X^n is a *(topological) stratified pseudomanifold* if $X_n = X^n - X^{n-1}$ is dense in X.

A space is called simply a *pseudomanifold* if it possesses a filtration with respect to which it is a stratified pseudomanifold.

Remark 2.4.2 It is much more common throughout the literature to also assume that a pseudomanifold must satisfy $X^{n-1} = X^{n-2}$, i.e. that X not have any codimension-one strata. It will be useful for us not to assume this. We will refer to a stratified pseudomanifold such that $X^{n-1} = X^{n-2}$ as a *classical stratified pseudomanifold.* A space is called a *classical pseudomanifold* if it possesses a filtration with respect to which it is a classical stratified pseudomanifold.

It is also often part of the definition to assume that each point of a stratified pseudomanifold has a link that is a stratified pseudomanifold; however we will show in Lemma 2.4.11 that this follows automatically from our definition. Note that when treating stratified pseudomanifolds, we will only consider links with this property.

Remark 2.4.3 Notice that, by definition, a stratified pseudomanifold always has regular strata; cf. Remarks 2.2.15 and 2.2.27. Therefore, the formal and topological dimensions of a stratified pseudomanifold are always the same.

The following lemma seems somewhat obvious, although Example 2.3.9 shows that some care is necessary.

Lemma 2.4.4 *Suppose X is an n-dimensional stratified pseudomanifold. Then $X_n = X^n - X^{n-1}$ is homeomorphic to a disjoint union of connected n-manifolds.*

Proof By definition, each connected component of the open set X_n is a manifold. Let $X_n = \amalg S_i$ be the *set-wise* decomposition of X_n into strata. We must show that a set W is open in X_n if and only if its restriction to each S_i is open in S_i. One direction is trivial: if W is open in X_n, then $W \cap S_i$ is open in S_i by definition of the subspace topology on S_i. Conversely, suppose $W \cap S_i$ is an open set in S_i and suppose $x \in W \cap S_i$. From the definitions, x must have a distinguished neighborhood N in X filtered homeomorphic to $U \times cL$, where U is an open neighborhood of x in S_i, and since S_i is already a top-dimensional stratum, it follows that $L = \emptyset$. Using the Euclidean topology of the distinguished neighborhood, any smaller Euclidean neighborhood of x in N is also a neighborhood of x in X_n and in S_i. Since $W \cap S_i$ is an open set in S_i in its subspace topology as a manifold, we can then choose a Euclidean neighborhood N' of x such that N' is a neighborhood of x in X_n and $N' \subset W \cap S_i$. It follows that $W \cap S_i$ is open in X_n. □

Since the disjoint union of manifolds is also a manifold, we thus see that an n-dimensional stratified pseudomanifold X is the closure of an n-dimensional manifold contained in X.

Example 2.4.5 The n-dimensional manifold M is a stratified pseudomanifold with the trivial filtration $\emptyset \subset M^n$. △

Example 2.4.6 Let N^n be a smooth submanifold of codimension > 0 embedded in a smooth manifold M^m. If we filter M in the obvious way as $N^n \subset M^m$ then M will be a stratified pseudomanifold. The locally cone-like property follows from the Tubular Neighborhood Theorem, so each point of N will have a trivially filtered sphere S^{m-n-1} as a link. Similarly, we obtain a stratified pseudomanifold if N and M are topological manifolds and the embedding is assumed to be locally flat, i.e. that each point $x \in N$ has a neighborhood pair that is homeomorphic to the standard Euclidean pair $(\mathbb{R}^m, \mathbb{R}^n)$. △

Example 2.4.7 The suspension of a compact manifold is a stratified pseudomanifold, as are the other stratified spaces in Example 2.3.4.

In fact, if we begin with a compact manifold and then engage in any finite iterated process of suspensions and taking products with other compact manifolds, the resulting space will be a stratified pseudomanifold. △

Example 2.4.8 If $X = X^2 = S^2 \vee S^1$ filtered by $\{x_0\} \subset S^1 \subset X$, where x_0 is the basepoint of the wedge, then X is a CS set but not a stratified pseudomanifold. In fact, no filtration of X will yield a stratified pseudomanifold, and so X is not a pseudomanifold. △

Example 2.4.9 Our next example is somewhat controversial as it does not yield a classical pseudomanifold: Let M^n be a compact n-dimensional ∂-manifold with boundary $\partial M \neq \emptyset$. We can filter M by $\partial M \subset M$. Then M is a stratified pseudomanifold. The interior points of M have distinguished neighborhoods of the form $\mathbb{R}^n \times c\emptyset$ and the boundary points have distinguished neighborhoods of the form $\mathbb{R}^{n-1} \times cL$, where L is a single point. Note that this is *not* a classical stratified pseudomanifold because the components of ∂M are strata of codimension one.

This example also illustrates the important point that the choice of filtration is critical. If M is considered as a space with the trivial filtration $\emptyset \subset M$, then M is *not* a stratified pseudomanifold; in fact it is not even a manifold stratified space, as M is not a manifold (manifolds with boundary are technically not manifolds!). △

Lemma 2.4.10 *An open subset U of an n-dimensional stratified pseudomanifold X filtered by the subspace filtration $U^i = U \cap X^i$ is an n-dimensional stratified pseudomanifold.*

Proof By Lemma 2.3.13, it is only necessary to verify that U_n is dense in U. Let $x \in U$. If W is any neighborhood of x in U, then W is also a neighborhood of x in X, as U is open in X. But since X_n is dense in X, each such neighborhood W intersects X_n. So every neighborhood of x in U intersects $U \cap X_n = U_n$. □

Lemma 2.4.11 *If X^n is a stratified pseudomanifold and L is a link of a point in X that is* *a recursive CS set, then L is a stratified pseudomanifold. Furthermore, if X^n is a classical stratified pseudomanifold, then so is L.*

Proof Suppose L is a link of the stratified pseudomanifold X so that there is a point $x \in X_i$ such that x has a distinguished neighborhood N filtered homeomorphic to $\mathbb{R}^i \times cL$ with $L = L^{n-i-1}$ a compact recursive CS set. We need only show that L_{n-i-1} is dense in L. Suppose $y \in L$ is a point such that y has a neighborhood W in L that does not intersect L_{n-i-1}. If we identify $N \subset X$ with $\mathbb{R}^i \times cL$ and think of L as embedded as, say, $\{0\} \times \{1/2\} \times L \subset R^i \times cL$, then $\mathbb{R}^i \times (0,1) \times W$ is a neighborhood of y in N that does not intersect $\mathbb{R}^i \times (0,1) \times L_{n-i-1}$. But $\mathbb{R}^i \times (0,1) \times L_{n-i-1} \cong N \cap X_n$ under the filtered homeomorphism. So the image of y in X has a neighborhood that does not intersect X_n, which contradicts X being a stratified pseudomanifold. Hence L_{n-i-1} is dense in L.

For the claim about classical stratified pseudomanifolds, we need only notice that if $L = L^{n-i-1}$ has a codimension-one stratum then $L_{n-i-2} \neq \emptyset$, so $\mathbb{R}^i \times (cL_{n-i-2} - \{v\})$ is non-empty and is a codimension-one stratum of $N \subset X$. So if any L is not classical, then X cannot be classical either, and our claim follows by the contrapositive. □

Corollary 2.4.12 *Every point in a stratified pseudomanifold has a link that is a stratified pseudomanifold.*

Proof This follows immediately from the definitions and the preceding lemma. □

Given Lemma 2.4.11 and its corollary, it is natural to formulate an alternative definition of topological stratified pseudomanifolds that does not directly refer to CS sets; this version of the definition is common in the literature (e.g. see [106]). The definition is recursive on dimension.

Definition 2.4.13 (Alternative definition of stratified pseudomanifold) A zero-dimensional *(topological) stratified pseudomanifold* is a discrete set of points.

For $n > 0$, an n-dimensional *(topological) stratified pseudomanifold* X^n is an n-dimensional filtered space such that:

1. Each connected component of $X^i - X^{i-1}$ is an i-dimensional manifold.
2. $X_n = X^n - X^{n-1}$ is dense in X.
3. For all i and for each $x \in X_i$, there is an open neighborhood U of x in X_i, a neighborhood N of x in X, a compact stratified pseudomanifold L (which may be empty), and a homeomorphism $h : U \times cL \to N$ such that $h(U \times c(L^k)) = X^{i+k+1} \cap N$.

We call an n-dimensional (topological) stratified pseudomanifold *classical* if $X^{n-1} = X^{n-2}$.

In what follows, whenever discussing stratified pseudomanifolds we will only consider links that are also stratified pseudomanifolds.

Remark 2.4.14 Let L be a link of a point $x \in X_i$ for the stratified pseudomanifold X, and let ℓ be a link of L. So L has a point y in some L_k such that y has a distinguished neighborhood in L filtered homeomorphic to $\mathbb{R}^k \times c\ell$. Then as observed in the proof of Lemma 2.4.11, the image of y under an embedding of L within a distinguished neighborhood in X has a neighborhood of the form $\mathbb{R}^i \times (0,1) \times \mathbb{R}^k \times c\ell \cong \mathbb{R}^{i+k+1} \times c\ell$. These homeomorphisms preserve the filtrations, and this demonstrates that ℓ is a link of y in X. In other words, a link in a link of a stratified pseudomanifold is a link in the stratified pseudomanifold.

2.5 PL Spaces and PL Pseudomanifolds

While our primary focus in this book will eventually concern singular intersection homology on filtered spaces and topological pseudomanifolds, another useful class of spaces, both for us as tools here and in practical application, are the piecewise linear filtered spaces and piecewise linear pseudomanifolds. In general, piecewise linear spaces, or *PL spaces*, are topological spaces with some extra structure. Roughly, these are the spaces that are homeomorphic to simplicial complexes, which the reader is likely to have encountered to one degree or another in most introductory algebraic topology texts, e.g. [220, Chapter 3], [125, Section 2.1], and vast swaths of [182]. Given a simplicial complex, it is possible to compute its simplicial homology groups algorithmically, so these are the most useful spaces for computational applications of homology. Furthermore, the extra rigidity afforded by such geometric structures is often enough to ensure a certain "tameness" in a space, preventing it from having bad properties, or at least making the properties it does have easier to prove. For example, the theory of PL manifolds is quite a bit more tractable than that of purely topological manifolds while they are still a bit more general than the category of smooth manifolds.

While a complete introduction to PL spaces would be beyond our purview, we recognize that they are not often covered in introductory topology courses. To complicate matters further, the standard references (Rourke and Sanderson [198], Hudson [130], Zeeman [254], and Stallings [222]) do not always approach the material from the same point of view or from a point of view that will be most useful for us in what follows. Therefore, we present in this section a good working definition of PL spaces and PL maps that seems to be well known in the literature (e.g. see [121, Section 1.3] or [2, Section 1]), and then in Appendix B we will review Hudson's broader development of the subject in further detail, though mostly without proofs. We will also show in the appendix that the category we define here in the main text is equivalent to the PL category implicit in [130]; see Section B.3. Furthermore, the appendix contains some fundamentals concerning simplicial complexes that the reader may want to review before proceeding.

In general, Hudson [130] will be our primary reference for PL topology, with occasional citations to Rourke and Sanderson [198] or to Munkres [182], the latter especially for results about simplicial complexes.

2.5.1 PL Spaces

The general idea of a PL space is that it is a space that can be triangulated, i.e. it is homeomorphic to a simplicial complex, but rather than fix a particular triangulation we think of the space as possessing a family of compatible triangulations. To make this more explicit, we need a sequence of definitions. If K is a simplicial complex, we let $|K|$ denote the underlying space of K as a topological space (see Appendix B for a review of simplicial complexes). A simplicial complex K is *locally finite* if every point $x \in |K|$ possesses a neighborhood U in $|K|$ that intersects only a finite number of simplices of K.

Definition 2.5.1 A *triangulation* T of a topological space X is a pair $T = (K, h)$, where K is a locally finite (possibly infinite) simplicial complex and $h : |K| \to X$ is a homeomorphism. A subdivision of $T = (K, h)$ is a pair $T' = (K', h)$, where K' is a *subdivision* of the simplicial complex K. We will say that $T = (K, h)$ and $S = (L, j)$ are *equivalent triangulations* if $j^{-1}h$ is a simplicial isomorphism[12]; it is easy to check that this is an equivalence relation. We say that T and S have a *common subdivision* if there are respective subdivisions T' and S' of T and S such that T' and S' are equivalent.

Definition 2.5.2 A *PL space* is a second-countable Hausdorff space X together with a family of triangulations $\mathcal{T}$ satisfying the following compatibility properties:

1. if $T \in \mathcal{T}$ and T' is any subdivision of T, then $T' \in \mathcal{T}$;
2. if $T, S \in \mathcal{T}$, then T and S have a common subdivision.

If $(X, \mathcal{T})$ is a PL space, we call the triangulations in $\mathcal{T}$ *admissible triangulations*. We often abbreviate notation by speaking of the "PL space X," leaving the family of triangulations tacit unless needed explicitly. Similarly, we sometimes say that "X is triangulated by K," leaving the precise homeomorphism tacit. Alternatively, once we have fixed a triangulation $T = (K, h)$ we will sometimes refer to a "simplex of T" without mentioning either K or h explicitly.

Example 2.5.3 If we start off with a simplicial complex K, we can think of $|K|$ as a PL space whose admissible triangulations are id : $|K| \to |K|$ and id : $|K'| \to |K|$ for all the subdivisions K' of K. To see that this family of triangulations satisfies the compatibility requirements, note that any two subdivisions of a simplicial complex have a common simplicial subdivision. Perhaps the easiest way to see this is to observe that if K' and K'' are any two subdivisions of K, then the intersection of any simplex of K' with a simplex of K'' is a polyhedron and together these polyhedra give $|K|$ the structure of a cell complex (see [198, Section 2.8]). It is then not too hard to subdivide a cell complex into a simplicial complex [198, Proposition 2.9]. Alternatively, this follows from Example B.2.21 and our other work in the appendix. △

We also need a definition of PL map:

[12] See Definitions B.1.14 and B.1.17 regarding simplicial maps and simplicial isomorphisms.

Definition 2.5.4 If $(X,\mathcal{T})$ and $(Y,\mathcal{S})$ are two PL spaces, a *PL map* $(X,\mathcal{T}) \to (Y,\mathcal{S})$ is a (topological) map $f : X \to Y$ such that if given any admissible triangulations (K,h) of X and (L,j) of Y there is a subdivision K' of K such that $j^{-1}fh$ takes each simplex of K' linearly into a simplex of L. Note that we are not saying that the map $g = j^{-1}fh : K' \to L$ is simplicial, only that the image of each simplex of K' is contained in some simplex of L and that the map itself is linear. If $\{v_i\}$ are the vertices of a simplex σ of K' and $x = \sum t_i v_i$ is a point of σ (using barycentric coordinates – see [125, Section 2.1] or [182, Section 1]), then $g(x) = \sum t_i g(v_i)$ using the linear structure of the simplex of L that contains $f(\sigma)$.

It might help to picture the definition diagrammatically:

$$\begin{array}{ccccc} & & X & \xrightarrow{\quad f \quad} & Y \\ & & {\scriptstyle h}\uparrow{\scriptstyle \cong} & & {\scriptstyle j}\uparrow{\scriptstyle \cong} \\ |K'| & \xrightarrow{\quad = \quad} & |K| & \xrightarrow{\ j^{-1}fh\ } & |L| \end{array}$$

The definition says that if we start with triangulations (K,h) and (L,j) of X and Y, then a map $f : X \to Y$ is PL if we can find a subdivision K' such that the map across the bottom of the diagram takes each simplex of K' linearly into a simplex of L.

Example 2.5.5 Let K and L be simplicial complexes and let $|K|$ and $|L|$ denote the PL spaces constructed in Example 2.5.3, e.g. the admissible triangulations of $|K|$ are K and its subdivisions. Let $f : K \to L$ be any simplicial map. Then f is PL. △

The reader might have expected a PL map to be defined as a map that is simplicial with respect to some choices of triangulations so that essentially every PL map would be as in the preceding example. But, it is not true that every PL map can be made simplicial in this way. It is not difficult to construct counterexamples; see Remark B.2.20 in Appendix B. However, this can be done if f is a proper[13] PL map. The following result is Theorem 3.6.C of [130], which we repeat also as Theorem B.2.19 in Appendix B:

Theorem 2.5.6 *If $f : X \to Y$ is a proper PL map, then there are triangulations (K,h) of X and (L,j) of Y such that $j^{-1}fh$ is simplicial.*

The next proposition gives us another example of a PL map and tells us that a PL structure on a space is determined by any one of its triangulations.

Proposition 2.5.7 *Let $(X,\mathcal{T})$ and $(X,\mathcal{S})$ be two PL spaces with the same underlying topological space X. Suppose that $\mathcal{T} \cap \mathcal{S} \neq \emptyset$, i.e. that $\mathcal{T}$ and $\mathcal{S}$ share a triangulation in common. Then the identity map* id $: X \to X$ *induces a PL homeomorphism $(X,\mathcal{T}) \to (X,\mathcal{S})$, i.e. a PL map with a PL inverse.*

[13] Recall that a map $f : X \to Y$ of topological spaces is called *proper* if for each compact set $K \subset Y$, the set $f^{-1}(K)$ is a compact subset of X.

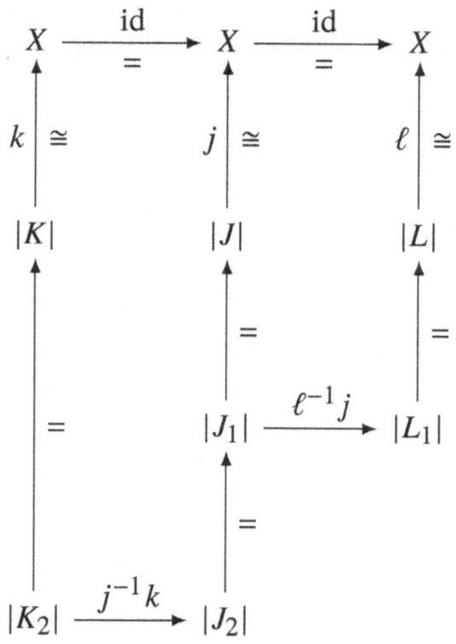

Figure 2.3

Proof To show that id gives a PL map $(X,\mathcal{T}) \to (X,\mathcal{S})$, we must show that for any $T = (K,k) \in \mathcal{T}$ and $S = (L,\ell) \in \mathcal{S}$ there is a subdivision $T' = (K',k)$ of T such that $\ell^{-1}k$ takes each simplex of K' linearly into a simplex of L. Let $R = (J,j) \in \mathcal{T} \cap \mathcal{S}$. Then as $S, R \in \mathcal{S}$, there is a common subdivision, consisting of triangulations (L_1,ℓ) and (J_1,j) with L_1 and J_1 subdivisions of L and J and with $\ell^{-1}j$ a simplicial isomorphism from J_1 to L_1. But since $(J,j) \in \mathcal{T}$, so is (J_1,j), and so there is a common subdivision of J_1 and K consisting of triangulations (J_2,j) and (K_2,k) with $j^{-1}k$ a simplicial isomorphism. We have visualized this data in Figure 2.3. Chasing this diagram, we see that $\ell^{-1}k = \ell^{-1}jj^{-1}k$, which can be identified with the path along the bottom of the diagram, takes each simplex of K_2 by a simplicial isomorphism to a simplex of J_2, which is contained in a simplex of J_1, which is mapped by a simplicial isomorphism to a simplex of L_2, which is contained in a simplex of L. Hence $\ell^{-1}k$ takes each simplex of K_2 linearly into a simplex of L. So id is piecewise linear.

The inverse of id is of course $\mathrm{id}^{-1} = \mathrm{id}$, which is also a piecewise linear map from $(X,\mathcal{S}) \to (X,\mathcal{T})$ by an analogous argument. So id is a PL homeomorphism. □

Next, let us briefly mention PL subspaces, again referring to Appendix B, particularly Section B.4, for a more thorough treatment.

Definition 2.5.8 If X is a PL space and Y is a subspace of X that is endowed with a PL structure in its own right, then Y is a *PL subspace* of X if the inclusion map $Y \hookrightarrow X$ is a PL map.

Example 2.5.9 By Example B.4.3 in the appendix, which utilizes [130, Lemma 3.7], if $T = (K,h)$ is an admissible triangulation of X, then the image of any subcomplex of K under

h is a PL subspace of X. Conversely, if Y is any *closed* PL subspace of X and $T = (K, h)$ is any triangulation of the PL space X, then there is a subdivision $T' = (K', h)$ such that $Y = h(K')$. So the closed PL subspaces of a PL space correspond to subcomplexes of triangulations. △

Example 2.5.10 It is also true that every open subset of a PL space is a PL subspace. This is much easier to see from the point of view in the appendix, so we refer the reader to Section B.4. We do note, however, that given any triangulation $T = (K, h)$ of X and any open subspace Y, there is a triangulation $S = (L, j)$ of Y that "subdivides" T in the sense that $h^{-1}j$ takes every simplex of L linearly and injectively into a simplex of K. See Example B.4.2. △

To complete this section, we observe that the PL spaces and maps constitute a category.

Definition 2.5.11 The PL spaces and maps as we have defined them here form a category $\mathcal{AT}$. We leave verification of the category axioms as an exercise for the reader. The name $\mathcal{AT}$ is chosen here to stand for "admissible triangulations" and to avoid confusion with the more traditional definition of the PL category, which we shall denote $\mathcal{PL}$ in the appendix. We show in Theorem B.3.7 of the appendix that $\mathcal{AT}$ and $\mathcal{PL}$ are equivalent categories.

2.5.2 Piecewise Linear and Simplicial Pseudomanifolds

We now consider filtered and stratified PL spaces.

Lemma 2.5.12 *Suppose X is a PL space and that X is filtered by a sequence of closed PL subspaces*

$$\emptyset = X^{-1} \subset X^0 \subset X^1 \subset X^2 \subset \cdots \subset X^{n-1} \subset X^n = X.$$

Then there is a triangulation $T = (K, h)$ of X with respect to which each of the X^i is the image under h of a subcomplex of K.

Proof Let $T_0 = (K_0, h)$ be any admissible triangulation of X. By Example 2.5.9, there is a subdivision $T_1 = (K_1, h)$ with respect to which there is a subcomplex K_1^1 of K_1 such that $X^{n-1} = h(K_1^1)$. But then there is similarly a subdivision $T_2 = (K_2, h)$ of T_1 with respect to which there is a subcomplex K_2^2 of K_2 such that $X^{n-2} = h(K_2^2)$. But if K_2^1 is the subdivision of K_1^1 induced by the subdivision from T_1 to T_2, then (K_2^1, h) is still a triangulation of X^{n-1}, just as (K_2, h) is still a triangulation of X. Continuing inductively in this manner, we reach a subdivision $T_n = (K_n, h)$ such that $h(K_n^i) = X^{n-i}$ for all i, $0 \leq i \leq n$, as desired. □

If X is a PL filtered space and $\mathcal{T}$ is the set of admissible triangulations of X, the proof of the lemma shows that any $T \in \mathcal{T}$ has a subdivision that is compatible with the filtration in the sense that every skeleton will be triangulated as a subcomplex. As this compatibility continues to hold under subdivision, the subset $\mathcal{T}_F \subset \mathcal{T}$ of triangulations compatible with the filtration is also a family of admissible triangulations, and in fact $(X, \mathcal{T})$ and $(X, \mathcal{T}_F)$ are

PL homeomorphic by Proposition 2.5.7. So in what follows we may safely limit ourselves to families of admissible triangulations that are compatible with filtrations, which we do without further comment.

We can now define PL filtered spaces, PL stratified spaces, PL manifold stratified spaces, PL CS sets, etc. All the definitions remain the same as in previous sections with the additional requirements that the filtration should be compatible with the PL structure, i.e. all subspaces are PL subspaces, all manifolds should be PL manifolds[14], and all structural homeomorphisms should be PL homeomorphisms.

For example, here is the full definition of a PL stratified pseudomanifold.

Definition 2.5.13 A *PL stratified pseudomanifold* X is a stratified pseudomanifold such that

1. X is a PL filtered space,
2. the strata of X are PL manifolds,
3. each point x has a distinguished neighborhood $N \cong \mathbb{R}^i \times cL$ such that the link L is a recursive PL CS set and the filtered homeomorphism $N \to \mathbb{R}^i \times cL$ is PL.

If X is a PL stratified pseudomanifold that has been given a fixed admissible triangulation such that each X^i is triangulated as a subcomplex, we will call X a *simplicial stratified pseudomanifold.*

We will call a PL space X a *PL pseudomanifold* if it possesses some filtration with respect to which it is a PL stratified pseudomanifold.

Remark 2.5.14 Again, our definition is uncommon in that it is usually assumed for a PL stratified pseudomanifold of dimension n that $X^{n-1} = X^{n-2}$. Once again we will refer to PL stratified pseudomanifolds satisfying this condition as *classical.*

Remark 2.5.15 By the same argument as in Lemma 2.4.11, the links of a PL stratified pseudomanifold will themselves be PL stratified pseudomanifolds, and the links of a classical PL stratified pseudomanifold will be classical PL stratified pseudomanifolds. As for topological stratified pseudomanifolds, we assume from now on that, whenever we refer to a link of a PL pseudomanifold, we have chosen a link that is itself a PL stratified pseudomanifold.

Once again, it is nice to have a direct definition that does not refer directly to CS sets, at the expense of the definition becoming recursive:

Definition 2.5.16 (Alternative definition of PL stratified pseudomanifold) A zero-dimensional *PL stratified pseudomanifold* is a discrete set of points.

For $n > 0$, an n-dimensional *PL stratified pseudomanifold* X^n is an n-dimensional filtered PL space such that:

[14] Being a PL n-manifold requires a bit more than being a topological n-manifold with a PL structure; it is also required that each point possess a neighborhood PL homeomorphic to the simplex Δ^n. See Definition B.2.4 or [130, Section I.5]. There are triangulations of topological manifolds that do not yield PL manifolds. This follows from the Double Suspension Theorem; see [73, Theorem 1].

1. Each connected component of $X^i - X^{i-1}$ is an i-dimensional PL manifold.
2. $X_n = X^n - X^{n-1}$ is dense in X.
3. For all i and for each $x \in X_i$, there is an open neighborhood U of x in X_i, a neighborhood N of x in X, a compact PL stratified pseudomanifold L (which may be empty), and a PL homeomorphism $h : U \times cL \to N$ such that $h(U \times c(L^k)) = X^{i+k+1} \cap N$.

We call an n-dimensional PL stratified pseudomanifold *classical* if $X^{n-1} = X^{n-2}$.

In a number of ways, PL spaces are more nicely behaved than their topological counterparts, which is not surprising given their additional structure. The next few results illustrate some of these niceties.

Our first lemma shows that every finite-dimensional PL space can be given the structure of a CS set. This is certainly not true in the topological category as we saw in Example 2.3.9.

Lemma 2.5.17 *Every finite-dimensional PL space has a filtration with respect to which it is a PL CS set.*

Proof Let X be a PL space, and fix an admissible triangulation $T = (K, h)$ of X. Let K^i be the simplicial skeleta of K. We claim the images $X^i = h(K^i)$ provide a CS set filtration of X. Indeed, the strata of X^i are then the images of the interiors of simplices of K, which are manifolds. Furthermore, by basic simplicial topology (see e.g. [182, 198]), if σ is a simplex of K, then σ has a (relative) star neighborhood $\overline{St}(\hat{\sigma})$ obtained by taking the union of all simplices in the barycentric subdivision[15] of the triangulation that include the barycenter $\hat{\sigma}$ of σ as a vertex. The simplicial link L of σ is then the union of the simplices in $\overline{St}(\hat{\sigma})$ that do not intersect σ. In this case, $\overline{St}(\hat{\sigma})$ is PL homeomorphic to the join $\sigma * L$ and the interior of $\overline{St}(\hat{\sigma})$ is then PL homeomorphic to $\mathring{\sigma} \times cL$, where $\mathring{\sigma}$ is the open simplex. Filtering L by its intersection with the skeleta of K provides the necessary PL filtration of L. □

As we saw in Example 2.3.6, the links in CS sets are not necessarily unique, and that example also applies to topological stratified pseudomanifolds. However, the extra rigidity in the PL setting does yield uniqueness results for links.

Lemma 2.5.18 *Let X be a PL CS set, and let S be a stratum of X. Then the links of any two points in S are PL homeomorphic.*

Proof Suppose that X is a PL CS set and that $x \in X$ is contained in an i-dimensional stratum S so that x has a neighborhood PL homeomorphic to $\mathbb{R}^i \times cL$. By basic PL topology (see [198, Exercise 2.24(3)] or the argument of [2, page 419]), $\mathbb{R}^i \times cL \cong c(S^{i-1} * L)$, where $S^{i-1} * L$ is the join of L with S^{i-1} or, equivalently, this is the ith suspension of L. In other words, $S^{i-1} * L$ is the *polyhedral link*[16] [198, Section 1.1] of x in X. If L'

15 See Example B.1.13 in the appendix for a review of barycentric subdivision.

16 In PL topology, the space $\mathrm{Lk}(x)$ such that x has a neighborhood of the form $c\mathrm{Lk}(x)$ is often called simply the "link," but since we have another meaning for "link," we use the expression "polyhedral link."

were another possible link for x, then we would similarly have a neighborhood of x that is PL homeomorphic to $c(S^{i-1} * L')$. But since polyhedral links are unique up to PL homeomorphism, we must have $S^{i-1} * L \cong S^{i-1} * L'$, and in the PL category, this implies[17] that $L \cong L'$ [179, Theorem 1]. So the link of x is unique.

It is also true that the links of any two points in the same stratum are PL homeomorphic. Since strata are connected, it suffices to show that the set of points in a stratum S with links homeomorphic to the link at a given point $z \in S$ is both open and closed in S. So let L be the link of z, let A be the set of points in S with link PL homeomorphic to L, and suppose $x \in A$. Then x has a neighborhood N in X that is PL homeomorphic to $\mathbb{R}^i \times L$ and where $\mathbb{R}^i \times \{v\}$ is taken by the homeomorphism to a neighborhood U of x in S. Any point in U also has N as a neighborhood in X, and so also has L as link. Therefore, A is open in S. Now, suppose x is in the closure of A in S. Then x has a neighborhood PL homeomorphic to $\mathbb{R}^i \times L'$ for some L'. But since x is in the closure of A, there is a point $y \in A$ that is in the image of $\mathbb{R}^i \times \{v\}$ under the homeomorphism. Hence the link of y is both L' and PL homeomorphic to L, and we must have $L' \cong L$ by the arguments of the preceding paragraph. Therefore, $x \in A$. So A is closed and open in S and so must be all of S. □

The preceding two lemmas concern CS sets, but in the PL category it is also much simpler to recognize which spaces are pseudomanifolds. The following proposition shows that any PL space which is *dimensionally homogeneous* can be stratified as a PL pseudomanifold.

Proposition 2.5.19 *Suppose X is any PL space of dimension n containing a closed PL subspace Σ of dimension $< n$ such that $X - \Sigma$ is an n-dimensional PL manifold that is dense in X. Then X is a PL pseudomanifold (i.e. there is some filtration with respect to which it is a PL stratified pseudomanifold). If, additionally,* $\dim(\Sigma) < n - 1$, *then X can be stratified as a classical PL stratified pseudomanifold.*

Proposition 2.5.19 will follow from Proposition 2.10.18, below, concerning intrinsic filtrations of PL spaces.

By [152] any semianalytic subset of a finite-dimensional affine space or of a countable real analytic manifold can be triangulated and in such a way that any closed semianalytic subspace can be triangulated as a subcomplex. This includes affine and projective complex analytic varieties, and so in particular affine and projective complex *algebraic* varieties. If such varieties are irreducible then they possess connected dense submanifolds [153, Corollary IV.2.8.3] and thus all such spaces can be stratified as classical PL stratified pseudomanifolds. These are important classes of spaces to which all of the most significant intersection homology results will apply!

Remark 2.5.20 In the setting of the proposition, it is tempting to attempt to define a stratification of X by fixing a triangulation of X with respect to which Σ is a subcomplex and $X - \Sigma$ is a PL manifold and then letting $X^n = X$ and letting X^i for $i < n$ be the union

[17] This fact is not true in the topological category. The Double Suspension Theorem again provides counterexamples as in Example 2.3.6.

of the i-simplices of Σ in the triangulation. This is the approach suggested, for example, in [121, Proposition 1.4]. However, it is not proven there, and it is not obvious to the author that the resulting links satisfy the required condition of being themselves PL stratified pseudomanifolds. Regardless, the approach of Proposition 2.10.18 via intrinsic filtrations is in many ways more natural as it does not depend on a choice of triangulation.

Corollary 2.5.21 *One has that X is an n-dimensional PL pseudomanifold if and only if X can be triangulated as a union of n-simplices, in which case every triangulation has this property. Furthermore, X can be filtered as a classical n-dimensional PL stratified pseudomanifold if and only if there is such a triangulation for which every $n-1$ simplex is the face of exactly two n-simplices, in which case every triangulation has this property.*

Proof If X is the union of n-simplices then the union of the interiors of those simplices constitutes a dense PL manifold, and so the proposition shows that X is a PL pseudomanifold.

Conversely, suppose X is a PL pseudomanifold. Then any simplex in any triangulation of X must be a face of an n-simplex in order for the condition to be fulfilled that X possesses a dense n-dimensional subspace. Hence every simplex is a face of an n-simplex, and since X is n-dimensional, X is a union of n-simplices as a simplicial complex.

Now, suppose that X is a union of n-simplices such that every $n-1$ simplex is the face of exactly two n-simplices. Then if we let Σ be the simplicial $n-2$ skeleton of X, Proposition 2.10.18 provides a filtration of X as a classical PL stratified pseudomanifold.

Again conversely, if X is a classical stratified PL pseudomanifold, then not only must any triangulation of X be a union of n-simplices, but every $(n-1)$-dimensional simplex must be the face of exactly two n-simplices. For if not, then any point of any $(n-1)$-simplex τ that is not the face of exactly two n-simplices cannot have a Euclidean neighborhood. Thus the entire $(n-1)$-dimensional interior of τ must be part of a stratum of dimension at least $n-1$ that is not contained in the n-manifold $X-\Sigma$. Therefore τ must be contained in some codimension-one stratum of X, a contradiction. □

(a) Classical Simplicial Pseudomanifolds

We pause in this section to discuss briefly the classical simplicial approach to pseudomanifolds, which preceded and motivates the more modern topological notions. Simplicial pseudomanifolds were originally defined as simplicial complexes such that (1) every simplex is the face of an n-simplex for some fixed n and (2) every $(n-1)$-simplex is the face of exactly two n-simplices. One also often sees a third condition, strong connectedness, which we will give below.

As motivation for this definition, consider the most primitive possible approach to constructing an n-dimensional triangulated manifold by gluing simplices together. Certainly one must begin with all simplices of the same dimension n since a manifold is dimensionally homogeneous – all points have neighborhoods of the same dimension. So there can be no simplices of dimension $> n$, and any simplices of dimension $< n$ would eventually have to be attached as faces of n-simplices. Next, there can be no free $(n-1)$-dimensional faces (since we are constructing manifolds, not manifolds with boundary), so each $(n-1)$-dimensional face must be the face of at least two n-simplices. But similarly an $(n-1)$-simplex cannot be the face of more than two n-simplices, since again we would not have a manifold otherwise. So, now suppose we have taken all our n-simplices and attached each $(n-1)$-face to exactly one other $(n-1)$-face. Do we have a manifold? We might or we might not. Any point in the interior of an n-simplex or the interior of an $(n-1)$-face now has a PL Euclidean neighborhood, but things can go wrong around lower-dimensional faces. The reader can verify that a suspended torus can also be constructed this way by gluing together two-dimensional faces of 3-simplices. We do see, however, that any non-manifold points must occur in the $(n-2)$-skeleton of the triangulation, and so in some sense things are not quite too bad. What we have constructed is a classical simplicial pseudomanifold, and by Corollary 2.5.21 there is some filtration with respect to which such a space is a PL stratified pseudomanifold.

In fact, classical simplicial pseudomanifolds are yet a bit more general than the discussion so far indicates because we may also make other gluings of our n-simplices along lower-dimensional faces. For example, if we have a PL n-sphere, $n \geq 2$, that we have constructed by gluing n-simplices along $n-1$ faces, we might yet glue two distinct vertices together and still have a classical simplicial pseudomanifold.

Nonetheless, despite these extra gluings, our classical simplicial pseudomanifold is not too far from being an n-manifold at least as far as the following property is concerned. Suppose that X is a compact simplicial pseudomanifold and that the manifold $X - X^{-2}$ is oriented, where X^{n-2} is the $(n-2)$-skeleton of the triangulation. This implies that it is possible to orient each of the n-simplices so that the chain $\Gamma = \sum_i \sigma_i$ is a cycle, i.e. its boundary as an element of the simplicial chain complex associated to the triangulation is 0. So Γ is a *fundamental class* for X in the same sense as for the homology of closed manifolds. And just as for manifolds, we can eliminate the orientation assumption if we are willing to work with $\mathbb{Z}_2$ coefficients. Precisely these fundamental classes will arise later when we consider intersection homology Poincaré duality for PL pseudomanifolds.

Another condition one typically sees in the definition of a simplicial n-dimensional pseudomanifold (e.g. [213, Section 24]) is that it should be possible to embed a path from any point in the interior of an n-simplex to any point in the interior of any other n-simplex such that the path only intersects interiors of n-simplices and $(n-1)$-simplices. This condition ensures that the PL manifold $X - \Sigma$ is connected and thus that $H_n(X) \cong \mathbb{Z}$, generated by Γ; more generally, $H_n(X)$ will be a direct sum of $\mathbb{Z}$ summands corresponding to connected components of $X - \Sigma$. This condition is sometimes called a "strong connectedness" condition. An example of a simplicial pseudomanifold that is connected but not strongly connected would be two n-spheres that are attached at a vertex.

Now, what about our PL stratified pseudomanifolds that do have $\dim(\Sigma) = \dim(X) - 1$? In this case, we are not requiring triangulations such that each $n-1$ simplex is the face of exactly two n-simplices. In this setting, we can still form the chain $\Gamma = \sum_i \sigma_i$, but it is clear that it will no longer be a cycle; it will only be a cycle in the relative chain group $C_*(X, \Sigma)$. In this sense, one might then expect that stratified pseudomanifolds, as we have defined them here, are more analogous to manifolds with boundary. As we will see below in discussing Poincaré duality results for intersection homology, this is not quite the case either, and, in fact, with the proper definitions they really do behave more like manifolds than like ∂-manifolds, or perhaps as something of a hybrid of the two. However, they do need to be handled with somewhat more general tools (see Chapter 6), which is why it is not unusual for some sources to restrict attention only to classical pseudomanifolds.

As we have observed above in Corollary 2.5.21, the simplicial pseudomanifolds that can be assembled from n-simplices can all be given filtrations making them PL stratified pseudomanifolds in the sense of Definitions 2.5.13 or 2.5.16 (though the filtration is not necessarily unique). We can then see how the definitions of topological stratified pseudomanifolds and CS sets constitute natural generalizations. In fact, we will even see below in Chapter 8 that topological stratified pseudomanifolds possess fundamental classes, though we must use singular chains rather than simplicial.

2.6 Normal Pseudomanifolds

Technical difficulties can sometimes arise in working with stratified pseudomanifolds that have points whose links are not connected. Pseudomanifolds whose links are connected are called *normal*, and such spaces also have the benefit that each connected component can contain only one regular stratum. It turns out that each stratified pseudomanifold X has a unique *normalization*, that is a normal stratified pseudomanifold $\tilde{X}$ together with a map $p : \tilde{X} \to X$ satisfying certain properties that we will soon describe. The motivation for normalization is that p induces isomorphisms of intersection homology groups, at least with some marginal further restrictions. So, historically, it has often been useful in proving results about the intersection homology of X to "resolve" first to $\tilde{X}$ and then work there; for an example of this, see [100]. For the purposes of this book, however, we have performed all proofs using X itself, and so we will not give normalization a completely thorough treatment. In particular, we will not prove the existence and uniqueness of normalizations for topological stratified pseudomanifolds, instead referring to Padilla [186] or Matthews [162]. We will discuss how intersection homology behaves with respect to normalizations below in Propositions 5.1.11 and 6.3.17.

We begin with an official statement of the definition of a normal pseudomanifold:

Definition 2.6.1 A (topological or PL) stratified pseudomanifold is called *normal* if the link of any point is connected.

Remark 2.6.2 Of course in the topological case a point does not have a unique link, but if one link is connected, they all must be. This is not completely obvious geometrically, but it follows, for example, from Corollary 5.3.14, which says that all links of a point have the same intersection homology groups, and from the fact that intersection homology can be used to detect connectivity, just as ordinary homology groups can. In fact, $I^{\bar{p}}H_0^{\mathrm{GM}}(L) \cong H_0(L)$

for a large enough perversity $\bar{p}$ (all of this will be explained once we get into intersection homology below).

The following lemma makes the case that, in some sense, normal stratified pseudomanifolds should play a role analogous to connected closed manifolds:

Lemma 2.6.3 *If X is a normal stratified pseudomanifold, then:*

1. *Every link of X is a normal stratified pseudomanifold.*
2. *If X is connected, then X has only one regular stratum.*

Proof The first statement follows from our earlier observation in Remark 2.4.14 that a link ℓ in a link L of a stratified pseudomanifold X is also a link in X.

We prove the second statement by induction. It is true for zero-dimensional pseudomanifolds. Assume for induction that the second statement of the lemma holds for normal stratified pseudomanifolds of dimension less than some $n > 0$. Suppose now that $X = X^n$ is connected and normal but that X has more than one regular stratum. Let S be one of the regular strata, and let T be the union of the other regular strata. As X is a stratified pseudomanifold $X = \overline{S \cup T}$, which is equal to $\bar{S} \cup \bar{T}$ by basic topology [247, Theorem 3.7]. If $\bar{S}$ and $\bar{T}$ are disjoint, that would make X disconnected, so there must be a point $z \in \bar{S} \cap \bar{T}$. Such a point z must be contained in a singular stratum of X. Consider a distinguished neighborhood $N \cong \mathbb{R}^i \times cL$ of z. The intersections of S and T with N have the respective forms $\mathbb{R}^i \times (cU - \{v\})$ and $\mathbb{R}^i \times (cV - \{v\})$, where $\{v\}$ is the cone point, and U and V are unions of regular strata of the link L. So L must have more than one regular stratum. But L is normal by the first part of the lemma and so this contradicts our induction hypothesis. Hence T must be empty. □

Unfortunately, different sources provide different technical definitions of *normalization*, but they all have essentially the same goal: to construct from a stratified pseudomanifold X a normal stratified pseudomanifold $\tilde{X}$, basically by "pulling apart" neighborhoods that are modeled on $\mathbb{R}^i \times c(\amalg L_k)$, for each L_k a connected component of the link L, into disjoint subsets that look like $\mathbb{R}^i \times \amalg c(L_k) = \amalg(\mathbb{R}^i \times cL_k)$. However, this must be done inductively over depth to ensure that each L_k is first itself a normal pseudomanifold. More specifically, we would like to construct such a normal $\tilde{X}$ together with a map $p : \tilde{X} \to X$ such that:

1. p is a proper surjection,
2. p maps i-dimensional strata of $\tilde{X}$ to i-dimensional strata of X (in fact these should be finite-sheeted covering maps),
3. the restriction of p to $\tilde{X} - \Sigma_{\tilde{X}}$ is a homeomorphism onto $X - \Sigma_X$, and
4. for any point $x \in \Sigma_X$, the set $p^{-1}(x)$ is a disjoint union of points and the number of such points is equal to the number of regular strata of any link of x.

Before giving a technical definition of normalization that implies these properties, let us present some examples to convey the basic idea.

Example 2.6.4 Suppose $\amalg_{k=1}^m M_k$ is a finite disjoint union of compact connected $(n-1)$-manifolds. Then $c(\amalg_{k=1}^m M_k)$ with the natural filtration is an n-dimensional stratified pseudomanifold, but it is not normal unless $m = 1$. The normalization is $\amalg_{k=1}^m (cM_k)$, with normalization map given by the quotient map that identifies the cone points together.

Similarly, something like $N \times c(\amalg_{k=1}^m M_k)$ normalizes to $\amalg_{k=1}^m (N \times cM_k)$. △

Example 2.6.5 If M is a compact n-manifold, $n > 0$, then the quotient map $M \to X$, where X is M with a finite number of points identified together, is a normalization map. More generally, if M is a smooth manifold containing disjoint smooth embeddings N_k of the same manifold N, then we can obtain a stratified pseudomanifold by gluing the N_k together. The quotient map is a normalization.

Even more generally, suppose X, Y are stratified pseudomanifolds with $\Sigma_X = \Sigma_Y$. Then the union of X and Y along the common Σ will be a non-normal pseudomanifold, and the quotient will be a normalization if X and Y are normal. △

Example 2.6.6 Suppose, as in the preceding example, that X is the quotient space of a finite union $\amalg M_k$ of compact n-manifolds identified along a finite set of points. Consider cX. Even if X is connected, cX will not be normal. In order to normalize X, we first need to normalize X. Then the normalization of cX will be $\amalg cM_k$, with its evident projection to cX. Note that in the normalization we have one cone point for each manifold M_k, and hence for each regular stratum of X. △

Example 2.6.7 Suppose X is a classical n-dimensional simplicial pseudomanifold with triangulation T. Then in [105], Goresky and MacPherson construct a normalization of X by taking the disjoint union of the n-simplices of T and then identifying the $(n-1)$-simplices as they are identified in X but without identifying any lower-dimensional simplices except as forced by identifying the $(n-1)$-simplices. As discussed in Section 2.5.2, this certainly gets us a simplicial pseudomanifold and with an obvious map to X obtained by making the remaining identifications. We leave the reader to consider why the $\tilde{X}$ constructed in this way must be normal and satisfy the other desired properties. △

For working with intersection homology, it is useful to give a more technical definition of normalization. We provide a version of the one in [162], which is a generalization of that in [186]. Analogously to the definition of pseudomanifold, the definition is inductive, assuming we know the definition already for stratified pseudomanifolds of lesser depth.

Definition 2.6.8 Let X be an n-dimensional stratified pseudomanifold. A *normalization* of X is an n-dimensional normal stratified pseudomanifold $\tilde{X}$ together with a proper surjective stratified map $p : \tilde{X} \to X$ such that:

1. the restriction $p : \tilde{X} - \Sigma_{\tilde{X}} \to X - \Sigma_X$ is a homeomorphism, and
2. if $x \in X$ has a distinguished neighborhood $N \cong U \times cL$, then there exists a commutative

diagram

$$\begin{array}{ccc} U \times \widetilde{cL} & \overset{\tilde{\phi}}{\hookrightarrow} & \tilde{X} \\ \downarrow p_0 & & \downarrow p \\ U \times cL & \overset{\phi}{\hookrightarrow} & X \end{array}$$

such that

(a) ϕ is a filtered homeomorphism from $U \times cL$ onto $N \subset X$,
(b) $\widetilde{cL} \cong \amalg cK_j$, where the K_j are the connected components of the normalization of[18] L, which we denote $p_L : \tilde{L} \to L$,
(c) p_0 is defined so that if $u \in U$ and $[t, k]$ with $t \in [0, 1)$ and $k \in K_j$ represents a point in cK_j, then $p_0(u, [t, k]) = (u, [t, p_L(k)])$, and
(d) $\tilde{\phi}$ is a filtered homeomorphism onto $p^{-1}(N)$.

It is a standard abuse of language to refer to $\tilde{X}$ alone as the normalization, leaving the map p tacit.

From the construction, we see that we must have $\tilde{X}^k = p^{-1}(X^k)$, and so p is compatible with the filtrations. Moreover, for each stratum $S \subset X$, we see that $p : p^{-1}(S) \to S$ is a locally trivial finite covering, and, applying Lemma 2.6.3, if $x \in X$ with link L, the cardinality of $|p^{-1}(x)|$ is equal to the number of regular strata of L. So the normalization map presented in the definition satisfies our desired properties.

The existence and uniqueness of normalizations is proven in[19] [162, 186].

2.7 Pseudomanifolds with Boundaries

Stratified pseudomanifolds constitute a generalization of manifolds. Since one also wants to consider the important class of "manifolds with boundary," it is reasonable to ask for "pseudomanifolds with boundary." This is provided by the following definition.

Definition 2.7.1 An n-dimensional *∂-stratified pseudomanifold* is a pair (X, B) together with a filtration on X such that:

1. $X - B$ with the induced filtration $(X - B)^i = (X - B) \cap X^i$ is an n-dimensional stratified pseudomanifold.

[18] So here we assume inductively that we have defined normalization of stratified pseudomanifolds of lesser depth than that of X. By condition (1), the normalization of a manifold is just the manifold itself with the identity map.

[19] In [186], Padilla assumes that his pseudomanifolds can be covered by an atlas of distinguished neighborhood charts such that, for a given stratum, the links of all points in the charts covering that stratum can be taken to be homeomorphic. Matthews [162] does not require such hypotheses.

2. B with the induced filtration $B^{i-1} = B \cap X^i$ is an $(n-1)$-dimensional stratified pseudomanifold.
3. B has an *open filtered collar neighborhood* in X, i.e. there exists a neighborhood N of B and a filtered homeomorphism $N \to [0,1) \times B$ (where $[0,1)$ is given the trivial filtration) that takes B to $\{0\} \times B$.

We call B the *boundary* of X, also denoted ∂X. We will often abuse notation by referring to the "∂-stratified pseudomanifold X," leaving B tacit.

A space is called simply a *∂-pseudomanifold* if it possesses a filtration with respect to which it is a ∂-stratified pseudomanifold. We will refer to a ∂-stratified pseudomanifold such that $X^{n-1} = X^{n-2}$ as a *classical ∂-stratified pseudomanifold*; a space is called a *classical ∂-pseudomanifold* if it possesses a filtration with respect to which it is a classical ∂-stratified pseudomanifold.

Example 2.7.2 A stratified pseudomanifold X is a ∂-stratified pseudomanifold with $\partial X = \emptyset$.

If X is a stratified pseudomanifold, then $X \times [0,1]$ is a ∂-stratified pseudomanifold with boundary $(X \times \{0\}) \cup (X \times \{1\})$.

If X is a compact stratified pseudomanifold, then the closed cone $\bar{c}X$ is a ∂-stratified pseudomanifold with $\partial(\bar{c}X) \cong X$. △

A critical point to observe is that the boundary of a ∂-stratified pseudomanifold depends upon the filtration. This is demonstrated by the following example.

Example 2.7.3 Let M be a paracompact n-dimensional ∂-manifold, and let P be its boundary (in the usual manifold-with-boundary sense). Suppose $P \neq \emptyset$.

1. Suppose we filter M trivially so that M itself is the only non-empty stratum. Then (M, P) is a ∂-stratified pseudomanifold. Note that all the conditions of Definition 2.7.1 are fulfilled: $M - P$ is an n-manifold, P is an $(n-1)$-manifold, and P is collared in M by classical manifold theory (see [125, Proposition 3.42]).
2. On the other hand, suppose X is the filtered space $P \subset M$. Then it is easy to check that X is a stratified pseudomanifold; the link of each point in P is a single point. But with this filtration, we cannot have $\partial X = P$ because condition (3) of Definition 2.7.1 would not be satisfied: P has a collared neighborhood in X but the collar homeomorphism does *not* preserve the filtration. With this filtration, $\partial X = \emptyset$.

△

So, unlike for manifolds, boundaries are not intrinsic to the topology of the space in general. However, our next result (from [100]) shows that when there are no codimension-one strata ∂X does depend only on the underlying space X and not on the choice of filtration (without codimension-one strata). Unfortunately, the proof is fairly technical and relies on several outside references concerning dimension theory.

Proposition 2.7.4 *Let (X, B) and (X', B') be ∂-stratified pseudomanifolds of dimension n with no codimension-one strata, and let $h : X \to X'$ be a homeomorphism (which is not required to be filtration-preserving). Then h takes B onto B'.*

Proof It suffices to show that h takes B to B', as the equivalent result for h^{-1} shows then that h takes B *onto* B'.

It further suffices to show that h takes the union of the regular strata of B to B', since the regular strata are dense in B and B' is closed. So let x be in a regular stratum of B and suppose that $h(x)$ is not in B'. Then there is a Euclidean neighborhood E of x in B such that $h(E) \subset X' - B'$. The existence of an open collar neighborhood of B shows that the local homology group $H_n(X, X - \{y\})$ is 0 for each $y \in E$, so by topological invariance of homology $h(E)$ must be contained in the singular set Σ' of $X' - B'$, for otherwise each point $h(y)$ would have a Euclidean neighborhood and $H_n(X', X' - h(y)) \cong \mathbb{Z}$.

Next we use the dimension theory of [37, Section II.16]. We will use the fact that each skeleton of a pseudomanifold (and in particular the singular set) is locally compact, as follows from Lemma 2.3.15 noting that each skeleton of a CS set is itself a CS set.

As defined in [37, Definition II.16.6], $\dim_{\mathbb{Z}} E$ is $n - 1$ by [37, Corollary II.16.28], so $\dim_{\mathbb{Z}} h(E)$ is also $n - 1$, and by [37, Theorem II.16.8] (using the fact that Σ' is locally compact) this implies that $\dim_{\mathbb{Z}} \Sigma'$ is $\geq n - 1$. To obtain a contradiction it suffices to show that $\dim_{\mathbb{Z}}$ of the i-skeleton of a stratified pseudomanifold is $\leq i$, as Σ' is the $(n-2)$-skeleton of $X' - B'$ due to the classical stratification.

So let Y be a pseudomanifold and assume by induction that $\dim_{\mathbb{Z}} Y^j \leq j$ for some j. This holds for $j = 0$ by [37, Corollary II.16.28]. Let c denote the family of compact supports, and let $\dim_{c,\mathbb{Z}}(Z)$ denote the least integer n (or ∞) such that $H_c^k(Z; \mathcal{A}) = 0$ for all sheaves $\mathcal{A}$ of abelian groups and all $k > n$ (see [37, Definition II.16.3]). Then $\dim_{\mathbb{Z}}$ is equal to $\dim_{c,\mathbb{Z}}$ for any locally compact space by [37, Definition II.16.6]. Since Y^i is a closed subset of Y^{i+1} and $Y^{i+1} - Y^i$ is a (possibly empty) $(i + 1)$-manifold, [37, Exercise II.11 and Corollary II.16.28] imply that $\dim_{c,\mathbb{Z}} Y^{i+1}$ is $\leq i + 1$ as required. □

The two cases of Example 2.7.3, together with the identity map of the underlying spaces, show that Proposition 2.7.4 is not true if codimension-one strata are allowed.

We conclude this section with some further observations concerning Definition 2.7.1.

Remark 2.7.5 When working with ∂-manifolds, the existence of collared boundaries is not usually part of the definition but is rather a theorem (at least for paracompact manifolds); see e.g. [125, Proposition 3.42 and remarks following]. For pseudomanifolds, however, it is necessary to make this desired property part of the definition.

For example, let M be an $(n - 1)$-dimensional ∂-manifold with $\partial M \neq \emptyset$ and so $n > 1$. Consider the closed cone $X = \bar{c}M$. If we filter X by $\{v\} \subset X$, where v is the cone vertex, then X is a stratified space, though it is not manifold stratified as $X - \{v\}$ is not a manifold. If we let $B = M \cup_{\partial M} \bar{c}(\partial M)$, then $X - B$ is an n-manifold homeomorphic to $M \times (0, 1)$, and B

is a stratified pseudomanifold with the cone point as a zero-dimensional stratum. However, B cannot have a filtered collar neighborhood as there are no one-dimensional strata of X.

Remark 2.7.6 The strata of a ∂-stratified pseudomanifold X will not necessarily be manifolds, and so ∂-stratified pseudomanifolds are not necessarily manifold stratified spaces. For example, a trivially filtered ∂-manifold with non-empty boundary is not a manifold, though it is a ∂-stratified pseudomanifold. However, the strata of a ∂-stratified pseudomanifold will be ∂-manifolds with the boundary of the stratum S consisting of $S \cap \partial X$.

On the other hand, if $x \in \partial X$ has a distinguished neighborhood in ∂X of the form $\mathbb{R}^i \times cL$, then thanks to the collar condition it also has a filtered neighborhood in X of the form $\left([0,1) \times \mathbb{R}^i\right) \times cL$. So it is useful to continue to refer to L as a link of x in X. Note that L is also a link in $X - \partial X$ of each point in the image of $(0,1) \times \{x\}$ under the collar homeomorphism.

Remark 2.7.7 When studying the intersection homology of ∂-stratified pseudomanifolds, we will make fairly regular use of the existence of filtered collars. However, to really see why these are so critical for us, see Example 8.3.11, which shows that they are needed for our coming Lefschetz duality results.

The following generalizes Lemma 2.4.10 to ∂-pseudomanifolds, though we must make some additional hypotheses about the underlying topology of the boundary. This lemma and its proof were provided by Jim McClure.

Lemma 2.7.8 *Let X be an n-dimensional ∂-stratified pseudomanifold, and suppose that ∂X is hereditarily paracompact*[20]. *If $U \subset X$ is an open subset filtered by the subspace filtration $U^i = U \cap X^i$, then U is an n-dimensional ∂-stratified pseudomanifold.*

Proof Applying Lemma 2.4.10, it's only necessary to give a filtered collar for $U \cap \partial X$ in U. Let $c : [0,1) \times \partial X \to X$ be a filtered collar of ∂X in X.

It suffices to find a continuous $f : U \cap \partial X \to (0,1)$ with $c\left([0, f(x)) \times \{x\}\right) \subset U$ for each $x \in U \cap \partial X$, because then we can let the collar for $U \cap \partial X$ be defined by $(t,x) \to c(f(x)t, x)$.

Using the standard basis for the product topology, for each $x \in U \cap \partial X$ there is a neighborhood V_x of x in $U \cap \partial X$ and a $d_x \in (0,1)$ such that $c([0,d_x) \times V_x) \subset U$. Since $U \cap \partial X$ is paracompact by the assumption that ∂X is hereditarily paracompact, the covering $\{V_x\}$ has a locally finite refinement $\{W_\alpha\}$, and for each α there is a d_α with $c([0,d_\alpha) \times W_\alpha) \subset U$. Let ϕ_α be a partition of unity subordinate to the cover $\{W_\alpha\}$ of $U \cap \partial X$. Then $\sum_\alpha d_\alpha \phi_\alpha$ is the desired function f. □

20 A space is hereditarily paracompact if every open subset is paracompact [37, page 21].

2.8 Other Species of Stratified Spaces

In this section we mention some other types of manifold stratified spaces. These spaces will not be used directly in the remainder of the book, so this section can be safely skipped. However, we do note that we will cite here results about our first two types of spaces, Whitney stratified and Thom–Mather stratified spaces, that imply that all irreducible algebraic and analytic varieties can be stratified as pseudomanifolds (as already discussed in Section 2.5.2), as can the connected orbit spaces of manifolds under smooth actions of compact Lie groups. Thus pseudomanifolds arise "in nature."

In contrast to Whitney stratified and Thom–Mather stratified spaces, which possess more structure than CS sets, we will also briefly consider manifold homotopically stratified spaces, which are not necessarily CS sets. Rather than satisfying a local cone-like condition, these spaces possess homotopy-theoretic conditions imposed on the interaction of strata.

For a more detailed survey of all of these spaces and others, see [133].

2.8.1 Whitney Stratified Spaces

The following geometric conditions on a manifold stratified space are due to Whitney [245, 246] and assume that our stratified space X is a closed subspace of a smooth manifold M. This will always be the case, for example, if one studies affine or projective algebraic or analytic varieties.

Definition 2.8.1 The stratified space $X \subset M$ is *Whitney stratified*:

1. If each stratum of X is a locally closed smooth submanifold of M.
2. (Whitney's Condition A) If $\{x_i\} \subset S'$ is a sequence of points in a k-dimensional stratum S' converging to a point x in a stratum $S \subset \bar{S}'$ and if the k-dimensional tangent spaces $T_{x_i}S'$ to S' at x_i converge to a k-dimensional subspace V of T_xM, the tangent plane to M at x, then V contains the tangent space to S at x, i.e. $T_xS \subset V$.
3. (Whitney's Condition B) If the hypotheses of Condition A hold and $\{y_i\} \subset S$ is a sequence of points also converging to x such that the sequence of secant lines between x_i and y_i converges to a line ℓ, then $\ell \subset V$.

To understand the condition on secant lines, one should choose a local coordinate chart for M around x. It can be shown that the condition is independent of the choice.

The definition turns out to be somewhat redundant, as it was shown by Mather [161] that Condition B implies Condition A.

Whitney's conditions were formulated with algebraic varieties in mind; however not every algebraic variety satisfies Whitney's conditions with its natural filtration, in which X^{i-1} is the set of singular (non-smooth) points of the subvariety X^i .

The standard example is the *Whitney umbrella* from [246, Example 18.7]; see Figure 2.4. Let $W = \{(x, y, z) \in \mathbb{R}^3 \mid x^2 = zy^2\}$, which is an irreducible real algebraic variety (see

[160, Example 5.1]). The space W is the union of the z-axis, Z, with the two-dimensional manifold that is the graph of the surface $z = x^2/y^2$ for $y \neq 0$. Notice that each point along the negative z-axis has a neighborhood whose intersection with W is equal to its intersection with the z-axis. By contrast, any slice of W determined by $z = c$, for c a positive constant, is the union of two lines. When $z = 0$, then also $x = 0$ and y can be arbitrary, so the intersection of W with the x–y plane is the y-axis. As $W - Z$ is a smooth two-dimensional manifold, the natural filtration of W is $Z \subset W$, since Z is the set of points at which W is not a smooth two-dimensional manifold, and Z itself is a smooth one-dimensional manifold. However, notice that this is not even a stratification as $Z \cap (\overline{W - Z}) \neq \emptyset$ but $Z \not\subset \overline{W - Z}$, so the Frontier Condition is not satisfied. Whitney's conditions are also violated as we see by letting $\{x_i\}$ be a sequence of points along the positive y-axis converging to the origin 0. These are points of the stratum $W - Z$, and their tangent spaces can all be identified with the x–y plane. Hence the limit of $T_{x_i}(W - Z)$ at the origin is also the x–y plane. But clearly this plane does not contain $T_0 Z$.

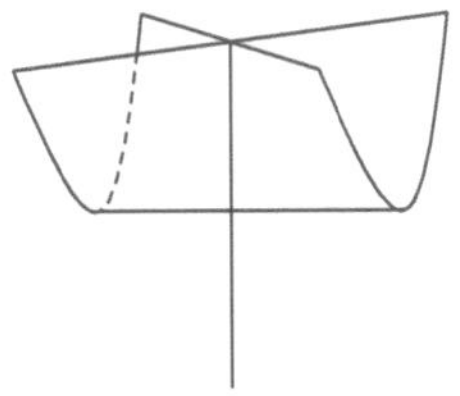

Figure 2.4 Whitney's umbrella $\{(x, y, z) \in \mathbb{R}^3 \mid x^2 = zy^2\}$.

Nonetheless, it is possible to choose a different filtration of W with respect to which it is a stratified space satisfying the Whitney conditions, namely $\{0\} \subset Z \subset W$. In fact all algebraic sets, semi-algebraic sets[21], analytic sets and semi-analytic sets[22], and sub-analytic sets[23] all can be filtered so as to possess Whitney stratifications. Expository references include [109, 215].

2.8.2 Thom–Mather Spaces

Thom–Mather spaces [161] often arise in settings where one wants to be able to make analytic arguments concerning stratified spaces. The idea is that each stratum should have an analogue of a tubular neighborhood but that the different tubes around the different strata should interact compatibly. We adapt our version of the definition from [133].

21 Semi-algebraic sets are finite unions of sets determined by finitely many polynomial equations or inequalities.
22 These are defined as for algebraic and semi-algebraic sets but using analytic functions rather than just polynomials.
23 We will not define these here; see [215].

Definition 2.8.2 For $0 \le k \le \infty$, the manifold stratified space X is a *Thom–Mather C^k stratified space* if:

1. Each stratum of X is a C^k manifold.
2. There is a tube system $\{T_i, \pi_i, \rho_i\}$ such that T_i is an open neighborhood of X_i in X (called a tubular neighborhood), $\pi_i : T_i \to X_i$ is a retraction (called the local retraction), and $\rho_i : T_i \to [0, \infty)$ is a map such that $\rho_i^{-1}(0) = X_i$.
3. For each pair X_i, X_j, if $T_{ij} = T_i \cap X_j$ and the restriction of π_i, ρ_i to T_{ij} are denoted π_{ij}, ρ_{ij}, then the map $(\pi_{ij}, \rho_{i,j}) : T_{ij} \to X_i \times (0, \infty)$ is a C^k submersion.
4. If $x \in T_{jk} \cap T_{ik} \cap \pi_{jk}^{-1}(T_{ij})$, then $\pi_{ij}\pi_{jk}(x) = \pi_{ik}(x)$ and $\rho_{ij}\pi_{jk}(x) = \rho_{ik}(x)$.

The idea here is that each π_i plays a role analogous to the projection of a tubular neighborhood to a submanifold in manifold theory, while each ρ_i is a measure of radial distance from a stratum. Condition 3 says that the π_i and ρ_i are not too wild as functions. The first equation of condition 4 says that the image of a point under two successive local retractions, from X_k to X_j and then from X_j to X_i, is the same as its image under the local retraction directly from X_k to X_i when the point is close enough to X_i and X_j to be contained in all the relevant tubes. The second equation of condition 4 says roughly that local retraction from X_k to X_j should not change the radial distance of a point from X_i, again when points are close enough to all relevant strata.

These sorts of conditions are relevant when one wants to study stratified spaces using techniques of global analysis; see for example [5].

By the work of Mather, Whitney stratified spaces always possess tube data, making them Thom–Mather spaces [161]. Conversely, Thom–Mather spaces can be embedded as Whitney stratified subanalytic sets in Euclidean space by Noirel [184], see also [185]. It then follows from theorems of Hardt [123] or Hironaka [127] that such spaces can all be triangulated; for compact Thom–Mather spaces triangulability was shown directly by Goresky [112]. Hence if X is an n-dimensional Thom–Mather space such that $X - \Sigma_X$ is dense, then X is a PL pseudomanifold by Proposition 2.5.19. Irreducible complex analytic varieties, and so in particular irreducible complex algebraic varieties, possess connected dense submanifolds [153, Corollary IV.2.8.3], and they can be Whitney stratified by complex submanifolds [109, Section I.1.7]. Thus such varieties can be stratified as classical PL stratified pseudomanifolds, as we saw already by a different argument in Section 2.5.2.

Whitney's umbrella shows that not every irreducible real variety will be a pseudomanifold, as it does not contain a dense manifold subset. However, any such variety that does possess a dense manifold subset will be a pseudomanifold. This condition is sometimes referred to as X being *equidimensional* or *pure*.

Another important class of spaces that can be given Thom–Mather stratifications are the orbit spaces of smooth manifolds under smooth actions of compact Lie groups. In fact, if G is a compact Lie group that acts smoothly on the manifold M, then M can be given a Whitney stratification whose strata are G-invariant submanifolds, and the strata of the orbit space M/G can be taken to be the images of the strata of M under the quotient map; see

Section II.4 and, particularly, [72, Theorems II.4.4 and II.4.5]. As M is a manifold, the union of regular strata of the Whitney stratification is dense, and if M is connected and the Whitney stratification can be taken to have no codimension-one strata then there must be just one regular stratum R. As R is dense in M, the image stratum R/G must be dense in M/G, and so M/G will be a pseudomanifold. In fact, even if M has a codimension-one stratum, so long as M/G is connected it follows from the Principal Orbit Theorem [36, Theorem IV.3.1] that the union of the regular strata M is dense in M and its image is connected in M/G, and so M/G must have a connected dense stratum. Therefore, for any smooth action of a compact Lie group on a manifold such that M/G is connected, M/G is a PL pseudomanifold.

2.8.3 Homotopically Stratified Spaces

CS sets impose fairly rigid local conditions; any point must possess a neighborhood of a given form. Hence one might wonder whether it is possible to work effectively with manifold stratified spaces that do not possess such conditions. This is indeed the case for a class of spaces we refer to as *manifold homotopically stratified spaces*.

These spaces were introduced by Quinn in [192] to provide "a setting for the study of purely topological stratified phenomena, particularly group actions on manifolds." Quinn's spaces are manifold stratified spaces that are not necessarily locally cone-like. Instead, they must satisfy certain homotopy-theoretic conditions concerning how the strata fit together. To explain these conditions, we will need to introduce some definitions.

If X is a filtered space, a map $f : Z \times A \to X$ is *stratum-preserving along* A if $f(z \times A)$ lies in a single stratum of X for each $z \in Z$. If $A = I = [0, 1]$, we call f a *stratum-preserving homotopy*. If $f : Z \times I \to X$ is only stratum-preserving when restricted to $Z \times [0, 1)$, we say f is *nearly stratum-preserving*.

If X is a filtered space, then $Y \subset X$ is *forward tame* in X if there is a neighborhood U of Y in X and a nearly stratum-preserving deformation retraction $R : U \times I \to X$ retracting U to Y relative to Y. So R is a strong deformation retraction that keeps each point in its original stratum until time 1 when everything collapses into Y.

The *stratified homotopy link* of Y in X, denoted $\operatorname{holink}_s(X, Y)$, is the space (with compact-open topology) of nearly stratum-preserving paths with their heads in Y and their tails in $X - Y$:

$$\operatorname{holink}_s(X, Y) = \{\omega \in X^I \mid \omega(1) \in Y, \omega([0, 1)) \subset \text{a single stratum of } X - Y\}.$$

The *holink evaluation* map takes a path $\omega \in \operatorname{holink}_s(X, Y)$ to $\omega(1)$. For $x \in X_i$, the *local holink*, denoted $\operatorname{holink}_s(X, x)$, is simply the subset of paths $\omega \in \operatorname{holink}_s(X, X_i)$ such that $\omega(1) = x$. Holinks inherit natural filtrations from their defining spaces, as in Example 2.2.6:

$$(\operatorname{holink}_s(X, Y))^j = \{\omega \in \operatorname{holink}(X, Y) \mid \omega(0) \in X^j\}.$$

Using these notions, we can now provide the definition of *manifold homotopically stratified spaces*:

Definition 2.8.3 A filtered space X is a *manifold homotopically stratified space (MHSS)* if the following conditions hold:

- X is locally compact, separable, and metric.
- Each X_i is an i-manifold and is locally closed in X.
- For each $k > i$, X_i is forward tame in $X_i \cup X_k$.
- For each $k > i$, the holink evaluation $\mathrm{holink}_s(X_i \cup X_k, X_i) \to X_i$ is a fibration.
- For each $x \in X$, there is a stratum-preserving homotopy

$$\mathrm{holink}(X,x) \times I \to \mathrm{holink}(X,x)$$

from the identity into a compact subset of $\mathrm{holink}(X,x)$[24].

While these spaces may seem complex, they have important applications. For example, manifold homotopically stratified spaces can arise as quotient spaces of manifolds under topological group actions and they have been utilized in this context by Yan [253], Beshears [27], and Weinberger and Yan [240, 241] to study topological group actions on manifolds. They also arise naturally in categories with more structure – for example, Cappell and Shaneson showed that they occur as mapping cylinders of maps between smoothly stratified spaces [48]. MHSSs even show up when simply studying manifolds and their submanifolds; for example a locally-flat topological submanifold of a higher-dimensional manifold may not possess a mapping cylinder neighborhood, but such a pair does satisfy the homotopy conditions required to constitute a manifold homotopically stratified space (it is also a stratified pseudomanifold); see Hughes, Taylor, Weinberger, and Williams, [132]. This does imply the existence of a certain type of neighborhood structure more general than a mapping cylinder, and such structures for MHSSs with more strata have been developed by Hughes [131]. There is even a surgery theory for MHSSs that has been developed by Weinberger [239]. A further survey of MHSSs in such geometric settings can be found in Hughes and Weinberger [133]. More recently, the homotopy properties of manifold homotopically stratified spaces have been studied in the work of Miller [172, 173] and Woolf [251].

Intersection homology of manifold homotopically stratified spaces has been studied by Quinn [191] and the author [83, 84, 86]. Unfortunately, providing a detailed treatment in this book would take us too far afield, so we simply mention that [191] provides a version for such spaces of the topological invariance theorem that occurs below as Theorem 5.5.1, while [86] contains a Poincaré Duality Theorem. The latter is proven via a combination of sheaf-theoretic axiomatics and chain-theoretic computations involving stratified homotopies.

[24] This condition, requiring *compactly dominated local holinks*, was not part of the original definition of Quinn [192]. It first appears in the work of Hughes leading towards his Approximate Tubular Neighborhood Theorem in [131].

2.9 Maps of Stratified Spaces

A central tenet of topology is that even if one is interested only in studying a specific space, it is important to be able to consider maps into and out of that space. When working with stratified spaces, it is natural to work with maps that are in some sense compatible with the stratification. For example, the definition of a distinguished neighborhood uses only homeomorphisms that preserve the filtration between the distinguished neighborhood N and its "model" $\mathbb{R}^i \times cL$. But of course it is too limiting to work only with homeomorphisms, and so it is necessary to define more general "stratified maps." There are various definitions in the literature. We will use the following ones.

The basic idea of a stratified map $f : X \to Y$ is that a stratum of X should not be mapped across multiple strata of Y.

Definition 2.9.1 If X, Y are filtered spaces and $f : X \to Y$ is a continuous function (map), we say that f is a *stratified map* if for each stratum $S \subset X$ there is a unique stratum $T \subset Y$ such that $f(S) \subset T$.

Example 2.9.2 If X is a filtered space and Y is filtered trivially as $\emptyset \subset Y$, then any map $f : X \to Y$ is a stratified map. In particular, any map between trivially filtered spaces is stratified (trivially).

If Z is a subset of a filtered space X endowed with the filtration $Z^i = Z \cap X^i$, then the inclusion $Z \hookrightarrow X$ is a stratified map. △

Suppose that $f : X \to Y$ is a stratified map that is also a homeomorphism and that f^{-1} is also stratified. If $f(S) \subset T$ for strata $S \subset X$ and $T \subset Y$, then f must take S homeomorphically onto T: The restriction of f to S is certainly injective, and if it were not surjective onto T then f^{-1} could not be a stratified map. It follows that f sets up a bijection between strata of S and strata of T. However, skeleta of X and Y might have different formal dimensions. For the purposes of intersection homology, what is really important in this setting is the preservation of codimension, so we make the following definition:

Definition 2.9.3 If X and Y are filtered spaces and $f : X \to Y$ is a homeomorphism such that f and f^{-1} are both stratified maps, we call f a *stratified homeomorphism* if for each stratum $S \subset X$ the codimension of the stratum $f(S)$ in Y is equal to the codimension of S in X.

Example 2.9.4 Suppose $X = X^n = \mathbb{R}^n$ filtered by $\{0\} \subset \mathbb{R}^n$ as a manifold stratified space (so $\{0\} = X^0$). Then X is not stratified homeomorphic to the trivially filtered $\mathbb{R}^n$. △

Example 2.9.5 Let X be a filtered space, and let Y be the filtered space with the same underlying space as X but such that $Y^i = X^{i-k}$ for some $k \geq 0$; this includes the assumption that if X has formal dimension n, then Y has formal dimension $n + k$. Then the identity map on the underlying space provides a stratified homeomorphism between X and Y that is not a filtered homeomorphism unless $k = 0$. △

Example 2.9.6 The filtered homeomorphisms of Definition 2.3.2 are stratified homeomorphisms that also preserve dimension.

If $f : X \to Y$ is a stratified homeomorphism of manifold stratified spaces then f must be a filtered homeomorphism because the formal dimension of each stratum must agree with its topological dimension as a manifold. △

Example 2.9.7 An important class of maps of stratified spaces is the *normally nonsingular maps*. As noted in Fulton and MacPherson [101], these are maps such that "the singularities of X at any point x are no better or worse than the singularities of Y at $f(x)$." Normally nonsingular inclusions will play a role in our discussion of L-classes in Section 9.4. We next define these maps. △

Often, normally nonsingular maps are defined without explicit reference to stratifications (see [106, Section 5.4], [101, Section 4.1], or [113, Section 8.6]), though we'll include stratification information here:

Definition 2.9.8 Suppose X is a filtered space and that $Z \subset X$ is a subspace. The inclusion $j : Z \hookrightarrow X$ is called a *normally nonsingular inclusion of codimension c* if there is a filtration on Z and a vector bundle $p : V \to Z$ with fibers $\mathbb{R}^c$ and filtered by $V^{i+c} = p^{-1}(Z^i)$ such that, identifying Z with the zero section of V, the embedding j extends to a filtered homeomorphism from V onto a neighborhood N of Z in X; see Figure 2.5. A normally nonsingular inclusion is a stratified map, and Z is called a *normally nonsingular subspace of codimension c* of X. Notice that Z is also a normally nonsingular subspace of N of codimension c and that N itself is a normally nonsingular subspace of X of codimension zero.

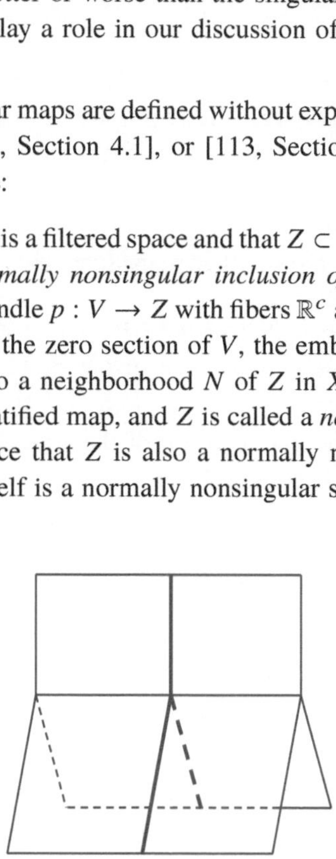

Figure 2.5 The bold subspace is a normally nonsingular subspace of codimension one with trivial normal bundle.

If Y is a filtered space and there is a vector bundle $\pi : W \to Y$ with fibers $\mathbb{R}^k$ and filtered by $W^{i+k} = p^{-1}(Y^i)$, then the projection $\pi : W \to Y$ is called a *normally nonsingular projection* of codimension $-k$. Normally nonsingular projections are also stratified maps. This definition is sometimes generalized to include bundles with manifold fibers as in [106, Section 5.4.2].

The composition of a normally nonsingular inclusion of codimension c and a normally nonsingular projection of codimension $-k$ is called a *normally nonsingular map of codimension $c - k$*.

Example 2.9.9 For any filtered space X, any of the inclusions $X \hookrightarrow cX$ determined by $x \to (t, x)$ for some fixed $t > 0$ is a normally nonsingular inclusion of codimension one. △

We also have a notion of stratified homotopy. For stratified homotopy equivalences, we will again want codimension to be preserved appropriately for our later applications to intersection homology.

Definition 2.9.10 Let X, Y be filtered spaces, and let I be the unit interval with the trivial filtration. Endow $I \times X$ with its product filtration. Then a stratified map $H : I \times X \to Y$ is called a *stratified homotopy*; in particular for each stratum $S \subset X$, we must have $H(I \times S)$ contained in a single stratum of Y. If $f = H|_{\{0\}\times X}$ and $g = H|_{\{1\}\times X}$, we say that f and g are *stratified homotopic* stratified maps.

If $f : X \to Y$ and $g : Y \to X$ are stratified maps such that

1. fg is stratified homotopic to id_Y and gf is stratified homotopic to id_X,
2. for each stratum $S \subset X$, the codimension of the stratum $f(S)$ in Y is equal to the codimension of S in X, and
3. for each stratum $T \subset Y$, the codimension of the stratum $g(T)$ in X is equal to the codimension of T in Y,

then we say that f and g are *stratified homotopy equivalences*, that f and g are *stratified homotopy inverses* to each other, and that X and Y are *stratified homotopy equivalent*.

Remark 2.9.11 As for stratified homeomorphisms, a stratified homotopy equivalence is possible if and only if there is a bijection between the set of strata of X and the set of strata of Y such that if $S \subset X$ and $T \subset Y$ correspond under the bijection then $f(S) \subset T$ and $g(T) \subset S$.

Remark 2.9.12 Some care must be taken not to confuse *stratified homotopies*, as just defined, with *stratum-preserving homotopies*, as defined in Section 2.8.3. Our stratum-preserving homotopies did not require the domains to be filtered spaces and only required that $H(\{x\} \times I)$ be contained in a single stratum of the codomain for each $x \in X$. As usual, the reader should be careful with terminology elsewhere in the literature as it might differ from what we utilize here.

Example 2.9.13 Suppose X is a filtered space, that $\mathbb{R}^n$ is given the trivial filtration, and that $\mathbb{R}^n \times X$ is given the product filtration as in Example 2.2.25. Then the inclusion $X \to \mathbb{R}^n \times X$ given by $x \to (0, x)$ is a stratified homotopy equivalence.

If cX is the cone on X with vertex v, filtered as in Example 2.2.11, then $cX - \{v\}$ is stratified homeomorphic to $\mathbb{R} \times X$ and so stratified homotopy equivalent to X. △

2.10 Advanced Topic: Intrinsic Filtrations

In order to prove some of our more advanced results later, we will need a deeper understanding of CS sets, including results about intrinsic filtrations of CS sets and of PL pseudomanifolds. These are certain filtrations inherent to the topology of the space and of which all other filtrations are refinements, in a sense to be made precise below. For reference purposes, this chapter is the most natural place to include such results, but we strongly urge the first-time reader to proceed on to our discussion of intersection homology in Chapter 3 and return here as needed. The first such necessity will be in our discussion of the invariance theorem (Theorem 5.5.1) in Section 5.5.

Our first lemma will be a general theorem of point-set topology concerning conical neighborhoods. In King [139], the theorem[25] is attributed to Stallings with references to [221] and [216]. The precise statement of the lemma does not seem to be contained in those references, though the proof we give is certainly a direct application of their techniques; in particular it is a slick combination of Stallings's "invertible cobordisms" with an infinite process trick.

Lemma 2.10.1 *Let X and Y be compact[26] topological spaces, and let v and w be the respective cone points of cX and cY. If there is a neighborhood U of the vertex v of cX such that $(U,v) \cong (cY,w)$, then $(cX,v) \cong (cY,w)$.*

Proof For $0 < t \leq 1$, let $c_tX = [0,t) \times X/\sim$, which we identify as a subset of $cX = c_1X = [0,1) \times X/\sim$. Each c_tX is a retraction of cX along its cone lines. Similarly, for $0 < t < 1$, let $\bar{c}_tX = [0,t] \times X/\sim \subset cX$. Note that c_tX is the interior of $\bar{c}_tX$.

We need to build a nested collection of neighborhoods of v. Since U is a neighborhood of v and X is compact, there is a $\delta \in (0,1)$ such that $\bar{c}_\delta X \subset U$. This follows from the Tube Lemma [181, Lemma 26.8]: Let $\pi : [0,1) \times X \to cL$ be the quotient map. By definition of the quotient topology, $U \subset cX$ is open if and only if $\pi^{-1}(U) \subset [0,1) \times X$ is open. In order for U to contain the cone point, $\pi^{-1}(U)$ must contain $\{0\} \times X$, and since X is compact, the Tube Lemma tells us that there is an open subset of $[0,1) \times X$ of the form $[0,s) \times X$ contained in $\pi^{-1}(U)$. But then if $\delta = s/2$, $\bar{c}_\delta X \subset U$. Similarly, using the homeomorphism $h : (cY,w) \to (U,v)$, there is a μ such that $h(\bar{c}_\mu Y)$ is contained in $c_\delta X$. Further applications of the argument then provide γ, ν such that $\bar{c}_\gamma X \subset h(c_\mu Y)$ and $h(\bar{c}_\nu Y) \subset c_\gamma X$. See Figure 2.6, where X_t denotes the image of $\{t\} \times X$ in cX and similarly for Y.

Let $P = \bar{c}_\gamma X - h(c_\nu Y)$, $Q = h(\bar{c}_\mu Y) - c_\gamma X$ and $R = \bar{c}_\delta X - h(c_\mu Y)$. Observe that P has disjoint boundary components homeomorphic to X and Y and that the boundary components can be taken to have disjoint collars; similar statements hold for Q and R.

We will write PQ to stand for the union of P and Q along their common boundary that

[25] Note that there is a typo in the conclusion of this theorem in its statement as Proposition 1 of [139]: in King's notation, the last symbol in the statement should be $(\mathring{C}Y, *)$.

[26] The assumption of compactness does not occur in [139]. I am not sure whether or not the lemma still holds without this assumption, but the compact case will be sufficient for our purposes.

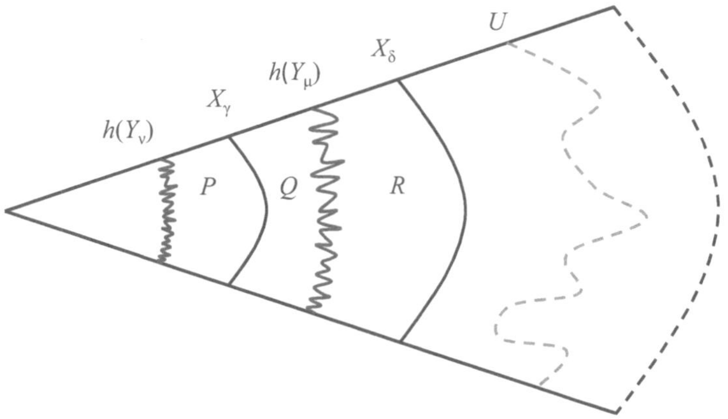

Figure 2.6 A diagram for the proof of Lemma 2.10.1.

is homeomorphic to X, and QR for the union of Q and R along the common boundary that is homeomorphic to Y. Notice that

$$PQ \cong h(\bar{c}_\mu Y) - h(c_\nu Y) \cong h(\bar{c}_\mu Y - c_\nu Y) \cong h([0,1] \times Y) \cong [0,1] \times Y$$

and

$$QR \cong \bar{c}_\delta X - c_\gamma X \cong [0,1] \times X.$$

We next claim that also $RQ \cong Y \times [0,1]$, where RQ is the union of Q and R, now along their common copy of X. To see this, we observe (using the collars of the boundaries and continuing to use concatenation to represent union along common boundaries) that

$$RQ \cong ([0,1] \times Y)RQ \cong (PQ)RQ \cong P(QR)Q \cong P([0,1] \times X)Q \cong PQ \cong [0,1] \times Y.$$

Now consider the infinite union

$$(\bar{c}_\gamma X)(QR)(QR)(QR) \cdots \cong (\bar{c}_\gamma X)([0,1] \times X)([0,1] \times X) \cdots \cong cX.$$

But if we regroup, this is the same as

$$\begin{aligned}((\bar{c}_\gamma X)Q)(RQ)(RQ) \cdots &\cong h(\bar{c}_\mu Y)([0,1] \times Y)([0,1] \times Y) \cdots \\ &\cong \bar{c}_\mu Y([0,1] \times Y)([0,1] \times Y) \cdots \cong cY.\end{aligned}$$

This completes the proof. □

Corollary 2.10.2 *Let X and X' be two CS set stratifications of the same underlying topological space $|X|$. Let $x \in |X|$, and let N, N' be distinguished neighborhoods of x in X and X', respectively. Then N and N' are homeomorphic as topological spaces.*

Proof Any distinguished neighborhood of x in some CS set stratification has the form $\mathbb{R}^k \times cL$ by definition. Suppose $N \cong \mathbb{R}^i \times cL$ and $N' \cong \mathbb{R}^j \times cL'$ are two such distinguished neighborhoods. Notice that

$$(N,x) \cong (\mathbb{R}^i \times cL, x) \cong (c(S^{i-1} * L), x)$$

and

$$(N',x) \cong (\mathbb{R}^j \times cL', x) \cong (c(S^{j-1} * L'), x),$$

where $*$ indicates the join of two spaces. By contracting along cone lines, we may assume, up to homeomorphism, that $N' \subset N$. But now, by the lemma, $(N,x) \cong (N',x)$. □

Although we will not need Lemma 2.10.1 or Corollary 2.10.2 until Section 5.5, they are closely related to our next goal, which is to construct from a CS set X a new CS set $\mathfrak{X}$ with the same underlying space and formal dimension as X but with the filtration of $\mathfrak{X}$ determined only by the topological properties of X. In other words, we will construct an *intrinsic filtration*. We provide a version of the argument of King [139], who credits Dennis Sullivan with the construction, though see also [122].

Definition 2.10.3 Let X be a CS set of formal dimension n, and define an equivalence relation $\sim$ on X such that two points $x_0, x_1 \in X$ are equivalent if they possess neighborhoods U_0, U_1 such that $(U_0, x_0) \cong (U_1, x_1)$ as topological space pairs (i.e. ignoring the filtrations). We sometimes abbreviate this condition by saying that x_0 and x_1 have "homeomorphic neighborhoods," although this isn't technically precise.

It is clear that this is an equivalence relation; we need one further property:

Lemma 2.10.4 *If x_0, x_1 are both in the same stratum of X, then $x_0 \sim x_1$.*

Proof Let S be a stratum of X of dimension i, and let $x_0 \in S$. We will show that the set W of points of S that are equivalent to x_0 is both open and closed in S. Since S is connected, it will follow that $W = S$, which will prove the lemma.

First, suppose $x \in W$ and notice that, by definition of CS sets, x has a neighborhood N homeomorphic to $\mathbb{R}^i \times cL$ with x corresponding to the point $0 \times v$ and with $S \cap N \cong \mathbb{R}^i \times \{v\}$. But since $(\mathbb{R}^i \times cL, 0 \times v) \cong (\mathbb{R}^i \times cL, z \times v)$ for any $z \in \mathbb{R}^i$, we see that we must have $x \sim y$ for any $y \in S \cap N$. Hence W is open in S.

Next, suppose $y \in \bar{W}$, where $\bar{W}$ is the closure of W in S. Then again y has a distinguished neighborhood N homeomorphic to $\mathbb{R}^i \times cL$ with y corresponding to the point $0 \times v$ and with $S \cap N \cong \mathbb{R}^i \times \{v\}$. Since $y \in \bar{W}$, there must be some $x \in W$ corresponding to $z \times v \in \mathbb{R}^i \times \{v\}$ for some $z \in \mathbb{R}^i$. But now again since $(\mathbb{R}^i \times cL, 0 \times v) \cong (\mathbb{R}^i \times cL, z \times v)$, we have $y \sim x \sim x_0$. So W is also closed, and it follows that $W = S$. □

Now, since any two points in a stratum of X are equivalent, it follows that the equivalence classes under $\sim$ must be unions of strata of the filtration. Let $\mathfrak{X}^i$ be the union of the equivalence classes that only contain strata of dimension $\leq i$. Then each $\mathfrak{X}^i$ and $\mathfrak{X}_i = \mathfrak{X}^i - \mathfrak{X}^{i-1}$

will be a union of equivalence classes of X and a union of strata of X of dimension $\leq i$. The $\mathfrak{X}^i$ also give a CS set filtration of the underlying space[27] $|X|$, and this new filtration is intrinsic to $|X|$ as a topological space.

Proposition 2.10.5 *Given a CS set X, let $\mathfrak{X}^i$ be the union of the equivalence classes that only contain strata of X of dimension $\leq i$. Suppose m is the dimension of the highest-dimensional non-empty stratum of X. Then for any integer[28] $k \geq m$, the subsets $\mathfrak{X}^i$ for $-1 \leq i \leq k$ filter X as a CS set of formal dimension k. Furthermore, the sets $\mathfrak{X}^i$ do not depend on the initial filtration of X as a CS set; in other words, if we begin with a different CS set stratification $\hat{X}$ of the underlying space $|X|$ and construct a filtration $\hat{\mathfrak{X}}^i$ analogously to the construction of the $\mathfrak{X}^i$, then $\mathfrak{X}^i = \hat{\mathfrak{X}}^i$.*

Before proving the proposition, let us consider a definition and some examples.

Definition 2.10.6 Given a CS set X of formal dimension n, let $\mathfrak{X}$ denote $|X|$ with the filtration constructed in Proposition 2.10.5 and with the same formal dimension n as X. This is called the *intrinsic filtration of X of formal dimension n.*

Remark 2.10.7 In general, if X is a filtered space then a coarsening of a filtration $\{X^i\}$ of X is a second filtration $\{Z^i\}$ of the same underlying space such that each stratum of the Z filtration is a union of strata of the X filtration. The proposition shows that $\mathfrak{X}$ is the *intrinsic coarsest filtration* of $|X|$ as a CS set in the sense that $\mathfrak{X}$ is also a CS set and no matter what stratification of $|X|$ as a CS set we begin with, $\mathfrak{X}$ is a coarsening of it.

Example 2.10.8 Suppose $X = X^n = M$ is a smooth n-dimensional manifold and that $V \subset M$ is a smooth submanifold so that X is filtered as $V \subset M$. Then every point $x \in M$ has a neighborhood homeomorphic to $(\mathbb{R}^n, x)$, so all points of M are equivalent and the intrinsic filtration is the trivial stratification on M. △

Example 2.10.9 The suspended torus ST^2 of Figure 1.1 on page 2 already has its intrinsic filtration. However, the twice suspended torus of Figure 1.2 on page 3 is not intrinsically filtered. As we leave the reader to show, all of the points of X^1 are equivalent. The coarsening $\emptyset \subset X^1 \subset X^4$ of that example is an intrinsic filtration. △

Now let us prove Proposition 2.10.5.

Proof of Proposition 2.10.5 First, we show that the $\mathfrak{X}^i$ are closed subsets of $|X|$. Suppose $x \in |X| - \mathfrak{X}^i$. So x must be equivalent to a point in a stratum of X of dimension greater than i. But then from the definitions every point of any distinguished neighborhood N of x is in $X - X^i$. So no point of N is contained in $\mathfrak{X}^i$, so $\mathfrak{X}^i$ is closed. Thus $\emptyset = \mathfrak{X}^{-1} \subset \mathfrak{X}^0 \subset \cdots \subset \mathfrak{X}^n$ is a filtration of $|X|$. In fact, if m is the dimension of the highest-dimensional non-empty stratum of X, then we have both $X^m = \cdots = X^n$ and $\mathfrak{X}^m = \cdots = \mathfrak{X}^n$, as every point of X is contained in a stratum of dimension at most m. Therefore, $\emptyset = \mathfrak{X}^{-1} \subset \mathfrak{X}^0 \subset \cdots \subset \mathfrak{X}^k$ is a filtration of $|X|$ for any $k \geq m$ and with $\mathfrak{X}^m = \cdots = \mathfrak{X}^k$.

[27] Recall that we use $|X|$ to denote the underlying space of X, disregarding any filtration information.
[28] Note that $\mathfrak{X}^i$ is well defined even if i exceeds the formal dimension of X.

Next we must show that points have distinguished neighborhoods with respect to the filtration by $\mathfrak{X}^i$. First we suppose $z \in X_i \cap \mathfrak{X}_i$. Then z has a distinguished neighborhood N in X that we identify with $\mathbb{R}^i \times cL$. We also identify L with $\{0\} \times \{1/2\} \times L \subset \mathbb{R}^i \times cL$. Let $Z^j = |N| \cap \mathfrak{X}^j$, and let $\ell^{j-i-1} = |L| \cap Z^j$. We claim that $Z^j = \mathbb{R}^i \times c\ell^{j-i-1}$. In other words, this says that if we filter $|L|$ by the ℓ^k and denote the resulting filtered space by ℓ, then $\mathbb{R}^i \times c\ell$ is filtered homeomorphic to $|N|$ filtered by the Z^j. So this will provide a distinguished neighborhood of z in $\mathfrak{X}$.

To verify the claim, we first observe that every point of N is contained in a stratum of X of dimension $\geq i$, so $|N|$ can only intersect nontrivially strata of $\mathfrak{X}$ of dimension $\geq i$. Let us show that $\mathbb{R}^i \times \{v\}$ must equal $|N| \cap \mathfrak{X}^i = Z^i$: We already know that $\mathbb{R}^i \times \{v\} = N \cap X_i$. As these points are all in the same stratum of X, they are all equivalent and so all must lie in $\mathfrak{X}_i$. Furthermore, the points in $|N| - X_i$ are all contained in strata of X of dimension $> i$ and so must be in strata of $\mathfrak{X}$ of dimension $> i$. Thus $|N| \cap \mathfrak{X}^i = |N| \cap X^i = \mathbb{R}^i \times \{v\}$. Next, for $j > i$, we have that $\ell^{j-i-1} \subset Z^j$ by definition. But if s is a point of $|L| = |\ell|$, then the points of $\mathbb{R}^i \times (0,1) \times \{s\} \subset |\mathbb{R}^i \times cL|$ all have homeomorphic neighborhoods. So the equivalence class of all the points in any $\mathbb{R}^i \times (0,1) \times \{s\}$ is determined completely by the equivalence class of $\{s\}$. It follows that we must have $Z^j = \mathbb{R}^i \times c\ell^{j-i-1}$, as desired. So z has a distinguished neighborhood in $\mathfrak{X}$.

Now, suppose x is any point in $\mathfrak{X}_i$. Then x is equivalent to a point z in an i-dimensional stratum of X since, by definition of $\mathfrak{X}^i$, the point x is in an equivalence class containing strata of X of dimension at most i, but if x is in an equivalence class containing only strata of dimension at most $i-1$, then x would be in $\mathfrak{X}^{i-1}$. Since z is equivalent to x, we also have $z \in \mathfrak{X}_i$, so $z \in \mathfrak{X}_i \cap X_i$. We have already shown that z possesses a distinguished neighborhood, and we will use this to construct a distinguished neighborhood for x. Since x is equivalent to z, they have homeomorphic neighborhood pairs (U, x) and (V, z). Let $h : V \to U$ be the homeomorphism. Let $\mathfrak{N} = \mathbb{R}^i \times c\ell$ be the distinguished neighborhood of z in $\mathfrak{X}$. By contracting $\mathbb{R}^i$ and $c\ell$ by homeomorphisms if necessary, we may assume $\mathbb{R}^i \times c\ell \subset V$. We claim that $h(\mathfrak{N})$ is then a distinguished neighborhood of x. For this we observe that, as h is a homeomorphism and as equivalence classes are determined entirely by local topology, h takes every point in $\mathfrak{N}$ to an equivalent point, i.e. y and $h(y)$ are equivalent for all $y \in \mathfrak{N}$. But then as the $\mathfrak{X}_j$ are unions of equivalence classes, h must take points of $\mathfrak{N} \cap \mathfrak{X}_j$ to points of $\mathfrak{X}_j$. So h induces a filtered homeomorphism from $\mathfrak{N}$ to its image, providing a distinguished neighborhood for x.

So we have shown that every point $x \in \mathfrak{X}_i$ has a neighborhood filtered homeomorphic to $\mathbb{R}^i \times c\ell$, and so we have a CS set.

Finally, let us demonstrate that the filtration by the $\mathfrak{X}^i$ does not depend on the initial filtration of X as a CS set. Let $\hat{X}$ denote an alternative CS set filtration of X, and let $\hat{\mathfrak{X}}^i$ denote the corresponding CS set filtration, i.e. each $\hat{\mathfrak{X}}^i$ is the union of the equivalence classes that only contain strata of $\hat{X}$ of dimension $\leq i$. Note that the definitions of the sets $\mathfrak{X}^i$ and $\hat{\mathfrak{X}}^i$ make sense for all integers $i \geq -1$, so we will allow that possibility for the rest of the argument.

We make the following observation: Suppose $x \in X$ has a neighborhood $|N|$ that is homeomorphic to $|\mathbb{R}^j \times cL|$ for some compact L (ignoring the filtration information). Then we claim x is equivalent to a point in some stratum of X of dimension at least j. Indeed, consider the points of $|N|$ contained in the homeomorphic image in $|X|$ of $|\mathbb{R}^j \times \{v\}|$. All the points in this j-dimensional set are clearly equivalent. We know that each equivalence class is a union of strata of X; so, if the equivalence class of x contained only strata of dimension $< j$, this would create a contradiction, as the union of strata of dimension $< j$ cannot cover a j-dimensional set due to the niceness of the local conical structures. Therefore, x must be equivalent to a point in a stratum of dimension $\geq j$. Now, suppose $x \in \mathfrak{X}^i$. By definition, x is not equivalent to any point in a stratum of X of dimension $> i$. Therefore, by our immediately preceding argument, x cannot have a neighborhood homeomorphic to $|\mathbb{R}^j \times cL|$ for any $j > i$. In particular, x must be contained in a stratum of $\hat{X}$ of dimension $\leq i$, and the same must be true of all points equivalent to x, i.e. all points equivalent to x are contained in $\hat{X}^i$. But this implies that $x \in \hat{\mathfrak{X}}^i$. So $\mathfrak{X}^i \subset \hat{\mathfrak{X}}^i$. The equivalent argument then shows that $\hat{\mathfrak{X}}^i \subset \mathfrak{X}^i$, and so $\mathfrak{X}^i = \hat{\mathfrak{X}}^i$ for all i. □

Lemma 2.10.10 *Let U be an open subset of the CS set X. Then U is a CS set with the filtration $U^i = U \cap X^i$ and the intrinsic filtration $\mathfrak{U}$ agrees with the restriction of the stratification of $\mathfrak{X}$ to $|U|$. In other words, $\mathfrak{U}^i = \mathfrak{X}^i \cap U$.*

Proof That U with the filtration coming from X is a CS set is Lemma 2.3.13. Therefore, U possesses an intrinsic filtration $\mathfrak{U}$ by Proposition 2.10.5. As the equivalence relation of Definition 2.10.3 is determined entirely by local conditions, we see that two points in U are equivalent if and only if they are equivalent in X. The lemma therefore follows from the definitions of $\mathfrak{U}$ and $\mathfrak{X}$. □

We will need one more lemma, which is a slight generalization of [139, Lemma 2].

Lemma 2.10.11 *Let M be a manifold (trivially filtered) and X a filtered space such that $M \times X$ is a CS set. Then the intrinsic filtration $(M \times X)^*$ of $|M \times X|$ is filtered homeomorphic to $M \times Z$, where Z is some coarsening of X (see Remark 2.10.7). Furthermore, if M_1, M_2 are two n-manifolds and Z_1, Z_2 are coarsenings of X such that $(M_i \times X)^*$ is filtered homeomorphic to $M_i \times Z_i$ for $i = 1, 2$, then $Z_1 = Z_2$. In other words, the coarsening Z of X depends only on the dimension of the manifold M.*

Proof Let $x \in X$, and notice that all points of $M \times \{x\}$ are equivalent under $\sim$. So $(M \times X)^*$ must have the form $M \times Z$ for some filtration Z of X. Furthermore, since the intrinsic filtration of $M \times X$ is coarser than that of $M \times X$, the filtration of Z must be coarser than that of X. The second claim of the lemma follows because the definition of the intrinsic filtration is local, and so the filtration Z depends only on the subspace $|\mathbb{R}^n \times X|$. □

Notice that the filtered space Z of the lemma is not necessarily a cone, and even if v is the cone point of cW, then $M \times \{v\}$ might not be a stratum of $|M \times cW|$ with its intrinsic filtration; it may be a subset of a larger stratum. For example, if $W = S^{n-1}$ and $M = \mathbb{R}^k$, then $M \times cW$ is homeomorphic to $\mathbb{R}^{k+n}$, whose intrinsic filtration is trivial and so stratified homeomorphic to $\mathbb{R}^{n+k} \cong \mathbb{R}^k \times \mathbb{R}^n$. Here the trivial filtration of $\mathbb{R}^n$ provides the intrinsic filtration of cW.

Remark 2.10.12 We will see below in Lemma 2.10.17 that, in the piecewise linear world, Lemma 2.10.11 can be strengthened to the statement that if M is a PL n-manifold and X a PL filtered space then the intrinsic filtration of $M \times X$ is PL homeomorphic to $M \times \mathfrak{X}$, where $\mathfrak{X}$ is the intrinsic filtration of X; no such claim is made in Lemma 2.10.11. The proof relies strongly on facts of PL topology, so it is not clear that there are versions of this available in the topological world, even with additional restrictions.

2.10.1 Intrinsic PL Filtrations

One can also consider intrinsic PL filtrations of PL spaces. These have been a historically important tool in the study of PL spaces (e.g. [2]), and they will be particularly important for us in our construction of L-classes in Section 9.4. They are also useful in studying bordism groups of PL pseudomanifolds and the resulting bordism homology theories; see [3, 94].

Recall that every PL space is a CS set with respect to some filtration; see Lemma 2.5.17. We can then set up an equivalence relation $\sim_{\text{PL}}$ analogous to that of Definition 2.10.3 but requiring PL homeomorphisms of neighborhoods; we will say that points satisfying the PL version of Definition 2.10.3 are *PL equivalent*. Then PL analogues of Lemma 2.10.4 and Proposition 2.10.5 hold with the identical proofs, assuming each homeomorphism is a PL homeomorphism. For reference purposes, we restate them in this context:

Lemma 2.10.13 *If X is a PL CS set and x_0, x_1 are both in the same stratum of X, then $x_0 \sim_{\text{PL}} x_1$.*

Proposition 2.10.14 *Given a PL CS set X, let $\mathfrak{X}^i_{\text{PL}}$ be the union of the PL equivalence classes that only contain strata of X of dimension $\leq i$. Suppose m is the dimension of the highest-dimensional non-empty stratum of X. Then for any integer $k \geq m$, the subsets $\mathfrak{X}^i_{\text{PL}}$ for $-1 \leq i \leq k$ filter X as a PL CS set of formal dimension k. The PL CS set filtration $\{\mathfrak{X}^i_{\text{PL}}\}$ does not depend on the initial filtration of X as a PL CS set.*

One extra observation is needed for the proposition: the intrinsic PL skeleta, say $\mathfrak{X}^i_{\text{PL}}$, are PL subsets, i.e. they will be subcomplexes in an admissible triangulation of X. To see this, let K be a triangulation of X, and let σ be a simplex of K. Then all points in the open simplex $\mathring{\sigma}$ have the same PL neighborhood, the open star simplicial star of σ in K. So if $\mathfrak{X}^i_{\text{PL}}$ intersects $\mathring{\sigma}$, it contains $\mathring{\sigma}$, and hence all of σ, since each $\mathfrak{X}^i_{\text{PL}}$ is closed by the PL analogue of the arguments of Proposition 2.10.5. Therefore, every $\mathfrak{X}^i_{\text{PL}}$ is a union of simplices, and so it is a subcomplex of the chosen triangulation.

Definition 2.10.15 If X is a PL set, let $\mathfrak{X}_{\text{PL}}$ denote the underlying space X filtered with skeleta $\{\mathfrak{X}^i_{\text{PL}}\}$ and with the same formal dimension as X. We call $\mathfrak{X}_{\text{PL}}$ the *intrinsic PL* *filtration* of X. If the PL context is understood, we may write simply $\mathfrak{X}$ and $\mathfrak{X}^i$, as we will do for the remainder of this section.

The PL version of Lemma 2.10.10 holds by the same arguments used to prove that lemma:

Lemma 2.10.16 *Let U be an open subset of the PL CS set X. Then U is a PL CS set with the filtration $U^i = U \cap X^i$ and the intrinsic PL filtration $\mathfrak{U}$ agrees with the restriction of the intrinsic PL filtration $\mathfrak{X}$ to $|U|$. In other words, $\mathfrak{U}^i = \mathfrak{X}^i \cap U$.*

In the PL setting, we also have a stronger version of Lemma 2.10.11:

Lemma 2.10.17 *Let M be a PL n-manifold (trivially filtered) and X a PL filtered space. Then the intrinsic filtration of $M \times X$ is PL filtered homeomorphic to $M \times \mathfrak{X}$, where $\mathfrak{X}$ is the intrinsic filtration of X.*

Proof Let $(M \times X)^*$ denote the intrinsic filtration of $M \times X$. Since the intrinsic filtration of a PL CS set is the coarsest PL CS set filtration, and since $(M \times X)^*$ and $M \times \mathfrak{X}$ are both PL CS sets, $(M \times X)^*$ must be a coarsening of $M \times \mathfrak{X}$. Suppose these filtrations are not the same. Then there must be two points, say (t, x) and (s, y), that are in the same stratum of $(M \times X)^*$ but different strata of $M \times \mathfrak{X}$. This implies that x and y are in different strata of $\mathfrak{X}$. Since (t, x) and (s, y) are in the same stratum of $(M \times X)^*$, they have PL homeomorphic neighborhoods. Let ℓ be the polyhedral link of x in X, i.e. x has a neighborhood $c\ell$ in X. The space ℓ is unique up to PL homeomorphism by basic PL topology [198, Lemma 2.19]. Owing to the product structure on $M \times X$ and basic PL topology, the point (t, x) then has a neighborhood of the form $c(S^{n-1} * \ell)$, where $S^{n-1} * \ell$ is the PL join of S^{n-1} with ℓ. Notice that $S^{n-1} * \ell$ is also the nth suspension of ℓ. Similarly, if y has polyhedral link ℓ' in X, then (s, y) has a neighborhood in $M \times X$ of the form $c(S^{n-1} * \ell')$. But since (t, x) and (s, y) have PL homeomorphic neighborhoods, by the uniqueness of polyhedral links, we must have $S^{n-1} * \ell \cong S^{n-1} * \ell'$, where $\cong$ denotes PL homeomorphism. But now we can invoke another basic result of PL topology to conclude that $\ell \cong \ell'$ [179, Theorem 1]. This implies that $c\ell \cong c\ell'$, so that x and y have PL homeomorphic neighborhoods in X. This is not quite enough yet to conclude that x and y are in the same stratum of $\mathfrak{X}$, as x and y could have homeomorphic neighborhoods but lie in different strata. However, now let us consider a path from (t, x) to (s, y) in $(M \times X)^*$ in the stratum containing the two points. This is possible because the stratum is a connected PL set. The same arguments we employed above apply to any two points along the path, so if (u, z) is such a point, then z is PL equivalent to x and y in X. Projecting the path to X provides a path between x and y consisting entirely of points that are PL equivalent to both x and y. Therefore, x and y must be in the same stratum of $\mathfrak{X}$. We have reached a contradiction, and so $(M \times X)^* = M \times \mathfrak{X}$. □

Notice that the proof of Lemma 2.10.17 leans heavily upon PL topology.

Thanks to the more structured behavior of the PL category, we are able to prove the following about intrinsic PL filtrations of PL pseudomanifolds.

Proposition 2.10.18 *Let X be an n-dimensional PL space containing a dense n-dimensional PL manifold M. Then $\mathfrak{X}$ is a PL stratified pseudomanifold. If $X - M$ has dimension $\leq n-2$, then $\mathfrak{X}$ is a classical PL stratified pseudomanifold.*

Proof We will prove the proposition by induction on the dimension of X. If X has dimension zero, then we must have that $X = \mathfrak{X}$ is a discrete set of points, and the proposition is immediate. Now, we assume we have proven the lemma for dimensions $< n$ and suppose that X is n-dimensional.

By Lemma 2.5.17, X can be filtered as a PL CS set, so, using that filtration to get started, we can apply the PL analogue of Proposition 2.10.5 to see that $\mathfrak{X}$ is a PL CS set. To show that $\mathfrak{X}$ is a stratified pseudomanifold as defined in Definition 2.5.13, we need to verify that the union of the regular strata of $\mathfrak{X}$ is dense in X and that the links in $\mathfrak{X}$ are themselves recursive PL CS sets.

For the density requirement, we have assumed that X possesses a dense PL manifold subset M. We claim that $M \subset \mathfrak{X}^n - \mathfrak{X}^{n-1}$ so that $\mathfrak{X}^n - \mathfrak{X}^{n-1}$ must also be dense in X. To verify the claim, let's utilize the filtration, say X', of X guaranteed by Lemma 2.5.17; this was simply the simplicial filtration with respect to some admissible triangulation K of X. Since $\mathfrak{X}$ coarsens all other PL CS set filtrations by the PL analogue of Remark 2.10.7, every skeleton of $\mathfrak{X}$ is a union of strata of X'. By definition, we take as the $(n-1)$-skeleton of $\mathfrak{X}$ the union of the PL equivalence classes of points of X that contain only strata of X' of dimension $\leq n-1$. So if $x \in \mathfrak{X}^{n-1}$, then x is contained in an $(n-1)$-simplex of K and x is not equivalent to any point in $(X')^n - (X')^{n-1}$. But $(X')^n - (X')^{n-1}$ is a union of open n-simplices, so this means that x cannot have an n-dimensional PL Euclidean neighborhood, so $x \notin M$. So $x \in M$ implies $x \in \mathfrak{X}^n - \mathfrak{X}^{n-1}$. This is the desired result. We also observe here that if $\dim(X - M) \leq n-2$, then $\dim(\mathfrak{X}^{n-1}) \leq n-2$, and so $\mathfrak{X}$ can then have no codimension-one strata, making it a classical PL stratified pseudomanifold once we have finished showing it is a PL stratified pseudomanifold.

Now we must consider the links of $\mathfrak{X}$ and show that, with the filtrations that are compatible with $\mathfrak{X}$, they are recursive PL CS sets.

Suppose $x \in \mathfrak{X}$ has a distinguished neighborhood U filtered PL homeomorphic to $\mathbb{R}^i \times cL$ for some PL filtered space L. Since $\mathfrak{X}$ is a CS set, every point has some such neighborhood, and this assumption implies that x is in an i-dimensional stratum of $\mathfrak{X}$. Let us identify L as the subset $\{0\} \times \{1/2\} \times L \subset \mathbb{R}^i \times cL$, and let us identify $\mathbb{R}^i \times cL$ with U so that we can think of L as embedded in $\mathfrak{X}$. We claim that the given filtration on L (the one compatible with it being a subset of $\mathfrak{X}$ in this way) is the intrinsic PL filtration of L, which exists by Proposition 2.10.14 because L is a PL space and every PL space can be filtered as a PL CS set. Notice that $\mathbb{R}^i \times (cL - \{v\}) \cong \mathbb{R}^{i+1} \times L$ is an open set of $\mathfrak{X}$ and so is intrinsically PL filtered by Lemma 2.10.16. Lemma 2.10.17 then implies that L must be intrinsically PL filtered. In particular, L is filtered as a PL CS set.

Next we observe that L possesses a dense PL $(n-i-1)$-manifold. For this, fix an admissible triangulation of L, and let M_L be the union of the interiors of the $(n-i-1)$-simplices of the triangulation. We claim that M_L is dense in L. By way of contradiction, assume that $x \in L$ is a point that has no neighborhood that intersects M_L; this implies that x is not a face of any $(n-i-1)$-simplex of L, so x has a neighborhood V in L that has dimension $< n-i-1$. But continuing to think of L as a subspace of X, if x has a neighborhood V of dimension $< n-i-1$ in L, then $0 \times 1/2 \times x$ has a neighborhood homeomorphic to $\mathbb{R}^{i+1} \times V$ in X that must have dimension $< n$. But this is a contradiction with X containing a dense n-manifold. Therefore, every point of L must be in a face of an $(n-i-1)$-simplex and M_L is dense in L. Thus, we can apply the induction hypothesis to conclude that L is a PL stratified pseudomanifold, and, in particular, a PL recursive CS set as desired. □

The following corollaries are immediate:

Corollary 2.10.19 *If X is an n-dimensional PL stratified pseudomanifold, then $\mathfrak{X}$ is an n-dimensional PL stratified pseudomanifold. If X is a classical PL stratified pseudomanifold, then $\mathfrak{X}$ is a classical PL stratified pseudomanifold.*

Corollary 2.10.20 *If X is a PL space with a triangulation in which every simplex is a face of an n-simplex, then $\mathfrak{X}$ is an n-dimensional PL stratified pseudomanifold and so $|X|$ is a PL pseudomanifold. If X is a PL space with a triangulation in which every simplex is a face of an n-simplex and such that every $(n-1)$-simplex is the face of exactly two n-simplices, then $\mathfrak{X}$ is a classical n-dimensional PL stratified pseudomanifold and $|X|$ is a classical PL pseudomanifold.*

For the latter corollary, the manifolds for Proposition 2.10.18 are respectively the union of the interiors of the n-simplices and the union of the interiors of the n- and $(n-1)$-simplices of the triangulations.

(a) Intrinsic Filtrations of PL Pseudomanifolds with Boundary

It is also useful to have a notion of an intrinsic filtration for a pseudomanifold with boundary. This is a bit more delicate, as we know from Example 2.7.3 that the notion of "boundary" itself can depend upon the filtration. Let us reconsider Example 2.7.3. There, we considered a ∂-manifold M with non-empty boundary (in the manifold sense) P. If we let X be the filtered space $P \subset M$, then we have a stratified pseudomanifold (without boundary!). In fact, we can easily verify that this is the intrinsic filtration of X. But if we instead think of M as trivially filtered it becomes a ∂-stratified pseudomanifold with boundary P. It is reasonable to ask for a version of intrinsic filtration that continues to "know" that there is a boundary present.

In fact, no intrinsic filtration, following our previous definitions, could have a boundary. The reason for this is that we have defined ∂-stratified pseudomanifolds so that the boundary has a filtered collar. In other words, if X is a ∂-stratified pseudomanifold, ∂X must have a neighborhood in X stratified homeomorphic to the product $[0,1) \times \partial X$, with $\{0\} \times \partial X$ being taken to $\partial X \subset X$ by the homeomorphism. In particular, then, if $x \in \partial X$ then all the points in[29] $[0,1) \times \{x\}$ live in a single stratum of X. However, the points (t,x), for $0 < t < 1$, will have identical neighborhoods in X, while $(0,x)$ will have a different neighborhood. Thus $(0,x)$ and the (t,x) for $t > 0$ cannot live in the same stratum of any intrinsic filtration. Thus no intrinsic filtration on X could have a non-empty boundary.

Nonetheless, there is still a way that we can usefully introduce intrinsic filtrations into the context of ∂-stratified pseudomanifolds.

Definition 2.10.21 Let X be a ∂-stratified pseudomanifold. We will say that X is *naturally filtered* if $X - \partial X$ and ∂X are intrinsically filtered pseudomanifolds with filtrations determined from X as in Definition 2.7.1, i.e. $(X-\partial X)^i = (X-\partial X)\cap X^i$ and $(\partial X)^{i-1} = (\partial X)\cap X^i$.

Example 2.10.22 Consider again a trivially filtered ∂-manifold M with $\partial M \neq \emptyset$. Then M is naturally stratified, as M and ∂M are both intrinsically stratified. △

Proposition 2.10.23 *Let X be a PL ∂-pseudomanifold. Then X can be naturally filtered. In other words, there exists a filtration $\check{X}$ of X such that:*

1. *$\check{X}$ is a PL ∂-stratified pseudomanifold,*
2. *$(\check{X}, \partial\check{X})$ and $(X, \partial X)$ have the same underlying PL space pairs, and*
3. *$\check{X} - \partial\check{X}$ and $\partial\check{X}$ are intrinsically stratified PL pseudomanifolds.*

Proof By the definition of ∂-stratified pseudomanifolds, $X - \partial X$ and ∂X are each PL stratified pseudomanifolds, and so by Corollary 2.10.19 each has an intrinsic filtration as a PL stratified pseudomanifold. Also by the definition of ∂-stratified pseudomanifolds, ∂X has a collar neighborhood in X that is PL homeomorphic to $[0,1)\times X$. Consider the subspace of X PL homeomorphic to $(0,1)\times\partial X$. If we let $(X-\partial X)^*$ denote the intrinsic filtration, then by Lemma 2.10.16, the restriction of this filtration to $(0,1)\times\partial X$ is intrinsically filtered, and by Lemma 2.10.17 it is $(0,1)\times(\partial X)^*$, where $(\partial X)^*$ is $|\partial X|$ with its intrinsic filtration. Therefore, if we take $(X-\partial X)^*$ and $[0,1)\times(\partial X)^*$, we can glue these spaces along their common PL filtered subset $(0,1)\times(\partial X)^*$. The resulting space is the desired $\check{X}$. Notice that $\check{X}$ does indeed meet the requirements to be a PL ∂-stratified pseudomanifold with $\partial\check{X} = (\partial X)^*$. □

Remark 2.10.24 Notice that the proof of Proposition 2.10.23 depends on Lemma 2.10.17, and, as noted in Remark 2.10.12, we do not necessarily have this available in the topological world. This thwarts our attempts to prove the existence of appropriate analogous natural filtrations of topological ∂-stratified pseudomanifolds in terms of intrinsically stratified CS sets.

Corollary 2.10.25 *Suppose X is a PL space possessing a triangulation such that:*

1. *every simplex is a face of an n-simplex,*
2. *every $(n-1)$-simplex is a face of either one or two n-simplices, and*

[29] Here we use the collar homeomorphism to provide coordinates for points in the collar.

3. *if B is the union of all $(n-1)$-simplices of X that are a face of only one n-simplex, then B has a collar, meaning that there is a PL embedding of $[0, 1) \times B$ into X taking $\{0\} \times B$ to $B \subset X$.*

Then X is an n-dimensional PL ∂-pseudomanifold. If each $(n-2)$-simplex of B is a face of exactly two $(n-1)$-simplices of B, then X is an n-dimensional classical ∂-pseudomanifold.

Proof If we let M denote the union of the interiors of the n-simplices of the triangulation and the interiors of the $(n-1)$-simplices of the triangulation that are not in B, then M is a manifold that is dense in the PL set $X - B$ and $(X - B) - M$ has dimension $\leq n - 2$. So by Proposition 2.10.18, $X - B$ is a classical PL pseudomanifold. Similarly, the interiors of the $(n-1)$-simplices of B are dense in B, so B is a PL pseudomanifold, also by Proposition 2.10.18. By assumption, B is collared in X, so by the same arguments as used in Proposition 2.10.23, we can glue together the intrinsic filtrations of $X - B$ and $[0, 1) \times B$ to obtain a PL ∂-stratified pseudomanifold with X as its underlying space. If B satisfies the extra condition, then both $X - B$ and B will be classical PL pseudomanifolds by Proposition 2.10.18, so X will be a classical PL ∂-stratified pseudomanifold. □

2.11 Advanced Topic: Products and Joins

This section contains proofs that products and joins of CS sets and stratified pseudomanifolds are again CS sets and stratified pseudomanifolds. We also show that the product of ∂-stratified pseudomanifolds is a ∂-stratified pseudomanifold and that the product of intrinsically stratified PL pseudomanifolds is intrinsically stratified. These are obviously desirable results, and, as for the previous section, this material is included here because it fits naturally with our chapter on stratified spaces. However, once again, the first-time reader is encouraged to skip this material for now in order to "get on with it" and to come back to this section as needed later on. In fact, most of the results of this section are the expected ones, so this section should serve more as a reference for the purposes of completeness.

As we observed in Example 2.2.25, if X and Y are filtered spaces, then $X \times Y$ has a natural *product filtration* such that

$$(X \times Y)^i = \bigcup_{j+k=i} X^j \times Y^k.$$

If X and Y have respective formal dimensions n and m, then this product has formal dimension $m + n$.

Lemma 2.11.1 *The strata of $X \times Y$ all have the form $S \times T$, where S is a stratum of X and T is a stratum of Y.*

Proof By a basic set-theoretic argument, which we leave to the reader,

$$(X \times Y)^i - (X \times Y)^{i-1} = \Big(\bigcup_{j+k=i} X^j \times Y^k\Big) - \Big(\bigcup_{j+k=i-1} X^j \times Y^k\Big)$$
$$= \bigcup_{j+k=i} \Big((X^j - X^{j-1}) \times (Y^k - Y^{k-1})\Big).$$

Now consider $S \times T$ for S a stratum of X and T a stratum of Y. To be specific, suppose S has formal dimension j and T has formal dimension k. Since S and T are connected, the set $S \times T$ is connected [181, Theorem 23.6]. We must show that each such $S \times T$ is in fact a connected component of $(X \times Y)^i - (X \times Y)^{i-1}$. As $S \times T$ is connected, it suffices to show that $S \times T$ is separated from each other $S' \times T'$ with S' a stratum of X, T' a stratum of y, $\dim(S') + \dim(T') = j + k = i$, and $S \times T \neq S' \times T'$. In other words, we show that $(S \times T) \cup (S' \times T')$ is not connected. By [181, Lemma 23.1], it suffices to show that neither of $S \times T$ or $S' \times T'$ contains a limit point of the other. The arguments are symmetric, so we will show that $S' \times T'$ cannot contain a limit point of $S \times T$.

Suppose (x, y) is a limit point of $S \times T$. If $(x, y) \in S \times T$, then $(x, y) \notin S' \times T'$ as $(S \times T) \cap (S' \times T')$ is non-empty only if $S = S'$ and $T = T'$. So, suppose $(x, y) \notin S \times T$. Then x is a limit point of S not contained in S or y is a limit point of T not contained in T. Suppose $x \notin S$. As X^j is closed, $x \in X^j$, but x cannot be contained in a j-dimensional stratum because S is a connected component of $X^j - X^{j-1}$. Therefore, $x \in X^{j-1}$. Applying the same argument to y, we have that either $x \in X^{j-1}$ or $y \in Y^{k-1}$, so $(x, y) \in (X \times Y)^{i-1}$. Thus (x, y) is not contained in any of the other strata of dimension i, in particular $S' \times T'$. □

We will see that taking products preserves other nice structure. For example the products of CS sets are CS sets and the products of pseudomanifolds are pseudomanifolds. In order to verify these claims, it is necessary to study not just products of filtered spaces, but also their joins, as the joins arise as links in product spaces.

We recall the construction of the join of two spaces X and Y; see for example [125, Chaper 0 and Additional Topic 4.G]. In all of our applications of joins, X and Y will be compact. Conceptually, the join $X * Y$ of two spaces is the union of all line segments connecting a point of X to a point of Y. A more constructive definition is that $X * Y$ is the quotient space of $X \times [0, 1] \times Y$ under the relations $(x, 0, y) \sim (x, 0, y')$ and $(x, 1, y) \sim (x', 1, y)$, where $x, x' \in X$ and $y, y' \in Y$. As for cones, it is convenient to parameterize points of the join with coordinates (x, t, y), noting that the coordinate system is degenerate when $t = 0$ or $t = 1$. We can observe that $X * Y$ contains canonical copies of X and Y as the respective images of $X \times \{0\} \times Y$ and $X \times \{1\} \times Y$ under the quotient map, and we will identify X and Y with these canonical copies. We have $X * Y - X \cong cX \times Y$, while $X * Y - Y \cong X \times cY$. Of course, $X * Y - (X \amalg Y) \cong X \times (0, 1) \times Y$. If we identify $X \times Y$ with the subset $X \times \{1/2\} \times Y \subset X * Y$, then we can also identify $X * Y$ as the union of closed subsets by

$$X * Y \cong (X \times \bar{c}Y) \cup_{X \times Y} (\bar{c}X \times Y),$$

where the cone parameter of $\bar{c}Y$ runs from 0 to 1/2 with the vertex at the 0 end and the cone parameter of $\bar{c}X$ runs from 1/2 to 1 with the vertex at the 1 end.

These descriptions allow us to introduce a filtration for $X * Y$ if X and Y are filtered. If the respective formal dimensions of X and Y are n and m, then $X * Y$ will have formal dimension $m + n + 1$. Looking at $X \times (0, 1) \times Y$, we can use the product filtration, letting $(0, 1)$ be filtered trivially, i.e.

$$(X \times (0,1) \times Y)^i = \bigcup_{j+k=i-1} X^j \cup (0,1) \cup Y^k.$$

On $X * Y - Y$, the product filtration on $X \times cY$ is $(X \times cY)^i = \bigcup_{j+k=i} X^j \times (cY)^k$, but each $X^j \times (cY)^k$ has the form $X^j \times c(Y^{k-1})$. Therefore, if we consider $X * Y - Y$ with this product filtration, the induced filtration on the subset $((X * Y) - Y) - X$ is consistent with the product filtration on $X \times (0, 1) \times Y$. Similarly, looking at the product filtration on $X * Y - X \cong cX \times Y$ with its product filtration, the induced filtration on the subset $((X * Y) - X) - Y$ is consistent with the product filtration on $X \times (0, 1) \times Y$. Therefore, assembling $X * Y$ as the union of $X \times cY$ and $cX \times Y$ with their product filtrations provides a natural *join filtration* on $X * Y$.

Even better, we can find a more explicit filtration of $X * Y$ by observing that the union in $X * Y$ of the set $X^j \times (cY^{k-1}) \subset X \times cY$ with the set $c(X^j) \times Y^{k-1} \subset cX \times Y$ is itself the join $X^j * Y^{k-1}$. So we can write the i-skeleton of $X * Y$ as $\bigcup_{a+b=i-1} X^a * Y^b$. Notice that the reason we have $a + b = i - 1$ instead of $a + b = i$ is that the $[0, 1]$ factor in the join adds a dimension that is only implicit in the notation. In particular, if $a = -1$, then $X^a = X^{-1} = \emptyset$ and $X^{-1} * Y^i = \emptyset * Y^i = Y^i$ is in the i-skeleton of $X * Y$. The equivalent observation holds if $b = -1$, so the i-skeleton of $X * Y$ includes the i-skeleta of X and Y. One can also compute that the set of i-dimensional strata of $X * Y$ comprises the i-strata of X, the i-strata of Y, and the i-strata of $X \times (0, 1) \times Y$.

Example 2.11.2 Let $M = M^m$ and $N = N^n$ be compact manifolds, filtered trivially, and suppose $m < n$. Then the smallest-dimensional non-empty skeleton of $M * N$ is $(M * N)^m = M^m * \emptyset^{-1} = M$. The next lowest-dimensional skeleton that is not equal to M is the n-skeleton $M \amalg N = (M^m * \emptyset^{-1}) \cup (\emptyset^{-1} * N^n)$. And the next skeleton that is not equal to the n-skeleton is the $(m + n + 1)$-skeleton $M * N = M^m * N^n$. △

Example 2.11.3 In Figure 2.7 we have the join $X * Y$ where X and Y are both intervals. Suppose first that we filter X trivially and think of Y as the interval I filtered as $\{y\} \subset I$ with y an interior point. We also assume all strata have their natural dimensions. Then we can notice that $X * Y - X \cong (cX) \times Y$ and $X * Y - Y$ is filtered homeomorphic to the filtered space $X \times c\{y\} \subset X \times cI$. We also observe $(X * Y)^0 = \{y\}$, $(X * Y)^1 = X \amalg Y$, $(X * Y)^2 = X \cup (X * \{y\}) \cup Y$, and $(X * Y)^3 = X * Y$.

We could also interpret Figure 2.7 so that X is given its simplicial filtration as a 1-simplex and so that Y is a simplicial complex with two 1-simplices, also filtered simplicially. In this case, $X * Y$ is simplicially filtered as a simplicial complex with two 3-simplices joined along a 2-simplex. △

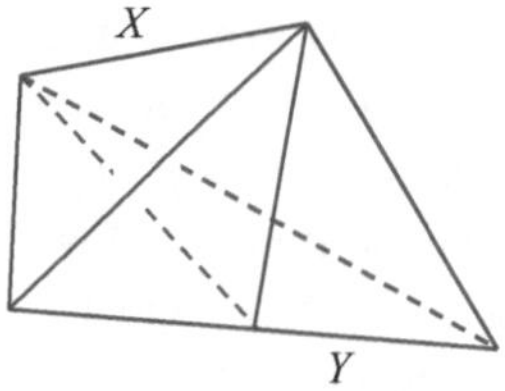

Figure 2.7 The join of Example 2.11.3.

Lemma 2.11.4 *If X and Y are (recursive) CS sets, then so is $X \times Y$ with the product filtration. If X and Y are compact (recursive) CS sets, then so is $X * Y$ with the join filtration.*

Proof Since the strata of $X \times Y$ have the form $S \times T$, where S is a stratum of X and T is a stratum of Y, the strata of $X \times Y$ are manifolds. Similarly, as the strata of $X * Y$ are strata of X, strata of Y, or have the form $S \times (0, 1) \times T$ where S and T are respective strata of X and Y, the strata of $X * Y$ are manifolds.

To verify the locally cone-like property, we will proceed by a simultaneous induction on $\dim(X) + \dim(Y)$ for $X \times Y$ and $X * Y$. We first dispense with some trivial cases by noting that if either X or Y is empty, then so is $X \times Y$, and if X is empty, then $X * Y = Y$, while if Y is empty, $X * Y = X$. So the result is established whenever $\dim(X)$ or $\dim(Y)$ is < 0. If $\dim(X) = \dim(Y) = 0$, then both X and Y are discrete unions of points or empty, and hence so is $X \times Y$, which is a recursive CS set. If $\dim(X) = \dim(Y) = 0$ and X and Y are compact (and so finite), then $X * Y$ is the union of all intervals between X and Y, with $(X * Y)^0$ consisting of the $\#X + \#Y$ points in X and Y and $(X * Y)^1$ consisting of $(\#X)(\#Y)$ open intervals. The link of each point of X is homeomorphic to Y, and the link of each point of Y is homeomorphic to X. So $X * Y$ is a recursive CS set.

We will also need to consider separately the case of $X \times Y$ with $\dim(X) + \dim(Y) = 1$. If either $\dim(X)$ or $\dim(Y)$ is < 0, the product is empty and there is nothing to prove. Otherwise, one of X or Y must be zero-dimensional. Choosing $\dim(X) = 0$ without loss of generality, we then have $X \times Y \cong \amalg_{\#X} Y$. In other words, $X \times Y$ is a disjoint collection of copies of Y, one for each point of X, and again the conclusion is trivial.

Now let A_n be the statement that if $\dim(X), \dim(Y) \geq 0$, and $\dim(X) + \dim(Y) \leq n$, then $X \times Y$ is a (recursive) CS set if X and Y are, and let B_n be the statement that if X and Y are compact, $\dim(X), \dim(Y) \geq 0$, and $\dim(X) + \dim(Y) \leq n$, then $X * Y$ is a (recursive) CS set if X and Y are. We will show that $A_{n+1} \Rightarrow B_n$ if $n \geq 0$ and that $B_{n-2} \Rightarrow A_n$ if $n - 2 \geq 0$. Thus we have the chain of implications

$$B_0 \Rightarrow A_2 \Rightarrow B_1 \Rightarrow A_3 \Rightarrow B_2 \Rightarrow A_4 \Rightarrow B_3 \Rightarrow \cdots .$$

Together with our low-dimensional cases, this will demonstrate the lemma.

First, we assume A_{n+1}, with $n \geq 0$, and consider $X * Y$ with X and Y compact and $\dim(X) + \dim(Y) = n \geq 0$. In our initial discussion, we observed that $X * Y$ is the union of the open subsets $X \times cY$ and $cX \times Y$. If X is a (recursive) CS set, then so is cX by Example 2.3.5, and similarly for Y. Thus $X \times cY$ and $cX \times Y$ are (recursive) CS sets by our assumption that A_{n+1} holds, as $\dim(X) + \dim(cY) = \dim(cX) + \dim(Y) = n + 1$. Since the (recursive) locally cone-like condition is a local condition, it follows then for all points in $X * Y$.

Next, let us assume B_{n-2}, with $n - 2 \geq 0$, and let $\dim(X) + \dim(Y) = n$. We will demonstrate A_n. Let $(x, y) \in X \times Y$ with x in a stratum $S \subset X$ and y in a stratum $T \subset Y$ with $\dim(S) = j$, $\dim(T) = k$. Then x has a distinguished neighborhood N in X with a filtered homeomorphism $h_N : U \times cL \to N$ such that $h_N(U \times cL^a) = X^{j+a+1} \cap N = N^{j+a+1}$ for all $a \geq -1$, and y has a distinguished neighborhood M in Y with a filtered homeomorphism $h_M : V \times cK \to M$ such that $h_M(V \times cK^b) = Y^{k+b+1} \cap M = M^{k+b+1}$ for all $b \geq -1$. By the definition of distinguished neighborhoods, we assume L and K are compact and that they are recursive if X and Y are. Then $N \times M$ is a neighborhood of (x, y) in $X \times Y$, and $(h_N \times h_M)^{-1}$ provides a filtered homeomorphism between $N \times M$ and $U \times cL \times V \times cK$. Ignoring filtrations for a moment, we then have

$$U \times cL \times V \times cK \cong U \times V \times cL \times cK \cong (U \times V) \times c(L * K),$$

where we have used the basic topological fact that $cL \times cK \cong c(L * K)$; see Figures 2.8 and 2.9. Since $\dim(L) \leq \dim(X) - 1$ and $\dim(K) \leq \dim(Y) - 1$, we have $\dim(L) + \dim(K) \leq \dim(X) + \dim(Y) - 2 = n - 2$. As L and K are compact filtered spaces then so is $L * K$ with its join filtration, and if L and K are recursive CS sets then so is $L * K$ by the hypothesis that B_{n-2} holds. So this provides a distinguished neighborhood of (x, y) of the desired form provided we verify the compatibility of the filtrations. In other words, we need to show that $U \times V \times cL \times cK \cong (U \times V) \times c(L * K)$ is a filtered homeomorphism.

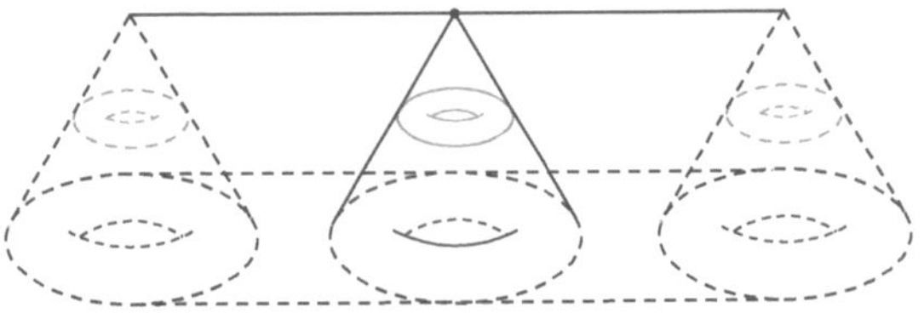

Figure 2.8 A schematic illustration of $cL \times cK \cong c(L * K)$. In this case we have $L = c(S^0) \cong (-1, 1)$, the open interval, and $K = c(T^2)$ so that $L * K \cong S^0 * T^2 \cong ST^2$, the suspension of the torus. We see the suspension here in the figure as the dashed (so unincluded) boundary of $c(L * K)$ as the union of the two cones on T^2 on the left and right with $I \times T^2$ along the bottom.

From the definitions, the product of the skeleta N^{j+a+1} and M^{k+b+1} are contained in the $(j + k + a + b + 2)$-skeleton of $N \times M$. Via $(h_N \times h_M)^{-1}$, this product corresponds to

$$U \times cL^a \times V \times cK^b \cong U \times V \times cL^a \times cK^b \cong U \times V \times c(L^a * K^b).$$

So we have

$$N^{j+a+1} \times M^{k+b+1} \cong U \times V \times c(L^a * K^b).$$

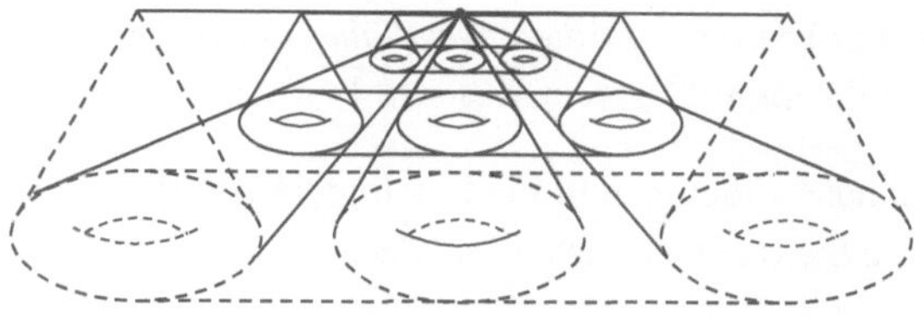

Figure 2.9 Figure 2.8 viewed more as a cone.

Taking the unions over all $a, b \geq -1$ such that $a + b = i$, we see that $(h_N \times h_M)^{-1}$ takes the $(j + k + i + 2)$-skeleton of $N \times M$ homeomorphically onto

$$U \times V \times c(\{(i + 1)\text{-skeleton of } K * L\}),$$

as desired. □

The following fact can be abstracted directly from the preceding proof:

Corollary 2.11.5 *If L and K are compact filtered spaces then $cL \times cK$, formed using the cone and product filtrations, is filtered homeomorphic to $c(L * K)$, formed using the join and cone filtrations.*

The next corollary is a version of Lemma 2.11.4 for stratified pseudomanifolds.

Corollary 2.11.6 *If X and Y are stratified pseudomanifolds then so is $X \times Y$. If X and Y are compact stratified pseudomanifolds then so is $X * Y$. Similarly, if X and Y are PL stratified pseudomanifolds then so is $X \times Y$ and, if X and Y are compact PL stratified pseudomanifolds then so is $X * Y$.*

Proof If U and V are respectively the unions of the regular strata of X and Y, then U is dense in X and V is dense in Y. The union of the regular strata of $X \times Y$ is $U \times V$, and this is dense in $X \times Y$ by basic point-set topology. Similarly, unless one of X or Y is empty, in which case the result is trivial, the union of the regular strata of $X \times Y$ is $U \times (0, 1) \times V$, which is again easily seen to be dense. So, by Lemma 2.11.4 and Definition 2.4.1, $X \times Y$ is a topological stratified pseudomanifold.

For the PL case, we note that products and joins of PL spaces are PL spaces by Section B.5. So we only need to note additionally that, using the definition of PL stratified pseudomanifolds, the homeomorphisms of the proof of Lemma 2.11.4 can all be taken to be PL, and, for PL spaces, $c(L) \times c(K) \cong c(L * K)$ piecewise linearly by [198, Exercise 2.24(3)] or the argument of [2, page 419]. □

It is also true that the product of ∂-stratified pseudomanifolds is a ∂-stratified pseudomanifold. Given Corollary 2.11.6, the main additional technicality is the need to demonstrate the collaring of the boundary.

Lemma 2.11.7 *If X and Y are ∂-stratified pseudomanifolds, then so is $X \times Y$. If X and Y are PL ∂-stratified pseudomanifolds, then so is $X \times Y$.*

Proof The proofs are nearly identical in the topological and PL cases, so we focus on the former. We will indicate the one place where it is not evident we have a PL map.

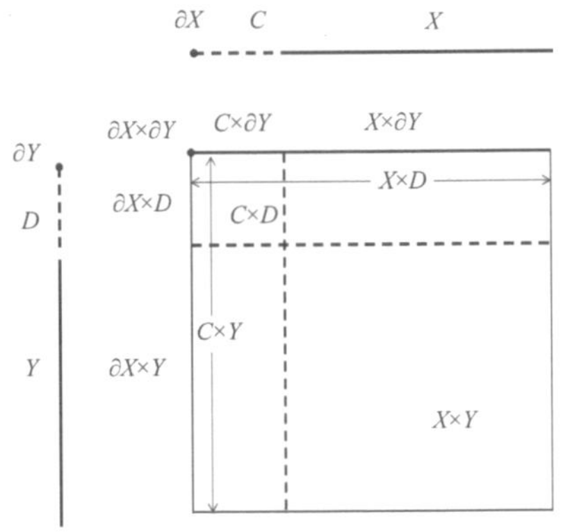

Figure 2.10 A schematic of the product $X \times Y$ showing the neighborhood of the boundary.

The interior of $X \times Y$ is $(X - \partial X) \times (Y - \partial Y)$, which is a stratified pseudomanifold by Corollary 2.11.6. The boundary of $X \times Y$ will be $(\partial X \times Y) \cup (X \times \partial Y)$ once we show that it is a pseudomanifold and that it is collared in $X \times Y$. So we will provisionally label this set $\partial(X \times Y)$; see Figure 2.10. Again by Corollary 2.11.6,

$$\partial(X \times Y) - \partial X \times \partial Y = (\partial X \times (Y - \partial Y)) \amalg ((X - \partial X) \times \partial Y)$$

is a stratified pseudomanifold, so to see that $\partial(X \times Y)$ is a stratified pseudomanifold, we only have to be careful near the "corner" $\partial X \times \partial Y$. Now, we have stratified collars that we can identify as $C = [0, 1) \times \partial X$ in X and $D = (-1, 0] \times Y$ in Y with $[0, 1)$ and $(-1, 0]$ trivially filtered; in the latter cases, the only difference from our standard conventions is a choice of a different parameterization on the collar, which will be useful below. So in $X \times \partial Y$, the subspace $\partial X \times \partial Y$ has a neighborhood $C \times \partial Y = [0, 1) \times \partial X \times \partial Y$ with the product filtration. Similarly, $\partial X \times \partial Y$ has a neighborhood

$$\partial X \times D = \partial X \times (-1, 0] \times \partial Y \cong (-1, 0] \times \partial X \times \partial Y$$

with the product filtration. So we then see that $\partial X \times \partial Y$ has a neighborhood in $\partial(X \times Y)$ that is filtered homeomorphic to

$$(\partial X \times D) \cup (C \times \partial Y) \cong (-1, 1) \times \partial X \times \partial Y,$$

which is a stratified pseudomanifold by Corollary 2.11.6. We have shown that $\partial(X \times Y)$ is covered by open sets that are stratified pseudomanifolds, and so $\partial(X\times Y)$ must be a stratified pseudomanifold.

Slightly more complex is the issue of demonstrating that $\partial(X \times Y)$ has a stratified collar. For this, it will be useful to assume that the closures of C and D have the form $\bar{C} \cong [0,1]\times\partial X$ and $\bar{D} \cong [-1,0] \times \partial Y$ by filtered homeomorphisms; this entails no loss of generality as we could, for example, form a new C from the subset $[0,1/2) \times \partial X$ of the original C. Then

$$N = (C \times Y) \cup (X \times D) \cong ([0,1) \times \partial X \times Y) \cup (X \times (-1,0] \times \partial Y)$$

is a neighborhood of $\partial(X \times Y)$ in $X \times Y$. Here, the product filtrations of the pieces of this neighborhood are consistent with those inherited from $X \times Y$. We need to show that $N \cong [0,1) \times \partial(X \times Y)$. For this, we notice that

$$\begin{aligned}(C \times Y) \cap (X \times D) = C \times D &\cong [0,1) \times \partial X \times (-1,0] \times \partial Y\\ &\cong [0,1) \times (-1,0] \times \partial X \times \partial Y,\end{aligned}$$

and the closure of this intersection in N has the form

$$(([0,1] \times [-1,0]) - (\{1\} \times \{-1\})) \times \partial X \times \partial Y.$$

In particular, we can think of N as consisting of three closed (in N) pieces: $C \times (Y - D)$, $(X - C) \times D$, and the closure in N of $C \times D$. These pieces are glued along

$$(\overline{C \times D}) \cap (C \times (Y - D)) \cong [0,1) \times \{-1\} \times \partial X \times \partial Y$$

and

$$(\overline{C \times D}) \cap ((X - C) \times D) \cong \{1\} \times (-1,0] \times \partial X \times \partial Y.$$

But $([0,1]\times[-1,0])-(\{1\}\times\{-1\})$ is simply a solid square with one corner point removed, and this space is piecewise linearly homeomorphic to $[0,1) \times [-1,1]$. We can see this using Figure 2.11. Thus the closure in N of $C \times D$ is PL homeomorphic to

$$[0,1) \times [-1,1] \times \partial X \times \partial Y.$$

Now, we form the collar neighborhood of $\partial(X \times Y)$ by gluing the pieces back together having "straightened" the closure of $C \times D$ and now identifying $\{1\} \times (-1,0] \times \partial X \times \partial Y$ in $(X - C) \times D$ with $[0,1) \times \{1\} \times \partial X \times \partial Y$ in the homeomorphed image of the closure of $C \times D$. See Figure 2.12. Together, we obtain a space that has the form $[0,1) \times \partial(X \times Y)$, using our earlier observation that

$$\partial(X \times Y) \cong (X - C) \cup (Y - D) \cup ([-1,1] \times \partial X \times \partial Y).$$

This is the desired collar. □

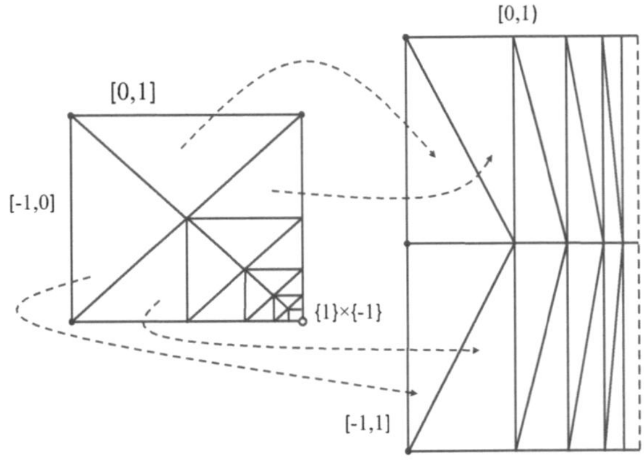

Figure 2.11 A PL homeomorphism $([0,1) \times [0,1)) - (\{1\} \times \{-1\}) \to [0,1) \times [-1,1]$. We see the first steps in triangulations of the spaces and a simplicial map that takes each L-shaped region with four triangles on the left to a rectangular region with four triangles on the right.

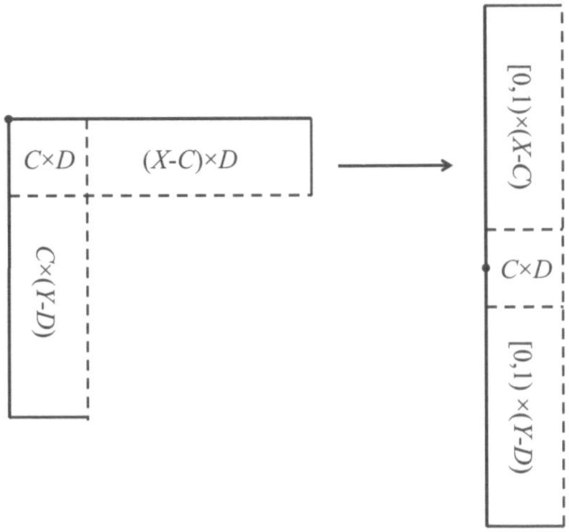

Figure 2.12 Straightening the collar neighborhood of $\partial(X \times Y)$.

2.11.1 Products of Intrinsic Filtrations

A reasonable question to ask is whether the products of intrinsically filtered spaces are intrinsically filtered in their product filtrations. In the topological category, results like the Double Suspension Theorem demonstrate that this is too much to ask. However, in the PL category, and even here with an extra assumption, we can have such results.

Proposition 2.11.8 *Let X and Y be two PL stratified pseudomanifolds with intrinsic filtrations $\mathfrak{X}$ and $\mathfrak{Y}$, and let $(X \times Y)^*$ denote the intrinsic filtration of $X \times Y$. If at least one of X, Y is a classical pseudomanifold, then $(X \times Y)^* = \mathfrak{X} \times \mathfrak{Y}$.*

Remark 2.11.9 The assumption that at least one of X or Y be classical is necessary, as demonstrated by the following example: Let $I = [0, 1]$ be filtered as $\{0, 1\} \subset I$. This is a PL stratified pseudomanifold filtration, and it is intrinsic. Then $I \times I$ is filtered as

$$\{(0,0), (0,1), (1,0), (1,1)\} \subset (\{0,1\} \times I) \cup (I \times \{0,1\}) \subset I \times I.$$

This is not an intrinsic filtration as all of the boundary points of the square $I \times I$ have PL homeomorphic relative neighborhoods. This follows from, among other arguments, the PL unbending procedures that we utilized in the proof of Lemma 2.11.7.

The proof of the proposition relies upon the following lemma:

Lemma 2.11.10 *A filtration of a PL pseudomanifold X is the PL intrinsic filtration if and only if no link of any point in X is PL homeomorphic to a suspension; i.e. if and only if for every link L in X there is no compact PL space Z such that L is PL homeomorphic to SZ. Furthermore, if X is a classical PL pseudomanifold, then no link of the intrinsic filtration is PL homeomorphic to a closed cone.*

Proof Throughout this proof, $\cong$ will denote PL homeomorphism without regard for filtrations. First, suppose L is a link of a point in an i-dimensional stratum of $\mathfrak{X}$, i.e. there is some $x \in \mathfrak{X}$ with a distinguished neighborhood filtered PL homeomorphic to $\mathbb{R}^i \times cL$. Suppose L is PL homeomorphic to a suspension, so that $L \cong SZ$ for some compact PL space Z. Then we have

$$\mathbb{R}^i \times cL \cong \mathbb{R}^i \times c(SZ) \cong \mathbb{R}^i \times \mathbb{R}^1 \times cZ \cong \mathbb{R}^{i+1} \times cZ,$$

using again that $cA \times cB \cong c(A * B)$ with $A = S^0$ and $B = Z$. Then if w is the cone vertex of cZ, all the points in $\mathbb{R}^{i+1} \times \{w\}$, including x, have PL homeomorphic neighborhoods, contradicting that x is contained in an i-dimensional stratum of $\mathfrak{X}$ (see the proof of Proposition 2.10.5). Thus L is not a suspension. So no link in $\mathfrak{X}$ can be a suspension.

Conversely, suppose X is a PL stratified pseudomanifold such that no link is a suspension. We claim that X is filtered by the intrinsic filtration. Suppose not, and let $\mathfrak{X}$ be the intrinsic filtration. Since $\mathfrak{X}$ is the coarsest filtration, there must be a stratum T of X that is contained in a stratum S of $\mathfrak{X}$ with $\dim(S) > \dim(T)$. Let $\dim(T) = i$ and $\dim(S) = j$, and suppose $x \in T$. Then x has a distinguished neighborhood in X of the form $\mathbb{R}^i \times cL \cong c(S^{i-1} * L)$, while x has a distinguished neighborhood in $\mathfrak{X}$ of the form $\mathbb{R}^j \times cL' \cong c(S^{j-1} * L')$. Note that if $i = 0$, we let $S^{-1} = \emptyset$ and $\emptyset * L = L$; similarly, the formulas apply if $L = \emptyset$. By the uniqueness of polyhedral links[30] [130, Corollary 1.15], this implies that $S^{i-1} * L \cong S^{j-1} * L'$, or, written in terms of iterated suspensions, $S^i L \cong S^j L'$. Since $i < j$, then $L \cong S^{j-i} L'$ by

[30] See footnote 16 on page 44.

[179, Theorem 1], so L is a suspension, contradicting the assumption. Therefore, X must actually be $\mathfrak{X}$.

Lastly, suppose X is a classical PL pseudomanifold and $\mathfrak{X}$ its intrinsic filtration. By Remark 2.5.15, each link L is also a classical PL pseudomanifold. By Corollary 2.5.21, this implies that if L has dimension k, then no triangulation of L can have a dimension $k-1$ face that does not bound exactly two k-simplices. This implies that L cannot have the structure of a closed cone $\bar{c}Z$ unless Z is empty. But if Z is empty, $\bar{c}Z$ is a point. However, since X is classical, it has a PL pseudomanifold filtration X' with no codimension-one strata, and since $\mathfrak{X}$ is coarser than X', $\mathfrak{X}$ also has no codimension-one strata. Therefore, L also cannot be a point. Therefore, L is not a closed cone. □

Proof of Proposition 2.11.8 By Lemma 2.11.10, it suffices to show that the links of $\mathfrak{X}\times\mathfrak{Y}$ are not suspensions. But every link of $\mathfrak{X}\times\mathfrak{Y}$ has the form $L*K$, where L and K are respective links of $\mathfrak{X}$ and $\mathfrak{Y}$, by our computations in the proof of Lemma 2.11.4. Furthermore, L and K cannot be suspensions, again by Lemma 2.11.10. We must show that $L*K$ is not a suspension.

In [179], Morton defines a compact polyhedron (PL space) to be *reduced* if it is not a closed cone or a suspension. By [179, Corollary to Theorems 1 and 2], every compact polyhedron (PL space) factors uniquely as the join of a ball or sphere to a reduced polyhedron. Since L and K are not suspensions, their unique factorizations must have the form $L \cong B^a * D$ and $K \cong B^b * E$, where D and E are reduced and B^a and B^b are balls of respective dimensions a and b such that $a, b \leq 0$ (letting $B^{-1} = \emptyset$). The last condition is due to the fact that any ball of dimension > 0 is a suspension, so if, for example, $L \cong B^a * D$ with $a > 0$, then $L \cong (S^0 * B^{a-1}) * D \cong S^0 * (B^{a-1} * D)$, presenting a contradiction. In fact, one of a or b must be < 0, as if $a = 0$ then L is a closed cone, and similarly if $b = 0$ then K is a closed cone. But we have assumed that one of X or Y is classical, and so by Lemma 2.11.10, it is not possible for both L and K to be closed cones.

So now we must have

$$L * K \cong B^a * D * B^b * E \cong B^{a+b+1} * D * E$$

with $a+b+1 \leq 0$. Furthermore, by [179, Lemma 3], the join of reduced polyhedra is reduced, so $D*E$ is reduced. So if $a+b+1 < 0$, then $L*K \cong D*E$ is reduced, and we are done. If $a+b+1 = 0$, then $L*K \cong B^0 * D * E$. Suppose this is a suspension, $B^0 * D * E \cong S^0 * F$ for some compact polyhedron F. Then, again by [179, Corollary to Theorems 1 and 2], we have either $F \cong B^c * G$ or $F \cong S^d * H$ with G and H reduced and $c, d \geq 0$. The reason c or d must be ≥ 0 is that otherwise F would be G or H, and so reduced, making $B^0 * D * E \cong S^0 * F$ impossible by the uniqueness part of Morton's corollary. But then $S^0 * F$ is PL homeomorphic to either $S^0 * B^c * G \cong B^{c+1} * G$ or $S^0 * S^d * H \cong S^{d+1} * H$, with $c+1, d+1 \geq 1$. These spaces are intended to be PL homeomorphic to $B^0 * D * E$ with $D*E$ reduced, so the uniqueness part of Morton's corollary again implies that both scenarios are impossible. Thus $B^0 * D * E \cong L*K$ cannot be a suspension, as desired. □

We also have a product result for naturally stratified PL ∂-stratified pseudomanifolds; see Definition 2.10.21.

Proposition 2.11.11 *Let X, Y be naturally stratified PL ∂-stratified pseudomanifolds such that at least one of X or Y is a classical PL ∂-stratified pseudomanifold. Then $X \times Y$ is naturally stratified.*

Proof By Lemma 2.11.7, $X \times Y$ is a ∂-stratified pseudomanifold. According to Definition 2.10.21, we must show that $X \times Y - \partial(X \times Y)$ and $\partial(X \times Y)$ are intrinsically stratified.

First, we see that

$$X \times Y - \partial(X \times Y) = (X - \partial X) \times (Y - \partial Y)$$

is intrinsically stratified by Proposition 2.11.8. By the same proposition, $(X - \partial X) \times \partial Y$, $\partial X \times (Y - \partial Y)$, and $\partial X \times \partial Y$ are intrinsically stratified. For this we note that if X is a classical PL stratified ∂-pseudomanifold, then ∂X is also classical, as if ∂X has codimension-one strata, then so does the collar neighborhood of ∂X in X; the same is, of course, true of Y. From the proof of Lemma 2.11.7, we observe that $\partial(X \times Y)$ is the union of the open subsets $(X - \partial X) \times \partial Y$, $\partial X \times (Y - \partial Y)$, and $(-1, 1) \times \partial X \times \partial Y$, where the latter space is also intrinsically filtered by Proposition 2.11.8. Piecing these spaces together to form $\partial(X \times Y)$ and using that intrinsic filtration is a local property, we see that $\partial(X \times Y)$ must be intrinsically filtered. □

3
Intersection Homology

In this chapter, we will define the *intersection homology groups* and compute some examples that demonstrate their most fundamental properties.

We begin in Section 3.1 by defining the *perversity parameters* whose values control how *intersection chains* are allowed to intersect with the strata of a space. Intersection homology itself is first defined in Section 3.2, beginning with simplicial intersection homology. We provide a number of examples there to develop the reader's geometric intuition.

In Section 3.3, we turn to piecewise linear (PL) intersection homology. PL chains can be thought of as simplicial chains but without requiring a fixed triangulation of the space in advance. In other words, a PL chain is represented by a simplicial chain chosen from any one of the triangulations admissible within the PL structure of the space. This outlook has some technical advantages; for example, all possible subdivisions are already built into the definition of the chain complex. However, as the reader is likely less familiar with PL homology than simplicial or singular homology, we develop the necessary background in Sections 3.3.1 and 3.3.2. PL intersection homology is then defined in Section 3.3.3 and its relation with the simplicial theory explored in Section 3.3.4.

Intersection homology defined with singular chains comes in Section 3.4.

This chapter is mostly concerned with definitions, examples, and the most fundamental properties of intersection homology. Further properties will be developed in the following chapters.

3.1 Perversities

The idea for defining intersection homology groups is that we should consider the simplices and chains that are ordinarily used to define homology groups but that we should place some limitations on how such chains are allowed to interact with different strata. In practice this is done by restricting the dimensions of the intersections of chains with strata. Of course there can be many different ways to do this: we could forbid a chain from intersecting a stratum altogether, we could pose no limitation with a given stratum, or we could make various choices in between. These choices are encoded in a *perversity parameter*, most often referred to simply as a *perversity*. As there are many choices for these parameters, there are many different kinds of intersection homology groups. In this section, we provide the precise definition of perversity.

In fact, the definition of perversity has evolved. We begin with the most general definition and then discuss some of the other limitations that were originally imposed.

Definition 3.1.1 Let X be a filtered space of formal dimension n, and let $\mathscr{S}$ be the set of strata of X. A *perversity* on X is a function

$$\bar{p} : \mathscr{S} \to \mathbb{Z}$$

such that $\bar{p}(S) = 0$ if $S \subset X - \Sigma_X$, i.e. if S is a regular stratum.

Remark 3.1.2 Given the generality of the definition, one might wonder whether we could do without the requirement that $\bar{p}(S) = 0$ if $S \subset X - \Sigma_X$, i.e. if S is a regular stratum. It will turn out that we could just as well require that $\bar{p}(S) \geq 0$ if S is a regular stratum. However, this would not change the intersection homology groups (or even the intersection chain complexes), and it is occasionally simpler in technical statements to have that $\bar{p}(S) = 0$ for such strata. On the other hand, if $\bar{p}(S) < 0$, the definition of intersection homology either becomes trivial or simply doesn't see the regular strata at all. We choose to avoid this degenerate case; see Remarks 3.2.9 and 3.4.5 for more details.

Definition 3.1.3 If $\bar{p}$ and $\bar{q}$ are perversities on a filtered space X such that $\bar{p}(S) \leq \bar{q}(S)$ for all singular strata S, then we will write $\bar{p} \leq \bar{q}$.

3.1.1 GM Perversities

The original definition of perversity was more complex, owing to the setting and properties of the original Goresky and MacPherson intersection homology groups. In particular, their intersection homology of classical PL stratified pseudomanifolds possesses a Poincaré Duality Theorem and is invariant to the filtration of the space as a pseudomanifold. We will see below in detail how these requirements force additional conditions on the perversity parameters. In fact, they will also place certain requirements on the space itself in that pseudomanifolds for which these properties hold simultaneously must be classical pseudomanifolds.

For now, we will simply provide an alternative definition for what we will call *Goresky–MacPherson perversities* or *GM perversities*. We will come to understand the additional requirements later on. We will sometimes refer to perversities as defined in Definition 3.1.1 as *general perversities* when we wish to distinguish them from GM perversities.

One further point worth mentioning before providing the definition is that GM perversities assign the same value to all strata of the same codimension. Hence given that perversities always evaluate to 0 on codimension-zero strata and that we will assume when using GM perversities that there are no codimension-one strata, it is standard to write GM perversities as functions of codimension with domain $\{2,3,4,\dots\}$.

Definition 3.1.4 A *Goresky–MacPherson perversity* (or *GM perversity*) is a function

$$\bar{p} : \{2,3,4,\dots\} \to \mathbb{Z}$$

such that

1. $\bar{p}(2) = 0$,
2. $\bar{p}(k) \le \bar{p}(k+1) \le \bar{p}(k) + 1$.

The conditions of the definition say that a perversity is a function defined on the integers ≥ 2 that "starts" at 0 and then for each transition in the domain from k to $k+1$ the perversity value either stays the same or increases by 1. So a GM perversity is sort of a "substep" function.

One convenient way to describe GM perversities is to think of them as sequences

$$[\bar{p}(2), \bar{p}(3), \bar{p}(4), \dots].$$

So, for example, a GM perversity might look like

$$\bar{p} = [0,1,1,2,3,3,3,4,5,5,\dots].$$

Remark 3.1.5 A GM perversity determines a general perversity in the following way. If $\mathfrak{p}$ is a GM perversity and X is a filtered space of formal dimension n with no codimension-one strata, then we can define an associated perversity $\bar{p}$ on X by $\bar{p}(S) = \mathfrak{p}(\operatorname{codim}(S))$ for any singular stratum S of X. In what follows, we will abuse notation by using the same symbol, typically $\bar{p}$, for a GM perversity and the general perversity it determines. All perversities should be assumed to be general perversities unless explicitly stated otherwise.

There are a few particular perversities that have special importance:

Example 3.1.6 The minimal GM perversity is the *zero perversity* $\bar{0}$, which takes the smallest possible values at each step by starting at 0 and then never increasing:

$$\bar{0} = [0,0,0,0,\dots].$$

On the other hand, the maximal GM perversity is the *top perversity* $\bar{t}$, which always takes the step up

$$\bar{t} = [0,1,2,3,\dots].$$

Both of these GM perversities can be extended to general perversities on arbitrary filtered spaces in somewhat obvious ways. The general zero perversity always takes the value 0 for all strata S of a filtered space X, i.e. $\bar{0}(S) = 0$. Similarly, the general top perversity can be defined by the function $\bar{t}(S) = \operatorname{codim}(S) - 2$ for all singular strata S. Note that we still always set $\bar{t}(S) = 0$ if S is a regular stratum. We shall see by our duality results that the definition we have given here is indeed the most reasonable extension of $\bar{t}$ for spaces with strata of codimension one. △

3.1.2 Dual Perversities

The general top perversity $\bar{t}$, defined so that $\bar{t}(S) = \text{codim}(S) - 2$ for all singular strata S, plays an especially important role in intersection homology Poincaré duality, as intersection homology groups dualize not only with respect to dimension index but also with respect to perversities, which is defined in terms of the top perversity.

Definition 3.1.7 Given a perversity $\bar{p}$, its *dual perversity* (or *complementary perversity*) is the perversity $D\bar{p}$ defined so that

$$D\bar{p}(S) = \bar{t}(S) - \bar{p}(S) = \text{codim}(S) - 2 - \bar{p}(S)$$

for all singular strata S, and $D\bar{p}(S) = 0$ if S is a regular stratum. We will often abbreviate this property by saying $D\bar{p} = \bar{t} - \bar{p}$ or $\bar{p} + D\bar{p} = \bar{t}$.

Example 3.1.8 We have $D\bar{0} = \bar{t}$ and $D\bar{t} = \bar{0}$. △

The following lemma, whose proof is immediate, shows that dualization acts as an involution on the set of perversities.

Lemma 3.1.9 *For any perversity,* $D(D\bar{p}) = \bar{p}$.

Manifold theory, and in particular the study of the cup product pairing or the intersection pairing, might lead us to expect that if duality between two objects is important, then objects that are dual to themselves are even more important. For example, the symmetric self-dual cup product pairing on $H^{2k}(M;\mathbb{Q})$ of a closed oriented $4k$-manifold yields signature invariants. This theme will be developed in detail in Chapter 9. To see now what sorts of perversities might have this property, we observe that since

$$\bar{p}(S) + D\bar{p}(S) = \text{codim}(S) - 2$$

for all singular strata S, it will only be possible to have $\bar{p}(S) = D\bar{p}(S)$ on strata that are of even codimension, in which case we would want $\bar{p}(S) = D\bar{p}(S) = (\text{codim}(S) - 2)/2$. We would be justified in calling this a *middle perversity* since it is halfway between the perversities $\bar{0}$ and $\bar{t}$, which are the extreme possibilities of GM perversities.

What about spaces that do possess odd-codimension strata? If our goal is to continue to have $\bar{p}(S)$ and $D\bar{p}(S)$ remain as close in value as possible, we would want $|\bar{p}(S) - D\bar{p}(S)| \leq 1$ for all S. This implies that one of the perversities must have value[1] $\lfloor(\text{codim}(S) - 2)/2\rfloor$ and the other one have value $\lceil(\text{codim}(S) - 2)/2\rceil$. Such perversities were defined by Goresky and MacPherson so that one would always take the higher value and the other would always take the lower value:

Definition 3.1.10 The *lower middle GM perversity* $\bar{m}$ and the *upper middle GM perversity*

[1] Here $\lfloor x \rfloor$ denotes the greatest integer less than or equal to the real number x, and $\lceil x \rceil$ denotes the least integer greater than or equal to x.

$\bar{n}$ are defined by

$$\bar{m} = [0, 0, 1, 1, 2, 2, 3, \dots], \qquad \bar{n} = [0, 1, 1, 2, 2, 3, 3, \dots].$$

These extend to general perversities with the definitions

$$\bar{m}(S) = \left\lfloor \frac{\operatorname{codim}(S) - 2}{2} \right\rfloor, \qquad \bar{n}(S) = \left\lceil \frac{\operatorname{codim}(S) - 2}{2} \right\rceil.$$

Notice that if X has no strata of odd codimension then $\bar{m}$ and $\bar{n}$ take the same value on all strata. In this case it is customary to use either symbol $\bar{m}$ or $\bar{n}$ to stand for a single "middle perversity."

Remark 3.1.11 The reader might well ask why we want to make a choice such that always $\bar{m}(S) \leq \bar{n}(S)$. For example, why is this better than two dual GM perversities such as $[0, 0, 1, 2, 2, \dots]$ and $[0, 1, 1, 1, 2, \dots]$ in which sometimes one and sometimes the other perversity is allowed to be the greater one. Beyond it being pleasing to have made a definite choice, we will see that it is possible to have maps between intersection homology groups of a given space of the form $I^{\bar{p}}H_*(X) \to I^{\bar{q}}H_*(X)$ when $\bar{p}(S) \leq \bar{q}(S)$ for all singular strata S. Thus it is possible to have such a comparison map $I^{\bar{m}}H_*(X) \to I^{\bar{n}}H_*(X)$, but we would not generally be able to do this for other pairs of dual perversity "near" the middle. This comparison map will be critical for finding intersection homology groups that are self-dual under the intersection pairing in Chapter 9.

3.2 Simplicial Intersection Homology

Now that we have defined perversities on stratified spaces, we are prepared to provide a first definition of intersection chains $I^{\bar{p}}C_*(X)$ and intersection homology $I^{\bar{p}}H_*(X)$. As for classical homology, there are in fact several different types of chain complexes – simplicial, singular, piecewise linear, etc.[2] – that lead to the same homology groups, at least with the proper assumptions. We will consider each of these in turn. However, the reader should also be aware that there are at least two competing definitions in another sense, reflecting certain challenges that arise when perversities take values that are "too high," meaning that they take values on strata that exceed the value of the top perversity $\bar{t}$, or when working with spaces with strata of codimension one. When either of these conditions arises, the definitions of intersection homology become incompatible. However, they do all agree, for example, when considering Goresky–MacPherson perversities and classical pseudomanifolds.

In order to ease the exposition, we will begin with the definitions of intersection homology closest to the original definition of Goresky and MacPherson [105], though the reader should be aware that this is not the definition that we will ultimately want when $\bar{p}(S) \geq \bar{t}(S)$ for

[2] This might be a good place to note that a good CW theory of intersection homology does not seem to have been worked out, perhaps for reasons that will become evident as we proceed.

some stratum or when discussing stratified pseudomanifolds with codimension-one strata (though it is still defined in those cases). Once we have gotten used to the basic ideas, we will proceed to discuss how the definition should be modified to obtain the best results in the general settings beginning in Chapter 8. Since we will eventually want to use the notations $I^{\bar{p}}C_*(X)$ and $I^{\bar{p}}H_*(X)$ for the modified definition we will present later, for now we use the notations $I^{\bar{p}}C_*^{\mathrm{GM}}(X)$ and $I^{\bar{p}}H_*^{\mathrm{GM}}(X)$. When we need to distinguish between the two theories, we will refer to "GM intersection homology" or "non-GM intersection homology."

We will begin with a simplicial[3] version of intersection homology. For this we let X be a *simplicial filtered space*, meaning that X is a filtered space with a fixed triangulation such that each skeleton of the filtration is a simplicial subcomplex (not necessarily a simplicial skeleton).

We assume the reader is familiar with standard simplicial homology built from oriented simplices as in e.g. [182, 220]. Let us briefly review simplicial chain complexes to fix our notation; see [182, Section 5] or [220, Section 4.1] for more details[4].

Recall that if K is a simplicial complex then an *oriented i-simplex* of K is an i-dimensional simplex of K together with an equivalence class of orderings of its vertices, where two orderings are equivalent if they differ by an even permutation. If we fix an orientation for each simplex, then the *simplicial chain group* $C_i(K)$ is the free abelian group generated by the oriented simplices. So if $\xi \in C_i(K)$, it can be written as a sum $\xi = \sum_j a_j\sigma_j$, where each σ_j is a unique simplex of K with its fixed orientation, each $a_j \in \mathbb{Z}$, and the sum is finite. We typically leave σ_j out of the sum if its coefficient a_j is 0, and we say that σ_j is a "simplex of ξ" if $a_j \neq 0$. The *support* of ξ, written $|\xi|$, is the union of the simplices of ξ (though if $\xi = \sigma$ for some oriented simplex σ, we will tend to write σ rather than $|\sigma|$). For each oriented simplex σ, we identify the oriented simplex that has the same underlying geometric simplex but the opposite orientation with $-\sigma \in C_i(K)$. If the vertices of σ are $\{v_j\}_{j=0}^i$, we can write $[v_0, \dots, v_i]$ to represent σ with the orientation corresponding to the indicated ordering. The boundary map $\partial : C_i(K) \to C_{i-1}(K)$ is defined to act on an oriented simplex by

$$\partial[v_0, \dots, v_i] = \sum_{k=0}^{i}(-1)^k[v_0, \dots, \hat{v}_k, \dots, v_i],$$

where $\hat{v}_k$ indicates that we remove the vertex v_k and so obtain an oriented $(i-1)$-simplex. Then we obtain $\partial : C_i(K) \to C_{i-1}(K)$ by extending linearly, with $\partial \circ \partial = 0$ so that $C_*(K)$ is a chain complex. The *ith simplicial homology group* is

$$H_i(K) = \frac{\ker(\partial : C_i(K) \to C_{i-1}(K))}{\operatorname{im}(\partial : C_{i+1}(K) \to C_i(K))}.$$

[3] The initial Goresky–MacPherson intersection homology of [105] is defined with respect to *PL* chains, meaning the direct limit complex of simplicial chains over all triangulations compatible with the stratification. We will address this version shortly.

[4] The simplicial homology treated by Hatcher in [125, Section 2.1] is a bit different, being built on *ordered* simplices as opposed to *oriented simplices*. We will need ordered simplices a bit later in Section 4.4.2 and will review them at that time. The two resulting chain complexes, oriented and ordered, are chain homotopy equivalent by [182, Theorem 13.6].

If X is a space triangulated by K, we will write $C_*(X)$ to mean $C_*(K)$ when the triangulation is understood.

Remark 3.2.1 An orientation on a simplex as just described also determines an orientation for it as a ∂-manifold if $i > 0$: By our Definition B.1.4, our simplicial complexes are all contained in Euclidean space, so every i-simplex σ is contained in a unique minimal i-dimensional affine space that can be identified with the tangent space at each point of σ. If σ is oriented by the ordering $[v_0, \dots, v_i]$, then the ordered set of vectors $[\overrightarrow{v_0v_1}, \dots, \overrightarrow{v_0v_i}]$ gives a basis for the tangent space at v_0, and hence fixes an orientation for σ, which is clearly orientable. This observation is useful if, for example, we have another i-simplex τ contained in σ, for it then makes sense to speak of orienting τ compatibly with σ: the orientation of σ as a ∂-manifold restricts to an orientation of τ as a ∂-manifold, and then we may choose an ordering of the vertices of τ consistent with this orientation.

We can now define the complex of *intersection chains* as a subcomplex of $C_*(X)$:

Definition 3.2.2 Let X be a simplicial filtered space endowed with a general perversity $\bar{p}$, and let $C_*(X)$ be the chain complex of oriented simplices of X.

We deem an i-simplex σ of X to be *$\bar{p}$-allowable* if, for each stratum $S \subset X$,

$$\dim(\sigma \cap S) \le i - \operatorname{codim}(S) + \bar{p}(S). \tag{3.1}$$

Here $\operatorname{codim}(S)$ is the formal codimension of S in the filtered space X, while $\dim(\sigma \cap S)$ is the topological dimension. As everything here is simplicial, this intersection will be a union of open faces of σ, so $\dim(\sigma \cap S)$ will be the highest dimension of such an open face. If $\sigma \cap S = \emptyset$, we let $\dim(\emptyset) = -\infty$.

If inequality (3.1) is satisfied for some σ and some S, we say that σ is *$\bar{p}$-allowable with respect to the stratum S*. If the perversity $\bar{p}$ has been fixed in advance, we will sometimes simply say that σ is *allowable*.

A chain $\xi \in C_i(X)$ is *$\bar{p}$-allowable* if all of the simplices of ξ and all of the simplices of $\partial\xi$ (with non-zero coefficient) are $\bar{p}$-allowable.

Let $I^{\bar{p}}C_*^{\mathrm{GM}}(X) \subset C_*(X)$ be the chain complex of $\bar{p}$-allowable chains, which we call the *perversity $\bar{p}$ intersection chain complex*. Let the *perversity $\bar{p}$ intersection homology groups* be the homology groups $H_*(I^{\bar{p}}C_*^{\mathrm{GM}}(X))$.

A number of observations are in order. First of all, we should note that each $I^{\bar{p}}C_*^{\mathrm{GM}}(X)$ is well defined as a chain complex. In particular, if ξ and η are $\bar{p}$-allowable chains, then every simplex in $\xi + \eta$, $-\xi$, $\partial(\xi + \eta) = \partial\xi + \partial\eta$, or $\partial(-\xi) = -\partial\xi$ must be a simplex already contained in ξ, η, $\partial\xi$, or $\partial\eta$ and so must be $\bar{p}$-allowable. Thus each $I^{\bar{p}}C_*^{\mathrm{GM}}(X)$ is a subgroup of $C_*(X)$. Each is also a well-defined chain complex by fiat, owing to the declaration that in order for ξ to be $\bar{p}$-allowable so must be all of the simplices of $\partial\xi$. Of course, as $I^{\bar{p}}C_*^{\mathrm{GM}}(X) \subset C_*(X)$, we always have $\partial(\partial\xi) = 0$, so $\partial\xi$ is a $\bar{p}$-allowable chain so long as it is composed of $\bar{p}$-allowable simplices.

We will discuss motivation for this definition in Section 3.2.2 after computing a few examples. The interested reader may feel free to skip ahead to that section now and then come back to the examples here.

3.2.1 First Examples

In order to get a feel for working with intersection homology, let us compute some elementary examples.

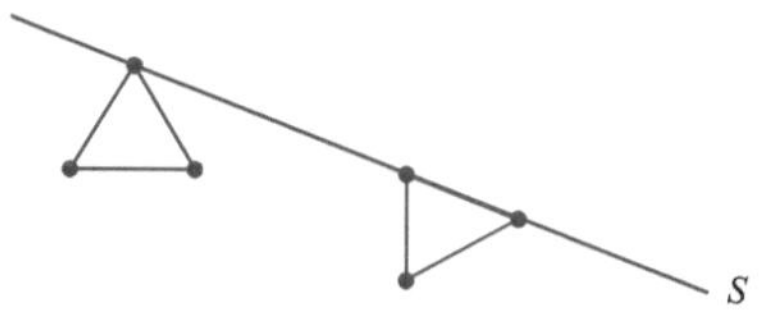

Figure 3.1 Two 2-simplices that intersect the stratum S.

Example 3.2.3 For the general idea in abstract, consider Figure 3.1, which shows two 2-simplices that intersect a stratum S. The intersection dimension on the left is 0 and on the right it is 1. If we use the perversity $\bar{t}$, then $2-\text{codim}(S)+\bar{t}(S) = 2-\text{codim}(S)+\text{codim}(S)-2 = 0$. So the simplex on the left is $\bar{t}$-allowable with respect to S, but the one on the right is not. △

Example 3.2.4 Let $X = X^0$ be a point. In this case there is only one stratum, X itself with codimension zero, and it is a regular stratum[5] so $\bar{p}(X) = 0$ for any perversity $\bar{p}$. There is also only one simplex to work with, a 0-simplex we shall denote v. The allowability condition then becomes that

$$\dim(v) = \dim(v \cap X) \leq \dim(v) - \text{codim}(S) + \bar{p}(S) = 0 - 0 + 0 = 0.$$

This is evidently true, so v is allowable, $\partial v = 0$ is allowable, and $I^{\bar{p}}C_*^{\text{GM}}(X) = C_*(X)$, the ordinary chain complex. So in this case intersection homology yields nothing new. △

Example 3.2.5 For a more interesting example, let X be the boundary of the simplex $[v_0, v_1, v_2]$; see Figure 3.2. Suppose X is filtered as $X^0 = \{v_0\} \subset X = X^1$. Then X has two strata: the regular stratum $X - \{v_0\}$ and the singular stratum $\{v_0\}$ of codimension one. For any perversity, we must have $\bar{p}(X - \{v_0\}) = 0$, but $\bar{p}(\{v_0\})$ could be any integer. Let us generically use the notation v for a 0-simplex and e for a 1-simplex. Then a 0-simplex v is

[5] Recall that writing $X = X^0$ lets us know that the formal dimension is zero.

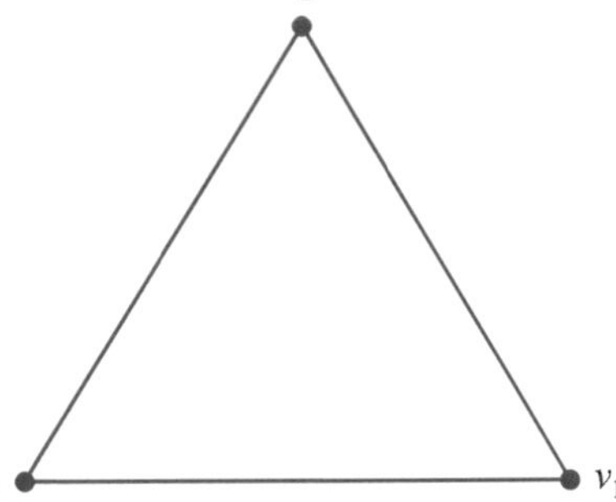

Figure 3.2 The boundary of the simplex $[v_0, v_1, v_2]$.

allowable if it satisfies the conditions

$$\begin{aligned}\dim(v \cap (X - \{v_0\})) &\leq \dim(v) - \operatorname{codim}(X - \{v_0\}) + \bar{p}(X - \{v_0\})\\ &= 0 - 0 + 0 = 0,\\ \dim(v \cap \{v_0\}) &\leq \dim(v) - \operatorname{codim}(\{v_0\}) + \bar{p}(\{v_0\}) = 0 - 1 + \bar{p}(\{v_0\})\\ &= \bar{p}(\{v_0\}) - 1.\end{aligned}$$

Since $\dim(v \cap (X - \{v_0\}))$ must always be ≤ 0 (as v is a 0-simplex), we see that any 0-simplex in $X - \{v_0\}$ is allowable. By contrast, the 0-simplex v_0 itself is allowable only if $\bar{p}(\{v_0\}) \geq 1$.

Similarly, for a 1-simplex e, the allowability conditions are

$$\begin{aligned}\dim(e \cap (X - \{v_0\})) &\leq \dim(e) - \operatorname{codim}(X - \{v_0\}) + \bar{p}(X - \{v_0\})\\ &= 1 - 0 + 0 = 1,\\ \dim(e \cap \{v_0\}) &\leq \dim(e) - \operatorname{codim}(\{v_0\}) + \bar{p}(\{v_0\}) = 1 - 1 + \bar{p}(\{v_0\})\\ &= \bar{p}(\{v_0\}).\end{aligned}$$

Again $\dim(e \cap (X - \{v_0\}))$ must always be ≤ 1 since e is a 1-simplex, and so the first condition always holds. Additionally, the 1-simplex $[v_1, v_2]$ does not intersect $\{v_0\}$, so it is allowable for any $\bar{p}$. The 1-simplices $[v_0, v_1]$ and $[v_0, v_2]$ both intersect $\{v_0\}$ with $\dim(e \cap \{v_0\}) = 0$, so they will be allowable only if $\bar{p}(\{v_0\}) \geq 0$.

So already we see that there are three distinct cases according to whether $\bar{p}(\{v_0\})$ is > 0, $= 0$, or < 0:

- If $\bar{p}(\{v_0\}) > 0$, then we have seen that all simplices will be allowable, and hence all chains will be allowable. In this case $I^{\bar{p}}C_*^{\mathrm{GM}}(X) = C_*(X)$, and we recover the standard simplicial chain complex, so $I^{\bar{p}}H_*^{\mathrm{GM}}(X) = H_*(X)$.

- If $\bar{p}(\{v_0\}) < 0$ then neither the 0-simplex $[v_0]$ nor either of the 1-simplices of X intersecting $\{v_0\}$ will be allowable, but the 1-simplex $[v_1, v_2]$ and the 0-simplices $[v_1], [v_2]$ are allowable. So $I^{\bar{p}}C_*^{\mathrm{GM}}(X) = C_*([v_1, v_2])$, the simplicial chain complex of the 1-simplex $[v_1, v_2]$. Correspondingly $I^{\bar{p}}H_*^{\mathrm{GM}}(X)$ is the homology of the interval.
- If $\bar{p}(\{v_0\}) = 0$, then our analysis concerning 0-simplices is the same as in the previous case: $[v_0]$ is not allowable, but $[v_1]$ and $[v_2]$ are. Now, however, all the 1-simplices are allowable. So what is the intersection chain complex $I^{\bar{p}}C_*^{\mathrm{GM}}(X)$? Here for the first time we must pay attention to boundaries. The 1-simplex $[v_0, v_1]$ is allowable *as a simplex*, but it is not allowable *as a chain* because its boundary $[v_1] - [v_0]$ contains a 0-simplex that is not allowable. However, the chain $[v_0, v_1] - [v_0, v_2]$ is allowable because its boundary is $[v_1] - [v_0] - ([v_2] - [v_0]) = [v_1] - [v_2]$, which is allowable. In fact, we can see that $I^{\bar{p}}C_1^{\mathrm{GM}}(X) \cong \mathbb{Z} \oplus \mathbb{Z}$ generated by $[v_1, v_2]$ and $[v_0, v_1] - [v_0, v_2]$. We have also seen that $I^{\bar{p}}C_0^{\mathrm{GM}}(X) \cong \mathbb{Z} \oplus \mathbb{Z}$, generated by $[v_1]$ and $[v_2]$. One can then compute by hand, using the boundary map, that $I^{\bar{p}}H_1^{\mathrm{GM}}(X) \cong \mathbb{Z}$, generated by $[v_0, v_1] - [v_0, v_2] + [v_1, v_2]$, just as for the standard homology, while $I^{\bar{p}}H_0^{\mathrm{GM}}(X) \cong \mathbb{Z}$ generated by either $[v_1]$ or $[v_2]$, which are homologous via the 1-simplex $[v_1, v_2]$. So ultimately it turns out that $I^{\bar{p}}H_*^{\mathrm{GM}}(X) \cong H_*(X)$ again here.

In the last case, namely $\bar{p}(\{v_0\}) = 0$, there are some shortcuts we could have used to compute the groups $I^{\bar{p}}H_*^{\mathrm{GM}}(X)$ without having to compute the complexes $I^{\bar{p}}C_*^{\mathrm{GM}}(X)$. For example, since $I^{\bar{p}}C_*^{\mathrm{GM}}(X) \subset C_*(X)$ and since we know from familiar homology computations that the only cycles in $C_1(X)$ are multiples of

$$\xi = [v_0, v_1] - [v_0, v_2] + [v_1, v_2],$$

these are also the only possible cycles in $I^{\bar{p}}C_*^{\mathrm{GM}}(X)$. Therefore, since $I^{\bar{p}}C_2^{\mathrm{GM}}(X) = 0$ trivially, to compute $I^{\bar{p}}H_1^{\mathrm{GM}}(X)$ we need only determine whether ξ is allowable. Once we have determined that it is, then we must have $I^{\bar{p}}H_1^{\mathrm{GM}}(X) \cong \mathbb{Z}$. Similarly, since all 0-simplices are cycles, once we have noticed that $[v_0]$ is not allowable but that $[v_1]$, $[v_2]$, and $[v_1, v_2]$ are allowable, we can quickly conclude that $I^{\bar{p}}H_0^{\mathrm{GM}}(X) \cong \mathbb{Z}$. Such computational techniques will prove very useful. △

Example 3.2.6 Let X again be the boundary of the 2-simplex $[v_0, v_1, v_2]$, but this time suppose the filtration is the simplicial filtration, i.e. $X^0 = \{[v_0], [v_1], [v_2]\}$ and $X^1 = X$. Suppose $\bar{p}(\{v_0\}) = \bar{p}(\{v_1\}) = \bar{p}(\{v_2\}) = 0$. Now, by the same analysis as in Example 3.2.5, none of the vertices are allowable but all of the edges are. Hence the cycle

$$[v_0, v_1] - [v_0, v_2] + [v_1, v_2]$$

is allowable and $I^{\bar{p}}H_1^{\mathrm{GM}}(X) \cong \mathbb{Z}$, but since no vertices are allowable, $I^{\bar{p}}H_0^{\mathrm{GM}}(X) = 0$. △

(a) Allowability with Respect to Regular Strata

One observation we might conjecture from our first examples is that the allowability condition is vacuous when it comes to regular strata. This is indeed the case as we formalize in the following lemma, which will help shorten the computations in our further examples.

Lemma 3.2.7 *Let σ be an i-simplex of a simplicial filtered space X and let S be a* regular *stratum of X. Then the allowability condition* (3.1) *is always satisfied.*

Proof Since σ is an i-simplex, for any subspace $Z \subset X$ it must be true that $\dim(\sigma \cap Z) \le i$, and since $\operatorname{codim}(S) = \bar{p}(S) = 0$, the right-hand side of the inequality (3.1) reduces to i. □

Example 3.2.8 Suppose X is an n-dimensional simplicial filtered space that is filtered trivially so that there are only regular strata; see Example 2.2.28. Then it follows from the preceding lemma that $I^{\bar{p}}C_*^{\mathrm{GM}}(X) = C_*(X)$. △

Remark 3.2.9 Lemma 3.2.7 allows us to provide some justification for setting $\bar{p}(S) = 0$ for all regular strata. We see from the lemma that with $\bar{p}(S) = 0$ all simplices are allowable with respect to all regular strata. Furthermore, if $\bar{p}(S) = m$ for any $m \ge 0$, then it is easy to see that the same conclusion will hold, so as mentioned in Remark 3.1.2, any choice of $\bar{p}(S) \ge 0$ for regular strata would provide the same intersection chains, but we choose $\bar{p}(S) = 0$ for definiteness and convenience.

By contrast if S is regular and $\bar{p}(S) \le -1$, then for an i-simplex to be allowable with respect to S, we would need

$$\dim(\sigma \cap S) \le i - \operatorname{codim}(S) + \bar{p}(S) = i - 0 + \bar{p}(S) \le i - 1.$$

This implies that the interior of σ cannot be completely contained in any regular stratum, and as the skeleta of the filtration are closed, this implies that $\sigma \subset \Sigma_X$. So if $\bar{p}(S) \le -1$ for any regular stratum, no simplex can intersect that stratum at all. In other words, $I^{\bar{p}}H_*^{\mathrm{GM}}(X)$ does not see that stratum, so it is equal to $I^{\bar{p}}H_*^{\mathrm{GM}}(X - S)$, noting that $X - S$ is a subcomplex of X so that this group is defined. Therefore, having regular strata with negative perversities is the same as working on spaces without those strata, and we could just as well have taken that view from the beginning and worked on a different space.

One could imagine there might be circumstances where it would nonetheless be useful to keep the regular stratum and use a negative perversity, but in fact these don't seem to come up meaningfully in applications. Altogether, we therefore believe it reasonable to always have $\bar{p}(S) = 0$ for regular strata.

See Remark 3.4.5 below for the analogous considerations for singular intersection chains.

(b) Effects of Subdivision

Next we explore the effects of subdivision on the computation of intersection homology groups.

Example 3.2.10 Let $X = X^1$ again be the boundary of the 2-simplex $[v_0, v_1, v_2]$ as in Example 3.2.6. Now consider X', the first barycentric subdivision of X (see Example

B.1.13), but with the same filtration as in Example 3.2.6 so that the 0-skeleton of the filtration is $\{v_0, v_1, v_2\}$; see Figure 3.3. We also continue to suppose $\bar{p}(\{v_0\}) = \bar{p}(\{v_1\}) = \bar{p}(\{v_2\}) = 0$. So as seen in the preceding example, these vertices are not allowable. But the barycenters of the edges will be allowable vertices, and they will be homologous via simplicial paths that cross through v_0, v_1, or v_2, as all 1-simplices are allowable. Thus $I^{\bar{p}}H_0^{\mathrm{GM}}(X') \cong \mathbb{Z}$. △

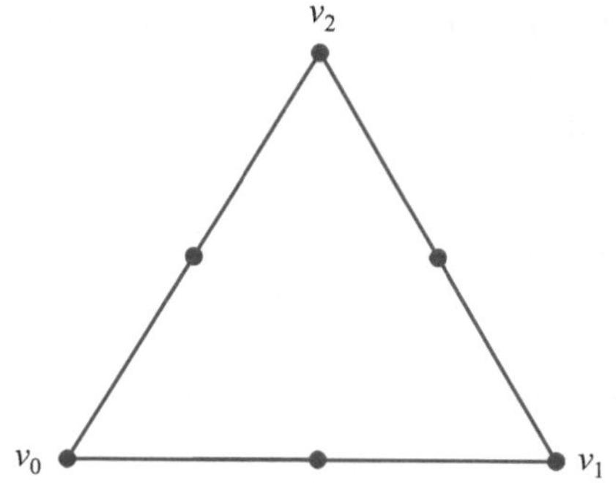

Figure 3.3 A simplicial complex X' with six 0-simplices and six 1-simplices. The filtration is $(X')^0 \subset (X')^1 = X'$ with $(X')^0 = \{v_0, v_1, v_2\}$.

Example 3.2.11 Let $Y = Y^2$ be the suspension of the boundary of the 2-simplex $[v_0, v_1, v_2]$, and let $Y^0 = Y^1 = \{v_0, v_1, v_2\}$; see Figure 3.4. Let $\bar{p}$ again be a perversity with $\bar{p}(\{v_0\}) = \bar{p}(\{v_1\}) = \bar{p}(\{v_2\}) = 0$. Let $\{n, s\}$ be the other two vertices. Then we can easily check that the two cone vertices $[n]$ and $[s]$ are the only allowable 0-simplices. Furthermore, as $1-2+\bar{p}(\{v_i\}) = -1$, no 1-simplex of Y is allowable, and we must have $I^{\bar{p}}H_0^{\mathrm{GM}}(Y) = \mathbb{Z}\oplus\mathbb{Z}$. Yet if Y' is the first barycentric subdivision of Y, keeping the same filtration, then there are allowable paths connecting any vertices and that don't contain any of the v_i, and so in this case $I^{\bar{p}}H_0^{\mathrm{GM}}(Y) = \mathbb{Z}$. △

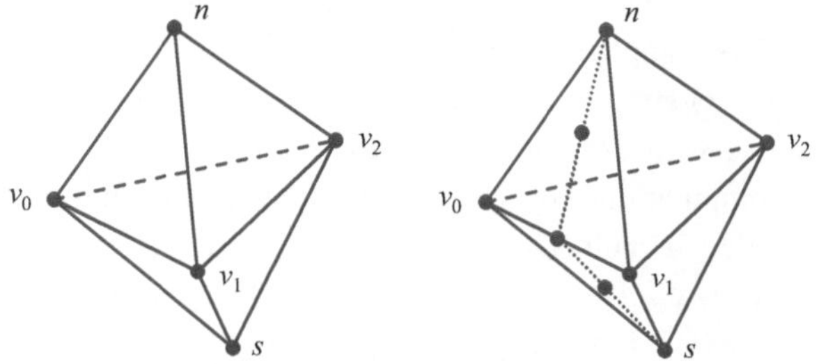

Figure 3.4 Left: the space Y. right: The dotted path indicates a homology between $[n]$ and $[s]$ in the barycentric subdivision.

These two examples show that the intersection homology groups are not independent of the triangulation. This might raise some reasonable concerns; however, we will show below in Theorem 3.3.20 that there is independence of the triangulation assuming some minor conditions. In particular, the groups will stabilize with respect to repeated barycentric subdivision.

(c) Some More Advanced Examples

The next examples involve computations of the intersection homology of stratified spaces built by coning off the boundary of a manifold. This example provides an intriguing first glimpse of the duality results we shall study later.

Example 3.2.12 Let M be a connected n-dimensional triangulated ∂-manifold with boundary $\partial M \neq \emptyset$. Let $\bar{c}(\partial M)$ be the simplicial cone on the boundary of M. In other words, for each simplex $[v_0, \dots, v_i]$ of ∂M, we add a simplex $[v, v_0, \dots, v_i]$; see [182, Section 8]. Let X be the space obtained by coning off the boundary of M:

$$X = X^n = M \cup_{\partial M} \bar{c}(\partial M).$$

If v is the cone vertex, let X be filtered as a manifold stratified space by $\{v\} \subset X$. We compute $I^{\bar{p}}H_*^{\mathrm{GM}}(X)$ for any perversity $\bar{p}$.

We have already seen that every simplex is allowable with respect to the regular stratum $X - \{v\}$, so we need only check which simplices are allowable with respect to $\{v\}$. This is only an issue for those simplices containing v, for which $\dim(\sigma \cap \{v\}) = 0$. So if σ is an i-simplex, we need

$$0 \leq i - \operatorname{codim}(\{v\}) + \bar{p}(\{v\}) = i - n + \bar{p}(\{v\}).$$

In other words, an i-simplex is allowed to contain v if and only if $i \geq n - \bar{p}(\{v\})$.

From this computation, we deduce that for $i < n - \bar{p}(\{v\})$, no allowable simplex, and hence no allowable chain, may contain v. Hence in this range every simplex must be a simplex in M itself and $I^{\bar{p}}C_i^{\mathrm{GM}}(X) = C_i(M)$. On the other hand, since every simplex is allowable for $i \geq n - \bar{p}(\{v\})$, every chain is allowable for $i > n - \bar{p}(\{v\})$, and so in this range $I^{\bar{p}}C_i^{\mathrm{GM}}(X) = C_i(X)$. The complicated case arises for $i = n - \bar{p}(\{v\})$. Now each i-simplex is allowable, but no $(i-1)$-simplex may contain v. So $I^{\bar{p}}C_{n-\bar{p}(\{v\})}^{\mathrm{GM}}(X)$ consists of all the $(n - \bar{p}(\{v\}))$-chains of X whose boundaries are in M.

Let us use these computations to compute the intersection homology groups. Since $I^{\bar{p}}C_{n-\bar{p}(\{v\})}^{\mathrm{GM}}(X)$ in particular includes all cycles of $C_{n-\bar{p}(\{v\})}(X)$, we have $I^{\bar{p}}H_i^{\mathrm{GM}}(X) = H_i(X)$ for $i \geq n - \bar{p}(\{v\})$. It also follows readily from the above discussion that $I^{\bar{p}}H_i^{\mathrm{GM}}(X) = H_i(M)$ for $i < n - \bar{p}(\{v\}) - 1$. In order to compute $I^{\bar{p}}H_{n-\bar{p}(\{v\})-1}^{\mathrm{GM}}(X)$, we observe that the cycles in $I^{\bar{p}}C_{n-\bar{p}(\{v\})-1}^{\mathrm{GM}}(X)$ are precisely the cycles in M, but they may be bounded by any chain in X. This is a description of the image group of the homomorphism $H_{n-\bar{p}(\{v\})-1}(M) \to H_{n-\bar{p}(\{v\})-1}(X)$ induced by the inclusion $M \hookrightarrow X$.

Summarizing, we have computed

$$I^{\bar{p}}H_i^{\mathrm{GM}}(X) \cong \begin{cases} H_i(X), & i \geq n - \bar{p}(\{v\}), \\ \mathrm{im}(H_i(M) \to H_i(X)), & i = n - \bar{p}(\{v\}) - 1, \\ H_i(M), & i < n - \bar{p}(\{v\}) - 1. \end{cases}$$

In particular, if $\bar{p}(\{v\}) \geq n$ then $I^{\bar{p}}H_*^{\mathrm{GM}}(X) \cong H_*(X)$, and if $\bar{p}(\{v\}) \leq -2$ then $I^{\bar{p}}H_*^{\mathrm{GM}}(X) \cong H_*(M)$. In fact, the latter isomorphism is also true when $\bar{p}(\{v\}) = -1$, since then

$$H_n(M) = \mathrm{im}(H_n(M) \to H_n(X)) = 0.$$

It is interesting to observe that if $i > 0$ then $H_i(X) \cong H_i(M, \partial M)$ by an easy homological argument (employ the exact sequence of the pair, the contractibility of cones, and excision), so that in this case we can identify $I^{\bar{p}}H_i^{\mathrm{GM}}(X)$ with $H_i(M, \partial M)$ if $i \geq n - \bar{p}(\{v\}) > 0$ and $I^{\bar{p}}H_{n-\bar{p}(\{v\})-1}^{\mathrm{GM}}(X)$ with

$$\mathrm{im}(H_{n-\bar{p}(\{v\})-1}(M) \to H_{n-\bar{p}(\{v\})-1}(M, \partial M))$$

if $n - \bar{p}(\{v\}) - 1 > 0$. In particular, if $n > \bar{p}(\{v\}) + 1$, we can reformulate our computation as

$$I^{\bar{p}}H_i^{\mathrm{GM}}(X) \cong \begin{cases} H_i(M, \partial M), & i \geq n - \bar{p}(\{v\}), \\ \mathrm{im}(H_i(M) \to H_i(M, \partial M)), & i = n - \bar{p}(\{v\}) - 1, \\ H_i(M), & i < n - \bar{p}(\{v\}) - 1. \end{cases}$$

This innocent-seeming computation is actually fairly remarkable. The intersection homology groups incorporate both the groups $H_*(M)$ and the groups $H_*(M, \partial M)$ with a transition point depending on the perversity. At the transition point, we have $\mathrm{im}(H_i(M) \to H_i(M, \partial M))$. This is already reminiscent of duality results, as Lefschetz duality provides pairings (in appropriate dimensions) between $H_*(M)$ and $H_*(M, \partial M)$ and self-pairings on $\mathrm{im}(H_*(M) \to H_*(M, \partial M))$. △

Example 3.2.13 The last example involved considering what happens at a cone vertex of a cone on a manifold. This next example, in which we compute the intersection homology of a suspension of a simplicial filtered space, is somewhat similar in terms of the involvement of cones, though the computations become more involved. We present this computation as an extended example of the kinds of computations that might arise when working with simplicial intersection homology. We note that such computations become much simpler once further tools are developed, such as singular intersection homology and the Mayer–Vietoris sequence. Singular intersection homology might give a different computation depending on the properties of the triangulation; however, singular and simplicial intersection homology will agree for suitably fine triangulations by Corollary 3.3.22 and Theorem 5.4.2, below. Thus the reader who doesn't want to get too bogged down in the following lengthy computation may safely skip ahead, perhaps simply noting the final results at the end of the example.

Let X be a compact $(m-1)$-dimensional simplicial filtered space, and let SX be the simplicial suspension of X obtained by adjoining two closed cones on X along X. In other words, we form SX from X by adding two new 0-simplices n and s and then for each simplex $[v_0, \dots, v_i]$ of X, we also now have simplices $[n, v_0, \dots, v_i]$ and $[s, v_0, \dots, v_i]$; see [182, Section 8, Exercise 1]. The suspension has a natural filtration given by $(SX)^{i+1} = S(X^i)$ for $i \geq 0$ and by letting $(SX)^0 = \{n, s\}$, the "north and south poles" given by the two cone vertices. The strata of SX are $\{n\}, \{s\}$, and the collection of $ST - \{n, s\}$ where T is a stratum of X.

Let $\bar{p}$ be a perversity on SX. For simplicity, we will compute the example for which $\bar{p}(\{n\}) = \bar{p}(\{s\})$, but we encourage the reader to consider the more general case as an instructive exercise. It will be useful to let $\bar{p}_X$ be the perversity on X defined by $\bar{p}_X(T) = \bar{p}(ST - \{n, s\})$.

To begin to compute $I^{\bar{p}}C_*^{\mathrm{GM}}(X)$, we first observe that since the codimension of a stratum T in X is the same as the codimension of $ST - \{n, s\}$ in SX, the condition for a simplex contained in X to be $\bar{p}_X$-allowable with respect to a stratum $T \subset X$ is exactly the same as the condition for the simplex to be $\bar{p}$-allowable with respect to $ST - \{n, s\}$ in SX.

Next we consider simplices containing n or s. Notice that any simplex of dimension > 0 containing one of these vertices can be written as a cone on a simplex contained in X, i.e. it has the form $[w, v_0, \dots, v_i]$ for $[v_0, \dots, v_i]$ an i-simplex of X and $w \in \{n, s\}$ (see [182, Section 8]). If $\sigma = [v_0, \dots, v_i]$ is a simplex of X, let $\bar{c}_n\sigma = [n, v_0, \dots, v_i]$ and $\bar{c}_s\sigma = [s, v_0, \dots, v_i]$. It is easy to check that $\bar{c}_n$ and $\bar{c}_s$ generate homomorphisms $C_*(X) \to C_{*+1}(SX)$ if we let $\bar{c}_n(0) = 0$ and $\bar{c}_s(0) = 0$, though these are not chain maps. In fact $\partial(\bar{c}_n\xi) = \xi - \bar{c}_n(\partial\xi)$ if ξ is a chain of dimension greater than zero; if $[v]$ is a 0-simplex, then $\partial(\bar{c}_n[v]) = [v] - [n]$. Similar formulas hold for $\bar{c}_s$. Conversely, if τ is a simplex containing n, then τ can be written (up to sign) as $[n, v_0, \dots, v_i]$, and we recognize τ as a cone on $\pm\sigma = \pm[v_0, \dots, v_i] \in X$. Again the analogous statement holds for simplices containing s. No simplex contains both n and s.

Now, let $\sigma = [v_0, \dots, v_i]$ be an i-simplex in X. Then σ is $\bar{p}_X$-allowable in X with respect to a stratum $T \subset X$ if and only if $\bar{c}_n\sigma$ and $\bar{c}_s\sigma$ are $\bar{p}$-allowable in SX with respect to $ST - \{n, s\}$. The argument is the same for each of $\bar{c}_n\sigma$ and $\bar{c}_s\sigma$ so we provide only the former. Clearly, $\dim(\bar{c}_n\sigma \cap (ST - \{n, s\})) = \dim(\sigma \cap T) + 1$, while $\operatorname{codim}_X(T) = \operatorname{codim}_{SX}(ST - \{n, s\})$, using the notation $\operatorname{codim}_Z(\cdot)$ to denote codimension within the space Z. Thus the inequality

$$\dim(\sigma \cap T) \leq i - \operatorname{codim}_X(T) + \bar{p}_X(T)$$

is equivalent to the inequality

$$\dim(\bar{c}_n\sigma \cap (ST - \{n, s\})) - 1 \leq i - \operatorname{codim}_{SX}(ST - \{n, s\}) + \bar{p}(ST - \{n, s\}),$$

which we can rewrite as

$$\dim(\bar{c}_n\sigma \cap (ST - \{n, s\})) \leq i + 1 - \operatorname{codim}_{SX}(ST - \{n, s\}) + \bar{p}(ST - \{n, s\}).$$

Hence as $\bar{c}_n\sigma$ is an $(i+1)$-simplex, we see that the conditions for the allowability of σ and $\bar{c}_n\sigma$ with respect to T and $ST - \{n, s\}$ are equivalent.

Next we look at allowability with respect to the strata $\{n\}$ and $\{s\}$. As in our previous computation, if a simplex contains the vertex n then $\dim(\sigma \cap \{n\}) = 0$. So for an i-simplex σ containing n to be allowable with respect to $\{n\}$, we need

$$0 \le i - \operatorname{codim}(\{n\}) + \bar{p}(\{n\}) = i - m + \bar{p}(\{n\}).$$

In other words, an i-simplex is allowed to contain n only if $i \ge m - \bar{p}(\{n\})$, and similarly for the vertex s.

So if $i < m - \bar{p}(\{n\})$, no allowable i-simplex, and hence no allowable chain, may contain n (or s), and so all such chains must be contained in X. But we have already noted that a simplex in X is $\bar{p}$-allowable in SX if and only if it is $\bar{p}_X$-allowable in X. So for $i < m - \bar{p}(\{n\})$, we have $I^{\bar{p}}C_i^{\mathrm{GM}}(SX) = I^{\bar{p}_X}C_i^{\mathrm{GM}}(X)$, and it follows that $I^{\bar{p}}H_i^{\mathrm{GM}}(SX) = I^{\bar{p}_X}H_i^{\mathrm{GM}}(X)$ for $i < m - \bar{p}(\{n\}) - 1$.

We also have that any cycle, say ξ, in $I^{\bar{p}}C^{\mathrm{GM}}_{m-\bar{p}(\{n\})-1}(SX)$ is contained in X, but chains in $I^{\bar{p}}C^{\mathrm{GM}}_{m-\bar{p}(\{n\})}(SX)$ may contain the suspension vertices. So given ξ, suppose that $m - \bar{p}(\{n\}) - 1 > 0$. Then $\xi \in I^{\bar{p}_X}C^{\mathrm{GM}}_{n-\bar{p}(\{n\})-1}(X)$ and $\bar{c}_n\xi \in I^{\bar{p}_X}C^{\mathrm{GM}}_{n-\bar{p}(\{n\})}(SX)$ will be allowable – we have seen above that the cone on the simplices of ξ, which are each $\bar{p}_X$-allowable, will be $\bar{p}$-allowable with respect to all strata $ST - \{n, s\}$, but we have also just seen allowability of simplices of this dimension with respect to $\{n\}$. Since $\partial(\bar{c}_n\xi) = \xi - \bar{c}_n(\partial\xi) = \xi$, the boundary of $\bar{c}_n\xi$ also consists of allowable simplices. So $\bar{c}_n\xi$ is allowable, and ξ represents 0 in intersection homology. Since this argument applies to any cycle, we have $I^{\bar{p}}H^{\mathrm{GM}}_{m-\bar{p}(\{n\})-1}(SX) = 0$.

If $m - \bar{p}(\{n\}) - 1 = 0$, we have a slightly more delicate situation, as, if v is a 0-simplex in X, then $\partial(\bar{c}_n[v]) = [v] - [n]$, and the 0-simplex $[n]$ is not allowable. However, for any two allowable 0-simplices $[v], [w]$ in X, the chain $\bar{c}_n([v] - [w])$ is an allowable 1-chain and has boundary $[v] - [w]$. Hence any two allowable 0-simplices in SX are homologous. So in this case $I^{\bar{p}}H_0^{\mathrm{GM}}(SX)$ is either $\mathbb{Z}$ or 0 according as there are or are not any allowable vertices in X.

Finally, we must compute the intersection homology for degrees $i \ge m - \bar{p}(\{n\})$. In this case all cycles are allowable with respect to $\{n, s\}$. We will see that, except in low-dimensional cases, these intersection homology groups are equal to the intersection homology groups of X in dimension $i - 1$. This agrees with what one might expect from the suspension formula for ordinary homology $\tilde{H}_i(SX) = \tilde{H}_{i-1}(X)$. Also as for ordinary homology, we will see that the isomorphism can be given by suspending chains in X. We will consider the possibilities $i = 0, 1$, which must be handled separately, below. For now assume that $i \ge m - \bar{p}(\{n\}) \ge 2$.

First suppose $\xi \in I^{\bar{p}}C_i^{\mathrm{GM}}(SX)$, with $i \ge m - \bar{p}(\{n\}) \ge 2$, is a cycle that does not intersect $\{n, s\}$. Then the same coning argument used above shows that ξ is the boundary of the allowable chain $\bar{c}_n\xi$. Similarly, suppose ξ is a cycle containing a simplex that includes, say

n, but not s. Then we can write ξ uniquely as $\xi = x + \bar{c}_n y$, where x and y are each contained in X; $\bar{c}_n y$ contains exactly those simplices of ξ containing n and then $x = \xi - \bar{c}_n y$. We do not assume that x or $c_n y$ are allowable chains, as their boundaries might not be allowable, though of course all simplices of x and $c_n y$ are allowable, as ξ is. This implies, as above, that all simplices of $\bar{c}_n x$ are allowable, and we next show that $\partial(\bar{c}_n x) = \xi$. Since ξ is a cycle, we have $\partial x = -\partial(\bar{c}_n y) = -y + \bar{c}_n(\partial y)$. As x and y are contained in X, so must be $\partial x + y = \bar{c}_n(\partial y)$, which is a contradiction unless $\partial y = 0$. So we must have $\partial y = 0$, and now we compute that

$$\begin{aligned}\partial(\bar{c}_n x) &= x - \bar{c}_n(\partial x) = x + \bar{c}_n(\partial(\bar{c}_n y))\\ &= x + \bar{c}_n(y - \bar{c}_n \partial y) = x + \bar{c}_n y\\ &= \xi.\end{aligned}$$

So we see that the only cycles that might not be trivial in intersection homology in this dimension range are those containing both n and s. We can write any such intersection cycle uniquely as $\xi = x + \bar{c}_n y - \bar{c}_s z$, where x, y, z are contained in X and composed of $\bar{p}$-allowable simplices, though their individual boundaries might not be. By arguments similar to those above, $\bar{c}_n x$ is composed of allowable simplices but now with boundary

$$\begin{aligned}\partial(\bar{c}_n x) &= x - \bar{c}_n(\partial x) = x + \bar{c}_n \partial(\bar{c}_n y - \bar{c}_s z)\\ &= x + \bar{c}_n(y - \bar{c}_n(\partial y) - z + \bar{c}_s(\partial z))\\ &= x + \bar{c}_n(y - z).\end{aligned}$$

For the last equation, we have used that

$$\partial x = -\partial(\bar{c}_n y) + \partial(\bar{c}_n z) = -y + \bar{c}_n(\partial y) + z - \bar{c}_s(\partial z)$$

and so, since $x, y, z \in X$, we must have $\partial y = \partial z = 0$, as in the argument above. Therefore,

$$\xi - \bar{c}_n z + \bar{c}_s z = x + \bar{c}_n y - \bar{c}_s z - \bar{c}_n z + \bar{c}_s z = x + \bar{c}_n(y - z) = \partial(\bar{c}_n x).$$

As the cycle z is allowable, so is the cycle $\bar{c}_n z - \bar{c}_s z$, which we denote Sz and call the suspension of z. This calculation also shows that $\partial \bar{c}_n x$ is allowable, and so $\bar{c}_n x$ is an allowable chain. Our last computation therefore shows that every cycle of $I^{\bar{p}}C_i^{\mathrm{GM}}(SX)$ for $i \geq m - \bar{p}(\{n\}) \geq 2$ is homologous in $I^{\bar{p}}C_*^{\mathrm{GM}}(SX)$ to a suspension of an allowable cycle of $I^{\bar{p}_X}C_{i-1}^{\mathrm{GM}}(X)$, i.e. suspension induces a surjective homomorphism $S : I^{\bar{p}_X}H_{i-1}^{\mathrm{GM}}(X) \to I^{\bar{p}}H_i^{\mathrm{GM}}(SX)$.

We next show that the intersection homology homomorphism S is also injective in this dimension range. Suppose z is an allowable cycle in $I^{\bar{p}}C_{i-1}^{\mathrm{GM}}(X)$ so that Sz is a cycle in $I^{\bar{p}}C_i^{\mathrm{GM}}(SX)$ for $i \geq m - \bar{p}(\{n\})$, and suppose Sz bounds an allowable $(i+1)$-chain Z. Again we can uniquely write $Z = \bar{c}_n A - \bar{c}_s B + D$ for chains A, B, D contained in X and composed of allowable simplices; and again A, B, D need not be allowable as chains, *a priori*. However, we have

$$\begin{aligned} Sz = \partial Z &= \partial(\bar{c}_n A - \bar{c}_s B + D) \\ &= A - \bar{c}_n(\partial A) - B + \bar{c}_s(\partial B) + \partial D, \end{aligned}$$

and since $Sz = \bar{c}_n z - \bar{c}_s z$, we can identify the subchains of Sz in each equation that contain n to obtain $\bar{c}_n z = -\bar{c}_n \partial A$. It follows that $z = -\partial A$. As A consists of allowable cycles, this shows that z bounds an intersection chain in X. So the suspension homomorphism is injective for $i \geq m - \bar{p}(\{n\}) \geq 2$.

To conclude we must consider the cases $i \geq m - \bar{p}(\{n\})$ and $i = 0, 1$. For $i = 0$, we note that $[n]$ and $[s]$ are now both allowable 0-cycles. If there is any allowable 0-simplex $[v]$ in X, then $\bar{c}_n$ and $\bar{c}_s$ are allowable and show that $[v]$, $[n]$, and $[s]$ are all intersection homologous so $I^{\bar{p}}H_0^{\mathrm{GM}}(SX) = \mathbb{Z}$. However, if there is no allowable 0-simplex $[v]$ in X, then $[n]$ and $[s]$ are not intersection homologous as any 1-chain with boundary $[n] - [s]$ would have to include an edge $[n, v]$ for some 0-simplex $[v]$ in X, and we know this will be allowable only if $[v]$ is allowable by our work way back at the beginning of the example. So in this case $I^{\bar{p}}H_0^{\mathrm{GM}}(SX) = \mathbb{Z} \oplus \mathbb{Z}$.

Finally, we consider $i = 1 \geq m - \bar{p}(\{n\})$. The main concern here is that cones and suspensions of 0-chains will be involved in the argument, but these must be treated carefully since, for example, if $[v]$ is a 0-simplex in X, then $\partial(\bar{c}_n[v]) = [v] - [n]$, and, similarly, $\partial(S[v]) = [s] - [n]$. So the suspension of a 0-chain is not necessarily a cycle. In fact, if $\eta = \sum a_j[v_j]$, then $\partial(S\eta) = \sum a_j([n] - [s])$, so that $S\eta$ is a cycle if and only if $\sum a_j = 0$, i.e. if the augmentation $\mathbf{a}(\eta)$ is 0. Similarly, $\partial(\bar{c}_n\eta) = \partial(\bar{c}_s\eta) = \eta$ if and only if $\mathbf{a}(\eta) = 0$. Thus we should consider $\tilde{S} : I^{\bar{p}_X}\tilde{C}_0^{\mathrm{GM}}(X) \to I^{\bar{p}}C_1^{\mathrm{GM}}(SX)$, where $I^{\bar{p}_X}\tilde{C}_0^{\mathrm{GM}}(X)$ is the kernel of the augmentation map $\mathbf{a} : I^{\bar{p}_X}C_0^{\mathrm{GM}}(X) \to \mathbb{Z}$ restricted from $\mathbf{a} : C_0(X) \to \mathbb{Z}$. Now if $\xi \in I^{\bar{p}}C_1^{\mathrm{GM}}(SX)$, and $\xi = x + \bar{c}_n y - \bar{c}s_z$, then y and z must have trivial augmentations or ξ could not be a cycle. With this observation, the above argument goes through to show that ξ is intersection homologous to a suspension of an allowable 0-cycle with trivial augmentation. Similarly, the above injectivity argument carries over for $\tilde{S}$, and it follows that $\tilde{S}$ induces an isomorphism $I^{\bar{p}_X}\tilde{H}_0^{\mathrm{GM}}(X) \to I^{\bar{p}}H_1^{\mathrm{GM}}(SX)$.

At last we can provide now the full formula[6]:

$$I^{\bar{p}}H_i^{\mathrm{GM}}(SX) = \begin{cases} I^{\bar{p}_X}\tilde{H}_{i-1}^{\mathrm{GM}}(X), & i \geq m - \bar{p}(\{n\}), \quad i \neq 0, \\ 0, & i = m - \bar{p}(\{n\}) - 1, \quad i \neq 0, \\ I^{\bar{p}_X}H_i^{\mathrm{GM}}(X), & i < m - \bar{p}(\{n\}) - 1, \\ \mathbb{Z} \oplus \mathbb{Z}, & i = 0 \geq m - \bar{p}(\{n\}), \quad X \text{ does not have an allowable 0-simplex}, \\ \mathbb{Z}, & i = 0 \geq m - \bar{p}(\{n\}) - 1, \quad X \text{ has an allowable 0-simplex}, \\ 0, & i = 0 = m - \bar{p}(\{n\}) - 1, \quad X \text{ does not have an allowable 0-simplex}. \end{cases}$$

[6] Note that, as for ordinary homology, we let $I^{\bar{p}_X}\tilde{H}_i^{\mathrm{GM}}(X) = I^{\bar{p}_X}H_i^{\mathrm{GM}}(X)$ if $i \neq 0$.

Suppose that X has regular strata and is sufficiently finely triangulated that there is a 0-simplex in a regular stratum. Then X will have $\bar{p}_X$-allowable 0-simplices. Furthermore, suppose that $\bar{p}(\{n\}) = \bar{p}(\{s\}) \leq m-2$, which will be the case if $\bar{p}$ is a GM perversity. Then our formula simplifies to a more manageable

$$I^{\bar{p}}H_i^{\mathrm{GM}}(SX) = \begin{cases} I^{\bar{p}_X}\tilde{H}_{i-1}^{\mathrm{GM}}(X), & i \geq m - \bar{p}(\{n\}), \quad i \neq 0, \\ 0, & i = m - \bar{p}(\{n\}) - 1, \quad i \neq 0, \\ I^{\bar{p}_X}H_i^{\mathrm{GM}}(X), & i < m - \bar{p}(\{n\}) - 1. \end{cases}$$

In this form, we can see more clearly that, except for some quirks in dimension zero in exceptional cases, $I^{\bar{p}}H_i^{\mathrm{GM}}(SX)$ agrees with $I^{\bar{p}_X}H_i^{\mathrm{GM}}(X)$ in lower dimensions and with $I^{\bar{p}_X}H_{i-1}^{\mathrm{GM}}(X)$ in higher dimensions; it is 0 at the transition dimension, which depends on the perversity.

In general, if $\bar{p}(\{n\}) = \bar{p}(\{s\}) \geq m$, then $I^{\bar{p}}H_i^{\mathrm{GM}}(SX)$ behaves like a suspension in ordinary homology in all dimensions except $i = 0$, by always being isomorphic to $I^{\bar{p}_X}H_{i-1}^{\mathrm{GM}}(X)$. For $\bar{p}(\{n\}) = \bar{p}(\{s\}) \leq -2$, we have $I^{\bar{p}}H_*^{\mathrm{GM}}(SX) = I^{\bar{p}_X}H_*^{\mathrm{GM}}(X)$, as if no suspension took place. △

3.2.2 Some Remarks on the Definition

In this section we briefly discuss the motivation for the definition of $\bar{p}$-allowability and a competing definition that we will not use.

(a) The Motivation for the Definition of Intersection Homology

We should next briefly discuss the curious allowability condition that $\dim(\sigma \cap S) \leq i - \mathrm{codim}(S) + \bar{p}(S)$ for an i-simplex σ. For this, recall that if M^m is a manifold with submanifolds N^n and P^p, then we say that N and P are in *general position* if $\dim(N \cap P) \leq n + p - m$. In particular, this is the case if N and P are actually *transverse*, meaning that at each point of $N \cap P$ we have a neighborhood U so that the triple $(U, N \cap U, P \cap U)$ is homeomorphic to the triple $(\mathbb{R}^m, \mathbb{R}^n \times \{0\}, \{0\} \times \mathbb{R}^p)$. In smooth manifold theory, transversality is often the more useful concept, but piecewise linear intersection theory requires only general position. We will discuss piecewise linear intersection products in Section 8.5. For our current purposes, we notice that since $\dim(P) = p$ and $\dim(M) = m$, then the codimension of P in M is $m - p$, so that the definition of general position can be rewritten $\dim(N \cap P) \leq n - \mathrm{codim}(P)$.

So now let ξ be an i-dimensional simplicial chain in M, which we continue to let be a manifold for the moment. We let $|\xi|$ be its support, i.e. the union of simplices of ξ. Furthermore, suppose M is filtered and that S is a manifold stratum. Then a requirement of the form $\dim(\sigma \cap S) \leq i - \mathrm{codim}(S)$ is simply the requirement that σ and S be in general position. If this condition holds for all σ in ξ, then $|\xi|$ and S are in general position.

Within a smooth manifold M, it is possible to manipulate smooth submanifolds by small isotopies to make them transverse; see, for example, [38, Section II.15]. Similarly, there are techniques for pushing polyhedra into general position; see [198, Chapter 5]. More generally, one can push chains into general position with respect to submanifolds or other chains by isotopies, with the technique varying slightly depending on what category we're in and what kinds of chains we're working with. And as far as homology is concerned, such isotopies don't change the homology class of a cycle. Consequently, if it suited our purposes, we might define the ordinary homology of a manifold using only chains in general position with respect to certain submanifolds or even more general subcomplexes.

Now, suppose X is a PL stratified pseudomanifold. If X is not a manifold, we no longer expect it to be possible to achieve general position with respect to subspaces. For example, let ξ be a 1-cycle that runs through the pinch point v of a pinched two-dimensional torus; see again Figure 1.4 on page 6. General position on a 2-manifold would require a 1-cycle and a 0-manifold to be disjoint, but there is no isotopy of the 1-cycle that can achieve that. So the idea of the $\bar{p}$-allowability condition is to provide a more flexible version of a general position constraint that might still allow that 1-cycle if we so desire.

For our i-simplex σ, the requirement

$$\dim(\sigma \cap S) \leq i - \operatorname{codim}(S) + \bar{p}(S)$$

differs from the general position condition only by the $\bar{p}(S)$ summand. So if $\bar{p}(S) = 0$, we recover general position, but having $\bar{p}(S) > 0$ lets us relax the general position requirement by a degree controlled by $\bar{p}$. By contrast, taking $\bar{p}(S) < 0$, which is less common, strengthens the general position requirement! As $\bar{p}$ is a function on strata, the perversity provides stratum-by-stratum control over how much deviation from general position we are willing to allow in defining the intersection chains. This is the origin of the term "perversity" – in some sense it is perverse that we are not requiring general position[7]!

Later, in Section 8.5, we will see how this loosening of general position requirements comes into play for constructing an intersection pairing that is well defined and nonsingular on pseudomanifolds, despite their singularities. This was the original setting for Poincaré duality on pseudomanifolds.

(b) Strata vs. Skeleta in the Definition of Intersection Chains

In some sources, notably including the original Goresky–MacPherson paper [105], the complex of intersection chains is defined not in terms of the dimensions of intersections of simplices with *strata* but rather in terms of the dimensions of intersections of simplices with *skeleta*. In particular, this alternative definition tends to arise in settings where one uses only GM perversities. Recall from Definition 3.1.4 that a GM perversity is a function $\bar{p} : \{2, 3, 4, \ldots\} \to \mathbb{Z}$ such that $\bar{p}(2) = 0$ and $\bar{p}(k) \leq \bar{p}(k+1) \leq \bar{p}(k) + 1$. Further, as noted in Remark 3.1.5, a GM perversity determines a general perversity on any filtered space X with no codimension-one strata by (abusing notation) letting $\bar{p}(S) = \bar{p}(\operatorname{codim}(S))$ for a singular stratum S.

[7] The historical survey [141] contains a bit more regarding the origins of the terminology.

Lemma 3.2.14 *Let X be a simplicial filtered space without codimension-one strata and of dimension n, and let $\bar{p}$ be a GM perversity. Let σ be an i-simplex of X. Then* $\dim(\sigma \cap S) \leq i - \operatorname{codim}(S) + \bar{p}(S)$ *for all singular strata S if and only if* $\dim(\sigma \cap X^k) \leq i - (n-k) + \bar{p}(n-k)$ *for all $k \leq n-2$.*

Proof Before beginning with the details, we remind the reader that X^k is the k-skeleton of the filtration on X, not the simplicial k-skeleton.

First, suppose that

$$\dim(\sigma \cap X^k) \leq i - (n-k) + \bar{p}(n-k)$$

for all $k \leq n-2$, and let S be a singular stratum of X. By definition, S is a connected component of $X^k - X^{k-1}$ for some k with $k \leq n-2$. So $\sigma \cap S \subset \sigma \cap X^k$. Therefore,

$$\dim(\sigma \cap S) \leq \dim(\sigma \cap X^k) \leq i - (n-k) + \bar{p}(n-k) = i - \operatorname{codim}(S) + \bar{p}(S).$$

Conversely, suppose $\dim(\sigma \cap S) \leq i - \operatorname{codim}(S) + \bar{p}(S)$ for all singular strata S, and choose some particular skeleton X^m, $m \leq n-2$. The skeleton X^m is a union of strata of dimension $\leq m$, and so of codimension $\geq n-m$. Suppose $X^m = \bigcup_j S_j$ for some collection of strata $\{S_j\}$ of codimension $\geq n-m$. Since σ is compact (or simply because it has a finite number of faces), it intersects only a finite number of the S_j, and $\dim(\sigma \cap X^m) = \max\{\dim(\sigma \cap S_j)\}$. So

$$\dim(\sigma \cap X^m) = \max\{\dim(\sigma \cap S_j)\} \leq \max\{i - \operatorname{codim}(S_j) + \bar{p}(S_j)\}.$$

As $\bar{p}(k) \leq \bar{p}(k+1) \leq \bar{p}(k) + 1$, the function $\bar{p}(k) - k$ is decreasing (non-strictly). So if we input to $i - \operatorname{codim}(S_j) + \bar{p}(S_j)$ strata of codimension $\geq n-m$, the maximum value will be $i - (n-m) + \bar{p}(n-m)$. Thus

$$\dim(\sigma \cap X^m) \leq i - (n-m) + \bar{p}(n-m),$$

as desired. □

This lemma implies that when X is a simplicial filtered space with no codimension-one strata and $\bar{p}$ is a GM perversity, we equally well could have used the dimensions of intersections of simplices with skeleta, rather than strata, to define intersection homology, and we would have obtained the same intersection chain complexes and intersection homology groups. In fact, this statement holds more generally, as the reader may note that the proof of the lemma really only used that the perversity was a nondecreasing function of the codimension. We invite the reader to think through the most general case where the two definitions are equivalent. But let us note that they are not equivalent in full generality.

Example 3.2.15 Here's a simple example. Let X be an n-simplex, $n \geq 2$, filtered by its simplicial skeleton. Let $\bar{p}$ be a perversity defined in terms of codimension such that $\bar{p}(n) = n$ and $\bar{p}(n-1) = 0$. Then for each vertex v, $\bar{p}(\{v\}) = n$ and $[v]$ is an allowable 0-simplex by our usual definition. However, it is not true that

$$0 = \dim(\{v\} \cap X^1) \leq 0 - (n-1) + \bar{p}(n-1) = 1 - n.$$

△

So, in general, it matters that we have used intersections with strata rather than intersections with skeleta to define intersection homology. As there are more strata than skeleta in general and as strata are smaller than skeleta, the stratum definition is the more flexible one and the one we maintain throughout the book, including as we move on to PL and singular intersection homology theories.

3.3 PL Intersection Homology

As one learns in an introductory text on algebraic topology, there are many benefits to working with simplicial homology. For example, unlike singular homology, if the space is compact then all of the chain groups are finitely generated. Additionally, there are no simplices of dimension higher than that of the space. Simplicial homology is also completely combinatorial, and so, at least in theory, computations can be done by a computer. Nonetheless, there are drawbacks to working with a fixed triangulation. Even given a triangulable manifold, there is not necessarily any canonical choice of triangulation, and, as we have seen, intersection homology might depend on the choice of triangulation. There are also theorems that become difficult to prove when locked into a specific triangulation.

In this section we turn to a homology theory that takes away the choice of triangulation but that still takes advantage of piecewise linear structure when it exists. This is piecewise linear (PL) homology. PL homology could be presented in any introductory algebraic topology text that treats simplicial homology, though usually it is not considered, as most texts have little reason not to jump straight to singular homology. Nonetheless, for the purposes of treating transversality and intersections in manifolds (or pseudomanifolds), which is a topic that will concern us briefly later in Section 8.5, PL chains come in handy. In fact, the language of PL chains was the original language in which intersection homology was formulated by Goresky and MacPherson in [105]; despite our limited consideration, the intersection pairing from the PL chain perspective was a motivating cornerstone for the entire intersection homology theory!

Additionally, PL intersection homology will help us prove theorems about simplicial intersection homology. This is not as big of an issue for ordinary homology where less care needs to be taken to make sure that simplicial maps don't inadvertently change the allowability of simplices with respect to strata. But even for ordinary homology the reader has likely had past opportunity to note that it is useful to have these multiple perspectives available for computing what often amount to the same homology groups. For example, CW homology is often the easiest to compute by hand, simplicial homology is perhaps the best for asking a computer to compute homology, while singular homology is often the simplest to prove certain theorems about (while of course also being defined on more general spaces). There does not seem to have developed a useful theory of intersection homology from the CW perspective, but we will treat simplicial, PL, and singular intersection homology throughout the first few chapters, ultimately demonstrating that the PL and singular theories agree on PL spaces, and that these also agree with the simplicial theory if we impose a few additional hypotheses. However, once we get to more sophisticated topics, it will be useful to focus primarily, and often exclusively, on the singular theory, which applies in more situations. One could also develop PL versions of these later results in detail and attempt to show that they are equivalent to the singular versions, but this book will be long enough as it is!

3.3.1 PL Homology

Let X be a PL space. Recall again that this means that X is endowed with a family of admissible locally finite triangulations such that any subdivision of an admissible triangulation is admissible and any two admissible triangulations have a common subdivision. The idea of the PL chain complex is that it should contain simplicial chains that are simplicial with respect to any admissible triangulation of X, and any two simplicial chains that differ just by a subdivision should represent the same PL chain.

To formalize this notion, let us first recall that in Definition 2.5.1 we defined two triangulations $T = (K, h)$ and $S = (L, j)$ of a PL space X to be *equivalent triangulations* if $j^{-1}h$ is a simplicial isomorphism. Further, recall that if $T = (K, h)$ is a triangulation then a subdivision of T is a triangulation of the form $T' = (K', h)$, where K' is a simplicial subdivision (not necessarily barycentric) of K as a simplicial complex. If $T = (K, h)$ and $S = (L, j)$ are equivalent triangulations, then any subdivision $T' = (K', h)$ induces an equivalent subdivision $S' = (L', j)$ by letting the vertices of K' determine the vertices of L' via $j^{-1}h$, using that $j^{-1}h$ is already linear on each simplex of K.

Now, let us observe that $\mathcal{T}$ can be given the structure of a directed set. Recall that a directed set has a relation $\leq$ that is reflexive and transitive and such that any two elements have a common upper bound, meaning that, given any elements a, b, there is an element c with $a \leq c$ and $b \leq c$. If $T = (K, h)$ and $S = (L, j)$ are two triangulations in $\mathcal{T}$, we will say that $T \leq S$ if S is equivalent to a subdivision of T; this includes the possibility that S is equivalent to T itself. This relation is clearly reflexive, and it satisfies the upper bound property because any two triangulations in $\mathcal{T}$ have a common subdivision, i.e. they have subdivisions that are equivalent (see Definitions 2.5.1 and 2.5.2). In particular, if T' and S' are equivalent subdivisions of T and S, respectively, then we have $T \leq T'$ and $S \leq T'$ (as well as $T \leq S'$ and $S \leq S'$). Finally, if $T \leq S \leq R$, with $T = (K, k)$, $S = (L, \ell)$, and $R = (J, j)$, then, by definition, there exist $T' = (K', k)$ and $S' = (L', \ell)$ such that $\ell^{-1}k : K' \to L$ and $j^{-1}\ell : L' \to J$ are simplicial isomorphisms. But L' induces via $k^{-1}\ell$ an equivalence between S' and a subdivision $T'' = (K'', k)$ of T', which is then also a subdivision of T. The composition $j^{-1}\ell\ell^{-1}k = j^{-1}k : K'' \to J$ is then a simplicial isomorphism, so $T \leq R$.

If $T = (K, h) \in \mathcal{T}$ is an admissible triangulation of X, we set $C_*^T(X) = C_*(K)$, the simplicial chain complex of K. If T is equivalent to $S = (L, j)$, the simplicial isomorphism $j^{-1}h$ induces an isomorphism $C_*^T(X) \to C_*^S(X)$. If $T' = (K', h)$ is a subdivision of T, then there is a subdivision chain map $C_*^T(X) \to C_*^{T'}(X)$ that takes each oriented i-simplex σ of K to the sum of the i-simplices of K' contained in σ with the compatible orientations, as per Remark 3.2.1; see Figure 3.5. The existence of subdivision chain maps typically appears in texts either only for barycentric subdivisions – where an easy precise inductive formula is available – or as a consequence of the fancier machinery of the Acyclic Model Theorem. Unfortunately, such accounts do not tend to include the more geometric description we have given for non-barycentric subdivisions; cf. [182, Section 17] and [220, Chapter 4.6]. So let us sketch a proof of how to draw our description out of, say, [182, Theorem 17.2], which says that if K' is a subdivision of K then there is a unique augmentation-preserving chain map $\lambda : C_*(K) \to C_*(K')$ such that $|\lambda(\sigma)| \subset \sigma$ for each σ.

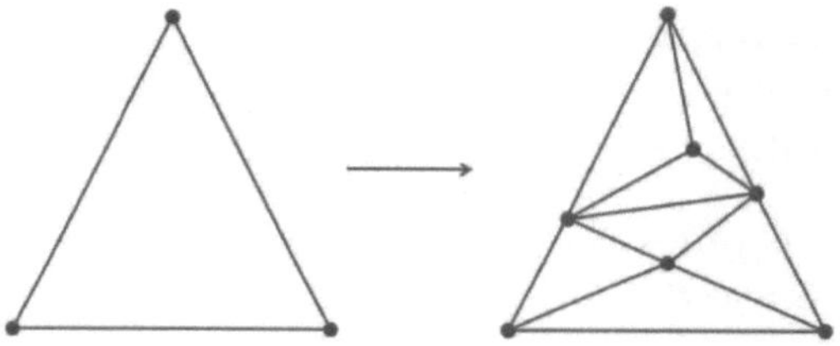

Figure 3.5 The subdivision chain map $C_*(K) \to C_*(K')$ takes each oriented i-simplex of K to the sum of the compatibly oriented i-simplices of K' that it contains.

Lemma 3.3.1 *Let K be a simplicial complex and K' a subdivision of K. Then the unique augmentation-preserving chain map $\lambda : C_*(K) \to C_*(K')$ such that $|\lambda(\sigma)| \subset \sigma$ for each oriented simplex σ of K satisfies $\lambda(\sigma) = \sum \tau_j$, where if σ is an i-simplex the sum is over all i-simplices of K' contained in σ, assuming each is oriented compatibly with σ.*

Proof Suppose that σ is an i-simplex of K and that in the subdivision K' the i-simplices contained in σ are denoted $\{\tau_j\}_{j\in\mathcal{J}}$. We can assume that σ and the τ_j are oriented compatibly; see Remark 3.2.1. Then [182, Theorem 17.2] implies that we must have $\lambda(\sigma) = \sum_{j\in\mathcal{J}} a_j\tau_j$ and that $\partial\lambda(\sigma) = \lambda(\partial\sigma)$ must be supported in $|\partial\sigma|$. Each $(i-1)$-simplex of K' contained in σ but not contained in $|\partial\sigma|$ must be the face of exactly two of the τ_j by [182, Corollary 63.3.b], noting that $(\sigma, \partial\sigma)$ is a relative homology i-manifold. So in order for $\partial\lambda(\sigma)$ to be contained in $|\partial\sigma|$, we must have that if τ_j and τ_k share such an $i-1$ face η then $a_j = \pm a_k$. But as τ_j and τ_k are compatibly oriented, we must in fact have $a_j = a_k$. This follows from the orientation on η being determined by the orientations of τ_j and τ_k as in manifold theory using the outward-pointing normals: as τ_j and τ_k have compatible orientations but induce oppositely directed outward-pointing normals on η, they induce opposite orientations on η so that the coefficient of η in $\partial\lambda(\sigma)$ is 0 only if the coefficients of τ_j and τ_k agree. Furthermore, as we know that $H_i(\sigma, |\partial\sigma|) \cong \mathbb{Z}$, an argument analogous to that of [182, Corollary 65.2] shows that we can move from any τ_j to any other τ_k by a sequence of i-simplices pairwise intersecting in $(i-1)$-simplices so that all of the coefficients of all of the τ_j must agree. If we call this common value a, we have $\lambda(\sigma) = a\big(\sum_{j\in\mathcal{J}} \tau_j\big)$.

So it remains to show that $a = 1$. For this, we use an induction argument. When $i = 0$, the condition that λ be augmentation-preserving guarantees that $a = 1$. Now suppose $i > 0$ and that the result has been proven for dimensions $< i$. As λ is a chain map, $\lambda(\partial\sigma) = \partial\lambda(\sigma) = \sum_{j\in\mathcal{J}} a\partial\tau_j$. If η is a particular $(i-1)$-face of σ given an arbitrary orientation then, by induction, $\lambda(\eta)$ is the sum over the $(i-1)$-simplices of K' contained in $|\eta|$ and compatibly oriented with η, each with coefficient 1. If δ is such an $(i-1)$-simplex, it follows that the coefficient of δ in $\lambda(\partial\sigma)$ is 1 or -1, depending on whether η appears with coefficient 1 or -1 in $\partial\sigma$. Furthermore, δ is the $(i-1)$-face of precisely one of the τ_j, say τ_0. So the coefficient of δ in $\sum_{j\in\mathcal{J}} a\partial\tau_j$ is a or $-a$ depending on whether δ appears with coefficient 1 or -1 in $\partial\tau_0$. But, as τ_0 is compatibly oriented with σ and as δ is compatibly oriented with η, it follows that the sign of the coefficient of η in $\partial\sigma$ must be the same as the coefficient of δ in $\partial\tau_0$, and so $a = 1$. □

Corollary 3.3.2 *If $\lambda : C_*(K) \to C_*(K')$ is the subdivision chain map and $\xi \in C_*(K)$, then $|\lambda(\xi)| = |\xi|$ and λ is injective.*

Proof The first claim follows directly from the lemma and definitions, recalling that if $\xi \in C_i(K)$ then $|\xi|$ is the union of the i-simplices of K, i.e. the σ_k with non-zero coefficient in the unique expression $\xi = \sum a_k\sigma_k$ for ξ. The injectivity of λ follows, as a chain has empty support if and only if it is the chain representing 0. □

From the description in the lemma, it is clear that subdivision chain maps commute with each other in the sense that if K has subdivisions K' and K'' with common subdivision K''', then the composition of subdivision maps $C_*(K) \to C_*(K') \to C_*(K''')$ and $C_*(K) \to C_*(K'') \to C_*(K''')$ agree: the image of an i-simplex in K under either composition is the sum over all i-simplices of K''' contained in σ and with the compatible orientation. Similarly, the chain maps induced by simplicial isomorphisms commute with each other, and they commute with the subdivision chain maps. So we obtain from the directed set $\mathcal{T}$ a directed system of chain complexes consisting of the $C_*^T(X)$ for $T \in \mathcal{T}$ with the maps of the directed system being the subdivision chain maps and maps induced by simplicial isomorphisms.

Definition 3.3.3 If X is a PL space and $\mathcal{T}$ is its family of admissible triangulations, the *PL chain complex* $\mathfrak{C}_*(X)$ is defined to be

$$\mathfrak{C}_*(X) = \varinjlim_{T\in\mathcal{T}} C_*^T(X).$$

So each chain in $\mathfrak{C}_*(X)$ can be described as a simplicial chain with respect to *some* admissible triangulation of X, and two such simplicial chains defined with respect to possibly different triangulations T, S represent the same PL chain if their images agree in some common subdivision of T and S (up to the simplicial isomorphism between those

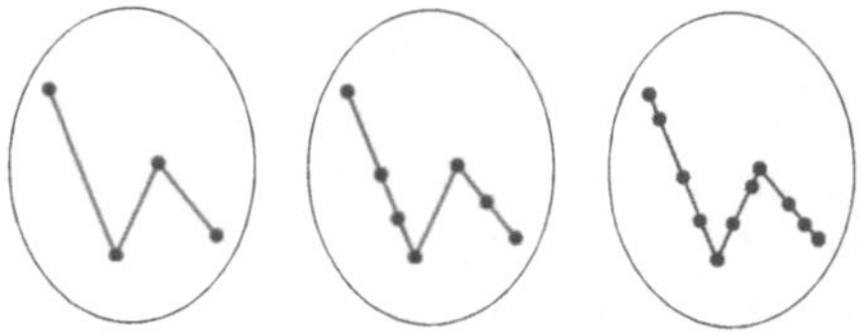

Figure 3.6 Simplicial 1-chains all representing the same PL chain.

subdivisions); see Figure 3.6. As a consequence of Corollary 3.3.2, each of the canonical maps $C_*^T(X) \to \mathfrak{C}_*(X)$ is injective, and we can let the support $|\xi|$ of a PL chain correspond to the support of any of its representatives, i.e. if $\xi \in C_i^T(K)$ with $T = (K,h)$ represents $[\xi] \in \mathfrak{C}_i(X)$, then $|[\xi]| = h(|\xi|)$ is the image of the union of the i-simplices of ξ.

We will tend to denote chains in $\mathfrak{C}_*(X)$ by letters like ξ. However, if we need to distinguish between an element of $\mathfrak{C}_*(X)$ and an element of some $C_*^T(X)$ representing it, as in the preceding paragraph, then we will tend to use ξ for the element of $C_*^T(X)$ and $[\xi]$ for the element of $\mathfrak{C}_*(X)$ that ξ represents. We also occasionally use $[\xi]$ to represent a homology class of the PL chain ξ. Which interpretation of $[\xi]$ is meant will be determined by context or by direct statement if confusion would otherwise result.

The following lemma might help simplify things a bit in thinking about PL chains.

Lemma 3.3.4 *Let X be a PL space with family of admissible triangulations $\mathcal{T}$. Let $T_0 = (K,h) \in \mathcal{T}$, and let $\mathcal{T}_0$ be the subset of $\mathcal{T}$ consisting of subdivisions of T_0. Then*

$$\mathfrak{C}_*(X) = \varinjlim_{T \in \mathcal{T}} C_*^T(X) \cong \varinjlim_{T \in \mathcal{T}_0} C_*^T(X).$$

Remark 3.3.5 Essentially, this lemma says that you can fix your favorite triangulation of X and then just work with the subdivisions of that triangulation. In what follows, we will generally use this model of $\mathfrak{C}_*(X)$ implicitly, leaving h tacit and simply saying that "K is a triangulation of X."

The proof of the lemma is really just a combination of the arguments that the family $\mathcal{T}_0$ is cofinal in $\mathcal{T}$ and that any cofinal subsystem of a direct system yields the same direct limit as the full direct system[8]. So any cofinal system of triangulations would do for the role of $\mathcal{T}_0$ in the lemma.

Proof of Lemma 3.3.4 As $\mathcal{T}_0$ is a subset of $\mathcal{T}$, we have a canonical map

$$\phi : \varinjlim_{T \in \mathcal{T}_0} C_*^T(X) \to \varinjlim_{T \in \mathcal{T}} C_*^T(X) = \mathfrak{C}_*(X).$$

If $[\xi] \in \mathfrak{C}_*(X)$, then $[\xi]$ can be represented by $\xi \in C_*(L)$ for some triangulation $S = (L, j)$

[8] Recall that a subset J_0 of a directed set J is *cofinal* if for any $y \in J$ there is a $z \in J_0$ with $y \leq z$; see [182, Section 73]. Furthermore, recall that if G_j is a direct system of abelian groups indexed by J then there is a canonical isomorphism $\varinjlim_{j \in J_0} G_j \to \varinjlim_{j \in J} G_j$; see [182, Lemma 73.1].

of X. But T_0 and S have some common subdivision given, say, by $T_0' = (K', h)$ and $S' = (L', j)$ with $j^{-1}h$ a simplicial isomorphism. So then $[\xi]$ is also represented by the image ξ' of ξ in $C_*(L')$ under the subdivision map. As K' and L' are isomorphic, it follows that ξ' is in the image of $C_*(K')$ in $C_*(L')$ under the map induced by the simplicial isomorphism, and so $[\xi]$ is in the image of $\varinjlim_{T \in \mathcal{T}_0} C_*^T(X)$. Thus ϕ is surjective.

Similarly, if $[\xi] \in \varinjlim_{T \in \mathcal{T}_0} C_*^T(X)$ is represented by some $\xi \in C_*(K')$ for some subdivision $T_0' = (K', h)$ of T_0 and if $\phi([\xi]) = 0$, then there is some $S = (L, j)$ with $T_0' \le S$ such that the image ξ' of ξ in $C_*^S(X) = C_*(L)$ is 0. But then the image of ξ must also be 0 in any common subdivision of T_0' and S, and it follows that $[\xi] = 0$ in $\varinjlim_{T \in \mathcal{T}_0} C_*^T(X)$. So ϕ is injective. □

The PL homology groups are defined to be the homology groups of the chain complex $\mathfrak{C}_*(X)$:

Definition 3.3.6 If X is a PL space, we define its PL homology groups to be

$$\mathfrak{H}_*(X) = H_*(\mathfrak{C}_*(X)).$$

As taking homology of a chain complex behaves well with respect to direct limits, these groups are really just the ordinary simplicial or singular homology groups of X:

Proposition 3.3.7 *If X is a PL space, then $\mathfrak{H}_*(X) \cong H_*(X)$, where $H_*(X)$ are the singular homology groups of X or the simplicial homology groups with respect to any triangulation.*

Proof The isomorphism of simplicial and singular homology groups is, of course, a well-known property: see [182, Theorem 34.3] or [125, Theorem 2.27]. But also, letting $\mathcal{T}$ be the admissible triangulations of X, we have

$$\begin{aligned}\mathfrak{H}_*(X) &= H_*(\mathfrak{C}_*(X)) \\ &= H_*\left(\varinjlim_{T \in \mathcal{T}} C_*^T(X)\right) \\ &\cong \varinjlim_{T \in \mathcal{T}} H_*\left(C_*^T(X)\right) \\ &\cong H_*(X).\end{aligned}$$

The two equalities come from the definitions. The first isomorphism is a standard algebraic property of direct limits (see [71, Proposition VIII.5.20]). The last isomorphism follows because the $H_*\left(C_*^T(X)\right)$ are all isomorphic to the simplicial homology of X with the subdivision maps inducing homology isomorphisms; so the direct system of homology groups is constant. □

PL Chains and PL Maps

We also need to know that PL maps induce maps of PL chain complexes, and so of PL homology. The following preliminary lemma will be useful. We refer to [182, Section 12] for background about the chain map between simplicial chain complexes induced by a simplicial map of simplicial complexes.

Lemma 3.3.8 *Let $f : K \to L$ be a simplicial map, and let K' and L' be subdivisions such that the induced map $f' : K' \to L'$ is also simplicial. Let $s_K : C_*(K) \to C_*(K')$ and $s_L : C_*(L) \to C_*(L')$ be the subdivision chain maps. Suppose $\xi \in C_*(K)$. Then $s_L f(\xi) = f' s_K(\xi)$.*

Proof Let σ be a (geometric) i-simplex of K. As f is simplicial, either it takes σ to a lower-dimensional simplex of L or it takes it linearly homeomorphically onto some i-simplex τ of L. Now suppose σ is oriented and consider how f acts on σ by the map of chain complexes $f : C_*(K) \to C_*(L)$. In the first case, $f(\sigma) = 0 \in C_*(L)$, and, in the second case, $f(\sigma)$ is $\pm\tau \in C_*(L)$. Next, consider the subdivision maps. The map s_K takes σ and subdivides it into a chain $s_K(\sigma) \in C_i(K')$, which is the sum over the compatibly oriented i-simplices in the subdivision σ' of σ determined by K', and s_L acts analogously on τ. In the case where f collapses σ to a lower-dimensional face, this must be true of every i-simplex in σ', and so $f'(s_K(\sigma)) = 0 = s_L f(\sigma)$. And if f takes σ homeomorphically to an i-simplex τ of L, then our assumptions imply that f must also take each simplex of σ' homeomorphically onto a simplex in the subdivision τ' of τ in L', and there is a bijection between simplices of σ' and the corresponding simplices of τ'. As all the simplices of σ' are oriented coherently with the orientation of σ and similarly all the simplices of τ' are oriented coherently with the orientation of τ, then we must have $f'(s_K(\sigma)) = \pm s_L\tau$, where the sign agrees with the sign in $f(\sigma) = \pm\tau$. So $f'(s_K(\sigma)) = s_L(f(\sigma))$, which implies the lemma. □

Lemma 3.3.9 *Let X and Y be PL spaces, and let $f : X \to Y$ be a PL map. Then f induces a chain map $f : \mathfrak{C}_*(X) \to \mathfrak{C}_*(Y)$ and so a map of PL homology $\mathfrak{H}_*(X) \to \mathfrak{H}_*(Y)$.*

Proof By Lemma 3.3.4, it suffices to limit the triangulations of X and Y that we consider to one family each, each of which consists of subdivisions of a single triangulation. It is therefore safe to leave the triangulation maps tacit and identify X and Y with the underlying spaces of simplicial complexes in Euclidean space and refer to f as the map between them. All simplicial complexes arising in the proof below will be assumed to derive from these fixed complexes by subdivision or as subcomplexes so that they are all compatible with the fixed PL structures on X and Y.

Let $[\xi] \in \mathfrak{C}_i(X)$. By the definition, this means that $[\xi]$ can be represented as a simplicial chain in $C_*(K)$, where K is some simplicial complex with $|K| = X$. Let ξ be such a simplicial chain, let $|\xi|$ be its support, and let J_0 be any finite subcomplex of K containing $|\xi|$. Let L be a simplicial complex with $|L| = Y$. As J_0 is compact, the restriction of f to J_0 is a proper map, and so by Theorem B.2.19 there are subdivisions J_1 of J_0 and L_1 of L with respect to which $f : J_1 \to L_1$ is simplicial. So then f induces a map of simplicial chain complexes $f_1 : C_*(J_1) \to C_*(L_1)$. Let ξ_1 be the image of ξ in the subdivision J_1. We can now set $f([\xi]) = [f_1(\xi_1)]$. To see that this is well defined, we need to verify that the construction does not depend on the choices of K, J_0, J_1, L, or L_1.

Suppose we instead choose a representative $\bar{\xi}$ for $[\xi]$ in a triangulation $\bar{K}$ of X with $|\bar{K}| = |K| = X$, and let $\bar{L}$ be an alternative triangulation for Y with $|\bar{L}| = |L| = Y$. Note that we must have $|\bar{\xi}| = |\xi|$. In fact, as ξ and $\bar{\xi}$ both represent the same PL chain, there is a common subdivision $K' = \bar{K}'$ of K and $\bar{K}$ such that the images of ξ and $\bar{\xi}$ in $C_*(K') = C_*(\bar{K}')$, say ξ' and $\bar{\xi}'$, must agree. Then by Corollary 3.3.2 we have $|\bar{\xi}| = |\bar{\xi}'| = |\xi'| = |\xi|$. Let $\bar{J}_0$ be any finite subcomplex of $\bar{K}$ containing $|\bar{\xi}|$. Let $\bar{J}_1$ and $\bar{L}_1$ be subdivisions of J_0 and L with respect to which f is simplicial, let $\bar{\xi}_1$ be the image of $\bar{\xi}$ in the subdivision $\bar{J}_1$, and let $\bar{f}_1 : C_*(\bar{J}_1) \to C_*(\bar{L}_1)$. We need to show that $f_1(\xi_1)$ and $\bar{f}_1(\bar{\xi})$ represent the same element of $\mathfrak{C}_*(Y)$.

The intersection $|J_0| \cap |\bar{J}_0| = |J_1| \cap |\bar{J}_1|$ is a closed PL subspace of both $|J_0| = J_1|$ and $|\bar{J}_0| = |\bar{J}_1|$, and so by Example B.4.3 there are subdivisions J_2 of J_1 and $\bar{J}_2$ of $\bar{J}_1$ with respect to which the intersection is triangulated by subcomplexes, say M_2 and $\bar{M}_2$. Let J_3 and L_3 be subdivisions of J_2 and L_1 such that $f : J_3 \to L_3$ is simplicial and induces $f_3 : C_*(J_3) \to C_*(L_3)$, and let ξ_3 denote the image of ξ under these subdivisions. By Lemma 3.3.8, $f_1(\xi_1)$ and $f_3(\xi_3)$ represent the same element in $\mathfrak{C}_*(Y)$. Furthermore, for the purposes of computing $f_3(\xi_3)$, we can consider the domain of f_3 to be restricted to the subcomplex of J_3 triangulating $|J_1| \cap |\bar{J}_1|$; call this subcomplex M_3. Analogously, we can perform a "barred" version of this construction, and the proof reduces to comparing $f_3(\xi_3)$, thinking of f_3 as restricted to M_3, with $\bar{f}_3(\bar{\xi}_3)$, constructed analogously by restricting f to $\bar{M}_3$. So far, we have simplified the problem in that now $|M_3| = |\bar{M}_3|$, whereas $|J_1|$ and $|\bar{J}_1|$ did not necessarily agree.

As the inclusion $|M_3| \hookrightarrow X$ is PL and $|M_3|$ is compact, there are subdivisions M_3' of M_3 and K_3 of K so that the inclusion is simplicial, i.e. M_3' is a subcomplex of K_3. Similarly, we can construct $\bar{M}_3'$ and $\bar{K}_3$. The complexes K_3 and $\bar{K}_3$ must have a common subdivision K_4, and as ξ and $\bar{\xi}$ represent the same PL chain, we can assume this is a subdivision in which their images under subdivision $\xi_4 \in C_i(K_4)$ and $\bar{\xi}_4 \in C_i(\bar{K}_4)$ agree. The restriction of this subdivision to $|M_3| = |\bar{M}_3|$ gives us a common subdivision $M_4 = \bar{M}_4$ of M_3 and $\bar{M}_3$ containing $\xi_4 = \bar{\xi}_4$. Let $L_4 = \bar{L}_4$ be a common subdivision of L_3 and $\bar{L}_3$. Then let $M_5 = \bar{M}_5$ and $L_5 = \bar{L}_5$ be respective subdivisions of $M_4 = \bar{M}_4$ and $L_4 = \bar{L}_4$ such that f is simplicial as a restricted map $f_5 : M_5 \to L_5$. We can define $\bar{f}_5 : \bar{M}_5 \to \bar{L}_5$ similarly, but as $M_5 = \bar{M}_5$, $L_5 = \bar{L}_5$, and f_5 and $\bar{f}_5$ are both induced by the same map f, we have $f_5 = \bar{f}_5$. Furthermore, as $\xi_4 = \bar{\xi}_4$, the images ξ_5 and $\bar{\xi}_5$ under subdivision remain equal. So, employing Lemma 3.3.8 again, we have $f_3(\xi_3) = f_5(\xi_5) = \bar{f}_5(\bar{\xi}_5) = \bar{f}_3(\bar{\xi}_3)$.

So we have shown that the definition of $f : \mathfrak{C}_*(X) \to \mathfrak{C}_*(Y)$ is independent of the choices made.

To see that $f : \mathfrak{C}_*(X) \to \mathfrak{C}_*(Y)$ so defined is a homomorphism, suppose that $[\xi], [\eta] \in \mathfrak{C}_*(X)$. If M and N are triangulations of X that contain simplicial representations of $[\xi]$ and $[\eta]$, we can let K be a common subdivision of M and N, and let ξ and η be the representatives of $[\xi]$ and $[\eta]$ in $C_*(K)$. Then we can choose J_0 in the above construction large enough to contain both $|\xi|$ and $|\eta|$, and so also $|\xi + \eta| \subset |\xi| \cup |\eta|$. As the subdivision chain map, the simplicial chain map induced by f_1, and the maps $C_*^T(Y) \to \mathfrak{C}_*(Y)$ are homomorphisms, it follows that

$$\begin{aligned} f([\xi + \eta]) &= [f_1((\xi + \eta)_1)] = [f_1(\xi_1 + \eta_1)] \\ &= [f_1(\xi_1) + f_1(\eta_1)] = [f_1(\xi_1)] + [f_1(\eta_1)] \\ &= f([\xi]) + f([\eta]). \end{aligned}$$

Similarly, to see that the map $f : \mathfrak{C}_*(X) \to \mathfrak{C}_*(Y)$ we have just defined is a chain map, we need only observe that each of the maps of simplicial complexes used in the above discussion are chain maps, as are the subdivision operators and the maps $C_*^T(X) \to \mathfrak{C}_*(X)$. So,

$$\begin{aligned} \partial(f[\xi]) &= \partial[f_1(\xi_1)] = [\partial f_1(\xi_1)] = [f_1\partial(\xi_1)] \\ &= [f_1((\partial\xi)_1)] = f([\partial\xi]) \\ &= f(\partial[\xi]). \end{aligned}$$

Of course, once we have a chain map, we have an induced map on homology. □

3.3.2 A Useful Alternative Characterization of PL Chains

There is an interesting alternative way to describe a PL chain that is sometimes very useful. The idea is that a PL i-chain ξ can be recovered from its support $|\xi|$, the support of its boundary $|\partial\xi|$, and a homology class in $H_i(|\xi|, |\partial\xi|)$. Basically, the supports tells us what simplices can be involved and the homology class carries the coefficient information. This observation was made by Goresky and MacPherson [105, Section 1.2] and has been utilized in [89, 119, 168] and with some generalizations in [95].

The main point comes from the following Useful Lemma:

Lemma 3.3.10 *Let X be a PL space and let $C \subset B \subset A$ be closed PL subspaces of X such that* $\dim(A) \le i$, $\dim(B) \le i - 1$, *and* $\dim(C) \le i - 2$. *Let*

$$\mathfrak{C}_i^{A,B} = \{\xi \in \mathfrak{C}_i(X) \mid |\xi| \subset A, |\partial\xi| \subset B\},$$

and define $\mathfrak{C}_{i-1}^{B,C}$ analogously. Then

1. *$\mathfrak{C}_i^{A,B} \cong \mathfrak{H}_i(A, B)$, and*
2. *the following diagram commutes.*

$$\begin{array}{ccc} \mathfrak{C}_i^{A,B} & \xrightarrow{\cong} & \mathfrak{H}_i(A,B) \\ \Big\downarrow \partial & & \Big\downarrow \partial_* \\ \mathfrak{C}_{i-1}^{B,C} & \xrightarrow{\cong} & \mathfrak{H}_{i-1}(B,C) \end{array}$$

Here all horizontal maps take chains (relative cycles) to the homology classes they represent.

Proof The isomorphism comes directly from the definitions: $\mathfrak{H}_i(A,B)$ is by definition the quotient of the relative PL i-cycles in A (relative to B) by the relative i-boundaries. But the set $\mathfrak{C}_i^{A,B}$ is precisely the group of relative i-cycles. And the group of relative boundaries is trivial by the dimension hypothesis: As $\dim(A) = i$, there can be no PL $(i+1)$-chains in A. This shows that $\mathfrak{C}_i^{A,B} \cong \mathfrak{H}_i(A,B)$, with a chain in $\mathfrak{C}_i^{A,B}$ representing a homology class in $\mathfrak{H}_i(A,B)$.

The second part of the lemma follows because we know from basic homology theory that if $\xi \in \mathfrak{C}_i^{A,B}$ represents the homology class $[\xi] \in \mathfrak{H}_i(A,B)$, then $\partial_*[\xi]$ is represented by $\partial\xi$. □

As a special case of the lemma, let ξ be any i-chain in $\mathfrak{C}_i(X)$. Then $\xi \in \mathfrak{C}_i^{|\xi|,|\partial\xi|}$, and the lemma tells us that ξ corresponds to an element of $\mathfrak{H}_i(|\xi|,|\partial\xi|)$. So ξ is completely determined by $|\xi|$, $|\partial\xi|$, and an element of $\mathfrak{H}_i(|\xi|,|\partial\xi|)$. Furthermore, every element of $\mathfrak{H}_i(|\xi|,|\partial\xi|)$ yields a specific chain supported in $|\xi|$ and whose boundary is supported in $|\partial\xi|$.

Example 3.3.11 It takes a bit of thinking about examples for this all to seem reasonable.

To see how a PL chain determines a homology class, suppose X is a PL space and that in some triangulation we have a subspace A consisting of two compatibly oriented 2-simplices σ_1, σ_2 that share a common 1-simplex; see Figure 3.7. Suppose the PL chain ξ of X is represented by $\sigma_1 + 2\sigma_2$. So the support of ξ is all of A, and the support of $\partial\xi$ is $|\partial\sigma_1| \cup |\partial\sigma_2|$. Using excision and the isomorphism between simplicial and PL homology, we have[9]

$$\mathfrak{H}_2(|\xi|,|\partial\xi|) = \mathfrak{H}_2(A,|\partial\sigma_1| \cup |\partial\sigma_2|) \cong \mathfrak{H}_2(\sigma_1,|\partial\sigma_1|) \oplus \mathfrak{H}_2(\sigma_2,|\partial\sigma_2|) \cong \mathbb{Z} \oplus \mathbb{Z},$$

and ξ is represented by the element $(1,2)$ in the sum.

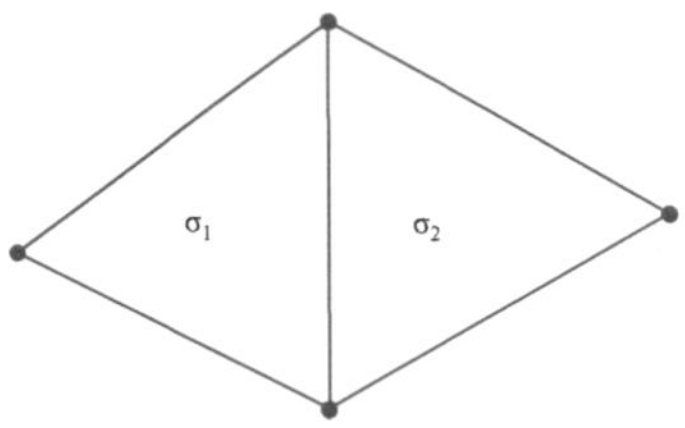

Figure 3.7 Two 2-simplices with a common 1-simplex.

Conversely, let's see in a more interesting case how a homology class determines a chain. Suppose that in $A \subset X$, we have a one-dimensional PL subspace Z that traverses the interior of A, breaking A into two PL subspaces each PL homeomorphic to the disk as in Figure 3.8. Then we still have $\mathfrak{H}_2(A, \partial A \cup Z) \cong \mathbb{Z} \oplus \mathbb{Z}$, and the lemma tells us that there is

[9] Recall that if σ is an oriented simplex then we write σ rather than $|\sigma|$ for its support, which is just the geometric simplex σ itself.

a PL chain corresponding to each such homology class. If we have any triangulation of the pair $(A, \partial A \cup Z)$ as in Figure 3.9, the class $(a, b) \in \mathbb{Z} \oplus \mathbb{Z}$ will correspond to a PL chain represented by a simplicial chain of the form $\sum a\tau_i + \sum b\eta_j$, where the τ_i are the 2-simplices in the first disk (compatibly oriented with A) and the η_j are the 2-simplices in the second disk (compatibly oriented with A). Notice that it is possible to have either a or b equal to 0 so that the support of the chain is not necessarily all of A.

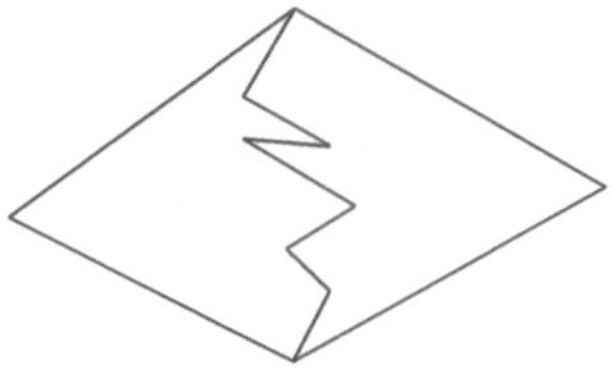

Figure 3.8 A different division of A, separated by Z.

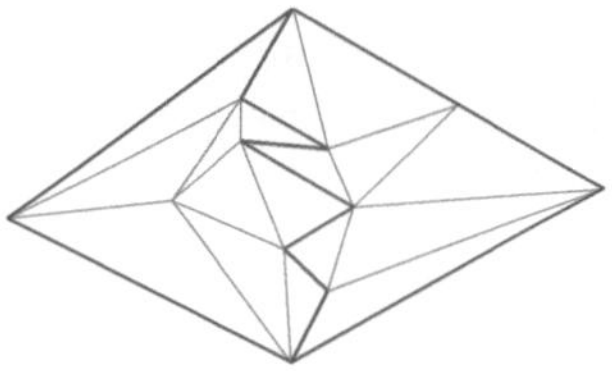

Figure 3.9 A triangulation compatible with A and Z.

To illustrate the boundary property of the lemma, let us continue to consider the pair (A, Z) within X. Then an element of $\mathfrak{H}_1(Z, \partial A \cap Z) \cong \mathbb{Z}$ determines a PL 1-chain supported in Z with boundary in $\partial A \cap Z$. And the boundary map

$$\partial_* : \mathfrak{H}_2(A, \partial A \cup Z) \to \mathfrak{H}_1(\partial A \cup Z, \partial A) \cong \mathfrak{H}_1(Z, \partial A \cap Z)$$

takes (a, b) to a 1-chain supported on Z such that each simplex in our triangulation has coefficient $a - b$ or $b - a$ depending on the choices of orientations. If we orient these 1-simplices consistently with an orientation of Z going from the bottom to the top of the diagram and if A is given the standard orientation of the plane, then $\partial_*(a, b)$ corresponds to $a - b$ times the fundamental class of Z in the triangulation. But the principal point of Lemma 3.3.10 is that this discussion is the same for *any* triangulation and so support and homology information are enough to determine PL chains and their boundaries. △

This description of chains via homology classes is somewhat reminiscent of the technical definition of the CW chain complex, though in general our spatial regions need not be cells. For example, if M is a compact oriented PL n-manifold, our isomorphism relates $\mathfrak{C}_n^{M,\emptyset}$ with $\mathfrak{H}_n(M)$, which is generated by the fundamental classes of the connected components.

(a) Adding Chains

If we fix closed PL subsets $B \subset A \subset X$ with $\dim(A) = i$ and $\dim(B) < i$ then each $\mathfrak{C}_i^{A,B} \cong \mathfrak{H}_i(A, B)$ is a group. But suppose we want to add two PL chains with different supports. For this we can use the following lemma:

Lemma 3.3.12 *Suppose we have PL subspace pairs of X given by $(A, B) \subset (C, D)$ with $\dim(A) \leq \dim(C) \leq i$ and $\dim(B) \leq \dim(D) \leq i - 1$. Then we have the following commutative diagram of inclusions/isomorphisms.*

$$\begin{array}{ccc} \mathfrak{C}_i^{A,B} & \xrightarrow{\cong} & \mathfrak{H}_i(A,B) \\ \big\downarrow & & \big\downarrow \\ \mathfrak{C}_i^{C,D} & \xrightarrow{\cong} & \mathfrak{H}_i(C,D) \end{array}$$

Proof The horizontal isomorphisms come from Lemma 3.3.10. We have an inclusion $\mathfrak{C}_i^{A,B} \hookrightarrow \mathfrak{C}_i^{C,D}$ because any PL chain in X supported in A and with boundary in B is also a PL chain in X supported in C and with boundary in D. The square evidently commutes by considering representative chains, and it follows that the right-hand vertical map is injective. □

So, suppose we are given two PL chains ξ_j, $j = 1, 2$, represented by classes $[\xi_j] \in \mathfrak{H}_i(A_j, B_j)$ with $\dim(A_j) \leq i$ and $\dim(B_j) < i$. To add them "homologically," we can consider their images under the injections $\mathfrak{H}_i(A_j, B_j) \hookrightarrow \mathfrak{H}_i(A_1 \cup A_2, B_1 \cup B_2)$ and then form the sum $[\xi_1] + [\xi_2]$ in the latter group. In fact, we could form the sum in any $\mathfrak{H}_i(C, D)$ with $|\xi_j| \subset C$ and $|\partial \xi_j| \subset D$ for $j = 1, 2$ with C and D satisfying the dimension conditions. In particular, given any PL i-chains ξ_1, ξ_2, we can represent $\xi_1 + \xi_2$ as an element of $\mathfrak{H}_i(|\xi_1| \cup |\xi_2|, |\partial\xi_1| \cup |\partial\xi_2|)$.

(b) Compatibility with PL Maps

The description of PL chains as homology classes via Lemma 3.3.10 is compatible with PL maps:

Lemma 3.3.13 *Let $f : X \to Y$ be a PL map of PL spaces, and let $\xi \in \mathfrak{C}_i(X)$ be represented by the homology class $[\xi] \in H_i(|\xi|, |\partial\xi|)$. Then $f(\xi) \in \mathfrak{C}_i(Y)$ is represented by the class $f([\xi]) \in H_i(f(|\xi|), f(|\partial\xi|))$.*

Proof By Lemma 3.3.9, the image $f(\xi) \in \mathfrak{C}_i(Y)$ can be found by triangulating X and Y by simplicial complexes K and L so that $|\xi|$ is triangulated as a subcomplex and the restriction of f to $|\xi|$ is a simplicial map. Then if ξ can be represented as $\sum a_j\sigma_j$ in the simplicial chain complex associated to K, we have $f(\xi)$ represented by $\sum a_j f(\sigma_j)$ using the triangulation L. By $f(\sigma_j)$, we mean that if f takes the vertices of σ_j to distinct vertices in L, then $f(\sigma_j)$ is the oriented simplex with those image vertices in the order of the orientation of σ_j, and otherwise it is 0 [182, Section 12]. But the map induced by f from $H_i(|\xi|, |\partial\xi|)$ to $H_i(f(|\xi|), f(|\partial\xi|))$ can be described by precisely the same simplicial map. Since the simplicial representative in $H_i(f(|\xi|), f(|\partial\xi|))$ determines the PL chain, the lemma follows. □

(c) Realization

Another natural question is the following: Suppose again that we are given closed PL subspaces A and B of a PL space X with $\dim(A) \le i$ and $\dim(B) \le i-1$ and an element $x \in \mathfrak{H}_i(A, B)$. By Lemma 3.3.10, this data determines a PL chain ξ with support in A and support of its boundary in B. It is then very reasonable to ask what chain ξ is. More specifically, if we have a triangulation T of X such that A and B are triangulated as subcomplexes, can we identify a simplicial chain in $C_i^T(X)$ that represents ξ? The answer is given by the following construction:

Let A_T^{i-1} denote the simplicial $(i-1)$-skeleton of A in the triangulation T. As B is triangulated as a subspace of A of dimension $< i$, we have $B \subset A_T^{i-1}$. So as in Lemma 3.3.12 we have the following diagram.

$$\begin{array}{ccc}
\mathfrak{C}_i^{A,B} & \xrightarrow{\cong} & \mathfrak{H}_i(A,B) \\
\big\downarrow & & \big\downarrow \\
\mathfrak{C}_i^{A,A_T^{i-1}} & \xrightarrow{\cong} & \mathfrak{H}_i(A, A_T^{i-1})
\end{array}$$

By excisions and homotopy equivalences $\mathfrak{H}_i(A, A_T^{i-1}) \cong \bigoplus_\sigma \mathfrak{H}_i(\sigma, |\partial\sigma|)$, where the sum is over the i-simplices of T. If we assign to each i-simplex σ of T an arbitrary orientation, this determines an isomorphism $\mathfrak{H}_i(\sigma, |\partial\sigma|) \cong \mathbb{Z}$. For $x \in \mathfrak{H}_i(A, B)$, let $a_\sigma(x)$ denote the image of x under projection to the $\mathbb{Z}$ summand corresponding to σ. Then we see that $\sum_\sigma a_\sigma(x)\sigma$ is a chain in $C_i^T(X)$ that represents the image of x in $\mathfrak{H}_i(A, A_T^{i-1})$. As an element of $C_i^T(X)$ that is supported in A, the sum $\sum_\sigma a_\sigma(x)\sigma$ also represents an element of $\mathfrak{C}_i^{A,A_T^{i-1}}$ that again maps to the image of x in $\mathfrak{H}_i(A, A_T^{i-1})$. Chasing the above diagram then reveals that our desired chain $\xi \in \mathfrak{C}_i^{A,B}$ corresponding to $x \in \mathfrak{H}_i(A, B)$ must map to the element represented by $\sum_\sigma a_\sigma(x)\sigma$ in $\mathfrak{C}_i^{A,A_T^{i-1}}$. But, again from the diagram chase, $\sum_\sigma a_\sigma(x)\sigma$ represents an element of $\mathfrak{C}_i^{A,A_T^{i-1}}$ in the image of $\mathfrak{C}_i^{A,B}$, so we must have $|\partial(\sum_\sigma a_\sigma(x)\sigma)| \subset B$, and $\sum_\sigma a_\sigma(x)\sigma$ represents ξ.

So we have shown that, given an $x \in \mathfrak{H}_i(A, B)$ and a triangulation of X that triangulates A and B as subcomplexes, the PL chain ξ corresponding to x under the isomorphism of Lemma 3.3.10 has the form $\sum_\sigma a_\sigma(x)\sigma$, where the sum is over the i-simplices of A in the triangulation with fixed orientations and $a_\sigma(x) \in \mathbb{Z}$ is the image of x under the maps $\mathfrak{H}_i(A, B) \to \mathfrak{H}_i(A, A_T^{i-1}) \to \mathfrak{H}_i(\sigma, |\partial\sigma|) \cong \mathbb{Z}$, with the second arrow being the projection to a summand and the isomorphism being determined by the orientation of σ.

Remark 3.3.14 A further useful observation is that if z is a point in the interior of an i-simplex σ in the triangulation of A, then by homotopy invariance and excision $\mathfrak{H}_i(\sigma, |\partial\sigma|) \cong \mathfrak{H}_i(\sigma, \sigma - \{z\}) \cong \mathfrak{H}_i(A, A - \{z\})$. So we can determine the coefficients of the simplices in a simplicial representation of a PL chain by looking at local homology groups.

3.3.3 PL Intersection Homology

Now, suppose that X is a PL filtered space so that each skeleton of the filtration is a subcomplex of any admissible triangulation; see Section 2.5.2. We would like to define the PL intersection chain complex as

$$I^{\bar{p}}\mathfrak{C}_*^{\mathrm{GM}}(X) = \varinjlim_{T\in\mathcal{T}} I^{\bar{p}}C_*^{\mathrm{GM},T}(X),$$

where $I^{\bar{p}}C_*^{\mathrm{GM},T}(X)$ is the simplicial intersection chain complex with respect to the triangulation T and $\mathcal{T}$ is a set of admissible triangulations of X compatible with the filtration. In other words, if $T = (K, h)$ and we filter $|K|$ by the $h^{-1}(X^i)$, which must be subcomplexes of K, and we similarly give $|K|$ the perversity with (abusing notation) $\bar{p}(h^{-1}(S)) = \bar{p}(S)$ for any stratum S of X, then we let $I^{\bar{p}}C_*^{\mathrm{GM},T}(X) = I^{\bar{p}}C_*^{\mathrm{GM}}(|K|)$. In order to ensure that this definition makes sense, we need to show that subdivision provides a well-defined chain map $I^{\bar{p}}C_*^{\mathrm{GM},T}(X) \to I^{\bar{p}}C_*^{\mathrm{GM},T'}(X)$ when T' is a subdivision of T; it is clear that this is true of the maps induced by simplicial isomorphisms.

Lemma 3.3.15 *For any perversity $\bar{p}$ and for any admissible triangulations T, T' of the PL filtered space X such that T' is a subdivision of T, the subdivision chain map $\bar{\nu} : C_*^T(X) \to C_*^{T'}(X)$ restricts to a chain map $\nu : I^{\bar{p}}C_*^{\mathrm{GM},T}(X) \to I^{\bar{p}}C_*^{\mathrm{GM},T'}(X)$.*

Proof Let $T = (K, h)$ and $T' = (K', h)$. As the subdivision map is already a chain map $C_*^T(X) = C_*(K) \to C_*^{T'}(X) = C_*(K')$, it is only necessary to show that if the i-simplex $\sigma \in K$ is allowable (with respect to the induced filtration and perversity on $|K|$), then so is each i-simplex $\sigma' \in K'$ such that σ' is contained in σ. But if σ is allowable, then for each singular stratum S,

$$\dim(\sigma \cap S) \leq i - \operatorname{codim}(S) + \bar{p}(S).$$

If σ' is contained in σ, we must have $\dim(\sigma' \cap S) \leq \dim(\sigma \cap S)$, so σ' is also allowable. □

Definition 3.3.16 Given the preceding lemma, we can now define

$$I^{\bar{p}}\mathfrak{C}_*^{\mathrm{GM}}(X) = \varinjlim_{T\in\mathcal{T}} I^{\bar{p}}C_*^{\mathrm{GM},T}(X)$$

and

$$I^{\bar{p}}\mathfrak{H}_*^{\mathrm{GM}}(X) = H_*(I^{\bar{p}}\mathfrak{C}_*^{\mathrm{GM}}(X)) \cong \varinjlim_{T\in\mathcal{T}} I^{\bar{p}}H_*^{\mathrm{GM},T}(X).$$

Remark 3.3.17 Analogously to the observation of Remark 3.3.5 and by a proof just like that of Lemma 3.3.4, rather than using all admissible triangulations of X to define $I^{\bar{p}}\mathfrak{C}_*^{\mathrm{GM}}(X)$ and hence $I^{\bar{p}}\mathfrak{H}_*^{\mathrm{GM}}(X)$, it suffices to stick with a smaller cofinal family of triangulations. In particular, it is useful to choose some particular admissible triangulation $T_0 = (K, h)$ and replace $\mathcal{T}$ with $\mathcal{T}_0$, the family of all subdivisions of T_0. There is then little harm in *identifying* X with $|K|$, keeping the homeomorphism h implicit, and then working entirely with subdivisions of K as the family of admissible triangulations, now redesignated as $\mathcal{T}$. In what follows, we will generally employ this identification without further comment, writing, for example, that "K is a triangulation of X" or "the PL chain ξ in X."

Lemma 3.3.18 *Let $\xi \in \mathfrak{C}_i(X)$, and let $|\xi| \subset X$ be the support of ξ, i.e. the union of the simplices with non-zero coefficient in some representation of ξ with respect to some triangulation. Then $\xi \in I^{\bar{p}}\mathfrak{C}_i^{\mathrm{GM}}(X)$ if and only if, for each stratum S of X,*

$$\dim(|\xi| \cap S) \le i - \operatorname{codim}(S) + \bar{p}(S)$$

and

$$\dim(|\partial\xi| \cap S) \le i - 1 - \operatorname{codim}(S) + \bar{p}(S).$$

Proof Suppose $\xi \in \mathfrak{C}_i(X) = \varinjlim_{T\in\mathcal{T}} C_i^T(X)$. Then ξ is represented by a chain ξ^T in $C_i^T(X)$ for some T, and if the given conditions on $|\xi|$ and $|\partial\xi|$ hold, then they clearly also hold on each simplex of ξ^T and $\partial\xi^T$. Thus $\xi^T \in I^{\bar{p}}C_i^{\mathrm{GM},T}(X)$, and it follows that $\xi \in I^{\bar{p}}\mathfrak{C}_i^{\mathrm{GM}}(X)$.

Conversely, suppose $\xi \in I^{\bar{p}}\mathfrak{C}_i^{\mathrm{GM}}(X) = \varinjlim_{T\in\mathcal{T}} I^{\bar{p}}C_i^{\mathrm{GM},T}(X)$. Then again ξ is represented by a chain $\xi^T \in I^{\bar{p}}C_*^{\mathrm{GM},T}(X)$ for some T, and each simplex of ξ^T and $\partial\xi^T$ must be allowable. But if

$$\dim(\sigma \cap S) \le i - \operatorname{codim}(S) + \bar{p}(S)$$

for each simplex σ of ξ^T, then $\sigma \cap S$ is contained in the $(i - \operatorname{codim}(S) + \bar{p}(S))$-skeleton of T, and therefore so will be $|\xi| \cap S = \bigcup_{\{\sigma \text{ in } \xi\}}(\sigma \cap S)$. Therefore,

$$\dim(|\xi| \cap S) \le i - \operatorname{codim}(S) + \bar{p}(S).$$

The same argument holds for $\partial\xi$ and shows that $|\xi|$ and $|\partial\xi|$ satisfy the required conditions. □

3.3.4 The Relation between Simplicial and PL Intersection Homology

For the ordinary PL chains, the homology of $\mathfrak{C}_*(X)$ agrees with the simplicial homology of X with respect to any triangulation because

$$H_i(\mathfrak{C}_*(X)) = H_*\left(\varinjlim_{T\in\mathcal{T}} C_*^T(X)\right) \cong \varinjlim_{T\in\mathcal{T}} H_*\left(C_*^T(X)\right) \cong H_*^T(X)$$

for any T, as homology commutes with direct limits and every map in the direct system $\varinjlim_{T\in\mathcal{T}} H_*(C_*^T(X))$ is an isomorphism given that simplicial homology is preserved by subdivision (see [182, Theorem 17.2]). However, in Example 3.2.10, we saw that intersection homology is not preserved by simplicial subdivision. In this section we will show that preservation of intersection homology under subdivision does hold provided we impose some reasonable conditions on the triangulations. It will follow that $I^{\bar{p}}\mathfrak{H}_*^{\mathrm{GM}}(X)$ does in fact agree with $H_*(I^{\bar{p}}C_*^{\mathrm{GM},T}(X))$ for "most" triangulations T.

Recall that a subcomplex $L \subset K$ of a simplicial complex is called a *full subcomplex* if for any simplex $\sigma \in K$ it is true that if the vertices of σ are all in L then σ itself is contained in L. If a subcomplex $L \subset K$ is full, it remains full under any further subdivisions $L' \subset K'$ such that $|K| = |K'|$, $|L| = |L'|$ [198, Lemma 3.3.b]. If T is an admissible triangulation of a PL filtered space X, we will say that T is a *full triangulation* if each skeleton of the filtration of X is triangulated as a full subcomplex.

Lemma 3.3.19 *If X is a PL filtered space, then X possesses a full triangulation.*

Proof By Lemma 2.5.12, we know that X has triangulations with respect to which all the skeleta of the filtration are subcomplexes. Suppose X is triangulated by such a triangulation K. By [198, Lemma 3.3.a], for every subcomplex L of K, there is a subdivision K' for which L is a full subcomplex (it is not even necessary to subdivide L!); by [198, Lemma 3.3.b], if J is a full subcomplex of K and K' is any subdivision of K, then the resulting complex J' remains a full subcomplex in K'. It follows that if X has finitely many skeleta then we can inductively subdivide K finitely many times until each skeleton is a full subcomplex. □

Notice that in Examples 3.2.10 and 3.2.11, in which we demonstrated the poor behavior of intersection homology under subdivision, we worked with triangulations that were not full. Full triangulations are much better behaved, as the following theorem shows.

Theorem 3.3.20 *Suppose T is a full triangulation of a PL filtered space and that T' is any subdivision of T. Then the maps induced by subdivision,*

$$I^{\bar{p}}H_*^{\mathrm{GM},T}(X) \to I^{\bar{p}}H_*^{\mathrm{GM},T'}(X)$$

are isomorphisms, as is the canonical map $I^{\bar{p}}H_*^{\mathrm{GM},T}(X) \to I^{\bar{p}}\mathfrak{H}_*^{\mathrm{GM}}(X)$.

The second assertion was first proven by Goresky and MacPherson in an appendix to [157]. Our proof of Theorem 3.3.20 is based upon an elaboration of their argument.

The theorem will depend on two key lemmas. We will first state the lemmas and prove immediate corollaries, which will provide the proof of the theorem. Then we will prove the lemmas, which will involve some fairly technical work.

The first lemma provides a slightly stronger statement of part of Theorem 3.3.20 and implies part of the theorem.

Lemma 3.3.21 *Suppose T is a full triangulation of a PL filtered space and that T' is any subdivision of T. Then the subdivision chain map $\nu : I^{\bar{p}}C_*^{\mathrm{GM},T}(X) \to I^{\bar{p}}C_*^{\mathrm{GM},T'}(X)$ has a* *left inverse chain map* $\mu : I^{\bar{p}}C_*^{\mathrm{GM},T'}(X) \to I^{\bar{p}}C_*^{\mathrm{GM},T}(X)$ *so that* $\mu\nu = \mathrm{id}$. *In particular,* ν *induces an injection on intersection homology.*

Corollary 3.3.22 *If* T *is a full triangulation of* X*, then the canonical map*

$$\beta : I^{\bar{p}}H_*^{\mathrm{GM},T}(X) \to I^{\bar{p}}\mathfrak{H}_*^{\mathrm{GM}}(X)$$

is injective.

Proof Since the subdivisions of T form a cofinal system $\mathcal{T}' \subset \mathcal{T}$ by the proof of Lemma 3.3.4, we can compute $I^{\bar{p}}\mathfrak{H}_*^{\mathrm{GM}}(X)$ as $\varinjlim_{T' \in \mathcal{T}'} I^{\bar{p}}H_*^{\mathrm{GM},T'}(X)$. Since T is a full triangulation, so will be any subdivision of T [198, Lemma 3.3.b]. So by Lemma 3.3.21 each map of this direct system of groups will be injective, and it follows that each map from any $I^{\bar{p}}H_*^{\mathrm{GM},T'}(X)$, including $T' = T$, to $\varinjlim_{T' \in \mathcal{T}'} I^{\bar{p}}H_*^{\mathrm{GM},T'}(X)$ will be injective (see [71, Proposition VIII.5.18.ii]). □

Unfortunately, the proof that each $\nu : I^{\bar{p}}H_*^{\mathrm{GM},T}(X) \to I^{\bar{p}}H_*^{\mathrm{GM},T'}(X)$ is surjective for full T will need to be a bit more roundabout and utilize the PL intersection homology as an intermediary.

Lemma 3.3.23 *If* T *is a full triangulation of* X*, then the canonical map* $\beta : I^{\bar{p}}H_*^{\mathrm{GM},T}(X) \to I^{\bar{p}}\mathfrak{H}_*^{\mathrm{GM}}(X)$ *is surjective.*

Corollary 3.3.24 *Suppose* T *is a full triangulation of a PL filtered space and that* T' *is any subdivision of* T*. Then the subdivision maps* $\nu : I^{\bar{p}}H_*^{\mathrm{GM},T}(X) \to I^{\bar{p}}H_*^{\mathrm{GM},T'}(X)$ *and* $\beta : I^{\bar{p}}H_*^{\mathrm{GM},T}(X) \to I^{\bar{p}}\mathfrak{H}_*^{\mathrm{GM}}(X)$ *are isomorphisms.*

Proof Map ν is injective by Lemma 3.3.21, and β is injective by Corollary 3.3.22. By Lemma 3.3.23, β is onto. To obtain surjectivity of ν, notice that β factors through ν, i.e. β is the composite

$$I^{\bar{p}}H_*^{\mathrm{GM},T}(X) \xrightarrow{\nu} I^{\bar{p}}H_*^{\mathrm{GM},T'}(X) \xrightarrow{\cong} I^{\bar{p}}\mathfrak{H}_*^{\mathrm{GM}}(X),$$

where the second map is also the canonical map to the direct limit and hence an isomorphism by Lemma 3.3.23 (replacing T with T'). Since β is surjective, so must be ν. □

This corollary includes the statement of Theorem 3.3.20 and so proves the theorem.

Now we turn to proving Lemmas 3.3.21 and 3.3.23. We will need one further small lemma before we begin.

Lemma 3.3.25 *Suppose* T *is a full triangulation of a PL filtered space. The interior of a simplex* σ *is contained in the stratum* S *if and only if*

1. *all vertices of* σ *are contained in the closure* $\bar{S}$ *(in particular every vertex is in a stratum* R *such that*[10] $R \prec S$*), and*
2. *at least one vertex of* σ *is contained in* S*.*

[10] Recall from Section 2.2.2 that $R \prec S$ if $R \subset \bar{S}$.

Proof Suppose the interior of σ is contained in S. Then the closure of σ (which includes all the vertices of σ) is contained in the closure of S. Furthermore, if no vertex of σ is contained in S, then every vertex is contained in some stratum R with $R < S$, $R \neq S$. So every vertex of σ is contained in some skeleton that does not contain S. But this is a contradiction, as T is a full triangulation.

Conversely, suppose the two conditions are met. Suppose $S \subset X^i - X^{i-1}$. Since every vertex of σ is contained in X^i, the fullness of T implies that σ is contained in X^i. Since σ has at least one vertex in S, this implies that σ is not contained in X^{i-1}. It follows that the interior must actually be in S, as a simplex cannot intersect multiple strata of the same formal dimension without violating connectedness of the simplex. □

Proof of Lemma 3.3.21 We have already seen in Lemma 3.3.15 that ν takes allowable chains to allowable chains. We must construct μ. We will first define μ as a chain map $C_*^{T'}(X) \to C_*^{T}(X)$, and then we will check that μ preserves allowability. Finally, we will see that it provides a left inverse to ν.

Suppose T triangulates X by the simplicial complex K and that K' is the subdivision corresponding to T'. We can assume that the vertices of K have been given a total order by the Well-Ordering Theorem.

Let σ be a simplex of K, and let S_σ be the stratum of X containing the interior of σ. By Lemma 3.3.25, there must be some vertex v of σ such that $v \in S_\sigma$. For each σ, let the vertex of σ in S_σ that is greatest in the order be called v_σ.

Now, each vertex w of K' is contained within the interior of some simplex σ_w of K. We define $\bar{\mu}$ on vertices by $\bar{\mu}(w) = v_{\sigma_w}$; see Figure 3.10. Since, by the definition of subdivision, each simplex of K' is contained within some simplex of K, this description of $\bar{\mu}$ on vertices is enough to extend $\bar{\mu}$ to a simplicial map $K' \to K$, and hence $\bar{\mu}$ induces a chain map $\bar{\mu} : C_*^{T'}(X) \to C_*^{T}(X)$.

We verify that $\bar{\mu}$ restricts to a well-defined chain map $\mu : I^{\bar{p}}C_*^{\mathrm{GM},T'}(X) \to I^{\bar{p}}C_*^{\mathrm{GM},T}(X)$.

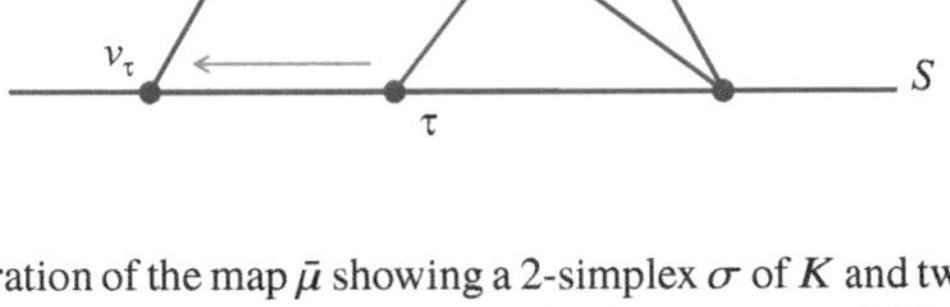

Figure 3.10 An illustration of the map $\bar{\mu}$ showing a 2-simplex σ of K and two smaller 2-simplices of K'. The map $\bar{\mu}$ takes each vertex of a simplex of K' to a vertex of K contained in the same stratum.

For this it suffices to show that the image of an allowable simplex in K' is an allowable simplex of K.

Suppose $\sigma' \in K'$. If $\bar{\mu}(\sigma')$ is degenerate, i.e. if $\bar{\mu}$ is not injective on the vertices of σ', then $\bar{\mu}(\sigma')$ represents 0 in the chain group and is automatically allowable. So assume $\bar{\mu}(\sigma')$ is nondegenerate. As T is already full, so is T', and we have shown in Lemma 3.3.25 that, for any simplex, the stratum containing the interior of that simplex is determined entirely by the vertices of the simplex. But by construction, if v is a vertex of σ', then v and $\bar{\mu}(v)$ are contained in the same stratum of X. Therefore the data assigning to each vertex the stratum containing it is the same for σ' and $\bar{\mu}(\sigma')$. Thus, the interiors of the faces (of all dimensions) of σ' and $\bar{\mu}(\sigma')$ are contained in corresponding strata, and so the dimension of intersection of $\bar{\mu}(\sigma')$ with each stratum of X must be the same as the dimension of intersection of σ with that stratum. So $\bar{\mu}(\sigma')$ must be allowable if σ' is, and the restriction of $\bar{\mu}$ to intersection chains provides the intersection chain map μ.

Lastly, we will verify that $\bar{\mu}\bar{\nu} = \text{id}$. This will imply that $\mu\nu = \text{id}$ upon restricting to the intersection chains. The argument will be inductive over dimension of simplices. In fact, we claim that if an i-simplex $\sigma \in K$ is triangulated as a complex K'_σ in K', then exactly one i-simplex of K_σ, call it η_σ, maps onto σ under μ (compatibly with orientation), while all other i-simplices of K_σ map to degenerate simplices and hence have trivial image under μ. This will suffice due to the description of the subdivision map given by Lemma 3.3.1.

To start the induction, it is evident by construction that $\bar{\mu}\bar{\nu}(v) = v$ for each vertex of K, and this is consistent with the claimed properties. Now, suppose we have verified the claim for all simplices of dimension up through $i-1$, and let σ be an i-simplex of K. Recall that v_σ is the vertex of σ that all vertices of K' in the interior of σ will map to under $\bar{\mu}$. Let τ be the $i-1$ face of σ that does not contain v_σ, and let η_τ be the $(i-1)$-simplex of K' contained in τ that maps onto τ under $\bar{\mu}$; such an η_τ exists by the induction hypothesis. The simplex η_τ must be an $i-1$ face of a unique i-simplex s of K'_σ, which we claim must be the desired η_σ. Let w be the vertex of s that is not contained in τ. We claim that w must map to v_σ, which would verify that s maps onto σ; it would also provide the necessary compatibility with orientation as, by assumption, η_τ and s are oriented compatibly with τ and σ and we have $\bar{\mu}$ orientation-preserving from η_τ to τ. Now, w certainly maps to v_σ if w is in the interior of σ or if $w = v_\sigma$. Otherwise, w is contained in another $i-1$ face of σ that is not τ. But since v_σ is contained in the stratum S_σ, in fact so must be at least all of $\sigma - \tau$. Since v_σ is the highest in the ordering of vertices among all vertices of σ contained in S_σ, then v_σ must also be the image of all vertices of K'_σ contained in $\sigma - \tau$ (as all faces of σ containing such vertices also contain v_σ). Thus $\bar{\mu}(w) = v_\sigma$.

It remains to show that no other i-simplex of K'_σ maps onto σ by $\bar{\mu}$. By the preceding argument, any simplex, say t, with more than one vertex in $\sigma - \tau$ must map multiple vertices to v_σ, and hence must be degenerate. But the only other possibility is to have all but at most one vertex in τ. Clearly no i-simplex t can have all of its vertices in τ, thus the only possibility is to have an $i-1$ face of t in τ. But now except for the η_σ we have constructed, no such simplex can have its $i-1$ face in τ be η_τ, and so for any other i-simplex t, the $i-1$ face in τ must degenerate under $\bar{\mu}$ by the induction hypothesis, and so $\bar{\mu}(\tau)=0$.

This completes the proof. □

Proof of Lemma 3.3.23 Let $[\xi]$ be an element of $I^{\bar{p}}\mathfrak{H}_i^{\mathrm{GM}}(X)$. The class $[\xi]$ can be represented by a cycle $\xi \in I^{\bar{p}}C_i^{\mathrm{GM},T'}(X)$ for some subdivision T' of T. Let ν be the subdivision map $I^{\bar{p}}C_*^{\mathrm{GM},T}(X) \to I^{\bar{p}}C_*^{\mathrm{GM},T'}(X)$, and let μ and $\bar{\mu}$ be the left inverses constructed in the proof of Lemma 3.3.21. We claim that the images of ξ and $\mu(\xi)$ are homologous in $I^{\bar{p}}\mathfrak{C}_*^{\mathrm{GM}}(X)$, which then demonstrates that $[\xi]$ is in the image of $\beta : I^{\bar{p}}H_*^{\mathrm{GM},T}(X) \to I^{\bar{p}}\mathfrak{H}_*^{\mathrm{GM}}(X)$. The reason we need to go into $I^{\bar{p}}\mathfrak{C}_*^{\mathrm{GM}}(X)$ to find the homology, rather than finding it directly in $I^{\bar{p}}C_*^{\mathrm{GM},T'}(X)$, will become clear from the construction.

Once again, suppose T triangulates X by the simplicial complex K and that K' is the subdivision corresponding to T'. We can assume that the vertices of K have been given a total order by the Well-Ordering Theorem.

Let $|\xi|$ be the support of ξ, and let I be the interval $[0,1]$. Consider the space $I \times |\xi|$, and provide it with the standard prism triangulation based on the triangulation of $|\xi|$ in K' (see, e.g. [125, Section 2.1] and Figure 3.11). In other words, suppose $\{0\}\times|\xi|$ and $\{1\}\times|\xi|$ are triangulated just as $|\xi|$ is triangulated in K', and for each simplex $\sigma=[v_0,\dots,v_i]$ of $|\xi|$, let $I\times\sigma$ be triangulated by the $(i+1)$-simplices of the form $[u_0,\dots,u_\ell,w_\ell,\dots,w_i]$ and their faces, where u_j, w_j are respectively the copies of v_j in $\{0\}\times\sigma$ and $\{1\}\times\sigma$. That this is indeed a triangulation is shown in [125] or follows from Lemma B.6.3, below. If the i-simplex $\sigma=[v_0,\dots,v_i]$ has coefficient m in ξ, then let the $(i+1)$-simplex $[u_0,\dots,u_\ell,w_\ell,\dots,w_i]$ have coefficient $(-1)^\ell m$ in order to obtain a simplicial chain Ξ on the space $I\times|\xi|$. Then $\partial\Xi=\{1\}\times\xi-\{0\}\times\xi$, recalling that ξ is a cycle; see the proof of [125, Theorem 2.10].

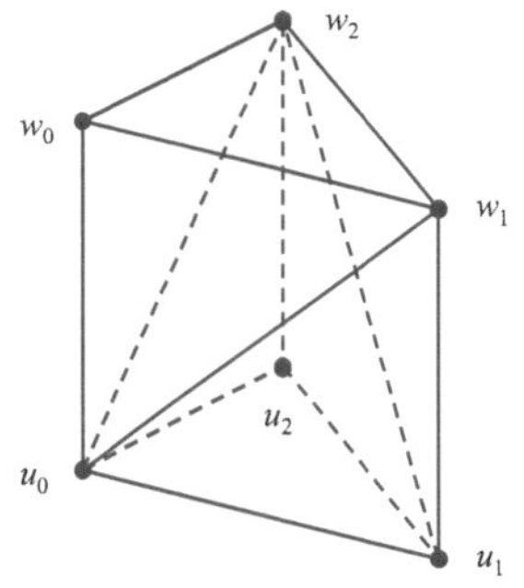

Figure 3.11 The prism triangulation of a 2-simplex.

Now we construct a piecewise linear map γ from $I\times|\xi|$ to X. For each vertex v of ξ, set $\gamma(\{0\}\times v)=v\in X$ and let $\gamma(\{1\}\times v)=\bar{\mu}(v)\in X$. Since every $I\times\tau$ gets mapped into a single simplex of K, this determines γ as a linear map on each simplex of $I\times|\xi|$, and so overall we obtain a piecewise linear map. The image of Ξ as a PL chain represents a chain

$[\gamma(\Xi)] \in \mathfrak{C}_{i+1}(X)$ such that

$$\partial[\gamma(\Xi)] = [\mu(\xi)] - [\xi] \in \mathfrak{C}_i(X),$$

where $[\cdot]$ denotes the class of a chain in the direct limit $\mathfrak{C}_*(X)$. We notice, by the way, that the reason we need to descend all the way to $\mathfrak{C}_*(X)$ is that γ will not necessarily be simplicial with respect to T or T'.

It only remains to show that $\gamma(\Xi)$ is allowable. In order to do this, it suffices by Lemma 3.3.18 to show that $|\gamma(\Xi)|$ satisfies the necessary inequalities. As we already know that $\partial[\gamma(\Xi)] = [\mu(\xi)] - [\xi]$ is allowable, we need only check the allowability condition with respect to $|\gamma(\Xi)|$, itself. For this, it will suffice to show that if σ is an $(i+1)$-simplex of $I \times |\xi|$ then $|[\gamma(\sigma)]|$ satisfies the allowability conditions in X, as then the required dimension conditions will also be true over a finite union of such σ. To do so, we claim that γ is stratum-preserving in the sense that if x is a point in $|\xi|$, then γ maps all of $I \times \{x\}$ to the same stratum of X. This will suffice to finish the proof as follows: Assume that τ is an i-simplex of ξ. We know that $\dim(\tau \cap S) \leq i - \operatorname{codim}(S) + \bar{p}(S)$ for any stratum S because τ is allowable. But assuming the claim, we will have $\dim(\gamma(I \times \tau) \cap S) \leq \dim(\tau \cap S) + 1$, as only points $(t, y) \in I \times \tau$ such that $y \subset S$ can map to S under γ. Thus, in particular, if σ is an $(i+1)$-simplex of $I \times \tau$, we have

$$\dim(\gamma(\sigma) \cap S) \leq \dim(\gamma(I \times \tau) \cap S) \leq \dim(\tau \cap S) + 1 \leq i + 1 - \operatorname{codim}(S) + \bar{p}(S).$$

So σ is an allowable $(i+1)$-simplex!

Now we must prove the claim. So, for any dimension k, let τ be a k-dimensional face of a simplex of $|\xi|$ in the triangulation K', and let $\pi : I \times |\xi| \to |\xi|$ be the projection. By Lemma 3.3.25, the interior of τ is contained in whatever stratum S_τ of X has the property that all vertices of τ are contained in $\bar{S}_\tau$, and at least one vertex of τ is contained in S_τ. Consider now the simplices of $I \times \tau \subset I \times |\xi|$ that intersect $\pi^{-1}(\mathring{\tau})$, where $\mathring{\tau}$ is the interior of τ. These are the $(k+1)$-simplices of the form $[u_0, \ldots, u_\ell, w_\ell, \ldots, w_k]$ and the k-simplices of the form $[u_0, \ldots, u_\ell, w_{\ell+1}, \ldots, w_k]$; all other simplices in the triangulation of $I \times \tau$ are contained in $\pi^{-1}(I \times |\partial\tau|)$ (the vertices of these other simplices all project to vertices of a proper face of τ). Now recall $\gamma(u_j)$ is simply the corresponding vertex v_j of τ in X and $\gamma(w_j) = \bar{\mu}(v_j)$, and by construction v_j and $\bar{\mu}(v_j)$ always lie in the same stratum of X. Therefore all of the $\gamma(u_j)$ and $\gamma(w_j)$ lie in $\bar{S}_\tau$ and, for at least one index m, $\gamma(w_m)$ and $\gamma(u_m)$ are contained in S_τ. Therefore, again by Lemma 3.3.25, the interiors of the $(k+1)$- and k-simplices that intersect $\pi^{-1}(I \times \mathring{\tau})$ are all contained in S_τ. Furthermore, the interiors of τ and $\mu(\tau)$ are contained in S_τ and thus all of $\gamma(I \times \mathring{\tau})$ is contained in S_τ. As the interiors of the faces (of all dimensions) of the simplices of $|\xi|$ partition $|\xi|$, the claim follows. □

3.4 Singular Intersection Homology

For ordinary homology, singular homology presents many advantages over simplicial homology, at the cost of trading a manageable number of simplices (finite on a compact simplicial space) for an uncountable number of simplices (on a space that is not a point) and thus of not being computable combinatorially. That said, many other properties, particularly homotopy properties, become much more transparent for singular homology, and of course singular homology applies to more general classes of spaces that might not even be triangulable. Singular intersection homology faces many of the same trade-offs. Singular intersection homology applies to more general spaces, and it will become easier to prove some theorems, at the expense of computation at first becoming more complicated. Ultimately, singular intersection homology will provide a setting for our most general duality results.

In this section, X will be any filtered space, not necessarily simplicial or PL. Let $S_*(X)$ be the complex of singular chains of X. Recall that a singular simplex $\sigma \in S_i(X)$ is a continuous function $\sigma : \Delta^i \to X$, where Δ^i is the standard geometric i-dimensional simplex. For definiteness, we could suppose that Δ^i is embedded in $\mathbb{R}^i$ with vertices

$$(0,\dots,0),(1,0,\dots,0),\dots,(0,\dots,0,1),$$

though we will typically use the more generic notation $[v_0,\dots,v_i]$ and think of Δ^i as an abstract space. In any case, we will always assume that Δ^i comes with a fixed ordering of its vertices. The boundary formula is then

$$\partial\sigma = \sum_{k=0}^{i}(-1)^k \sigma|_{[v_0,\dots,\hat{v}_k,\dots,v_i]},$$

where $\hat{v}_k$ indicates that we remove the vertex v_k so that $[v_0,\dots,\hat{v}_k,\dots,v_i]$ represents a geometric $(i-1)$-simplex and $\sigma|_{[v_0,\dots,\hat{v}_k,\dots,v_i]}$ represents a singular $(i-1)$-simplex[11]. Then we obtain $\partial : S_i(X) \to S_{i-1}(X)$ by extending linearly, with $\partial \circ \partial = 0$ so that $S_*(K)$ is a chain complex. The *ith singular homology group* is

$$H_i(K) = \frac{\ker(\partial : S_i(X) \to S_{i-1}(X))}{\operatorname{im}(\partial : S_{i+1}(X) \to S_i(X))}.$$

Note that we follow the fairly common practice of writing H_i for both simplicial and singular homology, which is justified by the isomorphism between the two when X is the underlying space of a simplicial complex [182, Section 34].

Just as in the simplicial case, we wish to define a subcomplex $I^{\bar{p}}S_*^{\mathrm{GM}}(X) \subset S_*(X)$ for each perversity $\bar{p}$. A little thought will make the reader leery of trying to use dimension of intersection to measure allowability since the images of singular simplices might now be

[11] If we really want to think of Δ^{i-1} as a fixed space, we could replace $\sigma|_{[v_0,\dots,\hat{v}_k,\dots,v_i]}$ with the composition of σ with the embedding $\Delta^{i-1} \hookrightarrow \Delta^i$ that takes Δ^{i-1} to the face of Δ^i spanned by $\{v_0,\dots,\hat{v}_k,\dots,v_i\}$ by an order-preserving map of vertices; see [182, Section 29]. We will usually leave these inclusion maps as implicit.

quite complex (think of pathological things like space-filling curves). Instead, we have the following pleasant adaptation of the simplicial notion of allowability introduced by Henry King[12] [139].

Definition 3.4.1 Let X be a filtered space endowed with a general perversity $\bar{p}$, and let $S_*(X)$ be the singular chain complex of X.

We deem a singular i-simplex $\sigma : \Delta^i \to X$ to be *$\bar{p}$-allowable* if, for all strata S of X,

$$\sigma^{-1}(S) \subset \{(i - \operatorname{codim}(S) + \bar{p}(S))\text{-skeleton of } \Delta^i\}. \tag{3.2}$$

Here $\operatorname{codim}(S)$ is the formal codimension of S in the filtered space X, while the skeleta of Δ^i are its simplicial skeleta. If inequality (3.2) is satisfied for some σ and some S, we say that σ is *$\bar{p}$-allowable with respect to the stratum S*. If the perversity $\bar{p}$ has been fixed in advance, we will sometimes simply say that σ is *allowable*.

A chain $\xi \in S_i(X)$ is *$\bar{p}$-allowable* if all of the simplices[13] in ξ and all of the simplices of $\partial\xi$ are $\bar{p}$-allowable.

Let $I^{\bar{p}}S^{\mathrm{GM}}_*(X) \subset S_*(X)$ be the chain complex of $\bar{p}$-allowable chains, which we call the *(perversity $\bar{p}$ singular) intersection chain complex*. Let the *(perversity $\bar{p}$ singular) intersection homology groups* be the homology groups $H_*(I^{\bar{p}}S^{\mathrm{GM}}_*(X))$. At the risk of some possible confusion with the simplicial homology groups, we will generally denote these also by $I^{\bar{p}}H^{\mathrm{GM}}_*(X)$. The analogous notation is always justified when working with ordinary homology because the simplicial and singular homology groups always agree on simplicial spaces. While we have already seen by Example 3.2.10 that this cannot always be the case here, we will have agreement with the simplicial intersection homology of "most" triangulations via Theorem 5.4.2, which says that singular and PL intersection homology are isomorphic, and Theorem 3.3.20, which said that PL and simplicial intersection homology agree for full triangulations. Between this fact and contextual clues, we hope the reader will not be too misled by the notation.

Notice that if X is a simplicial filtered space and the singular simplex $\sigma \to X$ is simply the inclusion of one of the i-simplices in the triangulation of X, then this definition of allowability corresponds exactly to our simplicial allowability conditions.

Let us compute some examples:

Example 3.4.2 Let $X = X^0$ be a point. In this case, there is only one stratum, X itself, and it is a regular stratum so $\bar{p}(X) = 0$ for any perversity $\bar{p}$. There is exactly one simplex in each dimension, the unique map $\sigma_i : \Delta^i \to X$. In this case, $\sigma_i^{-1}(X) = \Delta^i$, and the allowability

12 As in the original work of Goresky–MacPherson in [105, 106], King assumes that all strata of the same (co)dimension take the same perversity value. However, he dispenses with the other requirements of a Goresky and MacPherson perversity (except that it should be 0 on regular strata) and calls these *loose perversities*.

13 By saying that "σ is a simplex in ξ" or that "σ belongs to ξ," we mean that σ is a simplex in ξ with non-zero coefficient. In other words, if we write $\xi = \sum_j n_j\sigma_j$ for $n_j \in \mathbb{Z}$ and the σ_j are singular simplices such that $\sigma_j \neq \sigma_\ell$ if $j \neq \ell$, then we mean that $\sigma = \sigma_k$ for some k such that $n_k \neq 0$.

condition then becomes that

$$\Delta^i \subset \{(i - \text{codim}(X) + \bar{p}(X))\text{-skeleton of } \Delta^i\} = \{i\text{-skeleton of } \Delta^i\}.$$

So every singular simplex is allowable, and $I^{\bar{p}}S_*^{\text{GM}}(X) = S_*(X)$, the ordinary chain complex. △

As for simplicial intersection homology, we can observe that the allowability condition is vacuous when it comes to regular strata:

Lemma 3.4.3 *Let σ be a singular i-simplex of a filtered space X, and let S be a* regular *stratum of X. Then the allowability condition* (3.2) *is always satisfied.*

Proof In this case, the same computation as performed in Example 3.4.2 shows that the allowability requirement becomes that $\sigma^{-1}(S) \subset \{i\text{-skeleton of } \Delta^i\}$. This is satisfied trivially. □

Example 3.4.4 Suppose $X = X^n$ is a filtered space that is filtered trivially so that there are only regular strata; see Example 2.2.28. Then it follows from the preceding lemma that $I^{\bar{p}}S_*^{\text{GM}}(X) = S_*(X)$. △

Remark 3.4.5 Lemma 3.4.3 allows us to provide further justification for setting $\bar{p}(S) = 0$ for all regular strata. We see from the lemma that with $\bar{p}(S) = 0$ all simplices are allowable with respect to all regular strata. Furthermore, if $\bar{p}(S) = m$ for any $m \geq 0$, then it is easy to see that the same conclusion will hold, so as mentioned in Remark 3.1.2, any choice of $\bar{p}(S) \geq 0$ for regular strata would provide the same intersection chains, but we choose $\bar{p}(S) = 0$ for definiteness and convenience.

By contrast if S is regular and $\bar{p}(S) \leq -1$, then for an i-simplex to be allowable with respect to S, we would need

$$\sigma^{-1}(S) \subset \{(i + \bar{p}(S))\text{-skeleton of } \Delta^i\},$$

where $i + \bar{p}(S) \leq i - 1$. In other words, at most the $(i-1)$-skeleton of Δ^i could map to S (and less of it if $\bar{p}(S) < -1$). But since $X - \Sigma_X$ is an open subset of X, $\sigma^{-1}(X - \Sigma_X)$ must be an open subset of Δ^i, and so the allowability condition can only be satisfied if $\sigma^{-1}(X - \Sigma) = \emptyset$, i.e. if the image of σ is in Σ_X. In other words, $I^{\bar{p}}H_*^{\text{GM}}(X)$ would not see that regular stratum, so it is equal to $I^{\bar{p}}H_*^{\text{GM}}(X - S)$. Therefore having regular strata with negative perversities is the same as working on spaces without those strata, and we could just as well have taken that view from the beginning and worked on a different space.

Altogether, this makes it reasonable to always have $\bar{p}(S) = 0$ for regular strata.

Example 3.4.6 Suppose $\bar{p}(S) \leq \text{codim}(S) - 2$ for all singular strata S (this is a common condition to require for a perversity). Then $i - \text{codim}(S) + \bar{p}(S) \leq i - 2$ and so the allowability condition requires

$$\sigma^{-1}(S) \subset \{(i-2)\text{-skeleton of } \Delta^i\}$$

for all singular strata. So if $i = 0$ or 1, no i-simplex may intersect any singular stratum. Consequently, we must have that $I^{\bar{p}}H_0^{\mathrm{GM}}(X) \cong \mathbb{Z}^m$, where m is the number of path components of $X - \Sigma_X$. △

Example 3.4.7 Let M be a compact $(n-1)$-dimensional manifold, and let $X = X^n = cM$ be stratified by $\{v\} \subset X$, where v is the vertex of the cone. Since all simplices are allowable with respect to the regular stratum, the allowability condition for an i-simplex becomes

$$\sigma^{-1}(\{v\}) \subset \{(i - n + \bar{p}(\{v\}))\text{-skeleton of } \Delta^i\}.$$

If $i < n - \bar{p}(\{v\})$ then the image of σ cannot contain v at all, and so for[14] $i < n - \bar{p}(\{v\}) - 1$ we have $I^{\bar{p}}S_i^{\mathrm{GM}}(X) = S_i(X - \{v\})$, and so $I^{\bar{p}}H_i^{\mathrm{GM}}(X) = H_i(X - \{v\}) \cong H_i(M)$.

For $i \geq n - \bar{p}(\{v\})$, each i-simplex is allowed to map at least a vertex to v, and possibly more depending on dimension. To compute $I^{\bar{p}}H_i^{\mathrm{GM}}(X)$, suppose that σ is an allowable i-simplex. Let $\bar{c}\sigma : \Delta^{i+1} \to X$ be the (singular) cone on σ. This is defined as follows; see Figure 3.12: Since X is the cone $cX = [0,1) \times M/\sim$, every point in X can be described as a pair (t, z), where $t \in [0,1)$, $z \in M$ (with z non-unique if $t = 0$). In particular, if $x \in \Delta^i$, then $\sigma(x) = (\sigma_I(x), \sigma_M(x))$, where σ_I is the composition of σ with the projection to $[0,1)$ and similarly for σ_M, letting $\sigma_M(x)$ be arbitrary if $\sigma_I(x) = 0$. Now, think of Δ^{i+1} as the closed cone on Δ^i, i.e.

$$\Delta^{i+1} = \bar{c}\Delta^i = [0,1] \times \Delta^i/\sim,$$

and each point of Δ^{i+1} can be written (s, x) with $s \in [0,1]$ and $x \in \Delta^i$ (again non-uniquely if $s = 0$). Define the cone $\bar{c}\sigma$ so that $\bar{c}\sigma(s, x) = (s\sigma_I(x), \sigma_M(x))$. Then $\bar{c}\sigma(1, x) = (\sigma_I(x), \sigma_M(x)) = \sigma(x)$, and $\bar{c}\sigma(0, x) = (0, \sigma_M(x)) = v$. We also note that if $\sigma(x) = (0, \sigma_M(x)) = v$, then $\bar{c}\sigma(s, x) = (0, \sigma_M(x)) = v$ for all s. This map is readily seen to be continuous, and so $\bar{c}\sigma$ is a singular $(i+1)$-simplex.

We next claim that if σ is a $\bar{p}$-allowable i-simplex then so is $\bar{c}\sigma$, provided $i \geq n - \bar{p}(\{v\}) - 1$. The key issue, of course, is to compute $(\bar{c}\sigma)^{-1}(\{v\})$. This set certainly includes the cone vertex $(0, x) \in \Delta^{i+1}$. Otherwise, it consists of the points (s, x) such that $\sigma(x) = v$. Suppose x is contained in the j-skeleton of Δ^i. Then each point (s, x) is contained in at most the $(j+1)$-skeleton of Δ^{i+1}. So if $j \geq -1$ and $\sigma^{-1}(\{v\})$ is contained in the j-skeleton of Δ^i (letting the -1 skeleton be empty), then $(\bar{c}\sigma)^{-1}(\{v\})$ is contained in the $(j+1)$-skeleton of Δ^{i+1}. If σ is allowable, then

$$\sigma^{-1}(\{v\}) \subset \{(i - n + \bar{p}(\{v\}))\text{-skeleton of } \Delta^i\},$$

and we have just seen that if $i - n + \bar{p}(\{v\}) \geq -1$, which is precisely the case we're in, then

$$(\bar{c}\sigma)^{-1}(\{v\}) \subset \{(i + 1 - n + \bar{p}(\{v\}))\text{-skeleton of } \Delta^{i+1}\}.$$

But this is okay because $\bar{c}\sigma$ is an $(i+1)$-simplex! So the increase in the dimension of intersection with the skeleton of the model simplex is offset by the increase in the dimension of the simplex itself, and if the i-simplex σ is allowable then so is $\bar{c}\sigma$ when $i \geq n - \bar{p}(\{v\}) - 1$.

[14] The extra -1 is because homology in dimension i depends on chains in degrees i and $i + 1$.

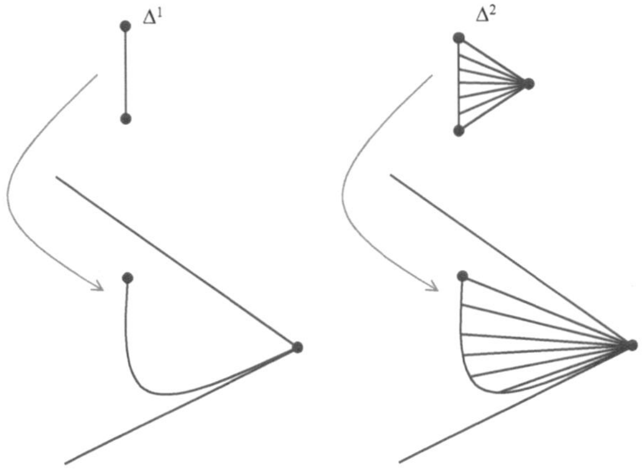

Figure 3.12 A singular 1-simplex and its singular cone.

Notice that we do need to be careful to have $i - n + \bar{p}(\{v\}) \geq -1$ since there is no difference between j-skeleta of Δ^i for $j < 0$; in fact in all these cases saying that $\sigma^{-1}(\{v\}) \subset \{j\text{-skeleton of } \Delta^i\}$ simply means that v isn't in the image of σ. However, in all these cases $(\bar{c}\sigma)^{-1}(\{v\})$ is contained in the 0-skeleton of Δ^{i+1}, regardless.

Now, let us see what this tells us about intersection homology. Let $\bar{c}$ act on chains as the linear extension of its action on simplices. Suppose $\xi \in I^{\bar{p}}S_i^{\mathrm{GM}}(X)$ is an allowable i-cycle for $i \geq n - \bar{p}(\{v\}) - 1$, which puts us in the situation $i - n + \bar{p}(\{v\}) \geq -1$. Thus $\bar{c}\xi$ is allowable. Furthermore, if $i > 0$ then since ξ is a cycle we will have $\partial(\bar{c}\xi) = \xi$. In fact, notice that on each i-simplex, $i > 0$, $\partial(\bar{c}\sigma) = \sigma - \bar{c}(\partial\sigma)$, just as for simplicial simplices, and so in general $\partial(\bar{c}\xi) = \xi - \bar{c}(\partial\xi)$. So if ξ is a cycle, it bounds $\bar{c}\xi$, and $I^{\bar{p}}H_i^{\mathrm{GM}}(X) = 0$, as we'd expect for a cone!

There is one last case to be careful about: when $i = 0$. This case is fundamentally different even for ordinary homology because, while the cone on an i-cycle ξ, $i > 0$, always has $\partial\xi = \xi$, and so $\bar{c}\xi$ provides a null-homology of ξ, this is not always true in the 0-cycle case, because if σ is a singular 0-simplex, then $\partial(\bar{c}\sigma) = \sigma - \sigma_v$, where σ_v is the singular 0-simplex with image v. So for ordinary homology we just get a homology from any singular simplex to σ_v. But we can't even do this in intersection homology because even if σ and $\bar{c}\sigma$ are allowable as *simplices*, the *chain* $\bar{c}\sigma$ might not be allowable, as σ_v might not be allowable. That said, if we continue to assume that $0 \geq n - \bar{p}(\{v\}) - 1$ so that $\bar{c}\sigma$ is allowable as a simplex, then if σ_1, σ_2 are any two allowable 0-simplices, the cone $\bar{c}(\sigma_2 - \sigma_1)$ will have allowable boundary $\sigma_2 - \sigma_1$, and so any two allowable 0-simplices are allowably homologous. Since there are allowable simplices in the regular stratum $X - \{v\}$, we have $I^{\bar{p}}H_0^{\mathrm{GM}}(X) \cong \mathbb{Z}$.

Altogether, we have computed the following:

$$I^{\bar{p}}H_i^{\mathrm{GM}}(X) \cong \begin{cases} 0, & i \geq n - \bar{p}(\{v\}) - 1, \quad i \neq 0, \\ \mathbb{Z}, & i \geq n - \bar{p}(\{v\}) - 1, \quad i = 0, \\ H_i(M), & i < n - \bar{p}(\{v\}) - 1. \end{cases}$$

△

Example 3.4.8 Let $X = X^1 = S^1$, the circle, and let $x_0 \in S^1$ be any point. Suppose X is filtered as $\{x_0\} \subset X$. Then X has two strata: the regular stratum $X - \{x_0\}$ and the singular stratum $\{x_0\}$. We wish to compute $I^{\bar{p}}H_*^{\mathrm{GM}}(X)$. As this computation will become much simpler once we have established some general properties of singular intersection homology in the next chapter, we limit ourselves here to the calculations that can be carried out in fairly short order.

We have already seen, in Lemma 3.4.3, that all simplices are allowable with respect to regular strata, so we have to check allowability at $\{x_0\}$, where $\bar{p}(\{x_0\})$ could be any fixed integer. An i-simplex $\sigma : \Delta^i \to X$ is allowable if it satisfies the conditions

$$\begin{aligned} \sigma^{-1}(\{x_0\}) &\subset \{(i - \operatorname{codim}(\{x_0\}) + \bar{p}(\{x_0\}))\text{-skeleton of } \Delta^i\} \\ &= \{(i - 1 + \bar{p}(\{x_0\}))\text{-skeleton of } \Delta^i\}. \end{aligned}$$

Already things have become much more complicated for singular intersection homology!

If $i - 1 + \bar{p}(\{x_0\}) \geq i$, i.e. if $\bar{p}(\{x_0\}) \geq 1$, then any simplex is allowable and so $I^{\bar{p}}S_*^{\mathrm{GM}}(X) = S_*(X)$ and $I^{\bar{p}}H_*^{\mathrm{GM}}(X) = H_*(X)$ in all degrees.

If $i - 1 + \bar{p}(\{x_0\}) < 0$, i.e. if $i < 1 - \bar{p}(\{x_0\})$, then no simplex whose image contains x_0 is allowable. In this case $I^{\bar{p}}S_i^{\mathrm{GM}}(X) = S_i(X - \{x_0\})$, which implies that $I^{\bar{p}}H_i^{\mathrm{GM}}(X) \cong H_i(X - \{x_0\})$ for i in the range $i < -\bar{p}(\{x_0\})$; we cannot draw this conclusion regarding $i = -\bar{p}(\{x_0\})$ because we do not know yet about $I^{\bar{p}}S_{1-\bar{p}(\{x_0\})}^{\mathrm{GM}}(X)$.

What about the other cases with $\bar{p}(\{x_0\}) \leq 0$? This is more complex, and we will defer the full computation until after we have developed some tools; see Example 4.4.22. △

As for ordinary singular homology, we have begun to see that singular intersection homology can be difficult to compute "by hand." Therefore, we would like to have some of the standard tools of homology available to us – long exact sequences, homotopy invariance, excision, etc. We will begin to explore these properties, and whether or not they carry over, in the next chapter.

Remark 3.4.9 Before moving on to investigate properties of intersection homology groups, it is useful to make an important observation about the chain complexes $I^{\bar{p}}S_*^{\mathrm{GM}}(X)$ that we will need to keep in mind. The ordinary singular chain groups $S_i(X)$ are free groups generated by the i-dimensional singular simplices. Since $I^{\bar{p}}S_i^{\mathrm{GM}}(X) \subset S_i(X)$ by definition, each $I^{\bar{p}}S_*^{\mathrm{GM}}(X)$ is also a free group; see [147, Theorem III.7.1]. However, $I^{\bar{p}}S_*^{\mathrm{GM}}(X)$ does not necessarily have a basis of singular simplices since we know that an allowable simplex of $S_i(X)$ is not necessarily allowable *as a chain* because its boundary might not be an allowable chain. Hence $I^{\bar{p}}S_*^{\mathrm{GM}}(X)$ has some basis of allowable chains, but in general we will not know what it is. This necessitates some care.

Similar remarks apply for simplicial intersection chain complexes. On the other hand, PL intersection chains are already more complex because the groups $\mathfrak{C}_*(X)$ are themselves only direct limits of free groups and so in general it is not clear whether they are free even in the non-intersection case.

4
Basic Properties of Singular and PL Intersection Homology

In this chapter we establish the basic properties of intersection homology. We will see that many of the axioms of ordinary homology persist, though often in modified forms that are suitable to the study of stratified spaces.

In treatments of classical homology, it is sufficient to develop properties for singular homology and then call upon the equivalence of singular and simplicial homology on simplicial spaces to transfer the properties to simplicial homology, sometimes with some minor modifications. Unfortunately, even establishing the equivalence of singular and simplicial/PL intersection homology will require knowing that certain properties hold for both theories. Thus we will have to develop these properties independently. Fortunately, however, the proof techniques in the two settings often complement each other, so we will be able to proceed in close parallel.

In the piecewise linear world, we will restrict our attention to PL intersection homology rather than simplicial intersection homology. This is justified by the isomorphism we have already established in Theorem 3.3.20 between PL and simplicial intersection homology for most triangulations, but it is also necessary due to the limitations of the simplicial theory. For example, a simplicial Mayer–Vietoris sequence akin to the one for ordinary simplicial homology as in, for example, [182, Theorem 25.1] would be problematic. To see this, suppose that K is a simplicial complex with subcomplexes K_1, K_2 such that $K = K_1 \cup K_2$. The usual surjectivity of the map $C_*(K_1) \oplus C_*(K_2) \to C_*(K)$ that occurs in establishing the simplicial homology Mayer–Vietoris sequence depends upon the ability to break up a chain in K into the sum of chains in K_1 and K_2. But breaking up chains creates new boundaries, and so with a fixed triangulation we cannot always break apart an allowable chain into a sum of allowable chains. By contrast, as PL intersection chains are not based on fixed triangulations and as open subsets of PL spaces are again PL spaces (by contrast with simplicial complexes), there is more flexibility, and it turns out the PL intersection homology can be treated more analogously to the singular theory. Consequently, there is a PL Mayer–Vietoris sequence in intersection homology that we develop in Section 4.4.

In Section 4.1, we study the behavior of intersection homology under stratified maps. Not every such map induces a well-defined function on intersection homology; rather, the particular perversities under consideration for the domain and codomain must play a role. But with the right assumptions, topological maps $f : X \to Y$ can induce homomorphisms $f : I^{\bar{p}}H^{\mathrm{GM}}_*(X) \to I^{\bar{q}}H^{\mathrm{GM}}_*(Y)$. Furthermore, the induced algebraic map will not generally be independent of the homotopy class of f, so intersection homology is not a true generalized homology theory. However, we will see that the algebraic maps induced by $f, g : X \to Y$ are the same if f and g are homotopic in an appropriately stratified sense.

In Section 4.2, we compute the intersection homology of a cone on a filtered space. This seemingly innocuous specific computation turns out to be of vital importance throughout all of intersection homology theory, as this is the archetype of a local computation on a CS set. There is a general precept that global homological properties follow from piecing together local homological properties, and the cone computation provides the local input data.

From here, we move on to other properties that one might expect from something called a homology theory: relative intersection homology and the long exact sequence of a pair in Section 4.3 and Mayer–Vietoris sequences and excision in Section 4.4.

4.1 Stratified Maps, Homotopies, and Homotopy equivalences

So far we have set up intersection homology of filtered spaces, but we have not yet considered morphisms of intersection homology groups induced by maps of spaces.

If we try to define a homomorphism $f_* : I^{\bar{p}}H^{\mathrm{GM}}_*(X) \to I^{\bar{p}}H^{\mathrm{GM}}_*(Y)$ for an arbitrary continuous map $f : X \to Y$ of filtered spaces, we immediately run into trouble. For one thing, as we have defined them, perversities are dependent upon the filtration of the space, so without further conditions it does not necessarily make sense to have the same perversity $\bar{p}$ defined on both X and Y. Even if we take $\bar{p}$ to be a GM perversity, so that it depends only on the codimensions of the strata and so can be applied to multiple spaces, there are still difficulties. As the simplest example, suppose X is a point with the trivial filtration so that there is one regular stratum and Y is any space with a non-empty singular stratum $S \subset Y = Y^n$. Let $f : X \to Y$ be any map that takes the point X into S. We have previously computed in Example 3.4.2 that any singular i-simplex σ is allowable in X with respect to any perversity. But there is no reason to expect that $f(\sigma) \in S_i(Y)$ is allowable with respect to $\bar{p}$ and S. In fact, this will only be possible if $i \le i - \mathrm{codim}(S) + \bar{p}(S)$, i.e. if $\bar{p}(S) \ge \mathrm{codim}(S)$. Thus in general it is not possible to set up a fully general functoriality with respect to any single perversity and map between filtered spaces.

Similarly, intersection homology will not be a homotopy invariant of spaces. For example, let $X = cM$ be the open cone on the $(n-1)$-manifold M. Then X is contractible to a point, and we have seen in Example 3.4.2 that every intersection homology group of the point (trivially filtered) is the same as the ordinary homology group of the point. However, as seen in Example 3.4.7, the intersection homology of a cone is not always the homology of a point.

That said, certainly one should be able to set up some reasonable situations in which one obtains homomorphisms of intersection homology groups, and that is what we turn to now.

Of course in the most general situation, one could simply consider all maps $f : X \to Y$ between filtered spaces with respective perversities $\bar{p}$ and $\bar{q}$ that take all allowable simplices to allowable simplices. This would certainly yield maps of intersection homology groups. In practice, however, there are more specific classes of maps that seem sufficiently useful for the required purposes.

Recall from Definition 2.9.1 that $f : X \to Y$ is a *stratified map* of filtered spaces if for each stratum $S \subset X$ there is a unique stratum $T \subset Y$ such that $f(S) \subset T$. We impose further limitations as follows:

Definition 4.1.1 A map $f : X \to Y$ is *GM stratified with respect to* $\bar{p}, \bar{q}$ (or $(\bar{p}, \bar{q})^{\mathrm{GM}}$-*stratified*) if:

1. the image of each stratum of X is contained in a single stratum of Y, i.e. if $T \subset Y$ is a stratum, then $f^{-1}(T)$ is a union of strata of X;
2. when the stratum $S \subset X$ maps to the stratum $T \subset Y$, then

$$\bar{p}(S) - \operatorname{codim}_X(S) \leq \bar{q}(T) - \operatorname{codim}_Y(T).$$

Remark 4.1.2 When we get to non-GM intersection homology in Chapter 6 we will need the further condition that $f(\Sigma_X) \subset \Sigma_Y$. At that point we will remove the "GM" decoration and call such maps $(\bar{p}, \bar{q})$-*stratified*. See Definition 6.3.2.

Here are two important examples:

Example 4.1.3 Let X be an open subset of the filtered space Y with the subspace filtration $X^i = X \cap Y^i$, and let $\bar{p}$ be the perversity on X inherited by the perversity $\bar{q}$ on Y. In other words, if S is a stratum of X and $S \subset T$ for a stratum T of Y, then $\bar{p}(S) = \bar{q}(T)$. Then the inclusion $X \hookrightarrow Y$ is GM stratified with respect to $\bar{p}, \bar{q}$. △

Example 4.1.4 Let $Y = X \times Z$, where Z is any trivially filtered space and Y has the product filtration. Then the strata of Y have the form $S \times Z$ for S a stratum of X, and for any such stratum $S \subset X$, the codimension of S in X equals the codimension of $S \times Z$ in Y. Let $f : X \to Y = X \times Z$ be the inclusion $f(x) = (x, z_0)$ for some fixed point $z_0 \in Z$, and suppose Y has the perversity $\bar{q}$ defined so that $\bar{q}(S \times Z) = \bar{p}(S)$ for any stratum S of X. Then f is $(\bar{p}, \bar{q})^{\mathrm{GM}}$-stratified. More generally, if f is a normally nonsingular inclusion (recall Definition 2.9.8), then the same considerations apply if the perversities are compatible in this way in a neighborhood of the image of X. △

Example 4.1.5 A stratified map $f : X \to Y$ is called *placid*[1] if for each stratum $T \subset Y$ we have $\operatorname{codim}_X(S) \geq \operatorname{codim}_Y(T)$ for each stratum $S \subset f^{-1}(T)$. If $\bar{p}$ is a GM perversity (see Section 3.1.1) and f is placid, then f is $(\bar{p}, \bar{p})^{\mathrm{GM}}$-stratified. For this we observe that if $\bar{p}$ is a GM perversity and if $\operatorname{codim}_Y(T) \leq \operatorname{codim}_X(S)$, then $\bar{p}(S) - \bar{p}(T) \leq \operatorname{codim}_X(S) - \operatorname{codim}_Y(T)$

[1] See [108, Section 4].

by the growth condition on GM perversities. So $\bar{p}(S) - \text{codim}_X(S) \leq \bar{p}(T) - \text{codim}_Y(T)$, as required. Δ

Proposition 4.1.6 *If X, Y are filtered spaces and $f : X \to Y$ is $(\bar{p}, \bar{q})^{\text{GM}}$-stratified, then f induces a chain map of singular intersection chain complexes*[2]

$$f : I^{\bar{p}}S_*^{\text{GM}}(X) \to I^{\bar{q}}S_*^{\text{GM}}(Y).$$

If, furthermore, X, Y are PL filtered spaces and f is a PL map that is $(\bar{p}, \bar{q})^{\text{GM}}$-stratified, then f induces a chain map $f : I^{\bar{p}}\mathfrak{C}_*^{\text{GM}}(X) \to I^{\bar{q}}\mathfrak{C}_*^{\text{GM}}(Y)$ of PL intersection chain complexes. In either case, we obtain corresponding maps of intersection homology groups.

Proof In both cases, there are the usual maps of chain complexes induced on the ordinary singular and PL chains by maps of spaces. Since the intersection chain complexes are subcomplexes, we need only check that allowability of simplices is preserved.

First consider the singular intersection chains. If $\sigma : \Delta^i \to X$ is a $\bar{p}$-allowable simplex, then $f(\sigma)$ is just the composition $f\sigma$. So we must consider $(f\sigma)^{-1}(T) = \sigma^{-1}f^{-1}(T)$ for singular strata T of Y. Now

$$\sigma^{-1}f^{-1}(T) \subset \bigcup_{\{S | f(S) \subset T\}} \sigma^{-1}(S).$$

So, as σ is $\bar{p}$-allowable, for each such S we have

$$\begin{aligned}\sigma^{-1}(S) &\subset \{(i - \text{codim}(S) + \bar{p}(S))\text{-skeleton of } \Delta^i\} \\ &\subset \{(i - \text{codim}(T) + \bar{q}(T))\text{-skeleton of } \Delta^i\},\end{aligned}$$

where we have used the definition of being $(\bar{p}, \bar{q})^{\text{GM}}$-stratified. But then the union of the $\sigma^{-1}(S)$ such that $S \subset f^{-1}(T)$ is also in the $(i - \text{codim}(T) + \bar{q}(T))$-skeleton of Δ^i, and so σ is allowable.

The PL version is perhaps even simpler. If σ is a $\bar{p}$-allowable simplex in some triangulation of X, then the image of σ under f is a PL subset of Y, though not necessarily a simplex because f might be simplicial only with respect to some subdivision of the triangulation with respect to which σ is given. But whatever the dimension of $\sigma \cap S$ is for a stratum $S \subset X$, $\dim(f(\sigma \cap S)) \leq \dim(\sigma \cap S)$ simply by the properties of PL maps. So

$$\begin{aligned}\dim(f(\sigma) \cap T) &\leq \dim\left(\bigcup_{\{S | f(S) \subset T\}} f(\sigma \cap S)\right) \leq \dim\left(\bigcup_{\{S | f(S) \subset T\}} \sigma \cap S\right) \\ &\leq \max_{\{S | f(S) \subset T\}} \{i + \bar{p}(S) - \text{codim}_X(S)\} \\ &\leq i - \text{codim}_Y(T) + \bar{q}(T),\end{aligned}$$

[2] We abuse notation by letting the same symbol f stand for maps of spaces and for the algebraic homomorphisms they induce. We hope context will reduce the confusion while this practice will slightly reduce notational clutter.

again utilizing the definitions. Thus each i-simplex of $f(\sigma)$ must be allowable, which suffices for the proof. □

Remark 4.1.7 Although we have not strictly set up a categorical structure, we remark that basic functorial properties do apply. In particular, if $f : X \to Y$ is $(\bar{p}, \bar{q})^{\mathrm{GM}}$-stratified and $g : Y \to Z$ is $(\bar{q}, \bar{r})^{\mathrm{GM}}$-stratified, then we easily verify that $gf : X \to Z$ is $(\bar{p}, \bar{r})^{\mathrm{GM}}$-stratified and that we can therefore also compose the resulting chain maps and maps of intersection homology groups to obtain composition maps that agree with those induced by gf. Similarly, the identity map $X \to X$ is $(\bar{p}, \bar{p})^{\mathrm{GM}}$-stratified for any perversity $\bar{p}$ on X and induces the identity map on intersection chains and intersection homology.

Given the proposition, there are evident corollaries, such as the following:

Corollary 4.1.8 *If $f : X \to Y$ is a stratified homeomorphism*[3] *and the perversities $\bar{p}$ on X and $\bar{q}$ on Y correspond (i.e. $\bar{p}(S) = \bar{q}(T)$ if $f(S) = T$), then $I^{\bar{p}}H_*^{\mathrm{GM}}(X) \cong I^{\bar{q}}H_*^{\mathrm{GM}}(Y)$. The corresponding fact holds for PL spaces, PL stratified homeomorphisms, and PL intersection homology.*

Proof In this case, the maps f and f^{-1} are respectively GM stratified with respect to the appropriate perversities, and since ff^{-1} and $f^{-1}f$ are identity maps, functoriality implies that f and f^{-1} induce isomorphisms of the intersection chain complexes. □

We now adapt Definition 2.9.10 concerning stratified homotopies to add in the perversity information.

Definition 4.1.9 Let X, Y be filtered spaces with respective perversities $\bar{p}, \bar{q}$. Let I be the trivially filtered unit interval, and let $I \times X$ be given the product filtration. Abusing notation, we also let $I \times X$ have the perversity $\bar{p}$ such that $\bar{p}(I \times S) = \bar{p}(S)$ for any stratum $S \subset X$.

1. A $(\bar{p}, \bar{q})^{\mathrm{GM}}$-stratified map $H : I \times X \to Y$ is called a *GM stratified homotopy (with respect to $\bar{p}, \bar{q}$)*.
2. If $f = H|_{\{0\}\times X}$ and $g = H|_{\{1\}\times X}$ then f and g are *GM stratified homotopic (with respect to $\bar{p}, \bar{q}$)* stratified maps.

We note that as the codimension of S in X is the same as the codimension of $I \times S$ in $I \times X$, the maps f and g of the definition are indeed $(\bar{p}, \bar{q})^{\mathrm{GM}}$-stratified.

The proof that $(\bar{p}, \bar{q})^{\mathrm{GM}}$-stratified homotopies induce the same maps of intersection homology is precisely the same as the proof for ordinary homology provided we can demonstrate allowability.

Proposition 4.1.10 *Suppose $f, g : X \to Y$ are $(\bar{p}, \bar{q})^{\mathrm{GM}}$-stratified homotopic $(\bar{p}, \bar{q})^{\mathrm{GM}}$-stratified maps. Then f and g induce chain homotopic chain maps $I^{\bar{p}}S_*^{\mathrm{GM}}(X) \to I^{\bar{q}}S_*^{\mathrm{GM}}(Y)$ and so $f = g : I^{\bar{p}}H_*^{\mathrm{GM}}(X) \to I^{\bar{q}}H_*^{\mathrm{GM}}(Y)$. If X, Y, f, g are PL and f, g are PL $(\bar{p}, \bar{q})^{\mathrm{GM}}$-stratified homotopic, then they induce chain homotopic chain maps $I^{\bar{p}}\mathfrak{C}_*^{\mathrm{GM}}(X) \to I^{\bar{q}}\mathfrak{C}_*^{\mathrm{GM}}(Y)$ and so $f = g : I^{\bar{p}}\mathfrak{H}_*^{\mathrm{GM}}(X) \to I^{\bar{q}}\mathfrak{H}_*^{\mathrm{GM}}(Y)$.*

[3] See Definition 2.9.3.

Proof It suffices to show that the identity homotopy $\mathrm{id} : I \times X \to I \times X$ between the inclusion maps $\mathrm{i}_0 : X \hookrightarrow \{0\} \times X \subset I \times X$ and $\mathrm{i}_1 : X \hookrightarrow \{1\} \times X \subset I \times X$ induces chain homotopy operators $P : I^{\bar{p}}S_i^{\mathrm{GM}}(X) \to I^{\bar{p}}S_{i+1}^{\mathrm{GM}}(I\times X)$ and $\mathfrak{P} : I^{\bar{p}}\mathfrak{C}_i^{\mathrm{GM}}(X) \to I^{\bar{p}}\mathfrak{C}_{i+1}^{\mathrm{GM}}(I\times X)$. For suppose that P is a chain homotopy operator, i.e. $\partial P = \mathrm{i}_1 - \mathrm{i}_0 - P\partial$. Then as H induces chain maps by Proposition 4.1.6, we have

$$\partial HP = H\partial P = H\mathrm{i}_1 - H\mathrm{i}_0 - HP\partial = g - f - HP\partial,$$

and so HP is a chain homotopy between f and g. Similarly $H\mathfrak{P}$ is a chain homotopy in the PL case.

Next, recall that the strata of $I \times X$ all have the form $I \times S$, where S is a stratum of X, and that $\mathrm{codim}_{I\times X}(I \times S) = \mathrm{codim}_X(S)$. Together with our convention $\bar{p}(I \times S) = \bar{p}(S)$, it is not difficult to check that i_0 and i_1 are $(\bar{p}, \bar{p})^{\mathrm{GM}}$-stratified. In fact, this is a special case of Example 4.1.4.

Now, as for ordinary homology, the idea is to construct a chain homotopy operator by a prism construction as in the proof of Lemma 3.3.23; see [125, proof of Theorem 2.10] for more details. Recall from Lemma 3.3.23 that if $\Delta^i = [v_0, \dots, v_i]$ then we can triangulate the prism $I \times \Delta^i$ so that the $(i + 1)$-simplices of the prism will be of the form $[u_0, \dots, u_\ell, w_\ell, \dots, w_i]$, where u_j, w_j are respectively the copies of v_j in $\{0\}\times\Delta^i$ and $\{1\}\times\Delta^i$. In fact, if $\eta = [v_{m_0}, \dots, v_{m_k}]$, with the v_{m_a} vertices of Δ^i, is a face of Δ^i, then $I\times\eta \subset I\times\Delta^i$ is triangulated as a subcomplex by the $(k + 1)$-simplices $[u_{m_0}, \dots, u_{m_b}, w_{m_b}, \dots, v_{m_k}]$ and their faces. A thorough proof of this will follow from Corollary B.6.5 in our discussion below of the more general Eilenberg–Zilber shuffle triangulation of $\Delta^p \times \Delta^q$. Consequently, we observe that if $(\Delta^i)^k$ is the simplicial k-skeleton of Δ^i, then $I \times (\Delta^i)^k \subset I \times \Delta^i$ is triangulated as a subcomplex of the $(k + 1)$-skeleton of the prism triangulation. In what follows, we always assume $I \times \Delta^i$ has this triangulation. We will construct P and $\mathfrak{P}$ as chain homotopy operators on the ordinary singular and PL chain complexes, obtaining the intersection versions by restriction to the allowable chains.

Let $\sigma : \Delta^i \to X$ be a singular simplex. We will define $P(\sigma)$, from which P is defined in general by extending linearly. Let $\tau_\ell : \Delta^{i+1} \to [u_0, \dots, u_\ell, w_\ell, \dots, w_i]$ be the simplicial homeomorphism determined by the order-preserving bijection of the vertices; this embeds Δ^{i+1} as one of the simplices of the prism triangulation of $I \times \Delta^i$. Let $P(\sigma)$ be the singular chain in $I \times X$ given by $P(\sigma) = \sum_j(\mathrm{id} \times \sigma)\tau_j$. Then it is not difficult to compute (or see [125]) that $\partial P = \mathrm{i}_1 - \mathrm{i}_0 - P\partial$, so P has the form of a chain homotopy operator. We must show that if σ is allowable then the $(i + 1)$-simplices of $P(\sigma)$ are allowable. This is sufficient, because if ξ is an allowable chain, then we will have $P(\xi)$ and $P(\partial\xi)$ consisting of allowable simplices, but also $\partial P(\xi) = \mathrm{i}_1(\xi) - \mathrm{i}_0(\xi) - P(\partial\xi)$ will consist of allowable simplices because i_0 and i_1 are $(\bar{p}, \bar{p})^{\mathrm{GM}}$-stratified. Therefore, we will have $P(\xi) \in I^{\bar{p}}S_{i+1}^{\mathrm{GM}}(I \times X)$.

So we must show that if σ is allowable then each $(\mathrm{id}\times\sigma)\tau_j : \Delta^{i+1} \to I\times X$ is an allowable singular simplex of $I \times X$ with its product filtration and the corresponding perversity as in Definition 4.1.9. As $\tau_j : \Delta^{i+1} \to I \times \Delta^i$ is a simplicial embedding, to check the allowability condition on $(\mathrm{id} \times \sigma)\tau_j$ it suffices to show for each stratum S of X that $(\mathrm{id} \times \sigma)^{-1}(I \times S)$ is contained in the $(i+1-\operatorname{codim}_{I\times X}(I\times S)+\bar{p}(I\times S))$-skeleton of the prism triangulation of $I\times\Delta^i$. Using the correspondences between perversities and codimensions of $I\times S$ in $I\times X$ and S in X, this means we must show that

$$(\mathrm{id}\times\sigma)^{-1}(I\times S)\subset\{(i+1-\operatorname{codim}_X(S)+\bar{p}(S))\text{-skeleton of } I\times\Delta^i\}.$$

Now observe that

$$(\mathrm{id}\times\sigma)^{-1}(I\times S)=I\times\sigma^{-1}(S).$$

And as σ is allowable, then $\sigma^{-1}(S)\subset\{(i-\operatorname{codim}(S)+\bar{p}(S))\text{-skeleton of }\Delta^i\}$, so that

$$(\mathrm{id}\times\sigma)^{-1}(I\times S)\subset I\times\{(i-\operatorname{codim}(S)+\bar{p}(S))\text{-skeleton of }\Delta^i\}.$$

But we have already noted that the product of I with the k-skeleton of Δ^i is contained in the $(k+1)$-skeleton of $I\times\Delta^i$.

This completes the proof for singular intersection homology.

For the PL case, the central idea is basically the same. We will start with a simplicial construction and show that it is compatible with passing to PL chains. So suppose that we choose some fixed admissible triangulation T of X and that $I\times X$ is given the prism triangulation based on the preceding construction over each simplex; see Theorem B.6.6. We will construct

$$P_T : I^{\bar{p}}C_i^{\mathrm{GM},T}(X)\to I^{\bar{p}}C_{i+1}^{\mathrm{GM},I\times T}(I\times X),$$

where $I\times T$ denotes the prism triangulation based on T. Analogously to the singular case, suppose $\sigma=[v_0,\dots,v_i]$ is a simplex of X, and let

$$\tau_j=[u_0,\dots,u_j,w_j,\dots,w_i]$$

in $I\times X$, where the u_j and w_j are the copies of v_j in $\{0\}\times X$ and $\{1\}\times X$. We let $P_T(\sigma)=\sum\tau_j$ in $I\times X$ and extend P_T linearly to simplicial chains. As above, we have $\partial P_T=\mathrm{j}_1-\mathrm{j}_0-P_T\partial$ so that P_T is a chain homotopy between j_0 and j_1. So if σ being allowable implies that the $(i+1)$-simplices of $P_T(\sigma)$ are allowable, then it follows just as for the argument in the singular case that P_T takes intersection chains to intersection chains.

If σ is allowable,

$$\dim(\sigma\cap S)\le i-\operatorname{codim}(S)+\bar{p}(S),$$

and so

$$\dim(I\times(\sigma\cap S))\le i+1-\operatorname{codim}(S)+\bar{p}(S).$$

But then certainly

$$\dim(\tau_j\cap(I\times S))\le\dim(I\times(\sigma\cap S))\le i+1-\operatorname{codim}(S)+\bar{p}(S).$$

So each τ_j is allowable, as desired.

Unfortunately, the operators P_T do not commute with the subdivision operators, as the reader can check with some easy examples. However, to obtain $\mathfrak{P}$ from the P_T we can show that if σ is a simplex in the triangulation T and σ' is a subdivision in the triangulation T' then the images of $P_T(\sigma)$ and $P_{T'}(\sigma')$ agree in the PL chain complex. For this we use Lemma 3.3.10, observing that the supports of $P_T(\sigma)$ and $P_{T'}(\sigma')$ are both $I \times |\sigma|$, the supports of $\partial P_T(\sigma)$ and $\partial P_{T'}(\sigma')$ are both

$$\partial(I \times |\sigma|) = (I \times |\partial\sigma|) \cup (\partial I \times |\sigma|),$$

and both chains represent the fundamental class generator of $H_{i+1}(I \times |\sigma|, \partial(I \times |\sigma|))$. So $P_T(\sigma)$ and $P_{T'}(\sigma')$ represent the same PL chains. □

Analogously to ordinary homology, the preceding result quickly implies the following corollary, which says that stratified homotopy equivalences (Definition 2.9.10) induce isomorphisms of intersection homology:

Corollary 4.1.11 *Suppose $f : X \to Y$ is a stratified homotopy equivalence and that $\bar{p}$ on X and $\bar{q}$ on Y agree in the sense that if S and T are strata of X and Y, respectively, then $\bar{p}(S) = \bar{q}(T)$ if $f(S) \subset T$. Then f induces an isomorphism $I^{\bar{p}}H_*^{\mathrm{GM}}(X) \cong I^{\bar{q}}H_*^{\mathrm{GM}}(Y)$. The analogous result holds in the PL category.*

Proof Recall that by Definition 2.9.10 a stratified homotopy equivalence $f : X \to Y$ is assumed to map each stratum of X to a stratum of Y of the same codimension. Furthermore, by Remark 2.9.11 such an f establishes a bijection between the strata of X and the strata of Y. So if S and T are strata of X and Y such that $f(S) \subset T$, then for any singular simplex $\Delta^i \to X$, we have

$$(f\sigma)^{-1}(T) = \sigma^{-1}(f^{-1}(T)) = \sigma^{-1}(S).$$

So using the correspondence of codimensions and perversities, $f\sigma$ is $\bar{q}$-allowable if σ is $\bar{p}$-allowable, and thus f is $(\bar{p}, \bar{q})^{\mathrm{GM}}$-stratified. Similarly, if $g : Y \to X$ is the stratified homotopy inverse of f, then g is $(\bar{q}, \bar{p})^{\mathrm{GM}}$-stratified. The compositions fg and gf are stratified homotopic to identity maps by assumption, and using the product filtrations and perversities as in Definition 4.1.9 on $I \times X$ and $I \times Y$, the homotopies are also GM stratified with respect to the perversities. So as the identity maps certainly induce isomorphisms on intersection homology, so do the compositions fg and gf by Proposition 4.1.10. Consequently each of f and g is an isomorphism. A similar argument shows the same thing in the PL category. □

Remark 4.1.12 In such situations, especially when X is a subset of Y, we will tend to abuse notation and use the same perversity symbol $\bar{p}$ for the perversities on both spaces. Then the result of the previous corollary would be written $I^{\bar{p}}H_*^{\mathrm{GM}}(X) \cong I^{\bar{p}}H_*^{\mathrm{GM}}(Y)$.

Example 4.1.13 As X is stratified homotopy equivalent to $\mathbb{R}^n \times X$ when $\mathbb{R}^n$ is given the trivial filtration and $\mathbb{R}^n \times X$ is given the product filtration, we have $I^{\bar{p}}H_*^{\mathrm{GM}}(\mathbb{R}^n \times X) \cong I^{\bar{p}}H_*^{\mathrm{GM}}(X)$, with the isomorphism induced either by inclusion $X \hookrightarrow \mathbb{R}^n \times X$, $x \to (z, x)$ for some fixed $z \in \mathbb{R}^n$, or by collapse $\mathbb{R}^n \times X \to X$, $(z, x) \to x$. △

4.2 The Cone Formula

In Example 3.4.7, we computed the intersection homology of the open cone on a manifold. In this section, we will extend this example to the cone on a filtered space. This computation turns out to be phenomenally important: we know that every point of a CS set has a neighborhood of the form $\mathbb{R}^k \times cL$, so once we know how to compute the intersection homology of a cone, the stratified homotopy invariance of Corollary 4.1.11 tells us how to compute the intersection homology of all these distinguished neighborhoods. A general principle of topology is that to understand something about a space it is often useful to study the pieces it is made of and how these pieces fit together; for example one sees this notion at work in the Mayer–Vietoris sequence. Another powerful example of this principle is at work in sheaf theory, which is precisely a machine for piecing local information together into global information. The intersection homology of distinguished neighborhoods of points constitutes the local information, and so this computation provides the foundation for the sheaf-theoretic approach to intersection homology. While we will not travel down the sheaf-theoretic road here, we will see similar ideas at play in our "Mayer–Vietoris arguments" of the next chapter. Via these, the intersection homology of cones is a critical stepping stone for almost all of our major theorems.

So, let $X = X^{n-1}$ be a compact $(n-1)$-dimensional filtered space, and consider the open cone cX with its cone filtration as in Example 2.2.11. The strata of cX will be the cone vertex v and strata of the form $(0,1) \times S$ for S a stratum of X. If $\bar{p}$ is a perversity on cX, define a perversity $\bar{p}_X$ on X such that if S is a stratum of X then $\bar{p}_X(S) = \bar{p}((0,1) \times S)$. For any fixed $t \in (0,1)$, the inclusion map $X \hookrightarrow cX$ that takes $x \in X$ to (x,t) preserves codimensions and perversities, and so it is $(\bar{p}_X, \bar{p})^{\mathrm{GM}}$-stratified. Therefore, it induces a map $I^{\bar{p}_X} S_*^{\mathrm{GM}}(X) \to I^{\bar{p}} S_*^{\mathrm{GM}}(cX)$.

We will demonstrate the following theorem, which turns out to be completely analogous to the computation when X is a manifold:

Theorem 4.2.1 *If $X = X^{n-1}$ is a compact filtered space of formal dimension $n-1$, then*

$$I^{\bar{p}} H_i^{\mathrm{GM}}(cX) \cong \begin{cases} 0, & i \geq n - \bar{p}(\{v\}) - 1,\ i \neq 0, \\ \mathbb{Z}, & i \geq n - \bar{p}(\{v\}),\ i = 0, \\ \mathbb{Z}, & i = n - \bar{p}(\{v\}) - 1,\ i = 0,\ I^{\bar{p}_X} H_0^{\mathrm{GM}}(X) \neq 0, \\ 0, & i = n - \bar{p}(\{v\}) - 1,\ i = 0,\ I^{\bar{p}_X} H_0^{\mathrm{GM}}(X) = 0, \\ I^{\bar{p}_X} H_i^{\mathrm{GM}}(X), & i < n - \bar{p}(\{v\}) - 1. \end{cases}$$

Furthermore, the isomorphisms of the last case are induced by inclusion. An equivalent conclusion holds for PL intersection homology when X is a compact PL filtered space.

If $\bar{p}(\{v\}) \leq n-2$, for example if $n > 1$ and $\bar{p}$ is a GM perversity, then $n - \bar{p}(\{v\}) - 1 \geq 1$ and the special behavior in low dimensions is avoided. In that case we obtain the much *simpler formula*

$$I^{\bar{p}}H_i^{\mathrm{GM}}(cX) \cong \begin{cases} 0, & i \geq n - \bar{p}(\{v\}) - 1, \\ I^{\bar{p}_X}H_i^{\mathrm{GM}}(X), & i < n - \bar{p}(\{v\}) - 1. \end{cases}$$

Before providing the proof, we make a few remarks.

Remark 4.2.2 The simpler formula in the case $\bar{p}(\{v\}) \leq n-2$ is consistent with what we will see later when we perform the analogous cone computations for non-GM intersection homology but arbitrary perversities. So the complication of the additional cases in Theorem 4.2.1 might be taken as a first indication that our present definition of intersection homology isn't quite right for arbitrary perversities.

Remark 4.2.3 The special case when $i = n - \bar{p}(\{v\}) - 1$, $i = 0$, and $I^{\bar{p}_X}H_0^{\mathrm{GM}}(X) = 0$ is not usually noted in the literature. Presumably this is because one is usually most interested in spaces that possess regular strata, and so this case does not arise, as regular strata must contain allowable 0-simplices, implying that $I^{\bar{p}_X}H_0^{\mathrm{GM}}(X) \neq 0$. So, for example, this case is unnecessary when working only with stratified pseudomanifolds by Remark 2.4.3. However, as noted in Remarks 2.2.15 and 2.2.27, spaces with no regular strata will be unavoidable for us in general; see Section 4.3, below, for more details.

It is easy to overlook this special case, and, indeed, the author is not aware of any prior reference to it, including in his own work or in [139], where singular intersection homology was first introduced.

Remark 4.2.4 Notice that the larger the value of $\bar{p}(v)$, the more chains in cX are allowable and so the more intersection homology groups are 0, thus agreeing with our expectations for the ordinary homology of a cone.

Proof of Theorem 4.2.1 The proof mirrors the argument of Example 3.4.7 nearly completely.

We begin by checking the allowability condition at $\{v\}$, for which the condition for an i-simplex becomes

$$\sigma^{-1}(\{v\}) \subset \{(i - n + \bar{p}(\{v\}))\text{-skeleton of } \Delta^i\}.$$

If $i < n - \bar{p}(\{v\})$ then the image of σ cannot contain v at all, and so in this range we have $I^{\bar{p}}S_i^{\mathrm{GM}}(cX) = I^{\bar{p}}S_i^{\mathrm{GM}}(cX - \{v\})$. Therefore, for $i < n - \bar{p}(\{v\}) - 1$, we obtain[4]

$$I^{\bar{p}}H_i^{\mathrm{GM}}(cX) = I^{\bar{p}}H_i^{\mathrm{GM}}(cX - \{v\}) \cong I^{\bar{p}}H_i^{\mathrm{GM}}((0,1) \times X).$$

Note that the extra -1 is because homology in dimension i depends on chains in dimensions i and $i+1$. But now by Corollary 4.1.11 and Example 4.1.13, the inclusion $X \to I \times X$ induces an isomorphism

$$I^{\bar{p}_X}H_i^{\mathrm{GM}}(X) \to I^{\bar{p}}H_i^{\mathrm{GM}}((0,1) \times X).$$

[4] Here we slightly abuse notation and use $\bar{p}$ also to stand for the perversity on $cX - \{v\}$ that evaluates on strata exactly as it would thinking of them as strata of cX; see Section 4.3, below, for a more general discussion of subsets of filtered spaces.

For $i \geq n - \bar{p}(\{v\}) - 1$, we again consider $\bar{c}\sigma$, the singular cone on σ; see Example 3.4.7. We claim that if σ is a $\bar{p}$-allowable i-simplex then so is $\bar{c}\sigma$, provided $i - n + \bar{p}(\{v\}) \geq -1$. Let us recall the definition of $\bar{c}\sigma$: We think of Δ^{i+1} as the closed cone on Δ^i, i.e.$\Delta^{i+1} = \bar{c}\Delta^i = [0,1] \times \Delta^i/\sim$, and write each point of Δ^{i+1} as (s,x) with $s \in [0,1]$ and $x \in \Delta^i$, non-uniquely if $s = 0$. Similarly, we give cX coordinates (t,z), where $t \in [0,1)$, $z \in X$, with z non-unique if $t = 0$. Then we can write $\sigma(x) = (\sigma_I(x), \sigma_X(x))$ and define $\bar{c}\sigma$ so that $\bar{c}\sigma(s,x) = (s\sigma_I(x), \sigma_X(x))$. We define $\bar{c}$ on chains by extending linearly, and, furthermore, this operator satisfies $\partial\bar{c}(\sigma) = \sigma - \bar{c}(\partial\sigma)$.

Now, consider the allowability of $\bar{\sigma}$ with respect to $\{v\}$. The argument here is identical to that of Example 3.4.7: The set $(\bar{c}\sigma)^{-1}(\{v\})$ includes the cone vertex $(0,x) \in \Delta^{i+1}$ and also the points (s,x) such that $\sigma(x) = v$. If x is contained in the j-skeleton of Δ^i. Then each point (s,x) is contained in at most the $(j+1)$-skeleton of Δ^{i+1}. So if $\sigma^{-1}(\{v\})$ is contained in the j-skeleton of Δ^i for $j \geq -1$, then $(\bar{c}\sigma)^{-1}(\{v\})$ is contained in the $(j+1)$-skeleton of Δ^{i+1} for $j \geq -1$. If σ is allowable, then

$$\sigma^{-1}(\{v\}) \subset \{(i - n + \bar{p}(\{v\}))\text{-skeleton of } \Delta^i\},$$

and we have just see that if $i - n + \bar{p}(\{v\}) \geq -1$ then

$$\sigma^{-1}(\{v\}) \subset \{(i + 1 - n + \bar{p}(\{v\}))\text{-skeleton of } \Delta^i\}.$$

As $\bar{c}\sigma$ is an $(i+1)$-simplex, $\bar{c}\sigma$ is allowable at $\{v\}$.

Next, let S be a stratum of X, and let us consider $(\bar{c}\sigma)^{-1}((0,1) \times S)$. Owing to the cone construction, a point (s,x), $s \neq 0$, of Δ^{i+1} maps to $(0,1) \times S$ if and only if $\sigma(x) \in S$. As we have already noted that $(\bar{c}\sigma)(0,x) = v$, it follows that

$$(\bar{c}\sigma)^{-1}((0,1) \times S) = (0,1] \times \sigma^{-1}(S) \subset \Delta^{i+1}.$$

But if σ is allowable, $\sigma^{-1}(S)$ lies in the $(i - \text{codim}(S) + \bar{p}(S))$-skeleton of Δ^i, so $(0,1) \times \sigma^{-1}(S)$ is contained in the $(1 + i - \text{codim}(S) + \bar{p}(S))$-skeleton of Δ^{i+1}. But this shows that $\bar{c}\sigma$ is allowable with respect to S. It now follows just as in Example 3.4.7 that if $i \geq n - \bar{p}(\{v\}) - 1$ and $i > 0$, then $I^{\bar{p}}H_i^{\text{GM}}(cX) = 0$ because if ξ is an allowable cycle then $\xi = \partial\bar{c}\xi$.

Finally, when $i = 0 \geq n - \bar{p}(\{v\}) - 1$, we again have to be careful, as in Example 3.4.7, because the cone on a singular 0-simplex generally has two boundary simplices. But again if σ_1, σ_2 are any two allowable 0-simplices, then the cone $\bar{c}(\sigma_2 - \sigma_1)$ will have allowable boundary $\sigma_2 - \sigma_1$, and so any two allowable 0-simplices are allowably homologous. So if there exists an allowable 0-simplex in cX, then we have $I^{\bar{p}}H_0^{\text{GM}}(X) \cong \mathbb{Z}$. This will occur if $\bar{p}(T) \geq \text{codim}_{cX}(T)$ for any stratum $T \subset cX$, in particular if X, and hence cX, has a regular stratum. There are two possibilities for this:

1. either $\bar{p}(\{v\}) \geq n$ (i.e. if $0 \geq n - \bar{p}(\{v\})$), in which case the unique 0-simplex with image $\{v\}$ is itself allowable; or
2. $\bar{p}(T) \geq \text{codim}_{cX}(T)$ for some $T = (0,1) \times S$, in which case $\bar{p}_X(S) \geq \text{codim}_X(S)$ and so S also allows 0-simplices in X and $I^{\bar{p}_X}H_0^{\text{GM}}(X) \neq 0$.

The last remaining possibility is if $0 = n - \bar{p}(\{v\}) - 1$ and $\bar{p}(T) < \text{codim}(T)$ for all strata of cX. Then neither cX nor X can have an allowable 0-simplex. Thus we must have $I^{\bar{p}}H_0^{\text{GM}}(cX) = I^{\bar{p}_X}H_0^{\text{GM}}(X) = 0$. This finishes the proof for singular intersection homology.

Now suppose X is a PL filtered space. The argument here is almost completely the same! In fact, except for minor modifications, the only difference is that we need to define a PL version of the cone operator $\bar{c}$. For this, first suppose we identify X with some simplicial complex in $\mathbb{R}^m$ via some triangulation, and then identify $\bar{c}X$ with a simplicial complex in $\mathbb{R}^{m+1}$ by embedding $\mathbb{R}^m$ as $\{0\} \times \mathbb{R}^m \subset \mathbb{R}^{m+1}$ and letting the point $(1, 0, \ldots, 0)$ be the cone vertex v. Then $\bar{c}X$ is obtained by coning off the simplices of X to v; see Lemma B.5.1, Corollary B.5.2, and their proofs. The open cone cX is the PL subspace $\bar{c}X - X$, and it has a triangulation T in which each simplex is contained linearly in some simplex of the triangulation of $\bar{c}X$ by Example B.4.2. Now, suppose σ is an oriented i-simplex in some subdivision of T. Every PL chain is a linear combination of such simplices. If we order the vertices of σ consistently with the orientation, then we can associate to σ some embedding $\mathfrak{j} : \Delta^i = [v_0, \ldots, v_i] \hookrightarrow cX$. Now extend $\mathfrak{j}$ linearly to a map $\bar{c}\mathfrak{j}$ from $[z, v_0, \ldots, v_i] = \Delta^{i+1} = \bar{c}\Delta^i$ to $\mathbb{R}^{m+1}$ by taking the new vertex z of $\bar{c}\Delta^i$ to the vertex v of $\bar{c}X$. As $\bar{c}\mathfrak{j}$ is linear on Δ^{i+1}, it is a PL map, using Definition B.2.15, the compactness of Δ^{i+1} and the local finiteness of the triangulations, and Definition B.1.24. Using Lemma 3.3.9, if $[\Delta]$ is the class of Δ^{i+1} as an element of $\mathfrak{C}_{i+1}(\Delta^{i+1})$, then $(\bar{c}\mathfrak{j})$ takes $[\Delta]$ to a PL chain we call $\bar{c}\sigma \in \mathfrak{C}_{i+i}(cX)$. Clearly, $\partial(\bar{c}\sigma) = \sigma - \bar{c}(\partial\sigma)$ as PL chains, where $\bar{c}(\partial\sigma)$ is defined by applying the same procedure to the simplices of $\partial\sigma$.

This gives us our PL cone operator $\bar{c}$, and from here the intersection homology computations are completely analogous to those for the singular chain case. We invite the reader to work through the details in this setting as a good exercise. □

4.3 Relative Intersection Homology

We now turn to relative intersection homology groups. For simplicity, we begin by discussing the singular chain case, though the same discussion applies in the PL setting.

Recall that if X is any space and $Y \subset X$ then the ordinary relative singular homology $H_*(X,Y)$ is defined to be the homology of the chain complex $S_*(X,Y) = S_*(X)/S_*(Y)$. This quotient makes sense, as each singular i-chain of Y is also an i-chain of X, and so each $S_i(Y)$ is naturally a subgroup of $S_i(X)$. Furthermore, the boundary map on X takes i-chains of Y to $(i-1)$-chains of Y, and so $S_*(Y)$ is a subcomplex of $S_*(X)$. It follows that $\partial : S_i(X,Y) \to S_{i-1}(X,Y)$ is well defined, yielding the chain complex $S_*(X,Y)$.

Suppose now that X is a filtered space with perversity $\bar{p}$ and that $Y \subset X$. We might then expect to define $I^{\bar{p}}S_*^{\text{GM}}(X,Y)$ as $I^{\bar{p}}S_*^{\text{GM}}(X)/I^{\bar{p}}S_*^{\text{GM}}(Y)$, but now there is a question of precisely what $I^{\bar{p}}S_*^{\text{GM}}(Y)$ should mean, as to define the $\bar{p}$-intersection chains on a space it must be a filtered space and the perversity should be defined with respect to that filtration. It turns out that there are two natural, though ultimately equivalent, ways to proceed:

1. Define a filtration on Y and a perversity $\bar{p}_Y$, obtained in a canonical way from the filtration on X and $\bar{p}$, so that $I^{\bar{p}_Y}S_*^{\mathrm{GM}}(Y)$ is defined and such that it can be identified with a subcomplex of $I^{\bar{p}}S_*^{\mathrm{GM}}(X)$. Then let
$$I^{\bar{p}}S_*^{\mathrm{GM}}(X,Y) = I^{\bar{p}}S_*^{\mathrm{GM}}(X)/I^{\bar{p}_Y}S_*^{\mathrm{GM}}(Y).$$
2. Rather than giving Y its own filtration and perversity information, we can define $I^{\bar{p}}S_*^{\mathrm{GM}}(Y \subset X)$ to be the intersection of complexes $I^{\bar{p}}S_*^{\mathrm{GM}}(X) \cap S_*(Y) \subset S_*(X)$. Another way to say this is that
$$I^{\bar{p}}S_*^{\mathrm{GM}}(Y \subset X) = \{\xi \in I^{\bar{p}}S_*^{\mathrm{GM}}(X) \mid |\xi| \subset Y\}.$$
Then we can let
$$I^{\bar{p}}S_*^{\mathrm{GM}}(X,Y) = I^{\bar{p}}S_*^{\mathrm{GM}}(X)/I^{\bar{p}}S_*^{\mathrm{GM}}(Y \subset X).$$

The appeal of the latter approach is that it does not require putting a filtration on Y that might not be intrinsically well suited to Y. However, both perspectives will be useful, and it turns out that there is a natural way to give Y a filtration and perversity $\bar{p}_Y$ such that $I^{\bar{p}_Y}S_*^{\mathrm{GM}}(Y) = I^{\bar{p}}S_*^{\mathrm{GM}}(Y \subset X)$. Before proceeding to the official definitions, let us consider an example that illustrates what "not intrinsically well suited" can mean.

Example 4.3.1 Suppose X is a filtered space with perversity $\bar{p}$ and that $x \in X$ is a point, which we can view as a subspace $\{x\} \subset X$. As a space in its own right, the most natural filtration for $\{x\}$ would be the trivial filtration as a zero-dimensional manifold. In that case, there is only a regular stratum, determining the perversity, say $\bar{q}$, with $\bar{q}(\{x\}) = 0$. Furthermore, then $I^{\bar{q}}S_*^{\mathrm{GM}}(\{x\}) = S_*(\{x\})$ as we know all simplices are allowable with respect to regular strata. But if x is contained in a singular stratum of X, then singular chains with image in $\{x\}$ might not be allowable in $I^{\bar{p}}S_*^{\mathrm{GM}}(X)$. Then we will not have $I^{\bar{q}}S_*^{\mathrm{GM}}(\{x\}) \subset I^{\bar{p}}S_*^{\mathrm{GM}}(X)$, and so we cannot define a relative intersection chain complex $I^{\bar{p}}S_*^{\mathrm{GM}}(X,\{x\})$ as a quotient.

By contrast, $I^{\bar{p}}S_*^{\mathrm{GM}}(\{x\} \subset X)$ is defined as a subcomplex of $I^{\bar{p}}S_*^{\mathrm{GM}}(X)$; it is generated in degree i by the singular simplex $\sigma_i : \Delta^i \to \{x\}$ if σ_i is $\bar{p}$-allowable in X. So if, as claimed, there is a filtration of $\{x\}$ and a perversity $\bar{p}_{\{x\}}$ such that $I^{\bar{p}_{\{x\}}}S_*^{\mathrm{GM}}(\{x\}) = I^{\bar{p}}S_*^{\mathrm{GM}}(\{x\} \subset X)$, it must not necessarily be the natural filtration as a 0-manifold. In particular, the formal dimension may be > 0, the only other filtration option for a single point. △

It turns out that while the filtration and perversity on Y that get us $I^{\bar{p}_Y}S_*^{\mathrm{GM}}(Y) = I^{\bar{p}}S_*^{\mathrm{GM}}(Y \subset X)$ might not be the most intrinsically natural to Y, they are the ones inherited from the information on X in the simplest possible way. We provide this in the next definition. In Section 4.3.1, we will discuss situations for which these inherited filtrations on subspaces do correspond to more natural inherent filtrations. Historically, such examples have been important.

Definition 4.3.2 Suppose X is a filtered space endowed with a perversity $\bar{p}$, and suppose $Y \subset X$ is an arbitrary subspace. Recall from Example 2.2.9 the subspace filtration $Y^i = Y \cap X^i$, which gives Y the same formal dimension as X. We define the *subspace perversity* $\bar{p}_Y$ on Y with the subspace filtration by $\bar{p}_Y(S) = \bar{p}(T)$ if S is a stratum of Y contained in the stratum $T \subset X$. We will also say that this filtration and perversity on Y are *inherited from* X, that the filtration and perversity are the *restrictions* of the filtration and perversity from X to Y, or that (X,Y) is a *filtered pair*. We will use these terminologies interchangeably.

Whenever we discuss a subspace $Y \subset X$, we will automatically assume it is endowed with the inherited filtration and perversity unless noted otherwise. We will also generally denote $\bar{p}_Y$ simply as $\bar{p}$ unless there is some danger of confusion, though we will continue to be rather pedantic in this section while first working out the details.

Remark 4.3.3 Suppose X is a filtered space and $Y \subset X$ is given the subspace filtration. Then

$$Y^i - Y^{i-1} = Y \cap X^i - Y \cap X^{i-1} = Y \cap (X^i - X^{i-1}).$$

So if S is a formally i-dimensional stratum of Y, then it is contained in a formally i-dimensional stratum, say T, of X. And as Y is given the same formal dimension as X in the subspace filtration, we then have $\operatorname{codim}_Y(S) = \operatorname{codim}_X(T)$, as well as $\bar{p}_Y(S) = \bar{p}(T)$ (which is just the definition of $\bar{p}_Y$).

Here is a first way in which the subspace filtration and perversity provide compatibility between the intersection chains of Y and those of X.

Lemma 4.3.4 *If X is a filtered space with perversity $\bar{p}$ and $Y \subset X$ is given the subspace filtration and perversity, then the inclusion $\jmath : Y \hookrightarrow X$ is $(\bar{p}_Y, \bar{p})^{\mathrm{GM}}$-stratified.*

Proof It is immediate from the definition of the subspace filtration and Remark 4.3.3 that $\jmath$ takes strata of Y into strata of X of the same codimension. Furthermore, by the definition of the subspace perversity, $\bar{p}_Y(S) = \bar{p}(T)$ if S is a stratum of Y contained in the stratum $T \subset X$. So for any such strata, we have $\bar{p}_Y(S) - \operatorname{codim}_Y(S) = \bar{p}(T) - \operatorname{codim}_X(T)$, thus satisfying the definition for $\jmath$ to be $(\bar{p}_Y, \bar{p})^{\mathrm{GM}}$-stratified. ∎

Corollary 4.3.5 *Suppose X is a filtered space with a perversity $\bar{p}$ and that $Y \subset X$ is given the subspace filtration and perversity. Then $I^{\bar{p}_Y} S^{\mathrm{GM}}_*(Y) = I^{\bar{p}} S^{\mathrm{GM}}_*(Y \subset X)$. Similarly, if X is PL and Y is a PL subspace, then $I^{\bar{p}_Y} \mathfrak{C}^{\mathrm{GM}}_*(Y) = I^{\bar{p}} \mathfrak{C}^{\mathrm{GM}}_*(Y \subset X)$, letting*

$$I^{\bar{p}} \mathfrak{C}^{\mathrm{GM}}_*(Y \subset X) = I^{\bar{p}} \mathfrak{C}^{\mathrm{GM}}_*(X) \cap \mathfrak{C}_*(Y) \subset \mathfrak{C}_*(X).$$

Proof By the preceding lemma and Proposition 4.1.6, the inclusion $\jmath$ induces maps of singular and PL intersection chain complexes, in fact commutative diagrams, as follows

$$
\begin{array}{ccc}
I^{\bar{p}_Y} S_*^{\mathrm{GM}}(Y) & \xrightarrow{\mathrm{j}} & I^{\bar{p}} S_*^{\mathrm{GM}}(X) \\
\big\downarrow & & \big\downarrow \\
S_*(Y) & \xhookrightarrow{\mathrm{j}} & S_*(X)
\end{array}
\qquad\qquad
\begin{array}{ccc}
I^{\bar{p}_Y} \mathfrak{C}_*^{\mathrm{GM}}(Y) & \xrightarrow{\mathrm{j}} & I^{\bar{p}} \mathfrak{C}_*^{\mathrm{GM}}(X) \\
\big\downarrow & & \big\downarrow \\
\mathfrak{C}_*(Y) & \xhookrightarrow{\mathrm{j}} & \mathfrak{C}_*(X).
\end{array}
$$

And so we can identify $I^{\bar{p}_Y} S_*^{\mathrm{GM}}(Y)$ with its image under j as a subcomplex of $I^{\bar{p}} S_*^{\mathrm{GM}}(X)$, and similarly in the PL case. In particular, the chains in the image of $I^{\bar{p}_Y} S_*^{\mathrm{GM}}(Y)$ under j are intersection chains of X and supported in Y, and so, suppressing the map j in the notation, $I^{\bar{p}_Y} S_*^{\mathrm{GM}}(Y) \subset I^{\bar{p}} S_*^{\mathrm{GM}}(Y \subset X)$ and, in the PL case, $I^{\bar{p}_Y} \mathfrak{C}_*^{\mathrm{GM}}(Y) \subset I^{\bar{p}} \mathfrak{C}_*^{\mathrm{GM}}(Y \subset X)$.

Conversely, suppose $\xi \in I^{\bar{p}} S_i^{\mathrm{GM}}(Y \subset X)$, i.e. that $\xi \in I^{\bar{p}} S_i^{\mathrm{GM}}(Y) \cap S_i(Y)$. The claim that $\xi \in I^{\bar{p}_Y} S_*^{\mathrm{GM}}(Y)$ is simply the claim that ξ satisfies its allowability requirements with respect to the inherited filtration and perversity $\bar{p}_Y$ on Y. But if S is a stratum of Y contained in a stratum T of X, then from the definition of $\bar{p}_Y$ and Remark 4.3.3, we see that the condition of being $\bar{p}_Y$-allowable with respect to S is precisely the condition of being $\bar{p}$-allowable with respect to T. And we know that all such allowabilities hold by the assumption that $\xi \in I^{\bar{p}} S_i^{\mathrm{GM}}(X)$. So $I^{\bar{p}} S_*^{\mathrm{GM}}(Y \subset X) \subset I^{\bar{p}_Y} S_*^{\mathrm{GM}}(Y)$ and, similarly in the PL case, $I^{\bar{p}} \mathfrak{C}_*^{\mathrm{GM}}(Y \subset X) \subset I^{\bar{p}_Y} \mathfrak{C}_*^{\mathrm{GM}}(Y)$. □

Remark 4.3.6 Earlier in this section, we made the observation that $I^{\bar{p}} S_*^{\mathrm{GM}}(Y \subset X)$ can also be described as $\{\xi \in I^{\bar{p}} S_*^{\mathrm{GM}}(X) \mid |\xi| \subset Y\}$. Of course similarly

$$I^{\bar{p}} \mathfrak{C}_*^{\mathrm{GM}}(Y \subset X) = \{\xi \in I^{\bar{p}} \mathfrak{C}_*^{\mathrm{GM}}(X) \mid |\xi| \subset Y\}$$

if Y is a PL subspace of the PL filtered space X. But notice now that the latter complex is still well defined if Y is *any* subspace of X, PL or not. Of course in this case it is possible that $\mathfrak{C}_*(Y)$, and hence $I^{\bar{p}_Y} \mathfrak{C}_*^{\mathrm{GM}}(Y)$, may not be defined, as Y might not even be triangulable. In this case, we will adopt the *definitions*

$$I^{\bar{p}_Y} \mathfrak{C}_*^{\mathrm{GM}}(Y) = I^{\bar{p}} \mathfrak{C}_*^{\mathrm{GM}}(Y \subset X) = \{\xi \in I^{\bar{p}} \mathfrak{C}_*^{\mathrm{GM}}(X) \mid |\xi| \subset Y\},$$

and Corollary 4.3.5 will not generally apply. But but we can still use $I^{\bar{p}} \mathfrak{C}_*^{\mathrm{GM}}(Y \subset X)$ to define relative chains and relative homology.

Definition 4.3.7 If X is a filtered space with perversity $\bar{p}$ and $Y \subset X$ with the subspace filtration and perversity, we define the relative singular intersection chain complex by

$$I^{\bar{p}} S_*^{\mathrm{GM}}(X,Y) = I^{\bar{p}} S_*^{\mathrm{GM}}(X)/I^{\bar{p}} S_*^{\mathrm{GM}}(Y \subset X) = I^{\bar{p}} S_*^{\mathrm{GM}}(X)/I^{\bar{p}_Y} S_*^{\mathrm{GM}}(Y).$$

If X is PL and Y is any subspace, we let

$$I^{\bar{p}} \mathfrak{C}_*^{\mathrm{GM}}(X,Y) = I^{\bar{p}} \mathfrak{C}_*^{\mathrm{GM}}(X)/I^{\bar{p}} \mathfrak{C}_*^{\mathrm{GM}}(Y \subset X) = I^{\bar{p}} \mathfrak{C}_*^{\mathrm{GM}}(X)/I^{\bar{p}_Y} \mathfrak{C}_*^{\mathrm{GM}}(Y).$$

We let $I^{\bar{p}} H_*^{\mathrm{GM}}(X,Y)$ and $I^{\bar{p}} \mathfrak{H}_*^{\mathrm{GM}}(X,Y)$ be the corresponding relative intersection homology groups.

We will continue with the $I^{\bar{p}}S_*^{\mathrm{GM}}(X)/I^{\bar{p}_Y}S_*^{\mathrm{GM}}(Y)$ and $I^{\bar{p}}\mathfrak{C}_*^{\mathrm{GM}}(X)/I^{\bar{p}_Y}\mathfrak{C}_*^{\mathrm{GM}}(Y)$ version of the notation in what follows, although, past this section, we will simplify the notation a bit further by writing $\bar{p}_Y$ simply as $\bar{p}$.

The following is now an immediate consequence of standard homological algebra; see [125, Theorem 2.16] or [182, Lemma 24.1]:

Theorem 4.3.8 *If X is a filtered space with perversity $\bar{p}$ and $Y \subset X$ with the subspace filtration and perversity, there is a short exact inclusion/quotient sequence*

$$0 \to I^{\bar{p}_Y}S_*^{\mathrm{GM}}(Y) \to I^{\bar{p}}S_*^{\mathrm{GM}}(X) \to I^{\bar{p}}S_*^{\mathrm{GM}}(X,Y) \to 0,$$

and, hence, a long exact intersection homology sequence

$$\cdots \to I^{\bar{p}_Y}H_i^{\mathrm{GM}}(Y) \to I^{\bar{p}}H_i^{\mathrm{GM}}(X) \to I^{\bar{p}}H_i^{\mathrm{GM}}(X,Y) \to \cdots.$$

Analogous statements hold for PL intersection chains and PL intersection homology.

4.3.1 Further Commentary on Subspace Filtrations

We have just seen that subspace filtrations on subspaces of filtered spaces lead naturally to relative intersection homology and long exact sequences of the pair. This justifies our use of subspace filtrations. But it also demonstrates our claims in Remarks 2.2.15, 2.2.27, and 4.2.3 that filtered spaces with no regular strata are unavoidable when treating subsets of more reasonable spaces:

Example 4.3.9 Consider the space $X = X^2 = S^2 \amalg S^1$ of Remark 2.2.15. This is a two-dimensional manifold stratified space filtered by $X^1 = S^1 \subset S^2 \amalg S^1 = X$. If we let $Y = X^1 = S^1 \subset X$ and wish to consider the relative intersection chain complex $I^{\bar{p}}S_*^{\mathrm{GM}}(X,Y)$ for some perversity $\bar{p}$ on X, then we will need the subcomplex $I^{\bar{p}_Y}S_*^{\mathrm{GM}}(Y)$, which consists of the chains of $I^{\bar{p}}S_*^{\mathrm{GM}}(X)$ contained in Y. But with the subspace filtration, Y is treated as a formally two-dimensional filtered space whose only non-empty stratum is the singular stratum S^1 with codimension one. With this filtration, S^1 has no regular strata.

It is similarly not hard to come up with less artificial, e.g. connected, examples. △

However, there are important situations in which such "bad" behavior does not occur or in which a subspace filtration gives intersection homology groups on the subspace that agree with those arising from a more natural filtration with more natural dimensions. The simplest of these situations occurs for open subsets of pseudomanifolds:

Example 4.3.10 By Lemma 2.4.10, an open subset U of an n-dimensional stratified pseudomanifold X is an n-dimensional stratified pseudomanifold with its inherited filtration.

More generally, suppose X is any n-dimensional manifold stratified space such that the union of its n-dimensional strata is dense in X, and let $U \subset X$ be an open set. In this case, the intersection of U with the i-dimensional strata of X will be i-dimensional manifolds, so U is also a manifold stratified space. The union of the n-dimensional strata of U will be dense in U, and so the filtration of U inherited from X is n-dimensional, topologically and not just formally.

In these examples, U has regular strata and the subspace dimensions are consistent with treating U as a pseudomanifold or manifold stratified space in its own right. △

Another important case with many applications is that of normally nonsingular subspaces:

Example 4.3.11 Normally nonsingular subspaces provide another useful class of subspaces for which intersection homology can often be computed using the intuitive dimensions of the subspace.

First, suppose Z is a k-dimensional filtered space, and consider the product $\mathbb{R}^m \times Z$ with the product filtration, where $\mathbb{R}^m$ is trivially filtered with one regular stratum of dimension m comprising all of $\mathbb{R}^m$. The formal dimension of $\mathbb{R}^m \times Z$ is $m + k$. Recall that, as in Example 2.2.25, there is a bijection between strata $S \subset Z$ and the strata $\mathbb{R}^m \times S \subset \mathbb{R}^m \times Z$; if S is a j-dimensional stratum of Z, then $\mathbb{R}^m \times S$ is a $(j + k)$-dimensional stratum of $\mathbb{R}^m \times Z$. It follows that the codimension of S in Z is equal to the codimension of $\mathbb{R}^m \times S$ in $\mathbb{R}^m \times Z$.

Now, suppose instead that we began with the product filtration on $\mathbb{R}^m \times Z$ and wanted to consider $\hat{Z} = \{0\} \times Z$ as a subset. In the subspace filtration we consider $\hat{Z}$ to have formal dimension $m + k$, and similarly the i-dimensional strata of $\hat{Z}$ will be the intersection of $\hat{Z}$ with the i-dimensional strata of $\mathbb{R}^m \times Z$. This is disconcerting: for example, if Z were a manifold stratified space, then an i-dimensional manifold stratum S in Z thought of as a stratum, say $\hat{S}$, in $\hat{Z}$ would have to have formal dimension $m + i$, and so $\hat{Z}$ could not be a manifold stratified space!

However, we notice that the codimension of a stratum S in Z is nonetheless the same as the codimension of the corresponding stratum $\hat{S}$ in $\hat{Z}$, namely $m + k - (m + i) = k - i$. Furthermore, suppose that $\bar{p}$ is a perversity on Z that we extend to a perversity $\bar{p}^\times$ on $\mathbb{R}^m \times Z$ so that $\bar{p}^\times(\mathbb{R}^m \times S) = \bar{p}(S)$. Then, as defined above, the restriction of $\bar{p}^\times$ to $\hat{Z}$, which we will denote $\bar{p}_{\hat{Z}}$, must have $\bar{p}_{\hat{Z}}(\hat{S}) = \bar{p}^\times(\mathbb{R}^m \times S) = \bar{p}(S)$. Given this correspondence of perversity values and codimensions, we see that Z is stratified homeomorphic[5] to $\hat{Z}$ and

$$I^{\bar{p}} S_*^{\mathrm{GM}}(Z) = I^{\bar{p}_{\hat{Z}}} S_*^{\mathrm{GM}}(\hat{Z}).$$

In fact, the inclusion map $Z \to \mathbb{R}^m \times Z$ is a stratified homotopy equivalence, so altogether we have

$$I^{\bar{p}^\times} H_*^{\mathrm{GM}}(\mathbb{R}^m \times Z) \cong I^{\bar{p}} H_*^{\mathrm{GM}}(Z) \cong I^{\bar{p}_{\hat{Z}}} H_*^{\mathrm{GM}}(\hat{Z}).$$

In other words, if we start with a filtered space Z with perversity $\bar{p}$ and then want to treat Z as a subspace of $\mathbb{R}^m \times Z$ with the subspace filtration and subspace perversity of the product perversity $\bar{p}^\times$, then this is equivalent, for intersection homology purposes, to working with Z and $\bar{p}$ themselves.

At first this example might seem somewhat artificial, but recall from Definition 2.9.8 that a normally nonsingular inclusion with trivial normal bundle is a stratified inclusion $\mathfrak{i} : Z \hookrightarrow X$

[5] See Definition 2.9.3.

such that, for some m, the map $\mathfrak{i}$ extends to a filtered homeomorphism $\bar{\mathfrak{i}}$, from $\mathbb{R}^m \times Z$ with the product filtration ($\mathbb{R}^m$ filtered trivially) onto some neighborhood of $\mathfrak{i}(Z)$. Let $Y = \mathfrak{i}(Z)$, the *normally nonsingular subspace* given the subspace filtration. If $\bar{p}$ is a perversity on X, $\bar{p}_Z$ is the perversity on Z such that $\bar{p}(T) = \bar{p}_Z(S)$ if $\mathfrak{i}(S) \subset T$, and $\bar{p}_Y$ is the subspace perversity on Y, then the preceding argument shows that $I^{\bar{p}_Y} S_*^{\mathrm{GM}}(Y) = I^{\bar{p}_Z} S_*^{\mathrm{GM}}(Z)$. In fact, as Y and Z are stratified homeomorphic, the only real differences between these two expressions are the formal dimensions of the spaces and the strata, though the codimensions of corresponding strata are the same. However, if, for example, Z is a stratified pseudomanifold, then it may make sense to study Z from that perspective, in which case $I^{\bar{p}_Z} S_*^{\mathrm{GM}}(Z)$ is the more natural chain complex. Our computation here shows that the two perspectives, Z as its own space or Z as a subspace, agree as far as intersection chains are concerned. While for simplicity we have here treated only the normally nonsingular inclusions with trivial bundle neighborhoods, the arguments extend directly to more general normally nonsingular inclusions.

We have already seen this example in play when computing the intersection homology of a cone in Theorem 4.2.1. Starting with $X = X^{n-1}$ as the initially given filtered space and forming the n-dimensional cone cX as in Example 2.2.11, we computed that if $\bar{p}$ is a perversity on cX then $I^{\bar{p}} H_i^{\mathrm{GM}}(cX) \cong I^{\bar{p}_X} H_i^{\mathrm{GM}}(X)$ in the degree range $i < n - \bar{p}(\{v\}) - 1$ where $\bar{p}_X(S) = \bar{p}((0,1) \times S)$ for S a stratum of X. We have a normally nonsingular inclusion

$$X \to \{t_0\} \times X \subset cX = [0,1) \times X/\sim,$$

for any choice of $t_0 \subset (0,1)$; see Example 2.9.9. Alternatively, we can let $\hat{X} = \{t_0\} \times X$ thought of with its subspace filtration, in which case $\dim(\hat{X}) = n$. So we have perspectives in which X can be considered to have dimension $n-1$ or dimension n. But nonetheless

$$I^{\bar{p}_X} S_*^{\mathrm{GM}}(X) = I^{\bar{p}_{\hat{X}}} S_*^{\mathrm{GM}}(\hat{X}),$$

and the isomorphism $I^{\bar{p}} H_*^{\mathrm{GM}}(cX) \cong I^{\bar{p}_X} H_*^{\mathrm{GM}}(X)$ (in the appropriate degrees) is induced by the equivalent inclusions of these subcomplexes into $I^{\bar{p}} S_*^{\mathrm{GM}}(X)$.

Consequently, in what follows we will tend to elide the distinction between what we here call X and $\hat{X}$ as we have seen that the distinction is not usually relevant for intersection homology purposes. △

Example 4.3.12 An important example combining both open subsets and normally nonsingular subspaces occurs when X is an n-dimensional stratified pseudomanifold (endowed with perversity $\bar{p}$) and U is a distinguished neighborhood of a point $x \in X_{n-k}$. Then U is itself an n-dimensional stratified pseudomanifold by Lemma 2.4.10. Furthermore, U is filtered homeomorphic to the n-dimensional stratified pseudomanifold $\mathbb{R}^{n-k} \times cL^{k-1}$, where L^{k-1} is a $(k-1)$-dimensional stratified pseudomanifold, which we can identify with a normally nonsingular subspace of $\mathbb{R}^{n-k} \times cL^{k-1}$. By the preceding examples, assuming we choose compatible perversities, the intersection chain complex of the link L thought of as a subspace is isomorphic to the intersection chain complex thinking of L as a $(k-1)$-

dimensional filtered space in its own right. In later sections, we will simply label these complexes as $I^{\bar{p}}S^{\mathrm{GM}}_*(L)$.

Treating L as a stratified pseudomanifold, all notions of dimension correspond with what we would expect topologically; there is no need for formal dimensions, except perhaps in the intermediate steps that we can now bury. △

Example 4.3.13 Finally, another useful example of a "reasonably behaved" subset occurs when $Y = \partial X$ is the boundary of an n-dimensional ∂-stratified pseudomanifold. This isn't quite a normally nonsingular subspace, as Y only has a collar neighborhood filtered homeomorphic to $[0,1) \times Y$ in X. However, exactly the same sorts of arguments apply as in the preceding examples and demonstrate that the intersection chain complex $I^{\bar{p}_{\partial X}}S^{\mathrm{GM}}_*(\partial X)$ obtained by thinking of ∂X as a subspace is equal to the intersection chain complex $I^{\bar{p}_Y}S^{\mathrm{GM}}_*(Y)$ obtained by thinking of $\partial X = Y$ as an $(n-1)$-dimensional stratified pseudomanifold in its own right with appropriately compatible perversities. This example is particular pleasing as the density of the union of the regular strata is one of the key defining properties of a stratified pseudomanifold, and so we would certainly rather think of ∂X as being $(n-1)$-dimensional than as inheriting the formal dimension n from X. △

4.3.2 Stratified Maps Revisited

Suppose (X, A) and (Y, B) are filtered pairs. So X and Y are filtered spaces and $A \subset X$ and $B \subset Y$ inherit the subspace filtration and subspace perversities from X and Y. Suppose that $f : X \to Y$ is a $(\bar{p}, \bar{q})^{\mathrm{GM}}$-stratified map that takes A into B. Using Proposition 4.1.6, f takes $\bar{p}$ intersection chains on X supported in A to $\bar{q}$ intersection chains on Y supported in B. Thus f induces maps $I^{\bar{p}}H^{\mathrm{GM}}_*(X, A) \to I^{\bar{q}}H^{\mathrm{GM}}_*(Y, B)$. In other words, we have the following relative version of Proposition 4.1.6:

Proposition 4.3.14 *If (X, A) and (Y, B) are filtered pairs, $f : X \to Y$ is $(\bar{p}, \bar{q})^{\mathrm{GM}}$-stratified, and $f(A) \subset B$, then f induces a chain map*

$$f : I^{\bar{p}}S^{\mathrm{GM}}_*(X, A) \to I^{\bar{q}}S^{\mathrm{GM}}_*(Y, B).$$

If, furthermore, X, Y are PL filtered spaces, A, B are PL subspaces, and f is a PL map that is $(\bar{p}, \bar{q})^{\mathrm{GM}}$-stratified, then f induces a chain map $f : I^{\bar{p}}\mathfrak{C}^{\mathrm{GM}}_*(X, A) \to I^{\bar{q}}\mathfrak{C}^{\mathrm{GM}}_*(Y, B)$. We thus obtain corresponding maps

$$f : I^{\bar{p}}H^{\mathrm{GM}}_*(X, A) \to I^{\bar{q}}H^{\mathrm{GM}}_*(Y, B) \quad \text{and} \quad f : I^{\bar{p}}\mathfrak{H}^{\mathrm{GM}}_*(X, A) \to I^{\bar{q}}\mathfrak{H}^{\mathrm{GM}}_*(Y, B).$$

From here, it is not difficult to modify the arguments of the various results of Section 4.1 so that they hold for such maps of relative intersection homology groups. Thus we have the following:

Corollary 4.3.15 *If $f : X \to Y$ is a stratified homeomorphism that is also a homeomorphism of pairs $f : (X, A) \to (Y, B)$ and the perversities $\bar{p}$ on X and $\bar{q}$ on Y correspond, then*

$I^{\bar{p}}H_*^{\mathrm{GM}}(X,A) \cong I^{\bar{q}}H_*^{\mathrm{GM}}(Y,B)$. The corresponding fact holds for PL spaces, PL stratified homeomorphisms, and PL intersection homology.

This follows from the naturality of long exact sequences, Corollary 4.1.8, and the Five Lemma.

Proposition 4.3.16 *Suppose (X,A) and (Y,B) are filtered pairs and $f,g : X \to Y$ are $(\bar{p},\bar{q})^{\mathrm{GM}}$-stratified maps that are $(\bar{p},\bar{q})^{\mathrm{GM}}$-stratified homotopic via a $(\bar{p},\bar{q})^{\mathrm{GM}}$-stratified homotopy taking the pair $(I \times X, I \times A)$ to (Y,B). Then f and g induce chain homotopic maps $I^{\bar{p}}S_*^{\mathrm{GM}}(X,A) \to I^{\bar{q}}S_*^{\mathrm{GM}}(Y,B)$ and so*

$$f = g : I^{\bar{p}}H_*^{\mathrm{GM}}(X,A) \to I^{\bar{q}}H_*^{\mathrm{GM}}(Y,B).$$

The analogous result holds in the PL category.

The proof here is the same as that of Proposition 4.1.10 by using prism operators and noting that if $|\sigma| \subset A$ then $|P(\sigma)| \subset I \times A$.

Corollary 4.3.17 *Suppose (X,A) and (Y,B) are filtered pairs and that $f : X \to Y$ is a stratified homotopy equivalence that restricts to a stratified homotopy equivalence $A \to B$. Suppose that the values of $\bar{p}$ on X and $\bar{q}$ on Y agree on corresponding strata. Then f induces an isomorphism $I^{\bar{p}}H_*^{\mathrm{GM}}(X,A) \cong I^{\bar{q}}H_*^{\mathrm{GM}}(Y,B)$. The analogous result holds in the PL category.*

The corollary follows from Corollary 4.1.11, the naturality of the long exact homology sequence, and the Five Lemma.

4.3.3 Reduced Intersection Homology and the Relative Cone Formula

Our main goal in this section is to compute the relative intersection homology groups $I^{\bar{p}}H_*^{\mathrm{GM}}(cX, cX - \{v\})$, which are an important complement to the cone intersection homology groups we computed in Theorem 4.2.1. For the relative groups, it is useful to have reduced intersection homology available.

(a) Reduced Intersection Homology

Definition 4.3.18 Let X be a filtered space with perversity $\bar{p}$. Let $\mathbf{a} : S_0(X) \to \mathbb{Z}$ be the augmentation homomorphism such that $\mathbf{a}(\sigma) = 1$ for each singular 0-simplex $\sigma : \Delta^0 \to X$. Let $\tilde{S}_*(X)$ be the augmented chain complex with $\tilde{S}_i(X) = S_i(X)$ for $i \geq 0$, $S_{-1}(X) = \mathbb{Z}$, and $\mathbf{a}$ serving as the boundary map $\tilde{S}_0(X) \to \tilde{S}_{-1}(X)$; see [182, Section 29] or [125, Section 2.1]. We define the *augmented intersection chain complex* $I^{\bar{p}}\tilde{S}_*^{\mathrm{GM}}(X) \subset \tilde{S}_*(X)$ to be the subcomplex of $\tilde{S}_*(X)$ with $I^{\bar{p}}\tilde{S}_i^{\mathrm{GM}}(X) = I^{\bar{p}}S_i^{\mathrm{GM}}(X)$ for $i \geq 0$, and $I^{\bar{p}}\tilde{S}_{-1}^{\mathrm{GM}}(X) = \mathbb{Z}$. We define the *reduced singular intersection homology groups* by $I^{\bar{p}}\tilde{H}_i^{\mathrm{GM}}(X) = H_*(I^{\bar{p}}\tilde{S}_*^{\mathrm{GM}}(X))$ for $i \geq 0$, and we set $I^{\bar{p}}\tilde{H}_{-1}^{\mathrm{GM}}(X) = 0$ by definition.

The *reduced PL intersection homology groups* $I^{\bar{p}}\tilde{\mathfrak{H}}_*^{\mathrm{GM}}(X) = 0$ are defined analogously.

Remark 4.3.19 The reason for declaring $I^{\bar{p}}\tilde{H}^{\mathrm{GM}}_{-1}(X) = 0$ by definition is to avoid having $I^{\bar{p}}\tilde{H}^{\mathrm{GM}}_{-1}(X) = \mathbb{Z}$ if there are no $\bar{p}$-allowable 0-simplices in X.

Proposition 4.3.20 *For any filtered space X and perversity $\bar{p}$, we have $I^{\bar{p}}H^{\mathrm{GM}}_i(X) \cong I^{\bar{p}}\tilde{H}^{\mathrm{GM}}_i(X)$ for $i > 0$ and $I^{\bar{p}}H^{\mathrm{GM}}_0(X) = I^{\bar{p}}\tilde{H}^{\mathrm{GM}}_0(X) \oplus \mathbb{Z}$ if $I^{\bar{p}}H^{\mathrm{GM}}_0(X) \neq 0$. Moreover, if $I^{\bar{p}}H^{\mathrm{GM}}_0(X) = 0$, then $I^{\bar{p}}\tilde{H}^{\mathrm{GM}}_0(X) \neq 0$. And similarly for PL intersection homology.*

Proof We notice that there is a surjective chain map $I^{\bar{p}}\tilde{S}^{\mathrm{GM}}_*(X)$ to $I^{\bar{p}}S^{\mathrm{GM}}_*(X)$ that is the identity map in non-negative degrees and with kernel complex K_* consisting only of the group $\mathbb{Z}$ in degree -1. The homology of K_* vanishes except in degree -1, so from the long exact homology sequence associated to our short exact sequence of chain complexes, $I^{\bar{p}}H^{\mathrm{GM}}_i(X) = I^{\bar{p}}\tilde{H}^{\mathrm{GM}}_i(X)$ for $i > 0$. In low degrees, we obtain the following exact sequence:

$$0 \longrightarrow I^{\bar{p}}\tilde{H}^{\mathrm{GM}}_0(X) \to I^{\bar{p}}H^{\mathrm{GM}}_0(X) \longrightarrow \mathbb{Z} \longrightarrow H_{-1}(I^{\bar{p}}\tilde{S}^{\mathrm{GM}}_*(X)) \longrightarrow 0.$$

Clearly if $I^{\bar{p}}H^{\mathrm{GM}}_0(X) = 0$ then $I^{\bar{p}}\tilde{H}^{\mathrm{GM}}_0(X) = 0$, and incidentally $H_{-1}(I^{\bar{p}}\tilde{S}^{\mathrm{GM}}_*(X)) \cong \mathbb{Z}$, as this is the case in which no 0-simplex is allowable so $I^{\bar{p}}S^{\mathrm{GM}}_0(X) = 0$. If $I^{\bar{p}}H^{\mathrm{GM}}_0(X) \neq 0$, then there is an allowable 0-simplex, say σ_0, and so a splitting $s : \mathbb{Z} \to I^{\bar{p}}H^{\mathrm{GM}}_0(X)$ such that $s(m) = m\sigma_0$. Therefore, $I^{\bar{p}}H^{\mathrm{GM}}_0(X) = I^{\bar{p}}\tilde{H}^{\mathrm{GM}}_0(X) \oplus \mathbb{Z}$. □

As for ordinary homology, if $Y \subset X$ is a filtered subspace, we can form the reduced pair $I^{\bar{p}}\tilde{S}^{\mathrm{GM}}_*(X,Y)$, which is identical to $I^{\bar{p}}\tilde{S}^{\mathrm{GM}}_*(X,Y)$ because the inclusion $I^{\bar{p}}\tilde{S}^{\mathrm{GM}}_*(Y) \to I^{\bar{p}}\tilde{S}^{\mathrm{GM}}_*(X)$ is the identity on $\mathbb{Z}$ in degree -1. We then get a corresponding long exact homology sequence, though note that due to our convention it is only exact at $I^{\bar{p}}H_0(X,Y)$ if $H_{-1}(I^{\bar{p}}\tilde{S}_*(Y)) = 0$, i.e. if there is an allowable 0-simplex in Y or, equivalently, if $I^{\bar{p}}H_0(Y) \neq 0$.

(b) The Relative Cone formula

Now let us compute the relative cone intersection homology groups.

Let X be a compact $(n-1)$-dimensional filtered space, and let cX be the open cone on X. In Theorem 4.2.1, we computed the intersection homology of cX. We will now compute the intersection homology groups $I^{\bar{p}}H^{\mathrm{GM}}_*(cX, cX - \{v\})$, where v is the cone point. As $cX - \{v\} \cong (0,1) \times X$, we have

$$I^{\bar{p}}H^{\mathrm{GM}}_*(cX - \{v\}) \cong I^{\bar{p}}H^{\mathrm{GM}}_*((0,1) \times X) \cong I^{\bar{p}}H^{\mathrm{GM}}_*(X),$$

by the preservation of intersection homology under stratified homeomorphisms and stratified homotopy equivalences. Notice that we have begun to embrace using $\bar{p}$ as the notation for all suitably compatible perversities as threatened in Definition 4.3.2. Via these isomorphisms, the long exact sequence of the pair is isomorphic to

$$\to I^{\bar{p}}H^{\mathrm{GM}}_i(X) \xrightarrow{\mathfrak{j}} I^{\bar{p}}H^{\mathrm{GM}}_i(cX) \to I^{\bar{p}}H^{\mathrm{GM}}_i(cX, cX - \{v\}) \to I^{\bar{p}}H^{\mathrm{GM}}_{i-1}(X) \to,$$

where $\mathfrak{j}$ stands for the inclusion map into a level set $x \to (t_0, x)$ for fixed $t_0 \in (0,1)$.

By Theorem 4.2.1 this inclusion is an isomorphism for $i < n - \bar{p}(\{v\}) - 1$, and so $I^{\bar{p}}H_i^{\mathrm{GM}}(cX, cX - \{v\}) = 0$ for $i < n - \bar{p}(\{v\}) - 1$.

For $i \geq n - \bar{p}(\{v\}) - 1$, $i > 0$, Theorem 4.2.1 tells us that $I^{\bar{p}}H_i^{\mathrm{GM}}(cX) = 0$. And so in these cases $I^{\bar{p}}H_i^{\mathrm{GM}}(X) \cong I^{\bar{p}}H_{i+1}^{\mathrm{GM}}(cX, cX - \{v\})$. Alternatively stated, for $i > n - \bar{p}(\{v\}) - 1$, $i > 1$, we have $I^{\bar{p}}H_i^{\mathrm{GM}}(cX, cX - \{v\}) \cong I^{\bar{p}}H_{i-1}^{\mathrm{GM}}(X)$.

Next we consider $I^{\bar{p}}H_i^{\mathrm{GM}}(cX, cX - \{v\})$ for $i = n - \bar{p}(\{v\}) - 1 > 0$. In this case, $I^{\bar{p}}H_{n-\bar{p}(\{v\})-1}^{\mathrm{GM}}(cX) = 0$ and $I^{\bar{p}}H_{n-\bar{p}(\{v\})-2}^{\mathrm{GM}}(X) \to I^{\bar{p}}H_{n-\bar{p}(\{v\})-2}^{\mathrm{GM}}(cX)$ is an isomorphism. Thus $I^{\bar{p}}H_{n-\bar{p}(\{v\})-1}^{\mathrm{GM}}(cX, cX - \{v\}) = 0$.

This leaves the following low-dimensional cases to check:

1. $I^{\bar{p}}H_1^{\mathrm{GM}}(cX, cX - \{v\})$, when $1 > n - \bar{p}(\{v\}) - 1$,
2. $I^{\bar{p}}H_0^{\mathrm{GM}}(cX, cX - \{v\})$, when $0 \geq n - \bar{p}(\{v\}) - 1$.

Observe that in both cases $0 \geq n - \bar{p}(\{v\}) - 1$.

In all of these remaining cases, $I^{\bar{p}}H_1^{\mathrm{GM}}(cX) = 0$, and so the tail of the exact sequence is

$$\begin{aligned} 0 &\to I^{\bar{p}}H_1^{\mathrm{GM}}(cX, cX - \{v\}) \to I^{\bar{p}}H_0^{\mathrm{GM}}(X) \to I^{\bar{p}}H_0^{\mathrm{GM}}(cX) \\ &\to I^{\bar{p}}H_0^{\mathrm{GM}}(cX, cX - \{v\}) \to 0. \end{aligned}$$

In the special case when $I^{\bar{p}}H_0^{\mathrm{GM}}(cX) = 0$, which can only happen if $0 = n - \bar{p}(\{v\}) - 1$ and $I^{\bar{p}}H_0^{\mathrm{GM}}(X) = 0$, we must have

$$I^{\bar{p}}H_1^{\mathrm{GM}}(cX, cX - \{v\}) = I^{\bar{p}}H_0^{\mathrm{GM}}(cX, cX - \{v\}) = 0.$$

Otherwise $I^{\bar{p}}H_0^{\mathrm{GM}}(cX) \cong \mathbb{Z}$. If the only allowable 0-simplex of cX is contained in $\{v\}$, which will happen if $0 \geq n - \bar{p}(\{v\})$ and $I^{\bar{p}}H_0^{\mathrm{GM}}(X) = 0$, then we must have $I^{\bar{p}}H_1^{\mathrm{GM}}(cX, cX - \{v\}) = 0$ and $I^{\bar{p}}H_0^{\mathrm{GM}}(cX, cX - \{v\}) \cong \mathbb{Z}$.

Finally, if $I^{\bar{p}}H_0^{\mathrm{GM}}(cX) \cong \mathbb{Z}$ but there are allowable 0-simplices in X, then any such 0-simplex is a generator so $I^{\bar{p}}H_0^{\mathrm{GM}}(X) \to I^{\bar{p}}H_0^{\mathrm{GM}}(cX)$ is a surjection. It follows that $I^{\bar{p}}H_0^{\mathrm{GM}}(cX, cX - \{v\}) = 0$. Furthermore, because $I^{\bar{p}}H_0^{\mathrm{GM}}(cX) \cong \mathbb{Z}$, we must have $I^{\bar{p}}\tilde{H}_0^{\mathrm{GM}}(cX) \cong 0$. So the reduced intersection homology long exact sequence becomes

$$0 \to I^{\bar{p}}H_1^{\mathrm{GM}}(cX, cX - \{v\}) \to I^{\bar{p}}\tilde{H}_0^{\mathrm{GM}}(X) \to 0,$$

and hence $I^{\bar{p}}H_1^{\mathrm{GM}}(cX, cX - \{v\}) \cong I^{\bar{p}}\tilde{H}_0^{\mathrm{GM}}(X)$.

We have now computed the following results for the special low-dimensional cases:

1. If $1 > n - \bar{p}(\{v\}) - 1$, then $I^{\bar{p}}H_1^{\mathrm{GM}}(cX, cX - \{v\}) \cong I^{\bar{p}}\tilde{H}_0^{\mathrm{GM}}(X)$. This also includes the two cases above in which $I^{\bar{p}}H_1^{\mathrm{GM}}(cX, cX - \{v\}) = 0$, since in both those cases $I^{\bar{p}}H_0^{\mathrm{GM}}(X) = 0$ and hence $I^{\bar{p}}\tilde{H}_0^{\mathrm{GM}}(X) = 0$.
2. If $0 \geq n - \bar{p}(\{v\}) - 1$, then $I^{\bar{p}}H_0^{\mathrm{GM}}(cX, cX - \{v\}) = 0 = I^{\bar{p}}\tilde{H}_{-1}^{\mathrm{GM}}(X)$ unless $0 \geq n - \bar{p}(\{v\})$ and $I^{\bar{p}}H_0^{\mathrm{GM}}(X) = 0$, in which case $I^{\bar{p}}H_0^{\mathrm{GM}}(cX, cX - \{v\}) \cong \mathbb{Z}$.

Assembling these results, we have shown the following.

Theorem 4.3.21 *If X is a compact $(n-1)$-dimensional filtered space then*

$$I^{\bar{p}}H_i^{\mathrm{GM}}(cX, cX - \{v\}) \cong \begin{cases} I^{\bar{p}}\tilde{H}_{i-1}^{\mathrm{GM}}(X), & i \geq n - \bar{p}(\{v\}), \quad i \neq 0, \\ I^{\bar{p}}\tilde{H}_{i-1}^{\mathrm{GM}}(X) = 0, & i \geq n - \bar{p}(\{v\}), \quad i = 0, \\ & \qquad I^{\bar{p}}H_0^{\mathrm{GM}}(X) \neq 0, \\ \mathbb{Z}, & i \geq n - \bar{p}(\{v\}), \quad i = 0, \\ & \qquad I^{\bar{p}}H_0^{\mathrm{GM}}(X) = 0, \\ 0, & i < n - \bar{p}(\{v\}). \end{cases}$$

If $\bar{p}(\{v\}) \leq n-1$, for example if $n > 1$ and $\bar{p}$ is a GM perversity, then $n - \bar{p}(\{v\}) \geq 1$ and the special behavior in low dimensions is avoided. In that case we obtain the much simpler formula

$$I^{\bar{p}}H_i^{\mathrm{GM}}(cX, cX - \{v\}) \cong \begin{cases} I^{\bar{p}}\tilde{H}_{i-1}^{\mathrm{GM}}(X), & i \geq n - \bar{p}(\{v\}), \\ 0, & i < n - \bar{p}(\{v\}). \end{cases}$$

Equivalent formulas holds for PL intersection homology when X is PL.

Remark 4.3.22 As for Theorem 4.2.1 (see Remark 4.2.3), the oddities in the low-dimensional cases $I^{\bar{p}_X}H_0^{\mathrm{GM}}(X) = 0$ do not seem to have been previously noticed in the literature. However, such cases do not arise if all spaces possess regular strata, for example when working only with stratified pseudomanifolds.

4.4 Mayer–Vietoris Sequences and Excision

As we have already seen many times, the properties of intersection homology can often be developed quite analogously to the corresponding properties for ordinary homology with just some extra care to ensure that allowability of chains is not compromised. In some sense this is also true when treating Mayer–Vietoris sequences and excision; however, in this case the extra care needed is a bit more subtle and complex, and we must be careful to avoid what might be called the *standard mistake of intersection homology*.

The issue is the following: Suppose $\xi \in S_i(X)$ is an ordinary singular i-chain. We may write $\xi = \sum_{j=1}^m c_j\sigma_j$ for some collection of singular simplices $\{\sigma_j\}$ and some coefficients $c_j \in \mathbb{Z}$. It is quite usual in chain arguments to break ξ into pieces, for example $\xi = (\sum_{j=1}^k c_j\sigma_j) + (\sum_{j=k+1}^m c_j\sigma_j)$. Suppose now that ξ is a chain that is allowable with respect to some perversity $\bar{p}$. By definition, each σ_j is an allowable simplex and, furthermore, each $(i-1)$-simplex of $\partial\xi$ is allowable. However, there is no reason to suppose that all the $(i-1)$-simplices of either $\partial(\sum_{j=1}^k c_j\sigma_j)$ or $\partial(\sum_{j=k+1}^m c_j\sigma_j)$ are allowable. There might be $(i-1)$-simplices of each of these that are not allowable but that cancel each other out in $\partial\xi$.

Here is a simple example: Consider the real line stratified as the one-dimensional manifold stratified space $\{0\} \subset \mathbb{R}$. Suppose $\bar{p}(\{0\}) = 0$. Let σ_1 be the orientation-preserving linear homeomorphism $\Delta^1 \to [-1,0] \subset \mathbb{R}$, and let σ_2 be the orientation-preserving linear homeomorphism $\Delta^1 \to [0,1] \subset \mathbb{R}$. Then for each singular simplex, $\sigma_j^{-1}(\{0\})$ lies in the 0-skeleton of Δ^1, and so each σ_j is allowable, as $\dim(\sigma_j)-\text{codim}(\{0\})+\bar{p}(\{0\}) = 1-1+0 = 0$. Furthermore, the *chain* $\sigma_1+\sigma_2$ is allowable since each simplex is allowable and $\partial(\sigma_1+\sigma_2) = \tau_1-\tau_{-1}$, where τ_α is the singular 0-simplex mapping Δ^0 to $\alpha \in \mathbb{R}$. The 0-simplices τ_1 and τ_{-1} have images in the regular stratum and so are allowable. However, neither σ_1 nor σ_2 are allowable *as chains* because each of their boundaries contains τ_0, which is not allowable, as in this case $\dim(\tau_0) - \text{codim}(\{0\}) + \bar{p}(\{0\}) = 0 - 1 + 0 = -1$.

This is clearly a difficulty when discussing excision. Suppose $K \subset U \subset X$ and the closure of K is contained in the interior of U. The idea behind the excision isomorphism $H_*(X,U) \cong H_*(X-K,U-K)$ in ordinary homology is that one can first perform subdivisions to make simplices of a chain as small as necessary and then "throw away" the simplices of the chain that intersect K; proofs of excision (e.g. in [125]) make this intuition precise. However, we must be careful when throwing away simplices not to leave exposed boundaries that are not allowable.

Similarly, in proving the existence of the Mayer–Vietoris exact sequence for a pair U,V with $U \cup V = X$, it is necessary to demonstrate that the inclusion $S_*(U) + S_*(V) \to S_*(X)$ induces an isomorphism on homology. Again the basic idea of the proof first involves subdividing chains of $S_*(X)$ to make them small enough so that every simplex fits inside one of U or V (which does not affect homology, which is preserved under subdivisions) and then showing that in fact $S_*(U) + S_*(V)$ is isomorphic to the complex of such chains of small simplices. But this requires showing that every chain made of small simplices can be written as the sum of a chain in U and a chain in V. For ordinary chains, there is no problem – just split the chain up into two chains, say one containing all the simplices that are contained completely in U and one containing all the rest (so all the simplices contained in both U and V get grouped into the chain in $S_*(U)$). But again intersection chains require much more care to make sure we are not creating unallowable boundary faces.

In this section we work through the intersection homology details. Ultimately, analogues of the ordinary homology arguments can be made to work out, but only with a good deal of care. We will first work through the PL intersection homology to get a feel for the arguments. Then we will turn to singular intersection homology, which will require a deeper investigation of singular subdivision.

4.4.1 PL Excision and Mayer–Vietoris

We begin with the PL theory. Throughout this section, X is a PL filtered space with perversity $\bar{p}$. By an allowable simplex of $\mathfrak{C}_*(X)$ we mean an allowable simplex with respect to some admissible triangulation of X and the perversity $\bar{p}$. Of course, as an element of $\mathfrak{C}_*(X)$, the simplex σ is identified with any chain obtained from σ via subdivision, but we will not mean by this language that σ is allowable *as a chain*, and so we cannot write $\sigma \in I^{\bar{p}}\mathfrak{C}_*^{\text{GM}}(X)$. If we want σ to be allowable *as a chain*, we will say so explicitly or we will write $\sigma \in I^{\bar{p}}\mathfrak{C}_*^{\text{GM}}(X)$.

The key to avoiding the aforementioned perils of breaking up chains is provided by the following lemma.

Lemma 4.4.1 *Let σ be an allowable i-simplex of X. Suppose τ is an $(i-1)$-simplex of some subdivision of σ such that for each face η of σ we have $\dim(\tau \cap \eta) < \dim(\eta)$; see Figure 4.1. Then τ is an allowable simplex.*

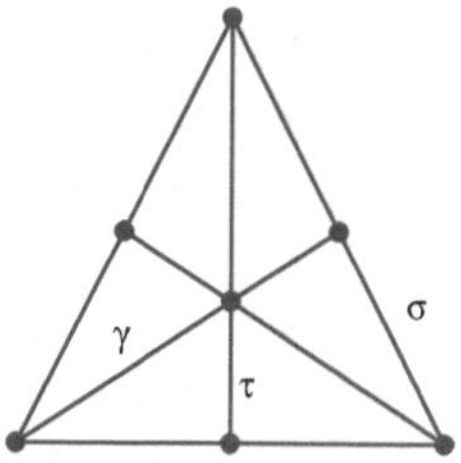

Figure 4.1 In the shown subdivision of the 2-simplex σ, the 1-simplex τ satisfies the hypotheses of the lemma, but the 1-simplex γ does not because γ has a zero-dimensional intersection with a vertex of σ.

Proof Let S be a stratum of X, and let $d = i - \operatorname{codim}(S) + \bar{p}(S)$. As σ is allowable, $\dim(\sigma \cap S) \leq d$. Since the filtration of X is compatible with the triangulation containing σ, this means that $\sigma \cap S$ must be contained in the simplicial d-skeleton of σ, i.e. the union of the faces of σ of dimension $\leq d$. We can label this simplicial skeleton σ^d. As $\tau \subset \sigma$, we then have $\tau \cap S \subset \sigma \cap \sigma^d$. In particular, $\tau \cap S \subset \tau \cap \sigma^d$. But, by assumption, the intersection of τ with any d-dimensional face of σ must have dimension $< d$, and so $\dim(\tau \cap S) < d$. In other words, $\dim(\tau \cap S) \leq i - 1 - \operatorname{codim}(S) + \bar{p}(S)$, so τ is allowable. □

So now the idea is to always break up chains in such a way that any new boundary face created has the form described in the lemma. Luckily, there are plenty such boundary faces due to our next lemma.

Lemma 4.4.2 *Let σ be a simplicial i-simplex, and let σ' be its barycentric subdivision. Let τ be an $(i-1)$-simplex of σ' that does not contain a vertex of σ. Then for each face η of σ, $\dim(\tau \cap \eta) < \dim(\eta)$.*

Proof Each such simplex must have the form $\tau = [\hat{\sigma}_1, \dots, \hat{\sigma}_i]$, where σ_j is a j-dimensional face of σ and $\hat{\sigma}_j$ is its barycenter (see e.g. [182, Section 15]). Thus $\tau \cap \sigma_k = [\hat{\sigma}_1, \dots, \hat{\sigma}_k]$, which has dimension $k - 1$, and if η is a face of σ such that $\eta \neq \sigma_i$ for any i, then $\tau \cap \eta = \emptyset$. □

We can now demonstrate PL excision. As for singular homology, the excision isomorphism will have the form $I^{\bar{p}}\mathfrak{H}^{\mathrm{GM}}_*(X - K, U - K) \xrightarrow{\cong} I^{\bar{p}}\mathfrak{H}^{\mathrm{GM}}_*(X, U)$ for subsets $K \subset U \subset X$

such that $\bar{K} \subset \mathring{U}$. The subspaces U and K do not necessarily have to be PL; see Remark 4.3.6.

Theorem 4.4.3 *Let X be a PL filtered space, and suppose $K \subset U \subset X$ are subsets such that $\bar{K} \subset \mathring{U}$. Then inclusion induces an isomorphism $I^{\bar{p}}\mathfrak{H}_*^{\mathrm{GM}}(X - K, U - K) \xrightarrow{\cong} I^{\bar{p}}\mathfrak{H}_*^{\mathrm{GM}}(X, U)$.*

Proof Let us first show that the map on intersection homology is surjective. Let $[\xi] \in I^{\bar{p}}\mathfrak{H}_i^{\mathrm{GM}}(X, U)$ be an allowable relative cycle. It will suffice to show that if ξ is a chain representing $[\xi]$ in some triangulation of X then there is a subdivision ξ''' of ξ such that $\xi''' = x + y$ with x, y allowable *chains* and such that x is supported in $X - K$ and y is supported in U. In this case, ξ and x represent the same element of $I^{\bar{p}}\mathfrak{H}_i^{\mathrm{GM}}(X, U)$ and x is in the image of $I^{\bar{p}}\mathfrak{H}_i^{\mathrm{GM}}(X - K, U - K)$.

Let us choose a triangulation T of X such that $[\xi]$ can be represented by a chain ξ in T. Suppose σ is an i-simplex of ξ. The simplex σ is covered by the two open sets $A = \sigma \cap \mathring{U}$ and $B = \sigma \cap (X - \bar{K})$. By a Lebesgue number argument (cf. the proof of [182, Theorem 16.1]), there exists a number $d(\sigma)$, such that if σ' is the $d(\sigma)$th barycentric subdivision of σ, then every i-simplex of σ' is contained in either A or B. Let $D = \max_{\sigma \in \xi}\{d(\sigma)\}$, where the maximum is taken over all i-simplices of ξ. Since ξ has only a finite number of such simplices, D is well defined.

Now, let us consider the Dth barycentric subdivision T' of T. Let ξ' denote the image of ξ under this subdivision; then $[\xi] = [\xi'] \in I^{\bar{p}}\mathfrak{H}_i^{\mathrm{GM}}(X, U)$. Let z consist of the simplices of ξ' (with their coefficients) in T' that intersect $\bar{K}$. By construction, z must be contained in $\mathring{U}$ and $\xi' - z$ is contained in $X - K$. So if we were looking to prove excision in ordinary homology, we'd be done; see Figure 4.2. However, we have no reason to expect that z and $\xi' - z$ will be allowable. This requires another level of work; see Figure 4.3.

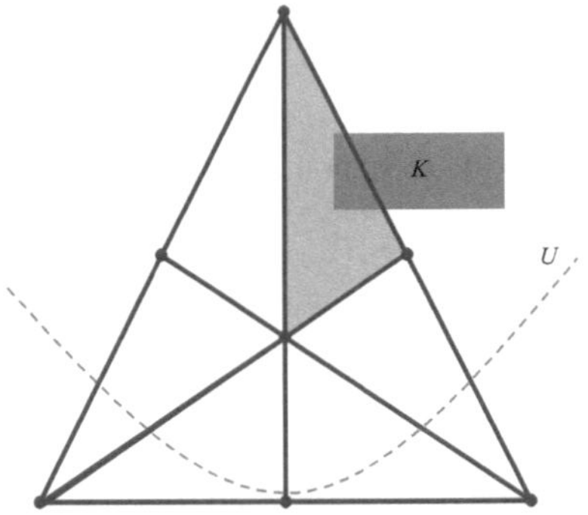

Figure 4.2 A subdivision of a 2-simplex into simplices contained in $\mathring{U}$ and simplices in $X - \bar{K}$. The simplex intersecting K is shaded. The 1-simplices bounding the shaded simplex might not all be allowable.

Let $|z|$ be the support of z, i.e. the union of the simplices of z. Since $|z| \subset \mathring{U}$, we can

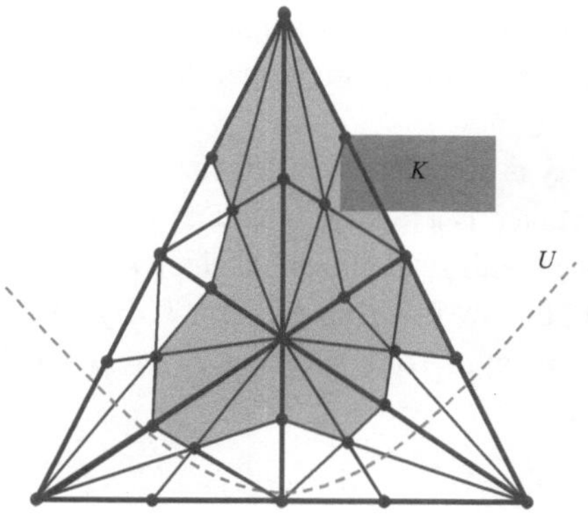

Figure 4.3 After further subdivisions, we form a chain with more shaded simplices, designed so that the new simplices in the boundary will satisfy the conditions of Lemma 4.4.2 and so be allowable. Note that possible neighboring 2-simplices are not shown, so some boundary simplices in the figure, such as the upper left 1-simplex, may not be allowable but, if not, will not be part of the boundary of the entire chain.

emulate our previous argument to obtain a further barycentric subdivision T'' such that every simplex of ξ'' is contained in $\mathring{U}$ or $X - |z|$. We will need to perform one more barycentric subdivision taking us from T'' to T'''.

Let z''' denote the images of z under the subdivision to T'''. Let y consist of the simplices (with coefficients) of ξ''' that intersect $|z| = |z'''|$. By the construction, every simplex of y intersects $|z|$ and so is not in $X - |z|$ and so is contained in $\mathring{U}$. Let $x = \xi''' - y$. Then $|x| \in X - |z| \subset X - K$. It therefore only remains to show that y is an allowable *chain*, as ξ''' is an allowable chain by Lemma 3.3.15 and so this will also imply that $x = \xi''' - y$ is an allowable chain.

Now, by the proof of Lemma 3.3.15, since each i-simplex of ξ''' is allowable, so is each i-simplex of y. So we need only check the simplices of ∂y. Some of the simplices of ∂y are simplices of $\partial\xi'''$, and so these are allowable. The only simplices of concern, then, are the simplices that are in ∂y but not in $\partial\xi'''$. In other words, these are simplices that must occur in canceling pairs in ∂y and ∂x. Let τ be such an $(i-1)$-simplex occurring in ∂y and ∂x. We claim that τ cannot contain a vertex of the triangulation T''. It will follow by Lemmas 4.4.2 and 4.4.1 that τ is allowable.

To prove the claim, we first note that τ cannot contain a vertex of T'' in $|z| = |z'''|$ because $|x| \cap |z| = \emptyset$ by construction. So suppose that τ contains a vertex of T'' not in $|z'''|$. Let σ''' be any i-simplex of T''' with τ as a face. Then σ''' also contains v. Suppose σ''' is contained in the i-simplex σ'' of T''. As $|z|$ is a subcomplex of T'' and $v \notin |z|$, the intersection of $|z|$ with σ'' must be in the opposite face of σ'' from v, i.e. the $i-1$ face spanned by the vertices of σ'' not including v. But since we are in a barycentric subdivision, σ''' cannot intersect both v and this opposite face, and so σ''' cannot be contained in y. But this contradicts τ being a simplex of ∂y. So τ cannot contain a vertex of T''. This completes the proof of the claim and hence the argument that the inclusion map induces a surjective map on intersection homology.

But the proof of injectivity is completely analogous! Suppose ξ represents an element of $I^{\bar{p}}\mathfrak{H}_i^{\mathrm{GM}}(X-K, U-K)$ and that ξ is a relative boundary in X, i.e. there is an allowable chain ζ in X such that $\partial\zeta = \xi + \rho$, with ρ an allowable chain supported in U. Suppose all these chains are represented simplicially in a triangulation T. By an argument analogous to that above, we can find a subdivision ζ''' of ζ such that $\zeta''' = \mu + \nu$, $|\nu| \subset \mathring{U}$, $\mu \subset X - K$, and μ and ν both allowable. Then

$$\partial\mu = \partial\zeta''' - \partial\nu = \xi''' + \rho''' - \partial\nu.$$

Both ρ''' and $\partial\nu$ are contained in U, and, in fact, since μ and ξ''' are contained in $X-K$, then so is $\rho''' - \partial\nu$, which must then be in $U-K$. So ξ''' is a relative boundary in $(X-K, U-K)$, and so represents 0 in $I^{\bar{p}}\mathfrak{H}_i^{\mathrm{GM}}(X-K, U-K)$. □

Similar arguments allow us to formulate a Mayer–Vietoris sequence.

Theorem 4.4.4 *Suppose $U, V \subset X$ such that $\mathring{U} \cup \mathring{V} = X$. Then there is an exact Mayer–Vietoris sequence*

$$\to I^{\bar{p}}\mathfrak{H}_i^{\mathrm{GM}}(U\cap V) \to I^{\bar{p}}\mathfrak{H}_i^{\mathrm{GM}}(U) \oplus I^{\bar{p}}\mathfrak{H}_i^{\mathrm{GM}}(V) \to I^{\bar{p}}\mathfrak{H}_i^{\mathrm{GM}}(X) \to I^{\bar{p}}\mathfrak{H}_{i-1}^{\mathrm{GM}}(U\cap V) \to .$$

Here U, V, and $U \cap V$ inherit their filtrations and perversities from X. There is similarly a Mayer–Vietoris sequence in reduced intersection homology.[6]

Proof The standard arguments (see e.g. [182, Section 33] or [125, Section 2.2]) demonstrate that there is a short exact sequence

$$0 \to I^{\bar{p}}\mathfrak{C}_i^{\mathrm{GM}}(U\cap V) \xrightarrow{\phi} I^{\bar{p}}\mathfrak{C}_i^{\mathrm{GM}}(U) \oplus I^{\bar{p}}\mathfrak{C}_i^{\mathrm{GM}}(V) \xrightarrow{\psi} I^{\bar{p}}\mathfrak{C}_i^{\mathrm{GM}}(U) + I^{\bar{p}}\mathfrak{C}_i^{\mathrm{GM}}(V) \to 0$$

To explain the notation, we let $I^{\bar{p}}\mathfrak{C}_*^{\mathrm{GM}}(U) + I^{\bar{p}}\mathfrak{C}_*^{\mathrm{GM}}(V)$ be the subcomplex of $I^{\bar{p}}\mathfrak{C}_*^{\mathrm{GM}}(X)$ generated by allowable chains supported in U or in V, and let $\mathfrak{i}_{A,B}$ stand for the inclusion map of spaces $A \hookrightarrow B$. Then we let

$$\phi(\xi) = \left(\mathfrak{i}_{U\cap V,U}(\xi), -\mathfrak{i}_{U\cap V,V}(\xi)\right)$$

and

$$\psi(\xi,\eta) = \mathfrak{i}_{U,X}(\xi) + \mathfrak{i}_{V,X}(\eta).$$

In the reduced case we simply extend this short exact sequence to degree -1, where the short exact sequence becomes

$$0 \to \mathbb{Z} \xrightarrow{\phi} \mathbb{Z}\oplus\mathbb{Z} \xrightarrow{\psi} \mathbb{Z} \to 0$$

[6] Due to our conventions about reduced intersection homology, the reduced sequence will only be exact at $I^{\bar{p}}\mathfrak{H}_0^{\mathrm{GM}}(X)$ if $I^{\bar{p}}\mathfrak{H}_0^{\mathrm{GM}}(U\cap V) \neq 0$; see Section 4.3.3.

with $\phi(a) = (a, -a)$ and $\psi(a, b) = a + b$.

The short exact sequence yields a long exact homology sequence. What needs to be shown is that the inclusion map $\psi : I^{\bar{p}}\mathfrak{C}_i^{\mathrm{GM}}(U) + I^{\bar{p}}\mathfrak{C}_i^{\mathrm{GM}}(V) \to I^{\bar{p}}\mathfrak{C}_i^{\mathrm{GM}}(X)$ yields an isomorphism on homology. The argument in the reduced case is the same, so we focus on the unreduced case.

The proof is basically the same as the argument we used to prove excision. Notice that if $x \in X$ is contained in the closure of $X - V$, it cannot be contained in the interior of V, so it must be contained in the interior of U. Therefore $\overline{X - V} \subset \mathring{U}$. Thus the argument of Theorem 4.4.3 shows how we can take an allowable cycle ξ in X, subdivide it to an appropriate ξ''', and then break it into two allowable pieces $\xi''' = x + y$, where y is contained in $\mathring{U}$ and x is contained in $X - \overline{X - V} \subset V$. So $y \in I^{\bar{p}}\mathfrak{C}_i^{\mathrm{GM}}(U)$ and $x \in I^{\bar{p}}\mathfrak{C}_i^{\mathrm{GM}}(V)$, and this shows that ψ is surjective on homology.

Similarly, if $x + y$ is a cycle in $I^{\bar{p}}\mathfrak{C}_i^{\mathrm{GM}}(U) + I^{\bar{p}}\mathfrak{C}_i^{\mathrm{GM}}(V)$ that bounds a chain ζ in X, then we can similarly split up an appropriate ζ''' as $\zeta''' = \mu + \nu$ and still have $\partial\zeta''' = \partial(\mu + \nu) = x''' + y'''$, so $x + y$ is homologically trivial in $I^{\bar{p}}\mathfrak{C}_i^{\mathrm{GM}}(U) + I^{\bar{p}}\mathfrak{C}_i^{\mathrm{GM}}(V)$. So ψ is injective on homology. □

4.4.2 Singular Subdivision, Excision, and Mayer–Vietoris

We now turn to the singular versions of excision and the Mayer–Vietoris sequence. The basic ideas are similar to those we have already explored in the PL case, but there are additional technicalities. For one thing, we have not yet discussed subdivision of singular simplices, which will be a necessary, though technically complex, component of the proof. We turn to that next, though the reader not particularly interested in the technical detail may just want to skim ahead, noting Theorem 4.4.18, which concerns excision, and Theorem 4.4.19, which concerns Mayer–Vietoris sequences. Following the proof of Theorem 4.4.19, we resume with some important examples that we encourage the reader not to skip.

Some of the results in this section are based on [85].

(a) Singular Subdivision

What should it mean for a singular chain to have a subdivision? The basic idea is that if $\sigma : \Delta^i \to X$ is a singular simplex, then we will build a singular chain $\hat{\sigma}$, which we shall call a *singular subdivision of* σ, patterned upon a simplicial subdivision $\hat{\Delta}^i$ of Δ^i. Roughly speaking, $\hat{\sigma}$ will be the sum over the restriction of σ to each of the i-simplices of $\hat{\Delta}^i$. Then if $\xi = \sum_a n_a\sigma_a$ is a singular chain, and if we think of each σ_a as a map $\sigma_a : \Delta^i_a \to X$, where Δ^i_a is simply an indexed copy of Δ^i, then we can construct a singular subdivision $\hat{\xi}$ of ξ via singular subdivisions $\hat{\sigma}_a$ of each σ_a, each possibly patterned on a different subdivision $\hat{\Delta}^i_a$ of Δ^i_a. However, we must take care to ensure that there is compatibility among simplices that share boundaries. This will be necessary so that the "boundary of the subdivision is the subdivision of the boundary."

To distinguish the domain polyhedron Δ^i from the singular simplices that will arise, we refer to Δ^i, or more generally any simplex in a simplicial complex, as a "geometric simplex." We will assume that the standard geometric simplex Δ^i is given an ordering of its vertices and write $\Delta^i = [v_0, \dots, v_i]$ with $v_j < v_k$ if and only if $j < k$.

There will be three steps involved in constructing a singular subdivision of a singular simplex. First we will perform a simplicial subdivision of Δ^i to a simplicial complex $\hat{\Delta}^i$, which yields a subdivision chain map $C_*(\Delta^i) \to C_*(\hat{\Delta}^i)$ as discussed in Section 3.3.1. Then we will use an ordering on the vertices of $\hat{\Delta}^i$ to construct a chain map[7] $C_*(\hat{\Delta}^i) \to S_*(|\hat{\Delta}^i|) = S_*(|\Delta^i|)$. We will define the *singular subdivision of* Δ^i to be the image of the canonical generator of $C_i(\Delta^i)$ under these chain maps. Finally, if $\sigma : |\Delta^i| \to X$ is a singular simplex, we will apply the induced chain map $\sigma : S_*(|\Delta^i|) \to S_*(X)$. The image of the singular subdivision of Δ^i under the chain map σ will be our $\hat{\sigma}$.

To carry out this program, we first need to discuss how an ordering on the vertices of a simplicial complex K provides a chain map $C_*(K) \to S_*(|K|)$.

Singular Chains from Ordered Simplicial Complexes. Suppose that K is a simplicial complex with an ordering on its vertices. Such an ordering can always be found by the Well-Ordering Theorem. In fact, we only need a partial ordering that restricts to a total ordering on each simplex of K, and in practice this is often what we will have. Recall from our review in Section 3.2 that the simplicial chain group $C_i(K)$ is generated by the i-simplices of K, each with a fixed orientation, where an orientation of a simplex of K is just an equivalence class of orderings on its vertices. Furthermore, if σ is such an oriented simplex, then we identify the oriented simplex that has the same underlying geometric simplex but the opposite orientation with $-\sigma$ as an element of $C_i(K)$. But now if each geometric simplex of K has a total ordering given on its vertices, that ordering provides a canonical orientation, and so we have canonical generators of $C_i(K)$. We can use these to define a chain map $\phi : C_*(K) \to S_*(|K|)$.

So now let $\tau = [w_0, \dots, w_i]$ be such a generator, i.e. τ is an oriented simplex of K with the unique vertex ordering such that $w_j < w_k$ if and only if $j < k$. Then we can let $\phi(\tau)$ be the singular i-simplex $\phi(\tau) : |\Delta^i| \to |K|$ that is the linear embedding determined by $\phi(\tau)(v_j) = w_j$ for all j. In other words, $\phi(\tau)$ is just the embedding that takes the standard i-simplex $|\Delta^i| = |[v_0, \dots, v_i]|$ linearly homeomorphically onto τ in K in a manner preserving the vertex ordering. This construction defines ϕ on the generators of $C_i(K)$, and, since $C_i(K)$ is free on such generators, extending linearly provides a homomorphism $C_i(K) \to S_i(|K|)$ for each i. But these homomorphisms are also compatible with the boundary maps: We

[7] We will be a bit careful here by writing $|\Delta^i|$ for the underlying space of Δ^i. This is because we also treat Δ^i as a simplicial complex within the current discussion.

know[8]

$$\partial\tau = \partial[w_0,\dots,w_i] = \sum_{k=0}^{i}(-1)^k[w_0,\dots,\hat{w}_k,\dots,w_i],$$

and each $[w_0,\dots,\hat{w}_k,\dots,w_i]$ is again oriented compatibly with the ordering of the vertices of K. But also by the definition of the singular chain complex we have $\partial(\phi(\tau)) = \sum_{k=0}^{i}(-1)^k\phi(\tau)\circ f_k$, where f_k is the face inclusion $f_k : \Delta^{i-1} \to \Delta^i$ that takes Δ^{i-1} to the face of Δ^i omitting v_k by a linear homeomorphism that is order-preserving on the vertices. Note that $\phi(\tau)\circ f_k$ is precisely $\phi([w_0,\dots,\hat{w}_k,\dots,w_i])$. So

$$\begin{aligned}\partial(\phi(\tau)) &= \sum_{k=0}^{i}(-1)^k\phi(\tau)\circ f_k\\ &= \sum_{k=0}^{i}(-1)^k\phi([w_0,\dots,\hat{w}_k,\dots,w_i]) = \phi\left(\sum_{k=0}^{i}(-1)^k[w_0,\dots,\hat{w}_k,\dots,w_i]\right)\\ &= \phi(\partial\tau).\end{aligned}$$

Thus ϕ is a chain map $C_*(K) \to S_*(|K|)$.

For later reference, we state our conclusion as a proposition:

Proposition 4.4.5 *Suppose that K is a simplicial complex with a partial ordering on its vertices that restricts to a total ordering on each simplex. Then there is a chain map $\phi : C_*(K) \to S_*(|K|)$ defined so that if $\tau = [w_0,\dots,w_i]$ is an i-simplex of K with vertices ordered as written and similarly $\Delta^i = [v_0,\dots,v_i]$, then $\phi(\tau) : |\Delta^i| \to |K|$ is the linear embedding onto τ in K determined by $\phi(\tau)(v_j) = w_j$ for all j.*

The map ϕ induces an isomorphism from simplicial homology to singular homology. We will not provide a proof here; see [220, Section 4.4 and Theorem 4.6.8].

While we will not need it quite yet, we observe that the map ϕ of Proposition 4.4.5 restricts to a chain map of intersection chain complexes:

Corollary 4.4.6 *Suppose that K is a* filtered *simplicial complex, i.e. that $|K|$ is filtered such that each skeleton of the filtration is a subcomplex of K. Suppose further that K possesses a partial ordering on its vertices that restricts to a total ordering on each simplex. Let $\bar{p}$ be a perversity on $|K|$. Then ϕ restricts to a chain map $\phi : I^{\bar{p}}C_*^{\mathrm{GM}}(K) \to I^{\bar{p}}S_*^{\mathrm{GM}}(|K|)$.*

Proof As we already know by Proposition 4.4.5 that ϕ is a chain map and as $I^{\bar{p}}C_*^{\mathrm{GM}}(K) \subset C_*(K)$, it suffices to verify that ϕ takes $\bar{p}$-allowable simplices to $\bar{p}$-allowable singular simplices. So suppose σ is a $\bar{p}$-allowable i-simplex of K. Then for any stratum S of $|K|$, we have $\dim(\sigma\cap S) \le i - \operatorname{codim}(S) + \bar{p}(S)$. As each skeleton of the filtration is a subcomplex of K, each stratum is a union of interiors of simplices of K, and so $\sigma\cap S$ is a union of interiors

[8] Be careful to recall that the hat in such expressions denotes omission, not subdivision. We apologize that the symbol ^ is used for two purposes in this section – to denote omission of vertices in boundary formulas but otherwise to denote subdivisions. We hope the contexts are sufficiently different to avoid confusion.

of faces of σ. Therefore, the assumption $\dim(\sigma \cap S) \le i - \operatorname{codim}(S) + \bar{p}(S)$ implies that $\sigma \cap S$ lies in some k-skeleton of σ with $k \le i - \operatorname{codim}(S) + \bar{p}(S)$. By definition, the singular simplex $\phi(\sigma)$ is the linear embedding of Δ^i onto σ determined by the vertex ordering. In particular, $\phi(\sigma)$ is a simplicial isomorphism onto σ. So $\phi(\sigma)^{-1}(S) = \phi(\sigma)^{-1}(\sigma \cap S)$ must be contained in the k-skeleton of Δ^i. Therefore, $\phi(\sigma)$ is $\bar{p}$-allowable. □

Singular Subdivision of Singular Simplices. Now, let us consider Δ^i as a simplicial complex with ordered vertices $[v_0, \dots, v_i]$, and let $\hat{\Delta}^i$ be a simplicial subdivision of Δ^i. We suppose a partial ordering on the vertices of $\hat{\Delta}^i$ such that the vertices of each simplex of $\hat{\Delta}^i$ are totally ordered. It will be convenient for what will come later to always choose our partial orderings so that they have the following additional properties:

1. The ordering on the vertices of Δ^i is preserved.
2. For each vertex w of $\hat{\Delta}^i$, let $d(w)$ be the smallest dimension of a face of Δ^i that contains w. This is equivalent to saying that w is contained in the interior of a face of dimension $d(w)$. We now require of our ordering of the vertices of $\hat{\Delta}^i$ that if $d(w_1) < d(w_2)$ then $w_1 < w_2$.

Definition 4.4.7 Let Δ^i be the standard i-simplex, and let $\hat{\Delta}^i$ be a simplicial subdivision of Δ^i given a partial ordering on its vertices satisfying the above conditions. Let $\lambda : C_*(\Delta^i) \to C_*(\hat{\Delta}^i)$ be the subdivision chain map as in Lemma 3.3.1, and let $\phi : C_*(\hat{\Delta}^i) \to S_*(|\hat{\Delta}^i|) = S_*(|\Delta^i|)$ be our chain map from simplicial to singular chains obtained using the ordering of the vertices. Let $\mathfrak{o}$ denote the generator of $C_i(\Delta^i)$ consistent with the orientation given by the ordering of the vertices of Δ^i. We will call $\phi(\lambda(\mathfrak{o}))$ a *singular subdivision of* Δ^i *based on the subdivision* $\hat{\Delta}^i$.

If $\sigma : |\Delta^i| \to X$ is a singular simplex, then we define the *singular subdivision* $\hat{\sigma}$ *of* σ *based on* $\hat{\Delta}^i$ to be the image of $\phi(\lambda(\mathfrak{o}))$ under the chain map $\sigma : S_*(|\Delta^i|) \to S_*(X)$ induced by σ; see Figure 4.4.

Suppose $\hat{\Delta}^i$ is a subdivision of Δ^i and that we let $\{\delta^i_j\}$ be the collection of i-simplices of $\hat{\Delta}^i$. Further, suppose we let $\mathfrak{i}_j : \Delta^i \to \Delta^i$ be the unique order-preserving linear embedding that takes Δ^i onto δ^i_j. Then from Lemma 3.3.1 and the definitions, we can describe the singular subdivision of Δ^i based on the subdivision $\hat{\Delta}^i$ explicitly as the singular chain $s = \sum \operatorname{sgn}(\mathfrak{i}_j)\mathfrak{i}_j$, where $\operatorname{sgn}(\mathfrak{i}_j)$ is 1 if the orientation of δ^i_j induced by the ordering of its vertices agrees with the orientation of Δ^i and -1 if it disagrees. If $\sigma : |\Delta^i| \to X$ is a singular simplex, the singular subdivision of σ based on $\hat{\Delta}^i$ is then $\hat{\sigma} = \sigma(s) = \sum \operatorname{sgn}(\mathfrak{i}_j)\sigma \circ \mathfrak{i}_j$.

Let f_k again be the order-preserving linear embedding $\Delta^{i-1} \to \Delta^i$ that takes Δ^{i-1} to the $i-1$ face of Δ^i omitting v_k. If $\sigma : |\Delta^i| \to X$ is a singular simplex, then the compositions $\tau_k = \sigma \circ f_k$ are the faces of σ. If $\hat{\Delta}^i$ is a subdivision of Δ^i, it determines a subdivision of each face of Δ^i by restriction, and for each k this induces a subdivision of Δ^{i-1} via the simplicial homeomorphism f_k between Δ^{i-1} and the kth face of Δ^i. Based on these subdivisions of Δ^{i-1}, we can form the singular subdivisions $\hat{\tau}_k$ just as $\hat{\sigma}$ was defined above.

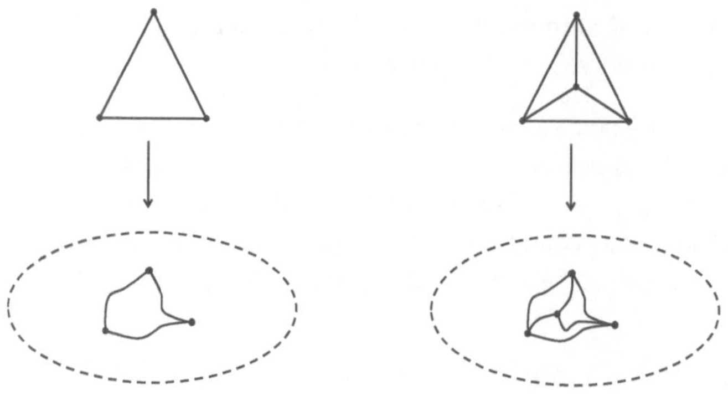

Figure 4.4 A singular simplex σ (left) and a singular subdivision $\hat{\sigma}$ (right). We think of the figure on the right as being the sum of compositions of σ with (signed) embeddings $\Delta^i \to \Delta^i$.

Lemma 4.4.8 *With the notation just established, we have* $\partial\hat{\sigma} = \sum_{k=0}^{i}(-1)^k \hat{\tau}_k$.

Proof We have the following commutative diagram using the subdivision $\hat{\Delta}^i$ of Δ^i for the top row and its restriction to the kth face for the bottom row:

$$\begin{array}{ccccccc}
C_*(\Delta^i) & \xhookrightarrow{\lambda} & C_*(\hat{\Delta}^i) & \xhookrightarrow{\phi} & S_*(|\hat{\Delta}^i|) & \xrightarrow{\sigma} & S_*(X) \\
\uparrow{\scriptstyle f_k} & & \uparrow{\scriptstyle f_k} & & \uparrow{\scriptstyle f_k} & & \uparrow{\scriptstyle =} \\
C_*(\Delta^{i-1}) & \xhookrightarrow{\lambda} & C_*(\hat{\Delta}^{i-1}) & \xhookrightarrow{\phi} & S_*(|\hat{\Delta}^{i-1}|) & \xrightarrow{\tau_k} & S_*(X).
\end{array}$$

Now let $\mathfrak{v}^i = [v_0, \ldots, v_i]$ be the generator of $C_i(\Delta^i)$ determined by the ordering of the vertices of Δ^i, and let $\mathfrak{v}_k^{i-1}$ be the analogous generator of $C_{i-1}(\Delta_k^{i-1})$, where Δ_k^{i-1} is an indexed copy of Δ^{i-1}. Then we have $f_k(\mathfrak{v}_k^{i-1}) = [v_0, \ldots, \hat{v}_k, \ldots, v_i]$, and so $\partial\mathfrak{v}^i = \sum_{k=0}^{i}(-1)^k f_k(\mathfrak{v}_k^{i-1})$. From the above diagram, the definitions, and the fact that λ, ϕ, and σ are all chain maps, we have

$$\begin{aligned}
\partial\hat{\sigma} &= \partial\sigma\phi\lambda(\mathfrak{v}^i) = \sigma\phi\lambda\partial(\mathfrak{v}^i) = \sigma\phi\lambda\left(\sum_{k=0}^{i}(-1)^k f_k(\mathfrak{v}_k^{i-1})\right) \\
&= \sum_{k=0}^{i}(-1)^k\sigma\phi\lambda f_k(\mathfrak{v}_k^{i-1}) = \sum_{k=0}^{i}(-1)^k\tau_k\phi\lambda(\mathfrak{v}_k^{i-1}) \\
&= \sum_{k=0}^{i}(-1)^k\hat{\tau}_k,
\end{aligned}$$

as desired. □

Singular Subdivision of Singular Chains. Next we want to extend from subdivision of singular simplices to subdivision of singular chains.

Definition 4.4.9 Let ξ be a singular i-chain of X given by $\xi = \sum_a n_a\sigma_a$ with $n_a \in \mathbb{Z}$ and each σ_a a singular i-simplex of X. For each σ_a, let Δ^i_a represent a copy of the standard i-simplex so that $\sigma_a : |\Delta^i_a| \to X$. We say that the subdivisions $\hat{\Delta}^i_a$ of Δ^i_a are *compatible with respect to* ξ if the following condition holds: Suppose that σ_a and σ_b are singular simplices of ξ and that they have $i-1$ faces $\tau_{a,j}$ and $\tau_{b,k}$ such that $\tau_{a,j} = \tau_{b,k}$ as singular simplices, i.e.

$$\tau_{a,j} : |\Delta^{i-1}| \overset{f_j}{\hookrightarrow} |\Delta^i| \xrightarrow{\sigma_a} X \quad \text{equals} \quad \tau_{b,k} : |\Delta^{i-1}| \overset{f_k}{\hookrightarrow} |\Delta^i| \xrightarrow{\sigma_b} X,$$

where the first map in each composition is the order-preserving embedding of the appropriate face. Then the singular subdivisions $\hat{\tau}_{a,j}$ and $\hat{\tau}_{b,k}$ based on the restrictions of the subdivisions $\hat{\Delta}^i_a$ and $\hat{\Delta}^i_b$ are required to be equal as chains. Note that a may equal b so this condition may impose compatibilities among the subdivisions of the faces of the same singular i-simplex.

When such compatible subdivisions of the Δ^i_a are given, we denote the associated singular subdivisions of the σ_a by $\hat{\sigma}_a$ and we call $\hat{\xi} = \sum_a n_a\hat{\sigma}_a$ a *singular subdivision* of ξ.

The reason for the compatibility requirement is as follows. Suppose that $\xi = \sum n_a\sigma_a$ is a singular i-chain. Then by Lemma 4.4.8 we have

$$\partial\hat{\xi} = \sum_a n_a\partial\hat{\sigma}_a = \sum_a n_a \sum_k (-1)^k \hat{\tau}_{a,k}.$$

On the other hand, suppose we consider

$$\partial\xi = \sum_a n_a\partial\sigma_a = \sum_a n_a \sum_k (-1)^k \tau_{a,k}.$$

We would like to be able to say that $\partial\hat{\xi} = (\partial\xi)\hat{}$, where $(\partial\xi)\hat{}$ indicates a singular subdivision of $\partial\xi$, by taking each $\tau_{a,k}$ to the singular subdivision $\hat{\tau}_{a,k}$, which will be an $(i-1)$-chain. What the compatibility condition tells us is that if $\tau_{a,k}$ is a face of the singular simplex σ_a of ξ, then the $\hat{\tau}_{a,k}$ obtained using the restriction of the subdivision $\hat{\Delta}^i_a$ to its kth face is the same as the singular subdivision that would be determined using the subdivision of any singular i-simplex of ξ that has the singular $(i-1)$-simplex $\tau_{a,k}$ (though labeled differently) as a face. So the compatibility lets us define the $\hat{\tau}_{a,k}$, and hence $(\partial\xi)\hat{}$, unambiguously and in such a way that $\partial\hat{\xi} = (\partial\xi)\hat{}$.

Note that we have only defined singular subdivision of a chain. We do not claim that anything we have done so far necessarily results in a chain map $S_*(X) \to S_*(X)$. Nonetheless, there are such examples:

Example 4.4.10 The standard example of singular subdivision is given by the *barycentric subdivision* of singular chains for which we let $\hat{\Delta}^i$ be the barycentric subdivision of Δ^i for all singular simplices of all dimensions; see Example B.1.13 or, more generally, [182, Section 31]. In this case, there is a natural partial ordering on the vertices of $\hat{\Delta}^i$: if v_τ denotes the barycenter of the face τ of Δ^i, then we let $v_{\tau_1} < v_{\tau_2}$ if $\dim(\tau_1) < \dim(\tau_2)$. For each fixed dimension d, each simplex of $\hat{\Delta}^i$ has at most one vertex that is the barycenter of a face of Δ^i of dimension d, so this partial ordering gives a total ordering on each simplex of $\hat{\Delta}^i$. The uniformity of the construction over all dimensions ensures compatibility among simplices in any chain, and so in this case we have a subdivision chain map $T : S_*(X) \to S_*(X)$. Iterations of T are then also subdivision chain maps. Similarly, we can find natural vertex orderings for generalized barycentric subdivisions, in which not every face is subdivided at each step (see [182, Section 16]), though in this case it takes more care to ensure compatibility among simplices. △

Remark 4.4.11 A slightly more general concept, which will be useful below, are chain maps $T : S_*(X) \to S_*(X)$ such that for each singular simplex σ the chain $T(\sigma)$ is a singular subdivision of σ. The reason this is more general is that for a chain ξ we do not require that $T(\xi)$ be a singular subdivision in the sense defined above. The difference is that we do not require here complete compatibility among all common faces of simplices. It is a good exercise to think through why such compatibility is not forced by T being a chain map!

We will see below in Corollary 4.4.15 that the restriction of such a chain map to $T : I^{\bar{p}}S^{\text{GM}}_*(X) \to I^{\bar{p}}S^{\text{GM}}_*(X)$ is chain homotopic to the identity. The advantage of our less general definition above of a singular subdivision of a chain is that it does not require the existence of such a subdivision chain map defined for all chains. We will show in Proposition 4.4.14 that an intersection cycle and any singular subdivision represent the same intersection homology class.

Remark 4.4.12 We note for future use that the idea of constructing a singular subdivision of a singular simplex $\sigma : |\Delta^i| \to X$ based on a subdivision $\hat{\Delta}^i$ of Δ^i can be extended to define singular chains starting with any simplicial complex K, a subdivision K' of K given a partial ordering on its vertices that restricts to a total ordering on each simplex, and a map $f : |K| \to X$. Given this information and an i-chain $\xi \in C_i(K)$, then we can apply analogues of λ and ϕ together with f to obtain a chain in $S_i(X)$, i.e. we use the composition

$$C_*(K) \xrightarrow{\lambda} C_*(K') \xrightarrow{\phi} S_*(|K'|) = S_*(|K|) \xrightarrow{f} S_*(X).$$

In particular, we will utilize below singular chains based on triangulations of prisms in order to create homologies.

Singular Subdivision of Intersection Chains. Let us now return to intersection homology. We would like the singular subdivision of a $\bar{p}$-allowable chain to be itself allowable:

Lemma 4.4.13 *Let $\hat{\xi}$ be a singular subdivision of the i-chain $\xi \in I^{\bar{p}}S^{\text{GM}}_i(X)$. Then $\hat{\xi} \in I^{\bar{p}}S^{\text{GM}}_i(X)$.*

Proof By assumption, for each σ in ξ and each stratum S of X, we have $\sigma^{-1}(S)$ contained in the $(i - \operatorname{codim}(S) + \bar{p}(S))$-skeleton of Δ^i. Similarly, for each $(i-1)$-simplex τ in $\partial\xi$, we have $\tau^{-1}(S)$ contained in the $(i - 1 - \operatorname{codim}(S) + \bar{p}(S))$-skeleton of Δ^{i-1}. Now $\hat{\xi}$ is composed of singular i-simplices of the form $\sigma\mathfrak{i}_j$ where $\mathfrak{i}_j : \Delta^i \to \Delta^i$ is a linear embedding. Let us denote the image $\mathfrak{i}_j(\Delta^i)$ by δ^i_j. Then the intersection of δ^i_j with an r-dimensional face F of Δ^i will be the face f of δ^i_j spanned by the vertices of δ^i_j in F. Clearly, we must have $\dim(f) \leq r$. Therefore, the inverse image under $\mathfrak{i}_j^{-1}$ of the r-skeleton of Δ^i must lie in the r-skeleton of the domain Δ^i. So now

$$\begin{aligned}(\sigma\mathfrak{i}_j)^{-1}(S) &= \mathfrak{i}_j^{-1}\sigma^{-1}(S)\\ &\subset \mathfrak{i}_j^{-1}(\{(i - \operatorname{codim}(S) + \bar{p}(S))\text{-skeleton of } \Delta^i\})\\ &\subset \{(i - \operatorname{codim}(S) + \bar{p}(S))\text{-skeleton of } \Delta^i\}.\end{aligned}$$

Thus each $\sigma\mathfrak{i}_j$ is allowable, and $\hat{\xi}$ is composed of allowable i-simplices. Similarly, the simplices in $\partial\hat{\xi}$ are allowable since we have seen that $\partial\hat{\xi}$ is a singular subdivision of $\partial\xi$, so the above arguments hold analogously for $\partial\xi$. □

Our goal now is to prove the following proposition. The proof, unfortunately, is a bit long and technical, and we wouldn't blame the reader for skipping it at a first pass.

Proposition 4.4.14 *Let ξ be a $\bar{p}$-allowable chain representing an element of $I^{\bar{p}}H_i^{\mathrm{GM}}(X,A)$, where A is a possibly empty subset of X. Then ξ is intersection homologous to any singular subdivision $\hat{\xi}$, so ξ and $\hat{\xi}$ represent the same element of $I^{\bar{p}}H_i^{\mathrm{GM}}(X,A)$.*

Proof Ultimately, we want to make a prism argument, so we need to construct the relevant prisms for the current situation. Therefore, as a first step in the proof, we will show that if $\hat{\Delta}^i$ is a subdivision of Δ^i then we can construct the following objects in a standardized way:

1. A triangulation K of the prism $[0,1] \times |\Delta^i|$ such that $\{0\} \times |\Delta^i|$ is triangulated as a subcomplex identical to Δ^i and $\{1\} \times |\Delta^i|$ is triangulated as a subcomplex identical to $\hat{\Delta}^i$.
2. For each face $F = [v_{j_0}, \ldots, v_{j_m}]$ of $\Delta^i = [v_0, \ldots, v_i]$ of dimension m, we define a chain $\Gamma_F \in C_{m+1}(K)$, which must satisfy the following property. Let $C_*(F)$ be the oriented simplicial chain complex associated to F, treating F as the simplicial complex consisting of F and its faces, and let $C_*(\hat{F})$ be the chain complex associated to the subdivision of F determined by restricting $\hat{\Delta}^i$ to F. We identify $C_*(F)$ and $C_*(\hat{F})$ with subcomplexes of $C_*(K)$ via the identifications in the definition of K. Let $\mathfrak{o}_F$ be the generator of $C_m(F)$ given by the ordering of the vertices of F. Let $\lambda_F : C_*(F) \to C_*(\hat{F})$ be the subdivision map as in Lemma 3.3.1. Let F_k be the face of F obtained by removing the kth vertex. Then the chain Γ_F must satisfy
$$\partial\Gamma_F = \lambda_F(\mathfrak{o}_F) - \mathfrak{o}_F - \sum_{k=0}^{m}(-1)^k\Gamma_{F_k}.$$

We will be most interested in the chain Γ_F where F is the unique top-dimensional face of Δ^i, though we need to consider all faces due to the inductive nature of the construction. In this case, stated roughly, what we have claimed is that we can triangulate the prism

$[0,1] \times |\Delta^i|$ and give it an $(i+1)$-chain whose boundary is the difference between the standard generator of $C_i(\Delta^i)$ at the bottom and its subdivision determined by the simplicial subdivision $\hat{\Delta}^i$ at the top, modulo a chain supported in $[0,1] \times |\partial\Delta^i|$. We are also claiming that we can perform the construction in a certain standardized way that will eventually lead us to compatibility along boundaries of simplices in chains.

So let us begin by constructing the triangulation K of $[0,1] \times |\Delta^i|$. We have already declared that $\{0\} \times |\Delta^i|$ will be triangulated as Δ^i, i.e. it is just the standard simplex (as a simplicial complex), and that we triangulate $\{1\} \times |\Delta^i|$ according to the subdivision $\hat{\Delta}^i$. So we have given a triangulation of the "top and bottom" of each prism $[0,1] \times |\Delta^i|$. We extend this triangulation inductively on the dimensions of faces of Δ^i:

1. For each 0-simplex v of Δ^i, we triangulate $[0,1] \times v$ as $\bar{c}[(\{0\} \times v) \amalg (\{1\} \times v)]$, where $\bar{c}$ represents the closed cone, with the cone vertex appearing at $\{1/2\} \times v$.
2. Now suppose for a face F of Δ^i we have inductively constructed a triangulation of $[0,1] \times |\partial F|$. Together with the trivial triangulation of $\{0\} \times |F|$ as a face of Δ^i and the triangulation of $\{1\} \times |F|$ as a subcomplex of $\hat{\Delta}^i$, we then have a triangulation L_F of $(\{0\} \times |F|) \cup (\{1\} \times |F|) \cup ([0,1] \times |\partial F|)$, which is the boundary of the prism $I \times |F|$. Now triangulate $[0,1] \times |F|$ by taking the closed cone on L_F with the new cone vertex positioned at $\{1/2\} \times \{b_F\}$, where b_F is the barycenter of F.

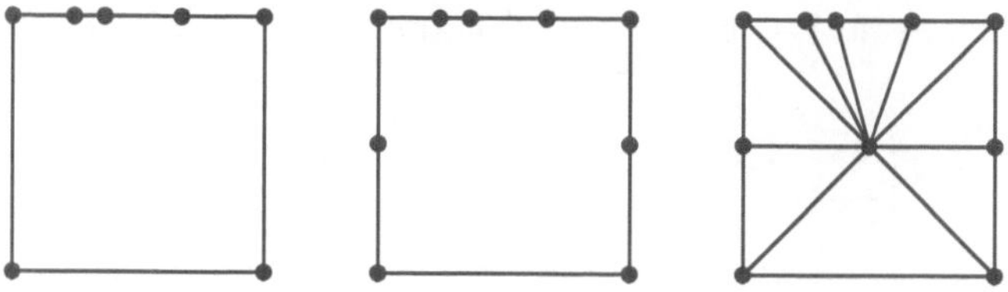

Figure 4.5 The stages in a triangulation of $[0,1] \times |\Delta^1|$ based on a subdivision $\hat{\Delta}^1$ of Δ^1 that can be seen at the top of each square.

This inductive procedure terminates with a triangulation of $[0,1] \times |\Delta^i|$ that depends only on knowing $\hat{\Delta}^i$. See Figure 4.5.

Now let us construct the chains Γ_F, which we also do inductively.

1. First suppose F is a 0-simplex of Δ^i. Let $(0,F)$ and $(1,F)$ be the copies of F in $\{0\} \times |\Delta^i|$ and $\{1\} \times |\Delta^i|$, respectively. Let u be the new vertex in our triangulation of $[0,1] \times |F|$. Let Γ_F be the simplicial chain $[(0,F),u] + [u,(1,F)]$. Then

$$\partial\Gamma_F = [u] - [(0,F)] + [(1,F)] - [u] = [(1,F)] - [(0,F)].$$

But $[(0,F)]$ is precisely $\mathfrak{o}_F$ in this case, and $[(1,F)] = \lambda(\mathfrak{o}_F)$. So Γ_F satisfies the desired formula as F has no faces.

2. Suppose now that we have defined Γ_F for all faces of dimension $< m$ and that F has dimension m. The simplicial chain

$$\zeta_F = \lambda(\mathfrak{o}_F) - \mathfrak{o}_F - \sum_{k=0}^{m}(-1)^k\Gamma_{F_k} \in C_m(K)$$

is thus defined and supported in the boundary of $[0,1]\times|F|$. Let u_F be the additional vertex added in defining the triangulation of $[0,1]\times|F|$ by coning off the triangulation of its boundary. We define Γ_F to be $\bar{c}\zeta_F$, the simplicial chain obtained by appending u_F as the new first vertex of each simplex; see Example 3.2.12 and [182, Section 8]. In other words, if $[w_0,\dots,w_m]$ is a simplex of ζ_F, then we let $\bar{c}([w_0,\dots,w_m]) = [u_F,w_0,\dots,w_m]$. Then $\bar{c}$ acts on the chain ζ_F by extending this construction linearly. With this definition, $\partial(\bar{c}\zeta_F) = \zeta_F - \bar{c}(\partial\zeta_F)$. But we claim that $\partial\zeta_F = 0$, and so $\partial\Gamma_F = \partial(\bar{c}\zeta_F) = \zeta_F$, which is precisely what we want. To see that $\partial\zeta_F = 0$, we first observe that $\partial\mathfrak{o}_F = \sum_{k=0}^{m}(-1)^k\mathfrak{o}_{F_k}$ by the definition of the simplicial boundary map, and we compute

$$\begin{aligned}
\partial\zeta_F &= \partial\left(\lambda(\mathfrak{o}_F) - \mathfrak{o}_F - \sum_{k=0}^{m}(-1)^k\Gamma_{F_k}\right)\\
&= \partial\lambda(\mathfrak{o}_F) - \partial\mathfrak{o}_F - \sum_{k=0}^{m}(-1)^k\partial\Gamma_{F_k}\\
&= \lambda(\partial\mathfrak{o}_F) - \partial\mathfrak{o}_F - \sum_{k=0}^{m}(-1)^k\left(\lambda(\mathfrak{o}_{F_k}) - \mathfrak{o}_{F_k} - \sum_{\ell=0}^{m-1}(-1)^\ell\Gamma_{F_{k,\ell}}\right)\\
&= \lambda\left(\sum_{k=0}^{m}(-1)^k\mathfrak{o}_{F_k}\right) - \sum_{k=0}^{m}(-1)^k\mathfrak{o}_{F_k} - \sum_{k=0}^{m}(-1)^k\lambda(\mathfrak{o}_{F_k})\\
&\qquad + \sum_{k=0}^{m}(-1)^k\mathfrak{o}_{F_k} + \sum_{k=0}^{m}(-1)^k\sum_{\ell=0}^{m-1}(-1)^\ell\Gamma_{F_{k,\ell}}\\
&= \sum_{k=0}^{m}(-1)^k\sum_{\ell=0}^{m-1}(-1)^\ell\Gamma_{F_{k,\ell}}.
\end{aligned}$$

Here $F_{k,\ell}$ is the ℓth face of the kth face of F. But then the form of the last sum is exactly the form of the sum obtained when we take the boundary of the boundary of a simplex. Hence the same cancellations occur, and the sum is zero.

Next, suppose that $\sigma : |\Delta^i| \to X$ is a singular simplex and that $\hat{\sigma}$ is a singular subdivision of σ based on $\hat{\Delta}^i$. Let us fix F as the top face of Δ^i, i.e. the unique i-dimensional face. We can choose a partial ordering on the vertices of the corresponding triangulation K of $[0,1]\times|\Delta^i|$ by letting each new vertex in the inductive construction be greater in the ordering than the previously added vertices. Let $\phi : C_*(K) \to S_*(|K|)$ be our map that uses the vertex ordering to assign to a canonically oriented simplex a singular simplex. Then let $p : [0,1]\times|\Delta^i| \to |\Delta^i|$ be the projection. We define a singular chain $P(\sigma) = \sigma p\phi(\Gamma_F)$, where p and σ act as chain maps:

$$S_*(|K|) \xrightarrow{p} S_*(|\Delta^i|) \xrightarrow{\sigma} S_*(X).$$

We compute $\partial P(\sigma)$.

$$\begin{aligned}\partial P(\sigma) &= \partial(\sigma p\phi(\Gamma_F)) = \sigma p\phi(\partial\Gamma_F)\\ &= \sigma p\phi\left(\lambda(\mathfrak{o}_F) - \mathfrak{o}_F - \sum_{k=0}^{m}(-1)^k\Gamma_{F_k}\right)\\ &= \sigma p\phi\lambda(\mathfrak{o}_F) - \sigma p\phi(\mathfrak{o}_F) - \sum_{k=0}^{m}(-1)^k\sigma p\phi(\Gamma_{F_k}).\end{aligned}$$

But thinking through the maps we see that $\sigma p\phi(\mathfrak{o}_F)$ is just the singular chain σ, and $\sigma p\phi\lambda(\mathfrak{o}_F)$ is our singular subdivision $\hat{\sigma}$ of σ. And if we let $\partial_k\sigma$ be the kth face of σ, then $\sigma p\phi(\Gamma_{F_k})$ is the same singular chain we would obtain by constructing $P(\partial_k\sigma)$ using the subdivision of the kth face of Δ^i obtained by restricting $\hat{\Delta}^i$. So we have

$$\partial P(\sigma) = \hat{\sigma} - \sigma - \sum_{k=0}^{m}(-1)^k P(\partial_k\sigma).$$

Next, let us check that if σ is a $\bar{p}$-allowable singular i-simplex then the $(i+1)$-simplices of $P(\sigma)$ will be allowable. Each such singular $(i+1)$-simplex η has the form $\sigma p\phi(\gamma)$, where γ is an $(i+1)$-simplex of K (with its canonical orientation). The singular simplex $\phi(\gamma)$ is a linear embedding $\mathfrak{j} : \Delta^{i+1} \to |K|$ that takes Δ^{i+1} homeomorphically onto γ; in fact $\mathfrak{j}$ is a simplicial isomorphism onto γ. So for the stratum S of X, we have $\eta^{-1}(S) = \mathfrak{j}^{-1}p^{-1}\sigma^{-1}(S)$. Since σ is allowable, $\sigma^{-1}(S)$ is in the $(i - \operatorname{codim}(S) + \bar{p}(S))$-skeleton of Δ^i, and so

$$p^{-1}\sigma^{-1}(S) \subset [0,1] \times \{(i - \operatorname{codim}(S) + \bar{p}(S))\text{-skeleton of } \Delta^i\},$$

which by construction must lie in the $(i-\operatorname{codim}(S)+\bar{p}(S)+1)$-skeleton of our simplicial complex K. In particular, $p^{-1}\sigma^{-1}(S)$ intersects γ in at most its $(i+1-\operatorname{codim}(S)+\bar{p}(S))$-skeleton. And since $\mathfrak{j}$ is just a simplicial isomorphism from Δ^{i+1} onto γ, we have $j^{-1}p^{-1}\sigma^{-1}(S)$ contained in the $(i+1-\operatorname{codim}(S)+\bar{p}(S))$-skeleton of Δ^{i+1}. Thus η is a $\bar{p}$-allowable simplex.

Equipped with these prismatic tools, we can now return to subdivision of singular chains. Suppose that $\xi \in I^{\bar{p}}S_i^{\mathrm{GM}}(X)$ is a chain with $|\partial\xi| \subset A$ and that $\hat{\xi}$ is a singular subdivision. We will construct a chain $\Xi \in I^{\bar{p}}S_{i+1}^{\mathrm{GM}}(X)$ such that $\partial\Xi = \hat{\xi} - \xi + \omega$, where $\omega \in I^{\bar{p}}S_i^{\mathrm{GM}}(X)$ and $|\omega| \subset A$. This will prove the proposition.

Suppose that $\xi = \sum_a n_a\sigma_a$, and let Δ_a^i be the domain simplex for σ_a with subdivision

$\hat{\Delta}^i_a$ and prism triangulation K_a. We let $\Xi = \sum_a n_a P(\sigma_a)$, and then we have

$$\begin{aligned}
\partial \Xi &= \partial\left(\sum_a n_a P(\sigma_a)\right) = \sum_a n_a \partial P(\sigma_a) \\
&= \sum_a n_a \left(\hat{\sigma}_a - \sigma_a - \sum_{k=0}^{m}(-1)^k P(\partial_k \sigma_a)\right) \\
&= \sum_a n_a \hat{\sigma}_a - \sum_a n_a \sigma_a - \sum_a n_a \sum_{k=0}^{m}(-1)^k P(\partial_k \sigma_a) \\
&= \hat{\xi} - \xi - \sum_a n_a \sum_{k=0}^{m}(-1)^k P(\partial_k \sigma_a).
\end{aligned}$$

Thus we have $\partial \Xi = \hat{\xi} - \xi + \omega$, if we set $\omega = -\sum_a n_a \sum_{k=0}^{m}(-1)^k P(\partial_k \sigma_a)$.

To see that $|\omega| \subset A$, we note that by the assumption that $\hat{\xi}$ is a singular subdivision of ξ, the subdivisions of the i-simplices of ξ are compatible along their boundaries. Furthermore, our construction of the triangulations was consistent in that if corresponding $i-1$ faces of Δ^i_s and Δ^i_t are subdivided in the same way in $\hat{\Delta}^i_s$ and $\hat{\Delta}^i_t$, then there will be correspondingly equal subdivisions of the subcomplexes of K_s and K_t over those $i-1$ faces. It follows that whatever cancellations of $(i-1)$-simplices occur when we take $\partial \xi$ are mirrored by cancellations of terms in ω, and so the only remaining terms in ω will similarly be those for which $\partial_k \sigma_a$ remains in $\partial \xi$. But by assumption $|\partial \xi| \subset A$ so the remaining $\partial_k \sigma_a$ are all in A. The map P clearly preserves supports; hence $|\omega| \subset A$.

It remains to check the allowability. Since ξ and $\hat{\xi}$ are allowable and since $\partial \omega = \partial \xi - \partial \hat{\xi}$, it remains only to check that the $(i+1)$-simplices of Ξ are allowable and that the i-simplices of ω are allowable. But as each σ_a is allowable, we have seen that this implies that each $(i+1)$-simplex of the corresponding $P(\sigma_a)$ is allowable. Similarly, we know that each i-simplex of ω is contained in a chain of the form $P(\partial_k \sigma_a)$, where $\partial_k \sigma_a$ is an $(i-1)$-simplex of $\partial \xi$ and so allowable. It follows again by our above argument that the i-simplices of each $P(\partial_k \sigma_a)$ are thus allowable. This completes the proof of Proposition 4.4.14. □

In the case where we actually have a chain map that produces singular subdivisions, we can say a bit more. The following corollary will be useful below in Section 7.1.

Corollary 4.4.15 *Suppose that $T : S_*(X) \to S_*(X)$ is a chain map that restricts to a singular subdivision on each singular simplex. Then the induced map $T : I^{\bar{p}}S^{\mathrm{GM}}_*(X, A) \to I^{\bar{p}}S^{\mathrm{GM}}_*(X, A)$ is chain homotopic to the identity for any subset $A \subset X$.*

Proof First, assume $A = \emptyset$. By the proof of Lemma 4.4.13, the image under T of each allowable simplex is allowable, and since T is a chain map, if $\xi \in I^{\bar{p}}S^{\mathrm{GM}}_*(X)$, then $T(\xi) \in I^{\bar{p}}S^{\mathrm{GM}}_*(X)$. The argument that T is chain homotopic to the identity follows from the construction of the proof of Proposition 4.4.14: For a simplex σ, recall the chain $P(\sigma)$ constructed in that proof. As T gives us a singular subdivision of every simplex, we can in this setting extend P to a homomorphism $S_i(X) \to S_{i+1}(X)$ for every i. Then, by construction, we have

$$\begin{aligned}\partial P(\sigma) &= T(\sigma) - \sigma - \sum_{k=0}^{m}(-1)^k P(\partial_k\sigma) = T(\sigma) - \sigma - P\left(\sum_{k=0}^{m}(-1)^k \partial_k\sigma\right)\\ &= T(\sigma) - \sigma - P(\partial\sigma).\end{aligned}$$

Thus for each chain $\xi \in S_i(X)$, we have $\partial P(\xi) = T(\xi) - \xi - P(\partial\xi)$. If $\xi \in I^{\bar{p}}S_i^{\mathrm{GM}}(X)$, then we know that all the simplices of ξ, $T(\xi)$, $P(\xi)$, and $P(\partial\xi)$ are allowable, using again Lemma 4.4.13 together with the allowability of the simplices of ξ and $\partial\xi$ and the properties established for P. It follows that $P(\xi) \in I^{\bar{p}}S_{i+1}^{\mathrm{GM}}(X)$. Therefore P provides a chain homotopy between the restriction of T to $I^{\bar{p}}S_*^{\mathrm{GM}}(X)$ and the identity.

If $A \neq \emptyset$, then since subdivision takes simplices supported in A to chains supported in A, the chain map T therefore induces a chain map from $I^{\bar{p}}S_*^{\mathrm{GM}}(X,A)$ to itself. Furthermore, since P also preserves (or reduces) support, the prism operator P is well defined as a map $I^{\bar{p}}S_*^{\mathrm{GM}}(X,A) \to I^{\bar{p}}S_{*+1}^{\mathrm{GM}}(X,A)$. The above boundary formula for P holds in this setting up to intersection chains supported in A, so P also induces a chain homotopy between $T : I^{\bar{p}}S_*^{\mathrm{GM}}(X,A) \to I^{\bar{p}}S_*^{\mathrm{GM}}(X,A)$ and the identity. □

(b) Excision

Now that we have established that subdivision of singular chains preserves intersection homology classes, we can demonstrate excision by using barycentric subdivision to ensure that chains are composed of small simplices and then breaking the chains into pieces, being careful to ensure that the pieces are each allowable chains. The proof is analogous to that in the PL setting, beginning with a version of Lemma 4.4.2, which provided a way to recognize boundary simplices as allowable if they are sufficiently "interior":

Definition 4.4.16 Suppose $\sigma : \Delta^i \to X$ is a singular simplex. Let $\hat{\Delta}^i$ be the barycentric subdivision of Δ^i. Let γ be an $(i-1)$-simplex of $\hat{\Delta}^i$ that does not contain any of the vertices of $\hat{\Delta}^i$. If we let $\mathfrak{i} : \Delta^{i-1} \to \Delta^i$ be the vertex-order-preserving embedding of γ, then we call the singular simplex $\sigma\mathfrak{i} : \Delta^{i-1} \to X$ a *completely interior simplex* of σ. More generally, we call $\sigma\mathfrak{i}$ *interior* if the corresponding γ is not contained in $\partial\Delta^i$.

Lemma 4.4.17 *If σ is an allowable singular i-simplex, and τ is a completely interior $(i-1)$-simplex of σ, then τ is allowable.*

Proof The proof is completely analogous to that of Lemma 4.4.1. By Lemma 4.4.2, the intersection of γ with every face η of Δ^i has dimension less than $\dim(\eta)$. So the intersection of γ with the k-skeleton of Δ^i must be contained in the $(k-1)$-skeleton of $\hat{\Delta}^i$, or in other words,

$$\mathfrak{i}^{-1}(k \text{ skeleton of } \Delta^i) \subset \{(k-1)\text{-skeleton of } \Delta^{i-1}\}.$$

If S is a stratum of X, then $\sigma^{-1}(S)$ is contained in the $(i - \operatorname{codim}(S) + \bar{p}(S))$-skeleton of Δ^i, by assumption, so it follows that $\tau^{-1}(S) = \mathrm{i}^{-1}\sigma^{-1}(S)$ is contained in the $(i - 1 - \mathrm{codim}(S) + \bar{p}(S))$-skeleton of Δ^i. So τ is allowable. □

Theorem 4.4.18 *Let X be a filtered space, and suppose $K \subset U \subset X$ such that $\bar{K} \subset \mathring{U}$. Then inclusion induces an isomorphism*

$$I^{\bar{p}}H_*^{\mathrm{GM}}(X - K, U - K) \xrightarrow{\cong} I^{\bar{p}}H_*^{\mathrm{GM}}(X, U).$$

Proof We first show that inclusion induces a surjection on intersection homology. Let $\xi \in I^{\bar{p}}S_i^{\mathrm{GM}}(X)$ be a chain representing an element of $I^{\bar{p}}H_i^{\mathrm{GM}}(X, U)$. The strategy is essentially the same as for PL subdivision. We first replace ξ by a subdivision $\hat{\xi}$ such that $\hat{\xi} = x + y$, where x and y are allowable singular *chains*, x is supported in $X - \bar{K}$ and y is supported in U. Then, applying Proposition 4.4.14, the chains x and ξ represent the same element of $I^{\bar{p}}H_i^{\mathrm{GM}}(X, U)$, but x is in the image of $I^{\bar{p}}H_i^{\mathrm{GM}}(X - K, U - K)$.

We will perform an iterated barycentric subdivision of ξ via the singular subdivision technique discussed above. Let β be the operator that replaces a chain with its singular barycentric subdivision, and let β^k denote the kth iteration of β. As ξ has only a finite number of simplices, and as simplices are compact, an easy Lebesgue number argument as in [182, Theorem 31.3] suffices to show that there is an m such that $\beta^m\xi$ consists entirely of singular simplices with image in $X - \bar{K}$ or $\mathring{U}$, which together constitute an open cover of X. If we were working with ordinary homology, this would be sufficient. However, we will need to employ techniques analogous to those we used to demonstrate PL excision, and so we need a slight buffer around the simplices that intersect K. Let Z be the union of the images of all the simplices of $\beta^m\xi$ whose images intersect $\bar{K}$; note that Z is compact and $Z \subset \mathring{U}$. We can now further subdivide $\beta^m\xi$ to obtain $\beta^M\xi$ such that every simplex of $\beta^M\xi$ lies in $X - (\bar{K} \cup Z)$ or $\mathring{U}$. We let $\hat{\xi} = \beta^{M+1}\xi$.

To explain our plan, note again that if we were working with ordinary homology, we'd be content to let y consist of the sum of all the simplices of $\beta^m\xi$ (with their coefficients) whose images intersect $\bar{K}$. The problem here is that this might create unallowable boundaries. So the purpose of the extra subdivisions is to make sure that we have enough extra singular simplices forming a "halo" around those touching $\bar{K}$, but still inside $\mathring{U}$, that we can cut the halo simplices along completely interior faces (see Definition 4.4.16) of one further subdivision, ensuring allowability of the new boundaries by Lemma 4.4.17. This is the program we now undertake in detail.

Let $\mathcal{A}$ be the set of singular simplices σ_j in $\beta^M\xi$ such that the image of σ_j intersects $\bar{K}$, and let $\mathcal{B} \supset \mathcal{A}$ be the set of singular simplices of $\beta^M\xi$ that share a singular vertex with a simplex in $\mathcal{A}$. By sharing a singular vertex, we mean that there is a point of X that is the common image of some vertex of each of the domain simplices. Since every simplex of $\mathcal{A}$ must be contained in a singular subdivision of a simplex with support in Z, the support of every simplex of $\mathcal{B}$ must intersect Z, and it follows from the construction that every simplex of $\mathcal{B}$ has image in $\mathring{U}$. Furthermore, every simplex of $\beta^M\xi$ not in $\mathcal{B}$ is contained in $X - \bar{K}$. Conceptually, $\mathcal{A}$ is the core of simplices that intersect $\bar{K}$, while $\mathcal{B} - \mathcal{A}$ is our "halo."

We now let y consist of the following simplices of $\hat{\xi} = \beta^{M+1}\xi$ (along with the coefficients they have in $\hat{\xi}$); see Figure 4.6:

1. If $\sigma \in \mathcal{A}$, then all i-simplices of the singular barycentric subdivision of σ are in y.
2. Suppose $\sigma \in \mathcal{B} - \mathcal{A}$ is an i-simplex with a vertex v that is shared with a simplex in $\mathcal{A}$ in the sense described above. Then we place every i-simplex of the singular barycentric subdivision of σ that contains the vertex v in y. Since there is some room for confusion, let us describe this in more detail. We are assuming that $\sigma : \Delta^i \to X$ is a simplex of $\beta^M\xi$ and that v is a vertex of Δ^i such that there is some other simplex $\sigma_2 : \Delta^i_2 \to X$ of $\beta^M\xi$ with $\sigma_2(\Delta^i_2) \cap \bar{K} \neq \emptyset$ and a vertex $v_2 \in \Delta^i_2$ such that $\sigma(v) = \sigma_2(v_2)$. This is what we abbreviate as "sharing the vertex v." Then we consider all the i-simplices in the barycentric subdivision of Δ^i that contain v, and we let the corresponding singular simplices in the singular barycentric subdivision of σ be in y. This does *not* mean that an i-simplex of the subdivision gets to be in y if one of its new vertices (one that arises from the subdivision of Δ^i and isn't an original vertex of Δ^i) gets mapped to the same image as some vertex of some simplex of $\beta^M\xi$.

In either case, each simplex σ' of $\hat{\xi}$ that qualifies for y is given the same coefficient it would have in $\hat{\xi}$. All simplices of y must have image that intersects Z, so $|y| \subset \mathring{U}$. Furthermore, it follows from the construction that y contains all the simplices of $\hat{\xi}$ that intersect $\bar{K}$. Since ξ is allowable, it follows from Lemma 4.4.13 that $\hat{\xi} = \beta^{M+1}\xi$ is allowable. So if y is allowable then $x = \hat{\xi} - y$ is also allowable and is contained in $X - \bar{K}$, so we will be finished with the proof of surjectivity.

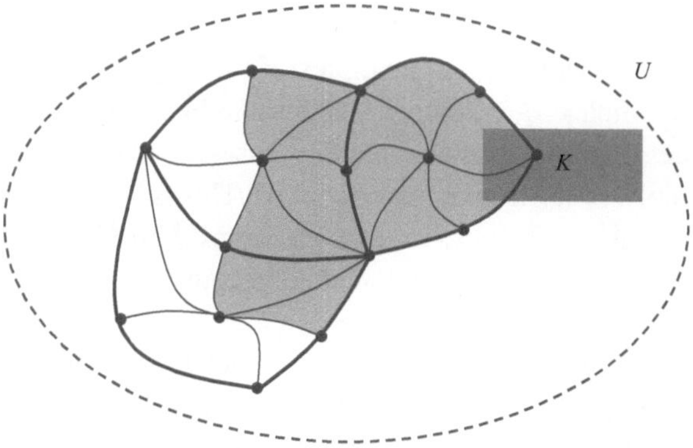

Figure 4.6 The construction of the chain y from the chain $\hat{\xi}$. The simplices with bold outlines are simplices of $\beta^M\xi$. The smaller simplices are from its barycentric subdivision $\hat{\xi} = \beta^{M+1}\xi$. The chains may have other simplices that are not shown. Among those simplices shown, the shaded ones are part of y because, for each one, the vertex it shares with the simplex from which it is subdivided is also shared by a simplex of $\beta^M\xi$ that intersects K.

We must show that y is allowable. As $\hat{\xi}$ is allowable, each i-simplex of y is allowable.

We need to check ∂y. Let τ be an $(i-1)$-simplex in ∂y. Then τ occurs in the boundary of a singular simplex of $\hat{\xi}$; of course, τ might occur as the boundary of multiple such simplices. The form of our argument will be to show either that τ is allowable or that, in fact, the coefficient of τ in ∂y is 0 so that τ is not in ∂y after all. There are three cases to consider. See Figure 4.7.

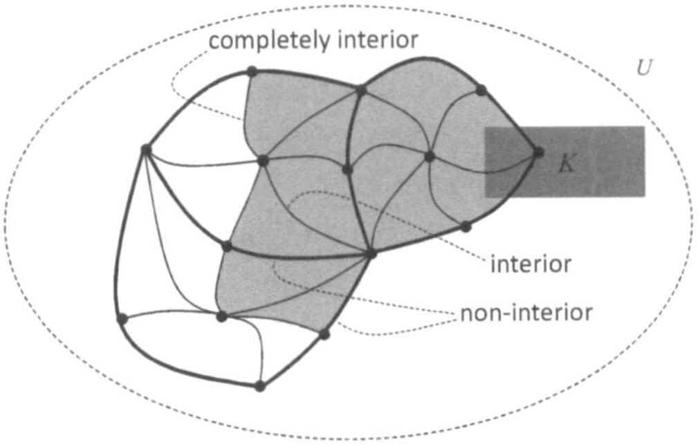

Figure 4.7 Demonstrating the three cases of $(i-1)$-simplices that we must consider for ∂y: completely interior, interior but not completely interior, and non-interior. The second case cannot actually contribute a simplex to ∂y due to cancellations. Similarly, any non-allowable non-interior $(i-1)$-simplices must be contained in faces that are non-allowable prior to the subdivision. Such faces must already cancel in $\beta^M\xi$, and they will contribute corresponding cancellations in ∂y due to the construction of y.

First, let σ'_τ be an i-simplex of y of which τ is a boundary face, and suppose that σ_τ is an i-simplex of $\beta^M\xi$ of which σ'_τ is a simplex of the singular subdivision $\beta\sigma_\tau$. If τ is a completely interior simplex to σ_τ, then τ is allowable by Lemma 4.4.17. Otherwise, we must have that τ shares a vertex v with σ_τ, and this must also be a vertex that σ_τ shares with a simplex in $\mathcal{A}$ (possibly σ_τ itself as σ_τ may be in $\mathcal{A}$). This is because an i-simplex δ in the barycentric subdivision $\hat{\Delta}^i$ of Δ^i shares exactly one vertex with Δ^i, and any $(i-1)$-simplex that does not share that vertex will be the domain of a completely interior singular simplex of $\sigma_\tau : \Delta^i \to X$.

Next, suppose that τ is interior to σ_τ but not completely interior. Then τ is a face of a σ'_τ that contains the vertex v of σ_τ. But all singular i-simplices in the subdivision of σ_τ containing v are in y, and so, τ being internal, there must actually be two such i-simplices in y, both in the singular subdivision of σ_τ and possessing τ as a common face. So these copies of τ cancel out in computing ∂y, and such a τ is not a concern for allowability of ∂y.

Finally, we must consider the case in which τ appears non-internally in the singular subdivision of some simplex σ_τ. In this case, τ is contained in some $i-1$ face F_τ of σ_τ. If F_τ is a simplex of $\partial\beta^M\xi$, then F_τ is allowable and hence so is τ. More generally, τ is allowable if F_τ is allowable for any reason, by Lemma 4.4.13. If F_τ is not allowable, then all the copies of F_τ must cancel in $\partial\beta^M\xi$. But since F_τ contains τ, F_τ must contain the vertex v that σ_τ shares with some simplex in $\mathcal{A}$. But then this implies that *every* simplex of $\beta^M\xi$ that has F_τ as a face is in $\mathcal{B}$. Since the coefficients of F_τ must cancel out to 0 in $\partial\beta^M\xi$, it follows also that all coefficients of τ arising from its appearance in subdivisions of F_τ cancel out (since all i-simplices containing v of the subdivisions of the i-simplices that have F_τ as a face appear in y). Considering then all possible faces F_τ in which τ appears, the same arguments show overall that, if τ is not allowable, its coefficient in ∂y must be 0.

This completes the proof of surjectivity. The proof of injectivity now follows from the proof of surjectivity, just as in the PL case in Theorem 4.4.3:

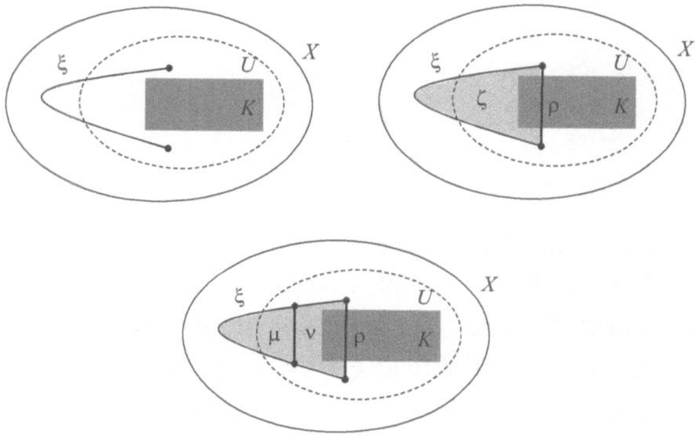

Figure 4.8 A schematic of the chains arising in the injectivity argument. We do not indicate the subdivisions in the picture or its labeling.

Suppose $\xi \in I^{\bar{p}}S_i^{\mathrm{GM}}(X-K)$ represents an element of $I^{\bar{p}}H_i^{\mathrm{GM}}(X-K, U-K)$ and that ξ is a relative boundary in X, i.e. there is an allowable chain ζ such that $\partial\zeta = \xi + \rho$, with ρ an allowable chain supported in U. We can now subdivide ζ as in the proof of surjectivity: construct analogous $\mathcal{A}$ and $\mathcal{B}$, and let v be the part of $\hat{\zeta} = \beta^{M+1}\zeta$ consisting of simplices that share a vertex with a simplex of $\beta^M\zeta$ in $\mathcal{A}$. Let $\mu = \hat{\zeta} - v$. Then by exactly the same arguments as above, μ and v are allowable, v is supported in U, and μ is supported in $X-K$; see Figure 4.8. Then

$$\partial\mu = \partial\hat{\zeta} - \partial v = \beta^{M+1}\xi + \beta^{M+1}\rho - \partial v.$$

Let us write

$$\partial\mu - \beta^{M+1}\xi = \beta^{M+1}\rho - \partial v. \tag{4.1}$$

Both $\beta^{M+1}\rho$ and ∂v are contained in U, and μ and $\beta^{M+1}\xi$ are contained in $X-K$. So both sides of (4.1) are in $U\cap(X-K) = U-K$. Putting this all together, we have $\mu \in I^{\bar{p}}S_{i+1}^{\mathrm{GM}}(X-K)$ with $\partial\mu = \beta^{M+1}\xi + (\beta^{M+1}\rho - \partial v)$, the term $\beta^{M+1}\xi$ representing the same class as ξ in $I^{\bar{p}}H_i^{\mathrm{GM}}(X-K, U-K)$ by Proposition 4.4.14, and the term $\beta^{M+1}\rho - \partial v$ allowable and supported in $U-K$. Thus $\xi = 0$ in $I^{\bar{p}}H_i^{\mathrm{GM}}(X-K, U-K)$. This completes the proof of Theorem 4.4.18. □

(c) Mayer–Vietoris

As in the PL setting, our hard work on excision also pays off with Mayer–Vietoris sequences:

Theorem 4.4.19 *Let X be a filtered space and suppose $X = U \cup V$, where U, V are subspaces such that $X = \mathring{U} \cup \mathring{V}$. Then there is an exact Mayer–Vietoris sequence*

$$\to I^{\bar{p}}H_i^{\mathrm{GM}}(U \cap V) \to I^{\bar{p}}H_i^{\mathrm{GM}}(U) \oplus I^{\bar{p}}H_i^{\mathrm{GM}}(V) \to I^{\bar{p}}H_i^{\mathrm{GM}}(X) \to I^{\bar{p}}H_{i-1}^{\mathrm{GM}}(U \cap V) \to \cdot.$$

There is also an analogous Mayer–Vietoris sequence in reduced intersection homology.[9]

Proof There is a short exact sequence

$$0 \to I^{\bar{p}}S_i^{\mathrm{GM}}(U \cap V) \xrightarrow{\phi} I^{\bar{p}}S_i^{\mathrm{GM}}(U) \oplus I^{\bar{p}}S_i^{\mathrm{GM}}(V) \xrightarrow{\psi} I^{\bar{p}}S_i^{\mathrm{GM}}(U) + I^{\bar{p}}S_i^{\mathrm{GM}}(V) \to 0,$$

and this yields a long exact homology sequence. Analogously to the PL case, we here have that $I^{\bar{p}}S_*^{\mathrm{GM}}(U) + I^{\bar{p}}S_*^{\mathrm{GM}}(V)$ is the subcomplex of $I^{\bar{p}}S_*^{\mathrm{GM}}(X)$ generated by allowable chains supported in U or in V, and, if we let $\mathfrak{i}_{A,B}$ stand for the inclusion map of spaces $A \hookrightarrow B$, then $\phi(\xi) = \big(\mathfrak{i}_{U\cap V,U}(\xi), -\mathfrak{i}_{U\cap V,V}(\xi)\big)$ and $\psi(\xi,\eta) = \mathfrak{i}_{U,X}(\xi) + \mathfrak{i}_{V,X}(\eta)$. In the reduced case we extend this short exact sequence to degree -1, where it is

$$0 \to \mathbb{Z} \xrightarrow{\phi} \mathbb{Z} \oplus \mathbb{Z} \xrightarrow{\psi} \mathbb{Z} \to 0$$

with $\phi(a) = (a, -a)$ and $\psi(a, b) = a + b$. What needs to be shown is that the inclusion map $\psi : I^{\bar{p}}S_*^{\mathrm{GM}}(U) + I^{\bar{p}}S_*^{\mathrm{GM}}(V) \to I^{\bar{p}}S_*^{\mathrm{GM}}(X)$ (extended by the identity $\mathbb{Z} \to \mathbb{Z}$ in degree -1 in the reduced case) yields an isomorphism on homology. We focus on the unreduced case, with the argument in the reduced case being the same.

The proof is basically the same as the argument we used to prove excision. The argument of Theorem 4.4.18 shows how we can take an allowable cycle ξ in X and subdivide it into a chain $\hat{\xi}$ representing the same intersection homology class and such that $\hat{\xi} = x + y$, with x, y allowable and such that y is contained in $\mathring{U}$ and x is contained in $X - (X - \mathring{V}) = \mathring{V}$, i.e. $y \in I^{\bar{p}}S_i^{\mathrm{GM}}(U)$ and $x \in I^{\bar{p}}S_i^{\mathrm{GM}}(V)$. This shows that ψ is surjective on homology.

Similarly, if $y + x$ is a cycle in $I^{\bar{p}}S_i^{\mathrm{GM}}(U) + I^{\bar{p}}S_i^{\mathrm{GM}}(V)$ that bounds a chain ζ in X, then we can split up a subdivision $\hat{\zeta}$ as $\hat{\zeta} = \nu + \mu$ with $\nu \in I^{\bar{p}}S_{i+1}^{\mathrm{GM}}(U)$ and $\mu \in I^{\bar{p}}S_{i+1}^{\mathrm{GM}}(V)$. Then $\partial\hat{\zeta} = \partial(\mu + \nu) = \hat{x} + \hat{y}$, where $\hat{x}, \hat{y}$ are the induced subdivisions of x and y. So $\hat{x} + \hat{y} = 0 \in H_*(I^{\bar{p}}S_i^{\mathrm{GM}}(U) + I^{\bar{p}}S_i^{\mathrm{GM}}(V))$. But we can show that $\hat{x} + \hat{y}$ represents the same homology class as $x + y$ in $H_*(I^{\bar{p}}S_i^{\mathrm{GM}}(U) + I^{\bar{p}}S_i^{\mathrm{GM}}(V))$ using the argument from the proof of Corollary 4.4.15. If we assume that we have been using barycentric subdivisions, then we have our prism operator P, which by construction preserves supports: $|P(\xi)| \subset |\xi|$. So

[9] Due to our conventions about reduced intersection homology, the reduced sequence will only be exact at $I^{\bar{p}}H_0^{\mathrm{GM}}(X)$ if $I^{\bar{p}}H_0^{\mathrm{GM}}(U \cap V) \neq 0$; see Section 4.3.3.

$|P(x)| \subset V$, $|P(y)| \subset U$, and

$$\begin{aligned}\partial(P(x)+P(y)) &= \partial P(x)+\partial P(y)\\ &= \hat{x}-x-P(\partial x)+\hat{y}-y-P(\partial y)\\ &= (\hat{x}+\hat{y})-(x+y)-P(\partial x+\partial y)\\ &= (\hat{x}+\hat{y})-(x+y),\end{aligned}$$

as $\partial x+\partial y=\partial(x+y)=0$ by assumption. So $x+y$ and $\hat{x}+\hat{y}$ represent the same class in $H_*(I^{\bar{p}}S_i^{\mathrm{GM}}(U)+I^{\bar{p}}S_i^{\mathrm{GM}}(V))$. Since we have shown that $\hat{x}+\hat{y}$ is trivial in $H_i(I^{\bar{p}}S_*^{\mathrm{GM}}(U)+I^{\bar{p}}S_*^{\mathrm{GM}}(V))$, so is $x+y$. Thus ψ is injective. □

Within the proof of the theorem, we demonstrated that the inclusion map $\psi: I^{\bar{p}}S_i^{\mathrm{GM}}(U)+I^{\bar{p}}S_i^{\mathrm{GM}}(V)\to I^{\bar{p}}S_i^{\mathrm{GM}}(X)$ induces an isomorphism on homology. Below, in Proposition 6.5.1 of Section 6.5, we will prove the stronger statement that if $\mathcal{V}$ is a covering of X such that the interiors of the elements of $\mathcal{V}$ constitute an open covering of X and if we let

$$I^{\bar{p}}S_*^{\mathrm{GM},\mathcal{V}}(X)=\sum_{V\in\mathcal{V}} I^{\bar{p}}S_*^{\mathrm{GM}}(V)\subset I^{\bar{p}}S_*^{\mathrm{GM}}(X),$$

then the inclusion $I^{\bar{p}}S_*^{\mathrm{GM},\mathcal{V}}(X)\hookrightarrow I^{\bar{p}}S_*^{\mathrm{GM}}(X)$ is a chain homotopy equivalence. For PL chains, the corresponding inclusion $I^{\bar{p}}\mathfrak{C}_*^{\mathrm{GM},\mathcal{V}}(X)\hookrightarrow I^{\bar{p}}\mathfrak{C}_*^{\mathrm{GM}}(X)$ will be shown to be an isomorphism.

(d) Examples

We now turn to some important applications of the tools we have now developed. First we compute the intersection homology of a suspension. Then we compute the intersection homology of a ∂-stratified pseudomanifold whose boundary has been coned off.

Example 4.4.20 Let us use the Mayer–Vietoris sequence to compute the intersection homology of the suspension of a compact filtered space. Let X be an $(n-1)$-dimensional compact filtered space, and let $SX=[-1,1]\times X/\sim$ be the suspension of X. We filter SX so that $(SX)^i=S(X^{i-1})$ and $(SX)^0=\{\mathfrak{n},\mathfrak{s}\}$, the north and south suspension vertices; in particular, SX has dimension n. Let $\bar{p}$ be a perversity on SX; for simplicity, let us also assume that $\bar{p}(\{\mathfrak{n}\})=\bar{p}(\{\mathfrak{s}\})=p$ and that $I^{\bar{p}}H_0^{\mathrm{GM}}(X)\neq 0$ (for example, if X, and hence also SX, possesses a regular stratum). We leave it as a fun exercise for the reader to consider the cases $\bar{p}(\{\mathfrak{n}\})\neq\bar{p}(\{\mathfrak{s}\})$ or $I^{\bar{p}}H_0^{\mathrm{GM}}(X)=0$. We will also use $\bar{p}$ to denote the perversity restricted to X.

We will use the reduced Mayer–Vietoris sequence for the two pieces $U=[-1,1)\times X/\sim\;\cong cX$ and $V=(-1,1]\times X/\sim\;\cong cX$ of the suspension. Then the intersection of these two pieces is $U\cap V\cong(-1,1)\times X$, and we know from stratified homotopy invariance (Corollary 4.1.11) that $I^{\bar{p}}H_*^{\mathrm{GM}}((-1,1)\times X)\cong I^{\bar{p}}H_*^{\mathrm{GM}}(X)$, induced by inclusion. We also know from

Theorem 4.2.1 that, since we've assumed X has regular strata and hence $I^{\bar p}H_0^{\mathrm{GM}}(X) \neq 0$,

$$I^{\bar p}H_i^{\mathrm{GM}}(cX) \cong \begin{cases} 0, & i \geq n-p-1, \quad i \neq 0, \\ \mathbb{Z}, & i = 0 \geq n-p-1, \\ I^{\bar p}H_i(X), & i < n-p-1, \end{cases}$$

where the isomorphisms in dimensions $i < n-p-1$ are induced by inclusions. It follows that the reduced intersection homology is

$$I^{\bar p}\tilde H_i^{\mathrm{GM}}(cX) \cong \begin{cases} 0, & i \geq n-p-1, \\ I^{\bar p}\tilde H_i(X), & i < n-p-1. \end{cases}$$

This gives us the computations of $I^{\bar p}\tilde H_*^{\mathrm{GM}}(U)$ and $I^{\bar p}\tilde H_*^{\mathrm{GM}}(V)$.

So for $i \geq n-p-1$, we see that $I^{\bar p}\tilde H_i^{\mathrm{GM}}(U) = I^{\bar p}\tilde H_i^{\mathrm{GM}}(V) = 0$, and so in this range from the reduced Mayer–Vietoris sequence we must have $I^{\bar p}\tilde H_i^{\mathrm{GM}}(X) \cong I^{\bar p}\tilde H_{i+1}^{\mathrm{GM}}(SX)$, via the Mayer–Vietoris boundary map and the stratified homotopy equivalence of $(-1,1)\times X$ with X. In other words, for $i > n-p-1$, we have $I^{\bar p}\tilde H_i^{\mathrm{GM}}(SX) \cong I^{\bar p}\tilde H_{i-1}^{\mathrm{GM}}(X)$. If $i = 0 \geq n-p-1$, we also get $I^{\bar p}\tilde H_0(SX) = 0$.

On the other hand, if $i < n-p-1$, the inclusion maps $I^{\bar p}\tilde H_i^{\mathrm{GM}}(U\cap V) \to I^{\bar p}\tilde H_i^{\mathrm{GM}}(U)$ and $I^{\bar p}\tilde H_i^{\mathrm{GM}}(U\cap V) \to I^{\bar p}\tilde H_i^{\mathrm{GM}}(V)$ are each isomorphisms (via the cone formula and stratified homotopy invariance), so the maps

$$\phi : I^{\bar p}\tilde H_i^{\mathrm{GM}}(U\cap V) \to I^{\bar p}\tilde H_i^{\mathrm{GM}}(U) \oplus I^{\bar p}H_i^{\mathrm{GM}}(V)$$

of the Mayer–Vietoris sequence just have the form of an "anti-diagonal" map $G \to G\oplus G$, $g \to (g,-g)$. So in this degree range, the Mayer–Vietoris sequence splits into short exact sequences of the form

$$0 \to G \xrightarrow{\phi} G\oplus G \xrightarrow{\psi} G \to 0$$

with $\phi(g) = (g,-g)$ and $\psi(g,h) = g+h$. Hence we have $I^{\bar p}\tilde H_i^{\mathrm{GM}}(SX) \cong I^{\bar p}\tilde H_i(X)$ for $i < n-p-1$, induced by inclusion.

Finally, we must compute $I^{\bar p}\tilde H_{n-p-1}^{\mathrm{GM}}(SX)$ for $n-p-1 > 0$. We have seen that

$$I^{\bar p}\tilde H_{n-p-2}^{\mathrm{GM}}(U\cap V) \to I^{\bar p}\tilde H_{n-p-2}^{\mathrm{GM}}(U) \oplus I^{\bar p}\tilde H_{n-p-2}^{\mathrm{GM}}(V)$$

is injective, and we have $I\tilde H_{n-p-1}^{\mathrm{GM}}(U) = I\tilde H_{n-p-1}^{\mathrm{GM}}(V) = 0$, so $I\tilde H_{n-p-1}^{\mathrm{GM}}(SX) = 0$ for $n-p-1 \geq 1$.

Altogether, we have shown the following:

$$I^{\bar p}\tilde H_i^{\mathrm{GM}}(SX) = \begin{cases} I^{\bar p}\tilde H_{i-1}^{\mathrm{GM}}(X), & i > n-p-1, \\ 0, & i = n-p-1 \\ I^{\bar p}\tilde H_i^{\mathrm{GM}}(X), & i < n-p-1. \end{cases}$$

Rewriting in terms of unreduced intersection homology for SX, and using the fact that

$I^{\bar{p}}H_0(SX) \neq 0$, as we have assumed there are allowable $\bar{0}$-simplices in X, we obtain the next theorem. △

Theorem 4.4.21 *If X is an $(n-1)$-dimensional compact filtered space with $I^{\bar{p}}H_0^{\mathrm{GM}}(X) \neq 0$ and $\bar{p}$ is a perversity on SX that takes the same value p at the two suspension points, then*

$$I^{\bar{p}}H_i^{\mathrm{GM}}(SX) \cong \begin{cases} I^{\bar{p}}\tilde{H}_{i-1}^{\mathrm{GM}}(X), & i > n-p-1, \quad i \neq 0, \\ 0, & i = n-p-1, \quad i \neq 0, \\ \mathbb{Z}, & i = 0 \geq n-p-1, \\ I^{\bar{p}}H_i^{\mathrm{GM}}(X), & i < n-p-1. \end{cases}$$

Notice that this agrees with the computation of Example 3.2.13, as we have assumed that there are always allowable singular 0-simplices.

It is interesting to compare this calculation to that for the suspension in ordinary homology. In that case, we have $\tilde{H}_i(SX) \cong \tilde{H}_{i-1}(X)$ for all i; see [182, Theorem 33.2] or [125, Section 2.2, Exercise 32]. So, for suspensions, intersection homology behaves like ordinary homology above a certain degree that depends on the perversity of the suspension points, shifting all the groups of the original space up by a degree. However, in intersection homology there is a transition degree at which the group is 0, and below that degree the groups are as though the suspension had not been done at all. The reason for some of the odd behavior in degree 0 is that when 0 should be the transition degree we can't quite manage $I^{\bar{p}}H_0^{\mathrm{GM}}(SX) = 0$, assuming there are regular strata, which then must contain allowable 0-cycles (when the transition degree is below 0, i.e. when $0 > n-p-1$, we will still have $I^{\bar{p}}H_0^{\mathrm{GM}}(SX) \cong \mathbb{Z}$, and we could just as well have written $I^{\bar{p}}\tilde{H}_0^{\mathrm{GM}}(SX) = I^{\bar{p}}\tilde{H}_{-1}^{\mathrm{GM}}(SX) = 0$ in our tabulation in the statement of the theorem). This defect in degree 0 will be corrected when we turn to non-GM intersection homology in Chapter 6; in particular, see Theorem 6.3.13.

Example 4.4.22 Here is another important example that includes, as a special case, a singular intersection homology version of Example 3.2.12. Let X be a compact n-dimensional ∂-stratified pseudomanifold with $\partial X \neq \emptyset$. As a special case, X may be a ∂-manifold as in Example 3.2.12, though now we do not assume a triangulation. Let $X^+ = X \cup_{\partial X} \bar{c}(\partial X)$. In other words, X^+ is X but with its boundary coned off. This space is filtered homeomorphic to the quotient space $X/\partial X$ in which we collapse ∂X to a point. Let v be the cone point. Let $\bar{p}$ be a perversity on X^+, which restricts to a perversity on X that we continue to call $\bar{p}$. We have an inclusion map of pairs $(X, \partial X) \hookrightarrow (X^+, \bar{c}(\partial X))$, which induces a map of long exact sequences:

$$\begin{array}{ccccccc}
\longrightarrow & I^{\bar p}H_i^{\mathrm{GM}}(\partial X) & \longrightarrow & I^{\bar p}H_i^{\mathrm{GM}}(X) & \longrightarrow & I^{\bar p}H_i^{\mathrm{GM}}(X,\partial X) & \longrightarrow \\
 & \downarrow & & \downarrow & & \downarrow \cong & \\
\longrightarrow & I^{\bar p}H_i^{\mathrm{GM}}(\bar c(\partial X)) & \longrightarrow & I^{\bar p}H_i^{\mathrm{GM}}(X^+) & \longrightarrow & I^{\bar p}H_i^{\mathrm{GM}}(X^+,\bar c(\partial X)) & \longrightarrow .
\end{array}$$

To see that the rightmost vertical map in the diagram is an isomorphism for all i, we use excision to excise a small cone neighborhood of the cone point in $I^{\bar p}H_i^{\mathrm{GM}}(X^+,\bar c(\partial X))$ and then apply a stratified homotopy equivalence to retract what remains of $\bar c(\partial X)$ back to ∂X.

We can now use the cone formula (Theorem 4.2.1) and the stratified homotopy equivalence of $\bar c(\partial X)$ with $c(\partial X)$ to compute $I^{\bar p}H_i^{\mathrm{GM}}(\bar c(\partial X))$. To simplify things somewhat, we will assume that $\bar p(\{v\}) \leq n-2$; see Remark 4.2.2. This allows us to avoid some messy cases when $i = 0, 1$, and it also provides a formula that's consistent with what we will get for non-GM intersection homology and arbitrary perversities later in Example 6.3.15. We encourage the reader to compute the more general cases as an exercise.

So, assuming $\bar p(\{v\}) \leq n-2$ (so that $n - \bar p(\{v\}) - 1 \geq 1$), Theorem 4.2.1 and Remark 4.2.2 give us

$$I^{\bar p}H_i^{\mathrm{GM}}(\bar c(\partial X)) \cong \begin{cases} 0, & i \geq n - \bar p(\{v\}) - 1, \\ I^{\bar p}H_i^{\mathrm{GM}}(\partial X), & i < n - \bar p(\{v\}) - 1. \end{cases}$$

From the long exact sequences, it follows immediately that

$$I^{\bar p}H_i^{\mathrm{GM}}(X^+) \cong I^{\bar p}H_i^{\mathrm{GM}}(X^+,\bar c(\partial X)) \cong I^{\bar p}H_i^{\mathrm{GM}}(X,\partial X)$$

for $i > n - \bar p(\{v\}) - 1$.

The cone formula also tells us that the isomorphism

$$I^{\bar p}H_i^{\mathrm{GM}}(\bar c(\partial X)) \cong I^{\bar p}H_i^{\mathrm{GM}}(\partial X)$$

when $i < n - \bar p(\{v\}) - 1$ is induced by inclusion, so we get

$$I^{\bar p}H_i^{\mathrm{GM}}(X) \cong I^{\bar p}H_i^{\mathrm{GM}}(X^+)$$

by the Five Lemma when $i < n - \bar p(\{v\}) - 1$.

This leaves the case $i = n - \bar p(\{v\}) - 1$. In this case, the diagram reduces to

$$\begin{array}{ccccccccc}
\longrightarrow & \mathrm{I}_i(\partial X) & \longrightarrow & \mathrm{I}_i(X) & \longrightarrow & \mathrm{I}_i(X,\partial X) & \xrightarrow{\partial_*} & \mathrm{I}_{i-1}(\partial X) & \longrightarrow \\
 & \downarrow & & \downarrow & & \downarrow \cong & & \downarrow \cong & \\
\longrightarrow & 0 & \longrightarrow & \mathrm{I}_i(X^+) & \longrightarrow & \mathrm{I}_i(X^+,\bar c(\partial X)) & \xrightarrow{\partial_*} & \mathrm{I}_{i-1}(\bar c(\partial X)) & \longrightarrow ,
\end{array}$$

where for clarity we have written I as shorthand for $I^{\bar{p}}H^{\mathrm{GM}}$. We can therefore identify $I^{\bar{p}}H^{\mathrm{GM}}_{n-\bar{p}(\{v\})-1}(X^+)$ with the kernel of the boundary map

$$\partial_* : I^{\bar{p}}H^{\mathrm{GM}}_{n-\bar{p}(\{v\})-1}(X,\partial X) \to I^{\bar{p}}H^{\mathrm{GM}}_{n-\bar{p}(\{v\})-2}(\partial X).$$

Equivalently, from the long exact sequences, this is the image of the inclusion-induced $I^{\bar{p}}H^{\mathrm{GM}}_{n-\bar{p}(\{v\})-1}(X) \to I^{\bar{p}}H^{\mathrm{GM}}_{n-\bar{p}(\{v\})-1}(X,\partial X)$.

So, summarizing, we have computed

$$I^{\bar{p}}H^{\mathrm{GM}}_i(X^+) \cong \begin{cases} I^{\bar{p}}H^{\mathrm{GM}}_i(X,\partial X), & i > n-\bar{p}(\{v\})-1, \\ \mathrm{im}(I^{\bar{p}}H^{\mathrm{GM}}_i(X) \to I^{\bar{p}}H^{\mathrm{GM}}_i(X,\partial X)), & i = n-\bar{p}(\{v\})-1, \\ I^{\bar{p}}H^{\mathrm{GM}}_i(X), & i < n-\bar{p}(\{v\})-1. \end{cases}$$

In the special case where X is a (trivially filtered) compact oriented n-dimensional ∂-manifold M with $\partial M \neq \emptyset$, and continuing to assume $\bar{p}(\{v\}) \leq n-2$, this reduces to

$$I^{\bar{p}}H^{\mathrm{GM}}_i(M^+) \cong \begin{cases} H_i(M,\partial M), & i > n-\bar{p}(\{v\})-1, \\ \mathrm{im}(H_i(M) \to H_i(M,\partial M)), & i = n-\bar{p}(\{v\})-1, \\ H_i(M), & i < n-\bar{p}(\{v\})-1. \end{cases}$$

This example provides a tantalizing glimpse of Poincaré duality results to come, as we know that

$$H_i(M;\mathbb{Q}) \cong \mathrm{Hom}(H_{n-i}(M,\partial M;\mathbb{Q}),\mathbb{Q})$$

and

$$\mathrm{im}(H_i(M;\mathbb{Q}) \to H_i(M,\partial M;\mathbb{Q})) \cong \mathrm{Hom}(\mathrm{im}(H_{n-i}(M;\mathbb{Q}) \to H_{n-i}(M,\partial M;\mathbb{Q})),\mathbb{Q})$$

as consequences of Lefschetz duality. Thus it would appear that, at least with field coefficients, finding a duality between $I^{\bar{p}}H^{\mathrm{GM}}_i(X)$ and $I^{\bar{q}}H^{\mathrm{GM}}_{n-i}(X)$ is a matter of choosing perversities such that the dual groups line up properly. This intuition will be validated below in Chapter 8.

The reader may now want to apply the computations of this example to complete Example 3.4.8. Recall that Example 3.4.8 dealt with $X = X^1 = S^1$, the circle, and a point $x_0 \in S^1$. The space X is filtered as $\{x_0\} \subset X$. Such a space arises as in our example here by letting our ∂-stratified pseudomanifold be I, the standard closed interval with its trivial filtration. In order to consider all the possible cases, the reader will want to explore the options $\bar{p}(\{v\}) > n-2$ as well. △

(e) Relative Mayer–Vietoris sequences

Once one has a Mayer–Vietoris sequence, it is not difficult to formulate a relative version.

Theorem 4.4.23 *Suppose $X = U \cup V$, where U, V are subspaces such that $X = \mathring{U} \cup \mathring{V}$. Let $A \subset X$, let $C = A \cap U$, and let $D = A \cap V$. Then there is an exact Mayer–Vietoris sequence*

$$\to I^{\bar{p}}H_i^{\mathrm{GM}}(U \cap V, C \cap D) \to I^{\bar{p}}H_i^{\mathrm{GM}}(U, C) \oplus I^{\bar{p}}H_i^{\mathrm{GM}}(V, D) \to I^{\bar{p}}H_i^{\mathrm{GM}}(X, A) \to .$$

Similarly, there is an analogous PL intersection homology sequence.

Proof As the proofs are equivalent, we focus on singular chains.

Let I_S denote $I^{\bar{p}}S_*^{\mathrm{GM}}$. Consider the following diagram:

$$\begin{array}{ccccccccc} 0 & \longrightarrow & \mathrm{I}_S(C \cap D) & \longrightarrow & \mathrm{I}_S(C) \oplus \mathrm{I}_S(D) & \longrightarrow & \mathrm{I}_S(C) + \mathrm{I}_S(D) & \longrightarrow & 0 \\ & & \downarrow & & \downarrow & & \downarrow & & \\ 0 & \longrightarrow & \mathrm{I}_S(U \cap V) & \longrightarrow & \mathrm{I}_S(U) \oplus \mathrm{I}_S(V) & \longrightarrow & \mathrm{I}_S(U) + \mathrm{I}_S(V) & \longrightarrow & 0. \end{array}$$

The top and bottom rows are Mayer–Vietoris short exact sequences of chain complexes, and each vertical map is an inclusion of complexes. Therefore, applying the Snake Lemma[10] [197, Corollary 6.12] in each degree yields a short exact sequence

$$0 \longrightarrow \mathrm{I}_S(U \cap V, C \cap D) \longrightarrow \mathrm{I}_S(U, C) \oplus \mathrm{I}_S(V, D) \longrightarrow \frac{\mathrm{I}_S(U) + \mathrm{I}_S(V)}{\mathrm{I}_S(C) + \mathrm{I}_S(D)} \longrightarrow 0$$

and a corresponding long exact homology sequence. It only remains to show that

$$H_i\left(\frac{\mathrm{I}_S(U) + \mathrm{I}_S(V)}{\mathrm{I}_S(C) + \mathrm{I}_S(D)}\right) \cong I^{\bar{p}}H_i^{\mathrm{GM}}(U \cap V, C \cap D).$$

But now we have yet another diagram of short exact sequences

$$\begin{array}{ccccccccc} 0 & \longrightarrow & \mathrm{I}_S(C) + \mathrm{I}_S(D) & \longrightarrow & \mathrm{I}_S(U) + \mathrm{I}_S(V) & \longrightarrow & \dfrac{\mathrm{I}_S(U) + \mathrm{I}_S(V)}{\mathrm{I}_S(C) + \mathrm{I}_S(D)} & \longrightarrow & 0 \\ & & \downarrow & & \downarrow & & \downarrow & & \\ 0 & \longrightarrow & \mathrm{I}_S(A) & \longrightarrow & \mathrm{I}_S(X) & \longrightarrow & \mathrm{I}_S(X, A) & \longrightarrow & 0. \end{array}$$

We have already established that the middle vertical map induces an isomorphism on homology in the proof of Theorem 4.4.19, and similarly the left-hand vertical map induces an isomorphism on homology using that the union of the interiors of C and D in A cover A (this is an easy exercise in point-set topology). Therefore, by the Five Lemma, the right-hand vertical map also induces homology isomorphisms. □

[10] Famously, a partial proof of the Snake Lemma opens the 1980 film *It's My Turn*. Unfortunately, the author knows of no film references to intersection homology.

5

Mayer–Vietoris Arguments and Further Properties of Intersection Homology

A basic question in topology is how to compute invariants of a space from invariants of subspaces. The standard tool for this purpose is the Mayer–Vietoris sequence. In the first section of this chapter, we examine a generalization that will be useful for proving many of the theorems that occur later in the book, including Künneth theorems and Poincaré duality. Following Bott and Tu [29], we refer to these techniques as "Mayer–Vietoris arguments."

The basic idea is that we want to know that if two homology theories agree on small pieces of a space, then they agree on the space as a whole. Roughly speaking, this is a generalization of the principle that if we know that $X = U \cup V$ and that two homology theories agree on U, V, and $U \cap V$, then they agree on X by applying the Five Lemma to a diagram of Mayer–Vietoris sequences (assuming sufficiently natural compatibility amongst the theories). Such arguments will provide a means to prove theorems about CS sets without the need to rely on sheaf theory or on acyclic model arguments, which are often used for ordinary homology but have no appropriate analogue for intersection homology.

After introducing the Mayer–Vietoris arguments in Section 5.1, we provide some easy first applications in Section 5.1.1 by showing that for sufficiently large perversities intersection homology is simply ordinary homology and that intersection homology behaves well with respect to normalization maps. For the remainder of the chapter, we turn to more complex applications.

In Section 5.2, we prove that there is a Künneth Theorem for the intersection homology of the product of a stratified space with a manifold; we will see a more general Künneth Theorem later in Chapter 6. This section contains an appendix that provides the technical details of the construction of the cross product using "Eilenberg–Zilber shuffles." In Section 5.3, we introduce intersection homology with coefficients. We will see that the Universal Coefficient Theorem does not hold in general but that it can be recovered by assuming that the space satisfies certain "locally torsion-free" hypotheses. Section 5.4 contains a proof that PL and singular intersection homology are isomorphic on PL CS sets, and Section 5.5 demonstrates that for certain perversities the intersection homology groups of a CS set do not depend upon the choice of stratification as a CS set. Finally, in Section 5.6, we consider finite generation of the intersection homology groups.

5.1 Mayer–Vietoris Arguments

Suppose X is a space and that O is the category whose objects are open subsets of X (so $\mathrm{Ob}(O)$ is the topology of X) and whose morphisms are inclusion maps of these subsets. Suppose that one has two functors F, G from O to some other category, such as the category of abelian groups, and that one wants to know whether these two functors are equivalent in the sense that $F(U) \cong G(U)$ for all $U \in O$. In intersection homology settings, one most often sees this question at work within sheaf theory (see [11, 28, 106], among others). Without getting into the details of sheaf theory, one often starts with data showing (roughly stated) that two types of sheaf cohomology agree in neighborhoods of each point of X and then uses this to conclude that more global sheaf cohomology groups must be isomorphic. Since we do not want to introduce sheaves, we will need Mayer–Vietoris arguments. For ordinary homology and cohomology, such techniques are put to much use for "good covers" of manifolds by Bott and Tu in [29]; see in particular [29, Section I.5] and also [38, Lemma V.9.5]. The first use of similar arguments for intersection homology seems to be in King [139], where arguments of this type were used to compute Künneth theorems (for which one factor is a manifold), to prove topological invariance of intersection homology (given certain conditions on perversities), and to provide a general comparison principle for intersection homology theories [139, Theorem 10]. Saralegi [205] later used these techniques to prove a de Rham theorem for intersection homology, and Friedman and McClure [100] made use of similar ideas in providing a non-sheaf-theoretic proof of Poincaré duality for singular intersection homology on pseudomanifolds patterned after the Poincaré duality proof for manifolds in Hatcher [125]. The following theorems are modifications of these prior arguments that we hope will prove to be more general; in particular, they incorporate modifications that will be necessary for our applications.

We begin with a Mayer–Vietoris argument for manifolds in Theorems 5.1.1 and 5.1.2. We then turn to a version for CS sets in Theorem 5.1.4.

Theorem 5.1.1 *Let $\mathcal{M}$ be the category whose objects are manifolds and whose morphisms are open inclusions, and let $\mathcal{A}b_*$ be the category of graded abelian groups. Let $F_*, G_* : \mathcal{M} \to \mathcal{A}b_*$ be covariant functors and let $\Phi : F_* \to G_*$ be a natural transformation. Suppose that Φ has the following three properties:*

(1) *$\Phi : F_*(U) \to G_*(U)$ is an isomorphism for U homeomorphic to $\mathbb{R}^n$ or $\emptyset$.*

(2) *F_* and G_* admit exact Mayer–Vietoris sequences, i.e. if U, V are open submanifolds of a manifold then there is an exact sequence*

$$\to F_i(U \cap V) \to F_i(U) \oplus F_i(V) \to F_i(U \cup V) \to F_{i-1}(U \cap V) \to,$$

and similarly for G_*, such that Φ induces a commutative diagram of such sequences.

(3) *If $\{U_\alpha\}$ is an increasing collection of open submanifolds of a manifold M (meaning that the indices α are taken from a totally ordered set and $\alpha < \beta$ implies $U_\alpha \subset U_\beta$) and Φ :*

$F_*(U_\alpha) \to G_*(U_\alpha)$ is an isomorphism for each α, then $\Phi : F_*\big(\bigcup_\alpha U_\alpha\big) \to G_*\big(\bigcup_\alpha U_\alpha\big)$ is an isomorphism.

Then $\Phi : F_*(M) \to G_*(M)$ is an isomorphism for every manifold M.

The theorem remains true using instead the category $\mathcal{M}_{\mathrm{PL}}$ of PL manifolds and inclusion of open subsets, using in condition (5.1.1) *the requirement that U be PL homeomorphic to $\mathbb{R}^n$.*

Proof For a manifold M, let $P(M)$ be the statement that $\Phi : F_*(M) \to G_*(M)$ is an isomorphism.

We will first demonstrate the conclusion of the theorem for manifolds that are open subsets of $\mathbb{R}^n$. Note that since $\mathbb{R}^n$ is a PL manifold, so are all such open submanifolds [198, Example 1.9].

So let M be an open subset of $\mathbb{R}^n$. M must possess a countable dense set, and taking open convex PL balls about the points of that dense set provides a countable covering $\mathcal{V}$ by open convex sets, each PL homeomorphic to $\mathbb{R}^n$. Furthermore, as the intersection of open PL convex sets is open PL convex, each non-empty finite intersection $V_{\beta_1} \cap \cdots \cap V_{\beta_m}$ of elements of $\mathcal{V}$ is PL homeomorphic[1] to $\mathbb{R}^n$.

By assumption (5.1.1), the statement $P(V_{\beta_1} \cap \cdots \cap V_{\beta_m})$ is true for each such intersection of elements of $\mathcal{V}$, and in particular $P(V_\beta)$ is true for each $V_\beta \in \mathcal{V}$.

Next we will show that $P(U)$ is true for any U that is the finite union of finite intersections of elements of $\mathcal{V}$, i.e. for $U = \bigcup_{i=1}^k U_i$, where each U_i has the form $U_i = \bigcap_{j=1}^{\ell_i} V_{i,j}$, $V_{i,j} \in \mathcal{V}$. As the base case, we have already seen that $P(U)$ is true when $k = 1$. Now assume that $P(U)$ is true for the union of fewer than k finite intersections of elements of $\mathcal{V}$, $k > 1$. We notice that

$$U_k \cap \left(\bigcup_{i=1}^{k-1} U_i\right) = \bigcup_{i=1}^{k-1} (U_k \cap U_i),$$

[1] It turns out, as of the time of writing, that even a proof that an open convex subset of $\mathbb{R}^n$ is homeomorphic to $\mathbb{R}^n$ is hard to find in the literature, so much so that math reference websites seem to comment on the obscurity [183]. A proof of topological homeomorphism written in 2012 appears as [103]. We require the stronger PL statement, which follows[2] from the yet stronger statement that every open star-shaped region of $\mathbb{R}^n$ is C^∞-diffeomorphic (and hence PL homeomorphic) to $\mathbb{R}^n$. This is also apparently a well-known folk theorem, though, as observed by Bruce Evans in a Mathematics Stack Exchange post [78], "Most books don't prove it. Some say that it is hard and others give it as an exercise." Evans outlines an argument in his post, while an explicit proof due to Stefan Born appears as [80, Theorem 237]. Since, for our purposes, it is sufficient to have only a C^1-diffeomorphism from a bounded convex open subset of $\mathbb{R}^n$ to $\mathbb{R}^n$, we can also cite Gromov [117, I.4.C1].

[2] This is yet another seemingly well-known fact that is difficult to pin down authoritatively. It seems to follow from J.H.C. Whitehead's proof of the existence and uniqueness of smooth triangulations of C^1-manifolds [243], an expository treatment of which can be found in Munkres's work [180]. Here is an argument: Let $f : M \to N$ be a diffeomorphism of C^1-manifolds. By [180, Theorem 10.6], M and N each possess C^1-triangulations, say via maps $k : |K| \to M$ and $\ell : |L| \to N$, where K and L are simplicial complexes. The composition fk provides another C^1-triangulation of N; see [180, Theorem 8.4]. Now, by [180, Theorem 10.5], since $fk : |K| \to N$ and $\ell : |L| \to N$ are two C^1-triangulations of N, there exist subdivisions of K and L that are "linearly isomorphic." Via the definitions given in [180, p. 70], this means precisely that $|K|$ and $|L|$ are PL homeomorphic, and hence so are M and N by Remark B.2.16.

and since each U_i is a finite intersection of elements of $\mathcal{V}$, the same is true of each $U_k \cap U_i$, $i < k$. So $P\big(U_k \cap \big(\bigcup_{i=1}^{k-1} U_i\big)\big)$ holds by induction, as does $P\big(\bigcup_{i=1}^{k-1} U_i\big)$. Since $P(U_k)$ holds by assumption (5.1.1), it follows now from assumption (5.1.1) and the Five Lemma that $P(U)$ is true.

Now, let $W_k = \bigcup_{i \le k} V_i$, where the indices now reflect that $\mathcal{V}$ is countable, so we can choose a bijection of $\mathcal{V}$ with the natural numbers to obtain an order. It follows from the previous paragraph that each $P(W_k)$ is true, and hence by assumption (5.1.1), $P(\bigcup W_k) = P(M)$ must be true. This completes the proof for the case where M is an open subset of $\mathbb{R}^n$.

Now, we let M be an arbitrary (non-empty) n-dimensional manifold (which we assume to be Hausdorff, but not necessarily second-countable). Let $\mathcal{U}$ be the collection of open sets of M for which $P(U)$ holds. Since every point of M has a neighborhood homeomorphic to $\mathbb{R}^n$, $\mathcal{U}$ is non-empty by condition (5.1.1). The set $\mathcal{U}$ is partially ordered by inclusion, and assumption (5.1.1) implies that every totally ordered collection $\{U_\alpha\}$ has an upper bound in $\mathcal{U}$, namely $\bigcup_\alpha U_\alpha$. By Zorn's Lemma, it follows that $\mathcal{U}$ has a maximal element. If this maximal element is M, then we are finished; so suppose that there is a maximal element W of $\mathcal{U}$ such that $W \neq M$. Suppose $x \in M - W$, and let V be a neighborhood of x homeomorphic to $\mathbb{R}^n$. Then $V \cap W$ is homeomorphic to an open subset of $\mathbb{R}^n$ (possibly empty) and so $P(V \cap W)$ holds by the argument above. Furthermore, $P(W)$ is true by assumption and $P(V)$ holds because $V \cong \mathbb{R}^n$. Therefore $P(V \cup W)$ holds by assumption (5.1.1) and the Five Lemma, contradicting the maximality of W. It follows that in fact $W = M$, and $P(M)$ is true.

The argument given in the last paragraph continues to hold in $\mathcal{M}_{\mathrm{PL}}$ if we assume M to be a (non-empty) PL manifold, using that every point has a neighborhood PL homeomorphic to $\mathbb{R}^n$. □

The technique of the proof yields the following variant of the theorem, which provides a useful alternative perspective:

Theorem 5.1.2 *Let $\mathcal{M}_M$ be the category whose objects are homeomorphic to open subsets of a given n-dimensional manifold M and whose morphisms are homeomorphisms and inclusions, and let $\mathcal{A}b_*$ be the category of graded abelian groups. Let $F_*, G_* : \mathcal{M}_M \to \mathcal{A}b_*$ be covariant functors and let $\Phi : F_* \to G_*$ be a natural transformation such that F_*, G_*, Φ satisfy the conditions of Theorem 5.1.1 with respect to $\mathcal{M}_M$. Then $\Phi : F_*(M) \to G_*(M)$ is an isomorphism.*

The theorem remains true using the category $\mathcal{M}_{M,PL}$ of open subsets of a given PL n-manifold M using in condition (5.1.1) *the requirement that U be PL homeomorphic to $\mathbb{R}^n$.*

Proof Since M is a manifold, it has an open subset homeomorphic to $\mathbb{R}^n$ and so all open subsets of $\mathbb{R}^n$ are homeomorphic to open subsets of M, so the first part of the proof of Theorem 5.1.2 goes through unchanged. Then we observe that the conclusion of Theorem 5.1.2 for the space M only utilizes open subsets of M. □

We will use Theorem 5.1.1 below to prove Theorem 5.2.25, which is a Künneth Theorem for intersection homology of spaces of the form $X \times M$, where X is a filtered space and M is an unfiltered manifold. The reader who is interested in seeing an immediate application of Theorem 5.1.1 could safely peek ahead to that theorem at this point.

We next provide a Mayer–Vietoris argument for CS sets. This theorem is a variation of a theorem of King [139, Theorem 10] that we have adapted a bit to suit the purposes for which we will need it.

Remark 5.1.3 The theorem does utilize the local structure available in CS sets, and so going forward to applications there will be things we can prove for CS sets that are not accessible for arbitrary filtered spaces by these techniques. However, it is sometimes possible to extend the applications to some spaces that are not CS sets, particularly to ∂-stratified pseudomanifolds, which are of particular interest. We will not provide these extensions every time, but for a good example of the typical arguments involved in such generalizations see Corollary 5.4.5.

Theorem 5.1.4 *Let $\mathcal{F}_X$ be the category whose objects are filtered homeomorphic to open subsets of a given CS set X and whose morphisms are filtered homeomorphisms and inclusions. Let $\mathcal{A}b_*$ be the category of graded abelian groups. Let $F_*, G_* : \mathcal{F}_X \to \mathcal{A}b_*$ be covariant functors, and let $\Phi : F_* \to G_*$ be a natural transformation such that F_*, G_*, Φ satisfy the conditions listed below.*

1. *F_* and G_* admit exact Mayer–Vietoris sequences as in Theorem 5.1.1 and Φ induces a commutative diagram of these sequences.*
2. *If $\{U_\alpha\}$ is an increasing collection of open subspaces of X and $\Phi : F_*(U_\alpha) \to G_*(U_\alpha)$ is an isomorphism for each α, then $\Phi : F_*\left(\bigcup_\alpha U_\alpha\right) \to G_*\left(\bigcup_\alpha U_\alpha\right)$ is an isomorphism.*
3. *If L is a compact filtered space such that X has an open subset filtered homeomorphic to $\mathbb{R}^i \times cL$ and $\Phi : F_*(\mathbb{R}^i \times (cL - \{v\})) \to G_*(\mathbb{R}^i \times (cL - \{v\}))$ is an isomorphism (where v is the cone vertex), then so is $\Phi : F_*(\mathbb{R}^i \times cL) \to G_*(\mathbb{R}^i \times cL)$.*
4. *If U is empty or an open subset of X contained within a single stratum and homeomorphic to Euclidean space[3], then $\Phi : F_*(U) \to G_*(U)$ is an isomorphism.*

Then $\Phi : F_*(X) \to G_*(X)$ is an isomorphism.

If X is a PL CS set and $\mathcal{F}_{X,PL}$ is the category whose objects are filtered PL homeomorphic to open subsets of X and whose morphisms are filtered PL homeomorphisms and inclusions, the theorem remains true if we replace the homeomorphisms in the conditions with PL homeomorphisms.

Proof Suppose X is a CS set. Let $\mathfrak{M}$ be the union of the strata of X with depth zero, i.e. the strata that are not contained in the closure of any other strata. Then we can think of $\mathfrak{M}$ as a disjoint union of (not necessarily connected) manifolds M^i, one for each dimension i such

[3] Note that it is possible for a CS set to have open Euclidean subsets of various dimensions; for example, let $X = S^2 \vee S^1$, filtered by $\{x_0\} \subset S^1 \subset X$, where x_0 is the basepoint of the wedge.

that X has non-empty strata of dimension i of depth 0. Each M^i itself is a disjoint union of strata of dimension i. Then every point in $\mathfrak{M}$ has a neighborhood homeomorphic to a Euclidean space. By assumption (4), for any such neighborhood U, the map $\Phi : F_*(U) \to G_*(U)$ is an isomorphism. It follows now from Theorem 5.1.2 and assumptions (2) and (1) that Φ is an isomorphism on each M^i, and in fact on any open submanifold of M^i. But since X must have finite formal dimension, an induction using property (1), the $\emptyset$ hypothesis of property (4), and the Five Lemma implies that Φ is an isomorphism on all of $\mathfrak{M}$, or in fact any open subspace of $\mathfrak{M}$.

The proof for arbitrary open $Y \subset X$ will now proceed by induction on the depth of Y (recall Definition 2.2.29). We have just established the theorem for all Y of depth 0, so we assume that we have verified the theorem for open subsets of X of depth $< K$ for some $K > 0$. We must show that this implies the theorem for Y of depth K. The proof will then be completed by induction up through the depth of X. For the remainder of the proof, let Y be an open subspace of X of depth K.

As in the proof of Theorem 5.1.1, the condition on unions of chains of subspaces allows us to conclude by Zorn's Lemma that there is a largest open subset W of Y on which Φ is an isomorphism. Using the induction assumption, if $Y_{\min}$ is the union of minimal strata in the partial ordering on strata (i.e. the strata S for which there do not exist strata T with $T < S$), then $Y - Y_{\min}$ has depth less than K, so Φ is an isomorphism on $Y - Y_{\min}$. This implies $Y - Y_{\min} \subset W$, since if not there would be a point $y \in Y - Y_{\min}$, $y \notin W$. But then y has an open neighborhood U in Y of depth $< K$ and $W \cap U$ then also has depth $< K$, so by the Mayer–Vietoris sequences and the Five Lemma, the map Φ would be an isomorphism on $W \cup U$, a contradiction.

Now we want to show that $W = Y$, again by a contradiction argument, assuming there is some $y \in Y_{\min}$, $y \notin W$. By the definition of a CS set, the point y has an open neighborhood N that is filtered homeomorphic to $\mathbb{R}^m \times cL$ if y is contained in a stratum of dimension m; note that $L \neq \emptyset$, as we have assumed y is in a stratum of depth > 0. Since $\mathbb{R}^m \times (cL - \{v\})$ has depth $< K$, the map Φ is also an isomorphism on it by the assumption on depth. But then Φ is an isomorphism on N by assumption (3). So if we can show that Φ is an isomorphism on $W \cap N$, then by the Mayer–Vietoris sequences and the Five Lemma, it will follow that Φ is an isomorphism on $W \cup N$, contradicting the maximality of W.

So we consider $W \cap N$. Let $V = Y_{\min} \cap W \cap N$. Since W includes all of $Y - Y_{\min}$, then $W \cap N$ is homeomorphic to the disjoint union of $\mathbb{R}^m \times (cL - \{v\})$ and V. We can also describe $W \cap N$ up to homeomorphism as the (not disjoint) union of $\mathbb{R}^m \times (cL - \{v\})$ with $V \times cL$. Both $\mathbb{R}^m \times (cL - \{v\})$ and the intersection

$$(\mathbb{R}^m \times (cL - \{v\})) \cap (V \times cL) \cong V \times (cL - \{v\})$$

are homeomorphic to open subsets of $Y - Y_{\min}$ and so have depth $< K$; thus Φ is an isomorphism on these sets by the induction hypothesis. So to use the Mayer–Vietoris sequences and the Five Lemma to show that Φ is an isomorphism on $W \cap N$, we only need to show that Φ is an isomorphism on $V \times cL$. For this we will use Theorem 5.1.2, with the manifold in the statement of the theorem being V and, for open $U \subset V$, the functors will be $\hat{F}_*(U) = F_*(U \times cL)$ and $\hat{G}_*(U) = G_*(U \times cL)$ and $\hat{\Phi}$ will be Φ restricted to sets of this form. The second and third hypotheses of Theorem 5.1.2 follow immediately from the corresponding statements for Φ, and the first hypothesis is satisfied since we have already seen in the preceding paragraph that Φ must be an isomorphism on any $U \times cL$ with $U \subset Y_{\min}$ homeomorphic to $\mathbb{R}^m$. Theorem 5.1.2 then provides that $\hat{\Phi}$ is an isomorphism on all of V, which is equivalent to the statement that $\Phi : F_*(V \times cL) \to G_*(V \times cL)$ is an isomorphism.

The proof in the PL case is identical. □

Remark 5.1.5 Theorem 5.1.4 implies, and in some ways generalizes, a result of King [139, Theorem 10]. Let us explain the relation.

The first main difference is that King assumes his functors are defined on a category of all filtered spaces whose maps are open inclusions and inclusions $0 \times X \to \mathbb{R}^k \times X$. His conclusion then holds for all CS sets. By contrast, our theorem essentially proves the theorem one CS set at a time. An immediate benefit is that we do not need to make sure our functors are defined for all filtered sets but only on spaces filtered homeomorphic to subsets of the particular CS set X. This does not necessarily weaken the conclusion since the arbitrariness of the X in the statement of Theorem 5.1.4 allows for the possibility that we might draw conclusions for all CS sets, provided F_*, G_*, and Φ possess the hypothesized properties in this generality. However, our version of the theorem is in some sense more flexible, since if one can produce F_*, G_*, and Φ that satisfy the hypothesized conditions on some particular class of CS set possessing a property that is preserved by taking open subsets (e.g. PL CS sets, pseudomanifolds, oriented CS sets, or locally torsion-free CS sets (see Definition 5.3.9)), then one can use Theorem 5.1.4 to draw conclusions just for the spaces in this class.

The second main difference is that in place of condition (3), King has the conditions

- the inclusion $0 \times X \hookrightarrow \mathbb{R}^k \times X$ induces isomorphisms on F_* and G_*,
- if L is a compact filtered space and Φ is an isomorphism on L, then Φ is an isomorphism on cL, and
- if Φ is an isomorphism on the filtered space L, then Φ is an isomorphism on $M \times L$ for a manifold M.

The first two of these properties can be used to imply our hypothesis (3), though not necessarily vice versa. The last property here would be used in King's version of the theorem to provide the last isomorphism of the proof of Theorem 5.1.4, but here we instead use (3), induction, and an appeal to Theorem 5.1.2.

The following lemma will be a useful way, in practice, to conclude that condition (2) of Theorem 5.1.4 holds.

Lemma 5.1.6 *Let $\mathcal{F}_X$ be the category whose objects are homeomorphic to open subsets* *of a given CS set X and whose morphisms are filtered homeomorphisms and inclusions. Let $\mathcal{A}b_*$ be the category of graded abelian groups. Let $F_*, G_* : \mathcal{F}_X \to \mathcal{A}b_*$ be functors, and let $\Phi : F_* \to G_*$ be a natural transformation. Suppose that if $\{U_\alpha\}$ is an increasing collection of open subspaces of X then the natural maps $\varinjlim_\alpha F_*(U_\alpha) \to F_*(\bigcup_\alpha U_\alpha)$ and $\varinjlim_\alpha G_*(U_\alpha) \to G_*(\bigcup_\alpha U_\alpha)$ are isomorphisms. Then if $\Phi : F_*(U_\alpha) \to G_*(U_\alpha)$ is an isomorphism for each α, the map $\Phi : F_*(\bigcup_\alpha U_\alpha) \to G_*(\bigcup_\alpha U_\alpha)$ is an isomorphism.*

Proof The naturality of Φ implies that we have a commutative diagram

$$\begin{array}{ccc} \varinjlim\limits_{\alpha} F_*(U_\alpha) & \longrightarrow & F_*\left(\bigcup\limits_\alpha U_\alpha\right) \\ \Big\downarrow {\scriptstyle \varinjlim_\alpha \Phi} & & \Big\downarrow {\scriptstyle \Phi} \\ \varinjlim\limits_{\alpha} G_*(U_\alpha) & \longrightarrow & G_*\left(\bigcup\limits_\alpha U_\alpha\right). \end{array}$$

By assumption, the horizontal maps are isomorphisms, and the left vertical map is an isomorphism since we have assumed each $\Phi : F_*(U_\alpha) \to G_*(U_\alpha)$ is. Hence the right-hand vertical arrow is also an isomorphism. □

To accompany this lemma, it is worth adding another lemma:

Lemma 5.1.7 *If X is a filtered space with perversity $\bar{p}$ and $\{U_\alpha\}$ is an increasing collection of open subspaces of X then the natural map $f : \varinjlim_\alpha I^{\bar{p}}H_*^{\mathrm{GM}}(U_\alpha) \to I^{\bar{p}}H_*^{\mathrm{GM}}(\bigcup_\alpha U_\alpha)$ is an isomorphism. Similarly, if X is PL then $\varinjlim_\alpha I^{\bar{p}}\mathfrak{H}_*^{\mathrm{GM}}(U_\alpha) \to I^{\bar{p}}\mathfrak{H}_*^{\mathrm{GM}}(\bigcup_\alpha U_\alpha)$ is an isomorphism.*

Proof This lemma is well known for ordinary homology and the proof for intersection homology is identical: If $[\xi] \in I^{\bar{p}}H_*^{\mathrm{GM}}(\bigcup_\alpha U_\alpha)$, then $[\xi]$ is represented by some specific cycle ξ, which has compact support. Hence ξ is contained in U_k for some k. It follows that the image of the element of $I^{\bar{p}}H_*^{\mathrm{GM}}(U_k)$ represented by ξ under the natural maps

$$I^{\bar{p}}H_*^{\mathrm{GM}}(U_k) \to \varinjlim_\alpha I^{\bar{p}}H_*^{\mathrm{GM}}(U_\alpha) \to I^{\bar{p}}H_*^{\mathrm{GM}}\left(\bigcup_\alpha U_\alpha\right)$$

represents $[\xi]$, so f is surjective.

Similarly, if $[\xi] \in \varinjlim_\alpha I^{\bar{p}}H_*^{\mathrm{GM}}(U_\alpha)$ and $f([\xi]) = 0$, then ξ is represented by a cycle ξ contained in some U_k, and ξ bounds some Ξ in $I^{\bar{p}}S_*^{\mathrm{GM}}(\bigcup_\alpha U_\alpha)$. But Ξ must also have compact support and so is contained in U_ℓ for some $\ell \geq k$. But it then follows that $[\xi] = 0 \in \varinjlim_\alpha I^{\bar{p}}H_*^{\mathrm{GM}}(U_\alpha)$.

The argument in the PL case is the same. □

5.1.1 First Applications: High Perversities and Normalization

Let us provide two fairly straightforward applications of Mayer–Vietoris arguments.

(a) High Perversities

As a first application of Mayer–Vietoris arguments on CS sets, we will prove the following proposition:

Proposition 5.1.8 *Let X be a CS set and $\bar{p}$ a perversity such that*

1. *every point in X has a neighborhood filtered homeomorphic to $\mathbb{R}^k \times cL$ such that $I^{\bar{p}}H_0^{\mathrm{GM}}(cL) \cong \mathbb{Z}$ and $I^{\bar{p}}H_i^{\mathrm{GM}}(cL) = 0$ for $i > 0$, and*
2. *the only strata of depth zero are regular strata.*

Then $I^{\bar{p}}H_*^{\mathrm{GM}}(X) \cong H_*(X)$, and similarly for PL intersection homology.

Before proving the proposition, we observe that the conditions required are not as extraordinary as they might at first seem. In fact, we have the following corollary, the first part of which was first demonstrated for PL stratified pseudomanifolds in [105].

Corollary 5.1.9 *The conditions of Lemma 5.1.8 hold if X is a* normal[4] *stratified pseudomanifold and $\bar{p}$ is the top perversity $\bar{t}$ such that $\bar{t}(S) = \operatorname{codim}(S) - 2$ for each singular stratum. So, in particular, if X is a normal stratified pseudomanifold, then $I^{\bar{t}}H_*^{\mathrm{GM}}(X) \cong H_*(X)$.*

Furthermore, the conditions hold for an arbitrary (not necessarily normal) stratified pseudomanifold if $\bar{p}(S) \geq \bar{t}(S)$ for all strata and $\bar{p}(S) > \bar{t}(S)$ for any stratum containing a point that has a link that is not connected.

Proof The condition that the only depth-zero strata are regular strata holds for all stratified pseudomanifolds by the definition, Definition 2.4.1, which requires that the union of the regular strata of X be dense in X.

For the first condition, assuming X is normal, we utilize the cone formula (Theorem 4.2.1). If X has dimension n and x is in a stratum of dimension i (hence codimension $n - i$), then L has dimension $n - i - 1$, so $I^{\bar{t}}H_*^{\mathrm{GM}}(\mathbb{R}^i \times cL) \cong I^{\bar{t}}H_*^{\mathrm{GM}}(cL)$ is trivial except when

$$* < n - i - \bar{t}(S) - 1 = n - i - (\operatorname{codim}(S) - 2) - 1 = n - i - (n - i) + 1 = 1.$$

Therefore, the only nontrivial group can be $I^{\bar{t}}H_0^{\mathrm{GM}}(cL) \cong I^{\bar{t}}H_0^{\mathrm{GM}}(L)$. As follows from Example 3.4.6, $I^{\bar{t}}H_0^{\mathrm{GM}}(L) \cong \mathbb{Z}^m$, where m is the number of regular strata of L. But by Lemma 2.6.3, L is itself a normal stratified pseudomanifold, and since it is connected, it has only one regular stratum, again by Lemma 2.6.3. Therefore, $m = 1$.

Finally, if there are strata for which the links are not connected, we need only observe that

- if $\bar{p}(S) > \bar{t}(S)$, then $0 \geq n - i - \bar{p}(S) - 1$, and
- $I^{\bar{p}}H_0^{\mathrm{GM}}(L) \neq 0$, as L is a stratified pseudomanifold by Lemma 2.4.11 and so possesses regular strata (and therefore allowable 0-simplices).

Thus again $I^{\bar{p}}H_0^{\mathrm{GM}}(L) \cong \mathbb{Z}$ by Theorem 4.2.1. □

[4] Recall from Definition 2.6.1 that this means that the link of any point is connected.

Proof of Proposition 5.1.8 We apply Theorem 5.1.4 with $F_*(U) = I^{\bar{p}}H_*^{\mathrm{GM}}(U)$, $G_*(U) = H_*(U)$, and Φ induced by the inclusion $I^{\bar{p}}S_*^{\mathrm{GM}}(U) \to S_*(U)$; the proof in the PL case is equivalent. The first condition of the theorem holds because the short exact Mayer–Vietoris sequence of intersection chain complexes maps to the corresponding ordinary short exact Mayer–Vietoris sequence of chain complexes, yielding a commutative diagram. This implies the commutative diagram of long exact homology sequences. Also, since the only strata of depth zero are regular strata, the only Euclidean open sets of X contained in a single stratum must all be subsets of regular strata, and so also the last condition is fulfilled tautologically, as all simplices in regular strata are allowable by Lemma 3.4.3. For the second condition, we invoke Lemmas 5.1.7 and 5.1.6, noting that the version of Lemma 5.1.7 for ordinary homology is standard (the proof is identical).

For the third condition, consider for a distinguished neighborhood the map

$$\Phi : I^{\bar{p}}H_*^{\mathrm{GM}}(\mathbb{R}^i \times cL) \to H_*(\mathbb{R}^i \times cL)$$

induced by inclusion, under the assumption that

$$I^{\bar{p}}H_*^{\mathrm{GM}}(\mathbb{R}^i \times (cL - \{v\})) \to H_*(\mathbb{R}^i \times (cL - \{v\}))$$

is an isomorphism. We wish to show that

$$I^{\bar{p}}H_*^{\mathrm{GM}}(\mathbb{R}^i \times cL) \to H_*(\mathbb{R}^i \times cL)$$

is an isomorphism; in fact, we will not even need the assumption. By homotopy invariance of homology, $H_*(\mathbb{R}^i \times cL)$ is trivial except for $H_0(\mathbb{R}^i \times cL) \cong \mathbb{Z}$. By stratified homotopy invariance of intersection homology $I^{\bar{p}}H_*^{\mathrm{GM}}(\mathbb{R}^i \times cL) \cong I^{\bar{p}}H_*^{\mathrm{GM}}(cL)$, and by assumption $I^{\bar{p}}H_0^{\mathrm{GM}}(cL) \cong \mathbb{Z}$ and $I^{\bar{p}}H_i^{\mathrm{GM}}(cL) = 0$ for $i > 0$. So abstractly

$$I^{\bar{p}}H_0^{\mathrm{GM}}(\mathbb{R}^i \times cL) \cong \mathbb{Z} \cong H_0(\mathbb{R}^i \times cL),$$

but it is now not difficult to observe that a generator for the former group, given by some 0-simplex, maps to a generator of the latter group. Technically, this is not quite enough yet to finish the proof, as Theorem 5.1.4 requires the third condition to hold for all neighborhoods of the form $\mathbb{R}^k \times cL$, while we have only assumed here that each point of X has at least one distinguished neighborhood satisfying the hypothesis. However, there are two solutions to this problem: One is to invoke Lemma 5.3.13, to be proven below, which states that every distinguished neighborhood of a point in a CS set has the same perversity $\bar{p}$ intersection homology. The other is to observe that, in fact, the proof of Theorem 5.1.4 only requires its condition (3) to hold for some distinguished neighborhood of each point, so in fact this hypothesis of the theorem could be weakened. □

Remark 5.1.10 Corollary 5.1.9 is not true for a general CS set with perversity $\bar{t}$, even for ones for which every point has a connected link. For example, let $X = S^2 \amalg S^1$, the disjoint union filtered by $S^1 \subset X$. Then by the computations of Example 3.4.6, no allowable 0- or

1-simplices may intersect S^1. It follows that we must have $I^{\bar{t}}H_1^{\text{GM}}(X) = 0$, while of course $H_1(X) \cong \mathbb{Z}$.

The normality condition is also critical, assuming perversity $\bar{t}$. For example, the cone X on the disjoint union $S^2 \amalg S^2$ is a stratified pseudomanifold. But $H_0(X) \cong \mathbb{Z}$, while $I^{\bar{t}}H_0^{\text{GM}}(X) \cong \mathbb{Z} \oplus \mathbb{Z}$.

(b) Normalization

In Section 2.6 we discussed normal pseudomanifolds and normalizations. We claimed there that intersection homology is preserved under normalization, at least with some minor restrictions. We here provide and prove the technical statement.

To see why we need some restriction on perversities, suppose that $X = S^2 \vee S^2$, the wedge of two spheres, filtered as $\{v\} \subset X$ with v the wedge point. In this case the normalization is $\tilde{X} = S^2 \amalg S^2$. If we choose perversities that are too high, then intersection homology will degenerate into ordinary homology by the preceding example, and we will have $I^{\bar{p}}H_0^{\text{GM}}(X) \cong \mathbb{Z}$ while $I^{\bar{p}}H_0^{\text{GM}}(\tilde{X}) \cong \mathbb{Z} \oplus \mathbb{Z}$. In fact, this problem will occur if any 1-simplices are allowed to intersect v at their endpoints, i.e. if $1 - \text{codim}(\{v\}) + \bar{p}(\{v\}) = -1 + \bar{p}(\{v\}) \geq 0$, or if $\bar{p}(\{v\}) \geq 1 = \text{codim}(\{v\}) - 1$. A generalization of this argument shows that to get the desired result in general we must limit ourselves to perversities such that $\bar{p}(S) \leq \text{codim}(S) - 2$ for all singular strata S, which was one of the original Goresky–MacPherson requirements.

Proposition 5.1.11 *Let X be a stratified pseudomanifold, and let $\pi : \tilde{X} \to X$ be a normalization. Suppose $\bar{p}$ is a perversity on X such that $\bar{p}(S) \leq \text{codim}(S) - 2$ for all singular strata S, and let $\tilde{p}$ be the perversity on $\tilde{X}$ defined so that $\tilde{p}(\tilde{S}) = \bar{p}(S)$ if $\pi(\tilde{S}) \subset S$. Then the map $\pi : I^{\tilde{p}}H_*^{\text{GM}}(\tilde{X}) \to I^{\bar{p}}H_*^{\text{GM}}(X)$ is an isomorphism.*

Proof If U is an open subset of X, let $G_*(U) = I^{\bar{p}}H_*^{\text{GM}}(U)$ and $F_*(U) = I^{\tilde{p}}H_*^{\text{GM}}(\pi^{-1}(U))$. We let $\Phi : F_* \to G_*$ be the map induced by p. The proposition will follow from Theorem 5.1.4 once we have verified the conditions on F_*, G_*, and Φ.

The Mayer–Vietoris sequences for F_* and G_* are just the intersection homology Mayer–Vietoris sequences in X or $\tilde{X}$, which we know exist from Theorem 4.4.19. The map π induces a natural transformation between the sequences because we have maps of the short exact Mayer–Vietoris sequences at the chain level. So the first condition of Theorem 5.1.4 holds. The fourth condition holds trivially because π is a homeomorphism when restricted to the regular strata. The second condition on increasing collections of open sets follows from Lemmas 5.1.6 and 5.1.7 applied to $I^{\bar{p}}H_*^{\text{GM}}(U_\alpha)$ and $I^{\tilde{p}}H_*^{\text{GM}}(\pi^{-1}(U_\alpha))$.

For the third condition of Theorem 5.1.4, we suppose that

$$p : I^{\tilde{p}}H_*^{\text{GM}}(\pi^{-1}(\mathbb{R}^k \times (cL - \{v\}))) \to I^{\bar{p}}H_*^{\text{GM}}(\mathbb{R}^k \times (cL - \{v\}))$$

is an isomorphism for each distinguished neighborhood $\mathbb{R}^k \times cL$ of a point x in a stratum S of X. From Definition 2.6.8 and stratified homotopy invariance, this is equivalent to assuming that the normalization map $\pi_L : \tilde{L} \to L$ on the link L induces an intersection homology isomorphism. Given this and recalling our notation from Definition 2.6.8 that the K_j are the connected components of $\tilde{L}$ and that $\pi^{-1}(\mathbb{R}^k \times cL) \cong \mathbb{R}^k \times \amalg cK_j$, the third condition reduces to showing that we have an isomorphism $\pi : I^{\tilde{p}}H_*^{\mathrm{GM}}(\mathbb{R}^k \times \amalg cK_j) \to I^{\bar{p}}H_*^{\mathrm{GM}}(\mathbb{R}^k \times cL)$. Up to isomorphisms induced by stratified homotopy equivalence, this reduces to considering the diagram

$$\begin{array}{ccc} I^{\tilde{p}}H_i^{\mathrm{GM}}(\amalg K_j) & \longrightarrow & I^{\tilde{p}}H_i^{\mathrm{GM}}(\amalg cK_j) \\ \downarrow & & \downarrow \\ I^{\bar{p}}H_i^{\mathrm{GM}}(L) & \longrightarrow & I^{\bar{p}}H_i^{\mathrm{GM}}(cL) \end{array}$$

with the horizontal maps induced by inclusions. The left-hand vertical map is precisely our assumed isomorphism $I^{\tilde{p}}H_i^{\mathrm{GM}}(\tilde{L}) \to I^{\bar{p}}H_i^{\mathrm{GM}}(L)$. From our assumptions about $\bar{p}$, if v is the cone point of cL, we have $\dim(L) - \bar{p}(v) = \mathrm{codim}(S) - 1 - \bar{p}(\{v\}) \geq 1$. So by the cone formula (Theorem 4.2.1), $I^{\bar{p}}H_i^{\mathrm{GM}}(cL)$ vanishes for $i \geq \dim(L) - p(v)$. Furthermore, from the definition of $\tilde{p}$, the groups $I^{\tilde{p}}H_i^{\mathrm{GM}}(\amalg cK_j)$ vanish in the same range. Also by the cone formula, the two horizontal maps, and so then also the right-hand vertical map, are isomorphisms in the complementary range $i < \dim(L) - p(v)$. Therefore, $\pi : I^{\tilde{p}}H_i^{\mathrm{GM}}(\mathbb{R}^k \times \amalg cK_j) \to I^{\bar{p}}H_i^{\mathrm{GM}}(\mathbb{R}^k \times cL)$ is an isomorphism for all i, which is what we needed to show. □

5.2 Cross Products and the Künneth Theorem with a Manifold Factor

In this section, we develop cross products for intersection chains. This will lead in Section 5.2.4 to a Künneth Theorem that holds for products $M \times X$, where X is a filtered space, M is an n-manifold with the trivial filtration, and $M \times X$ is given the product filtration $(M \times X)^i = M \times X^{i-n}$. We will show that the perversity $\bar{p}$ intersection homology of $M \times X$ is related by a Künneth formula to the ordinary homology of M and to the intersection homology of X with a corresponding perversity. The precise statement is in Theorem 5.2.25. Later, in Section 6.4.7, we will consider another, more general, Künneth Theorem for products of arbitrary CS sets, each with their own perversity.

We begin with the cross product for singular intersection chains in Section 5.2.1 and then treat the PL cross product in Section 5.2.2. Section 5.2.3 develops the basic properties of the cross products, and the Künneth Theorem involving a manifold factor comes in Section 5.2.4.

5.2.1 The Singular Chain Cross Product

For ordinary singular homology theory, the key morphism for topological Künneth theorems is the Eilenberg–Zilber cross product map

$$\varepsilon : S_*(X) \otimes S_*(Y) \to S_*(X \times Y).$$

In modern algebraic topology texts, the preference seems to be to construct this map abstractly using the method of acyclic models[5] (see [182, 220]). Unfortunately, this is insufficient for our purposes, as we have seen that even contractible spaces might have nontrivial intersection homology, depending on how they are filtered, and so we don't have the needed acyclicity. Luckily, there do exist concrete versions of the cross product, sometimes called Eilenberg–Zilber "shuffle products." Most sources (e.g. [155]) prefer to describe this map from the point of view of simplicial sets. However, a statement purely from the point of view of singular homology can be found as an exercise in Dold [71, Exercise VI.12.26.2].

The basic idea is analogous to that of the prism construction used in homotopy arguments: If $\sigma \in S_p(X)$ and $\tau \in S_q(Y)$ are singular simplices, then together they provide a product map

$$\sigma \times \tau : \Delta^p \times \Delta^q \to X \times Y.$$

If we have a suitable triangulation T of $\Delta^p \times \Delta^q$ with an ordering of the vertices, if $\{\delta_k^{p+q}\}$ is the collection of $(p+q)$-simplices of T, and if $\eta_k : \Delta^{p+q} \hookrightarrow \Delta^p \times \Delta^q$ is the vertex-order-preserving embedding of Δ^{p+q} onto δ_k, then this yields an element $\varepsilon(\sigma \otimes \tau)$ of $S_{p+q}(X \times Y)$ by

$$\varepsilon(\sigma \otimes \tau) = \sum \pm (\sigma \times \tau) \circ \eta_k.$$

Here the signs must be chosen appropriately, and of course the triangulations for various p and q must be chosen in such a way that the construction extends to a chain map. The standard such set of triangulations is described in terms of *shuffles*.

Let p, q be non-negative integers. Then a *(p,q)-shuffle* is a partition of the ordered set $[1, 2, \dots, p+q]$ into two disjoint ordered sets $\mu = [\mu_1, \dots, \mu_p]$ and $\nu = [\nu_1, \dots, \nu_q]$ with $\mu_i < \mu_{i+1}$ for each i and similarly for the ν_j. The idea is that this partition (μ, ν) tells us how to shuffle together two ordered sets, of respective cardinalities p and q, to form a new ordered set of cardinality $p+q$: the elements of the first set occupy the spots labeled by the μ and the elements of the second set are placed in the spots corresponding to the ν. So, for example, if we have ordered sets $[A, B, C]$ and $[\alpha, \beta]$, and a $(3,2)$-shuffle $([2,3,5],[1,4])$, then we can shuffle our sets by this prescription to get the ordered set $[\alpha, A, B, \beta, C]$. Note that the elements A, B, C go in spots $2, 3, 5$ as prescribed by the contents of μ, while α, β go in spots $1, 4$ as prescribed by the contents of ν.

[5] By contrast, the Alexander–Whitney map, which is a chain homotopy inverse for the cross product, is often constructed explicitly, but it will not be useful for constructing intersection homology and cohomology products. See Section 7.2 for a detailed discussion.

Another way to think of a (p,q)-shuffle is to imagine a walk on a $p \times q$ grid, where columns are labeled left to right by $\{0,\dots,p\}$ and the rows are labeled bottom to top by $\{0,\dots,q\}$. Then there is a bijection between (p,q)-shuffles and walks along the grid from $(0,0)$ to (p,q) in which each step must move one unit either up or to the right: on the ith step, if $i \in \mu$ we move to the right and if $i \in \nu$ we move up. Conversely, given such a path, if we move right on ith step then we put $i \in \mu$, and if we move up on the ith step, we put $i \in \nu$.

To see how shuffles give us triangulation of products, we label $\Delta^p = [u_0,\dots,u_p]$, $\Delta^q = [v_0,\dots,v_q]$, and $\Delta^{p+q} = [w_0,\dots,w_{p+q}]$, again as simplices with ordered vertices. Let $\eta^\mu : \Delta^{p+q} \to \Delta^p$ take the vertex $w_i \in \Delta^{p+q}$ to the vertex $u_j \in \Delta^p$ if $\mu_j \le i < \mu_{j+1}$ (letting $\mu_0 = 0$ and $\mu_{p+1} = p+q+1$), and define $\eta^\nu : \Delta^{p+q} \to \Delta^q$ analogously. We obtain a map

$$\eta_{\mu\nu} = (\eta^\mu, \eta^\nu) : \Delta^{p+q} \to \Delta^p \times \Delta^q$$

by extending linearly from what this map must do to vertices, and it is a linear embedding. We denote the image of $\eta_{\mu\nu}$ by $\delta_{\mu\nu}$, and we claim that the $\delta_{\mu\nu}$, as (μ,ν) ranges over all (p,q)-shuffles, are the $(p+q)$-simplices of a triangulation of $\Delta^p \times \Delta^q$.

To try to understand this triangulation even better, let us see explicitly where the vertices $\{w_i\}$ of Δ^{p+q} get mapped by $\eta_{\mu\nu}$. Since $\nu_0 = \mu_0 = 0$ by definition, w_0 gets mapped to (u_0, v_0). Now, if $1 \in \mu$, then w_1 gets mapped to (u_1, v_0), and if $1 \in \nu$, then w_1 gets mapped to (u_0, v_1). In general, if $w_i = w_{j+k}$ goes to (u_j, v_k), then w_{i+1} will go to either (u_{j+1}, v_k) or (u_j, v_{k+1}) depending respectively on whether $i+1$ is in μ or ν. In terms of walks on planar grids, if (μ,ν) is a shuffle, then the sequence of labels of the grid points of the corresponding walk gives the sequence of vertices of $\delta_{\mu\nu}$, with the point (j,k) on the grid indicating the vertex (u_j, v_k). This is the same principle we saw at work in constructing prisms for homotopy arguments, except those always have the form $\Delta^1 \times \Delta^q$, so there are only two choices for vertices of Δ^1 and so only one possible step "to the right."

Proving that this construction does indeed yield a triangulation of $\Delta^p \times \Delta^q$ and that it leads to a chain map is nontrivial, but it is also a bit of a diversion from our main development, so we include the details in the appendix in Section B.6 along with our more general development of PL spaces. More specifically, we show that if K and L are simplicial complexes with given vertex partial orderings that restrict to total orderings on each simplex, then there is a triangulation denoted $K \times L$ of $|K| \times |L|$ described locally in terms of shuffles. Furthermore, we define in Section B.6.5 a chain map of simplicial chain complexes

$$\bowtie : C_*(K) \otimes C_*(L) \to C_*(K \times L).$$

This *simplicial cross product* is defined so that if σ is a simplex of K oriented by the vertex ordering on K and if τ is similarly a vertex of L, then

$$\sigma \bowtie \tau = \sum \operatorname{sgn}(\mu,\nu)\delta_{\mu\nu},$$

where the sum is taken over all (p,q)-shuffles with $\delta_{\mu\nu}$ oriented according to the vertex ordering coming from the shuffle construction and with $\operatorname{sgn}(\mu,\nu)$ denoting the sign of the permutation from $[1,2,\dots,p+q]$ to $[\mu_1,\mu_2,\dots,\mu_p,\nu_1,\nu_2,\dots,\nu_q]$, i.e. 1 if the permutation is even and -1 if the permutation is odd. Furthermore, it is shown in Proposition B.6.8 that $\sigma \bowtie \tau$ is a fundamental class for the space $\sigma \times \tau$, i.e. it is the generator of $H_{p+q}(\sigma \times \tau, \partial(\sigma \times \tau))$ compatible with the product orientation.

To get from this to our singular cross product, let $\sigma_1 : \Delta^p \to X$ and $\sigma_2 : \Delta^q \to Y$ be singular simplices. Treating Δ^p and Δ^q as simplicial complexes, the shuffle product construction gives us a triangulation of $\Delta^p \times \Delta^q$ and an element of $C_{p+q}(\Delta^p \times \Delta^q)$ arising as the image under $\bowtie$ of the generators of $C_p(\Delta^p)$ and $C_q(\Delta^q)$ given by the vertex ordering (abusing notation, we also denote these generators as Δ^p and Δ^q). We again write

$$\Delta^p \bowtie \Delta^q = \sum \operatorname{sgn}(\mu, \nu)\delta_{\mu\nu} \in C_{p+q}(\Delta^p \times \Delta^q).$$

As the $\delta_{\mu\nu}$ also have vertex orderings, we can apply the chain map $\phi : C_*(\Delta^p \times \Delta^q) \to S_*(\Delta^p \times \Delta^q)$ of Proposition 4.4.5 that takes a k-simplex s of a simplicial complex to the embedding $\Delta^k \to s$ determined by the vertex orderings of Δ^k and s. In this setting, ϕ takes $\delta_{\mu\nu}$ to $\eta_{\mu\nu}$. Then we let

$$\varepsilon(\sigma_1 \otimes \sigma_2) = (\sigma_1 \times \sigma_2)\phi(\Delta^p \bowtie \Delta^q) = \sum \operatorname{sgn}(\mu, \nu)(\sigma_1 \times \sigma_2) \circ \eta_{\mu\nu}.$$

Proposition 5.2.1 *Suppose $\sigma_1 \in S_p(X)$ and $\sigma_2 \in S_q(Y)$. Then the sum over (p,q)-shuffles*

$$\varepsilon(\sigma_1 \otimes \sigma_2) = \sum \operatorname{sgn}(\mu, \nu)(\sigma_1 \times \sigma_2) \circ \eta_{\mu\nu}$$

extends linearly to a chain map

$$\varepsilon : S_*(X) \otimes S_*(Y) \to S_*(X \times Y).$$

Proof We need to show that ε commutes with boundaries. This is a consequence of the following observations:

1. The map $\sigma_1 \times \sigma_2 : S_*(\Delta^p \times \Delta^q) \to S_*(X \times Y)$ and the map $\phi : C_*(\Delta^p \times \Delta^q) \to S_*(\Delta^p \times \Delta^q)$ are both chain maps, as is $\bowtie$ by Proposition B.6.7.
2. Suppose F is a face of Δ^p or G is a face of Δ^q. The restriction of the shuffle triangulation of $\Delta^p \times \Delta^q$ to $F \times \Delta^q$ agrees with the shuffle product triangulation on $\Delta^{p-1} \times \Delta^q$, identifying Δ^{p-1} with F according to the vertex orderings, and the analogous statement holds for $\Delta^p \times G$. See Section B.6.4.

Putting these facts together with the definitions demonstrates that ε is a chain map. □

Definition 5.2.2 We refer to the map $\varepsilon : S_*(X) \otimes S_*(Y) \to S_*(X \times Y)$ as the *cross product*. We will also employ the notation $\varepsilon(x \otimes y) = x \times y$.

The following observation, which comes fairly directly from the definitions, will be useful below. In particular, we will demonstrate an intersection homology version in Corollary 5.2.16, which will be useful for proving the PL version of our generalized Künneth Theorem (Theorem 6.4.7).

Lemma 5.2.3 *Suppose K and L are simplicial complexes with given vertex partial orderings that restrict to total orderings on each simplex. Let $\phi_K : C_*(K) \to S_*(|K|)$ $\phi_L : C_*(L) \to S_*(|L|)$, and $\phi_{K\times L} : C_*(K \times L) \to S_*(|K \times L|)$ be the respective simplicial-to-singular chain maps of Proposition 4.4.5. Then there is a commutative diagram of chain maps:*

$$\begin{array}{ccc} C_*(K) \otimes C_*(L) & \xrightarrow{\bowtie} & C_*(K \times L) \\ \Big\downarrow{\scriptstyle \phi_K \otimes \phi_L} & & \Big\downarrow{\scriptstyle \phi_{K\times L}} \\ S_*(|K|) \otimes S_*(|L|) & \xrightarrow{\varepsilon} & S_*(|K \times L|). \end{array}$$

Proof The vertical arrows represent chain maps by Proposition 4.4.5, the top arrow is a chain map by Proposition B.6.7, and the bottom arrow is a chain map by Proposition 5.2.1. The complex $C_*(K) \otimes C_*(L)$ is generated by elements of the form $\sigma \otimes \tau$, with σ and τ respectively simplices of K and L with their given vertex orderings. For such elements, commutativity of the diagram follows directly from the definitions, identifying σ and τ with standard simplices Δ^p and Δ^q via linear embeddings into K and L and consequently identifying $\sigma \times \tau$ with $\Delta^p \times \Delta^q$ by a linear homeomorphism, inducing a simplicial isomorphism between the Eilenberg–Zilber shuffle triangulation of $\Delta^p \times \Delta^q$ and that of $\sigma \times \tau$. □

We now discuss the cross product for intersection chains.

Lemma 5.2.4 *If X and Y are filtered spaces, the cross product restricts to a chain map $\varepsilon : I^{\bar{p}}S_*^{\mathrm{GM}}(X) \otimes I^{\bar{q}}S_*^{\mathrm{GM}}(Y) \to I^{Q}S_*^{\mathrm{GM}}(X \times Y)$ if $Q(S \times T) \geq \bar{p}(S) + \bar{q}(T)$ for all strata $S \subset X$ and $T \subset Y$.*

Proof First recall that we know that $I^{\bar{p}}S_*^{\mathrm{GM}}(X)$ and $I^{\bar{q}}S_*^{\mathrm{GM}}(X)$ are free complexes, so flat; see Remark 3.4.9. Therefore $I^{\bar{p}}S_*^{\mathrm{GM}}(X) \otimes I^{\bar{q}}S_*^{\mathrm{GM}}(Y) \subset S_*(X) \otimes S_*(Y)$, so it makes sense to restrict the cross product to a map

$$I^{\bar{p}}S_*^{\mathrm{GM}}(X) \otimes I^{\bar{q}}S_*^{\mathrm{GM}}(Y) \to S_*(X \times Y).$$

Our claim is that if Q satisfies the given hypotheses then the image lies in $I^{Q}S_*^{\mathrm{GM}}(X \times Y)$.

Suppose $\sigma_1 \in S_i(X)$ and $\sigma_2 \in S_j(Y)$ are respectively $\bar{p}$- and $\bar{q}$-allowable simplices, and consider $\sigma_1 \times \sigma_2$. We want to show that each $(i+j)$-simplex of the chain $\sigma_1 \times \sigma_2$ is allowable (and then we will consider boundaries). Such a simplex corresponds to the composition

$$\Delta^{i+j} \xrightarrow{\eta_{\mu\nu}} \Delta^i \times \Delta^j \xrightarrow{\sigma_1 \times \sigma_2} X \times Y.$$

Now, if $S \subset X$ and $T \subset Y$ are strata, by the allowability assumptions,

$$\sigma_1^{-1}(S) \subset \{(i - \mathrm{codim}_X(S) + \bar{p}(S))\text{-skeleton of } \Delta^i\}$$

and

$$\sigma_2^{-1}(T) \subset \{(j - \text{codim}_Y(T) + \bar{q}(T))\text{-skeleton of } \Delta^j\}.$$

If we let $a = i - \text{codim}_X(S) + \bar{p}(S)$ and $b = j - \text{codim}_Y(T) + \bar{q}(T)$, then $\sigma_1^{-1} \times \sigma_2^{-1}(S \times T)$ lies in $(\Delta^i)^a \times (\Delta^j)^b$, where $(\Delta^i)^a$ is the a-skeleton of Δ^i and similarly for $(\Delta^j)^b$. But the triangulation of $\Delta^i \times \Delta^j$ coming from the cross product construction triangulates $(\Delta^i)^a \times (\Delta^j)^b$ as a subcomplex[6], which must have dimension $a + b$. Thus any $(i + j)$-simplex in our triangulation of $\Delta^i \times \Delta^j$ can intersect $(\Delta^i)^a \times (\Delta^j)^b$ in at most its $(a + b)$-skeleton. But this implies that $(\sigma_1 \times \sigma_2)^{-1}(S \times T)$ must lie in the $(a + b)$-skeleton of Δ^{i+j}, and

$$\begin{aligned} a + b &= i - \text{codim}_X(S) + \bar{p}(S) + j - \text{codim}_Y(T) + \bar{q}(T) \\ &= i + j - \text{codim}_{X \times Y}(S \times T) + \bar{p}(S) + \bar{q}(T). \end{aligned}$$

So $\sigma_1 \times \sigma_2$ is allowable with respect to any perversity Q such that $Q(S \times T) \geq \bar{p}(S) + \bar{q}(T)$.

This shows that the cross product of two allowable simplices is allowable with the given assumptions on perversities. Now, suppose ξ_1, ξ_2 are allowable *chains* in the respective spaces. Since $\xi_1 \otimes \xi_2$ can be written as a sum (with coefficients) of terms of the form $\sigma_1 \otimes \sigma_2$ in $S_*(X) \otimes S_*(Y)$ and since ε is the linear extension of how it acts on tensor products of simplices, the above argument implies that each singular simplex of $\xi_1 \times \xi_2$ is Q-allowable.

Next, consider that

$$\partial(\xi_1 \otimes \xi_2) = (\partial \xi_1) \otimes \xi_2 + (-1)^{|\xi_1|} \xi_1 \otimes (\partial \xi_2),$$

where $|\xi_1|$ occurring in an exponent denotes the degree of ξ_1. Since $\partial\xi_1$ and $\partial\xi_2$ are allowable, we see that $(\partial\xi_1) \otimes \xi_2$ and $\xi_1 \otimes (\partial\xi_2)$ are each contained in $I^{\bar{p}}S_*^{\text{GM}}(X) \otimes I^{\bar{q}}S_*^{\text{GM}}(Y)$. Therefore, by the preceding argument, each singular simplex of $(\partial\xi_1) \times \xi_2$ is Q-allowable, and similarly for $\xi_1 \times (\partial\xi_2)$. But since ε is a chain map,

$$\partial(\xi_1 \times \xi_2) = (\partial \xi_1) \times \xi_2 + (-1)^{|\xi_1|} \xi_1 \times (\partial \xi_2),$$

so $\partial(\xi_1 \times \xi_2)$ also consists of Q-allowable simplices. Therefore, we conclude that $\xi_1 \times \xi_2$ is a Q-allowable chain. □

Corollary 5.2.5 *Under the assumptions of Lemma 5.2.4, if also $A \subset X$ and $B \subset Y$, the cross product induces maps*

$$\begin{aligned} I^{\bar{p}}S_*^{\text{GM}}(X, A) \otimes I^{\bar{q}}S_*^{\text{GM}}(Y, B) &\longrightarrow \frac{I^Q S_*^{\text{GM}}(X \times Y)}{I^Q S_*^{\text{GM}}(A \times Y) + I^Q S_*^{\text{GM}}(X \times B)} \\ &\longrightarrow I^Q S_*^{\text{GM}}(X \times Y, (A \times Y) \cup (X \times B)). \end{aligned}$$

Proof The lemma implies that the cross product takes $I^{\bar{p}}S_*^{\text{GM}}(A) \otimes I^{\bar{q}}S_*^{\text{GM}}(Y)$ to $I^Q S_*^{\text{GM}}(A \times$

[6] We will see this in the detailed construction of the cross product in the appendix. See, in particular, Corollary B.6.5.

Y) and $I^{\bar{p}}S_*^{\mathrm{GM}}(X) \otimes I^{\bar{q}}S_*^{\mathrm{GM}}(B)$ to $I^Q S_*^{\mathrm{GM}}(X \times B)$. So by basic algebra and the multilinearity of the cross product, the image in

$$\frac{I^Q S_*^{\mathrm{GM}}(X \times Y)}{I^Q S_*^{\mathrm{GM}}(A \times Y) + I^Q S_*^{\mathrm{GM}}(X \times B)}$$

of the cross product of generators of $I^{\bar{p}}S_*^{\mathrm{GM}}(X, A)$ and $I^{\bar{q}}S_*^{\mathrm{GM}}(Y, B)$ is independent of the choice of coset representative. The second map is just a quotient map that is well defined because

$$I^Q S_*^{\mathrm{GM}}(A \times Y) + I^Q S_*^{\mathrm{GM}}(X \times B) \subset I^Q S_*^{\mathrm{GM}}((A \times Y) \cup (X \times B)). \qquad \square$$

Abusing notation, we will also use the symbol ε to refer to the map of the corollary or $\xi_1 \times \xi_2$ to refer to the image of $\xi_1 \otimes \xi_2$.

Remark 5.2.6 As for ordinary homology, the chain cross product induces a product on homology

$$\begin{aligned} I^{\bar{p}}H_*^{\mathrm{GM}}(X, A) \otimes I^{\bar{q}}H_*^{\mathrm{GM}}(Y, B) &\to H_*(I^{\bar{p}}S_*^{\mathrm{GM}}(X, A) \otimes I^{\bar{q}}S_*^{\mathrm{GM}}(Y, B)) \\ &\xrightarrow{\varepsilon} I^Q H_*^{\mathrm{GM}}(X \times Y, (A \times Y) \cup (X \times B)). \end{aligned}$$

The first map comes from basic algebra by noticing that if $x \in C_*$ and $y \in D_*$ are cycles, then $x \otimes y$ is a cycle in $C_* \otimes D_*$, while altering x and y in their homology classes does not alter the homology class of $x \otimes y$. In particular, $(\partial z) \otimes y = \partial(z \otimes y)$ if y is a cycle, and $x \otimes \partial z = (-1)^{|x|}\partial(x \otimes z)$ if x is a cycle. Therefore an element of $H_*(C_*) \otimes H_*(D_*)$ yields a well-defined element of $H_*(C_* \otimes D_*)$.

Remark 5.2.7 In the special case where the space Y is filtered trivially, then all strata are regular and there is only one perversity $\bar{q} = \bar{0}$ on Y. Thus $I^{\bar{q}}S_*^{\mathrm{GM}}(Y, B) = S_*(Y, B)$. In this case, the requirements of the form $Q(S \times T) \geq \bar{p}(S) + \bar{q}(T)$ become $Q(S \times Y) \geq \bar{p}(S)$. In particular, if we define Q on $X \times Y$ as the product perversity with $Q(S \times Y) = \bar{p}(S)$ and then abusively relabel Q to $\bar{p}$, we obtain from Corollary 5.2.5 cross products of the form

$$I^{\bar{p}}S_*^{\mathrm{GM}}(X, A) \otimes S_*(Y, B) \xrightarrow{\varepsilon} I^{\bar{p}}S_*^{\mathrm{GM}}(X \times Y, (A \times Y) \cup (X \times B)),$$

$$S_*(Y, B) \otimes I^{\bar{p}}S_*^{\mathrm{GM}}(X, A) \xrightarrow{\varepsilon} I^{\bar{p}}S_*^{\mathrm{GM}}(Y \times X, (B \times X) \cup (Y \times A)).$$

5.2.2 The PL Cross Product

If X and Y are PL spaces, then the product $X \times Y$ is a PL space by Proposition B.5.3 in Appendix B. Furthermore, given specific triangulations of X and Y and choices of vertex orderings for each, the Eilenberg–Zilber shuffle product gives a triangulation of $X \times Y$ by Theorem B.6.6 that is compatible with this PL structure. Also in Section B.6.5 of the appendix, as discussed in the preceding section, we construct a simplicial cross product $\bowtie: C_*(K) \otimes C_*(L) \to C_*(K \times L)$ for simplicial complexes K and L with chosen vertex orderings, and $\bowtie$ is shown to be a chain map in Proposition B.6.7. In this section, we discuss a PL cross product of the form $\mathfrak{C}_*(X) \otimes \mathfrak{C}_*(Y) \to \mathfrak{C}_*(X \times Y)$. Unfortunately, the shuffle product triangulations do not behave particularly well with respect to subdivision; see Figure B.2 on page 756. Therefore, we pursue a mixed approach, first defining the PL cross product by identifying PL chains with singular homology classes and then applying the singular homology cross product; this definition has the advantage of being independent of the choice of triangulation. Then we will show that this PL product does agree with the simplicial product when the triangulations are fixed, and this will allow us to see that the PL product is a chain map.

We begin with the definition via the singular product: If $\xi \in \mathfrak{C}_p(X)$ and $\eta \in \mathfrak{C}_q(Y)$, then by Lemma 3.3.10 we can identify ξ and η with elements $[\xi] \in H_p(|\xi|, |\partial\xi|)$ and $[\eta] \in H_q(|\eta|, |\partial\eta|)$, respectively. The singular homology cross product then produces an element

$$[\xi] \times [\eta] \in H_{p+q}(|\xi| \times |\eta|, (|\partial\xi| \times |\eta|) \cup (|\xi| \times |\partial\eta|)),$$

and we let the PL chain cross product $\xi \times \eta$ be the PL chain corresponding to this homology class, again via Lemma 3.3.10.

Definition 5.2.8 The PL cross product $\times : \mathfrak{C}_*(X) \otimes \mathfrak{C}_*(Y) \to \mathfrak{C}_*(X \times Y)$ is defined so that if $\xi \in \mathfrak{C}_p(X)$ and $\eta \in \mathfrak{C}_q(Y)$ then $\xi \times \eta$ is the PL chain corresponding to

$$[\xi] \times [\eta] \in H_{p+q}(|\xi| \times |\eta|, (|\partial\xi| \times |\eta|) \cup (|\xi| \times |\partial\eta|))$$

under the isomorphism of Lemma 3.3.10.

We should verify that this is a homomorphism and a chain map. To do so, we will show that this product can be described using the Eilenberg–Zilber shuffle construction if we choose triangulations of X and Y containing representatives of ξ and η and with orderings of their vertices. However, notice again that a benefit of Definition 5.2.8 is that it does not rely on any choices of triangulations and so is automatically independent of such choices.

Proposition 5.2.9 *Let K and L be simplicial complexes with partial orders on their vertices restricting to total orders on each simplex, and let $K \times L$ be the simplicial complex with $|K \times L| = |K| \times |L|$ given by the shuffle product triangulation. Let $\xi \in C_p(K)$ and $\eta \in C_q(L)$ represent PL chains in $|K|$ and $|L|$, respectively. Then the PL chain in $\mathfrak{C}_{p+q}(|K \times L|)$ represented by $\xi \bowtie \eta \in C_{p+q}(K \times L)$ is equal to the PL cross product $\xi \times \eta$.*

Proof Suppose $\xi = \sum a_i\sigma_i$ and $\eta = \sum b_j\tau_j$. Then by definition we can write the simplicial product $\xi \bowtie \sigma$ in the form

$$\sum_{ij} a_i b_j \sigma_i \bowtie \tau_j = \sum_{ij\mu\nu} a_i b_j \operatorname{sgn}(\mu, \nu)\delta^{ij}_{\mu\nu},$$

where the $\delta^{ij}_{\mu\nu}$ are the simplices determined by the (p, q)-shuffles in the product triangulation of $\sigma_i \times \tau_j$. By the proof of Proposition B.6.8, each $\operatorname{sgn}(\mu, \nu)\delta^{ij}_{\mu\nu}$ is oriented consistently with the product orientation of $\sigma_i \times \tau_j$ determined by the orientations of σ_i and τ_j coming from the vertex orderings.

Now, let us consider the PL product $\xi \times \eta$ represented by

$$[\xi] \times [\eta] \in H_{p+q}(|\xi| \times |\eta|, (|\partial\xi| \times |\eta|) \cup (|\xi| \times |\partial\eta|)).$$

The triangulation $K \times L$ contains $|\xi| \times |\eta|$ and $(|\partial\xi| \times |\eta|) \cup (|\xi| \times |\partial\eta|)$ as subcomplexes by Corollary B.6.5. Therefore, by our discussion in Section 3.3.2(c), the chain $\xi \times \eta$ can also be represented by a simplicial chain of the form $\sum_{ij\mu\nu} c^{ij}_{\mu\nu}\delta^{ij}_{\mu\nu}$, where the corresponding σ_i and τ_j range over just those simplices in ξ and η, respectively. By Remark 3.3.14, each $c^{ij}_{\mu\nu}$ is determined by the value of $\xi \times \eta$ in $H_{p+q}(|\xi| \times |\eta|, |\xi| \times |\eta| - \{z\})$, where z is contained in the interior of $\delta^{ij}_{\mu\nu}$ and we use the product orientation of $\sigma_i \times \tau_k$ to determine the sign. But if $x \in \sigma_i$ and $y \in \tau_j$ are such that $(x, y) = z \in \sigma_i \times \tau_j$, we have the commutative diagram

$$\begin{array}{ccc}
\begin{array}{c} H_p(|\xi|, |\partial\xi|) \otimes \\ H_q(|\eta|, |\partial\eta|) \end{array} & \xrightarrow{\times} & H_{p+q}(|\xi| \times |\eta|, (|\partial\xi| \times |\eta|) \cup (|\xi| \times |\partial\eta|)) \\
\downarrow & & \downarrow \\
\begin{array}{c} H_p(|\xi|, |\xi| - \{x\}) \otimes \\ H_q(|\eta|, |\eta| - \{y\}) \end{array} & \xrightarrow[\cong]{\times} & H_{p+q}(|\xi| \times |\eta|, |\xi| \times |\eta| - \{z\}) \\
\uparrow\cong & & \uparrow\cong \\
\begin{array}{c} H_p(\sigma_i, \sigma_i - \{x\}) \otimes \\ H_q(\tau_j, \tau_j - \{y\}) \end{array} & \xrightarrow[\cong]{\times} & H_{p+q}(\sigma_i \times \tau_j, \sigma_i \times \tau_j - \{z\}).
\end{array}$$

Here the lower vertical maps are isomorphisms by excision, and the bottom horizontal map is an isomorphism by the Künneth Theorem. As the coefficient of σ_i in ξ is a_i and the coefficient of τ_j in η is b_j, the coefficient $c^{ij}_{\mu\nu}$ is $\pm a_i b_j$ according as the orientation of $\delta^{ij}_{\mu\nu}$ agrees or disagrees with the product orientation of $\sigma_i \times \tau_j$. But, by the proof of Proposition B.6.8, the $\delta^{ij}_{\mu\nu}$ whose orientations agree with this product orientation are those whose signs are 1. So we can conclude that $c^{ij}_{\mu\nu} = \text{sgn}(\mu, \nu) a_i b_j$.

Therefore, the two cross products agree. □

Corollary 5.2.10 *The PL cross product is a chain map.*

Proof As the simplicial cross product is bilinear by definition, it follows from Proposition 5.2.9 that the PL cross product is a homomorphism; to add two chains we just work in a triangulation containing both. Similarly, Proposition 5.2.9 implies that the compatibility with boundaries proven for simplicial chains in Proposition B.6.7 carries over to the same conclusion for PL chains. So the PL cross product is a chain map. □

Corollary 5.2.11 *Let K and L be simplicial complexes with partial orders on their vertices restricting to total orders on each simplex, and let $K \times L$ be the simplicial complex with $|K \times L| = |K| \times |L|$ given by the shuffle product triangulation. Then there is a commutative diagram of chain maps*

$$\begin{array}{ccc} C_*(K) \otimes C_*(L) & \xrightarrow{\bowtie} & C_*(K \times L) \\ \downarrow & & \downarrow \\ \mathfrak{C}_*(|K|) \otimes \mathfrak{C}_*(|L|) & \xrightarrow{\times} & \mathfrak{C}_*(|K \times L|), \end{array}$$

with the vertical maps the canonical ones induced by taking simplicial chains to the PL chains they represent.

Proof Corollary 5.2.10 and Proposition B.6.7 show that the horizontal maps are chain maps. The commutativity follows immediately from Proposition 5.2.9 as $C_*(K) \otimes C_*(L)$ is generated by elements of the form $\xi \otimes \eta$. □

As for the singular cross product, the simplicial and PL cross products restrict to cross products of intersection chains, and we have the following versions of Lemma 5.2.4 and Corollary 5.2.5.

Lemma 5.2.12 *If X and Y are PL filtered spaces, the PL cross product restricts to a chain map*

$$I^{\bar{p}}\mathfrak{C}_*^{\mathrm{GM}}(X) \otimes I^{\bar{q}}\mathfrak{C}_*^{\mathrm{GM}}(Y) \xrightarrow{\times} I^{Q}\mathfrak{C}_*^{\mathrm{GM}}(X \times Y)$$

if $Q(S \times T) \geq \bar{p}(S) + \bar{q}(T)$ for all strata $S \subset X$ and $T \subset Y$. Similarly, if K and L are simplicial complexes triangulating X and Y with partial orders on their vertices restricting to total orders on each simplex, the simplicial cross product restricts to a chain map

$$I^{\bar{p}}C_*^{\mathrm{GM}}(K) \otimes I^{\bar{q}}C_*^{\mathrm{GM}}(L) \xrightarrow{\bowtie} I^{Q}C_*^{\mathrm{GM}}(K \times L)$$

under the same condition on perversities.

Corollary 5.2.13 *Under the assumptions of Lemma 5.2.12, if also $A \subset X$ and $B \subset Y$ are subsets, the PL cross product induces maps*

$$\begin{aligned} I^{\bar{p}}\mathfrak{C}_*^{\mathrm{GM}}(X,A) \otimes I^{\bar{q}}\mathfrak{C}_*^{\mathrm{GM}}(Y,B) &\longrightarrow \frac{I^{Q}\mathfrak{C}_*^{\mathrm{GM}}(X \times Y)}{I^{Q}\mathfrak{C}_*^{\mathrm{GM}}(A \times Y) + I^{Q}\mathfrak{C}_*^{\mathrm{GM}}(X \times B)} \\ &\longrightarrow I^{Q}\mathfrak{C}_*^{\mathrm{GM}}(X \times Y, (A \times Y) \cup (X \times B)). \end{aligned}$$

The proofs are essentially the same as for Lemma 5.2.4 and Corollary 5.2.5, although, since we don't know that the complexes $I\mathfrak{C}_*$ and $\mathfrak{C}_*$ are free, we need a different argument to see that

$$I^{\bar{p}}\mathfrak{C}_*^{\mathrm{GM}}(X) \otimes I^{\bar{q}}\mathfrak{C}_*^{\mathrm{GM}}(Y) \subset \mathfrak{C}_*(X) \otimes \mathfrak{C}_*(Y).$$

For this, we observe that these groups are torsion-free, and therefore they are flat $\mathbb{Z}$-modules, as $\mathbb{Z}$ is a Dedekind domain. Thus the desired inclusion holds. It will be useful to note for later that the same argument will apply for coefficients in any Dedekind domain; see Section A.4.2. Beyond this, the arguments that the perversities work out are completely analogous to the singular case.

Remark 5.2.14 Analogously to our observation in Remark 5.2.6, the chain cross products induce products on homology

$$\begin{aligned} I^{\bar p}\mathfrak{H}^{\mathrm{GM}}_*(X,A)\otimes I^{\bar q}\mathfrak{H}^{\mathrm{GM}}_*(Y,B) &\to H_*(I^{\bar p}\mathfrak{C}^{\mathrm{GM}}_*(X,A)\otimes I^{\bar q}\mathfrak{C}^{\mathrm{GM}}_*(Y,B)) \\ &\xrightarrow{\varepsilon} I^{Q}\mathfrak{H}^{\mathrm{GM}}_*(X\times Y,(A\times Y)\cup(X\times B)), \end{aligned}$$

and similarly in the simplicial case.

Using Lemma 5.2.12 to restrict the diagram of Corollary 5.2.11 to the intersection chain complexes, we obtain the following:

Corollary 5.2.15 *Let X and Y be PL filtered spaces, and suppose K and L are simplicial complexes triangulating X and Y with partial orders on their vertices restricting to total orders on each simplex. Let $\bar p$, $\bar q$, and Q be respective perversities on X, Y, and $X\times Y$ such that $Q(S\times T)\geq \bar p(S)+\bar q(T)$ for all strata $S\subset X$ and $T\subset Y$. Then there is a commutative diagram of chain maps*[7]

$$\begin{array}{ccc} I^{\bar p}C^{\mathrm{GM}}_*(K)\otimes I^{\bar q}C^{\mathrm{GM}}_*(L) & \xrightarrow{\bowtie} & I^{Q}C^{\mathrm{GM}}_*(K\times L) \\ \downarrow & & \downarrow \\ I^{\bar p}\mathfrak{C}^{\mathrm{GM}}_*(X)\otimes I^{\bar q}\mathfrak{C}^{\mathrm{GM}}_*(Y) & \xrightarrow{\times} & I^{Q}\mathfrak{C}^{\mathrm{GM}}_*(X\times Y), \end{array}$$

with the vertical maps the canonical ones induced by taking simplicial chains to the PL chains they represent.

Finally, this is a convenient time to observe a similar relationship between the simplicial and singular cross products:

Corollary 5.2.16 *Let X and Y be PL filtered spaces, and suppose K and L are simplicial complexes triangulating X and Y with partial orders on their vertices restricting to total orders on each simplex. Let $\bar p$, $\bar q$, and Q be respective perversities on X, Y, and $X\times Y$ such that $Q(S\times T)\geq \bar p(S)+\bar q(T)$ for all strata $S\subset X$ and $T\subset Y$. Then there is a commutative*

7 Here we identify $|K|$ with X via the triangulating homeomorphism, and similarly for the other spaces.

diagram of chain maps[8]

$$\begin{array}{ccc} I^{\bar p}C_*^{\mathrm{GM}}(K)\otimes I^{\bar q}C_*^{\mathrm{GM}}(L) & \xrightarrow{\bowtie} & I^{Q}C_*^{\mathrm{GM}}(K\times L) \\ \Big\downarrow{\scriptstyle \phi_K\otimes\phi_L} & & \Big\downarrow{\scriptstyle \phi_{K\times L}} \\ I^{\bar p}S_*^{\mathrm{GM}}(X)\otimes I^{\bar q}S_*^{\mathrm{GM}}(Y) & \xrightarrow{\varepsilon} & I^{Q}S_*^{\mathrm{GM}}(X\times Y), \end{array}$$

with the vertical maps being the simplicial-to-singular chain maps of Proposition 4.4.5 and Corollary 4.4.6.

Proof By Lemma 5.2.3, the diagram commutes for ordinary chains, and so it will also commute for the restriction to intersection chains if the maps are all well defined. But the vertical arrows are chain maps by Corollary 4.4.6, the top arrow is a chain map by Lemma 5.2.12, and the bottom arrow is a chain map by Lemma 5.2.4. □

5.2.3 Properties of the Cross Product

In this section, we will develop some of the basic properties of the singular and PL cross products. We will primarily make the arguments in the singular setting, as the PL arguments are analogous. When the PL case requires something different, we will indicate so; otherwise, we will not mention the PL proofs explicitly.

Proposition 5.2.17 (Naturality) *Let (X,A), (Y,B), (X',A'), and (Y',B') be pairs of filtered spaces and subsets. Let $f: X\to X'$ and $g: Y\to Y'$ be maps with $f(A)\subset A'$ and $f(B)\subset B'$. Suppose $\bar p,\bar q,\bar p',\bar q'$ are respective perversities on X,Y,X',Y' and that P and Q are respective perversities on $X\times Y$ and $X'\times Y'$ such that $P(S\times T)\geq \bar p(S)+\bar q(T)$ for all strata $S\subset X$ and $T\subset Y$ and $Q(S'\times T')\geq \bar p'(S')+\bar q'(T')$ for all strata $S'\subset X'$ and $T'\subset Y'$. Finally, suppose that f is $(\bar p,\bar p')^{\mathrm{GM}}$-stratified, g is $(\bar q,\bar q')^{\mathrm{GM}}$-stratified, and $f\times g$ is $(P,Q)^{\mathrm{GM}}$-stratified. Then the following diagram commutes:*

$$\begin{array}{ccc} I^{P}S_*^{\mathrm{GM}}(X\times Y,(A\times Y)\cup(X\times B)) & \xleftarrow{\varepsilon} & I^{\bar p}S_*^{\mathrm{GM}}(X,A)\otimes I^{\bar q}S_*^{\mathrm{GM}}(Y,B) \\ \Big\downarrow{\scriptstyle f\times g} & & \Big\downarrow{\scriptstyle f\otimes g} \\ I^{Q}S_*^{\mathrm{GM}}(X'\times Y',(A'\times Y')\cup(X'\times B')) & \xleftarrow{\varepsilon} & I^{\bar p'}S_*^{\mathrm{GM}}(X',A')\otimes I^{\bar q'}S_*(Y',B'). \end{array}$$

In other words, if $x\in I^{\bar p}S_*^{\mathrm{GM}}(X,A)$ and $y\in I^{\bar q}S_*^{\mathrm{GM}}(Y,B)$, then

$$f(x)\times g(y)=(f\times g)(x\times y).$$

Similarly for the PL cross product.

[8] Again, we here identify $|K|$ with X via the triangulating homeomorphism, and similarly for the other spaces.

Remark 5.2.18 Notice that it is possible in Proposition 5.2.17 to have $X = X'$ and $Y = Y'$ with the maps being identity maps. In this case, the lemma becomes a statement about naturality with respect to change of perversity.

Proof of Proposition 5.2.17 The hypotheses of the proposition guarantee that all the maps in the diagram are well defined. For commutativity, as all the maps involved are restrictions of the corresponding chain maps on complexes of ordinary singular chains, it suffices to verify commutativity of the diagram for ordinary singular chains. But $S_*(X,A) \otimes S_*(Y,B)$ is generated by elements represented by tensor products of singular simplices of the form $\sigma \otimes \tau$ with $\sigma : \Delta^i \to X$ and $\tau : \Delta^j \to Y$ for some i, j. But acting on $\sigma \otimes \tau$, the image of the maps left then down is represented by applying the spatial map $(f \times g)(\sigma \times \tau) = (f\sigma) \times (g\tau)$ to the singular chain version (i.e. the image under the map ϕ of Proposition 4.4.5) of the fundamental class $\Delta^p \bowtie \Delta^q$, while the map down then left similarly applies $(f\sigma) \times (f\tau)$ to $\phi(\Delta^p \bowtie \Delta^q)$.

In the PL case, it would be challenging to prove this lemma from the simplicial viewpoint, but using the homological description of PL chains and maps from Lemmas 3.3.10 and 3.3.13, the naturality for the PL chain cross product follows from the naturality of the singular homology cross product. □

Proposition 5.2.19 (Associativity) *Suppose X, Y, and Z are filtered spaces with respective perversities $\bar{p}$, $\bar{q}$, and $\bar{r}$ and subspaces $A \subset X$, $B \subset Y$, and $Z \subset C$. Suppose, furthermore, that*

1. *P is a perversity on $X \times Y$ such that $P(S \times S') \geq \bar{p}(S) + \bar{q}(S')$ for all strata $S \subset X$ and $S' \subset Y$,*
2. *Q is a perversity on $Y \times Z$ such that $Q(S' \times S'') \geq \bar{q}(S') + \bar{r}(S'')$ for all strata $S' \subset Y$ and $S'' \subset Z$, and*
3. *R is a perversity on $X \times Y \times Z$ such that $R(S \times S' \times S'') \geq P(S \times S') + \bar{r}(S'')$ and $R(S \times S' \times S'') \geq \bar{p}(S) + Q(S' \times S'')$ for all strata $S \subset X$, $S' \subset Y$, and $S'' \subset Z$.*

Then the following diagram commutes

$$\begin{array}{ccc}
\begin{array}{c} I^{\bar{p}}S_*^{\mathrm{GM}}(X,A) \otimes I^{\bar{q}}S_*^{\mathrm{GM}}(Y,B) \\ \otimes I^{\bar{r}}S_*^{\mathrm{GM}}(Z,C) \end{array} & \xrightarrow{\varepsilon \otimes \mathrm{id}} & \begin{array}{c} I^{P}S_*^{\mathrm{GM}}(X \times Y, (A \times Y) \cup (X \times B)) \otimes \\ I^{\bar{r}}S_*^{\mathrm{GM}}(Z,C) \end{array} \\
\Big\downarrow {\scriptstyle \mathrm{id} \otimes \varepsilon} & & \Big\downarrow {\scriptstyle \varepsilon} \\
\begin{array}{c} I^{\bar{p}}S_*^{\mathrm{GM}}(X,A) \otimes \\ I^{Q}S_*^{\mathrm{GM}}(Y \times Z, (B \times Z) \cup (Y \times C)) \end{array} & \xrightarrow{\varepsilon} & \begin{array}{c} I^{R}S_*^{\mathrm{GM}}(X \times Y \times Z, (A \times Y \times Z) \cup \\ (X \times B \times Z) \cup (X \times Y \times C)). \end{array}
\end{array}$$

In other words, if $x \in I^{\bar{p}}S_*^{\mathrm{GM}}(X,A)$, $y \in I^{\bar{q}}S_*^{\mathrm{GM}}(Y,B)$, $z \in I^{\bar{r}}S_*^{\mathrm{GM}}(Z,C)$, then

$$(x \times y) \times z = x \times (y \times z).$$

Similarly for the PL cross product.

Proof We first observe that all of the maps of the diagram are well defined by Lemma 5.2.4 and Corollary 5.2.5, noting that

$$\begin{aligned}&[A \times (Y \times Z)] \cup [X \times ((B \times Z) \cup (Y \times C))]\\&\quad = (A \times Y \times Z) \cup (X \times B \times Z) \cup (X \times Y \times C),\end{aligned}$$

and analogously for the terms on the right.

To verify the commutativity of the diagram, we recall that each group appearing in the diagram is a subgroup of the corresponding ordinary singular chain group, so it suffices to verify commutativity in that setting. There, $S_*(X, A) \otimes S_*(Y, B) \otimes S_*(Z, C)$ is generated by elements of the form $\sigma \otimes \tau \otimes \eta$, where $\sigma : \Delta^p \to X$, $\tau : \Delta^q \to Y$, and $\eta : \Delta^r \to Z$ are singular simplices. So the lemma reduces to showing that $(\sigma \times \tau) \times \eta = \sigma \times (\tau \times \eta)$.

Tracing through the definitions, the chain $(\sigma \times \tau) \times \eta$ is the singular chain that is obtained by applying the chain map corresponding to the spatial map $\sigma \times \tau \times \eta$ to the singular chain version (i.e. the image under the map ϕ of Proposition 4.4.5) of the fundamental class $(\Delta^p \bowtie \Delta^q) \bowtie \Delta^r$, which is made up of the $(p + q + r)$-simplices found by first applying the Eilenberg–Zilber shuffle triangulation to $\Delta^p \times \Delta^q$ and then applying the Eilenberg–Zilber shuffle triangulation to each $\delta^{p+q} \times \Delta^r$, where δ^{p+q} is one of the resulting $(p + q)$-simplices of the triangulation of $\Delta^p \times \Delta^q$. Similarly, $\sigma \times (\tau \times \eta)$ applies the chain map corresponding to the spatial map $\sigma \times \tau \times \eta$ to the singular chain version of the fundamental class $\Delta^p \bowtie (\Delta^q \bowtie \Delta^r)$ that comes by applying the Eilenberg–Zilber process first to $\Delta^q \times \Delta^r$. In both cases, by Proposition B.6.8 the fundamental classes are those corresponding to the product orientation of $\Delta^p \times \Delta^q \times \Delta^r$ with each factor given its standard orientation. Therefore, it suffices to verify that it does not matter in what order we perform the iterated Eilenberg–Zilber shuffle processes.

We must verify that both iterative procedures for triangulating $\Delta^p \times \Delta^q \times \Delta^r$ result in the same triangulation, which we can check by looking at the $(p + q + r)$-simplices of this triangulation. In our earlier descriptions of the triangulation process, we provided the vertices of the $(p + q + r)$-simplices of the triangulation. Looking at this prior description again, we know that if we first triangulate $\Delta^p \times \Delta^q$, each $(p+q)$-simplex will have its vertices determined by a (p, q)-shuffle. We have noted that there is a bijection between (p, q)-shuffles and walks in a $p \times q$ grid consisting of steps up and steps right and with the labels of the grid points determining the vertices of a $(p + q)$-simplex in $\Delta^p \times \Delta^q$. Now, the triangulation of $\Delta^p \times \Delta^q \times \Delta^r$ comes by taking each of the $(p + q)$-simplices δ of $\Delta^p \times \Delta^q$ and triangulating $\delta \times \Delta^r$ using $(p + q, r)$-shuffles. If we have fixed a (p, q)-shuffle and think of it as a path in a $p \times q$ grid, then the resulting $(p + q, r)$-shuffles corresponds to walks in a $p \times q \times r$ grid where each step is either a horizontal step dictated by the fixed (p, q)-shuffle or a vertical step. The corresponding triangulation of $\delta \times \Delta^r$ comes by considering the $(p + q + r)$-simplices with vertices labeled by the grid coordinates. As we work through all (p, q)-shuffles and all corresponding $(p + q, r)$-shuffles, we see that the collection of all $(p + q + r)$-simplices of the triangulation of $\Delta^p \times \Delta^q \times \Delta^r$ correspond to the collection of all paths on a $p \times q \times r$ grid in which each step increases a single coordinate by 1. An analogous argument beginning with (q, r)-shuffles and then, for each such shuffle, considering $(p, q + r)$-shuffles results in the same symmetrical description of what we should call (p, q, r)-shuffles. So we see that both procedures yield the same triangulation of $\Delta^p \times \Delta^q \times \Delta^r$. □

Proposition 5.2.20 (Commutativity) *Suppose X and Y are filtered spaces with respective perversities $\bar{p}$ and $\bar{q}$ and subspaces $A \subset X$ and $B \subset Y$. Suppose, furthermore, that P is a perversity on $X \times Y$ such that $P(S \times T) \geq \bar{p}(S) + \bar{q}(T)$ for all strata $S \subset X$ and $T \subset Y$ and that Q is a perversity on $Y \times X$ with $Q(T \times S) = P(S \times T)$. Then the following diagram commutes*

$$\begin{array}{ccc} I^{\bar{p}}S_*^{\mathrm{GM}}(X,A) \otimes I^{\bar{q}}S_*^{\mathrm{GM}}(Y,B) & \xrightarrow{\varepsilon} & I^{P}S_*^{\mathrm{GM}}(X \times Y, (A \times Y) \cup (X \times B)) \\ \downarrow{\scriptstyle \tau} & & \downarrow{\scriptstyle t} \\ I^{\bar{q}}S_*^{\mathrm{GM}}(Y,B) \otimes I^{\bar{p}}S_*^{\mathrm{GM}}(X,A) & \xrightarrow{\varepsilon} & I^{Q}S_*^{\mathrm{GM}}(Y \times X, (B \times X) \cup (Y \times A)), \end{array}$$

where τ is the standard (signed!) interchange map of tensor product factors and t is induced by the topological interchange map $t : X \times Y \to Y \times X$ given by $t(x, y) = (y, x)$. In other words, $t(x \times y) = (-1)^{|x||y|} y \times x$.

Similarly for the PL cross product.

Proof Notice that τ is a chain map: if $x \otimes y$ is a generator of the tensor product, then we have

$$\begin{aligned} \tau\partial(x \otimes y) &= \tau((\partial x) \otimes y + (-1)^{|x|} x \otimes \partial y) \\ &= (-1)^{(|x|-1)|y|} y \otimes \partial x + (-1)^{|x|+|x|(|y|-1)} (\partial y) \otimes x \\ &= (-1)^{(|x|-1)|y|} y \otimes \partial x + (-1)^{|x||y|} (\partial y) \otimes x \\ &= (-1)^{|x||y|} (\partial y) \otimes x + (-1)^{|x||y|+|y|} y \otimes \partial x \\ &= (-1)^{|x||y|} ((\partial y) \otimes x + (-1)^{|y|} y \otimes \partial x) \\ &= (-1)^{|x||y|} \partial(y \otimes x) \\ &= \partial(\tau(x \otimes y)). \end{aligned}$$

It is also clear that t takes P-allowable chains to Q-allowable chains and so induces a well-defined chain map.

Once again, all of the maps of the diagram are well defined by Lemma 5.2.4 and Corollary 5.2.5, and, as in the proof of Proposition 5.2.19, it suffices to verify commutativity for a tensor product of ordinary singular simplices $\sigma \otimes \eta$ with $\sigma : \Delta^p \to X$ and $\eta : \Delta^q \to Y$.

We know that ε acts on $\sigma \otimes \tau$ by applying the spatial map $\sigma \times \eta$ to the singular chain version (i.e. the image under the map ϕ of Proposition 4.4.5) of the fundamental class $\Delta^p \bowtie \Delta^q$ made up of simplices from the Eilenberg–Zilber (p, q)-shuffle triangulation. On the other hand, the map τ takes $\sigma \otimes \eta$ to $(-1)^{pq}\eta \otimes \sigma$, and then ε acts by applying $(-1)^{pq}\eta \times \sigma$ to the singular chain version of the fundamental class for $\Delta^q \times \Delta^p$ made up of simplices from the Eilenberg–Zilber (q, p)-shuffle triangulation. So we need to see that applying $t \circ (\sigma \times \eta)$ to the singular chain coming from the (p, q)-shuffle triangulation of $\Delta^p \times \Delta^q$ gives the same chain as applying $(-1)^{pq}\eta \times \sigma$ to the (q, p)-shuffle triangulation of $\Delta^q \times \Delta^p$.

We consider the following diagram, with $t_\Delta : \Delta^p \times \Delta^q \to \Delta^q \times \Delta^p$ denoting the interchange map with $t_\Delta(x, y) = (y, x)$:

$$\begin{array}{ccc} \Delta^p \times \Delta^q & \xrightarrow{\sigma\times\eta} & X \times Y \\ \downarrow{\scriptstyle t_\Delta} & & \downarrow{\scriptstyle t} \\ \Delta^q \times \Delta^p & \xrightarrow{\eta\times\sigma} & Y \times X \end{array}$$

The diagram clearly commutes, and, together with our above descriptions of $t\varepsilon(\sigma \otimes \eta)$ and $\varepsilon\tau(\sigma \otimes \eta)$, this implies that the lemma will follow if we can show that t_Δ takes the (p, q)-shuffle fundamental class of $\Delta^p \times \Delta^q$ to $(-1)^{pq}$ times the (q, p)-shuffle fundamental class of $\Delta^q \times \Delta^p$.

To prove this claim, we notice that for every (p, q)-shuffle there is a corresponding (q, p)-shuffle; in fact, if we identify (p, q)-shuffles as planar walks with only steps up or right on a $p \times q$ grid, then the corresponding (q, p)-shuffle is the walk on the $q \times p$ grid obtained by flipping the $p \times q$ grid along its southwest to northeast axis. Moreover, the map t_Δ takes the vertices of a $(p + q)$-simplex of $\Delta^p \times \Delta^q$ given by a (p, q)-shuffle to the vertices of $\Delta^q \times \Delta^p$ of the corresponding (q, p)-shuffle. As t_Δ is a linear map, it thus takes our (p, q)-shuffle triangulation of $\Delta^p \times \Delta^q$ to the (q, p)-shuffle triangulation of $\Delta^q \times \Delta^p$. It only remains to consider orientations. The singular triangulation of $\Delta^p \times \Delta^q$ obtained by the Eilenberg–Zilber shuffle process is constructed to conform to the product orientation of $\Delta^p \times \Delta^q$. Similarly, the triangulation of $\Delta^q \times \Delta^p$ conforms to the product orientation of $\Delta^q \times \Delta^p$. Identifying the two spaces via t_Δ, the orientations differ by a sign of $(-1)^{pq}$, owing to the interchange of factors, as desired. □

Proposition 5.2.21 (Unitality) *Suppose X is a filtered space with perversity $\bar{p}$ and subspace $A \subset X$. Let* pt *be the space with one point, and let $\sigma_0 : \Delta^0 \to \mathrm{pt}$ be the unique singular 0-simplex in $S_0(\mathrm{pt})$. Then if $\xi \in I^{\bar{p}}S_i^{\mathrm{GM}}(X, A)$, we have*

$$\begin{aligned} \sigma_0 \times \xi = \xi \times \sigma_0 = \xi \in I^{\bar{p}}S_i^{\mathrm{GM}}(\mathrm{pt} \times X, \mathrm{pt} \times A) &= I^{\bar{p}}S_i^{\mathrm{GM}}(X \times \mathrm{pt}, A \times \mathrm{pt}) \\ &= I^{\bar{p}}S_i^{\mathrm{GM}}(X, A). \end{aligned}$$

Similarly for the PL cross product.

Proof This follows immediately from the definitions, noting that the Eilenberg–Zilber triangulations yield $\Delta^p \times \Delta^0 = \Delta^0 \times \Delta^p = \Delta^p$ as simplicial complexes. □

Remark 5.2.22 The next property of cross products involves the boundary map ∂_* of long exact homology sequences. As this map lowers degree by one, we will treat it as a degree -1 map for the purposes of sign conventions. So, for example, if $\xi \in H_i(X)$ and $\eta \in H_j(Y,B)$, for appropriate spaces, then $(\mathrm{id} \otimes \partial_*)(\xi \otimes \eta) = (-1)^i \xi \otimes \partial_*(\eta) \in H_i(X) \otimes H_{j-1}(B)$. We note that this convention is not always followed in the literature; compare, for example, the following results with Statements VI.2.11, VI.2.12, and VI.2.13 in [71].

Proposition 5.2.23 (Stability) *Suppose X and Y are filtered spaces with respective perversities $\bar p$ and $\bar q$ and subspaces $A \subset X$ and $B \subset Y$. Suppose that Q is a perversity on $X \times Y$ such that $Q(S \times T) \geq \bar p(S) + \bar q(T)$ for all strata $S \subset X$ and $T \subset Y$. Then the following diagram commutes:*

$$\begin{array}{ccc}
I^{\bar p}H_i^{\mathrm{GM}}(X,A) \otimes I^{\bar q}H_j^{\mathrm{GM}}(Y,B) & \xrightarrow{\ \varepsilon\ } & I^{Q}H_{i+j}^{\mathrm{GM}}(X \times Y, (A \times Y) \cup (X \times B)) \\
\Big\downarrow{\scriptstyle \partial_* \otimes \mathrm{id}} & & \Big\downarrow{\scriptstyle \partial_*} \\
 & & I^{Q}H_{i+j-1}^{\mathrm{GM}}((A \times Y) \cup (X \times B)) \\
 & & \Big\downarrow \\
 & & I^{Q}H_{i+j-1}^{\mathrm{GM}}((A \times Y) \cup (X \times B), X \times B) \\
 & & \Big\uparrow \\
I^{\bar p}H_{i-1}^{\mathrm{GM}}(A) \otimes I^{\bar q}H_j^{\mathrm{GM}}(Y,B) & \xrightarrow{\ \varepsilon\ } & I^{Q}H_{i-1+j}^{\mathrm{GM}}(A \times Y, A \times B).
\end{array}$$

Here, the unlabeled maps of the diagram are induced by the inclusions

$$((A \times Y) \cup (X \times B), \emptyset) \to ((A \times Y) \cup (X \times B), X \times B)$$

and

$$(A \times Y, A \times B) \to ((A \times Y) \cup (X \times B), X \times B).$$

In other words, if $\xi \in I^{\bar p}H_i^{\mathrm{GM}}(X,A)$ and $\eta \in I^{\bar q}H_j^{\mathrm{GM}}(Y,B)$, then

$$(\partial_* \xi) \times \eta = \partial_*(\xi \times \eta) \in I^{Q}H_{i+j-1}^{\mathrm{GM}}((A \times Y) \cup (X \times B), X \times B).$$

Analogously, and via a similar diagram,

$$\xi \times \partial_* \eta = (-1)^i \partial_*(\xi \times \eta) \in I^{Q}H_{i+j-1}^{\mathrm{GM}}((A \times Y) \cup (X \times B), A \times Y).$$

Similarly for the PL cross product.

Proof Recall that, via the standard "zig-zag" construction of the boundary map of the long exact sequence of a pair, if ξ is a chain representing an element of $I^{\bar p}H_i^{\mathrm{GM}}(X,A)$, then the image of this element in $I^{\bar p}H_{i-1}^{\mathrm{GM}}(A)$ under the map ∂_* is represented by $\partial\xi$, and similarly for any other ∂_* map. Now, $I^{\bar p}H_i^{\mathrm{GM}}(X,A)\otimes I^{\bar q}H_j^{\mathrm{GM}}(Y,B)$ is generated by elements of the form $\xi\otimes\eta$ with ξ a chain representing an element of $I^{\bar p}H_i^{\mathrm{GM}}(X,A)$ and η a chain representing an element of $I^{\bar q}H_j^{\mathrm{GM}}(Y,B)$. The image of $\xi\otimes\eta$ in $I^{Q}H_{i+j-1}^{\mathrm{GM}}((A\times Y)\cup(X\times B),X\times B)$ working counterclockwise around the diagram is represented by $(\partial\xi)\times\eta$. The image working around the diagram clockwise is $\partial(\xi\times\eta)$, which, as the cross product is a chain map, is equal to $(\partial\xi)\times\eta+(-1)^i\xi\times\partial\eta$. But now $\xi\times\partial\eta$ is contained in $X\times B$. Therefore, $(\partial\xi)\times\eta$ and $(\partial\xi)\times\eta+(-1)^i\xi\times\partial\eta$ represent the same element in $I^{Q}H_{i+j-1}^{\mathrm{GM}}((A\times Y)\cup(X\times B),X\times B)$.

Analogously, $(\partial\xi)\times\eta$ is contained in $A\times Y$, and so $\partial(\xi\times\eta)$ and $(-1)^i\xi\times\partial\eta$ represent the same element in $I^{Q}H_{i+j-1}^{\mathrm{GM}}((A\times Y)\cup(X\times B),A\times Y)$. But the latter also represents $(-1)^i$ times the cross product of ξ with $\partial_*\eta$. □

Similarly, we have the following lemma:

Proposition 5.2.24 (Stability) *Suppose X and Y are filtered spaces with respective perversities $\bar p$ and $\bar q$ and subspaces $A\subset X$ and $B\subset Y$. Suppose that Q is a perversity on $X\times Y$ such that $Q(S\times T)\geq \bar p(S)+\bar q(T)$ for all strata $S\subset X$ and $T\subset Y$. Then the following diagram commutes:*

$$
\begin{array}{ccc}
I^{\bar p}H_i^{\mathrm{GM}}(X,A)\otimes I^{\bar q}H_j^{\mathrm{GM}}(Y,B) & \xrightarrow{\ \varepsilon\ } & I^{Q}H_{i+j}^{\mathrm{GM}}(X\times Y,(A\times Y)\cup(X\times B)) \\
 & & \downarrow{\scriptstyle \partial_*} \\
 & & I^{Q}H_{i+j-1}^{\mathrm{GM}}((A\times Y)\cup(X\times B)) \\
\downarrow{\scriptstyle (\partial_*\otimes\mathrm{id})+(\mathrm{id}\otimes\partial_*)} & & \downarrow \\
 & & I^{Q}H_{i+j-1}^{\mathrm{GM}}((A\times Y)\cup(X\times B),A\times B) \\
 & & \uparrow{\scriptstyle i_1+i_2} \\
\begin{array}{c}(I^{\bar p}H_{i-1}^{\mathrm{GM}}(A)\otimes I^{\bar q}H_j^{\mathrm{GM}}(Y,B))\oplus\\ (I^{\bar p}H_i^{\mathrm{GM}}(X,A)\otimes I^{\bar q}H_{j-1}^{\mathrm{GM}}(B))\end{array} & \xrightarrow{\ \varepsilon\oplus\varepsilon\ } & \begin{array}{c}I^{Q}H_{i-1+j}^{\mathrm{GM}}(A\times Y,A\times B)\oplus\\ I^{Q}H_{i-1+j}^{\mathrm{GM}}(X\times B,A\times B).\end{array}
\end{array}
$$

Here, the unlabeled map of the diagram is induced by the inclusions

$$((A\times Y)\cup(X\times B),\emptyset)\to((A\times Y)\cup(X\times B),A\times B)$$

and $i_1 + i_2$ denotes the sum of the two inclusion maps

$$(A \times Y, A \times B) \to ((A \times Y) \cup (X \times B), A \times B)$$

and

$$(X \times B, A \times B) \to ((A \times Y) \cup (X \times B), A \times B).$$

In other words, if $\xi \in I^{\bar{p}}H_i^{\mathrm{GM}}(X,A)$ and $\eta \in I^{\bar{q}}H_j^{\mathrm{GM}}(Y,B)$, then

$$\begin{aligned}\partial_*(\xi) \times \eta + (-1)^i \xi \times \partial_*(\eta) &= \partial_*(\xi \times \eta)\\ &\in I^Q H^{\mathrm{GM}}_{i+j-1}\big((A \times Y) \cup (X \times B), A \times B\big).\end{aligned}$$

Similarly for the PL cross product.

Proof As in the proof of Proposition 5.2.23, we consider a tensor product $\xi \otimes \eta$ of chains representing a generator of $I^{\bar{p}}H_i^{\mathrm{GM}}(X,A) \otimes I^{\bar{q}}H_j^{\mathrm{GM}}(Y,B)$. Chasing the diagram counterclockwise and using that $(\mathrm{id} \otimes \partial_*)(\xi \otimes \eta) = (-1)^i \xi \otimes \partial_* \eta$, the chain $\xi \otimes \eta$ gets taken first to $((\partial\xi) \otimes \eta) \oplus (-1)^i(\xi \otimes \partial\eta)$ then to $((\partial\xi) \times \eta) \oplus (-1)^i(\xi \times \partial\eta)$, and finally to

$$((\partial\xi) \times \eta) + (-1)^i(\xi \times \partial\eta) \in I^Q H^{\mathrm{GM}}_{i+j-1}((A \times Y) \cup (X \times B), A \times B).$$

Chasing the other way, and using that ε is a chain map, we get $\partial(\xi \times \eta) = (\partial\xi) \times \eta - (-1)^i \xi \times \partial\eta$. So the chain representatives agree. □

5.2.4 Künneth Theorem when One Factor Is a Manifold

Now that we have a cross product, in this section we will prove a Künneth Theorem for intersection homology in the case where one factor of the product is a trivially filtered manifold M and the other is a filtered space X. We will not be able to obtain a more general Künneth Theorem until we have redefined intersection homology slightly in Chapter 8, but this version of the Künneth Theorem nonetheless has important applications, including the proof of topological invariance of intersection homology with Goresky–MacPherson perversities; see Section 5.5.

Versions of the Künneth Theorem presented here seem to have been known quite early on; the special case for perversity $\bar{m}$, the space X a Witt space[9], and real coefficients[10] is a special case of Goresky–MacPherson [106, Section 6.3], while Siegel [218] has a proof in the PL category for Witt spaces with rational coefficients. A proof for singular intersection homology was provided by King [139]. We provide a different proof, though one that is very consonant with other techniques developed elsewhere in King's paper.

Theorem 5.2.25 *Suppose X is a filtered space with perversity $\bar{p}_X$ and that M is an n-dimensional manifold with its trivial filtration. Filter $M \times X$ with the product filtration so*

[9] See Section 9.1.1, below.
[10] See Section 5.3, below.

that $(M\times X)^i = M\times X^{i-n}$*, and define a perversity* $\bar p$ *on* $M\times X$ *whose value on the stratum* $R\times S$*, for* R *a connected component of* M*, is* $\bar p_X(S)$*. Then the cross product induces an isomorphism* $H_*(S_*(M)\otimes I^{\bar p_X}S_*^{\mathrm{GM}}(X)) \xrightarrow{\cong} I^{\bar p}H_*^{\mathrm{GM}}(M\times X)$*. If* X *is a PL filtered space and* M *is a PL manifold, then the same conclusion holds replacing singular chains with PL chains.*

Proof The singular and PL proofs are identical, so we will only provide the argument for singular intersection homology.

Before starting on the main body of the proof, we observe that for any $x_0\in M$ the inclusion of X into $M\times X$ by identifying it with $\{x_0\}\times X$ is a normally nonsingular inclusion; see Definition 2.9.8. In particular, the codimension of a stratum S in X is the same as the codimension of $M\times S$ in $M\times X$, so $I^{\bar p_X}H_*^{\mathrm{GM}}(X)$ is isomorphic to the intersection homology group $I^{\bar p}H_*^{\mathrm{GM}}(X)$ obtained by thinking of X as a subspace of $M\times X$ and using the inherited filtration and perversity. See Example 4.3.11 for a full discussion of this scenario. We will use this identification of intersection homology groups without further explicit mention.

Now, to prove the theorem, we will apply the Mayer–Vietoris argument, Theorem 5.1.1. We fix X and define functors from $\mathcal{M}$ to $\mathcal{A}b_*$ as follows: Let $F_*(M) = H_*(S_*(M)\otimes I^{\bar p_X}S_*^{\mathrm{GM}}(X))$, let $G_*(M) = I^{\bar p}H_*^{\mathrm{GM}}(M\times X)$, and let the natural transformation $\Phi: F_*\to G_*$ be induced by the cross product (see Remark 5.2.7). We must verify that the conditions of Theorem 5.1.1 are satisfied.

We first consider the cross product $S_*(\mathbb{R}^n)\otimes I^{\bar p_X}S_*^{\mathrm{GM}}(X)\to I^{\bar p}S_*^{\mathrm{GM}}(\mathbb{R}^n\times X)$. By the naturality of the cross product, we have the following commutative diagram

$$\begin{array}{ccc} S_*(\{0\})\otimes I^{\bar p_X}S_*^{\mathrm{GM}}(X) & \xrightarrow{\times} & I^{\bar p}S_*^{\mathrm{GM}}(\{0\}\times X) = I^{\bar p_X}S_*^{\mathrm{GM}}(X) \\ \downarrow & & \downarrow \\ S_*(\mathbb{R}^n)\otimes I^{\bar p_X}S_*^{\mathrm{GM}}(X) & \xrightarrow{\times} & I^{\bar p}S_*^{\mathrm{GM}}(\mathbb{R}^n\times X) \end{array}$$

in which the vertical maps are induced by inclusion. These vertical maps induce homology isomorphisms by stratified homotopy invariance and the algebraic Künneth Theorem[11]. Also, the top horizontal map induces a homology isomorphism because σ_0 is the generator of $H_*(\{0\}) = H_0(\{0\})\cong\mathbb{Z}$ and if $\xi\in I^{\bar p_X}S_i^{\mathrm{GM}}(X)$, then $\sigma_0\times\xi=\xi$ by Proposition 5.2.21. Therefore the bottom map also induces a homology isomorphism. These arguments are invariant up to homeomorphism, and so this computation also applies to subsets of M that are homeomorphic to $\mathbb{R}^n$.

Next suppose that $\{U_\alpha\}$ is a sequence of open subsets of M ordered by inclusion and

[11] To apply the algebraic Künneth Theorem, we observe that the chain complexes S_* and $\mathfrak{C}_*$ and their submodules are all torsion-free, and hence flat, $\mathbb{Z}$-modules; see [238, Theorem 3.6.3] for the Künneth Theorem and Section A.4.2 for the fact that torsion-free implies flat for Dedekind domains, including $\mathbb{Z}$.

such that $S_*(U_\alpha) \otimes I^{\bar{p}_X} S_*^{\mathrm{GM}}(X) \xrightarrow{\times} I^{\bar{p}} S_*^{\mathrm{GM}}(U_\alpha \times X)$ induces homology isomorphisms for each α. Consider the commuting diagrams of the form

$$\begin{array}{ccc} H_*(S_*(U_\beta) \otimes I^{\bar{p}_X} S_*^{\mathrm{GM}}(X)) & \xrightarrow{\times} & I^{\bar{p}} H_*^{\mathrm{GM}}(U_\beta \times X) \\ \downarrow & & \downarrow \\ H_*\left(S_*\left(\bigcup_\alpha U_\alpha\right) \otimes I^{\bar{p}_X} S_*^{\mathrm{GM}}(X)\right) & \xrightarrow{\times} & I^{\bar{p}} H_*^{\mathrm{GM}}\left(\bigcup_\alpha U_\alpha \times X\right), \end{array}$$

where U_β is a particular one of the $\{U_\alpha\}$. Suppose $[\xi] \in I^{\bar{p}} H_*^{\mathrm{GM}}(\bigcup_\alpha U_\alpha \times X)$, represented by the cycle ξ. Since ξ consists of a finite number of singular simplices, the union of the images of the simplices of ξ is compact and so must be contained in some $U_\beta \times X$. Hence $[\xi]$ is in the image of $I^{\bar{p}} H_*^{\mathrm{GM}}(U_\beta \times X)$. But the top line of the diagram is an isomorphism for any fixed β by assumption, and so it follows that the bottom map must be surjective.

Now assume $[\eta] \in H_*(S_*(\bigcup_\alpha U_\alpha) \otimes I^{\bar{p}_X} S_*^{\mathrm{GM}}(X))$ maps to 0 in $I^{\bar{p}} H_*^{\mathrm{GM}}(\bigcup_\alpha U_\alpha \times X)$. By definition, $\eta = \sum_j x_j \otimes \xi_j$ for some $x_j \in S_*(\bigcup_\alpha U_\alpha)$ and $\xi_j \in I^{\bar{p}} S_*^{\mathrm{GM}}(X)$, so we are assuming $\left[\sum x_j \times \xi_j\right] = 0$. Let ζ be a chain in $I^{\bar{p}} S_*^{\mathrm{GM}}(\bigcup_\alpha U_\alpha \times X)$ with $\partial\zeta = \sum x_j \times \xi_j$. Then, again by a compactness argument, there is a β such that all the x_j are supported in U_β and ζ is supported in $U_\beta \times X$. So then $\sum x_j \times \xi_j$ also represents 0 as an element of $I^{\bar{p}} H_*^{\mathrm{GM}}(U_\beta \times X)$. Since the top arrow is an isomorphism, there must be an element $\mu \in S_*(U_\beta) \otimes I^{\bar{p}_X} S_*^{\mathrm{GM}}(X)$ whose boundary is $\sum_j x_j \otimes \xi_j$. But then this must hold also under the inclusion of $S_*(U_\beta) \otimes I^{\bar{p}_X} S_*^{\mathrm{GM}}(X)$ into $S_*(\bigcup_\alpha U_\alpha) \otimes I^{\bar{p}} S_*^{\mathrm{GM}}(X)$ and so $[\eta] = 0$.

In sum, we have shown that if $S_*(U_\alpha) \otimes I^{\bar{p}_X} S_*^{\mathrm{GM}}(X) \xrightarrow{\times} I^{\bar{p}} S_*^{\mathrm{GM}}(U_\alpha \times X)$ induces homology isomorphisms for each α, then $H_*(S_*(\bigcup_\alpha U_\alpha) \otimes I^{\bar{p}_X} S_*^{\mathrm{GM}}(X)) \xrightarrow{\times} I^{\bar{p}} H_*^{\mathrm{GM}}(\bigcup_\alpha U_\alpha \times X)$ is an isomorphism.

Finally, consider the following diagram (where for clarity, we write I for $I^{\bar{p}} S_*^{\mathrm{GM}}$ and I_X for $I^{\bar{p}_X} S_*^{\mathrm{GM}}$):

$$\begin{array}{ccccccccc} 0 & \to & \begin{array}{c} S_*(U \cap V) \otimes \\ \mathrm{I}_X(X) \end{array} & \to & \begin{array}{c} (S_*(U) \otimes \mathrm{I}_X(X)) \oplus \\ (S_*(V) \otimes \mathrm{I}_X(X)) \end{array} & \to & \begin{array}{c} (S_*(U) + S_*(V)) \otimes \\ \mathrm{I}_X(X) \end{array} & \to & 0 \\ & & \downarrow \times & & \downarrow \times\oplus\times & & \downarrow \times & & \\ 0 & \to & \mathrm{I}((U \cap V) \times X) & \to & \begin{array}{c} \mathrm{I}(U \times X) \oplus \\ \mathrm{I}(V \times X) \end{array} & \to & \begin{array}{c} \mathrm{I}(U \times X) + \\ \mathrm{I}(V \times X) \end{array} & \to & 0. \end{array}$$

The bottom row is a Mayer–Vietoris short exact sequence, while the top row is obtained by tensoring the short exact Mayer–Vietoris sequence

$$0 \longrightarrow S_*(U \cap V) \longrightarrow S_*(U) \oplus S_*(V) \longrightarrow S_*(U) + S_*(V) \longrightarrow 0$$

with $I^{\bar{p}_X} S_*^{\mathrm{GM}}(X)$ then summing over degrees (recall $(A_* \otimes B_*)_i = \bigoplus_{j+k=i} A_j \otimes B_k$). Since each

$I^{\bar{p}\times}S_j^{\rm GM}(X)$ is $\mathbb{Z}$-torsion-free, it is a flat $\mathbb{Z}$-module, so tensoring with $I^{\bar{p}}S_j^{\rm GM}(X)$ preserves exactness. This diagram yields a map of long exact sequences in homology. By the proof of Theorem 4.4.19 (or Theorem 4.4.4 in the PL case), the inclusion

$$I^{\bar{p}}S_*^{\rm GM}(U\times X)+I^{\bar{p}}S_*^{\rm GM}(V\times X)\to I^{\bar{p}}S_*^{\rm GM}(U\times X)+I^{\bar{p}}S_*^{\rm GM}((U\cup V)\times X)$$

induces a homology isomorphism, so the bottom row has the form of a long exact Mayer–Vietoris sequence for G_*. Similarly, by the same arguments, or classically, $S_*(U)+S_*(V)\hookrightarrow S_*(U\cup V)$ induces a homology isomorphism, and it follows from the algebraic Künneth Theorem and the Five Lemma that the map

$$(S_*(U)+S_*(V))\otimes I^{\bar{p}\times}S_*^{\rm GM}(X)\to S_*(U\cup V)\otimes I^{\bar{p}\times}S_*^{\rm GM}(X)$$

induced by inclusion is also a homology isomorphism. So, again, substituting the homology of the latter expression for that of the former in the long exact sequence yields a long exact sequence of Mayer–Vietoris form for the functor F_*. We observe that the cross product continues to induce the isomorphism of exact sequences with these substitutions via the diagram

$$\begin{array}{ccc}
H_i((S_*(U)+S_*(V))\otimes I^{\bar{p}\times}S_*^{\rm GM}(X)) & \xrightarrow{\cong} & H_i(S_*(U\cup V)\otimes I^{\bar{p}\times}S_*^{\rm GM}(X)) \\
\downarrow{\scriptstyle\times} & & \downarrow{\scriptstyle\times} \\
H_i(I^{\bar{p}}S_*^{\rm GM}(U\times X)+I^{\bar{p}}S_*^{\rm GM}(V\times X)) & \xrightarrow{\cong} & I^{\bar{p}}H_i((U\cup V)\times X).
\end{array}$$

Thus the cross product induces a map of long exact Mayer–Vietoris sequences.

Given these verifications, Theorem 5.1.1 now implies the theorem. □

Corollary 5.2.26 *Under the assumptions of Theorem 5.2.25, if $A\subset X$ then the cross product induces an isomorphism*

$$H_*(S_*(M)\otimes I^{\bar{p}\times}S_*^{\rm GM}(X,A))\xrightarrow{\cong} I^{\bar{p}}H_*^{\rm GM}(M\times X,M\times A).$$

If X is a PL filtered space and M is a PL manifold, then the same conclusion holds replacing singular chains with PL chains.

Proof Consider the short exact sequence

$$0\longrightarrow I^{\bar{p}\times}S_*^{\rm GM}(A)\to I^{\bar{p}\times}S_*^{\rm GM}(X)\to I^{\bar{p}\times}S_*^{\rm GM}(X,A)\longrightarrow 0.$$

As $S_*(M)$ is torsion-free, and hence flat, tensoring with $S_*(M)$ preserves exactness, and, writing I for $I^{\bar{p}}S_*^{\mathrm{GM}}$ for clarity, we obtain a diagram of short exact sequences:

$$\begin{array}{ccccccccc}
0 & \longrightarrow & S_*(M)\otimes \mathrm{I}(A) & \longrightarrow & S_*(M)\otimes \mathrm{I}(X) & \longrightarrow & S_*(M)\otimes \mathrm{I}(X,A) & \longrightarrow & 0 \\
 & & \downarrow{\scriptstyle \times} & & \downarrow{\scriptstyle \times} & & \downarrow{\scriptstyle \times} & & \\
0 & \longrightarrow & \mathrm{I}(M\times A) & \longrightarrow & \mathrm{I}(M\times X) & \longrightarrow & \mathrm{I}(M\times X, M\times A) & \longrightarrow & 0.
\end{array}$$

It is not difficult to verify that this diagram commutes by looking at cross products of generators. The diagram then induces a map of long exact sequences, and the corollary follows from Theorem 5.2.25 and the Five Lemma. □

Remark 5.2.27 If we assume the subset A of Corollary 5.2.26 is open and that U is an open subset of the manifold M, then one can further show that the cross product induces an isomorphism

$$H_*(S_*(M,U)\otimes I^{\bar{p}x}S_*^{\mathrm{GM}}(X,A)) \xrightarrow{\cong} I^{\bar{p}}H_*^{\mathrm{GM}}(M\times X,(M\times A)\cup(U\times X)).$$

This follows from Theorem 5.2.25 by the same argument as that presented below to show that the more general relative Künneth Theorem for non-GM intersection homology (Theorem 6.4.13) follows from the absolute Künneth Theorem for non-GM intersection homology (Theorem 6.4.7). As the proof is a bit more involved than that for Corollary 5.2.26 alone, we refer the reader to that argument rather than reproduce the details here.

5.3 Intersection Homology with Coefficients and Universal Coefficient Theorems

So far, we have managed to show that intersection homology satisfies many of the key properties of ordinary homology. In this section, we will explore a topic whose translation to intersection homology is more problematic, namely intersection homology with coefficients. Part of the issue is that there are two possible competing definitions of intersection homology with coefficients.

5.3.1 Definitions of Intersection Homology with Coefficients

Perhaps the simplest approach to intersection homology with coefficients would be to consider $I^{\bar{p}}S_*^{\mathrm{GM}}(X)\otimes G$ for some abelian group G. Indeed, one could do this, and then of course the algebraic Universal Coefficient Theorem would give us split short exact sequences of the form

$$0\to I^{\bar{p}}H_i^{\mathrm{GM}}(X)\otimes G\to H_i(I^{\bar{p}}S_*^{\mathrm{GM}}(X)\otimes G)\to I^{\bar{p}}H_{i-1}^{\mathrm{GM}}(X)*G\to 0,$$

where $*$ denotes the torsion product $\mathrm{Tor}^1(\cdot,\cdot)$.

However, there is a more intriguing and ultimately more useful option, which is captured in the following definition:

Definition 5.3.1 Let G be an abelian group and X a filtered space with perversity $\bar{p}$. Define the complex of *intersection chains with coefficients in* G, denoted $I^{\bar{p}}S_*^{\mathrm{GM}}(X;G)$, to be the subcomplex of $S_*(X;G) = S_*(X) \otimes G$ such that $\xi \in I^{\bar{p}}S_*^{\mathrm{GM}}(X;G)$ if each simplex of[12] ξ is $\bar{p}$-allowable and each simplex of $\partial\xi$ is $\bar{p}$-allowable. In other words, we simply mirror the definition of the intersection chain complex $I^{\bar{p}}S_*^{\mathrm{GM}}(X)$ in terms of allowability, but using $S_*(X;G)$ as our starting point rather than $S_*(X)$.

If X is a PL filtered space, then we can similarly define for each admissible triangulation T the subcomplex $I^{\bar{p}}C_*^{CM,T}(X;G) \subset C_*^T(X;G) = C_*^T(X) \otimes G$ consisting of chains $\xi \in C_*^T(X;G)$ such that each simplex of ξ and $\partial\xi$ is $\bar{p}$-allowable. Subdivision $C_*^T(X) \to C_*^{T'}(X)$ induces a subdivision map $C_*^T(X;G) \to C_*^{T'}(X;G)$ by functoriality, and by the argument proving Lemma 3.3.15, each such map takes $I^{\bar{p}}C_*^{CM,T}(X;G)$ to $I^{\bar{p}}C_*^{CM,T'}(X;G)$. We can then let $I^{\bar{p}}\mathfrak{C}_*^{\mathrm{GM}}(X;G) = \varinjlim_{T\in\mathcal{T}} I^{\bar{p}}C_*^{\mathrm{GM},T}(X;G)$, where $\mathcal{T}$ is the set of admissible triangulations of X compatible with the filtration.

Remark 5.3.2 As the direct limit functor is exact and commutes with tensor products we have $I^{\bar{p}}\mathfrak{C}_*^{\mathrm{GM}}(X;G) \subset \mathfrak{C}_*(X;G) \cong \mathfrak{C}_*(X) \otimes G$. In particular, $I^{\bar{p}}\mathfrak{C}_*^{\mathrm{GM}}(X;G)$ is the subcomplex of chains in $\mathfrak{C}_*(X;G)$ representable by $\bar{p}$-allowable simplicial chains.

Remark 5.3.3 More generally, if M is an R-module for some commutative ring with unity R, then we can similarly define $I^{\bar{p}}S_*^{\mathrm{GM}}(X;M) \subset S_*(X;M)$, where we write $S_*(X;M) = S_*(X) \otimes_{\mathbb{Z}} M$. Here we use that an R-module is also a $\mathbb{Z}$-module via the unique ring morphism $\mathbb{Z} \to R$ taking 1 to the unity of R. We will not pursue this level of generality in what follows.

(a) Comparing the Options

The chain complexes $I^{\bar{p}}S_*^{\mathrm{GM}}(X;G)$ and $I^{\bar{p}}S_*^{\mathrm{GM}}(X) \otimes G$ do not yield the same homology groups! We will provide a concrete example in a moment, but let us first consider why this is plausible even though allowability of a simplex is no different in $I^{\bar{p}}S_*^{\mathrm{GM}}(X;G)$ than in $I^{\bar{p}}S_*^{\mathrm{GM}}(X) \otimes G$. The issue is what happens when we take boundaries.

For example, suppose that ξ is a singular chain in $S_*(X)$ and that $\partial\xi = 2\eta$, for some other chain η. Suppose every simplex of ξ is allowable, but that η contains simplices that are not allowable. So $\xi \notin I^{\bar{p}}S_*^{\mathrm{GM}}(X)$. Suppose that consequently[13] $\xi \otimes 1 \notin I^{\bar{p}}S_*^{\mathrm{GM}}(X) \otimes \mathbb{Z}_2$. However, $\xi \otimes 1$ will definitely be an element of $I^{\bar{p}}S_*^{\mathrm{GM}}(X;\mathbb{Z}_2)$: by assumption every simplex of ξ (and hence every simplex of $\xi \otimes 1$) is allowable, and now the boundary vanishes with $\mathbb{Z}_2$ coefficients. In fact, $\xi \otimes 1$ is a cycle in $I^{\bar{p}}S_*^{\mathrm{GM}}(X;\mathbb{Z}_2)$ and may well represent an intersection homology class.

[12] Analogously to the case with $\mathbb{Z}$ coefficients, we say that σ is a simplex of the chain $\xi \in S_*(X;G)$ if σ appears as one of the σ_i when we write ξ as a finite sum of the form $\sum_i g_i\sigma_i$ with each $g_i \in G$, $g_i \neq 0$, and no σ_i repeated.

[13] This is not automatic; for example, if all coefficients of ξ are even, then $\xi \otimes 1 = 0 \in I^{\bar{p}}S_*^{\mathrm{GM}}(X) \otimes \mathbb{Z}_2$ whether ξ is an intersection chain or not.

Conversely, suppose there are chains $\eta \in S_i(X), \xi \in S_{i+1}(X)$ such that $\xi \otimes 1, \eta \otimes 1 \in I^{\bar{p}}S_*^{\mathrm{GM}}(X;G)$ and $\partial(\xi \otimes 1) = \eta \otimes 1$. It might nonetheless be possible that ξ is not allowable as an element of $I^{\bar{p}}S_{i+1}^{\mathrm{GM}}(X)$; for example, perhaps G is 2-torsion and $\partial\xi = \eta + 2\zeta$ for some chain $\zeta \in S_i(X)$ containing non-allowable simplices. In this case, $\eta \otimes 1$ represents a trivial element of $I^{\bar{p}}H_i^{\mathrm{GM}}(X;G)$, but it might not be trivial in $H_i(I^{\bar{p}}S_*^{\mathrm{GM}}(X) \otimes G)$.

Let us look at an important concrete example where this latter situation occurs quite explicitly:

Example 5.3.4 Let $X = X^3$ be the cone $c(\mathbb{R}P^2)$ with the perversity $\bar{0}$ that assigns 0 to the cone vertex v, which is the only singular stratum. We first use homological tools to compute $I^{\bar{0}}H_*^{\mathrm{GM}}(c(\mathbb{R}P^2);\mathbb{Z}_2)$ and $H_*(I^{\bar{0}}S_*^{\mathrm{GM}}(c(\mathbb{R}P^2)) \otimes \mathbb{Z}_2)$.

The cone computation of Theorem 4.2.1 is easily generalized to include coefficient groups (see Theorem 5.3.5, below) to establish that

$$I^{\bar{0}}H_i^{\mathrm{GM}}(c(\mathbb{R}P^2);\mathbb{Z}_2) \cong \begin{cases} 0, & i \geq 2, \\ I^{\bar{0}}H_i^{\mathrm{GM}}(\mathbb{R}P^2;\mathbb{Z}_2), & i < 2. \end{cases}$$

Since $\mathbb{R}P^2$ is an unfiltered manifold, we have $I^{\bar{0}}H_*(\mathbb{R}P^2) = H_*(\mathbb{R}P^2)$. Therefore $I^{\bar{0}}H_0^{\mathrm{GM}}(c(\mathbb{R}P^2);\mathbb{Z}_2) \cong I^{\bar{0}}H_1^{\mathrm{GM}}(c(\mathbb{R}P^2);\mathbb{Z}_2) \cong \mathbb{Z}_2$ and $I^{\bar{0}}H_i^{\mathrm{GM}}(c(\mathbb{R}P^2);\mathbb{Z}_2) = 0$ otherwise.

On the other hand, with $\mathbb{Z}$ coefficients

$$I^{\bar{0}}H_i^{\mathrm{GM}}(c(\mathbb{R}P^2)) \cong \begin{cases} 0, & i \geq 2, \\ I^{\bar{0}}H_i^{\mathrm{GM}}(\mathbb{R}P^2), & i < 2, \end{cases}$$

and now from the ordinary homology computations it follows that $I^{\bar{0}}H_0^{\mathrm{GM}}(c(\mathbb{R}P^2)) \cong \mathbb{Z}$, $I^{\bar{0}}H_1^{\mathrm{GM}}(c(\mathbb{R}P^2)) \cong \mathbb{Z}_2$, and $I^{\bar{0}}H_i^{\mathrm{GM}}(c(\mathbb{R}P^2)) = 0$ otherwise. The algebraic Universal Coefficient Theorem then shows that

$$\begin{aligned} &H_0(I^{\bar{0}}S_*^{\mathrm{GM}}(c(\mathbb{R}P^2)) \otimes \mathbb{Z}_2) \cong \mathbb{Z}_2, \\ &H_1(I^{\bar{0}}S_*^{\mathrm{GM}}(c(\mathbb{R}P^2)) \otimes \mathbb{Z}_2) \cong \mathbb{Z}_2, \\ &H_2(I^{\bar{0}}S_*^{\mathrm{GM}}(c(\mathbb{R}P^2)) \otimes \mathbb{Z}_2) \cong \mathbb{Z}_2, \\ &H_i(I^{\bar{0}}S_*^{\mathrm{GM}}(c(\mathbb{R}P^2)) \otimes \mathbb{Z}_2) = 0, \qquad i > 2. \end{aligned}$$

This does not agree with the computation of $I^{\bar{0}}H_i^{\mathrm{GM}}(c(\mathbb{R}P^2);\mathbb{Z}_2)$ when $i = 2$.

So why the difference? Homologically, we can see the issue from the cone formula of Theorem 4.2.1 and its analogue with coefficients. From these cone formulas, we see that there is a sharp transition between degrees one and two. In degrees below two, the intersection homology of the cone on X agrees with the intersection homology of X itself, above that degree it is 0. But the Universal Coefficient Theorem not only references the homology in a given degree, but it reaches into a lower dimension to pull out torsion information. As a result, as we have seen in this example, the Universal Coefficient Theorem can then cause nontrivial homology to appear above the cutoff dimension where the homology of a cone should be zero. Since all CS sets are cones locally (or products of cones with Euclidean space), we can expect this local issue to percolate into a failure of the Universal Coefficient Theorem in general.

What is going on at the chain level? Let η be a cycle in $\mathbb{R}P^2$ representing the generator of $H_1(\mathbb{R}P^2) \cong \mathbb{Z}_2$. Identifying $\mathbb{R}P^2$ with $\{1/2\} \times \mathbb{R}P^2 \subset c(\mathbb{R}P^2)$, then η is also a generator of $I^{\bar{0}}H_1(c(\mathbb{R}P^2))$, while $\eta \otimes 1$ is a generator of both $I^{\bar{0}}H_1(c(\mathbb{R}P^2); \mathbb{Z}_2)$ and $H_1(I^{\bar{0}}S_*(c(\mathbb{R}P^2)) \otimes \mathbb{Z}_2)$. These groups are all isomorphic to $\mathbb{Z}_2$. Notice that η does not bound in $c(\mathbb{R}P^2)$ with any coefficients because it certainly does not bound in $\mathbb{R}P^2$, and it cannot bound in $c(\mathbb{R}P^2)$ because if we compute $2 - \text{codim}(\{v\}) - \bar{0}(\{v\}) = 2 - 3 - 0 = -1$, we see that no 2-chain can intersect the cone vertex.

Now, let $\xi \in S_2(\mathbb{R}P^2)$ be a chain such that $\partial\xi = 2\eta$. Then $\xi \otimes 1$ represents a cycle in $I^{\bar{p}}S_2^{\text{GM}}(c(\mathbb{R}P^2); \mathbb{Z}_2)$, but it bounds $\bar{c}\xi \otimes 1$ since we see from the dimension computation that a singular 3-cycle with a vertex at v *is* allowable. Thus $\xi \otimes 1$ represents 0 in $I^{\bar{p}}S_2^{\text{GM}}(c(\mathbb{R}P^2); \mathbb{Z}_2)$. But $\bar{c}\xi$ is *not* allowable as a chain in $I^{\bar{p}}S_2^{\text{GM}}(c(\mathbb{R}P^2))$, now with $\mathbb{Z}$ coefficients. This is because $\partial(\bar{c}\xi) = \xi - \bar{c}\partial\xi = \xi - 2\bar{c}\eta$. But $\bar{c}\eta$ is not allowable, as we have seen that 2-chains may not intersect the cone vertex. Therefore, $\xi \otimes 1$ does not bound $\bar{c}\xi \otimes 1$ in $I^{\bar{p}}S_*^{\text{GM}}(\mathbb{R}P^2) \otimes \mathbb{Z}_2$. Technically, this is not sufficient to demonstrate that ξ represents the nontrivial element of $H_2(I^{\bar{p}}S_*^{\text{GM}}(\mathbb{R}P^2) \otimes \mathbb{Z}_2)$, as we have only shown that $\xi\otimes$ does not bound $\bar{c}\xi \otimes 1$ and not that it can never bound. However, this example should give some idea of the additional intricacies of working with coefficients. △

Since $H_*(I^{\bar{p}}S_*^{\text{GM}}(X) \otimes G)$ can always be computed from $I^{\bar{p}}H_*^{\text{GM}}(X)$ via the universal coefficient theorem, the much more interesting groups are $I^{\bar{p}}H_*(X; G)$, and so we shall focus on them. From now on, "intersection homology with coefficients" will always mean $H_*(I^{\bar{p}}S_*(X; G))$ with $I^{\bar{p}}S_*(X; G)$ as defined in Definition 5.3.1.

(b) Basic Properties of Intersection Homology with Coefficients

As mentioned in Example 5.3.4, the entire argument of the proof of Theorem 4.2.1 can be copied nearly verbatim to establish the following generalization:

Theorem 5.3.5 *If X is a compact filtered space of formal dimension $n - 1$, then*

$$I^{\bar{p}}H_i^{\text{GM}}(cX; G) \cong \begin{cases} 0, & i \geq n - \bar{p}(\{v\}) - 1, \quad i \neq 0, \\ G, & i \geq n - \bar{p}(\{v\}), \quad i = 0, \\ G, & i = n - \bar{p}(\{v\}) - 1, \quad i = 0, \\ & \qquad I^{\bar{p}_X}H_0^{\text{GM}}(X; G) \neq 0, \\ 0, & i = n - \bar{p}(\{v\}) - 1, \quad i = 0, \\ & \qquad I^{\bar{p}_X}H_0^{\text{GM}}(X; G) = 0, \\ I^{\bar{p}}H_i^{\text{GM}}(X; G), & i < n - \bar{p}(\{v\}) - 1, \end{cases}$$

and similarly in the PL setting.

Likewise, all of our preceding work generalizes, mostly in evident ways. Our discussion of simplicial-versus-PL intersection homology, behavior under GM stratified maps, stratified homotopy invariance, relative intersection homology, the long exact sequences of pairs, subdivisions, Mayer–Vietoris sequences, and excision can be generalized to statements about $I^{\bar{p}}H_i^{\mathrm{GM}}(X;G)$ nearly verbatim, as can the obvious generalizations of Propositions 5.1.8 and 5.1.11. The one place where we must be slightly more careful is when considering cross products and the Künneth Theorem because there are places where we assumed we were working with free or flat chain complexes. In particular, we used freeness in Lemma 5.2.4 to argue that $I^{\bar{p}}S_*^{\mathrm{GM}}(X) \otimes I^{\bar{q}}S_*^{\mathrm{GM}}(Y) \subset S_*(X) \otimes S_*(Y)$, which is necessary to have the cross product defined. We also used flatness multiple times in the proof of Theorem 5.2.25, including where it is needed to invoke the algebraic Künneth Theorem, which itself does not hold for arbitrary rings. In order to extend to more general coefficients, we will need to put in place some restrictions.

First, notice that if R is a commutative ring with unity then $S_*(X;R) \cong S_*(X) \otimes_{\mathbb{Z}} R$, and we can extend the cross product to

$$\varepsilon : S_*(X;R) \otimes_R S_*(Y;R) \longrightarrow S_*(X \times Y;R)$$

by

$$\varepsilon((x \otimes_{\mathbb{Z}} r) \otimes_R (y \otimes_{\mathbb{Z}} s)) = (x \times y) \otimes_{\mathbb{Z}} rs.$$

Since $r, s \in R$ live in degree zero,

$$\begin{aligned}
&\partial\varepsilon((x \otimes r) \otimes_R (y \otimes s)) \\
&\quad = \partial((x \times y) \otimes_{\mathbb{Z}} rs) \\
&\quad = (\partial(x \times y)) \otimes_{\mathbb{Z}} rs \\
&\quad = ((\partial x) \times y + (-1)^{|x|} x \times (\partial y)) \otimes_{\mathbb{Z}} rs \\
&\quad = ((\partial x) \times y) \otimes_{\mathbb{Z}} rs + (-1)^{|x|}(x \times (\partial y)) \otimes_{\mathbb{Z}} rs \\
&\quad = \varepsilon(((\partial x) \otimes_{\mathbb{Z}} r) \otimes_R (y \otimes_{\mathbb{Z}} s) + (-1)^{|x|}(x \otimes_{\mathbb{Z}} r) \otimes_R ((\partial y) \otimes_{\mathbb{Z}} s)) \\
&\quad = \varepsilon(\partial((x \otimes_{\mathbb{Z}} r) \otimes_R (y \otimes_{\mathbb{Z}} s))),
\end{aligned}$$

where in these equations we have used ε in the sense defined here and $\times$ for the cross product with integer coefficients. This shows that our new, more general, ε is still a chain map. From here on, we will again use ε and $\times$ interchangeably to denote the cross product, letting context determine which coefficients are meant. Additionally, when fully in the setting of R coefficients, we will often write $\otimes$ rather than $\otimes_R$.

Now, to extend the cross product to intersection chains with coefficients, we assume that R is a Dedekind domain. When working with coefficient rings in what follows, we will often require them to be Dedekind domains due to the nice homological algebra properties they possess. Recall that a Dedekind domain is an integral domain with the property that every submodule of a projective R-module is projective[14]. In particular, principal ideal domains and fields are Dedekind domains. It is also true that any torsion-free module over a Dedekind domain is flat; in fact, this is true more generally of Prüfer domains[15] [146, Proposition 4.20], which could equally well be used below for the arguments where only this property of Dedekind domains is needed.

Since each $S_i(X;R)$ and $\mathfrak{C}_i(X;R)$ is an R-torsion-free R-module for any R, so will be their respective submodules $I^{\bar{p}}S_i^{\mathrm{GM}}(X;R)$ and $I^{\bar{p}}\mathfrak{C}_i^{\mathrm{GM}}(X;R)$. So if R is a Dedekind domain (or, more generally, a Prüfer domain), these R-modules will be flat. In fact, as $S_i(X;R)$ is a free R-module, $I^{\bar{p}}S_i^{\mathrm{GM}}(X;R)$ is projective, though in proofs for which we want to run parallel arguments for singular and PL chains, we will focus on the flatness. Therefore, for the purpose of short exact sequences of tensor products, $I^{\bar{p}}S_i^{\mathrm{GM}}(X;R)$ has the same properties we needed before when we used that $I^{\bar{p}}S_i^{\mathrm{GM}}(X)$ is flat as an abelian group. In particular, since we have an inclusion $I^{\bar{q}}S_j^{\mathrm{GM}}(Y;R) \hookrightarrow S_j(Y;R)$, tensoring with the flat module $I^{\bar{p}}S_i^{\mathrm{GM}}(X;R)$ yields an inclusion

$$I^{\bar{p}}S_i^{\mathrm{GM}}(X;R) \otimes_R I^{\bar{q}}S_j^{\mathrm{GM}}(Y;R) \hookrightarrow I^{\bar{p}}S_i^{\mathrm{GM}}(X;R) \otimes_R S_j(Y;R),$$

and similarly since $S_j(Y;R)$ is flat, tensoring it with the inclusion $I^{\bar{p}}S_i^{\mathrm{GM}}(X;R) \hookrightarrow S_i(X;R)$ yields an inclusion

$$I^{\bar{p}}S_i^{\mathrm{GM}}(X;R) \otimes_R S_j(Y;R) \hookrightarrow S_i(X;R) \otimes S_j(Y;R).$$

Summing over indices, we obtain once again

$$I^{\bar{p}}S_*^{\mathrm{GM}}(X;R) \otimes_R I^{\bar{q}}S_*^{\mathrm{GM}}(Y;R) \subset S_*(X;R) \otimes_R S_*(Y;R),$$

which allows us to restrict the cross product to intersection chain complexes. The properties of Section 5.2.3 follow.

We can now generalize Theorem 5.2.25 to the following:

Theorem 5.3.6 *Suppose X is a filtered space with perversity $\bar{p}_X$ and that M is an n-dimensional manifold with its trivial filtration. Filter $M \times X$ with the product filtration so that $(M \times X)^i = M \times X^{i-n}$, and define a perversity $\bar{p}$ on $M \times X$ whose value on $M \times S$ is $\bar{p}(S)$. Let R be a Dedekind domain. Then the cross product induces an isomorphism*

$$H_*(S_*(M;R) \otimes_R I^{\bar{p}_X}S_*^{\mathrm{GM}}(X;R)) \xrightarrow{\cong} I^{\bar{p}}H_*^{\mathrm{GM}}(M \times X;R).$$

If X is a PL filtered space and M is a PL manifold, then the same conclusion holds replacing singular chains with PL chains.

[14] This is essentially taken as the definition of a Dedekind domain in Cartan and Eilenberg [49, Section VII.5 and Theorem I.5.4]. Exercise 20 to Section 4 of Chapter VII of [30] shows that this property can be derived from other defining properties of Dedekind domains. A short literature search reveals that there are a very large number of equivalent definitions for Dedekind domains! See Appendix A.4.2 for more about Dedekind domains.

[15] Prüfer domains satisfy the more general property that submodules of *finitely generated* projective modules are projective. A module over a Prüfer domain is torsion-free if and only if it is flat [146, Proposition 4.20].

Proof The proof is essentially the same as that of Theorem 5.2.25. □

Remark 5.3.7 The assumption that R be Dedekind is not needed in the proof of Theorem 5.3.6 to define the cross product for singular chains since $S_*(M;R)$ is free for any manifold, which is sufficient to have

$$S_*(M;R) \otimes_R I^{\bar{p}}S_*^{\text{GM}}(X;R) \subset S_*(M;R) \otimes_R S_*(X;R)$$

for any ring.

Corollary 5.3.8 *Under the assumptions of Theorem 5.3.6, if $A \subset X$, then the cross product induces an isomorphism*

$$H_*(S_*(M;R) \otimes_R I^{\bar{p}}S_*^{\text{GM}}(X,A;R)) \xrightarrow{\cong} I^{\bar{p}}H_*^{\text{GM}}(M \times X, M \times A; R).$$

If X is a PL filtered space and M is a PL manifold, then the same conclusion holds replacing singular chains with PL chains.

The proof is the same as that for Corollary 5.2.26, using Theorem 5.3.6 in place of Theorem 5.2.25.

5.3.2 Universal Coefficient Theorems

It is natural to ask under what circumstances might it be true that $I^{\bar{p}}S_*^{\text{GM}}(X;G)$ and $I^{\bar{p}}S_*^{\text{GM}}(X) \otimes G$ have the same homology groups? This would be a useful property, for then we *could* use the Universal Coefficient Theorem for computations; at the same time, knowing when this property fails tells us when $I^{\bar{p}}H_*^{\text{GM}}(X;G)$ really is fundamentally different from $I^{\bar{p}}H_*^{\text{GM}}(X)$. Surprisingly enough, it turns out that the situation of Example 5.3.4 is essentially the only thing that can go wrong.

Recall that in Example 5.3.4 we saw a situation where

$$I^{\bar{p}}H_*^{\text{GM}}(cX;G) \ncong H_*(I^{\bar{p}}S_*^{\text{GM}}(cX) \otimes G)$$

because the way the Universal Coefficient Theorem blends information from two dimensions contradicts the strict truncation we see in the cone formula of Theorem 5.3.5. In particular, in the first dimension in which the cone formula tells us that $I^{\bar{p}}H_i^{\text{GM}}(cX;G)$ *must* be 0 (namely dimension $i = n - \bar{p}(\{v\}) - 1$ if X has dimension $n-1$), the universal coefficient computation reaches down to provide a possibly nontrivial term in $H_i(I^{\bar{p}}S_*^{\text{GM}}(cX) \otimes G)$ that comes from the torsion of $I^{\bar{p}}H_{i-1}^{\text{GM}}(cX)$. However, if $I^{\bar{p}}H_{i-1}^{\text{GM}}(cX) * G = 0$, we eliminate this problem. This discussion motivates the following definition, which is a slight generalization of that provided by Goresky and Siegel [111], who first considered this issue. We give the definition now for an arbitrary Dedekind domain R, though we will be mainly focused on the case $R = \mathbb{Z}$ until later chapters.

Definition 5.3.9 Let X be a CS set, R a Dedekind domain, and M an R-module. We say that X is *locally* $(\bar{p}, R; M)^{\mathrm{GM}}$*-torsion-free* if for each point $x \in X$ and for each link[16] L of x we have $I^{\bar{p}}H^{\mathrm{GM}}_{\dim(L)-\bar{p}(S)-1}(L; R) *_R M = 0$, where S is the stratum of X containing x and $*_R$ denotes the torsion product over R, i.e. $\mathrm{Tor}^1_R(\cdot,\cdot)$. If we only impose the condition for all points in a stratum $S \subset X$, we say that X is *locally* $(\bar{p}, R; M)^{\mathrm{GM}}$*-torsion-free along* S.

If $I^{\bar{p}}H^{\mathrm{GM}}_{\dim(L)-\bar{p}(S)-1}(L; R) *_R M = 0$ for all R-modules M, we simply say that X is *locally* $(\bar{p}, R)^{\mathrm{GM}}$*-torsion-free* – equivalent to asking that $I^{\bar{p}}H^{\mathrm{GM}}_{\dim(L)-\bar{p}(S)-1}(L; R)$ be flat as an R-module by [147, Theorem XVI.3.11]. In particular, this means that $I^{\bar{p}}H^{\mathrm{GM}}_{\dim(L)-\bar{p}(S)-1}(L)$ is torsion-free (as an R-module) by [146, Proposition 4.20].

Remark 5.3.10 Notice that since $\dim(L)+\dim(S)+1 = n$, the condition can be rewritten as $I^{\bar{p}}H^{\mathrm{GM}}_{\mathrm{codim}(S)-\bar{p}(S)-2}(L; R) * M = 0$, which more closely approximates the original definition in [111].

Remark 5.3.11 We could define locally torsion-free PL CS sets similarly, but, as we will soon see that PL and singular intersection homology are isomorphic for PL CS sets, this is unnecessary. A locally $(\bar{p}, R; M)^{\mathrm{GM}}$-torsion-free PL CS set is simply a CS set that is both a PL filtered space and locally $(\bar{p}, R; M)^{\mathrm{GM}}$-torsion-free in the sense above.

For the remainder of this section, we will concentrate mostly on $\mathbb{Z}$-modules, i.e. abelian groups, but we invite the reader to formulate the appropriate generalizations.

Example 5.3.12 Of course, if X is a manifold, trivially filtered, then every link is $\emptyset$ and so X is trivially locally $(\bar{p}, \mathbb{Z})^{\mathrm{GM}}$-torsion-free for any $\bar{p}$.

More generally[17], all CS sets are locally $(\bar{t}, \mathbb{Z})^{\mathrm{GM}}$-torsion-free, where $\bar{t}$ is the top perversity such that $\bar{t}(S) = \mathrm{codim}(S) - 2$ for any singular stratum S. To see this, we observe that if S is a stratum and L is a link of a point in S, then $\dim(X) = \dim(S) + \dim(L) + 1$, so $\mathrm{codim}(S) = \dim(L) + 1$. Therefore,

$$\begin{aligned}\dim(L) - \bar{t}(S) - 1 &= \dim(L) - (\mathrm{codim}(S) - 2) - 1 = \dim(L) - (\dim(L) + 1 - 2) - 1\\ &= 0.\end{aligned}$$

But by Example 3.4.6, we have that $I^{\bar{t}}H^{\mathrm{GM}}_0(L)$ is free abelian, and so $I^{\bar{t}}H^{\mathrm{GM}}_0(L) * G = 0$ for any abelian group G. This argument also carries over to coefficients in any Dedekind domain R and generalizes to any perversity $\bar{p}$ with $\bar{p} \geq \bar{t}$. △

As defined, the locally torsion-free condition seems to require checking the algebraic properties of $I^{\bar{p}}H^{\mathrm{GM}}_*(L)$ for all possible links of all points of X. The next lemma and its corollary show that these groups depend only on the stratum containing x and not on the specific choice of x or its distinguished neighborhood or link. Therefore, the condition of Definition 5.3.9 can be alternatively stated as requiring only that each stratum contains *some* point with *some* link such that $I^{\bar{p}}H^{\mathrm{GM}}_{\dim(L)-\bar{p}(S)-1}(L) * G = 0$.

[16] We will show below in Corollary 5.3.14 that $I^{\bar{p}}H^{\mathrm{GM}}_{\dim(L)-\bar{p}(S)-1}(L; R)$ depends only on the stratum S and not on the specific choices of $x \in S$ or link L of x.

[17] I first learned this from [56], but there may be earlier occurrences.

Lemma 5.3.13 *Let X be a CS set and $x \in X$. For $i = 1,2$, let $N_i \cong \mathbb{R}^k \times cL_i$ be distinguished neighborhoods of x. Then $I^{\bar{p}}H_*^{\mathrm{GM}}(L_1) \cong I^{\bar{p}}H_*^{\mathrm{GM}}(L_2)$ and $I^{\bar{p}}H_*^{\mathrm{GM}}(N_1) \cong I^{\bar{p}}H_*^{\mathrm{GM}}(N_2)$.*

Proof The second claim follows from the first because stratified homotopy invariance and the cone formula imply that the intersection homology of a distinguished neighborhood depends only on that of the link.

To prove $I^{\bar{p}}H_*^{\mathrm{GM}}(L_1) \cong I^{\bar{p}}H_*^{\mathrm{GM}}(L_2)$, first we claim that for any such distinguished neighborhoods N_i, there is another distinguished neighborhood $N_2' \subset N_2$ such that also $N_2' \subset N_1$ and $N_2' \hookrightarrow N_2$ is a stratified homotopy equivalence. We will prove this claim below. We note now though that if x is contained in the stratum S then the stratified homotopy equivalence $N_2' \hookrightarrow N_2$ of the claim restricts to a stratified homotopy equivalence $N_2' - (N_2' \cap S) \hookrightarrow N_2 - (N_2 \cap S)$ as we are simply removing corresponding strata from each space.

Next, we observe for $i = 1,2$ that $N_i - (N_i \cap S) \cong \mathbb{R}^k \times (cL_i - \{v_i\})$, which is stratified homotopy equivalent to L_i. So it suffices to show that $I^{\bar{p}}H_*(N_1 - (N_1 \cap S)) \cong I^{\bar{p}}H_*(N_2 - (N_2 \cap S))$.

Now, let $V = N_2 - (N_2 \cap S)$ and $U = N_1 - (N_1 \cap S)$. Replacing N_2 with N_2' and the corresponding $V' = N_2' - (N_2' \cap S)$, we have $V' \subset U$. But then by repeating the argument with the roles of the indices interchanged, we can assume there is a $U' \subset V'$ with $U' \hookrightarrow U$ a stratified homotopy equivalence. And then running the argument again, we have a $V'' \subset U'$ with $V'' \hookrightarrow V$ a stratified homotopy equivalence. In other words, we can have a sequence of spaces

$$V'' \overset{f}{\hookrightarrow} U' \overset{g}{\hookrightarrow} V' \overset{h}{\hookrightarrow} U,$$

such that the inclusions $hg : U' \hookrightarrow U$ and $gf : V'' \hookrightarrow V'$ are stratified homotopy equivalences and so induce isomorphisms of intersection homology. Therefore, the induced map $g : I^{\bar{p}}H_*^{\mathrm{GM}}(U') \to I^{\bar{p}}H_*^{\mathrm{GM}}(V')$ must be surjective and injective and so an isomorphism. But V' is stratified homotopy equivalent to V, which is stratified homotopy equivalent to L_2, and similarly U' is stratified homotopy equivalent to L_2. Hence, the intersection homology groups of L_1 and L_2 must be isomorphic.

To finish the proof, we must demonstrate our earlier claim about shrinking distinguished neighborhoods. Suppose x is contained in the stratum S. For simplicity of argument, via the given homeomorphisms let us identify N_2 canonically with $\mathbb{R}^k \times cL_2$, and let us assume that $x = (0, v)$, where 0 is the origin of $\mathbb{R}^k$. Then $N_1 \cap N_2 \cap S$ is an open subset of $\mathbb{R}^k$ and so contains a closed disk D_r of some radius r around the origin. N_2 is then stratified homotopy equivalent to $\mathring{D}_r \times cL_2$. Furthermore, since $N_1 \cap (D_r \times cL_2)$ must be a neighborhood of $D_r \times \{v\}$ in $D_r \times cL_2$ and, since D_r is compact, the Tube Lemma [181, Lemma 26.8] implies there must be a neighborhood W of v in cL_2 such that $D_r \times W \subset N_1 \cap (D_r \times cL_2)$. Furthermore, using the compactness of L_2, the definition of the quotient topology, and the Tube Lemma again, there is some s, $0 < s \leq 1$, such that $v \in cL_2$ has a neighborhood of the form

$$c_sL_2 = ([0,s) \times L_2/\sim) \subset cL_2 = ([0,1) \times L_2/\sim)$$

and such that $c_sL_2 \subset W$; see Figure 5.1. Thus $\mathring{D}_r \times c_sL_2 \subset N_1 \cap N_2$, and the inclusion $\mathring{D}_r \times c_sL_2 \hookrightarrow \mathbb{R}^k \times cL_2$ is a stratified homotopy equivalence (in the cone direction, we can retract along the cone lines). So we let $N_2' = \mathring{D}_r \times c_sL_2$. □

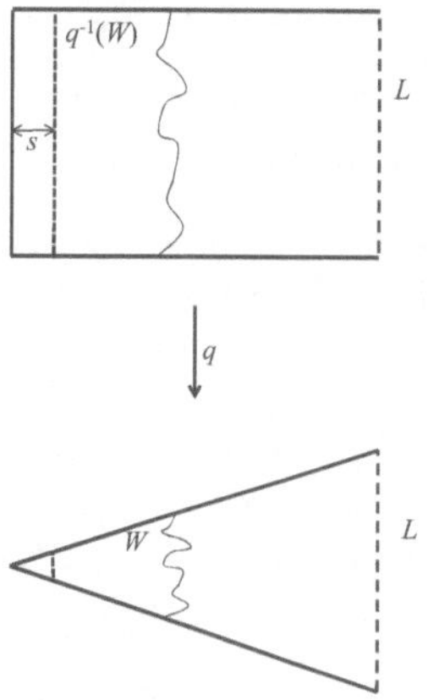

Figure 5.1 Given an open neighborhood W of the vertex of a cone with compact link L, we can always find a cone $c_sL \subset W$. If $q : [0,1) \times L \to cL$ is the quotient map, then $q^{-1}(W)$ is a neighborhood of $\{0\} \times L$ in $[0,1) \times L$. As L is compact, by the Tube Lemma there is an s such that $[0,s) \times L \subset q^{-1}(W)$. Then $q([0,s) \times L) = c_sL$ is a neighborhood of v in W.

Corollary 5.3.14 *Let X be a CS set. Then the intersection homology $I^{\bar{p}}H_*^{\rm GM}(L)$ of a link L of a point x in a stratum of S depends only on S. In other words, all links for any distinguished neighborhoods of any points in S have isomorphic intersection homology groups.*

Proof Let S be a stratum of X, and let $x_0 \in S$. The preceding lemma shows that all possible links of a given point in S have isomorphic intersection homology. Let W be the set of points of S whose links have intersection homology isomorphic to that of the links of x_0. We will show that W is both open and closed as a subset of S. Since S is connected, this will imply $W = S$.

Let x be any point in W, and let $N \cong \mathbb{R}^k \times cL$ be a distinguished neighborhood of x. Then the images under this homeomorphism of all points of the form $(z, v) \subset \mathbb{R}^k \times cL$, with v representing the cone vertex, share this distinguished neighborhood and hence have filtered homeomorphic choices of links. So each such point has a link whose intersection homology is isomorphic to that of a link of x, which is in turn isomorphic to the intersection homology of the link of x_0. This shows that W must be open.

Next, suppose y is a point in the closure of W, and let $N \cong \mathbb{R}^k \times cL$ be a distinguished neighborhood of y. The neighborhood N must contain a point $z \in W$. But then y and z share a distinguished neighborhood and hence a link. So the intersection homology of the links of y must agree with that of the links of z, which agree with the intersection homology of the links of x_0. So $y \in W$, and W must be closed. □

We will now use our Mayer–Vietoris argument (Theorem 5.1.4) to show that

$$I^{\bar{p}}H_*^{\mathrm{GM}}(cX;G) \cong H_*(I^{\bar{p}}S_*^{\mathrm{GM}}(X) \otimes G)$$

for locally $(\bar{p},\mathbb{Z};G)^{\mathrm{GM}}$-torsion-free CS sets. We first state the theorem and then prove some easy but important corollaries before moving on to the proof of the theorem.

Theorem 5.3.15 (Universal Coefficients) *Suppose X is a locally $(\bar{p},\mathbb{Z};G)^{\mathrm{GM}}$-torsion-free CS set. Then $I^{\bar{p}}H_*^{\mathrm{GM}}(X;G) \cong H_*(I^{\bar{p}}S_*^{\mathrm{GM}}(X) \otimes G)$. If X is also PL then $I^{\bar{p}}\mathfrak{H}_*^{\mathrm{GM}}(X;G) \cong H_*(I^{\bar{p}}\mathfrak{C}_*^{\mathrm{GM}}(X) \otimes G)$.*

Remark 5.3.16 While we state and prove the theorem here for a coefficient group G, it generalizes to the statement that $I^{\bar{p}}H_*(X;M) \cong H_*(I^{\bar{p}}S_*(X;R) \otimes_R M)$, and similarly in the PL setting, if X is a locally $(\bar{p},R;M)$-torsion-free CS set for a Dedekind domain R and R-module M.

Corollary 5.3.17 *For any CS set and any field F of characteristic zero, we have*

$$I^{\bar{p}}H_*^{\mathrm{GM}}(X;F) \cong I^{\bar{p}}H_*^{\mathrm{GM}}(X) \otimes_{\mathbb{Z}} F.$$

If X is also PL then

$$I^{\bar{p}}\mathfrak{H}_*^{\mathrm{GM}}(X;F) \cong I^{\bar{p}}\mathfrak{H}_*^{\mathrm{GM}}(X) \otimes_{\mathbb{Z}} F.$$

Proof As F is $\mathbb{Z}$-torsion-free, any space automatically satisfies the locally torsion-free condition with respect to F by [182, Theorem 54.4.c], so the result follows from Theorem 5.3.15 and the algebraic Universal Coefficient Theorem. □

Corollary 5.3.18 *For any CS set and any abelian group G, we have*

$$I^{\bar{t}}H_*^{\mathrm{GM}}(X;G) \cong H_*(I^{\bar{t}}S_*^{\mathrm{GM}}(X) \otimes G).$$

If X is also PL then $I^{\bar{t}}\mathfrak{H}_*^{\mathrm{GM}}(X;G) \cong H_*(I^{\bar{t}}\mathfrak{C}_*^{\mathrm{GM}}(X) \otimes G)$.

Proof This follows immediately from Example 5.3.12 and Theorem 5.3.15. □

Proof of Theorem 5.3.15 The singular and PL proofs are the same, so we give the singular proof.

We will use the Mayer–Vietoris argument from Theorem 5.1.4 with $F_*(U) = H_*(I^{\bar{p}}S_*^{\mathrm{GM}}(U) \otimes G)$ and $G_*(U) = I^{\bar{p}}H_*^{\mathrm{GM}}(U;G)$. The natural transformation $\Phi\colon F_*(U) \to G_*(U)$ is induced by the inclusion map $I^{\bar{p}}S_*^{\mathrm{GM}}(U) \otimes G \to I^{\bar{p}}S_*^{\mathrm{GM}}(U;G)$; note that if ξ is allowable in $I^{\bar{p}}S_*^{\mathrm{GM}}(U)$, then $\xi \otimes g$ will be so in $I^{\bar{p}}S_*^{\mathrm{GM}}(U;G)$, and $I^{\bar{p}}S_*^{\mathrm{GM}}(U) \otimes G$ is generated by terms of this form.

We must show that the conditions of Theorem 5.1.4 hold.

Condition 1: Notice that we have a commutative diagram with exact rows (where for clarity we write I for $I^{\bar{p}}S_*^{\mathrm{GM}}$):

$$\begin{array}{ccccccccc}
0 & \to & \mathrm{I}(U\cap V)\otimes G & \longrightarrow & (\mathrm{I}(U)\oplus \mathrm{I}(V))\otimes G & \longrightarrow & (\mathrm{I}(U)+\mathrm{I}(V))\otimes G & \to & 0\\
 & & \downarrow & & \downarrow & & \downarrow & & \\
0 & \to & \mathrm{I}(U\cap V;G) & \longrightarrow & \mathrm{I}(U;G)\oplus \mathrm{I}(V;G) & \longrightarrow & \mathrm{I}(U;G)+\mathrm{I}(V;G) & \to & 0.
\end{array}$$

To make sense of the sum terms, we identify $I^{\bar{p}}S_*^{\mathrm{GM}}(U) + I^{\bar{p}}S_*^{\mathrm{GM}}(V)$ as a subset of $I^{\bar{p}}S_*^{\mathrm{GM}}(U\cup V)$ and $I^{\bar{p}}S_*^{\mathrm{GM}}(U;G)+I^{\bar{p}}S_*^{\mathrm{GM}}(V;G)$ as a subset of $I^{\bar{p}}S_*^{\mathrm{GM}}(U\cup V;G)$. The top row is exact by tensoring the short exact sequence of free groups[18]

$$0 \to I^{\bar{p}}S_*^{\mathrm{GM}}(U\cap V)\to I^{\bar{p}}S_*^{\mathrm{GM}}(U)\oplus I^{\bar{p}}S_*^{\mathrm{GM}}(V)\to I^{\bar{p}}S_*^{\mathrm{GM}}(U)+I^{\bar{p}}S_*^{\mathrm{GM}}(V) \to 0$$

with G, and the bottom sequence is exact by our standard Mayer–Vietoris arguments. The left-hand map is induced by inclusion. Furthermore, by distributivity of tensor products over direct sums,

$$(I^{\bar{p}}S_*^{\mathrm{GM}}(U)\oplus I^{\bar{p}}S_*^{\mathrm{GM}}(V))\otimes G \cong (I^{\bar{p}}S_*^{\mathrm{GM}}(U)\otimes G)\oplus (I^{\bar{p}}S_*^{\mathrm{GM}}(V)\otimes G),$$

and so we may interpret the middle vertical map as a direct sum of inclusions, and so the middle map on homology corresponds to Φ. It is not difficult to check by hand that the left square commutes using that all groups are subgroups of the corresponding groups of the form[19] $S_*(W)\otimes G$. Finally, we have the commutative diagram

$$\begin{array}{ccc}
(I^{\bar{p}}S_*^{\mathrm{GM}}(U)+I^{\bar{p}}S_*^{\mathrm{GM}}(V))\otimes G & \longrightarrow & I^{\bar{p}}S_*^{\mathrm{GM}}(U\cup V)\otimes G\\
\downarrow & & \downarrow\\
I^{\bar{p}}S_*^{\mathrm{GM}}(U;G)+I^{\bar{p}}S_*^{\mathrm{GM}}(V;G) & \longrightarrow & I^{\bar{p}}S_*^{\mathrm{GM}}(U\cup V;G).
\end{array}$$

Commutativity is again easy to check by viewing all groups as subgroups of $S_*(U\cup V)\otimes G$. The top map induces homology isomorphisms by the proof of Theorem 4.4.4 and the algebraic Universal Coefficient Theorem, and the bottom map induces isomorphisms by the analogue of Theorem 4.4.4 with coefficients. So the resulting long exact Mayer–Vietoris homology sequences are compatible with Φ.

[18] If we were working with the various $I^{\bar{p}}S_*^{\mathrm{GM}}$ as R-modules for a Dedekind domain R, we would instead use here and elsewhere in the argument (in particular in invoking the algebraic Universal Coefficient Theorem [238, Theorem 3.6.1]) that these are all flat R-modules; see the discussion preceding Theorem 5.3.6.

[19] Note that if ξ is not allowable in $S_*(W)$, then no multiple of ξ can be allowable either, and so $S_*(W)/I^{\bar{p}}S_*^{\mathrm{GM}}(W)$ is torsion-free. Hence $S_*(W)/I^{\bar{p}}S_*^{\mathrm{GM}}(W)$ is flat, as torsion-free implies flat over any Dedekind domain; see Section A.4.2. Thus $S_*(W)/I^{\bar{p}}S_*^{\mathrm{GM}}(W) * G = 0$ [197, Theorem 7.2]. Hence, $I^{\bar{p}}S_*^{\mathrm{GM}}(W)\otimes G \subset S_*(W)\otimes G$ by the torsion/tensor product exact sequence; see [197, Corollary 7.3].

Condition 2: This property is satisfied for both F_* and G_* using Lemma 5.1.6, minor modifications of the arguments in the proof of Lemma 5.1.7, the Universal Coefficient Theorem, and the Five Lemma.

Condition 3: We must show that if L is a compact filtered space such that X has an open subset filtered homeomorphic to $\mathbb{R}^i \times cL$ and $\Phi : F_*(\mathbb{R}^i \times (cL - \{v\})) \to G_*(\mathbb{R}^i \times (cL - \{v\}))$ is an isomorphism, then so is $\Phi : F_*(\mathbb{R}^i \times cL) \to G_*(\mathbb{R}^i \times cL)$.

Using the stratified homotopy invariance of both functors F_* and G_*, this is equivalent to assuming that $H_*(I^{\bar{p}}S_*^{\mathrm{GM}}(L) \otimes G) \to I^{\bar{p}}H_*^{\mathrm{GM}}(L; G)$ is an isomorphism and needing to verify that consequently $H_*(I^{\bar{p}}S_*^{\mathrm{GM}}(cL) \otimes G) \to I^{\bar{p}}H_*^{\mathrm{GM}}(cL; G)$ is an isomorphism. By the assumption that X is locally $(\bar{p}, \mathbb{Z}; G)^{\mathrm{GM}}$-torsion-free, we must have $I^{\bar{p}}H^{\mathrm{GM}}_{\dim(L)-\bar{p}(S)-1}(L) * G = 0$.

Consider the following commutative diagram induced by inclusions:

$$\begin{array}{ccc} H_i(I^{\bar{p}}S_*^{\mathrm{GM}}(L) \otimes G) & \longrightarrow & I^{\bar{p}}H_i^{\mathrm{GM}}(L; G) \\ \downarrow & & \downarrow \\ H_i(I^{\bar{p}}S_*^{\mathrm{GM}}(cL) \otimes G) & \longrightarrow & I^{\bar{p}}H_i^{\mathrm{GM}}(cL; G). \end{array}$$

By assumption, the top horizontal map is an isomorphism for all i. Thanks to the cone formula (Theorem 5.3.5), the right-hand map is an isomorphism for all $i < \dim(L) - \bar{p}(S)$. Similarly, by Theorem 4.2.1, $I^{\bar{p}}H_i^{\mathrm{GM}}(L) \to I^{\bar{p}}H_i^{\mathrm{GM}}(cL)$ is an isomorphism in the same range, and hence this is also true of the left-hand map using the Universal Coefficient Theorem. Hence the bottom map is also an isomorphism in this range.

For $i \geq \dim(L) - \bar{p}(S)$, $i \neq 0$, by Theorem 5.3.5, $I^{\bar{p}}H_i^{\mathrm{GM}}(cL; G) = 0$. But this is also true of $H_i(I^{\bar{p}}S_*^{\mathrm{GM}}(cL) \otimes G)$ using Theorem 4.2.1, the Universal Coefficient Theorem, and the assumption that $I^{\bar{p}}H^{\mathrm{GM}}_{\dim(L)-\bar{p}(S)-1}(L) * G = 0$ (here is where the assumption that X is locally $(\bar{p}, \mathbb{Z}; G)^{\mathrm{GM}}$-torsion-free pays off!).

Finally, if $i = 0 \geq \dim(L) - \bar{p}(S)$, then $I^{\bar{p}}H_0^{\mathrm{GM}}(cL) \cong \mathbb{Z}$ or 0, so $H_0(I^{\bar{p}}S_*^{\mathrm{GM}}(cL) \otimes G) \cong G$ or 0, respectively, by the Universal Coefficient Theorem. By the same reasoning, $I^{\bar{p}}H_i^{\mathrm{GM}}(cL; G) \cong G$ or 0 in the corresponding situations by Theorem 5.3.5; note that $I^{\bar{p}}H_0^{\mathrm{GM}}(cL; G) = 0$ if and only if there are no allowable 0-simplices in cL, which is precisely when $I^{\bar{p}}H_0^{\mathrm{GM}}(cL) = 0$. In both nontrivial cases, the elements of the groups can be represented in the form $\sigma_0 \otimes g$, where σ_0 is any allowable 0-simplex and $g \in G$, so the bottom map of the diagram is again an isomorphism.

Thus $\Phi : H_i(I^{\bar{p}}S_*^{\mathrm{GM}}(cL) \otimes G) \to I^{\bar{p}}H_i^{\mathrm{GM}}(cL; G)$ is an isomorphism for all i.

Condition 4: If $U = \emptyset$, $F_*(U) = G_*(U) = 0$, so suppose $U \subset X$ is an open subset of X homeomorphic to Euclidean space and contained within a stratum S. Since the images of simplices of $S_*(U)$ are contained completely in S and cannot intersect other strata, the allowability condition for an i-simplex is that

$$\Delta^i \subset \{(i - \operatorname{codim}(S) + \bar{p}(S))\text{-skeleton of } \Delta^i\},$$

or, in other words, that $i \leq i - \operatorname{codim}(S) + \bar{p}(S)$. But this is simply the condition that $\bar{p}(S) \geq \operatorname{codim}(S)$, which is independent of i. So, depending on the value of $\bar{p}(S)$ and the codimension of S, either all simplices are allowable or none are! If none are, then $I^{\bar{p}}S_*^{\mathrm{GM}}(U) \otimes G = 0 = I^{\bar{p}}S_*^{\mathrm{GM}}(U;G)$, and if all are,

$$I^{\bar{p}}S_*^{\mathrm{GM}}(U) \otimes G = S_*(U) \otimes G = S_*(U;G) = I^{\bar{p}}S_*^{\mathrm{GM}}(U;G).$$

So either way Φ is an isomorphism on U. This completes the proof of Theorem 5.3.15. □

Corollary 5.3.19 *Suppose X is a locally $(\bar{p},\mathbb{Z};G)^{\mathrm{GM}}$-torsion-free CS set and that $A \subset X$ is also a locally $(\bar{p};G)^{\mathrm{GM}}$-torsion-free CS set, in particular if A is an open subset of X. Then $I^{\bar{p}}H_*^{\mathrm{GM}}(X,A;G) \cong H_*(I^{\bar{p}}S_*^{\mathrm{GM}}(X,A) \otimes G)$. If X is PL and A is a PL subset, then $I^{\bar{p}}\mathfrak{H}_*^{\mathrm{GM}}(X,A;G) \cong H_*(I^{\bar{p}}\mathfrak{C}_*^{\mathrm{GM}}(X,A) \otimes G)$.*

Proof Consider the diagram

$$\begin{array}{ccccccccc}
0 & \longrightarrow & I^{\bar{p}}S_*^{\mathrm{GM}}(A) \otimes G & \longrightarrow & I^{\bar{p}}S_*^{\mathrm{GM}}(X) \otimes G & \longrightarrow & I^{\bar{p}}S_*^{\mathrm{GM}}(X,A) \otimes G & \longrightarrow & 0 \\
 & & \downarrow & & \downarrow & & \downarrow & & \\
0 & \longrightarrow & I^{\bar{p}}S_*^{\mathrm{GM}}(A;G) & \longrightarrow & I^{\bar{p}}S_*^{\mathrm{GM}}(X;G) & \longrightarrow & I^{\bar{p}}S_*^{\mathrm{GM}}(X,A;G) & \longrightarrow & 0.
\end{array}$$

The top row is obtained by tensoring the short exact sequence of the pair (X,A) with G. Since $I^{\bar{p}}S_*^{\mathrm{GM}}(X,A)$ is a subgroup of $S_*(X,A)$, which is torsion-free, the group $I^{\bar{p}}S_*^{\mathrm{GM}}(X,A)$ is also torsion-free and hence flat. Therefore, tensoring with G preserves exactness [197, Corollary 7.3]. The bottom row is exact by the definition of $I^{\bar{p}}S_*^{\mathrm{GM}}(X,A;G)$ as the quotient $I^{\bar{p}}S_*^{\mathrm{GM}}(X;G)/I^{\bar{p}}S_*^{\mathrm{GM}}(A;G)$. The vertical maps are those of the proof of Theorem 5.3.15. It is straightforward that the left-hand vertical square commutes, and so the right-hand vertical map is induced as the quotient map; commutativity of the right-hand square follows.

The corollary now follows from the ensuing diagram of long exact sequences, Theorem 5.3.15, and the Five Lemma. □

Here is one final corollary for this section. It says that Bockstein maps exist when appropriate torsion-free conditions are met.

Corollary 5.3.20 *Suppose*

$$0 \to G_1 \to G_2 \to G_3 \to 0$$

is a short exact sequence of abelian groups and that X is a CS set that is locally $(\bar{p},\mathbb{Z};G_1)^{\mathrm{GM}}$-torsion-free and $(\bar{p},\mathbb{Z};G_3)^{\mathrm{GM}}$-torsion-free. Then there are Bockstein homomorphisms

$$\beta : I^{\bar{p}}H_i^{\mathrm{GM}}(X;G_3) \to I^{\bar{p}}H_{i-1}^{\mathrm{GM}}(X;G_1).$$

Similarly in the PL setting.

Proof As $I^{\bar{p}}S_*^{\mathrm{GM}}(X)$ is a complex of flat modules, tensoring the complex with the exact sequence of abelian groups yields a short exact sequence of chain complexes. The connecting map of the associated long exact homology sequences has the form

$$\beta : H_i(I^{\bar{p}}S_*^{\mathrm{GM}}(X) \otimes G_3) \to H_{i-1}(I^{\bar{p}}S_*^{\mathrm{GM}}(X) \otimes G_1).$$

But if X satisfies the given torsion-free conditions, Theorem 5.3.15 allows us to identify these groups with $I^{\bar{p}}H_i^{\mathrm{GM}}(X; G_3)$ and $I^{\bar{p}}H_{i-1}^{\mathrm{GM}}(X; G_1)$. □

Remark 5.3.21 It appears at first that we do not need X to be well behaved with respect to G_2 in order to define our Bockstein map, but the Snake Lemma implies that if $I^{\bar{p}}H^{\mathrm{GM}}_{\dim(L)-\bar{p}(S)-1}(L) * G_j = 0$ for $j = 1, 3$, then this also holds for $j = 2$. So for X to satisfy the hypotheses, it must also be locally $(\bar{p}, \mathbb{Z}; G_2)^{\mathrm{GM}}$-torsion-free.

5.4 Equivalence of PL and Singular Intersection Homology on PL CS Sets

In this section, we will show that for a PL CS set X, the PL intersection homology groups $I^{\bar{p}}\mathfrak{H}_*^{\mathrm{GM}}(X; G)$ are isomorphic to the singular intersection homology groups $I^{\bar{p}}H_*^{\mathrm{GM}}(X; G)$. This fact is certainly well known via sheaf theory [106]. A proof without sheaf theory is suggested in King [139], utilizing King's Theorem 10. However, on close examination, it was not completely clear to the author precisely what King had in mind for an "ordered PL theory," at least not without requiring some significant additional work to verify that such a theory would have the needed properties. So here we take a slightly different route (though no doubt this is not very far from what King had in mind).

We have already seen in Section 4.4.2 that there are ways to assign a singular chain to a simplicial chain, at least given additional vertex ordering information. Now we need to further build in the subdivisions inherent in the PL setting. It will be a bit easier to work with barycentric subdivisions, because then we can control in precisely what order we add new vertices, and we will see that we are free to work entirely with just the barycentric subdivision of some starting triangulation of our space X. We will also need to compute PL intersection homology groups of various open subsets U of X in order to employ Mayer–Vietoris arguments, and we will show that we can also compute the intersection homology of these subsets using only simplices coming from the barycentric subdivisions of our starting triangulation; this is the content of Lemma 5.4.1. With these tools, we will then describe a map from the PL intersection chains of U to an analogous limit complex of the singular chain complex. Then in Theorem 5.4.2, we will use Theorem 5.1.4 (which is itself similar to King's Theorem 10) to show that PL and singular intersection homology agree on PL CS sets. It seems reasonable to conjecture that such an isomorphism holds more generally for any PL filtered set, though we will not pursue such a result here.

The techniques of this section are not particular to the coefficients, so to keep the notation from getting even more cluttered than necessary we will use the notation for $\mathbb{Z}$ coefficients (meaning that we keep G tacit) in the proofs, though, for completeness, we provide the full notation in the statements of the theorems.

5.4.1 Barycentric Subdivisions and Maps from PL Chains to Singular Chains

Our first result is applicable to PL filtered spaces in general. Suppose $U \subset X$ is an open subset of a PL filtered space. Let T be a particular triangulation of X, and let T^i be the ith barycentric subdivision of T. The subdivision maps $C_*^{T^i}(X) \to C_*^{T^{i+1}}(X)$ form a direct system, whose limit we denote $\mathfrak{C}_*^T(X)$. Restricting to intersection chains, we similarly have $I^{\bar{p}}\mathfrak{C}_*^{\mathrm{GM},T}(X)$ as the limit of the $I^{\bar{p}}C_*^{\mathrm{GM},T^i}(X)$. We have chain maps, in fact inclusions, $\mathfrak{C}_*^T(X) \to \mathfrak{C}_*(X)$ and $I^{\bar{p}}\mathfrak{C}_*^{\mathrm{GM},T}(X) \to I^{\bar{p}}\mathfrak{C}_*^{\mathrm{GM}}(X)$. If we let $I^{\bar{p}}\mathfrak{C}_*^{\mathrm{GM},T}(U) \subset I^{\bar{p}}\mathfrak{C}_*^{\mathrm{GM},T}(X)$ be the subcomplex consisting of chains supported in U, then we also have a monomorphism $\theta : I^{\bar{p}}\mathfrak{C}_*^{\mathrm{GM},T}(U) \to I^{\bar{p}}\mathfrak{C}_*^{\mathrm{GM}}(U)$. Notice that elements of $I^{\bar{p}}\mathfrak{C}_*^{\mathrm{GM},T}(U)$ are not defined with respect to any fixed triangulation of U, but are rather defined with respect to triangulations of X and have their support in U.

Lemma 5.4.1 *For a PL filtered space X with open subset U, the map*

$$\theta : I^{\bar{p}}\mathfrak{C}_*^{\mathrm{GM},T}(U;G) \to I^{\bar{p}}\mathfrak{C}_*^{\mathrm{GM}}(U;G),$$

defined above, induces isomorphisms $\theta : I^{\bar{p}}\mathfrak{H}_*^{\mathrm{GM},T}(U;G) \to I^{\bar{p}}\mathfrak{H}_*^{\mathrm{GM}}(U;G)$.

Proof The proof utilizes the techniques used to prove Theorem 3.3.20.

We begin with surjectivity. Let ξ be a simplicial chain representing an element of $I^{\bar{p}}\mathfrak{H}_*^{\mathrm{GM}}(U)$. By the definition of PL chains, there is some triangulation T_1 of X such that ξ is represented as a simplicial chain with respect to T_1. Utilizing the compactness of the support of ξ, we can find an iterated barycentric subdivision T^i of our fixed starting triangulation T such that every simplex of T^i that intersects $|\xi|$ is contained in U. Let T_1' be a common subdivision of T^i and T_1, and let ξ' be the subdivision of ξ in the triangulation T_1'. Since we may assume that $i > 0$, we may assume without loss of generality that T^i is a full triangulation. The proof of Lemma 3.3.23 then shows how to construct an allowable homology in $I^{\bar{p}}\mathfrak{C}_*^{\mathrm{GM}}(X)$ from ξ' to an element of $I^{\bar{p}}C_*^{\mathrm{GM},T^i}(X)$. Furthermore, the argument shows that the homology will be supported in a subcomplex of T^i containing $|\xi|$. So, in particular, this homology is contained within U and demonstrates that θ must be surjective.

For injectivity, we similarly suppose $[\xi]$ is a cycle in $I^{\bar{p}}\mathfrak{C}_*^{\mathrm{GM},T}(U)$ whose image in $I^{\bar{p}}\mathfrak{C}_*^{\mathrm{GM}}(U)$ bounds an allowable PL chain $[\eta]$. Let T^i be a barycentric subdivision of T with i sufficiently large that $[\xi]$ can be represented by a simplicial chain in T^i and such that every simplex of T^i that intersects $|\eta|$ is contained in U. Let T_1 be a triangulation with respect to which $[\eta]$ can be represented by a simplicial chain η. Let T_1' be a common refinement of T^i and T_1 in which the images ξ' of ξ and η' of η satisfy $\partial\eta' = \xi'$. Then Lemma 3.3.21 provides a chain map $\mu : I^{\bar{p}}\mathfrak{C}_*^{\mathrm{GM},T_1'}(X) \to I^{\bar{p}}\mathfrak{C}_*^{\mathrm{GM},T^i}(X)$ which is a left inverse to the subdivision map; in particular it takes ξ' back to ξ in T^i. Once again, the proof of Lemma 3.3.21 shows that μ keeps the image of each simplex of T_1' within the simplex of T^i containing it, and so $|\mu(\eta')| \subset U$. Furthermore, as μ is a chain map, we have $\partial(\mu(\eta')) = \mu(\partial\eta') = \mu(\xi') = \xi$. This completes the proof of the lemma. □

We make the following observations concerning the intersection homology groups $I^{\bar{p}}\mathfrak{H}_*^{\mathrm{GM},T}(U)$ for a triangulation T of a PL filtered space X of which U is an open subset:

1. Using the arguments of Section 4.4.1, we have excision $I^{\bar{p}}\mathfrak{H}_*^{\mathrm{GM},T}(U,A) \cong I^{\bar{p}}\mathfrak{H}_*^{\mathrm{GM},T}(U-K, A-K)$ for $K \subset A \subset U$ with the closure of K in U contained in the interior of A in U (which is equal to the interior of A in X, since U is open in X). For example, if $[\xi] \in I^{\bar{p}}\mathfrak{H}_*^{\mathrm{GM},T}(U,A)$ is represented by a chain ξ in some subdivision of T that is also contained in U, then we can use the same construction as in the proof of Theorem 4.4.3 to find a further barycentric subdivision of T with respect to which we can split the image of ξ into two $\bar{p}$-allowable pieces, one contained in $U - \bar{K}$ and the other contained in A. The rest of the argument for excision is then exactly as in the proof of Theorem 4.4.3. Similarly, minor modifications to the proof of Theorem 4.4.4 establish that, for two open subsets $U, V \subset X$, there are Mayer–Vietoris sequences
$$\to I^{\bar{p}}\mathfrak{H}_*^{\mathrm{GM},T}(U\cap V) \to I^{\bar{p}}\mathfrak{H}_*^{\mathrm{GM},T}(U) \oplus I^{\bar{p}}\mathfrak{H}_*^{\mathrm{GM},T}(V) \to I^{\bar{p}}\mathfrak{H}_*^{\mathrm{GM},T}(U\cup V) \to .$$
2. Via the standard compactness of chains argument as in Lemma 5.1.7, if U_α is an increasing sequence of open subsets of X, then
$$\varinjlim I^{\bar{p}}\mathfrak{H}_*^{\mathrm{GM},T}(U_\alpha) \cong I^{\bar{p}}\mathfrak{H}_*^{\mathrm{GM},T}\left(\bigcup U_\alpha\right).$$

Now, continuing to assume we have a triangulation T of X, we may choose a total ordering of the vertices of T and use this to construct a chain map $\phi : C_*^T(X) \to S_*(X)$ by Proposition 4.4.5. Recall that if $\Delta^i = [v_0, \dots, v_i]$ is the standard simplex with fixed vertex order and if $\tau = [w_0, \dots, w_i]$ is an i-simplex of T with its vertex order, then $\phi(\tau) : \Delta^i \to X$ is just the linear map determined by taking each v_j to w_j. If T^1 is the first barycentric subdivision of T, we can partially order the vertices of T^1 consistently with the vertices of T by making each barycenter of a j-simplex smaller in order than a barycenter of a k-simplex if $j < k$; see Example 4.4.10. This is enough to prescribe a commutative diagram

$$\begin{array}{ccc} C_*^T(X) & \longrightarrow & S_*(X) \\ \downarrow \nu & & \downarrow \nu \\ C_*^{T^1}(X) & \longrightarrow & S_*(X), \end{array}$$

where the vertical maps are barycentric subdivision operators.

Continuing this process, we obtain a map $\mathfrak{C}_*^T(X) \to \mathfrak{S}_*(X)$, where $\mathfrak{S}_*(X)$ is the limit of

$S_*(X)$ under the barycentric subdivision maps, and similarly, applying Corollary 4.4.6, we obtain maps $\psi : I^{\bar{p}}\mathfrak{C}_*^{\mathrm{GM},T}(X) \to I^{\bar{p}}\mathfrak{S}_*^{\mathrm{GM}}(X)$, where $I^{\bar{p}}\mathfrak{S}_*^{\mathrm{GM}}(X)$ is the limit of the complexes $I^{\bar{p}}S_*^{\mathrm{GM}}(X)$ under subdivision. This is all well defined as the barycentric subdivisions of PL and singular intersection chains are allowable by Lemmas 3.3.15 and 4.4.13. If $U \subset X$ is open, then since the image of $I^{\bar{p}}\mathfrak{C}_*^{\mathrm{GM},T}(U) \subset I^{\bar{p}}\mathfrak{C}_*^{\mathrm{GM},T}(X)$ under ψ consists of chains supported in U, the map ψ restricts to a chain map $\psi : I^{\bar{p}}\mathfrak{C}_*^{\mathrm{GM},T}(U) \to I^{\bar{p}}\mathfrak{S}_*^{\mathrm{GM}}(U)$, where $I^{\bar{p}}\mathfrak{S}_*^{\mathrm{GM}}(U) \subset I^{\bar{p}}\mathfrak{S}_*^{\mathrm{GM}}(X)$ is the limit under barycentric subdivisions of $I^{\bar{p}}S_*^{\mathrm{GM}}(U)$. By Lemma 4.4.14, the barycentric subdivision map $I^{\bar{p}}S_*^{\mathrm{GM}}(U) \to I^{\bar{p}}S_*^{\mathrm{GM}}(U)$ induces the identity on homology, and so by the commutativity of direct limits with homology functors, $H_*(I^{\bar{p}}\mathfrak{S}_*^{\mathrm{GM}}(U)) \cong I^{\bar{p}}H_*^{\mathrm{GM}}(U)$.

5.4.2 The Isomorphism of PL and Singular Intersection Homology

We now proceed on to the main theorem of this section. In the statement, we don't assume that all of X is a CS set but only an open subset on which we will show that PL and singular intersection homology are isomorphic. This extra generality will be utilized below in the corollaries to the theorem.

Theorem 5.4.2 *Let X be a PL filtered space with triangulation T, and let $W \subset X$ be an open subset of X such that W is a PL CS set. Then the composition*

$$I^{\bar{p}}\mathfrak{H}_*^{\mathrm{GM}}(W;G) \xrightarrow{\theta^{-1}} I^{\bar{p}}\mathfrak{H}_*^{\mathrm{GM},T}(W;G) \xrightarrow{\psi} H_*(I^{\bar{p}}\mathfrak{S}_*^{\mathrm{GM}}(W;G))$$

is an isomorphism. In particular, $I^{\bar{p}}\mathfrak{H}_*^{\mathrm{GM}}(W;G) \cong I^{\bar{p}}H_*^{\mathrm{GM}}(W;G)$, and if X is a PL CS set then $I^{\bar{p}}\mathfrak{H}_*^{\mathrm{GM}}(X;G) \cong I^{\bar{p}}H_*^{\mathrm{GM}}(X;G)$.

Proof As the proof is the same for any coefficients, we work with $\mathbb{Z}$ coefficients for ease of notation.

The last sentence of the statement of the theorem follows from the preceding statements and our prior observation that $H_*(I^{\bar{p}}\mathfrak{S}_*^{\mathrm{GM}}(W)) \cong I^{\bar{p}}H_*^{\mathrm{GM}}(W)$.

The proof will use the Mayer–Vietoris argument of Theorem 5.1.4 with $\mathcal{F}_W$ being our domain category. Recall that this is the category whose objects are filtered homeomorphic to open subsets W and whose morphisms are filtered homeomorphisms and inclusions. Our functors on open sets $U \subset W$ are $F_*(U) = I^{\bar{p}}\mathfrak{H}_*^{\mathrm{GM}}(U)$ and $G_*(U) = H_*(I^{\bar{p}}\mathfrak{S}_*^{\mathrm{GM}}(U))$. We let $\Phi = \psi\theta^{-1}$. Notice that Φ is a natural transformation on the category of open subsets of W using the commutativity of diagrams of the form

$$\begin{array}{ccccc}
I^{\bar{p}}\mathfrak{H}_*^{\mathrm{GM}}(V) & \underset{\cong}{\xleftarrow{\theta}} & I^{\bar{p}}\mathfrak{H}_*^{\mathrm{GM},T}(V) & \xrightarrow{\psi} & H_*(I^{\bar{p}}\mathfrak{S}_*^{\mathrm{GM}}(V)) \\
\downarrow & & \downarrow & & \downarrow \\
I^{\bar{p}}\mathfrak{H}_*^{\mathrm{GM}}(U) & \underset{\cong}{\xleftarrow{\theta}} & I^{\bar{p}}\mathfrak{H}_*^{\mathrm{GM},T}(U) & \xrightarrow{\psi} & H_*(I^{\bar{p}}\mathfrak{S}_*^{\mathrm{GM}}(U))
\end{array}$$

for $V \subset U$. Similarly, we obtain a natural transformation of Mayer–Vietoris sequences. It is also not difficult to observe, in the usual way as in Lemma 5.1.7, that if $\{U_\alpha\}$ is an increasing collection of open subspaces of $X \in \mathcal{F}$ then the natural maps $\varinjlim_\alpha F_*(U_\alpha) \to F_*\left(\bigcup_\alpha U_\alpha\right)$ and $\varinjlim_\alpha G_*(U_\alpha) \to G_*\left(\bigcup_\alpha U_\alpha\right)$ are isomorphisms. Hence, using Lemma 5.1.6, conditions (1) and (2) of Theorem 5.1.4 are satisfied.

Next suppose U is an open subset of W that is contained in a single stratum S and PL homeomorphic to Euclidean space (or empty). By the same argument as in the proof[20] of Theorem 5.3.15, either every chain in U is allowable or none are, depending only on the codimension of S and $\bar{p}(S)$. If no chains can be allowable, then all homology groups are 0 and condition (4) of Theorem 5.1.4 holds trivially. If all chains are allowable, then, using Lemma 5.4.1, $I^{\bar{p}}\mathfrak{H}_*^{\mathrm{GM},T}(U) \cong I^{\bar{p}}\mathfrak{H}_*^{\mathrm{GM}}(U) \cong \mathfrak{H}_*(U)$ and $H_*(I^{\bar{p}}\mathfrak{S}_*^{\mathrm{GM}}(U)) \cong I^{\bar{p}}H_*^{\mathrm{GM}}(U) \cong H_*(U)$. Since U is PL homeomorphic to Euclidean space, these groups are all trivial except for $\mathfrak{H}_0(U) \cong H_0(U) \cong \mathbb{Z}$. In all cases the generator is represented by a single vertex (simplicial or singular), so it follows from the constructions that Φ is an isomorphism on U. Thus we have verified hypothesis (4) of Theorem 5.1.4.

Finally, we need to check condition (3) of Theorem 5.1.4. Let $N \cong \mathbb{R}^i \times cL$ be a distinguished neighborhood in W such that we have the diagram

$$\begin{array}{ccc}
I^{\bar{p}}\mathfrak{H}_*^{\mathrm{GM}}(\mathbb{R}^i \times (cL - \{v\})) & \xrightarrow{\Phi} & H_*(I^{\bar{p}}\mathfrak{S}_*^{\mathrm{GM}}(\mathbb{R}^i \times (cL - \{v\}))) \\
\downarrow & & \downarrow \\
I^{\bar{p}}\mathfrak{H}_*^{\mathrm{GM}}(\mathbb{R}^i \times cL) & \xrightarrow{\Phi} & H_*(I^{\bar{p}}\mathfrak{S}_*^{\mathrm{GM}}(\mathbb{R}^i \times cL)).
\end{array}$$

In order for Φ, which uses the triangulation T of X, to be well defined here, we are tacitly identifying $\mathbb{R}^i \times cL$ with an open subset N of W. We assume that the top map is an isomorphism in all dimensions. As we know that $H_*(I^{\bar{p}}\mathfrak{S}_*^{\mathrm{GM}}(V)) \cong I^{\bar{p}}H_*^{\mathrm{GM}}(V)$ for any V, the vertical maps can be computed from stratified homotopy invariance and the cone formula; for these computations, the triangulation of X is not involved. Furthermore, we know that these computations are identical for PL intersection homology and singular intersection homology. In particular, the map on the left is isomorphic to the map induced by inclusion $I^{\bar{p}}\mathfrak{H}_*^{\mathrm{GM}}(L) \to I^{\bar{p}}\mathfrak{H}_*^{\mathrm{GM}}(cL)$, and the same is true of the singular intersection homology. In the range of dimensions for which the cone formulas tell us that these vertical maps are isomorphisms, the commutativity of the diagram implies that the bottom map is also an isomorphism. In cases where the cone formula forces the bottom groups to be 0, the bottom map is an isomorphism trivially. Finally, in any of the special cases in which $I^{\bar{p}}\mathfrak{H}_0^{\mathrm{GM}}(\mathbb{R}^i \times cL) \cong I^{\bar{p}}\mathfrak{H}_0^{\mathrm{GM}}(N)$ and $H_0(I^{\bar{p}}\mathfrak{S}_*^{\mathrm{GM}}(\mathbb{R}^i \times cL)) \cong H_0(I^{\bar{p}}\mathfrak{S}_*^{\mathrm{GM}}(N))$ are forced to each be isomorphic to $\mathbb{Z}$, Lemma 5.4.1 shows that there must be an allowable 0-simplex in N that generates both $I^{\bar{p}}\mathfrak{H}_0^{\mathrm{GM},T}(N)$ and $I^{\bar{p}}\mathfrak{H}_0^{\mathrm{GM}}(N)$, and Φ then takes this 0-simplex to

[20] See the discussion of the proof of Condition (4) starting on page 232.

a singular 0-simplex generating $H_0(I^{\bar{p}}\mathfrak{S}_*^{\mathrm{GM}}(N))$. So the bottom map of the diagram is an isomorphism in all dimensions, verifying condition (3) of Theorem 5.1.4.

This completes our verification of the assumptions of Theorem 5.1.4, and the conclusion follows from that theorem. □

Two relative versions of Theorem 5.4.2 follow almost immediately:

Corollary 5.4.3 *Let X be a PL CS set, and let W be an open subset. Then $I^{\bar{p}}\mathfrak{H}_*^{\mathrm{GM}}(X,W;G) \cong I^{\bar{p}}H_*^{\mathrm{GM}}(X,W;G)$.*

Proof Consider the following diagram of short exact sequences, where T is a triangulation of X:

$$
\begin{array}{ccccccccc}
0 & \longrightarrow & I^{\bar{p}}\mathfrak{C}_*^{\mathrm{GM}}(W) & \longrightarrow & I^{\bar{p}}\mathfrak{C}_*^{\mathrm{GM}}(X) & \longrightarrow & I^{\bar{p}}\mathfrak{C}_*^{\mathrm{GM}}(X,W) & \longrightarrow & 0\\
 & & \uparrow{\scriptstyle\theta} & & \uparrow{\scriptstyle\theta} & & \uparrow{\scriptstyle\theta} & & \\
0 & \longrightarrow & I^{\bar{p}}\mathfrak{C}_*^{\mathrm{GM},T}(W) & \longrightarrow & I^{\bar{p}}\mathfrak{C}_*^{\mathrm{GM},T}(X) & \longrightarrow & I^{\bar{p}}\mathfrak{C}_*^{\mathrm{GM},T}(X,W) & \longrightarrow & 0\\
 & & \downarrow{\scriptstyle\psi} & & \downarrow{\scriptstyle\psi} & & \downarrow{\scriptstyle\psi} & & \\
0 & \longrightarrow & I^{\bar{p}}\mathfrak{S}_*^{\mathrm{GM}}(W) & \longrightarrow & I^{\bar{p}}\mathfrak{S}_*^{\mathrm{GM}}(X) & \longrightarrow & I^{\bar{p}}\mathfrak{S}_*^{\mathrm{GM}}(X,W) & \longrightarrow & 0.
\end{array}
$$

In each case row, the rightmost nontrivial group is defined to be the quotient under the evident inclusion of the leftmost group into the middle group.

This diagram induces a commutative diagram of long exact homology sequences. By Lemma 5.4.1 applied to W and X and by the Five Lemma, we obtain an isomorphism of the top two long exact sequences. It follows then that $\psi\theta^{-1}$ is well defined from the top long exact sequence to the bottom one, so now by Theorem 5.4.2 and the Five Lemma, $I^{\bar{p}}\mathfrak{H}_*^{\mathrm{GM}}(X,W) \cong H_*(I^{\bar{p}}\mathfrak{S}_*^{\mathrm{GM}}(X,W))$. But now by the exactness of the direct limit functor,

$$
\begin{aligned}
H_*(I^{\bar{p}}\mathfrak{S}_*^{\mathrm{GM}}(X,W)) &= H_*(I^{\bar{p}}\mathfrak{S}_*^{\mathrm{GM}}(X)/I^{\bar{p}}\mathfrak{S}_*^{\mathrm{GM}}(W))\\
&= H_*(\varinjlim I^{\bar{p}}S_*^{\mathrm{GM}}(X)/\varinjlim I^{\bar{p}}S_*^{\mathrm{GM}}(W))\\
&\cong H_*(\varinjlim (I^{\bar{p}}S_*^{\mathrm{GM}}(X)/I^{\bar{p}}S_*^{\mathrm{GM}}(W)))\\
&\cong \varinjlim H_*(I^{\bar{p}}S_*^{\mathrm{GM}}(X)/I^{\bar{p}}S_*^{\mathrm{GM}}(W))\\
&\cong I^{\bar{p}}H_*^{\mathrm{GM}}(X,W),
\end{aligned}
$$

as we know that the subdivision operator induces isomorphisms on intersection homology, and hence on relative intersection homology by another application of the Five Lemma. □

Corollary 5.4.4 *Let X be a PL CS set with closed PL subset A such that A is itself a PL CS set in its inherited filtration. Then $I^{\bar{p}}\mathfrak{H}_*^{\mathrm{GM}}(X,A;G) \cong I^{\bar{p}}H_*^{\mathrm{GM}}(X,A;G)$.*

Proof This time we assume that we begin with a triangulation T of X such that A is triangulated as a subcomplex. Once again, we have a diagram of short exact sequences

$$\begin{array}{ccccccccc}
0 & \longrightarrow & I^{\bar{p}}\mathfrak{C}_*^{\mathrm{GM}}(A) & \longrightarrow & I^{\bar{p}}\mathfrak{C}_*^{\mathrm{GM}}(X) & \longrightarrow & I^{\bar{p}}\mathfrak{C}_*^{\mathrm{GM}}(X,A) & \longrightarrow & 0 \\
 & & \uparrow{\scriptstyle\theta} & & \uparrow{\scriptstyle\theta} & & \uparrow{\scriptstyle\theta} & & \\
0 & \longrightarrow & I^{\bar{p}}\mathfrak{C}_*^{\mathrm{GM},T}(A) & \longrightarrow & I^{\bar{p}}\mathfrak{C}_*^{\mathrm{GM},T}(X) & \longrightarrow & I^{\bar{p}}\mathfrak{C}_*^{\mathrm{GM},T}(X,A) & \longrightarrow & 0 \\
 & & \downarrow{\scriptstyle\psi} & & \downarrow{\scriptstyle\psi} & & \downarrow{\scriptstyle\psi} & & \\
0 & \longrightarrow & I^{\bar{p}}\mathfrak{S}_*^{\mathrm{GM}}(A) & \longrightarrow & I^{\bar{p}}\mathfrak{S}_*^{\mathrm{GM}}(X) & \longrightarrow & I^{\bar{p}}\mathfrak{S}_*^{\mathrm{GM}}(X,A) & \longrightarrow & 0,
\end{array}$$

where, as before, $I^{\bar{p}}\mathfrak{C}_*^{\mathrm{GM},T}(A)$ denotes those chains of $I^{\bar{p}}\mathfrak{C}_*^{\mathrm{GM},T}(X)$ supported in A. Since the barycentric subdivisions of the restriction of T to A are compatible with restricting the barycentric subdivisions of T in all of X to A, the left and center vertical maps all induce isomorphisms on homology by Lemma 5.4.1 and Theorem 5.4.2. The corollary now follows from two applications of the Five Lemma to the resulting long exact sequences. □

Even though ∂-stratified pseudomanifolds are not technically CS sets, the following shows that the results of this section apply to them as well.

Corollary 5.4.5 *Suppose X is a PL ∂-stratified pseudomanifold. Then the composition*

$$I^{\bar{p}}\mathfrak{H}_*^{\mathrm{GM}}(X;G) \xrightarrow{\theta^{-1}} I^{\bar{p}}\mathfrak{H}_*^{\mathrm{GM},T}(X;G) \xrightarrow{\psi} H_*(I^{\bar{p}}\mathfrak{S}_*^{\mathrm{GM}}(X;G)) \xrightarrow{\cong} I^{\bar{p}}H_*^{\mathrm{GM}}(X;G)$$

is an isomorphism.

Proof Let $X - \partial X$ be the interior of X, and consider the following diagram in which the vertical maps are induced by inclusions:

$$\begin{array}{ccccccc}
I^{\bar{p}}\mathfrak{H}_*^{\mathrm{GM}}(X-\partial X) & \xleftarrow{\theta} & I^{\bar{p}}\mathfrak{H}_*^{\mathrm{GM},T}(X-\partial X) & \xrightarrow{\psi} & H_*(I^{\bar{p}}\mathfrak{S}_*^{\mathrm{GM}}(X-\partial X)) & \xleftarrow{\cong} & I^{\bar{p}}H_*^{\mathrm{GM}}(X-\partial X) \\
\downarrow & & \downarrow & & \downarrow & & \downarrow \\
I^{\bar{p}}\mathfrak{H}_*^{\mathrm{GM}}(X) & \xleftarrow{\theta} & I^{\bar{p}}\mathfrak{H}_*^{\mathrm{GM},T}(X) & \xrightarrow{\psi} & H_*(I^{\bar{p}}\mathfrak{S}_*^{\mathrm{GM}}(X)) & \xleftarrow{\cong} & I^{\bar{p}}H_*^{\mathrm{GM}}(X)
\end{array}$$

Commutativity can be seen at the chain level. The maps θ are both isomorphisms by Lemma 5.4.1, and the composition along the top is an isomorphism by Theorem 5.4.2. The leftmost and rightmost vertical maps are isomorphisms by stratified homotopy invariance (Corollary 4.1.11); the homotopy equivalence of the inclusion of $X - \partial X$ into X utilizes the collaring of the boundary[21]. It follows that the composition along the bottom of the diagram is an isomorphism, as desired. □

21 In fact, identifying the collar neighborhood of ∂X with $[0,1)\times\partial X$, we see that both X and $X-\partial X$ have stratified deformation retractions to $X-([0,1/2)\times\partial X)$.

While Theorem 5.4.2 and its corollaries provide isomorphisms between singular and PL intersection homology, in practice one prefers simplicial intersection homology as being the most directly computable. While PL techniques were necessary for our proofs in this section given their good functorial properties and the ability to work on open subsets of a larger PL space, our final results, together with our work from Section 3.3.4, allow us to recover a direct isomorphism between simplicial and singular intersection homology.

Corollary 5.4.6 *Let X be a PL CS set or a PL ∂-stratified pseudomanifold, and let T be a full triangulation of X compatible with the filtration and with an ordering on its vertices. Then the canonical chain map* $\phi : I^{\bar p}C_*^{\mathrm{GM},T}(X;G) \to I^{\bar p}S_*^{\mathrm{GM}}(X;G)$ *from simplicial to singular intersection chains determined using the ordering of the vertices induces an isomorphism on intersection homology.*

Proof The map ϕ is well defined by Proposition 4.4.5 and Corollary 4.4.6.

By thinking through the definitions of the maps, we have the following commutative diagram

$$\begin{array}{ccccc} & & I^{\bar p}C_*^{\mathrm{GM},T}(X;G) & \xrightarrow{\phi} & I^{\bar p}S_*^{\mathrm{GM}}(X;G) \\ & \swarrow & \downarrow & & \downarrow \\ I^{\bar p}\mathfrak{C}_*^{\mathrm{GM}}(X;G) & \xleftarrow{\theta} & I^{\bar p}\mathfrak{C}_*^{\mathrm{GM},T}(X;G) & \xrightarrow{\psi} & I^{\bar p}\mathfrak{S}_*^{\mathrm{GM}}(X;G) \end{array}$$

The map θ is a quasi-isomorphism (i.e. it induces homology isomorphisms) by Lemma 5.4.1, while the diagonal map is a quasi-isomorphism by Theorem 3.3.20. Therefore, the left-hand vertical map is a quasi-isomorphism by commutativity of the diagram. The right-hand vertical map is a quasi-isomorphism by our discussion preceding the proof of Theorem 5.4.2, while ψ is a quasi-isomorphism as a consequence of Theorem 5.4.2 or Corollary 5.4.5: the theorem or corollary demonstrates that $\theta^{-1}\psi$ is a homology isomorphism, but so is θ, hence ψ is as well. Thus three of the four sides of the square are quasi-isomorphisms, which implies that the top is. □

5.5 Topological Invariance

The definition of the intersection chains uses the stratification of the space to define which chains are allowable. The following fact is therefore quite remarkable: for certain perversities, the intersection homology groups do not depend on the stratification of a CS set, only on the underlying homeomorphism type of the space.

To even begin to make sense of this claim, we cannot work with our arbitrary perversities, which are also defined with reference to the stratification. Instead, we follow the original Goresky–MacPherson definition and use perversities that depend only on the codimension of the strata. Then it makes sense to apply such a perversity to multiple stratifications of the same space. Thus, throughout this section, we think of a perversity as a function $\bar{p} : \{1,2,3,\dots\} \to \mathbb{Z}$ with the input corresponding to the codimension of a stratum; as perversities are always zero on the codimension-zero (regular) strata, we omit mention of them or take $\bar{p}(0) = 0$ if needed explicitly. Additionally, since we must have a notion of codimension that depends only on the homeomorphism type of the space, we fold the formal dimension of the CS set into the data concerning its homeomorphism type.

Invariance of PL intersection homology under restratification of PL pseudomanifolds (keeping fixed the PL structure) was proven in [105], where Goresky and MacPherson introduced intersection homology. More general *topological invariance*, i.e. the dependence of intersection homology only on perversity and topological homeomorphism type, was first proven for pseudomanifolds by Goresky and MacPherson in [106] using the techniques of sheaf theory. That proof proceeds by finding axiomatic characterizations for sheaf complexes whose hypercohomology groups are isomorphic to the intersection homology groups. They first find such an axiomatic characterization in terms of a stratification, but then they show that this axiomatic characterization is equivalent to other ones that do not rely on the stratification. A good source for this material is [28, Section V], in which the details concerning *constructibility* of sheaf complexes are carried out a bit more carefully than in [106] (see, in particular, [28, Remark V.3.16]). The proof of topological invariance of singular intersection homology that we provide is a modification of that found by King [139] and applies more broadly to CS sets. Note that our CS sets follow the definition of Siebenmann [217] and so are more general than those of King [139], which assume dimensional homogeneity.

Given the equivalence of PL and singular intersection homology on PL spaces demonstrated in the preceding section, we will focus here only on singular homology, with the analogous PL result following from Theorem 5.4.2.

5.5.1 Which Perversities Work?

Let X be a CS set of a fixed formal dimension, and let $\bar{p}$ be a perversity $\bar{p} : \{1,2,\dots\} \to \mathbb{Z}$. If we want $I^{\bar{p}}H_*^{\mathrm{GM}}(X)$ to be a topological invariant, then $\bar{p}$ still cannot be arbitrary. To see this, consider a basic distinguished neighborhood of the form $\mathbb{R} \times cL$ for a compact filtered space L with regular strata. This space is homeomorphic to $c(SL)$, the cone on the suspension of L; see Figures 2.8 and 2.9. However, these two descriptions give rise to two different filtrations based on the filtration of L. The natural strata of $\mathbb{R} \times cL$ are $\mathbb{R} \times \{v\}$, where v is the cone vertex, and $\mathbb{R} \times (0,1) \times S$, where S is a stratum of L. For $c(SL)$, if we think of SL as a quotient of $[-1,1] \times L$ and cL as a quotient of $[0,1) \times L$, then there are three types of strata:

1. $\{w\}$, where w is the cone point of $c(SL)$,
2. $(0,1) \times \{v_{-1}\}$ and $(0,1) \times \{v_1\}$, where v_{-1}, v_1 are the vertices of the suspension SL, and
3. $(0,1) \times (-1,1) \times S$ for each stratum S of L.

Now, suppose L has dimension $k-2$. By Theorem 4.2.1, and using the stratified homotopy invariance, we have

$$I^{\bar{p}}H_i^{\mathrm{GM}}(\mathbb{R}\times cL)\cong\begin{cases}0, & i\geq k-2-\bar{p}(k-1),\quad i\neq 0,\\ \mathbb{Z}, & i=0\geq k-2-\bar{p}(k-1),\\ I^{\bar{p}}H_i(L), & i<k-2-\bar{p}(k-1).\end{cases}$$

Similarly,

$$I^{\bar{p}}H_i^{\mathrm{GM}}(c(SL))\cong\begin{cases}0, & i\geq k-1-\bar{p}(k),\quad i\neq 0,\\ \mathbb{Z}, & i=0\geq k-1-\bar{p}(k),\\ I^{\bar{p}}H_i(SL), & i<k-1-\bar{p}(k),\end{cases}$$

and by Theorem 4.4.21, if $I^{\bar{p}}H_0^{\mathrm{GM}}(L)\neq 0$ then

$$I^{\bar{p}}H_i^{\mathrm{GM}}(SL)=\begin{cases}I^{\bar{p}}\tilde{H}_{i-1}^{\mathrm{GM}}(L), & i>k-\bar{p}(k-1)-2,\quad i\neq 0,\\ 0, & i=k-\bar{p}(k-1)-2,\quad i\neq 0,\\ I^{\bar{p}}H_i^{\mathrm{GM}}(L), & i<k-\bar{p}(k-1)-2,\\ \mathbb{Z}, & i=0\geq k-\bar{p}(k-1)-2.\end{cases}$$

Now, what can we determine from this? Ignoring for the moment possible complications in low dimensions, we see that $I^{\bar{p}}H_i^{\mathrm{GM}}(\mathbb{R}\times cL)=0$ when $i\geq k-2-\bar{p}(k-1)$. Furthermore, $I^{\bar{p}}H_{k-3-\bar{p}(k-1)}^{\mathrm{GM}}(\mathbb{R}\times cL)\cong I^{\bar{p}}H_{k-3-\bar{p}(k-1)}^{\mathrm{GM}}(L)$. So at least assuming that $0\leq k-3-\bar{p}(k-1)\leq k-2$, it would not be hard to rig up an example where $I^{\bar{p}}H_{k-3-\bar{p}(k-1)}^{\mathrm{GM}}(\mathbb{R}\times cL)\neq 0$, or, for that matter, with $I^{\bar{p}}H_i^{\mathrm{GM}}(\mathbb{R}\times cL)\neq 0$ for all $i\leq k-3-\bar{p}(k-1)$. For example, suppose L is the product of $k-2$ circles. On the one hand, $I^{\bar{p}}H_i^{\mathrm{GM}}(c(SL))=0$ for $i\geq k-1-\bar{p}(k)$. So in order for topological invariance to hold, we should need $k-3-\bar{p}(k-1)<k-1-\bar{p}(k)$, or in other words, $\bar{p}(k)\leq\bar{p}(k-1)+1$.

On the other hand, again ignoring low-dimensional issues, we see that $I^{\bar{p}}H_i^{\mathrm{GM}}(c(SL))=0$ for $i\geq k-1-\bar{p}(k)$ and also for $i=k-\bar{p}(k-1)-2$, regardless of how $k-\bar{p}(k-1)-2$ compares to $k-1-\bar{p}(k)$. Once again, it is easy to choose L so that $I^{\bar{p}}H_i^{\mathrm{GM}}(c(SL))$ is non-zero for all other dimensions ≥ 0 and $<k$. Thus we will run into contradictions if there is a dimension j such that $k-\bar{p}(k-1)-2<j<k-1-\bar{p}(k)$, for then $I^{\bar{p}}H_j^{\mathrm{GM}}(c(SL))$ need not be 0, but $I^{\bar{p}}H_i^{\mathrm{GM}}(\mathbb{R}\times cL)=0$. So to avoid these contradictions we must have $k-\bar{p}(k-1)-2\geq k-1-\bar{p}(k)-1$, or in other words, $\bar{p}(k)\geq\bar{p}(k-1)$.

Together, these two arguments show that we must have

$$\bar{p}(k-1)\leq\bar{p}(k)\leq\bar{p}(k-1)+1. \tag{5.1}$$

This is one of the conditions for $\bar{p}$ to be a GM perversity (see Definition 3.1.4), and now we see the reason for it. It turns out that this condition is sufficient to obtain topological invariance so long as $\bar{p}(1)\geq 0$.

The reader might also have expected the condition (5.1) to place some stronger limitations on $\bar{p}(1)$ given that we have declared $\bar{p}(0)=0$. However, note that by Remark 3.4.5 we obtain identical intersection chain complexes by replacing $\bar{p}(0) = 0$ with $\bar{p}(0) = m$ for any $m \geq 0$. So if $\bar{p}(1) \geq 0$, we could for example set $\bar{p}(0) = \bar{p}(1)$ without changing the intersection chain complexes at all but now satisfying (5.1).

On the other hand, to see what goes wrong if $\bar{p}(1) < 0$ we first observe that it is not possible to have $\bar{p}(0) = 0$ and $\bar{p}(1) < 0$ together with (5.1). For a concrete example of this ruining topological invariance, the reader should work through the preceding discussion taking $L = \emptyset$ with $\dim(\emptyset) = -1$ (or, essentially the same thing, just consider some different filtrations of the real line).

But what if to adapt to $\bar{p}(1) < 0$ we were to break precedent and allow $\bar{p}(0) < 0$ in order to satisfy (5.1)? In this case (5.1) implies that $\bar{p}(k) < k$ for all k. Then if σ is an allowable i-simplex, we have $\sigma^{-1}(S) \subset \{(i - \operatorname{codim}(S) + \bar{p}(S))\text{-skeleton of } \Delta^i\}$ for all strata. But then if $\operatorname{codim}(S) = k$, we have $i - \operatorname{codim}(S) + \bar{p}(S) = i - k + \bar{p}(k) < i$ for all S. This means that no stratum, regular or singular, can intersect the image of the interior of Δ^i. So no allowable simplices can exist, so all intersection homology groups are trivial for any filtration! This might be topological invariance, but not a particularly useful one. So we'll stick with (5.1) and $\bar{p}(1) \geq 0$.

5.5.2 The Statement of the Theorem and Some Corollaries

So let's officially state the invariance theorem. Then, in the remainder of this section we make some observations and prove some corollaries. We will prove the theorem itself in the next section.

Theorem 5.5.1 *Suppose X is a CS set of formal dimension n and that $\bar{p} : \{1,2,\dots\} \to \mathbb{Z}$ is a perversity such that $\bar{p}(1) \geq 0$ and $\bar{p}(k-1) \leq \bar{p}(k) \leq \bar{p}(k-1) + 1$ for all $k \geq 2$. Let $\mathfrak{X}$ be $|X|$ with its intrinsic filtration*[22] *and the same formal dimension n. Then for any abelian group G the identity map of spaces induces an isomorphism $I^{\bar{p}}H^{\mathrm{GM}}_*(X;G) \to I^{\bar{p}}H^{\mathrm{GM}}_*(\mathfrak{X};G)$. It follows that $I^{\bar{p}}H^{\mathrm{GM}}_*(X;G)$ is independent (up to isomorphism) of the choice of stratification of X as a CS set of formal dimension n. In particular, if X' is another CS set of formal dimension n that is topologically homeomorphic to X (not necessarily filtered homeomorphic), then $I^{\bar{p}}H^{\mathrm{GM}}_*(X;G) \cong I^{\bar{p}}H^{\mathrm{GM}}_*(X';G)$.*

More generally, if A is an open subset of X and $(|X|,|A|) \cong (|X'|,|A'|)$ as topological spaces, then $I^{\bar{p}}H^{\mathrm{GM}}_*(X,A;G) \cong I^{\bar{p}}H^{\mathrm{GM}}_*(X',A';G)$.

Remark 5.5.2 The version of the invariance theorem in King [139] does not explicitly mention formal dimensions as it is implicit that each space is meant to be taken with its topological dimension. However, as we have remarked already, it will be useful for us to not maintain this assumption; see Remark 2.2.15.

Remark 5.5.3 The relative version of the theorem is stated only for open subsets. This is because the proof will rely upon intrinsic filtrations, and we know by Lemma 2.10.10 that

[22] See Section 2.10.

the restriction of an intrinsic filtration to an open subspace is the intrinsic filtration of the subspace. One could not necessarily expect such nice behavior for arbitrary subspaces.

We now turn to some corollaries of Theorem 5.5.1. The next four results are all about distinguished neighborhoods in CS sets and boundary collars in ∂-stratified pseudomanifolds. In this section we apply these results to deduce that the property of being locally torsion-free is independent of stratification. These will also be useful later in Chapter 8 in showing that homeomorphic spaces have isomorphic Poincaré and Lefschetz duality maps; for example, see Theorem 8.3.12.

Corollary 5.5.4 *Let X and X' be two n-dimensional CS set stratifications of the same underlying topological space, say $|X|$. Let $x \in |X|$, and let N, N' be distinguished neighborhoods of x in X and X', respectively. Suppose that $\bar{p} : \{1, 2, \ldots\} \to \mathbb{Z}$ is a perversity such that $\bar{p}(1) \geq 0$ and $\bar{p}(k-1) \leq \bar{p}(k) \leq \bar{p}(k-1)+1$ for $k \geq 2$. Then $I^{\bar{p}}H_*^{\mathrm{GM}}(N; F) \cong I^{\bar{p}}H_*^{\mathrm{GM}}(N'; G)$.*

This corollary follows immediately from the theorem and Corollary 2.10.2. It is interesting to compare this corollary with Lemma 5.3.13, which provides the same conclusion for arbitrary perversities but only when X and X' are the same stratification.

The next result is a more specific variant of the corollary that we will need later.

Lemma 5.5.5 *Let X and X' be CS sets and with $|X| = |X'|$, and suppose that $\bar{p} : \{1, 2, \ldots\} \to \mathbb{Z}$ is a perversity such that $\bar{p}(1) \geq 0$ and $\bar{p}(k-1) \leq \bar{p}(k) \leq \bar{p}(k-1)+1$ for $k \geq 2$. Take $x \in |X|$. Let U and V' be distinguished neighborhoods of x in X and X', respectively. Let U' denote $|U|$ with the filtration it inherits from X', let V denote V' in the filtration inherited from X, and let U^* and V^* be $|U|$ and $|V'|$ in their intrinsic filtrations. Suppose that $|V| \subset |U|$. Then inclusion induces isomorphisms $I^{\bar{p}}H_*^{\mathrm{GM}}(V; G) \to I^{\bar{p}}H_*^{\mathrm{GM}}(U; G)$, $I^{\bar{p}}H_*^{\mathrm{GM}}(V'; G) \to I^{\bar{p}}H_*^{\mathrm{GM}}(U'; G)$, and $I^{\bar{p}}H_*^{\mathrm{GM}}(V^*; G) \to I^{\bar{p}}H_*^{\mathrm{GM}}(U^*; G)$ for any abelian group G. Furthermore, the lemma remains true replacing the various spaces U, V, etc. with the deleted neighborhoods $U - \{x\}$, $V - \{x\}$, etc.*

Proof By shrinking U as in the proof of Lemma 5.3.13, there is a distinguished neighborhood U_1 of x in $|V|$ such that $U_1 \hookrightarrow U$ is a stratified homotopy equivalence, and then similarly there is a V_1' with $|V_1'| \subset |U_1|$ and $V_1' \hookrightarrow V'$ a stratified homotopy equivalence. We can form U_1' and V by again letting the primed spaces get their filtrations from X' while all those without primes get their filtrations from X. Similarly, asterisks denote the intrinsic filtration with all intrinsic filtrations inherited from $\mathfrak{X}$. Altogether, we have

$$|V_1| \subset |U_1| \subset |V| \subset |U|,$$

and this leads to the following commutative diagram:

$$\begin{array}{ccccccc}
I^{\bar{p}}H_*^{\mathrm{GM}}(V_1;G) & \longrightarrow & I^{\bar{p}}H_*^{\mathrm{GM}}(U_1;G) & \longrightarrow & I^{\bar{p}}H_*^{\mathrm{GM}}(V;G) & \longrightarrow & I^{\bar{p}}H_*^{\mathrm{GM}}(U;G) \\
\downarrow\cong & & \downarrow\cong & & \downarrow\cong & & \downarrow\cong \\
I^{\bar{p}}H_*^{\mathrm{GM}}(V_1^*;G) & \longrightarrow & I^{\bar{p}}H_*^{\mathrm{GM}}(U_1^*;G) & \longrightarrow & I^{\bar{p}}H_*^{\mathrm{GM}}(V^*;G) & \longrightarrow & I^{\bar{p}}H_*^{\mathrm{GM}}(U^*;G) \\
\uparrow\cong & & \uparrow\cong & & \uparrow\cong & & \uparrow\cong \\
I^{\bar{p}}H_*^{\mathrm{GM}}(V_1';G) & \longrightarrow & I^{\bar{p}}H_*^{\mathrm{GM}}(U_1';G) & \longrightarrow & I^{\bar{p}}H_*^{\mathrm{GM}}(V';G) & \longrightarrow & I^{\bar{p}}H_*^{\mathrm{GM}}(U';G).
\end{array}$$

The vertical maps are all isomorphisms by Theorem 5.5.1.

Now by stratified homotopy invariance the composition $I^{\bar{p}}H_*^{\mathrm{GM}}(U_1;G) \to I^{\bar{p}}H_*^{\mathrm{GM}}(U;G)$ is an isomorphism, as is $I^{\bar{p}}H_*^{\mathrm{GM}}(V_1';G) \to I^{\bar{p}}H_*^{\mathrm{GM}}(V';G)$. Together with the isomorphisms in the diagram, this is sufficient to conclude that

$$I^{\bar{p}}H_*^{\mathrm{GM}}(U_1^*;G) \to I^{\bar{p}}H_*^{\mathrm{GM}}(V^*;G)$$

is surjective and injective, so an isomorphism. It follows that the middle arrows in the top and bottom rows of the diagram are isomorphisms. Together with our two composite isomorphisms, this implies that $I^{\bar{p}}H_*^{\mathrm{GM}}(V;G) \to I^{\bar{p}}H_*^{\mathrm{GM}}(U;G)$ and $I^{\bar{p}}H_*^{\mathrm{GM}}(V_1';G) \to I^{\bar{p}}H_*^{\mathrm{GM}}(U_1';G)$ are isomorphisms, and it follows now that every map in the diagram is an isomorphism.

The proof of the last statement concerning the deleted neighborhoods is identical, noticing that we can choose our stratified homotopy equivalences in the preceding argument to fix x. □

For ∂-stratified pseudomanifolds, there are the following results about collars:

Lemma 5.5.6 *Let $\bar{p}: \{1,2,\dots\} \to \mathbb{Z}$ be a perversity such that $\bar{p}(1) \geq 0$ and $\bar{p}(k-1) \leq \bar{p}(k) \leq \bar{p}(k-1)+1$ for $k \geq 2$, and let X and X' be ∂-stratified pseudomanifolds with $(|X|,|\partial X|) = (|X'|,|\partial X'|)$. Suppose $|\partial X| = |\partial X'|$ is hereditarily paracompact. Let U and V' be filtered collar neighborhoods respectively of ∂X in X and of $\partial X'$ in X'. Let U' denote $|U|$ with the stratification it inherits from X', and let V denote V' in the stratification inherited from X. Suppose that $|V| \subset |U|$. Then inclusion induces isomorphisms $I^{\bar{p}}H_*^{\mathrm{GM}}(V;G) \to I^{\bar{p}}H_*^{\mathrm{GM}}(U;G)$ and $I^{\bar{p}}H_*^{\mathrm{GM}}(V';G) \to I^{\bar{p}}H_*^{\mathrm{GM}}(U';G)$.*

Note that we do not include statements about $\mathfrak{X}$ in the lemma because we do not have intrinsic filtrations in this setting; see Remark 2.10.24.

Proof The proof is very analogous to that of Lemma 5.5.5. Using Lemma 2.7.8, we can find filtered collars U_1 and V_1' so that again $|V_1| \subset |U_1| \subset |V| \subset |U|$, the inclusion $U_1 \hookrightarrow U$ is a stratified homotopy equivalence, etc. Let $\mathring{U}_1$ denote $U_1 - \partial X$, and similarly for the other sets. It remains true that the inclusion $\mathring{U}_1 \hookrightarrow \mathring{U}$ is a stratified homotopy equivalence and analogously for $\mathring{V}_1 \hookrightarrow \mathring{V}'$. As $\mathring{U}, \mathring{V}', \mathring{U}_1, \mathring{V}_1'$ are all CS sets, there are intrinsic filtrations, and repeating the diagram chase of the proof of Lemma 5.5.5 shows that the inclusions induce isomorphisms $I^{\bar{p}}H_*^{\mathrm{GM}}(\mathring{V}; G) \to I^{\bar{p}}H_*^{\mathrm{GM}}(\mathring{U}; G)$ and $I^{\bar{p}}H_*^{\mathrm{GM}}(\mathring{V}'; G) \to I^{\bar{p}}H_*^{\mathrm{GM}}(\mathring{U}'; G)$.

Finally, we have the following maps:

$$\begin{array}{ccc} I^{\bar{p}}H_*^{\mathrm{GM}}(\mathring{V}; G) & \longrightarrow & I^{\bar{p}}H_*^{\mathrm{GM}}(\mathring{U}; G) \\ \downarrow & & \downarrow \\ I^{\bar{p}}H_*^{\mathrm{GM}}(V; G) & \longrightarrow & I^{\bar{p}}H_*^{\mathrm{GM}}(U; G). \end{array}$$

We have just shown that the top map is an isomorphism, but so are the vertical maps by stratified homotopy invariance. The conclusion follows. □

Corollary 5.5.7 *Let $\bar{p} : \{1, 2, \ldots\} \to \mathbb{Z}$ be a perversity such that $\bar{p}(1) \geq 0$ and $\bar{p}(k-1) \leq \bar{p}(k) \leq \bar{p}(k-1) + 1$ for $k \geq 2$, and let X and X' be ∂-stratified pseudomanifolds with $(|X|, |\partial X|) = (|X'|, |\partial X'|)$ and $|\partial X|$ paracompact. Suppose $|\partial X| = |\partial X'|$ is compact. Let V' be a stratified collar neighborhood of $\partial X'$, and let V denote V' in the stratification inherited from X. Then inclusion induces isomorphisms $I^{\bar{p}}H_*^{\mathrm{GM}}(\partial X; G) \to I^{\bar{p}}H_*^{\mathrm{GM}}(V; G)$.*

Proof The set V is an open neighborhood of ∂X, so we can find a filtered collar U of ∂X in V by Lemma 2.7.8. Now consider the composition $I^{\bar{p}}H_*^{\mathrm{GM}}(\partial X; G) \to I^{\bar{p}}H_*^{\mathrm{GM}}(U; G) \to I^{\bar{p}}H_*^{\mathrm{GM}}(V; G)$. The first map is an isomorphism by stratified homotopy invariance, and the latter is an isomorphism by Lemma 5.5.7. The result follows. □

In the next example, we show that in the absence of condition (5.1) on the perversity then even whether or not a space is locally $(\bar{p}, R; M)^{\mathrm{GM}}$-torsion-free can depend on the filtration. We'll then prove that, in the presence of condition (5.1), being locally $(\bar{p}, R; M)^{\mathrm{GM}}$-torsion-free is a property of the space.

Example 5.5.8 Consider the space $X = X^5 = \mathbb{R} \times c(\mathbb{R}P^3)$ stratified as $\mathbb{R} \times \{v\} \subset X$. Then $\mathbb{R}P^3$ is a link of each point in the singular stratum $\mathbb{R} \times \{v\}$, and

$$I^{\bar{0}}H^{\mathrm{GM}}_{3-\bar{0}(S)-1}(\mathbb{R}P^3) = H_2(\mathbb{R}P^3) = 0.$$

So X is locally $(\bar{0}, \mathbb{Z})^{\mathrm{GM}}$-torsion-free. But now let's restratify this space as Y with $\{(0, v)\} \subset \mathbb{R} \times \{v\} \subset Y$, and let $\bar{p}$ be a perversity on Y that remains 0 on the one-dimensional strata. Then the suspension $S(\mathbb{R}P^3)$ is a link of $(0, v)$, and the cone points of the suspension lie in the one-dimensional strata, so, by Theorem 4.4.21,

$$I^{\bar{p}}H_i^{\mathrm{GM}}(S(\mathbb{R}P^3)) = \begin{cases} I^{\bar{p}}H_{i-1}^{\mathrm{GM}}(\mathbb{R}P^3), & i > 3, \\ 0, & i = 3, \\ I^{\bar{p}}H_i^{\mathrm{GM}}(\mathbb{R}P^3), & i < 3. \end{cases}$$

In particular, then $I^{\bar p}H_1^{\mathrm{GM}}(S(\mathbb{R}P^3)) \cong H_1(\mathbb{R}P^3) \cong \mathbb{Z}_2$. So if we take $\bar p(\{(0,v)\}) = 2$, we'll have $I^{\bar p}H^{\mathrm{GM}}_{\dim(S(\mathbb{R}P^3))-\bar p(\{(0,v)\})-1}(S(\mathbb{R}P^3)) \cong \mathbb{Z}_2$, and Y is not locally $(\bar p,\mathbb{Z})$-torsion-free. △

Proposition 5.5.9 *Let $\bar p : \{1,2,\dots\} \to \mathbb{Z}$ be a perversity such that $\bar p(1) \geq 0$ and $\bar p(k-1) \leq \bar p(k) \leq \bar p(k-1)+1$ for $k \geq 2$, and let X and X' be CS sets with $|X| = |X'|$. Then X is locally $(\bar p, R; M)^{\mathrm{GM}}$-torsion-free if and only if X' is.*

Proof We will show that X is locally $(\bar p, R; M)^{\mathrm{GM}}$-torsion-free if and only if $\mathfrak{X}$ is, where $\mathfrak{X}$ is $|X|$ with its intrinsic filtration (see Section 2.10). As X and X' have the same intrinsic filtration, the result will follow.

First, assume that X is locally $(\bar p, R; M)^{\mathrm{GM}}$-torsion-free. Recall that every stratum S of $\mathfrak{X}$ is a union of strata of X of dimension $\leq \dim(S)$. So let S be a stratum of $\mathfrak{X}$ of codimension ℓ, so that the dimensions of its links are $\ell - 1$, and let x be a point of X contained in a stratum T of X with $T \subset S$ and $\dim(S) = \dim(T)$. Let L be a link of x in X and let $\mathscr{L}$ be a link of x in $\mathfrak{X}$. As $\dim(S) = \dim(T)$, we have $\dim(L) = \dim(\mathscr{L}) = \ell - 1$. In this case, the locally torsion-free condition is a statement about intersection homology in degree $\ell - 1 - \bar p(\ell) - 1 = \ell - \bar p(\ell) - 2$. By the cone formula (Theorem 5.3.5), $I^{\bar p}H^{\mathrm{GM}}_{\ell-\bar p(\ell)-2}(cL;R) \cong I^{\bar p}H^{\mathrm{GM}}_{\ell-\bar p(\ell)-2}(L;R)$ and $I^{\bar p}H^{\mathrm{GM}}_{\ell-\bar p(\ell)-2}(c\mathscr{L};R) \cong I^{\bar p}H^{\mathrm{GM}}_{\ell-\bar p(\ell)-2}(\mathscr{L};R)$. It follows that if N and $\mathfrak{N}$ are distinguished neighborhoods of x in X and $\mathfrak{X}$, respectively, we have

$$I^{\bar p}H^{\mathrm{GM}}_{\ell-\bar p(\ell)-2}(L;R) \cong I^{\bar p}H^{\mathrm{GM}}_{\ell-\bar p(\ell)-2}(N;R) \cong I^{\bar p}H^{\mathrm{GM}}_{\ell-\bar p(\ell)-2}(\mathfrak{N};R)$$
$$\cong I^{\bar p}H^{\mathrm{GM}}_{\ell-\bar p(\ell)-2}(\mathscr{L};R),$$

using Corollary 5.5.4 for the middle isomorphism. As we have assumed that X is locally $(\bar p, R; M)^{\mathrm{GM}}$-torsion-free, the link L of x in X satisfies the required intersection homology torsion condition, and so $\mathscr{L}$ also satisfies the required torsion condition. Since the locally torsion-free condition is satisfied for a link at one point in S, it is satisfied at all points in S by Corollary 5.3.14. As S was an arbitrary stratum of $\mathfrak{X}$, it follows that $\mathfrak{X}$ is locally $(\bar p, R; M)^{\mathrm{GM}}$-torsion-free.

Conversely, suppose $\mathfrak{X}$ is locally $(\bar p, R; M)^{\mathrm{GM}}$-torsion-free. Let $x \in X$ be a point with distinguished neighborhood $N \cong \mathbb{R}^k \times cL$. Suppose $\dim(L) = \ell - 1$. As observed in the preceding paragraph, we have $I^{\bar p}H^{\mathrm{GM}}_{\ell-\bar p(\ell)-2}(L;R) \cong I^{\bar p}H^{\mathrm{GM}}_{\ell-\bar p(\ell)-2}(N;R)$. Now, let $\mathfrak{N}$ be a distinguished neighborhood of x in $\mathfrak{X}$. By Corollary 5.5.4, $I^{\bar p}H^{\mathrm{GM}}_{\ell-\bar p(\ell)-2}(N;R) \cong I^{\bar p}H^{\mathrm{GM}}_{\ell-\bar p(\ell)-2}(\mathfrak{N};R)$. But $\mathfrak{N} \cong \mathbb{R}^m \times c\mathscr{L}$ for some link $\mathscr{L}$ and some $\mathbb{R}^m$ with $m \geq k$, since the stratification of $\mathfrak{X}$ is coarser than that of X. Let $\dim(\mathscr{L}) = d - 1$. By stratified homotopy invariance, $I^{\bar p}H^{\mathrm{GM}}_{\ell-\bar p(\ell)-2}(\mathfrak{N};R) \cong I^{\bar p}H^{\mathrm{GM}}_{\ell-\bar p(\ell)-2}(c\mathscr{L};R)$. So, altogether, $I^{\bar p}H^{\mathrm{GM}}_{\ell-\bar p(\ell)-2}(L;R) \cong I^{\bar p}H^{\mathrm{GM}}_{\ell-\bar p(\ell)-2}(c\mathscr{L};R)$. If $m = k$, then $\dim(\mathscr{L}) = \ell - 1$ as well, and, by the cone formula again,

$$I^{\bar p}H^{\mathrm{GM}}_{\ell-\bar p(\ell)-2}(L;R) \cong I^{\bar p}H^{\mathrm{GM}}_{\ell-\bar p(\ell)-2}(\mathscr{L};R),$$

which shows that L satisfies the required torsion condition as $\mathscr{L}$ does by assumption. So suppose $m > k$, which implies that $d - 1 = \dim(\mathscr{L}) < \ell - 1$. We will show that

$I^{\bar{p}}H^{\mathrm{GM}}_{\ell-\bar{p}(\ell)-2}(c\mathscr{L};R)$ is isomorphic to 0, R, or $I^{\bar{p}}H^{\mathrm{GM}}_{d-\bar{p}(d)-2}(\mathscr{L};R)$. This will suffice, as we have assumed that $\mathfrak{X}$ is locally $(\bar{p},R;M)^{\mathrm{GM}}$-torsion-free and as R is free as a free module.

To prove our claim, it is sufficient, by the cone formula, to verify that $\ell-\bar{p}(\ell)-2 \geq d-\bar{p}(d)-2$, i.e. that $\ell-d \geq \bar{p}(\ell)-\bar{p}(d)$. But this is a consequence of having $\bar{p}(k-1) \leq \bar{p}(k) \leq \bar{p}(k-1)+1$ for all $k \geq 2$. □

5.5.3 Proof of Topological Invariance

In this section we prove Theorem 5.5.1. The proof is the same for all coefficient systems, so we provide the details only for $\mathbb{Z}$ coefficients.

We begin with one preliminary observation. Let X be a CS set, and recall the intrinsic filtration $\mathfrak{X}$ of $|X|$ constructed in Section 2.10. If X has formal dimension n, then we will assume that $\mathfrak{X}$ is also given formal dimension n. This allows us to construct a comparison map $I^{\bar{p}}H^{\mathrm{GM}}_*(X) \to I^{\bar{p}}H^{\mathrm{GM}}_*(\mathfrak{X})$. Indeed, the finer the stratification of a space, the more difficult it is for a simplex to be allowable, so we have the following lemma:

Lemma 5.5.10 *Suppose that X is a CS set and that X' is a coarsening of X, meaning that X and X' have the same underlying topological space but that each stratum of X' is a union of strata of X. Suppose X and X' have the same formal dimension, and let $\bar{p}$ be a perversity that depends only on codimension and such that $\bar{p}(k-1) \leq \bar{p}(k) \leq \bar{p}(k-1)+1$ for $k \geq 2$. Then $I^{\bar{p}}S^{\mathrm{GM}}_*(X) \subset I^{\bar{p}}S^{\mathrm{GM}}_*(X')$.*

Proof Suppose that S is a stratum of X of codimension k and that T is the stratum of X' of codimension $j \leq k$ containing S. If σ is an allowable i-simplex with respect to S, then $\sigma^{-1}(S) \subset \{(i-k+\bar{p}(k)_{\text{-skeleton of }\Delta^i}\}$. By the assumption on the perversities, $\bar{p}(k) \leq \bar{p}(j)+(k-j)$, so $i-k+\bar{p}(k) \leq i-k+\bar{p}(j)+(k-j) = i-j+\bar{p}(j)$, so σ is also allowable with respect to T. Hence there is an inclusion $I^{\bar{p}}S^{\mathrm{GM}}_*(X) \hookrightarrow I^{\bar{p}}S^{\mathrm{GM}}_*(X')$. □

Remark 5.5.11 This lemma implies that the identity map $\mathrm{id}: X \to X'$ is $(\bar{p},\bar{p})^{\mathrm{GM}}$-stratified with respect to the two filtrations.

Proof of Theorem 5.5.1 We will show that if $\mathfrak{X}$ is the intrinsic filtration of X, then the inclusion $I^{\bar{p}}S^{\mathrm{GM}}_*(X) \hookrightarrow I^{\bar{p}}S^{\mathrm{GM}}_*(\mathfrak{X})$ induces an isomorphism on homology. Then if X' is any other stratification of X (not necessarily a coarsening of X), we see that $I^{\bar{p}}H^{\mathrm{GM}}_*(X) \cong I^{\bar{p}}H^{\mathrm{GM}}_*(X')$ via the composite

$$I^{\bar{p}}H^{\mathrm{GM}}_*(X) \xrightarrow{\cong} I^{\bar{p}}H^{\mathrm{GM}}_*(\mathfrak{X}) \xleftarrow{\cong} I^{\bar{p}}H^{\mathrm{GM}}_*(X').$$

More generally, if X' is another n-dimensional CS set that is topologically homeomorphic to X, say by $h: |X'| \to |X|$, then h must yield a filtered homeomorphism $h: \mathfrak{X}' \to \mathfrak{X}$, as follows from the purely topological character of the definition of the intrinsic filtration. Therefore, we will have $I^{\bar{p}}H^{\mathrm{GM}}_*(X) \cong I^{\bar{p}}H^{\mathrm{GM}}_*(\mathfrak{X}) \cong I^{\bar{p}}H^{\mathrm{GM}}_*(\mathfrak{X}') \cong I^{\bar{p}}H^{\mathrm{GM}}_*(X')$.

Once we have shown that $I^{\bar{p}}S_*^{\mathrm{GM}}(X) \hookrightarrow I^{\bar{p}}S_*^{\mathrm{GM}}(\mathfrak{X})$ induces an isomorphism on homology, the claimed relative result will follow as we will have maps of short exact sequences:

$$\begin{array}{ccccccccc} 0 & \longrightarrow & I^{\bar{p}}S_*^{\mathrm{GM}}(A) & \longrightarrow & I^{\bar{p}}S_*^{\mathrm{GM}}(X) & \longrightarrow & I^{\bar{p}}S_*^{\mathrm{GM}}(X,A) & \longrightarrow & 0 \\ & & \downarrow & & \downarrow & & \downarrow & & \\ 0 & \longrightarrow & I^{\bar{p}}S_*^{\mathrm{GM}}(\mathfrak{A}) & \longrightarrow & I^{\bar{p}}S_*^{\mathrm{GM}}(\mathfrak{X}) & \longrightarrow & I^{\bar{p}}S_*^{\mathrm{GM}}(\mathfrak{X},\mathfrak{A}) & \longrightarrow & 0. \end{array}$$

The left side of the diagram commutes because the local nature of the definition of the intrinsic filtration implies that the intrinsic filtration $\mathfrak{A}$ of the open set $|A|$ is the restriction to $|A|$ of the intrinsic filtration of X. The commutativity of the right side then follows from the induced map on quotients. So once we have proven the theorem in the absolute case, the relative case will follow from the Five Lemma applied to the induced diagrams of long exact sequences. This then extends to the setting $(|A|,|X|) \cong (|A'|,|X'|)$ just as $I^{\bar{p}}H_*^{\mathrm{GM}}(X) \cong I^{\bar{p}}H_*^{\mathrm{GM}}(\mathfrak{X})$ implies $I^{\bar{p}}H_*^{\mathrm{GM}}(X) \cong I^{\bar{p}}H_*^{\mathrm{GM}}(X')$ whenever $|X| \cong |X'|$.

To prove Theorem 5.5.1 it remains to show that $I^{\bar{p}}S_*^{\mathrm{GM}}(X) \hookrightarrow I^{\bar{p}}S_*^{\mathrm{GM}}(\mathfrak{X})$ induces homology isomorphisms. The argument is by an induction on depth; see Definition 2.2.29. In fact, following King [139], the proof proceeds by an intertwined set of inductions on the following three statements:

$P(d)$. The comparison map $I^{\bar{p}}H_*^{\mathrm{GM}}(X) \longrightarrow I^{\bar{p}}H_*^{\mathrm{GM}}(\mathfrak{X})$ is an isomorphism for all CS sets X of depth $\leq d$.

$Q(d)$. The comparison map $I^{\bar{p}}H_*^{\mathrm{GM}}(X) \longrightarrow I^{\bar{p}}H_*^{\mathrm{GM}}(\mathfrak{X})$ is an isomorphism for all CS sets X filtered homeomorphic to $M \times cW$, where M is a trivially filtered manifold and W is a compact filtered space of depth $\leq d$.

$R(d)$. The comparison map $I^{\bar{p}}H_*^{\mathrm{GM}}(X) \longrightarrow I^{\bar{p}}H_*^{\mathrm{GM}}(\mathfrak{X})$ is an isomorphism for all CS sets X filtered homeomorphic to $\mathbb{R}^k \times cW$, where $\mathbb{R}^k$ is trivially filtered and W is a compact filtered space of depth $\leq d$.

We will show $P(d) \Rightarrow R(d)$, $R(d) \Rightarrow Q(d)$, and $P(d) \wedge Q(d) \Rightarrow P(d+1)$. So, for a CS set of depth n, it will follow from $P(n)$ that $I^{\bar{p}}H_*^{\mathrm{GM}}(X) \cong I^{\bar{p}}H_*^{\mathrm{GM}}(\mathfrak{X})$. To get the induction started, we notice that $P(0)$ is trivial, since if X has depth zero then $X = \mathfrak{X}$. We follow the arguments of King [139] closely for the first two implications. For the third, we follow King's basic idea but generalize using Zorn's Lemma so that we do not need to assume second-countability of our manifolds. To set notation throughout the argument, we declare that manifolds are assumed trivially filtered, unless stated otherwise, and cones and products are always given the respective cone or product filtrations. Furthermore, if Z is any filtered space, we write Z^* for $|Z|$ with its intrinsic filtration, recalling that $|X|$ is the topological space underlying the filtered space X. For example, spaces written $\mathbb{R}^m \times cL$ will always be filtered using the cone and product filtrations starting from a given filtration of

L and the trivial filtration of $\mathbb{R}^m$, while the intrinsic filtration of $|\mathbb{R}^m \times cL|$ will be written $(\mathbb{R}^m \times cL)^*$.

R(d) $\Rightarrow$ *Q(d)*. This is the simplest step of the argument. Let $\dim(M) = k$, and consider $M \times cW$, where W is a compact filtered space of depth $\le d$. By Lemma 2.10.11, there is a coarsening Z of cW such that $(M\times cW)^* \cong M\times Z$ and $(\mathbb{R}^k \times cW)^* \cong \mathbb{R}^k \times Z$. By assumption, $I^{\bar p}H_*^{\mathrm{GM}}(\mathbb{R}^k \times cW) \to I^{\bar p}H_*^{\mathrm{GM}}(\mathbb{R}^k \times Z)$ is an isomorphism, and using the stratified homotopy invariance of intersection homology, it follows that in fact we must have an isomorphism $I^{\bar p}H_*^{\mathrm{GM}}(cW) \to I^{\bar p}H_*^{\mathrm{GM}}(Z)$. But then it follows that $I^{\bar p}H_*^{\mathrm{GM}}(M \times cW) \to I^{\bar p}H_*^{\mathrm{GM}}(M \times Z)$ is an isomorphism using the Künneth Theorem (Theorem 5.2.25) and naturality of the isomorphisms involved.

P(d) $\Rightarrow$ *R(d)*. This is perhaps the most challenging part of the argument. Consider $\mathbb{R}^k \times cW$, where W is a compact filtered space of depth $\le d$. In this case, $\mathbb{R}^k \times cW$ has depth $\le d+1$. We need to show that $I^{\bar p}H_*^{\mathrm{GM}}(\mathbb{R}^k \times cW) \to I^{\bar p}H_*^{\mathrm{GM}}((\mathbb{R}^k \times cW)^*)$ is an isomorphism under the assumption that $I^{\bar p}H_*^{\mathrm{GM}}(X) \to I^{\bar p}H_*^{\mathrm{GM}}(\mathfrak{X})$ is an isomorphism for all *CS* sets X of depth $\le d$.

Let w be the cone point of cW, and let $y = (0, w) \in \mathbb{R}^k \times cW$. Note that $|\mathbb{R}^k \times \{w\}|$ is not necessarily a stratum of $(\mathbb{R}^k \times cW)^*$. However, since the intrinsic filtration $(\mathbb{R}^k \times cW)^*$ is a CS set, y will nonetheless have some distinguished neighborhood $\mathfrak{N} \subset (\mathbb{R}^k \times cW)^*$; let us suppose this distinguished neighborhood $\mathfrak{N}$ is filtered homeomorphic to $\mathbb{R}^m \times cL$ for some m and some compact filtered space L and that $y = (0, v)$, where v is the cone point of cL. Since $\mathbb{R}^k \times \{w\}$ is a stratum of $\mathbb{R}^k \times cW$ and $(\mathbb{R}^k \times cW)^*$ is coarser, the intersection of $|\mathbb{R}^k \times \{w\}|$ with $\mathfrak{N}$ must be contained in the stratum of $\mathfrak{N}$ homeomorphic to $\mathbb{R}^m \times \{v\}$.

Now treating $\mathbb{R}^j$ as cS^{j-1} with 0 as the cone point, up to topological homeomorphism we have $|\mathbb{R}^k \times cW| \cong |cS^{k-1} \times cW| \cong |c(S^{k-1} * W)|$ with y as the cone point, and similarly $|\mathfrak{N}| \cong |\mathbb{R}^m \times cL| \cong |c(S^{m-1} * L)|$; see Section 2.11. Since our neighborhood $\mathfrak{N}$ of y is contained in $|\mathbb{R}^k \times cW|$, we can conclude from Lemma 2.10.1 that in fact

$$|\mathbb{R}^k \times cW| \cong |c(S^{k-1} * W)| \cong |c(S^{m-1} * L)| \cong |\mathbb{R}^m \times cL|,$$

with each homeomorphism fixing y. Let $h : |\mathbb{R}^k \times cW| \to |\mathbb{R}^m \times cL|$ be the composite homeomorphism; see Figure 5.2.

Since the intrinsic filtration of a CS set is determined locally and purely topologically and since $\mathfrak{N}$ is an open subset of the intrinsic filtration $(\mathbb{R}^k \times cW)^*$, it follows that the images of the skeleta of $\mathbb{R}^k \times cL$ under h^{-1} also provide the intrinsic filtration of $|\mathbb{R}^k \times cW|$. In other words, h provides a filtered homeomorphism $(\mathbb{R}^k \times cW)^* \to \mathbb{R}^m \times cL$. So we have arrived at the following diagrams. The diagram of maps of spaces on the left leads to the diagram of intersection homology groups on the right taking filtrations into account; the dashed map is induced by the composite of the other two maps in the diagram as maps of intersection chains. The vertical homology map is an isomorphism because it is induced by a filtered

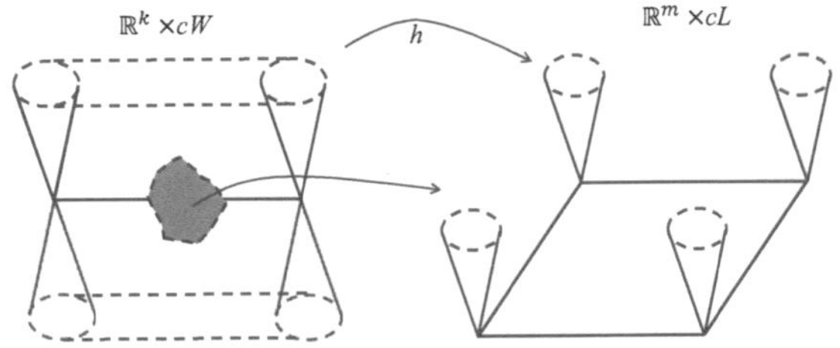

Figure 5.2 On the left, we have $\mathbb{R}^k \times cW$, which (topologically) contains the subspace $|\mathfrak{N}|$. In the intrinsic filtration $(\mathbb{R}^k \times cW)^*$, we have $\mathfrak{N} \cong \mathbb{R}^m \times cL$, shown on the right. But we also have a topological homeomorphism h from all of $|\mathbb{R}^k \times cW|$ to $|\mathbb{R}^m \times cL|$.

homeomorphism.

$$\begin{array}{ccc} |\mathbb{R}^k \times cW| & \xrightarrow{\text{id}} & |(\mathbb{R}^k \times cW)^*| \\ & \underset{\cong}{\overset{h}{\searrow}} & \Big\downarrow h\ \cong \\ & & |\mathbb{R}^m \times cL| \end{array} \qquad \begin{array}{ccc} I^{\bar p}H^{\mathrm{GM}}_*(\mathbb{R}^k \times cW) & \longrightarrow & I^{\bar p}H^{\mathrm{GM}}_*((\mathbb{R}^k \times cW)^*) \\ & \overset{h}{\searrow} & \Big\downarrow h\ \cong \\ & & I^{\bar p}H^{\mathrm{GM}}_*(\mathbb{R}^m \times cL) \end{array}$$

We need to show the maps on the right are all isomorphisms. Let us set $s = \dim W$ and $t = \dim L$. Then $k + s = m + t$, and since $m \geq k$ we have also $t \leq s$.

The easy case now is in the degree range $i \geq s - \bar{p}(s+1)$, $i \neq 0$. In this case, by the assumption on perversities and since $t \leq s$, we have $\bar{p}(s+1) - \bar{p}(t+1) \leq s - t$, so also $i \geq s - \bar{p}(t+1) + t - s = t - \bar{p}(t+1)$. So by stratified homotopy invariance and the cone formula, using that the codimensions of the cone points are respectively $s+1$ and $t+1$, we have in this range $I^{\bar p}H^{\mathrm{GM}}_i(\mathbb{R}^k \times cW) = 0$ and $I^{\bar p}H^{\mathrm{GM}}_i(\mathbb{R}^m \times cL) = 0$, so they must agree.

For the other cases, we will need to understand a bit more about how the strata of $\mathbb{R}^k \times cW$ and $\mathbb{R}^m \times cL$ interact via the map h. This will put us in a position to reframe the problem, as well as to apply the hypothesis $P(d)$ to the space $\mathbb{R}^k \times cW - \mathbb{R}^k \times \{w\} \cong \mathbb{R}^k \times (cW - \{w\}) \cong \mathbb{R}^{k+1} \times W$, which has depth d.

So let us begin by examining $h : |\mathbb{R}^k \times cW| \to |\mathbb{R}^m \times cL|$ a bit more. Our previous observation that the intersection of $|\mathbb{R}^k \times \{w\}| \subset |\mathbb{R}^k \times cW|$ with $|\mathfrak{N}|$ must be contained in the stratum of $\mathfrak{N}$ homeomorphic to $\mathbb{R}^m \times \{v\}$ now translates into the observation that $h(\mathbb{R}^k \times \{w\}) \subset \mathbb{R}^m \times \{v\}$; see Figure 5.3. Conversely, as $h^{-1}(\mathbb{R}^m \times cL)$ provides the intrinsic filtration of $|\mathbb{R}^k \times cW|$, the space $h^{-1}(\mathbb{R}^m \times \{v\})$ must be a union of strata of $\mathbb{R}^k \times cW$, including the stratum $\mathbb{R}^k \times \{w\}$. Since the skeleta of $\mathbb{R}^k \times cW$ all have the form $\mathbb{R}^k \times cW^j$ for some skeleton W^j of W (possibly empty), then $h^{-1}(\mathbb{R}^m \times \{v\})$ must have the form $\mathbb{R}^k \times cA$ for some closed set $A \subset W$ since $h^{-1}(\mathbb{R}^m \times \{v\})$ must be a closed subset and a union of strata including $\mathbb{R}^k \times \{w\}$. We note for later that as $\mathbb{R}^m - \{0\}$ has the homology of an $(m-1)$-sphere, this must also be true of $|\mathbb{R}^k \times cA| - \{(0, w)\}$, which is homotopy equivalent to $S^{k-1} * A$, the kth suspension of A. Since suspension increases the dimension of each reduced homology group by one, it follows that A must be a homology $(m-1-k)$-sphere.

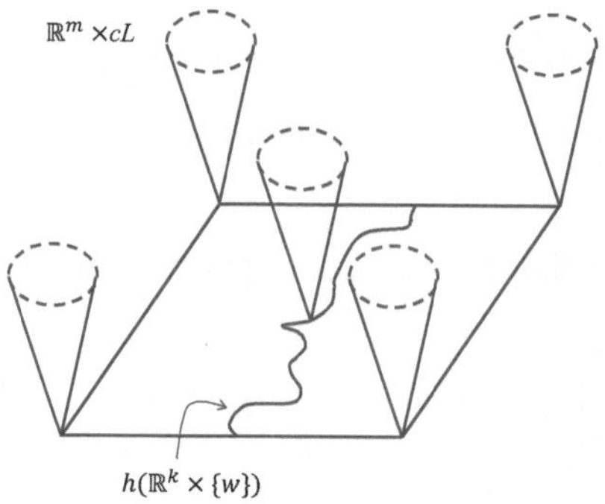

Figure 5.3 The image $h(\mathbb{R}^k \times \{w\})$ is contained in the stratum $\mathbb{R}^m \times \{v\}$.

Now we can return to homology groups. We next consider the following commutative diagram whose upper left horizontal map is the comparison map we want to be an isomorphism and whose vertical maps are induced by inclusions, and where for brevity we write $\mathbb{R}^k_{cW}$ for $\mathbb{R}^k \times cW$, $\mathbb{R}^k_w$ for $\mathbb{R}^k \times \{w\}$, and $\mathbb{R}^m_{cL}$ for $\mathbb{R}^m \times cL$:

$$\begin{array}{ccccc}
I^{\bar{p}}H^{\mathrm{GM}}_*(\mathbb{R}^k_{cW}) & \longrightarrow & I^{\bar{p}}H^{\mathrm{GM}}_*((\mathbb{R}^k_{cW})^*) & \xrightarrow[h]{\cong} & I^{\bar{p}}H^{\mathrm{GM}}_*(\mathbb{R}^m_{cL}) \\
\uparrow & & \uparrow & & \uparrow \\
I^{\bar{p}}H^{\mathrm{GM}}_*(\mathbb{R}^k_{cW} - \mathbb{R}^k w) & \xrightarrow{\cong} & I^{\bar{p}}H^{\mathrm{GM}}_*((\mathbb{R}^k_{cW} - \mathbb{R}^k_w)^*) & \xrightarrow[h]{\cong} & I^{\bar{p}}H^{\mathrm{GM}}_*(\mathbb{R}^m_{cL} - h(\mathbb{R}^k w)).
\end{array} \tag{5.2}$$

We have already seen that the top right horizontal map is an isomorphism as it is induced by a filtered homeomorphism. Similarly, the bottom right horizontal map is the filtered homeomorphism obtained by restricting the domain and codomain, so it also induces an isomorphism. The bottom left horizontal map is an isomorphism by the induction hypothesis $P(d)$, as $\mathbb{R}^k \times cW - \mathbb{R}^k \times \{w\} = \mathbb{R}^k \times (cW - \{w\})$ has depth d. The middle vertical map is well defined because it is induced by the open inclusion, and the intrinsic filtration of an open subset is the restriction of the intrinsic filtration of the whole space.

Let us now consider the degree range $i < s - \bar{p}(s+1)$. In this range, the left-hand vertical map of diagram (5.2) is an isomorphism by stratified homotopy invariance and the cone formula. So to show that the top left horizontal map is an isomorphism, which is our goal, it suffices to show the right-hand vertical map is an isomorphism. For this, we will analyze $I^{\bar{p}}H^{\mathrm{GM}}_*(\mathbb{R}^m \times cL - h(\mathbb{R}^k \times \{w\}))$ using the pair

$$(\mathbb{R}^m \times cL - h(\mathbb{R}^k \times \{w\}), \mathbb{R}^m \times cL - \mathbb{R}^m \times \{v\}).$$

If we excise $h(\mathbb{R}^k \times \{w\}) \times (cL - \{v\})$ from this pair, we remove everything else lying over $h(\mathbb{R}^k \times \{w\})$ in $\mathbb{R}^m \times cL$; see again Figure 5.3. The result is the pair

$$\begin{aligned}((\mathbb{R}^m \times \{v\} - h(\mathbb{R}^k \times \{w\})) \times cL, (\mathbb{R}^m \times \{v\} - h(\mathbb{R}^k \times \{w\})) \times (cL - \{v\}))\\ = ((\mathbb{R}^m \times \{v\} - h(\mathbb{R}^k \times \{w\})) \times (cL, cL - \{v\})).\end{aligned}$$

By the excision property, we thus have an isomorphism induced by inclusion

$$\begin{aligned}I^{\bar{p}}H_*^{\mathrm{GM}}((\mathbb{R}^m \times \{v\} - h(\mathbb{R}^k \times \{w\})) \times (cL, cL - \{v\}))\\ \to I^{\bar{p}}H_*^{\mathrm{GM}}(\mathbb{R}^m \times cL - h(\mathbb{R}^k \times \{w\}), \mathbb{R}^m \times cL - \mathbb{R}^m \times \{v\}).\end{aligned}$$

But via h, we have

$$\begin{aligned}|\mathbb{R}^m \times \{v\} - h(\mathbb{R}^k \times \{w\})| \cong |\mathbb{R}^k \times cA - \mathbb{R}^k \times \{w\}| = |\mathbb{R}^k \times (cA - \{w\})|\\ \cong |\mathbb{R}^{k+1} \times A|.\end{aligned}$$

As this is an open subset of $\mathbb{R}^m$, it is a manifold (trivially filtered). Abusing notation by letting $|\mathbb{R}^{k+1} \times A|$ denote this trivially filtered manifold, we obtain

$$\begin{aligned}I^{\bar{p}}H_*^{\mathrm{GM}}((\mathbb{R}^m \times \{v\} - h(\mathbb{R}^k \times \{w\})) \times (cL, cL - \{v\}))\\ \cong I^{\bar{p}}H_*^{\mathrm{GM}}(|\mathbb{R}^{k+1} \times A| \times (cL, (cL - \{v\}))).\end{aligned}$$

But A is a homology $(m-1-k)$-sphere, so it follows from the Künneth Theorem (Theorem 5.2.25) that

$$\begin{aligned}I^{\bar{p}}H_i^{\mathrm{GM}}(|\mathbb{R}^{k+1} \times A| \times (cL, cL - \{v\}))\\ \cong I^{\bar{p}}H_i^{\mathrm{GM}}(cL, cL - \{v\}) \oplus I^{\bar{p}}H_{i-m+1+k}^{\mathrm{GM}}(cL, cL - \{v\}),\end{aligned}$$

and by Theorem 4.3.21,

$$I^{\bar{p}}H_{i-m+1+k}^{\mathrm{GM}}(cL, cL - \{v\}) = 0 \quad \text{if} \quad i - m + 1 + k \le t - \bar{p}(t+1).$$

With these computations available, let us consider the commutative diagram of long exact sequences

$$\begin{array}{ccc}
\downarrow & & \downarrow \\
I^{\bar{p}}H_i^{\mathrm{GM}}(\mathbb{R}^m \times cL - \mathbb{R}^m \times \{v\}) & \longrightarrow & I^{\bar{p}}H_i^{\mathrm{GM}}(L - \{v\}) \\
\downarrow & & \downarrow \\
I^{\bar{p}}H_i^{\mathrm{GM}}(\mathbb{R}^m \times cL - h(\mathbb{R}^k \times \{w\})) & \longrightarrow & I^{\bar{p}}H_i^{\mathrm{GM}}(cL) \\
\downarrow & & \downarrow \\
I^{\bar{p}}H_i^{\mathrm{GM}}(\mathbb{R}^m \times cL - h(\mathbb{R}^k \times \{w\}), \mathbb{R}^m \times cL - \mathbb{R}^m \times \{v\}) & \longrightarrow & I^{\bar{p}}H_i^{\mathrm{GM}}(cL, cL - \{v\}) \\
\downarrow & & \downarrow
\end{array}$$

induced by the projection $\mathbb{R}^m \times cL \to \{z\} \times cL$ for some $z \notin h(\mathbb{R}^k \times \{w\})$. Such maps are well defined on intersection homology as they preserve codimension of strata. The top horizontal map is an isomorphism by stratified homotopy invariance. We have just seen that the bottom horizontal map is an isomorphism if $i - m + 1 + k \leq t - \bar{p}(t+1)$, using that our projection is compatible with the excision inclusion and that via the Künneth Theorem the elements of the $I^{\bar{p}}H_i^{\mathrm{GM}}(cL, cL - \{v\})$ summand are the cross products $\sigma_z \times \xi$, where $\sigma_z : \Delta^0 \to \{z\}$ is the unique singular 0-simplex and $\xi \in I^{\bar{p}}H_i^{\mathrm{GM}}(cL, cL - \{v\})$. In fact, by the same reasoning the map

$$I^{\bar{p}}H_i^{\mathrm{GM}}(\mathbb{R}^m \times cL - h(\mathbb{R}^k \times \{w\}), \mathbb{R}^m \times cL - \mathbb{R}^m \times \{v\}) \longrightarrow I^{\bar{p}}H_i^{\mathrm{GM}}(cL, cL - \{v\})$$

is a surjection in any degree. So using the general version of the Five Lemma[23], this

[23] Recall that the most general version of the Five Lemma only requires a diagram

$$\begin{array}{ccccccccc}
A & \longrightarrow & B & \longrightarrow & C & \longrightarrow & D & \longrightarrow & E \\
{\scriptstyle a}\downarrow & & {\scriptstyle b}\downarrow & & {\scriptstyle c}\downarrow & & {\scriptstyle d}\downarrow & & {\scriptstyle e}\downarrow \\
A' & \longrightarrow & B' & \longrightarrow & C' & \longrightarrow & D' & \longrightarrow & E'
\end{array}$$

is enough to establish that the middle horizontal map of the diagram is an isomorphism whenever $i - m + 1 + k \leq t - \bar{p}(t+1)$.

Now we look at

$$\begin{array}{ccc} I^{\bar{p}}H_*^{\mathrm{GM}}(\mathbb{R}^m \times cL) & & \\ \uparrow & \searrow & \\ & & I^{\bar{p}}H_*^{\mathrm{GM}}(\{z\} \times cL). \\ & \nearrow & \\ I^{\bar{p}}H_*^{\mathrm{GM}}(\mathbb{R}^m \times cL - h(\mathbb{R}^k \times \{w\})) & & \end{array}$$

The downward diagonal is an isomorphism by stratified homotopy invariance. If $i < s - \bar{p}(s+1)$, then $i-m+1+k < s-m+k+1-\bar{p}(s+1) \leq s-m+k+1-\bar{p}(t+1) = t+1-\bar{p}(t+1)$, the range in which we have just shown that the upward diagonal is an isomorphism. So the vertical arrow is an isomorphism when $i < s - \bar{p}(s+1)$. By our prior work, this establishes $I^{\bar{p}}H_*^{\mathrm{GM}}(\mathbb{R}^k \times cW) \to I^{\bar{p}}H_*^{\mathrm{GM}}((\mathbb{R}^k \times cW)^*)$ is an isomorphism in this range.

Finally, we need to consider the case $i = 0 \geq s - \bar{p}(s+1)$. In this case, we have $s \leq \bar{p}(s+1) \leq \bar{p}(t+1) + s - t$, so also $0 \geq t - \bar{p}(t+1)$. So, by the cone formula, $I^{\bar{p}}H_0^{\mathrm{GM}}(\mathbb{R}^k \times cW)$ and $I^{\bar{p}}H_0^{\mathrm{GM}}(\mathbb{R}^m \times cL)$ are each isomorphic to either $\mathbb{Z}$ or 0 depending on whether or not there is an allowable 0-simplex. If $I^{\bar{p}}H_0^{\mathrm{GM}}(\mathbb{R}^k \times cW) \cong \mathbb{Z}$, then there is an allowable 0-simplex σ_0 in $\mathbb{R}^k \times cW$. Since we identify $\mathbb{R}^m \times cL$ as a coarsening of $\mathbb{R}^k \times cW$ via h and since coarsening preserves allowability by Lemma 5.5.10, the simplex $h(\sigma_0)$ is allowable in $I^{\bar{p}}S_0^{\mathrm{GM}}(\mathbb{R}^m \times cL)$, and so also $I^{\bar{p}}H_0^{\mathrm{GM}}(\mathbb{R}^m \times cL) \cong \mathbb{Z}$. Thus h induces an isomorphism $I^{\bar{p}}H_0^{\mathrm{GM}}(\mathbb{R}^k \times cW) \to I^{\bar{p}}H_0^{\mathrm{GM}}(\mathbb{R}^m \times cL)$. Conversely, suppose there is an allowable 0-simplex σ_0 in $I^{\bar{p}}S_0^{\mathrm{GM}}(\mathbb{R}^m \times cL)$ so that $I^{\bar{p}}H_0^{\mathrm{GM}}(\mathbb{R}^m \times cL) \cong \mathbb{Z}$. Suppose the image of σ_0 lies in a stratum T of $\mathbb{R}^m \times cL$. Since allowability depends only on the stratum and the perversity, we see that then any singular 0-simplex $\Delta^0 \to T$ is also allowable. Now, since $\mathbb{R}^m \times cL$ is a coarsening of $\mathbb{R}^k \times cW$ via the homeomorphism h, it follows that $h^{-1}(T)$ is a union of strata of $\mathbb{R}^k \times cW$, and if T has dimension j, at least one of these strata of $\mathbb{R}^k \times cW$, say the stratum S, must also have dimension j, since T and $h^{-1}(T)$ are j-manifolds. But now $\mathrm{codim}(S) = \mathrm{codim}(T)$, so $\bar{p}(S) = \bar{p}(T)$, and we see that any 0-simplex with image in S must be allowable in $\mathbb{R}^k \times cW$. Therefore, $I^{\bar{p}}H_0^{\mathrm{GM}}(\mathbb{R}^k \times cW) \cong \mathbb{Z}$ and h induces the isomorphism $I^{\bar{p}}H_0^{\mathrm{GM}}(\mathbb{R}^k \times cW) \to I^{\bar{p}}H_0^{\mathrm{GM}}(\mathbb{R}^m \times cL)$.

So we have seen that $I^{\bar{p}}H_0^{\mathrm{GM}}(\mathbb{R}^k \times cW) \to I^{\bar{p}}H_0^{\mathrm{GM}}(\mathbb{R}^m \times cL)$ is an isomorphism in all cases. This finishes the proof that $P(d) \Rightarrow R(d)$.

$P(d)$ and $Q(d) \Rightarrow P(d+1)$. This final verification is essentially a Mayer–Vietoris argument. However, the setting doesn't quite fit the situation of Theorem 5.1.4, so we work through the details.

Suppose X is a CS set of depth $d+1$. By Lemma 2.10.11, if V is any open subset

with exact rows, b and d isomorphisms, a surjective, and e injective to conclude that c is an isomorphism; see the proof of the Five Lemma in [125, Section 2.1].

of X, then its intrinsic filtration V^* is the restriction to V of the intrinsic filtration of X; hence for any open subsets $V \subset Y \subset X$, the morphisms $I^{\bar{p}}H_*^{\mathrm{GM}}(V) \to I^{\bar{p}}H_*^{\mathrm{GM}}(V^*)$ and $I^{\bar{p}}H_*^{\mathrm{GM}}(Y) \to I^{\bar{p}}H_*^{\mathrm{GM}}(Y^*)$ are compatible, i.e. they form a commutative square with the inclusion maps. Let U be the largest open subset of X such that $I^{\bar{p}}H_*^{\mathrm{GM}}(U) \to I^{\bar{p}}H_*^{\mathrm{GM}}(U^*)$ is an isomorphism. Such a U exists by Zorn's Lemma, since if $\{U_\alpha\}$ is any increasing sequence of open sets such that $I^{\bar{p}}H_*^{\mathrm{GM}}(U_\alpha) \to I^{\bar{p}}H_*^{\mathrm{GM}}(U_\alpha^*)$ is an isomorphism for each α, then it also follows that $I^{\bar{p}}H_*^{\mathrm{GM}}\left(\bigcup_\alpha U_\alpha\right) \to I^{\bar{p}}H_*^{\mathrm{GM}}\left(\bigcup_\alpha U_\alpha^*\right) = I^{\bar{p}}H_*^{\mathrm{GM}}\left(\left(\bigcup_\alpha U_\alpha\right)^*\right)$, using Lemmas 5.1.6 and 5.1.7. Notice that U is non-empty since X must contain an open set of depth zero, and we have already observed that $P(0)$ is trivial.

We will show that $U = X$, which will complete the proof. We first observe that if $X(d+1)$ is the union of the strata of X of depth $d+1$, then $X - X(d+1) \subset U$. In fact, by $P(d)$, $I^{\bar{p}}H_*^{\mathrm{GM}}(N) \to I^{\bar{p}}H_*^{\mathrm{GM}}(N^*)$ is an isomorphism for any open subset N of $X - X(d+1)$. So if $x \in X - X(d+1)$, but $x \notin U$, then for any open neighborhood N_x of x in $X - X(d+1)$, both $I^{\bar{p}}H_*^{\mathrm{GM}}(N_x) \to I^{\bar{p}}H_*^{\mathrm{GM}}(N_x^*)$ and $I^{\bar{p}}H_*^{\mathrm{GM}}(N_x \cap U) \to I^{\bar{p}}H_*^{\mathrm{GM}}((N_x \cap U)^*)$ are isomorphisms. So using Mayer–Vietoris sequences, the Five Lemma, and the definition of U, also $I^{\bar{p}}H_*^{\mathrm{GM}}(N_x \cup U) \to I^{\bar{p}}H_*^{\mathrm{GM}}((N_x \cup U)^*)$ is an isomorphism, which would contradict the maximality of U. So $X - X(d+1) \subset U$.

Now suppose $x \in X - U$, so in particular $x \in X(d+1)$. Since X is a CS set, x has a neighborhood N filtered homeomorphic to $\mathbb{R}^k \times cL$ for some compact filtered L, which must have depth $\le d$. Let $Y = N \cap X(d+1) \cap U$; this is an open subset of $\mathbb{R}^k \times \{v\}$ and so a manifold with trivial filtration. Since U contains all of $X - X(d+1)$, the set Y has a neighborhood in U homeomorphic to $Y \times cL$, and in fact $N \cap U \cong (Y \times cL) \cup (\mathbb{R}^k \times (cL - \{v\}))$, where v is the cone vertex of cL. The intersection $(Y \times cL) \cap (\mathbb{R}^k \times (cL - \{v\})) \cong Y \times (cL - \{v\})$. By $P(d)$, we have $I^{\bar{p}}H_*^{\mathrm{GM}}(\mathbb{R}^k \times (cL - \{v\})) \cong I^{\bar{p}}H_*^{\mathrm{GM}}((\mathbb{R}^k \times (cL - \{v\}))^*)$ and $I^{\bar{p}}H_*^{\mathrm{GM}}(Y \times (cL - v)) \cong I^{\bar{p}}H_*^{\mathrm{GM}}((Y \times (cL - v))^*)$. By $Q(d)$, we have $I^{\bar{p}}H_*^{\mathrm{GM}}(Y \times cL) \cong I^{\bar{p}}H_*^{\mathrm{GM}}((Y \times cL)^*)$. So using the long exact Mayer–Vietoris sequences and the Five Lemma, also $I^{\bar{p}}H_*^{\mathrm{GM}}(N \cap U) \cong I^{\bar{p}}H_*^{\mathrm{GM}}((N \cap U)^*)$. But now $I^{\bar{p}}H_*^{\mathrm{GM}}(U) \cong I^{\bar{p}}H_*^{\mathrm{GM}}(U^*)$ by assumption and $I^{\bar{p}}H_*^{\mathrm{GM}}(N) \cong I^{\bar{p}}H_*^{\mathrm{GM}}(N^*)$ by $Q(d)$. So another Mayer–Vietoris and Five Lemma argument shows that $I^{\bar{p}}H_*^{\mathrm{GM}}(N \cup U) \cong I^{\bar{p}}H_*^{\mathrm{GM}}((N \cup U)^*)$.

But this contradicts the maximality of U if $U \neq X$, and so we must have $U = X$.

This completes the proof of Theorem 5.5.1. □

5.6 Finite Generation

It is often useful to know that the homology groups of certain "nice" compact spaces, such as manifolds, are finitely generated. We can obtain such a theorem for the intersection homology of CS sets assuming that the local intersection homology groups are finitely generated. By an inductive argument, this implies that the intersection homology groups of compact stratified pseudomanifolds are finitely generated. Using the results of Section

5.4, it suffices to limit our discussion in this section to singular intersection homology; by Theorem 5.4.2, the following results will also hold in the PL setting.

Here is the relevant local definition. Note its similarities to our locally torsion-free conditions.

Definition 5.6.1 A CS set X is called *locally* $(\bar{p}, \mathbb{Z}; G)^{\mathrm{GM}}$*-finitely generated* if G is a finitely generated abelian group and, for each point $x \in X$, there is a link L of X such that $I^{\bar{p}}H_i^{\mathrm{GM}}(L; G)$ is finitely generated for each i.

Arguments completely analogous to those of Lemma 5.3.13 show that the fact that $I^{\bar{p}}H_i^{\mathrm{GM}}(L; M)$ is finitely generated is really a property of the stratum containing x. In other words, if it is true of some link of some point in the stratum S, then it is true of all links of all points in S.

Remark 5.6.2 The following proposition only requires that every point have a distinguished neighborhood $N \cong \mathbb{R}^k \times cL$ such that each $I^{\bar{p}}H_i^{\mathrm{GM}}(N; G)$ is finitely generated. Thus, as $I^{\bar{p}}H_i^{\mathrm{GM}}(N; G) \cong I^{\bar{p}}H_i^{\mathrm{GM}}(L; G)$ only for some i, and is 0 otherwise, we do not need the full force of Definition 5.6.1 here. We will, however, need the stronger assumptions below when we discuss general Künneth Theorems; see Section 6.4.1.

Proposition 5.6.3 *Suppose X is a locally $(\bar{p}, \mathbb{Z}; G)^{\mathrm{GM}}$-finitely generated CS set. Suppose $U \subset W$ are open subsets of X, that $\bar{U} \subset W$, and that $\bar{U}$ is compact. Then the image of $I^{\bar{p}}H_i^{\mathrm{GM}}(U; G)$ in $I^{\bar{p}}H_i^{\mathrm{GM}}(W; G)$ is finitely generated. In particular, if X is compact, then each $I^{\bar{p}}H_i^{\mathrm{GM}}(X; G)$ is finitely generated.*

Proof The proof of this proposition is taken from [28, Theorem V.3.5].

We first observe that the last claim follows from the general situation by taking $U = W = X$. To prove the first statement, we will perform an induction on i. The proposition is clearly true when $i < 0$. So we suppose that the statement holds for all $i < k$ and consider degree k.

We assume that W is fixed and let $\mathcal{E}_W$ be the set of open subsets $U \subset W$ such that $\bar{U}$ is compact in W. Let $\mathcal{E}_W^i$ be the set of $U \in \mathcal{E}_W$ such that $\mathrm{im}(I^{\bar{p}}H_i^{\mathrm{GM}}(U; G) \to I^{\bar{p}}H_i^{\mathrm{GM}}(W; G))$ is finitely generated. We want to show that $\mathcal{E}_W^k = \mathcal{E}_W$ under the induction hypothesis $\mathcal{E}_W^i = \mathcal{E}_W$ for all $i < k$. Suppose that $V \in \mathcal{E}_W^k$ and that $U \subset V$. Then $\mathrm{im}(I^{\bar{p}}H_k^{\mathrm{GM}}(U; G) \to I^{\bar{p}}H_k^{\mathrm{GM}}(W; G)) \subset \mathrm{im}(I^{\bar{p}}H_k^{\mathrm{GM}}(V; G) \to I^{\bar{p}}H_k^{\mathrm{GM}}(W; G))$. So if $V \in \mathcal{E}_W^k$, then $U \in \mathcal{E}_W^k$, using that $\mathbb{Z}$ is a Noetherian ring so that any submodule of a finitely generated module over $\mathbb{Z}$ (i.e. of any finitely generated abelian group) is finitely generated; see [147, Chapter X]. Therefore, it suffices to show that every compact set $K \subset W$ has an open neighborhood $V \supset K$ with V in $\mathcal{E}_W^k$.

As X is a CS set, every point $x \in W$ has an open neighborhood N homeomorphic to $\mathbb{R}^j \times cL$ for some j and L that has compact closure in W; if necessary, take a distinguished neighborhood homeomorphic to $\mathbb{R}^j \times cL$ and then let N be a smaller open neighborhood $\mathring{D}_r \times c_s L$ within that (in the notation of Lemma 5.3.13) with closure homeomorphic to $D_r \times \bar{c}_s L$. We know from stratified homotopy invariance and the cone formula that $I^{\bar{p}}H_k^{\mathrm{GM}}(N; G) \cong$

$I^{\bar{p}}H_k^{\mathrm{GM}}(\mathbb{R}^j \times cL; G)$ is either 0 or isomorphic to $I^{\bar{p}}H_k^{\mathrm{GM}}(L; G)$ or G, all of which we have assumed to be finitely generated. So, again using that $\mathbb{Z}$ is Noetherian, the image of $I^{\bar{p}}H_k^{\mathrm{GM}}(N; G)$ in $I^{\bar{p}}H_k^{\mathrm{GM}}(W; G)$ is finitely generated [147, Proposition X.1.1]. Therefore, $N \in \mathcal{E}_W^k$. So now if K is any compact subset of W, there are $U_1, \dots, U_m \in \mathcal{E}_W^k$ that cover K. We want to show that there is a single $U \in \mathcal{E}_W^k$ covering K, so we will show that we can do an induction to decrease the number of elements of $\mathcal{E}_W^k$ needed to cover K.

Suppose we can show that any compact $K' \subset W$ that can be covered by two elements of $\mathcal{E}_W^k$ can be covered by one element of $\mathcal{E}_W^k$ and that K is covered by $U_1, \dots, U_m \in \mathcal{E}_W^k$. Then $K' = K - \cup_{j=1}^{m-2} U_j$ is compact and contained in $U_{m-1} \cup U_m$. So our assumption implies there is a $U'_{m-1} \in \mathcal{E}_W^k$ with $K' \subset U'_{m-1}$, and therefore $K \subset U_1 \cup \cdots \cup U_{m-2} \cup U'_{m-1}$. Thus, we could inductively reduce the number of elements of $\mathcal{E}_W^k$ needed to cover K down to one. So, it remains to show that we can reduce covers of compact subspaces by two elements of $\mathcal{E}_W^k$ to covers by one element.

So let $U_1, U_2 \in \mathcal{E}_W^k$ with $K \subset U_1 \cup U_2$. Let V_1 be a neighborhood of $K - U_2$ with $\bar{V}_1 \subset U_1$. As $K - U_2$ is compact and contained in U_1, such a V_1 exists by Corollary 2.3.18. Then $K \subset V_1 \cup U_2$, and we can let V_2 be an open neighborhood of $K - V_1$ such that $\bar{V}_2 \subset U_2$ for the same reasons; see Figure 5.4. Then $K \subset V_1 \cup V_2$, and we claim $V_1 \cup V_2 \subset \mathcal{E}_W^k$.

As $\bar{V}_1 \subset U_1$ and $\bar{V}_2 \subset U_2$, we have $\overline{V_1 \cup V_2} = \bar{V}_1 \cup \bar{V}_2 \subset \bar{U}_1 \cup \bar{U}_2$. As $U_1, U_2 \in \mathcal{E}_W^k$, it

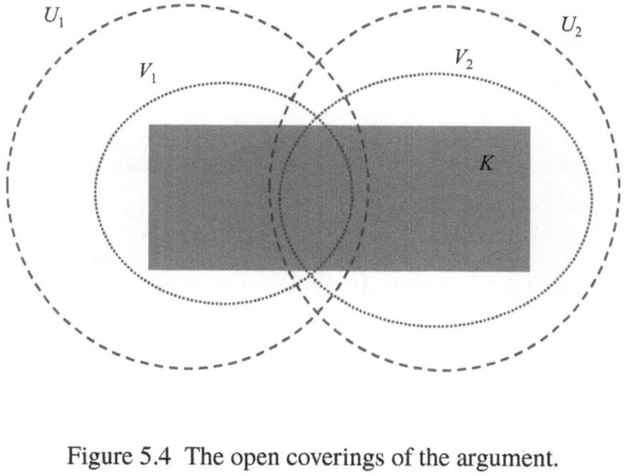

Figure 5.4 The open coverings of the argument.

follows that $\overline{V_1 \cup V_2}$ is compact. Now, consider the following diagram:

$$\begin{array}{ccccc}
\begin{array}{c} I^{\bar{p}}H_k^{\mathrm{GM}}(V_1;G) \oplus \\ I^{\bar{p}}H_k^{\mathrm{GM}}(V_2;G) \end{array} & \longrightarrow & I^{\bar{p}}H_k^{\mathrm{GM}}(V_1 \cup V_2;G) & \longrightarrow & I^{\bar{p}}H_{k-1}^{\mathrm{GM}}(V_1 \cap V_2;G) \\
\downarrow & & \downarrow \beta & & \downarrow \gamma \\
\begin{array}{c} I^{\bar{p}}H_k^{\mathrm{GM}}(U_1;G) \oplus \\ I^{\bar{p}}H_k^{\mathrm{GM}}(U_2;G) \end{array} & \xrightarrow{\alpha} & I^{\bar{p}}H_k^{\mathrm{GM}}(U_1 \cup U_2;G) & \xrightarrow{\delta} & I^{\bar{p}}H_{k-1}^{\mathrm{GM}}(U_1 \cap U_2;G) \\
& \searrow^{\mu} & & \swarrow_{\nu} & \\
& & I^{\bar{p}}H_k^{\mathrm{GM}}(W;G). & &
\end{array}$$

The rows of this diagram are from the Mayer–Vietoris sequences. Our claim is that $\mathrm{im}(\nu\beta)$ is finitely generated. The image of μ is finitely generated because $U_1, U_2 \in \mathcal{E}_W^k$, and so $\mathrm{im}(\nu\alpha)$ is finitely generated. As $\nu(\mathrm{im}(\beta) \cap \mathrm{im}(\alpha)) \subset \mathrm{im}(\nu\alpha)$, the subgroup $\nu(\mathrm{im}(\beta) \cap \mathrm{im}(\alpha))$ is finitely generated. We also have

$$\frac{\mathrm{im}(\beta)}{\mathrm{im}(\beta) \cap \mathrm{im}(\alpha)} \cong \mathrm{im}(\delta\beta) \subset \mathrm{im}(\gamma).$$

But we claim that $\mathrm{im}(\gamma)$ is finitely generated by the induction hypothesis. Notice that

$$\overline{V_1 \cap V_2} \subset \bar{V}_1 \cap \bar{V}_2 \subset U_1 \cap U_2 \subset \bar{U}_1 \cap \bar{U}_2.$$

So the closure of the open set $V_1 \cap V_2$ is contained in the open set $U_1 \cap U_2$. Also, as $U_1, U_2 \in \mathcal{E}_W$, the closures $\bar{U}_1$ and $\bar{U}_2$ are compact, so $\overline{V_1 \cap V_2}$ is compact as a closed subset of $\bar{U}_1 \cap \bar{U}_2$. Therefore, $\mathrm{im}(\gamma)$ is finitely generated in dimension $k-1$ by the induction assumption.

So, now $\nu(\mathrm{im}(\beta) \cap \mathrm{im}(\alpha))$ and $\frac{\mathrm{im}(\beta)}{\mathrm{im}(\beta)\cap\mathrm{im}(\alpha)}$ are finitely generated, and we can consider the diagram of short exact sequences:

$$\begin{array}{ccccccccc}
0 & \longrightarrow & \mathrm{im}(\beta) \cap \mathrm{im}(\alpha) & \longrightarrow & \mathrm{im}(\beta) & \longrightarrow & \dfrac{\mathrm{im}(\beta)}{\mathrm{im}(\beta) \cap \mathrm{im}(\alpha)} & \longrightarrow & 0 \\
& & \downarrow \nu & & \downarrow \nu & & \downarrow & & \\
0 & \longrightarrow & \nu(\mathrm{im}(\beta) \cap \mathrm{im}(\alpha)) & \longrightarrow & \nu(\mathrm{im}(\beta)) & \longrightarrow & \dfrac{\nu(\mathrm{im}(\beta))}{\nu(\mathrm{im}(\beta) \cap \mathrm{im}(\alpha))} & \longrightarrow & 0.
\end{array}$$

The commutativity of the left square induces the vertical map on the right, which is well defined and surjective. Therefore, the bottom right group is finitely generated as a quotient of a finitely generated group. And, lastly, the finite generation of the outer terms in the bottom short exact sequence implies that $\nu(\mathrm{im}(\beta)) = \mathrm{im}(\nu\beta)$ is finitely generated [147, Proposition X.1.2]. □

Corollary 5.6.4 *If G is a finitely generated abelian group and X is a compact recursive CS set, in particular if X is a compact stratified pseudomanifold, then $I^{\bar p}H_i^{\mathrm{GM}}(X;G)$ is finitely generated for all i.*

Proof By Proposition 5.6.3, $I^{\bar p}H_i^{\mathrm{GM}}(X;G)$ will be finitely generated if the intersection homology groups of the links are all finitely generated. But these must be lower depth compact recursive CS sets, so the result follows by induction on depth. In the base case, a depth zero compact CS sets is a compact manifold, so all the links must be empty and the proof of the proposition applies directly. □

Remark 5.6.5 Similar results to those in this section hold by replacing G with a finitely generated module M over a Noetherian ring R. Then analogous assumptions on $I^{\bar p}H_i^{\mathrm{GM}}(L;M)$ lead to analogous conclusions about $I^{\bar p}H_i^{\mathrm{GM}}(X;M)$. We treat these sorts of coefficients more explicitly in the setting of non-GM intersection homology; see Definition 6.3.38, Proposition 6.3.39, and Corollary 6.3.40.

6
Non-GM Intersection Homology

We now turn to "non-GM intersection homology," $I^{\bar{p}}H_*(X)$, a variant version of intersection homology that is better suited to working with completely general perversities and that will be necessary for our Poincaré duality results. While not the same as $I^{\bar{p}}H^{\mathrm{GM}}(X)$ in general, we do have $I^{\bar{p}}H_*(X) \cong I^{\bar{p}}H_*^{\mathrm{GM}}(X)$ when $\bar{p} \leq \bar{t}$, i.e. when $\bar{p}(S) \leq \operatorname{codim}(S) - 2$ for all singular strata S, so the two theories are the same in many important cases, including when $\bar{p}$ is a GM perversity (see Section 3.1.1).

We begin the chapter by explaining in detail why this variant is necessary before proceeding on to the official definitions in Section 6.2. Section 6.3 is about the properties of $IH_*(X)$, most of which mirror those of $IH_*^{\mathrm{GM}}(X)$. In fact, we largely cite our previous results, asking the reader to check that the proofs extend. For those instances that do require some modification of a previous argument, we indicate the needed changes. We also take the opportunity in Section 6.3.3 to provide a short discussion of intersection homology with local coefficient systems.

In Section 6.4, we prove a Künneth Theorem relating the intersection homology of the product of two CS sets to the intersection homology of the individual spaces. See Theorem 6.4.7 for details. As we will see in Remark 6.4.12, this Künneth Theorem is not true in general using GM intersection homology. This further motivates our introduction of non-GM intersection homology.

Section 6.5 contains some further technical propositions about splitting chains into pieces that will be needed in later chapters.

6.1 Motivation for Non-GM Intersection Homology

In order to proceed on to discuss intersection homology versions of Poincaré duality and a general Künneth Theorem, it is first necessary to modify the Goresky–MacPherson intersection homology, which is the theory we have been using thus far. Our modified PL and singular intersection chain complexes will be denoted simply $I^{\bar{p}}S_*$ and $I^{\bar{p}}\mathfrak{C}_*$, and the homology groups will be denoted $I^{\bar{p}}H_*$ and $I^{\bar{p}}\mathfrak{H}_*$. Note that we have dropped the GM from the notation. As we go along, we will explain why the modified definition is necessary and why we consider it to be the "right" definition for intersection homology.

Before providing the definition, we record the following facts that will be verified below:

1. For perversities $\bar{p}$ for which $\bar{p} \leq \bar{t}$, i.e. $\bar{p}(S) \leq \operatorname{codim}(S) - 2$ for all singular strata S, our new intersection homology groups $I^{\bar{p}}H_*$ will be identical to the GM intersection homology groups $I^{\bar{p}}H_*^{\mathrm{GM}}$ considered to this point. In particular, this will be true for any GM perversity on any pseudomanifold without codimension-one strata. In fact, we will show in Proposition 6.2.9 the stronger fact that if $\bar{p}(S) \leq \operatorname{codim}(S) - 2$ then $I^{\bar{p}}S_*(X; G) \cong I^{\bar{p}}S_*^{\mathrm{GM}}(X; G)$.
2. Versions of all our previous theorems will hold for $I^{\bar{p}}H_*$ with the exception of the topological invariance. If $\bar{p}$ is a perversity on a CS set X with $\bar{p}(S) > \operatorname{codim}(S) - 2$ for some singular stratum S, it will no longer be true in general that $I^{\bar{p}}H_*(X)$ is independent of the stratification of X.
3. If X is a compact oriented pseudomanifold, then we will see in Chapter 8 that there are intersection homology versions of Poincaré duality. With field coefficients F, for example, these can be formulated for $I^{\bar{p}}H_*(X; F)$ for any perversity $\bar{p}$ but not necessarily for $I^{\bar{p}}H_*^{\mathrm{GM}}(X; F)$ except in those cases where $I^{\bar{p}}H_*^{\mathrm{GM}}(X; F) \cong I^{\bar{p}}H_*(X; F)$ as in the preceding remark.

To illustrate this last point, let us provide a sample computation. Looking ahead to Theorem 8.5.11, if we let $F = \mathbb{Q}$, then one consequence of Poincaré duality on a compact oriented n-dimensional pseudomanifold X is an isomorphism

$$I^{\bar{p}}H_i(X; \mathbb{Q}) \cong \operatorname{Hom}(I^{D\bar{p}}H_{n-i}(X; \mathbb{Q}), \mathbb{Q}) \tag{6.1}$$

that corresponds to the nonsingularity of the intersection pairing on manifolds (see [71, Section VIII.13]). Recall that $D\bar{p}$ is defined so that

$$D\bar{p}(S) = \bar{t}(S) - \bar{p}(S) = \operatorname{codim}(S) - 2 - \bar{p}(S)$$

for singular strata S.

For our example, let ST be the suspension of the trivially filtered torus $T = S^1 \times S^1$. The only singular strata of ST are the two suspension points $\{\mathfrak{n}, \mathfrak{s}\}$. Let $\bar{p}$ be the perversity such that $\bar{p}(\{\mathfrak{n}\}) = \bar{p}(\{\mathfrak{s}\}) = 2$, which is one less than the codimension of the singular strata. Then $D\bar{p}(\{\mathfrak{n}\}) = D\bar{p}(\{\mathfrak{s}\}) = -1$.

Employing our suspension computation of Theorem 4.4.21, together with the Universal Coefficient Theorem for $\mathbb{Q}$ coefficients (Corollary 5.3.17), we have the following results:

$$I^{\bar{p}}H_i^{\mathrm{GM}}(ST; \mathbb{Q}) \cong \begin{cases} \mathbb{Q}, & i = 3, \\ \mathbb{Q} \oplus \mathbb{Q}, & i = 2, \\ 0, & i = 1, \\ \mathbb{Q}, & i = 0, \end{cases} \qquad I^{D\bar{p}}H_i^{\mathrm{GM}}(ST; \mathbb{Q}) \cong \begin{cases} 0, & i = 3, \\ \mathbb{Q}, & i = 2, \\ \mathbb{Q} \oplus \mathbb{Q}, & i = 1, \\ \mathbb{Q}, & i = 0. \end{cases}$$

Note the asymmetry that prevents (6.1) from being possible for all i. By contrast, we will see below in Theorem 6.3.13 that the following formulas hold:

$$I^{\bar{p}}H_i(ST;\mathbb{Q}) \cong \begin{cases} \mathbb{Q}, & i=3,\\ \mathbb{Q}\oplus\mathbb{Q}, & i=2,\\ \mathbb{Q}, & i=1,\\ 0, & i=0, \end{cases} \qquad I^{D\bar{p}}H_i(ST;\mathbb{Q}) \cong \begin{cases} 0, & i=3,\\ \mathbb{Q}, & i=2,\\ \mathbb{Q}\oplus\mathbb{Q}, & i=1,\\ \mathbb{Q}, & i=0, \end{cases}$$

and the restored symmetry is a reflection of Poincaré duality between $I^{\bar{p}}H_*$ and $I^{D\bar{p}}H_*$.

To get a first idea of where the problem lies with IH_*^{GM}, let us consider a related situation we studied in Example 4.4.22. There, we let M be a trivially filtered compact n-dimensional ∂-manifold with $\partial M \neq \emptyset$, and we studied the intersection homology of $M^+ = M \cup_{\partial M} \bar{c}(\partial M) \cong M/\partial M$. Under the assumption that $\bar{p}(\{v\}) \le n-2$, where v is the cone point, we computed that

$$I^{\bar{p}}H_i^{\mathrm{GM}}(M^+) \cong \begin{cases} H_i(M,\partial M), & i > n-\bar{p}(\{v\})-1,\\ \operatorname{im}(H_i(M)\to H_i(M,\partial M)), & i = n-\bar{p}(\{v\})-1,\\ H_i(M), & i < n-\bar{p}(\{v\})-1. \end{cases}$$

We also noted that this formula hints at the coming duality theorems because, assuming M is oriented, Lefschetz duality provides a nonsingular pairing of the groups $H_i(M,\partial M;F)$ and $H_{n-i}(M;F)$ with field coefficients F. But we see here that for large i the group $I^{\bar{p}}H_i^{\mathrm{GM}}(M^+)$ is not behaving like an absolute homology group: it is behaving like the *relative* homology group $H_i(M,\partial M)$. And the larger $\bar{p}(\{v\})$ is, the more degrees i for which this is the case. If we did let $\bar{p}(\{v\}) \ge n-1$, then we might expect to also see relative homology behavior in degree zero, but this is not what happens. In fact, if M is connected then $I^{\bar{p}}H_0^{\mathrm{GM}}(M^+) \cong \mathbb{Z}$ by Example 3.4.6. We could also arrive at this conclusion using the Mayer–Vietoris sequence and the cone formula as in the computation of Example 4.4.22.

In fact, the same "problem" can be observed in the cone formula itself, as we noted in Remark 4.2.2. Recall that if X is a compact $(n-1)$-dimensional filtered space and we assume that X has regular strata (so that $I^{\bar{p}}H_0^{\mathrm{GM}}(X) \neq 0$), then Theorem 4.2.1 gives

$$I^{\bar{p}}H_i^{\mathrm{GM}}(cX) \cong \begin{cases} 0, & i \ge n-\bar{p}(\{v\})-1, \quad i \neq 0,\\ \mathbb{Z}, & i = 0 \ge n-\bar{p}(\{v\})-1,\\ I^{\bar{p}}H_i^{\mathrm{GM}}(X), & i < n-\bar{p}(\{v\})-1. \end{cases}$$

Notice that "most" of the cases here follow a fairly simple formula: There is a cutoff dimension at $n-\bar{p}(\{v\})-1$. At this dimension and above, the intersection homology of the cone is 0. Below this dimension, we simply recover the intersection homology of X. The smaller the value of $\bar{p}(\{v\})$, the more of $I^{\bar{p}}H_i^{\mathrm{GM}}(X)$ we recover; the greater the value of $\bar{p}(\{v\})$, the more of $I^{\bar{p}}H_i^{\mathrm{GM}}(X)$ gets killed to 0. The discrepancy from this nice pattern arises as $\bar{p}(\{v\})$ gets so big as to be $\ge n-1$. Following the pattern, we would expect in this situation that all of the intersection homology groups of the cone should vanish, but rather we maintain the stable situation that $I^{\bar{p}}H_0^{\rm GM}(cX) \cong \mathbb{Z}$ no matter how large $\bar{p}(\{v\})$ gets beyond this point. Once again, this reflects intersection homology behaving like absolute homology and not relative homology, even though we have seen hints that it would be "better" for it to behave like relative intersection homology when the degree is high compared to a cutoff degree depending on the perversity[1].

So how do we modify our definitions to ensure that $I^{\bar{p}}H_0(cX) = 0$ for large enough $\bar{p}(\{v\})$? The idea is to replace $I^{\bar{p}}S_*^{\rm GM}$ with a chain complex that does behave a bit more like the relative chain complex, in particular something like $S_*(cX,\{v\}) = 0$, at least in low degrees when $\bar{p}(\{v\})$ is large enough. This gives $H_0(cX,\{v\}) = 0$ because all 0-simplices are homologous to the 0-simplex at v, which is trivial in the relative chain group. Unfortunately, however, for a general n-dimensional filtered space X, the solution is not quite as simple as replacing $I^{\bar{p}}S_*^{\rm GM}(X)$ with $I^{\bar{p}}S_*^{\rm GM}(X,\Sigma_X) \cong I^{\bar{p}}S_*^{\rm GM}(X)/I^{\bar{p}}S_*^{\rm GM}(\Sigma_X)$, where Σ_X is the singular locus of X (see Definition 2.2.13). By our definitions in Section 4.3, the most natural meaning for $I^{\bar{p}}S_*^{\rm GM}(\Sigma_X)$ in this context would be as the chain complex of $\bar{p}$-allowable chains on X that are supported in Σ_X. If, for example, $\bar{p}(S) = \operatorname{codim}(S) - 1$ for all singular S, which is one of the perversities we're concerned about, then this would mean that any allowable i-simplex must have $\sigma^{-1}(S)$ contained in the $(i - \operatorname{codim}(S) + \bar{p}(S))$-skeleton of Δ^i. But with our assumptions,

$$i - \operatorname{codim}(S) + \bar{p}(S) = i - \operatorname{codim}(S) + \operatorname{codim}(S) - 1 = i - 1,$$

which implies that any allowable simplex must map the interior of Δ^i to a regular stratum. So $I^{\bar{p}}S_*^{\rm GM}(\Sigma_X) = 0$! This is clearly no good, as we'd just have $I^{\bar{p}}S_*^{\rm GM}(X,\Sigma_X) = I^{\bar{p}}S_*^{\rm GM}(X)$ again. But we also don't want to quotient out by chains that are not contained in the singular set, because then we will start to lose the absolute homology behavior we need for lower degrees. What to do?

In the next section, we discuss two solutions[2] to defining $I^{\bar{p}}S_*(X)$ so that it provides the behavior that we want in all degrees. These were introduced independently by the author in [85] and by Saralegi in [205]. The two approaches look different at first, but they turn out to be essentially identical (though that of the author has a slightly broader applicability to local coefficient situations; see [85]). We will first present the author's original description, though in somewhat different language; then we will discuss Saralegi's version of the definition. Since coefficients will play an important role in what follows, we use coefficients in an abelian group G throughout the discussion.

[1] The bad behavior of $IH_*^{\rm GM}$ on cones also deviates from the Goresky–MacPherson sheaf-theoretic description of intersection homology. Lamentably, a discussion of this is beyond the purview of this book, but see [91] for an expository account.

[2] In the setting of subanalytic chains, an alternative solution obtained by throwing away the singular set altogether and working with locally finite chains on the regular strata was first hinted at in unpublished lecture notes of MacPherson's [156]. In this case, Σ_X becomes something like an "end" of $X - \Sigma_X$ off which boundaries of infinite chains vanish. However, once one begins using infinite chains, it changes the structure of the theory in other ways. This is not all bad, as this is in fact the key to connecting the chain theory and sheaf theory formalisms of intersection homology! But we will not pursue this here.

6.2 Definitions of Non-GM Intersection Homology

We have just seen in the preceding discussion that we wish to alter the intersection chain complex so that it yields, in the appropriate circumstances, something more like relative homology groups than absolute homology groups. To gain further insight, it is interesting to analyze just why it is that the absolute homology of a cone does not vanish in degree zero even as it does vanish in all other degrees. So suppose ξ is an i-cycle in cX with $i > 0$. Then we know that the singular cone $\bar{c}\xi$ (see Example 3.4.7) provides a null-homology of ξ – it is a chain whose boundary is ξ. However, suppose σ is a 0-simplex in $cX - \{v\}$. We can still form the cone $\bar{c}\sigma$, but now $\partial\bar{c}\sigma = \sigma - \sigma_v$, where σ_v is the 0-simplex whose image is v; we have already observed this phenomenon before when computing the intersection homology of the cone. The result is that $\bar{c}\sigma$ does not provide a null-homology, but simply a homology from σ to σ_v. Therefore every 0-simplex is homologous to the cone point, but neither is null-homologous. Hence $H_0(cX) \cong \mathbb{Z}$, due to this special property of 0-cycles compared to cycles of higher dimensions. On the other hand, $H_0(cX, \{v\})$ does vanish because $\bar{c}\sigma$ gives a homology from σ to σ_v, which is declared to represent 0 in $S_0(X, \{v\})$.

So, continuing to think about cones, we want something like the following: When the perversity at $\{v\}$ is large, we want the relative homology behavior, i.e. $\bar{c}\sigma$ should somehow provide a homology from σ to something trivial. When the perversity at $\{v\}$ is small, we want something more like the absolute behavior. We will first achieve the former goal by brute force: we will simply declare that boundary pieces of allowable simplices that land in the singular strata should vanish. Remarkably, this then works out in all perversity ranges because when the perversity is small the allowability requirements will "keep simplices away from the singular strata" so that our mandated vanishings don't come into play. But for high perversities, this extra vanishing of boundaries in Σ provides exactly the relative behavior we need.

6.2.1 First Definition of IH_*

Throughout this section, let X be an n-dimensional filtered space with perversity $\bar{p}$ and singular locus $\Sigma = \Sigma_X$, and let G be an abelian group.

There are essentially two pieces to our first definition of the non-GM intersection chain complex $I^{\bar{p}}S_*(G;X)$. Unlike the definition of the ordinary relative chain complex $S_*(X,\Sigma;G)$, this definition will not use quotient complexes, but we will impose some similar properties by hand. We do use quotient complexes in our other definitions, given below.

First, inspired by the behavior of $S_*(X,\Sigma;G)$, we want to make sure that no simplices are contained completely in Σ. To this end, we let $S_i^{\bar{p}}(X;G) \subset S_i(X;G)$ be generated by the $\bar{p}$-allowable i-simplices σ with support $|\sigma| \not\subset |\Sigma|$. We do not yet place a requirement on the boundaries of such chains because our second step will be to alter the definition of the boundary map.

To further mirror relative chain behavior, we need that any part of the boundary of a chain in $S_i^{\bar{p}}(X;G)$ that is contained in Σ must be treated as trivial. We do this by brute force by altering the boundary map. Suppose that σ is a $\bar{p}$-allowable i-simplex and that $\partial\sigma = \sum_{j=0}^{i}(-1)^j\sigma_j$ with σ_j the jth face of σ (i.e. $\sigma_j = \sigma|_{[0,\dots,\hat{j},\dots,i]}$). We define

$$\hat{\partial}\sigma = \sum_{|\sigma_j|\not\subset\Sigma_X} (-1)^j\sigma_j.$$

In other words, $\hat{\partial}\sigma$ is obtained from $\partial\sigma$ by *throwing out* the simplices with image in Σ; see Figure 6.1. If we extend $\hat{\partial}$ linearly to $\hat{\partial} : S_i^{\bar{p}}(X;G) \to S_{i-1}(X;G)$, we see that $\hat{\partial}\xi$ for $\xi \in S_i^{\bar{p}}(X;G)$ can also be described by taking $\partial\xi$ and then throwing out any simplices with support in Σ. So if we write $\partial\xi = \sum g_\tau\tau$ with each $g_\tau \in G$, then we can separate out this boundary chain into two pieces $\partial\xi = \sum_{|\tau|\subset\Sigma_X} g_\tau\tau + \sum_{|\tau|\not\subset\Sigma_X} g_\tau\tau$; then $\hat{\partial}\xi = \sum_{|\tau|\not\subset\Sigma_X} g_\tau\tau$.

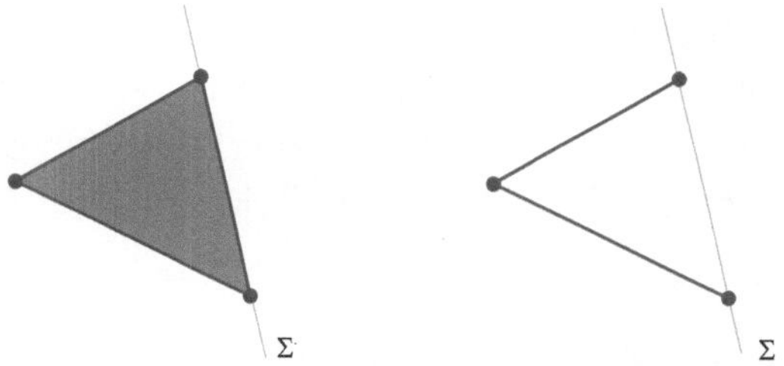

Figure 6.1 Left: a 2-simplex σ with one facet in Σ. Right: $\hat{\partial}\sigma$ omits the simplex contained in Σ.

If X is a PL filtered space with triangulation T, we can define simplicial chain groups $C_i^{T,\bar{p}}(X)$ with a modified boundary map $\hat{\partial}$ determined analogously by taking the usual boundary map and then throwing out any terms coming from simplices contained in Σ. We will show below that in both the singular and simplicial settings we have $\hat{\partial}\hat{\partial} = 0$ so that we can define the non-GM intersection chain complex as follows:

Definition 6.2.1 (Non-GM intersection homology, first definition) Let $I^{\bar{p}}S_i(X;G)$ consist of those chains in $S_i^{\bar{p}}(X;G)$ whose image under $\hat{\partial}$ is contained in $S_{i-1}^{\bar{p}}(X;G)$, i.e. such that each simplex of $\hat{\partial}\xi$ is $\bar{p}$-allowable. This gives a chain complex $I^{\bar{p}}S_*(X;G)$ with chain groups $I^{\bar{p}}S_i(X;G)$ and with boundary map $\hat{\partial}$. These are the *perversity $\bar{p}$ (non-GM) intersection chain complexes*, and the *(non-GM) intersection homology groups* are $I^{\bar{p}}H_*(X;G) = H_*(I^{\bar{p}}S_*(X;G))$.

Simplicial and PL chain complexes $I^{\bar{p}}C_*^T(X;G)$ and $I^{\bar{p}}\mathfrak{C}_*(X;G) = \varinjlim I^{\bar{p}}C_*^T(X;G)$ and homology groups $I^{\bar{p}}H_*^T(X) = H_*(I^{\bar{p}}C_*^T(X;G))$ and $I^{\bar{p}}\mathfrak{H}_*(X;G) = H_*(I^{\bar{p}}\mathfrak{C}_*(X;G))$ are defined analogously.

Remark 6.2.2 It is important to note that throughout this chapter we continue to use the notation $\hat{\partial}$ for the boundary map of the chain complex $I^{\bar{p}}S_*(X;G)$. However, once we have become more used to the notion, we will relapse to using the generic notation ∂ for this chain complex unless we specifically need to emphasize the distinction between different geometric boundary maps.

There is still some work to do to ensure that the preceding definitions all make sense. First, we verify that we always have $\hat{\partial}\hat{\partial} = 0$ so that we do have chain complexes. Then we'll check that subdivision does induce chain maps $I^{\bar{p}}C^T_*(X;G) \to I^{\bar{p}}C^{T'}_*(X;G)$ so that $I^{\bar{p}}\mathfrak{C}_*(X;G)$ is well defined.

Lemma 6.2.3 *Both $I^{\bar{p}}S_*(X;G)$ and $I^{\bar{p}}C^T_*(X;G)$ are chain complexes.*

Proof The argument is the same for both, so we focus on the singular chain case.

We first note that if $\xi, \eta \in S^{\bar{p}}_i(X)$, then so is $\xi + \eta$. Furthermore, if $\partial\xi = \sum_{|\sigma_j| \subset \Sigma_X} c_j\sigma_j + \sum_{|\sigma_j| \not\subset \Sigma_X} c_j\sigma_j$ and $\partial\eta = \sum_{|\tau_k| \subset \Sigma_X} d_k\tau_k + \sum_{|\tau_k| \not\subset \Sigma_X} d_k\tau_k$, then $\hat{\partial}\xi = \sum_{|\sigma_j| \not\subset \Sigma_X} c_j\sigma_j$ and $\hat{\partial}\eta = \sum_{|\tau_k| \not\subset \Sigma_X} d_k\tau_k$, so certainly $\hat{\partial}\xi + \hat{\partial}\eta = \hat{\partial}(\xi + \eta)$. It is also evident from the construction that the image of $\hat{\partial}$ applied to $I^{\bar{p}}S_i(X;G)$ lies in $I^{\bar{p}}S_{i-1}(X;G)$, so long as $\hat{\partial}\hat{\partial} = 0$.

Lastly, we observe that $\hat{\partial}^2 = 0$: By definition, $\partial\xi = \hat{\partial}\xi + \zeta$, where ζ consists of simplices in Σ_X, so we have $\hat{\partial}\xi = \partial\xi - \zeta$. Analogously then, $\hat{\partial}(\hat{\partial}\xi) = \partial(\hat{\partial}\xi) - \zeta_2 = \partial(\partial\xi - \zeta) - \zeta_2$, where ζ_2 contains the simplices of $\partial(\hat{\partial}\xi)$ contained in Σ_X. But then

$$\hat{\partial}(\hat{\partial}\xi) = \partial(\partial\xi - \zeta) - \zeta_2 = \partial\partial\xi - \partial\zeta - \zeta_2 = 0 - \partial\zeta - \zeta_2.$$

But since all simplices of $\partial\zeta$ and ζ_2 lie in Σ_X and since $\hat{\partial}(\hat{\partial}\xi)$ can contain no such simplices, we must have $-\partial\zeta - \zeta_2 = 0$. □

Lemma 6.2.4 *Subdivision induces chain maps $I^{\bar{p}}C^T_*(X;G) \to I^{\bar{p}}C^{T'}_*(X;G)$, and so $I^{\bar{p}}\mathfrak{H}_*(X;G) = \varinjlim H_*(I^{\bar{p}}C^T_*(X;G))$ is well defined.*

Proof Recall that all admissible triangulations of X are assumed compatible with the filtration.

If σ is an allowable i-simplex of the triangulation T and $|\sigma| \not\subset \Sigma$, then every i-simplex s in any subdivision of σ is allowable by Lemma 3.3.15 and also not contained in Σ (else Σ would intersect the interior of σ and thus contain σ). So subdivision takes allowable simplices not contained in Σ to chains composed of allowable simplices not contained in Σ.

It thus remains only to verify that the map $\hat{\partial}$ commutes with subdivision. Let σ be an i-simplex in the triangulation T and σ' the image of σ in the subdivision T' of T. On the one hand, the chain $\hat{\partial}\sigma'$ is obtained from $\partial\sigma'$ by throwing out the $(i-1)$-simplices of $\partial\sigma'$ that are contained in Σ. On the other, if $(\hat{\partial}\sigma)'$ is the subdivision of $\hat{\partial}\sigma$, then this chain differs from $(\partial\sigma)' = \partial\sigma'$ in that it is missing the $(i-1)$-simplices of $\partial\sigma'$ contained in the $(i-1)$-simplices of $\partial\sigma$ that are contained in Σ. So we show that the $(i-1)$-simplices of $\partial\sigma'$ that are contained in Σ are precisely the $(i-1)$-simplices of $\partial\sigma'$ contained in the $(i-1)$-simplices of $\partial\sigma$ that are contained in Σ.

If τ is an $(i-1)$-face of σ with $\tau \subset \Sigma$, then clearly any simplex of the induced subdivision (or any subdivision) τ' of τ is contained in Σ. Conversely, suppose some $(i-1)$-simplex t

of σ' is contained in Σ. If t is not contained in some $(i-1)$-face τ of σ, then Σ intersects the interior of σ and so $\sigma \subset \Sigma$, a contradiction, as $\sigma \in S_i^{\bar{p}}(X)$ implies that $|\sigma| \not\subset \Sigma$. If t is contained in some $(i-1)$-face τ of σ, then similarly all of τ must be contained in Σ. This completes the proof of the claim. □

6.2.2 Second Definition of IH_*

A more formal formulation of $I^{\bar{p}}S_*(X;G)$, closer to the definition of Saralegi [205], is as follows: Let $A^{\bar{p}}S_i(X;G)$ be the subgroup of $S_i(X;G)$ generated by the allowable simplices of X (with no assumption on boundaries or containment in Σ_X). Then $I^{\bar{p}}S_i^{\mathrm{GM}}(X;G)$ can be described as $A^{\bar{p}}S_i(X;G) \cap \partial^{-1}(A^{\bar{p}}S_{i-1}(X;G))$. Instead, let

$$I^{\bar{p}}S_i'(X;G) = \frac{(A^{\bar{p}}S_i(X;G) + S_i(\Sigma_X;G)) \cap \partial^{-1}(A^{\bar{p}}S_{i-1}(X;G) + S_{i-1}(\Sigma_X;G))}{S_i(\Sigma_X;G)}. \tag{6.2}$$

The idea is that we add to the allowable simplices all of the singular simplices living in Σ_X in order to quotient them out, but we do this in such a way as to maintain a chain complex. If we had just used $\frac{A^{\bar{p}}S_i(X;G)+S_i(\Sigma_X;G)}{S_i(\Sigma_X;G)}$, not only would this just be the same as the quotient of $A^{\bar{p}}S_i(X;G)$ by any allowable simplices supported in Σ_X, but we could also have some elements whose boundaries contain non-allowable simplices that are not contained in Σ_X, so we would not have a chain complex. The definition we have provided for $I^{\bar{p}}S_i'(X;G)$ clearly does not have this problem. Additionally, we note that $I^{\bar{p}}S_i'(X;G)$ is well defined because

$$S_i(\Sigma_X;G) \subset A^{\bar{p}}S_i(X;G) + S_i(\Sigma_X;G)$$

and also

$$S_i(\Sigma_X;G) \subset \partial^{-1}(A^{\bar{p}}S_{i-1}(X;G) + S_{i-1}(\Sigma_X;G)),$$

as every boundary of an element of $S_i(\Sigma_X;G)$ lies in $S_{i-1}(\Sigma_X;G)$.

We define $I^{\bar{p}}C_*^{T,'}(X;G)$ analogously, and let $I^{\bar{p}}\mathfrak{C}_*'(X;G) = \varinjlim(I^{\bar{p}}C_*^{T,'}(X;G))$. This is well defined as subdivision preserves each of the component groups in the simplicial analogue of (6.2).

Despite the different approaches, these definitions in fact yield isomorphic chain complexes to those defined previously:

Lemma 6.2.5 *One has $I^{\bar{p}}S_*(X;G) \cong I^{\bar{p}}S_*'(X;G)$, and similarly for the simplicial and PL versions.*

Proof As $I^{\bar{p}}S_*(X;G) \subset S_*^{\bar{p}}(X;G) \subset A^{\bar{p}}S_*(X;G)$, we will have canonical inclusion induced homomorphisms $\mathfrak{i}: I^{\bar{p}}S_i(X;G) \to I^{\bar{p}}S_i'(X;G)$ provided we also have

$$I^{\bar{p}}S_i(X;G) \subset \partial^{-1}(A^{\bar{p}}S_{i-1}(X;G) + S_{i-1}(\Sigma_X;G)).$$

But if $\xi \in I^{\bar{p}}S_*(X;G)$, then by definition $\partial\xi = \hat{\partial}\xi + \eta$, where $|\eta| \subset \Sigma_X$ and $\hat{\partial}\xi$ must be composed of allowable simplices by definition of $I^{\bar{p}}S_*(X;G)$. So indeed $\partial\xi \in A^{\bar{p}}S_{i-1}(X)+$

$S_{i-1}(\Sigma_X;G)$. The inclusion i is also a chain map since $\partial\mathrm{i}(\xi)$ will be represented by $\partial\xi = \hat{\partial}\xi + \eta$, which is equivalent in $I^{\bar{p}}S'_*(X;G)$ to $\hat{\partial}\xi$, because $|\eta| \subset S_{i-1}(X;G)$. But $\hat{\partial}\xi$ also represents $\mathrm{i}(\hat{\partial}\xi)$.

Next we observe that i is injective since by definition $I^{\bar{p}}S_*(X;G) \subset S^{\bar{p}}_*(X;G)$ and $S^{\bar{p}}_*(X;G)$ consists of simplices whose images do not lie completely in Σ_X.

Finally, we show that i is surjective. Let

$$x \in (A^{\bar{p}}S_i(X;G) + S_i(\Sigma_X;G)) \cap \partial^{-1}(A^{\bar{p}}S_{i-1}(X;G) + S_{i-1}(\Sigma_X;G)).$$

By definition, x can be written as a sum $\xi + \eta$ with $\xi \in A^{\bar{p}}S_i(X;G)$ and $\eta \in S_i(\Sigma_X;G)$. This decomposition is not necessarily unique, but we may assume that ξ contains no simplices contained completely in Σ_X, which will make the decomposition unique. We claim that in fact ξ represents an element of $I^{\bar{p}}S_i(X;G)$ so that $\mathrm{i}(\xi) = x$. We only need to show that $\hat{\partial}\xi \subset S^{\bar{p}}_{i-1}(X;G)$. But we know that

$$\partial x = \partial(\xi + \eta) \in A^{\bar{p}}S_{i-1}(X;G) + S_{i-1}(\Sigma_X;G),$$

hence any simplex of $\partial(\xi + \eta) = \partial\xi + \partial\eta$ that is not contained in Σ_X must be allowable. As η is contained in Σ, any simplex of $\partial(\xi + \eta)$ not contained in Σ must be part of $\partial\xi$, and this includes all of the simplices of $\hat{\partial}\xi$. Thus all simplices of $\hat{\partial}\xi$ are allowable, as required.

The proof for the simplicial/PL versions is the same. □

Remark 6.2.6 It will be useful for later to record the following facts from the proof of the lemma: Suppose x is a chain representing an element of $I^{\bar{p}}S'_*(X;G)$, and let ξ be the chain obtained by throwing away any terms in x coming from simplices contained in Σ_X. Then ξ and x represent the same element of $I^{\bar{p}}S'_*(X;G)$ and ξ also represents the corresponding element of $I^{\bar{p}}S_*(X;G)$. In particular, we can always represent an element of $I^{\bar{p}}S'_*(X;G)$ by an element of $I^{\bar{p}}S_*(X;G)$, and we typically assume that we do so.

Ordinarily, we might be concerned about creating new unallowable boundary terms when we start throwing away pieces of chains, but since any such new boundary simplices must be contained in Σ_X, they do not contribute to $\hat{\partial}$ and so cause no problems.

This lemma demonstrates that our two approaches to non-GM intersection chains via $I^{\bar{p}}S_*(X;G)$ and $I^{\bar{p}}S'_*(X;G)$ are equivalent, and hence yield the same intersection homology groups, which we will denote $I^{\bar{p}}H_*(X;G)$ in either setting. The author has some preference for the first definition since it avoids a complicated quotient formula and eliminates the ambiguities that arise when dealing with quotient groups. It is also suited for handling local coefficient systems that don't extend to Σ (see Section 6.3.3). Another benefit of the first definition is that it allows us to easily convert proofs we have already given for properties of $I^{\bar{p}}H^{\mathrm{GM}}_*(X;G)$ to proofs of properties of $I^{\bar{p}}H_*(X;G)$, which we shall do later in this chapter. However, the second definition certainly has the advantage of a clear mathematical formula. It also highlights that the non-GM intersection chain complexes can be thought of as subcomplexes of the *relative* chain complexes $S_*(X, \Sigma_X;G)$, though, once again, we will show shortly that this approach does recover $I^{\bar{p}}H^{\mathrm{GM}}_*$ for perversities with $\bar{p} \le \bar{t}$. In this case, the perversity conditions limit the behavior of chains near Σ_X sufficiently that the quotienting out of the chains in Σ becomes irrelevant for the homology computation.

As we proceed, we will feel free to use whichever of $I^{\bar{p}}S_*(X;G)$ or $I^{\bar{p}}S'_*(X;G)$ better suits the task at hand. Before moving on, however, we present a third equivalent definition, which is Saralegi's from [205]. Beyond the historical interest, this definition also has a very nice use in proving Proposition 6.2.9, below.

Remark 6.2.7 It is important to note that. analogously to our comments in Remark 6.2.2 concerning the notation $\hat{\partial}$, once we move on to later chapters in which the precise definition of the singular intersection chain complex is less relevant, we will revert to the notation $I^{\bar{p}}S_*(X;G)$ with boundary map written simply as ∂ except in those instances where it is necessary to utilize the distinction between $I^{\bar{p}}S_*(X;G)$ and $I^{\bar{p}}S'_*(X;G)$ explicitly.

6.2.3 Third Definition of IH_*

Next, we provide Saralegi's definition of intersection chains, which is a version of our second definition. The main difference is that instead of including and then quotienting out all of $S_*(\Sigma_X;G)$, instead we only include and quotient out the allowable chains on those strata for which the perversity is "too big." Following our cone computations above, this is reasonable, since these are the strata for which distinguished neighborhood computations will be affected by the "faulty" cone formula.

We will denote Saralegi's chain complex from [205] as[3] $I^{\bar{p}}S''_*(X;G)$. To define it, we continue to let $A^{\bar{p}}S_i(X;G)$ denote the groups generated by the $\bar{p}$-allowable i-simplices of X, and we let $X^{\bar{p}}$ denote the closure of the union of the singular strata S of X such that $\bar{p}(S) > \operatorname{codim}(S) - 2$. Then, for a perversity $\bar{q}$, let $A^{\bar{q}}S_i(X^{\bar{p}};G)$ be generated by the $\bar{q}$-allowable i-simplices with support in $X^{\bar{p}}$. We then define

$$\begin{aligned} &I^{\bar{p}}S''_i(X;G)\\ &= \frac{\left(A^{\bar{p}}S_i(X;G) + A^{\bar{p}+1}S_i(X^{\bar{p}};G)\right) \cap \partial^{-1}\left(A^{\bar{p}}S_{i-1}(X;G) + A^{\bar{p}+1}S_{i-1}(X^{\bar{p}};G)\right)}{A^{\bar{p}+1}S_i(X^{\bar{p}};G) \cap \partial^{-1}A^{\bar{p}+1}S_{i-1}(X^{\bar{p}};G)}. \end{aligned}$$

Here $\bar{p}+1$ is the perversity such that $(\bar{p}+1)(S) = \bar{p}(S)+1$ for all singular strata S, while of course we must have $(\bar{p}+1)(R) = 0$ for R a regular stratum. Once again, there are analogous simplicial and PL chain complexes.

We show that $I^{\bar{p}}S''_*(X;G)$ is also isomorphic to $I^{\bar{p}}S_*(X;G)$.

Lemma 6.2.8 *The chain map*

$$\mathfrak{i} : I^{\bar{p}}S''_*(X;G) \to I^{\bar{p}}S'_*(X;G)$$

induced by the inclusion $A^{\bar{p}+1}S_*(X^{\bar{p}};G) \hookrightarrow S_*(\Sigma)$ is an isomorphism. Hence,

$$I^{\bar{p}}S_*(X;G) \cong I^{\bar{p}}S'_*(X;G) \cong I^{\bar{p}}S''_*(X;G).$$

[3] Saralegi's original notation was $SC^{\bar{p}}_*(X, X_{\bar{q}};G)$, where $\bar{q} = D\bar{p}$.

Similarly for the simplicial/PL versions.

Proof The argument relies on the following two facts:

1. *If σ is a $\bar{p}$-allowable i-simplex, then each $(i-1)$-simplex of $\partial\sigma$ is $(\bar{p}+1)$-allowable*: By definition, for each singular stratum S the preimage $\sigma^{-1}(S)$ is contained in the $(i - \text{codim}(S) + \bar{p}(S))$-skeleton of Δ^i. But if F is any $i-1$ face of Δ^i, and $\sigma|_F$ is the restriction of σ to F, then $\sigma|_F^{-1}(S) = \sigma^{-1}(S) \cap F \subset \sigma^{-1}(S)$, and so this set must also be contained in the $(i - \text{codim}(S) + \bar{p}(S))$-skeleton of F. We now simply observe that $i - \text{codim}(S) + \bar{p}(S) = i - 1 - \text{codim}(S) + \bar{p}(S) + 1$. So $\sigma|_F$ is a $(\bar{p}+1)$-allowable $(i-1)$-simplex.
2. *If σ is a $\bar{p}$- or $(\bar{p}+1)$-allowable simplex with $|\sigma| \subset \Sigma$, then $|\sigma| \subset X^{\bar{p}}$*: As $\bar{p}$-allowable implies $(\bar{p}+1)$-allowable, it suffices to assume the latter. So suppose σ is a $(\bar{p}+1)$-allowable j-simplex, and let S be a singular stratum with $\bar{p}(S) \leq \text{codim}(S) - 2$. By definition, $\sigma^{-1}(S)$ is contained in the

$$j - \text{codim}(S) + \bar{p}(S) + 1 \leq j - \text{codim}(S) + \text{codim}(S) - 2 + 1 = j - 1$$

skeleton of Δ^j. So σ cannot take any points of the interior of Δ^j to S. As $|\sigma| \subset \Sigma$, it follows that σ must map the interior of Δ^j to $X^{\bar{p}}$, and hence it takes all of Δ^j to $X^{\bar{p}}$ as $X^{\bar{p}}$ is a closed set.

Now we prove that $\mathfrak{i}$ is injective.

Suppose the chain ξ represents an element of $I^{\bar{p}}S''_i(X;G)$ and that $\mathfrak{i}(\xi) = 0$. This means that $|\xi| \subset \Sigma$. We claim that this implies that $|\xi| \subset X^{\bar{p}}$. Indeed, by definition, ξ can be written as a sum $\xi = x + y$ with $x \in A^{\bar{p}}S_i(X;G)$ and $y \in A^{\bar{p}+1}S_i(X^{\bar{p}};G)$. Certainly $|y| \subset X^{\bar{p}} \subset \Sigma$ by definition, and so we also have $|x| = |\xi - y| \subset |\xi| \cup |y| \subset \Sigma$. So by fact (2), we have $|x| \subset X^{\bar{p}}$, and it follows that $|\xi| = |x+y| \subset |x| \cup |y| \subset X^{\bar{p}}$. Also, by definition, every simplex of x and y is $(\bar{p}+1)$-allowable, as $\bar{p}$-allowable implies $(\bar{p}+1)$-allowable, so $\xi \in A^{\bar{p}+1}S_i(X^{\bar{p}};G)$. Next, we have by definition, $\partial\xi \in A^{\bar{p}}S_{i-1}(X;G) + A^{\bar{p}+1}S_{i-1}(X^{\bar{p}};G)$, and our assumption that $|\xi| \subset \Sigma$ implies that $|\partial\xi| \subset \Sigma$. So by the same argument as for ξ itself, we must have $\partial\xi \in A^{\bar{p}+1}S_{i-1}(X^{\bar{p}};G)$. So

$$\xi \in A^{\bar{p}+1}S_i(X^{\bar{p}};G) \cap \partial^{-1}(A^{\bar{p}+1}S_{i-1}(X^{\bar{p}};G)),$$

which implies then that $\xi = 0 \in I^{\bar{p}}S''_*(X;G)$. This completes the proof that $\mathfrak{i}$ is injective.

For surjectivity, let $\xi \in I^{\bar{p}}S'_i(X;G)$ be represented by the sum $x+y$ with $x \in A^{\bar{p}}S_i(X;G)$ and $y \in S_i(\Sigma;G)$. We may assume that all simplices of $x+y$ that have image in Σ are part of y so that no simplex of x has image in Σ. As $y \in S_i(\Sigma;G)$ and $\partial y \in S_{i-1}(\Sigma;G)$, the chain y represents 0 in $I^{\bar{p}}S'_*(X;G)$. So ξ can be represented by $x \in A^{\bar{p}}S_i(X;G)$ alone. We also know $\partial\xi = \partial x \in A^{\bar{p}}S_{i-1}(X;G) + S_{i-1}(\Sigma;G)$. To show $\mathfrak{i}$ is surjective, we must verify that $\partial x \in A^{\bar{p}}S_{i-1}(X;G) + A^{\bar{p}+1}S_{i-1}(X^{\bar{p}};G)$, as this will imply that x represents an element of $I^{\bar{p}}S''_i(X;G)$. As every simplex of x is $\bar{p}$-allowable, fact (1) above tells us that ∂x consists of $(\bar{p}+1)$-allowable simplices, and we can write $\partial x = a + b$ with $a \in A^{\bar{p}}S_{i-1}(X;G)$ and

$b \in S_{i-1}(\Sigma; G)$. As ∂x and a consist of $(\bar{p}+1)$-allowable simplices, so must $b = \partial x - a$, and so each simplex of b is supported in $X^{\bar{p}}$ by fact (2). Thus ∂x decomposes as desired. □

6.2.4 Non-GM Intersection Homology below the Top Perversity

Saralegi's definition gives us what is perhaps the shortest route towards seeing that $I^{\bar{p}}S_*(X;G) \cong I^{\bar{p}}S_*^{\mathrm{GM}}(X;G)$ when $\bar{p}(S) \leq \bar{t}(S)$ for all singular strata S.

Proposition 6.2.9 *Let X be a filtered space. If $\bar{p}(S) \leq \bar{t}(S)$ for all singular strata S of X, then $I^{\bar{p}}S_*(X;G) \cong I^{\bar{p}}S_*^{\mathrm{GM}}(X;G)$, and similarly for the PL versions.*

Proof If $\bar{p}(S) \leq \bar{t}(S)$, then $\bar{p}(S) \leq \operatorname{codim}(S) - 2$. So there are no singular strata on which $\bar{p}(S) > \operatorname{codim}(S) - 2$, and so $X^{\bar{p}}$ is empty! So

$$I^{\bar{p}}S''_*(X;G) = A^{\bar{p}}S_*(X;G) \cap \partial^{-1}(A^{\bar{p}}S_{*-1}(X;G)),$$

which is precisely $I^{\bar{p}}S_*^{\mathrm{GM}}(X;G)$. The result now follows from Lemma 6.2.8. The PL argument is identical. □

Remark 6.2.10 Here is another way to prove Proposition 6.2.9, this time working with $I^{\bar{p}}S_*(X;G)$ directly. Recall that $S_*^{\bar{p}}(X;G)$ is the complex generated by the $\bar{p}$-allowable simplices. If $\bar{p} \leq \bar{t}$, then for every i, we have

$$i - \operatorname{codim}(S) + \bar{p}(S) \leq i - \operatorname{codim}(S) + \operatorname{codim}(S) - 2 = i - 2.$$

So, for an i-simplex σ to be allowable, $\sigma^{-1}(S)$ must be contained in the $(i-2)$-skeleton of Δ^i for any singular stratum S, and so $\sigma^{-1}(\Sigma_X) \subset \{(i-2)\text{-skeleton of } \Delta^i\}$. But this implies that no i- or $(i-1)$-dimensional face of σ can be contained completely in Σ_X. This in turn tells us that $\hat{\partial} = \partial$ when applied to elements of $S^{\bar{p}}(X;G)$. By definition, $I^{\bar{p}}S_*(X;G)$ consists of those elements $\xi \in S_*^{\bar{p}}(X;G)$ such that $\hat{\partial}\xi \in S_*^{\bar{p}}(X;G)$. But in this case, this means precisely that every simplex of ξ must be allowable and every simplex of $\partial\xi$ must be allowable. But this is precisely the definition of $I^{\bar{p}}S_*^{\mathrm{GM}}(X;G)$.

So when $\bar{p}(S) \leq \bar{t}(S)$ for all singular strata of X, we get nothing new. However, when $\bar{p}(S) > \bar{t}(S)$ for some S, we indeed get something different. One way to see this is to observe that when $\bar{p}(S) > \bar{t}(S)$ for some S, the groups $I^{\bar{p}}H_*(X;G)$ are not always topological invariants for a CS set even if the conditions of Theorem 5.5.1 hold; i.e. $I^{\bar{p}}H_*(X;G)$ may depend on the filtration.

Example 6.2.11 Let $\mathbb{R}$ be the unfiltered real line, and let $\bar{0}$ be the zero perversity. Then $I^{\bar{0}}S_*(\mathbb{R}) = I^{\bar{0}}S_*^{\mathrm{GM}}(\mathbb{R}) = S_*(\mathbb{R})$, so

$$I^{\bar{0}}H_0(\mathbb{R}) = I^{\bar{0}}H_0^{\mathrm{GM}}(\mathbb{R}) = H_0(\mathbb{R}) \cong \mathbb{Z}.$$

Now suppose we form the CS set X by filtering $\mathbb{R}$ instead as $\{0\} \subset \mathbb{R}$, keeping the perversity $\bar{0}$. The conditions of Theorem 5.5.1 hold, so

$$I^{\bar{0}}H_0^{\mathrm{GM}}(X) \cong I^{\bar{0}}H_0^{\mathrm{GM}}(\mathbb{R}) \cong H_0(\mathbb{R}) \cong \mathbb{Z}.$$

But now, let σ_x be the 0-simplex with image at $x \neq 0 \in \mathbb{R}$. Let e_x be the linear 1-simplex with boundary $\sigma_x - \sigma_0$. Then $e_x^{-1}(\{0\})$ is in the 0-skeleton of Δ^1, and $1 - \operatorname{codim}(\{0\}) + \bar{0}(\{0\}) = 1 - 1 + 0 = 0$, so e_x is allowable. Furthermore, $\hat{\partial} e_x = \sigma_x$. So e_x gives a null-homology of σ_x in $I^{\bar{0}}S_*(X)$, and therefore $I^{\bar{0}}H_0(X) = 0$. In particular, $I^{\bar{0}}H_0(X) \not\cong I^{\bar{0}}H_0^{\mathrm{GM}}(X)$, and so $I^{\bar{0}}H_0(X) \not\cong I^{\bar{0}}H_0(\mathbb{R})$. △

Remark 6.2.12 As we have seen that non-GM intersection homology can behave more like relative homology than absolute homology, and in particular that it can be trivial in degree zero, the non-GM intersection chain complex does not necessarily possess an augmentation map $\mathbf{a} : I^{\bar{p}}S_0(X; G) \to G$ so that

$$I^{\bar{p}}S_1(X; G) \xrightarrow{\partial} I^{\bar{p}}S_0(X; G) \xrightarrow{\mathbf{a}} G \to 0$$

is exact. Of course, if $\bar{p} \leq \bar{t}$, then $I^{\bar{p}}S_*(X; G) \cong I^{\bar{p}}S_*^{\mathrm{GM}}(X; G)$ and so there is an augmentation map in this case by Definition 4.3.18.

6.2.5 A New Cone Formula

Of course, the groups $I^{\bar{p}}S_*(X)$ should not equal $I^{\bar{p}}S_*^{\mathrm{GM}}(X)$ in all cases because the entire point of introducing $I^{\bar{p}}S_*(X)$ was to modify the cone formula. Let us now verify that we have done that successfully.

Theorem 6.2.13 *If X is a compact filtered space of formal dimension $n-1$, then*

$$I^{\bar{p}}H_i(cX; G) \cong \begin{cases} 0, & i \geq n - \bar{p}(\{v\}) - 1, \\ I^{\bar{p}}H_i(X; G), & i < n - \bar{p}(\{v\}) - 1. \end{cases}$$

Furthermore, the isomorphisms of the last case are induced by inclusion. An equivalent conclusion holds for PL intersection homology when X is a compact PL filtered space.

Proof The proof is nearly exactly the same as that of Theorems 4.2.1 and 5.3.5 except for the special computations that were required in dimension zero. We outline the arguments again, highlighting the necessary modifications.

If ξ is an i-cycle in $I^{\bar{p}}S_i(cX; G)$ for $i \geq n - \bar{p}(\{v\}) - 1$, $i > 0$, then, as in the proof of Theorem 4.2.1, we can check that $\bar{c}\xi$ is allowable and its boundary is ξ: As $\xi \in I^{\bar{p}}S_i(cX; G)$, it has no simplices in Σ_{cX} and therefore neither does $\bar{c}\xi$. As we assume that ξ is an i-cycle, we have $\hat{\partial}\xi = 0$. As a chain in $S_*(cX; G)$, we will have $\partial\xi = \hat{\partial}\xi + \eta$, where η is contained in Σ_{cX}. So then $\partial(\bar{c}\xi) = \xi - \bar{c}(\partial\xi) = \xi - \bar{c}(\hat{\partial}\xi + \eta)$. Since η is supported in Σ_{cX}, so is $\bar{c}\eta$, and we have assumed that $\hat{\partial}\xi = 0$. Therefore, $\hat{\partial}(\bar{c}\xi) = \xi$, using that no simplex of ξ is contained in Σ_{cX}. The allowability of the simplices of $\bar{c}\xi$ follows from the allowability of the simplices of ξ as in the proof of Theorem 4.2.1.

But now, remarkably, the argument of the preceding paragraph continues to hold even if $i = 0 \geq n - \bar{p}(\{v\}) - 1$, which wasn't the case in the proof of Theorem 4.2.1: With this perversity assumption, if σ is an allowable 0-simplex in cX not contained in Σ_{cX}, so that

$g\sigma \in I^{\bar{p}}S_0(cX;G)$, then $\partial(g\sigma) = g\partial\sigma = g(\sigma - \sigma_v)$, where σ_v is the 0-simplex with image in the cone vertex v. But then we have $\hat{\partial}(g\sigma) = g\sigma$ since σ_v has image in Σ_{cX}. Thus any $g\sigma \in I^{\bar{p}}S_0(cX;G)$ is null-homologous in $I^{\bar{p}}S_*(cX;G)$, and $I^{\bar{p}}H_0(cX;G) = 0$.

Finally, for $i < n - \bar{p}(\{v\}) - 1$, the chain $\bar{c}\xi$ is not allowable, and in fact no allowable simplex can intersect $\{v\}$ by the arguments in the proof of Theorem 4.2.1. So $I^{\bar{p}}H_i(cX;G) \cong I^{\bar{p}}H_i(cX - \{v\};G)$. We will see below in Corollary 6.3.8 that, just as for $I^{\bar{p}}H_*^{\mathrm{GM}}$, the groups $I^{\bar{p}}H_*$ are stratified homotopy invariants, and this completes the proof.

The PL argument is analogous using the modifications indicated in the proof of Theorem 4.2.1. □

6.2.6 Relative Non-GM Intersection Homology and the Relative Cone Formula

The relative intersection homology groups are defined just as we defined the relative GM intersection homology groups. If $Y \subset X$, we let $I^{\bar{p}}S_*(Y;G)$ be the subcomplex of $I^{\bar{p}}S_*(X;G)$ consisting of chains supported in Y or, equivalently, the complex defined natively in Y by using the filtration and perversity inherited from X (cf. Section 4.3). Then we let

$$I^{\bar{p}}S_*(X,Y;G) = I^{\bar{p}}S_*(X;G)/I^{\bar{p}}S_*(Y;G)$$

and

$$I^{\bar{p}}H_i(X,Y;G) = H_*(I^{\bar{p}}S_*(X,Y;G)).$$

Of course it follows that there is a long exact sequence of pairs for intersection homology:

$$\longrightarrow I^{\bar{p}}H_{i+1}(X,Y;G) \to I^{\bar{p}}H_i(Y;G) \to I^{\bar{p}}H_i(X;G) \to I^{\bar{p}}H_i(X,Y;G) \longrightarrow .$$

Remark 6.2.14 It will be useful below to notice that, for each i, the complex $I^{\bar{p}}S_*(X,Y;G)$ is a subcomplex of $S_*(X,Y;G)$. Indeed, there are evident maps

$$\begin{aligned} I^{\bar{p}}S_i(X,Y;G) = I^{\bar{p}}S_i(X;G)/I^{\bar{p}}S_i(Y;G) &\to S_i^{\bar{p}}(X;G)/S_i^{\bar{p}}(Y;G) \\ &\to S_i(X;G)/S_i(Y;G) \cong S_i(X,Y;G). \end{aligned}$$

To see that this composition is injective, we just need to observe that if $x \in I^{\bar{p}}S_i(X,Y;G)$ does not represent 0, then x has a representative chain containing a simplex that is not supported in Y, and so the image of x under this sequence of maps cannot be 0.

The following corollary is an immediate consequence of the long exact sequence of pairs.

Corollary 6.2.15 *If X is a compact $(n-1)$-dimensional filtered space then*

$$I^{\bar{p}}H_i(cX, cX - \{v\};G) \cong \begin{cases} I^{\bar{p}}H_{i-1}(X;G), & i > n - \bar{p}(\{v\}) - 1, \\ 0, & i \le n - \bar{p}(\{v\}) - 1. \end{cases}$$

An equivalent conclusion holds for PL intersection homology when X is a compact PL filtered space.

For some of our arguments below, it will be more useful to work with $I^{\bar{p}}S_*(X;G)$ in the form $I^{\bar{p}}S'_*(X;G)$. As these complexes are isomorphic, of course it follows that if we define $I^{\bar{p}}S'_*(X,Y;G)$ to be $I^{\bar{p}}S'_*(X;G)/I^{\bar{p}}S'_*(Y;G)$ then

$$I^{\bar{p}}S'_*(X,Y;G) = \frac{I^{\bar{p}}S'_*(X;G)}{I^{\bar{p}}S'_*(Y;G)} \cong \frac{I^{\bar{p}}S_*(X;G)}{I^{\bar{p}}S_*(Y;G)} = I^{\bar{p}}S_*(X,Y;G),$$

but it is also useful to have a better look at the form that $I^{\bar{p}}S'_*(X,Y;G)$ takes from the definitions. In other words, we have

$$I^{\bar{p}}S'_i(X,Y;G) = \frac{\dfrac{(A^{\bar{p}}S_i(X;G)+S_i(\Sigma_X;G))\cap\partial^{-1}(A^{\bar{p}}S_{i-1}(X;G)+S_{i-1}(\Sigma_X;G))}{S_i(\Sigma_X;G)}}{\dfrac{(A^{\bar{p}}S_i(Y;G)+S_i(\Sigma_Y;G))\cap\partial^{-1}(A^{\bar{p}}S_{i-1}(Y;G)+S_{i-1}(\Sigma_Y;G))}{S_i(\Sigma_Y;G)}},$$

where $\Sigma_Y = Y\cap\Sigma_X$.

Now, we notice that

$$\begin{aligned}&S_i(\Sigma_X;G)\cap[(A^{\bar{p}}S_i(Y;G)+S_i(\Sigma_Y;G))\cap\partial^{-1}(A^{\bar{p}}S_{i-1}(Y;G)+S_{i-1}(\Sigma_Y;G))]\\&\quad = S_i(\Sigma_Y;G).\end{aligned}$$

Indeed, $S_i(\Sigma_Y;G)$ is generated by the singular simplices in $\Sigma_Y = Y\cap\Sigma_X$, and these are certainly all contained both in Σ_X and in the expression in the square brackets. On the other hand, any chain in the left side of the expression is both a singular chain in Σ_X and a singular chain in Y (from the expression in brackets), so it must be in $S_i(\Sigma_Y;G)$. But now by basic group theory, if C, D, E are abelian groups, then

$$\frac{\dfrac{C}{D}}{\dfrac{E}{D\cap E}} \cong \frac{\dfrac{C}{D}}{\dfrac{D+E}{D}} \cong \frac{C}{D+E}.$$

So

$$I^{\bar{p}}S'_i(X,Y;G) = \frac{(A^{\bar{p}}S_i(X;G)+S_i(\Sigma_X;G))\cap\partial^{-1}(A^{\bar{p}}S_{i-1}(X;G)+S_{i-1}(\Sigma_X;G))}{S_i(\Sigma_X;G)+(A^{\bar{p}}S_i(Y;G)+S_i(\Sigma_Y;G))\cap\partial^{-1}(A^{\bar{p}}S_{i-1}(Y;G)+S_{i-1}(\Sigma_Y;G))}. \quad (6.3)$$

6.3 Properties of $\mathbf{I^{\bar{p}}H_*(X;G)}$

In this section, we establish for $I^{\bar{p}}H_*(X;G)$ versions of the properties we have already obtained for $I^{\bar{p}}H^{\mathrm{GM}}_*(X;G)$. In most cases, the proofs of these properties translate easily, and so we largely omit them except to focus on necessary modifications. More detailed proofs of most statements, written directly for $I^{\bar{p}}H_*$, can be found in [85]. When the arguments in the singular chain and PL chain cases are analogous, we give only the singular chain argument. We will also feel free to work with either chain complex $I^{\bar{p}}S_*(X;G)$ or $I^{\bar{p}}S'_*(X;G)$ according to whichever is more convenient at the time. At the end of this section, in Sections 6.3.2 and 6.3.3, we explore two other properties of non-GM intersection homology: dimensional homogeneity and intersection homology with local coefficients.

We begin with a general property that will be utilized often and that is most easily seen from the $I^{\bar{p}}S_*$ perspective.

Lemma 6.3.1 *Suppose X is a filtered space, $A \subset X$, and R is a Dedekind domain; in particular, this includes the case $R = \mathbb{Z}$. Then each $I^{\bar{p}}S_i(X,A;R) \cong I^{\bar{p}}S'_i(X,A;R)$ and any of their submodules are projective R-modules, and each $I^{\bar{p}}\mathfrak{C}_i(X,A;R) \cong I^{\bar{p}}\mathfrak{C}'_i(X,A;R)$ and any of their submodules are flat R-modules. If R is a field, then $I^{\bar{p}}S_i(X,A;R)$ and $I^{\bar{p}}\mathfrak{C}_i(X,A;R)$ are each free.*

Proof By construction we have $I^{\bar{p}}S_i(X;R) \subset S_i(X;R)$, and the inclusion induces a map $I^{\bar{p}}S_i(X,A;R) \to S_i(X,A;R)$. If the image of a chain $\xi \in I^{\bar{p}}S_i(X,A;R)$ is 0 in $S_i(X,A;R)$, then ξ must be contained in A, but this would force ξ to be 0 in $I^{\bar{p}}S_i(X,A;R)$. Therefore, $I^{\bar{p}}S_i(X,A;R) \subset S_i(X,A;R)$. But $S_i(X,A;R)$ is a free R-module, generated by the singular simplices not contained in A. Therefore, as R is Dedekind, $I^{\bar{p}}S_i(X,A;R)$ and any submodule of $I^{\bar{p}}S_i(X,A;R)$ is projective [49, Section VII.5 and Theorem I.5.4].

For $I^{\bar{p}}\mathfrak{C}_i(X,A;R)$, we cannot make the same argument, as $\mathfrak{C}_i(X,A;R)$ does not appear to be free in general. However, $I^{\bar{p}}\mathfrak{C}_i(X,A;R)$ is R-torsion-free, and so, as R is Dedekind, $I^{\bar{p}}\mathfrak{C}_i(X,A;R)$ is flat [146, Proposition 4.20]. Similarly, any submodule of $I^{\bar{p}}\mathfrak{C}_i(X,A;R)$ is torsion-free and so flat.

The claims when R is a field are immediate, as all vector spaces are free modules.

Of course, these properties extend to $I^{\bar{p}}S'_i(X,A;R)$ and $I^{\bar{p}}\mathfrak{C}'_i(X,A;R)$ by the isomorphisms of Lemma 6.2.5. □

6.3.1 Basic Properties

We now turn to adapting properties of GM intersection homology for non-GM intersection homology.

(a) Maps and Homotopies

We first define $(\bar{p},\bar{q})$-stratified maps. As mentioned in Remark 4.1.2, we need one additional condition beyond those for a map to be $(\bar{p},\bar{q})^{\text{GM}}$-stratified (see Definition 4.1.1), namely that such a map takes singular strata to singular strata. We will not have chain maps in general without such a condition, as the reader can verify by considering a manifold M, a point $x \in M$, and the identity map id from the filtered space $\{x\} \subset M$ to the unfiltered M. If we have an allowable simplex with non-zero boundary in $\{x\}$, then id will not be a chain map. Our additional condition will ensure that we do have chain maps, though this is still sometimes possible even when our new condition fails, as we will see below in Lemma 7.3.16.

So let us make an official definition:

Definition 6.3.2 A map $f : X \to Y$ is *stratified with respect to* $\bar{p}, \bar{q}$ (or $(\bar{p}, \bar{q})$*-stratified*) if

1. $f(\Sigma_X) \subset \Sigma_Y$;
2. the image of each stratum of X is contained in a single stratum of Y, i.e. if $T \subset Y$ is a stratum, then $f^{-1}(T)$ is a union of strata of X;
3. whenever the stratum $S \subset X$ maps to the stratum $T \subset Y$, then $\bar{p}(S) - \text{codim}(S) \leq \bar{q}(T) - \text{codim}(T)$.

Right away, we see that if a map of filtered pairs $f : (X, A) \to (Y, B)$ is a $(\bar{p}, \bar{q})$-stratified map, it induces a chain map $f : I^{\bar{p}}S'_*(X, A; G) \to I^{\bar{q}}S'_*(Y, B; G)$. This follows from the definitions of the $I^{\bar{p}}S'_*$ complexes because f takes Σ_X to Σ_Y and because Proposition 4.1.6 shows that the second and third conditions of the definition guarantee that $\bar{p}$-allowable simplices in X are taken to $\bar{q}$-allowable simplices in Y. Via the isomorphisms $IS_* \cong IS'_*$, this means that we also have a chain map that we briefly denote $\hat{f} : I^{\bar{p}}S_*(X, A; G) \to I^{\bar{q}}S_*(Y, B; G)$; once we have established this map, we will also call it f. We will show that these chain maps possess all of the expected properties, but first it will be useful to develop an explicit description of $\hat{f}$. This takes a bit of care because a simplex of $S^{\bar{p}}_*(X; G)$ might be taken under f into Σ_Y, and such simplices are not allowed in elements of $S^{\bar{q}}_*(Y; G)$. However, we can rectify this problem by just defining $\hat{f} : S^{\bar{p}}(X; G) \to S^{\bar{q}}(Y; G)$ so that $\hat{f}(\sigma) = 0$ if $|f\sigma| \subset \Sigma_Y$.

Lemma 6.3.3 *Suppose* $f : (X, A) \to (Y, B)$ *is a* $(\bar{p}, \bar{q})$*-stratified map. The map* $\hat{f}$ *is a well-defined chain map* $\hat{f} : I^{\bar{p}}S_*(X, A; G) \to I^{\bar{q}}S_*(Y, B; G)$ *that is equivalent to* $f : I^{\bar{p}}S'_*(X, A; G) \to I^{\bar{q}}S'_*(Y, B; G)$ *via the isomorphisms* $IS_* \cong IS'_*$ *of Lemma 6.2.5. The analogous result holds in the PL setting.*

Proof We first consider the case $A = B = \emptyset$ and verify that $\hat{f} : I^{\bar{p}}S_i(X; G) \to I^{\bar{q}}S_i(Y; G)$ is well defined for each i. Suppose $\xi \in I^{\bar{p}}S_i(X; G)$. As already noted, f takes $\bar{p}$-allowable simplices of X to $\bar{q}$-allowable simplices of Y by the proof of Proposition 4.1.6, and setting $\hat{f}(\sigma) = 0$ if $|f\sigma| \subset \Sigma_Y$ means that $\hat{f}(\xi) \in S^{\bar{q}}_i(Y; G)$. So we only need verify that $\hat{\partial}\hat{f}(\xi) \in S^{\bar{q}}_{i-1}(Y; G)$.

By assumption, we can write $\partial\xi = a + b$ with $|b| \subset \Sigma_Y$ and with each simplex of a being $\bar{p}$-allowable. Suppose $f(\xi) = x + y$ with $|y| \subset \Sigma_Y$ and no simplex of x supported in Σ_Y; by definition, $\hat{f}(\xi) = x$ and $\hat{\partial}\hat{f}(\xi)$ is obtained by taking ∂x and throwing away any simplices contained in Σ_Y. So it suffices to show that all simplices of ∂x not contained in Σ_Y are $\bar{q}$-allowable. But now

$$\partial x = \partial(f(\xi) - y) = f(\partial\xi) - \partial y = f(a + b) - \partial y = f(a) + [f(b) - \partial y].$$

The term on the right consists of simplices supported in Σ_Y, while every simplex of a is $\bar{p}$-allowable so that every simplex of $f(a)$ is $\bar{q}$-allowable, as required.

Now that we have shown that $\hat{f} : I^{\bar{p}}S_i(X;G) \to I^{\bar{q}}S_i(Y;G)$ is well defined, it is easy to verify that the following diagram commutes:

$$\begin{array}{ccc} I^{\bar{p}}S_i(X;G) & \xrightarrow[\cong]{\mathfrak{i}_X} & I^{\bar{p}}S_i'(X;G) \\ \Big\downarrow \hat{f} & & \Big\downarrow f \\ I^{\bar{q}}S_i(Y;G) & \xrightarrow[\cong]{\mathfrak{i}_Y} & I^{\bar{q}}S_i'(Y;G), \end{array}$$

where $\mathfrak{i}_X$ and $\mathfrak{i}_Y$ are the isomorphisms of Lemma 6.2.5. In particular, we have $\hat{f} = \mathfrak{i}_Y^{-1} f \mathfrak{i}_X$. As $\mathfrak{i}_Y^{-1} f \mathfrak{i}_X$ is a chain map, so is $\hat{f}$.

The relative version of $\hat{f}$ is then defined by taking quotients. □

Notation: Having established the chain map $\hat{f} : I^{\bar{p}}S_*(X,A;G) \to I^{\bar{q}}S_*(Y,B;G)$ and shown that it is equivalent to $f : I^{\bar{p}}S_*'(X,A;G) \to I^{\bar{q}}S_*'(Y,B;G)$, we can remove the hat and refer to both maps with the notation f.

Remark 6.3.4 In settings where we know that maps $f : X \to Y$ take strata of X to strata of Y of the same codimension, such as if f is a stratified homeomorphism or a stratified homotopy equivalence, then we have $f^{-1}(\Sigma_Y) = \Sigma_X$. In this case, simplices of $S_*^{\bar{p}}(X;G)$ cannot get taken into Σ_Y and so $f : I^{\bar{p}}S_*(X,A;G) \to I^{\bar{q}}S_*(Y,B;G)$ can be defined in the obvious way without having to kill any simplices manually. The reader would not lose much by taking this as an additional simplifying assumption in what is to come.

Noting that the same discussion carries over to the PL setting, we can conclude our discussion so far as follows:

Proposition 6.3.5 *If X,Y are filtered spaces, $f : X \to Y$ is $(\bar{p},\bar{q})$-stratified, and $A \subset X$ and $B \subset Y$ with $f(A) \subset B$, then f induces a chain map $f : I^{\bar{p}}S_*(X,A;G) \to I^{\bar{q}}S_*(Y,B;G)$. If, furthermore, X,Y are PL filtered spaces and f is a PL map that is $(\bar{p},\bar{q})$-stratified, then f induces a chain map $f : I^{\bar{p}}\mathfrak{C}_*(X,A;G) \to I^{\bar{q}}\mathfrak{C}_*(Y,B;G)$ of PL intersection chain complexes. In either case, we obtain corresponding maps of intersection homology groups.*

The following corollary is now immediate as in the GM setting; see Corollary 4.1.8.

Corollary 6.3.6 *If $f : X \to Y$ is a stratified homeomorphism that is also a homeomorphism of pairs $f : (X,A) \to (Y,B)$ and if the perversities $\bar{p}$ on X and $\bar{q}$ on Y correspond (i.e. $\bar{p}(S) = \bar{q}(T)$ if $f(S) = T$), then $I^{\bar{p}}H_*(X,A;G) \cong I^{\bar{q}}H_*(Y,B;G)$. The corresponding fact holds for PL spaces, PL stratified homeomorphisms, and PL intersection homology.*

Now we turn to homotopies.

Proposition 6.3.7 *Suppose $f,g : X \to Y$ are $(\bar{p},\bar{q})$-stratified maps that are $(\bar{p},\bar{q})$-stratified homotopic via a $(\bar{p},\bar{q})$-stratified homotopy $(I \times X, I \times A) \to (Y,B)$. Then f and g induce chain* *homotopic chain maps* $I^{\bar p}S_*(X,A;G) \to I^{\bar q}S_*(Y,B;G)$ *and so* $f = g : I^{\bar p}H_*(X,A;G) \to I^{\bar q}H_*(Y,B;G)$. *The analogous result holds in the PL setting.*

Proof The proofs of the analogous results in Propositions 4.1.10 and 4.3.16 for IH_*^{GM} followed the standard proofs (e.g. [125, Theorem 2.10]) that homotopic topological maps induce chain homotopic chain maps. In particular, we showed that there is a prism operator $P : I^{\bar p}S_i^{\mathrm{GM}}(X) \to I^{\bar p}S_{i+1}^{\mathrm{GM}}(I\times X)$ such that $\partial P = \mathrm{j}_1 - \mathrm{j}_0 - P\partial$, where $\mathrm{j}_0 : X \hookrightarrow \{0\}\times X \subset I\times X$ and $\mathrm{j}_1 : X \hookrightarrow \{1\} \times X \subset I \times X$ are the inclusion maps. We will argue that the same construction gives an analogous operator $P : I^{\bar p}S_i'(X) \to I^{\bar p}S_{i+1}'(I\times X;G)$, from which the rest of the argument is the same.

Recall that P is defined on simplices and then extended linearly, so the same argument works with any coefficient group. By construction, $|P(\sigma)| = I \times |\sigma| \subset I \times X$, and thus if $|\sigma| \subset \Sigma_X$ then $|P(\sigma)| \subset I \times \Sigma_X = \Sigma_{I\times X}$. Furthermore, we know from the proof of Proposition 4.1.10 that if σ is allowable then $P(\sigma)$ is composed of allowable simplices. So, P takes $A^{\bar p}S_i(X;G)$ to $A^{\bar p}S_{i+1}(I \times X;G)$ and $S_i(\Sigma_X;G)$ to $S_{i+1}(I \times \Sigma_X;G)$. Thus, by the definition of $I^{\bar p}S_*'(X;G)$, it only remains to show that if ξ is a chain representing an element of $I^{\bar p}S_i'(X;G)$ then $\partial P(\xi) \in A^{\bar p}S_i(I\times X;G) + S_i(\Sigma_{I\times X};G)$. But

$$\partial P(\xi) = \mathrm{j}_1(\xi) - \mathrm{j}_0(\xi) - P(\partial\xi).$$

As $\partial\xi \in A^{\bar p}S_{i-1}(X;G) + S_{i-1}(\Sigma_X;G)$ by assumption, we get $P(\partial\xi) \in A^{\bar p}S_i(I \times X;G) + S_i(\Sigma_{I\times X};G)$ by the arguments just given above. Meanwhile, j_0 and j_1 are $(\bar p,\bar p)$-stratified using our convention $\bar p(I \times S) = \bar p(S)$ (see the proof of Proposition 4.1.10) and that they take Σ_X to $I\times\Sigma_X = \Sigma_{I\times X}$. So j_0 and j_1 take $A^{\bar p}S_i(X;G)$ to $A^{\bar p}S_i(I\times X;G)$ and $S_i(\Sigma_X;G)$ to $S_i(\Sigma_{I\times X};G)$. Altogether then, $\partial P(\xi) \in A^{\bar p}S_i(I\times X;G) + S_i(I\times\Sigma_X;G)$. □

That stratified homotopy equivalences induce intersection homology isomorphisms now follows just as for Corollaries 4.1.11 and 4.3.17.

Corollary 6.3.8 *Suppose* (X,A) *and* (Y,B) *are filtered pairs and that* $f : X \to Y$ *is a stratified homotopy equivalence that restricts to a stratified homotopy equivalence* $A \to B$. *Suppose that the values of* $\bar p$ *on* X *and* $\bar q$ *on* Y *agree on corresponding strata. Then* f *induces an isomorphism* $I^{\bar p}H_*(X,A;G) \cong I^{\bar q}H_*(Y,B;G)$. *The analogous result holds in the PL category.*

(b) Subdivision, Excision, and Mayer–Vietoris

Next we turn to singular subdivision. Of course subdivision is built into the PL category, as verified in Lemma 6.2.4, and so does not require further work at this point.

Proposition 6.3.9 *Let* ξ *be a chain representing a cycle in* $I^{\bar p}S_i'(X,A;G)$. *Then* ξ *is intersection homologous to any singular subdivision* ξ', *so* ξ *and* ξ' *represent the same element of* $I^{\bar p}H_i(X,A;G)$. *Similarly, if* $\xi \in I^{\bar p}S_i(X,A;G)$ *is a cycle, then any chain obtained by subdividing* ξ *and then throwing out any terms corresponding to simplices supported in* Σ_X *represents the same element as* ξ *in* $I^{\bar p}H_i(X,A;G)$

Proof First, suppose $\xi \in I^{\bar{p}}S_i'(X,A;G)$ is a cycle. By assumption, ξ consists of simplices that are $\bar{p}$-allowable or contained in Σ_X. By Lemma 4.4.13, if σ is a $\bar{p}$-allowable simplex, then each simplex in any subdivision of σ is also $\bar{p}$-allowable, and clearly if $|\sigma| \subset \Sigma_X$ then the same is true of any simplex in any subdivision of σ. Furthermore, each simplex of $\partial\xi$ is either $\bar{p}$-allowable or contained in Σ_X, and so similarly, as singular subdivision commutes with boundaries, $\partial(\xi') = (\partial\xi)'$ also consists of simplices that are allowable or contained in Σ_X. So singular subdivisions of chains in $I^{\bar{p}}S_*'(X,A;G)$ are chains in $I^{\bar{p}}S_*'(X,A;G)$.

To see that ξ and ξ' represent the same homology class, recall that the non-GM argument of Proposition 4.4.14 involved a prism argument. Analogously to our observation in the proof of Proposition 6.3.7, such arguments carry over to the non-GM setting because we already know that such prisms over allowable simplices are composed of allowable simplices and because they preserve supports of chains so that a prism over a simplex in Σ is also contained in Σ. So the homology Ξ constructed as in the proof of Proposition 4.4.14 is an element of $I^{\bar{p}}S_{i+1}'(X,A;G)$ with $\partial\Xi = \xi' - \xi + \omega$, in this case with $\omega \subset A \cup \Sigma$ because $|\partial\xi| \subset A \cup \Sigma$ as ξ is a (relative) cycle.

For $\xi \in I^{\bar{p}}S_i(X,A;G)$ a cycle, we recall that ξ also represents an element of $I^{\bar{p}}S_i'(X,A;G)$. If ξ' is a subdivision, then we have just seen that ξ' is homologous to ξ in $I^{\bar{p}}S_i'(X,A;G)$. But $I^{\bar{p}}S_*(X,A;G)$ and $I^{\bar{p}}S_*'(X,A;G)$ are chain isomorphic by Lemma 6.2.5, and by Remark 6.2.6 the chains in $I^{\bar{p}}S_i(X,A;G)$ corresponding to ξ and ξ' are respectively ξ itself and the chain ξ'' obtained by throwing away from ξ' the terms corresponding to simplices supported in Σ. So ξ'' also and ξ represent the same element of $I^{\bar{p}}H_i(X,A;G)$. □

The following corollary derives from the tools of Proposition 6.3.9, just as Corollary 4.4.15 follows from the tools of Proposition 4.4.14.

Corollary 6.3.10 *Suppose that $T : S_*(X) \to S_*(X)$ is a chain map that restricts to a singular subdivision on each singular simplex. Then the induced map $T : I^{\bar{p}}S_*'(X,A;G) \to I^{\bar{p}}S_*'(X,A;G)$ is chain homotopic to the identity for any subset $A \subset X$. Similarly, the map $I^{\bar{p}}S_*(X,A;G) \to I^{\bar{p}}S_*(X,A;G)$ that applies T and then throws out simplices contained in Σ is chain homotopic to the identity.*

Using Proposition 6.3.9 and the properties of PL subdivision, singular and PL excision and Mayer–Vietoris sequences follow for non-GM intersection homology almost exactly as in the proofs of Theorems 4.4.3, 4.4.4, 4.4.18, and 4.4.19: One takes a sufficiently iterated barycentric subdivision of an i-chain and then splits it into pieces in such a way that the new boundary $(i-1)$-simplices carved out of the interiors of the allowable i-simplices are also allowable. That part of the argument is identical to those of Section 4.4. Simplices contained in Σ_X do not necessarily contribute allowable new boundary pieces when they are split, but any such new boundary simplex must also be contained in Σ_X and so can be thrown out when taking boundaries. We leave the reader to think through the prior arguments in greater detail and note that they still apply in the non-GM setting.

Theorem 6.3.11 *Let X be a filtered space, and suppose $K \subset U \subset X$ such that $\bar{K} \subset \mathring{U}$. Then inclusion induces an isomorphism*

$$I^{\bar{p}}H_*(X-K, U-K; G) \xrightarrow{\cong} I^{\bar{p}}H_*(X, U; G).$$

The equivalent results holds in the PL context.

Theorem 6.3.12 *Let X be a filtered space and suppose $X = U \cup V$, where U, V are subspaces such that $X = \mathring{U} \cup \mathring{V}$. Then there is an exact Mayer–Vietoris sequence*

$$\begin{aligned}\to I^{\bar{p}}H_i(U \cap V; G) \to I^{\bar{p}}H_i(U; G) \oplus I^{\bar{p}}H_i(V; G) \to I^{\bar{p}}H_i(U \cup V; G)\\ \to I^{\bar{p}}H_{i-1}(U \cap V; G) \to .\end{aligned}$$

The equivalent result holds in the PL context. There are also reduced intersection homology Mayer–Vietoris sequences and relative Mayer–Vietoris sequences analogous to those stated in Theorem 4.4.23.

Applying the Mayer–Vietoris sequence, we obtain the suspension formula for $I^{\bar{p}}H_*$:

Theorem 6.3.13 *If X is an $(n-1)$-dimensional compact filtered space and $\bar{p}$ is a perversity on SX that takes the same value p on the two suspension points, then*

$$I^{\bar{p}}H_i(SX; G) = \begin{cases} I^{\bar{p}}H_{i-1}(X; G), & i > n-p-1,\\ 0, & i = n-p-1,\\ I^{\bar{p}}H_i(X; G), & i < n-p-1,\end{cases}$$

and similarly in the PL setting.

Proof We can write SX as the union of two cones cX. The intersection is homeomorphic to $(-1,1) \times X$, which is stratified homotopy equivalent to X. By Theorem 6.2.13, the inclusion $X \hookrightarrow cX$ induces an intersection homology isomorphism for $i < n-p-1$. In the Mayer–Vietoris sequence, this becomes the anti-diagonal map

$$I^{\bar{p}}H_i(X; G) \hookrightarrow I^{\bar{p}}H_i(cX; G) \oplus I^{\bar{p}}H_i(cX; G) \cong I^{\bar{p}}H_i(X; G) \oplus I^{\bar{p}}H_i(X; G)$$

that takes ξ to $\xi \oplus -\xi$. Thus, by basic algebra, we have $I^{\bar{p}}H_i(SX; G) \cong I^{\bar{p}}H_i(X; G)$ in this degree range. For $i \geq n-p-1$, $I^{\bar{p}}H_i(cX; G) = 0$, so we get $I^{\bar{p}}H_i(SX; G) \cong I^{\bar{p}}H_{i-1}(X; G)$ for $i > n-p-1$, and $I^{\bar{p}}H_{n-p-1}(SX; G) = 0$. □

Remark 6.3.14 Notice that this formula is a little cleaner than that of Theorem 4.4.21 since no reduced homology groups are needed.

Example 6.3.15 In Example 4.4.22, we considered a compact n-dimensional ∂-stratified pseudomanifold X with $\partial X \neq \emptyset$ and defined $X^+ = X \cup_{\partial X} \bar{c}(\partial X)$. Under the assumption that $\bar{p}(\{v\}) \leq n-2$, we showed that

$$I^{\bar{p}}H_i^{\mathrm{GM}}(X^+) \cong \begin{cases} I^{\bar{p}}H_i^{\mathrm{GM}}(X,\partial X), & i > n-\bar{p}(\{v\})-1, \\ \mathrm{im}(I^{\bar{p}}H_i^{\mathrm{GM}}(X) \to I^{\bar{p}}H_i^{\mathrm{GM}}(X,\partial X)), & i = n-\bar{p}(\{v\})-1, \\ I^{\bar{p}}H_i^{\mathrm{GM}}(X), & i < n-\bar{p}(\{v\})-1. \end{cases}$$

The assumption about $\bar{p}(\{v\})$ was a simplification to avoid the special cases that can occur in computing $I^{\bar{p}}H_0^{\mathrm{GM}}(\bar{c}(\partial X))$ if $\bar{p}(\{v\})$ is allowed to be greater than $n-2$; see Theorem 4.2.1 and Remark 4.2.2. These special cases vanish, however, if we replace GM intersection homology with non-GM intersection homology. In this case, analogous arguments to those applied in Example 4.4.22, which can also be applied with arbitrary coefficients, yield

$$I^{\bar{p}}H_i(X^+;G) \cong \begin{cases} I^{\bar{p}}H_i(X,\partial X;G), & i > n-\bar{p}(\{v\})-1, \\ \mathrm{im}(I^{\bar{p}}H_i(X;G) \to I^{\bar{p}}H_i(X,\partial X;G)), & i = n-\bar{p}(\{v\})-1, \\ I^{\bar{p}}H_i(X;G), & i < n-\bar{p}(\{v\})-1, \end{cases}$$

for any perversity $\bar{p}$ on X^+. We also, once again, can take $X = M$ to be a trivially filtered manifold to get, for any perversity $\bar{p}$ on M^+,

$$I^{\bar{p}}H_i(M^+;G) \cong \begin{cases} H_i(M,\partial M;G), & i > n-\bar{p}(\{v\})-1, \\ \mathrm{im}(H_i(M;G) \to H_i(M,\partial M;G)), & i = n-\bar{p}(\{v\})-1, \\ H_i(M;G), & i < n-\bar{p}(\{v\})-1. \end{cases}$$

△

(c) Applications of Mayer–Vietoris Arguments

Using the $I^{\bar{p}}S_*$ incarnation of the non-GM intersection chain complex, the following version of Lemma 5.1.7 has an identical proof to that lemma and will be useful for Mayer–Vietoris arguments.

Lemma 6.3.16 *If X is a filtered space with perversity $\bar{p}$ and $\{U_\alpha\}$ is an increasing collection of open subspaces of X then the natural map $f : \varinjlim_\alpha I^{\bar{p}}H_*(U_\alpha;G) \to I^{\bar{p}}H_*\left(\bigcup_\alpha U_\alpha;G\right)$ is an isomorphism.*

Using Lemma 6.3.16, one can apply Mayer–Vietoris arguments for non-GM intersection homology to prove results such as the following generalization of Proposition 5.1.11 concerning normalization:

Proposition 6.3.17 *Let X be a stratified pseudomanifold, and let $\pi : \tilde{X} \to X$ be a normalization*[4]*. Suppose $\bar{p}$ is a perversity on X, and let $\tilde{p}$ be the perversity on $\tilde{X}$ defined so that $\tilde{p}(\tilde{S}) = \bar{p}(S)$ if $\pi(\tilde{S}) \subset S$. Then the map $\pi : I^{\tilde{p}}H_*(\tilde{X};G) \to I^{\bar{p}}H_*(X;G)$ is an isomorphism.*

[4] Recall Section 2.6.

The proof, using a Mayer–Vietoris argument, is the same as that of Proposition 5.1.11, using Lemma 6.3.16 and the non-GM cone formula (Theorem 6.2.13). In this case, we do not need to limit the perversities as we did for Proposition 5.1.11.

Our first application of Mayer–Vietoris arguments in Section 5.1.1 showed that $I^{\bar{p}}H_*^{\mathrm{GM}}(X) \cong H_*(X)$ when X is a normal stratified pseudomanifold and $\bar{p} \geq \bar{t}$. That result won't be true for non-GM intersection homology; rather one can prove the slightly less interesting result that if X is a stratified pseudomanifold then $I^{\bar{p}}H_*(X;G) \cong H_*(X,\Sigma_X;G)$ if $\bar{p} > \bar{t}$ (so in this case there is a stronger perversity requirement). The argument is the same as for Proposition 5.1.8 and Corollary 5.1.9, applying the Mayer–Vietoris argument of Theorem 5.1.4 together with the observation that the assumptions imply $I^{\bar{p}}H_*(cL;G) = H_*(cL,\Sigma_{cL};G) = 0$ when $L \neq \emptyset$ is a link of a singular stratum of X.

Mayer–Vietoris arguments also let us quickly prove the following related result that will be useful in our very brief discussion of local coefficients in Section 6.3.3. This proposition says that increasing the value of a perversity on a singular stratum S past $\bar{t}(S)+1 = \operatorname{codim}(S) - 1$ does not further alter the intersection homology groups. Note that the result would be trivial if we instead used $\bar{t}(S)+2 = \operatorname{codim}(S)$, as for this perversity value and above the allowability condition for S is always satisfied.

Proposition 6.3.18 *Let X be a CS set, and let $\bar{p}$ be a perversity on X. Define the perversity $\hat{p}$ on X so that for each singular stratum S we have*

$$\hat{p}(S) = \min\{\bar{p}(S), \operatorname{codim}(S) - 1\}.$$

Then the inclusion $I^{\hat{p}}S_*(X;G) \hookrightarrow I^{\bar{p}}S_*(X;G)$ induces an isomorphism $I^{\hat{p}}H_*(X;G) \cong I^{\bar{p}}S_*(X;G)$.

Proof The proof is by Mayer–Vietoris argument, Theorem 5.1.4. The only condition that does not follow immediately from the established properties of non-GM intersection homology is the condition about distinguished neighborhoods. But via stratified homotopy invariance, the hypothesis there amounts to the assumption that the proposition holds for the link. In this case, if L is the link of a stratum S for which $\bar{p}(S) = \hat{p}(S)$ then $I^{\bar{p}}H_*(cL;G) \cong I^{\hat{p}}H(cL;G)$ thanks to the cone formula (Theorem 6.2.13). Otherwise, we must have $\bar{p}(S) \geq \operatorname{codim}(S) - 1$ and then $I^{\bar{p}}H_*(cL;G) = I^{\hat{p}}H(cL;G) = 0$, again by the cone formula. The conclusion follows from Theorem 5.1.4. □

We call the perversity $\hat{p}$ of the proposition an *efficient* perversity.

(d) Cross Products

Next we turn to cross products and a version of Theorem 5.3.6, the Künneth Theorem with one factor a manifold. Cross products can be described equally well in the $I^{\bar{p}}S_*$ or $I^{\bar{p}}S'_*$ incarnations of non-GM intersection chains. We begin with the former.

If R is a commutative ring with unity, then we know from Lemma 5.2.4 and the discussion preceding Theorem 5.3.6 that the classical cross product

$$S_*(X;R) \otimes_R S_*(Y;R) \to S_*(X \times Y;R)$$

restricts to a cross product

$$I^{\bar{p}}S_*^{\mathrm{GM}}(X;R) \otimes_R I^{\bar{q}}S_*^{\mathrm{GM}}(Y;R) \to I^{Q}S_*^{\mathrm{GM}}(X \times Y;R),$$

if $Q(S \times T) \geq \bar{p}(S) + \bar{q}(T)$ for all strata $S \subset X$ and $T \subset Y$. We claim that similarly we have a cross product

$$I^{\bar{p}}S_*(X;R) \otimes_R I^{\bar{q}}S_*(Y;R) \to I^{Q}S_*(X \times Y;R).$$

As $I^{\bar{p}}S_i(X;R) \subset S_i(X;R)$ for all i, and similarly for Y, this cross product is certainly defined. We must check that it remains a chain map. For this, let $x \in I^{\bar{p}}S_i(X;R)$ and $y \in I^{\bar{q}}S_j(Y;R)$. This implies that no simplex of x or y is contained in Σ_X or Σ_Y, respectively. Suppose $\partial x = \hat{\partial} x + \xi$ and $\partial y = \hat{\partial} y + \eta$, where $|\xi| \subset \Sigma_X$ and $|\eta| \subset \Sigma_Y$. Then since the cross product is a chain map on ordinary chains, we have

$$\begin{aligned}
\partial(x \times y) &= (\partial x) \times y + (-1)^i x \times (\partial y) \\
&= (\hat{\partial} x + \xi) \times y + (-1)^i x \times (\hat{\partial} y + \eta) \\
&= (\hat{\partial} x) \times y + \xi \times y + (-1)^i x \times (\hat{\partial} y) + (-1)^i x \times \eta.
\end{aligned}$$

Since $|\xi| \subset \Sigma_X$ and $|\eta| \subset \Sigma_Y$, we have $|x \times \eta| \subset X \times \Sigma_Y \subset \Sigma_{X \times Y}$ and $|\xi \times y| \subset \Sigma_X \times Y \subset \Sigma_{X \times Y}$. Since no simplex of x, y, $\hat{\partial} x$, or $\hat{\partial} y$ is contained in Σ_X or Σ_Y, no simplex of $(\hat{\partial} x) \times y$ or $x \times (\hat{\partial} y)$ will be contained in $\Sigma_{X \times Y}$. This last fact can be seen by observing that in the singular subdivisions of $\Delta^i \times \Delta^j$ by shuffles used in the construction of the cross product each $(i + j)$-simplex of the triangulation of $\Delta^i \times \Delta^j$ projects onto both Δ^i and Δ^j by the standard projections; this is an immediate consequence of the definition of the shuffle product in Section 5.2. Therefore,

$$\hat{\partial}(x \times y) = (\hat{\partial} x) \times y + (-1)^i x \times (\hat{\partial} y),$$

showing that the cross product is indeed a chain map

$$I^{\bar{p}}S_*(X;R) \otimes_R I^{\bar{q}}S_*(Y;R) \to I^{Q}S_*(X \times Y;R)$$

if $Q(S \times T) \geq \bar{p}(S) + \bar{q}(T)$ for all strata $S \subset X$ and $T \subset Y$.

Analogously, if x and y are chains representing elements of $I^{\bar{p}}S'_i(X;R)$ and $I^{\bar{q}}S'_j(Y;R)$, respectively, then we can write $x = a_x + b_x$ and $y = a_y + b_y$ with $a_x \in A^{\bar{p}}S_i(X;R)$, $b_x \in S_i(\Sigma_X;R)$, $a_y \in A^{\bar{p}}S_j(Y;R)$, and $b_y \in S_j(\Sigma_Y;R)$. Then

$$x \times y = a_x \times a_y + b_x \times a_y + a_x \times b_y + b_x \times b_y,$$

and the first term on the right is in $A^{Q}S_{i+j}(X \times Y;R)$ while the others are in $S_{i+j}(\Sigma_{X \times Y};R)$. As each of ∂x and ∂y decomposes into allowable simplices and simplices in Σ_X or Σ_Y, the boundary $\partial(x \times y)$ decomposes similarly. Hence the cross product also is well defined $I^{\bar{p}}S'_*(X;R) \otimes_R I^{\bar{q}}S'_*(Y;R) \to I^{Q}S'_*(X \times Y;R)$.

These cross products extend to relative cross products

$$I^{\bar{p}}S_*(X,A;R)\otimes I^{\bar{q}}S_*(Y,B;R)\to I^{Q}S_*(X\times Y,(A\times Y)\cup(X\times B);R)$$

by the arguments of Corollary 5.2.5.

Similarly, following the development of Section 5.2.2, we obtain a non-GM simplicial cross product and a non-GM PL cross product

$$I^{\bar{p}}\mathfrak{C}_*(X,A;R)\otimes I^{\bar{q}}\mathfrak{C}_*(Y,B;R)\to I^{Q}\mathfrak{C}_*(X\times Y,(A\times Y)\cup(X\times B);R)$$

based upon the ordinary singular cross product via our Useful Lemma, Lemma 3.3.10. The argument that such PL products and their boundaries behave as desired is the same as that above for the singular product. Note, however, that when we apply Lemma 3.3.10 to represent an i-chain ξ as a homology class, it is the usual boundary ∂ (and not $\hat{\partial}$) that appears in the expression $H_i(|\xi|,|\partial\xi|)$.

We also have compatibility among the simplicial, singular, and PL cross products as in Corollaries 5.2.15 and 5.2.16. We will first need to discuss the relationship between these three types of homology more generally, below, so we postpone these results to Corollaries 6.3.36 and 6.3.37.

Recall by Lemma 6.3.1 that if R is a Dedekind domain then each $I^{\bar{p}}S_i(X,A;R)$ is a projective R-module and each $I^{\bar{p}}\mathfrak{C}_i(X,A;R)$ is a flat module. In this case, the properties of the cross product established in Section 5.2.3 hold for non-GM intersection homology $I\bar{H}_*$. The proofs are the same, although we note that in the non-GM analogues of the stability properties, Propositions 5.2.23 and 5.2.24, the boundary maps of the long exact sequences of the pair are determined on chains by applying the boundary maps of $I^{\bar{p}}S_*(X,A;R)$ or $I^{\bar{p}}S'_*(X,A;R)$ (or their PL versions), as appropriate. We state this all as a theorem:

Theorem 6.3.19 *If R is a Dedekind domain, then the properties of the cross product established in Section 5.2.3, including naturality, associativity, commutativity, unitality, and stability, hold for non-GM singular intersection homology with coefficients in R.*

The non-GM cross product yields the following Künneth Theorem:

Theorem 6.3.20 *Suppose X is a filtered space with perversity $\bar{p}_X$ and that M is an n-dimensional manifold with its trivial filtration. Filter $M\times X$ with the product filtration so that $(M\times X)^i=M\times X^{i-n}$, and define a perversity $\bar{p}$ on $M\times X$ whose value on the stratum $\mathcal{R}\times S$, for $\mathcal{R}$ a connected component of M, is $\bar{p}_X(S)$. Let R be a Dedekind domain. Then the cross product induces an isomorphism $H_*(S_*(M;R)\otimes I^{\bar{p}_X}S_*(X;R))\xrightarrow{\cong} I^{\bar{p}}H_*(M\times X;R)$ (or $H_*(S_*(M;R)\otimes I^{\bar{p}_X}S'_*(X;R))\xrightarrow{\cong} I^{\bar{p}}H_*(M\times X;R)$). If X is a PL filtered space and M is a PL manifold, then the same conclusion holds replacing singular chains with PL chains. The relative version of this theorem (analogous to Corollary 5.2.26) also holds.*

Just as for Theorems 5.2.25 and 5.3.6, this is an application of the Mayer–Vietoris argument (Theorem 5.1.1) with $F_*(M)=H_*(S_*(M;R)\otimes_R I^{\bar{p}_X}S_*(X;R))$, $G_*(M)=I^{\bar{p}}H_*(M\times X;R)$, and natural transformation $F_*\to G_*$ induced by the cross product. The relative version follows just as in Corollary 5.2.26.

(e) Coefficients

Now we turn to local torsion conditions and the Universal Coefficient Theorem in non-GM intersection homology, beginning with a non-GM version of Definition 5.3.9.

Definition 6.3.21 Let X be a CS set, R a Dedekind domain, and M an R-module. We say that X is *locally* $(\bar p, R; M)$*-torsion-free* if for each point $x \in X$ and for each link L of x we have $I^{\bar p}H_{\dim(L)-\bar p(S)-1}(L;R) *_R M = 0$, where S is the stratum of X containing x and $*_R$ denotes the torsion product over R, i.e. $\operatorname{Tor}^1_R(\cdot,\cdot)$. If we only impose the condition for all points in a stratum $S \subset X$, we say that X is *locally* $(\bar p, R; M)$*-torsion-free along* S.

If $I^{\bar p}H_{\dim(L)-\bar p(S)-1}(L;R) *_R M = 0$ for all R-modules M, we simply say that X is *locally* $(\bar p, R)$*-torsion-free*; this is equivalent to asking that $I^{\bar p}H_{\dim(L)-\bar p(S)-1}(L;R)$ be flat as an R-module by [147, Theorem XVI.3.11]. In particular, this means that $I^{\bar p}H_{\dim(L)-\bar p(S)-1}(L)$ is torsion-free (as an R-module) by [146, Proposition 4.20].

As noted in Remark 5.3.11, we do not define a separate PL version of the locally torsion-free condition as the fact that PL and singular intersection homology are isomorphic for PL CS sets makes it unnecessary. A locally $(\bar p, R; M)$-torsion-free PL CS set is simply a CS set that is both a PL filtered space and locally $(\bar p, R; M)$-torsion-free.

Example 6.3.22 As in Example 5.3.12, if X is a CS set with a perversity $\bar p$ such that $\bar p \geq \bar t$, then X is locally $(\bar p, R)$-torsion-free for any Dedekind domain R. This is because in this case $\dim(L) - \bar p(S) - 1 \leq 0$, and $I^{\bar p}H_0(L;R) \cong I^{\bar p}H_0^{\mathrm{GM}}(L;R)$ is always free. △

As for the GM locally torsion-free conditions, the non-GM condition does not depend on the choice of distinguished neighborhood by the following versions of Lemma 5.3.13 and Corollary 5.3.14.

Lemma 6.3.23 *Let X be a CS set and $x \in X$. For $i = 1, 2$, let $N_i \cong \mathbb{R}^k \times cL_i$ be distinguished neighborhoods of x. Then $I^{\bar p}H_*(L_1) \cong I^{\bar p}H_*(L_2)$ and $I^{\bar p}H_*(N_1) \cong I^{\bar p}H_*(N_2)$.*

Corollary 6.3.24 *Let X be a CS set. Then the intersection homology $I^{\bar p}H_*(L)$ of a link L of a point x in a stratum of S depends only on S. In other words, all links for any distinguished neighborhoods of any points in S have isomorphic intersection homology groups.*

The proofs are the same as for Lemma 6.3.23 and Corollary 6.3.24.

We then have a Universal Coefficient Theorem proven in the same way as Theorem 5.3.15; see also Remark 5.3.16.

Theorem 6.3.25 *Suppose X is a locally $(\bar p, R; M)$-torsion-free CS set for a Dedekind domain R and R-module M. Then $I^{\bar p}H_*(X;M) \cong H_*(I^{\bar p}S_*(X;R) \otimes_R M)$. If X is also PL then $I^{\bar p}\mathfrak{H}_*(X;M) \cong H_*(I^{\bar p}\mathfrak{C}_*(X;R) \otimes_R M)$. In particular, if G is an abelian group and X is a locally $(\bar p, \mathbb{Z}; G)$-torsion-free CS set, then $I^{\bar p}H_*(X;G) \cong H_*(I^{\bar p}S_*(X) \otimes G)$, and if X is also PL then $I^{\bar p}\mathfrak{H}_*(X;G) \cong H_*(I^{\bar p}\mathfrak{C}_*(X) \otimes G)$.*

Corollary 6.3.26 *For any CS set and any field F of characteristic zero, we have $I^{\bar p}H_*(X;F) \cong I^{\bar p}H_*(X) \otimes_{\mathbb{Z}} F$. If X is also PL then $I^{\bar p}\mathfrak{H}_*(X;F) \cong I^{\bar p}\mathfrak{H}_*(X) \otimes_{\mathbb{Z}} F$.*

Corollary 6.3.27 *Suppose X is a locally $(\bar{p}, R; M)$-torsion-free CS set for a Dedekind domain R and R-module M and that $A \subset X$ is also a locally $(\bar{p}, R; M)$-torsion-free CS set in particular if A is an open subset of X. Then $I^{\bar{p}}H_*(X, A; M) \cong H_*(I^{\bar{p}}S_*(X, A; R) \otimes_R M)$ If X is PL and A is a PL subset, then $I^{\bar{p}}\mathfrak{H}_*(X, A; G) \cong H_*(I^{\bar{p}}\mathfrak{C}_*(X, A) \otimes G)$.*

Example 6.3.28 The property of a space being locally torsion-free can depend on whether or not we are employing GM or non-GM intersection homology. Of course, if we use a perversity for which $\bar{p}(S) \leq \text{codim}(S) - 2$ then we know the two theories will be the same, but for a perversity that violates this condition, we can find spaces that are $\bar{p}$-torsion-free with respect to one type of intersection homology but not the other. Here we construct such an example.

Let $X = X^5 = S(\mathbb{R}P^2 \times S(S^1))$. Of course $S(S^1) \cong S^2$, but we give it the filtration induced by the suspension and label the suspension points $\{\mathfrak{n}, \mathfrak{s}\}$. We let $\mathbb{R}P^2$ have its trivial manifold filtration, give $\mathbb{R}P^2 \times S(S^1)$ the product filtration, and then we again stratify X itself as a suspension, with suspension points $\{\mathfrak{n}_0, \mathfrak{s}_0\}$. So ultimately there are three nontrivial skeleta:

$$X^0 = \{\mathfrak{n}_0, \mathfrak{s}_0\} \subset X^3 = S(\mathbb{R}P^2 \times \{\mathfrak{n}, \mathfrak{s}\}) \subset X^5 = X.$$

We will define a perversity $\bar{p}$ on X that depends only on codimension and such that $\bar{p}(2) = 1$. We will choose $\bar{p}(5)$ below. As usual, we use the same notation $\bar{p}$ for the restricted perversities on subspaces.

By the suspension computations, Theorem 4.4.21 and Theorem 6.3.13,

$$I^{\bar{p}}H_i^{\text{GM}}(S(S^1)) = \begin{cases} \mathbb{Z}, & i = 2, \\ 0, & i = 1, \\ \mathbb{Z}, & i = 0, \end{cases} \qquad I^{\bar{p}}H_i(S(S^1)) = \begin{cases} \mathbb{Z}, & i = 2, \\ \mathbb{Z}, & i = 1, \\ 0, & i = 0. \end{cases}$$

Next, let us apply the Künneth Theorems with one manifold factor (Theorems 5.2.25 and 6.3.20) to get

$$I^{\bar{p}}H_i^{\text{GM}}(\mathbb{R}P^2 \times S(S^1)) = \begin{cases} 0, & i = 4, \\ \mathbb{Z}_2, & i = 3, \\ \mathbb{Z}, & i = 2, \\ \mathbb{Z}_2, & i = 1, \\ \mathbb{Z}, & i = 0, \end{cases} \qquad I^{\bar{p}}H_i(\mathbb{R}P^2 \times S(S^1)) = \begin{cases} 0, & i = 4, \\ \mathbb{Z}_2, & i = 3, \\ \mathbb{Z} \oplus \mathbb{Z}_2, & i = 2, \\ \mathbb{Z}, & i = 1, \\ 0, & i = 0. \end{cases}$$

But now $\mathbb{R}P^2 \times S(S^1)$ is the link of the suspension points of X, and the locally torsion-free condition says that the intersection homology of these links must be torsion-free in degree $\dim(L) - \bar{p}(5) - 1 = 3 - \bar{p}(5)$. So then if we take $\bar{p}(5) = 1$, then X is locally $(\bar{p}, \mathbb{Z})^{\text{GM}}$-torsion-free but not locally $(\bar{p}, \mathbb{Z})$-torsion-free. And if we take $\bar{p}(5) = 2$, then X is locally $(\bar{p}, \mathbb{Z})$-torsion-free but not locally $(\bar{p}, \mathbb{Z})^{\text{GM}}$-torsion-free.

We could come up with orientable examples similarly using odd-dimensional projective spaces. △

(f) Agreement of Singular and PL Intersection Homology

All the material of Section 5.4 also carries over to provide an equivalence between PL and singular intersection homology on PL spaces.

We first note that the basic comparison maps are still well defined, generalizing Corollary 4.4.6:

Lemma 6.3.29 *Suppose that K is a* filtered *simplicial complex, i.e. $|K|$ is filtered such that each skeleton of the filtration is a subcomplex of K. Suppose further that K possesses a partial ordering on its vertices that restricts to a total ordering on each simplex. Let $\bar{p}$ be a perversity on $|K|$. Then the simplicial-to-singular chain map ϕ of Proposition 4.4.5 restricts to a chain map $\phi : I^{\bar{p}}C'_*(K;G) \to I^{\bar{p}}S'_*(|K|;G)$. Consequently, ϕ also induces a simplicial-to-singular chain map $\phi : I^{\bar{p}}C_*(K;G) \to I^{\bar{p}}S_*(|K|;G)$.*

Proof Adding in the coefficient groups, which do not alter the proofs of Proposition 4.4.5 or Corollary 4.4.6, the map $\phi : C_*(K;G) \to S_*(|K|;G)$ clearly takes simplices supported in Σ to simplices supported in Σ by definition, and the proof of Corollary 4.4.6 shows that it takes allowable simplices to allowable singular simplices. As $\phi : C_*(K;G) \to S_*(|K|;G)$ is a chain map, this is sufficient to show that ϕ induces a well-defined chain map $\phi : I^{\bar{p}}C'_*(K;G) \to I^{\bar{p}}S'_*(|K|;G)$. The last statement follows from the isomorphisms of Lemma 6.2.5. □

Given such comparison maps, the major results of Section 5.4 also hold in the non-GM setting. The proofs require also developing non-GM versions of the results in Section 3.3.4. These are left to the reader, noting that all subdivision maps, their inverses, and all of our prism constructions preserve allowability and take chains in Σ to chains in Σ. One arrives at the following non-GM version of Theorem 3.3.20:

Theorem 6.3.30 *Suppose T is a full triangulation of a PL filtered space and that T' is any subdivision of T. Then the maps induced by subdivision $I^{\bar{p}}H^T_*(X;G) \to I^{\bar{p}}H^{T'}_*(X;G)$ are isomorphisms, as is the canonical map $I^{\bar{p}}H^T_*(X;G) \to I^{\bar{p}}\mathfrak{H}_*(X;G)$.*

With these tools, one can obtain the following versions of Theorem 5.4.2 and Corollaries 5.4.3, 5.4.4, 5.4.5, and 5.4.6.

Theorem 6.3.31 *Let X be a PL CS set with triangulation T. Then the composition*[5]

$$I^{\bar{p}}\mathfrak{H}_*(W;G) \xrightarrow{\phi^{-1}} I^{\bar{p}}\mathfrak{H}^T_*(W;G) \xrightarrow{\psi} H_*(I^{\bar{p}}S_*(W;G))$$

is an isomorphism for any open set $W \subset X$. In particular, $I^{\bar{p}}\mathfrak{H}_*(X;G) \cong I^{\bar{p}}H_*(X;G)$.

Corollary 6.3.32 *Let X be a PL CS set, and let A be an open subset. Then $I^{\bar{p}}\mathfrak{H}_*(X,A;G) \cong I^{\bar{p}}H_*(X,A;G)$.*

[5] See Section 5.4 for the definition of these maps, which adapt readily to the non-GM setting.

Corollary 6.3.33 *Let X be a PL CS set with closed PL subset A such that A is itself a PL CS set in its inherited filtration. Then $I^{\bar p}\mathfrak{H}_*(X,A;G) \cong I^{\bar p}H_*(X,A;G)$.*

Corollary 6.3.34 *Suppose X is a PL ∂-stratified pseudomanifold. Then*

$$I^{\bar p}\mathfrak{H}_*(X;G) \xrightarrow{\cong} I^{\bar p}H_*^{\mathrm{GM}}(X;G).$$

Corollary 6.3.35 *Let X be a PL CS set or a PL ∂-stratified pseudomanifold, and let $T = (K,h)$ be a full triangulation of X compatible with the filtration and with an ordering on its vertices. Then the chain maps $\phi : I^{\bar p}C_*(K;G) \to I^{\bar p}S_*(X;G)$ and $\phi : I^{\bar p}C'_*(K;G) \to I^{\bar p}S'_*(X;G)$ of Lemma 6.3.29 induce isomorphisms on intersection homology.*

While we're discussing the relationships between simplicial, singular, and PL non-GM intersection homology, we note that we also have versions of Corollaries 5.2.15 and 5.2.16 relating the different types of cross products. We state the non-GM versions in a moment as Corollaries 6.3.36 and 6.3.37. For these, we already know that the non-GM products are well defined, as discussed earlier in this section starting on page 284. We have just seen in Lemma 6.3.29 that the simplicial-to-singular comparison maps carry over to the non-GM context, and the simplicial-to-PL maps were shown to exist in Lemma 6.2.4. So all the maps of the next two corollaries are known to be well-defined chain maps. The commutativity then follows from Corollary 5.2.11 and Lemma 5.2.3, respectively, as each non-GM intersection chain group is a subgroup of the corresponding ordinary chain group. Furthermore, none of the preceding arguments are disrupted by including coefficients in R.

Corollary 6.3.36 *Let X and Y be PL filtered spaces, and suppose K and L are simplicial complexes triangulating X and Y with partial orders on their vertices restricting to total orders on each simplex. Let $\bar p$, $\bar q$, and Q be respective perversities on X, Y, and $X \times Y$ such that $Q(S \times T) \geq \bar p(S) + \bar q(T)$ for all strata $S \subset X$ and $T \subset Y$. Then there is a commutative diagram of chain maps*

$$\begin{array}{ccc} I^{\bar p}C_*(K;R) \otimes I^{\bar q}C_*(L;R) & \xrightarrow{\bowtie} & I^{Q}C_*(K \times L;R) \\ \downarrow & & \downarrow \\ I^{\bar p}\mathfrak{C}_*(X;R) \otimes I^{\bar q}\mathfrak{C}_*(Y;R) & \xrightarrow{\times} & I^{Q}\mathfrak{C}_*(X \times Y;R), \end{array}$$

with the vertical maps the canonical ones induced by taking simplicial chains to the PL chains they represent.

Corollary 6.3.37 *Let X and Y be PL filtered spaces, and suppose K and L are simplicial complexes triangulating X and Y with partial orders on their vertices restricting to total orders on each simplex. Let $\bar p$, $\bar q$, and Q be respective perversities on X, Y, and $X \times Y$ such that $Q(S \times T) \geq \bar p(S) + \bar q(T)$ for all strata $S \subset X$ and $T \subset Y$. Then there is a commutative* *diagram of chain maps*

$$\begin{array}{ccc} I^{\bar{p}}C_*(K;R)\otimes I^{\bar{q}}C_*(L;R) & \xrightarrow{\bowtie} & I^{Q}C_*(K\times L;R) \\ \Big\downarrow{\scriptstyle \phi_K\otimes\phi_L} & & \Big\downarrow{\scriptstyle \phi_{K\times L}} \\ I^{\bar{p}}S_*(X;R)\otimes I^{\bar{q}}S_*(Y;R) & \xrightarrow{\varepsilon} & I^{Q}S_*(X\times Y;R), \end{array}$$

with the vertical maps being the simplicial-to-singular chain maps of Lemma 6.3.29.

(g) Finite Generation

Similarly, the results of Section 5.6 carry over, using the following definition. As noted in Remark 5.6.5, we expand here to finitely generated modules over Noetherian rings.

Definition 6.3.38 A CS set X is called *locally $(\bar{p},R;M)$-finitely generated* if R is a Noetherian ring, M is a finitely generated R-module, and, for each point $x\in X$, there is a link L of X such that $I^{\bar{p}}H_i(L;M)$ is finitely generated as an R-module for each i. When $M=R$, we will simply say that X is *locally $(\bar{p},R)$-finitely generated.* If the CS set X is locally $(\bar{p},R)$-finitely generated for all $\bar{p}$, we will simply say that X is *locally R-finitely generated.*

Arguments completely analogous to those of Lemmas 5.3.13 and 6.3.23 show that this definition is equivalent to requiring that each $I^{\bar{p}}H_i(L;M)$ be finitely generated for *every* link L in X.

Then we obtain the following proposition, noting that the proof of Proposition 5.6.3 goes through with only minor modifications. We also note that the only properties of abelian groups utilized in the proof of Proposition 5.6.3 are those arising from them being modules over the Noetherian ring $\mathbb{Z}$.

Proposition 6.3.39 *Let R be a Noetherian ring, and suppose X is a locally $(\bar{p},R;M)$-finitely generated CS set. Suppose $U\subset W$ are open subsets of X, that $\bar{U}\subset W$, and that $\bar{U}$ is compact. Then the image of $I^{\bar{p}}H_i(U;M)$ in $I^{\bar{p}}H_i(W;M)$ is finitely generated. In particular, if X is compact, then each $I^{\bar{p}}H_i(X;M)$ is finitely generated.*

Corollary 6.3.40 *If R is a Noetherian ring, M a finitely generated R-module, and X a compact recursive CS set, in particular if X is a compact stratified pseudomanifold, then $I^{\bar{p}}H_i(X;M)$ is finitely generated for all i.*

6.3.2 Dimensional Homogeneity

The trade-off for non-GM intersection homology being the theory that will provide a more general Künneth Theorem and Poincaré duality is that it does not "see" lower-dimensional pieces of a space. More precisely, suppose X is a CS set, and let $X^{\bullet}$ denote the closure of the union of the regular strata of X. In other words, $X^\bullet$ is the union of the strata S of X such that there exists a regular stratum $\mathcal{R}$ with $S < \mathcal{R}$. We will show that $I^{\bar p}H_*(X;G) \cong I^{\bar p}H_*(X^\bullet;G)$. In fact, more is true: the intersection homology of the closures of the regular strata do not interact. More precisely, we will see that if $\{\mathcal{R}_\alpha\}$ are the regular strata of X, then $I^{\bar p}H_*(X;G) \cong \bigoplus_\alpha I^{\bar p}H_*(\bar{\mathcal{R}}_\alpha;G)$ with the isomorphism being the sum of the maps induced by inclusion.

Example 6.3.41 Here are two examples illustrating the concept of $X^\bullet$.

1. Let $X = X^2$ be the one-point union (wedge product) of S^2 and S^1 with the filtration $\{v\} \subset S^1 \subset X$, where $\{v\}$ is the wedge point. Then $X^\bullet \cong S^2$ with filtration $\{v\} \subset S^2$.
2. If X is a CS set with formal dimension n but every non-empty stratum of X has dimension $< n$, then X has no regular strata and $X^\bullet = \emptyset$. For example, if $X = S^2$ but is given formal dimension six, then $X^\bullet = \emptyset$.

△

If X is a PL filtered space, the fact that $I^{\bar p}\mathfrak{H}_*(X;G) \cong I^{\bar p}\mathfrak{H}_*(X^\bullet;G)$ is not difficult to see. In fact, consider the inclusion map $I^{\bar p}\mathfrak{C}_*(X^\bullet;G) \to I^{\bar p}\mathfrak{C}_*(X;G)$. Since $X^\bullet$ is a closed union of strata of X, it is a subcomplex of any admissible triangulation of X compatible with the filtration. By definition, no simplex (with respect to some triangulation) σ of a chain in $I^{\bar p}\mathfrak{C}_*(X;G)$ can be contained in Σ_X, so every such simplex must have its interior in a regular stratum. This implies that σ is contained in $X^\bullet$. Thus the inclusion $I^{\bar p}\mathfrak{C}_*(X^\bullet;G) \to I^{\bar p}\mathfrak{C}_*(X;G)$ is also onto, so in fact $I^{\bar p}\mathfrak{C}_*(X^\bullet;G) = I^{\bar p}\mathfrak{C}_*(X;G)$ and then clearly $I^{\bar p}\mathfrak{H}_*(X^\bullet;G) = I^{\bar p}\mathfrak{H}_*(X;G)$. In fact, the same argument demonstrates that every allowable simplex must be contained in the closure of a single regular stratum $\mathcal{R}_\alpha$. So if $\xi \in I^{\bar p}\mathfrak{C}_i(X;G)$, there is a unique decomposition of ξ as $\xi = \sum_\alpha \xi_\alpha$, where ξ_α contains the simplices that intersect $\mathcal{R}_\alpha$. Furthermore, for each α, the boundary $\partial\xi_\alpha$ consists of simplices with interior in $\mathcal{R}_\alpha$ and simplices with interior in Σ. The simplices in Σ do not contribute to $\hat\partial$, while the simplices in $\mathcal{R}_\alpha$ must also be simplices of $\hat\partial\xi$, as $\mathcal{R}_\alpha \cap |\xi_\beta| = \emptyset$ for $\alpha \neq \beta$ and so no cancellations may occur among the different $\hat\partial\xi_\alpha$. As $\hat\partial\xi$ consists of allowable simplices, so then does each $\hat\partial\xi_\alpha$, and so $\xi_\alpha \in I^{\bar p}\mathfrak{C}_*(\bar{\mathcal{R}}_\alpha;G)$. Thus $I^{\bar p}\mathfrak{C}_*(X;G) = \bigoplus_\alpha I^{\bar p}\mathfrak{C}_*(\bar{\mathcal{R}}_\alpha;G)$ and $I^{\bar p}\mathfrak{H}_*(X;G) = \bigoplus_\alpha I^{\bar p}\mathfrak{H}_*(\bar{\mathcal{R}}_\alpha;G)$. We state this result as a lemma:

Lemma 6.3.42 *If X is a PL filtered space and $X^\bullet$ denotes the closure of the union of the regular strata of X, then $I^{\bar p}\mathfrak{C}_*(X^\bullet;G) = I^{\bar p}\mathfrak{C}_*(X;G)$ and so $I^{\bar p}\mathfrak{H}_*(X^\bullet;G) = I^{\bar p}\mathfrak{H}_*(X;G)$. Furthermore, if $\{\mathcal{R}_\alpha\}$ is the collection of regular strata of X, then $I^{\bar p}\mathfrak{C}_*(X;G) = \bigoplus_\alpha I^{\bar p}\mathfrak{C}_*(\bar{\mathcal{R}}_\alpha;G)$ and $I^{\bar p}\mathfrak{H}_*(X;G) = \bigoplus_\alpha I^{\bar p}\mathfrak{H}_*(\bar{\mathcal{R}}_\alpha;G)$.*

Example 6.3.43 Since the space in item (1) of Example 6.3.41 can be assumed to be PL, we see that $I^{\bar p}\mathfrak{H}_*(X;G) \cong I^{\bar p}\mathfrak{H}_*(S^2;G)$, where S^2 is filtered as above.

In both cases of item (2) of Example 6.3.41, $I^{\bar p}\mathfrak{H}_*(X;G) = 0$. △

The corresponding conclusion for singular intersection homology is more difficult to achieve. For one thing, a singular simplex that is not contained in Σ_X might nonetheless have its image intersect $X - X^\bullet$; for example, a singular 1-simplex might have both endpoints in regular strata of X but pass through strata of $X - X^\bullet$ in between. In order to draw upon local structure arguments, we must limit ourselves to CS sets, which will let us utilize Mayer–Vietoris arguments to get our desired results.

If X is a CS set (or, even more generally, a manifold stratified space) with n-dimensional regular strata, and we let $X^\bullet$ denote the closure of the union of the regular strata of X, then, by Proposition 2.2.20, $X^\bullet$ is itself a manifold stratified space whose strata comprise a subset of the strata of X. It turns out that if X is a CS set, then $X^\bullet$ is itself a CS set, which we will show below in Lemma 6.3.45. Furthermore, by definition, every point of $X^\bullet$ is contained in the closure of an n-dimensional stratum, but we can show something a bit more technical: if $X^\bullet \neq \emptyset$, each distinguished neighborhood in $X^\bullet$ is n-dimensional (using cohomological dimension as our dimension theory). This can be interpreted as a dimensional homogeneity property, akin to the sense in which an n-manifold is dimensionally homogeneous because every point has a neighborhood homeomorphic to n-dimensional Euclidean space. We will demonstrate this property below in Lemma 6.3.46. First, these lemmas motivate the following definition:

Definition 6.3.44 If X is a CS set, let $X^\bullet$ denote the CS set that is the closure of the union of the regular strata of X. We call $X^\bullet$ the *homogenization* of X. If $X^\bullet = X$, we say that X is *dimensionally homogeneous.*

After stating and proving Lemmas 6.3.45 and 6.3.46, we will show in Proposition 6.3.47 that if X is a CS set with homogenization $X^\bullet$, then $I^{\bar{p}}H_*(X;G) \cong I^{\bar{p}}H_*(X^\bullet;G)$.

Lemma 6.3.45 *If X is a CS set and $\mathfrak{R}$ is any union of regular strata of X, then the closure $\bar{\mathfrak{R}}$ is a CS set. In particular, the homogenization $X^\bullet$ is a CS set. If X is a stratified pseudomanifold, then so is $\bar{\mathfrak{R}}$.*

Proof The statement concerning $X^\bullet$ follows from the first statement by letting $\mathfrak{R}$ be the union of all the regular strata of X.

Since $\bar{\mathfrak{R}}$ is a union of strata of X by Proposition 2.2.20, $\bar{\mathfrak{R}}$ is automatically manifold stratified. So we must check that $\bar{\mathfrak{R}}$ is locally cone-like: Suppose $x \in \bar{\mathfrak{R}}$ has distinguished neighborhood N in X filtered homeomorphic to $\mathbb{R}^i \times cL$ by a filtered homeomorphism $h : N \to \mathbb{R}^i \times cL$. Since $\bar{\mathfrak{R}}$ is a union of strata of X, the set $N \cap \bar{\mathfrak{R}}$ is a union of strata of N and $h(N \cap \bar{\mathfrak{R}})$ must be a union of strata of $\mathbb{R}^i \times cL$. Since $x \in \bar{\mathfrak{R}}$, the entire stratum of N containing x must be in $\bar{\mathfrak{R}}$, so $h(N \cap \bar{\mathfrak{R}})$ must contain $\mathbb{R}^i \times \{v\}$; it now follows from the structure of the filtration of $\mathbb{R}^i \times cL$ that $h(N \cap \bar{\mathfrak{R}})$ must have the form $\mathbb{R}^i \times c\hat{L}$, where $\hat{L}$ is a union of strata of L. If we identify L with $\{0\} \times \{1/2\} \times L \subset \mathbb{R}^i \times cL$, then $\hat{L}$ is the intersection of L with $h(N \cap \bar{\mathfrak{R}})$ in $\mathbb{R}^i \times cL$. Since $\bar{\mathfrak{R}}$ is closed in X, the set $N \cap \bar{\mathfrak{R}}$ is closed in N and $h(N \cap \bar{\mathfrak{R}})$ is closed in $\mathbb{R}^i \times cL$; therefore, since L is compact, $\hat{L} = L \cap h(N \cap \bar{\mathfrak{R}})$ must also be compact. So $\bar{\mathfrak{R}}$ is a CS set.

Note that, in general, $\hat{L}$ is not required to be a CS set, as L itself is not required to be a CS set. However, suppose now that X is a stratified pseudomanifold, which implies that L is a stratified pseudomanifold by Lemma 2.4.11. Our subspace $\bar{\mathfrak{R}}$ is the closure of its regular strata, by definition, so we only need to show that the links $\hat{L}$ are also stratified pseudomanifolds. But a point-set argument shows that $\hat{L}$ is itself the closure of the union of the regular strata of L that h^{-1} takes into $\mathfrak{R}$. So it suffices to observe that the argument of the preceding paragraph holds replacing X with L, which has smaller depth than X. Repeating this argument iteratively, eventually we get to links that have depth zero, i.e. they're manifolds, and in this case it's clear that the closure of a union of connected components of a manifold is a manifold. □

We now state and prove Lemma 6.3.46. It shows that if X is an n-dimensional CS set and $X^\bullet$ is non-empty, then every distinguished neighborhood of $X^\bullet$ has topological dimension n. This justifies our claim that $X^\bullet$ is dimensionally homogeneous. This lemma is technical and uses some sheaf theory, including several citations to results in [37] that we will not explain in detail. Except for one later lemma, Lemma 8.1.9, the argument here will not be needed again and can be safely skipped by those not wishing to think too much about sheaf-theoretic dimension theory.

Lemma 6.3.46 *Suppose X is an n-dimensional CS set and that $\mathfrak{R}$ is a non-empty union of regular strata of X. Then every distinguished neighborhood N in $\bar{\mathfrak{R}}$ has* $\dim_{\mathbb{Z}} N = n$, *where* $\dim_{\mathbb{Z}} N$ *is the cohomological dimension with $\mathbb{Z}$ coefficients (see [37, Definition II.16.6]). In particular, if X has a non-empty regular stratum, then every distinguished neighborhood N in $X^\bullet$ has* $\dim_{\mathbb{Z}} N = n$.

Proof This argument is a straightforward generalization of that appearing for pseudomanifolds in [100, Proposition 7.3], which was due to Jim McClure.

We saw in Lemma 2.3.8 that the stratification of every CS set is locally finite, and so if $N \cong \mathbb{R}^i \times cL$ then L is a compact filtered space with finitely many strata. We also observe that since any open subset of a CS set is also a CS set, N is locally compact (every point has a neighborhood homeomorphic to $D^i \times \bar{c}L$, where D^i is the closed disk and $\bar{c}L$ is the closed cone on the compact space L), as is each skeleton of N, being a closed subset of a locally compact Hausdorff space [181, Corollary 29.3].

Since $N \subset \bar{\mathfrak{R}}$, we must have N intersect a regular stratum of X, and thus N must contain an open subspace M homeomorphic to a (non-empty) n-manifold. By [37, Corollary II.16.28], we have $\dim_{\mathbb{Z}}(M) = n$, and by [37, Theorem II.16.8], $\dim_{\mathbb{Z}}(N) \geq \dim_{\mathbb{Z}}(M) = n$ (here we use that N is locally compact and hence locally paracompact).

Next, we will show that $\dim_{\mathbb{Z}}(N) \leq \dim_{\mathbb{Z}}(M) = n$. In fact, we will show that if Y is any CS set, then the i-skeleton Y^i of Y satisfies $\dim_{\mathbb{Z}}(Y^i) \leq i$. As $N = N^n$, this will suffice. The proof will be by induction on i. In the case $i = 0$, the space Y^0 is a discrete collection of points, and so a 0-manifold, and we can again apply [37, Corollary II.16.28]. We now assume by induction hypothesis that $\dim_{\mathbb{Z}}(Y^{i-1}) \leq i-1$. If $Y^i - Y^{i-1}$ is empty, then we have $\dim_{\mathbb{Z}}(Y^i) = \dim_{\mathbb{Z}}(Y^{i-1}) \leq i-1$ by hypothesis, so we assume that $Y^i - Y^{i-1} \neq \emptyset$.

Let c denote the family of compact supports, and let $\dim_{c,\mathbb{Z}}$ be as before (see page 53). Then $\dim_{\mathbb{Z}}(Z)$ is equal to $\dim_{c,\mathbb{Z}}(Z)$ for any locally compact space Z by [37, Definition II.16.6], since c is paracompactifying for locally compact spaces (see [37, page 22]) and the extent $E(c)$ (i.e. the union of the elements of c; see [37, page 22]) is equal to Z, as every point has a compact neighborhood.

Now, by [181, Corollary 29.3], since Y^{i-1} and Y^i are closed in Y, they are each locally compact, and since $Y^i - Y^{i-1}$ is open in Y^i, we see $Y^i - Y^{i-1}$ is also locally compact. Thus, letting $c|A$ denote the compact sets in A, we have an equality

$$\dim_{c,\mathbb{Z}}(Y^i) = \max\{\dim_{c|Y^{i-1},\mathbb{Z}}(Y^{i-1}), \dim_{c|Y^i - Y^{i-1},\mathbb{Z}}(Y^i - Y^{i-1})\}$$

by [37, Exercise II.11, see also the solution on page 461 therein], utilizing again that c is paracompactifying on locally compact spaces. And using the discussion of the preceding paragraph, this becomes

$$\dim_{\mathbb{Z}}(Y^i) = \max\{\dim_{\mathbb{Z}}(Y^{i-1}), \dim_{\mathbb{Z}}(Y^i - Y^{i-1})\}.$$

We have assumed that $\dim_{\mathbb{Z}}(Y^{i-1}) \leq i - 1$ by induction, and, since $Y^i - Y^{i-1}$ is an i-manifold, we know $\dim_{\mathbb{Z}}(Y^i - Y^{i-1}) \leq i$ by [37, Corollary II.16.28]. Altogether, this shows that $\dim_{\mathbb{Z}}(Y^i) \leq i$, as desired. □

We are now ready for the main result of this section:

Proposition 6.3.47 *Let X be a CS set, and let $\{\mathcal{R}_\alpha\}$ be the regular strata of X. Then the inclusion maps induce isomorphisms*

$$\begin{array}{ccc} \bigoplus_\alpha I^{\bar{p}}H_*(\bar{\mathcal{R}}_\alpha; G) & \longrightarrow & I^{\bar{p}}H_*(X; G) \\ & \searrow \qquad \nearrow & \\ & I^{\bar{p}}H_*(X^\bullet; G). & \end{array}$$

The analogous result holds in the PL category.

Remark 6.3.48 As $I^{\bar{p}}H = I^{\bar{p}}H^{\mathrm{GM}}$ if $\bar{p} \leq \bar{t}$ by Proposition 6.2.9, Proposition 6.3.47 holds for GM intersection homology, as well, under this perversity restriction. However, Proposition 6.3.47 is not true of GM intersection homology in general. For example, suppose X has more than one regular stratum and that $\bar{p}$ is large enough on some stratum to allow a singular 1-simplex to run from one regular stratum to another, say from $\mathcal{R}_1$ to $\mathcal{R}_2$. Then

$$I^{\bar{p}}H_0^{\mathrm{GM}}(\overline{\mathcal{R}_1 \cup \mathcal{R}_2}) \cong I^{\bar{p}}H_0^{\mathrm{GM}}(\bar{\mathcal{R}}_1) \cong I^{\bar{p}}H_0^{\mathrm{GM}}(\bar{\mathcal{R}}_2) \cong \mathbb{Z}.$$

Proof of Proposition 6.3.47 The PL claim was already demonstrated in Lemma 6.3.42.

For singular intersection homology, we will apply Theorem 5.1.4 to get the isomorphisms $\oplus_\alpha I^{\bar{p}}H_*(\bar{\mathcal{R}}_\alpha; G) \to I^{\bar{p}}H_*(X; G)$ and $I^{\bar{p}}H_*(X^\bullet; G) \to I^{\bar{p}}H_*(X; G)$. The third isomorphism follows from the evident commutativity of the diagram.

For each open set $U \subset X$, let $D_*(U) = I^{\bar{p}}H_*(U; G)$, let $F_*(U) = I^{\bar{p}}H_*(U \cap X^\bullet; G)$, and let $E_*(U) = \oplus_\alpha I^{\bar{p}}H_*(U \cap \bar{\mathcal{R}}_\alpha; G)$. Let $\Phi : F_* \to D_*$ be induced by inclusion, and let $\Psi : E_* \to D_*$ be the sum of the maps induced by inclusion of the $\bar{\mathcal{R}}_\alpha$. Of course, D_* admits the standard Mayer–Vietoris sequence by Theorem 6.3.12, and if U, V are two open sets of X, then $U \cap X^\bullet$ and $V \cap X^\bullet$ are two open subsets of $X^\bullet$ with $(U \cap X^\bullet) \cap (V \cap X^\bullet) = (U \cap V) \cap X^\bullet$ and $(U \cap X^\bullet) \cup (V \cap X^\bullet) = (U \cup V) \cap X^\bullet$, so F_* admits a Mayer–Vietoris sequence coming from the intersection homology on $X^\bullet$. Compatibility of the short exact Mayer–Vietoris sequences at the chain level shows that Φ induces a map of homology Mayer–Vietoris sequences. For E_*, we have the direct sum of the Mayer–Vietoris sequences similarly associated to $U \cap \bar{\mathcal{R}}_\alpha$ and $V \cap \bar{\mathcal{R}}_\alpha$, and the direct sum of similarly associated maps to the Mayer–Vietoris sequence for D_*. This demonstrates condition (1) of Theorem 5.1.4.

Condition (2) of Theorem 5.1.4 follows from Lemma 5.1.6 using Lemma 6.3.16, applied on X for D_* and on $X^\bullet$ for F_*, noting that an increasing sequence of open subsets $\{U_\beta\}$ in X yields an increasing sequence of open subsets $\{U_\beta \cap X^\bullet\}$ in $X^\bullet$. Once again, the analogous argument is true for each summand of E_*, and so for the direct sum.

Next we look at condition (4) of Theorem 5.1.4. If U is empty, then $E_*(U) = F_*(U) = D_*(U) = 0$, trivially. If U is an open subset homeomorphic to Euclidean space and contained within a regular stratum of X, then U is contained in a single $\mathcal{R}_\alpha$, say $\mathcal{R}_0$, and $U \cap \mathcal{R}_0 = U \cap X^\bullet = U$, so Φ is certainly an isomorphism on such sets, as is Ψ because all summands will be zero except for $I^{\bar{p}}H_*(U \cap \bar{\mathcal{R}}_0; G) = I^{\bar{p}}H_*(U; G)$. If U is an open subset of X contained in any one singular stratum S, then we must have $S \cap \bar{\mathcal{R}}_\alpha = S \cap X^\bullet = \emptyset$ for all α, for otherwise S would have to be a stratum of some $\bar{\mathcal{R}}_\alpha$ (since we have seen that each $\bar{\mathcal{R}}_\alpha$, and so also $X^\bullet$, is a union of strata of X), in which case every neighborhood of every point of S would have to intersect some regular stratum of X, meaning that U could not be open. So, in this case, $U \cap \bar{\mathcal{R}}_\alpha = U \cap X^\bullet = \emptyset$ for all α and $E_*(U) = F_*(U) = 0$. But similarly, since $U \subset \Sigma_X$, we have $I^{\bar{p}}S_*(U; G) = 0$, as simplices of $I^{\bar{p}}S_*(X; G) = 0$ cannot be contained in Σ_X by definition, and therefore $D_*(U) = I^{\bar{p}}H_*(U; G) = 0$. Thus Φ and Ψ must again be isomorphisms in this case, trivially.

Finally, consider condition (3) of Theorem 5.1.4. Suppose we have an open subset U of X filtered homeomorphic to $\mathbb{R}^i \times cL$. It is convenient to identify U with $\mathbb{R}^i \times cL$ via the homeomorphism. Further suppose that $\Phi : F_*(\mathbb{R}^i \times (cL - \{v\})) \to D_*(\mathbb{R}^i \times (cL - \{v\}))$ and $\Psi : E_*(\mathbb{R}^i \times (cL - \{v\})) \to D_*(\mathbb{R}^i \times (cL - \{v\}))$ are isomorphisms (where v is the cone vertex). By definition, such an open set U is a distinguished neighborhood of all points of U contained in $\mathbb{R}^i \times \{v\}$. Let x be such a point in U, and first suppose $x \notin X^\bullet$, so also $x \notin \bar{\mathcal{R}}_\alpha$ for all α. Then the stratum containing x is not contained in $X^\bullet$ (or any $\bar{\mathcal{R}}_\alpha$), and, by the partial ordering on the strata of X and the fact that $X^\bullet$ (or $\bar{\mathcal{R}}_\alpha$) is closed, it follows that none of the strata that intersect U can be contained in $X^\bullet$ (or any $\bar{\mathcal{R}}_\alpha$), and therefore they do not intersect $X^\bullet$ (or $\bar{\mathcal{R}}_\alpha$). This means that $E_*(U) = F_*(U) = 0$, as $U \cap X^\bullet = U \cap \bar{\mathcal{R}}_\alpha = \emptyset$, but also $D_*(U) = 0$, as $U \cap X^\bullet = \emptyset$ implies that $U \subset \Sigma_X$. Thus Φ and Ψ are trivially isomorphisms on U.

Next, suppose that $x \in X^\bullet$, so the stratum of U containing the homeomorphic image of

$\mathbb{R}^i \times \{v\}$ is contained in $X^\bullet$. Consider the commutative diagram

$$\begin{array}{ccccc}
\bigoplus_\alpha I^{\bar{p}}H_*((\mathbb{R}^i_{\mathrm{A}}) \cap \bar{\mathcal{R}}_\alpha; G) & \xrightarrow{\cong} & I^{\bar{p}}H_*((\mathbb{R}^i_{\mathrm{A}}) \cap X^\bullet; G) & \xrightarrow{\cong} & I^{\bar{p}}H_*(\mathbb{R}^i_{\mathrm{A}}; G) \\
\downarrow & & \downarrow & & \downarrow \\
\bigoplus_\alpha I^{\bar{p}}H_*((\mathbb{R}^i_{\mathrm{B}}) \cap \bar{\mathcal{R}}_\alpha; G) & \longrightarrow & I^{\bar{p}}H_*((\mathbb{R}^i_{\mathrm{B}} \cap X^\bullet; G) & \longrightarrow & I^{\bar{p}}H_*(\mathbb{R}^i_{\mathrm{B}}; G),
\end{array}$$

where, for clarity, we have written $\mathbb{R}^i_{\mathrm{A}}$ for $\mathbb{R}^i \times (cL - \{v\})$, and $\mathbb{R}^i_{\mathrm{B}}$ for $\mathbb{R}^i \times cL$, and where the top maps are isomorphisms by assumption. By the arguments in the proof of Lemma 6.3.45, the space $(\mathbb{R}^i \times cL) \cap X^\bullet$ has the form $\mathbb{R}^i \times c\hat{L}$, where $\hat{L}$ is, roughly speaking, the intersection of L with $X^\bullet$. Similarly, each $\bar{\mathcal{R}}_\alpha$ that has a non-empty intersection with $\mathbb{R}^i \times cL$ intersects it in a space of the form $\mathbb{R}^i \times c\hat{L}_\alpha$. Of course it is possible for some α to have $\bar{\mathcal{R}}_\alpha \cap (\mathbb{R}^i \times cL) = \emptyset$, in which case the corresponding terms on the left side of the diagram are trivial.

Employing stratified homotopy invariance, the diagram is isomorphic to the diagram

$$\begin{array}{ccccc}
\bigoplus_\alpha I^{\bar{p}}H_*(\hat{L}_\alpha; G) & \xrightarrow{\cong} & I^{\bar{p}}H_*(\hat{L}; G) & \xrightarrow{\cong} & I^{\bar{p}}H_*(L; G) \\
\downarrow & & \downarrow & & \downarrow \\
\bigoplus_\alpha I^{\bar{p}}H_*(c\hat{L}_\alpha; G) & \longrightarrow & I^{\bar{p}}H_*(c\hat{L}; G) & \longrightarrow & I^{\bar{p}}H_*(cL; G).
\end{array}$$

But now we can employ the cone formula (Theorem 4.2.1). Recall that the cutoff dimension of the cone formula depends only on $\bar{p}(\{v\})$ and the codimension of the cone vertex. Since all strata inherit their formal dimensions from X, the codimension of the cone vertex in cL with the inherited filtration is the same as the codimension of the cone vertex in $c\hat{L}$, or in any $c\hat{L}_\alpha$, with the inherited filtrations. In particular, since the stratum containing x has dimension i (by our assumption that the distinguished neighborhood of x in X has the form $\mathbb{R}^i \times cL$), the cone vertex inherits codimension $n-i$ in cL, in $c\hat{L}$, and in each $c\hat{L}_\alpha$. The cone formula now says that in degrees $\geq n-i-1-\bar{p}(\{v\})$ the groups $I^{\bar{p}}H_*(cL;G)$, $I^{\bar{p}}H_*(c\hat{L};G)$, and all of the $I^{\bar{p}}H_*(c\hat{L}_\alpha;G)$ vanish, so the horizontal maps on the bottom of the diagram are trivially isomorphic. In degrees $< n-i-1-\bar{p}(\{v\})$, the other part of the cone formula tells us that all the vertical maps are isomorphisms (trivially so for the summands on the left for which $\bar{\mathcal{R}}_\alpha \cap (\mathbb{R}^i \times cL) = \emptyset$). Thus the bottom maps of the diagram are isomorphisms in this range as well. Therefore, $\Phi : F_*(U) \to D_*(U)$ and $\Psi : E_*(U) \to D_*(U)$ are isomorphisms in all degrees, and this establishes condition (3) of Theorem 5.1.4.

We have now verified all conditions of Theorem 5.1.4, which we can now invoke to conclude that Φ and Ψ are isomorphisms on X. □

The following corollary says that a version of the proposition carries down to the links, which we recall need not themselves be CS sets. This will be used below in the proof of Theorem 6.4.7. We leave the reader to formulate and prove a direct sum decomposition formula in this setting such as the one in the proposition, noting that the intersection of a regular stratum $\mathcal{R}_\alpha$ of X with an open set U may be the union of multiple disjoint regular strata of U.

Corollary 6.3.49 *Suppose X is a CS set and $x \in X^\bullet$. Let L be a link of x in X, and let $\hat{L}$ be the link of x in $X^\bullet$ as constructed in the proof of Lemma 6.3.45. Then $I^{\bar{p}}H_*(L;G) \cong I^{\bar{p}}H_*(\hat{L};G)$.*

Proof Let N be a distinguished neighborhood of x in X, which we identify with $\mathbb{R}^i \times cL$ via a filtered homeomorphism. Then, as shown in the proof of Lemma 6.3.45, we have $N \cap X^\bullet \cong \mathbb{R}^i \times c\hat{L}$ with $\hat{L} \subset L$ a union of strata of L. We claim that $(\mathbb{R}^i \times (cL - \{v\}))^\bullet \cong \mathbb{R}^i \times (c\hat{L} - \{v\})$. Then by the preceding proposition we have $I^{\bar{p}}H_*(\mathbb{R}^i \times (cL - \{v\}); G) \cong I^{\bar{p}}H_*(\mathbb{R}^i \times (c\hat{L} - \{v\}); G)$, which applying stratified homotopy invariance becomes $I^{\bar{p}}H_*(L;G) \cong I^{\bar{p}}H_*(\hat{L};G)$.

For the claim that $(\mathbb{R}^i \times (cL - \{v\}))^\bullet \cong \mathbb{R}^i \times (c\hat{L} - \{v\})$, we already know that $\mathbb{R}^i \times c\hat{L} \cong N \cap X^\bullet$. We also know that the regular strata of N are the connected components of the intersections of N with the regular strata of X. As every point in $\mathbb{R}^i \times c\hat{L}$ is in $X^\bullet$, each is in the closure of a regular stratum of X and so of a regular stratum of N, as N is open in X. Thus $\mathbb{R}^i \times (c\hat{L} - \{v\}) \subset (\mathbb{R}^i \times (cL - \{v\}))^\bullet$, noting that $\mathbb{R}^i \times (cL - \{v\})$ contains all the regular strata of N unless $L = \emptyset$, in which case the corollary holds vacuously. Conversely, if $x \in (\mathbb{R}^i \times (cL - \{v\}))^\bullet$, then x is in the closure of a regular stratum of $\mathbb{R}^i \times (cL - \{v\})$ and hence of a regular stratum of X. So $x \in X^\bullet$, which implies that $x \in \mathbb{R}^i \times c\hat{L}$. So $(\mathbb{R}^i \times (cL - \{v\}))^\bullet \subset \mathbb{R}^i \times (c\hat{L} - \{v\})$. We conclude $(\mathbb{R}^i \times (cL - \{v\}))^\bullet = \mathbb{R}^i \times (c\hat{L} - \{v\})$ as desired. □

To conclude this section, we provide one more lemma, which will be useful in the next section.

Lemma 6.3.50 *If X and Y are CS sets, then the product of the homogenizations $X^\bullet \times Y^\bullet$ is the homogenization $(X \times Y)^\bullet$ of $X \times Y$.*

Proof From the definition of the product stratification, the regular strata of $X \times Y$ are the products of the regular strata of X and Y. If either X or Y has no regular strata, then its corresponding homogenization is empty and so is the homogenization of $X \times Y$. So assume that all of the spaces have regular strata. If $x \in X^\bullet$ and $y \in Y^\bullet$, then every neighborhood U of x in X and every neighborhood V of y in Y intersect regular strata, so the product neighborhood $U \times V$ intersects a regular stratum of $X \times Y$. Since such neighborhoods are cofinal among neighborhoods of (x, y), the point (x, y) must be in $(X \times Y)^\bullet$.

Conversely, if $(x, y) \in (X \times Y)^\bullet$, then every neighborhood of (x, y) intersects some regular stratum of $X \times Y$, so in particular every neighborhood of the form $U \times V$ has this property, where U is a neighborhood of x in X and V is a neighborhood of $y \in Y$. But if $(u, v) \subset U \times V$

is contained in a regular stratum, so are $u \in U$ and $v \in V$. Therefore, x and y are contained respectively in the closures of regular strata of X and Y, so $(x, y) \in X^\bullet \times Y^\bullet$. □

6.3.3 Local coefficients

In this section, we pause to mention briefly an interesting feature of intersection homology theory that was observed already by Goresky and MacPherson in [105, Section 2.2] concerning intersection homology with local coefficients. Recall that ordinary homology with local coefficients over a space X can be defined by thinking of the coefficients of simplices as living not just in a group but in a bundle of groups over X, which allows for the coefficients to "twist" as we move around the space. The extra wrinkle that arises in intersection homology is that it is possible to define intersection homology with local coefficients even if the coefficient bundle is only defined over $X - \Sigma_X$. In particular, it need not extend to all of X.

To see the value of this, recall that if the underlying space of a classical stratified pseudomanifold X is actually a manifold and if $\bar{p}$ is a GM perversity, then $I^{\bar{p}}H_*^{\mathrm{GM}}(X) \cong H_*(X)$ by Theorem 5.5.1. But suppose now that $\mathcal{G}$ is a local coefficient system with fiber group G that is defined on $X - \Sigma$ but that does not extend to all of X. Then $H_*(X; \mathcal{G})$ is not defined, but $I^{\bar{p}}H_*^{\mathrm{GM}}(X; \mathcal{G})$ will be and it will not generally be isomorphic to $H_*(X; G)$. In this way, intersection homology gives us new stratification-dependent invariants of X, even when X is a manifold. This technique has been used, for example, to study versions of knot theory in [46, 82]. More generally, on both manifolds and pseudomanifolds, intersection homology with local coefficients allows for the definition of extended versions of many of the invariants to be studied below, including twisted signature and twisted L-classes [14, 20, 47]. In fact, intersection homology with local coefficients is ubiquitous, especially in the sheaf theoretic approach to intersection homology where it can be formulated more naturally into the basic definitions of the theory. See e.g. [28, Chapter V].

While intersection homology with local coefficients is important, we have chosen not to integrate it into our treatment throughout the book, which would have led to even more technical clutter. However, given the basics that we have developed, adding in local coefficients does not pose large technical challenges, and we leave it to the interested reader to fill in the necessary adaptations. Some of the details can be found in [81, Section 2] and [85, Section 2].

Turning to some specifics, we first provide a sketch of the needed background, referring the reader to [125, Additional Topic 3.H], [242, Chapter VI], or [67, Chapter 5] for more thorough treatments. Recall that there are (at least!) two ways to define homology with local coefficients on a space X. One way is principally algebraic: Let us suppose that X is path connected (or each component can be treated separately) with universal cover $\tilde{X}$, let $\pi = \pi_1(X)$, and let M be a left $\mathbb{Z}[\pi]$-module. Then the deck transformations of $\tilde{X}$ induce a right $\mathbb{Z}[\pi]$-module structure on $S_*(\tilde{X})$, and one can define the singular chain complex with coefficients in M to be $S_*(X; M) = S_*(\tilde{X}) \otimes_{\mathbb{Z}[\pi]} M$. While straightforward, this is not the definition that adapts most easily to intersection chains, and so we will turn to the more geometric construction. For some results and further applications on variants of intersection homology from this more algebraic perspective, see [98, 163].

Fixing an abelian group G, a *bundle of groups* (or *local system of coefficients*) $\mathcal{G}$ over X is a covering space $p : \mathcal{G} \to X$ such that all fibers $p^{-1}(x)$ are isomorphic to G. Each $x \in X$ is further assumed to have a neighborhood U and a homeomorphism $p^{-1}(U) \cong U \times G$ that takes each fiber over U isomorphically to G. In fancier language, $\mathcal{G}$ is a fiber bundle whose transition functions are automorphisms of G; the bundle $\mathcal{G}$ can also be considered to be a locally constant sheaf. Over a path connected and semi-locally simply connected space, the data of a bundle of groups is equivalent to the data of a group G with a $\pi_1(X)$ action (see [67, Lemma 4.7] or [242, Section VI.6]), and this is the link between the two definitions of homology with local coefficients.

Given a bundle of groups $p : \mathcal{G} \to X$ and a singular simplex $\sigma : \Delta^k \to X$, we define a coefficient of σ in $\mathcal{G}$ to be a lift $n : \Delta^k \to \mathcal{G}$, i.e. a map such that $pn = \sigma$. Since $\mathcal{G}$ is a covering space and Δ^k is contractible, to specify a coefficient for σ it is sufficient to specify a lift of σ over any point of Δ^k. Using the group structures on the fibers, it is possible to add coefficients fiberwise, and one can check that this yields a well-defined addition on coefficients. The singular chain complex $S_*(X;\mathcal{G})$ can then be defined to consist of the finite formal sums $\sum n_i\sigma_i$ with each n_i being a coefficient of σ_i. The addition on coefficients makes $S_*(X;\mathcal{G})$ into an abelian group in the evident way. Furthermore, if n_i is a coefficient of the singular k-simplex σ_i, then the restriction of n_i to any face τ of Δ^k is a coefficient for τ. This allows for the definition of boundary maps, and $S_*(X;\mathcal{G})$ becomes a chain complex with homology groups denoted $H_*(X;\mathcal{G})$, the *homology groups with coefficients in the local system* $\mathcal{G}$.

For intersection homology, we can define intersection chains with local coefficients over X just as above, limiting ourselves to chains composed of $\bar{p}$-allowable simplices whose boundaries are also composed of $\bar{p}$-allowable simplices. But, as noted above, what really makes this interesting is that, with minor restrictions on the perversities, this procedure makes sense even if $\mathcal{G}$ is only defined over $X - \Sigma$. In particular, we will assume that our perversities satisfy $\bar{p}(S) \leq \operatorname{codim}(S) - 1$ for each singular stratum S. As seen in Proposition 6.3.18, this is not really a limitation on the possible intersection homology groups with constant coefficients, but it will be useful here.

Consider a $\bar{p}$-allowable simplex $\sigma : \Delta^k \to X$. Since σ is allowable, for each singular stratum S we have that $\sigma^{-1}(S)$ is contained in the $(k - \operatorname{codim}(S) + \bar{p}(S) \leq k - 1)$-skeleton of Δ^k. In particular, σ takes the interior of Δ^k to $X - \Sigma$, and so specifying a lift to $\mathcal{G}$ of any point in the interior of Δ^k determines a lift of σ over all of $\sigma^{-1}(X - \Sigma)$, which is a contractible subset of Δ^k. We can let such a lift define a coefficient of σ. Notice that the allowability really is playing a role here, as, if $\mathcal{G}$ does not extend over Σ and if σ is a (not allowable) simplex with image in Σ, then we would not have a way to define a coefficient for Σ. Furthermore, even though the $k - 1$ faces of an allowable k-simplex may not be allowable, the restriction of a coefficient of σ to a $k - 1$ face τ does still give a partial lift of

τ over $\tau^{-1}(X - \Sigma)$, and these "partial coefficients" can still be used to compute boundaries. If $\tau^{-1}(X - \Sigma)$ is empty, then the image of τ is contained in Σ, and we can consider the coefficient to be 0 as in our first definition of non-GM intersection chains $I^{\bar{p}}S_*$. Thus, given a chain of allowable simplices and coefficients $\xi = \sum n_i\sigma_i$, the chain $\partial\xi$ is well defined from ξ as a sum over simplices with partial coefficient lifts, dropping the terms which go to zero because they come from boundary simplices contained in Σ. If all simplices of $\partial\xi$ are allowable, then each of those simplices will have a coefficient lift defined over its interior, and in this case we define ξ to be an *intersection chain with local coefficients in* $\mathcal{G}$. It is not hard to check that $\partial^2 = 0$, and we obtain a chain complex $I^{\bar{p}}S_*(X;\mathcal{G})$ and homology groups $I^{\bar{p}}H_*(X;\mathcal{G})$.

Given that we are willing to work with partial lifts, the reader might well wonder why we place any limitation on the perversities. This is probably not strictly necessary, but it does allow us to avoid having to work with some more serious pathologies. For example, if we allow perversities to be so high that they place no restrictions on the allowability of simplices, then it is possible for $\sigma^{-1}(X - \Sigma)$ to have infinite components: for example, let $X = \mathbb{R}$, let $\Sigma = \{0\}$, and let $\sigma : \Delta^1 = [0,1] \to X$ be given by $\sigma(t) = t\sin(1/t)$ for $t \neq 0$ and $\sigma(0) = 0$. Such simplices would have uncountably many partial coefficient lifts, even if the group G is finite. On the other hand, if $\sigma : \Delta^2 \to \mathbb{R}^2$ takes $\partial\Delta^2$ around the origin with non-zero winding number but $\mathcal{G}$ has nontrivial monodromy around the origin, then no continuous lifts of $\sigma|_{\sigma^{-1}(\mathbb{R}^2-\{0\})}$ exist except the one to the 0-element of G in each fiber and so such a σ cannot be a given any nontrivial coefficient at all.

Example 6.3.51 Let $X = S^2 = S(S^1)$, stratified as $\{\mathfrak{n},\mathfrak{s}\} \subset S^2$, with $\mathfrak{n},\mathfrak{s}$ being the suspension points. Let $\mathcal{G}$ be the bundle of coefficients over $S^2 - \{\mathfrak{n},\mathfrak{s}\} \cong (0,1) \times S^1$ with fiber $\mathbb{Z}$ but with nontrivial monodromy so that traveling along a generator of $\pi_1(S^2 - \{\mathfrak{n},\mathfrak{s}\}) \cong \pi_1(S^1)$ takes m in the fiber $\mathbb{Z}$ to $-m$. It is not difficult to verify that the basic properties of intersection homology continue to hold with local coefficients, and so by Theorem 6.3.13, if $\bar{p}(\{\mathfrak{n}\}) = \bar{p}(\{\mathfrak{s}\}) = p$, we have

$$I^{\bar{p}}H_i(X;\mathcal{G}) = \begin{cases} H_{i-1}(S^1;\mathcal{G}|_{S^1}), & i > 1-p, \\ 0, & i = 1-p, \\ H_i(S^1;\mathcal{G}|_{S^1}), & i < 1-p. \end{cases}$$

For example, if $p = 0$, we have $I^{\bar{p}}H_0(X;\mathcal{G}|_{S^1}) \cong H_0(S^1;\mathcal{G}|_{S^1}) \cong \mathbb{Z}_2$, while $I^{\bar{p}}H_2(X;\mathcal{G}|_{S^1}) \cong H_1(S^1;\mathcal{G}|_{S^1}) = 0$. These computations of $H_*(S^1;\mathcal{G}|_{S^1})$ can be done using CW homology with local coefficients [67, Chapter 5] by constructing S^1 as a 0-cell e_0 and a 1-cell e_1 and then observing that with these coefficients $\partial e_1 = 2e_0$.

Notice that such groups do not arise as intersection homology groups with any constant coefficients or from ordinary homology with local coefficients. In fact, as S^2 is simply connected, any coefficient system defined on the entire space must be constant. △

Example 6.3.52 As a more sophisticated example, let $X = S^n$, and let $K \subset S^n$ be a PL submanifold that is PL homeomorphic to S^{n-2}. Let Λ be the ring of rational Laurent polynomials $\Lambda = \mathbb{Q}[\mathbb{Z}] = \mathbb{Q}[t, t^{-1}]$, and let $\pi_1(S^n - K)$ act on Λ so that, if $g \in \Lambda$ and if a loop γ in $S^n - K$ has linking number ℓ with K then $\gamma(g) = t^\ell g$. This determines a bundle of coefficients Γ over $S^n - K$ with fiber Λ. If $\bar{p}$ is a GM perversity and K is locally flat, meaning that each point $x \in K$ has a neighborhood U in S^n such that $(U, U \cap K) \cong (\mathbb{R}^n, \mathbb{R}^{n-2})$, with the latter being the standard pair, then the Λ-modules $I^{\bar{p}}H_*(X; \Gamma)$ are isomorphic to the classical Alexander modules of the knot K with rational coefficients [151]. If K is not locally flat, then these are the *intersection Alexander modules* of the knot and have properties analogous to those of the classical Alexander modules. Analogous modules can be defined for links and hyperplane complements. For more, see [46, 82, 165]. △

Although we will not pursue this topic further in this book, there are also versions of intersection cohomology with local coefficient systems and versions of Poincaré duality. For a non-sheaf approach to some such results, see [163]. For sheaf-theoretic duality with local coefficients, see, for example, [28, 46, 106]; for applications of such duality to twisted signatures and L-classes see [14, 20, 46, 47]. For use of local coefficients in a "universal" duality theorem, see [98, 163]. Most of these results utilize coefficients in local systems of fields. A detailed study of the most general possible results over more general rings and using singular chain techniques remains to be written.

6.4 A General Künneth Theorem

We next turn toward establishing a more general Künneth Theorem that is not limited by the assumption that one factor of the product space must be a trivially filtered manifold. This will be possible using the non-GM intersection homology groups. Early Künneth Theorems focused primarily on attempts to determine for which GM perversities $\bar{p}$ it is true that

$$I^{\bar{p}}H_*(X \times Y; R) \cong H_*(I^{\bar{p}}S_*(X; R) \otimes_R I^{\bar{p}}S_*(Y; R)).$$

Recall that GM perversities are functions of codimension alone, so the formula makes sense. In [106], Goresky and MacPherson provided a sheaf-theoretic proof, based on the work of Cheeger [59] on L^2-cohomology, that such a formula holds for $\bar{p} = \bar{m}$, the lower middle perversity, using field coefficients and Witt spaces (see Definition 9.1.2 and Chapter 9, below). This result was generalized by Cohen, Goresky, and Ji [62], who, also using sheaf theory but now for coefficients in a principal ideal domain (PID) and for arbitrary compact pseudomanifolds, showed that such a formula holds whenever $\bar{p}$ satisfies the condition $\bar{p}(a) + \bar{p}(b) \leq \bar{p}(a + b) \leq \bar{p}(a) + \bar{p}(b) + 1$. They also show that this can be extended to $\bar{p}(a)+\bar{p}(b) \leq \bar{p}(a+b) \leq \bar{p}(a)+\bar{p}(b)+2$ provided one of X or Y is locally $(\bar{p}, R)$-torsion-free (see Definition 6.3.21).

In [87], the situation was generalized further to consider for what perversities $\bar{p}, \bar{q}, Q$ (not necessarily Goresky–MacPherson perversities) it is true that the cross product induces an isomorphism

$$H_*(I^{\bar{p}}S_*(X;R)\otimes_R I^{\bar{q}}S_*(Y;R)) \xrightarrow{\cong} I^Q H_*(X\times Y;R)$$

for stratified pseudomanifolds X,Y. The arguments in [87] were also sheaf-theoretic. We provide here a further generalization to CS sets using singular chains.

We will first consider a key example, the product of cones. This example will allow us to demonstrate what perversities Q may arise in the Künneth Theorem; the reader will also not be surprised by this point that cone computations will play an important role later in the proof. Then, we will state the Künneth Theorem, Theorem 6.4.7, and prove it using the computations of the key example and a Mayer–Vietoris argument.

Throughout this section we let R be a Dedekind domain, which includes the possibility that R is a PID or a field. Recall that this assumption implies that $I^{\bar{p}}S_*(X;R)$, $I^{\bar{p}}\mathfrak{C}_*(X;R)$, and any of their submodules are flat by Lemma 6.3.1. Once again, we will focus primarily on singular intersection homology and then show that a PL version of the theorem follows using the equivalence of singular and PL intersection homology.

6.4.1 A Key Example: the Product of Cones

We seek to understand what conditions are necessary and sufficient on the product space perversity Q in order for the map

$$H_*(I^{\bar{p}}S_*(X;R)\otimes I^{\bar{q}}S_*(Y;R)) \to I^Q H_*(X\times Y;R)$$

to be an isomorphism. We will do this by considering the key example of a product of cones $cX\times cY$, under the assumption that the Künneth Theorem already holds for product subsets of smaller depth. This situation will then play an important role in the inductive Mayer–Vietoris argument for the general case of the Künneth Theorem. We have already seen in the proof of Lemma 5.2.4 that we must in general have $Q(S\times S')\geq \bar{p}(S)+\bar{q}(S')$ in order for the cross product to be well defined. We will see that this condition, in addition to others, is imposed separately by other considerations in our example. The following discussion will be somewhat technically involved, so the reader more interested in the final answers might skip ahead to the statement of Lemma 6.4.3, or even Theorem 6.4.7 (our Künneth Theorem), and then continue on to later sections.

So to understand what Q may be, we consider $cX\times cY$, where X,Y are (non-empty) compact filtered sets of respective dimensions $n-1$ and $m-1$ and $cX\times cY$ is given the product filtration. Given that CS sets are locally products of this form (with additional Euclidean factors that do not influence the intersection homology), this is a good starting place.

Recall from Section 2.11 that if A,B are filtered spaces, then we define the *product filtration* on $A\times B$ so that $(A\times B)^i=\bigcup_{j+k=i}A^j\times B^k$. If A and B have respective formal dimensions n and m, then this product has formal dimension $n+m$. By Lemma 2.11.1, the strata of $A\times B$ each have the form $S\times T$, where S is a stratum of A and T is a stratum of B. We also recall from Corollary 2.11.5 that we have a filtered homeomorphism $cX \times cY \cong c(X*Y)$, where $X * Y$ is the join of X and Y; see Section 2.11 for more discussion of this fact and the definition of the filtration of a join. We suppose cX and cY have been endowed with respective perversities $\bar{p}$ and $\bar{q}$.

We need to compare $H_*(I^{\bar{p}}S_*(cX;R) \otimes I^{\bar{q}}S_*(cY;R))$ with $I^{Q}H_*(cX \times cY;R) \cong I^{Q}H_*(c(X*Y);R)$. We will assume as an induction hypothesis that we already have a Künneth isomorphism for any open subset of $cX \times cY - \{v \times w\}$ of the form $U \times V$, for open $U \subset cX$ and open $V \subset cY$. Here v and w are the respective cone points of cX and cY, and so any such product subspace has depth less than that of $cX \times cY$. We will discuss this assumption in detail later in the proof of Theorem 6.4.7 (in particular, see footnote 10 on page 316), but such an induction hypothesis is reasonable taking as base cases the products where at least one factor is a manifold.

With all of these assumptions, we will see that the cross product induces an isomorphism

$$H_*(I^{\bar{p}}S_*(cX;R) \otimes I^{\bar{q}}S_*(cY;R)) \xrightarrow{\cong} I^{Q}H_*(cX \times cY;R)$$

when we have the following conditions on Q:

$$\bar{p}(\{v\}) + \bar{q}(\{w\}) \leq Q(\{v \times w\}) \leq \bar{p}(\{v\}) + q(\{w\}) + 1. \qquad (6.4)$$

Furthermore, if the torsion product $I^{\bar{p}}H_{n-\bar{p}(\{v\})-2}(X;R) * I^{\bar{q}}H_{m-q(\{w\})-2}(Y;R) = 0$, then we can weaken this condition to

$$\bar{p}(\{v\}) + \bar{q}(\{w\}) \leq Q(\{v \times w\}) \leq \bar{p}(\{v\}) + q(\{w\}) + 2. \qquad (6.5)$$

We will also see that these conditions are necessary, in general.

(a) Necessity of Our Conditions on Q

Getting to work, we can easily write down $H_i(I^{\bar{p}}S_*(cX;R) \otimes I^{\bar{q}}S_*(cY;R))$ using the cone formula and the algebraic Künneth Theorem [126, Theorem V.2.1]:

$$\begin{aligned}
&H_i(I^{\bar{p}}S_*(cX;R) \otimes I^{\bar{q}}S_*(cY;R)) \\
&\quad\cong \bigoplus_{j+k=i} H_j(I^{\bar{p}}S_*(cX;R)) \otimes H_k(I^{\bar{q}}S_*(cY;R)) \\
&\qquad\quad \oplus \bigoplus_{j+k=i-1} H_j(I^{\bar{p}}S_*(cX;R)) * H_k(I^{\bar{q}}S_*(cY;R)) \\
&\quad\cong \bigoplus_{\substack{j+k=i \\ j<n-\bar{p}(\{v\})-1 \\ k<m-\bar{q}(\{w\})-1}} I^{\bar{p}}H_j(X;R) \otimes I^{\bar{q}}H_k(Y;R) \\
&\qquad\quad \oplus \bigoplus_{\substack{j+k=i-1 \\ j<n-\bar{p}(\{v\})-1 \\ k<m-\bar{q}(\{w\})-1}} I^{\bar{p}}H_j(X;R) * I^{\bar{q}}H_k(Y;R). \qquad (6.6)
\end{aligned}$$

A useful first observation here is that each tensor product term will be zero unless simultaneously $j \le n-\bar{p}(\{v\})-2$ and $k \le m-\bar{q}(\{w\})-2$, so $H_i(I^{\bar{p}}S_*(cX;R)\otimes I^{\bar{q}}S_*(cY;R))$ will have no tensor product terms if $i > n+m-\bar{p}(\{v\})-q(\{w\})-4$. Similarly, each torsion product term will be zero unless simultaneously $j \le n-\bar{p}(\{v\})-2$ and $k \le m-\bar{q}(\{w\})-2$, so $H_i(I^{\bar{p}}S_*(cX;R)\otimes I^{\bar{q}}S_*(cY;R))$ will have no torsion product terms if $i-1 > n+m-\bar{p}(\{v\})-q(\{w\})-4$, i.e. if $i > n+m-\bar{p}(\{v\})-q(\{w\})-3$. So, if $i \ge n+m-\bar{p}(\{v\})-q(\{w\})-2$, then $H_i(I^{\bar{p}}S_*(cX;R)\otimes I^{\bar{q}}S_*(cY;R)) = 0$.

The more challenging computation is the one for $I^Q H_*(cX \times cY;R) \cong I^Q H_*(c(X*Y);R)$. By the cone formula, we will have

$$I^Q H_i(cX\times cY;R) \cong I^Q H_i(c(X*Y);R) \cong \begin{cases} 0, & i \ge n+m-Q(\{v\times w\})-1,\\ I^Q H_i(X*Y;R), & i < n+m-Q(\{v\times w\})-1.\end{cases}$$

Given this computation, we can see why it is in general necessary to have

$$\bar{p}(\{v\})+q(\{w\}) \le Q(\{v\times w\}) \le \bar{p}(\{v\})+q(\{w\})+1$$

for there to be any hope of an isomorphism between $H_i(I^{\bar{p}}S_*(cX;R)\otimes I^{\bar{q}}S_*(cY;R))$ and $I^Q H_*(cX\times cY;R)$: We have just seen that $H_i(I^{\bar{p}}S_*(cX;R)\otimes I^{\bar{q}}S_*(cY;R)) = 0$ if $i \ge n+m-\bar{p}(\{v\})-q(\{w\})-2$; but it is also not hard to arrange for the term

$$H_{n+m-\bar{p}(\{v\})-q(\{w\})-3}(I^{\bar{p}}S_*(cX;R)\otimes I^{\bar{q}}S_*(cY;R))$$

to be non-zero. In particular, we have

$$\begin{aligned} &H_{n+m-\bar{p}(\{v\})-q(\{w\})-3}(I^{\bar{p}}S_*(cX;R)\otimes I^{\bar{q}}S_*(cY;R))\\ &\quad = I^{\bar{p}}H_{n-\bar{p}(\{v\})-2}(X;R) * I^{\bar{q}}H_{m-q(\{w\})-2}(Y;R),\end{aligned}$$

which could (generically) be non-zero, for example, by choosing X and Y to be manifolds with torsion in their homology modules of appropriate dimensions. For $i < n+m-\bar{p}(\{v\})-q(\{w\})-3$, we can obtain nontrivial non-torsion elements in all dimensions by letting X and Y be products of circles. On the other hand, we must have $I^Q H_i(c(X*Y);R) = 0$ if $i \ge n+m-Q(\{v\times w\})-1$. So we need to have $n+m-Q(\{v\times w\})-1 \ge n+m-\bar{p}(\{v\})-q(\{w\})-2$, i.e. that

$$Q(\{v\times w\}) \le \bar{p}(\{v\})+q(\{w\})+1.$$

If we happen to be in a situation where we know that

$$I^{\bar{p}}H_{n-\bar{p}(\{v\})-2}(X;R) * I^{\bar{q}}H_{m-q(\{w\})-2}(Y;R) = 0,$$

either by an assumption on the spaces or working with R being a field, then $H_i(I^{\bar{p}}S_*(cX;R)\otimes I^{\bar{q}}S_*(cY;R))$ will vanish when $i \ge n+m-\bar{p}(\{v\})-q(\{w\})-3$. So in this case we only need $n+m-Q(\{v\times w\})-1 \ge n+m-\bar{p}(\{v\})-q(\{w\})-3$, or $Q(\{v\times w\}) \le \bar{p}(\{v\})+q(\{w\})+2$. This provides the upper bounds on $Q(\{v\times w\})$ we claimed in conditions (6.4) and (6.5). We already know that the lower bound $\bar{p}(\{v\})+q(\{w\}) \le Q(\{v\times w\})$ is necessary in order for the cross product to be well defined; see also Remark 6.4.4 below. So, together, this establishes the general necessity of (6.4), or (6.5) if the torsion vanishing condition holds.

Remark 6.4.1 Thinking ahead to the sufficiency of conditions (6.4) or (6.5), one might expect that these still offer too much flexibility and that we would need to choose Q so that the dimension "cutoff" in the cone formula agrees with what we would expect from our computation of $H_i(I^{\bar{p}}S_*(cX;R) \otimes I^{\bar{q}}S_*(cY;R))$. However, we will see below that $I^QH_i(X * Y;R)$ turns out to be zero automatically in certain dimensions (reminiscent of the computation of the intersection homology of a suspension), and this provides the additional flexibility.

(b) Sufficiency of Our Conditions on Q

Now we turn to showing that our conditions (6.4) and (6.5) (with the torsion condition) on Q are sufficient to obtain a Künneth isomorphism. We have seen $H_i(I^{\bar{p}}S_*(cX;R) \otimes I^{\bar{q}}S_*(cY;R)) = 0$ for $i \geq n+m-\bar{p}(\{v\})-q(\{w\})-2$ and $I^QH_i(cX \times cY;R) = 0$ for $i \geq n+m-Q(\{v \times w\})-1$. Assuming $Q(\{v \times w\}) \geq \bar{p}(\{v\})+q(\{w\})$, we have $n+m-Q(\{v\times w\})-1 \leq n+m-\bar{p}(\{v\})-q(\{w\})-1$. Therefore, for $i \geq n+m-\bar{p}(\{v\})-q(\{w\})-1$ both $H_i(I^{\bar{p}}S_*(cX;R) \otimes I^{\bar{q}}S_*(cY;R))$ and $I^QH_i(cX \times cY;R)$ must be trivial, and we can focus on $i \leq n+m-\bar{p}(\{v\})-q(\{w\})-2$ for the rest of the discussion.

Note that when $Q(\{v\times w\}) = \bar{p}(\{v\})+q(\{w\})$, the range $i \leq n+m-\bar{p}(\{v\})-q(\{w\})-2$ is exactly the range where $I^QH_i(cX \times cY;R)$ is not forced to be zero by cone formula considerations alone. If $Q(\{v \times w\}) = \bar{p}(\{v\})+q(\{w\})+1$ then $I^QH_i(cX \times cY;R)$ is also automatically zero in dimension $i = n+m-\bar{p}(\{v\})-q(\{w\})-2$. If $Q(\{v \times w\}) = \bar{p}(\{v\})+q(\{w\})+2$ then $I^QH_i(cX \times cY;R)$ is also automatically 0 in dimension $i = n+m-\bar{p}(\{v\})-q(\{w\})-3$. Below these dimensions, and under our assumptions about Q, the module $I^QH_i(cX \times cY;R)$ is never automatically zero by the cone formula alone. To organize our thoughts going forward, it is helpful to separate out these observations into cases in terms of the degree i, given our conditions on Q:

1. If $i \geq n+m-\bar{p}(\{v\})-q(\{w\})-1$, then $H_i(I^{\bar{p}}S_*(cX;R)\otimes I^{\bar{q}}S_*(cY;R))$ and $I^QH_i(cX \times cY;R)$ are both zero. These are the only degrees in which $H_i(I^{\bar{p}}S_*(cX;R)\otimes I^{\bar{q}}S_*(cY;R))$ is always zero.
2. If $i \leq n+m-\bar{p}(\{v\})-q(\{w\})-4$, then $I^QH_i(cX \times cY;R)$ is never automatically zero from the cone formula alone under any of our conditions on Q.
3. If $i = n+m-\bar{p}(\{v\})-q(\{w\})-3$ then $I^QH_i(cX \times cY;R) = 0$ if $Q(\{v \times w\}) = \bar{p}(\{v\})+q(\{w\})+2$ but not necessarily otherwise.
4. If $i = n+m-\bar{p}(\{v\})-q(\{w\})-2$ then $I^QH_i(cX \times cY;R) = 0$ if $Q(\{v \times w\}) = \bar{p}(\{v\})+q(\{w\})+1$ or $Q(\{v\times w\}) = \bar{p}(\{v\})+q(\{w\})+2$ but not necessarily otherwise.

So we will need to compute $I^QH_i(cX \times cY;R)$ in those settings where it is not forced to be zero by the cone formula. In those cases, again by the cone formula, it will be isomorphic to $I^QH_i(X * Y;R) \cong I^QH_i(cX \times cY - \{v \times w\};R)$. So, we need to know something about these modules. But, we have

$$cX \times cY - \{v \times w\} \cong (cX \times (cY - \{w\})) \cup ((cX - \{v\}) \times cY),$$

while

$$(cX \times (cY - \{w\})) \cap ((cX - \{v\}) \times cY) \cong (cX - \{v\}) \times (cY - \{w\}).$$

As each of $cX \times (cY - \{w\})$, $(cX - \{v\}) \times cY$, and $(cX - \{v\}) \times (cY - \{w\})$ is a product of open subsets of depth less than that of $cX \times cY$, we can utilize our induction hypothesis that there is a Künneth isomorphism for these products. Furthermore, this decomposition of $cX \times cY - \{v \times w\}$ allows us to utilize a Mayer–Vietoris sequence. Therefore, we have the following diagram with the long exact Mayer–Vietoris sequence along the right side (coefficients tacit):

$$\begin{array}{ccc}
H_i(I^{\bar{p}}S_*(cX - \{v\}) \otimes I^{\bar{q}}S_*(cY - \{w\})) & \xrightarrow[\cong]{\varepsilon} & I^{Q}H_i((cX - \{v\}) \times (cY - \{w\})) \\
\downarrow & & \downarrow {\scriptstyle MV} \\
\begin{array}{c} H_i(I^{\bar{p}}S_*(cX) \otimes I^{\bar{q}}S_*(cY - \{w\})) \oplus \\ H_i(I^{\bar{p}}S_*(cX - \{v\}) \otimes I^{\bar{q}}S_*(cY)) \end{array} & \xrightarrow[\cong]{\varepsilon \oplus \varepsilon} & \begin{array}{c} I^{Q}H_i(cX \times (cY - \{w\})) \oplus \\ I^{Q}H_i((cX - \{v\}) \times (cY)) \end{array} \\
 & & \downarrow \\
 & & I^{Q}H_i(cX \times cY - \{v \times w\}) \\
 & & \downarrow
\end{array} \tag{6.7}$$

Note the map in the diagram that we have labeled MV.

Now, consider the terms in the exact sequence. The simplest term is the intersection term, $I^{Q}H_i((cX - \{v\}) \times (cY - \{w\}); R)$, which, by the induction assumption and the algebraic Künneth Theorem, is isomorphic via the cross product map to[6]

$$\begin{aligned}
&H_i(I^{\bar{p}}S_*(cX - \{v\}; R) \otimes I^{\bar{q}}S_*(cY - \{w\}; R)) \\
&\quad \cong \bigoplus_{j+k=i} H_j(I^{\bar{p}}S_*(cX - \{v\}; R)) \otimes H_k(I^{\bar{q}}S_*(cY - \{w\}; R)) \\
&\qquad \oplus \bigoplus_{j+k=i-1} H_j(I^{\bar{p}}S_*(cX - \{v\}; R)) * H_k(I^{\bar{q}}S_*(cY - \{w\}; R)).
\end{aligned} \tag{6.8}$$

In the Mayer–Vietoris sequence, this maps to the direct sum term $I^{Q}H_i((cX \times (cY - \{w\}); R)) \oplus I^{Q}H_*((cX - \{v\}) \times cY; R)$, which, again using the induction assumptions, is

[6] Of course, we could write e.g. $I^{\bar{p}}H_j(X; R)$ rather than $I^{\bar{p}}H_j(cX - \{v\}; R)$ in these formulas, but the latter forms will be better suited to the somewhat delicate argument we have coming up.

isomorphic to

$$H_i(I^{\bar{p}}S_*(cX;R) \otimes I^{\bar{q}}S_*(cY-\{w\};R)) \oplus H_i(I^{\bar{p}}S_*(cX-\{v\};R) \otimes I^{\bar{q}}S_*(cY;R)).$$

Using the algebraic Künneth theorem and the cone formula, this becomes

$$\begin{aligned}
&\bigoplus_{\substack{j+k=i\\ j<n-\bar{p}(\{v\})-1}} I^{\bar{p}}H_j(cX-\{v\};R) \otimes I^{\bar{q}}H_k(cY-\{w\};R)\\
&\oplus \bigoplus_{\substack{j+k=i-1\\ j<n-\bar{p}(\{v\})-1}} I^{\bar{p}}H_j(cX-\{v\};R) * I^{\bar{q}}H_k(cY-\{w\};R)\\
&\oplus \bigoplus_{\substack{j+k=i\\ k<m-\bar{q}(\{w\})-1}} I^{\bar{p}}H_j(cX-\{v\};R) \otimes I^{\bar{q}}H_k(cY-\{w\};R)\\
&\oplus \bigoplus_{\substack{j+k=i-1\\ k<m-\bar{q}(\{w\})-1}} I^{\bar{p}}H_j(cX-\{v\};R) * I^{\bar{q}}H_k(cY-\{w\};R). \qquad (6.9)
\end{aligned}$$

Notice that each summand here also occurs as a summand of (6.8). In fact, we claim that the maps

$$\begin{aligned}
&H_i(I^{\bar{p}}S_*(cX-\{v\};R) \otimes I^{\bar{q}}S_*(cY-\{w\};R))\\
&\qquad \to H_i(I^{\bar{p}}S_*(cX;R) \otimes I^{\bar{q}}S_*(cY-\{w\};R)),\\
&H_i(I^{\bar{p}}S_*(cX-\{v\};R) \otimes I^{\bar{q}}S_*(cY-\{w\};R))\\
&\qquad \to H_i(I^{\bar{p}}S_*(cX-\{v\};R) \otimes I^{\bar{q}}S_*(cY;R))
\end{aligned}$$

induced by the space inclusions correspond to the obvious projections onto summands. It turns out that this is really not so obvious[7], even given the nice topological situation, as the splittings guaranteed by the algebraic Künneth Theorem are not required to be preserved under morphisms (see [126, Section V.2]). Luckily, however, this naive expectation does turn out to hold. This will be the upshot of some technical work we will do below in Section 6.4.5, culminating in the following technical lemma. Although this lemma plays a critical role at several points in the remainder of our key example, its proof goes a bit far afield while we already have a few balls up in the air; hence, we put the proof aside until the end of Section 6.4.5, after we have finished the rest of our discussion of the Künneth Theorem.

Here is a statement of the lemma:

Lemma 6.4.2 *Given a Dedekind domain R and compact filtered sets $X = X^{n-1}$ and*

[7] This is an oversight in the proof of the Künneth Theorem in [87]; however, as we will see here, the result stated there does hold.

$Y = Y^{m-1}$, there are splittings of

$$\begin{aligned} &H_i(I^{\bar p}S_*(cX - \{v\}; R) \otimes I^{\bar q}S_*(cY - \{w\}; R)),\\ &H_i(I^{\bar p}S_*(cX; R) \otimes I^{\bar q}S_*(cY - \{w\}; R))),\\ &H_i(I^{\bar p}S_*(cX - \{v\}; R) \otimes I^{\bar q}S_*(cY; R)),\\ &H_i(I^{\bar p}S_*(cX; R) \otimes I^{\bar q}S_*(Y; R)) \end{aligned}$$

into direct sums of tensor products $I^{\bar p}H_j(cX - \{v\}; R) \otimes I^{\bar q}H_k(cY - \{w\}; R)$ and torsion products $I^{\bar p}H_{j-1}(cX - \{v\}; R) * I^{\bar q}H_k(cY - \{w\}; R)$, both with $j + k = i$, such that the maps in the diagram

$$\begin{array}{ccc} \begin{array}{c} H_i\big(I^{\bar p}S_*(cX - \{v\}; R) \otimes \\ I^{\bar q}S_*(Y - \{w\}; R)\big) \end{array} & \xrightarrow{\mathrm{i}\otimes\mathrm{id}} & \begin{array}{c} H_i\big(I^{\bar p}S_*(cX; R) \otimes \\ I^{\bar q}S_*(cY - \{w\}; R)\big) \end{array} \\ \Big\downarrow{\scriptstyle \mathrm{id}\otimes\bar{\mathrm{i}}} & & \Big\downarrow{\scriptstyle \mathrm{id}\otimes\bar{\mathrm{i}}} \\ \begin{array}{c} H_i\big(I^{\bar p}S_*(cX - \{v\}; R) \otimes \\ I^{\bar q}S_*(cY; R)\big) \end{array} & \xrightarrow{\mathrm{i}\otimes\mathrm{id}} & \begin{array}{c} H_i\big(I^{\bar p}S_*(cX; R) \otimes \\ I^{\bar q}S_*(cY; R)\big) \end{array} \end{array}$$

induced by the inclusions $\mathrm{i} : cX - \{v\} \hookrightarrow cX$ and $\bar{\mathrm{i}} : cY - \{w\} \hookrightarrow cY$ each restrict on each tensor or torsion product summand either to the zero map or to an isomorphism with the corresponding summand in the codomain. Furthermore, which of these options occurs is determined in the obvious way by the cone formula, Theorem 6.2.13; for example, the tensor product summand $I^{\bar p}H_j(cX - \{v\}; R) \otimes I^{\bar q}H_k(cY - \{w\}; R)$ maps to 0 in $H_i(I^{\bar p}S_*(cX; R) \otimes I^{\bar q}S_*(Y - \{w\}; R))$ when $j \geq n - \bar p(\{v\}) - 1$ and isomorphically to a corresponding summand otherwise.

So, given this lemma, let us examine the Mayer–Vietoris sequence of diagram (6.7) in dimensions $i \leq n + m - \bar p(\{v\}) - q(\{w\}) - 2$. It will be useful to adopt the following temporary notation: Let $T_{j,k}$, for $j + k = i$, denote the tensor product summand $I^{\bar p}H_j(cX - \{v\}; R) \otimes I^{\bar q}H_k(cY - \{w\}; R)$ of $H_i(I^{\bar p}S_*(cX - \{v\}; R) \otimes I^{\bar q}S_*(cY - \{w\}; R))$, and similarly let $\mathcal{T}_{j,k}$, for $j + k = i - 1$, denote the torsion product summand $I^{\bar p}H_j(cX - \{v\}; R) * I^{\bar q}H_k(cY - \{w\}; R)$. Lemma 6.4.2 says that under the maps from $H_i(I^{\bar p}S_*(cX - \{v\}; R) \otimes I^{\bar q}S_*(cY - \{w\}; R))$ to $H_i(I^{\bar p}S_*(cX; R) \otimes I^{\bar q}S_*(cY - \{w\}; R))$ and $H_i(I^{\bar p}S_*(cX - \{v\}; R) \otimes I^{\bar q}S_*(cY; R))$ induced by inclusion, each of the $T_{j,k}$ or $\mathcal{T}_{j,k}$ summands is taken either to zero or isomorphically to the appropriate corresponding summand, depending on whether or not the corresponding summand in the codomain vanishes due to the cone formula.

We claim that the restriction of the map MV of diagram (6.7) to each $T_{j,k}$ or $\mathcal{T}_{j,k}$ is injective if $i < n + m - \bar p(\{v\}) - q(\{w\}) - 2$, which will imply that, for all i in this range, the Mayer–Vietoris sequence breaks up into short exact sequences, each in the single degree i. Since the map MV takes each summand $T_{j,k}$ or $\mathcal{T}_{j,k}$ of (6.8) either to zero or to the corresponding term(s) in (6.9), we need only check that, in this range, each such summand appears in the expression (6.9). Each tensor product term $T_{j,k}$ with $i = j + k$ does not occur in (6.9) if and only if $j \geq n-\bar{p}(\{v\})-1$ and $k \geq m-\bar{q}(\{w\})-1$; in this case, we must have $i \geq m+n-\bar{p}(\{v\})-\bar{q}(\{w\})-2$. Each torsion product term $\mathcal{T}_{j,k}$ with $i-1=j+k$ does not occur in (6.9) if and only if $j \geq n-\bar{p}(\{v\})-1$ and $k \geq m-\bar{q}(\{w\})-1$; in this case, we must have $i \geq m+n-\bar{p}(\{v\})-\bar{q}(\{w\})-1$. So if $i < m+n-\bar{p}(\{v\})-\bar{q}(\{w\})-2$, each summand $T_{j,k}$ or $\mathcal{T}_{j,k}$ does appear in (6.9), and MV must be injective when restricted to the corresponding summand of (6.8). So MV is injective in the desired range $i < n+m-\bar{p}(\{v\})-q(\{w\})-2$.

So now for $i < m+n-\bar{p}(\{v\})-\bar{q}(\{w\})-2$, the Mayer–Vietoris sequence breaks into short exact sequences in each degree, and there are two possibilities for what happens to each summand $T_{j,k}$ or $\mathcal{T}_{j,k}$ of $H_i(I^{\bar{p}}S_*(cX-\{v\};R)\otimes I^{\bar{q}}S_*(cY-\{w\};R))$. If $T_{j,k}$ or $\mathcal{T}_{j,k}$ occur in both $H_i(I^{\bar{p}}S_*(cX;R)\otimes I^{\bar{q}}S_*(cY-\{w\};R))$ and $H_i(I^{\bar{p}}S_*(cX-\{v\};R)\otimes I^{\bar{q}}S_*(cY;R))$, then, up to isomorphism, MV restricts on the summand to the anti-diagonal map of the form $G \to G\oplus G$, $x \to (x,-x)$. However, if only one copy of $T_{j,k}$ or $\mathcal{T}_{j,k}$ occurs in the middle Mayer–Vietoris term, then MV takes the summand $T_{j,k}$ or $\mathcal{T}_{j,k}$ isomorphically to this copy. It follows that $I^QH_i(cX\times cY-\{v\times w\};R)$ will be isomorphic to the direct sum of those $T_{j,k}$ and $\mathcal{T}_{j,k}$ for which the first situation occurs, i.e. of those $T_{j,k}$ and $\mathcal{T}_{j,k}$ present in both $H_i(I^{\bar{p}}S_*(cX;R)\otimes I^{\bar{q}}S_*(cY-\{w\};R))$ and $H_i(I^{\bar{p}}S_*(cX-\{v\};R)\otimes I^{\bar{q}}S_*(cY;R))$.

Let us compute when this happens. From the cone formulas, we will have two copies of $T_{j,k}$ or $\mathcal{T}_{j,k}$ in (6.9) if both $j < n-\bar{p}(\{v\})-1$ and $k < m-\bar{q}(\{w\})-1$. So we see that for $i < m+n-\bar{p}(\{v\})-\bar{q}(\{w\})-2$, we have

$$\begin{aligned} &I^QH_i(cX\times cY-\{v\times w\};R)\\ &\cong \bigoplus_{\substack{j+k=i\\ j<n-\bar{p}(\{v\})-1\\ k<m-\bar{q}(\{w\})-1}} I^{\bar{p}}H_j(cX-\{v\};R)\otimes I^{\bar{q}}H_k(cY-\{w\};R)\\ &\oplus \bigoplus_{\substack{j+k=i-1\\ j<n-\bar{p}(\{v\})-1\\ k<m-\bar{q}(\{w\})-1}} I^{\bar{p}}H_j(cX-\{v\};R)*I^{\bar{q}}H_k(cY-\{w\};R). \end{aligned} \tag{6.10}$$

But this is exactly isomorphic to $H_i(I^{\bar{p}}S_*(cX;R)\otimes I^{\bar{q}}S_*(cY;R))$ as computed in (6.6), and we will show below, after cleaning up some other details, that this isomorphism is induced by the cross product.

Summing up, we can now update our list of facts beginning on page 306 as follows, again given the assumptions (6.4) or (6.5) concerning Q:

1. If $i \geq n+m-\bar{p}(\{v\})-q(\{w\})-1$, then $H_i(I^{\bar{p}}S_*(cX;R)\otimes I^{\bar{q}}S_*(cY;R))$ and $I^QH_i(cX\times cY\};R)$ are both zero.
2. If $i \leq n+m-\bar{p}(\{v\})-q(\{w\})-4$, then $I^QH_i(cX\times cY;R)\cong I^QH_i(cX\times cY-\{v\times w\};R)$, and this is isomorphic to $H_i(I^{\bar{p}}S_*(cX;R)\otimes I^{\bar{q}}S_*(cY;R))$ (we have yet to see that the isomorphism is induced by the cross product).
3. If $i=n+m-\bar{p}(\{v\})-q(\{w\})-3$ and $Q(\{v\times w\})\leq \bar{p}(\{v\})+\bar{q}(\{w\})+1$, then again $I^QH_i(cX\times cY;R)\cong I^QH_i(cX\times cY-\{v\times w\};R)$, and this is isomorphic to $H_i(I^{\bar{p}}S_*(cX;R)\otimes I^{\bar{q}}S_*(cY;R))$. If $Q(\{v\times w\})=\bar{p}(\{v\})+\bar{q}(\{w\})+2$, then $I^QH_i(cX\times$

$cY; R) \cong I^Q H_i(c(X * Y); R)$ must be zero from the cone formula, but, in this dimension, $H_i(I^{\bar{p}} S_*(cX; R) \otimes I^{\bar{q}} S_*(cY; R)) \cong I^{\bar{p}} H_{n-\bar{p}(\{v\})-2}(X; R) * I^{\bar{q}} H_{m-\bar{q}(\{w\})-2}(Y; R)$, so we will still have an isomorphism if this torsion term vanishes.

It remains only to consider $i = n + m - \bar{p}(\{v\}) - q(\{w\}) - 2$. We have already computed that $H_i(I^{\bar{p}} S_*(cX; R) \otimes I^{\bar{q}} S_*(cY; R)) = 0$ in this dimension. We will show that $I^Q H_i(cX \times cY - \{v \times w\}; R)$ also vanishes due to our assumptions about Q. Returning to our Mayer–Vietoris sequence of (6.7), since MV is injective in lower degrees, this group is the image of the term before it in the Mayer–Vietoris sequence. But if we look at (6.9), we see that unless $i = j + k \leq n - \bar{p}(\{v\}) - 2 + m - \bar{q}(\{w\}) - 2 = m + n - \bar{p}(\{v\}) - \bar{q}(\{w\}) - 4$, it is impossible to have two copies of a tensor product summand $T_{j,k}$ in (6.9). But we are assuming $i = m + n - \bar{p}(\{v\}) - \bar{q}(\{w\}) - 2$, so each tensor product summand $T_{j,k}$ occurs at most once in (6.9). Therefore, again using Lemma 6.4.2, the map MV takes $T_{j,k}$ in (6.8) either to zero or to the lone corresponding summand in (6.9), and so the summand does not survive into $I^Q H_{m+n-\bar{p}(\{v\})-\bar{q}(\{w\})-2}(cX \times cY - \{v \times w\}; R)$. Similarly, if $\mathcal{T}_{j,k}$ is a torsion product summand, it is impossible to have two copies of $\mathcal{T}_{j,k}$ in (6.9) unless $i - 1 = j + k \leq n - \bar{p}(\{v\}) - 2 + m - \bar{q}(\{w\}) - 2 = m + n - \bar{p}(\{v\}) - \bar{q}(\{w\}) - 4$. Since this is also not the case, the map of the corresponding $\mathcal{T}_{j,k}$ to (6.9) must again be zero or an isomorphism onto a lone summand, so there is again no contribution to $I^Q H_{m+n-\bar{p}(\{v\})-\bar{q}(\{w\})-2}(cX \times cY - \{v \times w\}; R)$. Thus $I^Q H_{m+n-\bar{p}(\{v\})-\bar{q}(\{w\})-2}(cX \times cY - \{v \times w\}; R) = 0$, as desired.

(c) The Cross Product

We have now shown, under the assumption (6.4) or (6.5), that the modules $H_i(I^{\bar{p}} S_*(cX; R) \otimes I^{\bar{q}} S_*(cY; R))$ and $I^Q H_i(cX \times cY; R) \cong I^Q H_i(c(X * Y); R)$ are isomorphic in all degrees. Let us confirm that this isomorphism is induced by the cross product.

The cases where the isomorphism is not trivial are those in degrees $i \leq m + n - \bar{p}(\{v\}) - \bar{q}(\{w\}) - 3$ with $Q(\{v \times w\}) \leq \bar{p}(\{v\}) + \bar{q}(\{w\}) + 1$. In this case, $I^Q H_i(cX \times cY; R) \cong I^Q H_i(cX \times cY - \{v \times w\}; R)$ due to the cone formula. In this range, our preceding discussion using the Mayer–Vietoris sequence of diagram (6.7) shows that up to isomorphism this group is isomorphic to the direct sum of some of the groups $T_{j,k}$ and $\mathcal{T}_{j,k}$. In particular, the groups $T_{j,k}$ and $\mathcal{T}_{j,k}$ that appear as summands in $I^Q H_i(cX \times cY - \{v \times w\}; R)$ are those that appear in both $I^Q H_i((cX - \{v\}) \times cY; R)$ and $I^Q H_i(cX \times (cY - \{w\}); R)$ and then survive to $I^Q H_i(cX \times cY - \{v \times w\}; R)$ in the quotient by the image of the anti-diagonal map. Thus all of $I^Q H_i(cX \times cY - \{v \times w\}; R)$ is in the image of the inclusion-induced map from $I^Q H_i((cX - \{v\}) \times cY; R)$.

By the naturality of the cross product (Proposition 5.2.17) we have a commutative diagram:

$$\begin{array}{ccc}
H_i(I^{\bar{p}}S_*(cX-\{v\};R)\otimes I^{\bar{q}}S_*(cY;R)) & \xrightarrow[\cong]{\varepsilon} & I^{Q}H_i((cX-\{v\})\times cY;R) \\
\Big\downarrow & & \Big\downarrow \\
 & & I^{Q}H_i(cX\times cY-\{v\times w\};R) \\
 & & \Big\downarrow{\cong} \\
H_i(I^{\bar{p}}S_*(cX;R)\otimes I^{\bar{q}}S_*(cY;R)) & \xrightarrow{\varepsilon} & I^{Q}H_i(cX\times cY;R).
\end{array}$$

Technically, Proposition 5.2.17 only gives us the commutativity of a diagram with the middle right term removed, but the inclusion-induced map $I^{Q}H_i((cX-\{v\})\times cY;R)\to I^{Q}H_i(cX\times cY;R)$ certainly factors through $I^{Q}H_i(cX\times cY-\{v\times w\};R)$. The surjectivity of the left vertical arrow comes from the cone formula and Lemma 6.4.2. The top horizontal arrow is an isomorphism by the induction hypothesis. The bottom right isomorphism comes from the cone formula in all degrees we are presently considering, and we saw in the preceding paragraph that the upper right arrow is surjective. It follows directly from the diagram that

$$\varepsilon : H_i(I^{\bar{p}}S_*(cX;R)\otimes I^{\bar{q}}S_*(cY;R))\to I^{Q}H_i(cX\times cY;R)$$

is surjective.

We must show that the bottom map ε is also injective. The top left module is a direct sum of terms $T_{j,k}=I^{\bar{p}}H_j(cX-\{v\};R)\otimes I^{\bar{q}}H_k(cY-\{w\};R)$ with $j+k=i$ and $\mathcal{T}_{j,k}=I^{\bar{p}}H_j(cX-\{v\};R)*I^{\bar{q}}H_k(cY-\{w\};R)$ with $j+k=i-1$, all with $k\leq m-\bar{q}(\{w\})-2$, as we here have only the full cone cY. By Lemma 6.4.2, the left vertical map is the projection to those summands also with $j\leq n-\bar{p}(\{v\})-2$. But we know from our preceding arguments with the Mayer–Vietoris sequence that the $T_{j,k}$ and $\mathcal{T}_{j,k}$ summands with both $j\leq n-\bar{p}(\{v\})-2$ and $k\leq m-\bar{q}(\{w\})-2$ are precisely the summands that survive the map from $I^{Q}H_i((cX-\{v\})\times cY;R)$ to $I^{Q}H_i(cX\times cY-\{v\times w\};R)$. So, in other words, every summand of $H_i(I^{\bar{p}}S_*(cX;R)\otimes I^{\bar{q}}S_*(cY;R))$ is the image of a summand of $H_i(I^{\bar{p}}S_*(cX-\{v\};R)\otimes I^{\bar{q}}S_*(cY;R))$ that maps isomorphically to a summand of $I^{Q}H_i(cX\times cY;R)$ by traveling right then down in the diagram. So the bottom horizontal map must be injective.

We can now wrap up the discussion thus far with a formal statement of our conclusions for our key example:

Lemma 6.4.3 *Let R be a Dedekind domain. Let X and Y be non-empty compact filtered spaces of respective dimensions $n-1, m-1$. Let $\bar{p}$ and $\bar{q}$ be respective perversities on cX and cY. Suppose a perversity Q is chosen on $cX\times cY$ such that for any open subset of $cX\times cY-\{v\times w\}$ of the form $U\times V$, with $U\subset cX$ and $V\subset cY$, we have an isomorphism* *induced by the cross product*

$$H_*(I^{\bar{p}}S_*(U;R) \otimes I^{\bar{q}}S_*(V;R)) \xrightarrow{\times} I^{Q}H_*(U \times V;R).$$

Then the cross product induces an isomorphism

$$H_*(I^{\bar{p}}S_*(cX;R) \otimes I^{\bar{q}}S_*(cY;R)) \xrightarrow{\times} I^{Q}H_*(cX \times cY;R)$$

if (and, in general, only if) $Q(\{v \times w\})$ equals $\bar{p}(\{v\}) + \bar{q}(\{w\})$ or $\bar{p}(\{v\}) + \bar{q}(\{w\}) + 1$. If $I^{\bar{p}}H_{n-\bar{p}(\{v\})-2}(X;R) * I^{\bar{q}}H_{m-\bar{q}(\{w\})-2}(Y;R) = 0$, then we can also use $Q(\{v \times w\}) = \bar{p}(\{v\}) + \bar{q}(\{w\}) + 2$.

Remark 6.4.4 As a side note, the Mayer–Vietoris sequence and Lemma 6.4.2 allow us to see that we need to have $Q(\{v \times w\}) \geq \bar{p}(\{v\}) + q(\{w\})$ for more reasons than just the well-definedness of the cross product: Indeed, if $Q(\{v \times w\}) < \bar{p}(\{v\}) + q(\{w\})$, then $n - m - \bar{p}(\{v\}) - q(\{w\}) - 1 < n + m - Q(\{v \times w\}) - 1$, so from the cone formula

$$\begin{aligned} &I^{Q}H_{n+m-\bar{p}(\{v\})-q(\{w\})-1}(cX \times cY;R) \\ &\quad \cong I^{Q}H_{n+m-\bar{p}(\{v\})-q(\{w\})-1}(cX \times cY - \{v \times w\};R). \end{aligned}$$

To study this group, let us consider the map MV in dimension $i = n+m-\bar{p}(\{v\})-q(\{w\})-2$. Suppose that $I^{\bar{p}}H_{n-\bar{p}(v)-1}(X;R) \otimes I^{\bar{q}}H_{m-\bar{q}(w)-1}(Y;R) \neq 0$, which is certainly a possibility (for example, if $\bar{p} = \bar{q} = \bar{0}$ and X, Y are oriented stratified pseudomanifolds, this could be the tensor product of the fundamental classes of X and Y; see Section 8.1, below). This tensor product corresponds to a summand in $H_i(I^{\bar{p}}S_*(cX - \{v\};R) \otimes I^{\bar{q}}S_*(cY - \{w\};R))$, but we see that its image must vanish in

$$H_i(I^{\bar{p}}S_*(cX;R) \otimes I^{\bar{q}}S_*(cY - \{w\};R)) \oplus H_i(I^{\bar{p}}S_*(cX - \{v\};R) \otimes I^{\bar{q}}S_*(cY;R)),$$

since $j = n - \bar{p}(v) - 1$ and $k = m - \bar{q}(w) - 1$ are both above the cutoff dimensions in the corresponding cone formulas. So $I^{Q}H_{n+m-\bar{p}(\{v\})-q(\{w\})-1}(cX \times cY;R)$ could be nonzero. But $H_{n+m-\bar{p}(\{v\})-q(\{w\})-1}(I^{\bar{p}}S_*(cX;R) \otimes I^{\bar{q}}S_*(cY;R))$ remains zero, as we computed previously directly from (6.6).

6.4.2 The Künneth Theorem

In this section, we produce our Künneth Theorem, relating $H_*(I^{\bar{p}}S_*(X;R) \otimes_R I^{\bar{q}}S_*(Y;R))$ with $I^{Q}H_*(X \times Y;R)$ for arbitrary CS sets X and Y with respective perversities $\bar{p}$ and $\bar{q}$. Our key example, studied in the preceding section, suggests that the perversity Q should be defined so that $Q(S \times T) = \bar{p}(S) + \bar{q}(T)$ or $Q(S \times T) = \bar{p}(S) + \bar{q}(T) + 1$ for singular strata $S \subset X$ and $T \subset Y$. If S is a regular stratum of X, then the Künneth Theorem with a manifold factor (Theorem 5.2.25) tells us that we need to be more restrictive and take $Q(S \times T) = \bar{p}(S) + \bar{q}(T) = \bar{q}(T)$. Similarly, if T is a regular stratum, we should use $Q(S \times T) = \bar{p}(S) + \bar{q}(T) = \bar{p}(S)$. We further expect that $Q(S \times T) = \bar{p}(S) + \bar{q}(T) + 2$ should also be acceptable for a product of singular strata so long as an appropriate torsion vanishing condition is met.

To simplify further statements, we introduce the following definition:

Definition 6.4.5 If Q is a perversity on $X \times Y$ satisfying the following properties, then we will say that Q is *$(\bar{p}, \bar{q})$-compatible*[8]:

1. if $S \subset X$ is a regular stratum and $T \subset Y$ is any stratum, then $Q(S \times T) = \bar{q}(T)$;
2. if $S \subset X$ is any stratum and $T \subset Y$ is a regular stratum, then $Q(S \times T) = \bar{p}(S)$;
3. if $S \subset X$ and $T \subset Y$ are both singular strata, then $Q(S \times T) = \bar{p}(S) + \bar{q}(T)$ or $Q(S \times T) = \bar{p}(S) + \bar{q}(T) + 1$;
4. if for each point $x \times y \in S \times T$ there are a distinguished neighborhood of x in X of the form $\mathbb{R}^a \times cL_1$ and a distinguished neighborhood of y in Y of the form $\mathbb{R}^b \times cL_2$ such that $I^{\bar{p}}H_{\dim(L_1)-\bar{p}(S)-1}(L_1; R) * I^{\bar{q}}H_{\dim(L_2)-\bar{q}(T)-1}(L_2; R) = 0$, then condition (3) on $Q(S \times T)$ may also include the possibility $Q(S \times T) = \bar{p}(S) + \bar{q}(T) + 2$. In particular, this is the case if X is locally $(\bar{p}, R)$-torsion-free along[9] the singular stratum S or Y is locally $(\bar{q}, R)$-torsion-free along the singular stratum T. Recall that by Lemma 6.3.24 this condition really depends only on S and T and not on the choices of x, y, L_1, or L_2.

Remark 6.4.6 If $S \subset X$ and $T \subset Y$ are both regular strata, then $\bar{p}(S) = \bar{q}(T) = 0$, and so both conditions (1) and (2) of the definition are consistent with the expectation that $Q(S \times T) = 0$ on the regular stratum $S \times T$ of $X \times Y$.

Furthermore, if $Y = M$ is an unfiltered manifold then $\bar{q}$ must be the trivial perversity that is 0 on each connected component $\mathcal{R}$ of M, and for any stratum $S \times \mathcal{R}$ of $X \times Y$, we must have $Q(S \times \mathcal{R}) = \bar{p}(S)$ for any $(\bar{p}, \bar{q})$-compatible perversity Q. This is consistent with the perversities appearing in the Künneth Theorem for which one factor is a manifold (Theorem 6.3.20).

It will turn out that this definition is precisely what works, and we obtain the following Künneth theorem:

Theorem 6.4.7 (Künneth Theorem) *Let X, Y be CS sets with respective perversities $\bar{p}, \bar{q}$, and let R be a Dedekind domain. Let Q be a $(\bar{p}, \bar{q})$-compatible perversity defined on $X \times Y$. Then the cross product induces an isomorphism*

$$H_*(I^{\bar{p}}S_*(X; R) \otimes_R I^{\bar{q}}S_*(Y; R)) \xrightarrow{\times} I^Q H_*(X \times Y; R).$$

The analogous result holds for PL intersection homology if X and Y are PL CS sets.

Remark 6.4.8 Using our observation about perversities in Remark 6.4.6, we see that this Künneth Theorem reduces to Theorem 6.3.20 if either of the factors is an unfiltered manifold.

Remark 6.4.9 The theorem applies as stated to stratified pseudomanifolds, which are CS sets, but it also applies to ∂-stratified pseudomanifolds: if X is a ∂-stratified pseudomanifold,

[8] Note that we leave the underlying spaces X and Y tacit in the notation, although the definitions of $\bar{p}$ and $\bar{q}$, as well as the local torsion properties, of course depend on the spaces.

[9] See Definition 6.3.21.

then X is stratified homotopy equivalent to the stratified pseudomanifold $X - \partial X$, and similarly for Y. So these interiors can be used as intersection homology substitutes for the full spaces in the following arguments. See Section 7.3.10 for more details.

Proof of Theorem 6.4.7 At first we will focus on singular intersection homology.

It will be simpler to work with dimensionally homogeneous CS sets, which can be done as follows: As observed in Lemma 6.3.50, $X^\bullet \times Y^\bullet = (X \times Y)^\bullet$, so by the naturality of the cross product we have a commutative diagram

$$\begin{array}{ccc} H_*(I^{\bar{p}}S_*(X^\bullet; R) \otimes_R I^{\bar{q}}S_*(Y^\bullet; R)) & \xrightarrow{\times} & I^{Q}H_*((X \times Y)^\bullet; R) \\ \downarrow & & \downarrow \\ H_*(I^{\bar{p}}S_*(X; R) \otimes_R I^{\bar{q}}S_*(Y; R)) & \xrightarrow{\times} & I^{Q}H_*(X \times Y; R) \end{array}$$

in which the vertical maps are induced by inclusions. By Proposition 6.3.47, the algebraic Künneth Theorem [238, Theorem 3.6.3] (which applies, as all intersection chain modules are flat by Lemma 6.3.1), and the Five Lemma, the vertical maps are both isomorphisms. Therefore, it will suffice to show that the top horizontal map is an isomorphism. Equivalently, as the homogenization of a CS set is a CS set by Lemma 6.3.45, we will assume in the following argument that all spaces are dimensionally homogeneous, and the result in general follows from this diagram. Note that, due to Corollary 6.3.49, our finite generation and torsion-free conditions on the links of X and Y carry over to the links of $X^\bullet$ and $Y^\bullet$.

The proof will proceed by an induction on the depth of $X \times Y$ using Mayer–Vietoris arguments.

If the depth of $X \times Y$ is zero, then X and Y both have depth zero, and the theorem reduces to the standard Künneth Theorem for ordinary homology.

If the depth of $X \times Y$ is one, then one of X or Y is a manifold, and the theorem reduces to the Künneth Theorem with a manifold factor (Theorem 6.3.20).

So now assume that the depth of $X \times Y$ is $K > 1$ and that we have proven the theorem in all cases where the depth of $X \times Y$ is $< K$.

Still assuming that $X \times Y$ has depth K, we will first prove the special case of the theorem in which Y is a CS set of the form $Y = \mathbb{R}^j \times c\mathcal{L}$ for some compact filtered $\mathcal{L}$. We will use a Mayer–Vietoris argument (Theorem 5.1.4) with functors defined on the open subsets of X. For $U \subset X$, we let $F_*(U) = H_*(I^{\bar{p}}S_*(U; R) \otimes_R I^{\bar{q}}S_*(Y; R))$ and $G_*(U) = I^{Q}H_*(U \times Y; R)$ with the natural transformation Φ corresponding to the cross product. The functor G_* admits Mayer–Vietoris sequences by Theorem 6.3.12. The argument that F_* admits Mayer–Vietoris sequences is essentially the same as that in the proof of Theorem 5.2.25: If we begin with the Mayer–Vietoris short exact sequence for $I^{\bar{p}}S_*(\cdot\,; R)$ for subsets of X, then we can tensor it with $I^{\bar{q}}S_*(Y; R)$, which is flat and so preserves exactness. The associated long exact sequence is the desired Mayer–Vietoris sequence for F_*. Also as in the proof of Theorem 5.2.25, the transformation Φ induces a map of short exact sequences, and hence a map of long exact sequences, even after replacing terms of the form $H_*((I^{\bar{p}}S_*(U;R) + I^{\bar{p}}S_*(V;R)) \otimes_R I^{\bar{q}}S_*(Y;R))$ with the isomorphic terms $H_*(I^{\bar{p}}S_*(U \cup V;R) \otimes_R I^{\bar{q}}S_*(Y;R))$

The second condition of Theorem 5.1.4 follows, as usual, from Lemmas 5.1.6 and 5.1.7 in the case of F_* using the algebraic Künneth Theorem and the commutativity of tensor products with direct limits and homology.

The last condition of Theorem 5.1.4 is trivial if U is empty. Otherwise, it is a consequence of Theorem 6.3.20, as the assumption that X is dimensionally homogeneous ensures that an open subset U of X can be contained in a single stratum of X only if U is an open subset of a regular stratum of X and so a trivially filtered manifold of the dimension of X.

Finally, we consider condition (3) of Theorem 5.1.4. By our induction on depth, the hypothesis that $\Phi : F_*(\mathbb{R}^i \times (cL - \{v\})) \to G_*(\mathbb{R}^i \times (cL - \{v\}))$ be an isomorphism is automatically fulfilled by the induction hypothesis whenever $\mathbb{R}^i \times cL$ is (homeomorphic to) a distinguished neighborhood of a point in a singular stratum of X. We need to show that $\Phi : F_*(\mathbb{R}^i \times cL) \to G_*(\mathbb{R}^i \times cL)$ is an isomorphism. But this is precisely the content of Lemma 6.4.3, up to the additional Euclidean space factors, which are not relevant to the computation due to the stratified homotopy invariance of intersection homology and naturality of the cross product. The hypotheses of the lemma are also satisfied due to the induction on depth[10]. And so Theorem 5.1.4 proves the special case $Y = \mathbb{R}^j \times c\mathscr{L}$.

Now we can move on to the general case of arbitrary CS sets $X \times Y$ of depth K. We again use Theorem 5.1.4, this time with functors defined on the open subsets of Y. For $U \subset Y$, we let $F_*(U) = H_*(I^{\bar{p}}S_*(X;R) \otimes_R I^{\bar{q}}S_*(U;R))$ and $G_*(U) = I^Q H_*(X \times U;R)$ with the functor Φ corresponding to the cross product. By the exact same reasoning as above, we have a commuting diagram of Mayer–Vietoris sequences, the condition on ascending chains of open sets holds, and the last condition of Theorem 5.1.4 is a consequence of Theorem 6.3.20, as the dimensional homogeneity assumption on Y again implies that if U is a non-empty open subset of Y contained in a single stratum, then U is a trivially-filtered manifold of the dimension of Y. For condition (3) of Theorem 5.1.4, the hypothesis that $\Phi : F_*(\mathbb{R}^i \times (cL - \{v\})) \to G_*(\mathbb{R}^i \times (cL - \{v\}))$ be an isomorphism (now for $\mathbb{R}^i \times cL$ a distinguished neighborhood of a point in a singular stratum of Y) is again automatically fulfilled by the induction hypothesis on depth, and we need to show that $\Phi : F_*(\mathbb{R}^i \times cL) \to G_*(\mathbb{R}^i \times cL)$ is an isomorphism. But this is exactly the special case of

[10] More precisely, if $U \subset cL$ and $V \subset c\mathscr{L}$ are such that $U \times V \subset cL \times c\mathscr{L} - \{v \times w\}$, then naturality of the cross product provides a diagram

$$\begin{array}{ccc} I^{\bar{p}}S_*(U;R) \otimes I^{\bar{q}}S_*(V;R) & \xrightarrow{\varepsilon} & I^Q S_*(U \times V;R) \\ \downarrow & & \downarrow \\ I^{\bar{p}}S_*(\mathbb{R}^i \times U;R) \otimes I^{\bar{q}}S_*(\mathbb{R}^j \times V;R) & \xrightarrow{\varepsilon} & I^Q S_*(\mathbb{R}^i \times U \times \mathbb{R}^j \times V;R). \end{array}$$

The vertical maps all induce homology isomorphisms by stratified homotopy invariance and the algebraic Künneth Theorem, and the bottom horizontal map induces a homology isomorphism by induction, using that all the spaces are (filtered homeomorphic to) open subsets of CS sets and so are themselves CS sets. Therefore, the top map is also a homology isomorphism, fulfilling the hypotheses of Lemma 6.4.3.

the theorem proven above for which the first factor is arbitrary but the second factor is the product of a Euclidean space and a cone. Since an open subset of a space Z has depth less than or equal to that of Z, our proof of the special case is allowable here. The theorem now follows for singular intersection homology from Theorem 5.1.4.

If X and Y are PL CS sets, suppose K and L are full simplicial complexes triangulating X and Y compatibly with their filtrations and with partial orders on their vertices restricting to total orders on each simplex. Then putting together the commutative diagrams of Corollaries 6.3.36 and 6.3.37, we get the commutative diagram

$$\begin{array}{ccc}
I^{\bar{p}}\mathfrak{C}_*(X;R)\otimes_R I^{\bar{q}}\mathfrak{C}_*(Y;R) & \xrightarrow{\times} & I^{Q}\mathfrak{C}_*(X\times Y;R)\\
\big\uparrow & & \big\uparrow\\
I^{\bar{p}}C_*(K;R)\otimes_R I^{\bar{q}}C_*(L;R) & \xrightarrow{\bowtie} & I^{Q}C_*(K\times L;R)\\
{\scriptstyle \phi_K\otimes\phi_L}\big\downarrow & & \big\downarrow{\scriptstyle \phi_{K\times L}}\\
I^{\bar{p}}S_*(X;R)\otimes_R I^{\bar{q}}S_*(Y;R) & \xrightarrow{\varepsilon} & I^{Q}S_*(X\times Y;R).
\end{array}$$

We know that maps of the form $I^{\bar{p}}C_*(K;R) \to I^{\bar{p}}\mathfrak{C}_*(X;R)$ induce isomorphisms on homology by Theorem 6.3.30, and the maps of the form $I^{\bar{p}}C_*(K;R) \to I^{\bar{p}}S_*(X;R)$ induce isomorphisms on homology by Corollary 6.3.35, identifying $|K|$ with X via the triangulating homeomorphism (and similarly for the other spaces). As all modules are flat, the vertical arrows on the left therefore also induce homology isomorphisms by the naturality of the algebraic Künneth Theorem and the Five Lemma. We have seen that the bottom map is an isomorphism on homology, thus the other horizontal maps are isomorphisms on homology as well. This completes the proof of Theorem 6.4.7. □

As a corollary, and applying also the algebraic Künneth Theorem, we obtain the following version of the theorem with coefficients in a field.

Corollary 6.4.10 *Let X,Y be CS sets with respective perversities $\bar{p},\bar{q}$. Let F be a field. Let Q be a $(\bar{p},\bar{q})$-compatible perversity defined on $X\times Y$. Then the cross product induces an isomorphism*

$$I^{\bar{p}}H_*(X;F)\otimes_F I^{\bar{q}}H_*(Y;F) \xrightarrow{\times} I^{Q}H_*(X\times Y;F).$$

The analogous result holds for PL intersection homology if X and Y are PL CS sets.

Example 6.4.11 Here is an example that will be useful below in Chapter 9 in our discussion of Witt signatures.

Let $\bar{n}$ be the upper-middle Goresky–MacPherson perversity defined by $\bar{n}(S) = \lceil \frac{1}{2}(\operatorname{codim}(S)-2)\rceil$ (Definition 3.1.10). As $\bar{n}$ depends only on codimension, we may also write simply $\bar{n}(k) = \lceil \frac{1}{2}(k-2) \rceil$. We claim that if X and Y are CS sets, each given the perversity $\bar{n}$, then $\bar{n}$ as a perversity on $X \times Y$ is $(\bar{n}, \bar{n})$-compatible.

Let $S \subset X$ and $T \subset Y$ be strata. If S is regular, it follows that $\text{codim}_{X\times Y}(S \times T) = \text{codim}_Y(T)$. If we call this common codimension ℓ, then both sides of the compatibility condition reduce to $\bar{n}(\ell)$. Things work out analogously if T is regular. So suppose S and T are both singular. If $\text{codim}_X(S) = k$ and $\text{codim}_Y(T) = \ell$, then it suffices to show $\bar{n}(k) + \bar{n}(\ell) \le \bar{n}(k+\ell) \le \bar{n}(k) + \bar{n}(\ell) + 1$. This can be seen from the following table:

k	ℓ	$\bar{n}(k)$	$\bar{n}(\ell)$	$\bar{n}(k)+\bar{n}(\ell)$	$\bar{n}(k+\ell)$
even	even	$\frac{k}{2}-1$	$\frac{\ell}{2}-1$	$\frac{k+\ell}{2}-2$	$\frac{k+\ell}{2}-1$
odd	even	$\frac{k+1}{2}-1$	$\frac{\ell}{2}-1$	$\frac{k+\ell+1}{2}-2$	$\frac{k+\ell+1}{2}-1$
even	odd	$\frac{k}{2}-1$	$\frac{\ell+1}{2}-1$	$\frac{k+\ell+1}{2}-2$	$\frac{k+\ell+1}{2}-1$
odd	odd	$\frac{k+1}{2}-1$	$\frac{\ell+1}{2}-1$	$\frac{k+\ell}{2}-1$	$\frac{k+\ell}{2}-1$

In each case $\bar{n}(k) + \bar{n}(\ell) \le \bar{n}(k+\ell) \le \bar{n}(k) + \bar{n}(\ell) + 1$, and so $\bar{n}$ is $(\bar{n}, \bar{n})$-compatible and the Künneth Theorem applies to provide an isomorphism

$$H_*(I^{\bar{n}}S_*(X;R) \otimes_R I^{\bar{n}}S_*(Y;R)) \xrightarrow{\times} I^{\bar{n}}H_*(X \times Y;R).$$

$\triangle$

Remark 6.4.12 This Künneth Theorem (Theorem 6.4.7) is not true if we replace IH_* with IH_*^{GM}. For example, let X and Y each be a cone on a torus $S^1 \times S^1$. Let $\bar{p}(\{v\}) = 5$ and $\bar{q}(\{w\}) = 0$. Using the cone formula (Theorem 4.2.1),

$$I^{\bar{p}}H_i^{\text{GM}}(X) \cong \begin{cases} \mathbb{Z}, & i = 0, \\ 0, & \text{otherwise,} \end{cases}$$

and

$$I^{\bar{q}}H_i^{\text{GM}}(Y) \cong \begin{cases} \mathbb{Z}, & i = 0, \\ \mathbb{Z} \oplus \mathbb{Z}, & i = 1, \\ 0, & \text{otherwise,} \end{cases}$$

and so also

$$H_i(I^{\bar{p}}S_*^{\text{GM}}(X) \otimes I^{\bar{q}}S_*^{\text{GM}}(Y)) \cong \begin{cases} \mathbb{Z}, & i = 0, \\ \mathbb{Z} \oplus \mathbb{Z}, & i = 1, \\ 0, & \text{otherwise.} \end{cases}$$

But if we take $Q(\{v \times w\}) = \bar{p}(\{v\}) + \bar{q}(\{w\}) = 5$, then, again by Theorem 4.2.1, we have

$$I^Q H_i^{\mathrm{GM}}(X \times Y; R) \cong \begin{cases} \mathbb{Z}, & i = 0, \\ 0, & \text{otherwise}. \end{cases}$$

Conceivably, there may be some way to modify the theorem so that Q provides some kind of additional truncation if perversities get "too big" for Theorem 4.2.1 to apply (which is essentially the problem here) and hence some way to extend the Künneth Theorem to IH_*^{GM}, but we will not pursue this here.

6.4.3 A Relative Künneth Theorem

Having established our general Künneth Theorem as Theorem 6.4.7, we now turn to proving a relative version of the theorem. We will not need to work from scratch; instead we will use Theorem 6.4.7 along with some homological algebra.

Theorem 6.4.13 (Künneth Theorem) *Let X,Y be CS sets with respective perversities $\bar{p},\bar{q}$, and let R be a Dedekind domain. Let $A \subset X$ and $B \subset Y$ be open*[11] *subspaces, and let Q be a $(\bar{p},\bar{q})$-compatible perversity defined on $X \times Y$. Then the cross product induces an isomorphism*

$$H_*(I^{\bar{p}}S_*(X,A;R) \otimes_R I^{\bar{q}}S_*(Y,B;R)) \xrightarrow{\times} I^Q H_*(X \times Y, (A \times Y) \cup (X \times B); R).$$

The analogous result holds for PL intersection homology if X and Y are PL CS sets.

Proof Consider the following diagram, in which we leave the R coefficients tacit:

$$\begin{array}{ccccccccc} 0 & \longrightarrow & \begin{array}{c} I^{\bar{p}}S_*(A) \otimes_R \\ I^{\bar{q}}S_*(Y) \end{array} & \longrightarrow & \begin{array}{c} I^{\bar{p}}S_*(X) \otimes_R \\ I^{\bar{q}}S_*(Y) \end{array} & \longrightarrow & \begin{array}{c} I^{\bar{p}}S_*(X,A) \otimes_R \\ I^{\bar{q}}S_*(Y) \end{array} & \longrightarrow & 0 \\ & & \downarrow{\scriptstyle \times} & & \downarrow{\scriptstyle \times} & & \downarrow{\scriptstyle \times} & & \\ 0 & \longrightarrow & I^Q S_*(A \times Y) & \longrightarrow & I^Q S_*(X \times Y) & \longrightarrow & I^Q S_*(X \times Y, A \times Y) & \longrightarrow & 0. \end{array}$$

The top row is the short exact $\bar{p}$-intersection chain sequence of the pair (X,A) tensored over R with $I^{\bar{q}}S_*(Y;R)$. Since each $I^{\bar{q}}S_i(Y;R)$ is flat, the sequence remains exact. The second row is the short exact Q-intersection chain sequence of the pair $(X \times Y, A \times Y)$. The vertical maps are all induced by the chain cross product, and the diagram commutes by naturality of the cross product. In the resulting diagram of long exact homology sequences, the cross product induces isomorphisms on the absolute homology terms by Theorem 6.4.7, observing that the links of A and B are all also links of X and Y. So

$$\times : H_*(I^{\bar{p}}S_*(X,A;R) \otimes_R I^{\bar{q}}S_*(Y;R)) \to I^Q H_*(X \times Y, A \times Y; R)$$

[11] It might be possible to prove this theorem in greater generality, but this will be a convenient assumption for us to utilize our previous results.

is also an isomorphism, by the Five Lemma.

Similarly, we now have the following diagram, again with coefficients tacit:

$$\begin{array}{ccccccccc}
0 & \to & I^{\bar p}S_*(X,A)\otimes_R I^{\bar q}S_*(B) & \to & I^{\bar p}S_*(X,A)\otimes_R I^{\bar q}S_*(Y) & \to & I^{\bar p}S_*(X,A)\otimes_R I^{\bar q}S_*(Y,B) & \to & 0\\
 & & \downarrow{\times} & & \downarrow{\times} & & \downarrow{\times} & & \\
0 & \to & I^{Q}S_*(X\times B, A\times B) & \to & I^{Q}S_*(X\times Y, A\times Y) & \to & \frac{I^{Q}S_*(X\times Y)}{(I^{Q}S_*(A\times Y)+I^{Q}S_*(X\times B))} & \to & 0
\end{array}$$

This time, the top row comes from tensoring the short exact $\bar q$-intersection chain sequence of the pair (Y,B) with $I^{\bar p}S_*(X,A;R)$. The sequence stays exact because $I^{\bar p}S_*(X,A;R)$ is also flat. The bottom row in the diagram is exact via some basic algebra: The inclusion $X\times B\hookrightarrow X\times Y$ induces an injection $I^{Q}S_*(X\times B,A\times B;R)\to I^{Q}S_*(X\times Y,A\times Y;R)$. Then

$$\begin{aligned}
&\frac{I^{Q}S_*(X\times Y,A\times Y;R)}{I^{Q}S_*(X\times B,A\times B;R)}\\
&\quad\cong\frac{I^{Q}S_*(X\times Y;R)/I^{Q}S_*(A\times Y;R)}{I^{Q}S_*(X\times B;R)/I^{Q}S_*(A\times B;R)}\\
&\quad\cong\frac{I^{Q}S_*(X\times Y;R)/I^{Q}S_*(A\times Y;R)}{(I^{Q}S_*(A\times Y;R)+I^{Q}S_*(X\times B;R))/I^{Q}S_*(A\times Y;R)}\\
&\quad\cong\frac{I^{Q}S_*(X\times Y;R)}{(I^{Q}S_*(A\times Y;R)+I^{Q}S_*(X\times B;R))},
\end{aligned}$$

using in the last isomorphism the Third Isomorphism Theorem and in the middle isomorphism the Second Isomorphism Theorem (see Theorems 2.13 and 2.12, respectively, of [197]), as

$$I^{Q}S_*(A\times Y;R)\cap I^{Q}S_*(X\times B;R)=I^{Q}S_*(A\times B;R).$$

The diagram of short exact sequences yields a diagram of long exact sequences in homology, and the maps corresponding to the first two vertical maps of the short exact sequence are isomorphisms by the case demonstrated above in which only one factor in the Künneth Theorem was a relative homology module. By the Five Lemma, we get isomorphisms

$$\begin{aligned}
&H_*(I^{\bar p}S_*(X,A;R)\otimes_R I^{\bar q}S_*(Y,B;R))\\
&\quad\xrightarrow{\times} H_*(I^{Q}S_*(X\times Y;R)/(I^{Q}S_*(A\times Y;R)+I^{Q}S_*(X\times B;R))).
\end{aligned}$$

Finally, there is a map of the short exact sequence

$$\begin{aligned}
&0\to I^{Q}S_*(A\times Y;R)+I^{Q}S_*(X\times B;R)\to I^{Q}S_*(X\times Y;R)\\
&\quad\to I^{Q}S_*(X\times Y;R)/(I^{Q}S_*(A\times Y;R)+I^{Q}S_*(X\times B;R))\to 0
\end{aligned}$$

to the short exact Q-intersection chain sequence of the pair $(X\times Y,(A\times Y)\cup(X\times B))$. The induced map

$$H_*(I^QS_*(A\times Y;R)+I^QS_*(X\times B;R))\to I^QH_*((A\times Y)\cup(X\times B);R)$$

is an isomorphism by the argument used to prove the Mayer–Vietoris sequence of Theorem 6.3.12; see the proof of Theorem 4.4.19. So, by the Five Lemma, we at last obtain the isomorphisms

$$\begin{aligned}&H_*(I^{\bar p}S_*(X,A;R)\otimes_R I^{\bar q}S_*(Y,B;R))\\&\xrightarrow{\times} H_*(I^QS_*(X\times Y;R)/(I^QS_*(A\times Y;R)+I^QS_*(X\times B;R)))\\&\cong I^QH_*(X\times Y,(A\times Y)\cup(X\times B);R).\end{aligned}$$

The argument for PL intersection homology is analogous. □

6.4.4 Applications of the Künneth Theorem

This section contains some rather immediate applications of the intersection homology Künneth Theorem.

One important consequence of Theorems 6.4.7 and 6.4.13, and one that will play a significant role in the next chapter, is that the singular chain cross product ε is a chain homotopy equivalence. As we use here that the singular intersection chain modules are projective (Lemma 6.3.1), we will not be able to extend this result to PL chains, which we only know to be flat.

Theorem 6.4.14 *Let X,Y be CS sets with respective perversities $\bar p,\bar q$, let $A\subset X$ and $B\subset Y$ be open subspaces, let R be a Dedekind domain, and let Q be a $(\bar p,\bar q)$-compatible perversity on $X\times Y$. Then the cross product*

$$\varepsilon: I^{\bar p}S_*(X,A;R)\otimes_R I^{\bar q}S_*(Y,B;R)\to I^QS_*(X\times Y,(A\times Y)\cup(X\times B);R)$$

is a chain homotopy equivalence.

Proof We first recall that each $I^{\bar p}S_i(X,A;R)$, $I^{\bar q}S_j(Y,B;R)$, and $I^QS_k(X\times Y,(A\times Y)\cup(X\times B);R)$ is a projective R-module; see Lemma 6.3.1. It follows that each $I^{\bar p}S_i(X,A;R)\otimes I^{\bar q}S_j(Y,B;R)$ is projective. In fact, if P and S are projective R-modules, then there are R-modules U and V such that $P\oplus U$ and $S\oplus V$ are each free (recall that a module is projective if and only if it is a direct summand of a free module; see Lemma A.4.1). Thus $(P\oplus U)\otimes(S\oplus V)$ is free [147, Corollary XVI.2.4]. But then

$$\begin{aligned}(P\oplus U)\otimes(S\oplus V)&\cong(P\otimes(S\oplus V))\oplus(U\otimes(S\oplus V))\\&\cong(P\otimes S)\oplus(P\otimes V)\oplus(U\otimes S)\oplus(U\times V),\end{aligned}$$

so $P\otimes S$ is a direct summand of a free module.

As each $I^{\bar p}S_i(X,A;R)\otimes I^{\bar q}S_j(Y,B;R)$ is projective, so then is $\bigoplus_{i+j=k}I^{\bar p}S_i(X,A;R)\otimes I^{\bar q}S_j(Y,B;R)$, and therefore $I^{\bar p}S_*(X,A;R)\otimes I^{\bar q}S_*(Y,B;R)$ is a complex of projectives. Under the assumptions of Theorem 6.4.13, the cross product induces a quasi-isomorphism of these complexes of projectives. This is sufficient for ε to be a chain homotopy equivalence. This last fact is well known, and we provide a proof as Lemma A.4.3 in Appendix A. □

For our next two applications of the Künneth Theorem, we examine when the product of locally torsion-free spaces is locally torsion-free in its own right (with respect to appropriate perversities) and similarly when the product of locally finitely generated spaces is locally finitely generated (recall Definition 6.3.38).

Proposition 6.4.15 *Suppose X and Y are CS sets with respective perversities $\bar{p}$ and $\bar{q}$. Let Q be a $(\bar{p},\bar{q})$-compatible perversity on $X \times Y$. Then $X \times Y$ is locally (Q,R)-torsion-free if and only if X is locally $(\bar{p},R)$-torsion-free and Y is locally $(\bar{q},R)$-torsion-free.*

Proof Let $S \times T$ be a stratum of $X \times Y$, let K be the link of S in X and let L be the link of T in Y. Then the join $K * L$ is the link of $S \times T$ in $X \times Y$. If S is a regular stratum of X, then $K = \emptyset$, and similarly for T and L. Recall that the join with the empty set is the identity construction, e.g. $K * \emptyset = K$.

By definition, $X \times Y$ is locally (Q,R)-torsion-free if for any such $S \times T$ the torsion product of $I^Q H_{\dim(K*L)-Q(S\times T)-1}(K * L;R)$ with any R-module vanishes, or, equivalently by [147, Theorem XVI.3.11], if this intersection homology module is flat. Recall that this expression makes sense as we can restrict the perversity Q on $X \times Y$ to the link $K * L$, which we can consider embedded in $X \times Y$. The locally torsion-free conditions for X and Y can be similarly restated.

If $K = \emptyset$, i.e. S is a regular stratum, then $Q(S\times T) = \bar{q}(T)$ and Q restricts to $\bar{q}$ on L, so the condition that $I^Q H_{\dim(K*L)-Q(S\times T)-1}(K * L;R)$ be flat reduces to $I^{\bar{q}} H_{\dim(L)-\bar{q}(T)-1}(L;R)$ being flat, which is the condition for Y to be locally $(\bar{q},R)$-torsion-free along T. Therefore, we see that if $X \times Y$ is locally (Q,R)-torsion-free, it follows that Y is locally $(\bar{q},R)$-torsion-free. The analogous argument holds when $L = \emptyset$. Thus if $X \times Y$ is locally (Q,R)-torsion-free, we must also have that X is locally $(\bar{p},R)$-torsion-free and Y is locally $(\bar{q},R)$-torsion-free. Conversely, if X is locally $(\bar{p},R)$-torsion-free and Y is a locally $(\bar{q},R)$-torsion-free, then $X \times Y$ is locally (Q,R)-torsion-free along strata $S \times T$ for which one of S or T is regular. It remains to show that if X and Y are each locally torsion-free then $X \times Y$ is locally torsion-free along products of singular strata.

So suppose that X and Y are both locally torsion-free and that neither K nor L is empty. To best mesh with our earlier computations, suppose $\dim(K) = m - 1$ and $\dim(L) = n - 1$. Then

$$\begin{aligned}\dim(K * L) - Q(S \times T) - 1 &= m + n - 1 - (\bar{p}(S) + \bar{q}(T) + C) - 1\\ &= m + n - \bar{p}(S) - \bar{q}(T) - 2 - C,\end{aligned}$$

where $C \in \{0, 1, 2\}$. Note that $C = 2$ is allowed because of the torsion-free hypotheses on X and Y. Now let us apply equation (6.10) and the computations below it on page 311, which hold by our assumption that Q is $(\bar{p},\bar{q})$-compatible.

First, we have computed that

$$I^{Q}H_{m+n-\bar{p}(S)-\bar{q}(T)-2}(K*L;R)=I^{Q}H_{m+n-\bar{p}(S)-\bar{q}(T)-2}(cK\times cL-\{v\times w\};R)=0,$$

where v, w are the respective cone vertices. This is certainly torsion-free.

Then, by equation (6.10) and deleting terms that must vanish, we obtain[12]

$$\begin{aligned}I^{Q}H_{m+n-\bar{p}(S)-\bar{q}(T)-4}(K*L;R)\cong&\left(I^{\bar{p}}H_{m-\bar{p}(S)-2}(K;R)\otimes_R I^{\bar{q}}H_{n-\bar{q}(T)-2}(L;R)\right)\\&\oplus\left(I^{\bar{p}}H_{m-\bar{p}(S)-3}(K;R)*_R I^{\bar{q}}H_{n-\bar{q}(T)-2}(L;R)\right)\\&\oplus\left(I^{\bar{p}}H_{m-\bar{p}(S)-2}(K;R)*_R I^{\bar{q}}H_{n-\bar{q}(T)-3}(L;R)\right)\end{aligned}$$

and

$$I^{Q}H_{m+n-\bar{p}(S)-\bar{q}(T)-3}(K*L;R)\cong I^{\bar{p}}H_{m-\bar{p}(S)-2}(K;R)*_R I^{\bar{q}}H_{n-\bar{q}(T)-2}(L;R).$$

Recalling that $\dim(K)=m-1$ and $\dim(L)=n-1$, we see that the locally torsion-free conditions on X and Y imply that $I^{\bar{p}}H_{m-\bar{p}(S)-2}(K;R)$ and $I^{\bar{q}}H_{n-\bar{q}(T)-2}(L;R)$ are both flat. It follows that all the torsion product terms vanish and that the tensor product term is flat as a tensor product of flat modules. Therefore, we obtain that $I^{Q}H_{m+n-\bar{p}(S)-\bar{q}(T)-C}(K*L;R)$ is flat for $C\in\{0,1,2\}$, as desired. □

Proposition 6.4.16 *Suppose X and Y are CS sets and that R is a Noetherian ring. Let X and Y have respective perversities $\bar{p}$ and $\bar{q}$, and let Q be a $(\bar{p},\bar{q})$-compatible perversity on $X\times Y$. Then $X\times Y$ is locally (Q,R)-finitely generated if and only if X is locally $(\bar{p},R)$-finitely generated and Y is locally $(\bar{q},R)$-finitely generated.*

Proof Let $S\times T$ be a stratum of $X\times Y$, let K be the link of S in X and let L be the link of T in Y. Then the join $K*L$ is the link of $S\times T$ in $X\times Y$.

By definition, $X\times Y$ is locally (Q,R)-finitely generated if, for any such $S\times T$ and each i, the modules $I^{Q}H_i(K*L;R)$ are all finitely generated. Recall that this expression makes sense as we can restrict the perversity Q on $X\times Y$ to the link $K*L$, which we can consider embedded in $X\times Y$.

If $K=\emptyset$, i.e. S is a regular stratum, then $Q(S\times T)=\bar{q}(T)$ and Q restricts to $\bar{q}$ on L, so the condition that $I^{Q}H_i(K*L;R)$ be finitely generated for all i reduces to $I^{\bar{q}}H_i(L;R)$ being finitely generated for all i, which is the condition for Y to be locally $(\bar{q},R)$-finitely generated along T. Therefore, we see that if $X\times Y$ is locally (Q,R)-finitely generated, it follows that Y is locally $(\bar{q},R)$-finitely generated. The analogous argument holds when $L=\emptyset$. Thus if $X\times Y$ is locally (Q,R)-finitely generated, we must also have that X is locally $(\bar{p},R)$-finitely generated and Y is locally $(\bar{q},R)$-finitely generated. Conversely, if X is locally $(\bar{p},R)$-finitely generated and Y is locally $(\bar{q},R)$-finitely generated, then $X\times Y$ is locally (Q,R)-finitely generated along strata $S\times T$ for which one of S or T is regular. It remains to show that if X and Y are each locally finitely generated then $X\times Y$ is locally finitely generated along products of singular strata.

[12] In this expression $K*L$ denotes the join of spaces and we use $*_R$ for the torsion product over R.

So, now, suppose that X and Y are both locally finitely generated and that neither K nor L is empty. We need for the intersection homology modules $I^QH_i(K * L; R)$ to be finitely generated for all i. But in our discussion in Section 6.4.1, we saw that $I^QH_i(K * L; R) \cong I^QH_i(cK \times cL - \{v \times w\}; R)$, where v, w are the respective cone vertices. Furthermore, $cK \times cL - \{v \times w\}$ is the union of $(cK - \{v\}) \times cL$ and $cK \times (cL - \{w\})$, which have intersection $(cK - \{v\}) \times (cL - \{w\})$, and there is a Mayer–Vietoris sequence involving these spaces. If $\dim(S) = s$, then $\mathbb{R}^s \times cK$ and $\mathbb{R}^s \times (cK - \{v\})$ are open sets of the CS set X (up to filtered homeomorphism), and so. up to Euclidean factors, which don't affect intersection homology computations, cK and $cK - \{v\}$ are open subsets of X and so can be treated as possessing the same local torsion properties as X, and analogously for Y, cL, and $cL - \{w\}$. In particular, the product perversity Q is $(\bar{p}, \bar{q})$-compatible on these products, and the Künneth Theorem (Theorem 6.4.7) applies to $I^QH_i((cK - \{v\}) \times cL; R)$, $I^QH_i(cK \times (cL - \{w\}); R)$, and $I^QH_i((cK - \{v\}) \times (cL - \{w\}); R)$ as in the argument of footnote 10 in the proof of Theorem 6.4.7. Therefore, using also the cone formula and stratified homotopy invariance, each of these modules is a finite direct sum of tensor and torsion products of modules of the form $I^{\bar{p}}H_j(K; R)$ or $I^{\bar{q}}H_k(L; R)$, each of which is finitely generated by hypothesis. Hence the terms on either side of $I^QH_i(K * L; R)$ in the Mayer–Vietoris sequence are finitely generated, and so, as R is Noetherian, $I^QH_i(K * L; R)$ must also be finitely generated, using the basic properties of Noetherian modules; see [147, Section X.1]. □

As a final application in this section, we prove the following lemma, which will be useful to have when we study Poincaré duality in Chapter 8. We could prove this lemma more geometrically using stratified homotopy invariance, but this is a nice application of the Künneth Theorem that lets us avoid the details of such an argument.

Lemma 6.4.17 *Let R be a Dedekind domain, let $D \subset C$ be compact convex subsets of $\mathbb{R}^k$, let L be a compact filtered space, and suppose cL is a CS set. For $0 \leq s < 1$, let $\bar{c}_s$ be the closed subcone $[0, s] \times L/\sim$ in $cL = [0, 1) \times L/\sim$; if $s = 0$ then $\bar{c}_0L = \{v\}$, the cone vertex. Then for $0 \leq t \leq s < 1$, the inclusion-induced map $I^{\bar{p}}H_i(\mathbb{R}^k \times cL, (\mathbb{R}^k \times cL) - (C \times \bar{c}_sL); R) \to I^{\bar{p}}H_i(\mathbb{R}^k \times cL, (\mathbb{R}^k \times cL) - (D \times \bar{c}_tL); R)$ is an isomorphism. In particular, if C contains the origin of $\mathbb{R}^k$, then $I^{\bar{p}}H_i(\mathbb{R}^k \times cL, (\mathbb{R}^k \times cL) - (C \times \bar{c}_sL); R) \to I^{\bar{p}}H_i(\mathbb{R}^k \times cL, (\mathbb{R}^k \times cL) - \{(0, v)\}; R)$ is an isomorphism.*

Proof Let B be a closed ball centered at a point $z \in D$ such that B has large enough positive radius to contain D in its interior $\mathring{B}$. Without loss of generality, let us suppose that z is the origin $0 \in \mathbb{R}^k$. Then there is a deformation retraction $r : I \times \mathbb{R}^k \to \mathbb{R}^k$ that retracts $\mathbb{R}^k - \{0\}$ to $\mathbb{R}^k - \mathring{B}$ by retracting outward along rays from the origin. As D is convex, any point in $\mathbb{R}^k - D$ must stay in $\mathbb{R}^k - D$ throughout the retraction, as if $x \in \mathbb{R}^k - D$ and $r(u, x) \in D$, then the entire line segment from 0 to $r(u, x)$ must be in D by convexity, but by construction x would also be in this line segment, a contradiction. Hence r restricts to a deformation retraction of $\mathbb{R}^k - D$ to $\mathbb{R}^k - \mathring{B}$, and so the inclusion $\mathbb{R}^k - \mathring{B} \hookrightarrow \mathbb{R}^k - D$ is a homotopy equivalence. We can choose B large enough to contain C as well, and so $\mathbb{R}^k - \mathring{B} \hookrightarrow \mathbb{R}^k - C$ is also a homotopy equivalence. From the diagram

$$\begin{array}{ccc} \mathbb{R}^k - \mathring{B} & \hookrightarrow & \mathbb{R}^k - C \\ & \searrow & \downarrow \\ & & \mathbb{R}^k - D, \end{array}$$

the inclusion $\mathbb{R}^k - C \hookrightarrow \mathbb{R}^k - D$ is also a homotopy equivalence. Similarly, the inclusion $cL - \bar{c}_s L \hookrightarrow cL - \bar{c}_t L$ is a stratified homotopy equivalence by retractions outward along the cone lines. So by (stratified) homotopy invariance, the inclusion-induced maps $H_*(\mathbb{R}^k, \mathbb{R}^k - C; R) \to H_*(\mathbb{R}^k, \mathbb{R}^k - D; R)$ and $I^{\bar{p}}H_*(cL, cL - \bar{c}_s L; R) \to I^{\bar{p}}H_*(cL, cL - \bar{c}_t L; R)$ are isomorphisms.

Next we observe that $(\mathbb{R}^k \times (cL - \bar{c}_s L)) \cup ((\mathbb{R}^k - C) \times cL) = (\mathbb{R}^k \times cL) - (C \times \bar{c}_s L)$, and similarly using D and $\bar{c}_t L$. By the intersection homology Künneth Theorem and the algebraic Künneth Theorem, we obtain a short exact sequence (coefficients tacit)

$$\begin{aligned} 0 &\to \bigoplus_{j+\ell=i} H_j(\mathbb{R}^k, \mathbb{R}^k - C) \otimes H_\ell(cL, cL - \bar{c}_s L) \\ &\to I^{\bar{p}}H_i(\mathbb{R}^k \times cL, (\mathbb{R}^k \times cL) - (C \times \bar{c}_s L)) \\ &\to \bigoplus_{j+\ell=i-1} H_j(\mathbb{R}^k, \mathbb{R}^k - C) * H_\ell(cL, cL - \bar{c}_s L) \to 0, \end{aligned}$$

and analogously for our other product space. But the cross product and the algebraic Künneth Theorem are natural in their inputs, so the inclusion

$$\big(\mathbb{R}^k \times cL, (\mathbb{R}^k \times cL) - (C \times \bar{c}_s L)\big) \hookrightarrow \big(\mathbb{R}^k \times cL, (\mathbb{R}^k \times cL) - (D \times \bar{c}_t L)\big)$$

induces a map of short exact sequences, and we have just seen that the maps of end terms are isomorphisms. The lemma therefore follows by the Five Lemma, with the last statement of the lemma being just a special case of the more general statement. □

6.4.5 Some Technical Stuff: the Proof of Lemma 6.4.2

In this section, we prove Lemma 6.4.2, which played a critical technical role in the proof of Lemma 6.4.3, the key example of the Künneth Theorem for the product of two cones. The proof of Lemma 6.4.2 uses explicitly that the singular intersection chain groups are projective R-modules for R a Dedekind domain, and so we do not provide a parallel PL development. We could demonstrate a PL version of the lemma following from the singular version using simplicial intersection homology as an intermediate, as we did for the proof of the Künneth Theorem itself, but as that theorem was our primary goal, we will leave PL versions of Lemma 6.4.2 for the interested reader to develop.

(a) Algebra of the Algebraic Künneth Theorem

What we will really need to prove Lemma 6.4.2 is some control over the cycles that represent homology classes in the homology of the tensor product of the intersection chain complexes of cones and deleted cones (cones with their vertices removed). To achieve this, we will need to develop a detailed understanding of the splitting of the short exact sequence guaranteed by the algebraic Künneth Theorem, and this in turn hinges upon a good understanding of the proof of that theorem. So we will begin by reviewing that proof. Then we discuss the splitting issues and prove a lemma, Lemma 6.4.19, that gives our desired result concerning the representation of homology classes; unfortunately, we won't be in a position to state this lemma precisely until we have developed some further notation. In Section 6.4.5(b) we will utilize Lemma 6.4.19 to prove Lemma 6.4.2.

Review of the Algebraic Künneth Theorem. Suppose that C_* and D_* are chain complexes of projective modules over the Dedekind domain R. Then the algebraic Künneth Theorem (e.g. [238, Theorem 3.6.3]) states that there is a short exact sequence, natural in the inputs C_* and D_*,

$$0 \to \bigoplus_{j+k=i} H_j(C_*) \otimes H_k(D_*) \to H_i(C_* \otimes D_*) \to \bigoplus_{j+k=i-1} H_j(C_*) * H_k(D_*) \to 0,$$

where $H_j(C_*) * H_k(D_*)$ denotes the torsion product. Furthermore, this sequence splits. However, the splitting is not natural.

Given an element $\sum [x_a] \otimes [y_a] \in H_j(C_*) \otimes H_k(D_*)$, where here $x_a \in C_j$ and $y_a \in D_k$ are cycles, the image of this element in $H_i(C_* \otimes D_*)$ is represented by $\sum x_a \otimes y_a$. Our current goal in reviewing the proof of the Künneth Theorem is to be able to say something about the cycles in $C_* \otimes D_*$ representing the elements of $\bigoplus_{j+k=i-1} H_j(C_*) * H_k(D_*)$ under some choice of splitting. In order to do this, we will review the proof of the algebraic Künneth Theorem. We will begin by following the proof in [126, Section V.2] with a few minor modifications in the details to pay closer attention to the splitting into summands and under the slightly different assumptions that R is a Dedekind domain and that C_* and D_* are chain complexes of projective R-modules. The discussion in [126] makes the more general assumption that C_* and D_* consist of flat modules but the stronger assumption that R is a PID; however, the argument there still goes through with our assumptions.

Proof of the Künneth Exact Sequence. Continuing to let C_* and D_* be chain complexes of projectives over a Dedekind domain, we begin with some notation. Let

$$Z_p = \ker(\partial_{C_*} : C_p \to C_{p-1}), \qquad B_p = \operatorname{im}(\partial_{C_*} : C_{p+1} \to C_p),$$
$$\bar{Z}_p = \ker(\partial_{D_*} : D_p \to D_{p-1}), \qquad \bar{B}_p = \operatorname{im}(\partial_{D_*} : D_{p+1} \to D_p).$$

In general, for any construction we make with C_*, the same notation occurring with a bar will denote the analogous construction for D_*. As submodules of projective modules over a Dedekind domain, the modules Z_p, $\bar{Z}_p$, B_p, and $\bar{B}_p$ are each projective for any p. We let Z_*

be the complex consisting of the modules Z_p and with trivial boundary maps (i.e. each is the zero homomorphism) and similarly for B_*. We also define a complex B'_* with $B'_p = B_{p-1}$ and all boundary maps trivial. The boundary map of C_* takes $x \in C_p$ to $\partial_{C_*}(x) \in B_{p-1} = B'_p$. This also determines a degree-zero chain map $\beta : C_* \to B'_*$. In fact, we have an exact sequence of degree-zero chain maps

$$0 \longrightarrow Z_* \xrightarrow{\ \mathrm{i}\ } C_* \xrightarrow{\ \beta\ } B'_* \longrightarrow 0, \tag{6.11}$$

with i the inclusion. The other advantage of introducing the modules B'_p is that, as each is projective, this short exact sequence splits in each degree (non-canonically and not necessarily compatibly among degrees), so that we can write $C_p \cong Z_p \oplus B'_p$ for each p. This will be important below when we get to splittings of the Künneth exact sequence.

We also have the evident inclusion maps

$$B'_{p+1} = B_p \overset{\mathrm{j}}{\hookrightarrow} Z_p \overset{\mathrm{i}}{\hookrightarrow} C_p.$$

Now, as D_* consists of projective modules, tensoring the sequence (6.11) with D_* preserves exactness, and we obtain a short exact sequence

$$0 \longrightarrow Z_* \otimes D_* \xrightarrow{\mathrm{i}\otimes\mathrm{id}} C_* \otimes D_* \xrightarrow{\beta\otimes\mathrm{id}} B'_* \otimes D_* \longrightarrow 0.$$

This short exact sequence yields a homology long exact sequence

$$\xrightarrow{\ \omega_i\ } H_i(Z_* \otimes D_*) \xrightarrow{\mathrm{i}\otimes\mathrm{id}} H_i(C_* \otimes D_*) \xrightarrow{\beta\otimes\mathrm{id}} H_i(B'_* \otimes D_*) \xrightarrow{\ \omega_{i-1}\ } . \tag{6.12}$$

Next we compute $H_i(B'_* \otimes D_*)$ and $H_i(Z_* \otimes D_*)$ a bit more explicitly. As B'_* has trivial boundary maps, the complex $B'_* \otimes D_*$ has boundary maps given by $\partial_{B'_*\otimes D_*}(b \otimes d) = (-1)^{|b|} b \otimes \partial_{D_*} d$, where $|b|$ is the degree of b in B'_*. We already know by definition that in each degree we have $(B'_* \otimes D_*)_i = \bigoplus_{j+k=i} B'_j \otimes D_k$, but this boundary computation shows that the restriction of the boundary map to $B'_j \otimes D_k$ has image in $B'_j \otimes D_{k-1}$. Thus, the complex $B'_* \otimes D_*$ is the direct sum of complexes $\mathcal{B}_{j,*}$ of the form

$$\longrightarrow B'_j \otimes D_{k+1} \longrightarrow B'_j \otimes D_k \longrightarrow B'_j \otimes D_{k-1} \longrightarrow ,$$

with $\mathcal{B}_{j,k} = B'_j \otimes D_k$ in degree $j + k$ and again with $\partial_{\mathcal{B}_{j,*}}(b \otimes d) = (-1)^j b \otimes \partial_{D_*} d$; note that the same sign $(-1)^j$ occurs in all degrees. In other words, up to the extra sign $(-1)^j$ in the boundary maps, the complex $\mathcal{B}_{j,*}$ is just the tensor product of the chain complex D_* with the *module* B'_j, treated as an object with degree j, or, equivalently, we can think of B'_j as representing a complex whose only non-zero module is B'_j in degree j. In any case, we have altogether $B'_* \otimes D_* \cong \bigoplus_j \mathcal{B}_{j,*}$.

Furthermore, as B'_j is projective, tensoring with it preserves kernels and images, and so

$$\ker\left(B'_j \otimes D_k \xrightarrow{\partial_{\mathcal{B}_{j,*}}} B'_j \otimes D_{k-1}\right) = B'_j \otimes \bar{Z}_k$$

$$\operatorname{im}\left(B'_j \otimes D_{k+1} \xrightarrow{\partial_{\mathcal{B}_{j,*}}} B'_j \otimes D_k\right) = B'_j \otimes \bar{B}_k$$

Now, as $H_k(D_*) = \bar{Z}_k/\bar{B}_k$ by definition, we have a short exact sequence

$$0 \longrightarrow \bar{B}_k \xrightarrow{\bar{\mathrm{i}}} \bar{Z}_k \xrightarrow{\bar{\eta}} H_k(D_*) \longrightarrow 0,$$

and tensoring with the projective B'_j gives the short exact sequence

$$0 \longrightarrow B'_j \otimes \bar{B}_k \xrightarrow{\mathrm{id}\otimes\bar{\mathrm{i}}} B'_j \otimes \bar{Z}_k \xrightarrow{\mathrm{id}\otimes\bar{\eta}} B'_j \otimes H_k(D_*) \longrightarrow 0.$$

But we have just identified $B'_j \otimes \bar{B}_k$ and $B_j \otimes \bar{Z}_k$ as the respective image and kernel of boundary maps of $\mathcal{B}_{j,*}$. So we see that

$$H_{j+k}(\mathcal{B}_{j,*}) \cong B'_j \otimes H_k(D_*).$$

Therefore,

$$H_i(B'_* \otimes D_*) \cong H_i\left(\bigoplus_j \mathcal{B}_{j,*}\right) \cong \bigoplus_j B'_j \otimes H_{i-j}(D_*) = \bigoplus_{j+k=i} B'_j \otimes H_k(D_*).$$

A similar argument shows that $H_i(Z_* \otimes D_*) \cong \bigoplus_{j+k=i} Z_j \otimes H_k(D_*)$. So the long exact sequence (6.12) becomes

$$\xrightarrow{\omega_i} \bigoplus_{j+k=i} Z_j \otimes H_k(D_*) \xrightarrow{\mathrm{i}\otimes\mathrm{id}} H_i(C_* \otimes D_*)$$
$$\xrightarrow{\beta\otimes\mathrm{id}} \bigoplus_{j+k=i} B'_j \otimes H_k(D_*) \xrightarrow{\omega_{i-1}} . \tag{6.13}$$

From the standard zig-zag computation of the boundary map of a long exact sequence, the connecting map ω_{i-1} corresponds to the direct sum of inclusions $\omega_{j-1,k} : B'_j \otimes H_k(D_*) \hookrightarrow Z_{j-1} \otimes H_k(D_*)$ induced by the maps $B'_j \overset{\mathrm{i}}{\hookrightarrow} Z_{j-1}$ (see the proof of [126, Theorem V.2.1] for more details, though this is also an easy diagram chase exercise). But we have an exact sequence

$$0 \longrightarrow B'_j \xrightarrow{\mathrm{i}} Z_{j-1} \xrightarrow{\eta} H_{j-1}(C_*) \longrightarrow 0,$$

so the complex whose only nontrivial terms are $B'_j \xrightarrow{\mathrm{i}} Z_{j-1}$ in degrees zero and one is a projective resolution of $H_{j-1}(C_*)$. So, tensoring with $H_j(D_*)$ and by the definition of the torsion product, we have exact sequences

$$0 \longrightarrow H_{j-1}(C_*) * H_k(D_*) \longrightarrow B'_j \otimes H_k(D_*) \xrightarrow{\omega_{j-1,k}}$$

$$Z_{j-1} \otimes H_k(D_*) \xrightarrow{\eta \otimes \mathrm{id}} H_{j-1}(C_*) \otimes H_k(D_*) \longrightarrow 0.$$

Taking direct sums and letting $\omega_{i-1} = \bigoplus_{j+k=i} \omega_{j-1,k}$, we see that

$$\mathrm{im}(\beta \otimes \mathrm{id}) = \ker(\omega_{i-1}) = \bigoplus_{j+k=i} \ker(\omega_{j-1,k}) \cong \bigoplus_{j+k=i} H_{j-1}(C_*) * H_k(D_*)$$

and

$$\ker(\beta \otimes \mathrm{id}) = \mathrm{im}(\imath \otimes \mathrm{id}) \cong \mathrm{cok}(\omega_i) \cong \bigoplus_{j+k=i} \mathrm{cok}(\omega_{j,k}) \cong \bigoplus_{j+k=i} H_j(C_*) \otimes H_k(D_*).$$

Hence the Künneth short exact sequence.

It is worth observing again here that if an element of $H_j(C_*) \otimes H_k(D_*)$ is represented by $[z] \otimes [\bar{z}]$ with z a cycle in C_* and $\bar{z}$ a cycle in D_* (and all generators of $H_j(C_*) \otimes H_k(D_*)$ have such a form), then the image of this element in $H_*(C_* \otimes D_*)$ is represented by $z \otimes \bar{z}$, as we see from chasing through our various sequences.

Splitting. Next we verify that the Künneth short exact sequence splits and examine the splittings. We have noted in considering diagram (6.11) that, as each B'_j is projective, there must exist (non-canonical) splittings $C_j \cong Z_j \oplus B'_j$ for each j, though these splittings will not necessarily be compatible across degrees. Let us fix a specific splitting for each j, and thus identify B'_j with a specific submodule of C_j; for the remainder of the discussion, we treat B'_j as a submodule of C_j via this splitting without further comment.

Remark 6.4.18 With this convention, the map $\jmath : B'_j \to Z_{j-1}$ can be identified with the restriction of the boundary map in C_* to the summand B'_j.

Let $\phi_j : C_j \to Z_j$ be the projections determined by our choices of splittings. By definition, with $\imath : Z_j \to C_j$ the inclusion, we have that $\phi_j \imath : Z_j \to Z_j$ is the identity, and we also have $\ker(\phi_j) = B'_j$. Similarly, we choose splittings that let us identify $D_k \cong \bar{Z}_k \oplus \bar{B}'_k$ and corresponding projection $\bar{\phi}_k : D_k \to \bar{Z}_k$. Using the distributivity of tensor products over direct sums, we have

$$\begin{aligned} C_j \otimes D_k &\cong (Z_j \oplus B'_j) \otimes (\bar{Z}_k \oplus \bar{B}'_k) \\ &\cong (Z_j \otimes \bar{Z}_k) \oplus (Z_j \otimes \bar{B}'_k) \oplus (B'_j \otimes \bar{Z}_k) \oplus (B'_j \otimes \bar{B}'_k), \end{aligned}$$

so that each $\phi_j \otimes \bar{\phi}_k$ is a projection to a summand of each $C_j \otimes D_k$, and putting these together, we obtain a splitting projection

$$\bigoplus_{j+k=i} \phi_j \otimes \bar{\phi}_k : \bigoplus_{j+k=i} C_j \otimes D_k \to \bigoplus_{j+k=i} Z_j \otimes \bar{Z}_k.$$

To obtain a homomorphism $H_i(C_* \otimes D_*) \to \bigoplus_{j+k=i} H_j(C_*) \otimes H_k(D_*)$ that splits our inclusion $\bigoplus_{j+k=i} H_j(C_*) \otimes H_k(D_*) \to H_i(C_* \otimes D_*)$ (given, as observed, by $[z] \otimes [\bar{z}] \to [z \otimes \bar{z}]$), we proceed as in the proof of [182, Lemma 58.1]: Let ρ_j be the composition $C_j \xrightarrow{\phi_j} Z_j \xrightarrow{\eta} H_j(C_*)$, with $\bar{\rho}_k$ defined similarly. If E_* is the chain complex with $H_j(C_*)$ in degree j and all boundary maps trivial, then the maps given by ρ_j in degree j give a chain map $\rho : C_* \to E_*$, as boundaries in C_* go to 0 in E_*. Define $\bar{E}_*$ and $\bar{\rho} : D_* \to \bar{E}_*$ similarly. These induce chain maps $\rho \otimes \bar{\rho} : C_* \otimes D_* \to E_* \otimes \bar{E}_*$ and thus homomorphisms $H_i(C_* \otimes D_*) \to H_i(E_* \otimes \bar{E}_*) = \bigoplus_{j+k=i} H_j(C_*) \otimes H_k(D_*)$, with the last equality due to the triviality of the boundary maps of E_* and $\bar{E}_*$. As $\rho \otimes \bar{\rho}$ takes $[z \otimes \bar{z}]$ to $[z] \otimes [\bar{z}]$, this is our desired projection, which demonstrates the splitting of the Künneth exact sequence.

Among other things, the arguments so far show that we can represent the homology classes in the summand $H_j(C_*) \otimes H_k(D_*)$ of $H_i(C_* \otimes D_*)$ (with $j + k = i$) by cycles contained in the summand $Z_j \otimes \bar{Z}_k$ of $C_* \otimes D_*$. The more interesting case is the torsion product terms, and we can now state our main lemma of this section:

Lemma 6.4.19 *Suppose $j + k = i$ and that we identify $H_{j-1}(C_*) * H_k(D_*)$ as a summand of $H_i(C_* \otimes D_*)$ via the splitting determined by $\rho \otimes \bar{\rho}$ as just above. Then any homology class in this summand can be represented by a cycle contained in $(B'_j \otimes \bar{Z}_k) \oplus (Z_{j-1} \otimes \bar{B}'_{k+1}) \subset (C_* \otimes D_*)_i$.*

Proof We already know that we have a surjection of the form

$$H_i(C_* \otimes D_*) \xrightarrow{\beta \otimes \mathrm{id}} \bigoplus_{j+k=i} H_{j-1}(C_*) * H_k(D_*)$$

from the Künneth Theorem and that the map

$$\rho \otimes \bar{\rho} : H_i(C_* \otimes D_*) \to \bigoplus_{j+k=i} H_j(C_*) \otimes H_k(D_*)$$

is a splitting of $H_i(C_* \otimes D_*)$ back to $\ker(\beta \otimes \mathrm{id}) \cong \bigoplus_{j+k=i} H_j(C_*) \otimes H_k(D_*)$. So, via this splitting, we can identify $\bigoplus_{j+k=i} H_{j-1}(C_*) * H_k(D_*)$ as $\ker(\rho \otimes \bar{\rho})$.

We first observe that any cycle of $C_* \otimes D_*$ contained in $(B'_j \otimes \bar{Z}_k) \oplus (Z_{j-1} \otimes \bar{B}'_{k+1}) \subset (C_* \otimes D_*)_i$ does indeed live in $\ker(\rho \otimes \bar{\rho})$, as ρ annihilates B'_j and $\bar{\rho}$ annihilates $\bar{B}'_{k+1}$ by the definitions. So any cycle in $(B'_j \otimes \bar{Z}_k) \oplus (Z_{j-1} \otimes \bar{B}'_{k+1})$ represents an element of the torsion product summand of $H_i(C_* \otimes D_*)$. It remains to show that every element of each $H_{j-1}(C_*) * H_k(D_*)$ can be represented by such a cycle.

Recall that the identification of the torsion product summand of $H_i(C_* \otimes D_*)$ as $\bigoplus_{j+k=i} H_{j-1}(C_*) * H_k(D_*)$ comes by considering the connecting map of the long exact sequence (6.13) and observing that

1. the map ω_{i-1} decomposes as a direct sum of maps $\omega_{j-1,k}$ with $j + k = i$, and
2. the kernel of each $\omega_{j-1,k}$ can be identified with $H_{j-1}(C_*) * H_k(D_*)$.

So, what we will do is the following: for each element $[\xi] \in H_{j-1}(C_*) * H_k(D_*)$ with $j + k = i$ represented as an element of $\ker(\omega_{j-1,k})$, we will show that there is a cycle

$\tilde{\xi} \in (B'_j \otimes \bar{Z}_k) \oplus (Z_{j-1} \otimes \bar{B}'_{k+1}) \subset (C_* \otimes D_*)_i$ that maps to a cycle representing $[\xi]$ under the map $\beta \otimes \mathrm{id}$ of (6.13).

Recall also that $\omega_{j-1,k}$ is just the map $B'_j \otimes H_k(D_*) \xrightarrow{\mathrm{j}\otimes\mathrm{id}} Z_{j-1} \otimes H_k(D_*)$. If we treat $B'_j \xrightarrow{\mathrm{j}} Z_{j-1}$ as a chain complex $P_{j-1,*}$ with nontrivial entries only in degrees zero and one, then its only nontrivial homology group is $H_0(P_{j-1,*}) \cong H_{j-1}(C_*)$, meaning that $P_{j-1,*}$ is a projective resolution of $H_{j-1}(C_*)$. The map $B'_j \otimes H_k(D_*) \xrightarrow{\mathrm{j}\otimes\mathrm{id}} Z_{j-1} \otimes H_k(D_*)$ can then be interpreted as the only nontrivial boundary map of $P_{j-1,*} \otimes H_k(D_*)$, whence $H_1(P_{j-1,*} \otimes H_k(D_*)) = \ker(\mathrm{j} \otimes \mathrm{id})$ is the torsion product $H_{j-1}(C_*) * H_k(D_*)$ by definition, and $H_0(P_{j-1,*} \otimes H_k(D_*)) = \mathrm{cok}(\mathrm{j} \otimes \mathrm{id}) \cong H_{j-1}(C_*) \otimes H_k(D_*)$ by the general theory of derived functors (see e.g. [126, Section IV.5]).

But now we will employ another general fact about the derived functors of the tensor product, namely that they are balanced. While we have used a projective resolution of $H_{j-1}(C_*)$ above to obtain $H_{j-1}(C_*) * H_k(D_*)$, we could just as well have used a projective resolution of $H_k(D_*)$, tensored it on the left with $H_{j-1}(C_*)$, and taken the first homology to obtain the same torsion product $H_{j-1}(C_*) * H_k(D_*)$ (up to isomorphism); see [126, Proposition IV.11.1]. Or, we can use both resolutions simultaneously! More precisely, let us choose the resolution $\bar{P}^{j-1}_{k,*}$ for $H_k(D_*)$ given by $\bar{B}'_{k+1} \xrightarrow{(-1)^{j-1}\bar{\mathrm{j}}} \bar{Z}_k$ in degrees zero and one. Then $H_{j-1}(C_*) * H_k(D_*)$ can also be computed as $H_1(P_{j-1,*} \otimes \bar{P}^{j-1}_{k,*})$. Furthermore, the isomorphism between $H_1(P_{j-1,*} \otimes \bar{P}^{j-1}_{k,*})$ and $H_1(P_{j-1,*} \otimes H_k(D_*))$ is induced by $\mathrm{id} \otimes \bar{\eta}$, where we extend $\bar{\eta} : \bar{Z}_k \to H_k(D_*)$ in the obvious way to a chain map from $\bar{P}^{j-1}_{k,*}$ to the complex whose only nontrivial entry is $H_k(D_*)$ in degree zero. This all follows from the proof of [238, Theorem 2.7.2], concerning balancing the Tor and Ext functors[13].

The reason for the sign in the definition of $\bar{P}^{j-1}_{k,*}$ is the following. Recalling that $P_{j-1,*}$ and $\bar{P}^{j-1}_{k,*}$ have nontrivial modules only in degrees zero and one, the degree-one module of $P_{j-1,*} \otimes \bar{P}^{j-1}_{k,*}$ is $(B'_j \otimes \bar{Z}_k) \oplus (Z_{j-1} \otimes \bar{B}'_{k+1})$. By the definition of the tensor product complex $P_{j-1,*} \otimes \bar{P}^{j-1}_{*}$, the boundary map takes elements $b \otimes \bar{z} \in B'_j \otimes \bar{Z}_k$ to

$$\partial(b \otimes \bar{z}) = \partial_{P_{j-1,*}}(b) \otimes \bar{z} - b \otimes \partial_{\bar{P}^{j-1}_{k,*}}(\bar{z}) = \mathrm{j}(b) \otimes \bar{z},$$

as b has degree one as an element of $P_{j-1,*}$. Similarly, the boundary of an element $z \otimes \bar{b} \in Z_{j-1} \otimes \bar{B}'_{k+1}$ is

$$\partial(z \otimes \bar{b}) = \partial_{P_{j-1,*}}(z) \otimes \bar{b} + z \otimes \partial_{\bar{P}^{j-1}_{k,*}}(\bar{b}) = (-1)^{j-1} z \otimes \bar{\mathrm{j}}(\bar{b}).$$

But as observed in Remark 6.4.18, treating B'_j as a submodule of C_j, the map j is just the restriction of ∂_{C_*} to the summand B'_j. So we have $\partial(b \otimes \bar{z}) = \partial_{C_*}(b) \otimes \bar{z}$ and $\partial(z \otimes \bar{b}) = (-1)^{j-1} z \otimes \partial_{D_*}(\bar{b})$. So the boundary map of $P_{j-1,*} \otimes \bar{P}^{j-1}_{k,*}$ acts on $(B'_j \otimes \bar{Z}_k) \oplus (Z_{j-1} \otimes \bar{B}')$

[13] Caution: in [238, Section 2.7], Weibel uses the notation $P \otimes Q$ to denote the double complex and $\mathrm{Tot}^{\oplus}(P \otimes Q)$ for the single complex, i.e. $\mathrm{Tot}^{\oplus}(P \otimes Q)_i = \bigoplus_{a+b=i} P_a \otimes Q_b$.

in exactly the same way that the boundary map of $C_* \otimes D_*$ acts on $(B'_j \otimes \bar{Z}_k) \oplus (Z_{j-1} \otimes \bar{B}')$ if we identify it as a submodule of $(C_* \otimes D_*)_i$. In particular, $(B'_j \otimes \bar{Z}_k) \oplus (Z_{j-1} \otimes \bar{B}')$ has the same cycles whether we think of it as a submodule of $(P_{j-1,*} \otimes \bar{P}^{j-1}_{k,*})_1$ or of $(C_* \otimes D_*)_i$

Therefore, let $[\xi] \in H_{j-1}(C_*) * H_k(D_*) = \ker(\omega_{j-1,k}) \subset B'_j \otimes H_k(D_*)$. Identifying $H_{j-1}(C_*) * H_k(D_*)$ with $H_1(P_{j-1,*} \otimes \bar{P}^{j-1}_{k,*})$ tells us that there is a cycle $\tilde{\xi} \in (B'_j \otimes \bar{Z}_k) \oplus (Z_j \otimes \bar{B}')$ representing $[\xi]$. Then identifying $(B'_j \otimes \bar{Z}_k) \oplus (Z_{j-1} \otimes \bar{B}')$ again as a submodule of $(C_* \otimes D_*)_i$, the cycle $\tilde{\xi}$ represents an element of $H_i(C_* \otimes D_*)$. It only remains to show that the image of the corresponding homology class under the map $\beta \otimes \mathrm{id}$ of diagram (6.13) is $[\xi]$.

As noted, the isomorphism between $H_1(P_{j-1,*} \otimes \bar{P}^{j-1}_{k,*})$ and $H_1(P_{j-1,*} \otimes H_k(D_*))$ (both representing $H_{j-1}(C_*) * H_k(D_*)$) is induced by $\mathrm{id} \otimes \bar{\eta}$. So, by definition, $\tilde{\xi}$ is a cycle in $(B'_j \otimes \bar{Z}_k) \oplus (Z_{j-1} \otimes \bar{B}'_{k+1})$ with the property that $(\mathrm{id} \otimes \bar{\eta})(\tilde{\xi})$ represents $[\xi]$. The map $\mathrm{id} \otimes \bar{\eta}$ acts trivially on the summand $Z_{j-1} \otimes \bar{B}'_{k+1}$ because $\bar{\eta}$ takes $\bar{B}'_{k+1}$ to 0. But on the summand $B'_j \otimes \bar{Z}_k$, we know that $\mathrm{id} \otimes \bar{\eta}$ is just the projection from the cycle module in degree $j + k = i$ of the complex $B'_j \otimes D_*$ to the homology module $H_i(B'_j \otimes D_*) \cong B'_j \otimes H_k(D_*)$. In particular, this tells us that $[\xi]$ is represented in $B'_j \otimes H_k(D_*)$ by the $B'_j \otimes \bar{Z}_k$ summand of $\tilde{\xi}$.

On the other hand, we must consider how $\beta \otimes \mathrm{id}$ acts on $(B'_j \otimes \bar{Z}_k) \oplus (Z_{j-1} \otimes \bar{B}')$ as a submodule of $(C_* \otimes D_*)_i$. In this case, the map β annihilates Z_{j-1}, so $\beta \otimes \mathrm{id}$ also annihilates $Z_{j-1} \otimes \bar{B}'_k$. But β restricts to the identity on B'_j, so the image of $\tilde{\xi}$ in $B'_j \otimes D_k$ under $\beta \otimes \mathrm{id}$ is also represented by the $B'_j \otimes \bar{Z}_k$ summand of $\tilde{\xi}$. As this agrees with the computation of the preceding paragraph, the cycle $\tilde{\xi}$ has the required properties. □

(b) Intersection Homology Products with Cones

We now apply the algebra we have been developing to the intersection homology of products. Let X and $\mathcal{Y}$ be filtered spaces, and suppose that X has dimension $n-1$ and that R is a Dedekind domain. We wish to consider $\mathrm{i} \otimes \mathrm{id} : I^{\bar{p}}S_*(cX - \{v\}; R) \otimes I^{\bar{q}}S_*(\mathcal{Y}; R) \to I^{\bar{p}}S_*(cX; R) \otimes I^{\bar{q}}S_*(\mathcal{Y}; R)$ induced by the inclusion $\mathrm{i} : cX - \{v\} \to cX$. The naturality of the algebraic Künneth Theorem gives us a map of short exact sequences (with R coefficients tacit):

$$\begin{array}{ccccc}
\bigoplus_{i=j+k} I^{\bar{p}}H_j(cX - \{v\}) \otimes I^{\bar{q}}H_k(\mathcal{Y}) & \hookrightarrow & H_i(I^{\bar{p}}S_*(cX - \{v\}) \otimes I^{\bar{q}}S_*(\mathcal{Y})) & \twoheadrightarrow & \bigoplus_{i=j+k} I^{\bar{p}}H_{j-1}(cX - \{v\}) * I^{\bar{q}}H_k(\mathcal{Y}) \\
\downarrow & & \downarrow & & \downarrow \\
\bigoplus_{i=j+k} I^{\bar{p}}H_j(cX) \otimes I^{\bar{q}}H_k(\mathcal{Y}) & \hookrightarrow & H_i(I^{\bar{p}}S_*(cX) \otimes I^{\bar{q}}S_*(\mathcal{Y})) & \twoheadrightarrow & \bigoplus_{i=j+k} I^{\bar{p}}H_{j-1}(cX) * I^{\bar{q}}H_k(\mathcal{Y}).
\end{array}$$

We know too that $cX - \{v\} \hookrightarrow cX$ induces an isomorphism $I^{\bar{p}}H_{j-1}(cX - \{v\}; R) \to I^{\bar{p}}H_{j-1}(cX; R)$ for $j - 1 < n - \bar{p}(\{v\}) - 1$ and that $I^{\bar{p}}H_{j-1}(cX; R) = 0$ for $j - 1 \geq$

$n - \bar{p}(\{v\}) - 1$. Therefore, naively, we expect that the map

$$H_i(I^{\bar{p}}S_*(cX - \{v\}; R) \otimes I^{\bar{q}}S_*(\mathcal{Y}; R)) \to H_i(I^{\bar{p}}S_*(cX; R) \otimes I^{\bar{q}}S_*(\mathcal{Y}; R))$$

should kill the summands

$$I^{\bar{p}}H_{j-1}(cX - \{v\}; R) \otimes I^{\bar{q}}H_k(\mathcal{Y}; R)$$

or

$$I^{\bar{p}}H_{j-1}(cX - \{v\}; R) * I^{\bar{q}}H_k(\mathcal{Y}; R)$$

with $j - 1 \geq n - \bar{p}(\{v\}) - 1$ and take the summands with $j - 1 < n - \bar{p}(\{v\}) - 1$ isomorphically to their counterparts in the codomain.

Unfortunately, the algebraic Künneth exact sequences do not split naturally, in general, and so it is conceivable for there to be unexpected subtleties. For example, one can construct maps $\phi \otimes \psi : C_* \otimes D_* \to C'_* \otimes D'_*$ for which, in the ensuing diagram of Künneth exact sequences, the torsion product summand of $H_i(C'_* \otimes D'_*)$ is zero and the tensor product summand for $H_i(C_* \otimes D_*)$ is zero, but for which the map $H_i(C_* \otimes D_*) \to H_i(C'_* \otimes D'_*)$ is not trivial; see [126, Section V.2]. Therefore, to verify our naive expectations, we cannot rely solely on algebra but must also utilize the topology involved.

Lemma 6.4.20 *Suppose $\mathcal{Y}$ is a filtered space, that X is a dimension $n - 1$ filtered space, and that R is a Dedekind domain. Consider*

$$\mathfrak{i} \otimes \mathrm{id} : I^{\bar{p}}S_*(cX - \{v\}; R) \otimes I^{\bar{q}}S_*(\mathcal{Y}; R) \to I^{\bar{p}}S_*(cX; R) \otimes I^{\bar{q}}S_*(\mathcal{Y}; R),$$

induced by the inclusion $\mathfrak{i} : cX - \{v\} \to cX$. The splittings and naturality of the Künneth Theorem induce the map (R coefficients tacit):

$$\begin{array}{ccc} H_i(I^{\bar{p}}S_*(cX - \{v\}) \otimes I^{\bar{q}}S_*(\mathcal{Y})) & \cong & \begin{array}{l} \bigoplus_{i=j+k} I^{\bar{p}}H_j(cX - \{v\}) \otimes I^{\bar{q}}H_k(\mathcal{Y}) \oplus \\ \bigoplus_{i=j+k} I^{\bar{p}}H_{j-1}(cX - \{v\}) * I^{\bar{q}}H_k(\mathcal{Y}) \end{array} \\ \downarrow {\scriptstyle \mathfrak{i} \otimes \mathrm{id}} & & \\ H_i(I^{\bar{p}}S_*(cX) \otimes I^{\bar{q}}S_*(\mathcal{Y})) & \cong & \begin{array}{l} \bigoplus_{i=j+k} I^{\bar{p}}H_j(cX) \otimes I^{\bar{q}}H_k(\mathcal{Y}) \oplus \\ \bigoplus_{i=j+k} I^{\bar{p}}H_{j-1}(cX) * I^{\bar{q}}H_k(\mathcal{Y}). \end{array} \end{array}$$

While naturality of the splittings is not a general property of the Künneth Theorem, in this setting the splittings of the Künneth exact sequence can be chosen so that each summand $I^{\bar{p}}H_j(cX - \{v\}; R) \otimes I^{\bar{q}}H_k(\mathcal{Y}; R)$ with $j \geq n - \bar{p}(\{v\}) - 1$ or $I^{\bar{p}}H_{j-1}(cX - \{v\}; R) * I^{\bar{q}}H_k(\mathcal{Y}; R)$ with $j - 1 \geq n - \bar{p}(\{v\}) - 1$ maps to 0 in $H_i(I^{\bar{p}}S_*(cX; R) \otimes I^{\bar{q}}S_*(\mathcal{Y}; R))$ and so that the other summands map to corresponding summands with identical chain representatives for homology classes in $H_i(I^{\bar{p}}S_*(cX; R) \otimes I^{\bar{q}}S_*(\mathcal{Y}; R))$.

Proof Recall the following facts from the proof of the cone formula in Theorem 4.2.1: If $\ell < n - \bar{p}(\{v\})$ then no allowable simplex can intersect $\{v\}$ at all, and so in this range we have $I^{\bar{p}}S_\ell(cX; R) = I^{\bar{p}}S_\ell(cX - \{v\}; R)$. On the other hand, if $\ell \geq n - \bar{p}(\{v\})$, then simplices may intersect the vertex. In fact, any cone $\bar{c}(\xi)$ on a chain $\xi \in I^{\bar{p}}S_\ell(cX - \{v\}; R)$ is allowable if either $\ell \geq n - \bar{p}(\{v\})$ or $\ell = n - \bar{p}(\{v\}) - 1$ with $\partial\xi = 0$ (the last condition ensures that $\partial(\bar{c}(\xi))$ remain allowable).

Let us see how this plays out for our map $H_i(I^{\bar{p}}S_*(cX - \{v\}; R) \otimes I^{\bar{q}}S_*(\mathcal{Y}; R)) \to H_i(I^{\bar{p}}S_*(cX; R) \otimes I^{\bar{q}}S_*(\mathcal{Y}; R))$, working one summand at a time.

First, let us consider the tensor product summands with domain $I^{\bar{p}}H_j(cX - \{v\}; R) \otimes I^{\bar{q}}H_k(\mathcal{Y}; R)$. By the naturality of the Künneth exact sequences, the tensor product summand $\bigoplus_{i=j+k} I^{\bar{p}}H_j(cX - \{v\}) \otimes I^{\bar{q}}H_k(\mathcal{Y})$ of $H_i(I^{\bar{p}}S_*(cX - \{v\}; R) \otimes I^{\bar{q}}S_*(\mathcal{Y}; R))$ gets taken to the tensor product summand $\bigoplus_{i=j+k} I^{\bar{p}}H_j(cX) \otimes I^{\bar{q}}H_k(\mathcal{Y})$ of $H_i(I^{\bar{p}}S_*(cX; R) \otimes I^{\bar{q}}S_*(\mathcal{Y}; R))$, so to understand the map on this summand, it is enough to look at each

$$\mathrm{i} \otimes \mathrm{id} : I^{\bar{p}}H_j(cX - \{v\}; R) \otimes I^{\bar{q}}H_k(\mathcal{Y}; R) \to \bigoplus_{i=j+k} I^{\bar{p}}H_j(cX) \otimes I^{\bar{q}}H_k(\mathcal{Y}).$$

But this is clear: Each tensor product summand $I^{\bar{p}}H_j(cX - \{v\}; R) \otimes I^{\bar{q}}H_k(\mathcal{Y}; R)$ is generated by elements of the form $[z] \otimes [y] \in I^{\bar{p}}H_j(cX - \{v\}; R) \otimes I^{\bar{q}}H_k(\mathcal{Y}; R)$, which map to the corresponding elements of $I^{\bar{p}}H_j(cX; R) \otimes I^{\bar{q}}H_k(\mathcal{Y}; R)$, and which project trivially onto the other summands $I^{\bar{p}}H_a(cX; R) \otimes I^{\bar{q}}H_b(\mathcal{Y}; R)$ with $a \neq j$ and $b \neq k$ via our splitting maps $\phi_a \otimes \bar{\phi}_b : \bigoplus_{r+s=i} C_r \otimes D_s \to Z_a \otimes \bar{Z}_b$ of the preceding subsection. So if $j \geq n - \bar{p}(\{v\}) - 1$, the summand $I^{\bar{p}}H_j(cX - \{v\}; R) \otimes I^{\bar{q}}H_k(\mathcal{Y}; R)$ maps to $I^{\bar{p}}H_j(cX; R) \otimes I^{\bar{q}}H_k(\mathcal{Y}; R) = 0$, and if $j < n - \bar{p}(\{v\}) - 1$, the summand maps isomorphically to the corresponding summand $I^{\bar{p}}H_j(cX; R) \otimes I^{\bar{q}}H_k(\mathcal{Y}; R)$. In fact, since in this latter case we have $j < n - \bar{p}(\{v\})$, we are in the range where $I^{\bar{p}}S_j(cX; R) = I^{\bar{p}}S_j(cX - \{v\}; R)$ and so the map takes summand generators to the precisely corresponding generators in the image.

Now, we must consider the torsion product summands. This is more subtle, as the lack of natural splitting in the Künneth exact sequences means that we cannot simply assume that the torsion product summand $\bigoplus_{i=j+k} I^{\bar{p}}H_{j-1}(cX - \{v\}) * I^{\bar{q}}H_k(\mathcal{Y})$ of $H_i(I^{\bar{p}}S_*(cX - \{v\}; R) \otimes I^{\bar{q}}S_*(\mathcal{Y}; R))$ maps only to the torsion product summand $\bigoplus_{i=j+k} I^{\bar{p}}H_{j-1}(cX) * I^{\bar{q}}H_k(\mathcal{Y})$ of $H_i(I^{\bar{p}}S_*(cX; R) \otimes I^{\bar{q}}S_*(\mathcal{Y}; R))$. By Lemma 6.4.19, each element of the summand

$$I^{\bar{p}}H_{j-1}(cX - \{v\}; R) * I^{\bar{q}}H_k(\mathcal{Y}; R) \subset H_i(I^{\bar{p}}S_*(cX - \{v\}) \otimes I^{\bar{q}}S_*(\mathcal{Y}))$$

can be represented as a cycle $\tilde{\xi}$ in $(B'_j \otimes \bar{Z}_k) \oplus (Z_{j-1} \otimes \bar{B}'_{k+1})$, where now we let Z_{j-1}, $\bar{Z}_k$, B'_j, and $\bar{B}'_{k+1}$ be cycle and boundary submodules as in the discussion of the preceding subsections, taking $C_* = I^{\bar{p}}S_*(cX - \{v\}; R)$ and $D_* = I^{\bar{q}}S_*(\mathcal{Y}; R)$. We here assume a choice of splitting so that B'_j is a summand of C_j and similarly for $\bar{B}'_{k+1}$ and D_{k+1}. In particular, then, $\tilde{\xi}$ must be of the form

$$\tilde{\xi} = \sum_\ell b_\ell \otimes \bar{z}_\ell + \sum_m z_m \otimes \bar{b}_m \tag{6.14}$$

with $b_\ell \in B'_j$, $\bar{z}_\ell \in \bar{Z}_k$, $z_m \in Z_{j-1}$, and $\bar{b}_m \in \bar{B}'_{k+1}$. As $\tilde{\xi}$ is a cycle, we must have

$$0 = \partial\tilde{\xi} = \sum_\ell \partial_{C_*}(b_\ell) \otimes \bar{z}_\ell + (-1)^{j-1} \sum_m z_m \otimes \partial_{D_*}(\bar{b}_m), \tag{6.15}$$

utilizing that the z_m and $\bar{z}_\ell$ are cycles. Let us see what homology class $[\tilde{\xi}] \in H_i(I^{\bar{p}}S_*(cX - \{v\}) \otimes I^{\bar{q}}S_*(\mathcal{Y}))$ maps to under $\mathfrak{i} \otimes \mathrm{id}$.

First, suppose that $j - 1 \geq n - \bar{p}(\{v\}) - 1$. Let

$$\zeta = \sum_\ell (\bar{c}(b_\ell)) \otimes \bar{z}_\ell + \sum_m (\bar{c}(z_m)) \otimes \bar{b}_m \in I^{\bar{p}}S_*(cX) \otimes I^{\bar{q}}S_*(\mathcal{Y}).$$

The chain ζ is allowable, by our above observations and recalling that the z_m are cycles. Then we have

$$\begin{aligned}
\partial\zeta &= \partial\left[\sum_\ell (\bar{c}(b_\ell)) \otimes \bar{z}_\ell + \sum_m (\bar{c}(z_m)) \otimes \bar{b}_m\right] \\
&= \sum_\ell \partial(\bar{c}(b_\ell)) \otimes \bar{z}_\ell + \sum_m \left(\partial(\bar{c}(z_m)) \otimes \bar{b}_m + (-1)^j \bar{c}(z_m) \otimes \partial\bar{b}_m\right) \\
&= \sum_\ell (b_\ell - \bar{c}(\partial b_\ell)) \otimes \bar{z}_\ell + \sum_m \left(z_m \otimes \bar{b}_m + (-1)^j \bar{c}(z_m) \otimes \partial\bar{b}_m\right) \\
&= \sum_\ell b_\ell \otimes \bar{z}_\ell + \sum_m z_m \otimes \bar{b}_m - \sum_\ell \bar{c}(\partial b_\ell) \otimes \bar{z}_\ell + (-1)^j \sum_m \bar{c}(z_m) \otimes \partial\bar{b}_m \\
&= \tilde{\xi} - \left[\sum_\ell \bar{c}(\partial b_\ell) \otimes \bar{z}_\ell + (-1)^{j-1} \sum_m (\bar{c}(z_m)) \otimes \partial\bar{b}_m\right],
\end{aligned}$$

identifying chains in $cX - \{v\}$ with their images under $\mathfrak{i}$ in cX. Observe now that $\tilde{\xi} - \partial\zeta$ looks just like our expression (6.15) above for $\partial\tilde{\xi}$ except that the first term in each tensor product now has a cone in the expression. In other words, $\tilde{\xi} - \partial\zeta$ is just the image of $\partial\tilde{\xi} = 0$ under the homomorphism $\bar{c} \otimes \mathrm{id}$, and so it is trivial. We conclude that the image of $\tilde{\xi}$ under $\mathfrak{i} \otimes \mathrm{id}$ is a boundary and so represents 0 in $H_i(I^{\bar{p}}S_*(cX; R) \otimes I^{\bar{q}}S_*(\mathcal{Y}; R))$. Thus, for $j - 1 \geq n - \bar{p}(\{v\}) - 1$, each of the $I^{\bar{p}}H_{j-1}(cX - \{v\}; R) * I^{\bar{q}}H_k(\mathcal{Y}; R)$, $j + k = i$, summands of $H_i(I^{\bar{p}}S_*(cX - \{v\}; R) \otimes I^{\bar{q}}S_*(\mathcal{Y}; R))$ maps to 0 in $H_i(I^{\bar{p}}S_*(cX; R) \otimes I^{\bar{q}}S_*(\mathcal{Y}; R))$.

Next, consider the torsion product summands $I^{\bar{p}}H_{j-1}(cX - \{v\}; R) * I^{\bar{q}}H_k(\mathcal{Y}; R)$ with $j - 1 < n - \bar{p}(\{v\}) - 1$. These are also represented by cycles $\tilde{\xi}$ of the form (6.14) and so are contained in summands of $I^{\bar{p}}S_*(cX - \{v\}; R) \otimes I^{\bar{q}}S_*(\mathcal{Y}; R)$ of the form

$$[I^{\bar{p}}S_j(cX - \{v\}; R) \otimes I^{\bar{q}}S_k(\mathcal{Y}; R)] \oplus [I^{\bar{p}}S_{j-1}(cX - \{v\}; R) \otimes I^{\bar{q}}S_{k+1}(\mathcal{Y}; R)].$$

But in this case $j < n - \bar{p}(\{v\})$, and we are in the range where $I^{\bar{p}}S_*(cX - \{v\}; R) = I^{\bar{p}}S_*(cX; R)$. In particular, we can assume that the splittings into cycles and boundaries are identical for $I^{\bar{p}}S_*(cX - \{v\}; R)$ and the identical $I^{\bar{p}}S_*(cX; R)$ in this degree range, while we choose arbitrary splittings into cycles and boundaries of $I^{\bar{p}}S_*(cX; R)$ in higher degrees. So, for $j < n - \bar{p}(\{v\})$, the submodules of the form $(B'_j \otimes \bar{Z}_k) \oplus (Z_{j-1} \otimes \bar{B}'_{k+1})$ are identical in $I^{\bar{p}}S_*(cX - \{v\}; R) \otimes I^{\bar{q}}S_*(\mathcal{Y}; R)$ and $I^{\bar{p}}S_*(cX; R) \otimes I^{\bar{q}}S_*(\mathcal{Y}; R)$, and they map under the corresponding $\beta \otimes \mathrm{id}$ to identical summands of the corresponding $\ker(\omega_{j-1,k})$. So the restriction of the map $\mathrm{i} \otimes \mathrm{id}$ to $I^{\bar{p}}H_{j-1}(cX - \{v\}; R) * I^{\bar{q}}H_k(\mathcal{Y}; R)$ followed by projection to $I^{\bar{p}}H_{j-1}(cX; R) * I^{\bar{q}}H_k(\mathcal{Y}; R)$ can be viewed as the identity isomorphism between corresponding summands. Furthermore, the image of $\tilde{\xi}$ in $I^{\bar{p}}S_*(cX; R) \otimes I^{\bar{q}}S_*(\mathcal{Y}; R)$ under $\mathrm{i} \otimes \mathrm{id}$ projects trivially to every other torsion summand $I^{\bar{p}}H_a(cX; R) * I^{\bar{q}}S_b(\mathcal{Y}; R)$ without $a = j - 1$, $b = k$, as, if we let $\mathscr{B}'_a$ be the boundary summands of $I^{\bar{p}}S_*(cX; R)$ (so $\mathscr{B}'_a = B'_a$ if $a < n - \bar{p}(\{v\})$), then the composition of the map $\beta \otimes \mathrm{id} : C_* \otimes D_* \to \bigoplus_{j+k=i} \mathscr{B}'_* \otimes D_*$ with the projection to $\mathscr{B}'_a \otimes D_b$ is 0 on elements of $(\mathscr{B}'_j \otimes \bar{Z}_k) \oplus (Z_{j-1} \otimes \bar{B}'_{k+1})$ unless $a = j - 1$. We also already know that a cycle of this form projects to 0 in any torsion summand under the splitting maps of the form $\bigoplus_{j+k=i} \phi_j \otimes \bar{\phi}_k$. Hence the homology class represented by $\tilde{\xi}$ maps only to a class in the summand $I^{\bar{p}}H_{j-1}(cX; R) * I^{\bar{q}}H_k(\mathcal{Y}; R)$ of $H_i(I^{\bar{p}}S_*(cX; R) \otimes I^{\bar{q}}S_*(\mathcal{Y}; R))$.

Altogether then, we can conclude that with our consistent choices of splittings the map $H_i(I^{\bar{p}}S_*(cX - \{v\}; R) \otimes I^{\bar{q}}S_*(\mathcal{Y}; R)) \to H_i(I^{\bar{p}}S_*(cX; R) \otimes I^{\bar{q}}S_*(\mathcal{Y}; R))$ is the naive one: it takes any summand $I^{\bar{p}}H_j(cX - \{v\}; R) \otimes I^{\bar{q}}H_k(\mathcal{Y}; R)$ to 0 if $j \geq n - \bar{p}(\{v\}) - 1$ and any summand $I^{\bar{p}}H_{j-1}(cX - \{v\}; R) * I^{\bar{q}}H_k(\mathcal{Y}; R)$ to 0 if $j - 1 \geq n - \bar{p}(\{v\}) - 1$; otherwise it takes each summand to an identically corresponding summand of $H_i(I^{\bar{p}}S_*(cX; R) \otimes I^{\bar{q}}S_*(\mathcal{Y}; R))$. □

Proof of Lemma 6.4.20. We can now prove Lemma 6.4.2, which we state again here for reference:

Lemma (6.4.20) *Given a Dedekind domain R and compact filtered sets $X = X^{n-1}$ and $Y = Y^{m-1}$, there are splittings of*

$$H_i(I^{\bar{p}}S_*(cX - \{v\}; R) \otimes I^{\bar{q}}S_*(cY - \{w\}; R)),$$
$$H_i(I^{\bar{p}}S_*(cX; R) \otimes I^{\bar{q}}S_*(cY - \{w\}; R))),$$
$$H_i(I^{\bar{p}}S_*(cX - \{v\}; R) \otimes I^{\bar{q}}S_*(cY; R)),$$
$$H_i(I^{\bar{p}}S_*(cX; R) \otimes I^{\bar{q}}S_*(Y; R))$$

into direct sums of tensor products $I^{\bar{p}}H_j(cX - \{v\}; R) \otimes I^{\bar{q}}H_k(cY - \{w\}; R)$ and torsion products $I^{\bar{p}}H_{j-1}(cX - \{v\}; R) * I^{\bar{q}}H_k(cY - \{w\}; R)$, both with $j + k = i$, such that the maps in the diagram

$$\begin{array}{ccc}
H_i(I^{\bar{p}}S_*(cX - \{v\}; R) \otimes I^{\bar{q}}S_*(Y - \{w\}; R)) & \xrightarrow{\mathrm{i} \otimes \mathrm{id}} & H_i(I^{\bar{p}}S_*(cX; R) \otimes I^{\bar{q}}S_*(cY - \{w\}; R)) \\
\downarrow{\scriptstyle \mathrm{id} \otimes \bar{\mathrm{i}}} & & \downarrow{\scriptstyle \mathrm{id} \otimes \bar{\mathrm{i}}} \\
H_i(I^{\bar{p}}S_*(cX - \{v\}; R) \otimes I^{\bar{q}}S_*(cY; R)) & \xrightarrow{\mathrm{i} \otimes \mathrm{id}} & H_i(I^{\bar{p}}S_*(cX; R) \otimes I^{\bar{q}}S_*(cY; R))
\end{array} \tag{6.16}$$

induced by the inclusions $\mathrm{i} : cX - \{v\} \hookrightarrow cX$ and $\bar{\mathrm{i}} : cY - \{w\} \hookrightarrow cY$ each restrict on each *tensor or torsion product summand either to the zero map or to an isomorphism with the corresponding summand in the codomain. Furthermore, which of these options occurs is determined in the obvious way by the cone formula, Theorem 6.2.13; for example, the tensor product summand* $I^{\bar{p}}H_j(cX-\{v\};R)\otimes I^{\bar{q}}H_k(cY-\{w\};R)$ *maps to* 0 *in* $H_i(I^{\bar{p}}S_*(cX;R)\otimes I^{\bar{q}}S_*(Y-\{w\};R))$ *when* $j\geq n-\bar{p}(\{v\})-1$ *and isomorphically to a corresponding summand otherwise.*

Proof The existence of such properties holds independently for each map in the diagram due to Lemma 6.4.20. However, we must also verify that the choices can be made compatibly.

First, consider the tensor product summands of the expressions in the diagram. The naturality of the Künneth Theorem and the arguments of the proof of Lemma 6.4.20 tell us that the tensor product summands always map to corresponding tensor product summands, and, in particular, the chain map $\mathfrak{i}\otimes\mathrm{id}$ induces the corresponding tensor product of homology maps. Therefore, the maps on the tensor product summands behave as expected.

Now we must consider the torsion product summands. For this, we can assume that we choose once and for all fixed splittings of each $I^{\bar{p}}S_\ell(cX-\{v\};R)$ and $I^{\bar{q}}S_\ell(cY-\{w\};R)$ into cycles and boundaries Z_ℓ and B'_ℓ, or $\bar{Z}_\ell$ and $\bar{B}'_\ell$, for all ℓ. As we know that $I^{\bar{p}}S_\ell(cX-\{v\};R)=I^{\bar{p}}S_\ell(cX;R)$ for $\ell<n-\bar{p}(\{v\})$ and $I^{\bar{q}}S_\ell(cY-\{w\};R)=I^{\bar{q}}S_\ell(cY;R)$ for $\ell<m-\bar{q}(\{w\})$, we can also assume the splittings for $I^{\bar{p}}S_\ell(cX;R)$ and $I^{\bar{q}}S_\ell(cY;R)$ are chosen to be the same in those degrees and arbitrary in higher degrees. Using these choices, we saw in the proof of Lemma 6.4.20 that each map of diagram (6.16) takes torsion summands in the domain to corresponding summands in the codomain. In particular, via those arguments, only the summands $I^{\bar{p}}H_{j-1}(cX-\{v\};R)*I^{\bar{q}}H_k(cY-\{w\};R)$ of $H_i(I^{\bar{p}}S_*(cX-\{v\};R)\otimes I^{\bar{q}}S_*(Y-\{w\};R))$ with $j-1<n-\bar{p}(\{v\})-1$ and $k<m-\bar{q}(\{w\})-1$ survive to the bottom right of the diagram, traveling in either direction, and these are represented in each module of the diagram by cycles contained in the submodules $(B'_j\otimes\bar{Z}_k)\oplus(Z_{j-1}\otimes\bar{B}'_{k+1})$, which are submodules of each chain complex in the diagram due to the assumptions on the ranges of j and k. So running any element of a summand $I^{\bar{p}}H_{j-1}(cX-\{v\};R)*I^{\bar{q}}H_k(cY-\{w\};R)$ of $H_i(I^{\bar{p}}S_*(cX-\{v\};R)\otimes I^{\bar{q}}S_*(Y-\{w\};R))$ to $H_i(I^{\bar{p}}S_*(cX;R)\otimes I^{\bar{q}}S_*(cY;R))$ by any route takes it to an element represented by the same cycle in the corresponding submodule $(B'_j\otimes\bar{Z}_k)\oplus(Z_{j-1}\otimes\bar{B}'_{k+1})$ of $I^{\bar{p}}S_*(cX;R)\otimes I^{\bar{q}}S_*(cY;R)$, and hence to the corresponding element of the corresponding homology summand. This is the claim of the lemma. □

6.5 Advanced Topic: Chain Splitting

In this section, we prove some technical results that will be needed in later chapters[14]. These all concern splitting intersection chains into small pieces and so are related to the arguments we made in proving the excision property and the existence of Mayer–Vietoris sequences. As in those arguments, the standard techniques from ordinary homology are not quite sufficient, as we need to be very careful to avoid the "standard mistake" of creating

[14] Specifically, see the proofs of Theorems 7.1.12 and 7.1.13, Proposition 7.3.59, and Lemmas 7.3.60 and 7.3.61.

faces that are not allowable in the boundaries when breaking a chain into pieces. The proofs here are based upon the arguments in [85] for demonstrating an analogue of Proposition 6.5.1. We state and prove these results for both GM and non-GM intersection chains, though the former will not be utilized below.

For the convenience of the reader who does not want to read through all of the details, we first state the main results of this section and then turn to the proofs. The proofs of the later statements all involve the constructions in the proof of our first statement, Proposition 6.5.1.

Proposition 6.5.1 *Let X be a filtered space with perversity $\bar{p}$. Let $\mathcal{V}$ be a covering of X such that the interiors of the elements of $\mathcal{V}$ constitute an open covering of X, let $A \subset X$, and let*

$$I^{\bar{p}}S_*^{\mathcal{V}}(X,A;G) = \sum_{V \in \mathcal{V}} I^{\bar{p}}S_*(V, A \cap V; G) \subset I^{\bar{p}}S_*(X,A;G).$$

Define $I^{\bar{p}}S_*^{\mathrm{GM},\mathcal{V}}(X,A;G)$, $I^{\bar{p}}\mathfrak{C}_*^{\mathcal{V}}(X,A;G)$, and $I^{\bar{p}}\mathfrak{C}_*^{\mathrm{GM},\mathcal{V}}(X,A;G)$ analogously. Then the inclusions $I^{\bar{p}}S_*^{\mathrm{GM},\mathcal{V}}(X,A;G) \hookrightarrow I^{\bar{p}}S_*^{\mathrm{GM}}(X,A;G)$ and $I^{\bar{p}}S_*^{\mathcal{V}}(X,A;G) \hookrightarrow I^{\bar{p}}S_*(X,A;G)$ are chain homotopy equivalences, and the corresponding inclusions $I^{\bar{p}}\mathfrak{C}_*^{\mathrm{GM},\mathcal{V}}(X,A;G) \hookrightarrow I^{\bar{p}}\mathfrak{C}_*^{\mathrm{GM}}(X,A;G)$ and $I^{\bar{p}}\mathfrak{C}_*^{\mathcal{V}}(X,A;G) \hookrightarrow I^{\bar{p}}\mathfrak{C}_*(X,A;G)$ are isomorphisms.

The same result holds as a statement about R-modules, replacing the abelian group G with an R-module over a commutative ring with unity R.

The following proposition first appears in this form in [23]:

Proposition 6.5.2 *Let X be a filtered space with perversity $\bar{p}$, and let $\mathcal{U}$ be an open cover of X. Suppose $\xi \in I^{\bar{p}}S_i^{\mathrm{GM}}(X;G) \cap S_i^{\mathcal{U}}(X;G)$, i.e. ξ is an intersection chain each of whose simplices is contained in an element of $\mathcal{U}$. Then $\xi \in I^{\bar{p}}S_i^{\mathrm{GM},\mathcal{U}}(X;G)$. Similarly, if $\xi \in I^{\bar{p}}S_i(X;G) \cap S_i^{\mathcal{U}}(X;G)$ then $\xi \in I^{\bar{p}}S_i^{\mathcal{U}}(X;G)$.*

The same result holds as a statement about R-modules, replacing the abelian group G with an R-module over a commutative ring with unity R.

Corollary 6.5.3 *Let A be an open subset of the filtered space X. Then the maps*

$$I^{\bar{p}}S_i^{\mathrm{GM}}(A;G) \to I^{\bar{p}}S_i^{\mathrm{GM}}(X;G) \quad \text{and} \quad I^{\bar{p}}S_i(A;G) \to I^{\bar{p}}S_i(X;G)$$

induced by inclusion are split inclusions.

The same result holds as a statement about R-modules, replacing the abelian group G with an R-module over a commutative ring with unity R.

Of course, we know this last result is true for $I^{\bar{p}}S_i^{\mathrm{GM}}(X;R)$ or $I^{\bar{p}}S_i(X;R)$ when R is a Dedekind domain, as in that case $I^{\bar{p}}S_i^{\mathrm{GM}}(X,A;R)$ or $I^{\bar{p}}S_i(X,A;R)$ are projective (see Lemma 6.3.1). This result is more general as it allows for any coefficients.

We now turn toward proving Propositions 6.5.1 and 6.5.2 and Corollary 6.5.3.

Proof of Proposition 6.5.1 As there is no difference in the proof between working with coefficients in an abelian group as opposed to coefficients in an R-module, we will stick with the former throughout the argument.

Recall that we assume $\mathcal{V}$ is a covering of X such that the interiors of the sets in $\mathcal{V}$ cover of X, and we let

$$I^{\bar{p}}S_*^{\mathcal{V}}(X,A;G) = \sum_{V\in\mathcal{V}} I^{\bar{p}}S_*(V,A\cap V;G) \subset I^{\bar{p}}S_*(X,A;G).$$

Notice that each $I^{\bar{p}}S_*(V,A\cap V;G)$ really does inject into $I^{\bar{p}}S_*(X,A;G)$: the only chains in the kernel of $I^{\bar{p}}S_*(V;G) \to I^{\bar{p}}S_*(X,A;G)$ are those that are supported in A and V, and those are zero in $I^{\bar{p}}S_*(V,A\cap V;G)$. Therefore, we can identify the image of $I^{\bar{p}}S_*(V,A\cap V;G)$ as a subgroup of $I^{\bar{p}}S_*(X,A;G)$, and the sum then makes sense. We also observe that $I^{\bar{p}}S_*^{\mathcal{V}}(X,A;G)$ consists of the elements of $\xi \in I^{\bar{p}}S_*(X,A;G)$ that can be represented as a finite sum of chains $\xi = \sum_{V\in\mathcal{V}} \xi_V$ with $\xi_V \in I^{\bar{p}}S_*(V;G)$. The complex $I^{\bar{p}}S_*^{\mathrm{GM},\mathcal{V}}(X,A;G)$ is defined analogously. Proposition 6.5.1 states that the inclusions $I^{\bar{p}}S_*^{\mathrm{GM},\mathcal{V}}(X,A;G) \hookrightarrow I^{\bar{p}}S_*^{\mathrm{GM}}(X,A;G)$ and $I^{\bar{p}}S_*^{\mathcal{V}}(X,A;G) \hookrightarrow I^{\bar{p}}S_*(X,A;G)$ are chain homotopy equivalences.

Our method of proof is to construct a singular subdivision map $T : S_*(X;G) \to S_*(X;G)$ satisfying certain properties that will let us show that it induces maps on $I^{\bar{p}}S_*^{\mathrm{GM}}(X,A;G)$ and $I^{\bar{p}}S_*(X,A;G)$ whose images lie in $I^{\bar{p}}S_*^{\mathrm{GM},\mathcal{V}}(X,A;G)$ and $I^{\bar{p}}S_*^{\mathcal{V}}(X,A;G)$, respectively. In fact, the induced map on $I^{\bar{p}}S_*^{\mathrm{GM}}(X,A;G)$ will simply be the (relative) restriction of T, and since its image lies in $I^{\bar{p}}S_*^{\mathrm{GM},\mathcal{V}}(X,A;G)$, we obtain a map that we denote by $\hat{T} : I^{\bar{p}}S_*^{\mathrm{GM}}(X,A;G) \to I^{\bar{p}}S_*^{\mathrm{GM},\mathcal{V}}(X,A;G)$. We will show that $\hat{T}$ is a chain homotopy inverse to the inclusion map.

For $I^{\bar{p}}S_*(X,A;G)$, the entire argument will be more complicated as a consequence of the fact that $I^{\bar{p}}S_*(X,A;G)$ is not a subcomplex of $S_*(X,A;G)$, as the boundary map $\hat{\partial}$ is not compatible with the boundary in $S_*(X,A;G)$; see Section 6.2.1. So as to most directly utilize the construction of T on $S_*(X,A;G)$, it is therefore more convenient to use instead the complex $I^{\bar{p}}S'_*(X,A;G)$, our alternative, though isomorphic (see Lemma 6.2.5), definition of non-GM intersection chains from Section 6.2.2. In particular, recall from formula (6.3) in Section 6.2.6 that the relative non-GM intersection chain groups can be written as

$$I^{\bar{p}}S'_i(X,A;G) = \frac{(A^{\bar{p}}S_i(X;G) + S_i(\Sigma_X;G)) \cap \partial^{-1}(A^{\bar{p}}S_{i-1}(X;G) + S_{i-1}(\Sigma_X;G))}{S_i(\Sigma_X;G) + (A^{\bar{p}}S_i(A;G) + S_i(\Sigma_A;G)) \cap \partial^{-1}(A^{\bar{p}}S_{i-1}(A;G) + S_{i-1}(\Sigma_A;G))},$$

where $A^{\bar{p}}S_i(X;G)$ is the subgroup of $S_i(X;G)$ generated by $\bar{p}$-allowable simplices, $A^{\bar{p}}S_i(A;G)$ is the subgroup of $S_i(A;G)$ generated by $\bar{p}$-allowable simplices, and $\Sigma_A = A \cap \Sigma_X$. Given $T : S_*(X : G) \to S_*(X;G)$, we can restrict T to

$$(A^{\bar{p}}S_i(X;G) + S_i(\Sigma_X;G)) \cap \partial^{-1}(A^{\bar{p}}S_{i-1}(X;G) + S_{i-1}(\Sigma_X;G)),$$

and we will see that the image of the restriction also lies in this group. Similarly,

$$(A^{\bar{p}}S_i(A;G) + S_i(\Sigma_A;G)) \cap \partial^{-1}(A^{\bar{p}}S_{i-1}(A;G) + S_{i-1}(\Sigma_A;G))$$

will be taken to itself, and, since subdivision maps preserve supports, T will furthermore take $S_i(\Sigma_X; G)$ to itself. Therefore we obtain an induced map $T' : I^{\bar{p}}S'_*(X,A;G) \to I^{\bar{p}}S'_*(X,A;G)$. As for the GM theory, we will show that the image of T' is contained in $I^{\bar{p}}S'^{,\mathcal{V}}_*(X,A;G)$, yielding a map $\bar{T} : I^{\bar{p}}S'_i(X,A;G) \to I^{\bar{p}}S'^{,\mathcal{V}}_*(X,A;G)$, which we will show is a chain homotopy inverse to the inclusion.

Before proceeding with this program, we should briefly discuss what exactly we mean by $I^{\bar{p}}S'^{,\mathcal{V}}_*(X,A;G)$. By Lemma 6.2.5, we know that $I^{\bar{p}}S_*(X;G) \cong I^{\bar{p}}S'_*(X;G)$ and similarly $I^{\bar{p}}S_*(A;G) \cong I^{\bar{p}}S'_*(A;G)$, and therefore the relative complexes are also isomorphic. Removing primes, we see that $I^{\bar{p}}S_*(V,A\cap V;G) \to I^{\bar{p}}S_*(X,A;G)$ is an injection because the only chains in the kernel of $I^{\bar{p}}S_*(V;G) \to I^{\bar{p}}S_*(X,A;G)$ are those that are supported in A and V, and those are zero in $I^{\bar{p}}S_*(V,A\cap V;G)$. Therefore, we can identify the image of $I^{\bar{p}}S_*(V,A\cap V;G)$ as a subgroup of $I^{\bar{p}}S_*(X,A;G)$. We then have the diagram

$$\begin{array}{ccc} I^{\bar{p}}S_*(V,A\cap V;G) & \hookrightarrow & I^{\bar{p}}S_*(X,A;G) \\ \downarrow \cong & & \downarrow \cong \\ I^{\bar{p}}S'_*(V,A\cap V;G) & \to & I^{\bar{p}}S'_*(X,A;G), \end{array}$$

so it follows that we can also regard each $I^{\bar{p}}S'_*(V,A\cap V;G)$ as a subgroup of $I^{\bar{p}}S'_*(X,A;G)$. The sum

$$I^{\bar{p}}S'^{,\mathcal{V}}_*(X,A;G) = \sum_{V\in\mathcal{V}} I^{\bar{p}}S'_*(V,A\cap V;G) \subset I^{\bar{p}}S'_*(X,A;G)$$

therefore makes sense.

In fact, we then obtain the diagram

$$\begin{array}{ccc} I^{\bar{p}}S^{\mathcal{V}}_*(X;G) & \hookrightarrow & I^{\bar{p}}S_*(X;G) \\ \downarrow & & \downarrow \cong \\ I^{\bar{p}}S'^{,\mathcal{V}}_*(X;G) & \hookrightarrow & I^{\bar{p}}S'_*(X;G). \end{array}$$

As the composition right then down is injective, it follows that the left vertical map must also be injective. It is also surjective, using the individual isomorphisms $I^{\bar{p}}S_*(V,A\cap V;G) \cong I^{\bar{p}}S'_*(V,A\cap V;G)$. So the left-hand vertical map is an isomorphism. It follows that the chain complex inclusions represented by the top and bottom horizontal maps of the diagram are isomorphic, and so to show that the top horizontal inclusion is a homotopy equivalence, it suffices to show that the bottom inclusion is.

The existence of the maps T, $\hat{T}$, and $\bar{T}$ is the subject of the following lemma:

Lemma 6.5.4 *Let $\mathcal{V}$ be a covering of X such that X is also covered by the interiors of the elements of $\mathcal{V}$. Then there exists a singular subdivision chain map $T : S_*(X;G) \to$*

$S_*(X;G)$ that induces chain maps $\hat{T} : I^{\bar{p}}S_*^{\mathrm{GM}}(X,A;G) \to I^{\bar{p}}S_*^{\mathrm{GM},\mathcal{V}}(X,A;G)$ and $\bar{T} : I^{\bar{p}}S'_*(X,A;G) \to I^{\bar{p}}S_*^{\prime,\mathcal{V}}(X,A;G)$.

The same result holds as a statement about R-modules, replacing the abelian group G with an R-module over a commutative ring with unity R.

The proof of Lemma 6.5.4 is provided below. First we demonstrate that the existence of such maps is sufficient to obtain the desired homotopy equivalences and so complete the proof of Proposition 6.5.1.

Consider $I^{\bar{p}}S_*^{\mathrm{GM}}(X)$. Let $\mathfrak{i}$ denote the inclusion $I^{\bar{p}}S_*^{\mathrm{GM},\mathcal{V}}(X;G) \hookrightarrow I^{\bar{p}}S_*^{\mathrm{GM}}(X;G)$. Then $\mathfrak{i}\hat{T}$ is simply the restriction of the $T : S_*(X;G) \to S_*(X;G)$ of the lemma to a map $I^{\bar{p}}S_*^{\mathrm{GM}}(X;G) \to I^{\bar{p}}S_*^{\mathrm{GM}}(X;G)$. By Corollary 4.4.15 (which applies just as well with coefficients in G), since this restriction of T is a singular subdivision map, the induced map $I^{\bar{p}}S_*^{\mathrm{GM}}(X,A;G) \to I^{\bar{p}}S_*^{\mathrm{GM}}(X,A;G)$ is chain homotopic to the identity. We will consider $\hat{T}\mathfrak{i}$ below.

Next, let $\bar{\mathfrak{i}} : I^{\bar{p}}S_*^{\prime,\mathcal{V}}(V,A\cap V;G) \hookrightarrow I^{\bar{p}}S'_*(X,A;G)$ be the inclusion and consider the singular subdivision map $T' = \bar{\mathfrak{i}}\bar{T} : I^{\bar{p}}S'_*(X,A;G) \to I^{\bar{p}}S'_*(X,A;G)$ induced by T. Once again, we know from the proof of Corollary 4.4.15 how to construct chain homotopies P (based on prism constructions) that show that singular subdivision operators are chain homotopic to the identity; we want to show that such a chain homotopy P descends to a well-defined chain homotopy operator $I^{\bar{p}}S'_*(X,A;G) \to I^{\bar{p}}S'_{*+1}(X,A;G)$. As constructed in Corollary 4.4.15, the chain homotopy P is defined on any singular simplex of X, it takes allowable simplices to sums of allowable simplices, and it preserves supports. Thus if $\xi \in A^{\bar{p}}S_i(X;G) + S_i(\Sigma_X;G)$, we will have $P(\xi) \in A^{\bar{p}}S_{i+1}(X;G) + S_{i+1}(\Sigma_X;G)$, while if also $\partial\xi \in A^{\bar{p}}S_{i-1}(X;G) + S_{i-1}(\Sigma_X;G)$, we will have

$$\partial P(\xi) = T(\xi) - \xi - P(\partial\xi) \in A^{\bar{p}}S_i(X;G) + S_i(\Sigma_X;G)$$

using that T also preserves allowability and supports. Similarly, P will take

$$(A^{\bar{p}}S_*(A;G) + S_*(\Sigma_A;G)) \cap \partial^{-1}(A^{\bar{p}}S_{*-1}(A;G) + S_{*-1}(\Sigma_A;G))$$

to itself and $S_i(\Sigma_X;G)$ to $S_{i+1}(\Sigma_X;G)$. Thus, P descends to a chain homotopy $I^{\bar{p}}S'_*(X,A;G) \to I^{\bar{p}}S'_{*+1}(X,A;G)$ between $T' = \bar{\mathfrak{i}}\bar{T}$ and the identity.

So we have now shown that $\mathfrak{i}\hat{T}$ and $\bar{\mathfrak{i}}\bar{T}$ are chain homotopic to identity maps. Next we consider $\hat{T}\mathfrak{i}$ and $\bar{T}\bar{\mathfrak{i}}$.

First, consider $T\mathfrak{i}$. Since $\mathfrak{i}$ is an inclusion map of a subcomplex, $\mathfrak{i}^{-1}$ is well defined on $\mathrm{im}(\mathfrak{i})$. We also observe that if ξ is a chain of $I^{\bar{p}}S_*^{\mathrm{GM}}(X;G)$ supported in some V, then the same is true of both $T(\xi)$ and $P(\xi)$. So if $\xi \in I^{\bar{p}}S_*^{\mathrm{GM},\mathcal{V}}(X,A;G)$ is represented by $\sum \xi_V$ with $\xi_V \in I^{\bar{p}}S_*^{\mathrm{GM}}(V;G)$, then $\mathfrak{i}^{-1}P\mathfrak{i}(\xi)$ is represented by $\sum_{V\in\mathcal{V}} \mathfrak{i}^{-1}P\mathfrak{i}(\xi_V)$, which represents an element in $I^{\bar{p}}S_*^{\mathrm{GM},\mathcal{V}}(X,A;G)$. Now we compute formally

$$\begin{aligned}\mathrm{id} - \hat{T}\mathfrak{i} &= \mathfrak{i}^{-1}\mathfrak{i}(\mathrm{id} - \hat{T}\mathfrak{i}) = \mathfrak{i}^{-1}\mathfrak{i} - \mathfrak{i}^{-1}\mathfrak{i}\hat{T}\mathfrak{i} = \mathfrak{i}^{-1}(\mathrm{id} - \mathfrak{i}\hat{T})\mathfrak{i}\\ &= \mathfrak{i}^{-1}(\partial P + P\partial)\mathfrak{i} = \mathfrak{i}^{-1}(\partial P)\mathfrak{i} + \mathfrak{i}^{-1}(P\partial)\mathfrak{i}\\ &= \partial\mathfrak{i}^{-1}P\mathfrak{i} + \mathfrak{i}^{-1}P\mathfrak{i}\partial.\end{aligned}$$

So $\hat{T}\mathfrak{i}$ is chain homotopic to the identity via the chain homotopy $\mathfrak{i}^{-1}P\mathfrak{i}$. Altogether, we have shown that $\mathfrak{i}$ is a chain homotopy equivalence, completing the proof for GM intersection chains. The argument for showing that $\bar{T}\bar{\mathfrak{i}}$ is homotopic to the identity is the same, recognizing that P preserves supports and takes chains $\xi_V \in S_*(V;G)$ representing elements of $I^{\bar{p}}S'_*(V, A\cap V;G) \subset I^{\bar{p}}S'_*(X,A;G)$ to chains that continue to represent elements in $I^{\bar{p}}S'_*(V,A\cap V;G)$. This follows from the same sorts of arguments applied just above to show that P descends to a chain homotopy on $I^{\bar{p}}S'_*(X,A;G)$.

This completes the proof for singular chains.

In the PL situation, we can draw the even stronger conclusion that the inclusion map $I^{\bar{p}}\mathfrak{C}^{\mathrm{GM},\mathcal{V}}_*(X,A;G) \hookrightarrow I^{\bar{p}}\mathfrak{C}^{\mathrm{GM}}_*(X,A;G)$ is an isomorphism. Once again this inclusion makes sense because the only chains in the kernel of $I^{\bar{p}}\mathfrak{C}_*(V;G) \to I^{\bar{p}}\mathfrak{C}_*(X,A;G)$ are those that are supported in A and V, and those are zero in $I^{\bar{p}}\mathfrak{C}_*(V,A\cap V;G)$. So $I^{\bar{p}}\mathfrak{C}^{\mathrm{GM},\mathcal{V}}_*(X,A;G) \subset I^{\bar{p}}\mathfrak{C}^{\mathrm{GM}}_*(X,A;G)$. So it suffices to demonstrate that the inclusion is a surjection. For this, let $[\xi] \in I^{\bar{p}}\mathfrak{C}^{\mathrm{GM}}_*(X,A;G)$, and suppose that we represent $[\xi]$ as a simplicial chain ξ in some triangulation. Choosing some ordering on the vertices of the triangulation, we can identify ξ with a singular chain as in Proposition 4.4.5. Now, we can apply the subdivision map T from our singular chain argument. As T subdivides simplices linearly, the resulting singular subdivision determines a simplicial subdivision of ξ, which we will also call $T(\xi)$. Furthermore, as the inclusion of $|\xi|$ into X is a proper PL embedding, by Theorem B.2.19 there is a triangulation of all of X such that some further subdivision, say $(T(\xi))'$, of $T(\xi)$ is a simplicial chain with respect to this triangulation. Via the properties of T, we can write $(T(\xi))' = \sum_{V\in\mathcal{V}}\xi_V$, with each ξ_V an element of $I^{\bar{p}}\mathfrak{C}^{\mathrm{GM}}_*(V;G)$. But in the PL setting, every chain is identified with its subdivisions, so in $I^{\bar{p}}\mathfrak{C}^{\mathrm{GM}}_*(X,A;G)$, we in fact have $[\xi] = \left[\sum_{V\in\mathcal{V}}\xi_V\right] = \sum_{V\in\mathcal{V}}[\xi_V]$.

The argument for $I^{\bar{p}}\mathfrak{C}'_*(X,A;G)$ is the same using T' and the fact that subdivisions preserve allowability and supports. This completes the proof of Proposition 6.5.1 subject to Lemma 6.5.4. □

Now we must prove Lemma 6.5.4.

Proof of Lemma 6.5.4 Once again, there is no difference in the proof between working with coefficients in an abelian group as opposed to an R-module, so we use the former.

We begin by constructing a singular subdivision chain map $T : S_*(X) \to S_*(X)$ with image in $S^{\mathcal{V}}_*(X)$, where $S^{\mathcal{V}}_*(X)$ is defined analogously with $I^{\bar{p}}S^{\mathrm{GM},\mathcal{V}}_*(X;G)$ but for ordinary singular chains and $\mathbb{Z}$-coefficients. In fact, the image of T will lie in $S^{\mathcal{U}}_*(X) \subset S^{\mathcal{V}}_*(X)$, where $\mathcal{U}$ is the set of interiors of the elements of $\mathcal{V}$. We later discuss the restriction of T to intersection chains and the generalization to other coefficients.

We first fix a well-ordering on the set $\mathcal{U}$, which is possible using the Well-Ordering Theorem; see [181, Section 10]. Suppose that $B \subset X$ is a subset that can be contained in some element U of $\mathcal{U}$. Then let $\psi(B) \in \mathcal{U}$ be the element of $\mathcal{U}$ that is least in the order among elements of $\mathcal{U}$ containing B. If $\sigma : \Delta^i \to X$ is a singular simplex such that the image $\sigma(\Delta^i)$ is contained in some $U \in \mathcal{U}$, then define $\psi(\sigma) = \psi(\sigma(\Delta^i))$.

The basic idea of the argument is as follows: We construct T such that, for each $\xi \in S_i(X)$, the chain $T(\xi)$ will be a sum of simplices each of which is supported in some element of $\mathcal{U}$. This alone would allow us to write $T(\xi) = \sum_{U \in \mathcal{U}} T_U(\xi)$ by letting $T_U(\xi)$ be the sum over those simplices σ of $T(\xi)$ (with their coefficients) such that $\psi(\sigma) = U$. In other words, if $T(\xi) = \sum n_i \sigma_i$, then $T_U(\xi) = \sum I_U(\sigma_i) n_i \sigma_i$, where the indicator function $I_U(\sigma_i)$ is 1 if $\psi(\sigma_i) = U$ and 0 otherwise. See Figure 6.2. Then $T_U(\xi)$ is supported in U. Such a T would be sufficient for working with ordinary singular chains (and this is the essence of such arguments in standard texts), but since our ultimate goal is to work with intersection chains, we must be a bit more subtle. In particular, if ξ is an intersection chain, we require that each of the $T_U(\xi)$ be an intersection chain, which means that we must be careful about the boundaries of the $T_U(\xi)$; as any i-simplex in a singular subdivision of a $\bar{p}$-allowable i-simplex is $\bar{p}$-allowable by the proof of Lemma 4.4.13, these boundaries are the only issue. So, as in our arguments concerning excision and Mayer–Vietoris sequences in Section 4.4, we must take some extra care to "shield" bad faces to ensure this doesn't happen.

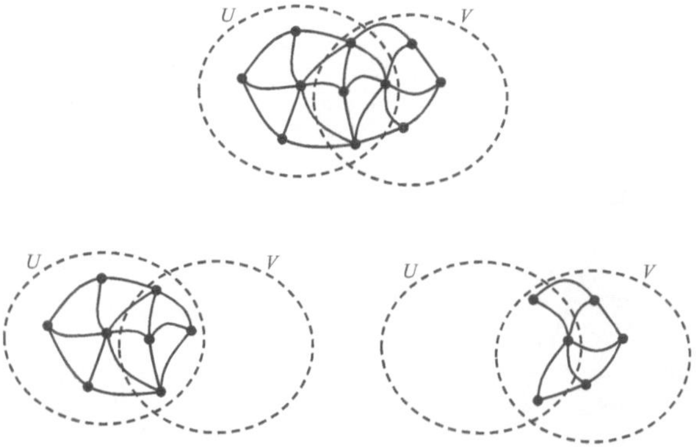

Figure 6.2 On the top, we see the singular chain $T(\xi)$ contained in $X = U \cup V$. If $U < V$ in the ordering of $\mathcal{U} = \{U, V\}$, then $T_U(\xi)$ is shown on the bottom left and $T_V(\xi)$ on the bottom right.

For this, we will construct T inductively to satisfy the properties in the following list. After giving this list, we will see why this is sufficient. Then we will see below that we can indeed construct a T with these properties.

1. T is a chain map $S_*(X) \to S_*(X)$.
2. For each singular simplex σ, the chain $T(\sigma)$ is a singular subdivision of σ as defined in Section 4.4.2. Recall that, roughly, this means that if $\sigma : \Delta^i \to X$, then there is some simplicial subdivision $\hat{\Delta}^i$ of Δ^i (with ordered vertices) such that $T(\sigma)$ is the sum of the restrictions of σ to each of the i-simplices of Δ^i. More technically speaking, we have $T(\sigma) = \sum \operatorname{sgn}(\mathfrak{i}_j) \sigma \circ \mathfrak{i}_j$, where each $\mathfrak{i}_j : \Delta^i \to \Delta^i$ is a linear homeomorphism of the standard i-simplex onto one of the i-simplices of $\hat{\Delta}^i$ with the inclusion map determined

by the ordering on the vertices and with $\mathrm{sgn}(\mathfrak{i}_j)$ equal to 1 or -1 according to whether the orientations of Δ^i and the image of $\mathfrak{i}_j$ agree or not.

3. The image of each simplex of $T(\sigma)$ is contained in some element of $\mathcal{U}$.
4. Suppose σ is a singular i-simplex and μ is a simplex (of any dimension) of the subdivision $\hat{\Delta}^i$ of Δ^i as in condition (2). Suppose further that there is some simplex η (of any dimension) of Δ^i such that $\mu \subset \eta$ and $\dim(\mu) = \dim(\eta)$ (in other words, μ is a top-dimensional simplex in the restricted subdivision of $\hat{\Delta}^i$ to the simplex η of Δ^i). Then $\psi(\sigma(\overline{St}(\mu, \hat{\Delta}^i))) = \psi(\sigma(\mu))$, where $\overline{St}(\mu, \hat{\Delta}^i)$ is the closed star of μ in $\hat{\Delta}^i$, which consists of all (closed) simplices of $\hat{\Delta}^i$ that have μ as a face (of any dimension). This condition says that the image under σ of every simplex of $\hat{\Delta}^i$ that has μ as a face has the same minimal containing element of $\mathcal{U}$ as μ itself does. This is the condition that will create the necessary "shielding." See Figure 6.3 for an illustration of this condition and Figure 6.5, below on page 351, for an illustration of the general idea of how these requirements allow us to split intersection chains into intersection chain pieces.

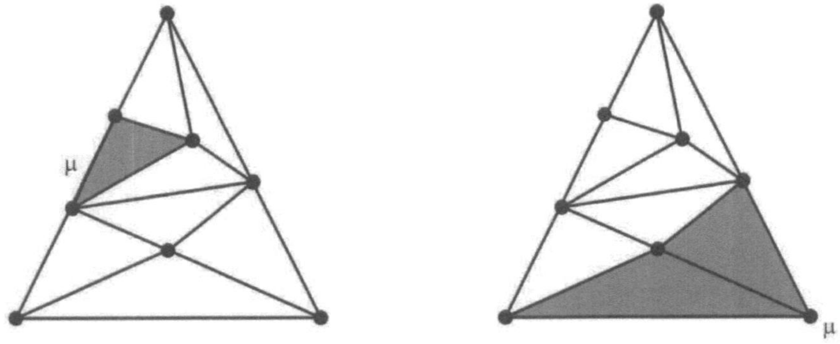

Figure 6.3 Two examples in a subdivision $\hat{\Delta}^2$ of Δ^2 of simplices μ satisfying the hypothesis of condition (4) (a 1-simplex on the left and a 0-simplex on the right). The shaded simplices indicate the closed stars $\overline{St}(\mu, \hat{\Delta}^i)$. The conclusion of the condition is that the least element of $\mathcal{U}$ containing $\sigma(\mu)$ must also contain the image under σ of the entire star.

Sublemma 6.5.5 *There exists a chain map T with these properties.*

Next, let us see that if we have a T with the listed properties and restrict it to $I^{\bar{p}}S_*^{\mathrm{GM}}(X) \subset S_*(X)$, then the image lies in $I^{\bar{p}}S_*^{\mathrm{GM},\mathcal{U}}(X)$. By Lemma 4.4.13, subdivisions preserve allowability, so, since T is a chain map, we certainly have that T induces a map $I^{\bar{p}}S_*^{\mathrm{GM}}(X) \to I^{\bar{p}}S_*^{\mathrm{GM}}(X) \subset S_*(X)$. Suppose $\xi \in I^{\bar{p}}S_i^{\mathrm{GM}}(X)$ and $U \in \mathcal{U}$. If $T(\sigma) = \sum n_j\tau_j$, for a singular simplex σ, let $T_U(\sigma) = \sum I_U(\tau_j)n_j\tau_j$, where I_U is the indicator function as defined above. Extending by linearity, we then have $T(\xi) = \sum_{U\in\mathcal{U}} T_U(\xi)$. Although $\mathcal{U}$ may contain infinite elements, this sum is necessarily finite as ξ consists of finitely many simplices, and hence so does the subdivision $T(\xi)$. Furthermore, by construction, $T_U(\xi)$ must consist of simplices supported in U. It remains to show that each $T_U(\xi)$ is an intersection chain under the assumption that $\xi \in I^{\bar{p}}S_i^{\mathrm{GM}}(X)$.

By Lemma 4.4.13, if $\xi \in I^{\bar{p}}S_i^{\mathrm{GM}}(X)$ then all of the i-simplices of each $T_U(\xi)$ will be allowable, so we are reduced to consider the allowability of the simplices of $\partial T_U(\xi)$. Let τ be an $(i-1)$-simplex of $\partial T_U(\xi)$. Each such τ can be identified as the restriction of σ to δ for some singular simplex $\sigma : \Delta^i \to X$ of ξ and some $(i-1)$-simplex δ in the subdivision $\hat{\Delta}^i$ of Δ^i. Technically, τ is the composition of σ with the inclusion of Δ^{i-1} into $\hat{\Delta}^i$ as δ, but we will abuse the notation slightly. If τ is allowable, there is no trouble, so let us assume that τ is not allowable and find a contradiction.

First, let us consider the possibility that τ is an $(i-1)$-simplex that comes from the interior of an i-simplex σ of ξ. In other words, the $(i-1)$-simplex δ in the the subdivision $\hat{\Delta}^i$ associated to $T(\sigma)$ is not contained in $\partial\Delta^i$. Suppose, furthermore, that for every face η of Δ^i it is true that $\dim(\delta \cap \eta) < \dim(\eta)$. Then by arguments completely analogous to those of Lemma 4.4.1 and Lemma 4.4.17, the simplex τ must in fact be allowable.

Thus, if τ is not allowable there is some face η of Δ^i for which it is true that $\dim(\delta \cap \eta) = \dim(\eta)$. This includes the possibility that δ is contained in the boundary of Δ^i, in which case $\dim(\delta) = \dim(\eta) = i - 1$. The intersection $\delta \cap \eta$ must be a (not necessarily proper) face of δ contained in the subdivision of η in $\hat{\Delta}^i$ and of the same dimension as η, and so $\delta \cap \eta = \mu$ is a simplex that satisfies the hypotheses of condition (4). Since T is assumed to satisfy condition (4), it follows that we have $\psi(\sigma(\overline{St}(\mu, \hat{\Delta}^i))) = \psi(\sigma(\mu))$. So, in particular, δ and any i-simplex γ of $\hat{\Delta}^i$ of which δ is a face, both of which contain μ as a face, have their images under σ contained in $\psi(\sigma(\mu))$. Furthermore, since δ and any such γ contain μ, their images under σ cannot be contained in an element of $\mathcal{U}$ that is smaller in the ordering because the image under σ of μ cannot be. Thus $\psi(\tau) = \psi(\sigma(\delta)) = \psi(\sigma(\gamma)) = \psi(\sigma(\mu))$. So, in particular, if γ is an i-simplex of $\hat{\Delta}^i$ of which δ is a face, then the i-simplex $\sigma|_\gamma$ is a simplex only of $T_{\psi(\tau)}(\xi)$ and not of any other $T_W(\xi)$, $W \in \mathcal{U}$, assuming it doesn't cancel with other simplices so that it does not appear in $T(\xi)$ at all. As we have assumed that τ occurs as a simplex in $\partial T_U(\xi)$, this also tells us that $\psi(\tau) = U$.

Now, consider all the ways that τ can arise as a face of an i-simplex of $T(\xi)$. As we assume that τ is not allowable, the discussion of the last paragraph holds, ranging across various possible simplices playing the roles of σ, η, δ, μ, and γ. In all such cases, the conclusion is that any i-simplex of $T(\xi)$ that contains τ as a boundary simplex must be included in $T_{\psi(\tau)}(\xi)$. But since we have assumed τ is not allowable, we know it does not occur in $\partial T(\xi)$. Therefore, there must be cancellations that occur among the boundaries of the i-simplices of $T(\xi)$ that contain τ as a boundary. Since all such simplices are contained in $T_{\psi(\tau)}(\xi) = T_U(\xi)$, the coefficients of τ must all cancel in $T_U(\xi)$. Therefore, the simplex τ cannot occur in any $\partial T_U(\xi)$, which was our desired contradiction. Therefore, we have $T_U(\xi) \in I^{\bar{p}}S_i^{\mathrm{GM}}(U)$ for each U.

It now follows that T restricts to a chain map $I^{\bar{p}}S_*^{\mathrm{GM}}(X) \to I^{\bar{p}}S_*^{\mathrm{GM},\mathcal{U}}(X)$. In fact, the exact same arguments extend immediately to chains with any coefficients to give a chain map $I^{\bar{p}}S_*^{\mathrm{GM}}(X;G) \to I^{\bar{p}}S_*^{\mathrm{GM},\mathcal{U}}(X;G)$.

So now consider the relative situation where we want to show that T induces $\hat{T}$: $I^{\bar{p}}S_*^{\mathrm{GM}}(X,A;G) \to I^{\bar{p}}S_*^{\mathrm{GM},\mathcal{V}}(X,A;G)$. Suppose an element of $I^{\bar{p}}S_*^{\mathrm{GM}}(X,A;G)$ is represented by $\xi + a$ with $\xi \in I^{\bar{p}}S_*^{\mathrm{GM}}(X;G)$ and $a \in I^{\bar{p}}S_*^{\mathrm{GM}}(A;G)$. We know that $T(\xi) = \sum T_U(\xi)$ with $T_U(\xi) \in I^{\bar{p}}S_*^{\mathrm{GM}}(U;G)$, and similarly we must have $T(a) = \sum T_U(a)$ with $T_U(a) \in I^{\bar{p}}S_*^{\mathrm{GM}}(A \cap U;G)$ because a subdivision of a chain in A will be in A. Therefore, $T(\xi + a) = \sum_U T_U(\xi + a) = \sum_U (T_U(\xi) + T_U(a))$ represents an element in $I^{\bar{p}}S_*^{\mathrm{GM},\mathcal{V}}(X,A;G)$. If we vary a within $I^{\bar{p}}S_*^{\mathrm{GM}}(A;G)$, we do not change the element that $T(\xi + a)$ represents in $I^{\bar{p}}S_*^{\mathrm{GM},\mathcal{V}}(X,A;G)$, so T induces a well-defined $\hat{T}$ with the desired properties.

Next, we consider

$$I^{\bar{p}}S_i'(X,A;G) = \frac{(A^{\bar{p}}S_i(X;G) + S_i(\Sigma_X;G)) \cap \partial^{-1}(A^{\bar{p}}S_{i-1}(X;G) + S_{i-1}(\Sigma_X;G))}{S_i(\Sigma_X;G) + (A^{\bar{p}}S_i(A;G) + S_i(\Sigma_A;G)) \cap \partial^{-1}(A^{\bar{p}}S_{i-1}(A;G) + S_{i-1}(\Sigma_A;G))};$$

recall again formula (6.3) in Section 6.2.6. We first show that T induces a chain map T' : $I^{\bar{p}}S_i'(X,A;G) \to I^{\bar{p}}S_i'(X,A;G)$. For convenience of notation, let us denote the "numerator" of the above expression by B. Since $B \subset S_*(X;G)$, the chain map T is defined on elements of B with image in $S_*(X;G)$; we must show that T takes elements of B to elements of B. So let $\xi \in B$. Then ξ is a linear combination of allowable simplices and simplices supported in Σ_X, and the same is true of $\partial\xi$. But the image of T on each allowable simplex is a chain composed of allowable simplices by Lemma 4.4.13, and the image of T on simplices supported in Σ_X is supported in Σ_X. This last statement is due to the fact T preserves (or reduces) supports because it is a subdivision map. Since T is a chain map, $\partial T(\xi) = T(\partial\xi)$, which then similarly must be composed of allowable simplices and simplices in Σ_X. By the same arguments, T must take simplices of

$$(A^{\bar{p}}S_i(A;G) + S_i(\Sigma_A;G)) \cap \partial^{-1}(A^{\bar{p}}S_{i-1}(A;G) + S_{i-1}(\Sigma_A;G))$$

back into this group, and we already know that it takes $S_i(\Sigma_X;G)$ to itself. So T induces a well-defined chain map $T' : I^{\bar{p}}S_*'(X,A;G) \to I^{\bar{p}}S_*'(X,A;G)$.

Now, we must observe that the image of $T' : I^{\bar{p}}S_*'(X,A;G) \to I^{\bar{p}}S_*'(X,A;G)$ is contained in $I^{\bar{p}}S_*^{\prime,\mathcal{U}}(X,A;G)$. In fact, if we represent an element of $I^{\bar{p}}S_i'(X,A;G)$ by a chain $\xi + s + a$ with $\xi \in B$, $s \in S_i(\Sigma_X;G)$, and

$$a \in (A^{\bar{p}}S_i(A;G) + S_i(\Sigma_A;G)) \cap \partial^{-1}(A^{\bar{p}}S_{i-1}(A;G) + S_{i-1}(\Sigma_A;G)),$$

then, as above, we have

$$T(\xi + s + a) = \sum_{U \in \mathcal{U}} T_U(\xi) + T_U(s) + T_U(a).$$

We must show that $T_U(\xi) + T_U(s) + T_U(a) \in I^{\bar{p}}S_*'(U, A \cap U;G)$. We first consider

$T_U(\xi)$. Preservation of allowability and supports shows that each $T_U(\xi)$ is contained in $A^{\bar{p}}S_*(U;G)+S_*(\Sigma_U;G)$, where $\Sigma_U = \Sigma_X \cap U$. Next we verify that $\partial T_U(\xi) \in A^{\bar{p}}S_{i-1}(U;G)+S_{i-1}(\Sigma_U;G)$. But the above arguments for $I^{\bar{p}}S_*^{\mathrm{GM}}(X;G)$ can be used again verbatim to show here that any simplex of $\partial T_U(\xi)$ not contained in Σ_X must be allowable. In particular, notice that if such a simplex τ is not contained in Σ_X, no simplex of $T_U(\xi)$ having τ as a face can be contained in Σ_X, so there is no disruption to our previous shielding arguments, using that we know that every simplex of $T(\xi)$ with τ as a face must be allowable. So

$$T_U(\xi) \in (A^{\bar{p}}S_i(U;G) + S_i(\Sigma_U;G)) \cap \partial^{-1}(A^{\bar{p}}S_{i-1}(U;G) + S_{i-1}(\Sigma_U;G)).$$

The same argument together with preservation of supports shows that

$$T_U(a) \in$$
$$(A^{\bar{p}}S_i(A \cap U;G) + S_i(\Sigma_A \cap U;G)) \cap \partial^{-1}(A^{\bar{p}}S_{i-1}(A \cap U;G) + S_{i-1}(\Sigma_A \cap U;G)).$$

Finally, preservation of supports tells us that $T_U(s) \in S_i(\Sigma_U;G)$. Therefore, each

$$T_U(\xi + s + a) = T_U(\xi) + T_U(s) + T_U(a)$$

represents an element of $I^{\bar{p}}S_i'(U, A \cap U;G)$. Furthermore, varying $s \in S_i(\Sigma_X;G)$ and a in its group does not change the element $T_U(\xi + s + a) \in I^{\bar{p}}S_i'(U, A \cap U;G)$. This proves our claim that T induces a well-defined map $\bar{T} : I^{\bar{p}}S_*'(X,A;G) \to I^{\bar{p}}S_*^{\prime,\mathcal{V}}(X,A;G)$.

This completes the proof of Lemma 6.5.4 subject to Sublemma 6.5.5. □

Proof of Sublemma 6.5.5 We will construct a subdivision chain map $T : S_*(X) \to S_*(X)$ that satisfies the conditions required. The construction will be inductive over the dimensions of chains. We first define $T : S_0(X) \to S_0(X)$ to be the identity map. Notice that this is consistent with the conditions we need to verify. Next, we assume that $T : S_j(X) \to S_j(X)$ has been constructed for $j < i$ satisfying the required properties. We will show how to extend T to $S_i(X)$.

Let $\sigma : \Delta^i \to X$ be a singular simplex of $S_i(X)$. We will define $T(\sigma)$. In general, T is extended to $S_i(X)$ by linearity. Since T will be a singular subdivision of σ, we seek to construct an appropriate subdivision $\hat{\Delta}^i$ of Δ^i (with vertex partial ordering), and this will determine the subdivision of σ; see Section 4.4.2 for a general discussion of singular subdivision. By assumption, $T(\partial\sigma)$ has already been defined, so we can assume a subdivision (with vertex partial ordering) of $\partial\Delta^i$ as given and denote it by $\partial\hat{\Delta}^i$. We begin by defining a subdivision[15] $\hat{\Delta}_1^i$ of Δ^i by letting $\hat{\Delta}_1^i$ be the cone on $\partial\hat{\Delta}^i$. In other words, we add the barycenter v of Δ^i as a vertex, and the i-simplices of $\hat{\Delta}_1^i$ will be simplices of the form $[w_0, \ldots, w_{i-1}, v]$, where $[w_0, \ldots, w_{i-1}]$ is an $(i-1)$-simplex of $\partial\hat{\Delta}^i$. See Figure 6.4. We also assume the new barycenter is placed in the order of vertices after the existing vertices of $\partial\hat{\Delta}^i$.

[15] The reason for this first subdivision $\hat{\Delta}_1^i$ is to ensure that we start the process to come with a legitimate simplicial subdivision of Δ^i that restricts to $\partial\hat{\Delta}^i$ on the boundary of Δ^i. If $\partial\hat{\Delta}^i = \partial\Delta^i$ then this step can be omitted by simply defining $\hat{\Delta}_1^i = \Delta^i$.

Next, we perform on $\hat{\Delta}^i_1$ iterated barycentric subdivisions relative to $\partial\hat{\Delta}^i$. The process of relative barycentric subdivision is described in detail in [182, Section 16], but here is the basic idea: Recall that barycentric subdivision of a simplicial complex K is performed inductively. The barycentric subdivision of the 0-skeleton of K is always just the 0-skeleton of K. Then assuming the barycentric subdivision K' of K has been constructed on the $(p-1)$-skeleton of K, one subdivides each p-simplex by coning off the barycentric subdivision of its boundary, analogously to our construction of $\hat{\Delta}^i_1$. To obtain ordered simplices in the subdivision, we let each successive barycenter come later in the order than those added at the previous stage of construction. For a *relative* barycentric subdivision, the difference is that one begins with a subcomplex $L \subset K$ and holds L fixed throughout the procedure. Again the subdivision of the 0-skeleton is just the 0-skeleton itself. Now assume we've constructed a relative barycentric subdivision up through the $(p-1)$-skeleton of K to obtain a $(p-1)$-dimensional complex K' with $L^{p-1} \subset K'$. Now let τ be a p-simplex of K. If τ is contained in L, then also $\partial\tau \subset L^{p-1} \subset K'$, and we add τ to K'. If τ is not contained in L, then we subdivide τ by taking the cone on the subdivision of $\partial\tau$ in K' that has already been constructed in the induction. Applying these procedures for all p-simplices of K, we obtain a p-skeleton for K' that contains L^p as a subcomplex. See Figure 6.4. Just as for ordinary barycentric subdivision, relative barycentric subdivision of K relative to L can be repeated iteratively.

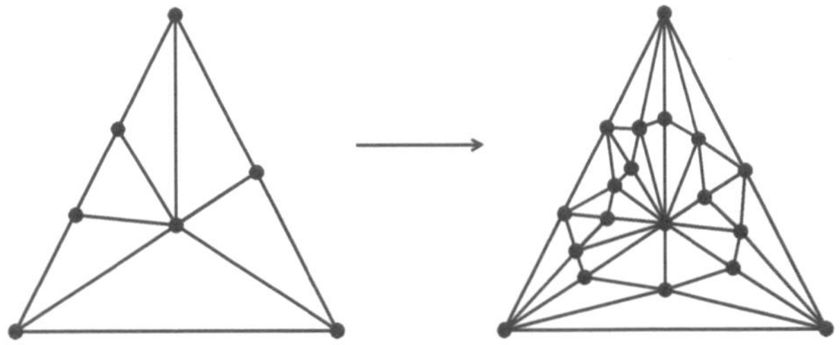

Figure 6.4 Relative barycentric subdivision on a complex of the form $\hat{\Delta}^2_1$ (left). Assuming that the subdivision $\partial\hat{\Delta}^2$ of $\partial\Delta^2$ is given, we form $\hat{\Delta}^2_1$ by taking the cone on that subdivision with the cone point at the barycenter of Δ^2. Then the relative barycentric subdivision does not subdivide any simplex of $\partial\hat{\Delta}^2$. Observe that the boundary is subdivided identically in the two complexes.

We will let $\hat{\Delta}^i$ be such an iterated barycentric subdivision of $\hat{\Delta}^i_1$ relative to $\partial\hat{\Delta}^i_1 = \partial\hat{\Delta}^i$, and $T(\sigma)$ will be the singular subdivision of σ based on this subdivision of Δ^i. The first two requirements for T, that it be a singular subdivision map and a chain map, will thus be satisfied by the construction. To obtain the other conditions, we must ensure that if we perform enough iterations of the relative barycentric subdivision then the other conditions become true. For this, we will prove a modified version of Lemma 16.3 of [182]; our argument (and some of our notation) will be modified versions of those found in [182].

Let $B = \partial\hat{\Delta}^i$, which we assume already constructed by induction. By the inductive argument, the image of each simplex of B under σ is contained in some element of the open covering $\mathcal{U}$. Furthermore, suppose μ is a j-dimensional simplex of B, $j < i$, that is contained in some j-dimensional face of Δ^i. By the inductive assumption, if F is any face of Δ^i containing μ (as a subset) and if $\hat{F}$ is the subdivision of F determined by the subdivision $\partial\hat{\Delta}^i$, then $\psi(\sigma(\overline{St}(\mu, \hat{F}))) = \psi(\sigma(\mu))$. Since this formula holds over all such F, we see that in fact $\psi(\sigma(\overline{St}(\mu, B))) = \psi(\sigma(\overline{St}(\mu, \partial\hat{\Delta}^i))) = \psi(\sigma(\mu))$. Now let $K = \hat{\Delta}^i_1$, and let $\mathrm{sd}^N(K/B)$ denote the Nth iterated barycentric subdivision of K relative to the subcomplex B. We will show that there is a sufficiently large N such that[16]

- for every μ in B satisfying the hypotheses of condition (4) on page 344, we have $\psi(\sigma(\overline{St}(\mu, \mathrm{sd}^N(K/B)))) = \psi(\sigma(\mu))$; and
- for every i-simplex ν of $\mathrm{sd}^N(K/B)$, then $\sigma(\nu)$ is contained in some element of $\mathcal{U}$.

Then if we let $\hat{\Delta}^i = \mathrm{sd}^N(K/B)$ and use this to define $T(\sigma)$, we will have satisfied all the conditions we need for T.

Here is where we use a slight variation of the method of argument of the proof of [182, Lemma 16.3]. We assume that $B = \partial\hat{\Delta}^i$ lies in some $\mathbb{R}^K \times \{0\} \subset \mathbb{R}^K \times \mathbb{R}$; in fact, since B is the boundary of an i-simplex, we can assume that it lies in $\mathbb{R}^i \times \{0\} \subset \mathbb{R}^{i+1}$. Then we let $p = (0, \ldots, 0, 1)$ and form the join (or, equivalently, the cone) $K = p * B$ inside $\mathbb{R}^{i+1}$ in this way; from now on, K will denote this specific simplicial complex in $\mathbb{R}^{i+1}$. Notice that K is still simplicially isomorphic to our $\hat{\Delta}^i_1$, and we identify K with Δ^i so that we may speak of $\sigma : K \to X$. Let μ be a simplex of B satisfying the hypotheses of condition (4). We observe that $\overline{St}(\mu, \mathrm{sd}^N(K/B)) \subset p * \overline{St}(\mu, B)$ for any N because any simplex of a relative subdivision having μ as a face must be contained within a subdivision of a simplex of $\hat{\Delta}^i_1$ that already has μ as a face. Consider $\sigma^{-1}(\psi(\sigma(\mu))) \subset K$, which is open in K and, from our previous observations, contains $\overline{St}(\mu, B)$. Since $p * \overline{St}(\mu, B) - \sigma^{-1}(\psi(\sigma(\mu)))$ is compact but does not intersect $\mathbb{R}^i \times \{0\}$, its orthogonal projection to $\{0\} \times \mathbb{R}$ has a positive minimum. In particular, there is an ϵ_μ such that any simplex contained in $(\mathbb{R}^i \times [0, \epsilon_\mu)) \cap (p * \overline{St}(\mu, B))$ is contained in $\sigma^{-1}(\psi(\sigma(\mu)))$. But now the arguments[17] of Step 1 of the proof of [182, Lemma 16.3] show precisely that, for any given ϵ_μ, there is a sufficiently large M_μ such that any simplex of $\mathrm{sd}^{M_\mu}(K/B)$ that intersects $\mathbb{R}^i \times \{0\}$ is contained in the strip $\mathbb{R}^i \times [0, \epsilon_\mu)$. It follows that, in $\mathrm{sd}^{M_\mu}(K/B)$, the star of μ is contained in $\sigma^{-1}(\psi(\sigma(\mu)))$, as desired. Since there are a finite number of such μ in B, it follows that there is an $M = \max_\mu\{M_\mu\}$ such that, in $\mathrm{sd}^M(K/B)$, the star of any μ satisfying the hypotheses of (4) is contained in

16 If these two conditions are already met for $\hat{\Delta}^i_1$ then no further subdivisions are necessary and we can let $\hat{\Delta}^i = \hat{\Delta}^i_1$.

17 Here is a sketch of the argument: Suppose K' is a subdivision of K relative to B such that any simplex of K' that intersects B is contained in $\mathbb{R}^i \times [0, m]$ for some m. There clearly exists such an $m \leq 1$. Let δ be a simplex of $\mathrm{sd}(K'/B)$ that intersects $\mathbb{R}^i \times \{0\}$. Then the vertices of δ are either vertices of B or barycenters of simplices of K' that intersect B. A computation with barycentric coordinates demonstrates that each of these barycenters must be contained in the strip $\mathbb{R}^i \times \left[0, (\frac{i}{i+1})m\right]$. Iterating the relative barycentric subdivision N times therefore results in all simplices of the iterated subdivision that intersect B being contained in $\mathbb{R}^i \times \left[0, (\frac{i}{i+1})^N m\right]$, and for a large enough N we have $(\frac{i}{i+1})^N m < \epsilon$.

$\sigma^{-1}(\psi(\sigma(\mu)))$. This is sufficient for condition (4) because if $Y \subset Z \subset X$ then $\psi(Y) \leq \psi(Z)$ in the ordering, and we have just shown that $\sigma(\overline{St}(\mu, \mathrm{sd}^M(K/B))) \subset \psi(\sigma(\mu))$ so that $\psi(\sigma(\overline{St}(\mu, \mathrm{sd}^M(K/B)))) \leq \psi(\sigma(\mu))$, while clearly $\sigma(\mu) \subset \sigma(\overline{St}(\mu, \mathrm{sd}^M(K/B)))$.

It remains to show that we can find an $N \geq M$ such that every simplex of $\mathrm{sd}^N(K/B)$ is contained in $\sigma^{-1}(U)$ for some $U \in \mathcal{U}$. By the preceding paragraph, every simplex of $\mathrm{sd}^M(K/B)$ that intersects B has this property. Let Q be the union of the simplices of $\mathrm{sd}^M(K/B)$ that do not intersect B, and let P be the union of the simplices of $\mathrm{sd}^N(K/B)$ that do intersect B. Then P and Q are finite complexes. As we perform further relative subdivisions, the simplices subdivided from the simplices in P continue to have the desired properties, while the simplices in Q, since they do not intersect B, undergo ordinary iterated barycentric subdivisions. But now we can appeal to the standard arguments: since Q is compact, it has a Lebesgue number [181, Lemma 27.5] with respect to the covering by $\sigma^{-1}(U)$, $U \in \mathcal{U}$, and by [182, Theorem 15.4], there is a finitely iterated barycentric subdivision of Q such that the diameters of the simplices of the subdivision are all less than the Lebesgue number. It follows that there is such an $\mathrm{sd}^N(K/B)$ as desired.

This completes the proof of Sublemma 6.5.5. □

We have at last finished proving Proposition 6.5.1. The tools we have just developed, however, can also be used to prove Proposition 6.5.2 and Corollary 6.5.3.

Proof of Proposition 6.5.2 The proof is the same with any coefficients (abelian groups or R-modules), so we provide the argument with $G = \mathbb{Z}$. The argument involves utilizing the constructions of the proof of Proposition 6.5.1, in particular the subdivision operators constructed in Lemma 6.5.4 and Sublemma 6.5.5.

First suppose $\xi \in I^{\bar{p}}S_i^{\mathrm{GM}}(X) \cap S_i^{\mathcal{U}}(X)$. As ξ is composed of a finite number of i-simplices, there is a finite set $\{U_1, \ldots, U_m\} \subset \mathcal{U}$ such that each simplex of ξ is contained in some U_j. We can choose our well-ordering of $\mathcal{U}$ such that the elements in $\{U_1, \ldots, U_m\}$ are the least elements and such that $U_j < U_k$ when $j < k$. In particular, we can assume that U_1 is the least element of $\mathcal{U}$. Let us decompose ξ in $S_i(X)$ as

$$\xi = \sum_{\{\sigma_j : |\sigma_j| \subset U_1\}} n_j \sigma_j + \sum_{\{\sigma_k : |\sigma_k| \not\subset U_1\}} n_k \sigma_k = \xi_1 + \xi_{>1}.$$

In other words, the chain ξ_1 consists of the simplices of ξ (with their coefficients) with image in U_1, and $\xi_{>1} = \xi - \xi_1$; neither ξ_1 nor $\xi_{>1}$ are necessarily intersection chains.

Now, let us apply the operator T_{U_1} from the proof of Lemma 6.5.4 to ξ. We have

$$T_{U_1}(\xi) = T_{U_1}(\xi_1 + \xi_{>1}) = T_{U_1}(\xi_1) + T_{U_1}(\xi_{>1}) \in I^{\bar{p}}S_i^{\mathrm{GM}}(U_1).$$

By the definition of T_{U_1} and the construction of T in the proof of Sublemma 6.5.5 (in particular see footnotes 15 on page 347 and 16 on page 349), as U_1 is the least element of $\mathcal{U}$ we may suppose that $T(\sigma) = \sigma$ for any σ with support in U_1. This then implies that $T_{U_1}(\xi_1) = \xi_1$. Also, each simplex of $T_{U_1}(\xi) - T_{U_1}(\xi_1) = T_{U_1}(\xi_{>1})$ must be a simplex from a singular subdivision of a simplex of $\xi_{>1}$; such a simplex must have support both in U_1

(in order to be a simplex of $T_{U_1}(\xi_{>1})$) and in some U_j with $j > 1$ (as each simplex of $\xi_{>1}$ is supported in some U_j with $j > 1$ and subdivision preserves or reduces supports). See Figure 6.5.

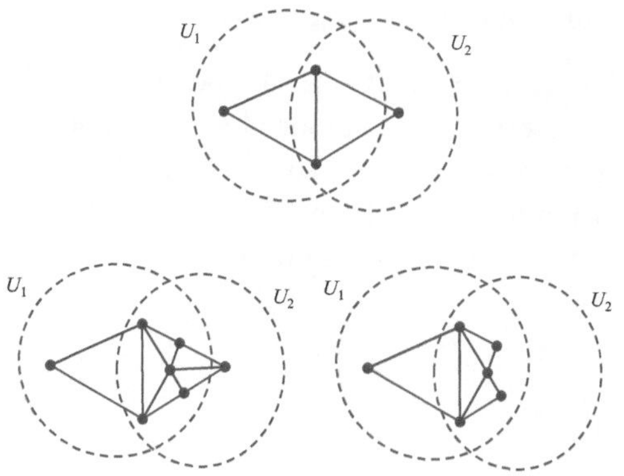

Figure 6.5 At the top we see an intersection chain ξ composed of two simplices, but the common face of the simplices may not be allowable, so we cannot split ξ along it. We construct the subdivision $T(\xi)$ (bottom left) using the subdivision operator constructed in the proof of Lemma 6.5.4. This subdivision preserves the simplex contained entirely in U_1 but subdivides the simplex in U_2 relative to its intersection with the simplex in U_1. Note that this subdivision satisfies all the properties required in Sublemma 6.5.5. On the bottom right, we see the resulting intersection chain $T_{U_1}(\xi)$.

Now consider

$$\begin{aligned}\xi = \xi_1 + \xi_{>1} &= \xi_1 + T_{U_1}(\xi_{>1}) - T_{U_1}(\xi_{>1}) + \xi_{>1}\\ &= T_{U_1}(\xi_1) + T_{U_1}(\xi_{>1}) - T_{U_1}(\xi_{>1}) + \xi_{>1}\\ &= T_{U_1}(\xi) + \xi_{>1} - T_{U_1}(\xi_{>1}).\end{aligned}$$

We know that $T_{U_1}(\xi) \in I^{\bar p}S_i^{\mathrm{GM}}(U_1) \subset I^{\bar p}S_i^{\mathrm{GM}}(X)$ by the proof of Lemma 6.5.4. So $\xi_{>1} - T_{U_1}(\xi_{>1}) = \xi - T_{U_1}(\xi)$ is also in $I^{\bar p}S_i^{\mathrm{GM}}(X)$. Additionally, we know that each simplex of $\xi_{>1} - T_{U_1}(\xi_{>1})$ is contained in an element of $\{U_2, \dots, U_m\}$, and so $\xi_{>1} - T_{U_1}(\xi_{>1})$ is subject to the same argument with a finite set of $\mathcal{U}$ with smaller cardinality. Continuing inductively, we see that we can decompose the chain ξ as desired.

If $\xi \in I^{\bar p}S_i(X)$, then ξ also represents a chain in $I^{\bar p}S_i'(X)$. The same argument just given still allows us to write $\xi = T_{U_1}(\xi) + \xi_{>1} - T_{U_1}(\xi_{>1})$, and by Lemma 6.5.4 the chain $T_{U_1}(\xi)$ represents an element of $I^{\bar p}S_i'(U_1) \subset I^{\bar p}S_i'(X)$. So $\xi - T_{U_1}(\xi) = \xi_{>1} - T_{U_1}(\xi_{>1})$ represents an element of $I^{\bar p}S_i'(X)$. Let $\phi : I^{\bar p}S_i'(X) \to I^{\bar p}S_i(X)$ be the isomorphism of Lemma 6.2.5. By Remark 6.2.6, if η is a chain representing an element of $I^{\bar p}S_i'(X)$, then $\phi(\eta)$ is obtained by throwing out from η any simplices contained in Σ_X. So $\phi(T_{U_1}(\xi_1)) \in I^{\bar p}S_i(U_1)$ and, for the same reasons as above, the image

$$\phi(\xi_{>1} - T_{U_1}(\xi_{>1})) = \phi(\xi - T_{U_1}(\xi)) = \xi - \phi(T_{U_1}(\xi)) \in I^{\bar{p}}S_i(X)$$

satisfies the hypotheses of the proposition with respect to a covering set of smaller cardinality. So, once again, an induction completes the argument. □

Proof of Corollary 6.5.3 Let $\mathcal{U} = \{A, X\}$ be an open cover of X, and let us order the cover so that $A < X$. As observed in the proof of Proposition 6.5.2, we can choose the subdivision operator T of the proof of Lemma 6.5.4 so that $T(\sigma) = \sigma$ for any σ supported in A. It then follows that the homomorphisms $\hat{T}_A : I^{\bar{p}}S_i^{\text{GM}}(X; G) \to I^{\bar{p}}S_i^{\text{GM}}(A; G)$ and $\bar{T}_A : I^{\bar{p}}S_i'(X; G) \to I^{\bar{p}}S_i'(A; G)$ provide the splittings. □

7
Intersection Cohomology and Products

We now turn to intersection cohomology and its properties and products. The reader should be cautioned that the term "intersection cohomology" can have a variety of meanings in the literature, including the following:

1. Sheaf theory almost always uses cohomological indexing, and so one speaks of "sheaf cohomology" (for the cohomology of a space with coefficients in a sheaf) or "sheaf hypercohomology" (for the cohomology of a space with coefficients in a complex of sheaves). In [106], a sheaf complex is introduced whose hypercohomology groups are the PL intersection homology groups on compact PL pseudomanifolds (see also [85, 90] for sheaves of singular chains). However, it soon became common to refer to intersection homology when constructed this way as intersection cohomology, and indeed that is the title of Borel's book [28]. On a non-compact space, these hypercohomology groups compute intersection homology "with closed supports," also called "homology of locally-finite chains" or "Borel–Moore homology." In this setting, a chain may be composed of infinite numbers of simplices so long as every point has a neighborhood intersecting only a finite number of them. The usual intersection homology groups can be recovered from the sheaf theory by considering "hypercohomology with compact supports."
2. In the PL setting, some sources (e.g. [110]) use the word "cochain" for what seems to refer to PL chains with closed support in the complementary dimension. In other words, they define $\mathfrak{C}^i(X)$ to be what we might call $\mathfrak{C}^{\infty}_{n-i}(X)$ on a PL space of dimension n. One sometimes sees cup and cap products given in this language (again, see [110]), but these seem to be versions of the Goresky–MacPherson intersection product that we will discuss in Section 8.5, with the cup product taking a pair of chains with closed support to a chain with closed support, and the cap product taking a chain with closed support and a chain with compact support to a chain with compact support.
3. The various flavors of intersection cohomology defined via differential forms are variants of de Rham cohomology.

By contrast to these other options, we shall utilize a definition analogous to that for ordinary singular cohomology by defining intersection cochains in terms of the Hom duals of the intersection chains. Such definitions have appeared previously, for example in [31, 100]. One can show using sheaf theory that this yields the same intersection cohomology groups as the sheaf-theoretic hypercohomology definition of [106] provided we use field coefficients [31, 99]. However, the intersection cohomology cone formula (Proposition 7.1.5) shows that these are not quite the same for more general coefficients, as these are evidently not just the intersection homology modules reindexed. We justify our current definition by its participation in the Poincaré Duality Theorems of the next chapter and by its parallel with the cohomology theory for manifolds.

After introducing these intersection cohomology groups in Section 7.1 and establishing their basic properties, we define singular cross, cup, and cap products in Section 7.2. Products of this form were first introduced in [100] with field coefficients, but we here generalize to coefficients in Dedekind domains. With an Alexander–Whitney construction unavailable for intersection cochains (as we will discuss below), the *algebraic diagonal* map that plays a critical role in the construction of the cup and cap products depends instead upon the Künneth Theorem of the preceding chapter. Section 7.3 is a long section developing the various properties of these products in analogy with the properties of the classical products; much of this material is developed for intersection cohomology here for the first time. For the reader more interested in the final statements than the arguments, a summary of these properties is provided in Section 7.3.9. Up to that point we consider the properties only for CS sets, which do not include pseudomanifolds with boundary. We explain how to extend the results to objects with boundary in Section 7.3.10. To conclude the chapter, we provide some results about intersection cohomology with compact supports in Section 7.4.

Unfortunately, we will see as we progress through the basic properties of intersection cohomology in Section 7.1 that it is more difficult to work with PL intersection cohomology than with singular intersection cohomology. The main trouble is that we know from Lemma 6.3.1 that the $I^{\bar{p}}S_i(X;R)$ are projective when R is a Dedekind domain, but we only know that the $I^{\bar{p}}\mathfrak{C}_i(X;R)$ are flat. So the PL chains do not behave as well with respect to dualization under Hom. Furthermore, the complex $I^{\bar{p}}\mathfrak{C}_i(X;R)$ is defined as a direct limit and so dualizes to an inverse limit, creating other difficulties. Meanwhile, as noted in Section 1.2, the advantage of the PL category in terms of allowing us to view chains geometrically and compute simplicially begins to fall away as we come to intersection cohomology, as we do not know of an Alexander–Whitney-type formula that would allow us to compute intersection cohomology products combinatorially. Consequently, though we briefly discuss properties of PL intersection cohomology alongside singular intersection cohomology in Section 7.1, beginning in Section 7.2 we turn to focusing solely on the singular intersection cohomology groups for the remainder of the book.

Remark 1 As per Remarks 6.2.2 and 6.2.7, for the remainder of the book we typically simply utilize the notation $(I^{\bar{p}}S_*(X;G),\partial)$ for the non-GM intersection chain complex and its boundary maps, reserving the notations $I^{\bar{p}}S'_*(X;G)$ or $\hat{\partial}$ for only those instances where making such distinctions explicit is critical.

7.1 Intersection Cohomology

So far, we have focused exclusively on intersection homology. It is time to introduce intersection cohomology. The definition works with coefficients in any commutative ring with unity R, though for certain results here and for many results as we progress we will need to assume that R is a Dedekind domain or a field. Recall that we have already seen such coefficient restrictions in Sections 5.3 and 6.4.

Definition 7.1.1 Let R be a commutative ring with unity, let (X, A) be a filtered space X with a subset A, and let $\bar{p}$ be a perversity on X. We define the *singular intersection cochain complex*

$$I_{\bar{p}}S^*(X, A; R) = \operatorname{Hom}_R(I^{\bar{p}}S_*(X, A; R), R).$$

We will denote the coboundary operator by d. Following the Koszul sign conventions (see appendix Section A.1), if $\alpha \in I_{\bar{p}}S^i(X, A; R)$ and $x \in I^{\bar{p}}S_{i+1}(X, A; R)$, then $(d\alpha)(x) = (-1)^{i+1}\alpha(\partial x)$.

The associated *intersection cohomology* modules are

$$I_{\bar{p}}H^i(X, A; R) = H^i(I_{\bar{p}}S^*(X, A; R)) = H^i(\operatorname{Hom}_R(I^{\bar{p}}S_*(X, A; R), R)).$$

Similarly, if X is a PL filtered space with PL subspace A, we let $I_{\bar{p}}\mathfrak{C}^*(X, A; R) = \operatorname{Hom}_R(I^{\bar{p}}\mathfrak{C}_*(X, A; R), R)$ and

$$I_{\bar{p}}\mathfrak{H}^i(X, A; R) = H^i(I_{\bar{p}}\mathfrak{C}^*(X, A; R)) = H^i(\operatorname{Hom}_R(I^{\bar{p}}\mathfrak{C}_*(X, A; R), R)).$$

Notice that in our notation both the degree index and the perversity index shift their subscript/superscript locations. Of course, shifting the degree index is standard; the shifting of the perversity index is simply meant as an additional visual cue.

Remark 7.1.2 We will not use it here, but for any R-module M one could just as easily define $I_{\bar{p}}S^*(X, A; M) = \operatorname{Hom}_R(I^{\bar{p}}S_*(X, A; R), M)$.

We could also define versions of intersection cohomology based on the complexes $I^{\bar{p}}S_*^{\mathrm{GM}}(X)$ and $I^{\bar{p}}\mathfrak{C}_*^{\mathrm{GM}}(X)$, but since our reason for introducing cohomology is to introduce cup products and as these require the Künneth Theorem and so non-GM intersection cohomology, we will not pursue separately the properties of GM intersection cohomology groups. Recall, though, that when $\bar{p} \le \bar{t}$, there is no difference between the GM and non-GM theories, by Proposition 6.2.9.

Remark 7.1.3 When the ground ring is understood, we will use the notations Hom and $\otimes$, rather than the more cumbersome Hom_R and $\otimes_R$.

When R is a Dedekind domain, the Universal Coefficient Theorem holds:

Theorem 7.1.4 *For a Dedekind domain R and for every i, there is a natural exact sequence*

$$0 \to \operatorname{Ext}(I^{\bar{p}}H_{i-1}(X, A; R), R) \to I_{\bar{p}}H^i(X, A; R) \to \operatorname{Hom}(I^{\bar{p}}H_i(X, A; R), R) \to 0.$$

This sequence splits, but not naturally. Additionally, if F is a field, then for PL intersection cohomology we have $I_{\bar p}\mathfrak{H}^i(X,A;F) \cong \mathrm{Hom}(I^{\bar p}\mathfrak{H}_i(X,A;F),F)$.

Proof This is just an application of the algebraic Universal Coefficient Theorem [238, Section 3.6], for which, for the singular chain case, we only need verify that each $I^{\bar p}S_i(X,A;R)$ and $\partial(I^{\bar p}S_i(X,A;R))$ are projective. But we know this from Lemma 6.3.1.

The difficulty with the PL case is that we do not know that $I^{\bar p}\mathfrak{C}_*(X;R)$ is projective, so we cannot invoke the needed algebraic results. Of course, if R is a field, then all modules are free, so in this case we can apply the algebraic theorem, with the Ext term vanishing. □

Notice that this Universal Coefficient Theorem for cohomology is not inconsistent with the general failure of the Universal Coefficient Theorem for homology, as discussed in Section 5.3.2. There, we introduced a complex $I^{\bar p}S_*(X;G)$ that is not of the form $C_* \otimes G$, and so the usual homological algebra does not apply. Here, by contrast, we define $I_{\bar p}S^*(X;R)$ by applying the $\mathrm{Hom}(\cdot,R)$ functor in the standard way.

Given Theorem 7.1.4 and the intersection homology cone formula (Theorem 6.2.13), it is not difficult to write down a cohomology version of the cone formula:

Proposition 7.1.5 *If X is a compact filtered space of formal dimension $n-1$ and R is a Dedekind domain, then*

$$I_{\bar p}H^i(cX;R) \cong \begin{cases} 0, & i > n-\bar p(\{v\})-1, \\ \mathrm{Ext}(I^{\bar p}H_{i-1}(X;R),R), & i = n-\bar p(\{v\})-1, \\ I_{\bar p}H^i(X;R), & i < n-\bar p(\{v\})-1. \end{cases}$$

Furthermore, the isomorphisms when $i < n-\bar p(\{v\})-1$ are induced by inclusion. This result extends to PL intersection cohomology if R is a field.

Proof The case $i > n-\bar p(\{v\})-1$ is immediate from the intersection homology cone formula and the Universal Coefficient Theorem. When $i < n-\bar p(\{v\})-1$, the naturality of the Universal Coefficient Theorem gives us the diagram

$$\begin{array}{ccccccccc} 0 \to & \mathrm{Ext}(I^{\bar p}H_{i-1}(cX;R),R) & \to & I_{\bar p}H^i(cX;R) & \to & \mathrm{Hom}(I^{\bar p}H_i(cX;R),R) & \to & 0 \\ & \downarrow & & \downarrow & & \downarrow & & \\ 0 \to & \mathrm{Ext}(I^{\bar p}H_{i-1}(X;R),R) & \to & I_{\bar p}H^i(X;R) & \to & \mathrm{Hom}(I^{\bar p}H_i(X;R),R) & \to & 0. \end{array}$$

The outer maps are isomorphisms by Theorem 6.2.13; therefore so is $I_{\bar p}H^i(cX;R) \to I_{\bar p}H^i(X;R)$ by the Five Lemma.

When $i = n-\bar p(\{v\})-1$, however, $I^{\bar p}H_i(cX;R) = 0$, but $I^{\bar p}H_{i-1}(cX;R)$ might be non-zero. So

$$I_{\bar p}H^i(cX;R) \cong \mathrm{Ext}(I^{\bar p}H_{i-1}(cX;R),R) \cong \mathrm{Ext}(I^{\bar p}H_{i-1}(X;R),R),$$

the last isomorphism again by the homology cone formula.

By Theorem 7.1.4, the argument can be extended to PL intersection cohomology when using field coefficients. □

Remark 7.1.6 Notice that the intersection cohomology version of the cone formula is not so clear cut as the homology version in Theorem 6.2.13. In homology, $I^{\bar{p}}H_i(cX;R)$ is either $I^{\bar{p}}H_i(X;R)$ or 0. But in cohomology $I_{\bar{p}}H^{n-\bar{p}(\{v\})-1}(cX;R)$ reaches down a degree to pluck out the torsion term $\operatorname{Ext}(I^{\bar{p}}H_{n-\bar{p}(\{v\})-2}(X;R),R)$. We have already seen this issue in Section 5.3 in our discussion of the Universal Coefficient Theorem in homology. This led to the definition of locally torsion-free spaces in Section 5.3.2. Similarly, the appearance of this one seemingly minor blip will force us to assume our spaces are locally torsion-free in order to get much of the intersection cohomology theory to work out right, including being able to define cup and cap products and, eventually, Poincaré duality.

Our next goal is to establish the basic properties of intersection cohomology, such as stratified homotopy invariance, excision, etc., utilizing the properties of intersection homology (though we reserve discussion of products, including cohomology cross products, to Section 7.2). Most follow almost directly from the intersection homology versions, in which case we will just briefly provide the relevant reasons. Some, however, require more elaboration, which we provide.

Proposition 7.1.7 *If X,Y are filtered spaces, $f : X \to Y$ is $(\bar{p},\bar{q})$-stratified, and $A \subset X$ and $B \subset Y$ with $f(A) \subset B$, then f induces a chain map $f^* : I_{\bar{q}}S^*(Y,B;R) \to I_{\bar{p}}S^*(X,A;R)$. If, additionally, X,Y are PL filtered spaces and f is a PL map, then f induces a chain map $f^* : I_{\bar{q}}\mathfrak{C}^*(Y,B;R) \to I_{\bar{p}}\mathfrak{C}^*(X,A;R)$ of PL intersection chain complexes. In either case, we obtain corresponding maps of intersection cohomology groups.*

Proof By Proposition 6.3.5, these are the conditions that allow the existence of maps of intersection chain complexes, and so the f^* here are just the resulting Hom duals. □

Corollary 7.1.8 *If $f : X \to Y$ is a stratified homeomorphism that is also a homeomorphism of pairs $f : (X,A) \to (Y,B)$ and the perversities $\bar{p}$ on X and $\bar{q}$ on Y correspond, then $I_{\bar{p}}H^*(X,A;R) \cong I_{\bar{q}}H^*(Y,B;R)$. The corresponding fact holds for PL spaces, PL stratified homeomorphisms, and PL intersection cohomology.*

Proof As for Corollaries 4.1.8 and 6.3.6, the maps induce isomorphisms of the intersection chain complexes, and so dually they induce isomorphisms of cochain complexes. □

Proposition 7.1.9 *Suppose $f,g : X \to Y$ are $(\bar{p},\bar{q})$-stratified maps that are $(\bar{p},\bar{q})$-stratified homotopic via a $(\bar{p},\bar{q})$-stratified homotopy taking the pair $(I\times X, I\times A)$ to (Y,B). Then f and g induce chain homotopic chain maps $I_{\bar{q}}S^*(Y,B;R) \to I_{\bar{p}}S^*(X,A;R)$ and so $f^* = g^* : I_{\bar{q}}H^*(Y,B;R) \to I_{\bar{p}}H^*(X,A;R)$. The analogous result holds in the PL category.*

Proof By Proposition 6.3.7, f and g induce chain homotopic chain maps of intersection chain complexes, and so we obtain the corresponding results about cochains by dualizing. □

The following corollary is an immediate consequence of the proposition and the fact that the duals of chain homotopies are (co)chain homotopies (Lemma A.2.2); compare Corollary 6.3.8.

Corollary 7.1.10 *Suppose (X, A) and (Y, B) are filtered pairs and that $f : X \to Y$ is a stratified homotopy equivalence that restricts to a stratified homotopy equivalence $A \to B$. Suppose that the values of $\bar{p}$ on X and $\bar{q}$ on Y agree on corresponding strata. Then f induces an isomorphism $I_{\bar{q}}H^*(Y, B; R) \cong I_{\bar{p}}H^*(X, A; R)$. The analogous result holds in the PL category.*

Theorem 7.1.11 *For a filtered space X and subspace A, if R is a Dedekind domain then there is a long exact sequence*

$$\longrightarrow I_{\bar{p}}H^i(X, A; R) \longrightarrow I_{\bar{p}}H^i(X; R) \longrightarrow I_{\bar{p}}H^i(A; R) \longrightarrow I_{\bar{p}}H^{i+1}(X, A; R) \longrightarrow .$$

The same is true of PL intersection cohomology using field coefficients.

Proof As R is Dedekind, $I^{\bar{p}}S_i(X, A; R)$ is projective by Lemma 6.3.1, and so the short exact sequence of intersection chain complexes of the pair splits in each dimension by Lemma A.4.2. Therefore, applying the functor $\mathrm{Hom}(\cdot, R)$ preserves exactness to yield the short exact sequence of the pair for intersection cochains. The existence of the long exact sequence of intersection cohomology follows by standard homological algebra.

In the PL case, the field coefficients ensure the exactness of the Hom functor. □

We now move on to excision and Mayer–Vietoris sequences. It is for these theorems that we need to utilize our chain splitting propositions of Section 6.5.

Theorem 7.1.12 (Excision) *Let X be a filtered space, and suppose $K \subset U \subset X$ such that $\bar{K} \subset \mathring{U}$. Then inclusion induces an isomorphism $I_{\bar{p}}H^*(X, U; R) \xrightarrow{\cong} I_{\bar{p}}H^*(X - K, U - K; R)$. The analogous statement holds for PL chains.*

Proof First consider singular chains. We will show that the inclusion $I^{\bar{p}}S_*(X - K, U - K; R) \to I^{\bar{p}}S_*(X, U; R)$ is a chain homotopy equivalence. Let $\mathcal{V} = \{X - K, U\}$ be a covering of X. Notice that the interiors of the sets of $\mathcal{V}$ continue to be a cover by the assumptions on K and U. By Proposition 6.5.1, the inclusion $\mathfrak{i} : I^{\bar{p}}S_*^{\mathcal{V}}(X; R) \to I^{\bar{p}}S_*(X; R)$ is a chain homotopy equivalence, and, as we will see in the proof of the proposition, we have a homotopy inverse T such that $\mathfrak{i}T$ is a singular subdivision map $\hat{T}$. In particular, $\hat{T}$ and T preserve or reduce supports. Similarly, the chain homotopies involved also preserve or reduce supports. It follows that $\mathfrak{i}$, T, and the chain homotopies descend to maps and chain homotopies relative to U, so that $\mathfrak{i}$ induces a chain homotopy equivalence

$$I^{\bar{p}}S_*^{\mathcal{V}}(X; R)/I^{\bar{p}}S_*(U; R) \to I^{\bar{p}}S_*(X; R)/I^{\bar{p}}S_*(U; R) = I^{\bar{p}}S_*(X, U; R).$$

But now, applying the second fundamental theorem of algebra and the basic definitions,

$$\begin{aligned}\frac{I^{\bar{p}}S_*^{\mathcal{V}}(X;R)}{I^{\bar{p}}S_*(U;R)} &= \frac{I^{\bar{p}}S_*(X-K;R)+I^{\bar{p}}S_*(U;R)}{I^{\bar{p}}S_*(U;R)}\\ &\cong \frac{I^{\bar{p}}S_*(X-K;R)}{I^{\bar{p}}S_*(X-K;R)\cap I^{\bar{p}}S_*(U;R)}\\ &\cong \frac{I^{\bar{p}}S_*(X-K;R)}{I^{\bar{p}}S_*(U-K;R)}\\ &\cong I^{\bar{p}}S_*(X-K,U-K;R).\end{aligned}$$

Therefore $I^{\bar{p}}S_*(X-K,U-K;R)$ and $I^{\bar{p}}S_*(X,U;R)$ are chain homotopy equivalent. Thinking through what happens to representative elements, we see that the chain homotopy equivalence is induced by the inclusion map.

The PL case is even more straightforward using the isomorphism given in Proposition 6.5.1 instead of homotopy equivalences. □

Theorem 7.1.13 (Mayer–Vietoris sequences) *Suppose that $X = U \cup V$, where U and V are subspaces such that $X = \mathring{U} \cup \mathring{V}$, and that R is a Dedekind domain. Then there is an exact Mayer–Vietoris sequence*

$$\begin{aligned}\to I_{\bar{p}}H^{i-1}(U\cap V;R) \to I_{\bar{p}}H^i(U\cup V;R) \to I_{\bar{p}}H^i(U;R)\oplus I_{\bar{p}}H^i(V;R)\\ \to I_{\bar{p}}H^i(U\cap V;R)\to .\end{aligned}$$

The equivalent results holds in the PL context if R is a field.

Proof First consider the short exact Mayer–Vietoris sequence of intersection chain complexes, analogous to that in the proof of Theorem 4.4.19. As R is Dedekind and $I^{\bar{p}}S_*(U;R)+I^{\bar{p}}S_*(V;R)$ is a submodule of $I^{\bar{p}}S_*(X;R)$, which is projective by Lemma 6.3.1, this submodule is also projective. So $\mathrm{Hom}(\cdot,R)$ preserves the exactness of the short exact sequence, and thus we obtain a short exact sequence of intersection cochain complexes and consequently a long exact sequence of cohomology modules. It remains to show that

$$H^i(\mathrm{Hom}(I^{\bar{p}}S_*(U;R)+I^{\bar{p}}S_*(V;R),R)) \cong I_{\bar{p}}H^i(X;R).$$

But the chain inclusion $I^{\bar{p}}S_*(U;R)+I^{\bar{p}}S_*(V;R)\hookrightarrow I^{\bar{p}}S_*(X;R)$ is, by Proposition 6.5.1, a chain homotopy equivalence, so the Hom dual is also a chain homotopy equivalence and induces an isomorphism on intersection cohomology.

For PL chains over a field, we use that all short exact sequences of vector spaces split and then apply the PL part of Proposition 6.5.1. □

The relative Mayer–Vietoris sequence requires a bit more work:

Theorem 7.1.14 *Suppose $X = U \cup V$, where U,V are open subspaces. Let $A \subset X$, let $C = A\cap U$, and let $D = A \cap V$. Let R be a Dedekind domain. Then there is an exact*

Mayer–Vietoris sequence

$$\to I_{\bar{p}}H^i(X,A;R) \to I_{\bar{p}}H^i(U,C;R) \oplus I_{\bar{p}}H^i(V,D;R) \to I_{\bar{p}}H^i(U \cap V, C \cap D;R) \to .$$

Proof As in the proof of Theorem 4.4.23, we have a diagram of exact sequences

$$\begin{array}{ccccccccc} 0 & \longrightarrow & I^{\bar{p}}S_*(C \cap D;R) & \longrightarrow & \begin{array}{c} I^{\bar{p}}S_*(C;R) \oplus \\ I^{\bar{p}}S_*(D;R) \end{array} & \longrightarrow & \begin{array}{c} I^{\bar{p}}S_*(C;R) + \\ I^{\bar{p}}S_*(D;R) \end{array} & \longrightarrow & 0 \\ & & \downarrow & & \downarrow & & \downarrow & & \\ 0 & \longrightarrow & I^{\bar{p}}S_*(U \cap V;R) & \longrightarrow & \begin{array}{c} I^{\bar{p}}S_*(U;R) \oplus \\ I^{\bar{p}}S_*(V;R) \end{array} & \longrightarrow & \begin{array}{c} I^{\bar{p}}S_*(U;R) + \\ I^{\bar{p}}S_*(V;R) \end{array} & \longrightarrow & 0, \end{array}$$

with the vertical maps all inclusions. This yields, by the Snake Lemma, a short exact sequence:

$$\begin{aligned} 0 &\longrightarrow I^{\bar{p}}S_*(U \cap V, C \cap D;R) \longrightarrow I^{\bar{p}}S_*(U,C;R) \oplus I^{\bar{p}}S_*(V,D;R) \\ &\longrightarrow \frac{I^{\bar{p}}S_*(U;R) + I^{\bar{p}}S_*(V;R)}{I^{\bar{p}}S_*(C;R) + I^{\bar{p}}S_*(D;R)} \longrightarrow 0. \end{aligned} \tag{7.1}$$

We claim that each $\frac{I^{\bar{p}}S_i(U;R)+I^{\bar{p}}S_i(V;R)}{I^{\bar{p}}S_i(C;R)+I^{\bar{p}}S_i(D;R)}$ is a submodule of $I^{\bar{p}}S_i(X,A;R)$ and so is projective, as the latter module is projective by Lemma 6.3.1. For this, we consider the evident map

$$\frac{I^{\bar{p}}S_i(U;R) + I^{\bar{p}}S_i(V;R)}{I^{\bar{p}}S_i(C;R) + I^{\bar{p}}S_i(D;R)} \to \frac{I^{\bar{p}}S_i(X;R)}{I^{\bar{p}}S_i(A;R)} = I^{\bar{p}}S_i(X,A;R)$$

induced by inclusions. If $\xi \in I^{\bar{p}}S_i(U;R) + I^{\bar{p}}S_i(V;R)$ maps to 0 in $I^{\bar{p}}S_i(X,A;R)$, then we must have $|\xi| \subset A$. But also every simplex of ξ is contained in U or V. So ξ can be written in terms of simplices in $A \cap U = C$ and $A \cap V = D$. Note that *a priori* if we are given a decomposition of ξ as $\xi = \xi_U + \xi_V$ with $\xi_U \in I^{\bar{p}}S_i(U;R)$ and $\xi_V \in I^{\bar{p}}S_i(V;R)$ then it is not necessarily the case that either ξ_U or ξ_V is contained in A, but the fact that $|\xi| \subset A$ means that any simplices of ξ_U or ξ_V not contained in A must cancel in ξ. Nonetheless, the assumptions of Proposition 6.5.2 are met (with $\mathcal{U} = \{C,D\}$, which is an open[1] cover of A), so we can conclude that there is a decomposition of ξ into $\xi_C + \xi_D$ with $\xi_C \in I^{\bar{p}}S_i(C;R)$ and $\xi_D \in I^{\bar{p}}S_i(D;R)$. Therefore, the chain ξ represents 0 in $\frac{I^{\bar{p}}S_i(U;R)+I^{\bar{p}}S_i(V;R)}{I^{\bar{p}}S_i(C;R)+I^{\bar{p}}S_i(D;R)}$. This proves the claim.

It now follows that the short exact sequence (7.1) splits in each degree and so remains exact after applying the functor $\mathrm{Hom}(\cdot, R)$. The associated long exact cohomology sequences

[1] The need to invoke Proposition 6.5.2 at this point is the reason we have assumed an open cover in our hypothesis, rather than the more general hypothesis of Theorem 7.1.13. In fact, it would be sufficient here for $U \cap A$ and $V \cap A$ to be an open cover of A, but it is unclear how to extend the argument to more general pairs $\{U,V\}$.

will be our Mayer–Vietoris sequence once we show that

$$H^i\left(\mathrm{Hom}\left(\frac{I^{\bar{p}}S_i(U;R)+I^{\bar{p}}S_i(V;R)}{I^{\bar{p}}S_i(C;R)+I^{\bar{p}}S_i(D;R)},R\right)\right)\cong I_{\bar{p}}H^i(X,A;R). \tag{7.2}$$

Here we again emulate the proof of Theorem 4.4.23. We have a diagram of exact sequences

$$\begin{array}{ccccccccc}
0 & \longrightarrow & \begin{array}{c} I^{\bar{p}}S_*(C;R)+\\ I^{\bar{p}}S_*(D;R)\end{array} & \to & \begin{array}{c} I^{\bar{p}}S_*(U;R)+\\ I^{\bar{p}}S_*(V;R)\end{array} & \to & \frac{I^{\bar{p}}S_*(U;R)+I^{\bar{p}}S_*(V;R)}{I^{\bar{p}}S_*(C;R)+I^{\bar{p}}S_*(D;R)} & \longrightarrow & 0\\
 & & \downarrow & & \downarrow & & \downarrow & & \\
0 & \longrightarrow & I^{\bar{p}}S_*(A;R) & \longrightarrow & I^{\bar{p}}S_*(X;R) & \longrightarrow & I^{\bar{p}}S_*(X,A;R) & \longrightarrow & 0,
\end{array}$$

each of which splits in each degree. So applying the $\mathrm{Hom}(\cdot,R)$ functor preserves exactness and yields a commutative diagram of long exact cohomology sequences. We know that the cohomology maps corresponding to the two left vertical arrows are isomorphisms from the proof of Theorem 7.1.13, so the isomorphism (7.2) follows by the Five Lemma. □

Applying the material of Section 5.4 provides an equivalence between PL and singular intersection cohomology on PL spaces, though we must again assume field coefficients.

Theorem 7.1.15 *Let F be a field, and let X be a PL CS set. Then $I_{\bar{p}}\mathfrak{H}^*(X;F)\cong I_{\bar{p}}H^*(X;F)$.*

Proof By the results of Section 5.4, which we have noted in Section 6.3 carry over to non-GM intersection homology, we have a diagram of maps

$$I^{\bar{p}}\mathfrak{S}_*(X;F)\xleftarrow{\psi} I^{\bar{p}}\mathfrak{C}^T_*(X;F)\xrightarrow{\phi} I^{\bar{p}}\mathfrak{C}_*(X;F),$$

in which each map induces isomorphisms on homology. We also have a map $I^{\bar{p}}S_*(X;F)\to I^{\bar{p}}\mathfrak{S}_*(X;F)$ that induces an isomorphism on homology. All chain complexes are bounded below and, since we work over a field, free. Therefore, all of the maps involved are in fact chain homotopy equivalences [182, Theorem 46.2]. Dualizing these homotopy equivalences provides homotopy equivalences of cochain complexes by Corollary A.2.3 and hence the desired intersection cohomology isomorphism. □

Remark 7.1.16 It seems reasonable to expect that the restriction that R be a field in Theorem 7.1.15 should not be necessary. However, it is not clear how to complete a more general proof, and so not clear that this is true. One method to pursue such a result would be to attempt to strengthen the results of Section 5.4 to make ϕ and ψ chain homotopy equivalences by constructing chain homotopy inverses. An alternative approach would be to attempt to find a way to show that $I^{\bar{p}}\mathfrak{C}_*(X;R)$ is always chain homotopy equivalent to a bounded-below complex of projective modules, say A_*; if R is Dedekind, then each $I^{\bar{p}}S_i(X;R)$ is already projective by Lemma 6.3.1. We would then have quasi-isomorphisms between A_* and $I^{\bar{p}}\mathfrak{S}_*(X;R)$, which would induce a chain homotopy equivalence between them by [238, Theorem 10.4.8]. Altogether, then, $I^{\bar{p}}\mathfrak{C}_*(X;R)$ and $I^{\bar{p}}S_*(X;R)$ would be chain homotopy equivalent.

Alternatively, a "more correct" way to proceed might be to work completely in the derived category and so define $I_{\bar{p}}S^*(X,A;R)$ instead using derived functors as $I_{\bar{p}}S^*(X,A;R) = \mathrm{RHom}_R(I^{\bar{p}}S_*(X,A;R),R)$ and similarly for $I_{\bar{p}}\mathfrak{C}^*$. This would have advantages at the expense of more sophistication. Since we do not expect the reader to be conversant with derived categories, and since we will not need to work much with PL intersection cohomology, we will not pursue this here.

We next extend the intersection homology results on topological invariance (Theorem 5.5.1) to cohomology. If we use field coefficients, this theorem implies a PL version via Theorem 7.1.15.

Theorem 7.1.17 *Suppose R is a Dedekind domain, X is a CS set of formal dimension n with no codimension-one strata, and $\bar{p}$ is a GM perversity. Then $I_{\bar{p}}H^*(X;R)$ is independent (up to isomorphism) of the choice of stratification of X as a CS set of formal dimension n. In particular, if X' is another CS set of formal dimension n that is topologically homeomorphic to X (not necessarily stratified homeomorphic), then $I_{\bar{p}}H^*(X;R) \cong I_{\bar{p}}H^*(X';R)$.*

More generally, if A is an open subset of X and $(X,A) \cong (X',A')$, then $I_{\bar{p}}H^*(X,A;R) \cong I_{\bar{p}}H^*(X',A';R)$.

Proof The hypotheses on the perversity ensure that $I^{\bar{p}}S_*(X;R) \cong I^{\bar{p}}S_*^{\mathrm{GM}}(X;R)$ by Proposition 6.2.9. The condition that there be no codimension-one strata is necessary to ensure both that $\bar{p} \leq \bar{t}$ and that $\bar{p}(k) \geq 0$ for all k, which is required for Theorem 5.5.1, which we can now invoke. Since R is Dedekind, each $I^{\bar{p}}S_i(X;R)$ is projective by Lemma 6.3.1, so the homology isomorphisms of the proof of Theorem 5.5.1 are, in fact, homotopy equivalences. This follows from [238, Theorem 10.4.8], noting that bounded-above cochain complexes are equivalent to bounded-below chain complexes. The $\mathrm{Hom}(\cdot,R)$ duals are then also chain homotopy equivalences by Corollary A.2.3, yielding isomorphisms on cohomology. The statement for relative intersection cohomology follows via a Five Lemma argument as in the proof of Theorem 5.5.1, using the arguments of Theorem 7.1.11 to obtain the diagrams of long exact sequences from the diagrams of short exact sequences of chain complexes. □

As a final observation, we note that there is an intersection cohomology version of Proposition 5.1.8 and Corollary 5.1.9, which told us that the perversity $\bar{t}$ intersection homology of a normal stratified pseudomanifold is isomorphic to ordinary homology. Looking ahead to our duality results, this will imply another result of [105], that for a compact normal stratified pseudomanifold of dimension n, we have $I^{\bar{0}}H_i(X) \cong H^{n-i}(X)$:

Proposition 7.1.18 *Let X be a CS set, R a Dedekind domain, and $\bar{p}$ a perversity such that*

1. *every point has a neighborhood stratified homeomorphic to $\mathbb{R}^k \times cL$ such that $I^{\bar{p}}H_0(cL;R) \cong R$ and $I^{\bar{p}}H_i(cL;R) = 0$ for $i > 0$, and*
2. *the only strata of depth zero are regular strata.*

Then $I_{\bar{p}}H^*(X;R) \cong H^*(X;R)$.

In particular, this is the case if X is a normal stratified pseudomanifold and $\bar{p}$ is the top perversity $\bar{t}$, i.e. in this case $I_{\bar{t}}H^*(X;R) \cong H^*(X;R)$.

This result holds for PL intersection cohomology using field coefficients.

Proof By Proposition 5.1.8 and Corollary 5.1.9, modified in the evident way for R coefficients, the assumptions imply that $I^{\bar{p}}H_*(X;R) \cong H_*(X;R)$. The result for cohomology is then a consequence of the Universal Coefficient Theorem and the Five Lemma. □

Let X be a normal, compact, and R-oriented n-dimensional stratified pseudomanifold. Then, as all CS sets are locally $\bar{t}$-torsion-free, we will have Poincaré duality for X by Theorem 8.2.4 below, implying that $I_{\bar{t}}H^{n-i}(X;R) \cong I^{\bar{0}}H_i(X;R)$. Putting this together with the preceding result yields $I^{\bar{0}}H_i(X;R) \cong H^{n-i}(X;R)$, and so also $I^{\bar{0}}\mathfrak{H}_i(X;R) \cong H^{n-i}(X;R)$ using Theorem 6.3.31. For $\mathbb{Z}$-coefficients on compact PL pseudomanifolds without codimension-one strata (in which case $I^{\bar{0}}\mathfrak{H}_i(X;R) \cong I^{\bar{0}}\mathfrak{H}_i^{\mathrm{GM}}(X;R)$ by Proposition 6.2.9), this result goes back to [105] and relates to other early results about computing cohomology using sufficiently transverse chains; see e.g. [40, 110, 113]. Together with Corollary 5.1.9, this observation motivates the notion that intersection homology, at least on normal pseudomanifolds, filters between ordinary cohomology and ordinary homology as the perversities range between $\bar{0}$ and $\bar{t}$.

7.2 Cup, Cap, and Cross Products

In this section, we introduce and study cup and cap products in intersection homology and cohomology, as well as an intersection cohomology cross product. Broadly, we follow the construction of intersection cup and cap products in [100], though only field coefficients are treated there. Here we generalize to allow coefficients in a Dedekind domain and provide a more comprehensive survey of properties than is found in [100].

We begin in Section 7.2.1 with a review of classical cohomology products together with a philosophical overview of the approach in the intersection context. The technical definitions follow in Section 7.2.2. Our extensive study of the properties of the products takes place in the next section, Section 7.3.

7.2.1 Philosophy

As for ordinary homology/cohomology theory, the advantage of working with cohomology over homology is that cohomology possesses an internal product. It is well known that singular cohomology always possesses a cup product, while homology only possesses a product in certain special situations, such as when we take the homology of a topological group or H space [71, Section VII.2] or when our space is a manifold, in which case there is an intersection product that is Poincaré dual to the cup product. In fact, the singular cup product can be defined at the level of cochains.

Unfortunately, we will not be able to define a cup product quite so broadly for intersection cohomology. We will mostly need to work at the level of cohomology (not cochains), and even when we do so the cup product will not generally be internal, meaning in this case that the cup product will take a pair of intersection cohomology classes with certain perversities to an intersection cohomology class with a third perversity. This last property is related to the formal structure of Poincaré duality for pseudomanifolds that we will discuss in the next chapter: in the intersection world, Poincaré duality pairs not just dual dimensions but *dual perversities*.

In fact, the first products in intersection homology theory were the intersection products introduced by Goresky and MacPherson [105] for PL stratified pseudomanifolds, generalizing the intersection product on PL manifolds. Among other results we shall discuss later, they showed that if $\bar{p}$ and $\bar{q}$ are dual GM perversities[2] and X is a closed connected oriented n-dimensional PL stratified pseudomanifold, then there is a nonsingular[3] intersection pairing

$$I^{\bar{p}}H_i^{\mathrm{GM}}(X;\mathbb{Q}) \otimes I^{\bar{q}}H_{n-i}^{\mathrm{GM}}(X;\mathbb{Q}) \xrightarrow{\pitchfork} \mathbb{Q}.$$

This was the original form of intersection homology Poincaré duality. When X satisfies these properties, our cup product pairing over $\mathbb{Q}$, discussed in Section 8.4, will be the Poincaré dual of this intersection product in cohomology. More generally, we will define our cup product over Dedekind domains on CS sets.

We will begin our discussion with a conceptual review of the cup product in ordinary singular homology. This will help us to see both the limitations and the possibilities for the intersection cohomology cup product.

We start by recalling perhaps the most familiar version of the cup product. If $\alpha \in S^i(X;R)$, $\beta \in S^j(X;R)$, and σ is an $(i+j)$-simplex, then we often see the cup product $\alpha \smile \beta$ defined by the formula[4]

$$(\alpha \smile \beta)(\sigma) = (-1)^{ij}\alpha(\sigma|_{[v_0,\dots,v_i]})\beta(\sigma|_{[v_i,\dots,v_{i+j}]}). \tag{7.3}$$

Here $\sigma|_{[v_0,\dots,v_i]}$ and $\sigma|_{[v_i,\dots,v_{i+j}]}$ are the singular i- and j-simplices obtained by restricting σ to the "front i-face" and "back j-face" of the standard model $(i+j)$-simplex $[v_0,\dots,v_{i+j}]$.

Already from the formula (7.3) we see what can go wrong for intersection cohomology. Suppose $\alpha \in I_{\bar{p}}S^i(X;R)$ and $\beta \in I_{\bar{q}}S^j(X;R)$. Then we would expect that $\alpha \smile \beta$ should

2 Recall Definition 3.1.7.

3 We review nonsingular pairings in Section 8.4.

4 See, for example, [125, Section 3.2] or [182, Section 48]. Unfortunately, many of the modern textbook sources for algebraic topology leave out the sign. Of course, provided one only cares about cup products $H^i(X;R) \times H^j(X;R) \to H^{i+j}(X;R)$ for fixed i and j, this sign doesn't really matter up to composition with the module isomorphism $x \to -x$. However, leaving out the sign ignores the Koszul sign conventions, which play a more important role when working at the level of complexes (as opposed to the level of groups or modules). We will err on the side of caution and attempt to maintain the Koszul conventions; see Section A.1. A treatment of cup products that includes the signs can be found in Section VII.8 of Dold [71].

be an element of $I_{\bar{r}}S^{i+j}(X;R)$ for some appropriate $\bar{r}$. Suppose now that $\xi \in I^{\bar{r}}S_{i+j}(X;R)$. If some sort of front/back formula were to hold, we would first expect that $(\alpha \smile \beta)(\xi)$ should be determined as a linear combination of terms $(\alpha \smile \beta)(\sigma)$, where σ is a simplex of ξ. This is problematic, as we know that σ being a simplex of ξ does not guarantee that $\sigma \in I^{\bar{r}}S_{i+j}(X;R)$, and so it's not clear that the expression $(\alpha \smile \beta)(\sigma)$ makes any sense, even abstractly. But even if σ is itself allowable as a chain, there is the further difficulty that we should not expect the $\bar{r}$-allowability of σ to tell us anything useful about the $\bar{p}$- and $\bar{q}$-allowability of its various faces. For example, suppose we would like $\bar{p}$, $\bar{q}$, and $\bar{r}$ to all be GM perversities; the intersection homology Poincaré duality of [105] makes this a not unreasonable request. Then we know that no 0- or 1-simplex that intersects Σ_X can be allowable (see Example 3.4.6). Therefore, if σ is any singular simplex that maps v_0 into Σ_X, the front 0-face cannot be allowable. Since it is not difficult, in general, to find allowable singular $(i+j)$-simplices that map v_0 into Σ_X, we see that the front face/back face formulation cannot be used to define $(\alpha \smile \beta)(\sigma)$ for $\alpha \in I_{\bar{p}}S^0(X;R)$.

Luckily, there are alternative ways to construct the cup product in cohomology, and in many ways the oft-used front face/back face formulation is not really the beginning of the cup product story but one of its nice end products that is simply unavailable to us in our intersection setting. So we turn to another well-known property of the cup product: If $\alpha \in S^i(X;R)$ and $\beta \in S^j(X;R)$, then

$$\alpha \smile \beta = \mathbf{d}^*(\alpha \times \beta), \tag{7.4}$$

where $\mathbf{d} : X \to X \times X$ is the diagonal map given by $\mathbf{d}(x) = (x,x)$ and $\alpha \times \beta$ here denotes the *cochain cross product*

$$S^i(X;R) \otimes S^j(X;R) \xrightarrow{\times} S^{i+j}(X \times X;R);$$

see e.g. [182, Theorem 61.3]. Equation (7.4) can be taken as an alternative definition of the cup product if we assume that we have first defined a cochain cross product. The cochain cross product can itself be defined in various ways, including in terms of the cup product as in [125, Section 3.2] or in terms of front and back faces as in [182, Sections 59–61]. But there is a more general construction, also consistent with the development in [182, Sections 59–61], which we now review.

First of all, if A, B are R-modules, then there is a natural map

$$\Theta : \mathrm{Hom}(A,R) \otimes \mathrm{Hom}(B,R) \to \mathrm{Hom}(A \otimes B, R)$$

defined so that if $\alpha \in \mathrm{Hom}(A,R)$, $\beta \in \mathrm{Hom}(B,R)$, $x \in A$, and $y \in B$, then

$$\Theta(\alpha \otimes \beta)(x \otimes y) = (-1)^{|\beta||x|}\alpha(x)\beta(y).$$

In particular, if

$$\alpha \in S^i(X;R) = \mathrm{Hom}(S_i(X;R),R) \quad \text{and} \quad \beta \in S^j(X;R) = \mathrm{Hom}(S_j(X;R),R),$$

then we obtain an element $\Theta(\alpha \otimes \beta) \in \text{Hom}(S_i(X;R) \otimes S_j(X;R), R)$. More generally, the map Θ extends to a chain map

$$\Theta : \text{Hom}(S_*(X;R), R) \otimes \text{Hom}(S_*(X;R), R) \to \text{Hom}(S_*(X;R) \otimes S_*(X;R), R)$$

by Lemma 7.2.1, which we will prove below. Secondly, recall that for ordinary homology and Dedekind domain R the chain complexes $S_*(X;R) \otimes S_*(X;R)$ and $S_*(X \times X;R)$ are chain homotopy equivalent; this is the Eilenberg–Zilber Theorem (see [220, Theorem 5.3.6] or [182, Theorem 59.2]). In the direction

$$\nu : S_*(X \times X;R) \to S_*(X;R) \otimes S_*(X;R),$$

the chain homotopy equivalences are sometimes called *Alexander–Whitney* maps. If we choose a specific ν, then we can define a cochain cross product by

$$\alpha \times \beta = \nu^*\Theta,$$

where ν^* is the $\text{Hom}(\cdot, R)$ dual of ν. If we choose a different Alexander–Whitney map, it will be chain homotopic to ν by the Acyclic Model Theorem (see [220, Theorem 4.2.8.b] or [182, Theorem 32.1.b]), which is used to prove the existence of ν in [220] and [182]. So, the cochain cross product is well defined up to chain homotopy, which implies that it induces a well-defined map after taking (co)homology.

Now, putting (7.4) together with our definition of the cochain cross product, we seem to be claiming that

$$\alpha \smile \beta = \mathbf{d}^*(\alpha \times \beta) = \mathbf{d}^*\nu^*\Theta(\alpha \otimes \beta).$$

This means that if σ is an $(i+j)$-simplex then we have

$$(\alpha \smile \beta)(\sigma) = \mathbf{d}^*\nu^*\Theta(\alpha \otimes \beta)(\sigma) = \Theta(\alpha \otimes \beta)(\nu\mathbf{d}(\sigma)).$$

Here $\mathbf{d}(\sigma) \in S_{i+j}(X \times X;R)$ and $\nu\mathbf{d}(\sigma) \in S_*(X;R) \otimes S_*(X;R)$. This composition

$$\nu\mathbf{d} : S_*(X;R) \to S_*(X;R) \otimes S_*(X;R)$$

is sufficiently useful in its own right that we will provide it with its own symbol $\bar{\mathbf{d}}$ and call it the *algebraic diagonal map*. Since ν is only defined up to chain homotopy, so is $\bar{\mathbf{d}}$. But if we choose a specific ν and write $\nu\mathbf{d}(\sigma) = \sum_k y_k \otimes z_k$, then we can compute explicitly

$$\begin{aligned}(\alpha \smile \beta)(\sigma) &= \Theta(\alpha \otimes \beta)(\nu\mathbf{d}(\sigma)) = \Theta(\alpha \otimes \beta)\Big(\sum_k y_k \otimes z_k\Big)\\ &= \sum_k (-1)^{|\beta||y_k|}\alpha(y_k)\beta(z_k).\end{aligned} \tag{7.5}$$

So what does all this have to do with the front face/back face formula from the beginning of our discussion? It turns out that a particular Alexander–Whitney map can be given explicitly by

$$\nu(\tau) = \sum_{k=0}^{i+j} \pi_1 \circ \tau|_{[v_0,\dots,v_k]} \otimes \pi_2 \circ \tau|_{[v_k,\dots,v_{i+j}]},$$

where τ is an $(i+j)$-simplex in $X \times X$ and $\pi_1, \pi_2 : X \times X \to X$ are the projections to the two factors (see [182, Theorem 59.5]). If $\alpha \in S^i(X;R)$ and $\beta \in S^j(X;R)$, then we have

$$\begin{aligned}\Theta(\alpha \otimes \beta)(\nu(\tau)) &= \Theta(\alpha \otimes \beta)\left(\sum_{k=0}^{i+j} \pi_1 \circ \tau|_{[v_0,\ldots,v_k]} \otimes \pi_2 \circ \tau|_{[v_k,\ldots,v_{i+j}]}\right)\\ &= \sum_{k=0}^{i+j} (-1)^{|\beta|k}\, \alpha\left(\pi_1 \circ \tau|_{[v_0,\ldots,v_k]}\right) \beta\left(\pi_2 \circ \tau|_{[v_k,\ldots,v_{i+j}]}\right)\\ &= (-1)^{ij}\, \alpha\left(\pi_1 \circ \tau|_{[v_0,\ldots,v_i]}\right) \beta\left(\pi_2 \circ \tau|_{[v_i,\ldots,v_{i+j}]}\right),\end{aligned}$$

using that α evaluates to zero on simplices not of degree i and that $|\beta| = j$. In case τ has the form $\tau = \mathbf{d}(\sigma)$, then using that $\pi_1\mathbf{d} = \pi_2\mathbf{d} = \mathrm{id}$ we have

$$\nu\mathbf{d}(\sigma) = \sum_{k=0}^{i+j} \pi_1\mathbf{d}\sigma|_{[v_0,\ldots,v_k]} \otimes \pi_2\mathbf{d}\sigma|_{[v_k,\ldots,v_{i+j}]} = \sum_{k=0}^{i+j} \sigma|_{[v_0,\ldots,v_k]} \otimes \sigma|_{[v_k,\ldots,v_{i+j}]},$$

and so

$$(\alpha \smile \beta)(\sigma) = \Theta(\alpha \otimes \beta)(\nu\mathbf{d}(\sigma)) = (-1)^{ij}\, \alpha\left(\sigma|_{[v_0,\ldots,v_i]}\right) \beta\left(\mathbf{d}\sigma|_{[v_i,\ldots,v_{i+j}]}\right).$$

And this is precisely the front face/back face cup product formula!

Let us emphasize again that it might be more appropriate to say that the front face/back face formulation of the cup product gives us *a* cup product and not *the* cup product. The point is that the front/back description of the cup product relies upon a particular choice of Alexander–Whitney map. If we choose another, chain homotopic, Alexander–Whitney map, we will obtain a different cup product formula at the cochain level. A formula of the form of (7.5) will still apply, but we might not have such nice explicit expressions for $\nu\mathbf{d}(\sigma)$. However, changing ν by a chain homotopy will not change the cup product at the *cohomology* level, since of course chain homotopic chain maps yield the same (co)homology morphisms. In other words, at the cohomology level, we can view the cup product as a well-defined composition[5]

$$\begin{aligned}&H^i(X;R) \otimes H^j(X;R) \xrightarrow{\Theta} H^{i+j}(\mathrm{Hom}(S_*(X;R) \otimes S_*(X;R), R))\\ &\xrightarrow{\quad \nu^* \quad} H^{i+j}(X \times X;R) \xrightarrow{\quad \mathbf{d}^* \quad} H^{i+j}(X;R).\end{aligned} \tag{7.6}$$

The composition of the first two maps is called the *cohomology cross product*.

With the composition (7.6) laid out, it is perhaps a good time to remind the reader that this formulation also demonstrates why we have a cup product in cohomology but not always an analogous internal product in homology: in homology we have the homology cross product $\varepsilon : H_i(X;R) \otimes H_j(X;R) \to H_{i+j}(X \times X;R)$, but the diagonal map points the wrong way,

[5] It is not difficult to show that Θ also makes sense as a map on cohomology $H^i(X;R) \otimes H^j(X;R) \xrightarrow{\Theta} H^{i+j}(\mathrm{Hom}(S_*(X;R) \otimes S_*(X;R), R))$; see Lemma 7.2.1, below.

$H_{i+j}(X \times X; R) \xleftarrow{\mathbf{d}} H_{i+j}(X; R)$. So in general there is no way to define a composition of the homology cross product with the diagonal!

Returning to the stratified world, we see that the composition (7.6) is our hope for defining a cup product in intersection cohomology. The maps Θ and $\mathbf{d}$ are canonical, but there is some flexibility in the choice of ν^*; we just need some chain homotopy equivalence ν to play the role of the Alexander–Whitney map. We have seen that we cannot hope for an intersection cochain analogue to the particular ν defined in terms of front and back faces, but Theorem 6.4.14 nonetheless promises that our intersection version of the Eilenberg–Zilber cross product map

$$\varepsilon : I^{\bar{p}}S_*(X; R) \otimes I^{\bar{q}}S_*(X; R) \to I^{Q}S_*(X \times X; R)$$

is a chain homotopy equivalence, given the proper assumptions on X, $\bar{p}$, $\bar{q}$, and Q, and so there are chain homotopy inverses, which we shall denote IAW for "intersection Alexander–Whitney" map. Although our IAW maps will be defined at the chain level only up to chain homotopy, it remains true that any two such IAW maps yield the same maps on cohomology[6]. So, while we lose the precision of having a specific nice Alexander–Whitney map given by a front face/back face formula at the cochain level, the general outline of the cohomology cup product construction still applies!

Historically, the suggestion by Jim McClure that one could obtain a cup product in intersection cohomology this way led to the author's work on the Künneth theorem in [87] (which extended previously known intersection homology Künneth Theorems from [62, 106]) and eventually to the construction of cup and cap products in Friedman and McClure [100], though in [100] we worked only with field coefficients.

Before moving on to the details of the intersection cohomology cup product, we note that the algebraic diagonal $\bar{\mathbf{d}}$ is the key not only to cup products but also to cap products, as well as slant products, which are less well known but sometimes useful. The cap product $H^j(X; R) \otimes H_{i+j}(X; R) \xrightarrow{\frown} H_i(X; R)$, $\alpha \otimes \xi \to \alpha \frown \xi$, is defined by

$$\alpha \frown \xi = \Phi(\mathrm{id} \otimes \alpha)\bar{\mathbf{d}}(\xi),$$

where Φ is the canonical isomorphism $S_*(X; R) \otimes R \to S_*(X; R)$. So if $\bar{\mathbf{d}}(\xi) = \sum_k y_k \otimes z_k$, then

$$\alpha \frown \xi = \sum_k (-1)^{|\alpha||y_k|} \alpha(z_k) y_k = \sum_k (-1)^{ij} \alpha(z_k) y_k,$$

as $\alpha(z_k) = 0$ unless $|z_k| = |\alpha| = j$, forcing $|y_k| = i$. See [71, Section VII.12].

[6] Our choice of IAW as a chain homotopy inverse to ε pins down a chain homotopy class of intersection Alexander–Whitney maps, but without the Acyclic Model Theorem available here, we do not guarantee that this is the only chain homotopy class of chain homotopy equivalences!

7.2.2 Intersection Homology Cup, Cap, and Cross Products

As seen in the preceding discussion, the three main ingredients needed to produce cross, cup, and cap products are the algebraic map Θ, the Alexander–Whitney map, and the diagonal map $\mathbf{d}$. We now turn to a careful consideration of each of these objects in the intersection homology setting, after which we will piece them together into the intersection homology and cohomology products. We will also want to have relative products available, so we develop the necessary tools in this generality.

(a) Hom of Tensor Products

First we consider the purely algebraic map Θ. Once again, if we have chain complexes of R-modules A_* and B_*, then we define

$$\Theta : \mathrm{Hom}(A_*, R) \otimes \mathrm{Hom}(B_*, R) \to \mathrm{Hom}(A_* \otimes B_*, R),$$

so that if $\alpha \in \mathrm{Hom}(A_*, R)$, $\beta \in \mathrm{Hom}(B_*, R)$, $x \in A_*$, and $y \in B_*$ then

$$\Theta(\alpha \otimes \beta)(x \otimes y) = (-1)^{|b||x|}\alpha(x)\beta(y).$$

We will show that Θ is a chain map.

Lemma 7.2.1 *Suppose we have chain complexes of R-modules C_* and D_*. Then $\Theta : \mathrm{Hom}(C_*, R) \otimes \mathrm{Hom}(D_*, R) \to \mathrm{Hom}(C_* \otimes D_*, R)$ is a chain map, and it induces a well-defined map $H^*(\mathrm{Hom}(C_*, R)) \otimes H^*(\mathrm{Hom}(D_*, R)) \to H^*(\mathrm{Hom}(C_* \otimes D_*, R))$.*

Proof We begin by checking that Θ is a chain map. Suppose[7] $\alpha \in \mathrm{Hom}^i(C_*, R) = \mathrm{Hom}(C_i, R)$ and $\beta \in \mathrm{Hom}^j(D_*, R) = \mathrm{Hom}(D_j, R)$. Then for any $x \otimes y \in C_* \otimes D_*$, we have

$$\begin{aligned}\Theta(d(\alpha \otimes \beta))(x \otimes y) &= \Theta((d\alpha) \otimes \beta + (-1)^i \alpha \otimes d\beta)(x \otimes y)\\ &= (-1)^{|x|j}((d\alpha)(x))\beta(y) + (-1)^{i+(j+1)|x|}\alpha(x)((d\beta)(y)),\end{aligned}$$

while

$$\begin{aligned}&d(\Theta(\alpha \otimes \beta))(x \otimes y)\\ &= (-1)^{i+j+1}\Theta(\alpha \otimes \beta)\partial(x \otimes y)\\ &= (-1)^{i+j+1}\Theta(\alpha \otimes \beta)((\partial x) \otimes y + (-1)^{|x|} x \otimes \partial y)\\ &= (-1)^{i+j+1+j(|x|-1)}\alpha(\partial x)\beta(y) + (-1)^{i+j+1+|x|+j|x|}\alpha(x)\beta(\partial y)\\ &= (-1)^{i+1+j|x|}\alpha(\partial x)\beta(y) + (-1)^{i+j+1+|x|+j|x|}\alpha(x)\beta(\partial y)\\ &= (-1)^{i+1+j|x|+i+1}((d\alpha)(x))\beta(y) + (-1)^{i+j+1+|x|+j|x|+j+1}\alpha(a)((d\beta)(y))\\ &= (-1)^{j|x|}((d\alpha)(x))\beta(y) + (-1)^{i+|x|+j|x|}\alpha(x)((d\beta)(y)).\end{aligned}$$

So $\Theta(d(\alpha\otimes\beta))$ and $d(\Theta(\alpha\otimes\beta))$ represent the same element of $\mathrm{Hom}(C_* \otimes D_*, R)$. Therefore, Θ is a chain map.

[7] See Section A.1.3 for a review of Hom complexes.

This is enough to show that Θ induces a cohomology map

$$H^*(\mathrm{Hom}(C_*,R)\otimes \mathrm{Hom}(D_*,R)) \to H^*(\mathrm{Hom}(C_*\otimes D_*,R)).$$

It remains to show that we have a well-defined map

$$H^*(\mathrm{Hom}(C_*,R))\otimes H^*(\mathrm{Hom}(D_*,R)) \to H^*(\mathrm{Hom}(C_*,R)\otimes \mathrm{Hom}(D_*,R)).$$

But this follows just as for the definition of the homology cross product in Remark 5.2.6. □

We will tend to abuse notation and refer to the induced map

$$H^*(\mathrm{Hom}(C_*,R))\otimes H^*(\mathrm{Hom}(D_*,R)) \to H^*(\mathrm{Hom}(C_*\otimes D_*,R))$$

also as Θ.

Remark 7.2.2 This is perhaps a good time to remind the reader that in contrast to the homological Künneth Theorem, there are extra conditions required to have an algebraic Künneth Theorem in cohomology. In particular, the map $H^*(\mathrm{Hom}(C_*,R))\otimes H^*(\mathrm{Hom}(D_*,R)) \to H^*(\mathrm{Hom}(C_*\otimes D_*,R))$ is not necessarily an isomorphism even when R is a field nor part of a Künneth short exact sequence when R is a Dedekind domain and C_* and D_* are complexes of projective modules. In order to obtain Künneth-like results in this Hom setting one needs additional assumptions, for example that $H_i(C_*)$ (or, symmetrically, $H_i(D_*)$) is finitely generated in each dimension. See [182, Section 60] or [71, Proposition VI.12.16] for further discussion and Section 7.3.8, below, for an intersection cohomology Künneth Theorem.

(b) Intersection Alexander–Whitney Maps

The existence of intersection Alexander–Whitney maps is a consequence of the Künneth Theorem (Theorem 6.4.7). In fact, we have already stated in Theorem 6.4.14 the conditions for the cross product

$$\varepsilon : I^{\bar p}S_*(X,A;R)\otimes I^{\bar q}S_*(Y,B;R) \to I^{Q}S_*(X\times Y,(A\times Y)\cup(X\times B);R)$$

to be a homotopy equivalence, in which case there exist homotopy inverses.

To review, if R is a Dedekind domain, X is a CS set with perversity $\bar p$ and open subset A, and Y is a CS set with perversity $\bar q$ and open subset B, then the above cross product ε induces a chain homotopy equivalence if Q is any $(\bar p,\bar q)$-compatible perversity on $X\times Y$. Recalling Definition 6.4.5, this means that Q satisfies the following conditions:

1. If $S\subset X$ is a regular stratum and $T\subset Y$ is any stratum, then $Q(S\times T)=\bar q(T)$.
2. If $S\subset X$ is any stratum and $T\subset Y$ is a regular stratum, then $Q(S\times T)=\bar p(S)$.
3. If $S\subset X$ and $T\subset Y$ are both singular strata, then $Q(S\times T)=\bar p(S)+\bar q(T)$ or $Q(S\times T)=\bar p(S)+\bar q(T)+1$.
4. If for each point $x\times y\in S\times T$ there are a distinguished neighborhood of x in X of the form $\mathbb{R}^a\times cL_1$ and a distinguished neighborhood of y in Y of the form $\mathbb{R}^b\times cL_2$ such that $I^{\bar p}H_{\dim(L_1)-\bar p(S)-1}(L_1;R) * I^{\bar q}H_{\dim(L_2)-\bar q(T)-1}(L_2;R)=0$, then condition (3)

on $Q(S \times T)$ may also include the possibility $Q(S \times T) = \bar{p}(S) + \bar{q}(T) + 2$. In particular, this is the case if X is locally $(\bar{p}, R)$-torsion-free along the singular stratum S or Y is locally $(\bar{q}, R)$-torsion-free along the singular stratum T. Recall that by Lemma 6.3.24 this condition really depends only on S and T and not on the choices of x, y, L_1, or L_2.

Definition 7.2.3 If the perversity Q on $X \times Y$ is $(\bar{p}, \bar{q})$-compatible then by Theorem 6.4.14 the map

$$\varepsilon : I^{\bar{p}}S_*(X, A; R) \otimes I^{\bar{q}}S_*(X, B; R) \to I^Q S_*(X \times Y, (A \times X) \cup (X \times B); R)$$

is a chain homotopy equivalence and so admits a chain homotopy inverse. We call such a chain homotopy inverse an *intersection Alexander–Whitney map* and label such chain maps by IAW. The map IAW is defined only up to chain homotopy, though we may fix a specific such map to bear that label when desired.

As we see from the conditions for Q to be $(\bar{p}, \bar{q})$-compatible, there may be many $(\bar{p}, \bar{q})$-compatible perversities on $X \times Y$. As for any set of perversities, these are partially ordered, and in fact there is a maximal element. It will be convenient in the arguments that follow to utilize this maximal compatible perversity, so we now establish a notation for it:

Definition 7.2.4 Let $\bar{p}$ be a perversity on the CS set X, let $\bar{q}$ be a perversity on the CS set Y, and let $\mathcal{Q}$ be the set of perversities on $X \times Y$ that are $(\bar{p}, \bar{q})$-compatible. Let $Q_{\bar{p},\bar{q}}$ be the perversity such that if $S \subset X$ and $T \subset Y$ are strata then

$$Q_{\bar{p},\bar{q}}(S \times T) = \max_{Q \in \mathcal{Q}} Q(S \times T).$$

Explicitly, this means that $Q_{\bar{p},\bar{q}}$ takes the following values:

1. if S and T are regular strata, then $Q_{\bar{p},\bar{q}}(S \times T) = \bar{p}(S) = \bar{q}(T) = 0$,
2. if S is a regular stratum, then $Q_{\bar{p},\bar{q}}(S \times T) = \bar{q}(T)$,
3. if T is a regular stratum, then $Q_{\bar{p},\bar{q}}(S \times T) = \bar{p}(S)$,
4. if neither S nor T are regular, if L_1 and L_2 are respective links in X and Y of points in S and T, and if

$$I^{\bar{p}}H_{\dim(L_1)-\bar{p}(S)-1}(L_1; R) * I^{\bar{q}}H_{\dim(L_2)-\bar{q}(T)-1}(L_2; R) = 0,$$

then

$$Q_{\bar{p},\bar{q}}(S \times T) = \bar{p}(S) + \bar{q}(T) + 2;$$

otherwise

$$Q_{\bar{p},\bar{q}}(S \times T) = \bar{p}(S) + \bar{q}(T) + 1.$$

By Lemma 6.3.24, the determination in this last condition depends only on S and T and not on the choices of L_1 and L_2.

Clearly $Q_{\bar{p},\bar{q}} \geq Q$ if Q is any other $(\bar{p}, \bar{q})$-compatible perversity on $X \times Y$.

Remark 7.2.5 If either X is locally $(\bar{p}, R)$-torsion-free or Y is locally $(\bar{q}, R)$-torsion-free then we have

$$Q_{\bar{p},\bar{q}}(S \times T) = \bar{p}(S) + \bar{q}(T) + 2$$

whenever S and T are both singular strata.

(c) The Diagonal Map

The topological diagonal map $\mathbf{d} : X \to X \times X$ is defined by $\mathbf{d}(x) = (x, x)$. We need to consider the maps it induces of the form

$$\mathbf{d} : I^{\bar{r}}S_*(X, A \cup B; R) \to I^{Q}S_*(X \times X, (A \times X) \cup (X \times B); R),$$

for which we need $\mathbf{d}$ to be $(\bar{r}, Q)$-stratified. Notice that if $a \in A$ then $\mathbf{d}(a) = (a, a) \in A \times X$, and if $b \in B$ then $\mathbf{d}(b) = (b, b) \in X \times B$. Also $\mathbf{d}$ takes the stratum S of X to the stratum $S \times S$ of $X \times X$ and thus, in particular, $\mathbf{d}(\Sigma_X) \subset \Sigma_{X \times X}$. So, by Definition 6.3.2, for $\mathbf{d}$ to be $(\bar{r}, Q)$-stratified we just need $\bar{r}(S) - \operatorname{codim}_X(S) \leq Q(S \times S) - \operatorname{codim}_{X \times X}(S \times S)$. As $\operatorname{codim}_{X \times X}(S \times S) = 2 \operatorname{codim}_X(S)$, this becomes the requirement that

$$\bar{r}(S) \leq Q(S \times S) - \operatorname{codim}_X(S). \tag{7.7}$$

We note that if $\bar{p}$ and $\bar{q}$ are two other perversities on X and if Q is $(\bar{p}, \bar{q})$-compatible, then the condition (7.7) also holds for $Q_{\bar{p},\bar{q}}$, as $Q_{\bar{p},\bar{q}} \geq Q$. Conversely, the larger $Q(S \times S)$ is, the larger $\bar{r}(S)$ may be, so that $\mathbf{d}$ induces allowable maps for a broader range of perversities $\bar{r}$. Hence we will typically work with $Q_{\bar{p},\bar{q}}$ when possible in order to maximize the possibilities. In fact, we will see that this is unavoidable in the most important cases, at least along the strata of the form $S \times S$.

Definition 7.2.6 Suppose that $\bar{p}$, $\bar{q}$, and $\bar{r}$ are perversities on the CS set X. We will say that $(\bar{p}, \bar{q}; \bar{r})$ is an *agreeable triple* if

$$\bar{r}(S) \leq Q_{\bar{p},\bar{q}}(S \times S) - \operatorname{codim}_X(S)$$

for each singular stratum $S \subset X$. More generally, if Q is any $(\bar{p}, \bar{q})$-compatible perversity on $X \times X$ such that condition (7.7) holds then we will say that $(\bar{p}, \bar{q}; \bar{r})$ is a *Q-agreeable triple*; so an agreeable triple is just a $Q_{\bar{p},\bar{q}}$-agreeable triple. Note that if $(\bar{p}, \bar{q}; \bar{r})$ is Q-agreeable for any Q then it is an agreeable triple as $Q_{\bar{p},\bar{q}}$ is maximal among $(\bar{p}, \bar{q})$-compatible perversities.

The definition is designed precisely to make the following statement hold:

Lemma 7.2.7 *Suppose that $(\bar{p}, \bar{q}; \bar{r})$ is a Q-agreeable triple of perversities on the CS set X, and that $A, B \subset X$ are open subsets. Then the diagonal map*

$$\mathbf{d} : I^{\bar{r}}S_*(X, A \cup B; R) \to I^{Q}S_*(X \times X, (A \times X) \cup (X \times B); R)$$

is $(\bar{r}, Q)$-stratified. In particular, if $(\bar{p}, \bar{q}; \bar{r})$ is agreeable then

$$\mathbf{d} : I^{\bar{r}}S_*(X, A \cup B; R) \to I^{Q_{\bar{p},\bar{q}}}S_*(X \times X, (A \times X) \cup (X \times B); R)$$

is $(\bar{r}, Q_{\bar{p},\bar{q}})$*-stratified.*

The following provides a useful alternative characterization of agreeable triples when X is appropriately locally torsion-free:

Lemma 7.2.8 *Suppose that X is a CS set with perversities $\bar{p}, \bar{q}, \bar{r}$ such that $Q_{\bar{p},\bar{q}}(S \times S) = \bar{p}(S) + \bar{q}(S) + 2$ for all singular strata S (for example, if X is either locally $(\bar{p}, R)$-torsion-free or locally $(\bar{q}, R)$-torsion-free for each S). Then $(\bar{p}, \bar{q}; \bar{r})$ is an agreeable triple if and only if*

$$D\bar{r} \geq D\bar{p} + D\bar{q}. \tag{7.8}$$

Proof Recall that the dual perversity $D\bar{p}$ of a perversity $\bar{p}$ is defined so that $D\bar{p}(S) = \operatorname{codim}(S) - 2 - \bar{p}(S)$ for S singular; see Definition 3.1.7.

By definition, the triple is agreeable if $\bar{r}(S) \leq Q_{\bar{p},\bar{q}}(S \times S) - \operatorname{codim}_X(S)$ for all singular strata $S \subset X$. If $Q_{\bar{p},\bar{q}}(S \times S) = \bar{p}(S) + \bar{q}(S) + 2$, this condition becomes $\bar{r}(S) \leq \bar{p}(S) + \bar{q}(S) + 2 - \operatorname{codim}_X(S)$. We now observe:

$$\begin{aligned}
&\bar{r}(S) \leq \bar{p}(S) + \bar{q}(S) + 2 - \operatorname{codim}_X(S) \\
&\quad \iff -\bar{r}(S) \geq -\bar{p}(S) - \bar{q}(S) - 2 + \operatorname{codim}_X(S) \\
&\quad \iff \operatorname{codim}_X(S) - 2 - \bar{r}(S) \geq \operatorname{codim}_X(S) - 2 - \bar{p}(S) + \operatorname{codim}_X(S) - 2 - \bar{q}(S) \\
&\quad \iff D\bar{r}(S) \geq D\bar{p}(S) + D\bar{q}(S).
\end{aligned}$$

□

In the form (7.8), the condition to be agreeable has a nice symmetry with conditions of the form $\bar{p} + \bar{q} \leq \bar{r}$ that arise when considering intersection pairings, which are dual to the cup product pairing on pseudomanifolds; see [105] and Section 8.5, below. This is also the form of the criterion used in [100] to define cup and cap products on locally torsion-free spaces.

Of course, if $Q_{\bar{p},\bar{q}}(S \times S)$ is not $\bar{p}(S) + \bar{q}(S) + 2$ for S singular, then we know that it must be $\bar{p}(S) + \bar{q}(S) + 1$. So by computations completely analogous to those in the proof of Lemma 7.2.8 we have the following more general corollary.

Corollary 7.2.9 *Suppose that X is a CS set with perversities $\bar{p}, \bar{q}, \bar{r}$. Without any further assumptions on X, the following statements hold:*

1. *If $(\bar{p}, \bar{q}; \bar{r})$ is an agreeable triple, then $D\bar{r} \geq D\bar{p} + D\bar{q}$.*
2. *If $D\bar{r} > D\bar{p} + D\bar{q}$, then $(\bar{p}, \bar{q}; \bar{r})$ is an agreeable triple.*

The most important triples of perversities have the form $(\bar{p}, D\bar{p}; \bar{0})$ as these are the triples that arise in the setting of the Poincaré Duality Theorem (Theorem 8.2.4). Lemma 7.2.8 allows us to identify them as agreeable under certain hypotheses:

Corollary 7.2.10 *Suppose that X is CS set that is locally $(\bar{p}, R)$-torsion-free or locally $(D\bar{p}, R)$-torsion-free. Then $(\bar{p}, D\bar{p}; \bar{0})$ is an agreeable triple. In particular, $(\bar{t}, \bar{0}; \bar{0})$ and $(\bar{0}, \bar{t}; \bar{0})$ are always agreeable.*

Proof The torsion-free conditions guarantee that $Q_{\bar{p},\bar{q}}(S \times S) = \bar{p}(S) + \bar{q}(S) + 2$ for all singular strata S. So by Lemma 7.2.8, the following verification is sufficient:

$$D\bar{0} = \bar{t} = \bar{p} + D\bar{p} = D(D\bar{p}) + D\bar{p} = D\bar{p} + D(D\bar{p}).$$

Here $\bar{t} = \bar{p} + D\bar{p}$ by the definition of $D\bar{p}$; see Definition 3.1.7. The last claim follows because all CS sets are locally $(\bar{t}, R)$-torsion-free by Example 6.3.22. □

Remark 7.2.11 Of course the hypotheses of Corollary 7.2.10 can be generalized somewhat, though we will see in Corollary 8.2.5 that if X is an R-orientable stratified pseudomanifold then the conditions of being locally $(\bar{p}, R)$-torsion-free and locally $(D\bar{p}, R)$-torsion-free are equivalent.

Here are two other special cases that will be useful in Sections 7.3.5 and 9.2, respectively:

Corollary 7.2.12 *Suppose that X is a CS set and R a Dedekind domain.*

1. *The triple $(\bar{p}, \bar{t}; \bar{p})$ is agreeable for any perversity $\bar{p}$.*
2. *If $\bar{n}$ is the upper-middle perversity (Definition 3.1.10) and X is locally $(\bar{n}; R)$-torsion-free, then $(\bar{n}, \bar{n}; \bar{0})$ is an agreeable triple.*

Proof All CS sets are locally $(\bar{t}; R)$-torsion-free by Example 5.3.12, so by Lemma 7.2.8 we need only observe for the first statement that

$$D\bar{p} \geq D\bar{p} + D\bar{t} = D\bar{p} + \bar{0} = D\bar{p}.$$

For the second statement, Lemma 7.2.8 says that we need $D\bar{0} \geq D\bar{n} + D\bar{n}$, which is equivalent to $\bar{t} \geq \bar{m} + \bar{m} = 2\bar{m}$, where $\bar{m}$ is the lower-middle perversity (Definition 3.1.10). By definition, we have $\bar{m}(S) = \lfloor \frac{1}{2}(\text{codim}(S) - 2) \rfloor$, which we abbreviate to $\bar{m}(k) = \lfloor \frac{1}{2}(k-2) \rfloor$ when k is the codimension of S. If k is even, then $\lfloor \frac{1}{2}(k-2) \rfloor = \frac{1}{2}(k-2)$, and so $2\bar{m}(k) = k-2 = \bar{t}(k)$. If k is odd, then $\lfloor \frac{1}{2}(k-2) \rfloor = \frac{1}{2}(k-3)$, and so $2\bar{m}(k) = k-3 = \bar{t}(k)-1 < \bar{t}(k)$. Thus $(\bar{n}, \bar{n}; \bar{0})$ is agreeable. □

(d) The Intersection Cup, Cap, and Cross Products

We can now put together our ingredients to construct intersection cohomology products.

Cross Products. We begin with the intersection cohomology cross product.

Definition 7.2.13 Let R be a Dedekind domain. Suppose that X is a CS set with perversity $\bar{p}$ and open subset A and that Y is a CS set with perversity $\bar{q}$ and open subset B. Let Q be a $(\bar{p}, \bar{q})$-compatible perversity on $X \times Y$. The *intersection cohomology cross product*

$$I_{\bar{p}}H^i(X, A; R) \otimes I_{\bar{q}}H^j(Y, B; R) \xrightarrow{\times} I_Q H^{i+j}(X \times Y, (A \times Y) \cup (X \times B); R)$$

is defined to be the composition

$$\begin{aligned} I_{\bar p}H^i(X,A;R)\otimes I_{\bar q}H^j(Y,B;R) &\xrightarrow{\Theta} H^{i+j}(\operatorname{Hom}(I^{\bar p}S_*(X,A;R)\otimes I^{\bar q}S_*(Y,B;R),R)) \\ &\xrightarrow{\mathrm{IAW}^*} I_QH^{i+j}(X\times Y,(A\times Y)\cup(X\times B);R), \end{aligned}$$

where IAW^* is the $\operatorname{Hom}(\cdot,R)$ dual of the intersection Alexander–Whitney map of Definition 7.2.3. So if $\alpha\in I_{\bar p}H^i(X,A;R)$ and $\beta\in I_{\bar q}H^j(Y,B;R)$ then

$$\alpha\times\beta=\mathrm{IAW}^*\Theta(\alpha\otimes\beta).$$

The intersection cohomology cross product is well defined by Lemma 7.2.1 and Definition 7.2.3.

Algebraic Diagonals. Next we can define the intersection chain algebraic diagonal map, which is used to define cup and cap products.

Definition 7.2.14 Let R be a Dedekind domain. Suppose that $(\bar p,\bar q;\bar r)$ is an agreeable triple of perversities on a CS set X and that $A,B\subset X$ are open subsets. Then the algebraic diagonal $\bar{\mathbf{d}}$ is defined up to chain homotopy to be the composition

$$\begin{aligned} I^{\bar r}S_*(X,A\cup B;R) &\xrightarrow{\mathbf{d}} I^{Q_{\bar p,\bar q}}S_*(X\times X,(A\times X)\cup(X\times B);R) \\ &\xrightarrow{\mathrm{IAW}} I^{\bar p}S_*(X,A;R)\otimes I^{\bar q}S_*(X,B;R). \end{aligned}$$

The algebraic diagonal is well defined up to chain homotopy thanks to Lemma 7.2.7 and Definition 7.2.3.

Remark 7.2.15 While we define the algebraic diagonal using $Q_{\bar p,\bar q}$, we would obtain an equivalent map using any Q such that $(\bar p,\bar q;\bar r)$ is Q-agreeable. To see this, we note that we have the commutative diagram

$$\begin{array}{ccccc} I^{\bar r}S_*(X,A\cup B;R) & \xrightarrow{\mathbf{d}} & \begin{array}{c} I^{Q}S_*(X\times X,(A\times X)\cup \\ (X\times B);R)\end{array} & \xleftarrow{\varepsilon} & \begin{array}{c} I^{\bar p}S_*(X,A;R)\otimes \\ I^{\bar q}S_*(X,B;R)\end{array} \\ & \searrow^{d} & \downarrow & \swarrow^{\varepsilon} & \\ & & I^{Q_{\bar p,\bar q}}S_*(X\times X,(A\times X)\cup(X\times B);R). & & \end{array}$$

The maps in the left-hand triangle exist by our assumptions about the perversities, and commutativity holds at the space level. The right-hand triangle commutes by the naturality of the cross product; see Proposition 5.2.17 and Theorem 6.3.19. It follows that we obtain a homotopy commutative diagram replacing each ε arrow with an IAW arrow pointing the opposite way.

Cup Products. Given the algebraic diagonal of Definition 7.2.14, we can now define cup and cap products completely analogously with the classical setting:

Definition 7.2.16 Let R be a Dedekind domain. Suppose that $(\bar{p}, \bar{q}; \bar{r})$ is an agreeable triple of perversities on a CS set X and that $A, B \subset X$ are open subsets. Then the *intersection cohomology cup product*

$$I_{\bar{p}}H^i(X, A; R) \otimes I_{\bar{q}}H^j(X, B; R) \xrightarrow{\smile} I_{\bar{r}}H^{i+j}(X, A \cup B; R)$$

is defined to be the composition

$$\begin{aligned} I_{\bar{p}}H^i(X, A; R) \otimes I_{\bar{q}}H^j(Y, B; R) &\xrightarrow{\Theta} H^{i+j}(\operatorname{Hom}(I^{\bar{p}}S_*(X, A; R) \otimes I^{\bar{q}}S_*(Y, B; R), R)) \\ &\xrightarrow{\bar{\mathbf{d}}^*} I_{\bar{r}}H^{i+j}(X, A \cup B; R). \end{aligned}$$

So if $\alpha \in I_{\bar{p}}H^i(X, A; R)$ and $\beta \in I_{\bar{q}}H^j(X, B; R)$ then

$$\alpha \smile \beta = \bar{\mathbf{d}}^*\Theta(\alpha \otimes \beta) = \mathbf{d}^*\mathrm{IAW}^*\Theta(\alpha \otimes \beta).$$

Even though $\bar{\mathbf{d}}$ is defined as a chain map only up to chain homotopy, the cup product is well defined on intersection cohomology by Lemma 7.2.1 and Definition 7.2.14.

The following observation is now immediate from the definitions:

Lemma 7.2.17 *Let R be a Dedekind domain. Suppose that $(\bar{p}, \bar{q}; \bar{r})$ is an agreeable triple of perversities on a CS set X and that $A, B \subset X$ are open subsets. If $\alpha \in I_{\bar{p}}H^i(X, A; R)$ and $\beta \in I_{\bar{q}}H^j(X, B; R)$ then*

$$\alpha \smile \beta = \mathbf{d}^*(\alpha \times \beta) \in I_{\bar{r}}H^{i+j}(X, A \cup B; R).$$

Cap Products. Now we define the cap product.

Definition 7.2.18 Let R be a Dedekind domain. Suppose that $(\bar{p}, \bar{q}; \bar{r})$ is an agreeable triple of perversities on a CS set X and that $A, B \subset X$ are open subsets. The *intersection cohomology cap product*

$$I_{\bar{q}}H^j(X, B; R) \otimes I^{\bar{r}}H_{i+j}(X, A \cup B; R) \xrightarrow{\frown} I^{\bar{p}}H_i(X, A; R)$$

is defined so that if $\beta \in I_{\bar{q}}H^j(X, B; R)$ and $\xi \in I^{\bar{r}}H_{i+j}(X, A \cup B; R)$ then

$$\beta \frown \xi = \Phi((\mathrm{id} \otimes \beta)\bar{\mathbf{d}}(\xi)), \tag{7.9}$$

where Φ is the canonical isomorphism $I^{\bar{p}}S_*(X, A; R) \otimes R \to I^{\bar{p}}S_*(X, A; R)$.

In other words, if $\mathbf{d}(\xi) = \sum_k x_k \otimes y_k \in I^{\bar{p}}S_*(X, A; R) \otimes I^{\bar{q}}S_*(X, B; R)$, then

$$\beta \frown \xi = \sum_k (-1)^{|\beta||x_k|}\beta(y_k)x_k = \sum_k (-1)^{ij}\beta(y_k)x_k.$$

Even though we have seen that **d** and IAW induce well-defined maps on cohomology, there is still a bit of work remaining to verify that the cap product is well defined, as it is not induced simply by applying (co)homology operators to chain maps. This requires some computations. We begin with a useful preliminary lemma.

Lemma 7.2.19 *Given the assumptions of Definition 7.2.18, suppose* $\beta \in I_{\bar{q}}S^j(X, B; R)$ *and* $\xi \in I^{\bar{r}}S_{i+j}(X, A \cup B; R)$*. If we fix a specific choice of* IAW *map, then we can define the cap product on the chain level via equation* (7.9)*. In this case the following chain-level formula holds:*

$$\partial(\beta \frown \xi) = (d\beta) \frown \xi + (-1)^{|\beta|}\beta \frown \partial\xi. \tag{7.10}$$

Proof Suppose that, given our choice of IAW map, we have $\bar{\mathbf{d}}(\xi) = \sum_k y_k \otimes z_k$. Since IAW and **d** are chain maps,

$$\bar{\mathbf{d}}(\partial\xi) = \partial\bar{\mathbf{d}}(\xi) = \partial\left(\sum_k y_k \otimes z_k\right) = \sum_k((\partial y_k) \otimes z_k + (-1)^{|y_k|}y_k \otimes (\partial z_k)).$$

Now, we compute using the definitions:

$$\begin{aligned}
(d\beta) \frown \xi &= \Phi(\mathrm{id} \otimes d\beta)\bar{\mathbf{d}}(\xi) \\
&= \Phi(\mathrm{id} \otimes d\beta)\sum_k y_k \otimes z_k = \sum_k(-1)^{(|\beta|+1)|y_k|}((d\beta)z_k)y_k \\
&= \sum_k(-1)^{(|\beta|+1)|y_k|+|\beta|+1}\beta(\partial z_k)y_k \\
&= \Phi(\mathrm{id} \otimes \beta)\sum_k(-1)^{(|\beta|+1)|y_k|+|\beta|+1+|\beta||y_k|}y_k \otimes \partial z_k \\
&= (-1)^{|\beta|+1}\Phi(\mathrm{id} \otimes \beta)\sum_k(-1)^{|y_k|}y_k \otimes \partial z_k \\
&= (-1)^{|\beta|+1}\Phi(\mathrm{id} \otimes \beta)\left(\sum_k \partial(y_k \otimes z_k) - \sum_k(\partial y_k) \otimes z_k\right) \\
&= (-1)^{|\beta|+1}\Phi(\mathrm{id} \otimes \beta)\partial\bar{\mathbf{d}}(\xi) + (-1)^{|\beta|}\Phi(\mathrm{id} \otimes \beta)\left(\sum_k(\partial y_k) \otimes z_k\right) \\
&= (-1)^{|\beta|+1}\Phi(\mathrm{id} \otimes \beta)\bar{\mathbf{d}}(\partial\xi) + (-1)^{|\beta|}\sum_k(-1)^{|\beta||\partial y_k|}\beta(z_k)\partial y_k \\
&= (-1)^{|\beta|+1}\beta \frown \partial\xi + (-1)^{|\beta|+|\beta|(|\xi|-|\beta|-1)}\sum_k \beta(z_k)\partial y_k \\
&= (-1)^{|\beta|+1}\beta \frown \partial\xi + (-1)^{|\beta||\xi|+|\beta|}\sum_k \beta(z_k)\partial y_k.
\end{aligned}$$

In the second to last line, we have used that all terms of the second summand vanish unless $|z_k| = |\beta|$, in which case $|\partial y_k| + |z_k| = |\partial\xi| = |\xi| - 1$ and so $|\partial y_k| = |\partial\xi| - |\beta| = |\xi| - 1 - |\beta|$.

By comparison, and using the same reasoning about degrees,

$$\begin{aligned}\partial(\beta \frown \xi) &= \partial\Phi(\mathrm{id}\otimes\beta)\bar{\mathbf{d}}(\xi) = \partial\Phi(\mathrm{id}\otimes\beta)\sum_k y_k\otimes z_k\\ &= \partial\sum_k(-1)^{|\beta|(|\xi|-|\beta|)}\beta(z_k)y_k\\ &= \sum_k(-1)^{|\beta||\xi|+|\beta|}\beta(z_k)\partial y_k.\end{aligned}$$

So, altogether, we see that

$$(d\beta)\frown\xi = (-1)^{|\beta|+1}\beta\frown\partial\xi + \partial(\beta\frown\xi),$$

or, equivalently,

$$\partial(\beta\frown\xi) = (d\beta)\frown\xi + (-1)^{|\beta|}\beta\frown\partial\xi. \qquad \square$$

Now we can show that the cap product induces a well-defined map on (co)homology, independent of choices.

Lemma 7.2.20 *Given the assumptions of Definition 7.2.18, the cap product*

$$I_{\bar q}H^j(X,B;R)\otimes I^{\bar r}H_{i+j}(X,A\cup B;R)\xrightarrow{\frown} I^{\bar p}H_i(X,A;R)$$

is well defined and independent of the choice of IAW *map.*

Proof Suppose $\beta\in I_{\bar q}S^j(X,B;R)$ is a cocycle and $\xi\in I^{\bar r}S_{i+j}(X,A\cup B;R)$ is a cycle. Let us first verify that $\beta\frown\xi$ is a cycle for any choice of IAW. We have just seen in Lemma 7.2.19 that

$$\partial(\beta\frown\xi) = (d\beta)\frown\xi + (-1)^{|\beta|}\beta\frown\partial\xi,$$

so if $d\beta=0$ and $\partial\xi=0$, we have $\partial(\beta\frown\xi)=0$.

Next, we must show that altering β and ξ within their (co)homology classes does not alter $\beta\frown\xi$. Equivalently, we must show that $\beta\frown\xi=0$ as an intersection homology class if β is a coboundary or ξ is a boundary. We continue to assume a fixed IAW map.

First, suppose $\xi=\partial\zeta$, continuing to assume β is a cocycle. Then, using equation (7.10), we have

$$\beta\frown\xi = \beta\frown\partial\zeta = (-1)^{|\beta|}\partial(\beta\frown\zeta) - (-1)^{|\beta|}(d\beta)\frown\zeta = (-1)^{|\beta|}\partial(\beta\frown\zeta).$$

So $\beta\frown\xi$ is a boundary.

Next, suppose $\beta=d\alpha$ and $\partial\xi=0$. Then, again using equation (7.10), we have

$$\beta\frown\xi = (d\alpha)\frown\xi = \partial(\alpha\frown\xi) - (-1)^{|\alpha|}\alpha\frown\partial\xi = \partial(\alpha\frown\xi).$$

So, again, $\beta\frown\xi$ is a boundary.

Summing up our computations thus far, we have seen that for a fixed choice of IAW the cap product takes a cohomology class and a homology class to a homology class. Next, we must show that altering IAW within its chain homotopy class does not affect the output.

To see this, we observe that if $\bar{\mathbf{d}}$ and $\bar{\mathbf{d}}'$ are two algebraic diagonals based on two different choices of IAW then we have

$$\bar{\mathbf{d}}(\xi) - \bar{\mathbf{d}}'(\xi) = D\partial\xi + \partial D\xi = \partial D\xi,$$

where D is the chain homotopy between $\bar{\mathbf{d}}$ and $\bar{\mathbf{d}}'$ induced by the chain homotopy between the two choices of IAW. So, changing either IAW within its chain homotopy class results in altering the cycle $\bar{\mathbf{d}}(\xi)$ by a boundary. Notice that a boundary in $I^{\bar{p}}S_*(X,A;R)\otimes I^{\bar{q}}S_*(X,B;R)$ has the form $\partial\left(\sum_\ell u_\ell \otimes v_\ell\right) = \sum_\ell((\partial u_\ell)\otimes v_\ell + (-1)^{|u_\ell|}u_\ell \otimes \partial(v_\ell))$.

But then we compute

$$\begin{aligned}
&(\Phi(\mathrm{id}\otimes\beta))\left(\partial\sum_\ell u_\ell\otimes v_\ell\right)\\
&= (\Phi(\mathrm{id}\otimes\beta))\left(\sum_\ell((\partial u_\ell)\otimes v_\ell + (-1)^{|u_\ell|}u_\ell\otimes\partial(v_\ell))\right)\\
&= \Phi\left(\sum_\ell(-1)^{j(|u_\ell|+1)}(\partial u_\ell)\otimes\beta(v_\ell) + \sum_\ell(-1)^{|u_\ell|+j|u_\ell|}u_\ell\otimes\beta(\partial(v_\ell))\right)\\
&= \sum_\ell(-1)^{j(|u_\ell|+1)}\beta(v_\ell)\partial u_\ell + \sum_\ell(-1)^{|u_\ell|+j|u_\ell|}\beta(\partial(v_\ell))u_\ell\\
&= \sum_\ell(-1)^{j(|u_\ell|+1)}\beta(v_\ell)\partial u_\ell + \sum_\ell(-1)^{|u_\ell|+j|u_\ell|}(-1)^{j+1}((d\beta)(v_\ell))u_\ell\\
&= \partial\left(\sum_\ell(-1)^{j(|u_\ell|+1)}\beta(v_\ell)u_\ell\right),
\end{aligned}$$

using that β is a cocycle. This term is a boundary, so we see that altering $\bar{\mathbf{d}}(\xi)$ by a boundary alters $\beta \frown \xi = \Phi(\mathrm{id}\otimes\beta)\bar{\mathbf{d}}(\xi)$ by a boundary, and therefore the homology class remains unchanged. □

Now that our cross, cup, and cap products are all defined, we turn to their properties in the next section.

7.3 Properties of Cup, Cap, and Cross Products.

In this (lengthy) section, we develop the various properties of cup, cap, and cross products in intersection homology and cohomology. Since we do not have a concrete Alexander–Whitney map to work with, but only one defined up to chain homotopy as a chain homotopy inverse of the intersection chain cross product, it is only at the level of homology and cohomology (as opposed to that of chains and cochains) that these properties can be formulated in a way that is independent of these choices. We will derive formulas reminiscent of the familiar ones from ordinary homology and cohomology theory, but statements will require some conditions and the proofs will require some care. The proofs of these properties in the standard textbook treatments generally either use the front face/back face formulas, which allow for very concrete computations at the chain level, or they rely on acyclic model arguments, which are also not available to us as we do not necessarily have acyclic generators of any of our chain complexes. What we rely on instead are the remarkable properties of the Eilenberg–Zilber shuffle product to show that certain diagrams commute on the nose; then we replace the Eilenberg–Zilber cross products with IAW maps going in the opposite directions to obtain homotopy commutative diagrams. These then have to be deployed in the right way in order to obtain the desired properties.

One nice feature of our program is that it demonstrates that acyclic models are unnecessary in the classical literature (which certainly must have been the original point of view when these tools were being developed), though after reading through our contortions below, the reader might well end up grateful for the Acyclic Model Theorem. We also note that many texts tend not to provide all the arguments for the following properties in detail. Therefore, we hope that we are providing a service for those who would like a complete modern reference to the standard properties even for ordinary (co)homology, to which intersection (co)homology reduces for trivial filtrations. One source that does provide a fairly thorough treatment is Dold's book [71] (though again relying on acyclic models to produce the initial homotopy commutative diagrams). We will not follow Dold precisely (though we do in many places), but it is a good reference for those wishing to see the treatment for ordinary (co)homology and a good place to find some additional properties that the reader might wish to generalize to the intersection setting.

There is a further point the reader should keep in mind when considering some of the more involved arguments below, especially the reader who will notice that the analogous arguments for intersection (co)homology in [100] are much simpler. The point is that everything would be much easier with coefficients in a field F. In that case, for any chain complex C_* over F, one has $H^*(\mathrm{Hom}(C_*, F)) \cong \mathrm{Hom}(H_*(C, F); F)$. So, if one wants to check, for example, that $\alpha \smile \beta = (-1)^{|\alpha||\beta|}\beta \smile \alpha$ in cohomology on a space X, it is only necessary to evaluate each expression in the claimed equality on cycles representing elements of $H_*(X; F)$. If the resulting evaluations are equal for arbitrary such representatives, then the equality is verified. Unfortunately, of course, in general $H^*(\mathrm{Hom}(C_*, R)) \ncong \mathrm{Hom}(H_*(C, R); R)$ when R is not a field, and, in particular, one cannot distinguish cohomology classes only by evaluating them on cycles – how they act on other chains is relevant. Thus, as already indicated, our main strategy for proving cohomological identities will be to show that two expressions are obtained by applying chain homotopic maps to a single cohomological expression. Since chain homotopic maps induce the same map on cohomology, we obtain identities in the image cohomology. Occasionally, in order for us to carry out this program, it will be necessary to perform some hands-on computations via evaluation of cochains on chains. However, it is important to note that such computations will be carried out at the level of *cochains* (not *cohomology*) and the evaluations will be applied to *chains*, not just cycles. Of course, this is an acceptable way to verify equalities at the cochain level.

Our pattern of attack for properties involving the cap product will tend to be a bit more irregular. As cap products involve both homology and cohomology, functoriality will not run in a single direction. Oddly enough, this will not cause a serious problem, and, in fact, proving properties involving cap products will often be easier than proving the analogous properties for cup products. In some sense, this is due to the fact that, in this setting, the relevant data on the homological side really does come in the form of homology classes (i.e. cycles) and not just arbitrary chains, as we will see.

As we proceed, we will group by topic, rather than by product type. For example, all of the associativity properties are discussed in a single section, as opposed to, say, all of the cup product properties being contained in a single section. As the properties of intersection cup, cap, and cross products are spread out over so many pages of proofs, we have provided a summary in Section 7.3.9.

We should also mention here Section 7.3.10. In most of the following sections, we will develop the properties of the cup, cap, and cross products while assuming that our spaces are CS sets. In Section 7.3.10, we will discuss how these results extend to ∂-stratified pseudomanifolds.

7.3.1 Naturality

The cup, cap, and cross products are all natural with respect to maps of spaces that satisfy enough conditions for all the relevant terms to be well defined. As we noted in Remark 5.2.18 concerning the naturality of the homology cross product, it is possible for the maps of spaces to be identity maps, in which case we obtain statements about naturality with respect to change of perversity; this observation applies just as well in this section.

We begin again with another algebraic lemma concerning Θ.

Lemma 7.3.1 *Let C_*, D_*, C'_*, D'_* be complexes of R-modules, and let $f : C_* \to C'_*$ and $g : D \to D'_*$ be degree zero chain maps. Then the following diagram commutes:*

$$\begin{array}{ccc}
\operatorname{Hom}(C_*, R) \otimes \operatorname{Hom}(D_*, R) & \xrightarrow{\Theta} & \operatorname{Hom}(C_* \otimes D_*, R) \\
\Big\uparrow {\scriptstyle f^* \otimes g^*} & & \Big\uparrow {\scriptstyle (f \otimes g)^*} \\
\operatorname{Hom}(C'_*, R) \otimes \operatorname{Hom}(D'_*, R) & \xrightarrow{\Theta} & \operatorname{Hom}(C'_* \otimes D'_*, R).
\end{array}$$

Consequently, there is a cohomology commutative diagram:

$$\begin{array}{ccc} H^*(\mathrm{Hom}(C_*,R)) \otimes H^*(\mathrm{Hom}(D_*,R)) & \xrightarrow{\Theta} & H^*(\mathrm{Hom}(C_* \otimes D_*,R)) \\ \uparrow{\scriptstyle f^*\otimes g^*} & & \uparrow{\scriptstyle (f\otimes g)^*} \\ H^*(\mathrm{Hom}(C'_*,R)) \otimes H^*(\mathrm{Hom}(D'_*,R)) & \xrightarrow{\Theta} & H^*(\mathrm{Hom}(C'_* \otimes D'_*,R)). \end{array}$$

Proof Let $\alpha \in \mathrm{Hom}(C'_*,R)$, $\beta \in \mathrm{Hom}(D'_*,R)$. To show that the diagram commutes, it suffices to apply both compositions to $\alpha \otimes \beta$ and check how the images act on generators $x \otimes y \in C_* \otimes D_*$. We have

$$\begin{aligned} [(f \otimes g)^*\Theta(\alpha \otimes \beta)](x \otimes y) &= \Theta(\alpha \otimes \beta)(f \otimes g)(x \otimes y) \\ &= \Theta(\alpha \otimes \beta)(f(x) \otimes g(y)) \\ &= (-1)^{|\beta||x|}\alpha(f(x))\beta(g(y)) \\ &= (-1)^{|\beta||x|}(f^*(\alpha))(x) \cdot (g^*(\beta))(y) \\ &= [\Theta((f^*\alpha) \otimes (g^*\beta))](x \otimes y) \\ &= [\Theta(f^* \otimes g^*)(\alpha \otimes \beta)](x \otimes y). \end{aligned}$$

Once again, the cohomology statement follows from the cochain-level commutativity as in Remark 5.2.6 by observing that if both inputs into Θ are cocycles and one is a coboundary then the output will be a coboundary; hence altering α and β within their cohomology classes does not alter the image in $H^*(\mathrm{Hom}(C_* \otimes D_*,R))$ by either route. □

(a) Naturality of the Cross Product

We now discuss naturality of the cross product.

Proposition 7.3.2 *Let R be a Dedekind domain, and let (X,A), (Y,B), (X',A'), and (Y',B') be pairs of CS sets and open subsets. Let $f : X \to X'$ and $g : Y \to Y'$ be maps with $f(A) \subset A'$ and $f(B) \subset B'$. Suppose $\bar p$, $\bar q$, $\bar p'$, $\bar q'$, Q, and Q' are respective perversities on X, Y, X', Y', $X \times Y$, and $X' \times Y'$ such that Q is $(\bar p,\bar q)$-compatible and Q' is $(\bar p',\bar q')$-compatible. Suppose that f is $(\bar p,\bar p')$-stratified, that g is $(\bar q,\bar q')$-stratified, and that $f \times g$ is (Q,Q')-stratified.*

Then if $\alpha \in I_{\bar p'}H^i(X',A';R)$ and $\beta \in I_{\bar q'}H^j(Y',B';R)$, we have

$$(f \times g)^*(\alpha \times \beta) = (f^*(\alpha)) \times (g^*(\beta)) \in I_Q H^{i+j}(X \times Y, (A \times Y) \cup (X \times B); R).$$

Proof Our assumptions imply that f^*, g^*, and $(f \times g)^*$ are well-defined intersection cohomology maps by Proposition 7.1.7. Furthermore, by Proposition 5.2.17 and Theorem 6.3.19 the following diagram commutes, and its horizontal maps are chain homotopy equivalences by Definition 7.2.3 and Theorem 6.4.14:

$$\begin{array}{ccc}
I^{Q}S_*(X\times Y,(A\times Y)\cup(X\times B);R) & \xleftarrow{\varepsilon} & I^{\bar p}S_*(X,A;R)\otimes I^{\bar q}S_*(Y,B;R)\\
\downarrow{\scriptstyle f\times g} & & \downarrow{\scriptstyle f\otimes g}\\
I^{Q'}S_*(X'\times Y',(A'\times Y')\cup(X'\times B');R) & \xleftarrow{\varepsilon} & I^{\bar p'}S_*(X',A';R)\otimes I^{\bar q'}S_*(Y',B';R).
\end{array}$$

So, reversing the horizontal arrows and replacing ε with IAW gives a homotopy commutative diagram.

Computing with cohomology classes gives

$$\begin{aligned}
&(f\times g)^*(\alpha\times\beta)\\
&= (f\times g)^*\mathrm{IAW}^*\Theta(\alpha\otimes\beta)\\
&= \mathrm{IAW}^*(f\otimes g)^*\Theta(\alpha\otimes\beta) && \text{by Proposition 5.2.17 and Theorem 6.3.19}\\
&= \mathrm{IAW}^*\Theta(f^*\otimes g^*)(\alpha\otimes\beta) && \text{by Lemma 7.3.1}\\
&= \mathrm{IAW}^*\Theta((f^*(\alpha))\otimes(g^*(\beta)))\\
&= (f^*(\alpha))\times(g^*(\beta)).
\end{aligned}$$

□

In the hypotheses of the preceding result we made assumptions not just that the maps f and g were stratified with respect to the perversities involved but also that $f\times g$ was (Q,Q')-stratified. In some situations the hypotheses on f and g are enough to tell us $f\times g$ is automatically stratified, at least if we use our maximal product perversities of the form $Q_{\bar p,\bar q}$. Here is one such result:

Lemma 7.3.3 *Suppose that $f : X\to X'$ and $g : Y\to Y'$ are maps of CS sets, that $\bar p,\bar q,\bar p',\bar q'$ are respective perversities on X,Y,X',Y', and that f is $(\bar p,\bar p')$-allowable and g is $(\bar q,\bar q')$-allowable. Furthermore, suppose X' is locally $(\bar p',R)$-torsion-free or Y' is locally $(\bar q',R)$-torsion-free. Then $f\times g$ is $(Q,Q_{\bar p',\bar q'})$-allowable for any $(\bar p,\bar q)$-compatible perversity Q on $X\times Y$.*

Proof As $Q\le Q_{\bar p,\bar q}$ for any $(\bar p,\bar q)$-compatible perversity Q, it suffices to prove the lemma with $Q=Q_{\bar p,\bar q}$. Recall from Definition 6.3.2 that f and g must take singular strata to singular strata, so $(f\times g)(\Sigma_{X\times Y})\subset\Sigma_{X'\times Y'}$. Therefore, if $S'\times T'$ is a regular stratum of $X'\times Y'$, then $(f\times g)^{-1}(S'\times T')$ is also regular (if non-empty), so the condition to be $(Q_{\bar p,\bar q},Q_{\bar p',\bar q'})$-stratified is trivial for such strata.

Next, suppose S' is a regular stratum of X' and that T' is a singular stratum of Y'. Then $f^{-1}(S')$ is a union of regular strata of X. Suppose $S\times T$ is a stratum of $X\times Y$ in $(f\times g)^{-1}(S'\times T')$, and so S is regular and T may be regular or singular. In this case, the definition of $(\bar p,\bar q)$-compatible and $(\bar p',\bar q')$-compatible perversities reduces the $(Q_{\bar p,\bar q},Q_{\bar p',\bar q'})$-stratified condition on such strata to the $(\bar q,\bar q')$-stratified condition for g.

Similarly, if S' is singular, T' is regular, and $S\times T \subset (f\times g)^{-1}(S'\times T')$ then the $(Q_{\bar{p},\bar{q}}, Q_{\bar{p}',\bar{q}'})$-stratified condition on such strata reduces to the $(\bar{p},\bar{p}')$-stratified condition for f.

Finally, suppose that $S' \subset X'$ and $T' \subset Y'$ are both singular and that $S \subset X$ and $T \subset Y$ are strata with $f(S\times T) \subset S' \times T'$. The locally torsion-free assumptions imply that $Q_{\bar{p}',\bar{q}'}(S'\times T') = \bar{p}'(S) + \bar{q}'(T) + 2$. So we have

$$\begin{aligned}
&Q_{\bar{p}',\bar{q}'}(S'\times T') - \operatorname{codim}_{X'\times Y'}(S'\times T')\\
&\quad = \bar{p}'(S') + \bar{q}'(T') + 2 - \operatorname{codim}_{X'}(S') - \operatorname{codim}_{Y'}(T')\\
&\quad = \bar{p}'(S') - \operatorname{codim}_{X'}(S') + \bar{q}'(T) - \operatorname{codim}_{Y'}(T') + 2\\
&\quad \geq p(S) - \operatorname{codim}_{X}(S) + \bar{q}(T) - \operatorname{codim}_{Y}(T) + 2\\
&\quad = p(S) + \bar{q}(T) + 2 - \operatorname{codim}_{X\times Y}(S\times T)\\
&\quad \geq Q_{\bar{p},\bar{q}}(S\times T) - \operatorname{codim}_{X\times Y}(S\times T).
\end{aligned}$$

Here the first inequality comes from the definitions of $(\bar{p},\bar{p}')$-stratified and $(\bar{q},\bar{q}')$-stratified and the second from the definition of $Q_{\bar{p},\bar{q}}$. Note that the last inequality works even if S or T is a regular stratum. The resulting overall inequality is the condition to be $(Q, Q_{\bar{p}',\bar{q}'})$-stratified. □

(b) Naturality of Cup and Cap Products

Next we turn toward cup and cap products. The following lemma provides the basis for the naturality of each of these by showing that algebraic diagonals behave naturally with respect to maps $f : X \to Y$. Notice that the lemma does not require any assumptions about the perversities on the product spaces, but rather provides a needed perversity on $X \times X$ under the assumptions that we have agreeable triples and that f is appropriately stratified with respect to the perversities in the triples.

Lemma 7.3.4 *Let R be a Dedekind domain, and let X and Y be CS sets with open subsets $A, B \subset X$ and $C, D \subset Y$. Let $f : X \to Y$ be a map with $f(A) \subset C$ and $f(B) \subset D$. Suppose $(\bar{p},\bar{q};\bar{r})$ is an agreeable triple of perversities on X and that $(\bar{u},\bar{v};\bar{s})$ is an agreeable triple of perversities on Y. Suppose that f is $(\bar{p},\bar{u})$-stratified, $(\bar{q},\bar{v})$-stratified, and $(\bar{r},\bar{s})$-stratified. Then there is a $(\bar{p},\bar{q})$-compatible perversity Q on $X\times X$ such that the following diagram commutes:*

$$\begin{array}{ccccc}
I^{\bar{r}}S_*(X, A\cup B; R) & \xrightarrow{\mathbf{d}} & \begin{array}{c} I^{Q}S_*(X\times X, (A\times X)\cup \\ (X\times B); R)\end{array} & \xleftarrow{\varepsilon} & \begin{array}{c} I^{\bar{p}}S_*(X, A; R)\otimes \\ I^{\bar{q}}S_*(X, B; R)\end{array}\\
\Big\downarrow f & & \Big\downarrow f\times f & & \Big\downarrow f\otimes f\\
I^{\bar{s}}S_*(Y, C\cup D; R) & \xrightarrow{\mathbf{d}} & \begin{array}{c} I^{Q_{\bar{u},\bar{v}}}S_*(Y\times Y, (C\times Y)\cup \\ (Y\times D); R)\end{array} & \xleftarrow{\varepsilon} & \begin{array}{c} I^{\bar{u}}S_*(Y, C; R)\otimes \\ I^{\bar{v}}S_*(Y, D; R).\end{array}
\end{array}$$

As a result, $\bar{\mathbf{d}}f$ is chain homotopic to $(f\otimes f)\bar{\mathbf{d}}$ as maps $I^{\bar{r}}S_*(X, A\cup B; R) \to I^{\bar{u}}S_*(Y, C; R)\otimes I^{\bar{v}}S_*(Y, D; R)$.

Proof We first demonstrate the existence of a perversity Q on $X \times X$ such that $(\bar{p}, \bar{q}; \bar{r})$ is Q-agreeable and $f \times f$ is $(Q, Q_{\bar{u},\bar{v}})$-stratified. As in the proof of Lemma 7.3.3, it is automatic that $f \times f$ takes singular strata to singular strata. We must define $Q(S \times T)$ for each stratum $S \times T \subset X \times X$ with S and T strata of X. Suppose that f takes S to the stratum S' of Y and T to the stratum T' of Y.

First, suppose that S' is a regular stratum; this implies that S is also regular. In this case we must have $Q(S \times T) = \bar{q}(T)$ and $Q_{\bar{u},\bar{v}}(S' \times T') = \bar{v}(T')$. But recalling that $\bar{p}(S) = \bar{u}(S') = 0$ for the regular strata, we can also write $Q(S \times T) = \bar{p}(S) + \bar{q}(T)$ and $Q_{\bar{u},\bar{v}}(S' \times T') = \bar{u}(S') + \bar{v}(T')$. So we have

$$\begin{aligned} Q_{\bar{u},\bar{v}}(S' \times T') - \operatorname{codim}_{Y \times Y}(S' \times T') &= \bar{u}(S') + \bar{v}(T') - \operatorname{codim}_Y(S') - \operatorname{codim}_Y(T') \\ &= \bar{u}(S') - \operatorname{codim}_Y(S') + \bar{v}(T') - \operatorname{codim}_Y(T') \\ &\geq \bar{p}(S) - \operatorname{codim}_X(S) + \bar{q}(T) - \operatorname{codim}_X(T) \\ &= Q(S \times T) - \operatorname{codim}_{X \times X}(S \times T). \end{aligned}$$

We have used here for the inequality the definition of $(\bar{p}, \bar{u})$- and $(\bar{q}, \bar{v})$-stratified maps, and the resulting inequality shows that $f \times f$ is $(Q, Q_{\bar{u},\bar{v}})$-stratified as far as these types of strata are concerned. An equivalent argument holds when T and T' are regular or all strata are regular.

This leaves the cases in which S' and T' are singular. First suppose that $S \neq T$, and let $Q(S \times T) = \bar{p}(S) + \bar{q}(T)$; note that this is consistent with the possibility that S or T is regular. We may have $S' = T'$, but in any case we know that $Q_{\bar{u},\bar{v}}(S' \times T') = \bar{u}(S') + \bar{v}(T') + C$, where $C \in \{1, 2\}$, depending on the local torsion behavior. In either case, the above computations continue to hold but now with $+C$ added to each of the expressions before the $\geq$.

Lastly, we must consider the case $S = T$, and in this case the perversity $\bar{r}$ plays a role, as we must have

$$\bar{r}(S) + \operatorname{codim}_X(S) \leq Q(S \times S) \tag{7.11}$$

in order for $\mathbf{d}$ to be $(\bar{r}, Q)$-allowable; see Definition 7.2.6. As we know that $(\bar{p}, \bar{q}; \bar{r})$ is agreeable, we do know that there exists some $(\bar{p}, \bar{q})$-compatible perversity P on $X \times X$ with

$$\bar{r}(S) + \operatorname{codim}_X(S) \leq P(S \times S) \leq Q_{\bar{p},\bar{q}}(S \times S).$$

In particular we must have

$$\bar{r}(S) + \operatorname{codim}_X(S) \leq Q_{\bar{p},\bar{q}}(S \times S) = \bar{p}(S) + \bar{q}(S) + K$$

for some $K \in \{0, 1, 2\}$. If $\bar{r}(S) + \operatorname{codim}_X(S) \leq \bar{p}(S) + \bar{q}(S)$, we can take $Q(S \times S) = \bar{p}(S) + \bar{Q}(S)$. Then (7.11) holds, and this is consistent with $f \times f$ being $(Q, Q_{\bar{u},\bar{v}})$-stratified, again by the same computations as just above. On the other hand, if $\bar{r}(S) + \operatorname{codim}_X(S) \not\leq \bar{p}(S) + \bar{q}(S)$, then we can take $Q(S \times S) = \bar{r}(S) + \operatorname{codim}_X(S)$. Once again, we know that $r(S) + \operatorname{codim}_X(S) \leq Q_{\bar{p},\bar{q}}(S \times S)$, so that we will have $\bar{p}(S) + \bar{q}(S) \leq Q(S \times S) \leq Q_{\bar{p},\bar{q}}(S \times S)$, which means that Q will remain $(\bar{p}, \bar{q})$-compatible. But also, as f is $(\bar{r}, \bar{s})$-stratified and $\mathbf{d}$ is

$(\bar{s}, \bar{Q}_{\bar{u},\bar{v}})$-stratified, we have

$$\begin{aligned}Q(S \times S) - \text{codim}_{X\times X}(S \times S) &= \bar{r}(S) + \text{codim}_X(S) - 2\text{codim}_X(S)\\ &= \bar{r}(S) - \text{codim}_X(S)\\ &\leq \bar{s}(S') - \text{codim}_Y(S')\\ &\leq Q_{\bar{u},\bar{v}}(S' \times S') - \text{codim}_{Y\times Y}(S' \times S').\end{aligned}$$

So we see that, as regards these strata, $f \times f$ is $(Q, Q_{\bar{u},\bar{v}})$-stratified, as desired.

Now that we have shown that we can find a Q that meets all requirements, we turn to the commutativity.

The commutativity of the left square of the diagram holds at the level of spaces, as $(f \times f)\mathbf{d}(x) = (f, f)(x, x) = (f(x), f(x)) = \mathbf{d}f(x)$. The square on the right is a special case of the diagram considered in Proposition 5.2.17 (and Theorem 6.3.19). Therefore, the diagram commutes. As the diagram commutes, the version of the diagram with each ε replaced by an IAW in the opposite direction homotopy commutes, and so $\bar{\mathbf{d}}f$ is chain homotopic to $(f \otimes f)\bar{\mathbf{d}}$, using the fact from Remark 7.2.15 that $\mathbf{d}$ does not depend on the choice of Q.

Since we will use analogous arguments often below, it is worth verifying this sort of claim in detail at least once, which we do here. By Proposition 5.2.17 and Theorem 6.3.19, we know that $(f \times f)\varepsilon = \varepsilon(f \otimes f)$ exactly. By applying the appropriate IAW maps to each side, we obtain that $\text{IAW}(f \times f)\varepsilon\text{IAW} = \text{IAW}\varepsilon(f \otimes f)\text{IAW}$. But now using that IAW and ε are chain homotopy inverses, $\text{IAW}(f \times f)\varepsilon\text{IAW} \sim \text{IAW}(f \times f)$ and $\text{IAW}\varepsilon(f \otimes f)\text{IAW} \sim (f \otimes f)\text{IAW}$. Thus $\text{IAW}(f \times f)$ and $(f \otimes f)\text{IAW}$ are chain homotopic. Therefore,

$$\bar{\mathbf{d}}f = \text{IAW}\mathbf{d}f = \text{IAW}(f \times f)\mathbf{d} \sim (f \otimes f)\text{IAW}\mathbf{d} = (f \otimes f)\bar{\mathbf{d}}. \qquad \square$$

With Lemma 7.3.4 in hand, we can now demonstrate the naturality of the cup and cap products.

Proposition 7.3.5 *Let R be a Dedekind domain, and let X and Y be CS sets with open subsets $A, B \subset X$ and $C, D \subset Y$. Let $f : X \to Y$ be a map with $f(A) \subset C$ and $f(B) \subset D$. Suppose $(\bar{p}, \bar{q}; \bar{r})$ is an agreeable triple of perversities on X and that $(\bar{u}, \bar{v}; \bar{s})$ is an agreeable triple of perversities on Y. Suppose that f is $(\bar{p}, \bar{u})$-stratified, $(\bar{q}, \bar{v})$-stratified, and $(\bar{r}, \bar{s})$-stratified.*

Then if $\alpha \in I_{\bar{u}}H^i(Y, C; R)$ and $\beta \in I_{\bar{v}}H^j(Y, D; R)$, we have

$$f^*(\alpha \smile \beta) = (f^*(\alpha)) \smile (f^*(\beta)) \in I_{\bar{r}}H^{i+j}(X, A \cup B; R).$$

Proof The conditions on f and the perversities ensure that all terms in the expression are well defined.

Now, we compute in cohomology using Lemmas 7.3.4 and 7.3.1:

$$\begin{aligned}
(f^*(\alpha)) \smile (f^*(\beta)) &= \bar{\mathbf{d}}^*\Theta((f^*(\alpha)) \otimes (f^*(\beta))) \\
&= \bar{\mathbf{d}}^*\Theta(f^* \otimes f^*)(\alpha \otimes \beta) \\
&= \bar{\mathbf{d}}^*(f \otimes f)^*\Theta(\alpha \otimes \beta) && \text{by Lemma 7.3.1} \\
&= f^*\bar{\mathbf{d}}^*\Theta(\alpha \otimes \beta) && \text{by Lemma 7.3.4} \\
&= f^*(\alpha \smile \beta).
\end{aligned}$$

□

Next we turn to naturality of the cap product, where the mixed functoriality makes the statement of naturality a bit more complex.

Proposition 7.3.6 *Let R be a Dedekind domain, and let X and Y be CS sets with open subsets $A, B \subset X$ and $C, D \subset Y$. Let $f : X \to Y$ be a map with $f(A) \subset C$ and $f(B) \subset D$. Suppose $(\bar{p}, \bar{q}; \bar{r})$ is an agreeable triple of perversities on X and that $(\bar{u}, \bar{v}; \bar{s})$ is an agreeable triple of perversities on Y. Suppose that f is $(\bar{p}, \bar{u})$-stratified, $(\bar{q}, \bar{v})$-stratified, and $(\bar{r}, \bar{s})$-stratified.*

Then if $\beta \in I_{\bar{v}}H^j(Y, D; R)$ and $\xi \in I^{\bar{r}}H_{i+j}(X, A \cup B; R)$, we have

$$\beta \frown f(\xi) = f(f^*(\beta) \frown \xi) \in I^{\bar{u}}H_i(Y, C; R).$$

Proof Once again, the conditions on f and the perversities ensure that all terms in the expression are well defined.

We compute

$$\begin{aligned}
\beta \frown f(\xi) &= \Phi(\mathrm{id} \otimes \beta)\bar{\mathbf{d}}f(\xi) \\
&= \Phi(\mathrm{id} \otimes \beta)(f \otimes f)\bar{\mathbf{d}}(\xi) && \text{by Lemma 7.3.4} \\
&= \Phi(f \otimes \beta f)\bar{\mathbf{d}}(\xi) \\
&= \Phi(f \otimes f^*\beta)\bar{\mathbf{d}}(\xi) \\
&= \Phi(f \otimes \mathrm{id})(\mathrm{id} \otimes f^*\beta)\bar{\mathbf{d}}(\xi) \\
&= f\Phi(\mathrm{id} \otimes f^*\beta)\bar{\mathbf{d}}(\xi) && \text{see below} \\
&= f(f^*(\beta) \frown \xi).
\end{aligned}$$

In the next to last equality, we have used that $\Phi(f \otimes \mathrm{id}) = f\Phi$. This is immediate in generality as, if $x \otimes 1 \in C_* \otimes R$ is a generator for some chain complex C_* and $f : C_* \to D_*$ is some chain map, then

$$\Phi(f \otimes \mathrm{id})(x \otimes 1) = \Phi(f(x) \otimes 1) = f(x) = f(\Phi(x \otimes 1)).$$

□

Remark 7.3.7 Recall from Examples 4.1.3 and 4.1.4 that inclusions of open subsets and normally nonsingular inclusions are what we might call $(\bar{p}, \bar{p})$-stratified maps, where the first perversity is the restricted perversity on the subset. Therefore, the naturality statements of this section apply to such inclusion maps. This is a nice fact that we will use below in proving Poincaré duality for stratified pseudomanifolds.

(c) Compatibility with Classical Products

As observed at the beginning of this section, it is possible for the maps of spaces to be identity maps, in which case the maps in our naturality lemmas are induced by the inclusions of the form $I^{\bar p}S_*(X,A;R) \hookrightarrow I^{\bar q}S_*(X,A;R)$ for $\bar p \leq \bar q$. There is a further observation we can make in this direction, which is that if $\bar p \leq \bar t$, then $I^{\bar p}S_*(X,A;R) \cong I^{\bar p}S_*^{\mathrm{GM}}(X,A;R)$ by Proposition 6.2.9, and so there is an inclusion $I^{\bar p}S_*(X,A;R) \hookrightarrow S_*(X,A;R)$. Of course all of our constructions of products mirror the classical constructions for ordinary homology and cohomology (e.g. [71, Chapter VII]), and in particular we know the Künneth Theorem holds for ordinary singular chains. So all of the arguments of this section go through replacing the codomain homology groups (or domain cohomology groups) with ordinary homology (or cohomology), so long as all perversities involved are $\leq \bar t$. For this we note that if $\bar p$ and $\bar q$ are both $\leq t$, then so is any $(\bar p, \bar q)$-compatible perversity on the product space; see the computation in the proof below of Lemma 7.3.11.

For example, we have the following version of Proposition 7.3.6.

Proposition 7.3.8 *Let R be a Dedekind domain, and let X be a CS set with open subsets $A, B \subset X$. Suppose $(\bar p, \bar q; \bar r)$ is an agreeable triple of perversities on X with $\bar p, \bar q, \bar r$ all $\leq \bar t$, and let $\omega_{\bar p} : I^{\bar p}S_*(X,A;R) \hookrightarrow S_*(X,A;R)$, $\omega_{\bar q} : I^{\bar q}S_*(X,B;R) \hookrightarrow S_*(X,B;R)$, and $\omega_{\bar r} : I^{\bar r}S_*(X,A\cup B;R) \hookrightarrow S_*(X,A\cup B;R)$.*

Then if $\beta \in H^j(X,B;R)$ and $\xi \in I^{\bar r}H_{i+j}(X,A\cup B;R)$, we have

$$\beta \frown \omega_{\bar r}(\xi) = \omega_{\bar p}(\omega_{\bar q}^*(\beta) \frown \xi) \in H_i(X,A;R).$$

We will utilize this lemma in Section 8.1.6, below, to discuss Goresky and MacPherson's observation in [105, Section 1.4] that the ordinary cap product with the fundamental class factors through the GM intersection homology groups. We will also use the following version of Proposition 7.3.6 in Section 8.5.3 to discuss the relationship of the factoring maps with cup and intersection products.

Proposition 7.3.9 *Let R be a Dedekind domain, and let X be a CS set with open subsets $A, B \subset X$. Suppose $(\bar p, \bar q; \bar r)$ is an agreeable triple of perversities on X with $\bar p, \bar q, \bar r$ all $\leq \bar t$, and let $\omega_{\bar p} : I^{\bar p}S_*(X,A;R) \hookrightarrow S_*(X,A;R)$, $\omega_{\bar q} : I^{\bar q}S_*(X,B;R) \hookrightarrow S_*(X,B;R)$, and $\omega_{\bar r} : I^{\bar r}S_*(X,A\cup B;R) \hookrightarrow S_*(X,A\cup B;R)$.*

Then if $\alpha \in I_{\bar p}H^i(X,A;R)$ and $\beta \in I_{\bar q}H^j(X,B;R)$, we have

$$\omega_{\bar r}^*(\alpha \smile \beta) = (\omega_{\bar p}^*(\alpha)) \smile (\omega_{\bar q}^*(\beta)) \in I_{\bar r}H^{i+j}(X,A\cup B;R).$$

(d) Topological Invariance

In later sections, particularly in Chapter 9, we will want to know in certain settings that our products and the invariants derived from them are topological invariants, i.e. independent of the choice of stratification up to canonical isomorphisms. Recall that we have established in Theorems 5.5.1 and 7.1.17 certain conditions under which the intersection homology and cohomology groups have such invariance. We will not pursue here the most general questions concerning when agreeability of a triple of perversities on a CS set X implies agreeability on other stratifications of $|X|$, and thus we will not pursue topological invariance of products in the greatest generality. Rather, we allow certain locally torsion-free assumptions, which greatly simplify the matter and are sufficient for our later purposes. We leave the more general questions for another day and will work toward Theorem 7.3.10, stated below.

So, suppose that X and X' are two CS set stratifications of the same underlying space $|X|$, neither of which possesses codimension-one strata. Let $\mathfrak{X}$ denote $|X|$ with its intrinsic filtration; see Section 2.10. Suppose that $\bar{p}$, $\bar{q}$, and $\bar{r}$ are GM perversities such that $D\bar{r} \geq D\bar{p} + D\bar{q}$. As $\bar{p}$, $\bar{q}$, and $\bar{r}$ are GM perversities, they depend only on the codimension of strata and so are defined on X, X', and $\mathfrak{X}$. As every GM perversity is below the top perversity, we have $I^{\bar{p}}H_*(X;R) \cong I^{\bar{p}}H_*^{\mathrm{GM}}(X;R)$ by Proposition 6.2.9, and similarly for intersection cohomology and for the other perversities and spaces. We assume that X is locally $(\bar{p},R)$-torsion-free, though we could instead take X to be locally $(\bar{q},R)$-torsion-free in the following discussion. By Proposition 5.5.9, the CS set X is locally $(\bar{p},R)$-torsion-free if and only if X' is, and if and only if $\mathfrak{X}$ is. So X, X', and $\mathfrak{X}$ are all locally $(\bar{p},R)$-torsion-free, which by Lemma 7.2.8 implies the triple $(\bar{p},\bar{q};\bar{r})$ is agreeable on all of them. So, as promised, the locally torsion-free assumption on X gets us a lot of mileage by allowing us an algebraic diagonal on all three spaces.

Next, we would like to apply the preceding lemmas of this section to establish naturality of products with respect to maps of the form $\mathrm{id} : X \to \mathfrak{X}$ (or $\mathrm{id} : X' \to \mathfrak{X}$, for which the arguments are the same). The place where we need some extra care is in noticing that these maps may take singular strata to regular strata and so they are not necessarily $(\bar{u},\bar{u})$-stratified for any of the perversities $\bar{u}$ we are considering, i.e. $\bar{u} \in \{\bar{p},\bar{q},\bar{r}\}$. But we have observed that as we are working with GM perversities we do have $I^{\bar{u}}S_* = I^{\bar{u}}S_*^{\mathrm{GM}}$ by Proposition 6.2.9, and so for id to induce intersection chain maps $I^{\bar{u}}S_*(X;R) \to I^{\bar{u}}S_*(\mathfrak{X};R)$ it is sufficient for id to be $(\bar{u},\bar{u})^{\mathrm{GM}}$-stratified, which it is by Remark 5.5.11. However, this is still not quite enough to extend Lemma 7.3.4, and hence the other naturality lemmas, as we must also consider the map on products $\mathrm{id} \times \mathrm{id} : I^{Q}S_*(X \times X;R) \to I^{Q_{\bar{p},\bar{q}}}S_*(\mathfrak{X} \times \mathfrak{X};R)$. In fact, we claim and will prove just below in Lemma 7.3.11 that $\mathrm{id} \times \mathrm{id} : I^{Q_{\bar{p},\bar{q}}}S_*(X \times X;R) \to I^{Q_{\bar{p},\bar{q}}}S_*(\mathfrak{X} \times \mathfrak{X};R)$ is a well-defined map of GM intersection chain complexes, which is also sufficient for the associated map of relative chain complexes. As we assume $D\bar{r} \geq D\bar{p} + D\bar{q}$ and a local torsion-free condition, the triple $(\bar{p},\bar{q};\bar{r})$ is $Q_{\bar{p},\bar{q}}$-agreeable on X and $\mathfrak{X}$, and thus we can utilize this $\mathrm{id} \times \mathrm{id}$ as the middle vertical map of the diagram of Lemma 7.3.4. From there, the rest of the proof of that lemma will hold, as Proposition 5.2.17 provides naturality of the cross product in the setting of GM-stratified maps.

So, continuing to assume that Lemma 7.3.11 holds and using the resulting modified version of Lemma 7.3.4, the naturality of cup and cap products follows just as argued in

Propositions 7.3.5 and 7.3.6. Thus, we obtain a diagram of the form

$$\begin{array}{ccc}
I_{\bar p}H^i(X,A;R)\otimes I_{\bar q}H^j(X,B;R) & \xrightarrow{\smile} & I_{\bar r}H^{i+j}(X,A\cup B;R)\\
\uparrow \cong & & \uparrow \cong\\
I_{\bar p}H^i(\mathfrak{X},\mathfrak{A};R)\otimes I_{\bar q}H^j(\mathfrak{X},\mathfrak{B};R) & \xrightarrow{\smile} & I_{\bar r}H^{i+j}(\mathfrak{X},\mathfrak{A}\cup\mathfrak{B};R)\\
\downarrow \cong & & \downarrow \cong\\
I_{\bar p}H^i(X',A';R)\otimes I_{\bar q}H^j(X',B';R) & \xrightarrow{\smile} & I_{\bar r}H^{i+j}(X',A'\cup B';R),
\end{array}$$

where $|A| = |\mathfrak{A}| = |A'|$ and $|B| = |\mathfrak{B}| = |B'|$. This demonstrates the topological invariance of the cup product.

Similarly, using the naturality of the cap product, we have the diagram[8]

$$\begin{array}{ccc}
I_{\bar q}H^j(X,B;R)\otimes I^{\bar r}H_{i+j}(X,A\cup B;R) & \xrightarrow{\frown} & I^{\bar p}H_i(X,A;R)\\
\downarrow \cong\ (\mathrm{id}^*)^{-1}\otimes\mathrm{id} & & \downarrow \cong\\
I_{\bar q}H^j(\mathfrak{X},\mathfrak{B};R)\otimes I^{\bar r}H_{i+j}(\mathfrak{X},\mathfrak{A}\cup\mathfrak{B};R) & \xrightarrow{\frown} & I^{\bar p}H_i(\mathfrak{X},\mathfrak{A};R)\\
\uparrow \cong\ (\mathrm{id}^*)^{-1}\otimes\mathrm{id} & & \uparrow \cong\\
I_{\bar q}H^j(X',B';R)\otimes I^{\bar r}H_{i+j}(X',A'\cup B';R) & \xrightarrow{\frown} & I^{\bar p}H_i(X',A';R).
\end{array}$$

So we have proven the following invariance theorem.

Theorem 7.3.10 *Let R be a Dedekind domain and suppose that $|X|$ is the underlying space of a CS set X with open subsets A and B and with no codimension-one strata. Suppose that $\bar p$, $\bar q$, and $\bar r$ are GM perversities such that $D\bar r \geq D\bar p + D\bar q$ and that X is locally $(\bar p, R)$-torsion-free or locally $(\bar q, R)$-torsion-free in some (and hence by Proposition 5.5.9 in every) CS set stratification with no codimension-one strata. Then, up to canonical isomorphisms, the cup and cap products*

$$I_{\bar p}H^i(X,A;R)\otimes I_{\bar q}H^j(X,B;R) \xrightarrow{\smile} I_{\bar r}H^{i+j}(X,A\cup B;R)$$

[8] Note that, in general, if $f : X \to Y$ induces homology and cohomology isomorphisms then we can rewrite the naturality of the cap product so that all of the functoriality is covariant using

$$f(\alpha \frown \xi) = f((f^*(f^*)^{-1}(\alpha)) \frown \xi) = (f^*)^{-1}(\alpha) \frown f(\xi).$$

and

$$I_{\bar{q}}H^j(X,B;R) \otimes I^{\bar{r}}H_{i+j}(X,A\cup B;R) \xrightarrow{\frown} I^{\bar{p}}H_i(X,A;R)$$

are independent of the choice of CS set stratification among such stratifications with no codimension-one strata.

We leave the reader to formulate and prove an analogous invariance statement concerning cross products and turn instead to finishing up with the promised lemma:

Lemma 7.3.11 *Suppose that X is a CS set and that $\mathfrak{X}$ is $|X|$ with its intrinsic filtration. Let $\bar{p}$ and $\bar{q}$ be GM perversities, and suppose X and $\mathfrak{X}$ are locally $(\bar{p},R)$-torsion-free. Then*

$$\mathrm{id}\times\mathrm{id} : I^{Q_{\bar{p},\bar{q}}}S_*(X\times X;R) \to I^{Q_{\bar{p},\bar{q}}}S_*(\mathfrak{X}\times\mathfrak{X};R)$$

is a well-defined chain map.

Proof For any strata S,T of X or $\mathfrak{X}$, we have

$$\begin{aligned} Q_{\bar{p},\bar{q}}(S\times T) &\le \bar{p}(S)+\bar{q}(T)+2 \le \bar{t}(S)+\bar{t}(T)+2\\ &= \mathrm{codim}_X(S)-2+\mathrm{codim}_X(T)-2+2 = \mathrm{codim}_{X\times X}(S\times T)-2\\ &= \bar{t}(S\times T), \end{aligned}$$

and so $I^{Q_{\bar{p},\bar{q}}}S_*(X\times X;R) \cong I^{Q_{\bar{p},\bar{q}}}S_*^{\mathrm{GM}}(X\times X;R)$ and $I^{Q_{\bar{p},\bar{q}}}S_*(\mathfrak{X}\times\mathfrak{X};R) \cong I^{Q_{\bar{p},\bar{q}}}S_*^{\mathrm{GM}}(\mathfrak{X}\times\mathfrak{X};R)$ by Theorem 6.3.19. Thus to have the desired map, we only need for $\mathrm{id}\times\mathrm{id}$ to be $(Q_{\bar{p},\bar{q}},Q_{\bar{p},\bar{q}})^{\mathrm{GM}}$-stratified.

As $\mathfrak{X}$ is always a coarser stratification of X, the map id takes all strata to strata of smaller codimension. In particular, it cannot take regular strata to singular strata. So, using the notation $f(S\times T)\subset S'\times T'$ established above, there are only the following cases to check.

- Regular × Regular ⟶ Regular × Regular:

$$Q_{\bar{p},\bar{q}}(S\times T)-\mathrm{codim}_{X\times X}(S\times T) = 0 = Q_{\bar{p},\bar{q}}(S'\times T')-\mathrm{codim}_{\mathfrak{X}\times\mathfrak{X}}(S'\times T').$$

- Singular × Regular ⟶ Regular × Regular (or equivalently, Regular × Singular ⟶ Regular × Regular):

$$\begin{aligned} Q_{\bar{p},\bar{q}}(S\times T)-\mathrm{codim}_{X\times X}(S\times T) &= \bar{p}(S)-\mathrm{codim}_X(S)\\ &\le \bar{t}(S)-\mathrm{codim}(S) = -2 \le 0\\ &= Q_{\bar{p},\bar{q}}(S'\times T')-\mathrm{codim}_{\mathfrak{X}\times\mathfrak{X}}(S'\times T'). \end{aligned}$$

- Singular × Regular ⟶ Singular × Regular (or equivalently, Regular × Singular ⟶ Regular × Singular):

$$\begin{aligned} Q_{\bar{p},\bar{q}}(S\times T)-\mathrm{codim}_{X\times X}(S\times T) &= \bar{p}(S)-\mathrm{codim}_X(S) \le \bar{p}(S)-\mathrm{codim}_{\mathfrak{X}}(S')\\ &= Q_{\bar{p},\bar{q}}(S'\times T')-\mathrm{codim}_{\mathfrak{X}\times\mathfrak{X}}(S'\times T'). \end{aligned}$$

- Singular × Singular → Regular × Regular:

$$\begin{aligned}
&Q_{\bar{p},\bar{q}}(S \times T) - \operatorname{codim}_{X\times X}(S \times T)\\
&\quad = \bar{p}(S) + \bar{q}(T) + 2 - \operatorname{codim}_X(S) - \operatorname{codim}_X(T)\\
&\quad \le \bar{t}(S) + \bar{t}(T) + 2 - \operatorname{codim}_X(S) - \operatorname{codim}_X(T) = -2 \le 0\\
&\quad = Q_{\bar{p},\bar{q}}(S' \times T') - \operatorname{codim}_{\mathfrak{X}\times\mathfrak{X}}(S' \times T').
\end{aligned}$$

- Singular × Singular → Singular × Regular (or equivalently, Singular × Singular → Regular × Singular):

$$\begin{aligned}
Q_{\bar{p},\bar{q}}(S \times T) - \operatorname{codim}_{X\times X}(S \times T) &= \bar{p}(S) + \bar{q}(T) + 2 - \operatorname{codim}_X(S) - \operatorname{codim}_X(T)\\
&\le \bar{p}(S) + \bar{t}(T) + 2 - \operatorname{codim}_X(S) - \operatorname{codim}_X(T)\\
&= \bar{p}(S) - \operatorname{codim}_X(S) \le \bar{p}(S') - \operatorname{codim}_{\mathfrak{X}}(S')\\
&= Q_{\bar{p},\bar{q}}(S' \times T') - \operatorname{codim}_{\mathfrak{X}\times\mathfrak{X}}(S' \times T').
\end{aligned}$$

- Singular × Singular → Singular × Singular:

$$\begin{aligned}
Q_{\bar{p},\bar{q}}(S \times T) - \operatorname{codim}_{X\times X}(S \times T) &= \bar{p}(S) + \bar{q}(T) + 2 - \operatorname{codim}_X(S) - \operatorname{codim}_X(T)\\
&\le \bar{p}(S') + \bar{q}(T') + 2 - \operatorname{codim}_{\mathfrak{X}}(S') - \operatorname{codim}_{\mathfrak{X}}(T')\\
&= Q_{\bar{p},\bar{q}}(S' \times T') - \operatorname{codim}_{\mathfrak{X}\times\mathfrak{X}}(S' \times T').
\end{aligned}$$

□

7.3.2 Commutativity

Next we turn to the (graded) commutativity of cup and cross products, starting with another algebraic lemma:

Lemma 7.3.12 *Suppose C_* and D_* are chain complexes of R-modules. The following diagram commutes:*

$$\begin{array}{ccc}
\operatorname{Hom}(C_*, R) \otimes \operatorname{Hom}(D_*, R) & \xrightarrow{\Theta} & \operatorname{Hom}(C_* \otimes D_*, R)\\
\Big\downarrow{\scriptstyle\tau} & & \Big\downarrow{\scriptstyle\tau^*}\\
\operatorname{Hom}(D_*, R) \otimes \operatorname{Hom}(C_*, R) & \xrightarrow{\Theta} & \operatorname{Hom}(D_* \otimes C_*, R).
\end{array}$$

Here both maps labeled τ are the maps that interchange factors, with appropriate signs.

Proof The map τ on the left is a chain map by the proof of Proposition 5.2.20. The map on the right is the Hom dual of the chain map $\tau : D_* \otimes C_* \to C_* \otimes D_*$.

We proceed by direct computation. If $y \otimes x$ is a generator of $D_* \otimes C_*$ and α, β are respective generators of $\mathrm{Hom}(C_*, R)$ and $\mathrm{Hom}(D_*, R)$, then

$$\begin{aligned}(\tau^*\Theta(\alpha\otimes\beta))(y\otimes x) &= \Theta(\alpha\otimes\beta)\tau(y\otimes x)\\ &= \Theta(\alpha\otimes\beta)(-1)^{|x||y|}x\otimes y\\ &= (-1)^{|x||y|+|\beta||x|}\alpha(x)\beta(y),\end{aligned}$$

while

$$\begin{aligned}\Theta(\tau(\alpha\otimes\beta))(y\otimes x) &= (-1)^{|\alpha||\beta|}\Theta(\beta\otimes\alpha)(y\otimes x)\\ &= (-1)^{|\alpha||\beta|+|\alpha||y|}\alpha(x)\beta(y).\end{aligned}$$

Now, both expressions will be zero unless $|y| = |\beta|$ and $|x| = |\alpha|$, and so both expressions are $\alpha(x)\beta(y)$. □

This lemma is enough to provide commutativity of the cross product:

Proposition 7.3.13 *Let R be a Dedekind domain. Suppose that X, Y are CS sets with respective perversities $\bar{p}, \bar{q}$ and that Q is a $(\bar{p}, \bar{q})$-compatible perversity on $X \times Y$. Define the perversity Q^τ on $Y \times X$ so that $Q^\tau(T \times S) = Q(S \times T)$ for strata $S \subset X$ and $T \subset Y$. Let $A \subset X$ and $B \subset Y$ be open subsets, and let $\alpha \in I_{\bar{p}}H^i(X, A; R)$ and $\beta \in I_{\bar{q}}H^j(Y, B; R)$. Then*

$$t^*(\alpha\times\beta) = (-1)^{ij}\beta\times\alpha \in I_{Q^\tau}H^{i+j}(Y\times X, (B\times X)\cup(Y\times A); R).$$

Proof Recall the topological map $t(x, y) = (y, x)$; it is straightforward to verify that

$$t : I^QS_*(X\times Y; (A\times Y)\cup(X\times B), R) \to I^{Q^\tau}S_*(Y\times X, (B\times X)\cup(Y\times A); R)$$

is (Q, Q^τ)-stratified, and so the arguments of Proposition 5.2.20 and Theorem 6.3.19 apply.

We compute

$$\begin{aligned}t^*(\alpha\times\beta) &= t^*\mathrm{IAW}^*\Theta(\alpha\otimes\beta)\\ &= \mathrm{IAW}^*\tau^*\Theta(\alpha\otimes\beta) &&\text{by Proposition 5.2.20 and Theorem 6.3.19}\\ &= \mathrm{IAW}^*\Theta\tau(\alpha\otimes\beta) &&\text{by Lemma 7.3.12}\\ &= (-1)^{ij}\mathrm{IAW}^*\Theta(\beta\otimes\alpha)\\ &= (-1)^{ij}\beta\times\alpha.\end{aligned}$$

In the second line, we have used that the strict commutativity of Proposition 5.2.20 becomes commutativity up to chain homotopy if we replace each map ε by a map IAW going in the opposite direction, and this implies that

$$t^*\mathrm{IAW}^* = (\mathrm{IAW}t)^* = (\tau\mathrm{IAW})^* = \mathrm{IAW}^*\tau^*$$

as a map on cohomology. □

Now we turn toward the cup product, starting with the next lemma.

Lemma 7.3.14 *Let R be a Dedekind domain. Suppose that $(\bar p, \bar q; \bar r)$ is an agreeable triple of perversities on a CS set X. Let $A, B \subset X$ be open subsets.*

Then $(\bar q, \bar p; \bar r)$ is also an agreeable triple of perversities and the following diagram commutes up to chain homotopy:

$$\begin{array}{ccc} I^{\bar r}S_*(X, A \cup B; R) & \xrightarrow{\bar{\mathbf{d}}} & I^{\bar p}S_*(X, A; R) \otimes I^{\bar q}S_*(X, B; R) \\ & \searrow^{\bar d} & \downarrow \tau \\ & & I^{\bar q}S_*(X, B; R) \otimes I^{\bar p}S_*(X, A; R). \end{array}$$

Here τ is the standard (signed!) interchange map of tensor product factors and the two algebraic diagonals are defined with respect to the appropriate ordering of perversities.

Proof It is immediate that $(\bar q, \bar p; \bar r)$ is an agreeable triple if $(\bar p, \bar q; \bar r)$ is, as $\bar p$ and $\bar q$ play symmetric roles in the definition of $(\bar p, \bar q)$-compatible perversities on $X \times X$.

We claim that the following diagram is commutative:

$$\begin{array}{ccccc} I^{\bar r}S_*(X, A \cup B; R) & \xrightarrow{\mathbf{d}} & I^{Q_{\bar p,\bar q}}S_*(X \times X; (A \times X) \cup (X \times B), R) & \xleftarrow{\varepsilon} & I^{\bar p}S_*(X, A; R) \otimes I^{\bar q}S_*(X, B; R) \\ & \searrow^{d} & \downarrow t & & \downarrow \tau \\ & & I^{Q_{\bar q,\bar p}}S_*(X \times X, (B \times X) \cup (X \times A); R) & \xleftarrow{\varepsilon} & I^{\bar q}S_*(X, B; R) \otimes I^{\bar p}S_*(X, A; R). \end{array}$$

Here the left-hand vertical arrow is induced by the topological map $t(x, y) = (y, x)$, which clearly satisfies $t\mathbf{d} = \mathbf{d}$ and takes $Q_{\bar p,\bar q}$-allowable chains to $Q_{\bar q,\bar p}$-allowable chains. The commutativity of the square follows from Proposition 5.2.20 and Theorem 6.3.19.

It now follows that replacing the maps ε with the homotopy inverses IAW (going in the opposite directions) will yield a diagram that is homotopy commutative. Commutativity of the diagram in the statement of the lemma now follows from the definition $\bar{\mathbf{d}} = \mathrm{IAW}\mathbf{d}$. □

Now we can show commutativity of the cup product:

Proposition 7.3.15 *Let R be a Dedekind domain. Suppose that $(\bar p, \bar q; \bar r)$ is an agreeable triple of perversities on the CS set X. Let $A, B \subset X$ be open subsets, and let $\alpha \in I_{\bar p}H^i(X, A; R)$ and $\beta \in I_{\bar q}H^j(X, B; R)$. Then $\alpha \smile \beta = (-1)^{ij}\beta \smile \alpha \in I_{\bar r}H^{i+j}(X, A \cup B; R)$.*

Proof Once again, the assumptions imply that all expressions are well defined. We compute

$$\begin{aligned}
\beta \smile \alpha &= \bar{\mathbf{d}}^*\Theta(\beta \otimes \alpha) \\
&= (\tau\bar{\mathbf{d}})^*\Theta(\beta \otimes \alpha) && \text{by Lemma 7.3.14} \\
&= \bar{\mathbf{d}}^*\tau^*\Theta(\beta \otimes \alpha) \\
&= \bar{\mathbf{d}}^*\Theta\tau(\alpha \otimes \beta) && \text{by Lemma 7.3.12} \\
&= (-1)^{ij}\bar{\mathbf{d}}^*\Theta(\beta \otimes \alpha) \\
&= (-1)^{ij}\beta \smile \alpha.
\end{aligned}$$

□

7.3.3 Unitality and Evaluation

The unital property of the standard cup product is the fact that $1 \smile \alpha = \alpha \smile 1 = \alpha$, where $\alpha \in H^*(X;R)$ and $1 \in H^0(X;R)$ is the element that evaluates each (positively oriented) singular 0-simplex to 1. Similarly, the unital property for cross products is that if $1 \in H^0(\text{pt}, R)$, then $\alpha \times 1 = 1 \times \alpha = \alpha$ in $H^*(X;R)$.

To have a unit for the cup product in intersection cohomology, we first need to see when it can be true that we have a cup product $I_{\bar{p}}H^*(X;R) \otimes I_{\bar{q}}H^0(X;R) \to I_{\bar{p}}H^*(X;R)$. From the definition, we require that $(\bar{p}, \bar{q}; \bar{p})$ be an agreeable triple. Corollary 7.2.12 shows us that $(\bar{p}, \bar{t}; \bar{p})$ is always agreeable on a CS set, and we will see that this is essentially the only choice. In particular, for $(\bar{p}, \bar{q}; \bar{p})$ to be agreeable then for any singular stratum $S \subset X$ we must have

$$\begin{aligned}
\bar{p}(S) - \operatorname{codim}_X(S) &\le Q_{\bar{p},\bar{q}}(S \times S) - \operatorname{codim}_{X\times X}(S \times S) \\
&\le \bar{p}(S) + \bar{q}(S) + 2 - 2\operatorname{codim}_X(S),
\end{aligned}$$

which implies that $\bar{q}(S) \ge \operatorname{codim}(S) - 2 = \bar{t}(S)$. So the condition $\bar{q} \ge \bar{t}$ is in general necessary for there to be an algebraic diagonal of the form $\bar{\mathbf{d}} : I^{\bar{p}}H_*(X;R) \to H_*(I^{\bar{p}}S_*(X;R) \otimes I^{\bar{q}}S_*(X;R))$.

However, the proof of Lemma 7.3.20, below, which demonstrates that $\bar{\mathbf{d}}$ is co-unital in the homotopy category (with appropriate conditions) and which leads to the unital property of cup products, requires consideration of projection maps $p_i : X \times X \to X$ with $p_1(x, y) = x$ and $p_2(x, y) = y$. For these spatial maps to induce maps on intersection chain complexes of the form $p_1 : I^Q S_*(X \times X; R) \to I^{\bar{p}}S_*(X;R)$ and $p_2 : I^Q S_*(X \times X; R) \to I^{\bar{q}}S_*(X;R)$, where here Q is a $(\bar{p}, \bar{q})$-compatible perversity on $X \times X$, we will see in Lemma 7.3.16 that in general we need to have $\bar{q} \le \bar{t}$. Together with the previous paragraph, this forces the top perversity $\bar{t}$ to play a privileged role in the unital property, and hence in intersection cohomology theory.

(a) Projection Maps

In order to proceed to results about cohomology products, we first need to study the behavior of intersection chains under projection maps.

Lemma 7.3.16 *Let X, Y be CS sets with respective perversities $\bar{p}$, $\bar{q}$, and let Q be a $(\bar{p},\bar{q})$-compatible perversity on $X \times Y$. The projection map $p_1 : X \times Y \to X$ induces a well-defined chain map $p_1 : I^Q S_*(X \times Y; R) \to I^{\bar{p}} S_*(X; R)$ if $\bar{q} \le \bar{t}$. Similarly, the projection map $p_2 : X \times Y \to Y$ induces a well-defined chain map $p_2 : I^Q S_*(X \times Y; R) \to I^{\bar{q}} S_*(Y; R)$ if $\bar{p} \le \bar{t}$. If these perversity requirements are not satisfied, then such chain maps do not exist in general.*

Proof We will demonstrate the lemma for p_1, the proof for p_2 being equivalent.

First, let us see when p_1 preserves allowability. By Definition 4.1.1, to show that p_1 takes allowable simplices to allowable simplices, we only need to check that $Q(S \times T) - \bar{p}(S) \le \operatorname{codim}_{X \times Y}(S \times T) - \operatorname{codim}_X(S)$ for each singular stratum S of X and arbitrary stratum T of Y. Notice that $\operatorname{codim}_{X \times Y}(S \times T) = \operatorname{codim}_X(S) + \operatorname{codim}_Y(T)$, so we need $Q(S \times T) - \bar{p}(S) \le \operatorname{codim}_Y(T)$. If T is a regular stratum then $Q(S \times T) = \bar{p}(S)$ and the inequality is satisfied. If T is singular and $\bar{q} \le \bar{t}$ then

$$\begin{aligned} Q(S \times T) - \bar{p}(S) &\le \bar{p}(S) + \bar{q}(T) + 2 - \bar{p}(S) \\ &= \bar{q}(T) + 2 \le \bar{t}(T) + 2 = \operatorname{codim}_Y(T) - 2 + 2 = \operatorname{codim}_Y(T). \end{aligned}$$

So, if $\bar{q} \le \bar{t}$, then p_1 takes allowable simplices to allowable simplices.

Now, as observed prior to Definition 6.3.2, preserving allowability is not in itself sufficient to guarantee a chain map of non-GM intersection chains. If Y possesses a singular stratum S and $\mathcal{R}$ is a regular stratum of X then p_1 takes $\mathcal{R} \times S$ to $\mathcal{R}$, and so it is not the case that $p_1(\Sigma_{X \times Y}) \subset \Sigma_X$. Therefore, p_1 is not $(Q, \bar{p})$-stratified, even when $\bar{q} \le \bar{t}$. Nonetheless, it turns out that p_1 does induce a chain map with these assumptions, but we still need a bit more work.

The critical observation is that if $\mathcal{R}$ is a regular stratum of X, then $p_1^{-1}(\mathcal{R})$ consists of strata of the form $\mathcal{R} \times S$, and, by the definition of Q, we have $Q(\mathcal{R} \times S) = \bar{q}(S)$. For an i-simplex in $X \times Y$ to be Q-allowable with respect to $\mathcal{R} \times S$, we must have that $\sigma^{-1}(\mathcal{R} \times S)$ is contained in the $(i - \operatorname{codim}_{X \times Y}(\mathcal{R} \times S) + Q(\mathcal{R} \times S))$-skeleton of Δ^i. But $\operatorname{codim}_X(\mathcal{R}) = 0$ and $Q(\mathcal{R} \times S) = \bar{q}(S)$, so if we assume that $\bar{q} \le \bar{t}$ then we obtain

$$\begin{aligned} i - \operatorname{codim}_{X \times Y}(\mathcal{R} \times S) + Q(\mathcal{R} \times S) &= i - \operatorname{codim}_Y(S) + \bar{q}(S) \le i - \operatorname{codim}_Y(S) + \bar{t}(S) \\ &= i - \operatorname{codim}_Y(S) + \operatorname{codim}_Y(S) - 2 \\ &= i - 2. \end{aligned}$$

It follows that if $\xi \in I^Q S_i(X \times Y; \mathcal{R})$ then any simplex of $\partial \xi$ that is contained completely in $\Sigma_{X \times Y}$ must in fact be contained in $p_1^{-1}(\Sigma_X)$, as the preceding argument shows that the interior of an $i-1$ face of an allowable simplex of ξ cannot intersect any singular stratum of the form $\mathcal{R} \times S$. So if τ is such an $i-1$ face that is contained in $\Sigma_{X \times Y}$, its interior must be contained in $p_1^{-1}(\Sigma_X)$, but this is a closed set, so all of it must be contained in $p_1^{-1}(\Sigma_X)$.

So now for $\xi \in I^Q S_i(X \times Y; R)$ let $\partial_1 \xi = \partial \xi - \hat{\partial} \xi$, where here we let ∂ denote the boundary of ξ as a chain in $S_*(X \times Y; R)$. Recall that $\hat{\partial} \xi$ consists, by definition, of the simplices of $\partial \xi$ (with their coefficients) that do not have image in $\Sigma_{X \times Y}$. Thus $\partial_1 \xi$ comprises those simplices of $\partial\xi$ that are contained in $\Sigma_{X\times Y}$, and, by the preceding paragraph, such simplices are actually contained in $p_1^{-1}(\Sigma_X)$. Since p_1 induces a chain map on ordinary chains, $\partial p_1(\xi) = p_1(\partial\xi) = p_1(\hat{\partial}\xi) + p_1(\partial_1\xi)$. As each simplex of $\partial_1\xi$ is contained in $p_1^{-1}(\Sigma_X)$, each simplex of $p_1(\partial_1\xi)$ is contained in Σ_X. On the other hand, since no simplex of $\hat{\partial}\xi$ is contained in $\Sigma_{X\times Y}$, it follows that no image of a simplex of $p_1(\hat{\partial}\xi)$ is contained in Σ_X. So, from the definition, we must have $\hat{\partial}p_1(\xi) = p_1(\hat{\partial}\xi)$, showing that p_1 is a chain map $p_1 : I^Q S_*(X\times Y;R) \to I^{\bar{p}}S_*(X;R)$ when $\bar{q} \le \bar{t}$.

Lastly, suppose $\bar{q} \not\le \bar{t}$. Let S be a stratum of Y such that $\bar{q}(S) > \operatorname{codim}_Y(S) - 2$, and let $\mathcal{U}$ be a regular stratum of Y with S in its closure. Let $\sigma : \Delta^i \to Y$ be a simplex with the image of one $i-1$ face in S and with the rest of Δ^i mapping into $\mathcal{U}$. Also, let x_0 be a point in a regular stratum $\mathcal{R}$ of X and let $\eta : \Delta^i \to X$ be the unique map with image x_0. Then the map $\xi = (\eta,\sigma) : \Delta^i \to X \times Y$ with $(\eta,\sigma)(z) = (\eta(z),\sigma(z))$ is a singular simplex with image in $p_1^{-1}(\mathcal{R})$. Furthermore, ξ is an allowable simplex, as the only singular stratum it intersects is $\mathcal{R}\times S$, which it intersects only in the image of the $(i-1)$-skeleton of Δ^i, and

$$i - \operatorname{codim}_{X\times Y}(\mathcal{R}\times S) + Q(\mathcal{R}\times S) = i - \operatorname{codim}_Y(S) + \bar{q}(S) \ge i - 1.$$

In fact, by an analogous computation for the boundary, ξ is allowable as a chain, and its boundary as an intersection chain is the sum (with signs) of the faces of ξ not contained in $\mathcal{R}\times S$. On the other hand, we have $p_1(\xi) = \eta$, which is contained in $\mathcal{R}$ and so an intersection chain, and $\hat{\partial}p_1(\xi) = \partial p_1(\xi) = \partial\eta$. This does not agree with $p_1(\hat{\partial}\xi)$, and thus p_1 is not a chain map of intersection chain complexes in this example. This shows that the condition that $\bar{q} \le \bar{t}$ is necessary provided Y has any singular strata. □

Corollary 7.3.17 *Let X,Y be CS sets with respective perversities $\bar{p}$, $\bar{q}$ and respective subspaces A,B. Let Q be a $(\bar{p},\bar{q})$-compatible perversity on $X\times Y$. Then the map p_1 induces a well-defined chain map $p_1 : I^Q S_*(X\times Y, A\times Y;R) \to I^{\bar{p}}S_*(X,A;R)$ if $\bar{q} \le \bar{t}$. Similarly, the map p_2 induces a well-defined chain map $p_2 : I^Q S_*(X\times Y, X\times B;R) \to I^{\bar{q}}S_*(Y,B;R)$ if $\bar{p} \le \bar{t}$. If the conditions on perversities are not satisfied, then such chain maps do not exist in general.*

Proof An equivalent argument to that in the proof of Lemma 7.3.16 demonstrates that p_1 induces a well-defined chain map $I^Q S_*(A\times Y;R) \to I^{\bar{p}}S_*(A;R)$. Therefore, $p_1 : I^Q S_*(X\times Y;R) \to I^{\bar{p}}S_*(X;R)$ induces a well-defined map on the quotient complexes $p_1 : I^Q S_*(X\times Y, A\times Y;R) \to I^{\bar{p}}S_*(X,A;R)$. The argument for p_2 is identical. □

The next lemma is similar in spirit to the last, though simpler. It will be used below in the proof of Lemma 7.3.20.

Lemma 7.3.18 *Let X be a filtered set with perversity $\bar{q}$, and let $p : X \to \mathrm{pt}$ be the map from X to a point. Then p induces a well-defined chain map $p : I^{\bar{q}}S_*(X;R) \to S_*(\mathrm{pt};R)$ if $\bar{q} \le \bar{t}$. If $\bar{q} \not\le \bar{t}$, then this is not true in general.*

Proof If $\bar{q} \le \bar{t}$, then $I^{\bar{q}}S_*(X;R) = I^{\bar{q}}S_*^{\mathrm{GM}}(X;R)$ by Proposition 6.2.9. All simplices are allowable in $S_*(\mathrm{pt}; R)$, so p takes allowable chains to allowable chains, and since $I^{\bar{q}}S_*^{\mathrm{GM}}(X; R) \subset S_*(X; R)$, the desired $p : I^{\bar{q}}S_*(X; R) \to S_*(\mathrm{pt}; R)$ is simply the restriction of the chain map $p : S_*(X; R) \to S_*(\mathrm{pt}; R)$. Therefore, p is a well-defined chain map when $\bar{q} \leq \bar{t}$.

If $\bar{q} \nleq \bar{t}$ and X has a singular stratum, then p will not necessarily be a chain map. In particular, let σ be as in the example at the end of the proof of Lemma 7.3.16. Then again $\hat{\partial}p(\sigma) = \partial p(\sigma) = p(\partial\sigma) \neq p(\hat{\partial}\sigma)$. □

(b) Unital Properties of Products

Once again, we will need an algebraic lemma:

Lemma 7.3.19 *Suppose C_* and D_* are chain complexes of R-modules. Let $\alpha \in \mathrm{Hom}(C_*, R)$ be a cochain and $\beta \in \mathrm{Hom}(D_*, R)$ a cocycle. Then*

$$(\mathrm{id} \otimes \beta)^*\Phi^*(\alpha) = \Theta(\alpha \otimes \beta) \in \mathrm{Hom}(C_* \otimes D_*, R),$$

where $\Phi : C_* \otimes R \to C_*$ is the standard isomorphism[9].

Proof It suffices to verify that both expressions in the claimed equality act identically on elements of $C_* \otimes D_*$. So let $x \otimes y$ be a generator of $C_* \otimes D_*$. Then

$$\begin{aligned}((\mathrm{id} \otimes \beta)^*\Phi^*(\alpha))(x \otimes y) &= \alpha(\Phi(\mathrm{id} \otimes \beta)(x \otimes y)) = (-1)^{|\beta||x|}\alpha(\Phi(x \otimes \beta(y)))\\ &= (-1)^{|\beta||x|}\alpha(\beta(y)x) = (-1)^{|\beta||x|}\beta(y)\alpha(x)\\ &= \Theta(\alpha \otimes \beta)(x \otimes y).\end{aligned}$$

□

Now, returning to the discussion at the beginning of this section, we have seen that in order to have both an agreeable triple of the form $(\bar{p}, \bar{q}; \bar{p})$ and projection-induced chain maps $p_1 : I^Q S_*(X \times X; R) \to I^{\bar{p}}S_*(X; R)$ and $p_2 : I^Q S_*(X \times X; R) \to I^{\bar{q}}S_*(X; R)$ with Q a $(\bar{p}, \bar{q})$-compatible perversity on $X \times X$, we must take $\bar{q} = \bar{t}$. This justifies the assumptions that appear in the following lemma, which states that with certain choices of perversity the algebraic diagonal is co-unital up to homotopy. Conveniently, the use of $\bar{q} = \bar{t}$ also makes a locally torsion-free condition automatic, as every CS set is locally torsion-free with respect to $\bar{t}$ by Example 5.3.12.

Lemma 7.3.20 *Let R be a Dedekind domain. Suppose that $\bar{p}$ is a perversity on a CS set X and that $A \subset X$ is an open subset. Then the compositions*

$$\begin{aligned}I^{\bar{p}}S_*(X, A; R) &\xrightarrow{\mathbf{d}} I^{\bar{t}}S_*(X; R) \otimes I^{\bar{p}}S_*(X, A; R) \xrightarrow{\mathbf{a}\otimes\mathrm{id}} R \otimes I^{\bar{p}}S_*(X, A; R)\\ &\underset{\cong}{\xrightarrow{\Phi}} I^{\bar{p}}S_*(X, A; R)\end{aligned}$$

[9] Here the expression on the left makes sense because a degree-i cocycle can be interpreted as a degree-$(-i)$ chain map $D_* \to R$, treating R as a chain complex with the module R in degree zero and all other modules trivial.

and

$$I^{\bar{p}}S_*(X,A;R) \xrightarrow{\bar{\mathbf{d}}} I^{\bar{p}}S_*(X,A;R) \otimes I^{\bar{t}}S_*(X;R) \xrightarrow{\mathrm{id}\otimes\mathbf{a}} I^{\bar{p}}S_*(S,A;R) \otimes R$$
$$\underset{\cong}{\xrightarrow{\Phi}} I^{\bar{p}}S_*(X,A;R),$$

in which **a** *is the augmentation map*[10]*, are each homotopic to the identity map* id : $I^{\bar{p}}S_*(X,A;R) \to I^{\bar{p}}S_*(X,A;R)$.

Proof We will demonstrate the claim regarding the first composition. The second argument is equivalent.

Every CS set is locally $(\bar{t}, R)$-torsion-free by Example 5.3.12, and using that $D\bar{t} = \bar{0}$ we have $D\bar{p} \geq D\bar{t} + D\bar{p}$. So $(\bar{p}, \bar{t}; \bar{p})$ is an agreeable triple by Lemma 7.2.8, and thus the algebraic diagonal is defined.

Now, consider the diagram

$$\begin{array}{ccc}
I^{Q_{\bar{p},\bar{t}}}S_*(X\times X, A\times X;R) & \xleftarrow{\varepsilon} & I^{\bar{p}}S_*(X,A;R)\otimes I^{\bar{t}}S_*(X;R) \\
\downarrow{\scriptstyle p_1} & & \downarrow{\scriptstyle \mathrm{id}\otimes p} \\
I^{\bar{p}}S_*(X,A;R) & \xleftarrow{\varepsilon} & I^{\bar{p}}S_*(X,A;R)\otimes S_*(\mathrm{pt};R).
\end{array} \tag{7.12}$$

The map p_1 is here induced by the projection to the first factor $X \times X \to X$, which is well defined by Corollary 7.3.17, and p is induced by the unique map $X \to \mathrm{pt}$ and is well defined by Lemma 7.3.18. The bottom cross product is that which occurs in the version of the Künneth Theorem for which one factor is a manifold. So each map of the diagram is a chain map, and it suffices to establish commutativity in each degree. For fixed degrees, each of the relevant chain modules is a submodule of either $S_k(X;R)$ or $S_k(X,A;R)$ for some k, and the maps are induced, at the level of modules (i.e. ignoring boundary maps), by the corresponding maps for the ordinary chain modules. Thus, to show that this diagram commutes, it suffices to show the commutavity of the following diagrams:

$$\begin{array}{ccc}
S_{i+j}(X\times X, A\times X;R) & \xleftarrow{\varepsilon} & S_i(X,A;R)\otimes S_j(X;R) \\
\downarrow{\scriptstyle p_1} & & \downarrow{\scriptstyle \mathrm{id}\otimes p} \\
S_{i+j}(X,A;R) = S_{i+j}(X\times \mathrm{pt}, A\times \mathrm{pt};R) & \xleftarrow{\varepsilon} & S_i(X,A;R)\otimes S_j(\mathrm{pt};R).
\end{array}$$

If σ is an i-simplex representing an element of $S_i(X,A;R)$ and $\tau : \Delta^j \to X$ is a j-simplex, then $\varepsilon(\sigma \otimes \tau)$ is defined by applying $\sigma \times \tau$ to the singular triangulation of $\Delta^i \times \Delta^j$ determined by the Eilenberg–Zilber shuffle process. So, proceeding left then down, we obtain the chain in $S_{i+j}(X,A;R)$ that comes from applying $p_1(\sigma\times\tau)$ to this singular triangulation.

[10] See Remark 6.2.12. Here we think of **a** as a degree 0 chain map **a** : $I^{\bar{t}}S_*(X;R) \to R$, treating the R on the right as the chain complex with the module R in degree 0 and the trivial module in all other degrees.

But if $(x,y) \in \Delta^i \times \Delta^j$, we have $p_1(\sigma \times \tau)(x,y) = p_1(\sigma(x), \tau(y)) = \sigma(x)$. If we let $\pi_1 : \Delta^i \times \Delta^j \to \Delta^i$ be the projection to the first factor, we similarly have $\sigma\pi_1(x,y) = \sigma(x)$, so $p_1(\sigma \times \tau) = \sigma\pi_1$. On the other hand, if $\pi : \Delta^j \to \mathrm{pt}$ is the unique map to a point, $\varepsilon(\mathrm{id} \otimes p)(\sigma \otimes \tau) = \varepsilon(\sigma \otimes p\tau)$ is given by applying $\sigma \times p(\tau) = \sigma \times \pi$ to the Eilenberg–Zilber singular triangulation of $\Delta^i \times \Delta^j$. But $\sigma\pi_1$ and $\sigma \times \pi$ agree up to identifying X with $X \times \mathrm{pt}$, demonstrating the commutativity.

It follows that if we replace each ε in Diagram (7.12) by a homotopy inverse IAW, we obtain a homotopy commutative diagram:

$$\begin{array}{ccc}
I^{Q_{\bar p,\bar t}}S_*(X \times X, A \times X; R) & \xrightarrow{\mathrm{IAW}} & I^{\bar p}S_*(X, A; R) \otimes I^{\bar t}S_*(X; R) \\
\downarrow{\scriptstyle p_1} & & \downarrow{\scriptstyle \mathrm{id} \otimes p} \\
I^{\bar p}S_*(X, A; R) & \xrightarrow{\mathrm{IAW}} & I^{\bar p}S_*(X, A; R) \otimes S_*(\mathrm{pt}; R).
\end{array}$$

Next, in the special case we are considering here, it is possible to write down an explicit map $\mathrm{IAW} : I^{\bar p}S_*(X, A; R) \xrightarrow{\mathrm{IAW}} I^{\bar p}S_*(X, A; R) \otimes S_*(\mathrm{pt}; R)$ that is a chain homotopy inverse to ε. In fact, we claim that the map $\nu : \xi \to \xi \otimes v_0$, where v_0 is the unique singular 0-simplex generating $S_0(\mathrm{pt}; R)$, is such a homotopy inverse. For this, we note that it follows from the definitions that $\varepsilon\nu = \mathrm{id}$ as chain maps (again identifying $X \times \mathrm{pt}$ with X). But this is enough to imply that ν is a chain homotopy inverse to ε by the following argument: Let g be a chain homotopy inverse to ε so that $g\varepsilon$ and εg are each homotopic to the appropriate identity; such a g exists by Theorem 6.4.14 using that $S_*(\mathrm{pt}; R) = I^{\bar q}S_*(\mathrm{pt}; R)$ for any $\bar q$ and that $Q_{\bar p,\bar q}$ agrees with $\bar p$ on $X \times \mathrm{pt} = X$ as pt has only a regular stratum. Then, as $g\varepsilon$ is homotopic to the identity, $g\varepsilon\nu$ is homotopic to ν, but $g\varepsilon\nu = g$. So ν is homotopic to g and so is a chain homotopy inverse for ε.

Now consider the larger diagram

$$\begin{array}{ccccc}
I^{\bar p}S_*(X, A; R) & \xrightarrow{\mathbf{d}} & I^{Q_{\bar p,\bar t}}S_*(X \times X, A \times X; R) & \xrightarrow{\mathrm{IAW}} & I^{\bar p}S_*(X, A; R) \otimes I^{\bar t}S_*(X; R) \\
& \searrow{\scriptstyle =} & \downarrow{\scriptstyle p_1} & & \downarrow{\scriptstyle \mathrm{id} \otimes p} \\
& & I^{\bar p}S_*(X, A; R) & \xrightarrow{\nu} & I^{\bar p}S_*(X, A; R) \otimes S_*(\mathrm{pt}; R) \\
& & & \nwarrow{\scriptstyle \Phi,\ \cong} & \downarrow{\scriptstyle \mathrm{id} \otimes \mathbf{a}} \\
& & & & I^{\bar p}S_*(X, A; R) \otimes R.
\end{array}$$

We have already seen that the square commutes up to homotopy; here we are letting ν be the map constructed in the preceding paragraph.

Commutativity of the upper left triangle is straightforward as $p_1\mathbf{d} = \mathrm{id}_X$. For the commutativity of the bottom right triangle, we need only observe that $\Phi(\mathrm{id} \otimes \mathbf{a})\nu(\xi) = \Phi(\mathrm{id}\otimes\mathbf{a})(\xi\otimes v_0) = \Phi(\xi\otimes 1) = \xi$, as $\mathbf{a}$ has degree zero and $\mathbf{a}(v_0) = 1$. Therefore, each component of the diagram homotopy commutes, and the composition counterclockwise around the outside of the diagram is homotopic to the identity. But, noting that the composition $I^{\bar{t}}S_*(X;R) \to S_*(\mathrm{pt};R) \xrightarrow{\mathbf{a}} R$ factors the direct augmentation $I^{\bar{t}}S_*(X;R) \xrightarrow{\mathbf{a}} R$, the path around the outside of the diagram is precisely $\Phi(\mathrm{id} \otimes \mathbf{a})\bar{\mathbf{d}}$. □

Proposition 7.3.21 *Let R be a Dedekind domain. Suppose that $\bar{p}$ is a perversity on a CS set X and that $A \subset X$ is an open subset. Let $\alpha \in I_{\bar{p}}H^i(X,A;R)$, and let $1 \in I_{\bar{t}}H^0(X;R)$ be represented by the cocycle that evaluates to 1 on each 0-simplex of $I^{\bar{t}}S_0(X;R)$. Then $1 \smile \alpha = \alpha \smile 1 = \alpha \in I_{\bar{p}}H^i(X,A;R)$.*

Proof We begin by observing that 1 really is a cocycle. For this, we need only show that if ξ is a 1-chain in $I^{\bar{t}}S_1(X;R)$, then $1(\partial\xi) = 0$. By Proposition 6.2.9, $I^{\bar{t}}S_*(X;R) = I^{\bar{t}}S_*^{\mathrm{GM}}(X;R)$, so $\partial\xi$ is the usual boundary of ξ, treating ξ as an element of $S_1(X;R)$. Since 1 is a cocycle in $S^0(X;R)$, it follows that $1(\partial\xi) = 0$. We also observe that if we treat the 0-cocycle $1 \in \mathrm{Hom}(I^{\bar{t}}S_0(X;R),R)$ as a degree-zero chain map $I^{\bar{t}}S_*(X;R) \to R$, then it is just the same as the augmentation map $\mathbf{a}$, utilized in the preceding lemma. So we will use the notation 1 when treating it as a cocycle and $\mathbf{a}$ when treating it as a chain map.

Now to verify the given property, we write $\mathrm{id} : I^{\bar{p}}S_*(X,A;R) \to I^{\bar{p}}S_*(X,A;R)$ and observe that on cohomology we have

$$\begin{aligned}\alpha &= \mathrm{id}^*\alpha \\ &= (\Phi(\mathrm{id}\otimes\mathbf{a})\bar{\mathbf{d}})^*\alpha \qquad \text{by Lemma 7.3.20} \\ &= \bar{\mathbf{d}}^*(\mathrm{id}\otimes\mathbf{a})^*\Phi^*(\alpha).\end{aligned}$$

And by Lemma 7.3.19, together with the relation between the cocycle 1 and the chain map $\mathbf{a}$ discussed above, $(\mathrm{id}\otimes\mathbf{a})^*\Phi^*(\alpha) = \Theta(\alpha\otimes 1)$. Therefore, $\alpha = \bar{\mathbf{d}}^*\Theta(\alpha\otimes 1) = \alpha \smile 1$. □

Proposition 7.3.22 *Let R be a Dedekind domain. Suppose that $\bar{p}$ is a perversity on a CS set X and that $A \subset X$ is an open subset. Let $\xi \in I^{\bar{p}}H_i(X,A;R)$, and let $1 \in I_{\bar{t}}H^0(X;R)$ be represented by the cocycle that evaluates to 1 on each 0-simplex of $I^{\bar{t}}S_0(X;R)$. Then $1 \frown \xi = \xi \in I^{\bar{p}}H_i(X,A;R)$.*

Proof We have just seen in the proof of Proposition 7.3.21 that $1 \in I_{\bar{t}}S^0(X;R)$ is a well-defined cocycle given by the augmentation map $\mathbf{a}$. Now we can compute as follows, again using Lemma 7.3.20:

$$1 \frown \xi = \Phi((\mathrm{id}\otimes 1)\bar{\mathbf{d}}(\xi)) = (\Phi(\mathrm{id}\otimes\mathbf{a})\bar{\mathbf{d}})(\xi) = \mathrm{id}(\xi) = \xi.$$ □

A unital property for cross products can be proven by reworking some of the pieces from Lemma 7.3.20.

Proposition 7.3.23 *Let R be a Dedekind domain. Suppose that $\bar{p}$ is a perversity on a CS set X and that $A \subset X$ is an open subset. Let $\alpha \in I_{\bar{p}}H^i(X, A; R)$, and let $1 \in H^0(\text{pt}; R)$ be represented by the cocycle that evaluates to 1 on the singular 0-simplex of $S_0(\text{pt}; R)$. Then $1 \times \alpha = \alpha \times 1 = \alpha$ in $I_{\bar{p}}H^i(\text{pt} \times X, \text{pt} \times A; R) = I_{\bar{p}}H^i(X \times \text{pt}, A \times \text{pt}; R) = I_{\bar{p}}H^i(X, A; R)$.*

Proof We will provide the argument for $\alpha \times 1$, the argument for $1 \times \alpha$ being equivalent.

By definition, $\alpha \times 1 = \text{IAW}^*\Theta(\alpha \otimes 1)$, and by Lemma 7.3.19, $\Theta(\alpha \otimes 1) = (\text{id} \otimes \mathbf{a})^*\Phi^*(\alpha)$, again using that the cocycle 1 corresponds to the chain map $\mathbf{a}$. Thus, $\alpha \times 1 = \text{IAW}^*(\text{id} \otimes \mathbf{a})^*\Phi^*(\alpha)$. But, we saw in the proof of Lemma 7.3.20 that $\Phi(\text{id} \otimes \mathbf{a})\text{IAW}$ is the identity (up to identifying $X \times \text{pt}$ with X) for a specific choice of IAW that we there labeled ν. Hence $\text{IAW}^*(\text{id} \otimes \mathbf{a})^*\Phi^*(\alpha) = \alpha$ at the level of cohomology. □

Proposition 7.3.24 *Let R be a Dedekind domain. Suppose that $\bar{p}$ is a perversity on a CS set X and that $A \subset X$ is an open subset. Let Y be a CS set with perversity $\bar{q} \leq \bar{t}$, and let Q be a $(\bar{p}, \bar{q})$-compatible perversity on $X \times Y$. Let $\alpha \in I_{\bar{p}}H^i(X, A; R)$, and let $1_Y \in I_{\bar{q}}H^0(Y; R)$ be represented by the cocycle that evaluates to 1 on each 0-simplex of $I^{\bar{q}}S_0(Y; R)$. Then $\alpha \times 1_Y = p_1^*(\alpha) \in I_QH^*(X \times Y, A \times Y; R)$, where $p_1 : X \times Y \to X$ is the projection.*

Proof By Lemma 7.3.17, p_1^* is well defined. Let $p : Y \to \text{pt}$ be the unique map, and observe that, by Lemma 7.3.18, p also induces a well-defined chain map $p : I^{\bar{q}}S_*(X; R) \to S_*(\text{pt}; R)$. We observe that if 1_{pt} denotes the element of $S^0(\text{pt}; R)$ that takes the unique 0-simplex to 1, then $p^*(1_{\text{pt}}) = 1_Y$. Thus $\alpha \times 1_Y = \alpha \times (p^*(1_{\text{pt}})) = (\text{id} \times p)^*(\alpha \times 1_{\text{pt}})$ by Proposition 7.3.2; technically, p and $\text{id} \times p = p_1$ are not stratified maps, but we know that they induce maps on intersection chains by Lemmas 7.3.17 and 7.3.18 and, given this, the arguments of Proposition 7.3.2 and Proposition 5.2.17 continue to apply. But $\alpha \times 1_{\text{pt}} = \alpha$ by Proposition 7.3.23 and $\text{id} \times p = p_1$ as maps (identifying $X \times \text{pt}$ with X, as usual). Therefore, $\alpha \times 1_Y = (\text{id} \times p)^*(\alpha \times 1_{\text{pt}}) = p_1^*(\alpha)$. □

(c) Products and Evaluations

Lemma 7.3.20 can also be used to show that the cap product corresponds to evaluation in the appropriate setting. Once again, we will need agreeable triples of the form $(\bar{p}, \bar{q}; \bar{p})$, forcing $\bar{q} \geq \bar{t}$ by our discussion at the beginning of this section. On the other hand, while we will not be using projection maps here, the use of the augmentation map $\mathbf{a}$ will force us to assume $\bar{q} \leq \bar{t}$; see Remark 6.2.12. So, once again, we must in fact use $\bar{q} = \bar{t}$, and $\bar{t}$ continues to play a special role.

Proposition 7.3.25 *Let R be a Dedekind domain, and suppose that $\bar{p}$ is a perversity on a CS set X and that $A \subset X$ is an open subset. Let $\alpha \in I_{\bar{p}}H^i(X, A; R)$ and $\xi \in I^{\bar{p}}H_i(X, A; R)$. Then $\mathbf{a}(\alpha \frown \xi) = \alpha(\xi) \in R$, where $\mathbf{a} : I^{\bar{t}}H_0(X; R) \to R$ is the augmentation map.*

Proof First, note that we have a well-defined cap product $I_{\bar{p}}H^i(X, A; R) \otimes I^{\bar{p}}H_i(X, A; R) \to I^{\bar{t}}H_0(X; R)$ as every CS set is locally $(\bar{t}, R)$-torsion-free and $D\bar{p} \geq D\bar{p} + D\bar{t} = D\bar{p} + \bar{0} = D\bar{p}$.

We next claim that if $\alpha \in I_{\bar{p}}S^i(X,A;R)$, $\mathbf{a} : I^{\bar{t}}S_*(X;R) \to R$ is the augmentation map, and $\Phi : R \otimes I^{\bar{t}}S_*(X;R) \to I^{\bar{t}}S_*(X;R)$ is the canonical isomorphism, then

$$\mathbf{a}\Phi(\mathrm{id} \otimes \alpha) = \alpha\Phi(\mathbf{a} \otimes \mathrm{id}) \in \mathrm{Hom}(I^{\bar{t}}S_*(X;R) \otimes I^{\bar{p}}S_*(X,A;R), R).$$

To check this, let $x \otimes y \in I^{\bar{t}}S_*(X;R) \otimes I^{\bar{p}}S_*(X,A;R)$ be a generator. Then we have

$$\begin{aligned}(\mathbf{a}\Phi(\mathrm{id} \otimes \alpha))(x \otimes y) &= (-1)^{i|x|}\mathbf{a}\Phi(x \otimes \alpha(y)) \; = (-1)^{i|x|}\mathbf{a}(\alpha(y)x) \\ &= (-1)^{i|x|}\alpha(y)\mathbf{a}(x) \; = \alpha(y)\mathbf{a}(x),\end{aligned}$$

where the last equality comes from the observation that $\mathbf{a}(x) = 0$ unless $|x| = 0$. Meanwhile,

$$(\alpha\Phi(\mathbf{a} \otimes \mathrm{id}))(x \otimes y) = (\alpha\Phi)(\mathbf{a}(x) \otimes y) \; = \alpha(\mathbf{a}(x)y) \; = \mathbf{a}(x)\alpha(y).$$

So, indeed, $a\Phi(\mathrm{id} \otimes \alpha) = \alpha\Phi(\mathbf{a} \otimes \mathrm{id})$.

Now, suppose $\alpha \in I_{\bar{p}}H^i(X,A;R)$ and $\xi \in I^{\bar{p}}H_i(X,A;R)$, and consider the evaluation $\alpha(\xi)$. Since evaluation is well defined at the level of (co)homology, we know that the element $\alpha(\xi) \in R$ is independent of the choice of intersection chain representing ξ. By Lemma 7.3.20, we can therefore replace ξ in this computation with $\Phi(\mathbf{a}\otimes\mathrm{id})\bar{\mathbf{d}}(\xi)$, where $\bar{\mathbf{d}}$ is defined with respect to some specific choice of IAW. Therefore, $\alpha(\xi) = \alpha(\Phi(\mathbf{a} \otimes \mathrm{id})\bar{\mathbf{d}}(\xi))$. But we have just seen that $\alpha\Phi(\mathbf{a} \otimes \mathrm{id}) = \mathbf{a}\Phi(\mathrm{id} \otimes \alpha)$, so $\alpha(\xi)$ becomes equal to $(\mathbf{a}\Phi(\mathrm{id} \otimes \alpha))(\bar{\mathbf{d}}\xi) = \mathbf{a}((\Phi(\mathrm{id} \otimes \alpha))\bar{\mathbf{d}}(\xi))$, which is precisely $\mathbf{a}(\alpha \frown \xi)$. □

Remark 7.3.26 There is an observation to be made concerning the proof of Proposition 7.3.25 that will be useful later in Section 8.4.3, though its utility isn't likely to be so apparent now. Let us continue to assume that $\alpha \in I_{\bar{p}}S^i(X,A;R)$ and $\xi \in I^{\bar{p}}S_i(X,A;R)$ but that they are not necessarily a cocycle and a cycle. We also continue to assume we have made a fixed choice of IAW with which to define the cap product. As Lemma 7.3.20 is stated at the chain level, it tells us that

$$\xi - \Phi(\mathbf{a} \otimes \mathrm{id})\bar{\mathbf{d}}(\xi) = D\partial\xi + \partial D\xi, \tag{7.13}$$

where D is a chain homotopy guaranteed by Lemma 7.3.20. The argument in the proof continues to imply that $\alpha(\Phi(\mathbf{a}\otimes\mathrm{id})\bar{\mathbf{d}}(\xi)) = \mathbf{a}(\alpha \frown \xi)$; so applying α to the entire expression (7.13) yields

$$\alpha(\xi) = \mathbf{a}(\alpha \frown \xi) + \alpha(D\partial\xi + \partial D\xi).$$

See the proof of Proposition 8.4.16 in Section 8.4.3 for our application of this formula.

Of course, there is also a nice formula for evaluation of the cohomology cross product on the homology cross product:

Proposition 7.3.27 *Let R be a Dedekind domain. Suppose that X,Y are CS sets with respective perversities $\bar{p}, \bar{q}$ and that Q is a $(\bar{p},\bar{q})$-compatible perversity on $X \times Y$. Let $A \subset X$ and $B \subset Y$ be open subsets. Let $\alpha \in I_{\bar{p}}H^a(X,A;R)$, $\beta \in I_{\bar{q}}H^b(Y,B;R)$, $\xi \in I^{\bar{p}}H_i(X,A;R)$, and $\eta \in I^{\bar{q}}H_j(Y,B;R)$. Then with $\alpha \times \beta \in I_QH^{a+b}(X \times Y, (A \times Y) \cup (X \times B); R)$ and $\xi \times \eta \in I^QH_{i+j}(X \times Y, (A \times Y) \cup (X \times B); R)$, we have $(\alpha \times \beta)(\xi \times \eta) = (-1)^{bi}\alpha(\xi)\beta(\eta) \in R$.*

Proof Given the assumptions, $\alpha\times\beta$ is well defined in $I_Q H^{a+b}(X\times Y,(A\times Y)\cup(X\times B);R)$ and $\xi\times\eta\in I^Q H_{i+j}(X\times Y,(A\times Y)\cup(X\times B);R)$. Now we have

$$\begin{aligned}(\alpha\times\beta)(\xi\times\eta) &= \left[\mathrm{IAW}^*\Theta(\alpha\otimes\beta)\right]\varepsilon(\xi\otimes\eta)\\ &= \Theta(\alpha\otimes\beta)\mathrm{IAW}\varepsilon(\xi\otimes\eta) = \Theta(\alpha\otimes\beta)(\xi\otimes\eta)\\ &= (-1)^{bi}\alpha(\xi)\beta(\eta).\end{aligned}$$

In the third line, we have used the fact that IAW and ε are chain homotopy inverses, so that

$$\mathrm{IAW}\varepsilon(\xi\otimes\eta) = \xi\otimes\eta \in H_{i+j}(I^{\bar p}S_*(X,A;R)\otimes I^{\bar q}S_*(Y,B;R)).$$ □

7.3.4 Associativity

We turn to the associativity of products. This is a nuisance. The problem is that we know that certain relations need to hold among perversities $\bar p$, $\bar q$, and $\bar r$ in order, for example, to have a cup product

$$I_{\bar p}H^i(X,A;R)\otimes I_{\bar q}H^j(X,B;R)\xrightarrow{\smile} I_{\bar r}H^{i+j}(X,A\cup B;R).$$

But in an associativity statement of the form

$$(\alpha\smile\beta)\smile\gamma = \alpha\smile(\beta\smile\gamma)$$

there are already four cup products, all of which need to be defined, and that's not even considering yet whether there may need to be some extra hypotheses about the perversities for the equality to hold.

To deal with the proliferation of necessary conditions, we will take a two-step process. First, we will prove the standard associativity results under very broad hypotheses of the form: "Suppose all the needed intermediary and product perversities exist to make all the maps in certain diagrams well defined and for the cross product maps to be chain homotopy equivalences." Then, in the second part of this section, we will show that such collections of perversities do indeed exist under some more easily stated, though less general, hypotheses.

(a) Associativity under Broad Assumptions

We can begin, as usual, with a purely algebraic lemma that does not involve any perversities:

Lemma 7.3.28 *Let C_*, D_*, E_* be chain complexes of R-modules. Then the following diagram commutes:*

$$\begin{array}{ccc}\begin{array}{c}\mathrm{Hom}(C_*,R)\otimes\mathrm{Hom}(D_*,R)\otimes\\ \mathrm{Hom}(E_*,R)\end{array} & \xrightarrow{\Theta\otimes\mathrm{id}} & \begin{array}{c}\mathrm{Hom}(C_*\otimes D_*,R)\otimes\\ \mathrm{Hom}(E_*,R)\end{array}\\ \Big\downarrow{\scriptstyle\mathrm{id}\otimes\Theta} & & \Big\downarrow{\scriptstyle\Theta}\\ \mathrm{Hom}(C_*,R)\otimes\mathrm{Hom}(D_*\otimes E_*,R) & \xrightarrow{\Theta} & \mathrm{Hom}(C_*\otimes D_*\otimes E_*,R).\end{array}$$

Proof Let $\alpha \in \operatorname{Hom}(C_*, R)$, $\beta \in \operatorname{Hom}(D_*, R)$, and $\gamma \in \operatorname{Hom}(E_*, R)$. To verify the commutativity, it suffices to check the evaluation of $\alpha \otimes \beta \otimes \gamma$ on a generator $x \otimes y \otimes z$ of $C_* \otimes D_* \otimes E_*$. We have

$$\begin{aligned}\Theta(\Theta \otimes \mathrm{id})(\alpha \otimes \beta \otimes \gamma)(x \otimes y \otimes z) &= \Theta(\Theta(\alpha \otimes \beta) \otimes \gamma)(x \otimes y \otimes z)\\ &= (-1)^{|\gamma|(|x|+|y|)}[\Theta(\alpha \otimes \beta)(x \otimes y)]\gamma(z)\\ &= (-1)^{|\gamma|(|x|+|y|)+|\beta||x|}\alpha(x)\beta(y)\gamma(z).\end{aligned}$$

Separately,

$$\begin{aligned}\Theta(\mathrm{id} \otimes \Theta)(\alpha \otimes \beta \otimes \gamma)(x \otimes y \otimes z) &= \Theta(\alpha \otimes \Theta(\beta \otimes \gamma))(x \otimes y \otimes z)\\ &= (-1)^{(|\beta|+|\gamma|)|x|}\alpha(x)\Theta(\beta \otimes \gamma)(y \otimes z)\\ &= (-1)^{(|\beta|+|\gamma|)|x|+|\gamma||y|}\alpha(x)\beta(y)\gamma(z).\end{aligned}$$

The two expressions are equal, completing the proof. □

Using the algebraic lemma, we have the following general associativity of the cross product.

Proposition 7.3.29 (Associativity) *Let R be a Dedekind domain. Suppose that $\bar p$, $\bar q$, $\bar r$ are perversities on CS sets X, Y, Z. Let $A \subset X$, $B \subset Y$, and $C \subset Z$ be open subsets. Suppose there are perversities Q_1 on $X \times Y$, Q_2 on $Y \times Z$, and Q_3 on $X \times Y \times Z$ such that*

1. *Q_1 is $(\bar p, \bar q)$-compatible,*
2. *Q_2 is $(\bar q, \bar r)$-compatible,*
3. *Q_3 is both $(\bar p, Q_2)$-compatible and $(Q_1, \bar q)$-compatible*[11].

Let $\alpha \in I_{\bar p}H^i(X, A; R)$, $\beta \in I_{\bar q}H^j(Y, B; R)$, and $\gamma \in I_{\bar r}H^k(Z, C; R)$. Then $(\alpha \times \beta) \times \gamma = \alpha \times (\beta \times \gamma)$ in

$$I_{Q_3}H^{i+j+k}(X \times Y \times Z, (A \times Y \times Z) \cup (X \times B \times Z) \cup (X \times Y \times C); R).$$

Proof From the conditions on the perversities, both iterated cross products are well defined and live in

$$I_{Q_3}H^{i+j+k}(X \times Y \times Z, (A \times Y \times Z) \cup (X \times B \times Z) \cup (X \times Y \times C); R).$$

Furthermore, leaving the R coefficients tacit, we have the following commutative diagram by Proposition 5.2.19; not that our assumptions about Q_1, Q_2, and Q_3 are sufficient to fulfill

[11] Such a collection Q_1, Q_2, Q_3 certainly exists: if $S \subset X$, $T \subset Y$, and $U \subset Z$ are strata (singular or regular), we can take $Q_1(S \times T) = \bar p(S) + \bar q(T)$, $Q_2(T \times U) = \bar q(S) + \bar r(T)$, and $Q_3(S \times T \times U) = \bar p(S) + \bar q(T) + \bar r(U)$.

the conditions there:

$$\begin{array}{ccc}
\begin{array}{c} I^{Q_3}S_*(X\times Y\times Z,(A\times Y\times Z)\cup \\ (X\times B\times Z)\cup(X\times Y\times C)) \end{array} & \xleftarrow{\varepsilon} & \begin{array}{c} I^{Q_1}S_*(X\times Y,(A\times Y)\cup(X\times B))\otimes \\ I^{\bar r}S_*(Z,C) \end{array} \\
\uparrow{\scriptstyle \varepsilon} & & \uparrow{\scriptstyle \varepsilon\otimes \mathrm{id}} \\
\begin{array}{c} I^{\bar p}S_*(X,A)\otimes \\ I^{Q_2}S_*(Y\times Z,(B\times Z)\cup(Y\times C)) \end{array} & \xleftarrow{\mathrm{id}\otimes\varepsilon} & \begin{array}{c} I^{\bar p}S_*(X,A)\otimes I^{\bar q}S_*(Y,B)\otimes \\ I^{\bar r}S_*(Z,C). \end{array}
\end{array}$$

Since each cross product is a chain homotopy equivalence by Theorem 6.4.14, if we replace each ε in the diagram with a choice of homotopy inverse IAW in the opposite direction we obtain a homotopy commutative diagram. Dualizing and letting $\sim$ denote chain homotopy we have

$$\mathrm{IAW}^*(\mathrm{IAW}\otimes \mathrm{id})^* \sim \mathrm{IAW}^*(\mathrm{id}\otimes \mathrm{IAW})^*, \tag{7.14}$$

with $\sim$ becoming equality if we act on cohomology.

We can now compute at the level of cohomology:

$$\begin{aligned}
(\alpha\times\beta)\times\gamma &= \mathrm{IAW}^*\Theta((\alpha\times\beta)\otimes\gamma) \\
&= \mathrm{IAW}^*\Theta(\mathrm{IAW}^*\Theta(\alpha\otimes\beta)\otimes\gamma) \\
&= \mathrm{IAW}^*\Theta(\mathrm{IAW}^*\Theta\otimes \mathrm{id})(\alpha\otimes\beta\otimes\gamma) \\
&= \mathrm{IAW}^*\Theta(\mathrm{IAW}^*\otimes \mathrm{id})(\Theta\otimes \mathrm{id})(\alpha\otimes\beta\otimes\gamma) \\
&= \mathrm{IAW}^*(\mathrm{IAW}\otimes \mathrm{id})^*\Theta(\Theta\otimes \mathrm{id})(\alpha\otimes\beta\otimes\gamma) && \text{by Lemma 7.3.1} \\
&= \mathrm{IAW}^*(\mathrm{IAW}\otimes \mathrm{id})^*\Theta(\mathrm{id}\otimes\Theta)(\alpha\otimes\beta\otimes\gamma) && \text{by Lemma 7.3.28} \\
&= \mathrm{IAW}^*(\mathrm{id}\otimes \mathrm{IAW})^*\Theta(\mathrm{id}\otimes\Theta)(\alpha\otimes\beta\otimes\gamma) && \text{by (7.14)} \\
&= \mathrm{IAW}^*\Theta(\mathrm{id}\otimes \mathrm{IAW}^*)(\mathrm{id}\otimes\Theta)(\alpha\otimes\beta\otimes\gamma) && \text{by Lemma 7.3.1} \\
&= \mathrm{IAW}^*\Theta(\mathrm{id}\otimes \mathrm{IAW}^*\Theta)(\alpha\otimes\beta\otimes\gamma) \\
&= \mathrm{IAW}^*\Theta(\alpha\otimes \mathrm{IAW}^*\Theta(\beta\otimes\gamma)) \\
&= \mathrm{IAW}^*\Theta(\alpha\otimes(\beta\times\gamma)) \\
&= \alpha\times(\beta\times\gamma).
\end{aligned}$$

□

We can now turn to the associativity properties of cup and cap products. Unfortunately, as noted in the introduction to this section, this is a point at which we will be unreasonably general, at least to begin. To see why, consider the following diagram (R coefficients implicit) whose commutativity will be required to obtain associativity identities of cup and cap products of the form $(\alpha \smile \beta) \smile \gamma = \alpha \smile (\beta \smile \gamma)$ and $(\alpha \smile \beta) \frown \xi = \alpha \frown (\beta \frown \xi)$:

$$\begin{array}{ccccc}
I^{\bar{s}}S_*(X, A \cup B \cup C) & \xrightarrow{\ \mathbf{d}\ } & I^{Q_4}S_*(X \times X, ((A \cup B) \times X) \cup (X \times C)) & \xleftarrow{\ \varepsilon\ } & I^{\bar{u}}S_*(X, A \cup B) \otimes I^{\bar{r}}S_*(X, C) \\
\Big\downarrow \mathbf{d} & & \Big\downarrow \mathbf{d} \times \mathrm{id} & & \Big\downarrow \mathbf{d} \otimes \mathrm{id} \\
I^{Q_5}S_*(X \times X, (A \times X) \cup (X \times (B \cup C))) & \xrightarrow{\ \mathrm{id} \times \mathbf{d}\ } & I^{Q_3}S_*(X \times X \times X, (A \times X \times X) \cup (X \times B \times X) \cup (X \times X \times C)) & \xleftarrow{\ \varepsilon\ } & I^{Q_1}S_*(X \times X, (A \times X) \cup (X \times B)) \otimes I^{\bar{r}}S_*(X, C) \\
\Big\uparrow \varepsilon & & \Big\uparrow \varepsilon & & \Big\uparrow \varepsilon \otimes \mathrm{id} \\
I^{\bar{p}}S_*(X, A) \otimes I^{\bar{v}}S_*(X, B \cup C) & \xrightarrow{\ \mathrm{id} \otimes \mathbf{d}\ } & I^{\bar{p}}S_*(X, A) \otimes I^{Q_2}S_*(X \times X, (B \times X) \cup (X \times C)) & \xleftarrow{\ \mathrm{id} \otimes \varepsilon\ } & I^{\bar{p}}S_*(X, A) \otimes I^{\bar{q}}S_*(X, B) \otimes I^{\bar{r}}S_*(X, C).
\end{array} \tag{7.15}$$

The commutativity itself is not particularly difficult. The problem is that even if we start with, say, $\alpha \in I_{\bar{p}}H^*(X; R)$, $\beta \in I_{\bar{q}}H^*(X; R)$, and $\gamma \in I_{\bar{r}}H^*(X; R)$ and want to consider $(\alpha \smile \beta) \smile \gamma$ and $\alpha \smile (\beta \smile \gamma)$ in $I_{\bar{s}}H^*(X; R)$ then we will additionally need in intermediate stages:

1. a triple $(\bar{p}, \bar{q}; \bar{u})$ that is Q_1-agreeable for some $(\bar{p}, \bar{q})$-compatible Q_1, and
2. a triple $(\bar{q}, \bar{r}; \bar{v})$ that is Q_2-agreeable for some $(\bar{q}, \bar{r})$-compatible Q_2,

such that

3. the triple $(\bar{u}, \bar{r}; \bar{s})$ must be Q_4-agreeable for some $(\bar{u}, \bar{r})$-compatible Q_4, and
4. the triple $(\bar{p}, \bar{v}; \bar{s})$ must be Q_5-agreeable for some $(\bar{p}, \bar{v})$-compatible Q_5.

And that is just to have both iterated cup products defined. To actually relate the two products requires a perversity Q_3 that fits in the middle of the diagram such that

5. Q_3 is $(Q_1, \bar{r})$- and $(\bar{p}, Q_2)$-compatible, and
6. the maps $\mathbf{d} \times \mathrm{id}$ and $\mathrm{id} \times \mathbf{d}$ of the diagram are respectively (Q_4, Q_3)- and (Q_5, Q_3)-stratified[12].

That is a lot to ask!

At first glance, it does not appear to be so easy to give a simple set of criteria in terms of $\bar{p}$, $\bar{q}$, $\bar{r}$, and $\bar{s}$ alone that will ensure that all these intermediaries exist and have the needed properties. If the values of $\bar{s}$, at least along the diagonal strata of the form $S \times S \times S$, can be made relatively small, then it is not so difficult to find perversities that work in the diagram. For example, we could choose the product perversities as in footnote 11 on page 405 and take $\bar{u}$ and $\bar{v}$ sufficiently small for the diagonal maps involving them to be stratified, at least so long as $\bar{s}$ is small enough to allow it. But as $\bar{s}$ gets larger, the constraints on the diagram begin to tighten, and one might want to use the largest possible values for the Q_i. But

[12] Note that indeed

$$(\mathrm{id} \times \mathbf{d})((A \times X) \cup (X \times (B \cup C))) \subset (A \times X \times X) \cup (X \times B \times X) \cup (X \times X \times C)$$

and analogously for $\mathbf{d} \times \mathrm{id}$.

here things get tricky. For example, the possibilities for Q_1 depend upon the intersection homology torsion in certain degrees of the links of X with respect to the perversities $\bar{p}$ and $\bar{q}$, but the possibilities for Q_3 depend upon things like the intersection homology torsion in certain degrees of the links of $X \times X$ with respect to Q_1. As we know, links in $X \times X$ are joins of links of X, and while we can actually compute their intersection homology in (at least some of) the relevant degrees using (6.10), we quickly see that torsion information from multiple degrees can start creeping in to our considerations, making general assertions quite complicated. We do have Proposition 6.4.15, which says that if X is locally $(\bar{p}, R)$-torsion-free and locally $(\bar{q}, R)$-torsion-free then $X \times X$ will be locally (Q_1, R)-torsion-free. But even if we have this, it seemingly provides no information about whether or not X is locally $(\bar{u}, R)$-torsion-free for any relevant $\bar{u}$.

Given these complications, it is remarkable that it is possible to make some organized statements, as we will see below culminating in Propositions 7.3.34 and 7.3.35. These will perhaps not be the most general statements, but they do provide readily verifiable criteria on $\bar{p}, \bar{q}, \bar{r}, \bar{u}, \bar{v}, \bar{s}$ that are sufficient to make the associativity work. Note that we will assume $\bar{u}$ and $\bar{v}$ as given in those statements, but this is not so unreasonable as presumably we either know or want to know something about where our intermediate cup and cap products live, and even if we do not want to make such assumptions, these hypotheses can be easily reinterpreted to read "if we have such $\bar{u}, \bar{v}$." Those results will also encompass a broader set of possibilities if we are willing to assume locally torsion-free conditions. In fact, with sufficient torsion-free assumptions, our final results imply that there can be associativity so long as the pleasing inequality $D\bar{s} \geq D\bar{p} + D\bar{q} + D\bar{r}$ holds. An application of Lemma 7.2.8 shows that this is the strongest statement we could hope for in terms of possibilities for $\bar{s}$.

As promised, we begin quite generally with the following lemma, which demonstrates a form of coassociativity.

Lemma 7.3.30 *Let R be a Dedekind domain. Suppose that $\bar{p}, \bar{q}, \bar{r}, \bar{s}$ are perversities on a CS set X. Let $A, B, C \subset X$ be open subsets. Suppose that perversities $\bar{u}, \bar{v}, Q_1, Q_2, Q_3, Q_4, Q_5$ exist so that conditions 1–6 above are satisfied.*

Then the following diagram commutes up to chain homotopy:

$$\begin{array}{ccc} I^{\bar{s}}S_*(X, A \cup B \cup C; R) & \xrightarrow{\bar{\mathbf{d}}} & I^{\bar{u}}S_*(X, A \cup B; R) \otimes I^{\bar{r}}S_*(X, C; R) \\ \Big\downarrow{\scriptstyle \bar{\mathbf{d}}} & & \Big\downarrow{\scriptstyle \bar{\mathbf{d}} \otimes \mathrm{id}} \\ I^{\bar{p}}S_*(X, A; R) \otimes I^{\bar{v}}S_*(X, B \cup C; R) & \xrightarrow{\mathrm{id} \otimes \bar{\mathbf{d}}} & I^{\bar{p}}S_*(X, A; R) \otimes I^{\bar{q}}S_*(X, B; R) \otimes I^{\bar{r}}S_*(X, C; R). \end{array}$$

Proof It will suffice to verify the commutativity of Diagram (7.15): The maps $\mathrm{id} \times \mathbf{d}$ and $\mathbf{d} \times \mathrm{id}$ are defined on spaces by $(\mathrm{id} \times \mathbf{d})(x, y) = (x, y, y)$ and $(\mathbf{d} \times \mathrm{id})(x, y) = (x, x, y)$. The assumptions of the lemma are that all the maps of Diagram (7.15) are well defined and imply that all cross product maps ε are chain homotopy equivalences by Theorem 6.4.14.

We will show that the squares in the diagram commute. Since each cross product is a chain homotopy equivalence, this implies that if we replace each ε with a map IAW going in the opposite direction, we obtain a diagram in which each square is homotopy commutative. Notice that the top and left of the diagram become algebraic diagonals $\bar{\mathbf{d}}$, while the right and bottom of the diagram take the form $\bar{\mathbf{d}} \otimes \mathrm{id}$ and $\mathrm{id} \otimes \bar{\mathbf{d}}$, assuming each symbol $\bar{\mathbf{d}}$ is interpreted with respect to the appropriate R-modules.

Since we know that all maps in the diagram are well-defined chain maps, it suffices to verify commutativity at the level of R-modules (as opposed to chain complexes). But we know that each R-module here is a submodule of the analogous ordinary singular chain R-modules, which are all free, generated by the singular simplices. So it suffices to verify commutativity on singular simplices.

The upper left square is induced by maps of spaces, and since $(\mathbf{d} \times \mathrm{id})\mathbf{d}(x) = (x, x, x) = (\mathrm{id} \times \mathbf{d})\mathbf{d}(x)$, this square commutes already at the space level. For the upper right square, consider a generator $\sigma \otimes \tau \in S_i(X; R) \otimes S_j(X; R)$. The map left then down yields the singular chain corresponding to applying $(\mathbf{d} \otimes \mathrm{id})(\sigma \times \tau) = (\mathbf{d}\sigma) \times \tau$ to the singular triangulation of $\Delta^i \times \Delta^j$ arising from the Eilenberg–Zilber shuffle procedure. On the other hand, the right-hand vertical map of the square takes $\sigma \otimes \tau$ to $(\mathbf{d}\sigma) \otimes \tau$, and then the bottom horizontal map takes this to $(\mathbf{d}\sigma) \times \tau$ applied to the singular triangulation of $\Delta^i \times \Delta^j$. So the two ways around the square agree. The argument for the lower left square is equivalent. This leaves the commutativity of the bottom right square, which is associativity of the chain cross product; see Proposition 5.2.19 and Theorem 6.3.19. □

We can now demonstrate the associativity of the cup and cap product under the assumptions that make Lemma 7.3.30 hold. Once again, we will state some nicer, but less general, versions of these results below as Propositions 7.3.34 and 7.3.35.

Lemma 7.3.31 (Associativity) *Let R be a Dedekind domain. Suppose that $\bar{p}, \bar{q}, \bar{r}, \bar{s}$ are perversities on a CS set X. Let $A, B, C \subset X$ be open subsets. Suppose that perversities $\bar{u}, \bar{v}, Q_1, Q_2, Q_3, Q_4, Q_5$ exist so that Lemma 7.3.30 holds. Let $\alpha \in I_{\bar{p}}H^i(X, A; R)$, $\beta \in I_{\bar{q}}H^j(X, B; R)$, and $\gamma \in I_{\bar{r}}H^k(X, C; R)$. Then $(\alpha \smile \beta) \smile \gamma$ and $\alpha \smile (\beta \smile \gamma)$ are well defined and equal elements of $I_{\bar{s}}H^{i+j+k}(X, A \cup B \cup C; R)$.*

Proof Representing α, β, and γ by cocycles and making specific choices of IAW maps, by definition we have

$$\begin{aligned}(\alpha \smile \beta) \smile \gamma = \bar{\mathbf{d}}^*\Theta((\alpha \smile \beta) \otimes \gamma) &= \bar{\mathbf{d}}^*\Theta((\bar{\mathbf{d}}^*\Theta(\alpha \otimes \beta)) \otimes \gamma) \\ &= \bar{\mathbf{d}}^*\Theta(\bar{\mathbf{d}}^*\Theta \otimes \mathrm{id})(\alpha \otimes \beta \otimes \gamma).\end{aligned}$$

Similarly,

$$\begin{aligned}\alpha \smile (\beta \smile \gamma) = \bar{\mathbf{d}}^*\Theta(\alpha \otimes (\beta \smile \gamma)) &= \bar{\mathbf{d}}^*\Theta(\alpha \otimes (\bar{\mathbf{d}}^*\Theta(\beta \otimes \gamma))) \\ &= \bar{\mathbf{d}}^*\Theta(\mathrm{id} \otimes \bar{\mathbf{d}}^*\Theta)(\alpha \otimes \beta \otimes \gamma).\end{aligned}$$

Notice here that, at the cochain level, these expressions depend on the choices of IAW maps but that they are well-defined expressions independent of these choices upon passing to cohomology, as all IAW maps are well defined up to chain homotopy. Now we compute at the level of cohomology:

$$\begin{aligned}(\alpha \smile \beta) \smile \gamma &= (\bar{\mathbf{d}}^*\Theta(\bar{\mathbf{d}}^*\Theta \otimes \mathrm{id}))(\alpha \otimes \beta \otimes \gamma) \\ &= \bar{\mathbf{d}}^*\Theta(\bar{\mathbf{d}}^* \otimes \mathrm{id})(\Theta \otimes \mathrm{id})(\alpha \otimes \beta \otimes \gamma) \\ &= \bar{\mathbf{d}}^*(\bar{\mathbf{d}} \otimes \mathrm{id})^*\Theta(\Theta \otimes \mathrm{id})(\alpha \otimes \beta \otimes \gamma) && \text{by Lemma 7.3.1} \\ &= \bar{\mathbf{d}}^*(\mathrm{id} \otimes \bar{\mathbf{d}})^*\Theta(\Theta \otimes \mathrm{id})(\alpha \otimes \beta \otimes \gamma) && \text{by Lemma 7.3.30} \\ &= \bar{\mathbf{d}}^*(\mathrm{id} \otimes \bar{\mathbf{d}})^*\Theta(\mathrm{id} \otimes \Theta)(\alpha \otimes \beta \otimes \gamma) && \text{by Lemma 7.3.28} \\ &= \bar{\mathbf{d}}^*\Theta(\mathrm{id} \otimes \bar{\mathbf{d}}^*)(\mathrm{id} \otimes \Theta)(\alpha \otimes \beta \otimes \gamma) && \text{by Lemma 7.3.1} \\ &= \bar{\mathbf{d}}^*\Theta(\mathrm{id} \otimes \bar{\mathbf{d}}^*\Theta)(\alpha \otimes \beta \otimes \gamma) \\ &= \alpha \smile (\beta \smile \gamma).\end{aligned}$$

□

Lemma 7.3.32 (Associativity) *Let R be a Dedekind domain. Suppose that $\bar{p}, \bar{q}, \bar{r}, \bar{s}$ are perversities on a CS set X. Let $A, B, C \subset X$ be open subsets. Suppose that perversities $\bar{u}, \bar{v}, Q_1, Q_2, Q_3, Q_4, Q_5$ exist so that Lemma 7.3.30 holds. Let $\alpha \in I_{\bar{q}}H^j(X, B; R)$, $\beta \in I_{\bar{r}}H^k(X, C; R)$, and $\xi \in I^{\bar{s}}H_{i+j+k}(X, A \cup B \cup C; R)$. Then $(\alpha \smile \beta) \frown \xi$ and $\alpha \frown (\beta \frown \xi)$ are well defined and equal elements of $I^{\bar{p}}H_i(X, A; R)$.*

Proof To demonstrate the equality, let us again assume that we have chosen fixed chain maps IAW and compute both expressions for given elements $\alpha \in I_{\bar{q}}S^j(X, B; R)$, $\beta \in I_{\bar{r}}S^k(X, C; R)$, and $\xi \in I^{\bar{s}}S_{i+j+k}(X, A \cup B \cup C; R)$:

$$\begin{aligned}(\alpha \smile \beta) \frown \xi &= \Phi(\mathrm{id} \otimes (\alpha \smile \beta))\bar{\mathbf{d}}(\xi) = \Phi(\mathrm{id} \otimes \bar{\mathbf{d}}^*\Theta(\alpha \otimes \beta))\bar{\mathbf{d}}(\xi) \\ &= \Phi(\mathrm{id} \otimes \Theta(\alpha \otimes \beta)\bar{\mathbf{d}})\bar{\mathbf{d}}(\xi) = \Phi(\mathrm{id} \otimes \Theta(\alpha \otimes \beta))(\mathrm{id} \otimes \bar{\mathbf{d}})\bar{\mathbf{d}}(\xi).\end{aligned}$$

The other computation is a bit more complicated. For it, we will want to assume we have fixed IAW maps, that $\bar{\mathbf{d}}(\xi) = \sum_\ell y_\ell \otimes z_\ell \in I^{\bar{u}}S_*(X, A \cup B; R) \otimes I^{\bar{r}}S_*(X, C; R)$, and that for each y_ℓ we have $\bar{\mathbf{d}}(y_\ell) = \sum_a u_{\ell a} \otimes v_{\ell a} \in I^{\bar{p}}S_*(X, A; R) \otimes I^{\bar{q}}S_*(X, B; R)$:

$$\begin{aligned}&\alpha \frown (\beta \frown \xi) \\ &= \Phi(\mathrm{id} \otimes \alpha)\bar{\mathbf{d}}(\beta \frown \xi) = \Phi(\mathrm{id} \otimes \alpha)\bar{\mathbf{d}}(\Phi(\mathrm{id} \otimes \beta))\bar{\mathbf{d}}(\xi) \\ &= \Phi(\mathrm{id} \otimes \alpha)\bar{\mathbf{d}}(\Phi(\mathrm{id} \otimes \beta))\Big(\sum_\ell y_\ell \otimes z_\ell\Big) = (-1)^{k|y_\ell|}\Phi(\mathrm{id} \otimes \alpha)\bar{\mathbf{d}}\sum_\ell \beta(z_\ell)y_\ell \\ &= (-1)^{k|y_\ell|}\Phi(\mathrm{id} \otimes \alpha)\sum_\ell \beta(z_\ell)\bar{\mathbf{d}}(y_\ell) = (-1)^{k|y_\ell|}\Phi(\mathrm{id} \otimes \alpha)\sum_\ell \beta(z_\ell)\sum_a u_{\ell a} \otimes v_{\ell a} \\ &= (-1)^{k|y_\ell|+j|u_{\ell a}|}\sum_\ell\sum_a \alpha(v_{\ell a})\beta(z_\ell)u_{\ell a} \\ &= (-1)^{k|y_\ell|+j|u_{\ell a}|+k|v_{\ell a}|}\Phi\Big(\sum_{\ell,a} u_{\ell a} \otimes \Theta(\alpha \otimes \beta)(v_{\ell a} \otimes z_\ell)\Big) \\ &= (-1)^{k|y_\ell|+j|u_{\ell a}|+k|v_{\ell a}|+(j+k)|u_{\ell a}|}\Phi(\mathrm{id} \otimes \Theta(\alpha \otimes \beta))\sum_{\ell,a} u_{\ell a} \otimes v_{\ell a} \otimes z_\ell \\ &= \Phi(\mathrm{id} \otimes \Theta(\alpha \otimes \beta))(\bar{\mathbf{d}} \otimes \mathrm{id})\bar{\mathbf{d}}(\xi).\end{aligned}$$

For the signs in the last line, we note that these expressions vanish unless $j = |\alpha| = |v_{\ell a}|$ and $k = |\beta| = |z_\ell|$, which leaves $|u_{\ell a}| = i$ in the nonvanishing terms. So,

$$\begin{aligned} k|y_\ell| + j|u_{\ell a}| + k|v_{\ell a}| + (j+k)|u_{\ell a}| &= k(i+j) + ij + jk + (j+k)i \\ &= ik + jk + ij + jk + ij + ik, \end{aligned}$$

which is even.

Suppose then that $\xi \in I^{\bar{s}}H_{i+j+k}(X, A \cup B \cup C; R)$. Since we know by Lemma 7.3.30 that $(\bar{\mathbf{d}} \otimes \mathrm{id})\bar{\mathbf{d}}$ and $(\mathrm{id} \otimes \bar{\mathbf{d}})\bar{\mathbf{d}}$ are chain homotopic chain maps, we have

$$\begin{aligned} &(\bar{\mathbf{d}} \otimes \mathrm{id})\bar{\mathbf{d}}(\xi) \\ &\quad = (\mathrm{id} \otimes \bar{\mathbf{d}})\bar{\mathbf{d}}(\xi) \in H_{i+j+k}(I^{\bar{p}}S_*(X, A; R) \otimes I^{\bar{q}}S_*(X, B; R) \otimes I^{\bar{r}}S_*(X, C; R)). \end{aligned}$$

Similarly, we know that if $\alpha \in I_{\bar{q}}H^j(X, B; R)$ and $\beta \in I_{\bar{r}}H^k(X, C; R)$, then $\Theta(\alpha \otimes \beta)$ is well defined in $H^*(\mathrm{Hom}(I^{\bar{p}}S_*(X, B; R) \otimes I^{\bar{q}}S_*(X, C; R), R))$. From here, the verification that

$$\Phi(\mathrm{id} \otimes \Theta(\alpha \otimes \beta))(\bar{\mathbf{d}} \otimes \mathrm{id})\bar{\mathbf{d}}(\xi) = \Phi(\mathrm{id} \otimes \Theta(\alpha \otimes \beta))(\mathrm{id} \otimes \bar{\mathbf{d}})\bar{\mathbf{d}}(\xi)$$

in homology follows exactly as in the proof of Lemma 7.2.20, where we showed that the cap product is independent of the choice of algebraic diagonal map up to chain homotopy. □

(b) Associativity in Some More Specific Settings

The following provides some reasonable conditions under which the hypotheses of Lemma 7.3.30 are fulfilled. As the careful reader will observe in the proof, there are various other such statements that could be made using different torsion-free hypotheses, but we will be content with this statement and its corollaries.

Lemma 7.3.33 *Suppose that $\bar{p}, \bar{q}, \bar{r}, \bar{s}$ are perversities on a CS set X. Suppose there are perversities $\bar{u}$ and $\bar{v}$ on X such that*

1. $D\bar{u} > D\bar{p} + D\bar{q}$,
2. $D\bar{v} > D\bar{q} + D\bar{r}$,
3. $D\bar{s} > D\bar{u} + D\bar{r}$, *and*
4. $D\bar{s} > D\bar{p} + D\bar{v}$.

Then there exist perversities Q_1, Q_2, Q_3, Q_4, Q_5 satisfying the requirements for Lemma 7.3.30. If X is locally $(\bar{p}, R)$-torsion-free and locally $(\bar{r}, R)$-torsion-free, then the conditions can all be replaced with non-strict inequalities, i.e. $D\bar{u} \geq D\bar{p} + D\bar{q}$, etc.

Proof By the computation in the proof of Lemma 7.2.8, the condition $D\bar{u} \geq D\bar{p} + D\bar{q}$ is equivalent to $\bar{u}(S) \leq \bar{p}(S) + \bar{q}(S) + 2 - \mathrm{codim}_X(S)$ for all singular strata $S \subset X$, and similarly $D\bar{u} > D\bar{p} + D\bar{q}$ is equivalent to $\bar{u}(S) \leq \bar{p}(S) + \bar{q}(S) + 1 - \mathrm{codim}_X(S)$ for all singular strata $S \subset X$. So if S and T are both singular strata, we take $Q_1(S \times T) = \bar{p}(S) + \bar{q}(T) + 2$ when we assume that X is locally $(\bar{p}, R)$-torsion-free and $Q_1(S \times T) = \bar{p}(S) + \bar{q}(T) + 1$ otherwise; then Q_1 will be $(\bar{p}, \bar{q})$-compatible, and $(\bar{p}, \bar{q}; \bar{u})$ will be a Q_1-agreeable triple by Definition

7.2.6. Similarly, if we define Q_2, Q_4, Q_5 analogously, we will have a Q_2-agreeable triple $(\bar{q}, \bar{r}; \bar{v})$, a Q_4-agreeable triple $(\bar{u}, \bar{r}; \bar{s})$, and a Q_5-agreeable triple $(\bar{p}, \bar{v}; \bar{s})$. So, assuming our hypotheses and with these choices, all arrows around the outside of Diagram (7.15) exist and satisfy the first four conditions following that diagram. Next we must define Q_3 and show that it satisfies the fifth and sixth conditions.

First, assume there are no torsion-free conditions. Let Q_3 be defined on strata $S \times T \times U$ so that if $S \times T \times U$ is not a regular stratum then

$$Q_3(S \times T \times U) = \bar{p}(S) + \bar{q}(T) + \bar{r}(U) + k - 1,$$

where k is the number of singular strata among S, T, U. Recalling that all perversities are assumed to evaluate to 0 on regular strata, it is then routine to verify that Q_3 is both $(Q_1, \bar{r})$- and $(\bar{p}, Q_2)$-compatible. For example, if S and T are singular and U is regular then

$$Q_3(S \times T \times U) = \bar{p}(S) + \bar{q}(T) + 1 = Q_1(S \times T) = \bar{p}(S) + Q_2(T \times U) + 1,$$

and if S, T, U are all singular then

$$\begin{aligned} Q_3(S \times T \times U) &= \bar{p}(S) + \bar{q}(T) + \bar{r}(U) + 2 = Q_1(S \times T) + \bar{r}(U) + 1 \\ &= \bar{p}(S) + Q_2(T \times U) + 1. \end{aligned}$$

The other necessary verifications are similar. If we assume X to be locally $(\bar{p}, R)$-torsion-free and locally $(\bar{r}, R)$-torsion-free, then we may take

$$Q_3(S \times T \times U) = \bar{p}(S) + \bar{q}(T) + \bar{r}(U) + 2(k - 1),$$

and again similar verifications hold.

Lastly, we must show that $\mathbf{d} \times \mathrm{id}$ and $\mathrm{id} \times \mathbf{d}$ are respectively (Q_4, Q_3)- and (Q_5, Q_3)-stratified. As the arguments are symmetric, we provide the first. It is evident that $\mathbf{d} \times \mathrm{id}$ takes points in $\Sigma_{X \times X}$ to points in $\Sigma_{X \times X \times X}$; in fact $(\mathbf{d} \times \mathrm{id})(S \times T) \subset S \times S \times T$. So we just need to verify the perversity condition. The computation is trivial if $S \times T \subset X \times X$ is a regular stratum. So we check the following three cases.

- S regular, T singular:

$$\begin{aligned} Q_4(S \times T) - \operatorname{codim}_{X \times X}(S \times T) &= \bar{r}(T) - \operatorname{codim}_X(T) \\ &= Q_3(S \times S \times T) - \operatorname{codim}_{X \times X \times X}(S \times S \times T). \end{aligned}$$

- S singular, T regular:

$$\begin{aligned} Q_4(S \times T) - \operatorname{codim}_{X \times X}(S \times T) &= \bar{u}(S) - \operatorname{codim}_X(S) \\ &\le \bar{p}(S) + \bar{q}(S) + C_{\bar{u}} - \operatorname{codim}_X(S) - \operatorname{codim}_X(S) \\ &= Q_3(S \times S \times T) - \operatorname{codim}_{X \times X \times X}(S \times S \times T), \end{aligned}$$

where $C_{\bar{u}} = 2$ if we assume that X is locally $(\bar{p}, R)$-torsion-free and $C_{\bar{u}} = 1$ otherwise.

- S singular, T singular:

$$\begin{aligned}
&Q_4(S \times T) - \operatorname{codim}_{X\times X}(S \times T)\\
&\quad = \bar{u}(S) + \bar{r}(T) + C_4 - \operatorname{codim}_{X\times X}(S \times T)\\
&\quad \le \bar{p}(S) + \bar{q}(S) + C_{\bar{u}} - \operatorname{codim}_X(S) + C_4 - \operatorname{codim}_{X\times X}(S \times T)\\
&\quad = Q_3(S \times S \times T) - \operatorname{codim}_{X\times X\times X}(S \times S \times T),
\end{aligned}$$

where $C_{\bar{u}} = C_4 = 2$ if we assume that X is locally $(\bar{p}, R)$-torsion-free and locally $(\bar{r}, R)$-torsion-free and $C_{\bar{u}} = C_4 = 1$ without these assumptions.

This demonstrates that $\mathbf{d} \times \mathrm{id}$ is (Q_4, Q_3)-stratified, as claimed. □

Using Lemmas 7.3.31, 7.3.32, and 7.3.33, and Corollary 7.2.9, we now have the following:

Proposition 7.3.34 (Associativity) *Let R be a Dedekind domain. Suppose that $\bar{p}, \bar{q}, \bar{r}, \bar{s}$ are perversities on a CS set X with open subsets A, B, C. Suppose there are perversities $\bar{u}$ and $\bar{v}$ on X such that*

1. $D\bar{u} > D\bar{p} + D\bar{q}$,
2. $D\bar{v} > D\bar{q} + D\bar{r}$,
3. $D\bar{s} > D\bar{u} + D\bar{r}$, *and*
4. $D\bar{s} > D\bar{p} + D\bar{v}$.

Let $\alpha \in I_{\bar{p}}H^i(X, A; R)$, $\beta \in I_{\bar{q}}H^j(X, B; R)$, $\gamma \in I_{\bar{r}}H^k(X, C; R)$, $\alpha \smile \beta \in I_{\bar{u}}H^{i+j}(X, A \cup B; R)$, and $\beta \smile \gamma \in I_{\bar{v}}H^{j+k}(X, B \cup C; R)$. Then $(\alpha \smile \beta) \smile \gamma = \alpha \smile (\beta \smile \gamma) \in I_{\bar{s}}H^{i+j+k}(X, A \cup B \cup C; R)$.

If X is locally $(\bar{p}, R)$-torsion-free and locally $(\bar{r}, R)$-torsion-free, then the conditions can all be replaced with non-strict inequalities, i.e. $D\bar{u} \ge D\bar{p} + D\bar{q}$, etc.

Proposition 7.3.35 (Associativity) *Let R be a Dedekind domain. Suppose that $\bar{p}, \bar{q}, \bar{r}, \bar{s}$ are perversities on a CS set X with open subsets A, B, C. Suppose there are perversities $\bar{u}$ and $\bar{v}$ on X such that*

1. $D\bar{u} > D\bar{p} + D\bar{q}$,
2. $D\bar{v} > D\bar{q} + D\bar{r}$,
3. $D\bar{s} > D\bar{u} + D\bar{r}$, *and*
4. $D\bar{s} > D\bar{p} + D\bar{v}$.

Let $\alpha \in I_{\bar{q}}H^j(X, B; R)$, $\beta \in I_{\bar{r}}H^k(X, C; R)$, $\xi \in I^{\bar{s}}H_{i+j+k}(X, A \cup B \cup C; R)$, $\alpha \smile \beta \in I_{\bar{v}}H^{j+k}(X, B \cup C; R)$, and $\beta \frown \xi \in I^{\bar{u}}H_{i+j}(X, A \cup B; R)$. Then $(\alpha \smile \beta) \frown \xi = \alpha \frown (\beta \frown \xi) \in I^{\bar{p}}H_i(X, A; R)$.

If X is locally $(\bar{p}, R)$-torsion-free and locally $(\bar{r}, R)$-torsion-free, then the conditions can all be replaced with non-strict inequalities, i.e. $D\bar{u} \ge D\bar{p} + D\bar{q}$, etc.

Remark 7.3.36 As promised above, we see that if we assume the torsion-free conditions in Propositions 7.3.34 and 7.3.35 and if we are willing to allow $\bar{u}$ and $\bar{v}$ so that $D\bar{u} = D\bar{p} + D\bar{q}$ and $D\bar{v} = D\bar{q} + D\bar{r}$, then the required relation among $\bar{p}$, $\bar{q}$, $\bar{r}$, and $\bar{s}$ for associativity relationships to hold becomes simply

$$D\bar{s} \geq D\bar{p} + D\bar{q} + D\bar{r}.$$

More generally, without assuming the torsion-free conditions, if we are willing to take $\bar{u}$ and $\bar{v}$ so that $D\bar{u} = D\bar{p} + D\bar{q} + 1$ and $D\bar{v} = D\bar{q} + D\bar{r} + 1$ (on singular strata), then we may have associativity whenever

$$D\bar{s} \geq D\bar{p} + D\bar{q} + D\bar{r} + 2.$$

7.3.5 Stability

We now turn to what Dold [71] refers to as "stability" properties of products. These are the properties that involve the connecting morphisms ∂_* and d^* in the long exact homology and cohomology sequences. Here we run into some additional difficulties because we do not have the fact from ordinary homology, which Dold achieves via acyclic model arguments, that the IAW maps are natural as chain maps with respect to maps of spaces; this naturality property plays a subtle role in the arguments of [71]. We do have such a fact for the chain cross product ε (see Proposition 5.2.17 and Theorem 6.3.19), but we cannot assume that this naturality continues to hold when we replace each ε with a chain homotopy inverse IAW. Therefore, our arguments will have to be more elaborate than those of Dold.

(a) Stability of Cap Products

We begin with the stability of cap products, which requires some big diagrams but fewer new techniques.

Proposition 7.3.37 *Let R be a Dedekind domain. Suppose X is a CS set and that $(\bar{p}, \bar{q}; \bar{r})$ is an agreeable triple of perversities on X. Let $A, B \subset X$ be open subsets with $\mathfrak{i} : B \hookrightarrow X$ the inclusion map. Suppose $\alpha \in I_{\bar{q}}H^j(B; R)$ and $\xi \in I^{\bar{r}}H_{i+j}(X, A \cup B; R)$. Then*

$$(d^*(\alpha)) \frown \xi = (-1)^{j+1}\mathfrak{i}(\alpha \frown e^{-1}\partial_*(\xi)) \in I^{\bar{p}}H_{i-1}(X, A; R),$$

where we have interpreted $\partial_*(\xi)$ as landing in $I^{\bar{r}}H_{i+j-1}(A \cup B, A; R)$ and where

$$e : I^{\bar{r}}H_{i+j-1}(B, A \cap B; R) \to I^{\bar{r}}H_{i+j-1}(A \cup B, A; R)$$

is the excision isomorphism.

In other words, the following diagram commutes[13]*:*

$$\begin{array}{ccccc}
\begin{array}{c} I_{\bar q}H^j(B;R)\otimes \\ I^{\bar r}H_{i+j}(X,A\cup B;R)\end{array} & \xrightarrow{d^*\otimes \mathrm{id}} & \begin{array}{c} I_{\bar q}H^{j+1}(X,B;R)\otimes \\ I^{\bar r}H_{i+j}(X,A\cup B;R)\end{array} & \xrightarrow{\frown} & I^{\bar p}H_{i-1}(X,A;R) \\
\downarrow{\scriptstyle -(\mathrm{id}\otimes\partial_*)} & & & & \uparrow{\scriptstyle \mathrm{i}} \\
\begin{array}{c} I_{\bar q}H^j(B;R)\otimes \\ I^{\bar r}H_{i+j-1}(A\cup B,A;R)\end{array} & \xleftarrow[\cong]{\mathrm{id}\otimes e} & \begin{array}{c} I_{\bar q}H^j(B;R)\otimes \\ I^{\bar r}H_{i+j-1}(B,A\cap B)\end{array} & \xrightarrow{\frown} & I^{\bar p}H_{i-1}(B,A\cap B;R).
\end{array}$$

Proof The proof will eventually use the following diagram, with R coefficients tacit:

$$\begin{array}{ccccc}
\begin{array}{c} H_{i+j}(I^{\bar p}S_*(X,A)\otimes \\ I^{\bar q}S_*(X,B))\end{array} & \xrightarrow{\varepsilon} & \begin{array}{c} I^{Q_{\bar p,\bar q}}H_{i+j}(X\times X, \\ (A\times X)\cup(X\times B))\end{array} & \xleftarrow{\mathbf{d}} & I^{\bar r}H_{i+j}(X,A\cup B) \\
 & & \downarrow{\scriptstyle \partial_*} & & \downarrow{\scriptstyle \partial_*} \\
 & & \begin{array}{c} I^{Q_{\bar p,\bar q}}H_{i+j-1}((A\times X) \\ \cup(X\times B))\end{array} & \xleftarrow{\mathbf{d}} & I^{\bar r}H_{i+j-1}(A\cup B) \\
\downarrow{\scriptstyle \partial_*} & & \downarrow & & \downarrow \\
 & & \begin{array}{c} I^{Q_{\bar p,\bar q}}H_{i+j-1}((A\times X) \\ \cup(X\times B),A\times X)\end{array} & \xleftarrow{\mathbf{d}} & I^{\bar r}H_{i+j-1}(A\cup B,A) \\
 & & \uparrow{\scriptstyle \cong\, e'} & & \uparrow{\scriptstyle \cong\, e} \\
\begin{array}{c} H_{i+j-1}(I^{\bar p}S_*(X,A)\otimes \\ I^{\bar q}S_*(B))\end{array} & \xrightarrow{\varepsilon} & \begin{array}{c} I^{Q_{\bar p,\bar q}}H_{i-1+j}(X\times B, \\ A\times B)\end{array} & \xleftarrow{\mathbf{d}} & I^{\bar r}H_{i+j-1}(B,A\cap B) \\
\uparrow{\scriptstyle \mathrm{i}\otimes\mathrm{id}} & & \uparrow{\scriptstyle \mathrm{i}\times\mathrm{id}} & \swarrow{\scriptstyle \mathbf{d}} & \\
\begin{array}{c} H_{i+j-1}(I^{\bar p}S_*(B,A\cap B)\otimes \\ I^{\bar q}S_*(B))\end{array} & \xrightarrow{\varepsilon} & \begin{array}{c} I^{Q_{\bar p,\bar q}}H_{i+j-1}(B\times B, \\ (A\cap B)\times B).\end{array} & &
\end{array} \tag{7.16}$$

(In the printed diagram, the diagonal map $\mathbf{d}$ goes from $I^{\bar r}H_{i+j-1}(B,A\cap B)$ to $I^{Q_{\bar p,\bar q}}H_{i+j-1}(B\times B,(A\cap B)\times B)$.)

Here, each of the diagonal maps is also composed with the evident inclusion. The maps i, e, and e' are induced by inclusion, and unlabeled maps are also induced by the evident inclusions or projections to quotient complexes. The map labeled ∂_* on the left is meant to be the boundary map in the long exact sequence associated to the short exact sequence

$$0\to \begin{array}{c} I^{\bar p}S_*(X,A;R)\otimes \\ I^{\bar q}S_*(B;R)\end{array} \to \begin{array}{c} I^{\bar p}S_*(X,A;R)\otimes \\ I^{\bar q}S_*(X;R)\end{array} \to \begin{array}{c} I^{\bar p}S_*(X,A;R)\otimes \\ I^{\bar q}S_*(X,B;R)\end{array} \to 0. \tag{7.17}$$

This is obtained by tensoring the exact sequence of $\bar q$ intersection chains of the pair (X,B)

[13] Recall that we treat ∂_* as a degree -1 map for sign purposes.

with the complex $I^{\bar{p}}S_*(X,A;R)$. The short exact sequence remains exact after tensoring, as $I^{\bar{p}}S_*(X,A;R)$ is projective by Lemma 6.3.1 and so flat.

We verify the commutativity of the diagram, starting with the upper left rectangle. Each element of $I^{\bar{p}}S_*(X,A;R) \otimes I^{\bar{q}}S_*(X,B;R)$ can be represented by a chain in $I^{\bar{p}}S_*(X;R) \otimes I^{\bar{q}}S_*(X;R)$, and if x is such a chain that is a cycle in $I^{\bar{p}}S_*(X,A;R) \otimes I^{\bar{q}}S_*(X,B;R)$, and so represents a homology class ξ, then by the standard zig-zag construction $\partial_*\xi \in H_*(I^{\bar{p}}S_*(X,A;R)\otimes I^{\bar{q}}S_*(B;R))$ is represented by ∂x. Therefore, $\varepsilon\partial_*(\xi) \in I^{Q_{\bar{p},\bar{q}}}H_{i-1+j}(X\times B, A\times B;R)$ is represented by $\varepsilon(\partial x)$, which also represents the image of this class in $I^{Q_{\bar{p},\bar{q}}}H_{i+j-1}((A\times X)\cup(X\times B), A\times X;R)$. But as ε is a chain map, $\varepsilon(\partial x) = \partial\varepsilon(x)$, which represents $\partial_*\varepsilon(\xi)$ in $I^{Q_{\bar{p},\bar{q}}}H_{i+j-1}((A\times X)\cup(X\times B);R)$ and so also its image in $I^{Q_{\bar{p},\bar{q}}}H_{i+j-1}((A\times X)\cup(X\times B), A\times X;R)$. So the upper left rectangle commutes.

The square on the bottom left commutes by the non-GM version of Proposition 5.2.17; see also Theorem 6.3.19. The triangle and the two bottom squares on the right commute at the space level because diagonal maps are natural, i.e. if $f : Z \to W$ is any map of spaces and $\mathbf{d}_Z$ and $\mathbf{d}_W$ are the respective diagonal maps then $\mathbf{d}_W f = (f\times f)\mathbf{d}_Z$. For the upper square on the right, let x be a chain in $I^{\bar{r}}S_{i+j}(X;R)$ representing an element $\xi \in I^{\bar{r}}H_{i+j}(X, A\cup B;R)$ and recall that ∂_* can be represented by taking the boundary of x. So $\mathbf{d}_{A\cup B}\partial_*(\xi)$ is represented by including $\mathbf{d}_{A\cup B}(\partial x)$ into $(A\times X)\cup(X\times B)\subset X\times X$. On the other hand, $\partial_*\mathbf{d}(\xi)$ is represented by $\partial\mathbf{d}_X(x) = \mathbf{d}_X\partial x$, as $\mathbf{d}$ is a chain map. But again using the naturality of $\mathbf{d}$, these are the same chain. Therefore, both images of ξ are represented by the same chain.

Let us also verify that the two maps labeled e and e', which are meant to indicate excision, really are isomorphisms. Since we have assumed A and B to be open subsets of X, the pair $\{A,B\}$ is an open cover of $A\cup B$. Consider now the maps that factor e:

$$\frac{I^{\bar{r}}S_*(B;R)}{I^{\bar{r}}S_*(A\cap B;R)} \to \frac{I^{\bar{r}}S_*(A;R)+I^{\bar{r}}S_*(B;R)}{I^{\bar{r}}S_*(A;R)} \to \frac{I^{\bar{r}}S_*(A\cup B;R)}{I^{\bar{r}}S_*(A;R)}.$$

The first map is an isomorphism by the Second Isomorphism Theorem, using $I^{\bar{r}}S_*(A;R)\cap I^{\bar{r}}S_*(B;R) = I^{\bar{r}}S_*(A\cap B;R)$, while the second map induces a homology isomorphism by the non-GM version of the arguments in the proof of Theorem 4.4.23. The map e' is similarly a homology isomorphism by the same arguments, replacing A with $A\times X$ and B with $X\times B$.

Next we recall how d^* works. Suppose α represents an element of $I_{\bar{q}}H^j(B;R)$. Then the zig-zag construction of d^* shows that $d^*\alpha \in I_{\bar{q}}H^{j+1}(X,B;R)$ is represented by $d\bar{\alpha}$, where $\bar{\alpha} \in I_{\bar{q}}S^j(X;R)$ restricts to α over B.

Now, suppose that $\xi \in I^{\bar{r}}H_{i+j}(X, A\cup B;R)$. Then $\bar{\mathbf{d}}(\xi)$ is obtained by going left across the top row of the diagram, using that ε is an isomorphism by the Künneth Theorem. Suppose we choose chain maps IAW and that $\bar{\mathbf{d}}(\xi)$ is then represented by $\sum_k y_k \otimes z_k \in I^{\bar{p}}S_*(X;R)\otimes I^{\bar{q}}S_*(X;R)$, noting that every element of $I^{\bar{p}}S_*(X,A;R)\otimes I^{\bar{q}}S_*(X,B;R)$ has such representatives. By definition, we then have

$$\begin{aligned}(d^*\alpha) \frown \xi &= \Phi(\mathrm{id} \otimes d^*\alpha)(\bar{\mathbf{d}}\xi) = \Phi(\mathrm{id} \otimes d\bar{\alpha})\Big(\sum\nolimits_k y_k \otimes z_k\Big)\\ &= \sum\nolimits_k (-1)^{(j+1)|y_k|} y_k d\bar{\alpha}(z_k) = \sum\nolimits_k (-1)^{(j+1)|y_k|+j+1} y_k \bar{\alpha}(\partial z_k)\\ &= \Phi(\mathrm{id} \otimes \bar{\alpha})\Big(\sum\nolimits_k (-1)^{(j+1)|y_k|+j+1+j|y_k|} y_k \otimes \partial z_k\Big)\\ &= \Phi(\mathrm{id} \otimes \bar{\alpha})\Big(\sum\nolimits_k (-1)^{|y_k|+j+1} y_k \otimes \partial z_k\Big)\\ &= (-1)^{j+1}\Phi(\mathrm{id} \otimes \bar{\alpha})\Big(\sum\nolimits_k (-1)^{|y_k|} y_k \otimes \partial z_k\Big),\end{aligned}$$

which we know must be a cycle in $I^{\bar{p}}S_{i-1}(X, A; R)$.

Notice that the expression

$$\sum_k (-1)^{|y_k|} y_k \otimes \partial z_k \in I^{\bar{p}}S_*(X, A; R) \otimes I^{\bar{q}}S_*(X, B; R)$$

is a piece of

$$\partial\Big(\sum\nolimits_k y_k \otimes z_k\Big) = \sum((\partial y_k) \otimes z_k + (-1)^{|y_k|} y_k \otimes \partial z_k).$$

So

$$\begin{aligned}(d^*\alpha) \frown \xi &= (-1)^{j+1}\Phi(\mathrm{id} \otimes \bar{\alpha})\left(\sum_k (-1)^{|y_k|} y_k \otimes \partial z_k\right)\\ &= (-1)^{j+1}\Phi(\mathrm{id} \otimes \bar{\alpha})\left(\partial\Big(\sum_k y_k \otimes z_k\Big) - \sum(\partial y_k) \otimes z_k\right)\\ &= (-1)^{j+1}\Phi(\mathrm{id} \otimes \bar{\alpha})\partial\left(\sum_k y_k \otimes z_k\right) - \sum_k (-1)^{j+1+j|\partial y_k|}\bar{\alpha}(z_k)\partial y_k.\end{aligned}$$

The terms of the second summand on the right are all boundaries in X, and so represent 0 in $I^{\bar{p}}H_{i-1}(X, A; R)$. Therefore, the element $(d^*\alpha) \frown \xi \in I^{\bar{p}}H_{i-1}(X, A; R)$ is also represented by $(-1)^{j+1}\Phi(\mathrm{id} \otimes \bar{\alpha})\partial(\sum_k y_k \otimes z_k)$.

Now, recall that $\sum_k y_k \otimes z_k \in I^{\bar{p}}S_*(X; R) \otimes I^{\bar{q}}S_*(X; R)$ was chosen so that its image under

$$I^{\bar{p}}S_*(X; R) \otimes I^{\bar{q}}S_*(X; R) \to I^{\bar{p}}S_*(X, A; R) \otimes I^{\bar{q}}S_*(X, B; R)$$

is a cycle representing $\bar{\mathbf{d}}(\xi)$. Of course, this map factors through $I^{\bar{p}}S_*(X, A; R) \otimes I^{\bar{q}}S_*(X; R)$, and so $\sum_k y_k \otimes z_k$ also represents a chain there. Moreover, since the image in $I^{\bar{p}}S_*(X, A; R) \otimes I^{\bar{q}}S_*(X, B; R)$ is a cycle, it follows from the zig-zag construction of the connecting morphism ∂_* of the homology exact sequence associated to (7.17) that the image of $\sum_k y_k \otimes z_k$ under the boundary map

$$\partial : (I^{\bar{p}}S_*(X, A; R) \otimes I^{\bar{q}}S_*(X; R))_{i+j} \to (I^{\bar{p}}S_*(X, A; R) \otimes I^{\bar{q}}S_*(X; R))_{i+j-1}$$

must actually be contained in the submodule $(I^{\bar{p}}S_*(X, A; R) \otimes I^{\bar{q}}S_*(B; R))_{i+j-1}$. So $\partial(\sum_k y_k \otimes$

z_k) represents the image of ξ after traveling all the way left and then down one step in diagram (7.16). We also observe that $\partial(\sum_k y_k \otimes z_k)$ being contained in $I^{\bar p}S_*(X,A;R) \otimes I^{\bar q}S_*(B;R)$ means that

$$\Phi(\mathrm{id} \otimes \bar\alpha)\partial\Big(\sum\nolimits_k y_k \otimes z_k\Big) = \Phi(\mathrm{id} \otimes \alpha)\partial\Big(\sum\nolimits_k y_k \otimes z_k\Big),$$

as $\bar\alpha$ restricts to α for chains in B, by definition.

At this point, we can fully employ Diagram (7.16) by using the commutativity to observe that $\partial(\sum_k y_k \otimes z_k) \in H_{i+j-1}(I^{\bar p}S_*(X,A;R) \otimes I^{\bar q}S_*(B;R))$ equals $(\mathrm{i} \otimes \mathrm{id})\bar{\mathbf{d}}e^{-1}\partial_*(\xi)$. By the argument in the proof of Lemma 7.2.20 showing that the cap product is well defined, we thus have that

$$\Phi(\mathrm{id} \otimes \alpha)\partial\Big(\sum\nolimits_k y_k \otimes z_k\Big) = \Phi(\mathrm{id} \otimes \alpha)(\mathrm{i} \otimes \mathrm{id})\bar{\mathbf{d}}e^{-1}\partial_*(\xi).$$

So

$$\begin{aligned}(d^*\alpha) \frown \xi &= (-1)^{j+1}\Phi(\mathrm{id} \otimes \alpha)(\mathrm{i} \otimes \mathrm{id})\bar{\mathbf{d}}e^{-1}\partial_*(\xi)\\ &= (-1)^{j+1}\Phi(\mathrm{i} \otimes \alpha)\bar{\mathbf{d}}e^{-1}\partial_*(\xi) = (-1)^{j+1}\mathrm{i}\Phi(\mathrm{id} \otimes \alpha)\bar{\mathbf{d}}e^{-1}\partial_*(\xi)\\ &= (-1)^{j+1}\mathrm{i}(\alpha \frown e^{-1}\partial_*(\xi)),\end{aligned}$$

which is what we needed to show.

For the third equality in the preceding sequence, we note that in general for any $u \otimes v \in C_* \otimes D_*$, $f : C_* \to C'_*$, and $\gamma \in \mathrm{Hom}(D_*, R)$ we have

$$\begin{aligned}\Phi(f \otimes \gamma)(u \otimes v) &= (-1)^{|\gamma||u|}\Phi(f(u) \otimes \gamma(v))\\ &= (-1)^{|\gamma||u|}\gamma(v)f(u) = f((-1)^{|\gamma||u|}\gamma(v)u)\\ &= f(\Phi(\mathrm{id} \otimes \gamma)(u \otimes v)).\end{aligned}$$

□

Proposition 7.3.38 *Let R be a Dedekind domain. Suppose X is a CS set and that $(\bar p, \bar q; \bar r)$ is an agreeable triple of perversities on X. Let $A, B \subset X$ be open subsets with $\mathrm{i} : A \hookrightarrow X$ the inclusion map. Suppose $\alpha \in I_{\bar q}H^j(X,B;R)$ and $\xi \in I^{\bar r}H_{i+j}(X, A \cup B; R)$. Then*

$$\partial_*(\alpha \frown \xi) = (-1)^j(\mathrm{i}^*(\alpha)) \frown (e^{-1}\partial_*(\xi)) \in I^{\bar p}H_{i-1}(A;R),$$

where we interpret $\partial_*(\xi)$ as landing in $I^{\bar r}H_{i+j-1}(A \cup B, B; R)$ and

$$e : I^{\bar r}H_{i+j-1}(A, A \cap B; R) \to I^{\bar r}H_{i+j-1}(A \cup B, B; R)$$

is the excision isomorphism.

In other words, the following diagram commutes[14]*:*

$$
\begin{array}{ccccc}
\begin{matrix} I_{\bar q}H^j(X,B;R)\otimes \\ I^{\bar r}H_{i+j}(X,A\cup B;R)\end{matrix} & & \xrightarrow{\quad\frown\quad} & & I^{\bar p}H_i(X,A;R) \\
\Big\downarrow{\scriptstyle i^*\otimes\partial_*} & & & & \Big\downarrow{\scriptstyle \partial_*} \\
\begin{matrix} I_{\bar q}H^j(A,A\cap B;R)\otimes \\ I^{\bar r}H_{i+j-1}(A\cup B,B;R)\end{matrix} & \xleftarrow[\cong]{\mathrm{id}\otimes e} & \begin{matrix} I_{\bar q}H^j(A,A\cap B;R)\otimes \\ I^{\bar r}H_{i+j-1}(A,A\cap B)\end{matrix} & \xrightarrow{\frown} & I^{\bar p}H_{i-1}(A;R).
\end{array}
$$

Proof Consider the following diagram, with R coefficients tacit:

$$
\begin{array}{ccccc}
\begin{matrix} H_{i+j}(I^{\bar p}S_*(X,A)\otimes \\ I^{\bar q}S_*(X,B))\end{matrix} & \xrightarrow{\varepsilon} & \begin{matrix} I^{Q_{\bar p,\bar q}}H_{i+j}(X\times X, \\ (A\times X)\cup(X\times B))\end{matrix} & \xleftarrow{\mathbf{d}} & I^{\bar r}H_{i+j}(X,A\cup B) \\
\Big\downarrow & & \Big\downarrow{\scriptstyle\partial_*} & & \Big\downarrow{\scriptstyle\partial_*} \\
 & & \begin{matrix} I^{Q_{\bar p,\bar q}}H_{i+j-1}((A\times X)\cup \\ (X\times B))\end{matrix} & \xleftarrow{\mathbf{d}} & I^{\bar r}H_{i+j-1}(A\cup B) \\
{\scriptstyle\partial_*}\Big\downarrow & & \Big\downarrow & & \Big\downarrow \\
 & & \begin{matrix} I^{Q_{\bar p,\bar q}}H_{i+j-1}((A\times X)\cup \\ (X\times B),X\times B)\end{matrix} & \xleftarrow{\mathbf{d}} & I^{\bar r}H_{i+j-1}(A\cup B,B) \\
\Big\downarrow & & \Big\uparrow{\scriptstyle\cong\, e'} & & \Big\uparrow{\scriptstyle\cong\, e} \\
\begin{matrix} H_{i+j-1}(I^{\bar p}S_*(A)\otimes \\ I^{\bar q}S_*(X,B))\end{matrix} & \xrightarrow{\varepsilon} & I^{Q_{\bar p,\bar q}}H_{i-1+j}(A\times X,A\times B) & \xleftarrow{\mathbf{d}} & I^{\bar r}H_{i+j-1}(A,A\cap B) \\
\Big\uparrow{\scriptstyle\mathrm{id}\otimes i} & & \Big\uparrow{\scriptstyle\mathrm{id}\times i} & \swarrow{\scriptstyle\mathbf{d}} & \\
\begin{matrix} H_{i+j-1}(I^{\bar p}S_*(A)\otimes \\ I^{\bar q}S_*(A,A\cap B))\end{matrix} & \xrightarrow{\varepsilon} & \begin{matrix} I^{Q_{\bar p,\bar q}}H_{i+j-1}(A\times A, \\ A\times(A\cap B)).\end{matrix} & &
\end{array}
$$

This is the same as the diagram in Proposition 7.3.37 except that the roles of (X,A) and (X,B) have been reversed. In particular, the ∂_* on the left is now the boundary map in the long exact sequence associated to the short exact sequence

$$
0 \longrightarrow \begin{matrix} I^{\bar p}S_*(A;R)\otimes \\ I^{\bar q}S_*(X,B;R)\end{matrix} \longrightarrow \begin{matrix} I^{\bar p}S_*(X;R)\otimes \\ I^{\bar q}S_*(X,B;R)\end{matrix} \longrightarrow \begin{matrix} I^{\bar p}S_*(X,A;R)\otimes \\ I^{\bar q}S_*(X,B;R)\end{matrix} \longrightarrow 0.
$$

The arguments for commutativity, however, remain the same.

Suppose that $\xi \in I^{\bar r}H_{i+j}(X,A\cup B;R)$ and that we have chosen fixed IAW maps. Then $\bar{\mathbf{d}}(\xi)$ is obtained by going left across the top row of the diagram, using that ε is an isomorphism

[14] Recall that we treat ∂_* as a degree-(-1) map for sign purposes.

by the Künneth Theorem. Suppose $\bar{\mathbf{d}}(\xi)$ is represented by $\sum_k y_k \otimes z_k \in I^{\bar{p}}S_*(X;R) \otimes I^{\bar{q}}S_*(X;R)$. By definition, $\alpha \frown \xi$ is represented by

$$\alpha \frown \xi = \Phi(\mathrm{id} \otimes \alpha)\bar{\mathbf{d}}(\xi) = \Phi(\mathrm{id} \otimes \alpha)\left(\sum_k y_k \otimes z_k\right) = \sum_k (-1)^{j|y_k|}\alpha(z_k)y_k,$$

which is an element of $I^{\bar{p}}S_i(X;R)$ representing a cycle in $I^{\bar{p}}S_i(X,A;R)$. Therefore, by the zig-zag construction, $\partial_*(\alpha \frown \xi)$ is represented by

$$\begin{aligned}\partial_*(\alpha \frown \xi) &= \partial\left(\sum_k (-1)^{j|y_k|}\alpha(z_k)y_k\right) = \sum_k (-1)^{j|y_k|}\alpha(z_k)\partial y_k\\ &= \Phi(\mathrm{id} \otimes \alpha)\sum (-1)^{j|y_k|+j(|y_k|-1)}(\partial y_k) \otimes z_k\\ &= (-1)^j\Phi(\mathrm{id} \otimes \alpha)\sum (\partial y_k) \otimes z_k,\end{aligned}$$

which must be a cycle in $I^{\bar{p}}S_{i-1}(A;R) \subset I^{\bar{p}}S_{i-1}(X;R)$.

As in the proof of Proposition 7.3.37, we recognize that $\sum(\partial y_k) \otimes z_k$ is a piece of $\partial(\sum_k y_k \otimes z_k) = \sum((\partial y_k) \otimes z_k + (-1)^{|y_k|}y_k \otimes (\partial z_k))$. But since $\alpha \in I_{\bar{q}}S^j(X,B;R)$ is a cocycle, for any term of the form $y_k \otimes (\partial z_k)$ we have

$$\Phi(\mathrm{id} \otimes \alpha)(y_k \otimes (\partial z_k)) = \pm y_k \otimes (\alpha(\partial z_k)) = \pm y_k \otimes ((d\alpha)(z_k)) = 0.$$

Thus we have

$$\begin{aligned}\partial_*(\alpha \frown \xi) &= (-1)^j\Phi(\mathrm{id} \otimes \alpha)\sum (\partial y_k) \otimes z_k\\ &= (-1)^j\Phi(\mathrm{id} \otimes \alpha)\left(\sum ((\partial y_k) \otimes z_k + (-1)^{|y_k|}y_k \otimes (\partial z_k))\right)\\ &= (-1)^j\Phi(\mathrm{id} \otimes \alpha)\partial\left(\sum_k y_k \otimes z_k\right).\end{aligned}$$

Analogously to the proof of Proposition 7.3.37, we can use the diagram to interpret $\partial\,(\sum_k y_k \otimes z_k)$ as a cycle in $I^{\bar{p}}S_*(A;R) \otimes I^{\bar{q}}S_*(X,B;R)$ representing the homology class $(\mathrm{id} \otimes \mathfrak{i})\bar{\mathbf{d}}e^{-1}\partial_*(\xi)$. So then

$$\begin{aligned}\partial_*(\alpha \frown \xi) &= (-1)^j\Phi(\mathrm{id} \otimes \alpha)\partial\left(\sum_k y_k \otimes z_k\right)\\ &= (-1)^j\Phi(\mathrm{id} \otimes \alpha)(\mathrm{id} \otimes \mathfrak{i})\bar{\mathbf{d}}e^{-1}\partial_*(\xi)\\ &= (-1)^j\Phi(\mathrm{id} \otimes \alpha\mathfrak{i})\bar{\mathbf{d}}e^{-1}\partial_*(\xi)\\ &= (-1)^j\Phi(\mathrm{id} \otimes \mathfrak{i}^*(\alpha))\bar{\mathbf{d}}e^{-1}\partial_*(\xi)\\ &= (-1)^j(\mathfrak{i}^*(\alpha)) \frown (e^{-1}\partial_*(\xi)).\end{aligned}$$

□

(b) Algebra of Shifts and Mapping Cones

The stability formulas for the cup and cross products are a little trickier. Part of the problem is that the maps ∂_* and d^*, in their usual constructions, aren't described via chain maps, which makes it difficult to apply our previous results as tools. To get around this difficulty, we utilize the algebraic mapping cone construction, which lets us replace quotient complexes by algebraic mapping cones (up to chain homotopy equivalence) and the connecting morphisms in long exact homology sequences by morphisms induced by chain maps. Having done this, we will be able to establish our desired stability formulas for cup and cross products, but first we must review the necessary algebra.

A brief review of shifts and mapping cones is provided in Appendix A.3. We recall the basic definitions here as well and develop a few more details.

Shifts. Recall Appendix A.3.1. If C_* is a homologically indexed chain complex with boundary map ∂_{C_*}, then the chain complex $C[k]_*$ is defined so that $C[k]_i = C_{i-k}$ and $\partial_{C[k]_*} = (-1)^k \partial_{C_*}$. In this section we only need to consider $k = 1$, and so we make that specialization in the rest of this section, i.e. we consider only $C[1]_*$ with $C[1]_i = C_{i-1}$ and $\partial_{C[1]_*} = -\partial_{C_*}$. The shift map $\mathfrak{s} : C[1]_* \to C_*$ is defined to take $C[1]_i$ identically to the corresponding module C_{i-1}. This is a (homological) degree -1 chain map. If x is an element of C_{i-1}, we also write $\bar{x}$ for $\mathfrak{s}^{-1}(x)$ and so $\mathfrak{s}(\bar{x}) = x$.

The assignment $C_* \to C[1]_*$ is functorial: Suppose $f : C_* \to D_*$ is a degree-0 chain map of chain complexes. Then we can define $f[1] : C[1]_* \to D[1]_*$ so that if $\bar{x} \in C[1]_*$ then $f[1](\bar{x}) = \overline{f(x)}$. Alternatively, $f[1]$ is defined so that the following diagram commutes, which can be done as $\mathfrak{s}$ is an isomorphism:

$$\begin{array}{ccc} C_* & \xrightarrow{\ f\ } & D_* \\ \uparrow{\scriptstyle \mathfrak{s}} & & \uparrow{\scriptstyle \mathfrak{s}} \\ C[1]_* & \xrightarrow{\ f[1]\ } & D[1]_*. \end{array}$$

It also follows that $f[1]$ is a degree-0 chain map. It is then clear that $(\cdot)[1]$ is a functor.

We will also need later the observation that if C_* and D_* are two chain complexes then there is a canonical (degree-0) isomorphism $\mathfrak{t} : C[1]_* \otimes D_* \cong (C_* \otimes D_*)[1]_*$. Indeed, we can realize $\mathfrak{t}$ as the composition

$$C[1]_* \otimes D_* \xrightarrow{\mathfrak{s} \otimes \mathrm{id}} C_* \otimes D_* \xrightarrow{\mathfrak{s}^{-1}} (C_* \otimes D_*)[1]_*.$$

This takes a generator $\bar{x} \otimes y$ to $\overline{x \otimes y}$.

Algebraic Mapping Cones. Recall Appendix A.3.2. Suppose $f : C_* \to D_*$ is a chain map of chain complexes. We let E_*^f (or simply E_* if there's no ambiguity) denote the algebraic mapping cone of f with $E_i = D_i \oplus C_{i-1} = D_i \oplus C[1]_i$ and $\partial(x, y) = (f(y) + \partial x, -\partial y)$. As verified in Appendix A.3.2, this is indeed a chain complex.

There is a short exact sequence of chain complexes

$$0 \longrightarrow D_* \xrightarrow{\ \mathfrak{e}\ } E_* \xrightarrow{\ \mathfrak{b}\ } C[1]_* \longrightarrow 0 \tag{7.18}$$

with $\mathfrak{e}(x) = (x, 0)$ and $\mathfrak{b}(x, y) = \bar{y}$, where $\bar{y}$ uses our notation for shifted elements from just above. The maps $\mathfrak{e}$ and $\mathfrak{b}$ are both degree-0 chain maps, recalling that the boundary map for $C[1]_*$ is the negative of that for C_*. It is not true in general that $E_* = D_* \oplus C[1]_*$ as chain complexes, since the boundary map of E_* is not a direct sum of the boundary maps of the summands.

The mapping cone construction is also functorial in the following sense: Suppose we have a commutative diagram of chain maps:

$$\begin{array}{ccc} C_* & \xrightarrow{f} & D_* \\ \downarrow g & & \downarrow h \\ C'_* & \xrightarrow{f'} & D'_*. \end{array}$$

Then there is an induced map $k : E_*^f \to E_*^{f'}$ with $k(x, y) = (h(x), g(y))$. We check that this is a degree-0 chain map:

$$\begin{aligned} k\partial(x, y) &= k(f(y) + \partial x, -\partial y) = (hf(y) + h(\partial x), -g(\partial y)) \\ &= (f'g(y) + \partial h(x), -\partial g(y)) = \partial(h(x), g(y)) \\ &= \partial k(x, y). \end{aligned}$$

We also obtain the commutative diagram

$$\begin{array}{ccccccccc} 0 & \longrightarrow & D_* & \xrightarrow{\mathfrak{e}} & E_* & \xrightarrow{\mathfrak{b}} & C[1]_* & \longrightarrow & 0 \\ & & \downarrow h & & \downarrow k & & \downarrow g[1] & & \\ 0 & \longrightarrow & D'_* & \xrightarrow{\mathfrak{e}} & E'_* & \xrightarrow{\mathfrak{b}} & C'[1]_* & \longrightarrow & 0. \end{array}$$

We leave the easy verification of commutativity to the reader.

Now, suppose that $\mathfrak{i} : C_* \to D_*$ is an inclusion and consider $E_*^{\mathfrak{i}}$. In this case, we will simply write $\partial(x, y) = (mfi(y) + \partial x, -\partial y) = (y + \partial x, -\partial y)$, as elements of C_* can also be considered elements of D_*. In this setting, there are useful interactions between the long exact homology and cohomology sequences associated with (7.18) and the usual long exact homology sequence of the pair (D_*, C_*).

Lemma 7.3.39 *Suppose that $\mathfrak{i} : C_* \to D_*$ is an inclusion of chain complexes with algebraic mapping cone E_*. There is a diagram of long exact homology sequences*

$$
\begin{array}{ccccccccc}
\longrightarrow H_{i+1}(C[1]_*) & \xrightarrow{\partial_*} & H_i(D_*) & \xrightarrow{\mathfrak{e}} & H_i(E_*) & \xrightarrow{\mathfrak{b}} & H_i(C[1]_*) & \longrightarrow \\
\Big\downarrow \mathfrak{s} & 1 & \Big\| & 1 & \Big\downarrow \mathfrak{q} & -1 & \Big\downarrow \mathfrak{s} & \\
\longrightarrow H_i(C_*) & \xrightarrow{\mathfrak{i}} & H_i(D_*) & \xrightarrow{p} & H_i(D_*/C_*) & \xrightarrow{\partial_*} & H_{i-1}(C_*) & \longrightarrow ,
\end{array}
$$

which commutes up to the signs indicated in each square. Here the top sequence is the long exact sequence associated to the short exact sequence (7.18) *and the bottom sequence is the long exact sequence associated to the short exact sequence*

$$0 \longrightarrow C_* \xrightarrow{\mathfrak{i}} D_* \xrightarrow{p} D_*/C_* \longrightarrow 0. \qquad (7.19)$$

Consequently, $\mathfrak{q} : H_*(E) \to H_*(D_*/C_*)$ *is an isomorphism.*

Proof We first observe that, via $\mathfrak{s}$, the cycles of $C[1]_i$ are taken bijectively to the cycles of C_{i-1} and the boundaries of $C[1]_i$ are taken bijectively to the boundaries of C_{i-1}, so $H_i(C[1]_*) \cong H_{i-1}(C_*)$. The vertical maps $\mathfrak{s}$ in the diagram involving C_* represent simply these canonical isomorphisms.

Next, let us define $\mathfrak{q} : E_* \to D_*/C_*$. If $(x, y) \in E_*$, we let $\mathfrak{q}(x, y) = x$, where we let x also represent the class of x in D_*/C_*. This is evidently a homomorphism, and we have $\mathfrak{q}(\partial(x, y)) = \mathfrak{q}((y + \partial x, -\partial y)) = y + \partial x$, which represents the same class as ∂x in D_*/C_*. So $\mathfrak{q}$ is a chain map. We can also see immediately from the definitions that $\mathfrak{q}\mathfrak{e} = p$, so the middle square in the diagram commutes.

Next, let us check that the other squares commute. Each of these includes one map that is the boundary map of a long exact sequence. Let us first suppose that (x, y) represents a cycle in $H_i(E_*)$. Then $\mathfrak{q}(x, y)$ is represented by x, and by the standard zig-zag definition of the boundary map in a long exact homology sequence, $\partial_*\mathfrak{q}(x, y)$ is represented by a cycle in C_{i-1} that maps to ∂x under the injection $\mathfrak{i}$. On the other hand $\mathfrak{b}(x, y) = \bar{y}$, so $\mathfrak{s}\mathfrak{b}(x, y) = y$. But we have stipulated that (x, y) is a cycle so that $(0, 0) = \partial(x, y) = (y + \partial x, -\partial y)$, and it follows that $\partial x = -y$. So the square commutes up to a sign of -1.

Now, suppose $\bar{y} \in C[1]_{i+1}$ is a cycle representing a homology class. Again we use the standard zig-zag construction to compute $\partial_*(\bar{y})$. We notice that $(0, y) \in E_{i+1}$ satisfies $\mathfrak{b}(0, y) = \bar{y}$, and then consider $\partial(0, y) = (y, -\partial y) = (y, 0) = \mathfrak{e}(y)$. Therefore, y represents $\partial_*(\bar{y})$, and of course $\mathfrak{i}\mathfrak{s}(\bar{y}) = y$, so this shows that the left square of the diagram commutes.

Finally, since the vertical maps of the diagram involving C_* and D_* are isomorphisms, the map $\mathfrak{q}$ must also induce a homology isomorphism by the Five Lemma. Technically, we do not have a strictly commutative diagram, but if we choose a fixed degree i and change the signs of the two vertical maps to the right of $\mathfrak{q} : H_i(E_*) \to H_i(D_*/C_*)$, then we obtain a strictly commutative diagram in a large enough vicinity of this particular $\mathfrak{q}$ to apply the Five Lemma and conclude that this $\mathfrak{q}$ is an isomorphism. But of course the same argument works for any i. □

Lemma 7.3.40 *Suppose $\mathfrak{i} : C_* \to D_*$ is a chain map of chain complexes of projective R-modules such that each module of C_*/D_* is also projective. Let E_* be the algebraic mapping cone of $\mathfrak{i}$. Then there is a diagram of long exact cohomology sequences*

$$\begin{array}{ccccccccc}
\longleftarrow H^{i+1}(\mathrm{Hom}(C[1]_*,R)) & \overset{d^*}{\longleftarrow} & H^i(\mathrm{Hom}(D_*,R)) & \overset{\mathfrak{e}^*}{\longleftarrow} & H^i(\mathrm{Hom}(E_*,R)) & \overset{\mathfrak{b}^*}{\longleftarrow} & H^i(\mathrm{Hom}(C[1]_*,R)) \longleftarrow \\
\Big\uparrow \mathfrak{s}^* & -1 & \Big\uparrow = & 1 & \Big\uparrow \mathfrak{q}^* & 1 & \Big\uparrow \mathfrak{s}^* \\
\longleftarrow H^i(\mathrm{Hom}(C_*,R)) & \overset{\mathfrak{i}^*}{\longleftarrow} & H^i(\mathrm{Hom}(D_*,R)) & \overset{p^*}{\longleftarrow} & H^i(\mathrm{Hom}(D_*/C_*,R)) & \overset{d^*}{\longleftarrow} & H^{i-1}(\mathrm{Hom}(C_*,R)) \longleftarrow ,
\end{array}$$

which commutes up to the signs indicated in each square. Here the bottom sequence is the long exact sequence associated to the short exact sequence

$$0 \longleftarrow \mathrm{Hom}(C_*,R) \overset{\mathfrak{i}^*}{\longleftarrow} \mathrm{Hom}(D_*,R) \longleftarrow \mathrm{Hom}(D_*/C_*,R) \longleftarrow 0$$

and the top sequence is the long exact sequence associated to the dual short exact sequence (7.18).

In particular, $\mathfrak{q}^* : H^*(\mathrm{Hom}(D_*/C_*,R)) \to H^*(\mathrm{Hom}(E_*,R))$ is an isomorphism.

Proof First notice that the assumption that all modules be projective implies that the Hom duals of the short exact sequences (7.18) and (7.19) remain exact and so generate long exact cohomology sequences.

Next, recall that if $\alpha \in \mathrm{Hom}(C_i,R)$ and $\bar{x} \in C[1]_{i+1}$, we have $\mathfrak{s}^*(\alpha)(\bar{x}) = (-1)^i\alpha(\mathfrak{s}(\bar{x})) = (-1)^i\alpha(x)$; the sign is due to the Koszul convention[15] as $\mathfrak{s}$ is a degree -1 map and α is a degree $-i$ map. Since $\mathfrak{s}$ is a (degree -1) chain map that is the identity on modules up to indexing, the same is true of $\mathfrak{s}^*$ up to sign. In particular, $\mathfrak{s}^*$ takes cocycles in $\mathrm{Hom}(C_i,R) = \mathrm{Hom}^i(C_*,R)$ bijectively to cocycles of $\mathrm{Hom}^{i+1}(C[1]_*,R)$ and coboundaries in $\mathrm{Hom}(C_i,R) = \mathrm{Hom}^i(C_*,R)$ bijectively to coboundaries of $\mathrm{Hom}^{i+1}(C[1]_*,R)$, so $H^{i+1}(\mathrm{Hom}(C[1]_*,R)) \cong H^i(\mathrm{Hom}(C_*,R))$. The vertical maps $\mathfrak{s}^*$ in the diagram involving C_* represent simply this canonical isomorphism.

We saw in the proof of Lemma 7.3.39 that $\mathfrak{q}\mathfrak{e} = p$, so $\mathfrak{e}^*\mathfrak{q}^* = p^*$, so the middle square in the diagram commutes.

Next, let us check that the other squares commute. Each of these includes one map that is the coboundary map of a long exact sequence.

First, suppose $\alpha \in \mathrm{Hom}^i(D_*,R)$ is a cocycle representing an element of $H^i(\mathrm{Hom}(D_*,R))$. By the zig-zag construction, $d^*\alpha \in H^{i+1}(\mathrm{Hom}(C[1]_*,R))$ is represented by choosing a cochain $\bar{\alpha} \in \mathrm{Hom}(E_i,R)$ that restricts to α on D_*, taking $d\bar{\alpha}$, and then restricting $d\bar{\alpha}$ to $\mathrm{Hom}(C[1]_{i+1},R)$. We are free to choose $\bar{\alpha}$ to be $\bar{\alpha} = (\alpha,0) \in \mathrm{Hom}(E_i,R) = \mathrm{Hom}(D_i \oplus C_{i-1},R) = \mathrm{Hom}(D_i,R) \oplus \mathrm{Hom}(C_{i-1},R)$. So then $d(\bar{\alpha})$ acts on a chain $(0,y) \in E_{i+1} = D_{i+1} \oplus C_i$ by

$$(d\bar{\alpha})(0,y) = (-1)^{i+1}\bar{\alpha}\partial(0,y) = (-1)^{i+1}(\alpha,0)(y,-\partial y) = (-1)^{i+1}\alpha(y).$$

[15] See Section A.1.

On the other hand, we have

$$\mathfrak{s}^*\mathfrak{i}^*(\alpha)(\bar{y}) = (-1)^i\mathfrak{i}^*\alpha(y) = (-1)^i\alpha(\mathfrak{i}(y)) = (-1)^i\alpha(y).$$

So the left square commutes up to the sign -1.

Next, suppose that $\alpha \in \mathrm{Hom}^{i-1}(C_*, R) = \mathrm{Hom}(C_{i-1}, R)$ represents an element of $H^{i-1}(\mathrm{Hom}(C_*, R))$. Then $d^*\alpha$ is represented by choosing a cochain $\bar{\alpha} \in \mathrm{Hom}(D_{i-1}, R)$ that restricts to α on C_{i-1} and then taking $d\bar{\alpha}$ and restricting it to act on D_*/C_*. If $(x,y) \in E_i$, we have

$$\mathfrak{q}^*d\bar{\alpha}(x,y) = (d\bar{\alpha})\mathfrak{q}(x,y) = (-1)^i\bar{\alpha}(\partial\mathfrak{q}(x,y)) = (-1)^i\bar{\alpha}(\partial x). \tag{7.20}$$

On the other hand, noting that $\mathfrak{b}$ is a degree-0 chain map,

$$\mathfrak{b}^*\mathfrak{s}^*\alpha(x,y) = (\mathfrak{s}^*\alpha)(\mathfrak{b}(x,y)) = (\mathfrak{s}^*\alpha)(\bar{y}) = (-1)^{i-1}\alpha(\mathfrak{s}(\bar{y})) = (-1)^{i-1}\alpha(y). \tag{7.21}$$

Consider, however, $(\bar{\alpha}, 0) \in \mathrm{Hom}(D_{i-1}, R) \oplus \mathrm{Hom}(C_{i-2}, R) = \mathrm{Hom}(D_{i-1} \oplus C_{i-2}, R) = \mathrm{Hom}(E_{i-1}, R)$. So we can compute

$$\begin{aligned}
d(\bar{\alpha},0)(x,y) &= (-1)^i(\bar{\alpha},0)\partial(x,y)\\
&= (-1)^i(\bar{\alpha},0)(y+\partial x, -\partial y)\\
&= (-1)^i(\bar{\alpha}(y)+\bar{\alpha}(\partial x))\\
&= (-1)^i\bar{\alpha}(y) + (-1)^i\bar{\alpha}(\partial x)\\
&= (-1)^i\alpha(y) + \mathfrak{q}^*d\bar{\alpha}(x,y) && \text{by (7.20)}\\
&= -\mathfrak{b}^*\mathfrak{s}^*\alpha(x,y) + \mathfrak{q}^*d\bar{\alpha}(x,y) && \text{by (7.21).}
\end{aligned}$$

In the fifth line, we have also used that $\alpha(y) = \bar{\alpha}(y)$, as $y \in C_*$. So $\mathfrak{b}^*\mathfrak{s}^*\alpha$ and $\mathfrak{q}^*d\bar{\alpha}$ represent the same cohomology class.

Finally, since the vertical maps of the diagram involving C_* and D_* are isomorphisms, the map $\mathfrak{q}^*$ must also induce a cohomology isomorphism by the Five Lemma. Technically, we do not have a strictly commutative diagram, but for any fixed i, we can change the signs of nearby vertical maps to obtain a strictly commutative diagram in a large enough vicinity of this particular $\mathfrak{q}^*$ to apply the Five Lemma and conclude that this $\mathfrak{q}^*$ is an isomorphism. □

(c) Stability of Cross Products and Cup Products

Now we can return to establishing stability formulas for cup and cross products. Of course, these come in pairs such as $(d^*\alpha) \times \beta$ and $\alpha \times (d^*\beta)$. We only prove one result for each such pair, as the other can be obtained using the commutative properties.

Various excision isomorphisms come into play in stating and proving the stability formulas. We denote these generically with an e, specifying in more detail where relevant.

We begin with some needed lemmas.

Lemma 7.3.41 *Suppose that $\bar{p}, \bar{q}$ are perversities on filtered sets X, Y and that Q is a $(\bar{p},\bar{q})$-compatible perversity on $X \times Y$. Let $A \subset X$ and $B \subset Y$ be open subsets. Let F_* denote* *the algebraic mapping cone of the inclusion* $\mathfrak{i} : I^{\bar{p}}S_*(A;R) \to I^{\bar{p}}S_*(X;R)$*, let* E_* *be the algebraic mapping cone of the map* $\mathfrak{i}\times \mathrm{id} : I^{Q}S_*(A\times Y, A\times B;R) \to I^{Q}S_*(X\times Y, X\times B;R)$*, and let* G_* *be the algebraic mapping cone of the inclusion* $I^{Q}S_*((A\times Y)\cup(X\times B), X\times B;R) \to I^{Q}S_*(X\times Y, X\times B;R)$*. The following diagram (with implicit R coefficients) commutes:*

$$\begin{array}{ccccc}
I^{Q}S_*(X\times Y,(A\times Y)\cup(X\times B)) & \xleftarrow{=} & I^{Q}S_*(X\times Y,(A\times Y)\cup(X\times B)) & \xleftarrow{\varepsilon} & I^{\bar{p}}S_*(X,A)\otimes I^{\bar{q}}S_*(Y,B) \\
\uparrow{\scriptstyle\cong} & & & & \\
\dfrac{I^{Q}S_*(X\times Y, X\times B)}{I^{Q}S_*((A\times Y)\cup(X\times B), X\times B)} & & \uparrow{\scriptstyle\mathfrak{q}'} & & \uparrow{\scriptstyle\mathfrak{q}\otimes\mathrm{id}} \\
\uparrow{\scriptstyle\mathfrak{q}} & & & & \\
G_* & \xleftarrow{k} & E_* & \xleftarrow{(\varepsilon,\varepsilon)} & F_*\otimes I^{\bar{q}}S_*(Y,B) \\
\downarrow{\scriptstyle\mathfrak{b}} & & \downarrow{\scriptstyle\mathfrak{b}} & & \downarrow{\scriptstyle\mathfrak{b}\otimes\mathrm{id}} \\
I^{Q}S_*((A\times Y)\cup(X\times B), X\times B)[1] & \xleftarrow{\mathfrak{e}[1]} & I^{Q}S_*(A\times Y, A\times B)[1] & \xleftarrow{\bar{\varepsilon}} & I^{\bar{p}}S_*(A)[1]\otimes I^{\bar{q}}S_*(Y,B).
\end{array}$$

Proof Let us first verify that all the maps make sense. The unlabeled maps are induced by the obvious inclusions and/or quotients. The upper left vertical map is an isomorphism by the Third Isomorphism theorem. Unlike F_* and G_*, the mapping cone E_* is not based on an inclusion of chain complexes. However, the map $E_* \to G_*$ makes sense by the functoriality of the algebraic mapping cone construction, and we define the map labeled $\mathfrak{q}'$ to act on

$$(x,y) \in E_i = I^{Q}S_i(X\times Y, X\times B;R) \oplus I^{Q}S_{i-1}(A\times Y, A\times B;R)$$

by taking (x,y) to the class of x in $I^{Q}S_i(X\times Y,(A\times Y)\cup(X\times B);R)$. This is a chain map because $\mathfrak{q}'\partial(x,y) = \mathfrak{q}'(y+\partial x, -\partial y) = y+\partial x$, but y here is represented by a chain supported in $A\times Y$, so $y+\partial x$ and ∂x represent the same element in $I^{Q}S_*(X\times Y,(A\times Y)\cup(X\times B))$. It also follows readily from this definition that the upper left rectangle commutes.

For the map labeled $\bar{\varepsilon}$, we let this be the composition of the canonical isomorphism $\mathfrak{t}$ (see page 421) and the shifted cross product $\varepsilon[1]$.

Similarly, for the map labeled $(\varepsilon,\varepsilon)$, recall that, as *modules*, we have

$$E_i = I^{Q}S_i(X\times Y, X\times B;R) \oplus I^{Q}S_{i-1}(A\times Y, A\times B;R)$$

and

$$F_j = I^{\bar{p}}S_j(X;R) \oplus I^{\bar{p}}S_{j-1}(A;R).$$

Thus $(F_* \otimes I^{\bar{q}}S_*(Y,B;R))_i$ is the direct sum over $j+k=i$ of terms of the form

$$\begin{aligned}&(I^{\bar{p}}S_j(X;R) \oplus I^{\bar{p}}S_{j-1}(A;R)) \otimes I^{\bar{q}}S_k(Y,B;R)\\&\quad = (I^{\bar{p}}S_j(X;R) \otimes I^{\bar{q}}S_k(Y,B;R)) \oplus (I^{\bar{p}}S(A;R)_{j-1} \otimes I^{\bar{q}}S_k(Y,B;R)).\end{aligned}$$

Such modules are generated by elements of the form $(x,a) \otimes y = (x \otimes y, a \otimes y)$, and we let $(\varepsilon,\varepsilon)$ act in the obvious way by $(\varepsilon,\varepsilon)(x \otimes y, a \otimes y) = (\varepsilon(x \otimes y), \varepsilon(a \otimes y))$. Let us check that this gives us a chain map, keeping $|x| = j$ and $|y| = k$, so that $|a| = j-1$ and $|(x,a)| = j$:

$$\begin{aligned}&(\varepsilon,\varepsilon)(\partial((x,a)\otimes y))\\
&\quad = (\varepsilon,\varepsilon)(\partial(x,a)\otimes y + (-1)^j(x,a)\otimes \partial y)\\
&\quad = (\varepsilon,\varepsilon)((\partial x + \mathfrak{i}(a), -\partial a)\otimes y + (-1)^j(x,a)\otimes \partial y)\\
&\quad = (\varepsilon,\varepsilon)((\partial x\otimes y + \mathfrak{i}(a)\otimes y, -(\partial a)\otimes y) + (-1)^j(x\otimes \partial y, a\otimes \partial y))\\
&\quad = (\varepsilon,\varepsilon)((\partial x\otimes y + \mathfrak{i}(a)\otimes y + (-1)^j x\otimes \partial y, -(\partial a)\otimes y + (-1)^j a\otimes \partial y))\\
&\quad = (\varepsilon,\varepsilon)(\partial(x\otimes y) + \mathfrak{i}(a)\otimes y, -\partial(a\otimes y))\\
&\quad = (\varepsilon(\partial(x\otimes y) + \mathfrak{i}(a)\otimes y), -\varepsilon(\partial(a\otimes y)))\\
&\quad = (\partial\varepsilon(x\otimes y) + (\mathfrak{i}\times \mathrm{id})\varepsilon(a\otimes y), -\partial\varepsilon(a\otimes y))\\
&\quad = \partial(\varepsilon(x\otimes y), \varepsilon(a\otimes y)).\end{aligned}$$

In the next to last line, we have used that ε is a chain map and that it is natural by Proposition 5.2.17 and Theorem 6.3.19.

For commutativity of the top right square of the diagram, we compute using representatives:

$$\begin{aligned}\mathfrak{q}'(\varepsilon,\varepsilon)((x,a)\otimes y) &= \mathfrak{q}'(\varepsilon,\varepsilon)(x\otimes y, a\otimes y) = \mathfrak{q}'(\varepsilon(x\otimes y), \varepsilon(a\otimes y))\\
&= \varepsilon(x\otimes y) = \varepsilon(\mathfrak{q}(x,a)\otimes y)\\
&= \varepsilon(\mathfrak{q}\otimes \mathrm{id})((x,a)\otimes y).\end{aligned}$$

Similarly, for the bottom right square, we have

$$\begin{aligned}\mathfrak{b}(\varepsilon,\varepsilon)((x,a)\otimes y) &= \mathfrak{b}(\varepsilon,\varepsilon)(x\otimes y, a\otimes y) = \mathfrak{b}(\varepsilon(x\otimes y), \varepsilon(a\otimes y))\\
&= \overline{\varepsilon(a\otimes y)} = \varepsilon[1]\overline{a\otimes y} = \varepsilon[1]\mathfrak{t}(\bar{a}\otimes y)\\
&= \bar{\varepsilon}(\bar{a}\otimes y) = \bar{\varepsilon}(\mathfrak{b}(x,a)\otimes y)\\
&= \bar{\varepsilon}(\mathfrak{b}\otimes \mathrm{id})((x,a)\otimes y).\end{aligned}$$

The commutativity of the bottom left square is a consequence of the naturality properties of the mapping cone construction. □

Corollary 7.3.42 *Let R be a Dedekind domain. Suppose that $\bar{p},\bar{q}$ are perversities on CS sets X,Y and that Q is a $(\bar{p},\bar{q})$-compatible perversity on $X\times Y$. Let $A\subset X$ and $B\subset Y$ be open subsets.*

The diagram below commutes up to homotopy:

$$
\begin{array}{ccccc}
I^{Q}S_*(X\times Y,(A\times Y)\cup(X\times B)) & \xrightarrow{=} & I^{Q}S_*(X\times Y,(A\times Y)\cup(X\times B)) & \xrightarrow{\mathrm{IAW}} & I^{\bar p}S_*(X,A)\otimes I^{\bar q}S_*(Y,B)\\
\downarrow{\scriptstyle\cong} & & & & \downarrow{\scriptstyle \mathfrak{q}^{-1}\otimes\mathrm{id}}\\
\dfrac{I^{Q}S_*(X\times Y,X\times B)}{I^{Q}S_*((A\times Y)\cup(X\times B),X\times B)} & & & & \\
\downarrow{\scriptstyle \mathfrak{q}^{-1}} & & & & \\
G_* & & \longrightarrow & & F_*\otimes I^{\bar q}S_*(Y,B)\\
\downarrow{\scriptstyle \mathfrak{b}} & & & & \downarrow{\scriptstyle \mathfrak{b}\otimes\mathrm{id}}\\
I^{Q}S_*((A\times Y)\cup(X\times B),X\times B)[1] & \xrightarrow{(e[1])^{-1}} & I^{Q}S_*(A\times Y,A\times B)[1] & \xrightarrow{\overline{\mathrm{IAW}}} & I^{\bar p}S_*(A)[1]\otimes I^{\bar q}S_*(Y,B)\\
\downarrow{\scriptstyle \mathfrak{s}} & & \downarrow{\scriptstyle \mathfrak{s}} & & \downarrow{\scriptstyle \mathfrak{s}\otimes\mathrm{id}}\\
I^{Q}S_*((A\times Y)\cup(X\times B),X\times B) & \xrightarrow{(e)^{-1}} & I^{Q}S_*(A\times Y,A\times B) & \xrightarrow{\mathrm{IAW}} & I^{\bar p}S_*(A)\otimes I^{\bar q}S_*(Y,B).
\end{array}
$$

Proof Everything except the bottom row of this diagram is obtained from the diagram of Lemma 7.3.41 by replacing some maps (or compositions of maps) with their chain homotopy inverses. In particular, by Lemma 7.3.39, the $\mathfrak{q}$ maps are quasi-isomorphisms of chain complexes of projective modules, so they are chain homotopy equivalences by Lemma A.4.3 (noting that it is sufficient in the argument for the complexes simply to both be bounded below). We let $\mathfrak{q}^{-1}$ denote the chain homotopy inverse. It follows that $\mathfrak{q}\otimes\mathrm{id}$ is also a homotopy equivalence. We know that ε is a chain homotopy inverse by Theorem 6.4.14, and IAW is our label for its inverse. It follows that the composition in the middle horizontal row of the diagram of Lemma 7.3.40 is also a homotopy equivalence. The map $\overline{\mathrm{IAW}}$ is the chain homotopy inverse for $\bar\varepsilon=\varepsilon[1]\mathfrak{t}$, which is a composition of an isomorphism and a shifted homotopy equivalence.

The inclusion-induced map

$$I^{Q}S_*(A\times Y,A\times B)\to I^{Q}S_*((A\times Y)\cup(X\times B),X\times B)$$

is also a quasi-isomorphism, and so a homotopy equivalence, as all modules are projective.

We can see this noting that the map factors as

$$\frac{I^Q S_*(A\times Y;R)}{I^Q S_*(A\times B;R)} = \frac{I^Q S_*(A\times Y;R)}{I^Q S_*((A\times Y)\cap(X\times B);R)}$$
$$\to \frac{I^Q S_*(A\times Y;R)+I^Q S_*(X\times B;R)}{I^Q S_*(X\times B;R)} \to \frac{I^Q S_*((A\times Y)\cup(X\times B);R)}{I^Q S_*(X\times B;R)}.$$

The first map in the second line is an isomorphism by the Second Isomorphism Theorem, noting that $I^Q S_*(A\times Y;R)\cap I^Q S_*(X\times B;R) = I^Q S_*(A\times B;R)$, while the second map induces a homology isomorphism by the non-GM version of the arguments in the proof of Theorem 4.4.23.

Therefore, appropriately reversing the arrows in the diagram of Lemma 7.3.41 results in a homotopy commutative diagram as labeled, except for the bottom row, which we have added on. The bottom left square commutes by the functoriality of shifting. The bottom right square is the following square with its horizontal maps inverted up to homotopy:

$$\begin{array}{ccc} I^Q S_*(A\times Y, A\times B)[1] & \xleftarrow{\bar\varepsilon} & I^{\bar p} S_*(A)[1]\otimes I^{\bar q} S_*(Y,B) \\ \downarrow{\scriptstyle \mathfrak{s}} & & \downarrow{\scriptstyle \mathfrak{s}\otimes \mathrm{id}} \\ I^Q S_*(A\times Y, A\times B) & \xleftarrow{\varepsilon} & I^{\bar p} S_*(A)\otimes I^{\bar q} S_*(Y,B). \end{array}$$

This commutes because

$$\begin{aligned}\mathfrak{s}\bar\varepsilon(\bar x\otimes y) &= \mathfrak{s}\varepsilon[1]\mathfrak{t}(\bar x\otimes y) = \mathfrak{s}\varepsilon[1]\overline{x\otimes y} = \mathfrak{s}(\overline{\varepsilon(x\otimes y)})\\ &= \varepsilon(x\otimes y) = \varepsilon(\mathfrak{s}(\bar x)\otimes y)\\ &= \varepsilon(\mathfrak{s}\otimes \mathrm{id})(\bar x\otimes y).\end{aligned}$$

□

Proposition 7.3.43 *Let R be a Dedekind domain. Suppose X and Y are CS sets with respective perversities $\bar p$ and $\bar q$ and open subspaces $A\subset X$ and $B\subset Y$. Let Q be a $(\bar p,\bar q)$-compatible perversity on $X\times Y$. Let $\alpha\in I_{\bar p}H^i(A;R)$ and $\beta\in I_{\bar q}H^j(Y,B;R)$. Then*

$$(d^*(\alpha))\times\beta = d^*(e^{-1})^*(\alpha\times\beta)\in I_Q H^{i+j+1}(X\times Y,(A\times Y)\cup(X\times B);R),$$

where

$$e: I^Q H_*(A\times Y, A\times B;R)\to I^Q H_*((A\times Y)\cup(X\times B), X\times B;R)$$

is an excision isomorphism and where we interpret the right-hand d^* as a map

$$I_Q H^{i+j}((A\times Y)\cup(X\times B),X\times B;R)\to I^Q H^{i+j+1}(X\times Y,(A\times Y)\cup(X\times B);R).$$

Proof By Lemma 7.3.40, we have $d^*(\alpha) = (\mathfrak{q}^*)^{-1}\mathfrak{b}^*\mathfrak{s}^*\alpha$. As observed in Corollary 7.3.42, since all of our modules are projectives, the quasi-isomorphism $\mathfrak{q}$ (see Lemma 7.3.39) is in fact a chain homotopy equivalence, so we can replace $(\mathfrak{q}^*)^{-1}$ with $(\mathfrak{q}^{-1})^*$, which is well defined up to chain homotopy. This lets us write $d^*(\alpha) = (\mathfrak{q}^{-1})^*\mathfrak{b}^*\mathfrak{s}^*\alpha = (\mathfrak{s}\mathfrak{b}\mathfrak{q}^{-1})^*\alpha$. So, from the definition of the cross product, we have

$$\begin{aligned}
(d^*\alpha) \times \beta &= \mathrm{IAW}^*\Theta((d^*(\alpha)) \otimes \beta)\\
&= \mathrm{IAW}^*\Theta(((\mathfrak{s}\mathfrak{b}\mathfrak{q}^{-1})^*(\alpha)) \otimes \beta)\\
&= \mathrm{IAW}^*\Theta((\mathfrak{s}\mathfrak{b}\mathfrak{q}^{-1})^* \otimes \mathrm{id})(\alpha \otimes \beta)\\
&= \mathrm{IAW}^*((\mathfrak{s}\mathfrak{b}\mathfrak{q}^{-1}) \otimes \mathrm{id})^*\Theta(\alpha \otimes \beta) && \text{by Lemma 7.3.1}\\
&= (e^{-1}\mathfrak{s}\mathfrak{b}\mathfrak{q}^{-1})^*\mathrm{IAW}^*\Theta(\alpha \otimes \beta) && \text{by Corollary 7.3.42}\\
&= d^*(e^{-1})^*(\alpha \times \beta) && \text{by Lemma 7.3.40.}
\end{aligned}$$

□

Proposition 7.3.44 *Let R be a Dedekind domain. Suppose X is a CS set with open subsets A and B, that $\mathfrak{i} : A \to X$ is the inclusion map, and that $(\bar{p}, \bar{q}; \bar{r})$ is an agreeable triple of perversities on X. Let $\alpha \in I_{\bar{p}}H^i(A; R)$ and $\beta \in I_{\bar{q}}H^j(X, B; R)$. Then*

$$(d^*(\alpha)) \smile \beta = d^*(e^{-1})^*(\alpha \smile \mathfrak{i}^*(\beta)) \in I_{\bar{r}}H^{i+j+1}(X, A \cup B; R),$$

where $e : I^{\bar{r}}H_*(A, A \cap B; R) \to I^{\bar{r}}H_*(A \cup B, B; R)$ is induced by inclusion and where we interpret the right-hand d^* as a map $I_{\bar{r}}H^{i+j}(A \cup B, B; R) \to I_{\bar{r}}H^{i+j+1}(X, A \cup B; R)$.

Proof Consider the diagram with tacit coefficients in Figure 7.1. Here D_* is the algebraic mapping cone of the inclusion

$$I^{\bar{r}}S_*(A \cup B, B; R) \to I^{\bar{r}}S_*(X, B; R)$$

and, as in Lemma 7.3.41 but with $X = Y$, we let G_* be the algebraic mapping cone of the inclusion

$$I^{Q_{\bar{p},\bar{q}}}S_*((A \times X) \cup (X \times B), X \times B; R) \to I^{Q_{\bar{p},\bar{q}}}S_*(X \times X, X \times B; R).$$

The bottom quadrilateral commutes by the naturality of the chain cross product (Proposition 5.2.17 and Theorem 6.3.19). The triangle and the square with the e maps commute at the level of maps of pairs of spaces. The isomorphisms in the top square come from the Third Isomorphism Theorem, and this square also commutes by looking at representative elements. The square involving $\mathfrak{s}$ is the naturality of shifts. So let us consider the squares involving the mapping cones. By the naturality properties of the mapping cone construction, the map labeled $(\mathbf{d}, \mathbf{d})$ is a chain map and the square involving the $\mathfrak{b}$ maps commutes. Furthermore, if $(x, y) \in D_i = I^{\bar{r}}S_i(X, B; R) \oplus I^{\bar{r}}S_{i-1}(A \cup B, B; R)$, we have

$$\mathbf{d}\mathfrak{q}(x, y) = \mathbf{d}(x) = \mathfrak{q}(\mathbf{d}(x), \mathbf{d}(y)) = \mathfrak{q}(\mathbf{d}, \mathbf{d})(x, y),$$

so the second square from the top commutes.

Therefore, the full diagram commutes. Furthermore, $\mathfrak{q}$ is again a quasi-isomorphism by Lemma 7.3.39, and it is a homotopy equivalence since we work with all projective modules. Also, our new e on the left of the diagram is a quasi-isomorphism, and hence a chain

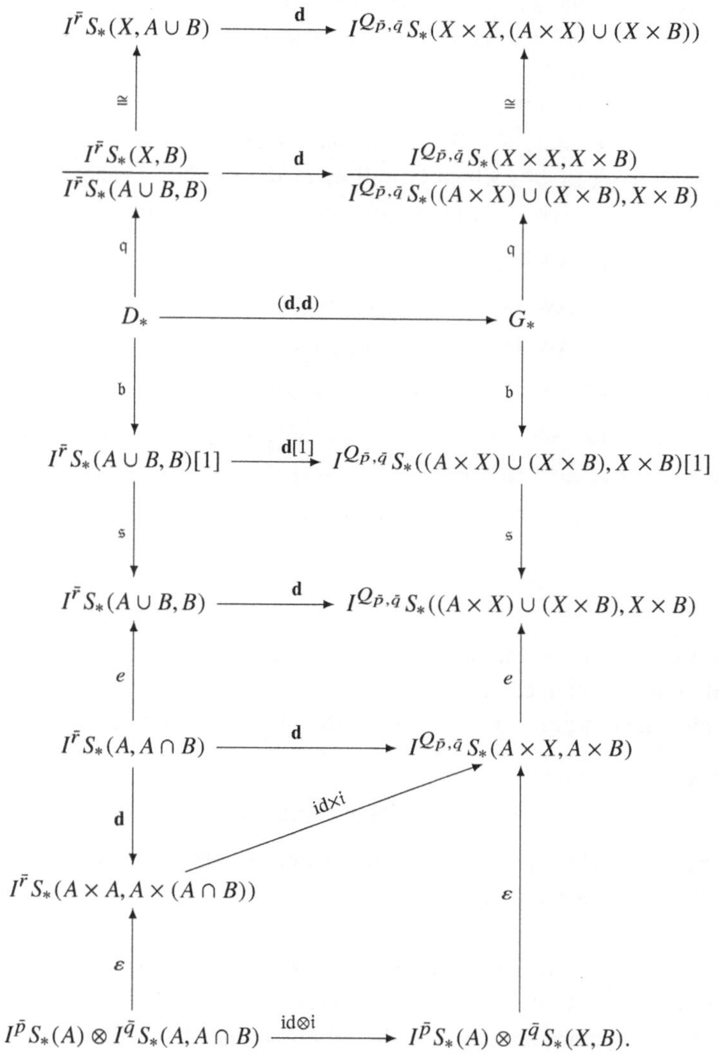

Figure 7.1

homotopy equivalence of complexes of projective modules, since we can factor it as

$$\frac{I^{\bar r}S_*(A;R)}{I^{\bar r}S_*(A\cap B;R)} \to \frac{I^{\bar r}S_*(A;R)+I^{\bar r}S_*(B;R)}{I^{\bar r}S_*(B;R)} \to \frac{I^{\bar r}S_*(A\cup B;R)}{I^{\bar r}S_*(B;R)}.$$

The first map is an isomorphism by the Second Isomorphism Theorem, noting that $I^{\bar r}S_*(A)\cap I^{\bar r}S_*(B) = I^{\bar r}S_*(A\cap B;R)$, while the second map induces a homology isomorphism by the non-GM version of the arguments in the proof of Theorem 4.4.23. Therefore, we can invert homotopy equivalences and adjoin this diagram to part of the diagram of Corollary 7.3.42 to get the diagram in Figure 7.2 that commutes up to homotopy.

Now we can compute. By Lemma 7.3.40, we get $d^*(\alpha) = (\mathfrak{q}^*)^{-1}\mathfrak{b}^*\mathfrak{s}^*\alpha \in I_{\bar{p}}H^{i+1}(X, A; R)$. As observed in Corollary 7.3.42, since all of our modules are projectives, the quasi-isomorphisms $\mathfrak{q}$ (see Lemma 7.3.39) are in fact chain homotopy equivalences, so we can replace $(\mathfrak{q}^*)^{-1}$ with $(\mathfrak{q}^{-1})^*$, which is well defined up to homotopy. This lets us write $d^*(\alpha) = (\mathfrak{q}^{-1})^*\mathfrak{b}^*\mathfrak{s}^*(\alpha) = (\mathfrak{s}\mathfrak{b}\mathfrak{q}^{-1})^*(\alpha)$. So, from the definition of the cup product, we have

$$
\begin{aligned}
(d^*(\alpha)) \smile \beta &= \mathbf{d}^*\mathrm{IAW}^*\Theta(d^*(\alpha) \otimes \beta) \\
&= \mathbf{d}^*\mathrm{IAW}^*\Theta(((\mathfrak{s}\mathfrak{b}\mathfrak{q}^{-1})^*(\alpha)) \otimes \beta) \\
&= \mathbf{d}^*\mathrm{IAW}^*\Theta((\mathfrak{s}\mathfrak{b}\mathfrak{q}^{-1})^* \otimes \mathrm{id})(\alpha \otimes \beta) \\
&= \mathbf{d}^*\mathrm{IAW}^*((\mathfrak{s}\mathfrak{b}\mathfrak{q}^{-1}) \otimes \mathrm{id})^*\Theta(\alpha \otimes \beta) && \text{by Lemma 7.3.1} \\
&= (e^{-1}\mathfrak{s}\mathfrak{b}\mathfrak{q}^{-1})^*\mathbf{d}^*\mathrm{IAW}^*(\mathrm{id} \otimes \mathfrak{i})^*\Theta(\alpha \otimes \beta) && \text{by the diagram} \\
&= d^*(e^{-1})^*\mathbf{d}^*\mathrm{IAW}^*(\mathrm{id} \otimes \mathfrak{i})^*\Theta(\alpha \otimes \beta) && \text{by Lemma 7.3.40} \\
&= d^*(e^{-1})^*\mathbf{d}^*\mathrm{IAW}^*\Theta(\mathrm{id} \otimes \mathfrak{i}^*)(\alpha \otimes \beta) && \text{by Lemma 7.3.1} \\
&= d^*(e^{-1})^*\mathbf{d}^*\mathrm{IAW}^*\Theta(\alpha \otimes \mathfrak{i}^*(\beta)) && \text{by Lemma 7.3.1} \\
&= d^*(e^{-1})^*(\alpha \smile \mathfrak{i}^*(\beta)).
\end{aligned}
$$

For the equality labeled "by the diagram," the compositions $(\mathrm{id} \otimes \mathfrak{i})\mathrm{IAW}\mathbf{d}e^{-1}\mathfrak{s}\mathfrak{b}\mathfrak{q}^{-1}$ and $(\mathfrak{s}\mathfrak{b}\mathfrak{q}^{-1} \otimes \mathrm{id})\mathrm{IAW}\mathbf{d}$ are obtained in the diagram by starting in the upper left and proceeding each way around the outside to the term one above the bottom right corner. □

7.3.6 Criss-Crosses

There are a variety of relations involving combinations of the cup, cap, and cross products, some of which we have already seen. We discuss the rest here.

(a) The Relation between Cup and Cross Products

As in ordinary cohomology theory, the cup and cross products can each be defined in terms of the other. In fact, we know directly from the definitions that the cup product is the pullback of the cross product by the diagonal map:

Proposition 7.3.45 *Let R be a Dedekind domain. Suppose that $(\bar{p}, \bar{q}; \bar{r})$ is a Q-agreeable triple of perversities on a CS set X. If $\alpha \in I_{\bar{p}}H^i(X, A; R)$, $\beta \in I_{\bar{q}}H^j(X, B; R)$, and $\alpha \times \beta \in I_QH^{i+j}(X \times Y, (A \times Y) \cup (X \times B); R)$, then*

$$\mathbf{d}^*(\alpha \times \beta) = \alpha \smile \beta \in I_{\bar{r}}H^{i+j}(X, A \cup B; R).$$

The next lemma demonstrates that we can also recover the cross product in terms of a cup product.

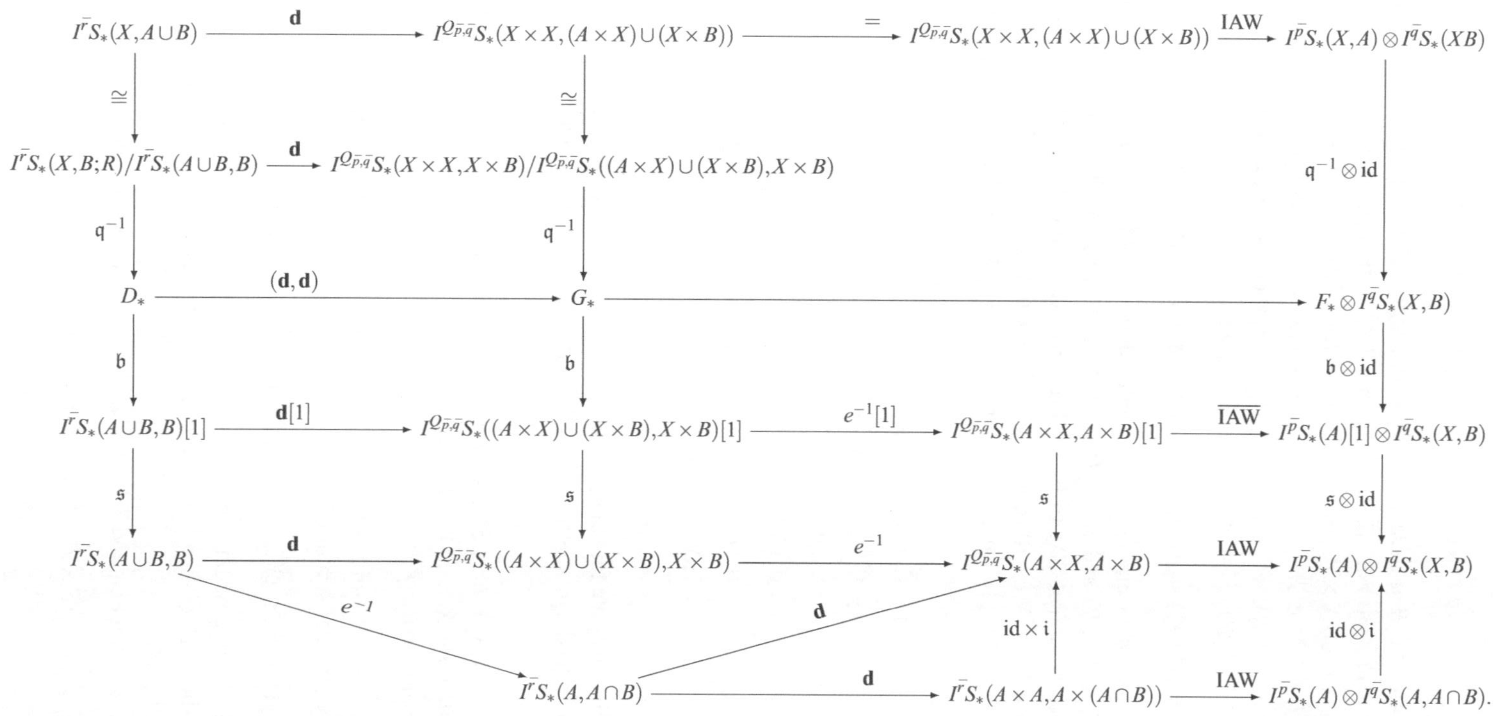
$I^{\bar{r}}S_*(X, A \cup B)$
$\mathbf{d}$
$I^{Q_{\bar{p},\bar{q}}}S_*(X \times X, (A \times X) \cup (X \times B))$
$=$
$I^{Q_{\bar{p},\bar{q}}}S_*(X \times X, (A \times X) \cup (X \times B))$
IAW
$I^{\bar{p}}S_*(X, A) \otimes I^{\bar{q}}S_*(XB)$
$\cong$
$\cong$
$I^{\bar{r}}S_*(X, B; R)/I^{\bar{r}}S_*(A \cup B, B)$
$\mathbf{d}$
$I^{Q_{\bar{p},\bar{q}}}S_*(X \times X, X \times B)/I^{Q_{\bar{p},\bar{q}}}S_*((A \times X) \cup (X \times B), X \times B)$
$\mathfrak{q}^{-1} \otimes \mathrm{id}$
$\mathfrak{q}^{-1}$
$\mathfrak{q}^{-1}$
D_*
$(\mathbf{d}, \mathbf{d})$
G_*
$F_* \otimes I^{\bar{q}}S_*(X, B)$
$\mathfrak{b}$
$\mathfrak{b}$
$\mathfrak{b} \otimes \mathrm{id}$
$I^{\bar{r}}S_*(A \cup B, B)[1]$
$\mathbf{d}[1]$
$I^{Q_{\bar{p},\bar{q}}}S_*((A \times X) \cup (X \times B), X \times B)[1]$
$e^{-1}[1]$
$I^{Q_{\bar{p},\bar{q}}}S_*(A \times X, A \times B)[1]$
$\overline{\mathrm{IAW}}$
$I^{\bar{p}}S_*(A)[1] \otimes I^{\bar{q}}S_*(X, B)$
$\mathfrak{s}$
$\mathfrak{s}$
$\mathfrak{s}$
$\mathfrak{s} \otimes \mathrm{id}$
$I^{\bar{r}}S_*(A \cup B, B)$
$\mathbf{d}$
$I^{Q_{\bar{p},\bar{q}}}S_*((A \times X) \cup (X \times B), X \times B)$
e^{-1}
$I^{Q_{\bar{p},\bar{q}}}S_*(A \times X, A \times B)$
IAW
$I^{\bar{p}}S_*(A) \otimes I^{\bar{q}}S_*(X, B)$
e^{-1}
$\mathbf{d}$
$\mathrm{id} \times \mathfrak{i}$
$\mathrm{id} \otimes \mathfrak{i}$
$I^{\bar{r}}S_*(A, A \cap B)$
$\mathbf{d}$
$I^{\bar{r}}S_*(A \times A, A \times (A \cap B))$
IAW
$I^{\bar{p}}S_*(A) \otimes I^{\bar{q}}S_*(A, A \cap B).$

Figure 7.2

Proposition 7.3.46 *Let R be a Dedekind domain. Suppose that X is a CS set with perversity $\bar{p}$, Y is a CS set with perversity $\bar{q}$, and $A \subset X$, $B \subset Y$ are open sets. Let Q be a $(\bar{p},\bar{q})$-compatible perversity on $X \times Y$. Let $p_1 : X \times Y \to X$ and $p_2 : X \times Y \to Y$ be the projection maps, and $p_1^* : I_{\bar{p}}H^*(X,A;R) \to I_{Q_{\bar{p},\bar{t}_Y}}H^*(X \times Y, A \times Y;R)$ and $p_2^* : I_{\bar{q}}H^*(Y,B;R) \to I_{Q_{\bar{t}_X,\bar{q}}}H^*(X \times Y, X \times B;R)$, where $\bar{t}_X$ and $\bar{t}_Y$ are the respective top perversities on X and Y. Then if $\alpha \in I_{\bar{p}}H^i(X,A;R)$ and $\beta \in I_{\bar{q}}H^j(X,B;R)$, we have*

$$\alpha \times \beta = (p_1^*(\alpha)) \smile (p_2^*(\beta)) \in I_Q H^{i+j}(X \times Y, (A \times Y) \cup (X \times B); R).$$

Proof The maps p_1^* and p_2^* are well defined with respective images in $I_{Q_{\bar{p},\bar{t}_Y}}H^*(X \times Y, A \times Y;R)$ and $I_{Q_{\bar{t}_X,\bar{q}}}H^*(X \times Y, X \times B;R)$ by Corollary 7.3.17. For the cup product $(p_1^*(\alpha)) \smile (p_2^*(\beta))$ to be well defined, we need $(Q_{\bar{p},\bar{t}_Y}, Q_{\bar{t}_X,\bar{q}}; Q)$ to be an agreeable triple on $X \times Y$. We will show this just below in Lemma 7.3.47. Therefore, the cup product of $p_1^*(\alpha)$ and $p_2^*(\beta)$ is well defined with image in $I_Q H^{i+j}(X \times Y, (A \times Y) \cup (X \times B); R)$.

Now we compute

$$\begin{aligned}
(p_1^*(\alpha)) \smile (p_2^*(\beta)) &= \bar{\mathbf{d}}^*\Theta((p_1^*(\alpha)) \otimes (p_2^*(\beta))) \\
&= \bar{\mathbf{d}}^*\Theta(p_1^* \otimes p_2^*)(\alpha \otimes \beta) \\
&= \bar{\mathbf{d}}^*(p_1 \otimes p_2)^*\Theta(\alpha \otimes \beta) && \text{by Lemma 7.3.1} \\
&= \mathbf{d}^*\mathrm{IAW}^*(p_1 \otimes p_2)^*\Theta(\alpha \otimes \beta) \\
&= \mathbf{d}^*(p_1 \times p_2)^*\mathrm{IAW}^*\Theta(\alpha \otimes \beta) && \text{by Proposition 5.2.17} \\
&= \mathrm{IAW}^*\Theta(\alpha \otimes \beta) && \text{see below} \\
&= \alpha \times \beta.
\end{aligned}$$

The next to last line uses the fact that, at the level of spaces, $(p_1 \times p_2)d : X \times Y \to X \times Y$ is the identity map. Indeed, $(p_1 \times p_2)d(x,y) = (p_1 \times p_2)((x,y),(x,y)) = (p_1(x,y), p_2(x,y)) = (x,y)$. □

Lemma 7.3.47 *Let X,Y be CS sets with respective perversities $\bar{p},\bar{q}$. Let $\bar{t}_X$ and $\bar{t}_Y$ be the top perversities on X and Y. Then $(Q_{\bar{p},\bar{t}_Y}, Q_{\bar{t}_X,\bar{q}}; Q)$ is an agreeable triple for any $(\bar{p},\bar{q})$-agreeable perversity Q.*

Proof As $Q \leq Q_{\bar{p},\bar{q}}$, it suffices to prove the lemma just for $Q_{\bar{p},\bar{q}}$. This means we must show that for any singular stratum $S \times T \subset X \times Y$ we have

$$Q_{\bar{p},\bar{q}}(S \times T) \leq Q_{Q_{\bar{p},\bar{t}_Y},Q_{\bar{t}_X,\bar{q}}}(S \times T \times S \times T) - \operatorname{codim}_{X \times Y}(S \times T).$$

The tricky part is dealing with the fact that we know the various Q perversities incorporate summands of either $+1$ or $+2$ with the precise value depending on the torsion information for the individual strata.

First, suppose S is singular and T is regular. Then $Q_{\bar{p},\bar{q}}(S \times T) = \bar{p}(S)$, while

$$\begin{aligned}
Q_{Q_{\bar{p},\bar{t}_Y},Q_{\bar{t}_X,\bar{q}}}(S \times T \times S \times T) &= Q_{\bar{p},\bar{t}_Y}(S \times T) + Q_{\bar{t}_X,\bar{q}}(S \times T) + C \\
&= \bar{p}(S) + \bar{t}_X(S) + C.
\end{aligned}$$

Here we know that $C \in \{1,2\}$, depending on a local torsion computation that we now perform. Let K be a link of S in X, and suppose $\dim(K) = m-1$, consistent with our conventions in Section 6.4. As T is assumed to be a regular stratum, K is also the link of $S\times T$ in $X\times Y$. The relevant computation is then that of the torsion product of $I^{Q_{\bar{p},\bar{t}_Y}}H_{m-Q_{\bar{p},\bar{t}_Y}(S\times T)-2}(K;R)$ with $I^{Q_{\bar{t}_X,\bar{q}}}H_{m-Q_{\bar{t}_X,\bar{q}}(S\times T)-2}(K;R)$. But

$$Q_{\bar{t}_X,\bar{q}}(S\times T) = \bar{t}_X(S) = \operatorname{codim}_X(S) - 2 = m-2,$$

recalling that as K is a link of S we have $\dim(K)+\dim(S)+1 = \dim(X)$, and so $\operatorname{codim}_X(S) = \dim(K)+1 = m$. Therefore

$$m - Q_{\bar{t}_X,\bar{q}}(S\times T) - 2 = 0.$$

But $I^{Q_{\bar{t}_X,\bar{q}}}H_0(K;R)$ must be torsion-free, and so the torsion product vanishes and we have $C = 2$. Consequently,

$$\begin{aligned} Q_{\bar{p},\bar{q}}(S\times T) &= \bar{p}(S) \\ &= \bar{p}(S) + \operatorname{codim}_X(S) - 2 + 2 - \operatorname{codim}_X(S) \\ &= \bar{p}(S) + \bar{t}_X(S) + 2 - \operatorname{codim}_X(S) \\ &= Q_{Q_{\bar{p},\bar{t}_Y},Q_{\bar{t}_X,\bar{q}}}(S\times T\times S\times T) - \operatorname{codim}_{X\times Y}(S\times T), \end{aligned}$$

which suffices.

Clearly we have a similar argument if S is regular and T is singular.

So now suppose S and T are both singular with respective links K, L of respective dimensions $m-1, n-1$. Then we have

$$Q_{\bar{p},\bar{q}}(S\times T) = \bar{p}(S) + \bar{q}(T) + C_1,$$

with $C_1 \in \{1,2\}$, the case $C_1 = 2$ occurring if and only if

$$I^{\bar{p}}H_{m-\bar{p}(S)-2}(K;R) * I^{\bar{q}}H_{n-\bar{q}(T)-2}(L;R) = 0.$$

Yet more complex, using that all spaces are locally torsion-free with respect to $\bar{t}$ by Example 5.3.12, we have

$$\begin{aligned} &Q_{Q_{\bar{p},\bar{t}_Y},Q_{\bar{t}_X,\bar{q}}}(S\times T\times S\times T) \\ &\quad= Q_{\bar{p},\bar{t}_Y}(S\times T) + Q_{\bar{t}_X,\bar{q}}(S\times T) + C_2 \\ &\quad= \bar{p}(S) + \bar{t}_Y(T) + 2 + \bar{t}_X(S) + \bar{q}(T) + 2 + C_2 \\ &\quad= \bar{p}(S) + \operatorname{codim}_Y(T) - 2 + 2 + \operatorname{codim}_X(S) - 2 + \bar{q}(T) + 2 + C_2 \\ &\quad= \bar{p}(S) + \bar{q}(T) + \operatorname{codim}_X(S) + \operatorname{codim}_Y(T) + C_2. \end{aligned}$$

As $S\times T$ has link $K * L$ of dimension $m+n-1$, here C_2 depends on the torsion product of $I^{Q_{\bar{p},\bar{t}_Y}}H_{m+n-Q_{\bar{p},\bar{t}_Y}(S\times T)-2}(K * L;R)$ with $I^{Q_{\bar{t}_X,\bar{q}}}H_{m+n-Q_{\bar{t}_X,\bar{q}}(S\times T)-2}(K * L;R)$. Amazingly, we can say something about these modules. We work with the first, as the second is analogous. Using again that all CS sets are locally torsion-free with respect to $\bar{t}$, we have $Q_{\bar{p},\bar{t}_Y}(S\times T)=\bar{p}(S)+\bar{t}_Y(T)+2$, and so

$$I^{Q_{\bar{p},\bar{t}_Y}}H_{m+n-Q_{\bar{p},\bar{t}_Y}(S\times T)-2}(K*L;R)=I^{Q_{\bar{p},\bar{t}_Y}}H_{m+n-\bar{p}(S)-\bar{t}_Y(T)-4}(K*L;R).$$

As in the proof of Proposition 6.4.15, we can apply equation (6.10) and the computations below it on page 311 to compute

$$\begin{aligned}
&I^{Q_{\bar{p},\bar{t}_Y}}H_{m+n-Q_{\bar{p},\bar{t}_Y}(S\times T)-2}(K*L;R)\\
&\quad=I^{Q_{\bar{p},\bar{t}_Y}}H_{m+n-\bar{p}(S)-\bar{t}_Y(T)-4}(K*L;R)\\
&\quad\cong\left(I^{\bar{p}}H_{m-\bar{p}(S)-2}(K;R)\otimes_R I^{\bar{t}_Y}H_{n-\bar{t}_Y(T)-2}(L;R)\right)\\
&\qquad\oplus\left(I^{\bar{p}}H_{m-\bar{p}(S)-3}(K;R)*_R I^{\bar{t}_Y}H_{n-\bar{t}_Y(T)-2}(L;R)\right)\\
&\qquad\oplus\left(I^{\bar{p}}H_{m-\bar{p}(S)-2}(K;R)*_R I^{\bar{t}_Y}H_{n-\bar{t}_Y(T)-3}(L;R)\right).
\end{aligned}$$

But $\bar{t}_Y(T)=\operatorname{codim}_Y(T)-2=n-2$. So $I^{\bar{t}_Y}H_{n-\bar{t}_Y(T)-3}(L;R)=I^{\bar{q}}H_{-1}(L;R)=0$ and $I^{\bar{t}_Y}H_{n-\bar{t}_Y(T)-2}(L;R)=I^{\bar{t}_Y}H_0(L;R)\cong I^{\bar{t}_Y}H_0^{\mathrm{GM}}(L;R)$, which is free. So the two torsion product terms vanish, and the tensor product term is a direct sum of copies of $I^{\bar{p}}H_{m-\bar{p}(S)-2}(K;R)$. In particular, $I^{Q_{\bar{p},\bar{t}_Y}}H_{m+n-Q_{\bar{p},\bar{t}_Y}(S\times T)-2}(K*L;R)$ has the same torsion as $I^{\bar{p}}H_{m-\bar{p}(S)-2}(K;R)$. Analogously, $I^{Q_{\bar{t}_X,\bar{q}}}H_{m+n-Q_{\bar{t}_X,\bar{q}}(S\times T)-2}(K*L;R)$ has the same torsion as $I^{\bar{q}}H_{n-\bar{q}(T)-2}(L;R)$. But this means that $C_1=C_2$.

So now we can compute

$$\begin{aligned}
&Q_{\bar{p},\bar{q}}(S\times T)\\
&\quad=\bar{p}(S)+\bar{q}(T)+C_1\\
&\quad=\bar{p}(S)+\bar{q}(T)+\operatorname{codim}_X(S)+\operatorname{codim}_Y(T)+C_1-\operatorname{codim}_X(S)-\operatorname{codim}_Y(T)\\
&\quad=Q_{Q_{\bar{p},\bar{t}_Y},Q_{\bar{t}_X,\bar{q}}}(S\times T\times S\times T)-\operatorname{codim}_{X\times Y}(S\times T),
\end{aligned}$$

which suffices to complete the proof. □

(b) Interchange Identities under Broad Assumptions

We would next like a statement of the form

$$(\alpha\times\gamma)\smile(\beta\times\delta)=(-1)^{|\gamma||\beta|}(\alpha\smile\beta)\times(\gamma\smile\delta).$$

For ordinary cohomology, the proof is very simple, utilizing the ordinary cohomology version of Proposition 7.3.46. It runs like this:

$$\begin{aligned}
(\alpha\times\gamma)\smile(\beta\times\delta)&=(p_1^*(\alpha)\smile p_2^*(\gamma))\smile(p_1^*(\beta)\smile p_2^*(\delta))\\
&=(-1)^{|\gamma||\beta|}p_1^*(\alpha)\smile p_1^*(\beta)\smile p_2^*(\gamma)\smile p_2^*(\delta)\\
&=(-1)^{|\gamma||\beta|}p_1^*(\alpha\smile\beta)\smile p_2^*(\gamma\smile\delta)\\
&=(-1)^{|\gamma||\beta|}(\alpha\smile\beta)\times(\gamma\smile\delta).
\end{aligned}$$

However, in order to apply this argument here, of course we need to make a number of assumptions; in particular, we need to make sure all the products in these formulas are defined. Rather than look into the above approach directly, we instead develop some more general technical lemmas that we will be able to apply both to the interchange of cup and cross products and to the interchange of cap and cross products. Thus we take a slightly different route than, for example, Dold [71].

As we did for the associative identities in Section 7.3.4, we will begin with very general assumptions about relationships among various perversities in order to have all products defined, then later we will show some conditions in which the general assumptions are satisfied.

We start with yet another algebraic lemma:

Lemma 7.3.48 *For chain complexes of R-modules C_*, D_*, E_*, F_*, the following diagram commutes:*

$$\begin{array}{ccccc}
\begin{matrix}\operatorname{Hom}(C_*,R)\otimes\operatorname{Hom}(D_*,R)\otimes\\ \operatorname{Hom}(E_*,R)\otimes\operatorname{Hom}(F_*,R)\end{matrix} & \xrightarrow{\Theta\otimes\Theta} & \begin{matrix}\operatorname{Hom}(C_*\otimes D_*,R)\otimes\\ \operatorname{Hom}(E_*\otimes F_*,R)\end{matrix} & \xrightarrow{\Theta} & \begin{matrix}\operatorname{Hom}(C_*\otimes D_*\otimes\\ E_*\otimes F_*,R)\end{matrix}\\
\Big\downarrow{\scriptstyle \mathrm{id}\otimes\tau\otimes\mathrm{id}} & & & & \Big\downarrow{\scriptstyle (\mathrm{id}\otimes\tau\otimes\mathrm{id})^*}\\
\begin{matrix}\operatorname{Hom}(C_*,R)\otimes\operatorname{Hom}(E_*,R)\otimes\\ \operatorname{Hom}(D_*,R)\otimes\operatorname{Hom}(F_*,R)\end{matrix} & \xrightarrow{\Theta\otimes\Theta} & \begin{matrix}\operatorname{Hom}(C_*\otimes E_*,R)\otimes\\ \operatorname{Hom}(D_*\otimes F_*,R)\end{matrix} & \xrightarrow{\Theta} & \begin{matrix}\operatorname{Hom}(C_*\otimes E_*\otimes\\ D_*\otimes F_*,R).\end{matrix}
\end{array}$$

Proof Let $\alpha \in \operatorname{Hom}(C_i,R)$, $\beta \in \operatorname{Hom}(D_j,R)$, $\gamma \in \operatorname{Hom}(E_k,R)$, $\delta \in \operatorname{Hom}(F_\ell,R)$, $x \in C_i$, $y \in D_j$, $z \in E_k$, and $w \in F_\ell$. Then

$$\begin{aligned}
&[(\mathrm{id}\otimes\tau\otimes\mathrm{id})^*\Theta(\Theta\otimes\Theta)(\alpha\otimes\beta\otimes\gamma\otimes\delta)](x\otimes z\otimes y\otimes w)\\
&\quad= (-1)^{jk}[\Theta(\Theta\otimes\Theta)(\alpha\otimes\beta\otimes\gamma\otimes\delta)](x\otimes y\otimes z\otimes w)\\
&\quad= (-1)^{jk}[\Theta(\Theta(\alpha\otimes\beta)\otimes\Theta(\gamma\otimes\delta))](x\otimes y\otimes z\otimes w)\\
&\quad= (-1)^{jk+(k+\ell)(i+j)}(\Theta(\alpha\otimes\beta)(x\otimes y))(\Theta(\gamma\otimes\delta)(z\otimes w))\\
&\quad= (-1)^{jk+(k+\ell)(i+j)+ij+k\ell}\alpha(x)\beta(y)\gamma(z)\delta(w)\\
&\quad= (-1)^{jk+ik+i\ell+jk+j\ell+ij+k\ell}\alpha(x)\beta(y)\gamma(z)\delta(w)\\
&\quad= (-1)^{ik+i\ell+j\ell+ij+k\ell}\alpha(x)\beta(y)\gamma(z)\delta(w).
\end{aligned}$$

On the other hand,

$$
\begin{aligned}
&[\Theta(\Theta \otimes \Theta)(\mathrm{id} \otimes \tau \otimes \mathrm{id})(\alpha \otimes \beta \otimes \gamma \otimes \delta)](x \otimes z \otimes y \otimes w) \\
&\quad = (-1)^{jk}[\Theta(\Theta \otimes \Theta)(\alpha \otimes \gamma \otimes \beta \otimes \delta)](x \otimes z \otimes y \otimes w) \\
&\quad = (-1)^{jk}[\Theta(\Theta(\alpha \otimes \gamma) \otimes \Theta(\beta \otimes \delta))](x \otimes z \otimes y \otimes w) \\
&\quad = (-1)^{jk+(j+\ell)(i+k)}(\Theta(\alpha \otimes \gamma)(x \otimes z))(\Theta(\beta \otimes \delta)(y \otimes w)) \\
&\quad = (-1)^{jk+(j+\ell)(i+k)+ik+j\ell}\alpha(x)\gamma(z)\beta(y)\delta(w) \\
&\quad = (-1)^{jk+ij+jk+i\ell+k\ell+ik+j\ell}\alpha(x)\gamma(z)\beta(y)\delta(w) \\
&\quad = (-1)^{ij+i\ell+k\ell+ik+j\ell}\alpha(x)\gamma(z)\beta(y)\delta(w).
\end{aligned}
$$

These two expressions are equal. With any other combination of degrees (i.e. if α and x do not have corresponding degrees), then each expression is zero. □

Now we turn to diagrams of intersection chains, verifying the commutativity while putting off for later the question of whether a sufficient collection of perversities can be found for the maps of the diagram to all make sense.

Lemma 7.3.49 *Let R be a Dedekind domain. Suppose that X and Y are CS sets with subsets $A, B \subset X$ and $C, D \subset Y$. Let $(\bar p, \bar q; \bar r)$ be a Q_1-agreeable triple of perversities on X and $(\bar u, \bar v; \bar s)$ a Q_2-agreeable triple of perversities on Y. Suppose perversities Q_3, Q_4, Q_5, Q_6, Q_7 exist so that all the maps in the following diagrams are well defined. Then the diagrams commute*[16] *(R coefficients tacit).*

$$
\begin{array}{ccc}
\begin{array}{c} I^{Q_5}S_*(X \times Y; ((A \cup B) \times Y) \cup \\ (X \times (C \cup D))) \end{array} & \xrightarrow{\ \mathbf{d}\ } & \begin{array}{c} I^{Q_6}S_*(X \times Y \times X \times Y; \\ (((A \times Y) \cup (X \times C)) \times (X \times Y)) \cup \\ ((X \times Y) \times ((B \times Y) \cup (X \times D)))) \end{array} \\
 & & \uparrow{\scriptstyle \mathrm{id} \times t \times \mathrm{id}} \\
\uparrow{\scriptstyle \varepsilon} & & \begin{array}{c} I^{Q_7}S_*(X \times X \times Y \times Y; \\ (((A \times X) \cup (X \times B)) \times (Y \times Y)) \cup \\ ((X \times X) \times ((C \times Y) \cup (Y \times D)))) \end{array} \\
 & & \uparrow{\scriptstyle \varepsilon} \\
\begin{array}{c} I^{\bar r}S_*(X, A \cup B) \otimes \\ I^{\bar s}S_*(Y, C \cup D) \end{array} & \xrightarrow{\ \mathbf{d} \otimes \mathbf{d}\ } & \begin{array}{c} I^{Q_1}S_*(X \times X, (A \times X) \cup (X \times B)) \otimes \\ I^{Q_2}S_*(Y \times Y, (C \times Y) \cup (Y \times D)) \end{array}
\end{array}
$$

16 Notice that the diagrams fit together side by side to make one large diagram, but it wouldn't fit on the page!

$$
\begin{array}{ccc}
\begin{array}{c} I^{Q_6}S_*(X\times Y\times X\times Y; \\ (((A\times Y)\cup(X\times C))\times(X\times Y))\cup \\ ((X\times Y)\times((B\times Y)\cup(X\times D)))) \end{array} & \xleftarrow{\varepsilon} & \begin{array}{c} I^{Q_3}S_*(X\times Y;(A\times Y)\cup \\ (X\times C))\otimes I^{Q_4}S_*(X\times Y; \\ (B\times Y)\cup(X\times D))) \end{array} \\
\Big\uparrow{\scriptstyle \mathrm{id}\times t\times\mathrm{id}}\ \cong & & \Big\uparrow{\scriptstyle \varepsilon\otimes\varepsilon} \\
\begin{array}{c} I^{Q_7}S_*(X\times X\times Y\times Y; \\ (((A\times X)\cup(X\times B))\times(Y\times Y))\cup \\ ((X\times X)\times((C\times Y)\cup(Y\times D)))) \end{array} & & \begin{array}{c} I^{\bar p}S_*(X,A)\otimes I^{\bar u}S_*(Y,C)\otimes \\ I^{\bar q}S_*(X,B)\otimes I^{\bar v}S_*(Y,D) \end{array} \\
\Big\uparrow{\scriptstyle \varepsilon} & & \Big\uparrow{\scriptstyle \mathrm{id}\otimes\tau\otimes\mathrm{id}}\ \cong \\
\begin{array}{c} I^{Q_1}S_*(X\times X,(A\times X)\cup(X\times B))\otimes \\ I^{Q_2}S_*(Y\times Y,(C\times Y)\cup(Y\times D)) \end{array} & \xleftarrow{\varepsilon\otimes\varepsilon} & \begin{array}{c} I^{\bar p}S_*(X,A)\otimes I^{\bar q}S_*(X,B)\otimes \\ I^{\bar u}S_*(Y,C)\otimes I^{\bar v}S_*(Y,D) \end{array}
\end{array}
$$

Proof Notice that

$$
\begin{aligned}
(A\times Y)\cup(X\times C)\cup(B\times Y)\cup(X\times D) &= (A\times Y)\cup(B\times Y)\cup(X\times C)\cup(X\times D)\\
&= ((A\cup B)\times Y)\cup(X\times(C\cup D))
\end{aligned}
$$

so that the subspaces are correct for the top diagonal map.

We also need to show that the map labeled $\mathrm{id}\times t\times\mathrm{id}$ makes sense. At the space level, this map interchanges the second and third coordinates. We can rewrite

$$(((A\times X)\cup(X\times B))\times(Y\times Y))\cup((X\times X)\times((C\times Y)\cup(Y\times D)))$$

as

$$(A\times X\times Y\times Y)\cup(X\times B\times Y\times Y)\cup(X\times X\times C\times Y)\cup(X\times X\times Y\times D).$$

Applying $\mathrm{id}\times t\times\mathrm{id}$, which is a homeomorphism, we get the space

$$(A\times Y\times X\times Y)\cup(X\times Y\times B\times Y)\cup(X\times C\times X\times Y)\cup(X\times Y\times X\times D),$$

and this is equal to

$$(((A\times Y)\cup(X\times C))\times(X\times Y))\cup((X\times Y)\times((B\times Y)\cup(X\times D))).$$

So $\mathrm{id}\times t\times\mathrm{id}$ is a homeomorphism of pairs of spaces, and we assume it is well defined as a map of intersection chains.

We now turn to verifying the commutativity. It might be helpful to compare the following argument with the proof of Proposition 5.2.20.

As in some of our previous arguments in this section (for example, in the proof of Lemma 7.3.20), it now suffices to verify commutativity of the analogous diagram of ordinary singular chain complexes. So, suppose σ is a singular a-simplex of X and τ is a singular b-simplex of Y. So $\sigma\otimes\tau$ represents an element of $S_a(X,A\cup B;R)\otimes S_*(Y,C\cup D;R)$. Then $\varepsilon(\sigma\otimes\tau)$ is the singular chain obtained by applying the product map $\sigma\times\tau:\Delta^a\times\Delta^b\to X\times Y$ to the Eilenberg–Zilber shuffle triangulation of $\Delta^a\times\Delta^b$, and $\mathbf{d}\varepsilon(\sigma\otimes\tau)$ applies

$(\sigma \times \tau, \sigma \times \tau)$ to the Eilenberg–Zilber shuffle triangulation of $\Delta^a \times \Delta^b$. Note that here we write $\sigma \times \tau : \Delta^a \times \Delta^b \to X \times Y$ and

$$(\sigma \times \tau, \sigma \times \tau) : \Delta^a \times \Delta^b \to (X \times Y) \times (X \times Y).$$

In other words, each point $(x, y) \in \Delta^a \times \Delta^b$ gets taken to $(\sigma(x), \tau(y), \sigma(x), \tau(y)) \in X \times Y \times X \times Y$.

On the other hand,

$$(\mathbf{d} \otimes \mathbf{d})(\sigma \otimes \tau) = \mathbf{d}(\sigma) \otimes \mathbf{d}(\tau) = (\sigma, \sigma) \otimes (\tau, \tau).$$

Then $\varepsilon(\mathbf{d} \otimes \mathbf{d})(\sigma \otimes \tau)$ applies $(\sigma, \sigma) \times (\tau, \tau)$ to the Eilenberg–Zilber shuffle singular triangulation of $\Delta^a \times \Delta^b$, and so each point $(x, y) \in \Delta^a \times \Delta^b$ gets taken to

$$(\sigma(x), \sigma(x), \tau(y), \tau(y)) \in X \times Y \times X \times Y.$$

Including the map $\mathrm{id} \times t \times \mathrm{id}$ therefore results in the top diagram commuting.

Next, suppose $\sigma, \tau, \upsilon, \omega$ are singular simplices such that $\sigma \otimes \tau \otimes \upsilon \otimes \omega$ represents an element of

$$S_a(X, A; R) \otimes S_b(X, B; R) \otimes S_c(Y, C; R) \otimes S_d(Y, D; R).$$

Then

$$(\mathrm{id} \otimes \tau \otimes \mathrm{id})(\sigma \otimes \tau \otimes \upsilon \otimes \omega) = (-1)^{bc} \sigma \otimes \upsilon \otimes \tau \otimes \omega.$$

So we see that, up to our sign $(-1)^{bc}$, the chain

$$\varepsilon(\varepsilon \otimes \varepsilon)(\mathrm{id} \otimes \tau \otimes \mathrm{id})(\sigma \otimes \tau \otimes \upsilon \otimes \omega)$$

is obtained by applying $\sigma \times \upsilon \times \tau \times \omega$ to a singular triangulation of $\Delta^a \times \Delta^c \times \Delta^b \times \Delta^d$ obtained by taking the Eilenberg–Zilber shuffle triangulations of $\Delta^a \times \Delta^c$ and $\Delta^b \times \Delta^d$ and then the Eilenberg–Zilber shuffle triangulation of the product of simplicial complexes $(\Delta^a \times \Delta^c) \times (\Delta^b \times \Delta^d)$.

On the other hand,

$$(\mathrm{id} \times t \times \mathrm{id})\varepsilon(\varepsilon \otimes \varepsilon)$$

is similarly defined, up to sign, by applying $(\mathrm{id} \times t \times \mathrm{id})(\sigma \times \tau \times \upsilon \times \omega)$ to a singular triangulation of $\Delta^a \times \Delta^b \times \Delta^c \times \Delta^d$ obtained by taking the Eilenberg–Zilber shuffle triangulations of $\Delta^a \times \Delta^b$ and $\Delta^c \times \Delta^d$ and then the Eilenberg–Zilber shuffle triangulation of the product of simplicial complexes $(\Delta^a \times \Delta^b) \times (\Delta^c \times \Delta^d)$. But we have seen in the proofs of Proposition 7.3.29 and Proposition 5.2.20 that, ignoring orientations, the Eilenberg–Zilber triangulation process is strictly associative and commutative, implying that if we apply $\mathrm{id} \times t \times \mathrm{id}$ to our triangulation of $\Delta^a \times \Delta^b \times \Delta^c \times \Delta^d$ in this paragraph, we obtain our triangulation of $\Delta^a \times \Delta^c \times \Delta^b \times \Delta^d$ from the preceding paragraph. However, since the orientation of the singular subdivision chain depends on the orientation of the underlying space, the signs of the chains in the two singular triangulations will differ by a factor of $(-1)^{bc}$, accounting for the interchange of Δ^b with Δ^c. Now since

$$(\mathrm{id} \times t \times \sigma)(\sigma \times \tau \times \upsilon \times \omega) = (\sigma \times \upsilon \times \tau \times \omega)(\mathrm{id} \times t \times \sigma),$$

where the t on the left is the one in the diagram and the one on the right acts on $\Delta^b \times \Delta^c$, we see that, also accounting for the sign $(-1)^{bc}$ that comes from τ, the rectangle on the right side of the diagram commutes exactly. □

We can now state and prove our desired identities in general form:

Lemma 7.3.50 *Let R be a Dedekind domain. Suppose that X and Y are CS sets with open subsets $A, B \subset X$ and $C, D \subset Y$. Let $(\bar{p}, \bar{q}; \bar{r})$ be a Q_1-agreeable triple of perversities on X and $(\bar{u}, \bar{v}; \bar{s})$ a Q_2-agreeable triple of perversities on Y. Suppose perversities Q_3, Q_4, Q_5, Q_6, Q_7 exist so that all the maps in the diagrams of Lemma 7.3.49 exist and so that all of the ε maps (except perhaps the vertical one shared by both diagrams) are chain homotopy equivalences.*

Let $\alpha \in I_{\bar{p}}H^i(X, A; R)$, $\beta \in I_{\bar{q}}H^j(X, B; R)$, $\gamma \in I_{\bar{u}}H^k(Y, C; R)$, and $\delta \in I_{\bar{v}}H^\ell(Y, D; R)$. Suppose $\alpha \smile \beta \in I_{\bar{r}}H^{i+j}(X, A \cup B; R)$, $\gamma \smile \delta \in I_{\bar{s}}H^{k+\ell}(Y, C \cup D; R)$, $\alpha \times \gamma \in I_{Q_3}H^{i+k}(X \times Y, (A \times Y) \cup (X \times C); R)$, and $\beta \times \delta \in I_{Q_4}H^{j+\ell}(X \times Y, (B \times Y) \cup (X \times D); R)$. Then

$$(\alpha \smile \beta) \times (\gamma \smile \delta) = (-1)^{jk}(\alpha \times \gamma) \smile (\beta \times \delta)$$
$$\in I_{Q_5}H^{i+j+k+\ell}(X \times Y, ((A \cup B) \times Y) \cup (X \times (C \cup D)); R).$$

Proof By assumption, we can utilize the diagrams of Lemma 7.3.49 with all ε arrows (except perhaps the vertical one shared by both diagrams) reversed and replaced with IAW maps to obtain a diagram that is homotopy commutative (putting the two diagrams of the lemma together).

Now, we compute[17]:

$$
\begin{aligned}
&(\alpha \smile \beta)\times(\gamma \smile \delta)\\
&= \mathrm{IAW}^*\Theta((\alpha \smile \beta)\otimes(\gamma \smile \delta))\\
&= \mathrm{IAW}^*\Theta(\bar{\mathbf{d}}^*\Theta(\alpha\otimes\beta)\otimes\bar{\mathbf{d}}^*\Theta(\gamma\otimes\delta))\\
&= \mathrm{IAW}^*\Theta(\bar{\mathbf{d}}^*\Theta\otimes\bar{\mathbf{d}}^*\Theta)(\alpha\otimes\beta\otimes\gamma\otimes\delta)\\
&= \mathrm{IAW}^*\Theta(\bar{\mathbf{d}}^*\otimes\bar{\mathbf{d}}^*)(\Theta\otimes\Theta)(\alpha\otimes\beta\otimes\gamma\otimes\delta)\\
&= \mathrm{IAW}^*(\bar{\mathbf{d}}\otimes\bar{\mathbf{d}})^*\Theta(\Theta\otimes\Theta)(\alpha\otimes\beta\otimes\gamma\otimes\delta) \quad \text{by Lemma 7.3.1}\\
&= \mathrm{IAW}^*(\mathrm{IAW}\mathbf{d}\otimes\mathrm{IAW}\mathbf{d})^*\Theta(\Theta\otimes\Theta)(\alpha\otimes\beta\otimes\gamma\otimes\delta)\\
&= [(\mathrm{IAW}\otimes\mathrm{IAW})(\mathbf{d}\otimes\mathbf{d})\mathrm{IAW}]^*\Theta(\Theta\otimes\Theta)(\alpha\otimes\beta\otimes\gamma\otimes\delta)\\
&= [(\mathrm{id}\otimes\tau\otimes\mathrm{id})(\mathrm{IAW}\otimes\mathrm{IAW})\mathrm{IAW}\mathbf{d}]^*\Theta(\Theta\otimes\Theta)(\alpha\otimes\beta\otimes\gamma\otimes\delta)\\
&\qquad\qquad\qquad\qquad\qquad\qquad\qquad\qquad\qquad\qquad \text{by Lemma 7.3.49}\\
&= \mathbf{d}^*\mathrm{IAW}^*(\mathrm{IAW}\otimes\mathrm{IAW})^*(\mathrm{id}\otimes\tau\otimes\mathrm{id})^*\Theta(\Theta\otimes\Theta)(\alpha\otimes\beta\otimes\gamma\otimes\delta)\\
&= \mathbf{d}^*\mathrm{IAW}^*(\mathrm{IAW}\otimes\mathrm{IAW})^*\Theta(\Theta\otimes\Theta)(\mathrm{id}\otimes\tau\otimes\mathrm{id})(\alpha\otimes\beta\otimes\gamma\otimes\delta)\\
&\qquad\qquad\qquad\qquad\qquad\qquad\qquad\qquad\qquad\qquad \text{by Lemma 7.3.48}\\
&= (-1)^{jk}\bar{\mathbf{d}}^*(\mathrm{IAW}\otimes\mathrm{IAW})^*\Theta(\Theta\otimes\Theta)(\alpha\otimes\gamma\otimes\beta\otimes\delta)\\
&= (-1)^{jk}\bar{\mathbf{d}}^*\Theta(\mathrm{IAW}^*\otimes\mathrm{IAW}^*)(\Theta\otimes\Theta)(\alpha\otimes\gamma\otimes\beta\otimes\delta) \quad \text{by Lemma 7.3.1}\\
&= (-1)^{jk}\bar{\mathbf{d}}^*\Theta(\mathrm{IAW}^*\Theta\otimes\mathrm{IAW}^*\Theta)(\alpha\otimes\gamma\otimes\beta\otimes\delta)\\
&= (-1)^{jk}\bar{\mathbf{d}}^*\Theta(\mathrm{IAW}^*\Theta(\alpha\otimes\gamma)\otimes\mathrm{IAW}^*\Theta(\beta\otimes\delta))\\
&= (-1)^{jk}\bar{\mathbf{d}}^*\Theta((\alpha\times\gamma)\otimes(\beta\times\delta))\\
&= (-1)^{jk}(\alpha\times\gamma)\smile(\beta\times\delta). \qquad \square
\end{aligned}
$$

Next we look at the interaction of the cap and cross products. Because we do not need the Künneth Theorem to form the cross product of chains (as opposed to cochains), there are fewer requirements needed on the ε maps.

Lemma 7.3.51 *Let R be a Dedekind domain. Suppose that X and Y are CS sets with open subsets $A, B \subset X$ and $C, D \subset Y$. Let $(\bar{p},\bar{q};\bar{r})$ be a Q_1-agreeable triple of perversities on X and $(\bar{u},\bar{v};\bar{s})$ a Q_2-agreeable triple of perversities on Y. Suppose perversities Q_3, Q_4, Q_5, Q_6, Q_7 exist so that all the maps in the diagrams of Lemma 7.3.49 exist and so that all of the* horizontal *ε maps are chain homotopy equivalences.*

Let $\alpha \in I_{\bar{q}}H^j(X,B;R)$, $x \in I^{\bar{r}}H_{i+j}(X, A\cup B;R)$, $\beta \in I_{\bar{v}}H^{\ell}(Y,D;R)$, and $y \in I^{\bar{s}}H_{k+\ell}(Y, C\cup D;R)$. Suppose $\alpha\times\beta \in I_{Q_4}H^{j+\ell}(X\times Y,(B\times Y)\cup(X\times D);R)$, $x\times y \in I^{Q_5}H_{i+j+k+\ell}(X{\times}Y,((A\cup B){\times}Y)\cup(X{\times}(C\cup D));R)$, $\alpha\frown x \in I^{\bar{p}}H_i(X,A;R)$ and

[17] Notice that, abusing notation, $\mathrm{id}^* = \mathrm{id}$, $\tau^{-1} = \tau$ and $\tau^* = \tau$. These expressions make perfect sense if each symbol is interpreted with the correct domain and codomain.

$\beta \frown y \in I^{\bar{u}}H_k(Y,C;R)$. *Then*

$$\begin{aligned}(\alpha \times \beta) \frown (x \times y) &= (-1)^{\ell(i+j)}(\alpha \frown x) \times (\beta \frown y)\\ &\in I^{Q_3}H_{i+k}(X \times Y, (A \times Y) \cup (X \times C); R).\end{aligned}$$

Proof The assumptions allow us to invoke the diagrams of Lemma 7.3.49, replacing the horizontal ε maps with IAW maps in the opposite directions to obtain a diagram that is homotopy commutative. Notice that IAW $\otimes$ IAW is indeed a chain homotopy inverse of $\varepsilon \otimes \varepsilon$ by Corollary A.2.6 and its proof.

Now we begin to compute using the definitions. Let us make specific choices of IAW maps and suppose that $\bar{\mathbf{d}}(x) = \sum_a u_a \otimes v_a \in I^{\bar{p}}S_*(X,A;R) \otimes I^{\bar{q}}S_*(X,B;R)$ and that $\bar{\mathbf{d}}(y) = \sum_b w_b \otimes z_b \in I^{\bar{u}}S_*(Y,C;R) \otimes I^{\bar{v}}S_*(Y,D;R)$.

Then we have

$$\begin{aligned}
&(\alpha \frown x) \times (\beta \frown y)\\
&\quad = [\Psi(\mathrm{id} \otimes \alpha)\bar{\mathbf{d}}(x)] \times [\Psi(\mathrm{id} \otimes \beta)\bar{\mathbf{d}}(y)]\\
&\quad = \left[\Psi(\mathrm{id} \otimes \alpha)\left(\sum_a u_a \otimes v_a\right)\right] \times \left[\Psi(\mathrm{id} \otimes \beta)\left(\sum_b w_b \otimes z_b\right)\right]\\
&\quad = \left[\left(\sum_a (-1)^{ij} u_a \alpha(v_a)\right)\right] \times \left[\left(\sum_b (-1)^{k\ell} w_b \beta(z_b)\right)\right]\\
&\quad = \sum_{a,b} (-1)^{ij+k\ell} \alpha(v_a)\beta(z_b) u_a \times w_b\\
&\quad = \sum_{a,b} (-1)^{ij+k\ell+j\ell} \left[(\alpha \times \beta)(v_a \times z_b)\right] u_a \times w_b \qquad \text{see Proposition 7.3.27}\\
&\quad = \sum_{a,b} (-1)^{ij+k\ell+j\ell+(j+\ell)(i+k)} \Psi(\mathrm{id} \otimes (\alpha \times \beta))((u_a \times w_b) \otimes (v_a \times z_b))\\
&\quad = (-1)^{j\ell+i\ell+jk} \Psi(\mathrm{id} \otimes (\alpha \times \beta)) \sum_{a,b} ((u_a \times w_b) \otimes (v_a \times z_b))\\
&\quad = (-1)^{j\ell+i\ell+jk} \Psi(\mathrm{id} \otimes (\alpha \times \beta)) \sum_{a,b} (\varepsilon(u_a \otimes w_b) \otimes \varepsilon(v_a \otimes z_b))\\
&\quad = (-1)^{j\ell+i\ell+jk} \Psi(\mathrm{id} \otimes (\alpha \times \beta)) \sum_{a,b} (\varepsilon \otimes \varepsilon)(u_a \otimes w_b \otimes v_a \otimes z_b)\\
&\quad = (-1)^{j\ell+i\ell+jk+jk} \Psi(\mathrm{id} \otimes (\alpha \times \beta)) \sum_{a,b} (\varepsilon \otimes \varepsilon)(\mathrm{id} \otimes \tau \otimes \mathrm{id})(u_a \otimes v_a \otimes w_b \otimes z_b)\\
&\quad = (-1)^{j\ell+i\ell} \Psi(\mathrm{id} \otimes (\alpha \times \beta))(\varepsilon \otimes \varepsilon)(\mathrm{id} \otimes \tau \otimes \mathrm{id})(\bar{\mathbf{d}}(x) \otimes \bar{\mathbf{d}}(y)).
\end{aligned}$$

Now, using the commutativity of the diagrams in Lemma 7.3.49 and the homotopy commutativity that ensues if we replace the horizontal ε maps in those diagrams with IAW maps in the opposite directions, we observe (starting from the bottom left corner of the diagram) that $(\varepsilon \otimes \varepsilon)(\mathrm{id} \otimes \tau \otimes \mathrm{id})(\bar{\mathbf{d}}(x) \otimes \bar{\mathbf{d}}(y))$ represents the same homology class as

$\bar{\mathbf{d}}\varepsilon(x \otimes y) = \bar{\mathbf{d}}(x \times y)$. So, using our argument from the proof of Lemma 7.2.20, where we showed that the cap product is well defined, we can replace $(\varepsilon \otimes \varepsilon)(\mathrm{id} \otimes \tau \otimes \mathrm{id})(\bar{\mathbf{d}}(x) \otimes \bar{\mathbf{d}}(y))$ with the homologous $\bar{\mathbf{d}}(x \times y)$ without altering the homology class of the full expression. Therefore, we have

$$\begin{aligned}
&(\alpha \frown x) \times (\beta \frown y) \\
&\quad = (-1)^{j\ell + i\ell}\Psi(\mathrm{id} \otimes (\alpha \times \beta))(\varepsilon \otimes \varepsilon)(\mathrm{id} \otimes \tau \otimes \mathrm{id})(\bar{\mathbf{d}}(x) \otimes \bar{\mathbf{d}}(y)) \\
&\quad = (-1)^{j\ell + i\ell}\Psi(\mathrm{id} \otimes (\alpha \times \beta))\bar{\mathbf{d}}(x \times y) \\
&\quad = (-1)^{\ell(i+j)}(\alpha \times \beta) \frown (x \times y).
\end{aligned}$$

□

(c) Interchange Identities in Some More Specific Settings

Our goal now is to show that there exist reasonable (or at least somewhat reasonable) hypotheses that guarantee the existence of Q_i perversities satisfying the hypotheses of Lemmas 7.3.49, 7.3.50, and 7.3.51. Once again, we will not pursue all possible scenarios but will limit ourselves to some that are simple to state, especially those that arise if we make enough locally torsion-free assumptions. If nothing else, we are then assured to have some identities when working with field coefficients or spaces that are nice enough. In fact, we will obtain two sets of results. The first is in some sense a stronger pair of results, providing a final product in a relatively large perversity. We will then need to scale that result back a bit to lower perversities that arise in later applications.

To begin, it is simplest to work with product perversities for which we always know whether we have a +1 or +2 summand. So let us define the following for use in this section:

Definition 7.3.52 If $\bar{p}$ and $\bar{q}$ are perversities on spaces X and Y and $a \in \mathbb{Z}$, let us define $\hat{Q}^a_{\bar{p},\bar{q}}$ so that

1. if $T \subset Y$ is regular, then $\hat{Q}^a_{\bar{p},\bar{q}}(S \times T) = \bar{p}(S)$,
2. if $S \subset X$ is regular, then $\hat{Q}^a_{\bar{p},\bar{q}}(S \times T) = \bar{q}(T)$,
3. if $S \subset X$ and $T \subset Y$ are both singular, then $\hat{Q}^a_{\bar{p},\bar{q}}(S \times T) = \bar{p}(S) + \bar{q}(T) + a$.

Principally we will be concerned with the cases $a \in \{1, 2\}$ of the definition.

Using these perversities, we have that the cross products of the form

$$I^{\bar{p}}S_*(X, A; R) \otimes I^{\bar{q}}S_*(Y, B; R) \longrightarrow I^{\hat{Q}^a_{\bar{p},\bar{q}}}S_*(X \times Y; (A \times Y) \cup (X \times B); R)$$

are always defined for $a \geq 0$; see the discussion preceding Theorem 6.3.19.

Lemma 7.3.53 *If*

- $D\bar{r} > D\bar{p} + D\bar{q}$, $D\bar{s} > D\bar{u} + D\bar{v}$, *and* $a = 1$, *or if*
- $D\bar{r} \geq D\bar{p} + D\bar{q}$, $D\bar{s} \geq D\bar{u} + D\bar{v}$, *and* $a = 2$,

then the hypotheses of Lemma 7.3.49 hold with

1. $Q_1 = \hat{Q}^a_{\bar{p},\bar{q}}$,

2. $Q_2 = \hat{Q}^a_{\bar{u},\bar{v}}$,
3. $Q_3 = \hat{Q}^a_{\bar{p},\bar{u}}$,
4. $Q_4 = \hat{Q}^a_{\bar{q},\bar{v}}$,
5. $Q_5 = \hat{Q}^a_{\bar{r},\bar{s}}$,
6. $Q_6 = \hat{Q}^a_{\hat{Q}^a_{\bar{p},\bar{u}},\hat{Q}^a_{\bar{q},\bar{v}}}$,
7. $Q_7 = \hat{Q}^a_{\hat{Q}^a_{\bar{p},\bar{q}},\hat{Q}^a_{\bar{u},\bar{v}}}$.

Proof From our discussion preceding the statement of the lemma, these choices make all the cross product maps in the diagrams of Lemma 7.3.49 well defined. So we only need to consider the other maps.

The map $\mathrm{id} \otimes \tau \otimes \mathrm{id}$ certainly makes sense, as do the maps

$$I^{\bar{r}}S_*(X, A \cup B) \xrightarrow{\mathbf{d}} I^{\hat{Q}^a_{\bar{p},\bar{q}}}S_*(X \times X, (A \times X) \cup (X \times B)),$$

$$I^{\bar{s}}S_*(Y, C \cup D) \xrightarrow{\mathbf{d}} I^{\hat{Q}^a_{\bar{u},\bar{v}}}S_*(Y \times Y, (C \times Y) \cup (Y \times D)),$$

using the basic computations concerning agreeability. So we need to consider only the diagonal map $\mathbf{d}$ in the top diagram of Lemma 7.3.49 and the map $\mathrm{id} \times t \times \mathrm{id}$.

We begin by considering the map $\mathbf{d}$ at the top of the first diagram of Lemma 7.3.49. From our discussion of agreeable triples preceding Definition 7.2.6, we must show that

$$\hat{Q}^a_{\bar{r},\bar{s}}(S \times T) \leq \hat{Q}^a_{\hat{Q}^a_{\bar{p},\bar{u}},\hat{Q}^a_{\bar{q},\bar{v}}}(S \times T \times S \times T) - \mathrm{codim}_{X \times Y}(S \times T). \tag{7.22}$$

As seen in the proof of Lemma 7.2.8, the condition $D\bar{r} > D\bar{p} + D\bar{q}$ is equivalent to

$$\bar{r}(S) < \bar{p}(S) + \bar{q}(S) + 2 - \mathrm{codim}_X(S)$$

for $S \subset X$ a singular stratum, and we have the same equivalences replacing all strict inequalities with non-strict inequalities. We can express this by saying that

$$\bar{r}(S) \leq \bar{p}(S) + \bar{q}(S) + a - \mathrm{codim}_X(S),$$

where $a \in \{1,2\}$ is chosen consistently with the hypotheses of the lemma. Similarly we have, for $T \subset Y$ singular,

$$\bar{s}(T) \leq \bar{u}(T) + \bar{v}(T) + a - \mathrm{codim}_Y(T).$$

The inequality (7.22) holds trivially if S and T are both regular strata. Suppose T is regular and S is singular. Then, plugging in $S \times T$ we have $\hat{Q}^a_{\bar{r},\bar{s}}(S \times T) = \bar{r}(S)$, while

$$\begin{aligned}
&\hat{Q}^a_{\hat{Q}^a_{\bar{p},\bar{u}},\hat{Q}^a_{\bar{q},\bar{v}}}(S \times T \times S \times T) - \mathrm{codim}_{X \times Y}(S \times T)\\
&= \hat{Q}^a_{\bar{p},\bar{u}}(S \times T) + \hat{Q}^a_{\bar{q},\bar{v}}(S \times T) + a - \mathrm{codim}_{X \times Y}(S \times T)\\
&= \bar{p}(S) + \bar{q}(S) + a - \mathrm{codim}_X(S).
\end{aligned}$$

So as we know $\bar{r}(S) \leq \bar{p}(S) + \bar{q}(S) + a - \mathrm{codim}_X(S)$ with either hypothesis, we see that the inequality (7.22) holds in this case. A similar computation verifies (7.22) if S is regular and T is singular.

Now, suppose S and T are both singular. Then, $\hat{Q}^a_{\bar{r},\bar{s}}(S\times T) = \bar{r}(S) + \bar{s}(T) + a$, while

$$\begin{aligned}
&\hat{Q}^a_{\hat{Q}^a_{\bar{p},\bar{u}},\hat{Q}^a_{\bar{q},\bar{v}}}(S\times T\times S\times T) - \operatorname{codim}_{X\times Y}(S\times T)\\
&\quad = \hat{Q}^a_{\bar{p},\bar{u}}(S\times T) + \hat{Q}^a_{\bar{q},\bar{v}}(S\times T) + a - \operatorname{codim}_{X\times Y}(S\times T)\\
&\quad = \bar{p}(S) + \bar{u}(T) + a + \bar{q}(S) + \bar{v}(T) + a + a - \operatorname{codim}_{X\times Y}(S\times T)\\
&\quad = \bar{p}(S) + \bar{u}(T) + \bar{q}(S) + \bar{v}(T) + 3a - \operatorname{codim}_{X\times Y}(S\times T).
\end{aligned}$$

So putting these together with the consequences of the hypotheses we have

$$\begin{aligned}
&\hat{Q}^a_{\bar{r},\bar{s}}(S\times T)\\
&\quad = \bar{r}(S) + \bar{s}(T) + a\\
&\quad \le \bar{p}(S) + \bar{q}(S) + a - \operatorname{codim}_X(S) + \bar{u}(T) + \bar{v}(T) + a - \operatorname{codim}_Y(T) + a\\
&\quad = \bar{p}(S) + \bar{q}(S) + \bar{u}(T) + \bar{v}(T) + 3a - \operatorname{codim}_{X\times Y}(S\times T)\\
&\quad = \hat{Q}^a_{\hat{Q}^a_{\bar{p},\bar{u}},\hat{Q}^a_{\bar{q},\bar{v}}}(S\times T\times S\times T) - \operatorname{codim}_{X\times Y}(S\times T).
\end{aligned}$$

So, again, the inequality (7.22) holds.

This completes our verification concerning the diagonal map at the top of the first diagram of Lemma 7.3.49.

Next, we need to check that $\mathrm{id}\times t\times \mathrm{id}$ is well defined as a map of intersection chains. In particular, we check that perversities of corresponding strata agree, at which point it becomes clear that allowable chains are taken to allowable chains. In other words, if S,T are strata of X and U,V are strata of Y, we need to know that

$$\hat{Q}^a_{\hat{Q}^a_{\bar{p},\bar{u}},\hat{Q}^a_{\bar{q},\bar{v}}}(S\times U\times T\times V) = \hat{Q}^a_{\hat{Q}^a_{\bar{p},\bar{q}},\hat{Q}^a_{\bar{u},\bar{v}}}(S\times T\times U\times V).$$

Surprisingly enough, this is true! We compute the relevant values in Table 7.1 depending on whether each of S,T,U,V is regular (denoted by an r) or singular (denoted by an s). The last column shows the common value of the perversity evaluations. The other two columns are meant to indicate the intermediate steps in each case, eliminating terms that evaluate directly to zero. □

The preceding lemma implies that Lemmas 7.3.50 and 7.3.51 hold if we can use the appropriate $\hat{Q}$ perversities in Lemma 7.3.49 and also have the ε maps required by Lemmas 7.3.50 and 7.3.51 be chain homotopy equivalences. This requires having $\hat{Q}^a_{\bar{p},\bar{q}}$ be $(\bar{p},\bar{q})$-compatible and similarly for the other cross products we wish to invert up to homotopy. We know that such compatibility will hold in general if $a = 1$, which we can use if we assume that $D\bar{r} > D\bar{p} + D\bar{q}$ and $D\bar{s} > D\bar{u} + D\bar{v}$. But if we only want to assume $D\bar{r} \ge D\bar{p} + D\bar{q}$ and $D\bar{s} \ge D\bar{u} + D\bar{v}$, which are situations that will be important when we consider duality, then we must use $a = 2$, which requires us to also make sure that enough local torsion conditions

Table 7.1

S	T	U	V	$\hat{Q}^a_{\hat{Q}^a_{\bar{p},\bar{q}},\hat{Q}^a_{\bar{u},\bar{v}}}(S\times T\times U\times V)$	$\hat{Q}^a_{\hat{Q}^a_{\bar{p},\bar{u}},\hat{Q}^a_{\bar{q},\bar{v}}}(S\times U\times T\times V)$	common value
r	r	r	r	0	0	0
s	r	r	r	$\hat{Q}^a_{\bar{p},\bar{q}}(S\times T)$	$\hat{Q}^a_{\bar{p},\bar{q}}(S\times U)$	$\bar{p}(S)$
r	s	r	r	$\hat{Q}^a_{\bar{p},\bar{q}}(S\times T)$	$\hat{Q}^a_{\bar{q},\bar{v}}(T\times V)$	$\bar{q}(T)$
r	r	s	r	$\hat{Q}^a_{\bar{u},\bar{v}}(U\times V)$	$\hat{Q}^a_{\bar{p},\bar{u}}(S\times U)$	$\bar{u}(U)$
r	r	r	s	$\hat{Q}^a_{\bar{u},\bar{v}}(U\times V)$	$\hat{Q}^a_{\bar{q},\bar{v}}(T\times V)$	$\bar{v}(V)$
s	s	r	r	$\hat{Q}^a_{\bar{p},\bar{q}}(S\times T)$	$\hat{Q}^a_{\bar{p},\bar{u}}(S\times U)+\hat{Q}^a_{\bar{q},\bar{v}}(T\times V)+a$	$\bar{p}(S)+\bar{q}(T)+a$
s	r	s	r	$\hat{Q}^a_{\bar{p},\bar{q}}(S\times T)+\hat{Q}^a_{\bar{u},\bar{v}}(U\times V)+a$	$\hat{Q}^a_{\bar{p},\bar{u}}(S\times U)$	$\bar{p}(S)+\bar{u}(U)+a$
s	r	r	s	$\hat{Q}^a_{\bar{p},\bar{q}}(S\times T)+\hat{Q}^a_{\bar{u},\bar{v}}(U\times V)+a$	$\hat{Q}^a_{\bar{p},\bar{u}}(S\times U)+\hat{Q}^a_{\bar{q},\bar{v}}(T\times V)+a$	$\bar{p}(S)+\bar{v}(V)+a$
r	s	s	r	$\hat{Q}^a_{\bar{p},\bar{q}}(S\times T)+\hat{Q}^a_{\bar{u},\bar{v}}(U\times V)+a$	$\hat{Q}^a_{\bar{p},\bar{u}}(S\times U)+\hat{Q}^a_{\bar{q},\bar{v}}(T\times V)+a$	$\bar{q}(T)+\bar{u}(U)+a$
r	s	r	s	$\hat{Q}^a_{\bar{p},\bar{q}}(S\times T)+\hat{Q}^a_{\bar{u},\bar{v}}(U\times V)+a$	$\hat{Q}^a_{\bar{q},\bar{v}}(T\times V)$	$\bar{q}(T)+\bar{v}(V)+a$
r	r	s	s	$\hat{Q}^a_{\bar{u},\bar{v}}(U\times V)$	$\hat{Q}^a_{\bar{p},\bar{u}}(S\times U)+\hat{Q}^a_{\bar{q},\bar{v}}(T\times V)+a$	$\bar{u}(U)+\bar{v}(V)+a$
s	s	s	r	$\hat{Q}^a_{\bar{p},\bar{q}}(S\times T)+\hat{Q}^a_{\bar{u},\bar{v}}(U\times V)+a$	$\hat{Q}^a_{\bar{p},\bar{u}}(S\times U)+\hat{Q}^a_{\bar{q},\bar{v}}(T\times V)+a$	$\bar{p}(S)+\bar{q}(T)+\bar{u}(U)+2a$
s	s	r	s	$\hat{Q}^a_{\bar{p},\bar{q}}(S\times T)+\hat{Q}^a_{\bar{u},\bar{v}}(U\times V)+a$	$\hat{Q}^a_{\bar{p},\bar{u}}(S\times U)+\hat{Q}^a_{\bar{q},\bar{v}}(T\times V)+a$	$\bar{p}(S)+\bar{q}(T)+\bar{v}(V)+2a$
s	r	s	s	$\hat{Q}^a_{\bar{p},\bar{q}}(S\times T)+\hat{Q}^a_{\bar{u},\bar{v}}(U\times V)+a$	$\hat{Q}^a_{\bar{p},\bar{u}}(S\times U)+\hat{Q}^a_{\bar{q},\bar{v}}(T\times V)+a$	$\bar{p}(S)+\bar{u}(U)+\bar{v}(V)+2a$
r	s	s	s	$\hat{Q}^a_{\bar{p},\bar{q}}(S\times T)+\hat{Q}^a_{\bar{u},\bar{v}}(U\times V)+a$	$\hat{Q}^a_{\bar{p},\bar{u}}(S\times U)+\hat{Q}^a_{\bar{q},\bar{v}}(T\times V)+a$	$\bar{q}(T)+\bar{u}(U)+\bar{v}(V)+2a$
s	s	s	s	$\hat{Q}^a_{\bar{p},\bar{q}}(S\times T)+\hat{Q}^a_{\bar{u},\bar{v}}(U\times V)+a$	$\hat{Q}^a_{\bar{p},\bar{u}}(S\times U)+\hat{Q}^a_{\bar{q},\bar{v}}(T\times V)+a$	$\bar{p}(S)+\bar{q}(T)+\bar{u}(U)+\bar{v}(V)+3a$

are met. The following statements contain enough torsion conditions to ensure we have all the needed compatibilities, at the expense of a lot of assumptions. The assumptions are worse in the analogue of Lemma 7.3.50, as we require more of the ε maps to be chain homotopy equivalences for that lemma.

Proposition 7.3.54 *Let R be a Dedekind domain. Suppose that X is a CS set with perversities $\bar{p}$, $\bar{q}$, $\bar{r}$ with $D\bar{r} > D\bar{p} + D\bar{q}$ and that Y is a CS set with perversities $\bar{u}$, $\bar{v}$, $\bar{s}$ with $D\bar{s} > D\bar{u} + D\bar{v}$. Let $A, B \subset X$ be open subsets and $C, D \subset Y$ be open subsets.*

Let

$$\alpha \in I_{\bar{p}}H^i(X,A;R), \quad \beta \in I_{\bar{q}}H^j(X,B;R), \quad \gamma \in I_{\bar{u}}H^k(Y,C;R), \quad \delta \in I_{\bar{v}}H^\ell(Y,D;R).$$

Suppose $\alpha \smile \beta \in I_{\bar{r}}H^{i+j}(X, A \cup B; R)$, $\gamma \smile \delta \in I_{\bar{s}}H^{k+\ell}(Y, C \cup D; R)$, $\alpha \times \gamma \in I_{\hat{Q}^a_{\bar{p},\bar{u}}}H^{i+k}(X\times Y, (A\times Y)\cup(X\times C); R)$, and $\beta\times\delta \in I_{\hat{Q}^a_{\bar{q},\bar{v}}}H^{j+\ell}(X\times Y, (B\times Y)\cup(X\times D); R)$, with $a = 1$.

Then

$$\begin{aligned}(\alpha \smile \beta) \times (\gamma \smile \delta) &= (-1)^{jk}(\alpha \times \gamma) \smile (\beta \times \delta)\\ &\in I_{\hat{Q}^a_{\bar{r},\bar{s}}}H^{i+j+k+\ell}(X \times Y, ((A \cup B) \times Y) \cup (X \times (C \cup D)); R).\end{aligned}$$

Under the assumptions, $D\bar{r} \geq D\bar{p} + D\bar{q}$ and $D\bar{s} \geq D\bar{u} + D\bar{v}$ then the above equalities hold with $a = 2$ assuming the following conditions are satisfied[18]*:*

- *X is locally $(\bar{p}, R)$-torsion-free or locally $(\bar{q}, R)$-torsion-free,*
- *Y is locally $(\bar{u}, R)$-torsion-free or locally $(\bar{v}, R)$-torsion-free,*
- *X is locally $(\bar{p}, R)$-torsion-free or Y is locally $(\bar{u}, R)$-torsion-free,*
- *X is locally $(\bar{q}, R)$-torsion-free or Y is locally $(\bar{v}, R)$-torsion-free,*
- *$X \times Y$ is locally $(\hat{Q}^2_{\bar{p},\bar{u}}, R)$-torsion-free or locally $(\hat{Q}^2_{\bar{q},\bar{v}}, R)$-torsion-free,*
- *X is locally $(\bar{r}, R)$-torsion-free or Y is locally $(\bar{s}, R)$-torsion-free.*

Proposition 7.3.55 *Let R be a Dedekind domain. Suppose that X is a CS set with perversities $\bar{p}$, $\bar{q}$, $\bar{r}$ with $D\bar{r} > D\bar{p} + D\bar{q}$ and that Y is a CS set with perversities $\bar{u}$, $\bar{v}$, $\bar{s}$ with $D\bar{s} > D\bar{u} + D\bar{v}$. Let $A, B \subset X$ and $C, D \subset Y$ be open subsets.*

Let

$$\begin{aligned}\alpha &\in I_{\bar{q}}H^j(X,B;R), & x &\in I^{\bar{r}}H_{i+j}(X, A \cup B; R),\\ \beta &\in I_{\bar{v}}H^\ell(Y,D;R), & y &\in I^{\bar{s}}H_{k+\ell}(Y, C \cup D; R).\end{aligned}$$

Suppose $\alpha\times\beta \in I_{\hat{Q}^a_{\bar{q},\bar{v}}}H^{j+\ell}(X\times Y, (B\times Y)\cup(X\times D); R)$, $x\times y \in I^{\hat{Q}^a_{\bar{r},\bar{s}}}H_{i+j+k+\ell}(X\times Y, ((A\cup B) \times Y) \cup (X \times (C \cup D)); R)$, $\alpha \frown x \in I^{\bar{p}}H_i(X, A; R)$, and $\beta \frown y \in$

[18] These conditions are not all independent; see, in particular, Proposition 6.4.15.

$I^{\bar{u}}H_k(Y,C;R)$, with $a=1$. Then

$$(\alpha\times\beta)\frown(x\times y)=(-1)^{\ell(i+j)}(\alpha\frown x)\times(\beta\frown y)$$
$$\in I^{\hat{Q}^a_{\bar{p},\bar{u}}}H_{i+k}(X\times Y,(A\times Y)\cup(X\times C);R).$$

Under the assumptions $D\bar{r}\geq D\bar{p}+D\bar{q}$ and $D\bar{s}\geq D\bar{u}+D\bar{v}$ then the above equalities hold with $a=2$ assuming the following conditions are satisfied[19]*:*

- *X is locally $(\bar{p},R)$-torsion-free or locally $(\bar{q},R)$-torsion-free,*
- *Y is locally $(\bar{u},R)$-torsion-free or locally $(\bar{v},R)$-torsion-free,*
- *$X\times Y$ is locally $(\hat{Q}^2_{\bar{p},\bar{u}},R)$-torsion-free or locally $(\hat{Q}^2_{\bar{q},\bar{v}},R)$-torsion-free.*

Unfortunately, we need one more iteration of this, as Propositions 7.3.54 and 7.3.55 are actually a little "too strong." For example, the end result of Proposition 7.3.54 ends up in

$$I_{\hat{Q}^a_{\bar{r},\bar{s}}}H^{i+j+k+\ell}(X\times Y,((A\cup B)\times Y)\cup(X\times(C\cup D));R).$$

The perversity $\hat{Q}^a_{\bar{r},\bar{s}}$ is designed to be on the larger end of $(\bar{r},\bar{s})$-compatible perversities, and we might want our result to live in one of the smaller $(\bar{r},\bar{s})$-compatible perversities. Of course, we can pull back an intersection cohomology module to one with a smaller perversity, but we might also want to weaken where we expect some of the intermediate products to live. This could all be done by applying the appropriate naturality statements, but to get desired versions of both Propositions 7.3.54 and 7.3.55, it is perhaps simplest now to revisit the diagrams of Lemma 7.3.49. To simplify, we will use the perversities as stand-ins for the modules, and we use the perversities of Lemma 7.3.53. We also make the necessary assumption to invert the indicated ϵ maps to IAW maps, and the resulting diagram commutes only up to homotopy.

$$\begin{array}{ccccc}
\hat{Q}^a_{\bar{r},\bar{s}} & \xrightarrow{\mathbf{d}} & \hat{Q}^a_{\hat{Q}^a_{\bar{p},\bar{u}},\hat{Q}^a_{\bar{q},\bar{v}}} & \xrightarrow{\text{IAW}} & \hat{Q}^a_{\bar{p},\bar{u}}\otimes\hat{Q}^a_{\bar{q},\bar{v}} \\
 & & \uparrow{\scriptstyle \text{id}\times t\times\text{id}\ \cong} & & \downarrow{\scriptstyle \text{IAW}\otimes\text{IAW}} \\
\downarrow{\scriptstyle \text{IAW}} & & \hat{Q}^a_{\hat{Q}^a_{\bar{p},\bar{q}},\hat{Q}^a_{\bar{u},\bar{v}}} & & \bar{p}\otimes\bar{u}\otimes\bar{q}\otimes\bar{v} \\
 & & \uparrow{\scriptstyle \varepsilon} & & \uparrow{\scriptstyle \text{id}\otimes\tau\otimes\text{id}\ \cong} \\
\bar{r}\otimes\bar{s} & \xrightarrow{\mathbf{d}\otimes\mathbf{d}} & \hat{Q}^a_{\bar{p},\bar{q}}\otimes\hat{Q}^a_{\bar{u},\bar{v}} & \xrightarrow{\text{IAW}\otimes\text{IAW}} & \bar{p}\otimes\bar{q}\otimes\bar{u}\otimes\bar{v}.
\end{array}$$

Now let us introduce the following additional perversities:

1. let $P_{\bar{r},\bar{s}}$ be $(\bar{r},\bar{s})$-compatible on $X\times Y$ with $P_{\bar{r},\bar{s}}\leq\hat{Q}^a_{\bar{r},\bar{s}}$,

[19] These conditions are not all independent; see, in particular, Proposition 6.4.15.

2. let $P_{\bar{p},\bar{u}}$ be $(\bar{p},\bar{u})$-compatible on $X \times Y$ with $P_{\bar{u},\bar{v}} \leq \hat{Q}^a_{\bar{u},\bar{v}}$,
3. let $P_{\bar{q},\bar{v}}$ be $(\bar{q},\bar{v})$-compatible on $X \times Y$ with $P_{\bar{v},\bar{v}} \leq \hat{Q}^a_{\bar{v},\bar{v}}$.

Furthermore, suppose that $(P_{\bar{p},\bar{u}}, P_{\bar{q},\bar{v}}; P_{\bar{r},\bar{s}})$ is $\mathfrak{P}$-agreeable with $\mathfrak{P} \leq \hat{Q}^a_{\hat{Q}^a_{\bar{p},\bar{u}},\hat{Q}^a_{\bar{q},\bar{v}}}$.

Then we can augment the preceding diagram to

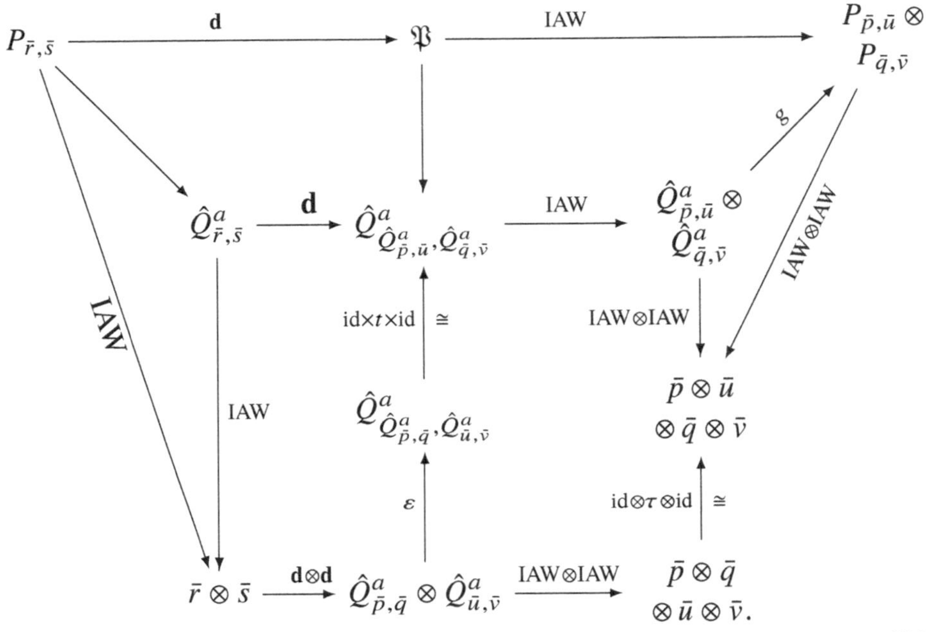

(7.23)

The unlabeled maps are induced by the identity maps on spaces, and they are allowable, as we have defined our perversities so that the domain perversities are less than or equal to the target perversities. The map labeled g is the homotopy inverse of such a map induced by an identity map; we are permitted to invert this identity map up to chain homotopy, as the two other maps in the top right triangle are chain homotopy equivalences. The top row is well defined by our agreeable assumption, and all of our new polygons commute by naturality of diagonal maps and cross products.

By chasing the diagram, one can now check that the two paths around the outside of the diagram from the top left to the bottom right commute up to homotopy. Thus the proof of Lemma 7.3.50 goes through. Similarly, replacing the vertical and diagonal IAW maps with their chain homotopy inverse ε maps, the two paths around the outside of the diagram from the bottom left to the top right right commute up to homotopy. This is what is needed in the proof of Lemma 7.3.51. This allows us to conclude with the following versions of the interchange properties.

Proposition 7.3.56 *Let R be a Dedekind domain. Suppose that X is a CS set with perversities $\bar{p}$, $\bar{q}$, $\bar{r}$ with $D\bar{r} > D\bar{p} + D\bar{q}$ and that Y is a CS set with perversities $\bar{u}$, $\bar{v}$, $\bar{s}$*

with $D\bar{s} > D\bar{u} + D\bar{v}$. Let $A, B \subset X$ be open subsets and $C, D \subset Y$ be open subsets. Suppose on $X \times Y$ with $a = 1$ that $P_{\bar{r},\bar{s}} \leq \hat{Q}^a_{\bar{r},\bar{s}}$ is $(\bar{r},\bar{s})$-compatible, $P_{\bar{p},\bar{u}} \leq \hat{Q}^a_{\bar{p},\bar{u}}$ is $(\bar{p},\bar{u})$-compatible, $P_{\bar{q},\bar{v}} \leq \hat{Q}^a_{\bar{q},\bar{v}}$ is $(\bar{q},\bar{v})$-compatible, and $(P_{\bar{p},\bar{u}}, P_{\bar{q},\bar{v}}; P_{\bar{r},\bar{s}})$ is $\mathfrak{P}$-agreeable with $\mathfrak{P} \leq \hat{Q}^a_{\hat{Q}^a_{\bar{p},\bar{u}},\hat{Q}^a_{\bar{q},\bar{v}}}$.

Let $\alpha \in I_{\bar{p}}H^i(X, A; R)$, $\beta \in I_{\bar{q}}H^j(X, B; R)$, $\gamma \in I_{\bar{u}}H^k(Y, C; R)$, and $\delta \in I_{\bar{v}}H^\ell(Y, D; R)$. Suppose $\alpha \smile \beta \in I_{\bar{r}}H^{i+j}(X, A \cup B; R)$, $\gamma \smile \delta \in I_{\bar{s}}H^{k+\ell}(Y, C \cup D; R)$, $\alpha \times \gamma \in I_{P_{\bar{p},\bar{u}}}H^{i+k}(X \times Y, (A \times Y) \cup (X \times C); R)$, and $\beta \times \delta \in I_{P_{\bar{q},\bar{v}}}H^{j+\ell}(X \times Y, (B \times Y) \cup (X \times D); R)$.

Then

$$\begin{aligned}(\alpha \smile \beta) \times (\gamma \smile \delta) &= (-1)^{jk}(\alpha \times \gamma) \smile (\beta \times \delta)\\ &\in I_{P_{\bar{r},\bar{s}}}H^{i+j+k+\ell}(X \times Y, ((A \cup B) \times Y) \cup (X \times (C \cup D)); R).\end{aligned}$$

Under the assumptions $D\bar{r} \geq D\bar{p} + D\bar{q}$ and $D\bar{s} \geq D\bar{u} + D\bar{v}$ then the above statement holds with $a = 2$ assuming the following conditions are satisfied[20]:

- *X is locally $(\bar{p}, R)$-torsion-free or locally $(\bar{q}, R)$-torsion-free,*
- *Y is locally $(\bar{u}, R)$-torsion-free or locally $(\bar{v}, R)$-torsion-free,*
- *X is locally $(\bar{p}, R)$-torsion-free or Y is locally $(\bar{u}, R)$-torsion-free,*
- *X is locally $(\bar{q}, R)$-torsion-free or Y is locally $(\bar{v}, R)$-torsion-free,*
- *$X \times Y$ is locally $(\hat{Q}^2_{\bar{p},\bar{u}}, R)$-torsion-free or locally $(\hat{Q}^2_{\bar{q},\bar{v}}, R)$-torsion-free,*
- *X is locally $(\bar{r}, R)$-torsion-free or Y is locally $(\bar{s}, R)$-torsion-free.*

Proposition 7.3.57 *Let R be a Dedekind domain. Suppose that X is a CS set with perversities $\bar{p}$, $\bar{q}$, $\bar{r}$ with $D\bar{r} > D\bar{p} + D\bar{q}$ and that Y is a CS set with perversities $\bar{u}$, $\bar{v}$, $\bar{s}$ with $D\bar{s} > D\bar{u} + D\bar{v}$. Let $A, B \subset X$ and $C, D \subset Y$ be open subsets. Suppose on $X \times Y$ with $a = 1$ that $P_{\bar{r},\bar{s}} \leq \hat{Q}^a_{\bar{r},\bar{s}}$ is $(\bar{r},\bar{s})$-compatible, $P_{\bar{p},\bar{u}} \leq \hat{Q}^a_{\bar{p},\bar{u}}$ is $(\bar{p},\bar{u})$-compatible, $P_{\bar{q},\bar{v}} \leq \hat{Q}^a_{\bar{q},\bar{v}}$ is $(\bar{q},\bar{v})$-compatible, and $(P_{\bar{p},\bar{u}}, P_{\bar{q},\bar{v}}; P_{\bar{r},\bar{s}})$ is $\mathfrak{P}$-agreeable with $\mathfrak{P} \leq \hat{Q}^a_{\hat{Q}^a_{\bar{p},\bar{u}},\hat{Q}^a_{\bar{q},\bar{v}}}$.*

Let $\alpha \in I_{\bar{q}}H^j(X, B; R)$, $x \in I^{\bar{r}}H_{i+j}(X, A \cup B; R)$, $\beta \in I_{\bar{v}}H^\ell(Y, D; R)$, and $y \in I^{\bar{s}}H_{k+\ell}(Y, C \cup D; R)$. Suppose $\alpha \times \beta \in I_{P_{\bar{q},\bar{v}}}H^{j+\ell}(X \times Y, (B \times Y) \cup (X \times D); R)$, $x \times y \in I^{P_{\bar{r},\bar{s}}}H_{i+j+k+\ell}(X \times Y, ((A \cup B) \times Y) \cup (X \times (C \cup D)); R)$, $\alpha \frown x \in I^{\bar{p}}H_i(X, A; R)$, and $\beta \frown y \in I^{\bar{u}}H_k(Y, C; R)$. Then

$$\begin{aligned}(\alpha \times \beta) \frown (x \times y) &= (-1)^{\ell(i+j)}(\alpha \frown x) \times (\beta \frown y)\\ &\in I^{P_{\bar{p},\bar{u}}}H_{i+k}(X \times Y, (A \times Y) \cup (X \times C); R).\end{aligned}$$

Under the assumption $D\bar{r} \geq D\bar{p} + D\bar{q}$ and $D\bar{s} \geq D\bar{u} + D\bar{v}$ then the above equalities hold with $a = 2$ assuming the following conditions are satisfied[21]:

- *X is locally $(\bar{p}, R)$-torsion-free or locally $(\bar{q}, R)$-torsion-free,*

[20] These conditions are not all independent; see, in particular, Proposition 6.4.15.

[21] These conditions are not all independent; see, in particular, Proposition 6.4.15. Also note that we need more torsion-free conditions here than in Proposition 7.3.55 because our argument here requires us to be able to obtain the map labeled g in Diagram (7.23) as a homotopy inverse.

- *Y is locally $(\bar{u}, R)$-torsion-free or locally $(\bar{v}, R)$-torsion-free,*
- *X is locally $(\bar{p}, R)$-torsion-free or Y is locally $(\bar{u}, R)$-torsion-free,*
- *X is locally $(\bar{q}, R)$-torsion-free or Y is locally $(\bar{v}, R)$-torsion-free,*
- *$X \times Y$ is locally $(\hat{Q}^2_{\bar{p},\bar{u}}, R)$-torsion-free or locally $(\hat{Q}^2_{\bar{q},\bar{v}}, R)$-torsion-free.*

Example 7.3.58 The following example will be utilized below in the proof of Theorem 9.3.17.3, which concerns the multiplicativity of Witt signatures. Suppose R is a field so that all torsion-free conditions hold automatically and we can use $a = 2$. We claim that the two interchange propositions hold taking $\bar{p} = \bar{q} = \bar{u} = \bar{v} = P_{\bar{p},\bar{u}} = P_{\bar{q},\bar{v}} = \bar{n}$ and $\bar{r} = \bar{s} = P_{\bar{r},\bar{s}} = \bar{0}$. Here $\bar{n}$ is the upper middle Goresky–MacPherson perversity (Definition 3.1.10). As $\bar{n}$ and $\bar{0}$ depend only on codimension, they are well defined on any space. We need only check that the perversities satisfy the required conditions.

We have $D\bar{0} \geq D\bar{n} + D\bar{n}$ by the proof of Corollary 7.2.12, while $\bar{n}$ is $(\bar{n},\bar{n})$-compatible by Example 6.4.11. It is easy to see that $\bar{0}$ is $(\bar{0},\bar{0})$-compatible right from the definition (Definition 6.4.5). It is also clear from Definition 7.3.52 that $\bar{0} \leq \hat{Q}^a_{\bar{0},\bar{0}}$.

To see that $\bar{n} \leq \hat{Q}^2_{\bar{n},\bar{n}}$, we note that we saw in the proof of Example 6.4.11 that on a stratum of $X \times Y$ of the form $S \times T$ with $\operatorname{codim}(S) = k > 0$ and $\operatorname{codim}(S) = \ell > 0$ we have

$$\bar{n}(S \times T) = \bar{n}(k + \ell) \leq \bar{n}(k) + \bar{n}(\ell) + 1 \leq \bar{n}(k) + \bar{n}(\ell) + 2 = \hat{Q}^2_{\bar{n},\bar{n}}(S \times T).$$

Similarly, if S is regular we have

$$\bar{n}(S \times T) = \bar{n}(\ell) = \hat{Q}^2_{\bar{n},\bar{n}}(S \times T),$$

and analogously if T is regular.

Finally, we need a $\mathfrak{P}$ such that $(\bar{n}, \bar{n}; \bar{0})$ is $\mathfrak{P}$-agreeable with $\mathfrak{P} \leq \hat{Q}^2_{\hat{Q}^2_{\bar{n},\bar{n}}, \hat{Q}^2_{\bar{n},\bar{n}}}$. Continuing to assume field coefficients, we know that $\hat{Q}^2_{\bar{n},\bar{n}}$ is the maximal $(\bar{n},\bar{n})$-compatible perversity and so the agreeability of $(\bar{n}, \bar{n}; \bar{0})$ means that it will be $\hat{Q}^2_{\bar{n},\bar{n}}$-agreeable. Let us see that $\hat{Q}^2_{\bar{n},\bar{n}} \leq \hat{Q}^2_{\hat{Q}^2_{\bar{n},\bar{n}}, \hat{Q}^2_{\bar{n},\bar{n}}}$. For this we can employ Table 7.2 applied to strata $S \times T \times U \times V \subset X \times Y \times X \times Y$, with r denoting a regular stratum and s denoting a singular stratum. In principle there are 16 possibilities, but we omit those cases that are redundant due to the evident symmetries.

As we know that $\bar{n}(k + \ell) \leq \bar{n}(k) + \bar{n}(\ell) + 1$, where k, ℓ represent codimensions, we see that indeed $\hat{Q}^2_{\bar{n},\bar{n}} \leq \hat{Q}^2_{\hat{Q}^2_{\bar{n},\bar{n}}, \hat{Q}^2_{\bar{n},\bar{n}}}$. △

7.3.7 Locality

There is another desirable property of the traditional cap product in ordinary homology that we will need to approximate. If we form the singular chain cap product $\alpha \frown \xi$ using the front face/back face formula, then every simplex of $\alpha \frown \xi$ will be a face of a simplex of ξ. This shows at the level of chains that the support of $\alpha \frown \xi$ will be a subset of the support of ξ itself, which one can imagine is a useful property. For example, it is utilized in the

Table 7.2

S	T	U	V	$\hat{Q}^2_{\bar{n},\bar{n}}((S\times T)\times(U\times V))$	$\hat{Q}^2_{\hat{Q}^2_{\bar{n},\bar{n}},\hat{Q}^2_{\bar{n},\bar{n}}}(S\times U\times T\times V)$
r	r	r	r	0	0
s	r	r	r	$\bar{n}(S)$	$\bar{n}(S)$
s	s	r	r	$\bar{n}(S\times T)$	$\bar{n}(S)+\bar{n}(T)+2$
s	r	s	r	$\bar{n}(S)+\bar{n}(U)+2$	$\bar{n}(S)+\bar{n}(U)+2$
s	s	s	r	$\bar{n}(S\times T)+\bar{n}(U)+2$	$[\bar{n}(S)+\bar{n}(T)+2+\bar{n}(U)]+2$
s	s	s	s	$\bar{n}(S\times T)+\bar{n}(U\times V)+2$	$[\bar{n}(S)+\bar{n}(T)+2+$ $\bar{n}(U)+\bar{n}(V)+2]+2$

proof that Poincaré duality isomorphisms are compatible with Mayer–Vietoris sequences in Hatcher [125, Lemma 3.36]. We will need a version of this fact for intersection homology below in Lemma 7.4.8. However, there is no reason to suppose in the intersection world that the support of $\alpha \frown \xi$ will be contained in the support of ξ, and we will not be able to recreate this property exactly. However, we can show that acting on an intersection chain by a cap product "doesn't move it too far" in a sense made precise in the following proposition. The arguments in this section are based closely on those in [98, 100].

Proposition 7.3.59 *Let R be a Dedekind domain. Suppose X is a CS set with an agreeable triple of perversities $(\bar{p},\bar{q};\bar{r})$. Let A,B be open subsets of X. Let $\mathcal{U}$ be an open covering of X. Then the image of*

$$\bar{\mathbf{d}} : I^{\bar{r}}H_*(X,A\cup B;R) \to H_*(I^{\bar{p}}S_*(X,A;R)\otimes I^{\bar{q}}S_*(X,B;R))$$

is contained in the image of

$$\kappa : H_*\left(\sum_{U\in\mathcal{U}} I^{\bar{p}}S_*(U,U\cap A;R)\otimes I^{\bar{q}}S_*(U,U\cap B;R)\right)$$
$$\to H_*(I^{\bar{p}}S_*(X,A;R)\otimes I^{\bar{q}}S_*(X,B;R)),$$

where κ is induced by inclusions.

Notice that $\sum_{U\in\mathcal{U}} I^{\bar{p}}S_*(U,U\cap A;R)\otimes I^{\bar{q}}S_*(U,U\cap B;R)$ makes sense because each $I^{\bar{p}}S_*(U,U\cap A;R)\otimes I^{\bar{q}}S_*(U,U\cap B;R)$ is a submodule of $I^{\bar{p}}S_*(X,A;R)\otimes I^{\bar{q}}S_*(X,B;R)$, using the fact that the intersection chain modules are all projective by Lemma 6.3.1.

The point of Proposition 7.3.59 is that, if we apply ε to an element of $\sum_{U\in\mathcal{U}} I^{\bar{p}}S_*(U;R)\otimes I^{\bar{q}}S_*(U;R)$, then the image will live in $\bigcup_{\mathcal{U}} U\times U$, which, depending on $\mathcal{U}$, can be considered a small neighborhood of the diagonal $\mathbf{d}(X)\subset X\times X$. This doesn't really ensure that $\bar{\mathbf{d}}(\xi)$

consists of chains "near $\mathbf{d}(\xi)$," but it does mean that $\bar{\mathbf{d}}(\xi)$ is a sum of tensor products of chains that are near each other, which will be sufficient for Lemma 7.4.8, below.

Proof of Proposition 7.3.59 Consider the following diagram, in which the unlabeled map and the maps labeled λ and κ are induced by inclusions and μ is induced by the chain cross product ε.

$$\begin{array}{ccccc}
I^{\bar r}S_*(X, A\cup B; R) & \xrightarrow{\mathbf{d}} & \begin{array}{c} I^{Q_{\bar p,\bar q}}S_*(X\times X, \\ (A\times X)\cup(X\times B); R)\end{array} & \xleftarrow{\varepsilon} & \begin{array}{c} I^{\bar p}S_*(X,A;R)\otimes \\ I^{\bar q}S_*(X,B;R)\end{array} \\
& \searrow^{\bar{\mathbf{d}}} & \uparrow & & \uparrow \kappa \\
& & \begin{array}{c} I^{Q_{\bar p,\bar q}}S_*\big(\bigcup_{\mathcal{U}} U\times U, \\ \bigcup_{\mathcal{U}}((A\cap U)\times U) \\ \cup(U\times(B\cap U)); R\big)\end{array} & & \\
& & \lambda\uparrow & & \\
& & \begin{array}{c} \sum_{\mathcal{U}} I^{Q_{\bar p,\bar q}}S_*\big(U\times U, \\ ((A\cap U)\times U)\cup \\ (U\times(B\cap U)); R\big)\end{array} & \xleftarrow{\mu} & \begin{array}{c} \sum_{\mathcal{U}} I^{\bar p}S_*(U, A\cap U; R)\otimes \\ I^{\bar q}S_*(U, B\cap U; R)\end{array}
\end{array}$$

The diagram commutes. In particular, the triangle commutes because the underlying map of spaces commutes, while the rectangle commutes using Proposition 5.2.17 and Theorem 6.3.19.

We know that ε induces a homology isomorphism by the Künneth Theorem (Theorem 6.4.7). We will show that λ and μ induce homology isomorphisms, as well. The lemma then follows by applying H_* to the diagram and reversing the arrows of the quasi-isomorphisms.

The map λ is an isomorphism on homology by Proposition 6.5.1. To see that this proposition applies, we note that

$$\begin{aligned}
(U\times U)\cap[(A\times X)\cup(X\times B)] &= [(U\times U)\cap(A\times X)]\cup[(U\times U)\cap(X\times B)] \\
&= [(U\cap A)\times(U\cap X)]\cup[(U\cap X)\times(U\cap B)] \\
&= [(U\cap A)\times U]\cup[U\times(U\cap B)],
\end{aligned}$$

and clearly this is contained in $\bigcup_{\mathcal{U}}((A\cap U)\times U)\cup(U\times(B\cap U))$. Therefore, the intersection of $U\times U$ with $\bigcup_{\mathcal{U}}((A\cap U)\times U)\cup(U\times(B\cap U))$ is $[(U\cap A)\times U]\cup[U\times(U\cap B)]$.

The proof that μ is a quasi-isomorphism is presented as Lemma 7.3.62, below. □

We now work toward stating and proving Lemma 7.3.62 in order to finish the proof of Proposition 7.3.59.

Lemma 7.3.60 *Let X be a filtered space, let B be an open subset, and let G be an abelian group. Suppose $\{U_j\}_{j=1}^k$ is a finite collection of open subsets of X. Then*

$$I^{\bar p}S_*(U_k, B\cap U_k; G)\cap\sum_{j=1}^{k-1} I^{\bar p}S_*(U_j, B\cap U_j; G) = \sum_{j=1}^{k-1} I^{\bar p}S_*(U_j\cap U_k, B\cap U_j\cap U_k; G)$$

as subgroups of $I^{\bar{p}}S_*(X,B;G)$.

Notice that the analogous lemma would be straightforward for ordinary singular chains, using a basis represented by singular simplices. However, it is not totally obvious in the intersection world where we do not have complete freedom in how to break chains apart into pieces. For example, it is certainly possible that there might be chains $x_1 \in I^{\bar{p}}S_*(U_1;G)$ and $x_2 \in I^{\bar{p}}S_*(U_2;G)$ such that $x_1 + x_2$ is supported in U_3 but such that x_1 is not supported in $U_1 \cap U_3$. If x_1 and x_2 were ordinary chains, that would mean that they share some simplices that cancel in $x_1 + x_2$, and we could throw away these simplices to be left with y_1 and y_2 such that $y_1 + y_2 = x_1 + x_2$ but with y_j supported in $U_j \cap U_3$. As usual, we cannot throw away simplices so cavalierly in the setting of intersection chains, so more argument is needed. Luckily, the groundwork has already been lain. In fact, this lemma could be proven as a nearly immediate application of Proposition 6.5.2. However, we will next need a version of this lemma for tensor products, which will be Lemma 7.3.61, just below. To work toward the proof of that lemma, we prove Lemma 7.3.60 instead by using Corollary 6.5.3, which we have seen is closely related to Proposition 6.5.2, both being consequences of the techniques used to prove Proposition 6.5.1, which we have already used in the proof of Proposition 7.3.59.

Proof of Lemma 7.3.60 First, observe that the expressions in the lemma make sense, as inclusion induces injections $I^{\bar{p}}S_*(U_j, B \cap U_j;G) \hookrightarrow I^{\bar{p}}S_*(X,B;G)$ for each j: for example, the only chains in the kernel of $I^{\bar{p}}S_*(U_j;G) \to I^{\bar{p}}S_*(X,B;G)$ are those that are supported in B and U_j, and those are zero in $I^{\bar{p}}S_*(U_j, B\cap U_j;G)$. Therefore, we can identify the image of $I^{\bar{p}}S_*(U_j, B \cap U_j;G)$ as a subgroup of $I^{\bar{p}}S_*(X,B;G)$, and the sum on the left then makes sense. The same argument holds for the terms involving further intersections.

Next, notice that we only need to prove the lemma for groups, i.e. that

$$I^{\bar{p}}S_i(U_k, B \cap U_k;G) \cap \sum_{j=1}^{k-1} I^{\bar{p}}S_i(U_j, B \cap U_j;G) = \sum_{j=1}^{k-1} I^{\bar{p}}S_i(U_j \cap U_k; B \cap U_j \cap U_k;G)$$

for each i, as each side of the equality takes care of itself as a chain complex. It is also obvious that $I^{\bar{p}}S_i(U_j \cap U_k, B\cap U_j \cap U_k;G)$ injects into $I^{\bar{p}}S_i(U_k, B\cap U_k;G)$ and $I^{\bar{p}}S_i(U_j, B\cap U_j;G)$ so that, altogether,

$$I^{\bar{p}}S_i(U_k, B \cap U_k;G) \cap \sum_{j=1}^{k-1} I^{\bar{p}}S_i(U_j, B \cap U_j;G) \supset \sum_{j=1}^{k-1} I^{\bar{p}}S_i(U_j \cap U_k, B \cap U_j \cap U_k;G).$$

So suppose $x \in I^{\bar{p}}S_i(U_k, B \cap U_k;G) \cap \sum_{j=1}^{k-1} I^{\bar{p}}S_i(U_j, B \cap U_j;G)$. In particular, this means that there is an intersection chain ξ in $I^{\bar{p}}S_i(U_k;G) \subset I^{\bar{p}}S_i(X;G)$ that represents x and intersection chains $\xi_j \in I^{\bar{p}}S_i(U_j;G)$ so that $\sum_j \xi_j \in \sum_{j=1}^{k-1} I^{\bar{p}}S_i(U_j;G) \subset I^{\bar{p}}S_i(X;G)$ represents x. In other words, the chains ξ and $\sum_j \xi_j$ both represent x as elements of $I^{\bar{p}}S_i(X,B;G)$. We want to show that x can also be represented by a chain in $\sum_{j=1}^{k-1} I^{\bar{p}}S_i(U_j \cap$

$U_k; G) \subset I^{\bar{p}}S_i(X; G)$, which of course also represents an element of $\sum_{j=1}^{k-1} I^{\bar{p}}S_i(U_j \cap U_k; B \cap U_j \cap U_k; G) \subset I^{\bar{p}}S_i(X, B; G)$.

Consider now Corollary 6.5.3, which says that if A is an open subset of X then the map $\mathfrak{i} : I^{\bar{p}}S_i(A; G) \to I^{\bar{p}}S_i(X; G)$ induced by space inclusion is a *split* inclusion. In our case, let $A = U_k$ and let $P : I^{\bar{p}}S_i(X; G) \to I^{\bar{p}}S_i(U_k; G)$ be the splitting such that $P\mathfrak{i} = \mathrm{id}$. Of course all the (non-relative) groups we have been considering are subgroups of $I^{\bar{p}}S_i(X; G)$, so we can restrict P to any of these subgroups. In the proof of Corollary 6.5.3, the map P is constructed by taking subdivisions of chains and then throwing away some of the subdivisions. As a result, for any chain ζ, the support of $P(\zeta)$ is contained in the support of ζ. So P takes chains in B to chains in B and is therefore also well defined: $P : I^{\bar{p}}S_i(X, B; G) \to I^{\bar{p}}S_i(U_k, B \cap U_k; G)$. Now, since ξ and $\sum_j \xi_j$ (now thought of as elements of $I^{\bar{p}}S_i(X; G)$) represent the same element in $I^{\bar{p}}S_i(X, B; G)$, so do $\mathfrak{i}P(\xi)$ and $\mathfrak{i}P\left(\sum_j \xi_j\right) = \sum_j \mathfrak{i}P(\xi_j)$. As $|\xi| \subset U_k$, we have, in fact, $\mathfrak{i}P(\xi) = \xi \in I^{\bar{p}}S_i(X; G)$, so $\mathfrak{i}P(\xi)$ and $\mathfrak{i}P\left(\sum_j \xi_j\right)$ also represent x. Furthermore, by the support properties of P, we have $|\mathfrak{i}P(\xi_j)| \subset U_j$ for each j, as $|\xi_j| \subset U_j$. Therefore,

$$\mathfrak{i}P(\xi_j) \in I^{\bar{p}}S_i(U_k; G) \cap I^{\bar{p}}S_i(U_j; G) = I^{\bar{p}}S_i(U_k \cap U_j; G) \subset I^{\bar{p}}S_i(X; G),$$

and so

$$\mathfrak{i}P\left(\sum_j \xi_j\right) = \sum_j \mathfrak{i}P(\xi_j) \in \sum_{j=1}^{k-1} I^{\bar{p}}S_i(U_j \cap U_k; G).$$

As we know that $\mathfrak{i}P\left(\sum_j \xi_j\right)$ represents x as an element of $I^{\bar{p}}S_i(X, B; G)$, we therefore see that the image $\mathfrak{i}P\left(\sum_j \xi_j\right)$ in

$$\sum_{j=1}^{k-1} I^{\bar{p}}S_i(U_j \cap U_k, B \cap U_j \cap U_k; G) \subset I^{\bar{p}}S_i(X, B; G)$$

represents x. This completes the proof. $\square$

Lemma 7.3.61 *Let X be a filtered space, let A and B be open subsets, and let R be a Dedekind domain. Suppose $\{U_j\}_{j=1}^k$ is a finite collection of open subsets of X. Then*

$$\begin{aligned}&\left[I^{\bar{p}}S_*(U_k, A \cap U_k; R) \otimes I^{\bar{q}}S_*(U_k, U_k \cap B; R)\right] \cap \\ &\quad \sum_{j=1}^{k-1} \left[I^{\bar{p}}S_*(U_j, A \cap U_j; R) \otimes I^{\bar{q}}S_*(U_j, U_j \cap B; R)\right] \\ &= \sum_{j=1}^{k-1} I^{\bar{p}}S_*(U_j \cap U_k, A \cap U_j \cap U_k; R) \otimes I^{\bar{q}}S_*(U_j \cap U_k, U_j \cap U_k \cap B; R)\end{aligned}$$

as submodules of $I^{\bar{p}}S_*(X, A; R) \otimes I^{\bar{q}}S_*(X, B; R)$.

Proof The proof is essentially the same as that for Lemma 7.3.60. First, recall from that proof that each $I^{\bar{p}}S_*(U_j, A \cap U_j; G) \hookrightarrow I^{\bar{p}}S_*(X, A; G)$ is an injection and similarly for the analogous terms. But we know that each of the individual groups involved in the expressions is a projective R-module by Lemma 6.3.1, and so also each

$$I^{\bar{p}}S_*(U_j, A \cap U_j; R) \otimes I^{\bar{q}}S_*(U_j, U_j \cap B; R)$$

injects into

$$I^{\bar{p}}S_*(X, A; R) \otimes I^{\bar{q}}S_*(X, B; R)$$

and so can be considered a submodule, and similarly for the

$$I^{\bar{p}}S_*(U_j \cap U_k, A \cap U_j \cap U_k; R) \otimes I^{\bar{q}}S_*(U_j \cap U_k, U_j \cap U_k \cap B; R).$$

It also follows that the inclusion $\supset$ holds for the expression in the statement of the lemma.

The elements of

$$I^{\bar{p}}S_*(U_k, A \cap U_k; R) \otimes I^{\bar{q}}S_*(U_k, U_k \cap B; R)$$

can be represented by chains of the form $\sum_\ell \xi_\ell \otimes \xi'_\ell$, where $\xi_\ell \in I^{\bar{p}}S_*(U_k; R)$ and $\xi'_\ell \in I^{\bar{q}}S_*(U_k; R)$. Similarly, elements of

$$I^{\bar{p}}S_*(U_j, A \cap U_j; R) \otimes I^{\bar{q}}S_*(U_j, U_j \cap B; R)$$

for $1 \leq j \leq k-1$ can be represented by chains $\sum_\ell \eta_{j\ell} \otimes \eta'_{j\ell}$ with $\eta_{j\ell} \in I^{\bar{p}}S_*(U_j; R)$ and $\eta'_{j\ell} \in I^{\bar{q}}S_*(U_j; R)$. Technically, the indexing sets for the ℓ should depend on j, but by including some zero terms, we can assume that the indexing sets for the ℓ are all the same. So, suppose we have an element x from the left-hand side of the expression in the statement of the lemma. Then there are choices of ξ_ℓ, ξ'_ℓ, $\eta_{j\ell}$, $\eta'_{j\ell}$ so that both $\sum_\ell \xi_\ell \otimes \xi'_\ell$ and $\sum_{j=1}^{k-1} \sum_\ell \eta_{j\ell} \otimes \eta'_{j\ell}$ represent x in $I^{\bar{p}}S_*(X, A; R) \otimes I^{\bar{q}}S_*(X, B; R)$.

Consider again the splitting map $P : I^{\bar{p}}S_*(X; R) \to I^{\bar{p}}S_*(U_k; R)$ of Lemma 7.3.60 that exists due to Corollary 6.5.3. This is not necessarily a chain map, but that will not be necessary. Similarly, we have a splitting map $P' : I^{\bar{q}}S_*(X; R) \to I^{\bar{q}}S_*(U_k; R)$, and we let $\mathfrak{i} : U_k \to X$ be the inclusion. As P and P' preserve (or reduce) supports, we have

$$P : I^{\bar{p}}S_*(X, A; R) \to I^{\bar{p}}S_*(U_k, U_k \cap A; R)$$

and

$$P' : I^{\bar{q}}S_*(X, B; R) \to I^{\bar{q}}S_*(U_k, U_k \cap B; R),$$

so we can apply $\mathfrak{i}P \otimes \mathfrak{i}P'$ to our chains $\sum_\ell \xi_\ell \otimes \xi'_\ell$ and $\sum_j^{k-1} \sum_\ell \eta_{j\ell} \otimes \eta'_{j\ell}$, and their images will represent the same element in $I^{\bar{p}}S_*(X, A; R) \otimes I^{\bar{q}}S_*(X, B; R)$. As each ξ_ℓ and ξ'_ℓ is already supported in U_k, the chain

$$(\mathfrak{i}P \otimes \mathfrak{i}P')\left(\sum_\ell \xi_\ell \otimes \xi'_\ell\right) = \sum_\ell \mathfrak{i}P(\xi_\ell) \otimes \mathfrak{i}P'(\xi'_\ell) = \sum_\ell \xi_\ell \otimes \xi'_\ell,$$

so $P \otimes P'$ acts by the identity on these elements. On the other hand,

$$(\mathrm{i}P \otimes \mathrm{i}P')\left(\sum_j^{k-1}\sum_\ell \eta_{j\ell} \otimes \eta'_{j\ell}\right) = \sum_j^{k-1}\sum_\ell \mathrm{i}P(\eta_{j\ell}) \otimes \mathrm{i}P'(\eta'_{j\ell}).$$

By the properties of P and P', as we saw in Lemma 7.3.60, each $P(\eta_{j\ell})$ or $P(\eta'_{j\ell})$ will be contained in $I^{\bar{p}}S_*(U_k \cap U_j; R)$ or $I^{\bar{q}}S_*(U_k \cap U_j; R)$, so

$$\sum_j^{k-1}\sum_\ell P(\eta_{j\ell}) \otimes P(\eta'_{j\ell}) \in \sum_{j=1}^{k-1} I^{\bar{p}}S_*(U_j \cap U_k; R) \otimes I^{\bar{q}}S_*(U_j \cap U_k; R).$$

Therefore, we have shown that every element of $I^{\bar{p}}S_*(X, A; R) \otimes I^{\bar{q}}S_*(X, B; R)$ in

$$I^{\bar{p}}S_*(U_k, A \cap U_k; R) \otimes I^{\bar{q}}S_*(U_k, U_k \cap B; R) \cap \sum_{j=1}^{k-1}\left[I^{\bar{p}}S_*(U_j, A \cap U_j; R) \otimes I^{\bar{q}}S_*(U_j, U_j \cap B; R)\right]$$

can be represented by an element in

$$\sum_{j=1}^{k-1} I^{\bar{p}}S_*(U_j \cap U_k; R) \otimes I^{\bar{q}}S_*(U_j \cap U_k; R)$$

and so also by an element of

$$\sum_{j=1}^{k-1} I^{\bar{p}}S_*(U_j \cap U_k, A \cap U_j \cap U_k; R) \otimes I^{\bar{q}}S_*(U_j \cap U_k, U_j \cap U_k \cap B; R).$$

This completes the proof of Lemma 7.3.61. □

We can now present Lemma 7.3.62, which will complete the proof of Proposition 7.3.59.

Lemma 7.3.62 *Let R be a Dedekind domain. Suppose X is a CS set with an agreeable triple of perversities $(\bar{p}, \bar{q}; \bar{r})$. Let A, B be open subsets of X. Let $\mathcal{V}$ be a collection of open subsets of X.*

Then the map

$$\mu : H_*\left(\sum_{\mathcal{V}} I^{\bar{p}}S_*(V, A \cap V; R) \otimes I^{\bar{q}}S_*(V, B \cap V; R)\right)$$
$$\to H_*\left(\sum_{\mathcal{V}} I^{Q_{\bar{p},\bar{q}}}S_*\left(V \times V, ((A \cap V) \times V) \cup (V \times (B \cap V)); R\right)\right)$$

induced by the chain cross product ε is an isomorphism.

Proof We will proceed by induction on the size of $\mathcal{V}$. If $|\mathcal{V}| = 1$, the result follows by the Künneth Theorem (Theorem 6.4.7). So now suppose that $|\mathcal{V}| = k > 1$, i.e. that $\mathcal{V} = \{V_i\}_{i=1}^k$, $k > 1$. Let $\mathcal{V}' = \{V_i\}_{i=1}^{k-1}$, and let $\mathcal{V}'' = \{V_i \cap V_k\}_{i=1}^{k-1}$.

Now we consider the following diagram of short exact sequences with coefficients tacit. Here we use the notation $(Y, Z) \times (C, D)$ to represent $(Y \times C, (Y \times D) \cup (Z \times C))$.

$$
\begin{array}{ccc}
0 & & 0 \\
\downarrow & & \downarrow \\
\begin{array}{c}\Sigma_{i=1}^{k-1} I^{\bar{p}} S_*(V_i \cap V_k, A \cap V_i \cap V_k) \otimes \\ I^{\bar{q}} S_*(V_i \cap V_k, B \cap V_i \cap V_k)\end{array} & \xrightarrow{\varepsilon} & \begin{array}{c}\Sigma_{i=1}^{k-1} I^{Q_{\bar{p},\bar{q}}} S_*[(V_i \cap V_k, A \cap V_i \cap V_k) \times \\ (V_i \cap V_k, B \cap V_i \cap V_k)]\end{array} \\
\downarrow & & \downarrow \\
\begin{array}{c}[I^{\bar{p}} S_*(V_k, A \cap V_k) \otimes \\ I^{\bar{q}} S_*(V_k, B \cap V_k)] \oplus \\ \Sigma_{i=1}^{k-1} [I^{\bar{p}} S_*(V_i, A \cap V_i) \otimes \\ I^{\bar{q}} S_*(V_i, B \cap V_i)]\end{array} & \xrightarrow{\varepsilon \oplus \varepsilon} & \begin{array}{c}[I^{Q_{\bar{p},\bar{q}}} S_*((V_k, A \cap V_k) \times \\ (V_k, B \cap V_k))] \oplus \\ \Sigma_{i=1}^{k-1} I^{Q_{\bar{p},\bar{q}}} S_*((V_i, A \cap V_i) \times \\ (V_i, B \cap V_i))\end{array} \\
\downarrow & & \downarrow \\
\begin{array}{c}\Sigma_{i=1}^{k} I^{\bar{p}} S_*(V_i, A \cap V_i) \otimes \\ I^{\bar{q}} S_*(V_i, B \cap V_i)\end{array} & \xrightarrow{\varepsilon} & \begin{array}{c}\Sigma_{i=1}^{k} I^{Q_{\bar{p},\bar{q}}} S_*[(V_i, A \cap V_i) \times \\ (V_i, B \cap V_i)]\end{array} \\
\downarrow & & \downarrow \\
0 & & 0
\end{array}
$$

Although this looks horrendous, we claim that each vertical sequence has the standard Mayer–Vietoris form

$$0 \longrightarrow C_* \cap D_* \longrightarrow C_* \oplus D_* \longrightarrow C_* + D_* \longrightarrow 0$$

and so is exact. In fact, it is clear that the transitions from the middle terms to the bottom terms have the form $B_* \oplus C_* \to B_* + C_*$, and it follows from Lemmas 7.3.60 and 7.3.61, together with some careful manipulation of set identities, that the top nontrivial term in each column is the intersection of the two summands of the middle term, as desired. The diagram also commutes thanks to the naturality of the cross product.

The map of short exact sequences now induces a map of long exact homology sequences. By the induction hypothesis, and by the fact that the homology map induced on a direct sum is a direct sum of homology maps, we have homology isomorphisms on every two out of three terms, so the Five Lemma completes the proof for $|\mathcal{V}| = k$.

By induction, the lemma is now proven for any finite collection $\mathcal{V}$.

Suppose now that $\mathcal{V}$ is not necessarily finite, and suppose

$$\xi \in H_i\left(\sum_{\mathcal{V}} I^{Q_{\bar{p},\bar{q}}} S_*\left(V \times V, ((A \cap V) \times V) \cup (V \times (B \cap V)); R\right)\right).$$

By definition, every such element can be represented as a finite sum $\xi = \sum_{j=1}^{k} \xi_j$, with

$$\xi_j \in I^{Q_{\bar{p},\bar{q}}} S_* \left(V_j \times V_j, ((A \cap V_j) \times V_j) \cup (V_j \times (B \cap V_j)); R\right)$$

for $\{V_j\}_{j=1}^{k} \subset \mathcal{V}$. Now consider the diagram

$$\begin{array}{ccc} H_i\big(\sum_{j=1}^{k} I^{\bar{p}} S_*(V_j, A \cap V_j; R) \otimes I^{\bar{q}} S_*(V_j, B \cap V_j; R)\big) & \xrightarrow{\mu} & H_i\big(\sum_{j=1}^{k} I^{Q_{\bar{p},\bar{q}}} S_*[V_j \times V_j, ((A \cap V_j) \times V_j) \cup (V_j \times (B \cap V_j)); R]\big) \\ \downarrow & & \downarrow \\ H_i\big(\sum_{\mathcal{V}} I^{\bar{p}} S_*(V, A \cap V; R) \otimes I^{\bar{q}} S_*(V, B \cap V; R)\big) & \xrightarrow{\mu} & H_i\big(\sum_{\mathcal{V}} I^{Q_{\bar{p},\bar{q}}} S_*[V \times V, ((A \cap V) \times V) \cup (V \times (B \cap V)); R]\big). \end{array}$$

By assumption, ξ is in the image of the right-hand vertical map, and the μ in the top line is an isomorphism by the arguments above. Therefore, ξ is also in the image of the bottom μ, and thus the bottom μ is surjective.

Next, assume that

$$\zeta \in H_i\left(\sum_{\mathcal{V}} I^{\bar{p}} S_*(V, A \cap V; R) \otimes I^{\bar{q}} S_*(V, B \cap V; R)\right)$$

with $\mu(\zeta) = 0$. Once again, by definition, we can represent ζ by a chain contained in a finite sum of terms, say over the sets $\{W_j\}_{j=\ell}^{k'} \subset \mathcal{V}$. As $\mu(\zeta) = 0$, this means that there must be an element

$$Z \in \sum_{\mathcal{V}} I^{Q_{\bar{p},\bar{q}}} S_* \left(V \times V, ((A \cap V) \times V) \cup (V \times (B \cap V)); R\right)$$

whose boundary is $\mu(\zeta)$, and again there must be some $\{U_m\}_{m=1}^{k''} \subset \mathcal{V}$ such that

$$\eta \in \sum_{m=1}^{k''} I^{Q_{\bar{p},\bar{q}}} S_* \left(U_m \times U_m, ((A \cap U_m) \times U_m) \cup (U_m \times (B \cap U_m)); R\right).$$

The collection $\{W_j\}_{j=\ell}^{k'} \cup \{U_m\}_{m=1}^{k''} \subset \mathcal{V}$ is finite, so let us relabel this as $\{V_j\}_{j=1}^{k}$ and again consider a diagram of the form just above but as a diagram of chain complexes, not homology. We start with the class that represents ζ in the upper left corner. By our assumptions, the composition down then right yields $\mu(\zeta)$, which we know is trivial in homology and bounds a chain Z in the image of the chain complex in the upper right. Therefore, as the vertical maps of the diagram are injective at the chain level, the chain $\mu(\zeta)$ must represent the homology class 0 already in the upper right. But as the upper μ is an isomorphism on homology, ζ must already represent the trivial homology class in the upper left, and therefore in the bottom left. This shows that the bottom μ is injective in homology.

Altogether now, we have shown that the bottom μ is a homology isomorphism. This completes the proof of Lemma 7.3.62. □

7.3.8 The Cohomology Künneth Theorem

As one final property of intersection (co)homology products, we can prove a cohomology Künneth Theorem.

Theorem 7.3.63 *Let R be a Dedekind domain. Suppose that X is a CS set with perversity $\bar{p}$, that Y is a CS set with perversity $\bar{q}$, and that Q is a $(\bar{p},\bar{q})$-compatible perversity on $X \times Y$. Furthermore, suppose that either $I^{\bar{p}}H_i(X,A;R)$ is finitely generated for each i or $I^{\bar{q}}H_j(Y,B;R)$ is finitely generated for each j. Then there is a natural exact sequence*

$$\begin{aligned} 0 \to \bigoplus_{i+j=k} I_{\bar{p}}H^i(X,A;R) \otimes I_{\bar{q}}H^j(Y,B;R) &\xrightarrow{\times} I_Q H^k(X \times Y,(A \times Y) \cup (X \times B);R) \\ \to \bigoplus_{i+j=k+1} I_{\bar{p}}H^i(X,A;R) * I_{\bar{q}}H^j(Y,B;R) &\to 0 \end{aligned}$$

that splits (non-naturally).

In particular, if X is a manifold $X = M$ (trivially stratified) and the finite generation hypotheses are satisfied then we have an exact sequence

$$\begin{aligned} 0 \to \bigoplus_{i+j=k} H^i(M,A;R) \otimes I_{\bar{q}}H^j(Y,B;R) &\xrightarrow{\times} I_{\bar{q}}H^k(M \times Y,(A \times Y) \cup (M \times B);R) \\ \to \bigoplus_{i+j=k+1} H^i(M,A;R) * I_{\bar{q}}H^j(Y,B;R) &\to 0, \end{aligned}$$

where we use $\bar{q}$ to denote both a perversity on Y and the corresponding perversity on $M \times Y$ whose value is $\bar{q}(T)$ on any stratum $R \times T \subset M \times Y$ with $R \subset M$ and $T \subset Y$.

Before the proof, we provide a standard important example:

Example 7.3.64 By taking $(X,A) = (\mathbb{R}^k,\mathbb{R}^k - \{0\})$ and recalling $H^k(\mathbb{R}^k,\mathbb{R}^k - \{0\};R) \cong R$ and $H^i(\mathbb{R}^k,\mathbb{R}^k - \{0\};R) = 0$ for $i \neq k$, we see that the cohomology cross product produces an isomorphism

$$\begin{aligned} I_{\bar{q}}H^j(Y,B;R) &\cong R \otimes I_{\bar{q}}H^j(Y,B;R) \\ &\cong H^k(\mathbb{R}^k,\mathbb{R}^k - \{0\};R) \otimes I_{\bar{q}}H^j(Y,B;R) \\ &\xrightarrow{\times} I_{\bar{q}}H^{j+k}(\mathbb{R}^k \times Y,((\mathbb{R}^k - \{0\}) \times Y) \cup (\mathbb{R}^k \times B);R). \end{aligned}$$ △

Proof of Theorem 7.3.63 We first note that the last statement, for which $X = M$ is a trivially-filtered manifold, follows from the more general statement of the theorem, using that any perversity $\bar{p}$ on M is trivial and gives $I_{\bar{p}}H^*(M;R) = H^*(M;R)$. Furthermore, if Q is $(\bar{p},\bar{q})$-compatible on $M \times Y$, then for any stratum $R \times T$ of $M \times Y$ we must have $Q(R \times T) = \bar{q}(T)$.

For the main statement of the theorem we follow the standard procedure that can be found, for example, in [182, Section 60] or [71, Proposition VI.12.16]. These sources, however, only consider R a PID, so we will indicate the proof and the necessary generalizations over a Dedekind domain. The basic idea is to treat $I_{\bar{p}}S^i(X,A;R)$ and $I_{\bar{q}}S^j(Y,B;R)$ as if they were homologically indexed and then we can apply the standard algebraic Künneth Theorem to obtain the short exact top row in a diagram of the form

$$\begin{array}{ccccc}
\bigoplus_{i+j=k} \begin{matrix} I_{\bar p}H^i(X,A;R)\otimes \\ I_{\bar q}H^j(Y,B;R) \end{matrix} & \hookrightarrow & \begin{matrix} H^k(I_{\bar p}S^*(X,A;R)\otimes \\ I_{\bar q}S^*(Y,B;R)) \end{matrix} & \twoheadrightarrow & \bigoplus_{i+j=k+1} \begin{matrix} I_{\bar p}H^i(X,A;R)* \\ I_{\bar q}H^j(Y,B;R) \end{matrix} \\
 & & \downarrow \cong & & \\
 & & \begin{matrix} H^k(\mathrm{Hom}(I^{\bar p}S_*(X,A;R) \\ \otimes I^{\bar q}S_*(Y,B;R),R)) \end{matrix} & & \\
 & & \mathrm{IAW}^* \downarrow \cong & & \\
 & & \begin{matrix} I_QH^k(X\times Y,(A\times Y)\cup \\ (X\times B);R). \end{matrix} & &
\end{array} \tag{7.24}$$

The first horizontal map composed with the vertical maps is the intersection cohomology cross product by definition, and replacing the middle term of the short exact sequence with its claimed isomorphic image $I_QH^k(X\times Y,(A\times Y)\cup(X\times B);R)$ will provide the claimed sequence of the theorem.

We already know that IAW is a chain homotopy equivalence by definition, and so IAW* is an isomorphism. What remains is to verify that $I_{\bar p}S^*(X,A;R)$ and $I_{\bar q}S^*(Y,B;R)$ satisfy the necessary conditions for the algebraic Künneth Theorem to hold and that the top vertical map is an isomorphism.

Suppose we have chain homotopy equivalences $f: C_*\to I^{\bar p}S_*(X,A;R)$ and $g: D_*\to I^{\bar q}S_*(Y,B;R)$ such that C_* and D_* are bounded-below complexes of projectives and such that each C_i is finitely generated if each $I^{\bar p}H_i(X,A;R)$ is finitely generated and each D_i is finitely generated if each $I^{\bar q}H_i(Y,B;R)$ is finitely generated. For simplicity, we will assume that it is C_* that satisfies the finiteness condition. Lemma A.4.4 in Appendix A shows that such chain homotopy equivalences can be constructed because the intersection chain complexes are complexes of projective modules. Then, we have the following short exact sequence:

$$\begin{array}{ccccc}
\bigoplus_{i+j=k} \begin{matrix} H^i(\mathrm{Hom}(C_*,R))\otimes \\ H^j(\mathrm{Hom}(D_*,R)) \end{matrix} & \hookrightarrow & \begin{matrix} H^k(\mathrm{Hom}(C_*,R)\otimes \\ \mathrm{Hom}(D_*,R)) \end{matrix} & \twoheadrightarrow & \bigoplus_{i+j=k+1} \begin{matrix} H^i(\mathrm{Hom}(C_*,R))* \\ H^j(\mathrm{Hom}(D_*,R)) \end{matrix} \\
f^*\otimes g^* \uparrow \cong & & f^*\otimes g^* \uparrow \cong & & f^* * g^* \uparrow \cong \\
\bigoplus_{i+j=k} \begin{matrix} I_{\bar p}H^i(X,A;R)\otimes \\ I_{\bar q}H^j(Y,B;R) \end{matrix} & & \begin{matrix} H^k(I_{\bar p}S^*(X,A;R)\otimes \\ I_{\bar q}S^*(Y,B;R);R) \end{matrix} & & \bigoplus_{i+j=k+1} \begin{matrix} I_{\bar p}H^i(X,A;R)* \\ I_{\bar q}H^j(Y,B;R). \end{matrix}
\end{array} \tag{7.25}$$

As f and g are chain homotopy equivalences, the vertical maps here are all isomorphisms: in the middle we are using that $f^*\otimes g^*$ as a chain map is a chain homotopy equivalence and so induces an isomorphism in (co)homology (see Section A.2); on the left, f^* and g^* are (co)homology maps that are isomorphisms and so their tensor product is an isomorphism by the functoriality of $\otimes$, and similarly on the right by the functoriality of the torsion product. The top line here is a split short exact sequence by the standard algebraic Künneth Theorem[22] [238, Theorem 3.6.3], up to the switch between homological and cohomological indexing. To cite this theorem, we need to know that each $\mathrm{Hom}(C_i, R)$ and $d(\mathrm{Hom}(C_i, R))$ is flat. We will show that $\mathrm{Hom}(C_i, R)$ is projective, which will suffice, as a submodule of a projective module over a Dedekind domain is projective: As C_i is finitely generated, there is a finitely generated free R-module F_i with $F_i \to C_i$ surjective and with kernel $K_i \subset F_i$. As C_i is projective, $F_i \cong K_i \oplus C_i$ by Lemma A.4.2, and as F_i is finitely generated free, $F_i = R^{m_i}$ for some m_i. So $\mathrm{Hom}(F_i, R) \cong \mathrm{Hom}(R^{m_i}, R) \cong (\mathrm{Hom}(R, R))^{m_i}$ is free, and, furthermore, $\mathrm{Hom}(F_i, R) \cong \mathrm{Hom}(K_i \oplus C_i, R) \cong \mathrm{Hom}(K_i, R) \oplus \mathrm{Hom}(C_i, R)$. Hence $\mathrm{Hom}(C_i, R)$ is a direct summand of a free module and is therefore projective.

So, we have now seen that the top row of Diagram (7.25) is short exact split by the algebraic Künneth Theorem, and, via the isomorphisms in the diagram we obtain a split short exact sequence with the same modules as in the top row of Diagram (7.24).

Next, we want to show that we have a commutative diagram:

$$\begin{array}{ccccc}
\displaystyle\bigoplus_{\substack{i+j\\=k}} \begin{array}{c} H^i(\mathrm{Hom}(C_*, R)) \otimes \\ H^j(\mathrm{Hom}(D_*, R)) \end{array} & \xhookrightarrow{\theta} & \begin{array}{c} H^k(\mathrm{Hom}(C_*, R) \otimes \\ \mathrm{Hom}(D_*, R)) \end{array} & \xrightarrow{\cong} & H^k(\mathrm{Hom}(C_* \otimes D_*, R)) \\
{\scriptstyle f^*\otimes g^*}\Big\uparrow{\scriptstyle \cong} & & {\scriptstyle f^*\otimes g^*}\Big\uparrow{\scriptstyle \cong} & & {\scriptstyle (f\otimes g)^*}\Big\uparrow{\scriptstyle \cong} \\
\displaystyle\bigoplus_{\substack{i+j\\=k}} \begin{array}{c} I_{\bar{p}}H^i(X, A; R) \\ \otimes I_{\bar{q}}H^j(Y, B; R) \end{array} & \xrightarrow{\theta} & \begin{array}{c} H^k(I_{\bar{p}}S^*(X, A; R) \\ \otimes I_{\bar{q}}S^*(Y, B; R); R) \end{array} & \xrightarrow{\cong} & \begin{array}{c} H^k(\mathrm{Hom}(I^{\bar{p}}S_*(X, A; R) \\ \otimes I^{\bar{q}}S_*(Y, B; R), R)). \end{array}
\end{array} \tag{7.26}$$

Notice that the first part of this diagram is compatible with the first part of Diagram (7.25), while the bottom line consists of the left-hand horizontal map and the first vertical map in Diagram (7.24). So, if we show that this diagram commutes and that the maps labeled as such are isomorphisms, it will follow that we have achieved a diagram of the form of (7.24), in which we know that the row is exact, that the top vertical map is an isomorphism, and that the composition right then one step down is Θ, so that the full map right then all the way down is the cohomology cross product. This will prove the theorem.

The commutativity of the square on the right follows from Lemma 7.3.1. The commutativity on the left is trivial as, following Remark 5.2.6, if $\alpha \in I_{\bar{p}}S^i(X, A; R)$ and $\beta \in I_{\bar{q}}S^j(Y, B; R)$ are cocycles representing cohomology classes, the map labeled θ simply takes $\alpha \otimes \beta$, as a tensor product of cohomology classes, to the cohomology class represented by $\alpha \otimes \beta$. The commutativity then comes by seeing what happens to such cochain repre-

[22] Curiously, in [238, Theorem 3.6.3], Weibel only discusses the (non-natural) splitting of the Künneth exact sequence when the base ring is $\mathbb{Z}$. Hilton and Stammbach [126, Theorem V.2.1] give a complete proof of the algebraic Künneth Theorem, including the splitting, when R is a PID. The proof given in [126] works over a Dedekind domain replacing "free" with "projective" throughout the argument.

sentatives. We have already observed that the two leftmost vertical maps are isomorphisms. The rightmost vertical map is similarly an isomorphism as the tensor products and duals of chain homotopy equivalences are chain homotopy equivalences; see Section A.2.

Lastly, we need to see that

$$H^k(\mathrm{Hom}(C_*,R)\otimes \mathrm{Hom}(D_*,R)) \to H^k(\mathrm{Hom}(C_*\otimes D_*,R))$$

is an isomorphism, which will imply that the bottom right horizontal map of Diagram (7.26) is also an isomorphism. In fact, $\mathrm{Hom}(C_*,R)\otimes \mathrm{Hom}(D_*,R) \to \mathrm{Hom}(C_*\otimes D_*,R)$ is an isomorphism of chain complexes with the given hypotheses. This is shown just below as Lemma 7.3.65.

This completes the proof of Theorem 7.3.63. □

Lemma 7.3.65 *Let R be a Dedekind domain. Suppose C_* and D_* are bounded-below complexes of projective R-modules and that one of C_* or D_* consists entirely of finitely generated modules. Then*

$$\Theta : \mathrm{Hom}(C_*,R)\otimes \mathrm{Hom}(D_*,R) \to \mathrm{Hom}(C_*\otimes D_*,R)$$

is an isomorphism.

Proof For any fixed total degree k, we have

$$\mathrm{Hom}^k(C_*\otimes D_*,R) = \mathrm{Hom}\left(\bigoplus_{i+j=k} C_i\otimes D_j, R) \cong \bigoplus_{i+j=k} \mathrm{Hom}(C_i\otimes D_j, R\right),$$

as there are a finite number of non-zero summands by the bounded-below conditions. For any fixed i and j, if $f\in \mathrm{Hom}(C_i,R)$ and $g\in \mathrm{Hom}(D_j,R)$, then $\Theta(f\otimes g)$ acts trivially on elements of $C_{i'}\otimes D_{j'}$ unless $i=i'$ and $j=j'$. So Θ takes $\mathrm{Hom}(C_i,R)\otimes\mathrm{Hom}(D_j,R)\to \mathrm{Hom}(C_i\otimes D_j,R)$. In other words, it preserves the direct sum structure. So it suffices to prove that if A and B are fixed projective R-modules one of which is finitely generated then $\Theta : \mathrm{Hom}(A,R)\otimes\mathrm{Hom}(B,R)\to\mathrm{Hom}(A\otimes B,R)$ is an isomorphism.

The proof is standard when A is free and finitely generated (see e.g. [220, Lemma 5.5.6]): In this case, $A\cong R^m$ for some m. Let $\{a_i\}_{i=1}^m$ be a basis for R^m, let $\{a_i^*\}_{i=1}^m$ be the dual basis (i.e. $a_i^*(a_j)=\delta_{i,j}$), and let $\langle a_i\rangle$ be the summand of R^m generated by a_i. If $\beta\in\mathrm{Hom}(B,R)$, then we see that $\Theta(a_i^*\otimes\beta)$ acts trivially on the summands $\langle a_j\rangle\otimes B$ for $i\neq j$. So Θ takes $\mathrm{Hom}(\langle a_i\rangle,R)\otimes\mathrm{Hom}(B,R)$ to $\mathrm{Hom}(\langle a_i\rangle\otimes B,R)$, and again the map splits into a direct sum of simpler maps. Restricting to each summand, it is easy to see that each $\Theta:\mathrm{Hom}(R,R)\otimes\mathrm{Hom}(B,R)\to\mathrm{Hom}(R\otimes B,R)$ is an isomorphism taking $\mathrm{id}\otimes\beta$ to β in $\mathrm{Hom}(R\otimes B,R)\cong\mathrm{Hom}(B,R)$. This completes the case for A free and finitely generated.

Suppose now A is finitely generated projective and that $A\oplus A'$ is a free module. We can find such a finitely generated A'. To see this, we observe that as A is finitely generated there is a finitely generated free R-module F with $F\to A$ surjective and with kernel $A'\subset F$. As Dedekind domains are Noetherian [30, Theorem VII.2.2.1], F and A' are thus Noetherian (see [147, Section X.1]), and so A' is finitely generated. The above proof for free modules says that Θ induces an isomorphism from

$$\begin{aligned}&\operatorname{Hom}(A \oplus A', R) \otimes \operatorname{Hom}(B, R)\\ &\quad \cong (\operatorname{Hom}(A, R) \otimes \operatorname{Hom}(B, R)) \oplus (\operatorname{Hom}(A', R) \otimes \operatorname{Hom}(B, R))\end{aligned}$$

to

$$\operatorname{Hom}((A \oplus A') \otimes B, R) \cong \operatorname{Hom}(A \otimes B, R) \oplus \operatorname{Hom}(A' \otimes B, R).$$

But Θ also preserves these splittings, by an argument just as above, and it follows that this larger Θ restricts to the desired isomorphism $\Theta : \operatorname{Hom}(A, R) \otimes \operatorname{Hom}(B, R) \to \operatorname{Hom}(A \otimes B, R)$ by the following argument. The injectivity for the free modules restricts to injectivity on each summand. Furthermore, as we have surjectivity in the free case, any element of $\operatorname{Hom}(A \otimes B, R) \subset \operatorname{Hom}((A \oplus A') \otimes B, R)$ is in the image of $\operatorname{Hom}(A \oplus A', R) \otimes \operatorname{Hom}(B, R)$. But we know the summand $\operatorname{Hom}(A', R) \otimes \operatorname{Hom}(B, R)$ of $\operatorname{Hom}(A \oplus A', R) \otimes \operatorname{Hom}(B, R)$ maps trivially to the summand $\operatorname{Hom}(A \otimes B, R)$ of $\operatorname{Hom}(A \otimes B, R) \oplus \operatorname{Hom}(A' \otimes B, R)$ and so all of $\operatorname{Hom}(A \otimes B, R)$ must be in the image of $\operatorname{Hom}(A, R) \otimes \operatorname{Hom}(B, R)$. □

7.3.9 Summary of Properties

Given the large number of properties of cup, cross, and cap products we have now seen and given how spread out through the proofs the statements have been, we here present for the benefit of the reader a quick summary of these properties and the required conditions necessary for the properties to hold. We include also the properties of the homology cross product from Section 5.2.3, which hold for non-GM intersection homology by Theorem 6.3.19. Throughout this section, spaces labeled X, Y, etc. are CS sets (unless noted otherwise), subspaces labeled A, B, etc. are open subspaces (unless noted otherwise), and R is a Dedekind domain. The perversity $\bar{t}_X$ is the top perversity on the space X.

(a) Naturality

- **Conditions:** All spaces only filtered
  Maps: $f : (X, A) \to (X', A')$ is $(\bar{p}, \bar{p}')$-stratified, $g : (Y, B) \to (Y', B')$ is $(\bar{q}, \bar{q}')$-stratified, and $f \times g$ is (P, Q)-stratified; $P(S \times T) \geq \bar{p}(S) + \bar{q}(T)$, $Q(S' \times T') \geq \bar{p}'(S') + \bar{q}'(T')$
  Terms: $\xi \in I^{\bar{p}}S_*(X, A; R)$, $\eta \in I^{\bar{q}}S_*(Y, B; R)$, $\xi \times \eta \in I^{P}H_*(X \times Y, (A \times Y) \cup (X \times B); R)$
  Property:
  $f(\xi) \times g(\eta) = (f \times g)(\xi \times \eta) \in I^{Q}S_*(X' \times Y', (A' \times Y') \cup (X' \times B'); R)$
  Location: Proposition 5.2.17
- **Conditions:** Q on $X \times Y$ is $(\bar{p}, \bar{q})$-compatible, Q' on $X' \times Y'$ is $(\bar{p}', \bar{q}')$-compatible

Maps: $f:(X,A)\to(X',A')$ is $(\bar p,\bar p')$-stratified, $g:(Y,B)\to(Y',B')$ is $(\bar q,\bar q')$-stratified, $f\times g$ is (Q,Q')-stratified

Terms: $\alpha\in I_{\bar p'}H^i(X',A';R)$, $\beta\in I_{\bar q'}H^j(Y',B';R)$, $\alpha\times\beta\in I_{Q'}H^{i+j}(X'\times Y',(A'\times Y')\cup(X'\times B');R)$

Property:

$$(f\times g)^*(\alpha\times\beta)=(f^*(\alpha))\times(g^*(\beta))\in I_QH^{i+j}(X\times Y,(A\times Y)\cup(X\times B);R)$$

Location: Proposition 7.3.2

- **Conditions:** $(\bar p,\bar q;\bar r)$ agreeable on X, $(\bar u,\bar v;\bar s)$ agreeable on Y

  Maps: $f:(X;A,B)\to(Y;C,D)$ is $(\bar p,\bar u)$-stratified, $(\bar q,\bar v)$-stratified, and $(\bar r,\bar s)$-stratified

  Terms: $\alpha\in I_{\bar u}H^i(Y,C;R)$, $\beta\in I_{\bar v}H^j(Y,D;R)$, $\alpha\smile\beta\in I_{\bar s}H^{i+j}(Y,C\cup D;R)$

  Property:

  $$f^*(\alpha\smile\beta)=(f^*(\alpha))\smile(f^*(\beta))\in I_{\bar r}H^{i+j}(X,A\cup B;R)$$

  Location: Proposition 7.3.5

- **Conditions:** $(\bar p,\bar q;\bar r)$ agreeable on X, $(\bar u,\bar v;\bar s)$ agreeable on Y

  Maps: $f:(X;A,B)\to(Y;C,D)$ is $(\bar p,\bar u)$-stratified, $(\bar q,\bar v)$-stratified, and $(\bar r,\bar s)$-stratified

  Terms: $\beta\in I_{\bar v}H^j(Y,D;R)$, $\xi\in I^{\bar r}H_{i+j}(X,A\cup B;R)$, $f^*(\beta)\frown\xi\in I^{\bar p}H_i(X,A;R)$

  Property:

  $$\beta\frown f(\xi)=f(f^*(\beta)\frown\xi)\in I^{\bar u}H_i(Y,C;R).$$

  Location: Proposition 7.3.6

(b) Associativity

- **Conditions:** All spaces only filtered; $P(S\times S')\geq\bar p(S)+\bar q(S')$ on $X\times Y$, $Q(S'\times S'')\geq\bar q(S')+\bar r(S'')$ on $Y\times Z$, $T(S\times S'\times S'')\geq P(S\times S')+\bar r(S'')$ and $T(S\times S'\times S'')\geq\bar p(S)+Q(S'\times S'')$ on $X\times Y\times Z$

  Terms: $x\in I^{\bar p}S_*(X,A;R)$, $y\in I^{\bar q}S_*(Y,B;R)$, $z\in I^{\bar r}S_*(Z,C;R)$

  Property:

  $$(x\times y)\times z=x\times(y\times z)\in I^TS_*(X\times Y\times Z,(A\times Y\times Z)\cup(X\times B\times Z)\cup(X\times Y\times C);R)$$

  Location: Proposition 5.2.19

- **Conditions:** Q_1 is $(\bar p,\bar q)$-compatible, Q_2 is $(\bar q,\bar r)$-compatible, Q_3 is both $(\bar p,Q_2)$-compatible and $(Q_1,\bar q)$-compatible

  Terms: $\alpha\in I_{\bar p}H^i(X,A;R)$, $\beta\in I_{\bar q}H^j(Y,B;R)$, $\gamma\in I_{\bar r}H^k(Z,C;R)$

  Property:

  $$(\alpha\times\beta)\times\gamma=\alpha\times(\beta\times\gamma)\quad\text{in}$$
  $$I_{Q_3}H^{i+j+k}(X\times Y\times Z,(A\times Y\times Z)\cup(X\times B\times Z)\cup(X\times Y\times C);R)$$

  Location: Proposition 7.3.29

- **Conditions:** $D\bar{u} > D\bar{p} + D\bar{q}$, $D\bar{v} > D\bar{q} + D\bar{r}$, $D\bar{s} > D\bar{u} + D\bar{r}$, and $D\bar{s} > D\bar{p} + D\bar{v}$; conditions can be replaced with non-strict inequalities if X is locally $(\bar{p}, R)$-torsion-free and locally $(\bar{r}, R)$-torsion-free[23]
  Terms: $\alpha \in I_{\bar{p}}H^i(X, A; R)$, $\beta \in I_{\bar{q}}H^j(X, B; R)$, $\gamma \in I_{\bar{r}}H^k(X, C; R)$, $\alpha \smile \beta \in I_{\bar{u}}H^{i+j}(X, A \cup B; R)$, $\beta \smile \gamma \in I_{\bar{v}}H^{j+k}(X, B \cup C; R)$
  Property:
  $$(\alpha \smile \beta) \smile \gamma = \alpha \smile (\beta \smile \gamma) \in I_{\bar{s}}H^{i+j+k}(X, A \cup B \cup C; R)$$
  Location: Proposition 7.3.34

(c) Commutativity

- **Conditions:** All spaces only filtered; $P(S \times S') \geq \bar{p}(S) + \bar{q}(S')$ on $X \times Y$, $Q(S' \times S) = P(S \times S')$ on $Y \times X$
  Maps: $t : X \times Y \to Y \times X$ such that $t(x, y) = (y, x)$
  Terms: $x \in I^{\bar{p}}S_*(X, A; R)$, $y \in I^{\bar{q}}S_*(Y, B; R)$
  Property:
  $$t(x \times y) = (-1)^{|x||y|}y \times x \in I^Q S_*(Y \times X, (B \times X) \cup (Y \times A); R)$$
  Location: Proposition 5.2.20

- **Conditions:** Q is $(\bar{p}, \bar{q})$-compatible on $X \times Y$, $Q^\tau(T \times S) = Q(S \times T)$ for $T \times S \subset Y \times X$
  Maps: $t : X \times Y \to Y \times X$ such that $t(x, y) = (y, x)$
  Terms: $\alpha \in I_{\bar{p}}H^i(X, A; R)$, $\beta \in I_{\bar{q}}H^j(Y, B; R)$
  Property:
  $$t^*(\alpha \times \beta) = (-1)^{ij}\beta \times \alpha \in I_{Q^\tau}H^{i+j}(Y \times X, (B \times X) \cup (Y \times A); R)$$
  Location: Proposition 7.3.13

- **Conditions:** $(\bar{p}, \bar{q}; \bar{r})$ agreeable
  Terms: $\alpha \in I_{\bar{p}}H^i(X, A; R)$, $\beta \in I_{\bar{q}}H^j(X, B; R)$
  Property:
  $$\alpha \smile \beta = (-1)^{ij}\beta \smile \alpha \in I_{\bar{r}}H^{i+j}(X, A \cup B; R)$$
  Location: Proposition 7.3.15

(d) Unital properties

- **Conditions:** X filtered
  Terms: $\sigma_0 : \Delta^0 \to \mathrm{pt}$ the unique singular 0-simplex, $\xi \in I^{\bar{p}}S_i(X, A; R)$
  Property:
  $$\begin{aligned}\sigma_0 \times \xi = \xi \times \sigma_0 = \xi &\in I^{\bar{p}}S_i(\mathrm{pt} \times X, \mathrm{pt} \times A; R)\\ &= I^{\bar{p}}S_i(X \times \mathrm{pt}, A \times \mathrm{pt}; R) = I^{\bar{p}}S_i(X, A; R).\end{aligned}$$
  Location: Proposition 5.2.21

- **Conditions:** None

[23] See also Lemma 7.3.31 for a more general associativity property.

Terms: $\alpha \in I_{\bar{p}}H^i(X,A;R)$, $1 \in I_{\bar{t}}H^0(X;R)$

Property:

$$1 \smile \alpha = \alpha \smile 1 = \alpha \in I_{\bar{p}}H^i(X,A;R)$$

Location: Proposition 7.3.21

- **Conditions:** None

  Terms: $\xi \in I^{\bar{p}}H_i(X,A;R)$, $1 \in I_{\bar{t}}H^0(X;R)$

  Property:

  $$1 \frown \xi = \xi \in I^{\bar{p}}H_i(X,A;R)$$

  Location: Proposition 7.3.22

- **Conditions:** None

  Terms: $\alpha \in I_{\bar{p}}H^i(X,A;R)$, $1 \in H^0(\text{pt};R)$

  Property:

  $$1 \times \alpha = \alpha \times 1 = \alpha \in I_{\bar{p}}H^i(\text{pt} \times X, \text{pt} \times A;R) = I_{\bar{p}}H^i(X \times \text{pt}, A \times \text{pt};R) = I_{\bar{p}}H^i(X,A;R)$$

  Location: Proposition 7.3.23

- **Conditions:** $\bar{q} \leq \bar{t}_Y$, Q is $(\bar{p},\bar{q})$-compatible on $X \times Y$.

  Terms: $\alpha \in I_{\bar{p}}H^i(X,A;R)$, $1_Y \in I_{\bar{q}}H^0(Y;R)$, $p_1 : X \times Y \to X$ the projection

  Property:

  $$\alpha \times 1_Y = p_1^*(\alpha) \in I_QH^*(X \times Y, A \times Y;R)$$

  Location: Proposition 7.3.24

(e) Evaluations

- **Conditions:** None

  Maps: $\mathbf{a} : I^{\bar{t}}H_0(X;R) \to R$ the augmentation map

  Terms: $\alpha \in I_{\bar{p}}H^i(X,A;R)$, $\xi \in I^{\bar{p}}H_i(X,A;R)$

  Property:

  $$\mathbf{a}(\alpha \frown \xi) = \alpha(\xi) \in R$$

  Location: Proposition 7.3.25

- **Conditions:** Q is $(\bar{p},\bar{q})$-compatible

  Terms: $\alpha \in I_{\bar{p}}H^a(X,A;R)$, $\beta \in I_{\bar{q}}H^b(Y,B;R)$, $\xi \in I^{\bar{p}}H_i(X,A;R)$, $\eta \in I^{\bar{q}}H_j(Y,B;R)$,
  $\alpha \times \beta \in I_QH^{a+b}(X \times Y, (A \times Y) \cup (X \times B);R)$,
  $\xi \times \eta \in I^QH_{i+j}(X \times Y, (A \times Y) \cup (X \times B);R)$

  Property:

  $$(\alpha \times \beta)(\xi \times \eta) = (-1)^{bi}\alpha(\xi)\beta(\eta) \in R$$

  Location: Proposition 7.3.27

(f) Stability

- **Conditions:** All spaces filtered, $Q(S \times S') \geq \bar{p}(S) + \bar{q}(S')$ on $X \times Y$
  Terms: $\xi \in I^{\bar{p}}H_i(X, A; R)$, $\eta \in I^{\bar{q}}H_j(Y, B; R)$
  Property:
  $$(\partial_* \xi) \times \eta = \partial_*(\xi \times \eta) \in I^Q H_{i+j-1}((A \times Y) \cup (X \times B), X \times B; R)$$
  Location: Proposition 5.2.23
- **Conditions:** All spaces filtered, $Q(S \times S') \geq \bar{p}(S) + \bar{q}(S')$ on $X \times Y$
  Terms: $\xi \in I^{\bar{p}}H_i(X, A; R)$, $\eta \in I^{\bar{q}}H_j(Y, B; R)$
  Property:
  $$\begin{aligned}&\partial_*(\xi) \times \eta + (-1)^i \xi \times \partial_*(\eta)\\&\quad = \partial_*(\xi \times \eta) \in I^Q H_{i+j-1}((A \times Y) \cup (X \times B), A \times B; R)\end{aligned}$$
  Location: Proposition 5.2.24
- **Conditions:** $(\bar{p}, \bar{q}; \bar{r})$ agreeable
  Maps: $\mathrm{i} : B \hookrightarrow X$ the inclusion map,
  $e : I^{\bar{r}}H_{i+j-1}(B, A \cap B; R) \to I^{\bar{r}}H_{i+j-1}(A \cup B, A; R)$ the excision isomorphism
  Terms: $\alpha \in I_{\bar{q}}H^j(B; R)$, $\xi \in I^{\bar{r}}H_{i+j}(X, A \cup B; R)$, $\partial_*(\xi) \in I^{\bar{r}}H_{i+j-1}(A \cup B, A; R)$
  Property:
  $$(d^*(\alpha)) \frown \xi = (-1)^{j+1}\mathrm{i}(\alpha \frown e^{-1}\partial_*(\xi)) \in I^{\bar{p}}H_{i-1}(X, A; R),$$
  Location: Proposition 7.3.37
- **Conditions:** $(\bar{p}, \bar{q}; \bar{r})$ agreeable
  Maps: $\mathrm{i} : A \hookrightarrow X$ the inclusion map,
  $e : I^{\bar{r}}H_{i+j-1}(A, A \cap B; R) \to I^{\bar{r}}H_{i+j-1}(A \cup B, B; R)$ the excision isomorphism
  Terms: $\alpha \in I_{\bar{q}}H^j(X, B; R)$, $\xi \in I^{\bar{r}}H_{i+j}(X, A \cup B; R)$, $\partial_*(\xi) \in I^{\bar{r}}H_{i+j-1}(A \cup B, B; R)$
  Property:
  $$\partial_*(\alpha \frown \xi) = (-1)^j(\mathrm{i}^*(\alpha)) \frown (e^{-1}\partial_*(\xi)) \in I^{\bar{p}}H_{i-1}(A; R)$$
  Location: Proposition 7.3.38
- **Conditions:** Q is $(\bar{p}, \bar{q})$-compatible on $X \times Y$
  Maps: $e : I^Q H_*(A \times Y, A \times B; R) \to I^Q H_*((A \times Y) \cup (X \times B), X \times B; R)$ an excision isomorphism,
  $d^* : I_Q H^{i+j}((A \times Y) \cup (X \times B), X \times B; R) \to I^Q H^{i+j+1}(X \times Y, (A \times Y) \cup (X \times B); R)$
  Terms: $\alpha \in I_{\bar{p}}H^i(A; R)$, $\beta \in I_{\bar{q}}H^j(Y, B; R)$
  Property:
  $$(d^*\alpha) \times \beta = d^*(e^{-1})^*(\alpha \times \beta) \in I_Q H^{i+j+1}(X \times Y, (A \times Y) \cup (X \times B); R)$$
  Location: Proposition 7.3.43
- **Conditions:** $(\bar{p}, \bar{q}; \bar{r})$ agreeable
  Maps: $\mathrm{i} : A \to X$ is the inclusion map, $e : I^{\bar{r}}H_*(A, A \cap B; R) \to I^{\bar{r}}H_*(A \cup B, B; R)$ an excision isomorphism, $d^* : I_{\bar{r}}H^{i+j}(A \cup B, B; R) \to I_{\bar{r}}H^{i+j+1}(X, A \cup B; R)$
  Terms: $\alpha \in I_{\bar{p}}H^i(A; R)$, $\beta \in I_{\bar{q}}H^j(X, B; R)$

Property:

$$(d^*\alpha) \smile \beta = d^*(e^{-1})^*(\alpha \smile \mathfrak{i}^*(\beta)) \in I_{\bar{r}}H^{i+j+1}(X, A \cup B; R)$$

Location: Proposition 7.3.44

(g) Combinations – properties that involve multiple types of products

- **Conditions:** $D\bar{u} > D\bar{p} + D\bar{q}$, $D\bar{v} > D\bar{q} + D\bar{r}$, $D\bar{s} > D\bar{u} + D\bar{r}$, and $D\bar{s} > D\bar{p} + D\bar{v}$; conditions can be replaced with non-strict inequalities if X is locally $(\bar{p}, R)$-torsion-free and locally $(\bar{r}, R)$-torsion-free[24]

  Terms: $\alpha \in I_{\bar{q}}H^j(X, B; R)$, $\beta \in I_{\bar{r}}H^k(X, C; R)$, $\xi \in I^{\bar{s}}H_{i+j+k}(X, A \cup B \cup C; R)$, $\alpha \smile \beta \in I_{\bar{v}}H^{j+k}(X, B \cup C; R)$, $\beta \frown \xi \in I^{\bar{u}}H_{i+j}(X, A \cup B; R)$

  Property:

  $$(\alpha \smile \beta) \frown \xi = \alpha \frown (\beta \frown \xi) \in I^{\bar{p}}H_i(X, A; R)$$

  Location: Proposition 7.3.35

- **Conditions:** $(\bar{p}, \bar{q}; \bar{r})$ is Q-agreeable

  Maps: $\mathbf{d} : X \to X \times X$ the diagonal map

  Terms: $\alpha \in I_{\bar{p}}H^i(X, A; R)$, $\beta \in I_{\bar{q}}H^j(X, B; R)$, $\alpha \times \beta \in I_Q H^{i+j}(X \times Y, (A \times Y) \cup (X \times B); R)$

  Property:

  $$\mathbf{d}^*(\alpha \times \beta) = \alpha \smile \beta \in I_{\bar{r}}H^{i+j}(X, A \cup B; R)$$

  Location: Proposition 7.3.45

- **Conditions:** Q is $(\bar{p}, \bar{q})$-compatible on $X \times Y$

  Maps: $p_1 : X \times Y \to X$ and $p_2 : X \times Y \to Y$ the projection maps, $p_1^* : I_{\bar{p}}H^*(X, A; R) \to I_{Q_{\bar{p},\bar{t}_Y}}H^*(X \times Y, A \times Y; R)$ and $p_2^* : I_{\bar{q}}H^*(Y, B; R) \to I_{Q_{\bar{t}_X,\bar{q}}}H^*(X \times Y, X \times B; R)$

  Terms: $\alpha \in I_{\bar{p}}H^i(X, A; R)$, $\beta \in I_{\bar{q}}H^j(X, B; R)$

  Property:

  $$\alpha \times \beta = (p_1^*(\alpha)) \smile (p_2^*(\beta)) \in I_Q H^{i+j}(X \times Y, (A \times Y) \cup (X \times B); R)$$

  Location: Proposition 7.3.46

- **Conditions:** 1. $D\bar{r} > D\bar{p} + D\bar{q}$ on X, $D\bar{s} > D\bar{u} + D\bar{v}$ on Y, $a = 1$, OR
  2. $D\bar{r} \geq D\bar{p} + D\bar{q}$ on X, $D\bar{s} \geq D\bar{u} + D\bar{v}$ on Y, $a = 2$, X is locally $(\bar{p}, R)$-torsion-free or locally $(\bar{q}, R)$-torsion-free, Y is locally $(\bar{u}, R)$-torsion-free or locally $(\bar{v}, R)$-torsion-free, X is locally $(\bar{p}, R)$-torsion-free or Y is locally $(\bar{u}, R)$-torsion-free, X is locally $(\bar{q}, R)$-torsion-free or Y is locally $(\bar{v}, R)$-torsion-free, $X \times Y$ is locally $(\hat{Q}^2_{\bar{p},\bar{u}}, R)$-torsion-free[25] or locally $(\hat{Q}^2_{\bar{q},\bar{v}}, R)$-torsion-free, X is locally $(\bar{r}, R)$-torsion-free or Y is locally $(\bar{s}, R)$-torsion-free

[24] See also Lemma 7.3.32 for a more general associativity property.

[25] See Definition 7.3.52 for the definition of $\hat{Q}^a_{\bar{p},\bar{q}}$.

Terms: $\alpha \in I_{\bar{p}}H^i(X,A;R)$, $\beta \in I_{\bar{q}}H^j(X,B;R)$, $\gamma \in I_{\bar{u}}H^k(Y,C;R)$, $\delta \in I_{\bar{v}}H^\ell(Y,D;R)$, $\alpha \smile \beta \in I_{\bar{r}}H^{i+j}(X,A\cup B;R)$, $\gamma \smile \delta \in I_{\bar{s}}H^{k+\ell}(Y,C\cup D;R)$, $\alpha\times\gamma \in I_{\hat{Q}^a_{\bar{p},\bar{u}}}H^{i+k}(X\times Y,(A\times Y)\cup(X\times C);R)$, $\beta\times\delta \in I_{\hat{Q}^a_{\bar{q},\bar{v}}}H^{j+\ell}(X\times Y,(B\times Y)\cup(X\times D);R)$

Property:

$$(\alpha\smile\beta)\times(\gamma\smile\delta) = (-1)^{jk}(\alpha\times\gamma)\smile(\beta\times\delta) \quad \text{in} \quad I_{\hat{Q}^a_{\bar{r},\bar{s}}}H^{i+j+k+\ell}(X\times Y,((A\cup B)\times Y)\cup(X\times(C\cup D));R)$$

Location: Proposition 7.3.54; see also Lemma 7.3.50 for a more general version of this property

- **Conditions:** 1a. $D\bar{r} > D\bar{p} + D\bar{q}$ on X, $D\bar{s} > D\bar{u} + D\bar{v}$ on Y, $a = 1$, OR
  1b. $D\bar{r} \geq D\bar{p} + D\bar{q}$ on X, $D\bar{s} \geq D\bar{u} + D\bar{v}$ on Y, $a = 2$, X is locally $(\bar{p},R)$-torsion-free or locally $(\bar{q},R)$-torsion-free, Y is locally $(\bar{u},R)$-torsion-free or locally $(\bar{v},R)$-torsion-free, X is locally $(\bar{p},R)$-torsion-free or Y is locally $(\bar{u},R)$-torsion-free, X is locally $(\bar{q},R)$-torsion-free or Y is locally $(\bar{v},R)$-torsion-free, $X\times Y$ is locally $(\hat{Q}^2_{\bar{p},\bar{u}},R)$-torsion-free[26] or locally $(\hat{Q}^2_{\bar{q},\bar{v}},R)$-torsion-free, X is locally $(\bar{r},R)$-torsion-free or Y is locally $(\bar{s},R)$-torsion-free
  2. On $X\times Y$: $P_{\bar{r},\bar{s}} \leq \hat{Q}^a_{\bar{r},\bar{s}}$ is $(\bar{r},\bar{s})$-compatible, $P_{\bar{p},\bar{u}} \leq \hat{Q}^a_{\bar{p},\bar{u}}$ is $(\bar{p},\bar{u})$-compatible, $P_{\bar{q},\bar{v}} \leq \hat{Q}^a_{\bar{q},\bar{v}}$ is $(\bar{q},\bar{v})$-compatible, $(P_{\bar{p},\bar{u}},P_{\bar{q},\bar{v}};P_{\bar{r},\bar{s}})$ is $\mathfrak{P}$-agreeable with $\mathfrak{P} \leq \hat{Q}^a_{\hat{Q}^a_{\bar{p},\bar{u}},\hat{Q}^a_{\bar{q},\bar{v}}}$

Terms: $\alpha \in I_{\bar{p}}H^i(X,A;R)$, $\beta \in I_{\bar{q}}H^j(X,B;R)$, $\gamma \in I_{\bar{u}}H^k(Y,C;R)$, $\delta \in I_{\bar{v}}H^\ell(Y,D;R)$, $\alpha \smile \beta \in I_{\bar{r}}H^{i+j}(X,A\cup B;R)$, $\gamma \smile \delta \in I_{\bar{s}}H^{k+\ell}(Y,C\cup D;R)$, $\alpha\times\gamma \in I_{P_{\bar{p},\bar{u}}}H^{i+k}(X\times Y,(A\times Y)\cup(X\times C);R)$, $\beta\times\delta \in I_{P_{\bar{q},\bar{v}}}H^{j+\ell}(X\times Y,(B\times Y)\cup(X\times D);R)$

Property:

$$(\alpha\smile\beta)\times(\gamma\smile\delta) = (-1)^{jk}(\alpha\times\gamma)\smile(\beta\times\delta) \quad \text{in} \quad I_{P_{\bar{r},\bar{s}}}H^{i+j+k+\ell}(X\times Y,((A\cup B)\times Y)\cup(X\times(C\cup D));R)$$

Location: Proposition 7.3.56; see also Lemma 7.3.50

- **Conditions:** 1. $D\bar{r} > D\bar{p} + D\bar{q}$ on X, $D\bar{s} > D\bar{u} + D\bar{v}$ on Y, $a = 1$, OR
  2. $D\bar{r} \geq D\bar{p} + D\bar{q}$ on X, $D\bar{s} \geq D\bar{u} + D\bar{v}$ on Y, $a = 2$, X is locally $(\bar{p},R)$-torsion-free or locally $(\bar{q},R)$-torsion-free, Y is locally $(\bar{u},R)$-torsion-free or locally $(\bar{v},R)$-torsion-free, $X\times Y$ is locally $(\hat{Q}^2_{\bar{p},\bar{u}},R)$-torsion-free[27] or locally $(\hat{Q}^2_{\bar{q},\bar{v}},R)$-torsion-free.

Terms: $\alpha \in I_{\bar{q}}H^j(X,B;R)$, $x \in I^{\bar{r}}H_{i+j}(X,A\cup B;R)$, $\beta \in I_{\bar{v}}H^\ell(X,D;R)$, $y \in I^{\bar{s}}H_{k+\ell}(X,C\cup D;R)$, $\alpha\times\beta \in I_{\hat{Q}^a_{\bar{q},\bar{v}}}H^{j+\ell}(X\times Y,(B\times Y)\cup(X\times D);R)$,

[26] See Definition 7.3.52 for the definition of $\hat{Q}^a_{\bar{p},\bar{q}}$.
[27] See Definition 7.3.52 for the definition of $\hat{Q}^a_{\bar{p},\bar{q}}$.

$x \times y \in I^{\hat{Q}^a_{\bar{r},\bar{s}}} H_{i+j+k+\ell}(X \times Y, ((A \cup B) \times Y) \cup (X \times (C \cup D)); R)$,
$\alpha \frown x \in I^{\bar{p}} H_i(X, A; R)$, and $\beta \frown y \in I^{\bar{u}} H_k(Y, C; R)$

Property:

$$(\alpha \times \beta) \frown (x \times y) = (-1)^{\ell(i+j)} (\alpha \frown x) \times (\beta \frown y) \quad \text{in} \quad I^{\hat{Q}^a_{\bar{p},\bar{u}}} H_{i+k}(X \times Y, (A \times Y) \cup (X \times C); R)$$

Location: Proposition 7.3.55; see also Lemma 7.3.51 for a more general version of this property.

- **Conditions:** 1a. $D\bar{r} > D\bar{p} + D\bar{q}$ on X, $D\bar{s} > D\bar{u} + D\bar{v}$ on Y, $a = 1$, OR
  1b. $D\bar{r} \geq D\bar{p} + D\bar{q}$ on X, $D\bar{s} \geq D\bar{u} + D\bar{v}$ on Y, $a = 2$, X is locally $(\bar{p}, R)$-torsion-free or locally $(\bar{q}, R)$-torsion-free, Y is locally $(\bar{u}, R)$-torsion-free or locally $(\bar{v}, R)$-torsion-free, X is locally $(\bar{p}, R)$-torsion-free or Y is locally $(\bar{u}, R)$-torsion-free, X is locally $(\bar{q}, R)$-torsion-free or Y is locally $(\bar{v}, R)$-torsion-free, $X \times Y$ is locally $(\hat{Q}^2_{\bar{p},\bar{u}}, R)$-torsion-free[28] or locally $(\hat{Q}^2_{\bar{q},\bar{v}}, R)$-torsion-free, X is locally $(\bar{r}, R)$-torsion-free or Y is locally $(\bar{s}, R)$-torsion-free
  2. On $X \times Y$: $P_{\bar{r},\bar{s}} \leq \hat{Q}^a_{\bar{r},\bar{s}}$ is $(\bar{r}, \bar{s})$-compatible, $P_{\bar{p},\bar{u}} \leq \hat{Q}^a_{\bar{p},\bar{u}}$ is $(\bar{p}, \bar{u})$-compatible, $P_{\bar{q},\bar{v}} \leq \hat{Q}^a_{\bar{q},\bar{v}}$ is $(\bar{q}, \bar{v})$-compatible, $(P_{\bar{p},\bar{u}}, P_{\bar{q},\bar{v}}; P_{\bar{r},\bar{s}})$ is $\mathfrak{P}$-agreeable with $\mathfrak{P} \leq \hat{Q}^a_{\hat{Q}^a_{\bar{p},\bar{u}}, \hat{Q}^a_{\bar{q},\bar{v}}}$

  Terms: $\alpha \in I_{\bar{q}} H^j(X, B; R)$, $x \in I^{\bar{r}} H_{i+j}(X, A \cup B; R)$, $\beta \in I_{\bar{v}} H^\ell(X, D; R)$, $y \in I^{\bar{s}} H_{k+\ell}(X, C \cup D; R)$, $\alpha \times \beta \in I_{P_{\bar{q},\bar{v}}} H^{j+\ell}(X \times Y, (B \times Y) \cup (X \times D); R)$, $x \times y \in I^{P_{\bar{r},\bar{s}}} H_{i+j+k+\ell}(X \times Y, ((A \cup B) \times Y) \cup (X \times (C \cup D)); R)$, $\alpha \frown x \in I^{\bar{p}} H_i(X, A; R)$, and $\beta \frown y \in I^{\bar{u}} H_k(Y, C; R)$

  Property:

  $$(\alpha \times \beta) \frown (x \times y) = (-1)^{\ell(i+j)} (\alpha \frown x) \times (\beta \frown y) \quad \text{in} \quad I^{P_{\bar{p},\bar{u}}} H_{i+k}(X \times Y, (A \times Y) \cup (X \times C); R)$$

  Location: Proposition 7.3.57; see also Lemma 7.3.51.

(h) Locality

- **Conditions:** $(\bar{p}, \bar{q}; \bar{r})$ agreeable, $\mathcal{U}$ a covering of X

  Maps: $\kappa : H_*(\Sigma_{U \in \mathcal{U}} I^{\bar{p}} S_*(U, U \cap A; R) \otimes I^{\bar{p}} S_*(U, U \cap B; R)) \to H_*(I^{\bar{p}} S_*(X, A; R) \otimes I^{\bar{p}} S_*(X, B; R))$ induced by inclusions

  Property:

  $$\text{im}(\bar{\mathbf{d}} : I^{\bar{r}} H_*(X, A \cup B; R) \to H_*(I^{\bar{p}} S_*(X, A; R) \otimes I^{\bar{p}} S_*(X, B; R))) \subset \text{im}(\kappa)$$

  Location: Proposition 7.3.59

(i) Cohomology Künneth Theorem

- **Conditions:** Q is $(\bar{p}, \bar{q})$-compatible on $X \times Y$, all $I^{\bar{p}} H_i(X, A; R)$ finitely generated or all $I^{\bar{q}} H_j(Y, B; R)$ finitely generated

[28] See Definition 7.3.52 for the definition of $\hat{Q}^a_{\bar{p},\bar{q}}$.

Property: There is a natural exact sequence

$$0 \to \bigoplus_{i+j=k} I_{\bar{p}}H^i(X,A;R) \otimes I_{\bar{q}}H^j(Y,B;R)$$
$$\xrightarrow{\times} I_Q H^k(X \times Y, (A \times Y) \cup (X \times B); R)$$
$$\to \bigoplus_{i+j=k+1} I_{\bar{p}}H^i(X,A;R) * I_{\bar{q}}H^j(Y,B;R) \to 0$$

that splits (non-naturally)

Location: Theorem 7.3.63

7.3.10 Products on ∂-pseudomanifolds

Throughout this chapter, we have developed the definitions and properties of cap, cup, and cross products under the assumptions that all spaces are CS sets and that all subsets are open subsets. This has largely been so that we can invoke the Künneth Theorem, usually the version stated in Theorem 6.4.14, which says that the chain cross product is a chain homotopy equivalence. This is used, in turn, to provide our IAW maps, which allow us to define the cup, cap, and cohomology cross products. In order to develop Lefschetz duality, we will need to work with products on ∂-pseudomanifolds and their boundaries. However, ∂-stratified pseudomanifolds are not technically CS sets, as their strata may be ∂-manifolds, not manifolds, and their boundaries are not open subsets. So our preceding results do not directly apply. Nonetheless, we will show in this section that most of these preceding results do apply by using that ∂-stratified pseudomanifold pairs $(X, \partial X)$ are stratified deformation retract pairs of CS sets with open subsets.

We can actually work a bit more generally than with pairs $(X, \partial X)$, so we adopt the following definition.

Definition 7.3.66 Suppose that X is a ∂-stratified pseudomanifold and that $\partial X = A \cup B$ with A and B themselves ∂-stratified pseudomanifolds with $A \cap B = \partial A = \partial B$. Let us call A and B satisfying these properties *partial boundaries*.

If A is a partial boundary of X, we will call the pair (X, A) a *partial boundary pair*.

Example 7.3.67 If X is a ∂-stratified pseudomanifold, then both ∂X and $\emptyset$ are partial boundaries. If (X, A) is a partial boundary pair, then so are $(X, \emptyset)$, $(A, \emptyset)$, and $(A, \partial A)$. △

Example 7.3.68 If X and Y are ∂-stratified pseudomanifolds, then $(\partial X) \times Y$ and $X \times \partial Y$ are partial boundaries of $X \times Y$ according to the proof of Lemma 2.11.7. Furthermore, we have $((\partial X) \times Y) \cup (X \times \partial Y) = \partial(X \times Y)$, so $(X \times Y, ((\partial X) \times Y) \cup (X \times \partial Y))$ is a partial boundary pair. △

We can now show that products exist for partial boundary pairs. Note that the notions of $(\bar{p}, \bar{q})$-compatible product perversities and agreeable triples of perversities extend to this context, as ∂-stratified pseudomanifolds have links.

Proposition 7.3.69 *Let R be a Dedekind domain. Suppose (X,A) and (Y,B) are partial boundary pairs and that Q is a $(\bar{p},\bar{q})$-compatible perversity on $X\times Y$. Then the cross product*

$$I^{\bar{p}}S_*(X,A;R)\otimes I^{\bar{q}}S_*(Y,B;R)\xrightarrow{\varepsilon} I^{Q}S_*(X\times Y,(A\times Y)\cup(X\times B);R)$$

is a chain homotopy equivalence and so admits chain homotopy inverse IAW *maps.*

Proof We begin by modeling our spaces as CS sets. Let N_1 be a filtered open collar of ∂X in X, which is guaranteed to exist from the definition of a ∂-stratified pseudomanifold (Definition 2.7.1). Let $N_2 \cong [0,1)\times\partial X$ be an external collar, which we can glue on to X to form the stratified pseudomanifold $X' = X\cup_{\partial X} N_2$. Then X' has a stratified deformation retraction to X by retracting the collar N_2. Let A^c be the complementary partial boundary such that $\partial X = A\cup A^c$ with $A\cap A^c = \partial A = \partial A^c$. Let A_1 be the union of A with the filtered collar $[0,1)\times\partial A$ in A^c; if $A=\partial X$, then $A_1 = A = \partial X$. Then A_1 is a CS set in ∂X that has a stratified deformation retraction to A. Identifying $N_1\cup N_2$ with $(-1,1)\times\partial X$, we can form the subspace $A' = (-1,1)\times A_1$. Then A' has a stratified deformation retraction to A, first by retracting A' to A_1 and then A_1 to A. See Figure 7.3. We also define the pair (Y',B') analogously. The perversities $\bar{p},\bar{q}$ extend in the obvious way to X' and Y', and Q extends to a $(\bar{p},\bar{q})$-compatible perversity on $X'\times Y'$.

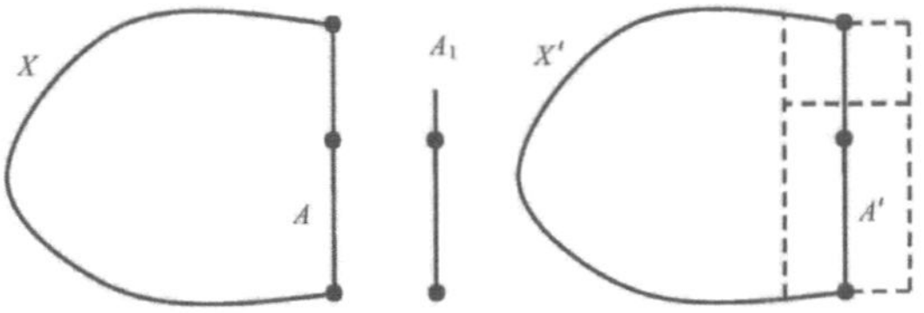

Figure 7.3 Schematics of: ∂-stratified pseudomanifold X with partial boundary A (left), extension of A to A_1 (middle), and the CS pair (X', A') obtained by adding collars (right).

Now let $f:(X,A)\hookrightarrow(X',A')$ and $g:(Y,B)\hookrightarrow(Y',B')$ be the inclusions and consider the diagram

$$\begin{array}{ccc}
I^{\bar{p}}S_*(X,A;R)\otimes I^{\bar{q}}S_*(Y,B;R) & \xrightarrow{\varepsilon} & I^{Q}S_*(X\times Y,(A\times Y)\cup(X\times B);R)\\
\Big\downarrow{\scriptstyle f\otimes g} & & \Big\downarrow{\scriptstyle f\times g}\\
I^{\bar{p}}S_*(X',A';R)\otimes I^{\bar{q}}S_*(Y',B';R) & \xrightarrow{\varepsilon} & I^{Q}S_*(X'\times Y',(A'\times Y')\cup(X'\times B');R).
\end{array} \tag{7.27}$$

The diagram commutes by Proposition 5.2.17 and Theorem 6.3.19. The induced maps $f: I^{\bar{p}}H_*(X,A;R)\to I^{\bar{p}}H_*(X',A';R)$ and $g: I^{\bar{q}}H_*(Y,B;R)\to I^{\bar{q}}H_*(Y',B';R)$ are isomorphisms by stratified homotopy invariance (applied to the inclusions $X\hookrightarrow X'$, $Y\hookrightarrow Y'$,

$A \hookrightarrow A'$, and $B \hookrightarrow B'$), the long exact sequences of the pairs, and the Five Lemma. Furthermore, the map

$$I^{\bar{p}}S_*(X',A';R) \otimes I^{\bar{q}}S_*(Y',B';R) \xrightarrow{\varepsilon} I^QS_*(X'\times Y',(A'\times Y')\cup(X'\times B');R)$$

induces a homology isomorphism by the Künneth Theorem (Theorem 6.4.13). So it will suffice to show that

$$f\times g : I^QH_*(X\times Y,(A\times Y)\cup(X\times B);R) \to I^QH_*(X'\times Y',(A'\times Y')\cup(X'\times B');R)$$

is also an isomorphism. For if so, then the bottom and sides of the diagram will all be chain homotopy equivalences, as all the modules involved are projective; see the proof of Theorem 6.4.14. It will follow that the top map of the diagram is also a chain homotopy equivalence.

To show that $f\times g$ induces the desired homology isomorphism, we note that the product of the stratified deformation retractions gives a stratified deformation retraction of $X'\times Y'$ to $X\times Y$ and so $f\times g : I^QH_*(X\times Y;R) \to I^QH_*(X'\times Y';R)$ is an isomorphism. So, again via the stratified homotopy invariance, the exact sequences of the pairs, and the Five Lemma, it suffices to show that we have an isomorphism

$$I^QH_*((A\times Y)\cup(X\times B);R) \to I^QH_*((A'\times Y')\cup(X'\times B');R)$$

induced by $f\times g$.

For this, we utilize the following diagram (coefficients tacit):

$$\begin{array}{ccccccccc}
 & & I^QH_*(A\times B) & & \begin{array}{c} I^QH_*(A\times Y)\oplus \\ I^QH_*(X\times B) \end{array} & & \begin{array}{c} I^QH_*((A\times Y)\cup \\ (X\times B)) \end{array} & & \\
 & & \Big\downarrow a & & \Big\downarrow b & & \Big\downarrow e & & \\
\longrightarrow & & \begin{array}{c} I^QH_*((A\times B')\cup \\ (A'\times B)) \end{array} & \longrightarrow & \begin{array}{c} I^QH_*((A\times Y')\cup \\ (A'\times B))\oplus \\ I^QH_*((A\times B')\cup \\ (X'\times B)) \end{array} & \longrightarrow & \begin{array}{c} I^QH_*((A\times Y')\cup \\ (X'\times B)) \end{array} & \longrightarrow & \\
 & & \Big\downarrow c & & \Big\downarrow d & & \Big\downarrow h & & \\
\longrightarrow & & I^QH_*(A'\times B') & \longrightarrow & \begin{array}{c} I^QH_*(A'\times Y')\oplus \\ I^QH_*(X'\times B') \end{array} & \longrightarrow & \begin{array}{c} I^QH_*((A'\times Y')\cup \\ (X'\times B')) \end{array} & \longrightarrow &
\end{array}$$

The bottom row is the long exact Mayer–Vietoris sequence of the pair $\{A'\times Y', X'\times B'\}$, each of which is an open subset of $(A'\times Y')\cup(X'\times B') = X'\times Y'$. The middle row is the Mayer–Vietoris sequence of the pair $\{(A\times Y')\cup(A'\times B),(A\times B')\cup(X'\times B)\}$. Each of these sets is open in their union $(A\times Y')\cup(X'\times B)$, so this is also a valid Mayer–Vietoris sequence. The modules in the top row do not automatically fit into a Mayer–Vietoris sequence, as the pair $\{A\times Y, X\times B\}$ is not necessarily excisive (at least not evidently so). We will show that the vertical arrows, each of which is induced by spatial inclusion, are all isomorphisms; in particular, then, the composition map he is an isomorphism, which is what we need to show.

First, as A' stratified deformation retracts to A and B' stratified deformation retracts to B, the product $A' \times B'$ has a stratified deformation retraction to $A \times B$, so the composition ca is an isomorphism. Similarly, starting with $(A \times B') \cup (A' \times B)$, we can first use the stratified deformation retraction of $A \times B'$ to $A \times B$ to perform a stratified deformation retraction of $(A \times B') \cup (A' \times B)$ to $A' \times B$ (holding $A' \times B$ fixed), and then we perform a stratified deformation retraction from $A' \times B$ to $A \times B$. So a is an isomorphism, and it follows that c is an isomorphism. In the middle column, we have product stratified deformation retractions of $A' \times Y'$ to $A \times Y$ and of $X' \times B'$ to $X \times B$, so db is an isomorphism. As $B \subset Y$, the stratified deformation retraction of Y' to Y can be used to stratified deformation retract $(A \times Y') \cup (A' \times B)$ to $(A \times Y) \cup (A' \times B)$, which then stratified deformation retracts to $(A \times Y) \cup (A \times B) = A \times Y$. Analogously, $(A \times B') \cup (X' \times B)$ stratified deformation retracts to $X \times B$. So the middle vertical maps are isomorphisms.

It follows now from the Five Lemma that h is an isomorphism. Lastly, again using that $B \subset Y$ and $A \subset X$, we can hold $X' \times B$ fixed and perform a stratified deformation retraction of $(A \times Y') \cup (X' \times B)$ to $(A \times Y) \cup (X' \times B)$; then holding $A \times Y$ fixed, $(A \times Y) \cup (X' \times B)$ stratified deformation retracts to $(A \times Y) \cup (X \times B)$. So e is an isomorphism. At last we conclude that he is an isomorphism, as claimed.

This completes the proof of Proposition 7.3.69. □

Remark 7.3.70 The key requirement in the above argument is that (X', A') be a CS set X' with open subset A' such that X' has a stratified deformation retraction to X and A' has a stratified deformation retraction to A, and similarly for the pair (Y, B). Hence the above argument, and those that follow, can be generalized to include other pairs with such "CS models." For example, we will need briefly below in the proof of Theorem 8.3.12 to work with a pair of the form (X, A), where X is a ∂-stratified pseudomanifold and A is a filtered open collar neighborhood of ∂X. In this case, we can form an appropriate (X', A') by attaching the same external collar to X and A.

The following corollary is now immediate from the preceding proposition and the definitions of the cross, cup, and cap products:

Corollary 7.3.71 *Let R be a Dedekind domain, and let X and Y be ∂-stratified pseudomanifolds with respective perversities $\bar{p}$ and $\bar{q}$ and partial boundaries $A \subset \partial X$ and $B \subset \partial Y$. If Q is a $(\bar{p}, \bar{q})$-compatible perversity on $X \times Y$ then there is a well-defined cohomology cross product*

$$I_{\bar{p}}H^i(X, A; R) \otimes I_{\bar{q}}H^j(Y, B; R) \xrightarrow{\times} I_Q H^{i+j}(X \times Y, (A \times Y) \cup (X \times B); R).$$

If X is a ∂-stratified pseudomanifold with an agreeable triple of perversities $(\bar{p}, \bar{q}; \bar{r})$ and partial boundaries $A, B \subset \partial X$, then there are a well-defined cup product

$$I_{\bar{p}}H^i(X, A; R) \otimes I_{\bar{q}}H^j(X, B; R) \xrightarrow{\smile} I_{\bar{r}}H^{i+j}(X, A \cup B; R)$$

and a well-defined cap product

$$I_{\bar{q}}H^j(X,B;R) \otimes I^{\bar{r}}H_{i+j}(X,A\cup B;R) \xrightarrow{\frown} I^{\bar{p}}H_i(X,A;R).$$

In particular, we have cup products of the form

$$I_{\bar{p}}H^i(X;R) \otimes I_{\bar{q}}H^j(X,\partial X;R) \xrightarrow{\smile} I_{\bar{r}}H^{i+j}(X,\partial X;R),$$
$$I_{\bar{p}}H^i(X,\partial X;R) \otimes I_{\bar{q}}H^j(X,\partial X;R) \xrightarrow{\smile} I_{\bar{r}}H^{i+j}(X,\partial X;R),$$

and cap products of the form

$$I_{\bar{q}}H^j(X,\partial X;R) \otimes I^{\bar{r}}H_{i+j}(X,\partial X;R) \xrightarrow{\frown} I^{\bar{p}}H_i(X;R),$$
$$I_{\bar{q}}H^j(X;R) \otimes I^{\bar{r}}H_{i+j}(X,\partial X;R) \xrightarrow{\frown} I^{\bar{p}}H_i(X,\partial X;R),$$
$$I_{\bar{q}}H^j(X,\partial X;R) \otimes I^{\bar{r}}H_{i+j}(X,\partial X;R) \xrightarrow{\frown} I^{\bar{p}}H_i(X,\partial X;R).$$

Beyond the existence of products for ∂-pseudomanifolds, we would like for our various properties to hold. Those whose statements and proofs depend only on having IAW maps of the form guaranteed by Proposition 7.3.69 will still hold, and this includes most of the properties we've developed. However, there could be some additional difficulties with some of the stability properties in Section 7.3.5 and locality properties in Section 7.3.7. The main issue with Section 7.3.5 is that there are certain excision maps involved in the statements and proofs of stability that do rely on our subsets being open subsets, or at least on having certain excisive couples. We will discuss some examples where this still holds in more detail below. The results of Section 7.3.7 are more problematic as, for example, Lemma 7.3.62 involves not just the global cross product but the cross products over small subsets. No doubt a version of the results of this section could be recovered with enough careful hypotheses, but, as we will not need such results below, we will not pursue this.

So, looking through the statements and proofs of the properties of cross, cup, and cap products to see where IAW maps come in, we obtain the following:

Theorem 7.3.72 *The following 14 properties of cup, cap, and cohomology cross products on spaces continue to hold if*

- *each space pair (X,A) or (Y,B) for cross products or*
- *each pair (X,A) or (X,B) for cup and cap products*

consists of a ∂-stratified pseudomanifold with a partial boundary subset (i.e. each is a partial boundary pair):

1. *all properties of homology cross products*[29]*,*
2. *existence of cohomology cross products, cup products, and cap products as defined in Section 7.2.2,*

[29] In fact, intersection homology cross products have already been presented for arbitrary filtered spaces and none of them require IAW maps. So we mention this here only for completeness.

3. *naturality of cross, cup, and cap products (Propositions 7.3.2, 7.3.5, and 7.3.5), noting that it is acceptable if some of the spaces are partial boundary pairs and some of the pairs are CS sets with open subsets,*
4. *commutativity of cross and cup products (Propositions 7.3.13 and 7.3.15),*
5. *unital property of cross products and cup products (Propositions 7.3.24, 7.3.23, and 7.3.21),*
6. *the evaluation properties of cross and cap products (Propositions 7.3.27 and 7.3.25),*
7. *associativity of cross, cup, and cap products (Propositions 7.3.29, 7.3.34, and 7.3.35 and Lemmas 7.3.31 and 7.3.32), assuming that* (X,A), (Y,B), (Z,C), $(X\times Y,(A\times Y)\cup(X\times B))$ *and* $(Y\times Z,(B\times Z)\cup(Y\times C))$ *are all partial boundary pairs (with* $X=Y=Z$ *for the cup and cap products),*
8. *stability of cross products (Proposition 7.3.43), assuming* $e: I^QH_*(A\times Y,A\times B;R)\to I^QH_*((A\times Y)\cup(X\times B),X\times B;R)$ *is an isomorphism,*
9. *stability of cup products (Proposition 7.3.44), assuming* $(A,A\cap B)$ *is a partial boundary pair and that*
$$e: I^{\bar r}H_*(A,A\cap B;R)\to I^{\bar r}H_*(A\cup B,B;R)$$
and
$$e: I^{Q_{\bar p,\bar q}}H_*(A\times X,A\times B;R)\to I^{Q_{\bar p,\bar q}}H_*((A\times X)\cup(X\times B),X\times B)$$
are isomorphisms,
10. *stability of cap products of Proposition 7.3.37 if* $(B,A\cap B)$ *is a partial boundary pair and the maps*
$$e: I^{\bar r}H_*(B,A\cap B;R)\to I^{\bar r}H_*(A\cup B,A;R)$$
and
$$e': I^{Q_{\bar p,\bar q}}H_*(X\times B,A\times B;R)\to I^{Q_{\bar p,\bar q}}H_*((A\times X)\cup(X\times B),A\times X;R)$$
are isomorphisms,
11. *stability of cap products of Proposition 7.3.38, assuming* $(A,A\cap B)$ *is a partial boundary pair and the maps*
$$e: I^{\bar r}H_*(A,A\cap B;R)\to I^{\bar r}H_*(A\cup B,B;R)$$
and
$$e': I^{Q_{\bar p,\bar q}}H_*(A\times X,A\times B;R)\to I^{Q_{\bar p,\bar q}}H_*((A\times X)\cup(X\times B),X\times B;R)$$
are isomorphisms,
12. *the relation between cross and cup products of Proposition 7.3.45,*
13. *the relation between cross and cup products of Proposition 7.3.46 if* $(X\times Y,A\times Y)$ *and* $(X\times Y,X\times B)$ *are partial boundary pairs,*

14. *the interchange properties (Lemmas 7.3.50 and 7.3.51 and Propositions 7.3.54 and 7.3.55), assuming that* (X,A), (X,B), (Y,C), (Y,D), $(X,A\cup B)$, $(Y,C\cup D)$, $(X\times X,(A\times X)\cup(X\times B))$, $(Y\times Y,(C\times Y)\cup(Y\times D))$, $(X\times Y,(A\times Y)\cup(X\times C))$, *and* $(X\times Y,(B\times Y)\cup(X\times D))$ *are all partial boundary pairs.*

To close this section, we demonstrate a useful condition for ensuring the existence of the excision maps necessary for the stability conditions of Propositions 7.3.37 and 7.3.38. For Proposition 7.3.37, these are the maps

$$e : I^{\bar{r}}H_*(B,A\cap B;R)\to I^{\bar{r}}H_*(A\cup B,A;R)$$

and

$$e' : I^{\bar{r}}H_*(X\times B,A\times B;R)\to I^{\bar{r}}H_*((A\times X)\cup(X\times B),A\times X;R).$$

The maps for Proposition 7.3.38 are equivalent, with the roles of A and B reversed. We will apply this later in our proof of Corollary 8.3.10.

Lemma 7.3.73 *Suppose X is a filtered space with closed subspaces $A,B\subset X$. Let $\bar{r}$ be any perversity.*

1. *If $A\cap B$ has an open neighborhood U in A such that $A\cap B$ is a stratified deformation retract of U and B is a stratified deformation retract of $U\cup B$, then the map induced by inclusion $e : I^{\bar{r}}H_*(B,A\cap B;R)\to I^{\bar{r}}H_*(A\cup B,A;R)$ is an isomorphism.*
2. *If B has an open neighborhood N in X such that B is a stratified deformation retract of N, then the map induced by inclusion $e' : I^{\bar{r}}H_*(X\times B,A\times B;R)\to I^{\bar{r}}H_*((A\times X)\cup(X\times B),A\times X;R)$ is an isomorphism.*

Proof The proof of the first statement is by excision and stratified homotopy invariance: Note that $A-U$ is closed in A, which is closed in X, so $A-U$ is closed in X and hence in $A\cup B$. Furthermore, $A-U$ is contained in $(A\cup B)-B$, which is a subset of A and open in $A\cup B$. So $A-U$ is closed and contained in the interior of A in $A\cup B$; thus excision applies and we have an isomorphism $I^{\bar{r}}H_*(U\cup B,U;R)\to I^{\bar{r}}H_*(A\cup B,A;R)$ with excised set $A-U$. Next, the inclusion $I^{\bar{r}}H_*(B,A\cap B;R)\to I^{\bar{r}}H_*(U\cup B,U;R)$ is an isomorphism using the Five Lemma and noting that the inclusions $A\cap B\hookrightarrow U$ and $B\hookrightarrow U\cup B$ are stratified homotopy equivalences. As e is the composition of these inclusions, it induces an intersection homology isomorphism.

The proof of the second statement is similar. In this case, we have an excision isomorphism

$$I^{\bar{r}}H_*((A\times N)\cup(X\times B),A\times N;R)\to I^{\bar{r}}H_*((A\times X)\cup(X\times B),A\times X;R)$$

that excises $A\times(X-N)$, which is closed and contained in $A\times(X-B)$, an open subset of $A\times X$. Then the stratified deformation retraction of N to B induces stratified deformation retractions from $A\times N$ to $A\times B$ and from $(A\times N)\cup(X\times B)$ to $(A\times B)\cup(X\times B)=X\times B$. So, again employing the Five Lemma, we have an isomorphism

$$I^{\bar{r}}H_*(X\times B,A\times B;R)\to I^{\bar{r}}H_*((A\times N)\cup(X\times B),A\times N;R).$$

Also once again, the map e' is the composition of these inclusions, so it induces an intersection homology isomorphism. □

Example 7.3.74 Suppose that X is a ∂-stratified pseudomanifold and that $A, B \subset X$ are complementary partial boundaries, i.e. $\partial X = A \cup B$ and $A \cap B = \partial A = \partial B$. Then this collection of spaces satisfies the hypotheses of the lemma taking U to be a filtered collar of ∂A in A and N to be $[0,1) \times (B \cup U)$ within the filtered collar $[0,1) \times \partial X$ of ∂X. Also in this case we have $(A, A \cap B) = (A, \partial A)$ and $(B, A \cap B) = (B, \partial B)$, so these are partial boundary pairs. Thus Propositions 7.3.37 and 7.3.38 hold in this setting. △

7.4 Intersection Cohomology with Compact Supports

Cohomology with compact supports plays an important role in ordinary cohomology theory, particularly in the statement and proof of the Poincaré Duality Theorem, and the same remains true for intersection cohomology. In this section we define intersection cohomology with compact supports and also discuss how, with proper assumptions, we obtain a cap product from compactly supported intersection cohomology to intersection homology. Moreover, we discuss compatibility of this cap product with Mayer–Vietoris sequences. We will assume familiarity with the direct limit functor; see [182, Section 73] for a review.

Recall (e.g. from [125, Section 3.3]) that for a space X and coefficient group G the ordinary cohomology groups with compact supports $H^i_c(X;G)$ can be defined as direct limits over the directed set of compact subsets of X. In fact, the compact subsets of X form a directed set if we define the relation $\leq$ so that $K \leq L$ if and only if $K \subset L$. Then we can easily check the properties required of a directed set (see e.g. [182, Section 73]):

1. $K \leq K$,
2. if $K \leq L$ and $L \leq J$ then $K \leq J$, and
3. given compact K and L there is a compact J with $K \leq J$ and $L \leq J$ (just take $J = K \cup L$).

Now, if $K \leq L$ then $X - L \subset X - K$, so the restriction $H^i(X, X-K; G) \to H^i(X, X-L; G)$ is well defined for each such pair. So the $H^i(X, X-K; G)$ form a direct system of groups, and we let

$$H^i_c(X;G) = \varinjlim H^i(X, X-K; G).$$

Roughly speaking, we can think of elements of the direct limit $H^i_c(X;G)$ as being represented by cocycles that annihilate chains in X sufficiently "close to infinity."

The idea for intersection cohomology is identical:

Definition 7.4.1 Let X be a filtered space with perversity $\bar{p}$, and let R be a commutative ring with unity. The *intersection cohomology modules with compact supports*, $I_{\bar{p}}H^i_c(X;R)$, are defined to be $\varinjlim I_{\bar{p}}H^i(X, X-K; R)$, where the limit is over all compact subsets of X.

Example 7.4.2 Let X be a compact filtered space. We claim[30] that

$$I_{\bar{p}}H^i_c(cX;R) = I_{\bar{p}}H^i(cX, cX - \{v\}; R).$$

Indeed, recall that we have $cX = ([0,1) \times X)/\sim$, and, for $0 < r < 1$, let $\bar{c}_r X$ be the subspace $([0,r] \times X)/\sim$. Then the compact sets $\bar{c}_r X$ are cofinal among the compact subsets of cX. In this case, this means that every compact subset of cX is contained within one of the $\bar{c}_r X$. Recall that for computing direct limits it is sufficient to restrict to any cofinal directed subset [182, Lemma 73.1]. But then if $r > s$ the restriction $I_{\bar{p}}H^i(cX, cX - \bar{c}_s X; R) \to I_{\bar{p}}H^i(cX, cX - \bar{c}_r X; R)$ is an isomorphism via stratified homotopy equivalence. Therefore, $I_{\bar{p}}H^i_c(cX;R)$ is isomorphic to any of these $I_{\bar{p}}H^i(cX, cX - \bar{c}_r X; R)$, and they are all isomorphic to $I_{\bar{p}}H^i(cX, cX - \{v\}; R)$, again using stratified homotopy equivalence. △

The functoriality of cohomology with compact supports runs opposite to the standard functoriality of cohomology and requires some additional assumptions. For example, suppose U is an open subset of X. Then for each compact $K \subset U$ excision guarantees that restriction induces an isomorphism $I_{\bar{p}}H^*(X, X - K; R) \to I_{\bar{p}}H^*(U, U - K; R)$. Consider now the following diagram with $K \subset L \subset U$, the sets K and L compact:

$$\begin{array}{ccc}
I_{\bar{p}}H^*(U,U-K;R) & \longrightarrow & I_{\bar{p}}H^*(U,U-L;R)\\
 & \searrow \quad \varinjlim_{J\subset U} I_{\bar{p}}H^*(U,X-J;R) \quad \swarrow & \\
\uparrow\cong & \downarrow & \uparrow\cong\\
 & \nearrow \quad \varinjlim_{J\subset X} I_{\bar{p}}H^*(X,X-J;R) \quad \nwarrow & \\
I_{\bar{p}}H^*(X,X-K;R) & \longrightarrow & I_{\bar{p}}H^*(X,X-L;R)
\end{array} \qquad (7.28)$$

The outside rectangle commutes by naturality with respect to the inclusion maps. The diagonal maps are all the canonical maps to the direct limits, letting $\varinjlim_{J\subset U}$ represent the limit over compact subsets of U and $\varinjlim_{J\subset X}$ the limit over compact subsets of X, and so the triangles commute. As this commutativity holds for all $K \subset L \subset U$ we therefore have compatible maps from the direct system of modules $I_{\bar{p}}H^*(U, U - J; R)$ (over compact $J \subset U$) to $\varinjlim_{J\subset X} I_{\bar{p}}H^*(X, X - J; R)$. Therefore, the dashed arrow exists by the universal property of direct limits. We thus obtain a map

$$I_{\bar{p}}H^*_c(U;R) = \varinjlim_{K\subset U} I_{\bar{p}}H^*(U, U-K;R) \to \varinjlim_{K\subset X} I_{\bar{p}}H^*(X, X-K;R) = I_{\bar{p}}H^*_c(X;R).$$

The following immediate consequence of this discussion is useful:

[30] Unfortunately, we have two different "c"s here: the c for compact supports and the c for the cone construction. However, context should make clear which is meant in each case.

Lemma 7.4.3 *Suppose $U \subset X$ is an open subset and that $L \subset U$ is compact. Then we have the commutative diagram*

$$\begin{array}{ccc} I_{\bar{p}}H^*(U, U - L; R) & \xleftarrow{\cong} & I_{\bar{p}}H^*(X, X - L; R) \\ \downarrow & & \downarrow \\ I_{\bar{p}}H^*_c(U; R) & \longrightarrow & I_{\bar{p}}H^*_c(X; R). \end{array}$$

Here the horizontal maps are induced by inclusion of U into X, the bottom map being that defined just above. The vertical maps are those taking an element of a direct system to its image in the direct limit.

Our next lemma is a cohomology version of Lemma 6.3.16, concerning limits over increasing collections of open sets. Like that lemma, this one is useful in combination with Lemma 5.1.6 for verifying the limit condition in Mayer–Vietoris arguments.

Lemma 7.4.4 *Suppose R is any commutative ring with unity. If X is a filtered space with perversity $\bar{p}$ and $\{U_\alpha\}$ is an increasing collection of open subspaces of X then the canonical map $f : \varinjlim_\alpha I_{\bar{p}}H^*_c(U_\alpha; R) \to I_{\bar{p}}H^*_c\left(\bigcup_\alpha U_\alpha; R\right)$ is an isomorphism.*

Proof First, recall that for any open $U \subset V \subset X$, we do have maps $I_{\bar{p}}H^*_c(U; R) \to I_{\bar{p}}H^*_c(V; R)$. These are discussed in Section 7.4. Additionally, if $U \subset W \subset V$, then

$$\begin{array}{ccc} I_{\bar{p}}H^*_c(U; R) & \longrightarrow & I_{\bar{p}}H^*_c(W; R) \\ & \searrow & \downarrow \\ & & I_{\bar{p}}H^*_c(V; R) \end{array}$$

commutes. Taking U and W from among the U_α and letting $V = \bigcup U_\alpha$, it follows from the universal property of limits that our map f is defined.

Now, suppose $a \in I_{\bar{p}}H^*_c\left(\bigcup_\alpha U_\alpha; R\right)$. For convenience, denote $\bigcup_\alpha U_\alpha$ by $\mathscr{U}$. By definition, a is represented by an element $a_K \in I_{\bar{p}}H^*(\mathscr{U}, \mathscr{U} - K; R)$, for some compact $K \subset \mathscr{U}$. In fact, we can think of a as the image of a_K under the canonical map

$$I_{\bar{p}}H^*(\mathscr{U}, \mathscr{U} - K; R) \to \varinjlim_{K \subset \mathcal{U}} I_{\bar{p}}H^*(\mathscr{U}, \mathscr{U} - K; R) = I_{\bar{p}}H^*_c(\mathscr{U}; R).$$

As K is compact and the collection $\{U_\alpha\}$ is increasing, there is some U_α, say U_0, with

$K \subset U_0$. By Lemma 7.4.3, we have a commutative diagram

$$\begin{CD}
I_{\bar p}H^*(U_0, U_0 - K; R) @<<< I_{\bar p}H^*(\mathcal{U}, \mathcal{U} - K; R)\\
@VVV @VVV\\
I_{\bar p}H_c^*(U_0; R) @>>> I_{\bar p}H_c^*(\mathcal{U}; R).
\end{CD}$$

It follows that $a \in I_{\bar p}H_c^*(\mathcal{U}; R)$ must be in the image of $I_{\bar p}H_c^*(U_0; R)$. Therefore, a is in the image of $\varinjlim_\alpha I_{\bar p}H_c^*(U_\alpha; R)$. So f is surjective.

For the proof of injectivity, it will be useful to refer to the following diagram whose terms we define as they arise:

$$\begin{CD}
I_{\bar p}H_c^*(U_0; R) @<<< I_{\bar p}H^*(U_0, U_0 - K; R) @<\cong<< I_{\bar p}H^*(\mathcal{U}, \mathcal{U} - K; R)\\
@VVV @A{\cong}AA @A{=}AA\\
I_{\bar p}H_c^*(U_1; R) @<<< I_{\bar p}H^*(U_1, U_1 - K; R) @<\cong<< I_{\bar p}H^*(\mathcal{U}, \mathcal{U} - K; R)\\
@V{=}VV @VVV @VVV\\
I_{\bar p}H_c^*(U_1; R) @<<< I_{\bar p}H^*(U_1, U_1 - K'; R) @<\cong<< I_{\bar p}H^*(\mathcal{U}, \mathcal{U} - K'; R)
\end{CD}$$

All the maps labeled as isomorphisms are excision maps.

Now suppose that $a \in \varinjlim_\alpha I_{\bar p}H_c^*(U_\alpha; R)$ and $f(a) = 0 \in I_{\bar p}H_c^*(\mathcal{U}; R)$. By definition, a is represented by an element $a_0 \in I_{\bar p}H_c^*(U_0; R)$ for some U_0, and, furthermore, a_0 must be the image of some $a_0^K \in I_{\bar p}H^*(U_0, U_0 - K; R)$ for some compact $K \subset U_0$. The excision isomorphism $I_{\bar p}H^*(U_0, U_0 - K; R) \cong I_{\bar p}H^*(\mathcal{U}, \mathcal{U} - K; R)$ then takes a_0^K to an element $a_0^{K,\mathcal{U}}$ that represents $f(a)$. But if $f(a) = 0$, this implies that there is some compact K' with $K \subset K'$ such that $a_0^{K,\mathcal{U}}$ maps to 0 in $I_{\bar p}H^*(\mathcal{U}, \mathcal{U} - K'; R)$. As K' is compact, there must be some U_1 with $K \subset K' \subset U_1$ and $U_0 \subset U_1$. We have now introduced all the spaces in the diagram. The top left square commutes by Lemma 7.4.3; the remaining squares commute more evidently. The diagram shows that the image of a_0 is trivial in $I_{\bar p}H_c^*(U_1; R)$, as we have seen that running clockwise around the outside of the diagram takes a_0^K through the image of $a_0^{K,\mathcal{U}}$ in $I_{\bar p}H^*(\mathcal{U}, \mathcal{U} - K'; R)$, which is 0. As a_0 maps to 0 in $I_{\bar p}H_c^*(U_1; R)$, it follows that it represents the trivial element of $\varinjlim_\alpha I_{\bar p}H_c^*(U_\alpha; R)$. This implies that $a = 0$. □

We also have the following Mayer–Vietoris sequence for cohomology with compact supports (compare [125, Lemma 3.36]).

Lemma 7.4.5 *Let X be a CS set*[31] *with perversity $\bar p$, and let R be a commutative ring with*

[31] The assumption that X is a CS set, not just an arbitrary filtered space, will be used here to cite Corollary

unity. Suppose $X = U \cup V$ for U, V open subsets. Then there is an exact Mayer–Vietoris sequence

$$\longrightarrow I_{\bar{p}}H^i_c(U \cap V; R) \longrightarrow I_{\bar{p}}H^i_c(U; R) \oplus I_{\bar{p}}H^i_c(V; R) \longrightarrow I_{\bar{p}}H^i_c(X; R) \longrightarrow .$$

Proof Suppose $K \subset U$ and $L \subset V$ are compact sets; see Figure 7.4. Then, by Theorem 7.1.13, we have the exact Mayer–Vietoris sequence

$$\begin{aligned}\to I_{\bar{p}}H^i(X, X - (K \cap L); R) &\to I_{\bar{p}}H^i(X, X - K; R) \oplus I_{\bar{p}}H^i(X, X - L; R)\\ &\to I_{\bar{p}}H^i(X, X - (K \cup L); R) \to ;\end{aligned}$$

note that $(X - K) \cap (X - L) = X - (K \cup L)$ and $(X - K) \cup (X - L) = X - (K \cap L)$. This sequence is natural in K and L, so we can take the direct limits with respect to the directed system γ of pairs (K, L) such that K is compact in U and L is compact in V, letting $(K, L) \leq (K', L')$ if $K \subset K'$ and $L \subset L'$. Taking direct limits preserves exactness (see e.g. [38, Theorem D.4]), so we will show that the resulting direct limit modules are the desired modules of the statement of the lemma.

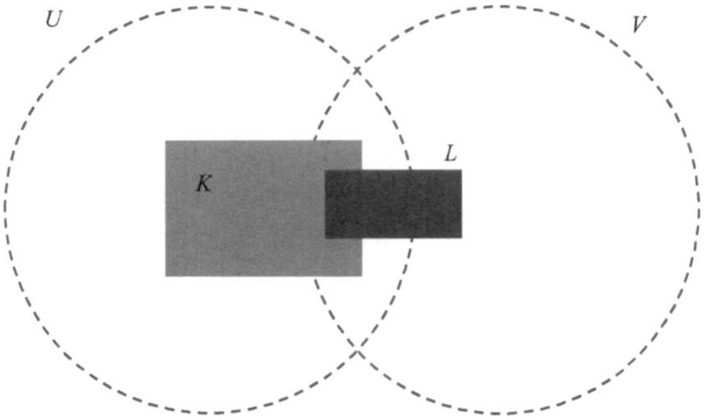

Figure 7.4 Compact $K \subset U$ and $L \subset V$.

We first consider the middle term of the exact sequence. By functoriality, direct limits distribute over direct sums, so we can consider the summands separately. The first summand is constant in L, so using that $\varinjlim_{\gamma} \cong \varinjlim_K \varinjlim_L \cong \varinjlim_L \varinjlim_K$ (see [38, Theorem D5]) we

2.3.17 in the argument below. However, as we see from the proof of that corollary, it would be sufficient to assume that X is a filtered space that is locally compact Hausdorff.

compute

$$\begin{aligned}\varinjlim_{\gamma} I_{\bar p}H^i(X, X-K; R) &\cong \varinjlim_{\gamma} I_{\bar p}H^i(U, U-K; R)\\ &\cong \varinjlim_{K\subset U}\varinjlim_{L\subset V} I_{\bar p}H^i(U, U-K; R)\\ &\cong \varinjlim_{K\subset U} I_{\bar p}H^i(U, U-K; R)\\ &\cong I_{\bar p}H^i_c(U; R).\end{aligned}$$

The first isomorphism here comes from the naturality in this setting of excision isomorphisms as in Diagram (7.28). The argument for the other summand is equivalent.

For $\varinjlim_{\gamma} I_{\bar p}H^i(X, X-(K\cap L); R)$, this direct sequence is isomorphic to

$$\varinjlim_{\gamma} I_{\bar p}H^i(U\cap V, U\cap V-(K\cap L); R),$$

again by excision. Then we have the canonical compatible maps from each $I_{\bar p}H^i(U\cap V, U\cap V-(K\cap L); R)$ to

$$I_{\bar p}H^i_c(U\cap V; R) = \varinjlim_{J\subset U} I_{\bar p}H^i(U\cap V, U\cap V-J; R),$$

with J running over the compact subsets of $U\cap V$, so there is a map

$$\phi : \varinjlim_{\gamma} I_{\bar p}H^i(U\cap V, U\cap V-(K\cap L); R) \to I_{\bar p}H^i_c(U\cap V; R).$$

If an element of $I_{\bar p}H^i_c(U\cap V; R)$ is represented by $\alpha \in I_{\bar p}H^i(U\cap V, U\cap V-J; R)$, then we can let $K = L = J$ and obtain the commutative diagram

$$\begin{array}{ccc} I_{\bar p}H^i(U\cap V, U\cap V-(J\cap J); R) & \longrightarrow & \varinjlim_{\gamma} I_{\bar p}H^i(U\cap V, U\cap V-(K\cap L); R)\\ \Big\downarrow{=} & & \Big\downarrow{\phi}\\ I_{\bar p}H^i(U\cap V, U\cap V-J; R) & \longrightarrow & I_{\bar p}H^i_c(U\cap V; R),\end{array}$$

so ϕ is surjective. Similarly, if we have an element $[\alpha] \in \varinjlim_{\gamma} I_{\bar p}H^i(U\cap V, U\cap V-(K\cap L); R)$ represented by some particular $\alpha \in I_{\bar p}H^i(U\cap V, U\cap V-(K\cap L); R)$ for some specific K, L and if $\phi([\alpha]) = 0$, then the image of α in some $I_{\bar p}H^i(U\cap V, U\cap V-J; R)$ with $K\cap L\subset J\subset U\cap V$ must be 0. But then the image of α in $I_{\bar p}H^i(U\cap V, U\cap V-(J\cap J); R)$ must be 0, so $[\alpha] = 0$.

The argument for $\varinjlim_{\gamma} I_{\bar p}H^i(X, X-(K\cup L); R)$ is similar in spirit but requires some minor additional work. Again we have an evident map $\phi : \varinjlim_{\gamma} I_{\bar p}H^i(X, X-(K\cup L); R) \to I_{\bar p}H^i_c(X; R) = \varinjlim_{J} I_{\bar p}H^i(X, X-J; R)$, now with J running over the compact subsets of

X. Suppose $[\alpha] \in I_{\bar{p}}H^i_c(X;R)$ represented by $\alpha \in I_{\bar{p}}H^i(X, X - J; R)$ for some specific J. By arguments similar to those of the last paragraph, it suffices for surjectivity of ϕ to show that we can write J as $J = J_1 \cup J_2$ with $J_1 \subset U$ and $J_2 \subset V$; see Figure 7.5. For this, let us consider the disjoint closed sets $J - (J \cap V)$ and $X - U$. The subspace $J - (J \cap V)$ is compact, so we can find disjoint open sets W_1, W_2 with $J - (J \cap V) \subset W_1$ and $X - U \subset W_2$ by Corollary 2.3.17. Now, let J_1 be the closure of $J \cap W_1$. As W_1 and W_2 are disjoint, $J_1 \cap W_2 = \emptyset$, which implies that $J_1 \cap (X - U) = \emptyset$, so $J_1 \subset U$. Also J_1 is a closed subset of J and so is compact. Let $J_2 = J - (J \cap W_1) = J \cap (X - W_1)$, which is also compact as a closed subset of J. Furthermore, $J \cap W_1$ contains $J - (J \cap V)$, i.e. all points of J outside of V are contained in $J \cap W_1$, so all the points left in J_2 must be contained in V. Finally, every point of J is either in $J \cap W_1$, and so in J_1, or in $J - (J \cap W_1)$, and so in J_2. Therefore, $J = J_1 \cup J_2$ with $J_1 \subset U$ and $J_2 \subset V$. This provides the surjectivity of ϕ.

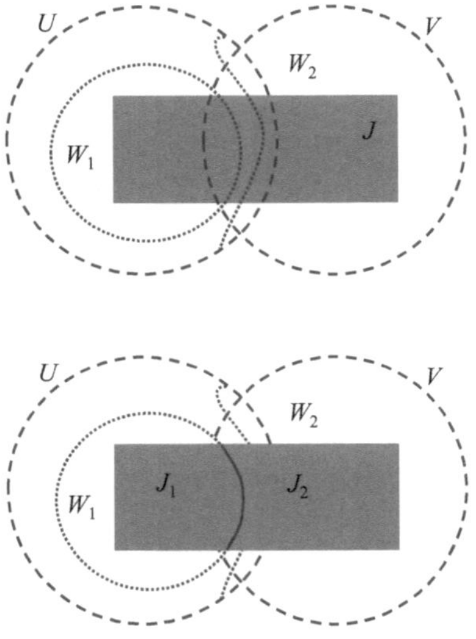

Figure 7.5 Writing J as the union of compact sets in U and V in the proof of Lemma 7.4.5.

Finally, we consider injectivity of the map ϕ of the preceding paragraph. Now we suppose we have $[\alpha] \in \varinjlim_{\gamma} I_{\bar{p}}H^i(X, X - (K \cup L); R)$ with $\phi([\alpha]) = 0$. Let $\alpha \in I_{\bar{p}}H^i(X, X - (K \cup L); R)$ for some specific K and L represent $[\alpha]$. Because $\phi([\alpha]) = 0$, there is some compact $J \subset X$ with $K \cup L \subset J$ and the image of α in $I_{\bar{p}}H^i(X, X - J; R)$ trivial. As in the preceding paragraph, let us choose compact J_1 and J_2 such that $J = J_1 \cup J_2$ and $J_1 \subset U$, $J_2 \subset V$. We do not know that either $K \subset J_1$ or $L \subset J_2$. However, we do have $K \cup J_1 \subset U$ and $L \cup J_2 \subset V$, so we can consider the image of α in $I_{\bar{p}}H^i(X, X - ((K \cup J_1) \cup (L \cup J_2)); R)$. We have $K \cup L \subset J \subset (K \cup J_1) \cup (L \cup J_2)$, and so the map from $I_{\bar{p}}H^i(X, X - (K \cup L); R)$

to $I_{\bar{p}}H^i(X, X - ((K \cup J_1) \cup (L \cup J_2)); R)$ factors through $I_{\bar{p}}H^i(X, X - J; R)$. Therefore, the image of α must be trivial in $\varinjlim_{\gamma} I_{\bar{p}}H^i(X, X - (K \cup L); R)$, and hence $[\alpha] = 0$. □

Let us turn to cap products. Here we will also utilize inverse limits; see [125, Additional Topic 3.F] or [197, Section 5.2].

Suppose X is a CS set with agreeable perversities $(\bar{p}, \bar{q}; \bar{r})$ and that R is a Dedekind domain so that we have a well-defined intersection (co)homology cap product. Suppose that for every compact $K \subset X$ we have a class $\xi_K \in I^{\bar{r}}H_{i+j}(X, X - K; R)$ such that if $K \subset L$ and $i_{K,L} : (X, X - L) \hookrightarrow (X, X - K)$ is the inclusion then $i_{K,L}(\xi_L) = \xi_K$. In other words, the collection $\{\xi_K\}$ represents an element of the inverse limit $\varprojlim I^{\bar{r}}H_{i+j}(X, X - K; R)$. Once again the relevant directed set is that of compact subsets of X, but now we have an inverse system of modules and maps of the form $I^{\bar{r}}H_{i+j}(X, X - L; R) \to I^{\bar{r}}H_{i+j}(X, X - K; R)$ for $K \leq L$.

Let $\alpha \in I_{\bar{q}}H^j_c(X; R)$. By the definition of the direct limit, the class α can be represented by some $\alpha_K \in I_{\bar{q}}H^j(X, X-K; R)$, and we can consider the cap product $\alpha_K \frown \xi_K \in I^{\bar{p}}H_i(X, R)$. Suppose now $K \subset L$. Let $\alpha_L = i^*_{K,L}\alpha_K \in I_{\bar{q}}H^j(X, X - L; R)$, so α_L also represents α in $I_{\bar{q}}H^j_c(X; R)$. Using the naturality property of the cap product (Proposition 7.3.6) in the second line below, we have the following computation:

$$\begin{aligned}\alpha_K \frown \xi_K &= \alpha_K \frown i_{K,L}(\xi_L)\\ &= i_{K,L}((i^*_{K,L}(\alpha_K)) \frown \xi_L)\\ &= i_{K,L}(\alpha_L \frown \xi_L).\end{aligned}$$

In this case we treat $i_{K,L}$ as a map of triples $(X; \emptyset, X - L) \to (X; \emptyset, X - K)$ so that in the last line $i_{K,L}$ is the identity map of $I^{\bar{p}}H_i(X; R)$. So we have $\alpha_K \frown \xi_K = i_{K,L*}(\alpha_L \frown \xi_L) = \alpha_L \frown \xi_L$, which tells us that we obtain an element of $I^{\bar{p}}H_i(X; R)$ depending only on α and ξ, not K or L. Altogether then, we have demonstrated the following lemma:

Lemma 7.4.6 *Suppose X is a CS set with an agreeable triple of perversities $(\bar{p}, \bar{q}; \bar{r})$ and that R is a Dedekind domain. Then the cap product induces a well-defined map*

$$I_{\bar{q}}H^j_c(X; R) \otimes \varprojlim I^{\bar{r}}H_{i+j}(X, X - K; R) \xrightarrow{\frown} I^{\bar{p}}H_i(X; R).$$

In our treatment of Poincaré duality, below, the element of $\varprojlim I^{\bar{r}}H_{i+j}(X, X - K; R)$ will correspond to the fundamental class of an R-oriented stratified pseudomanifold X, and so, up to some sign issues we will discuss later, this will be the cap product that induces Poincaré duality.

Next, we consider the following more general property of inverse limits, though stated in the context we will need below:

Lemma 7.4.7 *Suppose X is a CS set and $W \subset X$ is an open subset. Then there is a canonical map $\varprojlim_{K \subset X} I^{\bar{p}}H_i(X, X - K; R) \to \varprojlim_{K \subset W} I^{\bar{p}}H_i(W, W - K; R)$. We denote the* *image of* $\xi \in \varprojlim_{K \subset X} I^{\bar{p}}H_i(X, X-K;R)$ *by* ξ^W. *Furthermore, for any specific compact* $L \subset W$, *let* ξ^W_L *denote the canonical image of* ξ^W *in* $I^{\bar{p}}H_i(W, W-L;R)$.

Suppose $W \subset W' \subset X$ *are open subsets and* $L' \subset L$ *are compact subsets of* W. *Then* $\xi^W_L \in I^{\bar{p}}H_i(W, W-L;R)$ *maps to* $\xi^{W'}_{L'} \in I^{\bar{p}}H_i(W', W'-L';R)$ *under the map induced by inclusion.*

Proof First, we observe that the map $\xi \to \xi^W$ is well defined for any open $W \subset X$. Let $\xi \in \varprojlim I^{\bar{p}}H_i(X, X-K;R)$. By definition ξ determines elements $\xi_K \in I^{\bar{p}}H_i(X, X-K;R)$ for every $K \subset X$ with the property that if $L \subset K$ then the image of ξ_K in $I^{\bar{p}}H_i(X, X-L;R)$ under inclusion is ξ_L; conversely, any such compatible collection of ξ_K determines ξ. Now, suppose we restrict our attention to the compact K with $K \subset W$. By excision, $I^{\bar{p}}H_i(X, X-K;R) \cong I^{\bar{p}}H_i(W, W-K;R)$, and for all $L \subset K \subset W$, we have commutative diagrams

$$\begin{array}{ccc} I^{\bar{p}}H_i(X, X-K;R) & \longrightarrow & I^{\bar{p}}H_i(X, X-L;R) \\ \uparrow \cong & & \uparrow \cong \\ I^{\bar{p}}H_i(W, W-K;R) & \longrightarrow & I^{\bar{p}}H_i(W, W-L;R). \end{array}$$

Thus $\xi \in \varprojlim I^{\bar{p}}H_i(X, X-K;R)$ determines compatible elements $\xi^W_K \in I^{\bar{p}}H_i(W, W-K;R)$ for all compact $K \subset W$ and so an element $\xi^W \in \varprojlim I^{\bar{p}}H_i(W, W-K;R)$. So we do have a well-defined map $\varprojlim I^{\bar{p}}H_i(X, X-K;R) \to \varprojlim I^{\bar{p}}H_i(W, W-K;R)$, and the image of ξ^W in $I^{\bar{p}}H_i(W, W-K;R)$ is ξ^W_K by construction.

The last statement of the theorem now follows from the commutative diagram below for $W \subset W'$ and $L' \subset L \subset W$:

$$\begin{array}{cccccccc} & & I^{\bar{p}}H_i(X, X-L;R) & \xleftarrow{\cong} & I^{\bar{p}}H_i(W', W'-L;R) & \xleftarrow{\cong} & I^{\bar{p}}H_i(W, W-L;R) \\ & \nearrow & \downarrow & & \downarrow & & \downarrow \\ \varprojlim I^{\bar{p}}H_i(X, X-K;R) & & & & & & \\ & \searrow & I^{\bar{p}}H_i(X, X-L';R) & \xleftarrow{\cong} & I^{\bar{p}}H_i(W', W'-L';R) & \xleftarrow{\cong} & I^{\bar{p}}H_i(W, W-L';R). \end{array}$$

□

In the remainder of this section, we will prove the following lemma, which is critical to the proof of Poincaré duality. We base our treatment on Proposition 6.7 and Lemma 6.8 of [100], which itself is based on Hatcher [125, Lemma 3.36], though with some modification necessitated by the lack of exact control on supports of intersection chains under cap products (see Section 7.3.7). As observed in Hatcher, the proof is surprisingly nontrivial. We also include some necessary details that were overlooked in [100] (and are also not treated in full detail in the printed version of [125], though see the online errata).

Lemma 7.4.8 *Let* X *be a CS set with agreeable triple of perversities* $(\bar{p}, \bar{q}; \bar{r})$, *let* $U, V \subset$

X be two open subsets with $U \cup V = X$, let R be a Dedekind domain, and let $\xi \in \varprojlim I^{\bar{r}}H_{i+j}(X, X - K; R)$. If $W \subset X$ is open, let ξ^W denote the image of ξ under the map $\varprojlim_{K \subset X} I^{\bar{r}}H_{i+j}(X, X - K; R) \to \varprojlim_{K \subset W} I^{\bar{r}}H_{i+j}(W, W - K; R)$. Let $D^W : I_{\bar{q}}H^j_c(W; R) \to I^{\bar{p}}H_i(W; R)$ be given by $D^W(\alpha) = \alpha \frown \xi^W$. Then the following diagram of Mayer–Vietoris sequences commutes up to signs:

$$
\begin{array}{ccccccc}
\longrightarrow & I_{\bar{q}}H^j_c(U \cap V; R) & \longrightarrow & \begin{array}{c} I_{\bar{q}}H^j_c(U; R) \oplus \\ I_{\bar{q}}H^j_c(V; R) \end{array} & \longrightarrow & I_{\bar{q}}H^j_c(X; R) & \longrightarrow \\
& \Big\downarrow{\scriptstyle D^{U \cap V}} & & \Big\downarrow{\scriptstyle D^U \oplus -D^V} & & \Big\downarrow{\scriptstyle D^X} & \\
\longrightarrow & I^{\bar{p}}H_i(U \cap V; R) & \longrightarrow & \begin{array}{c} I^{\bar{p}}H_i(U; R) \oplus \\ I^{\bar{p}}H_i(V; R) \end{array} & \longrightarrow & I^{\bar{p}}H_i(X; R) & \longrightarrow .
\end{array}
$$

Proof As in Lemma 7.4.7 we let ξ^W_K denote the image of ξ in $I^{\bar{p}}H_i(W, W - K; R)$, and we let $D^W_K : I_{\bar{q}}H^j(W, W - K; R) \to I^{\bar{p}}H_i(W; R)$ be given by $D^W_K(\alpha) = \alpha \frown \xi^W_K$.

To demonstrate the commutativity up to sign of the diagram, we will demonstrate the signed commutativity of the three subdiagrams corresponding to the three squares (up to index shift) in the diagram of the lemma.

First Square. Suppose $K \subset U$ and $L \subset V$ are compact. We first consider the diagram:

$$
\begin{array}{ccc}
I_{\bar{q}}H^j(X, X - K \cap L; R) & \longrightarrow & I_{\bar{q}}H^j(X, X - K; R) \oplus I_{\bar{q}}H^j(X, X - L; R) \\
\Big\downarrow{\scriptstyle \cong} & & \Big\downarrow{\scriptstyle \cong} \\
I_{\bar{q}}H^j(U \cap V, U \cap V - K \cap L; R) & & I_{\bar{q}}H^j(U, U - K; R) \oplus I_{\bar{q}}H^j(V, V - L; R) \\
\Big\downarrow{\scriptstyle D^{U \cap V}_{K \cap L}} & & \Big\downarrow{\scriptstyle D^U_K \oplus -D^V_L} \\
I^{\bar{p}}H_i(U \cap V; R) & \longrightarrow & I^{\bar{p}}H_i(U; R) \oplus I^{\bar{p}}H_i(V; R).
\end{array}
$$

The direct limit over the compact pairs $(K, L) \subset (U, V)$ of the top half of this diagram determines the map $I_{\bar{q}}H^j_c(U \cap V; R) \to I_{\bar{q}}H^j_c(U; R) \oplus I_{\bar{q}}H^j_c(V; R)$ in the Mayer–Vietoris sequence, as we saw in the proof of Lemma 7.4.5. Furthermore, the lower vertical maps are precisely the cap products given by Lemma 7.4.6. So, once we show this diagram commutes, taking the limit of such diagrams over pairs (K, L) will provide the commutativity of the first square of Lemma 7.4.8.

For commutativity, we can work with the two summands on the right independently, with analogous arguments. So consider

$$\begin{array}{ccccc}
I_{\bar q}H^j(X, X-K\cap L;R) & & \longrightarrow & & I_{\bar q}H^j(X,X-K;R) \\
\downarrow\cong & & \searrow & & \downarrow\cong \\
I_{\bar q}H^j(U\cap V,\; U\cap V - K\cap L;R) & \longleftarrow & I_{\bar q}H^j(U,U-K\cap L;R) & \longrightarrow & I_{\bar q}H^j(U,U-K;R) \\
\downarrow D^{U\cap V}_{K\cap L} & & & \searrow D^U_{K\cap L} & \downarrow D^U_K \\
I^{\bar p}H_i(U\cap V;R) & & \longrightarrow & & I^{\bar p}H_i(U;R).
\end{array}$$

The maps in the upper half of the diagram are all induced by inclusion and commute at the space level. The map pointing to the left is an isomorphism by excision; the excised subspace is the closed subset $U - U\cap V$ of U, which is contained in the open subset $U - K\cap L$ of U. The map $D^{U\cap V}_{K\cap L}$ on the left is the cap product with the image $\xi^{U\cap V}_{K\cap L}$ of ξ in $I^{\bar r}H_{i+j}(U\cap V, U\cap V - K\cap L;R)$, while the $D^U_{K\cap L}$ on the right is the cap product with the image $\xi^U_{K\cap L}$ of ξ in $I^{\bar r}H_{i+j}(U,U-K\cap L;R)$. If we let $\mathfrak{i} : (U\cap V;\emptyset, U\cap V - K\cap L) \to (U;\emptyset, U-K\cap L)$ denote the inclusion of the triple, then $\mathfrak{i}(\xi^{U\cap V}_{K\cap L}) = \xi^U_{K\cap L}$ by Lemma 7.4.7. So the bottom left quadrilateral commutes by the naturality of the cap product given in Proposition 7.3.6 (compare the argument for Lemma 7.4.6). Similarly, if we let $\mathfrak{j} : (U;\emptyset, U-K) \to (U;\emptyset, U-K\cap L)$ denote another inclusion and ξ^U_K the image of ξ in $I^{\bar r}H_{i+j}(U,U-K;R)$, then $\mathfrak{j}(\xi^U_K) = \xi^U_{K\cap L}$ by Lemma 7.4.7, and the bottom right triangle also commutes by Proposition 7.3.6.

Finally, notice that the analogous square involving V acquires a sign in $-D^V_L$ to counter the negative sign of the second Mayer–Vietoris inclusion $I^{\bar p}H_i(U\cap V;R) \to I^{\bar p}H_i(V;R)$; see "Algebraic topology" notation item 7 on page xxi.

This establishes the commutativity of the first square in the diagram of the lemma.

Second Square. Next, we consider the diagram

$$\begin{array}{ccc}
I_{\bar q}H^j(X,X-K;R)\oplus I_{\bar q}H^j(X,X-L;R) & \longrightarrow & I_{\bar q}H^j(X,X-K\cup L;R) \\
\downarrow\cong & & \\
I_{\bar q}H^j(U,U-K;R)\oplus I_{\bar q}H^j(V,V-L;R) & & \downarrow D_{K\cup L} \\
\downarrow D^U_K\oplus -D^V_L & & \\
I^{\bar p}H_i(U;R)\oplus I^{\bar p}H_i(V;R) & \longrightarrow & I^{\bar p}H_i(X;R).
\end{array}$$

Once again, in the limit, the top part of the diagram corresponds to the map of the Mayer–Vietoris sequence of Lemma 7.4.5, this time as the inverse of the upper left vertical isomorphism composed with the upper horizontal map. Also once again, to show that this diagram commutes, we can consider the summands separately. So consider

$$\begin{array}{ccc}
I_{\bar q}H^j(X,X-K;R) & \longrightarrow & I_{\bar q}H^j(X,X-K\cup L;R) \\
\downarrow \cong & & \\
I_{\bar q}H^j(U,U-K;R) & \searrow^{D_K} & \downarrow D_{K\cup L} \\
\downarrow D^U_K & & \\
I^{\bar p}H_i(U;R) & \longrightarrow & I^{\bar p}H_i(X;R).
\end{array}$$

The lower left D^U_K is the cap product with ξ^U_K, the diagonal D_K is the cap product with ξ_K, and if $\mathfrak{i}:(U;\emptyset,U-K)\to(X;\emptyset,X-K)$ is the inclusion, then $\mathfrak{i}(\xi^U_K)=\xi_K$ by Lemma 7.4.7. So the left triangle commutes by naturality of the cap product (Proposition 7.3.6).

Similarly, if $\mathfrak{j}:(X;\emptyset,X-K\cup L)\to(X;\emptyset,X-K)$ is the inclusion, then $\mathfrak{j}(\xi_{K\cup L})=\xi_K$ by Lemma 7.4.7, and the right triangle commutes by Proposition 7.3.6.

In the analogous version of the diagram for V, the sign on $-D^V_L$ counteracts the negative sign from $I_{\bar q}H^j(X,X-L;R)\to I_{\bar q}H^j(X,X-K\cup L;R)$ that the inclusion map acquires in the definition of the Mayer–Vietoris cohomology sequence (dualized from the homology sequence).

This demonstrates the commutativity of the second square.

Third Square. This is the “hidden square” in the diagram that involves the boundary maps of the Mayer–Vietoris sequence. Here it is revealed:

$$\begin{array}{ccc}
I_{\bar q}H^j_c(X;R) & \xrightarrow{d^*} & I_{\bar q}H^{j+1}_c(U\cap V;R) \\
\downarrow D^X & & \downarrow D^{U\cap V} \\
I^{\bar p}H_i(X;R) & \xrightarrow{\partial_*} & I^{\bar p}H_{i-1}(U\cap V;R).
\end{array}$$

Commutativity of this diagram will follow from that of the diagram

$$\begin{array}{ccc}
I_{\bar{q}}H^{j}(X,X-K\cup L;R) & \xrightarrow{d^*} & I_{\bar{q}}H^{j+1}(X,X-K\cap L;R) \\
 & & \downarrow \cong \\
\Big\downarrow D_{K\cup L} & & I_{\bar{q}}H^{j+1}(U\cap V,U\cap V-K\cap L;R) \\
 & & \downarrow D^{U\cap V}_{K\cap L} \\
I^{\bar{p}}H_{i}(X;R) & \xrightarrow{\partial_*} & I^{\bar{p}}H_{i-1}(U\cap V;R).
\end{array} \tag{7.29}$$

Once again, these maps become those of the preceding diagram under the direct limits, and these are the maps in the Mayer–Vietoris diagram.

The proof utilizes a concrete realization of

$$\bar{\mathbf{d}}(\xi_{K\cup L}) \in H_{i+j}(I^{\bar{p}}S_*(X;R)\otimes I^{\bar{q}}S_*(X,X-K\cup L;R)).$$

In fact, let's take $\xi_{K\cup L} \in I^{\bar{r}}H_{i+j}(X,X-K\cup L;R)$. By Proposition 7.3.59, the image $\bar{\mathbf{d}}(\xi_{K\cup L})$ can be realized as the image under inclusion of an element of

$$H_*\left(\sum_{W\in\mathcal{W}} I^{\bar{p}}S_*(W,W\cap A;R)\otimes I^{\bar{q}}S_*(W,W\cap B;R)\right),$$

where $\mathcal{W}$ is an open covering of X and A,B are open subsets of X, in this case with $A=\emptyset$ and $B=X-K\cup L$. We will use the following specific covering of X: Let $W_1=U-U\cap L$, $W_2=U\cap V$, and $W_3=V-V\cap K$; see Figure 7.6.

As $L\subset V$ and $K\subset U$, the set W_2 contains all the points of U that are removed to form W_1 and all the points of V that are removed to form W_3, so $\mathcal{W}=\{W_1,W_2,W_3\}$ is a covering of X. Then

$$W_1\cap B=(U-U\cap L)\cap(X-K\cup L)=U-U\cap(K\cup L),$$
$$W_2\cap B=(U\cap V)\cap(X-K\cup L)=U\cap V-(U\cap V)\cap(K\cup L),$$

and

$$W_3\cap B=(V-V\cap K)\cap(X-K\cup L)=V-V\cap(K\cup L).$$

To simplify notation, we will abbreviate $U-U\cap(K\cup L)$ as $U-K\cup L$, and similarly for the others.

Thus, applying Proposition 7.3.59, the class $\bar{\mathbf{d}}(\xi_{K\cup L})$ can be represented by a cycle in

$$\begin{aligned}
&(I^{\bar{p}}S_*(U-L;R)\otimes I^{\bar{q}}S_*(U-L,U-K\cup L;R))\\
&\quad\oplus(I^{\bar{p}}S_*(U\cap V;R)\otimes I^{\bar{q}}S_*(U\cap V,U\cap V-K\cup L;R))\\
&\quad\oplus(I^{\bar{p}}S_*(V-K;R)\otimes I^{\bar{q}}S_*(V-K,V-K\cup L;R)).
\end{aligned}$$

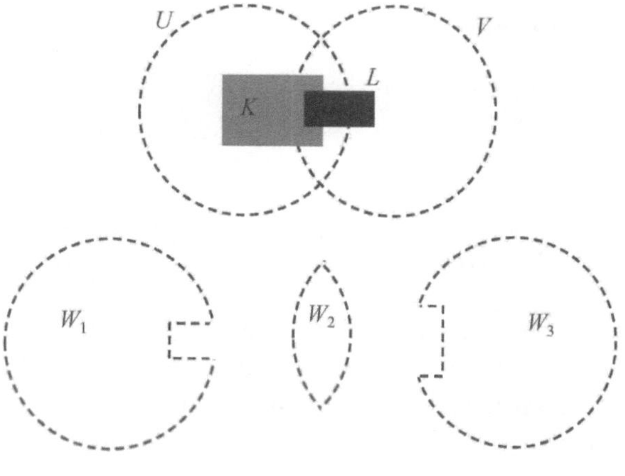

Figure 7.6 A covering of $X = U \cup VS$.

Following [100], let us therefore represent $\bar{\mathbf{d}}(\xi_{K\cup L})$ by a chain $\eta = \eta_{U-L} + \eta_{U\cap V} + \eta_{V-K}$ with[32] $\eta_{U-L} \in I^{\bar{p}}S_*(U-L;R) \otimes I^{\bar{q}}S_*(U-L;R)$, $\eta_{U\cap V} \in I^{\bar{p}}S_*(U\cap V;R) \otimes I^{\bar{q}}S_*(U\cap V;R)$, and $\eta_{V-K} \in I^{\bar{p}}S_*(V-K;R) \otimes I^{\bar{q}}S_*(V-K;R)$.

The argument of the proof of Proposition 7.3.59 is natural with respect to an inclusion $B \to B'$; we take $B' = X - K\cap L$. As $\xi_{K\cap L}$ is the image of $\xi_{K\cup L}$ under the inclusion $(X;\emptyset, X-K\cup L) \to (X;\emptyset, X-K\cap L)$, we obtain that the image of $\bar{\mathbf{d}}(\xi_{K\cap L})$ in $H_{i+j}(I^{\bar{p}}S_*(X;R) \otimes I^{\bar{q}}S_*(X, X-K\cap L;R))$ is also represented by $\eta_{U-L} + \eta_{U\cap V} + \eta_{V-K}$. However, observe that $(U-L)\cap(X-K\cap L) = U-L$ and $(V-K)\cap(X-K\cap L) = V-K$, so

$$\begin{aligned}
&(I^{\bar{p}}S_*(U-L;R) \otimes I^{\bar{q}}S_*(U-L, U-K\cap L;R))\\
&\quad \oplus (I^{\bar{p}}S_*(U\cap V;R) \otimes I^{\bar{q}}S_*(U\cap V, U\cap V-K\cap L;R))\\
&\quad \oplus (I^{\bar{p}}S_*(V-K;R) \otimes I^{\bar{q}}S_*(V-K, V-K\cap L;R))\\
&= I^{\bar{p}}S_*(U\cap V;R) \otimes I^{\bar{q}}S_*(U\cap V, U\cap V-K\cap L;R),
\end{aligned}$$

and so $\bar{\mathbf{d}}(\xi_{K\cap L})$ can be represented simply by $\eta_{U\cap V}$.

Now, let

$$\alpha \in I_{\bar{q}}H^j(X, X-K\cup L;R) = I_{\bar{q}}H^j(X, (X-K)\cap(X-L);R).$$

Let us find a cochain representing

$$d^*(\alpha) \in I_{\bar{q}}H^{j+1}(X, X-K\cap L;R) = I_{\bar{q}}H^j(X, (X-K)\cup(X-L);R).$$

Treating α as a cochain, $d^*(\alpha)$ is determined by the output of a zig-zag chase in the

32 The idea here is that, for example, $\eta_{U-L} \in I^{\bar{p}}S_*(U-L;R) \otimes I^{\bar{q}}S_*(U-L;R)$ is a precise choice of element representing an element of $I^{\bar{p}}S_*(U-L;R) \otimes I^{\bar{q}}S_*(U-L, U-K\cup L;R)$.

Mayer–Vietoris short exact sequence

$$0 \to I_{\bar q}S^*(X,(X-K)+(X-L);R) \xrightarrow{\mathfrak{d}} I_{\bar q}S^*(X,X-K;R) \oplus I_{\bar q}S^*(X,X-L;R)$$
$$\to I_{\bar q}S^*(X,(X-K)\cap(X-L);R) \to 0.$$

Here

$$I_{\bar q}S^*(X,(X-K)+(X-L);R) = \operatorname{Hom}\left(\frac{I^{\bar q}S_*(X;R)}{I^{\bar q}S_*(X-K;R)+I^{\bar q}S_*(X-L;R)}, R\right).$$

So these are intersection cochains that vanish on intersection chains in $I^{\bar q}S_*(X-K;R)$ or $I^{\bar q}S_*(X-L;R)$. Of course, as usual for Mayer–Vietoris sequences, we already know that the cohomology of this cochain complex is isomorphic to $I_{\bar q}H^*(X,(X-K)\cup(X-L);R)$, which is the identification we always tacitly use in Mayer–Vietoris cohomology sequences.

The zig-zag argument tells us that $\alpha \in I_{\bar q}S^j(X,(X-K)\cap(X-L);R)$ must be the image of some $\alpha_K \oplus \alpha_L \in I_{\bar q}S^j(X,X-K;R) \oplus I_{\bar q}S^j(X,X-L;R)$; in fact, pulling back by the Mayer–Vietoris inclusion map gives us $\alpha = \alpha_K - \alpha_L$. Then we take $d(\alpha_K \oplus \alpha_L) = d\alpha_K \oplus d\alpha_L$, and, α being a cocycle, $d\alpha_K \oplus d\alpha_L$ is in the image of the map labeled $\mathfrak{d}$. In fact, the map $\mathfrak{d}$ is the diagonal direct sum inclusion of the form $\mathfrak{d}(a) = a \oplus a$ (up to restrictions), as we can verify from $\mathfrak{d}$ being the dual of the map that adds two chains (up to inclusions). Therefore, $d^*(\alpha)$ is represented by $d\alpha_K$. Well, almost. Remember that $d\alpha_K \in I_{\bar q}H^{j+1}(X,(X-K)+(X-L);R)$, while we want an element of $I_{\bar q}H^{j+1}(X,(X-K)\cup(X-L);R) = I_{\bar q}H^{j+1}(X,X-K\cap L;R)$. The isomorphism between these modules is induced by the dual of the inclusion

$$\psi : I^{\bar q}S_*(X,(X-K)+(X-L);R) \to I^{\bar q}S_*(X,X-K\cap L;R).$$

Therefore, let $\beta \in I^{\bar q}S^{j+1}(X,X-K\cap L;R)$ be a cochain such that $\psi^*(\beta) = d\alpha_K$ in $I_{\bar q}H^{j+1}(X,(X-K)+(X-L);R)$, say by $\psi^*(\beta) - d\alpha_K = d\theta$. Then β represents $d^*(\alpha)$.

So the composition right then down in diagram (7.29) takes the class of α to the class of β, then restricts it to act on chains in $U \cap V$, and finally forms $\beta \frown \xi^{U\cap V}_{K\cap L}$. Note, here and in what follows, for simplicity of notation we will leave certain inclusion- and restriction-induced maps tacit; so, for example, the β in $\beta \frown \xi^{U\cap V}_{K\cap L}$ is really the restriction of β to $U \cap V$. This is not unreasonable, as both β and its restriction act the same way on chains; the context should be clear throughout, so this should not cause too much confusion.

Now, using our above observation that $\bar{\mathbf{d}}(\xi_{K\cap L})$ can be represented by $\eta_{U\cap V} \in I^{\bar p}S_*(U\cap V;R) \otimes I^{\bar q}S_*(U\cap V;R)$, we obtain that the image of the composition right then down in diagram (7.29) is represented by $\Phi(\mathrm{id}\otimes\beta)\eta_{U\cap V}$. In fact, recall $\eta_{U\cap V}$ represents a chain in

$$I^{\bar p}S_*(U\cap V;R) \otimes I^{\bar q}S_*(U\cap V, U\cap V - K\cup L;R),$$

and so also chains in

$$I^{\bar p}S_*(U\cap V;R) \otimes I^{\bar q}S_*(U\cap V,(U\cap V-K)+(U\cap V-L);R)$$

and

$$I^{\bar p}S_*(U\cap V;R) \otimes I^{\bar q}S_*(U\cap V, U\cap V - K\cap L;R)$$

via inclusion. So, invoking naturality, $\Phi(\mathrm{id}\otimes\beta)\eta_{U\cap V}$ and $\Phi(\mathrm{id}\otimes\psi^*(\beta))\eta_{U\cap V}$ represent exactly the same element in $I^{\bar p}H_{i-1}(U\cap V;R)$, so we can just as well use $\Phi(\mathrm{id}\otimes\psi^*(\beta))\eta_{U\cap V}$ as our representative for the composition in the diagram.

We next want to get back to an expression involving $d\alpha_K$, rather than β. For this, we recall that $\eta=\eta_{U-L}+\eta_{U\cap V}+\eta_{V-K}$ represents a cycle in $I^{\bar p}S_*(X;R)\otimes I^{\bar q}S_*(X,X-K\cup L;R)$. Using the fact that all our modules are projective gives

$$I^{\bar p}S_*(X;R)\otimes I^{\bar q}S_*(X,X-K\cup L;R)\cong\frac{I^{\bar p}S_*(X;R)\otimes I^{\bar q}S_*(X;R)}{I^{\bar p}S_*(X;R)\otimes I^{\bar q}S_*(X-K\cup L;R)},$$

so η being a cycle implies that

$$\partial\eta=\partial\eta_{U-L}+\partial\eta_{U\cap V}+\partial\eta_{V-K}\in I^{\bar p}S_*(X;R)\otimes I^{\bar q}S_*(X-K\cup L;R).$$

But we also have

$$\eta_{U-L}\in I^{\bar p}S_*(U-L;R)\otimes I^{\bar q}S_*(U-L;R)\subset I^{\bar p}S_*(X;R)\otimes I^{\bar q}S_*(X-L;R),$$
$$\eta_{V-K}\in I^{\bar p}S_*(V-K;R)\otimes I^{\bar q}S_*(V-K;R)\subset I^{\bar p}S_*(X;R)\otimes I^{\bar q}S_*(X-K;R).$$

Therefore, $\partial\eta_{U\cap V}=\partial\eta-\partial\eta_{U-L}-\partial\eta_{V-K}$ is a chain in $I^{\bar p}S_*(X;R)\otimes(I^{\bar q}S_*(X-K;R)+I^{\bar q}S_*(X-L;R))$. In fact, as all the tensor product terms of $\partial\eta_{U\cap V}$ are supported in $U\cap V$, it is a chain, therefore, in

$$I^{\bar p}S_*(U\cap V;R)\otimes(I^{\bar q}S_*(U\cap V-K;R)+I^{\bar q}S_*(U\cap V-L;R)).$$

Now, recall that $\psi^*(\beta)-d\alpha_K=d\theta$ in $I_{\bar q}S^{j+1}(X,(X-K)+(X-L);R)$, and this relation remains under the restriction to $I_{\bar q}S^{j+1}(U\cap V,(U\cap V-K)+(U\cap V-L);R)$. Suppose

$$\eta_{U\cap V}=\sum y_k\otimes z_k\in I^{\bar p}S_*(U\cap V;R)\otimes I^{\bar q}S_*(U\cap V;R).$$

Then we can compute

$$\begin{aligned}
&(\mathrm{id}\otimes\psi^*(\beta))\eta_{U\cap V}\\
&\quad=(\mathrm{id}\otimes(d\alpha_K+d\theta))\eta_{U\cap V}\\
&\quad=(\mathrm{id}\otimes d\alpha_K)\eta_{U\cap V}+(\mathrm{id}\otimes(d\theta))\eta_{U\cap V}\\
&\quad=(\mathrm{id}\otimes d\alpha_K)\eta_{U\cap V}+(\mathrm{id}\otimes(d\theta))\sum\nolimits_k y_k\otimes z_k\\
&\quad=(\mathrm{id}\otimes d\alpha_K)\eta_{U\cap V}+(-1)^{(j+1)(i-1)}\sum\nolimits_k y_k\otimes(d\theta)(z_k)\\
&\quad=(\mathrm{id}\otimes d\alpha_K)\eta_{U\cap V}+(-1)^{(j+1)(i-1)}\sum\nolimits_k y_k\otimes(-1)^{j+1}\theta(\partial z_k)\\
&\quad=(\mathrm{id}\otimes d\alpha_K)\eta_{U\cap V}+(-1)^{(j+1)(i-1)+j+1}\sum\nolimits_k y_k\otimes\theta(\partial z_k)\\
&\quad=(\mathrm{id}\otimes d\alpha_K)\eta_{U\cap V}+(-1)^{(j+1)(i-1)+j+1+j(i-1)}(\mathrm{id}\otimes\theta)\sum\nolimits_k y_k\otimes\partial z_k\\
&\quad=(\mathrm{id}\otimes d\alpha_K)\eta_{U\cap V}+(-1)^{i+j}(\mathrm{id}\otimes\theta)\sum\nolimits_k y_k\otimes\partial z_k\\
&\quad=(\mathrm{id}\otimes d\alpha_K)\eta_{U\cap V}+(-1)^{i+j+(i-1)}(\mathrm{id}\otimes\theta)\sum\nolimits_k(-1)^{|y_k|}y_k\otimes\partial z_k\\
&\quad=(\mathrm{id}\otimes d\alpha_K)\eta_{U\cap V}+(-1)^{j-1}(\mathrm{id}\otimes\theta)\sum\nolimits_k(\partial(y_k\otimes z_k)-(\partial y_k)\otimes z_k)\\
&\quad=(\mathrm{id}\otimes d\alpha_K)\eta_{U\cap V}+(-1)^{j-1}(\mathrm{id}\otimes\theta)\Big(\partial\eta_{U\cap V}-\sum\nolimits_k(\partial y_k)\otimes z_k\Big)\\
&\quad=(\mathrm{id}\otimes d\alpha_K)\eta_{U\cap V}+(-1)^{j}(\mathrm{id}\otimes\theta)\sum\nolimits_k(\partial y_k)\otimes z_k.
\end{aligned}$$

Let us explain all this. In the fourth line, we have used that $(d\theta)(z_k) = 0$ unless $|z_k| = |d\theta| = j+1$, in which case the corresponding y_k has $|y_k| = i-1$. The next few lines are just computations and simplifications. In the fourth line from the bottom, we again use that all the summands on the right are trivial unless $|y_k| = i-1$, and this allows us to include the $(-1)^{|y_k|}$ in all terms balanced off by $(-1)^{i-1}$ outside the sum. In the last line, we have used that θ kills elements of $I^{\bar{q}}S_*(X-K;R) + I^{\bar{q}}S_*(X-L;R)$, and we have seen that these are all that occur in the second tensor factors of $\partial\eta_{U\cap V}$.

Now, applying Φ to both sides of this computation, we get

$$\begin{aligned}\Phi(\mathrm{id}\otimes\psi^*(\beta))\eta_{U\cap V} &= \Phi(\mathrm{id}\otimes d\alpha_K)\beta_{U\cap V} + (-1)^j\Phi(\mathrm{id}\otimes\theta)\sum_k(\partial y_k)\otimes z_k\\ &= \Phi(\mathrm{id}\otimes d\alpha_K)\eta_{U\cap V} + (-1)^{j+(i-1)j}\sum_k\theta(z_k)\partial y_k.\end{aligned}$$

Therefore, $\Phi(\mathrm{id}\otimes\psi^*(\beta))\eta_{U\cap V}$ and $\Phi(\mathrm{id}\otimes d\alpha_K)\eta_{U\cap V}$ are homologous in $I^{\bar{p}}S_*(U\cap V;R)$. So, the composition right then down in the diagram is represented by $\Phi(\mathrm{id}\otimes d\alpha_K)\eta_{U\cap V}$.

Now that we've gotten $d\alpha_K$ back in the picture, we need to massage this just a bit more to fit with what we'll get going around the diagram the other way. Continue to suppose $\eta_{U\cap V} = \sum y_k\otimes z_k$. We notice that

$$\begin{aligned}&\Phi(\mathrm{id}\otimes\alpha_K)\partial\eta_{U\cap V}\\ &\quad= \Phi(\mathrm{id}\otimes\alpha_K)\partial\sum y_k\otimes z_k\\ &\quad= \Phi(\mathrm{id}\otimes\alpha_K)\sum((\partial y_k)\otimes z_k + (-1)^{|y_k|}y_k\otimes\partial z_k)\\ &\quad= \sum(-1)^{j|\partial y_k|}\alpha_K(z_k)\partial y_k + \sum(-1)^{i-1+j(i-1)}\alpha_K(\partial z_k)y_k\\ &\quad= \partial\left(\sum(-1)^{j(i-1)}\alpha_K(z_k)y_k\right) + \sum(-1)^{i-1+j(i-1)+j+1}((d\alpha_K)(z_k))y_k.\\ &\quad= \sum(-1)^{j(i-1)+ij}\partial\Phi(\mathrm{id}\otimes\alpha_K)(y_k\otimes z_k)\\ &\qquad\qquad+\sum(-1)^{i+j(i-1)+j+(j+1)(i-1)}\Phi(\mathrm{id}\otimes d\alpha_K)y_k\otimes z_k.\\ &\quad= (-1)^j\partial\Phi(\mathrm{id}\otimes\alpha_K)\eta_{U\cap V} + (-1)^{j+1}\Phi(\mathrm{id}\otimes d\alpha_K)\eta_{U\cap V}.\end{aligned}\tag{7.30}$$

Here, we have again used that $\eta_{U\cap V} = \sum y_k\otimes z_k$ is an $(i+j)$-chain, that α is a j-cochain, and that the expressions above will vanish unless a cochain acts on a chain of the same degree. So, we see that up to signs[33] that depend only on the fixed i and j, $\Phi(\mathrm{id}\otimes\alpha_K)\partial\eta_{U\cap V}$ and $\Phi(\mathrm{id}\otimes d\alpha_K)\eta_{U\cap V}$ together bound; therefore, $\Phi(\mathrm{id}\otimes\alpha_K)\partial\eta_{U\cap V}$ also represents the composition right then down in diagram (7.29), up to sign. At last, this is the final form that we want for this element.

Next, we consider the other way around the diagram (7.29). We first take the cap product of α with $\xi_{K\cup L}$, which is

$$\begin{aligned}\alpha\frown\xi_{K\cup L} &= \Phi(\mathrm{id}\otimes\alpha)\bar{\mathbf{d}}(\xi_{K\cup L}) = \Phi(\mathrm{id}\otimes\alpha)(\eta_{U-L}+\eta_{U\cap V}+\eta_{V-K})\\ &= \Phi(\mathrm{id}\otimes\alpha)\eta_{U-L} + \Phi(\mathrm{id}\otimes\alpha)\eta_{U\cap V} + \Phi(\mathrm{id}\otimes\alpha)\eta_{V-K}.\end{aligned}$$

33 These signs disagree with Hatcher [125, Lemma 3.36] because Hatcher's version of the cap product has the chain on the left and the cochain on the right.

The first of these chains is supported in U while the other two are supported in V. Therefore, the zig-zag construction of the map ∂_* in the homology Mayer–Vietoris sequence can proceed by pulling our chain representative for $\alpha \frown \xi_{K\cup L}$ back to

$$\Phi(\mathrm{id}\otimes\alpha)\eta_{U-L} \oplus (\Phi(\mathrm{id}\otimes\alpha)\eta_{U\cap V} + \Phi(\mathrm{id}\otimes\alpha)\eta_{V-K}) \in I^{\bar{p}}S_i(U;R) \oplus I^{\bar{p}}S_i(V;R),$$

then taking its boundary under $\partial\oplus\partial$, and then finally arrive at a preimage in $I^{\bar{p}}S_i(U\cap V;R)$ under the map $(\mathfrak{i}_U, -\mathfrak{i}_V)$, with the $\mathfrak{i}$ denoting the inclusion maps. In this case, the preimage is represented by $\partial(\Phi(\mathrm{id}\otimes\alpha)\eta_{U-L})$. Now, using computations identical to those in (7.30), but with α in place of α_K and η_{U-L} in place of $\eta_{U\cap V}$, we have that, up to signs,

$$\partial(\Phi(\mathrm{id}\otimes\alpha)\eta_{U-L}) = \pm\Phi(\mathrm{id}\otimes\alpha)\partial\eta_{U-L} \pm \Phi(\mathrm{id}\otimes d\alpha)\eta_{U-L}.$$

But α is a cocycle, so this becomes $\partial(\Phi(\mathrm{id}\otimes\alpha)\eta_{U-L}) = \pm\Phi(\mathrm{id}\otimes\alpha)\partial\eta_{U-L}$. Now, recall that η_{U-L}, and so also its boundary, is in $I^{\bar{p}}S_*(U-L;R)\otimes I^{\bar{q}}S_*(U-L;R)$, and that $\alpha = \alpha_K - \alpha_L$, as above. But α_L kills chains outside of L, so $(\mathrm{id}\otimes\alpha)\partial\eta_{U-L} = (\mathrm{id}\otimes\alpha_K)\partial\eta_{U-L}$. Next, once again, as $\bar{\mathbf{d}}(\xi_{K\cup L})$ is a cycle in $I^{\bar{p}}S_*(X;R)\otimes I^{\bar{q}}S_*(X, X-K\cup L;R)$, we have that $\partial\eta$, which represents $\partial\bar{\mathbf{d}}(\xi_{K\cup L})$, is in $I^{\bar{p}}S_*(X;R)\otimes I^{\bar{q}}S_*(X-K\cup L;R)$. Now, apply $\mathrm{id}\otimes\alpha_K$ to the expression $\partial\eta = \partial\eta_{U-L} + \partial\eta_{U\cap V} + \partial\eta_{V-K}$ and observe that $\mathrm{id}\otimes\alpha_K$ must kill $\partial\eta_{V-K}$ and $\partial\eta$, which are made of chains supported outside of K in the second tensor factor. Therefore,

$$\Phi(\mathrm{id}\otimes\alpha_K)\partial\eta_{U-L} = -\Phi(\mathrm{id}\otimes\alpha_K)\partial\eta_{U\cap V}.$$

So our representative for the image of α down then right in diagram (7.29) is, up to sign, $\Phi(\mathrm{id}\otimes\alpha_K)\partial\eta_{U\cap V}$. And, again up to sign, this is the same expression we obtained earlier for running right then down in diagram (7.29).

This completes the third square and so the proof of Lemma 7.4.8. Whew! □

8
Poincaré Duality

In this chapter, we prove intersection homology Poincaré duality for pseudomanifolds. Our approach follows the modern theory for manifolds in which the duality isomorphism is given by the cap product with a fundamental class. Much of our particular exposition is modeled on the treatment in Hatcher [125]. For intersection homology, this proof is quite different from the original techniques of Goresky and MacPherson in [105] and [106]; the former proceeds using piecewise linear intersection pairings and a filtration argument using "basic sets," while the latter utilizes the axiomatics of the derived category of sheaf complexes. We will not provide the details of either of these original proofs here, although in Section 8.5 we do provide an exposition of the PL intersection pairing.

We begin in Section 8.1 by discussing orientations and fundamental classes on pseudomanifolds. This is followed by our proof of Poincaré duality in Section 8.2 and Lefschetz duality for ∂-pseudomanifolds in Section 8.3. In Section 8.4.1, we derive in detail how these duality theorems yield nonsingular cup product and torsion pairings. In Section 8.4.5 we also discuss what we call "image pairings"; these are the intersection homology generalizations of the duality pairings on the image groups $\text{im}(H^*(M,\partial M) \to H^*(M))$ for a ∂-manifold M. We conclude with our discussion of intersection pairings in Section 8.5.

8.1 Orientations and Fundamental Classes

In this section, we continue to head toward an intersection homology version of Poincaré duality by constructing orientations and fundamental classes for stratified pseudomanifolds. Notice that we are here restricting ourselves from the larger generality of CS sets down to spaces with a bit more structure. Stratified pseudomanifolds are required to be *recursive* CS sets, and the union of the regular strata must be dense. This latter condition is necessary to have something that is dimensionally homogeneous, which we need in order for all points to be able to carry anything like an orientation in the proper degree. Such a restriction is not completely necessary for Poincaré duality as, by Proposition 6.3.47, non-GM intersection homology does not detect strata outside the homogenization[1] of a CS set. But this is also an argument that we might as well restrict our attention to the homogeneous CS sets. The idea for using recursive CS sets is that orientation properties are local and so we will need to demonstrate the proper homological properties on distinguished neighborhoods, and these properties are completely controlled by the links via stratified homotopy invariance and the cone formula. One alternative would be to simply make the necessary homological assumptions about the links, which could perhaps be done. However, a pleasant feature of stratified pseudomanifolds is that their links are also stratified pseudomanifolds by Lemma 2.4.11, and so the local homological properties will exist inductively via the fundamental class of the link.

The principal arguments in this section are based on those of [100, Section 5], which are themselves based on the arguments for manifolds in Section 3.3 of Hatcher [125]. The primary difference from [100] is that there we first developed the results for normal stratified pseudomanifolds and then made additional arguments to obtain them for arbitrary pseudomanifolds using the properties of normalization maps. Here we take a more direct route, treating arbitrary stratified pseudomanifolds throughout.

In Section 8.1.1 we review some material about orientations and fundamental classes for manifolds, then in Section 8.1.2 we discuss orientations of CS sets, including behavior with respect to changing the filtration. Section 8.1.3 contains the main theorems about fundamental classes for pseudomanifolds. In Section 8.1.4 we explain why fundamental classes are only defined for perversities that don't take negative values, and Section 8.1.5 considers the behavior of fundamental classes under changes of perversity or stratification. In Section 8.1.6 we revisit an observation of Goresky and MacPherson from [105] by showing that the cap product with the fundamental class in singular homology factors through the intersection homology cap product. Finally, we consider orientations and fundamental classes of product spaces in Section 8.1.7.

WARNING: In this section, we require some elementary sheaf theory at a variety of points. While we attempt to provide an overview of the relevant notions where necessary, the reader should be aware that not all of this section will be self-contained given our development thus far. Good references for most of what we need can be found in the first few chapters of Swan [230] or Bredon [37], each of which develops sheaf theory considerably more than we will need here.

8.1.1 Orientation and Fundamental Classes of Manifolds

Let us first briefly review in this subsection the principal definitions and results concerning orientation and fundamental classes for manifolds. One good reference, and the one we will mostly follow in our treatment below for pseudomanifolds, can be found in [125, Section 3.3]. We also assume the reader is familiar with bundles of groups as in, for example, [125,

[1] Recall Definition 6.3.44 in Section 6.3.2.

Additional Topic 3.H]; see also the discussion of orientations for manifolds in [125, Section 3.3]. We will mostly be interested in bundles of R-modules, but the basic ideas are the same.

Recall that for an n-dimensional manifold M and for any coefficient ring R we have an orientation bundle $\mathcal{O}$ with fiber (stalk) $\mathcal{O}_x = H_n(M, M - \{x\}; R) \cong R$ at $x \in M$. At any moment, we will work with a fixed base ring, so we will omit it from the notation for the orientation bundle. The bundle structure is determined by noting that every point in M has an open ball neighborhood B such that $H_n(M, M - B; R) \cong R$ and for any two $x, y \in B$ we have canonical isomorphisms induced by inclusion $H_n(M, M - \{x\}; R) \xleftarrow{\cong} H_n(M, M - B; R) \xrightarrow{\cong} H_n(M, M - \{y\}; R)$. Hence every point of M has a neighborhood on which we have canonical identification between fibers, and this determines the bundle $\mathcal{O}$.

Definition 8.1.1 The manifold M is *R-orientable* if $\mathcal{O}$ has a global section $\mathfrak{o}$ that restricts to a generator of R over each point; this is equivalent to assuming that the bundle $\mathcal{O}$ is trivial. In particular, every manifold is $\mathbb{Z}_2$-orientable. If M is R-orientable, an *R-orientation* is a choice of such a global section $\mathfrak{o}$.

Of course there are homological results about manifolds that are closely intertwined with their orientation properties. The following lemma, which is a restatement of [125, Lemma 3.27], lays the principal cornerstone:

Lemma 8.1.2 *Let M be an n-dimensional manifold and $K \subset M$ a compact set. Then the following hold:*

1. *$H_i(M, M - K; R) = 0$ for $i > n$, and a class in $H_n(M, M - K; R)$ is zero if and only if its image in $H_n(M, M - x; R)$ is zero for all $x \in K$.*
2. *Given a section $\mathfrak{s}$ of $\mathcal{O}$ over M, there is a unique class $\gamma_K \in H_n(M, M - K; R)$ whose image in $H_n(M, M - \{x\}; R)$ is $\mathfrak{s}(x)$ for any $x \in K$. In particular, if M is R-oriented then there is a unique class $\Gamma_K \in H_n(M, M - K; R)$ whose image in $H_n(M, M - \{x\}; R)$ is $\mathfrak{o}(x)$ for any $x \in K$.*

This lemma leads to the following theorem, which is two-thirds of [125, Theorem 3.26]:

Theorem 8.1.3 *Let M be a closed (compact with empty boundary) connected n-manifold. Then we have the following:*

1. *If M is R-orientable, then $H_n(M; R) \longrightarrow H_n(M, M - \{x\}; R) \cong R$ is an isomorphism for all $x \in M$.*
2. *$H_i(M; R) = 0$ for $i > n$.*

The last statement of the theorem follows immediately from the first statement of the lemma, taking $K = M$. The first statement of the theorem follows from the second statement of the lemma as follows: As M is connected and R-oriented, the orientation bundle $\mathcal{O}$ is the trivial bundle with fiber R, and its module of sections $\Gamma(M, \mathcal{O})$ is therefore isomorphic to R

via the evaluation map that takes $\mathfrak{s} \in \Gamma(M,\mathcal{O})$ to

$$\mathfrak{s}(x) \in \mathcal{O}_x = H_n(M, M - \{x\}; R) \cong R$$

for any $x \in M$. In particular, $\mathfrak{o}$ is a generator of $\Gamma(M,\mathcal{O}) \cong R$. Now, any element of $\xi \in H_n(M;R)$ determines a global section $\mathfrak{s}_\xi$ of $\mathcal{O}$ by letting $\mathfrak{s}_\xi(x)$ be the image of ξ in $H_n(M, M - \{x\}; R)$. The map $H_n(M;R) \to \Gamma(M,\mathcal{O})$ so described is both surjective and injective by the second statement of the lemma, taking $K = M$. Thus we have isomorphisms

$$H_n(M;R) \xrightarrow{\cong} \Gamma(M,\mathcal{O}) \xrightarrow{\cong} H_n(M, M - \{x\}; R) \cong R$$

for all $x \in M$.

Definition 8.1.4 We call the element of $H_n(M;R)$ corresponding to a given R-orientation $\mathfrak{o}$ of M the *fundamental class* $\Gamma_M \in H_n(M;R)$.

The proof of Lemma 8.1.2 requires some work. We will not discuss this here, but rather we refer again to [125, Section 3.3].

8.1.2 Orientation of CS Sets

Now, we turn to pseudomanifolds. In fact, the basic definitions concerning orientation are applicable more generally to CS sets; we will only need to restrict to pseudomanifolds when considering homological results in the next subsection, so we will work in the greater generality until then. We do assume in this section that our CS sets have regular strata.

Definition 8.1.5 Let X be an n-dimensional CS set with non-empty regular strata. We say that X is *R-orientable* if the n-manifold $X - X^{n-1} = X - \Sigma_X$ is R-orientable, and we say that X is *R-oriented* if a particular R-orientation for $X - X^{n-1} = X - \Sigma_X$ has been chosen.

Before launching into the homological properties related to orientability in the next subsection, we will first discuss the relationship between orientations of different CS set filtrations of a single space. We begin with a simple observation.

Lemma 8.1.6 *Suppose X is an n-dimensional CS set with non-empty regular strata and that X' is an n-dimensional CS set with the same underlying space $|X|$ and with a finer stratification, i.e. each stratum of X' is contained in a stratum of X. Then if X is R-orientable so is X', and any R-orientation of X determines a unique R-orientation of X'.*

Proof The assumption that the stratification of X' is finer than that of X implies immediately that $X' - \Sigma_{X'} \subset X - \Sigma_X$. So if the orientation bundle over $X - \Sigma_X$ is trivial, it restricts to a trivial bundle over $X' - \Sigma_{X'}$, and any choice of global section of generators similarly restricts. □

The converse to Lemma 8.1.6 is not true in general, as the following examples demonstrate.

Example 8.1.7 Let M be the unfiltered open Mobius band, and let M' be $|M|$ filtered as $\Sigma \subset M$, where Σ is an arc running widthwise across M. In other words, let M be formed from $[0,1] \times (0,1)$ by identifying $\{0\} \times (0,1)$ with $\{1\} \times (0,1)$ by $(0,t) \sim (1, 1-t)$, and let Σ be the image of $\{0,1\} \times (0,1)$. Of course M is not $\mathbb{Z}$-orientable, but $M' - \Sigma_{M'} = M - \Sigma$ is homeomorphic to the open disk, so it is $\mathbb{Z}$-orientable. △

Example 8.1.7 shows that it is possible to have one filtration of a CS set be R-orientable while another is not. The next example shows that even if two filtrations yield R-orientable CS sets, an R-orientation on a finer stratification does not necessarily determine an R-orientation on a coarser stratification.

Example 8.1.8 Let $\mathbb{R}'$ consist of the space $\mathbb{R}$ filtered by $\{0\} \subset \mathbb{R}$. Then $\mathbb{R}'$ has two regular strata, corresponding to the positive and negative real numbers, and we can orient these submanifolds, and hence $\mathbb{R}'$, in a way that is not compatible with a single orientation on all of the CS set $\mathbb{R}$. △

The trouble in both of these examples is caused by the addition of a stratum of codimension one. It turns out that Lemma 8.1.6 does have a converse if we forbid the addition of "new" codimension-one strata. The two following lemmas utilize some sheaf-theoretic notions of dimension theory that go even a bit further beyond the elementary treatment of sheaves that we will need more seriously below. The reader unacquainted with such notions should still be able to follow the general idea of the proof, though the reader willing to believe the result can safely skip the argument, which won't be needed again later.

Lemma 8.1.9 *Suppose X is an n-dimensional CS set with non-empty regular strata and that X' is an n-dimensional CS set with the same underlying space $|X|$ and with a finer stratification, i.e. each stratum of X' is contained in a stratum of X. Suppose further that any codimension-one stratum of X' is contained in a codimension-one stratum of X. Then if X' is R-orientable so is X, and any R-orientation of X' determines a unique R-orientation of X.*

Proof Let $M = X - \Sigma_X$ and $M' = X' - \Sigma_{X'}$. The spaces M and M' are n-dimensional manifolds, and as X' is stratified more finely than X, we have $M' \subset M$. We claim that M' is an open dense subset of M such that $M - M'$ has codimension at least two, utilizing the sheaf-theoretic $\dim_{\mathbb{Z}}$ as our notion of dimension (see Lemma 6.3.46, above, and [37, Section II.16] for a full treatment). This will imply the lemma using a result about bundle theory that we provide below.

It is clear that M' is open in M as M and M' are both open subsets of the underlying space $|X|$. For the issue of codimension, as M is an n-manifold, $\dim_{\mathbb{Z}}(M) = n$ by [37, Corollary II.16.28]. To see that the complement of M' in M has dimension $\leq n-2$, suppose $x \in M - M'$. Then x must be contained in a singular stratum of X', and we claim it is not a codimension-one singular stratum because, by assumption, if x is contained in a codimension-one stratum of X' then it is contained in a codimension-one stratum of X, which would contradict $x \in M$. Therefore, $M - M'$ is a subset of the $(n-2)$-skeleton of

X'. As $M - M'$ is the intersection of M with $\Sigma_{X'}$ by definition, it now follows that $M - M'$ is the intersection of M with the $(n-2)$-skeleton of X', which is thus an open subset of the $(n-2)$-skeleton of X'. We showed above in the proof of Lemma 6.3.46 that the i-skeleton of a CS set has $\mathbb{Z}$-dimension $\leq i$, and any open subset of such a skeleton also has $\mathbb{Z}$-dimension $\leq i$ by [37, Theorem II.16.8] (this also uses that $|X|$ is locally compact by Lemma 2.3.15). Thus $\dim_{\mathbb{Z}}(M - M') \leq n - 2$. Similarly, M' is dense in M because if $x \in M$ then any neighborhood of x contains an open Euclidean neighborhood of dimension n, and such an open neighborhood cannot be contained in the $(n-1)$-skeleton of X' for dimension reasons, again by [37, Theorem II.16.8]; so any neighborhood of x intersects M'. So we have shown that M' is an open dense subset of M such that $M - M'$ has codimension ≥ 2.

The lemma now follows from a basic result about bundle theory, which we present as Lemma 8.1.10, below. According to that result, if we have two bundles defined on a manifold then any bundle isomorphism between them defined on an open dense subset of codimension at least two has a unique extension to the entire manifold. So in our setting Lemma 8.1.10 says that an isomorphism of bundles over M' must extend uniquely over all of M. In particular, suppose O_M is the R-orientation bundle over M and that $\mathfrak{R}_M$ is the trivial bundle over M with stalk R. If X' is R-orientable, there is an isomorphism $\mathfrak{R}_M|_{M'} \to O_M|_{M'} = O_{M'}$, and an R-orientation corresponds to a specific choice of isomorphism (given a fixed identification of $\mathfrak{R}_M$ with $M \times R$, the R-orientation is the image section under the bundle isomorphism of the section of $\mathfrak{R}_M|_{M'}$ that takes each $x \in M'$ to $1 \in R$). Lemma 8.1.10 guarantees a unique extension of this isomorphism to all of M, demonstrating that M is orientable and extending uniquely the chosen R-orientation. □

The following lemma is basic to bundle theory, though the techniques required for the proof are a bit beyond the required background of most of this book. We'll provide some details of the argument from [28, Lemma V.4.11.a], though, once again, the reader willing to believe the result can safely skip the proof, which won't be needed again later.

Lemma 8.1.10 *Suppose M is an n-dimensional manifold and that U is a dense open subset of M whose complement has codimension at least two. Then if $\mathcal{E}$ and $\mathcal{F}$ are bundles of finitely generated R-modules on M and $\phi : \mathcal{E}|_U \to \mathcal{F}|_U$ is a bundle morphism, then there exists a unique bundle morphism $\psi : \mathcal{E} \to \mathcal{F}$ that extends ϕ. Furthermore, if ϕ is an isomorphism then so is ψ.*

Proof Without loss of generality, we can assume M is path connected; otherwise we can argue on each path component separately. It follows that U must also be path connected. The basic idea is that because $M - U$ has codimension ≥ 2 any path can be altered by a homotopy to avoid $M - U$. If we were working entirely with smooth objects, this would follow from general position arguments, but as our objects are purely topological, this is not completely straightforward. We will use the fact that if Y is a locally compact Hausdorff space and W is an open subspace then there is a long exact sequence

$$\longrightarrow \mathbb{H}^i_c(W;\mathbb{Z}_2) \to \mathbb{H}^i_c(Y;\mathbb{Z}_2) \to \mathbb{H}^i_c(Y - W;\mathbb{Z}_2) \longrightarrow ,$$

where $\mathbb{H}^i$ denotes sheaf cohomology. The existence of such a long exact sequence is not so evident using the singular cohomology definition of H^i_c, but it follows for $\mathbb{H}^i_c$ from basic sheaf cohomology theory[2] [37, Section II.10.3]. Now, $\dim_{\mathbb{Z}}(M-U) \leq n-2$ by assumption, and as $M-U$ is locally compact, it is locally paracompact, so from [37, Proposition II.16.15] we also have $\dim_{\mathbb{Z}_2}(M-U) \leq n-2$. By [37, Definition II.16.6], $\dim_{\mathbb{Z}_2}(M-U) = \dim_{c,\mathbb{Z}_2}(M-U)$, and so we have $\mathbb{H}^i_c(M-U;\mathbb{Z}_2) = 0$ for $i = n-1$ and $i = n$ by [37, Theorem II.16.4]. Thus, from the exact sequence, $\mathbb{H}^n_c(U;\mathbb{Z}_2) \cong \mathbb{H}^n_c(M;\mathbb{Z}_2)$. Finally, by Poincaré duality [37, Example IV.2.9], as U and M are n-manifolds and as the $\mathbb{Z}_2$-orientation sheaf of any space is constant, we have $\mathbb{H}^n_c(U;\mathbb{Z}_2) \cong H_0(U;\mathbb{Z}_2)$ and $\mathbb{H}^n_c(M;\mathbb{Z}_2) \cong H_0(M;\mathbb{Z}_2)$. So, if M is path connected we have $H_0(U;\mathbb{Z}_2) \cong H_0(M;\mathbb{Z}_2) \cong \mathbb{Z}_2$, and U is also path connected.

The same argument works locally to show that if B is an open ball neighborhood of a point in M then $U \cap B$ is path connected and dense in B. This observation is enough to imply that, for any basepoint $x_0 \in U$, the map $\mathrm{i}_* : \pi_1(U,x_0) \to \pi_1(M,x_0)$ is surjective, where i is the inclusion $U \hookrightarrow M$. Indeed, this is a basic exercise (that we leave for the reader) in locally modifying paths by homotopies within small balls.

Now, as M is path connected, the category of bundles on M whose stalks are finitely generated R-modules is equivalent to the category of finitely generated $\pi_1(M,x_0)$-modules for any $x_0 \in M$ (see [125, Additional Topic 3.H]), and similarly for U. Furthermore, restriction of a bundle to U corresponds to the "change of scalars" induced by $\mathrm{i}_* : \pi_1(U,x_0) \to \pi_1(M,x_0)$. This means that if E is a $\pi_1(M,x_0)$-module and $z \in E$, then we obtain a corresponding module, say E_U, with the same elements as E (though we will denote the version of z in E_U as z_U) and with action of $\gamma \in \pi_1(U,x_0)$ on E_U given by $\gamma z_U = (\mathrm{i}_*(\gamma)z)_U$.

The hypothesis of the lemma is equivalent to assuming that we have two $\pi_1(M,x_0)$-modules, say E, F, and a $\pi_1(U,x_0)$-morphism $\phi : E_U \to F_U$, where E_U and F_U denote E and F as modules after the restriction of scalars. We must show that there is a unique morphism $\psi : E \to F$ that induces ϕ. But $E = E_U$ and $F = F_U$ as groups, so ϕ certainly provides a function ψ determined by $(\psi(z))_U = \phi(z_U)$ for $z \in E$. We must show that ψ is a morphism of $\pi_1(M,x_0)$-modules by showing that $\psi(mz) = m\psi(z)$ for any $m \in \pi_1(M,x_0)$. As i_* is surjective we can choose $\bar{m} \in \pi_1(U,x_0)$ such that $\mathrm{i}_*(\bar{m}) = m$. Then, using that ϕ is a map of $\pi_1(U,x_0)$-modules, we have

$$\begin{aligned}(\psi(mz))_U = \phi((mz)_U) &= \phi((\mathrm{i}_*(\bar{m})z)_U) = \phi(\bar{m}z_U) = \bar{m}\phi(z_U) = \bar{m}(\psi(z))_U \\ &= (m\psi(z))_U.\end{aligned}$$

As the assignment $z \to z_U$ is bijective, we thus see that ψ is a $\pi_1(M,x_0)$-module morphism, as desired. It is also clearly the unique such morphism compatible with ϕ, and it is bijective if and only if ϕ is bijective. This completes the proof. □

[2] Note that the exact sequence in Bredon [37] is stated for any paracompactifying family of supports Φ. In our case $\Phi = c$, the family of compact supports, and the restrictions $\Phi|W$ and $\Phi|Y-W$ are also c, as follows directly from [37, Definition II.6.3]. In each case, c is paracompactifying by the observations of [37, page 22] as all of the spaces here are locally compact. Note that open and closed subspaces of locally compact Hausdorff spaces are locally compact [181, Corollary 29.3].

Corollary 8.1.11 *Suppose X is a CS set with non-empty regular strata, all of whose codimension-one strata are contained in the codimension-one strata of $\mathfrak{X}$, the intrinsic filtration of X (recall Definition 2.10.6); in particular, this will be the case if X has no codimension-one strata. Then any R-orientation of X determines a unique R-orientation for any CS set filtration X' of the underlying space $|X|$.*

Proof Recall that $\mathfrak{X}$ is a coarsening of any CS set stratification of $|X|$; see Remark 2.10.7. Therefore, by Lemma 8.1.9 and the assumptions, an R-orientation of X determines a unique R-orientation of $\mathfrak{X}$, which determines a unique R-orientation on any other stratification X' of $|X|$ by Lemma 8.1.6. □

Remark 8.1.12 As just observed in the proof of Corollary 8.1.11, it follows from Lemma 8.1.6 that any R-orientation of $\mathfrak{X}$ determines an R-orientation on any CS set filtration of $|X|$. As the CS set filtration $\mathfrak{X}$ is intrinsic to the underlying space, this provides a sense in which the notion of R-orientability is independent of choice of CS set filtration, assuming that $\mathfrak{X}$ is R-orientable. However, Example 8.1.7, just above, demonstrates that we might still have R-orientability of X even when $\mathfrak{X}$ is not R-orientable.

8.1.3 Homological Properties of Orientable Pseudomanifolds

In this section, we look at intersection homology versions of Lemma 8.1.2 and Theorem 8.1.3; in particular we construct intersection homology fundamental classes for oriented stratified pseudomanifolds. An immediate question is what perversity we should use. This will turn out to be immaterial so long as $\bar{p} \geq \bar{0}$, i.e. if $\bar{p}(S) \geq 0$ for all singular strata S. In this section we will proceed simultaneously with all such perversities, and in the next we will show that, in fact, the top-dimensional intersection homology groups behave identically with respect to any of these perversities, so there is no real distinction. In later sections, we will work with $\bar{0}$, which is initial among all these perversities and so provides a canonical fundamental class. It is the cap product with this $\bar{0}$-perversity fundamental class that takes $\bar{p}$-allowable intersection cochains to $D\bar{p}$-allowable intersection chains, demonstrating again why the notion of dual perversities is so relevant. On the other hand, if $\bar{p}(S) < 0$ for some singular stratum S, then we will see below, as a consequence of Proposition 8.1.24, that it is not possible for X to have a (global) fundamental class.

(a) The Orientation Sheaf

We begin with the preliminary observation that if X is an n-dimensional CS set and $\bar{p}$ is a perversity satisfying $\bar{p} \geq \bar{0}$ then the assignment $x \to I^{\bar{p}}H_n(X, X - \{x\}; R)$ will no longer be locally constant; in fact, at singular points of X we do not necessarily have $I^{\bar{p}}H_n(X, X - \{x\}; R) \cong R$. Therefore, we cannot talk about an orientation *bundle* over all of X, though we do still have such a bundle over $X - \Sigma_X$. The appropriate object on all of X is a *sheaf*. Sheaves generalize bundles of coefficients by allowing the possibility of different modules over different points. We will only need a bare minimum of material about sheaves, but since we do not assume the reader is necessarily familiar with any sheaf theory, we provide the basic idea. A readable elementary account from the following point of view can be found in the early chapters of Swan [230]. Other good basic introductions are [233] and the early sections of [37].

There are actually multiple equivalent definitions of sheaves, but for our purposes the simplest is the following: Like a bundle of coefficients, a sheaf $\mathcal{S}$ of R-modules over a space Y is a space $\mathcal{S}$ together with a local homeomorphism $\pi : \mathcal{S} \to Y$, meaning that for each $z \in \mathcal{S}$ the map π takes some open neighborhood V of z in $\mathcal{S}$ homeomorphically onto an open neighborhood of $\pi(z)$ in Y. Additionally, for each $x \in Y$, the preimage $\pi^{-1}(x)$, which is also denoted $\mathcal{S}_x$ and called the *stalk* of the sheaf at x, must be an R-module with the discrete topology. It is not required that the various $\mathcal{S}_x$ be isomorphic to each other. Finally, there is also a requirement that algebraic operations should be continuous; in other words, the map $\{(y,z) \in \mathcal{S} \times \mathcal{S} \mid \pi(y) = \pi(z) \in V\} \to \mathcal{S}$ given by $(y,z) \to y + z$ must be continuous, and an analogous statement holds for scalar multiplication. A bundle of coefficients with fiber module F is simply a sheaf for which each point in Y has a neighborhood V on which $\pi^{-1}(V) \cong V \times F$ with π corresponding to the projection onto V. In general, however, while each stalk of a sheaf $\mathcal{S}$ must inherit the discrete topology as a subspace of $\mathcal{S}$, the overall topology of $\mathcal{S}$ might be quite complicated and is very often non-Hausdorff.

In our particular case, the orientation sheaf $\mathcal{O}^{\bar{p}}$ on X will have stalk $\mathcal{O}^{\bar{p}}_x = I^{\bar{p}}H_n(X, X - \{x\}; R)$ over the point $x \in X$, and the topology on $\mathcal{O}^{\bar{p}}$ is given so that if $U \subset X$ is any open subset of X and $\xi \in I^{\bar{p}}H_n(X, X - \bar{U}; R)$ is any homology class, then the union of images of ξ in $I^{\bar{p}}H_n(X, X - \{x\}; R)$ as x runs over all points of U is an open subset of $\mathcal{O}^{\bar{p}}$. One can then check that this collection of subsets of $\mathcal{O}^{\bar{p}}$ generates a topology such that the projection $\pi : \mathcal{O}^{\bar{p}} \to X$ is a local homeomorphism; see[3] [37, Example I.1.11] and, more generally, [37, Section I.1]. We call $\mathcal{O}^{\bar{p}}$ the *R-orientation sheaf on X with perversity $\bar{p}$.*

If $\pi : \mathcal{S} \to Y$ is a sheaf over a space Y, then a *section* of Y over the subset $V \subset Y$ is a map $\mathfrak{s} : V \to Y$ such that $\pi\mathfrak{s}(y) = y$ for all $y \in V$. The sections over V constitute an R-module

[3] For the reader either familiar with some sheaf theory or who wishes to compare our rough description here with the details in [37], our sheaf $\mathcal{O}^{\bar{p}}$ is really the sheafification of the presheaf[4] $U \to I^{\bar{p}}H_n(X, X - \bar{U}; R)$. The stalk of this presheaf at $x \in X$ is $\varinjlim_{x \in U} I^{\bar{p}}H_n(X, X - \bar{U}; R)$, but this module is isomorphic to $I^{\bar{p}}H_n(X, X - \{x\}; R)$. To see this, recall that we are free to replace this direct limit with one over a cofinal system of distinguished neighborhoods of x. We can choose such a cofinal sequence of distinguished neighborhoods, $\cdots \supset N_i \supset N_{i+1} \supset \cdots$, so that the inclusion-induced maps

$$I^{\bar{p}}H_n(X, X - \bar{N}_i; R) \to I^{\bar{p}}H_n(X, X - \bar{N}_{i+1}; R) \to I^{\bar{p}}H_n(X, X - \{x\}; R)$$

are all isomorphisms. In fact, if x has a distinguished neighborhood $N \cong \mathbb{R}^k \times cL$ then we can choose a sequence of neighborhoods N_i of the form $B_{r_i} \times c_{s_i}L$ with B_{r_i} a ball of radius i centered at $0 \in \mathbb{R}^k$ and $c_{s_i}L$ the subcone $[0, s_i) \times L/\sim$ of cL. Then $\bar{N}_i = \bar{B}_{r_i} \times \bar{c}_{s_i}L$, and we can choose the r_i and s_i to be decreasing sequences converging to zero. This gives a cofinal system of neighborhoods of x, identifying x with $(0, v)$, and, up to excision isomorphisms, each $I^{\bar{p}}H_n(X, X - \bar{N}_i; R) \to I^{\bar{p}}H_n(X, X - \bar{N}_{i+1}; R)$ and each $I^{\bar{p}}H_n(X, X - \bar{N}_i; R) \to I^{\bar{p}}H_n(X, X - \{x\}; R)$ is an isomorphism by Lemma 6.4.17. So the direct system $I^{\bar{p}}H_n(X, X - \bar{N}_i; R)$ is isomorphic to the constant direct system with module $I^{\bar{p}}H_n(X, X - \{x\}; R)$ for each index, and consequently the direct limit is isomorphic to $I^{\bar{p}}H_n(X, X - \{x\}; R)$.

[4] We use the closure $\bar{U}$ here because it will be useful in later arguments for $X - \bar{U}$ to be an open set, but just to define $\mathcal{O}^{\bar{p}}$ we could have used $U \to I^{\bar{p}}H_n(X, X - U; R)$ instead.

via the algebraic operations on the sheaf. If $\mathcal{O}^{\bar{p}}$ is the orientation sheaf over the CS set X and $U \subset X$ is an open subset, then an element $\xi \in I^{\bar{p}}H_n(X, X - \bar{U}; R)$ determines a section $\mathfrak{s}_\xi$ of $\mathcal{O}^{\bar{p}}$ by the assignment that takes $x \in U$ to the image of ξ in $I^{\bar{p}}H_n(X, X - \{x\}; R)$: it is immediate that $\pi\mathfrak{s}_\xi$ is the identity on U. Furthermore, by the construction discussed in the preceding paragraph, the set of points $\mathfrak{s}_\xi(U)$ is an open neighborhood of $\mathfrak{s}_\xi(z)$ for any z in U. Suppose we fix $z \in U$. As π is a local homeomorphism, it follows that π restricts to a homeomorphism from some possibly smaller open neighborhood $W \subset \mathfrak{s}_\xi(U)$ of $\mathfrak{s}_\xi(z)$ onto $\pi(W)$, which is an open neighborhood of z in U. One can check that the restriction of $\mathfrak{s}_\xi$ to $\pi(W)$ is the inverse homeomorphism to the restriction of π to W, and from this one deduces that $\mathfrak{s}_\xi$ is continuous on $\pi(W)$. But z was arbitrary, so $\mathfrak{s}_\xi$ is continuous on U.

Remark 8.1.13 A useful consequence of the local homeomorphism property of the sheaf map $\pi : \mathcal{S} \to Y$ is that if $\mathfrak{s}$ and $\mathfrak{t}$ are any two sections of $\mathcal{S}$ defined on an open set $U \subset Y$ and if $\mathfrak{s}(y) = \mathfrak{t}(y)$ for some $y \in U$, then $\{z \in U \mid \mathfrak{s}(z) = \mathfrak{t}(z)\}$ is an *open* subset of U; see [230, Section II.2.1]. This takes a bit of getting used to for those of us who generally work with Hausdorff topologies!

In particular, for our sheaf $\mathcal{O}^{\bar{p}}$ over a CS set X, this has the following consequence: Let us fix $x \in X$, and suppose that $\mathfrak{t}$ is a section of $\mathcal{O}^{p}$ over some open set containing x. By definition, $\mathfrak{t}(x)$ is an element of $I^{\bar{p}}H_n(X, X - \{x\}; R)$. Let $\xi \in I^{\bar{p}}S_n(X; R)$ be a chain so that $[\xi] = \mathfrak{t}(x) \in I^{\bar{p}}H_n(X, X - \{x\}; R)$. Let W_x be a neighborhood of x and W_∂ a neighborhood of $|\partial\xi|$ with $W_1 \cap W_\partial = \emptyset$; such a W_x and W_∂ exist by Corollary 2.3.17. Now ξ also represents a class in $I^{\bar{p}}H_n(X, X - \bar{W}_x; R)$ and so an image class in $I^{\bar{p}}H_n(X, X - \{z\}; R)$ for any $z \in W_x$. By the discussion just above, the assignment $z \to [\xi] \in I^{\bar{p}}H_n(X, X - \{z\}; R)$ determines a continuous section $\mathfrak{s}_\xi$ of $\mathcal{O}^{\bar{p}}$ over the open set W_x. But by definition $\mathfrak{s}_\xi(x) = \mathfrak{t}(x)$, and so, by the observation of the preceding paragraph, $\mathfrak{t}$ and $\mathfrak{s}_\xi$ are equal on some neighborhood of x. This argument shows that for any point $x \in X$ and any section $\mathfrak{t}$ of $\mathcal{O}^{\bar{p}}$ defined in a neighborhood of x there is a possibly smaller neighborhood V_x of x and a class $[\xi] \in I^{\bar{p}}H_n(X, X - \bar{V}_x; R)$ such that for all $z \in V_x$ the section value $\mathfrak{t}(z)$ is equal to the image of $[\xi]$ in $I^{\bar{p}}H_n(X, X - \{z\}; R)$. This fact will be useful below, for example in the proof of Lemma 8.1.14.

(b) Homological Theorems

Now, we turn to a version of Lemma 8.1.2 in the stratified case, which will be Lemma 8.1.16, below. The proof of Lemma 8.1.16, which concerns global intersection homology classes, in particular fundamental classes, will be somewhat intertwined with another lemma, Lemma 8.1.14, which is concerned with properties of the orientation sheaf and so with local intersection homology. We will perform an induction on depth that requires Lemma 8.1.16 at depth $d - 1$ to prove Lemma 8.1.14 at depth d, which is needed for Lemma 8.1.16 at depth d, and so on. Thus, we will state both results together, as well as the theorem to which they lead, which will be our stratified version of Theorem 8.1.3; then we will move on to the proofs, followed by some corollaries.

We remark that even though Lemma 8.1.2 has some parts that do not require orientability, our main interest here is in orientable pseudomanifolds, and so we will make that assumption throughout in order to avoid making our web of intertwined arguments any more complicated than it already is. Such an assumption will also be useful in our induction steps. Additionally, while our construction of fundamental classes will require perversities $\bar{p}$ with $\bar{p} \geq 0$, we will also prove some results concerning intersection homology in degrees greater than $\dim(X)$ for arbitrary perversities. Throughout the proofs in this section, we will use $\bar{q}$ to denote an arbitrary perversity and $\bar{p}$ to denote a perversity with $\bar{p} \geq 0$.

Here is our lemma concerning properties of the orientation sheaf $\mathcal{O}^{\bar{p}}$:

Lemma 8.1.14 *Let R be a Dedekind domain, and let X be an R-oriented n-dimensional stratified pseudomanifold. Let $\bar{p}$ be a perversity with $\bar{p} \geq \bar{0}$. Then the following statements hold:*

1. *For all $x \in X$ and all $i > n$, $I^{\bar{q}}H_i(X, X - \{x\}; R) = 0$ for* any *perversity $\bar{q}$.*
2. *Any section defined over $X - \Sigma_X$ of the sheaf $\mathcal{O}^{\bar{p}}$ extends uniquely to a section of $\mathcal{O}^{\bar{p}}$ on all of X. In particular, there is a unique global section $\mathfrak{o}^{\bar{p}}$ of the sheaf $\mathcal{O}^{\bar{p}}$ that restricts to the given R-orientation on $X - \Sigma_X$, and if a section $\mathfrak{s}$ of $\mathcal{O}^{\bar{p}}$ is such that $\mathfrak{s}(x) = 0$ for all $x \in X - \Sigma_X$, then $\mathfrak{s}(x) = 0$ for all $x \in X$.*

Definition 8.1.15 We call the section $\mathfrak{o}^{\bar{p}}$ of $\mathcal{O}^{\bar{p}}$ the *orientation section* of X with respect to the perversity $\bar{p}$.

Now we have our lemma concerning fundamental classes:

Lemma 8.1.16 *Let R be a Dedekind domain, and let X be an R-oriented n-dimensional stratified pseudomanifold. Let $\bar{p}$ be a perversity with $\bar{p} \geq \bar{0}$. Let $K \subset X$ be a compact subset. Then the following hold:*

1. *$I^{\bar{q}}H_i(X, X - K; R) = 0$ for $i > n$ and for* any *perversity $\bar{q}$.*
2. *Given a section $\mathfrak{s}$ of $\mathcal{O}^{\bar{p}}$ over X, there is a unique class $\gamma \in I^{\bar{p}}H_n(X, X - K; R)$ whose image in $I^{\bar{p}}H_n(X, X - \{x\}; R)$ is $\mathfrak{s}(x)$ for any $x \in K$. In particular:*
   (a) *if $\mathfrak{o}^{\bar{p}}$ is an R-orientation section for X then there is a unique class $\Gamma_K^{\bar{p}} \in I^{\bar{p}}H_n(X, X - K; R)$ whose image in $I^{\bar{p}}H_n(X, X - \{x\}; R)$ is $\mathfrak{o}^{\bar{p}}(x)$, for any $x \in K$, and*
   (b) *if $\gamma \in I^{\bar{p}}H_n(X, X - K; R)$ then γ restricts to 0 in $I^{\bar{p}}H_n(X, X - \{x\}; R)$ for all $x \in K$ if and only if $\gamma = 0$.*

Definition 8.1.17 We call the class Γ_K of Lemma 8.1.16 the *fundamental class of X over K* with respect to the chosen R-orientation.

These results lead to the following important theorem:

Theorem 8.1.18 *Let R be a Dedekind domain, and let X be a* compact *R-oriented n-dimensional stratified pseudomanifold with perversity $\bar{p} \geq \bar{0}$. Then the following statements hold:*

1. *$I^{\bar{q}}H_i(X;R) = 0$ for $i > n$ and for* any *perversity $\bar{q}$.*
2. *There is a unique class $\Gamma_X^{\bar{p}} \in I^{\bar{p}}H_n(X;R)$ whose image in $I^{\bar{p}}H_n(X, X - \{x\}; R)$, for any $x \in X$, corresponds to the image of the orientation section $\mathfrak{o}^{\bar{p}}(x)$.*
3. *If $\{x_j\}_{j=1}^m$ is a collection of points of X, one in each regular stratum, then $I^{\bar{p}}H_n(X;R) \cong \bigoplus_j I^{\bar{p}}H_n(X, X - \{x_j\}; R) \cong R^m$ via the map that takes an element of $I^{\bar{p}}H_n(X;R)$ to the direct sum of its images in $I^{\bar{p}}H_n(X, X - \{x_j\}; R)$.*

Definition 8.1.19 We call the class $\Gamma_X^{\bar{p}}$ of Lemma 8.1.16 the *fundamental class of X with respect to the chosen R-orientation and perversity $\bar{p}$.* If $\bar{p} = \bar{0}$, we write simply Γ_X and call Γ_X the *fundamental class of X with respect to the chosen R-orientation.*

Proof of Theorem 8.1.18, assuming Lemmas 8.1.14 and 8.1.16 The first two statements follow immediately from Lemma 8.1.16 by taking $K = X$ and by using Lemma 8.1.14 to guarantee the existence of the orientation section $\mathfrak{o}^{\bar{p}}$.

For the last statement, we know by Proposition 6.3.47 that if $\{\mathcal{R}_j\}$ are the regular strata of X then $I^{\bar{p}}H_n(X;R) \cong \bigoplus_j I^{\bar{p}}H_n(\bar{\mathcal{R}}_j;R)$. Let $\Gamma_{X,j}^{\bar{p}}$ denote the image of $\Gamma_X^{\bar{p}}$ in the summand $I^{\bar{p}}H_n(\bar{\mathcal{R}}_j;R)$. As $\Gamma_{X,j}^{\bar{p}}$ is supported in $\bar{\mathcal{R}}_j$, and as the image of $\Gamma_X^{\bar{p}}$ in $I^{\bar{p}}H_n(X, X - \{x_j\}; R)$ is the generator $\mathfrak{o}^{\bar{p}}(x_j)$, it follows that the image of $\Gamma_{X,j}^{\bar{p}}$ in $I^{\bar{p}}H_n(X, X - \{x_j\}; R)$ is also $\mathfrak{o}^{\bar{p}}(x_j)$, while the image of $\Gamma_{X,j}^{\bar{p}}$ in $I^{\bar{p}}H_n(X, X - \{x_k\}; R)$ is 0 for $k \neq j$. Therefore, the map $I^{\bar{p}}H_n(X;R) \to \bigoplus_j I^{\bar{p}}H_n(X, X - \{x_j\}; R)$ that takes an element to the sum of its images in $I^{\bar{p}}H_n(X, X - \{x_j\}; R)$ is surjective, with $\Gamma_{X,j}^{\bar{p}}$ mapping onto the element of R^m that is a generator in the jth slot and 0 in the other slots.

For injectivity we observe that, for any $\xi \in I^{\bar{p}}H_n(X;R)$ there is a section $\mathfrak{s}_\xi$ of $\mathcal{O}^{\bar{p}}$ given by letting $\mathfrak{s}_\xi(x)$ be the image of ξ in $I^{\bar{p}}H_n(X, X - \{x\}; R)$. Suppose the image of some ξ is 0 in each $I^{\bar{p}}H_n(X, X - \{x_j\}; R)$ so that $\mathfrak{s}_\xi(x_j) = 0$ for each j. As $\mathcal{O}^{\bar{p}}$ is the trivial bundle over each regular stratum, due to X being R-oriented by assumption, we must have $\mathfrak{s}_\xi(x) = 0$ for all $x \in X - \Sigma_X$. By item (2) of Lemma 8.1.14, the section $\mathfrak{s}_\xi$ is thus the 0 section on all of X. So, by item (2) of Lemma 8.1.16, the class ξ must be 0 in $I^{\bar{p}}H_n(X;R)$. This demonstrates that our map $I^{\bar{p}}H_n(X;R) \to \bigoplus_j I^{\bar{p}}H_n(X, X - \{x_j\}; R)$ is injective and so an isomorphism. □

We now turn to proving the two lemmas, Lemma 8.1.14 and Lemma 8.1.16. As noted above, the proofs are inductively intertwined. They also utilize that the lemmas, at depth d, together prove Theorem 8.1.18 at depth d. To start the induction, we observe that both lemmas and the theorem are true for stratified pseudomanifolds of depth zero, i.e. manifolds, by [125, Theorem 3.26 and Lemma 3.27]. We now turn to showing that Lemma 8.1.14 holds for an X of depth $d > 0$ under the assumption that Lemmas 8.1.14 and 8.1.16 and Theorem 8.1.18 hold on all stratified pseudomanifolds of depth $< d$.

Proof of Lemma 8.1.14, given the induction assumptions We begin with the first statement of the lemma, noting that it is immediate when $x \in X - \Sigma_X$, as, in this case, x has a

Euclidean neighborhood. So, by excision, $I^{\bar{q}}H_i(X, X - \{x\}; R) \cong I^{\bar{q}}H_i(\mathbb{R}^n, \mathbb{R}^n - \{x\}; R) \cong H_i(\mathbb{R}^n, \mathbb{R}^n - \{x\}; R)$; of course, this is isomorphic to R if $i = n$ and 0 otherwise.

Next, suppose x has a distinguished neighborhood of the form $\mathbb{R}^k \times cL$ with $L \neq \emptyset$ a compact stratified pseudomanifold of dimension $n-k-1$. By excision, $I^{\bar{q}}H_i(X, X-\{x\}; R) \cong I^{\bar{q}}H_i(\mathbb{R}^k \times cL, \mathbb{R}^k \times cL - \{x\}; R)$. But this pair is homeomorphic to the product of the two pairs of spaces $(\mathbb{R}^k, \mathbb{R}^k - \{0\})$ and $(cL, cL - \{v\})$, using the convention $(A, B) \times (C, D) = (A \times C, (A \times D) \cup (B \times C))$. So, by the Künneth Theorem with one term being a manifold (Theorem 6.3.19), we have

$$I^{\bar{q}}H_i(\mathbb{R}^k \times cL, \mathbb{R}^k \times cL - \{x\}; R) \cong H_k(\mathbb{R}^k, \mathbb{R}^k - \{0\}; R) \otimes I^{\bar{q}}H_{i-k}(cL, cL - \{v\}; R)$$
$$\cong I^{\bar{q}}H_{i-k}(cL, cL - \{v\}; R).$$

We now apply the cone formula (Corollary 6.2.15), by which $I^{\bar{q}}H_{i-k}(cL, cL - \{v\}; R) \cong I^{\bar{q}}H_{i-k-1}(L; R)$ if $i - k > n - k - \bar{q}(\{v\}) - 1$, i.e. if $i > n - \bar{q}(\{v\}) - 1$, and is 0 otherwise.

If L is R-oriented, or even R-orientable, $I^{\bar{q}}H_j(L; R) = 0$ for $j > n-k-1$ by the induction assumption and Theorem 8.1.18. As $i > n$ implies $i - k - 1 > n - k - 1$, this would imply the first statement of the lemma, independent of which case of the cone formula applies for our particular choice of j. To see that L must be R-orientable if X is, we note that any orientation on $X - \Sigma_X$ restricts to an R-orientation on the union of regular strata of $\mathbb{R}^k \times cL$. But, by the local structure for CS sets, this union of regular strata must be isomorphic to $\mathbb{R}^k \times (0, 1) \times (L - \Sigma_L)$. We now invoke the fact that a product manifold is orientable if and only if all its manifold factors are[5]. So $L - \Sigma_L$ is R-orientable and thus L is R-orientable, which is enough to draw the necessary conclusion from Theorem 8.1.18. This completes the proof of the first statement of the lemma.

We turn to the second statement of the lemma. For this case, we care about $i = n$, and we are now in the setting where we have assumed that $\bar{p} \geq 0$. It is thus true that $n > n - \bar{p}(\{v\}) - 1$, and therefore, $I^{\bar{p}}H_n(X, X - \{x\}; R) \cong I^{\bar{p}}H_{n-k-1}(L; R)$ in this case.

The second sentence of the second statement of the lemma follows directly from the first sentence and from the fact that every sheaf has a zero section [37, page 4]. We must show that any section $\mathfrak{s}$ of $\mathcal{O}^{\bar{p}}$ over $X - \Sigma_X$ extends uniquely to all of X. So, let $\mathfrak{s}$ be such a section,

[5] This is the first exercise in Section VI.7 of Bredon's book [38]. Here's a sketch of the proof: Suppose $M_1^{m_1} \times M_2^{m_2}$ is a product of manifolds. Letting B_i be a local Euclidean ball in M_i, the local isomorphisms

$$H_{m_1}(M_1, M_1 - B_1; R) \otimes H_{m_2}(M_2, M_2 - B_2; R) \xrightarrow{\times} H_{m_1+m_2}(M_1 \times M_2, M_1 \times M_2 - B_1 \times B_2; R)$$

induce an isomorphism of bundles (locally constant sheaves) $\mathcal{O}_1 \hat{\otimes} \mathcal{O}_2 \to \mathcal{O}_\times$, where $\mathcal{O}_i$ is the orientation bundle of M_i and $\mathcal{O}_\times$ is the orientation bundle of the product. The "total tensor product" $\mathcal{O}_1 \hat{\otimes} \mathcal{O}_2$ has fiber $\mathcal{O}_{1,x_1} \otimes \mathcal{O}_{2,x_2}$ at $(x_1, x_2) \in M_1 \times M_2$; it can be formally defined as the tensor product over $M_1 \times M_2$ of the pullback bundles $\pi_1^*\mathcal{O}_1 \otimes \pi_2^*\mathcal{O}_2$, where $\pi_i : M_1 \times M_2 \to M_i$ is the projection. If $M_1 \times M_2$ is R-orientable, $\mathcal{O}_\times$ is the constant bundle and has a global orientation section $\mathfrak{o}_\times$. Consider now the subspace $\{x_1\} \times M_2$ for some fixed $x_1 \in M_1$. The restriction to this subspace of $\mathcal{O}_1 \hat{\otimes} \mathcal{O}_2$ has the constant bundle with fiber R in the first factor, and so the restriction is isomorphic to $\mathcal{O}_2$, living on a copy of M_2. But, as $\mathcal{O}_\times$ was constant, this restricted bundle is also constant, so $\mathcal{O}_2$ is constant. Therefore, M_2 is R-orientable. The argument for M_1 is identical. Conversely, if $\mathfrak{o}_i$ is a global R-orientation section in $\mathcal{O}_i$, then $\mathfrak{o}_1 \hat{\otimes} \mathfrak{o}_2$, which takes values $\mathfrak{o}_1(x_1) \otimes \mathfrak{o}_2(x_2)$ at (x_1, x_2), is an R-orientation section over $M_1 \times M_2$.

and let $x \in \Sigma_X$. We must define $\mathfrak{s}(x)$ for each such x and show that, overall, we obtain a well-defined global section. We choose a distinguished neighborhood N of x, which we identify with $\mathbb{R}^k \times cL$ via a filtered homeomorphism. We can assume that $x \in N$ has the coordinates $(0, v) \in \mathbb{R}^k \times cL$.

We must assign to x an element of $I^{\bar{p}}H_n(X, X - \{x\}; R) \cong I^{\bar{p}}H_n(\mathbb{R}^k \times cL, \mathbb{R}^k \times cL - \{(0, v)\}; R)$. Within N, we have a smaller compact neighborhood of x of the form $\bar{B}_r \times \bar{c}_s$, where $\bar{B}_r$ is a closed Euclidean ball of radius r in $\mathbb{R}^k$ and where $\bar{c}_s L = [0, s] \times L/\sim$, for $0 < s < 1$, is a smaller closed cone on L within $cL = [0, 1) \times L/\sim$. Let $\bar{B}_r \times \bar{c}_s L$ be denoted by $\bar{N}'$. We can also let $N' = B_r \times c_s L$ be the interior of $\bar{N}'$. Suppose that $\{\mathcal{L}_\alpha\}$ is the collection of the regular strata of L, which is finite as L is compact, and that $\mathcal{R}_\alpha = \mathbb{R}^k \times (0, 1) \times \mathcal{L}_\alpha$ are the corresponding regular strata of N. For each α, let x_α be some point in $\mathcal{R}_\alpha \cap \bar{N}'$. We claim that there are isomorphisms

$$I^{\bar{p}}H_n(X, X - \{x\}; R) \xleftarrow{\cong} I^{\bar{p}}H_n(X, X - \bar{N}'; R) \xrightarrow{\cong} \bigoplus_\alpha I^{\bar{p}}H_n(X, X - \{x_\alpha\}; R) \cong R^m. \quad (8.1)$$

Here, the map to the left is induced by inclusion, and the map to the right is the direct sum of maps induced by inclusions. The claim will be sufficiently useful later that we separate it out as a lemma in its own right, Lemma 8.1.20, which we prove just below. For now, we finish the proof of Lemma 8.1.14 assuming this claim.

By (8.1), there is a unique $\xi \in I^{\bar{p}}H_n(X, X - \bar{N}'; R)$ that restricts to the given $\mathfrak{s}(x_\alpha)$ in each $I^{\bar{p}}H_n(X, X - \{x_\alpha\}; R)$. In fact, as we vary through all $z \in N'$, the image of ξ in $I^{\bar{p}}H_n(X, X - \{z\}; R)$, $z \in N'$, determines a section $\mathfrak{s}_\xi$ of $\mathcal{O}^{\bar{p}}$ over N'. As $\mathcal{O}^{\bar{p}}$ is constant over the regular strata of X and as there is a representative x_α in each regular stratum of N', the section $\mathfrak{s}_\xi$ must agree with $\mathfrak{s}$ at all regular stratum points of N'. Thus $\mathfrak{s}_\xi$ extends $\mathfrak{s}$ to a section over $(X - \Sigma_X) \cup N'$. Let us show this is a unique extension. Suppose $\mathfrak{t}$ is any other section defined on a neighborhood of x that extends $\mathfrak{s}$ over a neighborhood of x. By Remark 8.1.13, there must be some open neighborhood V of x and some $\zeta \in I^{\bar{p}}H_n(X, X - \bar{V}; R)$ such that, for each $z \in V$, the section value $\mathfrak{t}(z)$ is the image of ζ in $I^{\bar{p}}H_n(X, X - \{z\}; R)$. But now let N_1 be an even smaller distinguished neighborhood of x inside $V \cap N'$ and with its own smaller $\bar{N}_1'$. Then ξ and ζ both map to elements of $I^{\bar{p}}H_n(X, X - \bar{N}_1'; R)$, and they each have the same images in $I^{\bar{p}}H_n(X, X - \{z\}; R)$ for all z in regular strata of N_1' by assumption. Therefore, applying Lemma 8.1.20 (or, equivalently, equation (8.1)) again, this time utilizing N_1 and N_1' as the neighborhoods in the lemma, we see that ξ and ζ must represent the same element of $I^{\bar{p}}H_n(X, X - \{x\}; R)$. Therefore, $\mathfrak{s}_\xi(x) = \mathfrak{t}(x)$. This shows that there is only one possible value of $\mathcal{O}^{\bar{p}}_x = I^{\bar{p}}H_n(X, X - \{x\}; R)$ that extends $\mathfrak{s}$ from the regular strata to some neighborhood of x.

Finally, we must show that we obtain a global section that extends $\mathfrak{s}$ to all of X. We have seen that for every $x \in X$ there is a section $\mathfrak{s}_x$ defined on a neighborhood U_x of x and such that $\mathfrak{s}_x$ agrees with $\mathfrak{s}$ at all points in regular strata of U_x. If U_x and U_y are two such neighborhoods, then the uniqueness of the extensions, proven in the preceding paragraph, implies that $\mathfrak{s}_x = \mathfrak{s}_y$ on the overlap $U_x \cap U_y$: for any point $z \in U_x \cap U_y$, the sections $\mathfrak{s}_x$ and

$\mathfrak{s}_y$ are both defined in neighborhoods of z and, by assumption, extend $\mathfrak{s}$ from the regular strata to a neighborhood of z. So, by the result of the preceding paragraph, $\mathfrak{s}_x(z) = \mathfrak{s}_y(z)$. It is now another fundamental property of sheaf theory that, given such local sections that agree on overlaps, they can be patched together to provide a global section; see [37, Section I.1]. In fact, this property is sometimes used to define sheaves.

This completes the proof of Lemma 8.1.14, given the induction assumptions. □

As we have seen, the following lemma was needed in the proof of Lemma 8.1.14, but we will use it several more times in this section. For Lemma 8.1.20, we will use a slightly more general $\bar{N}'$, which will be needed below.

Lemma 8.1.20 *Let R be a Dedekind domain, and let X be an R-oriented n-dimensional stratified pseudomanifold. Suppose $x \in \Sigma_X$ has a distinguished neighborhood $N \cong \mathbb{R}^k \times cL$. Suppose x is contained in a compact subset of $\mathbb{R}^k \times cL$ of the form $\bar{N}' = C \times \bar{c}_s L \subset \mathbb{R}^k \times cL$, where C is a compact convex subset of $\mathbb{R}^k$ and $\bar{c}_s L = [0,s] \times L/\sim$, for $0 < s < 1$, is a smaller closed cone on L within $cL = [0,1) \times L/\sim$. Let $\{\mathcal{R}_\alpha\}_{\alpha=1}^m$ denote the regular strata of $\bar{N}'$, which are bijective with the regular strata of L, and let $\{x_\alpha\}_{\alpha=1}^m$ be any collection of points with $x_\alpha \in \mathcal{R}_\alpha$. Then for any perversity $\bar{p}$ with $\bar{p} \geq \bar{0}$ the inclusion maps induce isomorphisms*

$$I^{\bar{p}}H_n(X, X - \{x\}; R) \xleftarrow{\cong} I^{\bar{p}}H_n(X, X - \bar{N}'; R) \xrightarrow{\cong} \bigoplus_\alpha I^{\bar{p}}H_n(X, X - \{x_\alpha\}; R) \cong R^m.$$

Therefore, if $\xi \in I^{\bar{p}}H_n(X, X - \bar{N}'; R)$ and $\mathfrak{s}_\xi$ is the section of $\mathcal{O}^{\bar{p}}$ over $\bar{N}'$ such that $\mathfrak{s}_\xi(z)$ is the image of ξ in $I^{\bar{p}}H_n(X, X - \{z\}; R)$ for any $z \in \bar{N}'$, then the value $\mathfrak{s}_\xi(x)$ is completely determined by the collection of values $\{s_\xi(x_\alpha)\}$, and the value $\mathfrak{s}_\xi(x)$ determines the values $\{s_\xi(x_\alpha)\}$ for any *collection of x_α satisfying the hypotheses.*

Proof For the proof, we continue to assume we are within the overall inductive scenario of this section, so that Lemmas 8.1.14 and 8.1.16 and Theorem 8.1.18 are all available for spaces of depth less than that of X, which we assume has depth d. We observe quickly that the "therefore" statement of this lemma follows directly from the existence of the stated isomorphisms of the lemma. Further, before getting into the details, we remind the reader that when we speak of the regular strata of $\bar{N}'$ we mean its intersections with the regular strata of N. In this case, the regular strata of $\bar{N}'$ have the form $C \times (0,s] \times \mathcal{L}_\alpha$, where the $\mathcal{L}_\alpha$ are the regular strata of L.

For the map to the left in the diagram, by excision it is isomorphic to the map

$$I^{\bar{p}}H_n(\mathbb{R}^k \times cL, \mathbb{R}^k \times cL - \bar{N}; R) \to I^{\bar{p}}H_n(\mathbb{R}^k \times cL, \mathbb{R}^k \times cL - \{x\}; R)$$

induced by inclusion. Without loss of generality, we can write $x = (0, v)$, and so this intersection homology map can be written in terms of the inclusion map from the product of pairs[6] $(\mathbb{R}^k, \mathbb{R}^k - C) \times (cL, cL - \bar{c}_s L)$ to the product of pairs $(\mathbb{R}^k, \mathbb{R}^k - \{0\}) \times (cL, cL - \{v\})$. The intersection homology map is thus an isomorphism by Lemma 6.4.17.

[6] Recall that $(A, B) \times (E, F)$ means $(A \times E, (A \times F) \cup (B \times E))$.

For the maps to the right in the claim of the lemma, which is the direct sum of the maps induced by the inclusions, we consider the following diagram:

$$\begin{array}{ccccc} I^{\bar{p}}H_{n-k-1}(L;R) & \xleftarrow{\cong} & \bigoplus I^{\bar{p}}H_{n-k-1}(\bar{\mathcal{L}}_\alpha;R) & \xrightarrow[\cong]{\phi} & \begin{array}{c}\bigoplus I^{\bar{p}}H_n(\mathbb{R}^k \times c\bar{\mathcal{L}}_\alpha, \\ \mathbb{R}^k \times c\bar{\mathcal{L}}_\alpha - \{x_\alpha\};R)\end{array} \\ \Big\downarrow{\scriptstyle\cong} & & & & \Big\downarrow{\scriptstyle\cong} \\ I^{\bar{p}}H_n(N,N-\bar{N}';R) & & \longrightarrow & & \bigoplus_\alpha I^{\bar{p}}H_n(N,N-\{x_\alpha\};R). \end{array} \tag{8.2}$$

By excisions, the bottom map is isomorphic to the map

$$I^{\bar{p}}H_n(X,X-\bar{N}';R) \xrightarrow{\cong} \bigoplus_\alpha I^{\bar{p}}H_n(X,X-\{x_\alpha\};R)$$

of the claim, so it suffices to show that this diagram commutes and that the other maps of the diagram are all isomorphisms.

We let the left-hand vertical map be the composition of the inverse of the isomorphism

$$\begin{aligned} I^{\bar{p}}H_{n-k}(cL,cL-\bar{c}_sL;R) \xrightarrow{\cong} I^{\bar{p}}H_{n-k}(cL,cL-\{v\};R) &\xrightarrow{\partial_*} I^{\bar{p}}H_{n-k-1}(cL-\{v\};R) \\ &\cong I^{\bar{p}}H_{n-k-1}(L;R), \end{aligned}$$

which is an isomorphism from the relative cone formula (Corollary 6.2.15 and using that $\bar{p} \geq 0$) and stratified homotopy invariance, with the Künneth isomorphism

$$\begin{aligned} I^{\bar{p}}H_{n-k}(cL,cL-\bar{c}_sL;R) &\cong R \otimes I^{\bar{p}}H_{n-k}(cL,cL-\bar{c}_sL;R) \\ &\cong H_k(\mathbb{R}^k,\mathbb{R}^k-C;R) \otimes I^{\bar{p}}H_{n-k}(cL,cL-\bar{c}_sL;R) \\ &\xrightarrow{\cong} I^{\bar{p}}H_{n-k}(\mathbb{R}^k \times cL;\mathbb{R}^k \times cL - \bar{N}';R). \end{aligned}$$

Tracing through these isomorphisms, the left-hand vertical map thus takes an element $\zeta \in I^{\bar{p}}H_{n-k}(L;R)$ to an element of

$$I^{\bar{p}}H_n(N,N-\bar{N}';R) \cong I^{\bar{p}}H_n(\mathbb{R}^k \times cL, \mathbb{R}^k \times cL - C \times \bar{c}_sL;R)$$

that can be represented by the product chain $\eta \times \mathcal{Z}$, where η is a generator of $H_k(\mathbb{R}^k,\mathbb{R}^k - C;R) \cong R$ and where $\mathcal{Z}$ is some allowable chain in cL whose boundary is a copy of ζ that we can suppose lives in a copy of L at the coordinates $\{0\} \times \{s'\} \times L \subset \mathbb{R}^k \times cL$ with $s < s' < 1$.

For convenience below, we will now construct an explicit $\mathcal{Z}$ as described in the preceding paragraph. For this, let us write each x_α as $x_\alpha = (y_\alpha, t_\alpha, z_\alpha)$ with $y_\alpha \in \mathbb{R}^k$, $t_\alpha \in (0,1)$, and $z_\alpha \in \mathcal{L}_\alpha$. So this provides coordinates for x_α in the distinguished neighborhood $N \cong \mathbb{R}^k \times cL$. Let s' be chosen as above with $s < s' < 1$, and let s'' be such that $0 < s'' < \min\{t_\alpha\}$. Then for all α we have a 1-simplex generator of $H_1((0,1),(0,1)-\{t_\alpha\};R)$ given by the linear homeomorphism $e : [0,1] \to [s'',s'] \subset (0,1)$ with $e(0) = s''$ and $e(1) = s'$. For a chain ζ in L, let ζ_t denote ζ thought of as living in a copy of L at cone coordinate t in cL. We propose to take $\mathcal{Z} = \bar{c}\zeta_{s''} + e \times \zeta$, where $\bar{c}\zeta_{s''}$ is the singular cone on $\zeta_{s''}$ (see the construction of Example 3.4.7) and where $e \times \zeta$ is the cross product using that $(0,1) \times L$ is filtered homeomorphic to the subspace $cL - \{v\}$ of cL. Then $\partial(\bar{c}\zeta_{s''} + e \times \zeta) = \zeta_{s''} + \zeta_{s'} - \zeta_{s''} = \zeta_{s'}$, as desired. Admittedly, $\bar{c}\zeta_{s''} + e \times \zeta$ is essentially the same chain as $\bar{c}\zeta_{s'}$, but it will be useful in what follows to have a part of this chain with a definitive product structure.

Returning now to Diagram (8.2), we let the right-hand vertical map be a direct sum of excision isomorphisms, excising $\mathbb{R}^k \times (c(L - \bar{\mathcal{L}}_\alpha) - \{v\})$, whose closure is in the complement of $\{x_\alpha\}$. So this map is an isomorphism. The map to the left in the diagram, also induced by inclusions, is an isomorphism by Proposition 6.3.47. For the map labeled ϕ, if $\zeta^\alpha \in I^{\bar{p}}H_{n-k-1}(\bar{\mathcal{L}}_\alpha; R)$, and if we continue to think of L as embedded in N at the coordinates $\{0\} \times \{s'\} \times L \subset \mathbb{R}^k \times cL$ with $s < s' < 1$, then we let this map take ζ^α to $\eta \times \mathcal{Z}^\alpha \in I^{\bar{p}}H_n(\mathbb{R}^k \times c\bar{\mathcal{L}}_\alpha, \mathbb{R}^k \times c\bar{\mathcal{L}}_\alpha - \{x_\alpha\}; R)$, where $\mathcal{Z}^\alpha$ is defined analogously to the earlier $\mathcal{Z}$ and η is a generator of $H_k(\mathbb{R}^k, \mathbb{R}^k - C; R)$, also as above. We observe that, for support reasons, the chain $\eta \times \mathcal{Z}^\alpha$ is automatically zero in each $I^{\bar{p}}H_n(\mathbb{R}^k \times c\bar{\mathcal{L}}_\beta, \mathbb{R}^k \times c\bar{\mathcal{L}}_\beta - \{x_\beta\}; R)$ with $\beta \neq \alpha$. With this definition, the reader can check that it follows from our constructions that the diagram commutes. What remains then is to show that ϕ is an isomorphism.

Consider next the diagram

$$\begin{array}{ccc}
\bigoplus I^{\bar{p}}H_{n-k-1}(\bar{\mathcal{L}}_\alpha; R) & \xrightarrow{\ \phi\ } & \bigoplus I^{\bar{p}}H_n(\mathbb{R}^k \times c\bar{\mathcal{L}}_\alpha, \mathbb{R}^k \times c\bar{\mathcal{L}}_\alpha - \{x_\alpha\}; R) \\
\downarrow \cong & & \uparrow \cong \\
\bigoplus I^{\bar{p}}H_{n-k-1}(\bar{\mathcal{L}}_\alpha, \bar{\mathcal{L}}_\alpha - \{z_\alpha\}; R) & \xrightarrow{\ \cong\ } & \begin{array}{c}\bigoplus I^{\bar{p}}H_n(\mathbb{R}^k \times (0,1) \times \bar{\mathcal{L}}_\alpha, \\ \mathbb{R}^k \times (0,1) \times \bar{\mathcal{L}}_\alpha - \{x_\alpha\}; R)\end{array}
\end{array}$$

The left-hand vertical map, induced by inclusions, is an isomorphism as each $\bar{\mathcal{L}}_\alpha$ is a stratified pseudomanifold (by Lemma 6.3.45) with only one regular stratum and of depth $< d$, so Theorem 8.1.18 applies via our induction assumptions. The right-hand vertical map is a direct sum of excision isomorphisms, in each case excising $\mathbb{R}^k \times \{v\}$. The bottom map comes from the Künneth Theorem (Theorem 6.3.20); in particular, it is the direct sum of the isomorphisms

$$\begin{aligned}
& I^{\bar{p}}H_{n-k-1}(\bar{\mathfrak{L}}_\alpha, \bar{\mathfrak{L}}_\alpha - \{z_\alpha\}; R) \\
&\quad \cong R \otimes I^{\bar{p}}H_{n-k-1}(\bar{\mathfrak{L}}_\alpha, \bar{\mathfrak{L}}_\alpha - \{z_\alpha\}; R) \\
&\quad \cong H_k(\mathbb{R}^k, \mathbb{R}^k - \{y_\alpha\}; R) \otimes H_1((0,1), (0,1) - \{t_\alpha\}; R) \\
&\qquad \otimes I^{\bar{p}}H_{n-k-1}(\bar{\mathfrak{L}}_\alpha, \bar{\mathfrak{L}}_\alpha - \{z_\alpha\}; R) \\
&\quad \cong I^{\bar{p}}H_n(\mathbb{R}^k \times (0,1) \times \bar{\mathcal{L}}_\alpha, \mathbb{R}^k \times (0,1) \times \bar{\mathcal{L}}_\alpha - \{x_\alpha\}; R)
\end{aligned}$$

defined so that if $\zeta^\alpha \in I^{\bar{p}}H_{n-k-1}(\bar{\mathfrak{L}}_\alpha, \bar{\mathfrak{L}}_\alpha - \{z_\alpha\}; R)$ then ζ^α goes to $\eta \times e \times \zeta^\alpha$, where η and e are as above. It remains to see that this last diagram commutes, which will imply that ϕ is an isomorphism. But, by our constructions, each $\phi(\zeta^\alpha)$ is represented by $\eta \times \mathcal{Z}^\alpha =$

$\eta \times (\bar{c}\zeta^{\alpha}_{s''} + e \times \zeta^{\alpha})$, which is equal to $\eta \times e \times \zeta^{\alpha}$ in $I^{\bar{p}}H_n(\mathbb{R}^k \times c\bar{\mathcal{L}}_{\alpha}, \mathbb{R}^k \times c\bar{\mathcal{L}}_{\alpha} - \{x_{\alpha}\}; R)$, as $\bar{c}\zeta^{\alpha}_{s''}$ is supported in $\mathbb{R}^k \times c\bar{\mathcal{L}}_{\alpha} - \{x_{\alpha}\}$. This proves the commutativity and so finishes the proof of Lemma 8.1.20. □

Remark 8.1.21 The computations in the above proof demonstrate one of the problems if we allow $\bar{p}(S) < 0$ for some singular stratum S. In particular, if $x \in S$ and S has dimension k, the computations in the proof show that $I^{\bar{p}}H_i(X, X - \{x\}; R) \cong I^{\bar{p}}H_{i-k}(cL, cL - \{v\}; R)$ which, by the cone formula (Corollary 6.2.15), is zero if $i - k \le n - k - \bar{p}(\{v\}) - 1$, i.e. if $i \le n - \bar{p}(\{v\}) - 1$. But if $\bar{p}(\{v\}) < 0$, this scenario will include $i = n$. This still results in a unique extension of sections of $\mathcal{O}^{\bar{p}}$ to S, as any section will be forced to be zero at each point on S. However, the isomorphism on the right in Lemma 8.1.20 will no longer hold, and this would cause some of the arguments we will see in the proof of Lemma 8.1.16, below, to fall apart.

Proof of Lemma 8.1.16, given the induction assumptions We have already proved that Lemma 8.1.20 holds at depth d under the assumptions that Lemma 8.1.20, Lemma 8.1.16, and Theorem 8.1.18 all hold at depth $d - 1$. So, we may use Lemma 8.1.20 at depth d and the other two results at depth $d - 1$ in the following argument.

We also note that parts (2a) and (2b) of the lemma's statement follow immediately from the main assertion of part (2). For the rest of the proof, we will broadly follow the pattern of the proof of [125, Lemma 3.27], which consists of a few separate steps.

Mayer–Vietoris Reduction Step. As a first step, suppose K, L are compact subsets of X such that the claims of Lemma 8.1.16 hold with respect to K, L, and $K \cap L$. We will show that they then hold for $K \cup L$. Consider the following portion of the long exact Mayer–Vietoris: sequence

$$\begin{aligned} &I^{\bar{q}}H_{n+1}(X, X - (K \cap L); R) \to I^{\bar{q}}H_n(X, X - (K \cup L); R) \\ &\xrightarrow{\Phi} I^{\bar{q}}H_n(X, X - K; R) \oplus I^{\bar{q}}H_n(X, X - L; R) \xrightarrow{\Psi} I^{\bar{q}}H_n(X, X - (K \cap L); R). \end{aligned}$$

Given the assumptions on K, L, and $K \cap L$, we have $I^{\bar{q}}H_i(X, X - K; R) = I^{\bar{q}}H_i(X, X - L; R) = I^{\bar{q}}H_i(X, X - (K \cap L); R) = 0$ for $i > n$. It thus follows from the portion of the sequence further to the left that $I^{\bar{q}}H_i(X, X - (K \cup L); R) = 0$ for $i > n$. We also have that the map labeled Φ is injective.

Continuing to assume the lemma on K, L, and $K \cap L$ and now assuming $\bar{p} \ge 0$ (and replacing $\bar{q}$ with $\bar{p}$ in our Mayer–Vietoris sequence), suppose that $\xi \in I^{\bar{p}}H_n(X, X - (K \cup L); R)$ maps to 0 in $I^{\bar{p}}H_n(X, X - \{x\}; R)$ for all $x \in K \cup L$. Then the image of ξ under Φ in the summand $I^{\bar{p}}H_n(X, X - K; R)$ must map to 0 in each $I^{\bar{p}}H_n(X, X - \{x\}; R)$ for $x \in K$, and analogously for the L summand. By the assumption, this implies that $\Phi(\xi) = 0$, so $\xi = 0$. This implies that any two elements $\xi, \xi' \in I^{\bar{p}}H_n(X, X - (K \cup L); R)$ that map to the same section of $\mathcal{O}^{\bar{p}}$ over $K \cup L$ must be equal, as $\xi - \xi' = 0$ by this argument. This provides the uniqueness part of the second statement of the lemma for $K \cup L$.

Finally, suppose we are given a section $\mathfrak{s}$ of $\mathcal{O}^{\bar{p}}$, and suppose $\gamma_K \in I^{\bar{p}}H_n(X, X - K; R)$

and $\gamma_L \in I^{\bar{p}}H_n(X, X - L; R)$ restrict to this section at points of K or L, respectively. Then $\Psi(\gamma_K, -\gamma_L)$, which is represented by $\gamma_K - \gamma_L$, restricts to $0 \in I^{\bar{p}}H_n(X, X - \{x\}; R)$ for each $x \in K \cap L$. But, again assuming the lemma holds for $K \cap L$, this means that $\Psi(\gamma_K, -\gamma_L) = 0$, and so from the exact sequence there exists a $\gamma_{K \cup L} \in I^{\bar{p}}H_n(X, X - (K \cup L); R)$, with $\Phi(\gamma_{K \cup L}) = (\gamma_K, -\gamma_L)$. As $\Phi(\xi)$ is represented by $(\xi, -\xi)$ in general, we see that $\gamma_{K \cup L}$ must restrict to the values $\mathfrak{s}(x)$ at each $I^{\bar{p}}H_n(X, X - \{x\}; R)$ for $x \in K \cup L$.

Distinguished Neighborhood Reduction Step. Suppose K is any compact subset of X. For each $x \in X$ there is a distinguished neighborhood N_x of x filtered homeomorphic to $\mathbb{R}^k \times cL$, with k and L depending on x. Within this neighborhood, x has smaller compact neighborhoods, say of the form $\bar{N}'_x \cong \bar{B}_r \times \bar{c}_s L$, where $\bar{B}_r$ is a closed ball in $\mathbb{R}^k$ and $\bar{c}_s L$ is the compact subcone of cL up to radius s; let $N'_x \cong B_r \times c_s L$ be the interior of $\bar{N}'_x$. As K is compact, it can be covered by a finite number of the N'_x, and therefore K is the union of a finite number of the compact spaces $\bar{N}'_x \cap K$, with each of these compact spaces being contained in the corresponding distinguished neighborhood N_x.

Suppose we can prove the lemma for any such compact set that is contained in some distinguished neighborhood. For the purposes of this step, let us call such compact sets *distinguished compact sets*. Then, we claim that the lemma will hold for any finite union of such distinguished compact sets by induction and the Mayer–Vietoris discussion above, and hence it will hold for K. To verify the claim, let $\{J_\alpha\}$ be any collection of distinguished compact sets. We have assumed the lemma holds for each J_α. Suppose, as induction hypothesis, that it holds for any union of $\ell - 1$ such sets, and let $\{J_1, \dots, J_\ell\}$ be a collection of ℓ such sets, $\ell > 1$. Then $J_\ell \cap \left(\bigcup_{i=1}^{\ell-1} J_i\right) = \bigcup_{i=1}^{\ell-1}(J_\ell \cap J_i)$. As each $J_\ell \cap J_i$ is a distinguished compact set and as $\bigcup_{i=1}^{\ell-1}(J_\ell \cap J_i)$ is a union of $\ell - 1$ such sets, the lemma holds for $J_\ell \cap \left(\bigcup_{i=1}^{\ell-1}\right)$ by the induction hypothesis. It also holds for J_ℓ and $\bigcup_{i=1}^{\ell-1} J_i$ by our assumptions, so it holds for $\bigcup_{i=1}^{\ell} J_i$ by the Mayer–Vietoris reduction step.

Therefore, as we have shown that K is the union of finitely many distinguished compact sets, the lemma will hold for K, provided that we show that the lemma holds for any single distinguished compact set.

Proof for PM-Convex Sets. We have shown that it now suffices to prove the lemma for any distinguished compact set in X, where a distinguished compact set is a compact set contained within a distinguished neighborhood of some point in x. If K is such a distinguished compact set in X contained in the distinguished neighborhood N, then we have $I^{\bar{p}}H_*(X, X - K; R) \cong I^{\bar{p}}H_*(N, N - K; R)$, and both the orientation bundle $\mathcal{O}^{\bar{p}}$ of X and the hypothesized section $\mathfrak{s}$ restrict to an orientation bundle and section over N. As the requirements of the lemma for K only concern the values of $\mathfrak{s}$ at points in K, it therefore suffices to prove the lemma with N in place of X. In the remaining steps, we are therefore free to assume that $X = \mathbb{R}^k \times cL$. We can also assume that $L \neq \emptyset$, otherwise we know the claimed result holds from the manifold case [125].

Following [100], we next prove the lemma for the case where K is a *PM-convex set* in $X = \mathbb{R}^k \times cL$. These are defined to be subsets K of $\mathbb{R}^k \times cL$ of either of the following forms:

1. $K = C \times \bar{c}_s L$, where $C \subset \mathbb{R}^k$ is a compact convex set and, as usual, $\bar{c}_s L = [0, s] \times L/\sim$ is a closed subcone of cL, or
2. $K = C \times [r, s] \times D$, where $C \subset \mathbb{R}^k$ is a compact convex set, $[r, s]$ is an interval "along the cone line" with $0 < r \le s < 1$, and $D \subset L$ is a compact subset.

As the set-theoretic intersection of products is the product of the intersections and as the intersection of two compact convex subsets of $\mathbb{R}^k$ is a compact convex subset, we see that the intersection of two PM-convex sets is also a PM-convex set. We also observe that every point in X has a PM-convex neighborhood. In fact, by some basic point-set topology, given any $x \in U \subset X$ with U open, there is a PM-convex neighborhood of x contained in U.

We now prove Lemma 8.1.16 in the case where K is a PM-convex set.

First, suppose that K is of the second type, $K = C \times [r, s] \times D$. Then K does not intersect $\mathbb{R}^k \times \{v\} \subset \mathbb{R}^k \times cL = X$, so, by excision,

$$I^{\bar{q}}H_*(X, X - K; R) \cong I^{\bar{q}}H_*(X - (\mathbb{R}^k \times \{v\}), X - (K \cup (\mathbb{R}^k \times \{v\})); R).$$

As $X - (\mathbb{R}^k \times \{v\})$ must have depth $< d$, where d continues to denote the depth of X, the lemma must hold on $X - (\mathbb{R}^k \times \{v\})$ by our induction assumptions. So, for $i > n$,

$$I^{\bar{q}}H_i(X, X - K; R) \cong I^{\bar{q}}H_i(X - (\mathbb{R}^k \times \{v\}), X - (K \cup (\mathbb{R}^k \times \{v\})); R) = 0.$$

Also, for $\bar{p} \ge 0$, any section $\mathfrak{s}$ of $\mathcal{O}^{\bar{p}}$ over X restricts to a section over $X - (\mathbb{R}^k \times \{v\})$. By induction, there is an element of $I^{\bar{p}}H_n(X - (\mathbb{R}^k \times \{v\}), X - (K \cup (\mathbb{R}^k \times \{v\})); R)$ that is compatible with this section over K, and the excision isomorphism therefore yields such an element of $I^{\bar{p}}H_n(X, X - K; R)$.

Next, suppose K is of the first type, $K = C \times \bar{c}_s L$. Notice that this is precisely the form of the sets $\bar{N}'$ of Lemma 8.1.20. If $y \in C$, then $I^{\bar{q}}H_i(X, X - K; R) \cong I^{\bar{q}}H_i(X, X - \{(y, v)\}; R)$ by the proof that the leftward map in Lemma 8.1.20 is an isomorphism. But we already know from Lemma 8.1.14 and the induction assumptions that $I^{\bar{q}}H_i(X, X - \{(y, v)\}; R) = 0$ for $i > n$.

Now let $\mathfrak{s}$ be a section of $\mathcal{O}^{\bar{p}}$ on X. For $i = n$ and $\bar{p} \ge \bar{0}$, we know by Lemma 8.1.20 that if we are given a collection of points $\{x_\alpha\}$ consisting of one point in each regular stratum of $\bar{N}'$ then there is a unique element $\xi \in I^{\bar{p}}H_n(X, X - \bar{N}'; R)$ whose image at each x_α is $\mathfrak{s}(x_\alpha)$. We must verify that the image of ξ in $I^{\bar{p}}H_n(X, X - \{z\}; R)$ is $\mathfrak{s}(z)$ for all $z \in \bar{N}'$; this will provide the unique element of $I^{\bar{p}}H_n(X, X - K; R)$ promised by the lemma. For this, it is simpler to work with a slightly larger open set that contains $\bar{N}'$. So let B_r be an open ball in $\mathbb{R}^k$ containing C, let $s < s'' < 1$, let $N'' = B_r \times c_{s''}L$, and let $\bar{N}'' = \bar{B}_r \times \bar{c}_{s''}L$. Then $\bar{N}' \subset N''$, and the inclusion $I^{\bar{p}}H_n(X, X - \bar{N}''; R) \to I^{\bar{p}}H_n(X, X - \bar{N}'; R)$ is an isomorphism by two applications of the proof that the leftward map in Lemma 8.1.20 is an isomorphism. Let $\xi'' \in I^{\bar{p}}H_n(X, X - \bar{N}''; R)$ be the unique element that maps to $\xi \in I^{\bar{p}}H_n(X, X - \bar{N}'; R)$. Then ξ'' determines a section $\mathfrak{s}_{\xi''}$ over N'', and again $\mathfrak{s}_{\xi''}(x_\alpha) = \mathfrak{s}(x_\alpha)$; in fact, $\mathfrak{s}_{\xi''}(z)$ must agree with the image of ξ in $I^{\bar{p}}H_n(X, X - \{z\}; R)$ for each $z \in \bar{N}'$. As N'' is a stratified pseudomanifold with orientation sheaf obtained by restricting $\mathcal{O}^{\bar{p}}$ from X, the restriction of $\mathfrak{s}$ to N'' and the section $\mathfrak{s}_{\xi''}$ must each be constant over each regular stratum of N''. Thus, the two sections must agree at all points of regular strata of N'', as they agree over one point in each regular stratum (each regular stratum $B_r \times (0, s'') \times \mathcal{L}_\alpha$ of N'' contains a regular stratum $C \times (0, s] \times \mathcal{L}_\alpha$ of $\bar{N}'$, the $\mathcal{L}_\alpha$ being the regular strata of L, and so each regular stratum of N'' contains one of the x_α). But, again using that N'' is a stratified pseudomanifold with orientation bundle restricted from that on X, Lemma 8.1.14 states that the restriction of $\mathfrak{s}$ over the regular strata extends uniquely to a section over all of N''. As $\mathfrak{s}_{\xi''}$ and the restriction of $\mathfrak{s}$ to N'' are both defined on all of N'' and are equal on the regular strata, this uniqueness means that $\mathfrak{s}(z) = \mathfrak{s}_{\xi''}(z)$ for all $z \in N''$. Thus, the image of ξ'' in $I^{\bar{p}}H_n(X, X - \{z\}; R)$ is $\mathfrak{s}(z)$ for all $z \in N''$, and so the image of ξ must then be $\mathfrak{s}(z)$ for each $z \in \bar{N}'$, as desired.

Proof for Arbitrary $K \subset \mathbb{R}^k \times cL$. Finally, suppose $K \subset X = \mathbb{R}^k \times cL$ is an arbitrary compact set. We will first verify the second statement of the lemma.

Let $\mathfrak{s}$ be a section of $O^{\bar{p}}$ on X. Any compact K must be contained in some PM-convex set, say Q. Let $\gamma_Q \in I^{\bar{p}}H_n(X, X - Q; R)$ be the unique element guaranteed by the previous step such that the image of γ_Q in $I^{\bar{p}}H_n(X, X - \{x\}; R)$ is $\mathfrak{s}(x)$ for each $x \in Q$. Then the image $\gamma_K \in I^{\bar{p}}H_n(X, X - K; R)$ of γ_Q must have image $\mathfrak{s}(x) \in I^{\bar{p}}H_n(X, X - \{x\}; R)$ for each $x \in K$. We must show that γ_K is the unique such element of $I^{\bar{p}}H_n(X, X - K; R)$.

Suppose $\gamma'_K \in I^{\bar{p}}H_n(X, X - K; R)$ also has image $\mathfrak{s}(x) \in I^{\bar{p}}H_n(X, X - \{x\}; R)$ for each $x \in K$. We will show that $\gamma_K - \gamma'_K = 0$. Let ξ be a relative cycle representing $\gamma_K - \gamma'_K \in I^{\bar{p}}H_n(X, X - K; R)$. Let $|\partial\xi|$ be the support of $\partial\xi$. Then $|\partial\xi| \cap K = \emptyset$, and ξ determines a section $\mathfrak{s}_\xi$ of $O^{\bar{p}}$ over $U = X - |\partial\xi|$ that must be zero at each $x \in K$. Now, choose any $x \in K$. By Remark 8.1.13, as $\mathfrak{s}_\xi(x) = 0$, the section $\mathfrak{s}_\xi$ must be identically zero in an open neighborhood V_x of x, and we may suppose $V_x \subset U$. Let $\bar{A}_x$ be a PM-convex neighborhood of x in V_x, and let A_x be the interior of $\bar{A}_x$. As we let x run through all the elements of K, the open subsets A_x cover K. Since K is compact, there is a finite subcollection $\{A_{x_j}\}_{j=1}^m$ that covers K, and the corresponding union $P = \cup_{j=1}^m \bar{A}_{x_j}$ is a compact subset of U that contains K. By construction, the image of ξ in $I^{\bar{p}}H_n(X, X - \{z\}; R)$ is zero for each $z \in P$; in other words, the images of ξ agree with the zero section of $O^{\bar{p}}$ over P. But P is a finite union of PM-convex sets, and we have proven the lemma for PM-convex sets. So, by the Mayer–Vietoris step and the induction argument in the distinguished neighborhood reduction step, the lemma holds for P. Therefore, $0 \in I^{\bar{p}}H_n(X, X - P; R)$ is the unique element that agrees with the zero section at each point of P, and so $\xi = 0 \in I^{\bar{p}}H_n(X, X - P; R)$. But then ξ must also equal zero under the inclusion-induced composition

$$I^{\bar{p}}H_n(X, X - U; R) \to I^{\bar{p}}H_n(X, X - P; R) \to I^{\bar{p}}H_n(X, X - K; R),$$

which is what we needed to show.

Lastly, we need to see that $I^{\bar{q}}H_i(X, X - K; R) = 0$ for $i > n$ and any $\bar{q}$. Suppose $\xi \in I^{\bar{q}}H_i(X, X - K; R)$ for $i > n$. Once again, let $U = X - |\partial\xi|$, and, for each $x \in K$, let $\bar{A}_x$ be a PM-convex neighborhood of x in U, although this time we impose no additional conditions on $\bar{A}_x$. By the same argument as in the preceding paragraph, there is a union P of a finite number of the $\bar{A}_x$ with $K \subset P$. But now, for each $\bar{A}_x$ we have

$I^{\bar q}H_i(X, X - \bar A_x; R) = 0$, because we have already proven Lemma 8.1.16 for PM-convex sets. Also once again, the Mayer–Vietoris reduction step, together with the induction argument in the distinguished neighborhood reduction step, now shows that $I^{\bar q}H_i(X, X - P; R) = 0$. So ξ must represent zero in $I^{\bar q}H_i(X, X - P; R)$, and so also zero under the inclusion-induced map $I^{\bar q}H_n(X, X - P; R) \to I^{\bar q}H_n(X, X - K; R)$. But ξ was an arbitrary element of $I^{\bar q}H_i(X, X - K; R)$, so $I^{\bar q}H_i(X, X - K; R) = 0$.

This completes the proof of Lemma 8.1.16 given the induction assumptions. □

So, just to catch up, we have now seen that assuming Lemma 8.1.14, Lemma 8.1.16, and Theorem 8.1.18 for depths $< d$ implies both lemmas and the theorem for depth d. So, by induction, our proof of all these results is complete.

(c) Useful Corollaries

In the second statement of Lemma 8.1.16, the hypothesis calls for a section $\mathfrak{s}$ of $\mathcal{O}^{\bar p}$ defined on all of X. However, now that we have proven the lemma we can provide a slight strengthening that only requires $\mathfrak{s}$ to be defined on a neighborhood of the compact set K. This observation will be useful in the proof of Proposition 8.1.25, below.

Corollary 8.1.22 *Let R be a Dedekind domain, and let X be an R-oriented n-dimensional stratified pseudomanifold. Let $\bar p$ be a perversity with $\bar p \geq \bar 0$. Let $K \subset X$ be a compact subset. Then given a section $\mathfrak{s}$ of $\mathcal{O}^{\bar p}$ defined over a neighborhood of K there is a unique class $\gamma \in I^{\bar p}H_n(X, X - K; R)$ whose image in $I^{\bar p}H_n(X, X - \{x\}; R)$ is $\mathfrak{s}(x)$ for any $x \in K$.*

Proof Let $U \subset X$ be an open subset containing K and on which $\mathfrak{s}$ is defined. As an open subset of an R-oriented stratified pseudomanifold, the space U is itself an R-oriented stratified pseudomanifold. We notice via the excision isomorphisms $I^{\bar p}H_n(X, X - \{x\}; R) \cong I^{\bar p}H_n(U, U - \{x\}; R)$ for $x \in U$ that the restriction $\mathcal{O}^{\bar p}|_U$ of $\mathcal{O}^{\bar p}$ to U is isomorphic[7] to the orientation sheaf for U. So $\mathfrak{s}$ determines a global section of the orientation sheaf over U (which we will also just call $\mathfrak{s}$) and thus by Lemma 8.1.16 a unique class $\gamma_U \in$

[7] Technically, this is not enough to show that these orientation sheaves are isomorphic, but for any open sets $W' \subset W \subset U$ we have a commutative diagram

$$\begin{array}{ccc} I^{\bar p}H_n(U, U - \bar W; R) & \longrightarrow & I^{\bar p}H_n(X, X - \bar W; R) \\ \downarrow & & \downarrow \\ I^{\bar p}H_n(U, U - \bar W'; R) & \longrightarrow & I^{\bar p}H_n(X, X - \bar W'; R), \end{array}$$

and the horizontal maps are excision isomorphisms whenever $\bar W \subset U$. This is sufficient by some elementary sheaf theory. In particular, this says we have a map of presheaves over U that induces a map of sheaves between the orientation sheaf over U and the restriction to U of the orientation sheaf over X; furthermore, the sheaf map is an isomorphism, as every point $x \in U$ has neighborhoods whose closures are contained in U (using that X is locally compact by Lemma 2.3.15) and so the maps $I^{\bar p}H_n(U, U - \bar W; R) \to I^{\bar p}H_n(X, X - \bar W; R)$ are all isomorphisms for "sufficiently small" W containing x. See [37, Section I.2] for the relevant sheaf theory background.

$I^{\bar{p}}H_n(U, U - K; R)$ whose image in $I^{\bar{p}}H_n(U, U - \{x\}; R)$ is $\mathfrak{s}(x)$ for any $x \in K$. But now we have a commutative diagram of excisions

$$\begin{array}{ccc} I^{\bar{p}}H_n(U, U - K; R) & \xrightarrow{\cong} & I^{\bar{p}}H_n(X, X - K; R) \\ \downarrow & & \downarrow \\ I^{\bar{p}}H_n(U, U - \{x\}; R) & \xrightarrow{\cong} & I^{\bar{p}}H_n(X, X - \{x\}; R), \end{array}$$

showing that the image of γ_U in $I^{\bar{p}}H_n(X, X - \{x\}; R)$ also maps to $\mathfrak{s}(x)$ for each $x \in K$. This image is the desired γ. Furthermore, the class γ is unique, as we see that if $\gamma' \in I^{\bar{p}}H_n(X, X - K; R)$ maps to $\mathfrak{s}(x)$ for each $x \in K$ then the image of γ' in $I^{\bar{p}}H_n(U, U - K; R)$ has the same property. But by the uniqueness of Lemma 8.1.16, the only such class in $I^{\bar{p}}H_n(U, U - K; R)$ is γ_U. □

Here is one more corollary, concerning the structure of fundamental classes at singular points. It will be used in the proof of the Poincaré Duality Theorem (Theorem 8.2.4).

Corollary 8.1.23 *Let R be a Dedekind domain, and let L be an $(n-k-1)$-dimensional compact R-oriented stratified pseudomanifold with fundamental class $\Gamma_L \in I^{\bar{0}}H_{n-k-1}(L; R)$. Giving $\mathbb{R}^k$ and $(0,1)$ their standard orientations, let $\mathbb{R}^k \times cL$ be oriented by the product orientation on $\mathbb{R}^k \times cL - \Sigma_{\mathbb{R}^k \times cL} \cong \mathbb{R}^k \times (0,1) \times (L - \Sigma_L)$. Let $\eta \in H_0(\mathbb{R}^k, \mathbb{R}^k - \{0\}; R)$ be the generator consistent with the orientation of $\mathbb{R}^k$. Let $\mathfrak{l} : L \hookrightarrow cL$ be the inclusion of L at a fixed cone coordinate, and let $\bar{c}\mathfrak{l}(\Gamma_L)$ be the singular cone[8] on Γ_L. Then $\eta \times \bar{c}\mathfrak{l}(\Gamma_L) \in I^{\bar{0}}H_n(\mathbb{R}^k \times cL, \mathbb{R}^k \times cL - \{(0,v)\}; R)$ represents the orientation class $\mathfrak{o}((0,v))$ of $\mathbb{R}^k \times cL$ over $\{(0,v)\}$.*

Proof Let $\bar{B}_r$ be a closed ball of radius r centered at the origin in $\mathbb{R}^k$ and choose a chain representative for η such that η is also a generator of $H_k(\mathbb{R}^k, \mathbb{R}^k - \bar{B}_r; R)$. For some s, $0 < s < 1$, let $\bar{N}' = \bar{B}_r \times \bar{c}_s L \subset \mathbb{R}^k \times cL$, and let x_α be a set of points, one in each regular stratum of $\bar{N}'$. As in the proof of Lemma 8.1.20, let x_α have coordinates $(y_\alpha, t_\alpha, z_\alpha)$, choose s', s'' with $s < s' < 1$ and $0 < s'' < \min\{t_\alpha\}$, and let $e : [0,1] \to [s'', s']$ be the linear isomorphism with $e(0) = s''$ and $e(1) = s'$. Now let $\zeta = \Gamma_L$, and also as in the proof of Lemma 8.1.20 let $\mathcal{Z}$ be the chain $\bar{c}\zeta_{s''} + e \times \zeta$, where $\zeta_{s''}$ is the image of ζ under the embedding of L at the s'' cone coordinate. We consider $\eta \times \mathcal{Z}$, which represents a homology class in $I^{\bar{0}}H_n(\mathbb{R}^k \times cL, \mathbb{R}^k \times cL - \bar{N}'; R)$.

By our choices of coordinates, for each x_α the chains $\eta \times \mathcal{Z}$ and $\eta \times e \times \Gamma_L$ represent the same class in $I^{\bar{0}}H_n(\mathbb{R}^k \times cL, \mathbb{R}^k \times cL - \{x_\alpha\}; R)$, which by excision is isomorphic to $I^{\bar{0}}H_n(\mathbb{R}^k \times (0,1) \times L, \mathbb{R}^k \times (0,1) \times L - \{(y_\alpha, t_\alpha, z_\alpha)\}; R)$. But from the assumptions, the class $\eta \times e \times \Gamma_L$ is the product of local orientation classes at $(y_\alpha, t_\alpha, z_\alpha)$ consistent with

[8] Recall Example 3.4.7.

the given orientation. So the image of $\eta \times \mathcal{Z}$ represents the orientation class $\mathfrak{o}(x_\alpha)$ in each $I^{\bar{0}}H_n(\mathbb{R}^k \times cL, \mathbb{R}^k \times cL - \{x_\alpha\}; R)$.

Now, by Lemma 8.1.16, there is a unique class $\Gamma_{\bar{N}'} \in I^{\bar{0}}H_n(\mathbb{R}^k \times cL, \mathbb{R}^k \times cL - \bar{N}'; R)$ whose image in $I^{\bar{0}}H_n(\mathbb{R}^k \times cL, \mathbb{R}^k \times cL - \{z\}; R)$ is $\mathfrak{o}(z)$ for each $z \in \bar{N}'$. So, in particular, $\Gamma_{\bar{N}'}$ maps to $\mathfrak{o}(x_\alpha)$ for each x_α. But by Lemma 8.1.20, there is a unique class in $I^{\bar{0}}H_n(\mathbb{R}^k \times cL, \mathbb{R}^k \times cL - \bar{N}'; R)$ with this last property, and we have already shown that $\eta \times \mathcal{Z}$ has the property. So we must have

$$\Gamma_{\bar{N}'} = \eta \times \mathcal{Z} \in I^{\bar{0}}H_n(\mathbb{R}^k \times cL, \mathbb{R}^k \times cL - \bar{N}'; R),$$

and so $\eta \times \mathcal{Z}$ must also represent $\mathfrak{o}((0, v)) \in I^{\bar{0}}H_n(\mathbb{R}^k \times cL, \mathbb{R}^k \times cL - \{(0, v)\}; R)$. But now if we consider the image of $\eta \times \mathcal{Z}$ in this last module, the piece $\eta \times e \times \Gamma_L$ is contained in the complement of $\{(0, v)\}$, and so

$$\eta \times \mathcal{Z} = \eta \times \bar{c}\zeta_{s''} = \eta \times \bar{c}(\Gamma_L)_{s''} \in I^{\bar{0}}H_n(\mathbb{R}^k \times cL, \mathbb{R}^k \times cL - \{(0, v)\}; R).$$

Thus $\mathfrak{o}((o, v))$ is represented by $\eta \times \bar{c}(\Gamma_L)_{s''}$.

Lastly, we observe that the class $\bar{c}(\Gamma_L)_{s''}$ in $I^{\bar{0}}H_{n-k}(cL, cL - \{v\}; R)$ is independent of the coordinate s'' for $0 < s'' < 1$, and so we can replace $(\Gamma_L)_{s''}$ with $\mathfrak{l}(\Gamma_L)$ for our original choice of $\mathfrak{l}$ in the statement of the lemma. This shows that $\mathfrak{o}((0, v))$ can be represented by a chain of the claimed form. □

8.1.4 Lack of Global Fundamental Classes for Subzero Perversities

We are now in a position to investigate a bit further what happens if $\bar{p}$ is a perversity on X with $\bar{p}(S) < 0$ for some singular stratum S. The following proposition shows that $I^{\bar{p}}H_*(X - S; R) \to I^{\bar{p}}H_*(X; R)$ is an isomorphism, so putting negative perversities on some strata is equivalent to leaving such strata out of the space altogether. Among other consequences, this shows that Theorem 8.1.18 cannot possibly hold for X, as $X - S$ will be non-compact (it also might not be a stratified pseudomanifold). In fact, given this isomorphism, any global fundamental class $\Gamma_X^{\bar{p}} \in I^{\bar{p}}H_n(X; R)$ can be represented by a chain in $X - S$, and this chain must have compact support. The chain's image in $I^{\bar{p}}H_n(X, X - \{x\}; R)$ has to be zero for any x outside this support. But, S must have a neighborhood that does not intersect the support of the chain and, using that $X - \Sigma_X$ is dense, there are therefore points of $X - \Sigma_X$ not contained in the support of the chain. This contradicts $\Gamma_X^{\bar{p}}$ being a fundamental class.

Proposition 8.1.24 *Let R be a Dedekind domain, and let X be a stratified pseudomanifold with perversity $\bar{p}$. Suppose $S \subset X$ is a singular stratum with $\bar{p}(S) < 0$. Then inclusion induces an isomorphism $I^{\bar{p}}H_*(X - S; R) \to I^{\bar{p}}H_*(X; R)$.*

Proof We will use a Mayer–Vietoris argument, and so we will invoke Theorem 5.1.4. In this case, for any open $U \subset X$, let $F_*(U) = I^{\bar{p}}H_*(U - (U \cap S); R)$ and $G_*(U) = I^{\bar{p}}H_*(U; R)$, and let $\Phi : F_* \to G_*$ be induced by inclusion. We must check the conditions of Theorem 5.1.4.

The functors F_* and G_* admit Mayer–Vietoris sequences by Theorem 6.3.12, and Φ induces a map between them. The direct limit condition follows from Lemmas 5.1.6 and 6.3.16.

Skipping to the fourth condition of Theorem 5.1.4, if U is empty or contained in a stratum other than S, then $U - (U \cap S) = U$, so Φ is the identity map for this U. If $U \subset S$ is an open subset homeomorphic to Euclidean space and if σ is an i-simplex in U, then σ takes all of Δ^i into S, and the $\bar{p}$-allowability condition for σ becomes $i \le i - \operatorname{codim}(S) + \bar{p}(S)$, i.e. $\operatorname{codim}(S) \le \bar{p}(S)$. But this is impossible if $\bar{p}(S) < 0$, so no simplices are allowable and $I^{\bar{p}}H_*(U) = 0 = I^{\bar{p}}H_*(U - (U \cap S))$, as $U - (U \cap S)$ is the empty set.

For the remaining condition of Theorem 5.1.4, we must show that if $\Phi : F_*(\mathbb{R}^k \times (cL - \{v\})) \to G_*(\mathbb{R}^k \times (cL - \{v\}))$ is an isomorphism, then so is $\Phi : F_*(\mathbb{R}^k \times cL) \to G_*(\mathbb{R}^k \times cL)$. Here, $\mathbb{R}^k \times cL$ is a distinguished neighborhood of a point $x \in X$. Let us make the inductive assumption that we have verified the proposition already for stratified pseudomanifolds of depth less than that of X. The proposition is trivial when the depth of X is zero, as in this case S must be empty. Therefore, if we can verify the condition for our X under the inductive assumption, the proposition will be proven on X by Theorem 5.1.4. As X was arbitrary of its depth, the entire proposition will follow by induction.

So, consider now the diagram:

$$\begin{array}{ccc} F_*(\mathbb{R}^k \times (cL - \{v\})) & \xrightarrow{\Phi} & G_*(\mathbb{R}^k \times (cL - \{v\})) \\ \downarrow & & \downarrow \\ F_*(\mathbb{R}^k \times cL) & \xrightarrow{\Phi} & G_*(\mathbb{R}^k \times cL). \end{array}$$

We are assuming the top horizontal map is an isomorphism. The vertical arrows are induced by inclusion. There are two cases to consider, depending upon whether $x \in S$. First, suppose $x \notin S$. If x is not contained in the closure of S, then the distinguished neighborhood cannot intersect S, so the bottom map in this diagram is trivially an isomorphism. Suppose x is in the closure of S. This means that L has a stratum that is the intersection of L with S, thinking of some copy of L as being embedded in X by the distinguished neighborhood homeomorphism. Therefore, $L \cap S$ is a union of strata of L, on each of which $\bar{p} < 0$, and the top horizontal arrow is an isomorphism and stratified homotopy invariance and induction. The vertical arrow on the right is basically (up to stratified homotopy equivalence) the inclusion of a link into its cone, and so, by the cone formula (Theorem 6.2.13), the codomain is zero for $* \ge \dim(L) - \bar{p}(\{v\})$ and the map is an isomorphism otherwise. The vertical map on the left is similarly, up to stratified homotopy equivalence, the inclusion of $L - (L \cap S)$ into $c(L - (L \cap S))$. Our cone formula as stated in Theorem 6.2.13 doesn't precisely apply because $L - (L \cap S)$ is not compact. One of our main concerns with having compact links was to avoid weird topologies on cones, which can happen in the quotient topology if the space being coned is not compact. In this case, we can avoid this problem by letting $c(L-(L\cap S))$ have the subspace topology from cL, and with this assumption the argument of Theorem 6.2.13 goes through without any trouble. So, for $*\geq \dim(L)-\bar{p}(\{v\})$, both modules on the bottom line of the diagram are trivial, and otherwise the top and sides of the diagram are isomorphisms, implying that the bottom map is also.

Next, suppose $x\in S$. In this case, the diagram reduces to

$$\begin{array}{ccc} I^{\bar p}H_*(\mathbb{R}^k\times(cL-\{v\});R) & \xrightarrow[=]{\Phi} & I^{\bar p}H_*(\mathbb{R}^k\times(cL-\{v\});R) \\ \Big\downarrow{=} & & \Big\downarrow \\ I^{\bar p}H_*(\mathbb{R}^k\times(cL-\{v\});R) & \xrightarrow{\Phi} & I^{\bar p}H_*(\mathbb{R}^k\times cL;R), \end{array}$$

so this case reduces to showing that the inclusion-induced $I^{\bar p}H_*(\mathbb{R}^k\times(cL-\{v\});R)\to I^{\bar p}H_*(\mathbb{R}^k\times cL;R)$ is an isomorphism in all dimensions. By the cone formula (and stratified homotopy invariance), this is true for $*<\dim(L)-\bar{p}(\{v\})$, and $I^{\bar p}H_*(\mathbb{R}^k\times cL;R)=0$ for $*\geq\dim(L)-\bar{p}(\{v\})$, so we must demonstrate that $I^{\bar p}H_*(\mathbb{R}^k\times(cL-\{v\});R)\cong I^{\bar p}H_*(L;R)$ is also zero in this range. But, by assumption, $\bar{p}(\{v\})<0$, so the dimension range $*\geq\dim(L)-\bar{p}(\{v\})$ only includes dimensions that are greater than $\dim(L)$. But, as L is a stratified pseudomanifold, $I^{\bar p}H_*(L;R)=0$ for $*>\dim(L)$ by Theorem 8.1.18. □

8.1.5 Invariance of Fundamental Classes

In this section, we will show that the fundamental classes we constructed in the previous section are invariants in two different senses. First, we will show that they are essentially independent of the choice of perversity $\bar p$ such that $\bar p\geq\bar 0$. The precise meaning of this claim can be found in the statements of Proposition 8.1.25 and Corollary 8.1.26. Then we will go on to see in what sense our fundamental classes are invariant to the choice of stratification.

(a) Fundamental Classes under Change of Perversity

We begin with change of perversity. Recall that if X is a filtered space with perversities $\bar p$ and $\bar q$ satisfying $\bar p\leq\bar q$ then the identity map $X\to X$ is $(\bar p,\bar q)$-stratified. We will denote the induced map by $\tau_{\bar p,\bar q}: I^{\bar p}S_*(X;R)\to I^{\bar q}S_*(X;R)$. On stratified pseudomanifolds, such maps then induce morphisms of orientation sheaves $\mathcal{O}^{\bar p}\to\mathcal{O}^{\bar q}$.

Proposition 8.1.25 *Let R be a Dedekind domain, and let X be an R-oriented n-dimensional stratified pseudomanifold. Suppose $\bar p$, and $\bar q$ are perversities on X with $\bar 0\leq\bar p\leq\bar q$. Then the following hold:*

1. *The canonical morphism of sheaves $\mathcal{O}^{\bar p}\to\mathcal{O}^{\bar q}$ is an isomorphism. In particular, every*

$O^{\bar{p}}$ is isomorphic to $O^{\bar{0}}$, which we can simply denote O and call the R-orientation sheaf over X. *We will also generally identify all the $O^{\bar{p}}$ via these isomorphisms.*

2. *If K is a compact subset of X then the canonical map $\tau_{\bar{p},\bar{q}} : I^{\bar{p}}H_n(X, X - K; R) \to I^{\bar{q}}H_n(X, X - K; R)$ is an isomorphism. Furthermore, given a section $\mathfrak{s}$ of O over K, if $\gamma^{\bar{p}} \in I^{\bar{p}}H_n(X, X - K; R)$ and $\gamma^{\bar{q}} \in I^{\bar{q}}H_n(X, X - K; R)$ each restrict to $\mathfrak{s}(x)$ in $I^{\bar{p}}H_n(X, X - \{x\}; R) \cong I^{\bar{q}}H_n(X, X - \{x\}; R)$ for each $x \in K$, then $\tau_{\bar{p},\bar{q}}(\gamma^{\bar{p}}) = \gamma^{\bar{q}}$. In particular, if $\mathfrak{s} = \mathfrak{o}^{\bar{p}} = \mathfrak{o}^{\bar{q}}$, then $\tau_{\bar{p},\bar{q}}(\Gamma_K^{\bar{p}}) = \Gamma_K^{\bar{q}}$.*

We will prove the proposition just below. As a corollary, we have the following statement about global fundamental classes on compact stratified pseudomanifolds. This corollary demonstrates the sense in which fundamental classes do not depend on the choice of perversity, provided we don't allow perversities with negative values.

Corollary 8.1.26 *Let R be a Dedekind domain, and let X be a* compact *R-oriented n-dimensional stratified pseudomanifold. Then, for any $\bar{p}$ and $\bar{q}$ that are both $\geq \bar{0}$, the diagram*

$$I^{\bar{p}}H_n(X; R) \xleftarrow{\tau_{\bar{0},\bar{p}}} I^{\bar{0}}H_n(X; R) \xrightarrow{\tau_{\bar{0},\bar{q}}} I^{\bar{q}}H_n(X; R)$$

consists of isomorphisms, and the composition left to right takes $\Gamma_X^{\bar{p}}$ to $\Gamma_X^{\bar{q}}$.

Proof The corollary follows directly from the second statement of the proposition taking $K = X$ and using $\bar{0}$ as an intermediary. □

Definition 8.1.27 It follows from Proposition 8.1.25 that every $\Gamma_K^{\bar{p}}$, $\bar{p} \geq \bar{0}$, is the image of $\Gamma_K^{\bar{0}}$, which we will simply denote Γ_K and call *the fundamental class of X over K* with respect to the R-orientation. If X is compact, then we call Γ_X *the fundamental class of X* with respect to the R-orientation.

Proof of Proposition 8.1.25 We will induct on the depth d of X. If $d = 0$, then X is a manifold and all intersection homology groups reduce to ordinary homology, making the statements trivial. So, suppose now that we have proven the proposition up through depth $d - 1$ and that X has depth d.

For the first statement, assume $\bar{0} \leq \bar{p} \leq \bar{q}$ so that we have maps

$$\tau_{\bar{p},\bar{q}} : I^{\bar{p}}H_*(X, A; R) \to I^{\bar{q}}H_*(X, A; R)$$

for any subset A. In particular, if $x \in X$ is any point and U is any neighborhood of x, we have maps $\tau_{\bar{p},\bar{q}} : I^{\bar{p}}H_n(X, X - \bar{U}; R) \to I^{\bar{q}}H_n(X, X - \bar{U}; R)$. Such maps induce a morphism of sheaves $O^{\bar{p}} \to O^{\bar{q}}$; see[9] [37, Section I.2]. To show that this morphism is an isomorphism, it suffices to demonstrate that the induced map of stalks $I^{\bar{p}}H_n(X, X - \{x\}; R) \to I^{\bar{q}}H_n(X, X - \{x\}; R)$ is an isomorphism for all x. If x is contained in a regular stratum of X, this map is isomorphic to the identity map $H_n(\mathbb{R}^n, \mathbb{R}^n - \{0\}; R) \to H_n(\mathbb{R}^n, \mathbb{R}^n - \{0\}; R)$ by excision. So suppose x has a distinguished neighborhood $\mathbb{R}^k \times cL$, $L \neq \emptyset$. Then, applying excision, the

9 We are using here that the collection of maps $I^{\bar{p}}H_n(X, X - \bar{U}; R) \to I^{\bar{q}}H_n(X, X - \bar{U}; R)$ constitute a map of presheaves, which induces a map of sheaves.

naturality of the Künneth Theorem (as used in the proof of Lemma 6.4.17), the naturality of the boundary map of the long exact sequence of the pair, the isomorphism of the cone formula (Corollary 6.2.15), and stratified homotopy invariance, this map reduces to the map $I^{\bar p}H_{n-k-1}(L;R) \to I^{\bar q}H_{n-k-1}(L;R)$. By the arguments of the proof of Lemma 8.1.14, the link L is orientable, so this map is an isomorphism by the induction assumption, as L must have smaller depth than X.

For the second statement, we first make an observation: If $\xi \in I^{\bar p}S_n(X;R)$ is an n-chain in X, let U be the open set $X - |\partial\xi|$; the set U is itself an R-oriented stratified pseudomanifold. Then ξ determines a class in $I^{\bar p}H_n(X,|\partial\xi^{\bar p}|;R)$ and the images of this class in $I^{\bar p}H_n(X,X-\{x\};R)$ as x varies over U gives a section $\mathfrak{s}$ of the orientation sheaf $\mathcal{O}^{\bar p}$ over U. Furthermore, suppose $K \subset U$ is a compact subset. Then for all $x \in K$ the factorization

$$I^{\bar p}H_n(X,|\partial\xi^{\bar p}|;R) \to I^{\bar p}H_n(X,X-K;R) \to I^{\bar p}H_n(X,X-\{x\};R)$$

shows that the class represented by ξ in $I^{\bar p}H_n(X,X-K;R)$ maps to the values of $\mathfrak{s}(x) \in I^{\bar p}H_n(X,X-\{x\};R)$ for each $x \in K$.

Now suppose that we have a class $[\xi^{\bar q}] \in I^{\bar q}H_n(X,X-K;R)$ represented by the chain $\xi^{\bar q}$, and let U be the open set $U = X - |\partial\xi^{\bar q}|$. Then, as just described, $\xi^{\bar q}$ determines a section $\mathfrak{s}^{\bar q}$ over U, and so in particular over a neighborhood of K. If we let $\mathfrak{s}^{\bar p}$ be the image of the section $\mathfrak{s}^{\bar q}$ under the isomorphism of orientation sheaves $\mathcal{O}^{\bar p} \cong \mathcal{O}^{\bar q}$, then $\mathfrak{s}^{\bar p}$ is also defined over a neighborhood of K. So by Corollary 8.1.22 there is a unique class $[\xi^{\bar p}] \in I^{\bar p}H_n(X,X-K;R)$ whose image in $I^{\bar p}H_n(X,X-\{x\};R)$ is $\mathfrak{s}^{\bar p}(x)$ for all $x \in K$. For each such $x \in K$ we have a diagram

$$\begin{array}{ccc} I^{\bar p}H_n(X,X-K;R) & \longrightarrow & I^{\bar p}H_n(X,X-\{x\};R) \\ \Big\downarrow \tau_{\bar p,\bar q} & & \Big\downarrow \tau_{\bar p,\bar q}\ \cong \\ I^{\bar q}H_n(X,X-K;R) & \longrightarrow & I^{\bar q}H_n(X,X-\{x\};R), \end{array} \tag{8.3}$$

with the right-hand vertical map an isomorphism by the first part of the proof of the proposition. In fact, this isomorphism takes the stalks of $\mathcal{O}^{\bar p}$ to the corresponding stalks of $\mathcal{O}^{\bar q}$. So, by the commutativity of the diagram, the class $\tau_{\bar p,\bar q}([\xi^{\bar p}])$ also maps to $\mathfrak{s}^{\bar q}(x)$ in $I^{\bar q}H_n(X,X-\{x\};R)$ for all $x \in K$, and thus by the uniqueness part of Corollary 8.1.22 we must have $\tau_{\bar p,\bar q}([\xi^{\bar p}]) = [\xi^{\bar q}]$. So the left-hand vertical map of the diagram is surjective.

Similarly, suppose we start with a class $[\xi^{\bar p}] \in I^{\bar p}H_n(X,X-K;R)$ represented by a chain $\xi^{\bar p}$. Suppose that $\tau_{\bar p,\bar q}([\xi^{\bar p}]) = 0 \in I^{\bar q}H_n(X,X-K;R)$. Now let $U = X - |\partial\xi^{\bar p}|$. For $x \in U$, we have seen that the images of $\xi^{\bar p}$ in $I^{\bar p}H_n(X,X-\{x\};R)$ give a section $\mathfrak{s}$ of the orientation sheaf $\mathcal{O}^{\bar p}$ over U, which is a neighborhood of K. Now suppose $x \in K$. Then we have also seen that $\mathfrak{s}(x)$ is the image of $\xi^{\bar{p}}$ down the left side of the following commutative diagram

$$\begin{array}{ccc} I^{\bar{p}}H_n(X,|\partial\xi^{\bar{p}}|;R) & & \\ \downarrow & & \\ I^{\bar{p}}H_n(X,X-K;R) & \xrightarrow{\tau_{\bar{p},\bar{q}}} & I^{\bar{q}}H_n(X,X-K;R) \\ \downarrow & & \downarrow \\ I^{\bar{p}}H_n(X,X-\{x\};R) & \underset{\cong}{\xrightarrow{\tau_{\bar{p},\bar{q}}}} & I^{\bar{q}}H_n(X,X-\{x\};R). \end{array}$$

But $[\xi^{\bar{p}}]$ is the image of $\xi^{\bar{p}}$ in $I^{\bar{p}}H_n(X,X-K;R)$, and this maps to 0 in $I^{\bar{q}}H_n(X,X-K;R)$ by assumption. So it follows from the diagram, and in particular the isomorphism at the bottom, that $\mathfrak{s}$ must be zero over K. So now by Corollary 8.1.22 again, there is a unique class in $I^{\bar{p}}H_n(X,X-K;R)$ whose image is $\mathfrak{s}(x) = 0$ in $I^{\bar{p}}H_n(X,X-\{x\};R)$ for each $x \in K$, and this must be the class zero. Thus $[\xi^{\bar{p}}] = 0$. So we have shown that $\tau_{\bar{p},\bar{q}} : I^{\bar{p}}H_n(X,X-K;R) \to I^{\bar{q}}H_n(X,X-K;R)$ is injective, and therefore it is an isomorphism.

Finally, for the last claim of the proposition, suppose $\gamma^{\bar{p}}$ and $\gamma^{\bar{q}}$ are as given. Employing Diagram (8.3) once again, the hypotheses state that the classes $\tau_{\bar{p},\bar{q}}(\gamma^{\bar{p}})$ and $\gamma^{\bar{q}}$ in $I^{\bar{q}}H_n(X,X-K;R)$ must each map to the same values in $I^{\bar{q}}H_n(X,X-\{x\};R)$ for each x in K, namely $\mathfrak{s}(x)$. Furthermore, using images of a chain representative of $\gamma^{\bar{q}}$ as above, we see that $\mathfrak{s}$ can be extended to a section on a neighborhood of K. This allows us to use Corollary 8.1.22 again to conclude that there is a unique element of $I^{\bar{q}}H_n(X,X-K;R)$ that maps to $\mathfrak{s}(x)$ for each $x \in K$. So we must have $\tau_{\bar{p},\bar{q}}(\gamma^{\bar{p}}) = \gamma^{\bar{q}}$.

This completes the proof of Proposition 8.1.25. □

(b) Fundamental Classes under Change of Stratification

Next we consider how fundamental classes behave under change of stratification. We first consider the simpler case of two stratifications that are related by coarsening/refinement before generalizing to arbitrary stratifications. So suppose X and X' are stratified pseudomanifolds with $|X| = |X'|$ and with X' refining X. Of course, two different stratifications of a single underlying space must each be equipped with its own perversity. In general, we could consider perversities $\bar{p}$ and $\bar{p}'$ on X and X', respectively, such that the set-theoretic identity map $X' \to X$ is $(\bar{p}',\bar{p})$-stratified. However, given the "independence of perversity" results just above, it is simpler, and no loss of generality, to consider only the zero perversities, which we can label as $\bar{0}$ on both X and X'.

Next, we will need some further restrictions on X and X'. The issue is that we want the set-theoretic identity map id : $X' \to X$ to be $(\bar{0},\bar{0})$-stratified; note that id : $X \to X'$ can never be stratified unless $X = X'$ as each stratum of the domain must map into just one stratum of the codomain by Definition 6.3.2. On the other hand, the map id : $X' \to X$ can only be stratified for non-GM perversities if $\Sigma_{X'} \subset \Sigma_X$. We could proceed with such an assumption, but, as we will also soon want to bring in the topological invariance properties of intersection homology, it makes sense instead to require that $\bar{0} \le \bar{t}$ on X and X' so that $I^{\bar{0}}H_n(X;R) \cong I^{\bar{0}}H_n^{\mathrm{GM}}(X;R)$ and $I^{\bar{0}}H_n(X';R) \cong I^{\bar{0}}H_n^{\mathrm{GM}}(X';R)$ by Proposition 6.2.9. This allows us to invoke topological invariance (Theorem 5.5.1), using that $\bar{0}$ is a GM perversity, but it also lets id : $X' \to X$ be $(\bar{0},\bar{0})^{\mathrm{GM}}$-stratified as Definition 4.1.1 does not require singular strata to map to singular strata. As $\bar{t}(S) = \operatorname{codim}(S) - 2$, the requirement $\bar{0} \le \bar{t}$ simply means that we forbid codimension-one strata[10].

Notationally, let $|X| = |X'| = |\mathfrak{X}|$ denote the common underlying topological space, while $\mathfrak{X}$, as always, is the intrinsic CS set stratification. As for our other statements concerning fundamental classes, there will be a compact subset, which we denote $K \subset |X|$. Of course, K can inherit different filtrations depending on whether we think of it as contained in X, X', or $\mathfrak{X}$, but for simplicity, and as our main interest will be in *removing* K to consider spaces such as $X - K$ or $X' - K$, we will abuse notation a bit and just use K, rather than $|K|$, throughout. Also, although we will be in the setting where GM and non-GM intersection homology agree, as discussed in the preceding paragraph, and so using that the id maps are GM-stratified, we will nonetheless continue to use the non-GM notation throughout.

Proposition 8.1.28 *Let R be a Dedekind domain, let X be an n-dimensional stratified pseudomanifold, and let X' be an n-dimensional stratified pseudomanifold refining the stratification of X. Suppose that X and X' have no codimension-one strata and that X and X' are compatibly R-oriented in the sense that the R-orientation on X induces that on X' as in Lemma 8.1.6. Let K be a compact subset of $|X|$, and let $\Gamma_K \in I^{\bar{0}}H_n(X, X-K;R)$ and $\Gamma'_K \in I^{\bar{0}}H_n(X', X'-K;R)$ be the fundamental classes over K given the orientation, as guaranteed by Lemma 8.1.16. Let $\phi : I^{\bar{0}}H_n(X', X'-K;R) \to I^{\bar{0}}H_n(X, X-K;R)$ be induced by the identity map* id : $X' \to X$. *Then $\phi\left(\Gamma'_K\right) = \Gamma_K$.*

Proof Let $\mathfrak{o}$ and $\mathfrak{o}'$ be the global orientation sections on X and X', respectively, as determined by the orientations (see Lemma 8.1.14). By the uniqueness clause of Lemma 8.1.16, it suffices to show that $\phi\left(\Gamma'_K\right)$, which is still represented by the chain Γ'_K, restricts to $\mathfrak{o}(x) \in I^{\bar{0}}H_n(X, X-\{x\};R)$ for each $x \in K$.

So, suppose $x \in K$, and let U' be a distinguished neighborhood of x in X' with $|U'| \subset |X' - |\partial\Gamma'_K||$. As Γ'_K has image $\mathfrak{o}'(x) \in I^{\bar{0}}H_n(X', X'-\{x\};R)$ by assumption, it follows from Lemma 8.1.20 that there is a possibly smaller distinguished neighborhood[11] V' of x in U' such that, at every z in the regular strata of V', the chain Γ'_K represents $\mathfrak{o}'(z) \in I^{\bar{0}}H_n(X, X-\{z\};R)$. Now, let W be a distinguished neighborhood of x in X such that

[10] In general, the modules $I^{\bar{0}}H_n(X;R)$ and $I^{\bar{0}}H_n(X';R)$ are not necessarily isomorphic if there are codimension-one strata. For example, let $X = \mathbb{R}$ be the unfiltered real line, and let X' be the finer filtration given by $\{0,1\} \subset \mathbb{R}$. Then $I^{\bar{0}}H_1(X;R) = 0$ but $I^{\bar{0}}H_1(X';R) \cong R$ (exercise!).

[11] We avoid using the notation $\bar{N}'$ from Lemma 8.1.20 to avoid a clash with the "prime" notation we are using here.

$|W| \subset |V'|$. As the regular strata of V' are dense in V', every regular stratum of W contains a point that is also in a regular stratum of V'. As $|W| \subset |V'|$, at any such point y we continue to have that Γ'_K represents $\mathfrak{o}'(y) \in I^{\bar{0}}H_n(X', X' - \{y\}; R) \cong H_n(\mathbb{R}^n, \mathbb{R}^n - \{0\}; R)$. But, on the regular strata of X', which are subsets of the regular strata of X, the orientation bundle of $X' - \Sigma_{X'}$ is the restriction of the orientation bundle over $X - \Sigma_X$, so it follows that $\phi(\Gamma'_K)$ also represents $\mathfrak{o}(y) \in I^{\bar{0}}H_n(X, X - \{y\}; R) \cong H_n(\mathbb{R}^n, \mathbb{R}^n - \{0\}; R)$ for any such y. In other words, each regular stratum of W has a point y at which $\phi(\Gamma'_K)$ represents $\mathfrak{o}(y)$. But the orientation bundle is constant on the regular strata, and so, in fact, $\phi(\Gamma'_K)$ represents $\mathfrak{o}(y)$ at any point y in a regular stratum of W. It now follows again from Lemma 8.1.20 that $\phi(\Gamma'_K)$ must represent $\mathfrak{o}(x)$ in $I^{\bar{0}}H_n(X, X - \{x\}; R)$. □

Unfortunately, we cannot quite next use Proposition 8.1.28 to parallel our treatment of orientations in Corollary 8.1.11 by allowing us to compare fundamental classes for *arbitrary* pseudomanifold stratifications X and X' of the underlying space $|X|$ by comparing with a fundamental class of $\mathfrak{X}$. The problem is that our previous work only guarantees that $\mathfrak{X}$ will be a CS set, even if X is itself a stratified pseudomanifold, and our preceding work on fundamental classes requires a pseudomanifold stratification. Nonetheless, we will be able to work around this difficulty using the isomorphisms

$$I^{\bar{0}}H_n(X', X' - K; R) \xrightarrow{\cong} I^{\bar{0}}H_n(\mathfrak{X}, \mathfrak{X} - K; R) \xleftarrow{\cong} I^{\bar{0}}H_n(X, X - K; R).$$

Proposition 8.1.29 *Let R be a Dedekind domain, and let X and X' be* any *two n-dimensional stratified pseudomanifolds with the same underlying space $|X|$ and without codimension-one strata. Suppose X and X' are compatibly R-oriented in the sense of Corollary 8.1.11. Let K be a compact subset of $|X|$, and let $\Gamma_K \in I^{\bar{0}}H_n(X, X - K; R)$ and $\Gamma'_K \in I^{\bar{0}}H_n(X', X' - K; R)$ be the fundamental classes over K given the R-orientation. Then Γ_K and Γ'_K correspond under the canonical isomorphisms*

$$I^{\bar{0}}H_n(X', X' - K; R) \xrightarrow{\cong} I^{\bar{0}}H_n(\mathfrak{X}, \mathfrak{X} - K; R) \xleftarrow{\cong} I^{\bar{0}}H_n(X, X - K; R).$$

In particular, if $|X|$ is compact then the fundamental classes Γ_X and $\Gamma_{X'}$ correspond in this manner.

Proof The idea of the proof is essentially the same as that of Proposition 8.1.28, though we must make some modifications. Here, we cannot use the uniqueness clause of Lemma 8.1.16 for $\mathfrak{X}$, but topological invariance provides a workaround. Suppose we start with Γ'_K. We will show that the image of Γ'_K under the isomorphisms is Γ_K. For this, we can invoke the uniqueness of Lemma 8.1.16 for X.

As topological invariance implies that

$$I^{\bar{0}}H_n(X', A'; R) \cong I^{\bar{0}}H_n(\mathfrak{X}, \mathfrak{A}; R) \cong I^{\bar{0}}H_n(X, A; R)$$

for any open $|A| \subset |X|$ with $|A| = |A'| = |\mathfrak{A}|$, we see that X, X', and $\mathfrak{X}$ all share a common orientation sheaf (up to canonical isomorphisms) and so the compatible R-orientations on these spaces come with a common orientation section $\mathfrak{o}$, which restricts to the given compatible manifold orientations on the regular strata of each filtration.

Now, let's take a chain representing Γ'_K. As K is compact, we can find disjoint open sets U' and U'_1 with $|K| \subset |U'|$ and $|\partial\Gamma'_K| \subset |U'|$ by Corollary 2.3.17. Then the chain Γ'_K also represents an element of $I^{\bar{0}}H_n(X, X - |\bar{U}'|; R)$ that maps onto $\Gamma'_K \in I^{\bar{0}}H_n(X', X' - K; R)$. Arguing as in the proof of Proposition 8.1.28, Lemma 8.1.20 implies that for each $x \in K$ there is a neighborhood V' of x in X' with $|V'| \subset |U'|$ on which Γ'_K represents $\mathfrak{o}(z) \in I^{\bar{0}}H_n(X', X' - \{z\}; R)$ for each z in each regular stratum of V'. Letting W be a distinguished neighborhood of x in X with $|\bar{W}| \subset |V'|$, and letting $y \in W$ be a point contained simultaneously in regular strata of X and X' (and so also of $\mathfrak{X}$, as $\mathfrak{X}$ coarsens both X and X'), we have a diagram

$$\begin{array}{ccccc}
I^{\bar{0}}H_n(X', X' - |\bar{U}'|; R) & \xrightarrow{\cong} & I^{\bar{0}}H_n(\mathfrak{X}, \mathfrak{X} - |\bar{U}'|; R) & \xleftarrow{\cong} & I^{\bar{0}}H_n(X, X - |\bar{U}'|; R) \\
\downarrow & & \downarrow & & \downarrow \\
I^{\bar{0}}H_n(X', X' - |\bar{W}|; R) & \xrightarrow{\cong} & I^{\bar{0}}H_n(\mathfrak{X}, \mathfrak{X} - |\bar{W}|; R) & \xleftarrow{\cong} & I^{\bar{0}}H_n(X, X - |\bar{W}|; R) \\
\downarrow & & \downarrow & & \downarrow \\
I^{\bar{0}}H_n(X', X' - \{y\}; R) & \xrightarrow{\cong} & I^{\bar{0}}H_n(\mathfrak{X}, \mathfrak{X} - \{y\}; R) & \xleftarrow{\cong} & I^{\bar{0}}H_n(X, X - \{y\}; R).
\end{array}$$

The isomorphisms of the diagram come from Theorem 5.5.1. So, suppose we let γ denote the image of the element of $I^{\bar{0}}H_n(X, X - |\bar{U}'|; R)$ represented by the chain Γ'_K under the maps across the top row, so $\gamma \in I^{\bar{0}}H_n(X', X' - |\bar{U}|; R)$. By our choice of W, the chain Γ'_K represents $\mathfrak{o}(y) \in I^{\bar{0}}H_n(X', X' - \{y\}; R)$, and so it follows from the diagram and the compatibility of the orientation sections (especially at points of regular strata) that γ also maps to $\mathfrak{o}(y) \in I^{\bar{0}}H_n(X, X - \{y\}; R)$. But y was an arbitrary point in a regular stratum of both X and X' in $|W|$. As the unions of regular strata in X and X' are dense in X and X', respectively, every regular stratum of W contains a point y that is also in a regular stratum of X. So every regular stratum of $W \subset X'$ has a point at which γ represents $\mathfrak{o}(y)$ in $I^{\bar{0}}H_n(X, X - \{y\}; R)$, and it follows from Lemmas 8.1.20 and 8.1.16 that γ must map to $\mathfrak{o}(x)$ in $I^{\bar{0}}H_n(X, X - \{x\}; R)$. But x was arbitrary in K, so the class γ must map to $\Gamma_K \in I^{\bar{0}}H_n(X, X - K; R)$, again by Lemma 8.1.16.

Finally, we have a diagram

$$\begin{array}{ccccc}
I^{\bar{0}}H_n(X', X' - |\bar{U}'|; R) & \xrightarrow{\cong} & I^{\bar{0}}H_n(\mathfrak{X}, \mathfrak{X} - |\bar{U}'|; R) & \xleftarrow{\cong} & I^{\bar{0}}H_n(X, X - |\bar{U}'|; R) \\
\downarrow & & \downarrow & & \downarrow \\
I^{\bar{0}}H_n(X', X' - K; R) & \xrightarrow{\cong} & I^{\bar{0}}H_n(\mathfrak{X}, \mathfrak{X} - K; R) & \xleftarrow{\cong} & I^{\bar{0}}H_n(X, X - K; R),
\end{array}$$

and as the class of Γ'_K in $I^{\bar{0}}H_n(X', X' - |\bar{U}'|; R)$ maps to the fundamental class $\Gamma'_K \in I^{\bar{0}}H_n(X', X' - K; R)$ and to $\gamma \in I^{\bar{0}}H_n(X, X - |\bar{U}'|; R)$, and as γ maps to $\Gamma_K \in I^{\bar{0}}H_n(X, X - K; R)$, it follows that the composition along the bottom of the diagram takes Γ'_K to Γ_K. □

Remark 8.1.30 It follows from Proposition 8.1.29 that if X is a compact n-dimensional R-oriented stratified pseudomanifold without codimension-one strata, then the fundamental class Γ_X is a topological invariant in the following sense: Suppose that Y is another compact R-oriented stratified pseudomanifold without codimension-one strata and that $f : |X| \to |Y|$ is a *topological homeomorphism*, i.e. that it is a homeomorphism of the underlying spaces without regard to the filtrations. Then X induces an image pseudomanifold stratification, say Y', on Y, and an image R-orientation on Y' via the pointwise isomorphisms $I^{\bar{0}}H_n(X, X - \{x\}; R) \cong I^{\bar{0}}H_n(Y', Y' - \{f(x)\}; R)$ induced by the *filtered* homeomorphism $X \to Y'$. Suppose that f is orientation-preserving in the sense that this image R-orientation is compatible with the given R-orientation on Y in the sense of Corollary 8.1.11. Then applying Proposition 8.1.29 it must also be the case that $f(\Gamma_X) \in I^{\bar{0}}H_n(Y'; R)$ corresponds to Γ_Y under the canonical isomorphisms

$$I^{\bar{0}}H_n(Y'; R) \xrightarrow{\cong} I^{\bar{0}}H_n(\mathfrak{Y}; R) \xleftarrow{\cong} I^{\bar{0}}H_n(Y; R),$$

with $\mathfrak{Y}$ being the intrinsic filtration of $|Y|$.

This is essentially the content of [100, Corollary 5.23], although there it is mistakenly asserted that $f(\Gamma_X) \in I^{\bar{0}}H_n(Y; R)$.

8.1.6 Intersection Homology Factors the Cap Product

Our discussion of fundamental classes sets us up to make an interesting, and still somewhat mysterious, observation that goes back to the first paper on intersection homology by Goresky and MacPherson [105, Section 1.4]. There, working with classical compact oriented n-dimensional PL stratified pseudomanifolds and with GM perversities $\bar{p}$, Goresky and MacPherson noticed that for any such $\bar{p}$ the cap product with the fundamental class in ordinary (co)homology $H^{n-i}(X) \xrightarrow{\frown \Gamma_X} H_i(X)$ factors as[12]

$$H^{n-i}(X) \xrightarrow{\alpha_{\bar{p}}} I^{\bar{p}}H_i^{\mathrm{GM}}(X) \xrightarrow{\omega_{\bar{p}}} H_i(X),$$

where $\omega_{\bar{p}}$ is induced by the inclusion $I^{\bar{p}}S_*^{\mathrm{GM}}(X) \hookrightarrow S_*(X)$ and the other pieces of this composition will be described below.

The fundamental class is not constructed in detail in [105] but rather taken to be "the unique class ... [in] $H_n(X)$ which restricts to the local orientation class in $H_n(X, X - p)$ for every 'nonsingular point' $p \in X - \Sigma$." Such a fundamental class Γ_X can be found by fixing a triangulation and taking the sum over all n-simplices, compatibly oriented with the

12 We should probably write the Goresky–MacPherson statement in terms of PL homology and cohomology, but in a moment we will generalize to the singular chain setting.

orientation of X. This is evidently a cycle and satisfies the local condition on $X - X^{n-2}$. It is also easy to check that this cycle is allowable with respect to any GM perversity $\bar{s}$, and then an argument using Theorem 8.1.18 shows that Γ is a fundamental class in $I^{\bar{s}}H_n(X)$, which is isomorphic to $I^{\bar{s}}H_n^{\mathrm{GM}}(X)$ by Proposition 6.2.9. The image of Γ under $\omega_{\bar{s}}$ in $H_n(X)$ then must satisfy the stated local homology property at all points of $X - \Sigma$.

The construction of the map $\alpha_{\bar{p}}$ in [105] is more complicated, and two descriptions are provided there. The first is in terms of representations of cohomology classes by "mock bundles," a theory developed by Buoncristiano, Rourke, and Sanderson in [40] for describing cohomology theories geometrically. The second is a more hands-on construction using the cap product and the dual cells in a simplicial complex (for which a detailed exposition can be found in [182, Section 64]). Rather than pursue these simplicial constructions, let us instead utilize Proposition 7.3.8 to provide a version of the Goresky–MacPherson result for singular intersection homology on topological pseudomanifolds.

For this, let X be a compact R-oriented n-dimensional topological stratified pseudomanifold for R a Dedekind domain. Suppose further that $\bar{p}$ is any perversity with $\bar{0} \le \bar{p} \le \bar{t}$ and such that X is locally $(\bar{p}; R)$-torsion-free. This implies that X must be a classical pseudomanifold (no strata of codimension one), and $I^{\bar{p}}S_*^{\mathrm{GM}}(X;R) \cong I^{\bar{p}}S_*(X;R)$ by Proposition 6.2.9. If $\Gamma \in I^{\bar{0}}H_n(X;R)$ is the fundamental class on X guaranteed by Theorem 8.1.18 then $\omega_{\bar{0}}(\Gamma) \in H_n(X;R)$ is also a cycle, and the image of $\omega_{\bar{0}}(\Gamma)$ in $H_n(X - \{x\};R)$ is a compatibly oriented generator for each $x \in X - \Sigma$. So analogously to Definition 8.1.27, we can also abuse notation and refer to $\omega_{\bar{0}}(\Gamma) \in H_n(X;R)$ as the fundamental class Γ in ordinary homology.

As we have assumed X to be locally $(\bar{p}; R)$-torsion-free, Corollary 7.2.10 gives us a cap product $I_{D\bar{p}}H^{n-i}(X;R) \xrightarrow{\frown\Gamma} I^{\bar{p}}H_i(X;R)$. By Proposition 7.3.8, we therefore have that the following diagram always commutes:

$$\begin{array}{ccc} H^{n-i}(X;R) & \xrightarrow{\frown\Gamma} & H_i(X;R) \\ \Big\downarrow{\scriptstyle \omega^*_{D\bar{p}}} & & \Big\uparrow{\scriptstyle \omega_{\bar{p}}} \\ I_{D\bar{p}}H^{n-i}(X;R) & \xrightarrow{\frown\Gamma} & I^{\bar{p}}H_i(X;R), \end{array} \qquad (8.4)$$

where the bottom Γ technically lives in $I^{\bar{0}}H_n(X;R)$ and the top Γ lives in $H_n(X;R)$. So this demonstrates the following proposition:

Proposition 8.1.31 *Let R be a Dedekind domain, let X be a compact oriented n-dimensional topological stratified pseudomanifold, and let $\bar{p}$ be any perversity with $\bar{0} \le \bar{p} \le \bar{t}$ such that X is locally $(\bar{p}; R)$-torsion-free. Let Γ be a fundamental class for X. Then* *the cap product* $H^{n-i}(X;R) \xrightarrow{\frown\Gamma} H_i(X;R)$ *factors as*

$$H^{n-i}(X;R) \xrightarrow{\omega^*_{D\bar{p}}} I_{D\bar{p}}H^{n-i}(X;R) \xrightarrow{\frown\Gamma} I^{\bar{p}}H_i(X;R) \xrightarrow{\omega_{\bar{p}}} H_i(X;R).$$

Definition 8.1.32 Motivated by Proposition 8.1.31 and continuing its hypotheses, we can define a singular intersection homology analogue of the Goresky–MacPherson map $\alpha_{\bar{p}}$ to be the composition

$$H^{n-*}(X;R) \xrightarrow{\omega^*_{D\bar{p}}} I_{D\bar{p}}H^{n-*}(X;R) \xrightarrow{\mathcal{D}} I^{\bar{p}}H_*(X;R),$$

where $\mathcal{D}(\beta) = (-1)^{|\beta|n}\beta \frown \Gamma$.

Remark 8.1.33 The map $\mathcal{D}$, which at this point incorporates a somewhat unexpected sign into the cap product with the fundamental class, will become our standard Poincaré duality map below in Section 8.2. The reason for the sign is so that the duality map will be induced by an appropriately graded degree-n chain map at the chain level. We introduce this here to make our map $\alpha_{\bar{p}}$ also induced by a degree-n chain map at the (co)chain level. See Remark 8.2.2 for more details.

Below in Section 8.5.3, we will see that $\alpha_{\bar{p}}$ has some nice compatibility properties with respect to cup, cap, and intersection products. These properties were first recognized in [105].

(a) More General Factorizations

Definition 8.1.32 gives us a map $\alpha_{\bar{p}}$ that factors the ordinary singular duality map (which is not in general an isomorphism) through $I^{\bar{p}}H_*(X;R)$ when X is locally $(\bar{p};R)$-torsion-free. But the Goresky–MacPherson map $\alpha_{\bar{p}}$ is defined on compact oriented PL stratified pseudomanifolds with $\mathbb{Z}$ coefficients for *any* GM perversity $\bar{p}$. The torsion-free condition is not required! How can this be achieved in singular intersection homology?

One attempt might be based on the following observations:

1. By Example 6.3.22, all CS sets are locally torsion-free with respect to $\bar{t}$, and so $\mathfrak{L}$: $I_{\bar{t}}H^{n-*}(X;R) \to I^{\bar{0}}H_*(X;R)$ is always defined.
2. The map $\omega_{\bar{0}} : I^{\bar{0}}S_*(X;R) \hookrightarrow S_*(X;R)$ factors through every perversity with $\bar{0} \le \bar{p} \le \bar{t}$ via the maps $\tau_{\bar{0},\bar{p}}$ defined in Section 8.1.5, i.e. the composition

$$I^{\bar{0}}S_*(X;R) \xhookrightarrow{\tau_{\bar{0},\bar{p}}} I^{\bar{p}}S_*(X;R) \xhookrightarrow{\omega_{\bar{p}}} S_*(X;R)$$

is equal to $\omega_{\bar{0}}$.

So, in fact, there is a commutative diagram

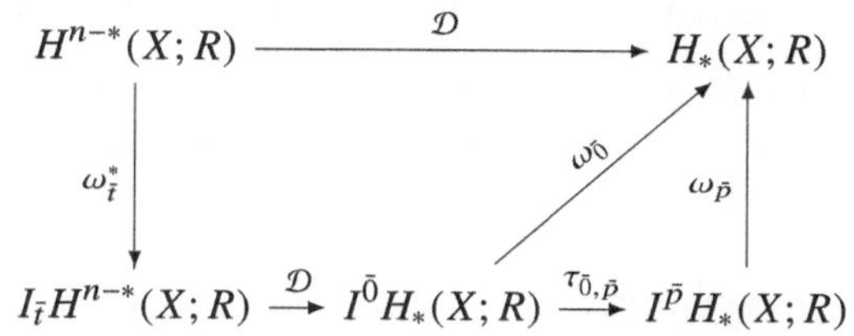

(8.5)

for *any* $\bar{p}$ with $\bar{0} \le \bar{p} \le \bar{t}$. Here the commutativity of the left-hand quadrilateral is just a special case of Proposition 8.1.31 by our first observation. So this construction demonstrates that the singular cap product factors through all intersection homology groups with $\bar{0} \le \bar{p} \le \bar{t}$, and so it looks like a reasonable alternative definition for $\alpha_{\bar{p}}$ in these more general cases might be as the composition $\alpha'_{\bar{p}}$ of the maps

$$H^{n-*}(X;R) \xrightarrow{\omega_{\bar{t}}^*} I_{\bar{t}}H^{n-*}(X;R) \xrightarrow{\mathcal{D}} I^{\bar{0}}H_*(X;R) \xrightarrow{\tau_{\bar{0},\bar{p}}} I^{\bar{p}}H_*(X;R).$$

However, the reader will see that it is not clear that one can extend the identities of Section 8.5.3 to $\alpha'_{\bar{p}}$. Of course, even if such identities do not hold in general, one would still like to know that $\alpha_{\bar{p}}$ and $\alpha'_{\bar{p}}$ at least agree when the former is defined. But even this is problematic! In particular, to have $\alpha_{\bar{p}} = \alpha'_{\bar{p}}$ when X is locally $(\bar{p};R)$-torsion-free, we would need to show that the following diagram commutes:

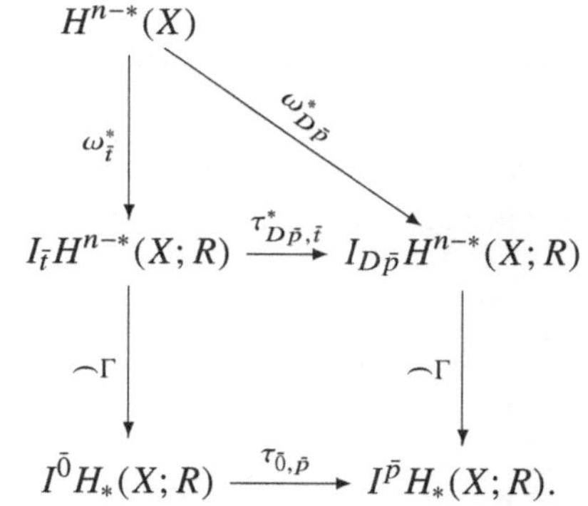

(8.6)

The top triangle certainly commutes, as the composite inclusion

$$I^{D\bar{p}}S_*(X;R) \xhookrightarrow{\tau_{D\bar{p},\bar{t}}} I^{\bar{t}}S_*(X;R) \xhookrightarrow{\omega_{\bar{t}}} S_*(X;R)$$

is equal to the inclusion $\omega_{D\bar{p}}$. But looking at the square, the top arrow points the wrong way for this to commute by naturality. And a "by-hands" attempt at a proof shows us the conundrum. By definition, the left cap product is defined using the algebraic diagonal that produces an element of $I^{\bar{0}}S_*(X;R) \otimes I^{\bar{t}}S_*(X;R)$, while the cap product on the right has its algebraic diagonal in $I^{\bar{p}}S_*(X;R) \otimes I^{D\bar{p}}S_*(X;R)$. As $\bar{0} \le \bar{p}$, but $\bar{t} \ge D\bar{p}$, there is no evident map to use to compare these two products. So it currently remains unclear how to define a map $\alpha_{\bar{p}}$ in singular intersection homology for arbitrary $\bar{p}$ with $\bar{0} \le \bar{p} \le \bar{t}$ that is satisfactorily compatible with the Goresky–MacPherson construction.

Remark 8.1.34 Symmetrically, we observe that for X a compact R-oriented stratified pseudomanifold the map $H^{n-*}(X;R) \xrightarrow{\frown \Gamma} H_*(X;R)$ also filters through all the intersection cohomology groups with $\bar{0} \le \bar{p} \le \bar{t}$ by the diagram

$$\begin{array}{ccccc}
H^{n-*}(X;R) & & \xrightarrow{\quad\mathcal{D}\quad} & & H_*(X;R) \\
\downarrow{\scriptstyle \omega^*_{\bar{p}}} & \searrow{\scriptstyle \omega^*_{\bar{0}}} & & & \uparrow{\scriptstyle \omega_{\bar{t}}} \\
I_{\bar{p}}H^{n-*}(X;R) & \xrightarrow{\tau^*_{\bar{0},\bar{p}}} & I^{\bar{0}}H_*(X;R) & \xrightarrow{\mathcal{D}} & I^{\bar{t}}H_*(X;R).
\end{array}$$

Here we use that such spaces are all locally $(\bar{0}, R)$-torsion-free by Example 6.3.22 and Corollary 8.2.5, below.

Remark 8.1.35 Chataur, Saralegi-Aranguren, and Tanré in [54] have constructed a diagram of the form of Diagram (8.4) for an arbitrary perversity $\bar{p}$ by replacing $I_{D\bar{p}}H^{n-i}(X;R)$ with an alternative construction they call Thom–Whitney cohomology, denoted $H^{n-*}_{TW,\bar{p}}(X;R)$. They use this to construct alternative factorizations of the singular cap product through $I^{\bar{p}}H_*(X;R)$. They also demonstrate that the corresponding version of $\alpha_{\bar{p}}$ satisfies a version of property (8.23) of Proposition 8.5.13, below, relating it to cup products and intersection products.

Remark 8.1.36 Another interesting observation of Chataur, Saralegi-Aranguren, and Tanré in [54], which we here adapt to our own setting, is the following. Suppose again that R is a Dedekind domain and that X is now a *normal* compact R-oriented n-dimensional stratified pseudomanifold. Consider once again Diagram (8.5) but with $\bar{p} = \bar{t}$ so that we have

$$\begin{array}{ccccc}
H^{n-*}(X;R) & & \xrightarrow{\quad\mathcal{D}\quad} & & H_*(X;R) \\
\downarrow{\scriptstyle \omega^*_{\bar{t}}} & & & \nearrow{\scriptstyle \omega_{\bar{0}}} & \uparrow{\scriptstyle \omega_{\bar{t}}} \\
I_{\bar{t}}H^{n-*}(X;R) & \xrightarrow{\mathcal{D}} & I^{\bar{0}}H_*(X;R) & \xrightarrow{\tau_{\bar{0},\bar{t}}} & I^{\bar{t}}H_*(X;R).
\end{array}$$

As X is assumed to be normal, Corollary 5.1.9 (together with the details of its proof) implies that $\omega_{\bar{t}}$ is an isomorphism, and similarly, by Proposition 7.1.18, the map $\omega^*_{\bar{t}}$ is an isomorphism. We will see below in Theorem 8.2.4 that $\mathcal{D} : I_{\bar{t}}H^{n-*}(X;R) \to I^{\bar{0}}H_*(X;R)$ is a Poincaré duality isomorphism. We thus obtain the following fact (cf. [54, Theorem B]):

Proposition 8.1.37 *Let R be a Dedekind domain and suppose that X is a* normal *compact R-oriented n-dimensional stratified pseudomanifold. Then the Poincaré duality map in ordinary singular (co)homology (given by the signed cap product with the fundamental* *class) is an isomorphism* $\mathfrak{D} : H^{n-i}(X;R) \xrightarrow{\mathfrak{D}} H_i(X;R)$ *if and only if the canonical map* $\tau_{\bar{0},\bar{t}} : I^{\bar{0}}H_i(X;R) \to I^{\bar{t}}H_i(X;R)$ *is an isomorphism.*

8.1.7 Product Spaces

We include here some results on product orientations that we shall need below.

Lemma 8.1.38 *Suppose* X_1 *and* X_2 *are stratified pseudomanifolds. Then* $X_1 \times X_2$ *is* R*-orientable if and only if* X_1 *and* X_2 *are* R*-orientable.*

Proof If $M_i = X_i - \Sigma_{X_i}$, then $X_1 \times X_2 - \Sigma_{X_1 \times X_2} = M_1 \times M_2$. For manifolds, we have already recalled that a product of manifolds is orientable if and only if the factors are; see footnote 5 on page 510. But by Definition 8.1.5 a stratified pseudomanifold X is orientable if and only if $X - \Sigma_X$ is orientable. The lemma follows. □

Turning to specific orientations, if M_1, M_2 are R-oriented manifolds of dimensions n_1, n_2, then $M_1 \times M_2$ has a natural R-orientation. In fact, recalling again footnote 5 on page 510, suppose $\mathcal{O}_1, \mathcal{O}_2$ are the orientation bundles over M_1, M_2, respectively, and that $\mathcal{O}$ is the orientation bundle for $M_1 \times M_2$. Let $\mathcal{O}_1 \hat{\otimes} \mathcal{O}_2$ be the bundle over $M_1 \times M_2$ whose fiber at $(x_1, x_2) \in M_1 \times M_2$ is $H_{n_1}(M_1, M_1 - \{x_1\}; R) \otimes H_{n_2}(M_2, M_2 - \{x_2\}; R) \cong H_{n_1+n_2}(M_1 \times M_2, M_1 \times M_2 - \{(x_1, x_2)\}; R) \cong R$. Then $\mathcal{O}_1 \hat{\otimes} \mathcal{O}_2$ is isomorphic[13] to $\mathcal{O}$. Furthermore, if $\mathfrak{o}_1, \mathfrak{o}_2$ are the global orientation sections of the respective orientation bundles $\mathcal{O}_1, \mathcal{O}_2$, then $\mathfrak{o}_1 \hat{\otimes} \mathfrak{o}_2 \in \mathcal{O}_1 \hat{\otimes} \mathcal{O}_2$ provides an orientation $\mathfrak{o}$ of $\mathcal{O}$ via the isomorphism with $\mathcal{O}_1 \hat{\otimes} \mathcal{O}_2$, as the cross product of generators of $H_{n_1}(M_1, M_1 - \{x_1\}; R)$ and $H_{n_2}(M_2, M_2 - \{x_2\}; R)$ is a generator of $H_{n_1+n_2}(M_1 \times M_2, M_1 \times M_2 - \{(x_1, x_2)\}; R)$. Locally, if $\mathfrak{o}_1(x_1)$ and $\mathfrak{o}_2(x_2)$ are represented by chains $\xi_1 \in H_{n_1}(M_1, M_1 - \{x_1\}; R)$ and $\xi_2 \in H_{n_2}(M_2, M_2 - \{x_2\}; R)$, then $\mathfrak{o}((x_1, x_2))$ is represented by $\xi_1 \times \xi_2 \in H_{n_1+n_2}(M_1 \times M_2, M_1 \times M_2 - \{(x_1, x_2)\}; R)$.

By the same reasoning, if X_1 and X_2 are R-oriented stratified pseudomanifolds, then $\mathcal{O}_1 \hat{\otimes} \mathcal{O}_2$ is isomorphic to $\mathcal{O}$ over the regular strata of $X_1 \times X_2$, and so by Lemma 8.1.14 the image under the cross product of $\mathfrak{o}_1 \hat{\otimes} \mathfrak{o}_2$ in $\mathfrak{o}$ is an orientation section of $X_1 \times X_2$. We also have an analogous statement to that for manifolds concerning chain representatives.

In the case of pseudomanifolds, we therefore have the following result:

Proposition 8.1.39 *Let* R *be a Dedekind domain. Suppose that* X_1 *and* X_2 *are* R*-oriented stratified pseudomanifolds of respective dimensions* n_1 *and* n_2*. Let* $K_i \subset X_i$ *be compact subsets. Let* Γ_{K_i} *be chains representing the fundamental classes of* X_i *over* K_i*.*

[13] In slightly more detail, the cross product sets up a map of (partially defined) presheaves such that if $U = U_1 \times U_2$ is a product open subset of $M_1 \times M_2$, then we have the cross product map

$$H_{n_1}(M_1, M_1 - \bar{U}_1; R) \otimes H_{n_2}(M_2, M_2 - \bar{U}_2; R) \xrightarrow{\times} H_{n_1+n_2}(M_1 \times M_2, M_1 \times M_2 - \bar{U}_1 \times \bar{U}_2; R).$$

For each $(x_1, x_2) \in M_1 \times M_2$ there is a cofinal system of such product neighborhoods with each U_i a Euclidean ball, in which case the cross product is an isomorphism. Taking direct limits, this induces an isomorphism of sheaves.

Then $\Gamma_{K_1} \times \Gamma_{K_2} \in I^{\bar{0}}H_{n_1+n_2}(X_1 \times X_2, (X_1 \times X_2) - (K_1 \times K_2); R)$ *is the fundamental class of* $X_1 \times X_2$ *over* $K_1 \times K_2$ *with respect to the product orientation on* $X_1 \times X_2$.

Proof The proof is similar in spirit to other proofs in the preceding sections, so we will be a little sketchy in the details here.

Suppose $(x_1, x_2) \in K_1 \times K_2$. We must show that for any such (x_1, x_2) the chain $\Gamma_{K_1} \times \Gamma_{K_2}$ represents

$$\mathfrak{o}((x_1, x_2)) \in I^{\bar{0}}H_{n_1+n_2}(X_1 \times X_2, (X_1 \times X_2) - \{(x_1, x_2)\}; R).$$

But suppose U_1 and U_2 are distinguished neighborhoods of x_1 and x_2 in X_1 and X_2, respectively. We can assume that $U_1 \subset X_1 - |\partial \Gamma_{K_1}|$ and $U_2 \subset X_2 - |\partial \Gamma_{K_2}|$. Then, we know by Lemma 8.1.20 that, assuming U_1 and U_2 are sufficiently small, the chain Γ_{K_1} represents $\mathfrak{o}_1(z_1)$ for each z_1 contained in a regular stratum of U_1 and that Γ_{K_2} represents $\mathfrak{o}(z_2)$ for each z_2 contained in a regular stratum of U_2. But, via the isomorphism between $\mathcal{O}$ and $\mathcal{O}_1 \hat{\otimes} \mathcal{O}_2$ over the regular strata that we discussed just above, this implies that $\Gamma_{K_1} \times \Gamma_{K_2}$ must represent

$$\mathfrak{o}((z_1, z_2)) \in I^{\bar{0}}H_{n_1+n_2}(X_1 \times X_2, (X_1 \times X_2) - \{(z_1, z_2)\}; R).$$

So, by applying Lemma 8.1.20 again to a small enough distinguished neighborhood of (x_1, x_2) in $X_1 \times X_2$, the chain $\Gamma_{K_1} \times \Gamma_{K_2}$ must indeed represent $\mathfrak{o}((x_1, x_2))$. □

8.2 Poincaré Duality

At last, in this section, we come to the *raison d'être* for intersection homology: Poincaré duality on stratified pseudomanifolds.

8.2.1 The Duality Map

By Theorem 8.1.18, we know that if R is a Dedekind domain and X is a compact R-oriented stratified pseudomanifold of dimension n, then X admits a fundamental class $\Gamma_X \in I^{\bar{0}}H_n(X; R)$ that is consistent with the R-orientation in the sense that, for any $x \in X$, the image of Γ_X in $I^{\bar{p}}H_n(X, X - \{x\}; R)$ takes the value of the orientation section $\mathfrak{o}(x)$. By Corollary 7.2.10, the cap product with this class,

$$\frown \Gamma_X : I_{\bar{p}}H^i(X; R) \to I^{D\bar{p}}H_{n-i}(X; R),$$

is defined if X is locally $(\bar{p}; R)$-torsion-free[14]. The intersection (co)homology version of Poincaré duality states that this map is an isomorphism. Of course, if X is an unfiltered manifold, and hence automatically locally torsion-free, this reduces to precisely the statement of classical Poincaré duality.

[14] The requirement for the existence of this cap product is actually that X be either locally $(\bar{p}; R)$-torsion-free or locally $(D\bar{p}; R)$-torsion-free, but it will follow as a consequence of Poincaré duality that an R-oriented stratified pseudomanifold is locally $(\bar{p}; R)$-torsion-free if and only if it is locally $(D\bar{p}; R)$-torsion-free. See Corollary 8.2.5

More generally, if X is any R-oriented locally $(\bar{p}; R)$-torsion-free n-dimensional stratified pseudomanifold, we have a Poincaré duality isomorphism

$$\mathcal{D} : I_{\bar{p}}H^i_c(X; R) \to I^{D\bar{p}}H_{n-i}(X; R),$$

where X is no longer assumed to be compact. For this, recall that if X is an R-oriented stratified pseudomanifold of dimension n (not necessarily compact) and if $K \subset X$ is a compact subset, we have shown in Lemma 8.1.16 that we have a fundamental class $\Gamma_K \in I^{\bar{0}}H_n(X, X - K; R)$ that is consistent with the R-orientation in the sense that, for any $x \in K$, the image of Γ_K in $I^{\bar{p}}H_n(X, X - \{x\}; R)$ takes the value of the orientation section $\mathfrak{o}(x)$. Suppose now that $K' \subset X$ is another compact subset with $K \subset K'$. If $\Gamma_{K'} \in I^{\bar{0}}H_n(X, X - K'; R)$ is the corresponding fundamental class, then the image of Γ' in $I^{\bar{0}}H_n(X, X - K; R)$ must be Γ_K: the map between homology groups is induced by inclusion so the image of $\Gamma_{K'}$ and its image in $I^{\bar{0}}H_n(X, X - K; R)$ can be represented by the same chain. But, by definition, this chain represents $\mathfrak{o}(x)$ for each $x \in K'$, and so, in particular, at each $x \in K$. It follows that the chain therefore represents Γ_K by the uniqueness of Lemma 8.1.16. This argument shows that the collection $\{\Gamma_K\}$, as K varies over the compact subsets of X, constitutes an element of $\varprojlim I^{\bar{0}}H_n(X, X - K; R)$. By Lemma 7.4.6, the cap product induces a map

$$\frown : I_{\bar{p}}H^i_c(X; R) \otimes \varprojlim I^{\bar{0}}H_n(X, X - K; R) \to I^{D\bar{p}}H_{n-i}(X; R).$$

Definition 8.2.1 Let R be a Dedekind domain and X an R-oriented locally $(\bar{p}; R)$-torsion-free n-dimensional stratified pseudomanifold. Let $\Gamma \in \varprojlim I^{\bar{0}}H_n(X, X-K; R)$ corresponding to the collection of fundamental classes $\{\Gamma_K\}$ as above. We define the *duality map* $\mathcal{D}$ to be the map

$$\begin{aligned} I_{\bar{p}}H^i_c(X; R) &\to I^{D\bar{p}}H_{n-i}(X; R), \\ \alpha &\to (-1)^{in}\alpha \frown \Gamma. \end{aligned}$$

In other words, if $\alpha \in I_{\bar{p}}H^i_c(X; R)$, then

$$\mathcal{D}(\xi) = (-1)^{in}\alpha \frown \Gamma.$$

When X is compact, the module $I^{\bar{0}}H_n(X; R)$ is initial among the $I^{\bar{0}}H_n(X, X - K; R)$, so $\varprojlim I^{\bar{0}}H_n(X, X - K; R) \cong I^{\bar{0}}H_n(X; R)$, and, in this case, $\mathcal{D}$ reduces, up to sign, to the standard cap product with Γ.

Remark 8.2.2 Wait a minute – where did that sign come from? Notice that we have defined $\mathcal{D}(\alpha)$ as $(-1)^{|\alpha|n}\alpha \frown \Gamma$ and not simply as $\alpha \frown \Gamma$. This deserves some explanation. The issue is that there are many circumstances in geometric topology where it is desirable to think of the Poincaré duality map as a chain map from cochains to chains. Of course, in our treatment here, we have generally only pursued cap products as operators on homology and cohomology elements, but, in classical algebraic topology, if $\xi \in S_k(X)$ is a fixed chain and we have chosen a fixed definition of the cap product[15], then $\frown \xi$ induces a function of chain complexes $S^*(X) \to S_{k-*}(X)$. In particular, if M is a compact manifold, then one would like the Poincaré duality map determined by some sort of cap product with a fundamental class to provide a map $S^*(M) \to S_{k-*}(M)$ in the appropriate category of (co)chain complexes. This is important, for example, in surgery theory.

The problem is that $\frown \Gamma$ is not a chain map because it does not obey the proper sign conventions. To explain, first let us recall that if we have a map of (homologically indexed) chain complexes $f : C_* \to D_*$ that raises degrees by k, i.e. f restricts to homomorphisms $f_i : C_i \to D_{i+k}$, then f is considered a chain map of degree k if $\partial f = (-1)^k f\partial$; see Appendix A.1.4 or [71, Section VI.10]. Most of the chain maps we have considered so far have been degree-zero chain maps, and so the sign is invisible. Next, recall that we can consider cohomologically indexed complexes to be equivalent to homologically indexed complexes via the identification $C^* = C_{-*}$. Unfortunately, this clashes with the standard topological use whereby, say, $S_*(X)$ and $S^*(X)$ are not reindexings of the same complex, but rather different complexes; so, for clarity in the remainder of this remark, we replace $S^*(X)$ with $\mathrm{Hom}^*(S_*(X); R) = \mathrm{Hom}_{-*}(S_*(X); R)$. This last identification allows us to treat cochain complexes as homologically indexed.

Now, fixing an element $\xi \in S_k(X)$, the function $\alpha \to \alpha \frown \xi$ provides homomorphisms $\mathrm{Hom}_{-*}(S_*(X); R)$ to $S_{k-*}(M)$, and so raises the (homological) degree by k. However, by Lemma 7.2.19, fixing a particular cap product at the chain level yields

$$\partial(\alpha \frown \xi) = (d\alpha) \frown \xi + (-1)^{|\alpha|}\alpha \frown \partial\xi.$$

If $\partial\xi = 0$, we obtain $\partial(\alpha \frown \xi) = (d\alpha) \frown \xi$, so $\alpha \to \alpha \frown \xi$ is not signed as a chain map

Continuing to assume ξ is a cycle, consider now the map $\mathcal{D}_\xi$ such that $\mathcal{D}_\xi(\alpha) = (-1)^{|\alpha||\xi|}\alpha \frown \xi$. Then we have

$$\begin{aligned}\mathcal{D}_\xi(d\alpha) &= (-1)^{(|\alpha|+1)|\xi|}(d\alpha) \frown \xi \\ &= (-1)^{(|\alpha|+1)|\xi|}\partial(\alpha \frown \xi) = (-1)^{|\xi|}\partial((-1)^{|\alpha||\xi|}\alpha \frown \xi) \\ &= (-1)^{|\xi|}\partial(\mathcal{D}_\xi(\alpha)).\end{aligned}$$

Therefore, $\mathcal{D}_\xi$ is a chain map. So, in particular, if M is a compact oriented manifold and Γ is the fundamental class, then $\mathcal{D} = \mathcal{D}_\Gamma$ is a chain map from cochains to chains, as desired. Similarly, if X is a compact oriented stratified pseudomanifold, then $\mathcal{D} = \mathcal{D}_\Gamma$ is a chain map, assuming we have fixed a particular choice[16] of $\bar{\mathbf{d}}$ so that we can speak of the cap product at the chain level.

Of course, these signs are not critical for obtaining our Poincaré duality isomorphisms in each fixed degree, but they allow us to stay consistent with the necessary properties

[15] As we have discussed, there is some flexibility in the definition of the cap product at the chain level due to the need to choose a specific Alexander–Whitney map, but the ambiguities can be removed by passing to the appropriate homotopy category of (co)chain complexes.

[16] In the manifold case, we can assume that we are using the traditional Alexander–Whitney diagonal.

of Poincaré duality as a chain map in other sources. In particular, this is consistent with [98, 100]. For more about this sign convention, see [89, Section 4.1].

Before proceeding on to the proof of Poincaré duality, we next present an example that demonstrates the necessity of the torsion-free condition if we hope to have a Poincaré duality. Of course we don't know how to define the cap product of the duality map if the locally torsion-free condition fails, but this example shows that we can't necessarily have duality isomorphisms by any means.

Example 8.2.3 Let X be the suspension of $\mathbb{R}P^3$, i.e. $X = S(\mathbb{R}P^3)$ with the standard suspension filtration as in Example 2.3.4, assuming $\mathbb{R}P^3$ is filtered trivially. So X has two singular points corresponding to the suspension points $\{\mathfrak{n}, \mathfrak{s}\}$. As $\mathbb{R}P^3$ is $\mathbb{Z}$-orientable, so is $X - \Sigma_X \cong (-1,1) \times \mathbb{R}P^3$. Let us choose a perversity $\bar{p}$ on X so that $\bar{p}(\{\mathfrak{n}\}) = \bar{p}(\{\mathfrak{s}\}) = 1$. As the codimension of the suspension points is four, the value of $D\bar{p}$ on these points is then

$$D\bar{p}(\{\mathfrak{n}\}) = \bar{t}(\{\mathfrak{n}\}) - \bar{p}(\{\mathfrak{n}\}) = \operatorname{codim}(\{\mathfrak{n}\}) - 2 - \bar{p}(\{\mathfrak{n}\}) = 4 - 2 - 1 = 1,$$

and similarly for $\mathfrak{s}$, so $D\bar{p} = \bar{p}$ in this situation. We also have $\bar{p} \leq \bar{t}$, so we even have $I^{\bar{p}}H_*(X) \cong I^{\bar{p}}H_*^{\mathrm{GM}}(X)$. Applying the suspension computation of Theorem 6.3.13, we have

$$I^{\bar{p}}H_i(X) \cong \begin{cases} I^{\bar{p}}\tilde{H}_{i-1}(\mathbb{R}P^3), & i > 2, \\ 0, & i = 2, \\ I^{\bar{p}}H_i(\mathbb{R}P^3), & i < 2. \end{cases}$$

So $I^{\bar{p}}H_4(X) \cong I^{\bar{p}}H_0(X) \cong \mathbb{Z}$ and $I^{\bar{p}}H_1(X) \cong \mathbb{Z}_2$, and the other groups are zero.

Similarly, as X is compact, we have

$$I_{\bar{p}}H^i_c(X) = I_{\bar{p}}H^i(X) \cong \operatorname{Hom}(I^{\bar{p}}H_i(X);\mathbb{Z}) \oplus \operatorname{Ext}(I^{\bar{p}}H_{i-1}(X);\mathbb{Z}),$$

using Theorem 7.1.4. So, applying our above computation for $I^{\bar{p}}H_*(X)$, we obtain the following:

$$\begin{array}{lcl} I_{\bar{p}}H^4(X) & \cong & \mathbb{Z}, \\ I_{\bar{p}}H^3(X) & \cong & 0, \\ I_{\bar{p}}H^2(X) & \cong & \mathbb{Z}_2, \\ I_{\bar{p}}H^1(X) & \cong & 0, \\ I_{\bar{p}}H^0(X) & \cong & \mathbb{Z}, \end{array} \qquad \begin{array}{lcl} I^{\bar{p}}H_0(X) & \cong & \mathbb{Z}, \\ I^{\bar{p}}H_1(X) & \cong & \mathbb{Z}_2, \\ I^{\bar{p}}H_2(X) & \cong & 0, \\ I^{\bar{p}}H_3(X) & \cong & 0, \\ I^{\bar{p}}H_4(X) & \cong & \mathbb{Z}. \end{array}$$

Comparing across the rows, we see that Poincaré duality fails. This is a manifestation of our observation in Remark 7.1.6 of the clash between the "cleanly truncated" cone formula in intersection homology (Theorem 6.2.13), and the cone formula in intersection cohomology (Proposition 7.1.5), which contains that extra torsion term. Here this contrast results in a failure of Poincaré duality.

We do observe, however, that if we replaced our $\mathbb{Z}$ coefficients with, say, coefficients in $\mathbb{Q}$, then similar computations would result in identical answers except with $\mathbb{Z}$ terms replaced with $\mathbb{Q}$ terms and $\mathbb{Z}_2$ terms replaced with zero. Then we have

$$\begin{array}{lclclcl} I_{\bar p}H^4(X;\mathbb{Q}) & \cong & \mathbb{Q}, & & I^{\bar p}H_0(X;\mathbb{Q}) & \cong & \mathbb{Q},\\ I_{\bar p}H^3(X;\mathbb{Q}) & \cong & 0, & & I^{\bar p}H_1(X;\mathbb{Q}) & \cong & 0,\\ I_{\bar p}H^2(X;\mathbb{Q}) & \cong & 0, & & I^{\bar p}H_2(X;\mathbb{Q}) & \cong & 0,\\ I_{\bar p}H^1(X;\mathbb{Q}) & \cong & 0, & & I^{\bar p}H_3(X;\mathbb{Q}) & \cong & 0,\\ I_{\bar p}H^0(X;\mathbb{Q}) & \cong & \mathbb{Q}, & & I^{\bar p}H_4(X;\mathbb{Q}) & \cong & \mathbb{Q}. \end{array}$$

In this case, we see that the corresponding groups are isomorphic, although we have not yet shown that this isomorphism is via the duality map. △

8.2.2 The Poincaré Duality Theorem

We now turn to the official statement and proof of Poincaré duality.

Theorem 8.2.4 (Poincaré duality) *Suppose R is a Dedekind domain, and let X be an n-dimensional R-oriented locally $(\bar p;R)$-torsion-free*[17] *stratified pseudomanifold. Then the duality map*

$$\mathcal{D} : I_{\bar p}H^i_c(X;R) \to I^{D\bar p}H_{n-i}(X;R)$$

is an isomorphism for all i. In particular, if X is compact, then the cap product

$$\frown \Gamma_X : I_{\bar p}H^i(X;R) \to I^{D\bar p}H_{n-i}(X;R)$$

is an isomorphism.

Proof Although we have been careful to define our duality map $\mathcal{D}$ to account for the signs needed for $\mathcal{D}$ to be a chain map when thought of at the chain/cochain level, we can safely ignore these signs and work with the cap product in fixed degrees for the purposes of proving the theorem.

The proof will be by induction on depth. The base case is that for which X is an unfiltered manifold, in which case this is classical Poincaré duality. The reader can find a proof of manifold duality in, for example, [125, Theorem 3.35], though it will not be difficult for us to prove this case along with the others. For the manifold base case, we will use the Mayer–Vietoris argument (Theorem 5.1.2), while the inductive step uses the Mayer–Vietoris argument for CS sets (Theorem 5.1.4). We will check the conditions to apply these theorems for the base and inductive steps in parallel. In the inductive case, we suppose X has depth $d \geq 1$ and that the theorem has been proven for stratified pseudomanifolds of depth $< d$. In fact, as we go through the argument we will complete the proof of the base case before we ever need to use the induction assumption.

To apply Theorems 5.1.2 and 5.1.4, we define our functors as follows. If $U \subset X$ is an open set, let $F_*(U) = I_{\bar p}H^*_c(U;R)$, let $G_*(U) = I^{D\bar p}H_{n-*}(U;R)$, and let $\Phi = \mathcal{D}^U$, where $\mathcal{D}^U$ is defined as in Lemma 7.4.8. Namely, if Γ^U is the image of Γ under the canonical map

[17] Or, equivalently, X can be locally $(D\bar p;R)$-torsion-free; see footnote 14 on page 536 and Corollary 8.2.5.

$\varprojlim_{K\subset X} I^{\bar{0}}H_n(X, X-K;R) \to \varprojlim_{K\subset U} I^{\bar{0}}H_n(U, U-K;R)$ of Lemma 7.4.7, then the map $\mathcal{D}^U$ is the signed cap product, in the sense of Lemma 7.4.6, with Γ^U. Furthermore, by the second paragraph of Lemma 7.4.7, the image Γ_K^U of Γ^U in $I^{\bar{0}}H_n(U, U-K;R)$ is also the image of $\Gamma_K^X = \Gamma_K$ under the excision isomorphism $I^{\bar{0}}H_n(X, X-K;R) \to I^{\bar{0}}H_n(U, U-K;R)$. So, by the commutative diagram

$$\begin{array}{ccc} I^{\bar{0}}H_n(X, X-K;R) & \xleftarrow{\cong} & I^{\bar{0}}H_n(U, U-K;R) \\ \downarrow & & \downarrow \\ I^{\bar{0}}H_n(X, X-\{x\};R) & \xleftarrow{\cong} & I^{\bar{0}}H_n(U, U-\{x\};R), \end{array}$$

the image of Γ_K^U in $I^{\bar{0}}H_n(U, U-\{x\};R)$ is just the value of the orientation section on U compatible by excision with the restriction of the orientation section over X. Therefore, the map $\mathcal{D}^U$ is really just the same as the duality map on U determined by restricting the orientation from X.

We must verify that the conditions of Theorems 5.1.2 and 5.1.4 are satisfied.

Mayer–Vietoris Step. The functors F_* and G_* admit Mayer–Vietoris sequences with Φ inducing a map between them by Lemma 7.4.8. Actually, the diagram of Lemma 7.4.8 only commutes up to signs, but this is sufficient for the arguments in Theorems 5.1.2 and 5.1.4, as we can change some signs of maps to still invoke the Five Lemma where it is used in the proofs of those theorems. Incidentally, the commutativity of the diagram of Lemma 7.4.8, appropriately restricted to each summand in the middle term, demonstrates that Φ is indeed a natural transformation (up to signs).

Limit Step. For the limit condition in Theorem 5.1.4, we aim to employ Lemma 5.1.6, for which we need to verify that the maps $\varinjlim_\alpha F_*(U_\alpha) \to F_*\left(\bigcup_\alpha U_\alpha\right)$ and $\varinjlim_\alpha G_*(U_\alpha) \to G_*\left(\bigcup_\alpha U_\alpha\right)$ are isomorphisms. For G_*, this is the content of Lemma 6.3.16; for F_* this is Lemma 7.4.4.

Euclidean Neighborhood Step. If U is empty, then Φ is trivially an isomorphism, and if U is an open subset of X contained in a single stratum and homeomorphic to Euclidean space, then U must be contained in a regular stratum, as X is a stratified pseudomanifold. In this case, the map Φ is the Poincaré duality map for Euclidean space.

Let $\bar{B}_r$ be the closed ball of radius r centered at the origin in $\mathbb{R}^n$. Such balls are cofinal among all compact subspaces of $\mathbb{R}^n$, and so $H_c^*(\mathbb{R}^n;R) \cong \varinjlim H^*(\mathbb{R}^n, \mathbb{R}^n - \bar{B}_r;R)$ and $\varprojlim H_*(\mathbb{R}^n, \mathbb{R}^n - K;R) \cong \varprojlim H_*(\mathbb{R}^n, \mathbb{R}^n - \bar{B}_r;R)$. But for $r \le s$ the maps $H^*(\mathbb{R}^n, \mathbb{R}^n - \bar{B}_r;R) \to H^*(\mathbb{R}^n, \mathbb{R}^n - \bar{B}_s;R)$ and $H_*(\mathbb{R}^n, \mathbb{R}^n - \bar{B}_s;R) \to H_*(\mathbb{R}^n, \mathbb{R}^n - \bar{B}_r;R)$ are all isomorphisms by the homotopy invariance of homology. Therefore $H_c^*(\mathbb{R}^n;R) \cong H^*(\mathbb{R}^n, \mathbb{R}^n - \bar{B}_s;R)$ and $\varprojlim H_*(\mathbb{R}^n, \mathbb{R}^n - \bar{B}_r;R) \cong H_*(\mathbb{R}^n, \mathbb{R}^n - \bar{B}_s;R)$ for any fixed s. It thus follows from the discussion preceding Lemma 7.4.6 that the map $H_c^*(\mathbb{R}^n;R) \xrightarrow{\frown\Gamma} H_{n-*}(\mathbb{R}^n;R)$ is isomorphic to the map $H^*(\mathbb{R}^n,\mathbb{R}^n - \bar{B}_s;R) \xrightarrow{\frown\Gamma_{\bar{B}_s}} H_{n-*}(\mathbb{R}^n;R)$ for any s. By elementary computations, the modules $H^*(\mathbb{R}^n,\mathbb{R}^n - \bar{B}_s;R)$ and $H_*(\mathbb{R}^n,\mathbb{R}^n - \bar{B}_s;R)$ are trivial unless $* = n$, in which case each is isomorphic to R. Furthermore, we know that $\Gamma_{\bar{B}_s}$ is a generator of $H_n(\mathbb{R}^n,\mathbb{R}^n - \bar{B}_s;R)$. By the Universal Coefficient Theorem, we can compute

$$H^n(\mathbb{R}^n,\mathbb{R}^n - \bar{B}_s;R) \cong \mathrm{Hom}(H_n(\mathbb{R}^n,\mathbb{R}^n - \bar{B}_s;R),R) \cong \mathrm{Hom}(R,R) \cong R,$$

as $H_{n-1}(\mathbb{R}^n,\mathbb{R}^n - \bar{B}_s;R) = 0$. Let $\alpha \in H^n(\mathbb{R}^n,\mathbb{R}^n - \bar{B}_s;R)$ be the generator such that $\alpha(\Gamma_{\bar{B}_s}) = 1$. Then by Proposition 7.3.25 we have

$$1 = \alpha(\Gamma_{\bar{B}_s}) = \mathbf{a}(\alpha \frown \Gamma_{\bar{B}_s}).$$

Consequently, we must have that $\alpha \frown \Gamma_{\bar{B}_s}$ represents a generator of $H_0(\mathbb{R}^n) \cong R$. This shows that $H^*(\mathbb{R}^n,\mathbb{R}^n - \bar{B}_s;R) \xrightarrow{\frown\Gamma_{\bar{B}_s}} H_{n-*}(\mathbb{R}^n;R)$ is an isomorphism, and hence so is $H_c^*(\mathbb{R}^n;R) \xrightarrow{\frown\Gamma} H_{n-*}(\mathbb{R}^n;R)$.

Wrapping Up the Base Case; On to Induction. We have now demonstrated (or referred out to Lemma 7.4.4) all the conditions needed to invoke Theorem 5.1.2 to demonstrate Poincaré duality for unfiltered manifolds. This completes the base case. So for the remainder of the argument we may assume the depth of X is $d > 0$ while, by the induction hypothesis, the theorem holds for depths less than d. So far we have not used the induction hypothesis, but it will be needed for the next step.

Distinguished Neighborhood Step. Finally, for the last remaining condition of Theorem 5.1.4, we must show that if $U \cong \mathbb{R}^k \times cL^{n-k-1}$ is a distinguished neighborhood in X and $\Phi : F_*(\mathbb{R}^k \times (cL - \{v\})) \to G_*(\mathbb{R}^k \times (cL - \{v\}))$ is an isomorphism, then so is $\Phi : F_*(\mathbb{R}^k \times cL) \to G_*(\mathbb{R}^k \times cL)$. As has often been the case in our Mayer–Vietoris arguments, we will show directly that $\Phi : F_*(\mathbb{R}^k \times cL) \to G_*(\mathbb{R}^k \times cL)$ is always an isomorphism, relying on our induction assumptions more so than the supposition that Φ is an isomorphism $F_*(\mathbb{R}^k \times (cL - \{v\})) \to G_*(\mathbb{R}^k \times (cL - \{v\}))$.

There are essentially two cases to consider, the first being the range of indices in which all the modules are zero: By stratified homotopy invariance and the cone formula (Theorem 6.2.13),

$$G_i(\mathbb{R}^k \times cL) = I^{D\bar{p}}H_{n-i}(\mathbb{R}^k \times cL;R) \cong I^{D\bar{p}}H_{n-i}(cL;R) = 0$$

if $n - i \geq n - k - D\bar{p}(v) - 1$, i.e. if $i \leq k + D\bar{p}(\{v\}) + 1$. But now

$$D\bar{p}(\{v\}) = \bar{t}(\{v\}) - \bar{p}(\{v\}) = \mathrm{codim}(\{v\}) - 2 - \bar{p}(\{v\}) = n - k - 2 - \bar{p}(\{v\}).$$

So, altogether, $G_i(\mathbb{R}^k \times cL) = 0$ if $i \leq n - \bar{p}(\{v\}) - 1$.

Next, consider $F_i(\mathbb{R}^k \times cL) = I_{\bar{p}}H_c^i(\mathbb{R}^k \times cL;R)$. To compute the compactly supported intersection cohomology, we can choose a cofinal collection of compact subsets of the form

$K_{r,s} = \bar{B}_r \times \bar{c}_s L$, where $\bar{B}_r$ is the closed ball of radius r in $\mathbb{R}^k$ and $\bar{c}_s L$ is our closed subcone out to s in the cone coordinate. By Lemma 6.4.17, the direct system $I_{\bar{p}}H^i(\mathbb{R}^k \times cL, \mathbb{R}^k \times cL - K_{r,s}; R)$ is constant, with all terms being isomorphic to $I_{\bar{p}}H^i(\mathbb{R}^k \times cL, \mathbb{R}^k \times cL - \{(0,v)\}; R)$. Therefore, applying the Universal Coefficient Theorem (Theorem 7.1.4) and the Künneth Theorem (Theorem 6.3.20) applied to

$$(\mathbb{R}^k \times cL, \mathbb{R}^k \times cL - \{(0,v)\}) \cong (\mathbb{R}^k, \mathbb{R}^k - \{0\}) \times (cL, cL - \{v\}),$$

we have

$$\begin{aligned}
&F_i(\mathbb{R}^k \times cL)\\
&\quad = I_{\bar{p}}H^i_c(\mathbb{R}^k \times cL; R) \;\cong I_{\bar{p}}H^i(\mathbb{R}^k \times cL, \mathbb{R}^k \times cL - \{(0,v)\}; R)\\
&\quad \cong \operatorname{Hom}(I^{\bar{p}}H_i(\mathbb{R}^k \times cL, \mathbb{R}^k \times cL - \{(0,v)\}; R), R)\\
&\qquad \oplus \operatorname{Ext}(I^{\bar{p}}H_{i-1}(\mathbb{R}^k \times cL, \mathbb{R}^k \times cL - \{(0,v)\}; R), R)\\
&\quad \cong \operatorname{Hom}(I^{\bar{p}}H_{i-k}(cL, cL - \{v\}; R), R) \oplus \operatorname{Ext}(I^{\bar{p}}H_{i-k-1}(cL, cL - \{v\}; R), R).
\end{aligned}$$

By the relative cone formula (Corollary 6.2.15), $I^{\bar{p}}H_j(cL, cL - \{v\}; R) = 0$ for $j \le n - k - \bar{p}(\{v\}) - 1$. So, the first summand of $F_i(\mathbb{R}^k \times cL)$ is zero when $i - k \le n - k - \bar{p}(\{v\}) - 1$, i.e. when $i \le n - \bar{p}(\{v\}) - 1$. Similarly, the second summand vanishes when $i - k - 1 \le n - k - \bar{p}(\{v\}) - 1$, i.e. when $i \le n - \bar{p}(\{v\})$. Therefore, $F_i(\mathbb{R}^k \times cL) = 0$ for $i \le n - \bar{p}(\{v\}) - 1$, so that $F_i(\mathbb{R}^k \times cL) = 0 = G_i(\mathbb{R}^k \times cL)$ for $i \le n - \bar{p}(\{v\}) - 1$, and this must be induced by Φ as there is a unique map between trivial modules.

Next, we must consider $i \ge n - \bar{p}(\{v\})$. This will involve the following diagram, in which η is the fundamental class of $H_k(\mathbb{R}^k, \mathbb{R}^k - \{0\}; R)$ consistent with the standard orientation:

$$\begin{array}{ccc}
I_{\bar{p}}H^{i-k-1}(L; R) & \xrightarrow[\cong]{\frown \Gamma_L} & I^{D\bar{p}}H_{n-i}(L; R)\\
\uparrow{\scriptstyle \mathfrak{l}^*}\ \cong & & \downarrow{\scriptstyle \mathfrak{l}}\ \cong\\
I_{\bar{p}}H^{i-k-1}(cL - \{v\}; R) & \xrightarrow{\frown \mathfrak{l}(\Gamma_L)} & I^{D\bar{p}}H_{n-i}(cL - \{v\}; R)\\
\downarrow{\scriptstyle d^*}\ \cong & & \downarrow\ \cong\\
I_{\bar{p}}H^{i-k}(cL, cL - \{v\}; R) & \xrightarrow{\frown \bar{c}\mathfrak{l}(\Gamma_L)} & I^{D\bar{p}}H_{n-i}(cL; R)\\
\downarrow{\scriptstyle \eta^* \times}\ \cong & & \downarrow\ \cong\\
I_{\bar{p}}H^i(\mathbb{R}^k \times cL, \mathbb{R}^k \times cL - \{(0,v)\}; R) & \xrightarrow{\frown (\eta \times \bar{c}\mathfrak{l}(\Gamma_L))} & I^{D\bar{p}}H_{n-i}(\mathbb{R}^k \times cL; R).
\end{array} \tag{8.7}$$

If we can show that this diagram commutes up to sign and that all the maps so labeled are isomorphisms if $i \geq n - \bar{p}(\{v\})$, then it will follow that all of the maps are isomorphisms, and so, in particular, the map at the bottom,

$$I_{\bar{p}}H^i(\mathbb{R}^k \times cL, \mathbb{R}^k \times cL - \{(0,v)\}; R) \xrightarrow{\frown(\eta \times \bar{c}\mathrm{I}(\Gamma_L))} I^{D\bar{p}}H_{n-i}(\mathbb{R}^k \times cL; R),$$

is an isomorphism. Furthermore, by Corollary 8.1.23 this bottom horizontal map is the cap product with the fundamental class of $\mathbb{R}^k \times cL$ over $\{(0,v)\}$, and we claim that this map being an isomorphism suffices to prove that Φ is an isomorphism, completing the proof of the theorem.

For this last claim, taking $K = \bar{B}_r \times \bar{c}_s L$ and $K' = \bar{B}_{r'} \times \bar{c}_{s'} L$ with $r < r'$ and $s < s' < 1$, we have diagrams of the form

$$\begin{array}{ccccc}
\begin{matrix} I_{\bar{p}}H^i(\mathbb{R}^k \times cL, \\ \mathbb{R}^k \times cL - \{(0,v)\}; R) \end{matrix} & \xrightarrow{\cong} & \begin{matrix} I_{\bar{p}}H^i(\mathbb{R}^k \times cL, \\ \mathbb{R}^k \times cL - K; R) \end{matrix} & \xrightarrow{\cong} & \begin{matrix} I_{\bar{p}}H^i(\mathbb{R}^k \times cL, \\ \mathbb{R}^k \times cL - K'; R) \end{matrix} \\
\downarrow{\scriptstyle \frown\Gamma_{\{(0,v)\}}} & & \downarrow{\scriptstyle \frown\Gamma_K} & & \downarrow{\scriptstyle \frown\Gamma_{K'}} \\
I_{D\bar{p}}H_{n-i}(\mathbb{R}^k \times cL; R) & \xleftarrow{=} & I_{D\bar{p}}H_{n-i}(\mathbb{R}^k \times cL; R) & \xleftarrow{=} & I_{D\bar{p}}H_{n-i}(\mathbb{R}^k \times cL; R).
\end{array}$$

The horizontal maps are isomorphisms by Lemma 6.4.17, and we let Γ'_K, Γ_K, and $\Gamma_{\{(0,v)\}}$ here denote the fundamental classes in $\mathbb{R}^k \times cL$. Then $\Gamma_{\{(0,v)\}}$ is the image of Γ_K, which is the image of $\Gamma_{K'}$, all under the relevant inclusion maps by the discussion establishing the duality map in Section 8.2.1. Commutativity of the diagram is due to naturality of the cap product (Proposition 7.3.6). Via the horizontal isomorphisms, taking the direct limit over all such K of the form $\bar{B}_r \times \bar{c}_s L$ shows that the map

$$I_{\bar{p}}H^i(\mathbb{R}^k \times cL, \mathbb{R}^k \times cL - \{(0,v)\}; R) \xrightarrow{\frown\Gamma_{\{(0,v)\}}} I_{D\bar{p}}H_{n-i}(\mathbb{R}^k \times cL; R)$$

is isomorphic (up to sign) to the duality map $\Phi : F_*(\mathbb{R}^k \times cL) \to G_*(\mathbb{R}^k \times cL)$.

So it remains to show that Diagram (8.7) commutes with all the vertical maps being isomorphisms.

The top square. Turning to our claims about Diagram (8.7), the top horizontal map $I_{\bar{p}}H^{i-k-1}(L; R) \xrightarrow{\frown\Gamma_L} I^{D\bar{p}}H_{n-i}(L; R)$ is an isomorphism by our induction assumption. For this, we observe that L is a stratified pseudomanifold by Lemma 2.4.11, and it is R-orientable as a link of the R-orientable X by the arguments in the proof of Lemma 8.1.14. Additionally, L is locally $(\bar{p}; R)$-torsion-free because, by Remark 2.4.14, its links are all also links of X, which is locally $(\bar{p}; R)$-torsion-free, and the locally torsion-free vanishing condition can be stated in terms of the dimensions of the links themselves, without reference to the ambient space. Therefore, as L has lower depth than X, the induction hypothesis applies. We can here choose a particular orientation of L such that if we give $\mathbb{R}^k$ and $(0,1)$ their standard orientations then the product orientation on $\mathbb{R}^k \times (0,1) \times L$ agrees with the given orientation inherited from $\mathbb{R}^k \times cL \subset X$.

The map I is the inclusion $\mathrm{I} : L \hookrightarrow cL - \{v\}$ of L into the cone at some fixed cone coordinate, so the two top vertical maps of Diagram (8.7) are isomorphisms by stratified homotopy invariance. Furthermore, I is a normally nonsingular inclusion, so the top square commutes by Proposition 7.3.6 and Remark 7.3.7.

The center square. We first show that the vertical maps are isomorphisms. By the cone formula, if $i \geq n - \bar{p}(\{v\})$ then we are in the range where we have isomorphisms

$$\begin{array}{ccc} I^{D\bar{p}}H_{n-i}(L;R) & \xrightarrow{\cong} & I^{D\bar{p}}H_{n-i}(cL - \{v\};R) \\ & \searrow^{\cong} & \downarrow{\cong} \\ & & I^{D\bar{p}}H_{n-i}(cL;R), \end{array}$$

treating L as a subspace of cL at some fixed cone coordinate.

For cohomology, we note again that by the Universal Coefficient Theorem $I_{\bar{p}}H^j(cL;R) \cong \operatorname{Hom}(I^{\bar{p}}H_j(cL;R),R) \oplus \operatorname{Ext}(I^{\bar{p}}H_{j-1}(cL;R),R)$, while $I^{\bar{p}}H_a(cL;R) = 0$ for $a \geq n - k - \bar{p}(\{v\}) - 1$ by the cone formula. So $I_{\bar{p}}H^j(cL;R) = 0$ for $j > n - k - \bar{p}(\{v\}) - 1$. If $j = n - k - \bar{p}(\{v\}) - 1$, then

$$I_{\bar{p}}H^j(cL;R) \cong \operatorname{Ext}(I^{\bar{p}}H_{n-k-\bar{p}(\{v\})-2}(cL;R),R) \cong \operatorname{Ext}(I^{\bar{p}}H_{n-k-\bar{p}(\{v\})-2}(L;R),R),$$

using the cone formula. But we have assumed that X is locally $(\bar{p};R)$-torsion-free, so by Definition 6.3.21, $I^{\bar{p}}H_{n-k-\bar{p}(\{v\})-2}(L;R)$ is flat. This module is also finitely generated by Corollary 6.3.40, as Dedekind domains are Noetherian [30, Theorem VII.2.2.1]. Furthermore, finitely generated flat modules over Noetherian rings are projective [146, Theorem 4.38], so $\operatorname{Ext}\left(I^{\bar{p}}H_{n-k-\bar{p}(\{v\})-2}(L;R),R\right) = 0$. Therefore, we have $I_{\bar{p}}H^j(cL;R) = 0$ for $j \geq n-k-\bar{p}(\{v\})-1$, and so, by the long exact sequence of the pair, $I_{\bar{p}}H^j(cL-\{v\};R) \xrightarrow{d^*} I_{\bar{p}}H^{j+1}(cL, cL-\{v\};R)$ is an isomorphism for $j \geq n-k-\bar{p}(\{v\})-1$. Taking $j = i-k-1$, this provides the isomorphisms $I_{\bar{p}}H^{i-k-1}(cL - \{v\};R) \xrightarrow{d^*} I_{\bar{p}}H^{i-k}(cL, cL - \{v\};R)$ for $i - k - 1 \geq n - k - \bar{p}(\{v\}) - 1$, i.e. for $i \geq n - \bar{p}(\{v\})$, as desired. Notice the role that the locally torsion-free condition plays in this argument!

For the commutativity of the center square of Diagram (8.7), we let $\bar{c}\mathrm{I}(\Gamma_L)$ be the class of the singular cone on the chain $\mathrm{I}(\Gamma_L)$ (see Example 3.4.7). In particular, this means that $\bar{c}\mathrm{I}(\Gamma_L)$ maps to $\mathrm{I}(\Gamma_L)$ under the isomorphism $\partial_* : I^{\bar{0}}H_{n-k}I(cL, cL - \{v\};R) \to I^{\bar{0}}H_{n-k-1}(cL - \{v\};R)$; this is an isomorphism by the relative cone formula (Corollary 6.2.15). For the commutativity of this square, up to sign, we can apply Proposition 7.3.37. Comparing that lemma to our setting, the X of the lemma is cL, the subspace B is $cL - \{v\}$, and $A = \emptyset$. The chain ξ of the lemma is our $\bar{c}\mathrm{I}(\Gamma_L) \in I^{\bar{0}}H_{n-k}(cL, cL - \{v\};R)$, and, as $A = \emptyset$, the map e in the lemma is the identity map. With these identifications, the lemma applies to demonstrate that this square commutes up to sign.

The bottom square. For commutativity of the bottom square of Diagram (8.7), we will apply Proposition 7.3.55 so that if $\alpha \in I_{\bar{p}}H^{i-k}(cL, cL - \{v\}; R)$ then

$$(\eta^* \times \alpha) \frown (\eta \times \bar{c}\mathfrak{l}(\Gamma_L)) = \pm(\eta^* \frown \eta) \times (\alpha \frown \bar{c}\mathfrak{l}(\Gamma_L)). \tag{8.8}$$

For the proposition to apply, we must check that the hypotheses are satisfied. As $\mathbb{R}^k$ is a manifold, all possible torsion-free conditions are satisfied. The space $\mathbb{R}^k \times cL$ is an open set of X, and so it is locally $(\bar{p}; R)$-torsion-free as this is a local condition. For cL, all links of cL are links of X by Remark 2.4.14 and so cL is also locally $(\bar{p}; R)$-torsion-free; here we continue our standard abuse of notation to let $\bar{p}$ also denote the perversity on L or cL induced by the perversity $\bar{p}$ on X. With this abuse, and as $\mathbb{R}^k$ is trivially filtered, we can identify $\bar{p}$ on $\mathbb{R}^k \times cL$ with the product perversity $\hat{Q}^2(\bar{0}, \bar{p})$ in the statement of Proposition 7.3.55, letting $\bar{0}$ stand in for the unique perversity on $\mathbb{R}^k$. This verifies all the requirements to apply the proposition, so that the equality (8.8) holds.

Now, as η^* is dual to η, we have using Proposition 7.3.25 that $\eta^*(\eta) = \mathbf{a}(\eta^* \frown \eta) = 1$, and so $\eta^* \frown \eta$ is a generator of $H_0(\mathbb{R}^k; R)$. Thus $\eta^* \frown \eta$ can be represented by a single 0-simplex, say σ_y, with image $y \in \mathbb{R}^k$. So

$$(\eta^* \times \alpha) \frown (\eta \times \bar{c}\mathfrak{l}(\Gamma_L)) = \pm\sigma_y \times (\alpha \frown \bar{c}\mathfrak{l}(\Gamma_L)).$$

But now if the vertical map on the right of the bottom square of Diagram (8.7) is the inclusion of cL as $\{y\} \times cL$, then $\sigma_y \times (\alpha \frown \bar{c}\mathfrak{l}(\Gamma_L))$ represents the image of $\alpha \frown \bar{c}\mathfrak{l}(\Gamma_L) \in I^{D\bar{p}}H_{n-i}(cL; R)$ under this inclusion by Theorem 6.3.19 (compare Proposition 5.2.21). Thus the bottom square commutes up to sign.

Conclusion. Now that we have shown commutativity of Diagram (8.7), implying that Φ is an isomorphism over $\mathbb{R}^k \times cL$, we may invoke Theorem 5.1.4 to complete the induction step, and thus the proof of the Poincaré Duality Theorem (Theorem 8.2.4) is complete. □

8.2.3 Duality of Torsion-Free Conditions

As an immediate consequence of Poincaré duality, we prove the fact, alluded to in footnote 14 on page 536, that an oriented stratified pseudomanifold is locally $(\bar{p}; R)$-torsion-free if and only if it is locally $(D\bar{p}; R)$-torsion-free.

Corollary 8.2.5 *Suppose R is a Dedekind domain and that X is an n-dimensional R-oriented stratified pseudomanifold (or ∂-stratified pseudomanifold). Then X is locally $(\bar{p}; R)$-torsion-free if and only if X is locally $(D\bar{p}; R)$-torsion-free.*

Proof Let L be a link of a point x contained in the singular stratum S; suppose $\dim(L) = \ell$. Then L is a stratified pseudomanifold by Lemma 2.4.11, and it is R-orientable as a link of the R-orientable X by the arguments in the proof of Lemma 8.1.14. By Remark 2.4.14, the links of L are links of X, and the definition of locally torsion-free (Definition 6.3.21) shows that only dimensions (and not codimensions) are involved in the locally torsion-free condition. Therefore, if X is locally $(\bar{q}; R)$-torsion-free for some perversity $\bar{q}$, then so is L.

Suppose now that X, and so L, is locally $(\bar{p}; R)$-torsion-free. So the Poincaré Duality Theorem (Theorem 8.2.4) applies to L. Noting that L is compact by definition, recalling that $D(D\bar{p}) = \bar{p}$, and applying the Universal Coefficient Theorem (Theorem 7.1.4), we have

$$\begin{aligned} I^{D\bar{p}}H_{\ell-D\bar{p}(S)-1}(L;R) &\cong I_{D(D\bar{p})}H^{D\bar{p}(S)+1}(L;R) \\ &\cong \operatorname{Hom}(I^{\bar{p}}H_{D\bar{p}(S)+1}(L;R),R) \oplus \operatorname{Ext}(I^{\bar{p}}H_{D\bar{p}(S)}(L;R),R). \end{aligned}$$

Next, using that $\dim(X) = \ell + \dim(S) + 1$, so that $\operatorname{codim}_X(S) = \ell + 1$, we compute

$$\begin{aligned} D\bar{p}(S) &= \bar{t}(S) - \bar{p}(S) = \operatorname{codim}_X(S) - 2 - \bar{p}(S) = \ell + 1 - 2 - \bar{p}(S) \\ &= \ell - \bar{p}(S) - 1. \end{aligned}$$

By assumption, $I^{\bar{p}}H_{\ell-\bar{p}(S)-1}(L;R)$ is a flat R-module (see Definition 6.3.21); it is also finitely generated by Corollary 6.3.40, using that Dedekind domains are Noetherian [30, Theorem VII.2.2.1]. But finitely generated flat modules over Noetherian rings are projective [146, Theorem 4.38], so $\operatorname{Ext}(I^{\bar{p}}H_{D\bar{p}(S)}(L;R),R) = 0$. Thus we have reduced our formula for $I^{D\bar{p}}H_{\ell-D\bar{p}(S)-1}(L;R)$ to

$$I^{D\bar{p}}H_{\ell-D\bar{p}(S)-1}(L;R) \cong \operatorname{Hom}(I^{\bar{p}}H_{D\bar{p}(S)+1}(L;R),R).$$

Furthermore, for any R-module A, the module $\operatorname{Hom}(A,R)$ is torsion-free. This is an elementary fact: if $f \in \operatorname{Hom}(A,R)$ and $f \neq 0$, then there is some $x \in A$ such that $f(x) \neq 0$. But then if $r \in R$ with $r \neq 0$, we have $(rf)(x) = r \cdot f(x) \neq 0$, as R is an integral domain. So $f \neq 0$ implies $rf \neq 0$. Thus $I^{D\bar{p}}H_{\ell-D\bar{p}(S)-1}(L;R)$ is R-torsion-free and so flat, as R is a Dedekind domain (see Section A.4.2). This concludes the proof that locally $(\bar{p}; R)$-torsion-free implies locally $(D\bar{p}; R)$-torsion-free.

For the other direction, if X is locally $(D\bar{p}; R)$-torsion-free, we can use $D\bar{p}$ in place of $\bar{p}$ in the part of the corollary already proven to conclude that X is locally $(D(D\bar{p}); R)$-torsion-free. But $D(D\bar{p}) = \bar{p}$, completing the proof. □

8.2.4 Topological Invariance of Poincaré Duality

We demonstrated in Proposition 8.1.29 that the orientation class of a stratified pseudomanifold is invariant under appropriate changes of stratification. We also know that intersection homology, in general, is invariant to the stratification of X as a CS set, assuming that X has no codimension-one strata and that $\bar{p}$ is a GM perversity; this follows from Theorem 5.5.1 and Proposition 6.2.9, recalling that GM perversities $\bar{p}$ are defined and satisfy $\bar{0} \leq \bar{p} \leq \bar{t}$ if there are no codimension-one strata. Similarly, with the same hypotheses, we have invariance of intersection cohomology by Theorem 7.1.17 and of the cap product by Theorem 7.3.10. Putting these results together, we see that the Poincaré duality isomorphism is independent of the stratification in the following sense:

Theorem 8.2.6 *Suppose R is a Dedekind domain and that $\bar{p}$ is a GM perversity. Let X and X' be two n-dimensional compatibly R-oriented compact stratified pseudomanifold stratifications with no codimension-one strata of the same underlying space $|X|$. Suppose*

X is locally $(\bar{p};R)$-torsion-free, which implies X' and $\mathfrak{X}$ are locally $(\bar{p};R)$-torsion-free by Proposition 5.5.9. Then there are canonical isomorphisms

$$\begin{array}{ccc} I_{\bar p}H^i(X;R) & \xrightarrow{\mathcal{D}} & I^{D\bar p}H_{n-i}(X;R)\\ \Big\uparrow\cong & & \Big\downarrow\cong \\ I_{\bar p}H^i(\mathfrak{X};R) & \xrightarrow{\mathcal{D}} & I^{D\bar p}H_{n-i}(\mathfrak{X};R)\\ \Big\downarrow\cong & & \Big\uparrow\cong\\ I_{\bar p}H^i(X';R) & \xrightarrow{\mathcal{D}} & I^{D\bar p}H_{n-i}(X';R). \end{array}$$

Technically, we haven't defined the $\mathcal{D}$ in the middle row, as $\mathfrak{X}$ is not guaranteed to be a stratified pseudomanifold, only a CS set. However, as guaranteed by Proposition 8.1.29, we can use the image of the fundamental class of X (or X') as a stand-in to define the duality map in the middle row.

Remark 8.2.7 Analogously to what we saw in Remark 8.1.30 concerning fundamental classes, it follows from such invariance results that Poincaré duality is topologically invariant in the following broader sense: Suppose X and Y are compact n-dimensional R-oriented stratified pseudomanifolds without codimension-one strata, and suppose that $f : |X| \to |Y|$ is a *topological homeomorphism*, i.e. that it is a homeomorphism of the underlying spaces without regard to the stratifications. Then X induces an image stratification, say Y', on Y, and an image R-orientation on Y'. Suppose that f is orientation-preserving in that the image R-orientation is compatible with the given R-orientation on Y in the sense of Corollary 8.1.11. Then employing Remark 8.1.30, Theorem 8.2.6, and naturality, we arrive at a canonical diagram of isomorphisms of the following form:

$$\begin{array}{ccc} I_{\bar p}H^i(X;R) & \xrightarrow{\mathcal{D}} & I^{D\bar p}H_{n-i}(X;R)\\ f^*\Big\uparrow\cong & & f\Big\downarrow\cong \\ I_{\bar p}H^i(Y';R) & \xrightarrow{\mathcal{D}} & I^{D\bar p}H_{n-i}(Y';R)\\ \Big\downarrow\cong & & \Big\downarrow\cong\\ I_{\bar p}H^i(Y;R) & \xrightarrow{\mathcal{D}} & I^{D\bar p}H_{n-i}(Y;R). \end{array}$$

8.3 Lefschetz Duality

In this section, we will extend our duality results to compact orientable ∂-stratified pseudomanifolds. The reader might want to look back at Section 2.7 for the definitions and details concerning these spaces. There appear to be more general versions of manifold duality that do not require compactness (for example, see [125, Section 3.3, Exercise 25]), but it does not seem to be as straightforward, for example, to construct the relevant fundamental classes without compactness. With X compact, and using that ∂X must have a stratified collar in X by Definition 2.7.1, it is relatively straightforward to derive Lefschetz duality as a consequence of Poincaré duality, and we will be content with this case.

8.3.1 Orientations and Fundamental Classes

We first consider orientations and fundamental classes for ∂-stratified pseudomanifolds.

Definition 8.3.1 Let X be an n-dimensional ∂-stratified pseudomanifold. We say that X is *R-orientable* if and only if the stratified pseudomanifold $X - \partial X$ is R-orientable. Equivalently, by Definition 8.1.5, X is R-orientable if and only if the manifold $(X - \Sigma_X) - \partial(X - \Sigma_X)$ is R-orientable. An *R-orientation* of X is an R-orientation of $X - \partial X$.

Lemma 8.3.2 *If X is an R-orientable ∂-stratified pseudomanifold, then so is ∂X.*

Proof By definition, if X is an R-orientable ∂-stratified pseudomanifold then $(X - \Sigma_X) - \partial(X-\Sigma_X)$ is R-orientable. But $(X-\Sigma_X)$ is a ∂-manifold. Therefore, its boundary $\partial(X-\Sigma_X)$ is R-orientable by classical manifold theory; see e.g. [71, Proposition VIII.2.19]. Lastly, we observe that $\partial(X - \Sigma_X) = \partial X - \Sigma_{\partial X}$, so ∂X is R-orientable. □

It follows from Lemma 8.1.14 and the rest of Section 8.1.3 that if the ∂-stratified pseudomanifold X is R-oriented then there is an orientation sheaf $\mathcal{O}^{\bar p}$ over $X - \partial X$ and a unique global section $\mathfrak{o}^{\bar p}$ determined by the orientation. We then have the following analogue of Theorem 8.1.18.

Theorem 8.3.3 *Let R be a Dedekind domain, and let X be a* compact *R-oriented n-dimensional ∂-stratified pseudomanifold with perversity $\bar p \geq \bar 0$. Then the following hold:*

1. *$I^{\bar q}H_i(X;R) = I^{\bar q}H_i(X,\partial X;R) = 0$ for $i > n$ and for* any *perversity $\bar q$.*
2. *There is a unique class $\Gamma_X^{\bar p} \in I^{\bar p}H_n(X,\partial X;R)$ such that, for any $x \in X - \partial X$, the image of $\Gamma_X^{\bar p}$ under the composition induced by inclusion and excision, $I^{\bar p}H_n(X,\partial X;R) \to I^{\bar p}H_n(X,X-\{x\};R) \cong I^{\bar p}H_n(X-\partial X,(X-\partial X)-\{x\};R)$, corresponds to the value of the orientation section $\mathfrak{o}^{\bar p}(x)$.*
3. *If $\{x_j\}_{j=1}^m$ is a collection of points of $X - \partial X$, one in each regular stratum, then $I^{\bar p}H_n(X,\partial X;R) \cong \bigoplus_j I^{\bar p}H_n(X,X-\{x_j\};R) \cong R^m$ via the map that takes an element of $I^{\bar p}H_n(X,\partial X;R)$ to the direct sum of its images in the $I^{\bar p}H_n(X,X-\{x_j\};R)$.*

Proof Recall that ∂X has a collar neighborhood N in X filtered homeomorphic to $[0,1)\times\partial X$ by Definition 2.7.1. Then the inclusion $\partial X \hookrightarrow N$ is a stratified homotopy equivalence, so it follows from the long exact sequences and the Five Lemma that $I^{\bar q}H_*(X,N;R) \cong I^{\bar q}H_*(X,\partial X;R)$.

By Lemma 8.3.2, we know ∂X is R-orientable, so $I^{\bar q}H_i(N;R) \cong I^{\bar q}H_i(\partial X;R) = 0$ for $i > n-1$ by Theorem 8.1.18. Furthermore, by excision, $I^{\bar q}H_i(X,N;R) \cong I^{\bar q}H_i(X-\partial X, N-\partial X;R)$. As $X-\partial X$ is an R-orientable stratified pseudomanifold and as $X-N = (X-\partial X)-(N-\partial X)$ is compact, $I^{\bar q}H_i(X,N;R) = 0$ for $i > n$ by Lemma 8.1.16. Thus $I^{\bar q}H_i(X,\partial X;R) = 0$ for $i > n$. Together with $I^{\bar q}H_i(\partial X;R) = 0$ for $i > n-1$, the long exact sequence of the pair $(X,\partial X)$ shows that $I^{\bar q}H_i(X;R) = 0$ for $i > n$.

Now, let $\bar p \geq 0$, and let $K = X - N$, which is compact as X is compact. Note that $(X-\partial X)-K = N-\partial X$. Let $\Gamma^{\bar p}_K \in I^{\bar p}H_n(X-\partial X, N-\partial X;R)$ be the fundamental class of $X-\partial X$ over K, as guaranteed by Lemma 8.1.16. By excision and homotopy equivalence,

$$I^{\bar p}H_n(X-\partial X, N-\partial X;R) \cong I^{\bar p}H_n(X,N;R) \cong I^{\bar p}H_n(X,\partial X;R).$$

We let $\Gamma^{\bar p}_X$ be the image of $\Gamma^{\bar p}_K$ under these isomorphisms. We will show that $\Gamma^{\bar p}_X$ has the desired properties and that it is independent of the choice of N.

First, suppose $x \in K = X - N$. We have a commutative diagram

$$\begin{array}{ccccc} I^{\bar p}H_n(X-\partial X, N-\partial X;R) & \xrightarrow{\cong} & I^{\bar p}H_n(X,N;R) & \xleftarrow{\cong} & I^{\bar p}H_n(X,\partial X;R)\\ \downarrow & & \downarrow & & \downarrow \\ I^{\bar p}H_n(X-\partial X, X-(\partial X\cup\{x\});R) & \xrightarrow{\cong} & I^{\bar p}H_n(X,X-\{x\};R) & \xleftarrow{=} & I^{\bar p}H_n(X,X-\{x\};R) \end{array}$$

with the horizontal maps being isomorphisms by excisions and stratified homotopy invariance. As $\Gamma^{\bar p}_K$ maps to a generator of $I^{\bar p}H_n(X-\partial X, X-(\partial X\cup\{x\});R)$, we see from the diagram and the definition that $\Gamma^{\bar p}_X$ maps to the corresponding generator of $I^{\bar p}H_n(X,X-\{x\};R)$, as desired.

Now, suppose that N' is another collar neighborhood of ∂X within N, and let $K' = X-N'$. Then we have the commutative diagram

$$\begin{array}{ccccc} I^{\bar p}H_n(X-\partial X, N-\partial X;R) & \xrightarrow{\cong} & I^{\bar p}H_n(X,N;R) & \xleftarrow{\cong} & I^{\bar p}H_n(X,\partial X;R)\\ \uparrow & & \uparrow & & \uparrow{=} \\ I^{\bar p}H_n(X-\partial X, N'-\partial X;R) & \xrightarrow{\cong} & I^{\bar p}H_n(X,N';R) & \xleftarrow{\cong} & I^{\bar p}H_n(X,\partial X;R). \end{array}$$

By the uniqueness properties of Lemma 8.1.16, the fundamental class $\Gamma^{\bar p}_{K'} \in I^{\bar p}H(X-\partial X, N'-\partial X;R)$ must map to $\Gamma^{\bar p}_K \in I^{\bar p}H(X-\partial X, N-\partial X;R)$. It therefore follows from the diagram that N and N' both yield the same $\Gamma^{\bar{p}}_X$. If now N'' is *any* other collar of ∂X (so not necessarily contained in N), then there is a collar N''' of ∂X in $N \cap N''$; as ∂X is compact, this follows from the Tube Lemma [181, Theorem 26.8]. Using the preceding argument twice, we see that the corresponding $\Gamma^{\bar{p}}_K$, $\Gamma^{\bar{p}}_{K''}$, and $\Gamma^{\bar{p}}_{K'''}$ all map to the same $\Gamma^{\bar{p}}_X$. So $\Gamma^{\bar{p}}_X$ is independent of the choice of collar. As every $x \in X - \partial X$ lies outside of some collar of ∂X, it follows that $\Gamma^{\bar{p}}_X$ restricts as desired for every $x \in X - \partial X$. Uniqueness of $\Gamma^{\bar{p}}$ follows from the uniqueness of the $\Gamma^{\bar{p}}_K$.

For the last part of the theorem, we may suppose $\partial X \neq \emptyset$, or the result follows immediately from Theorem 8.1.18. Let N be a collar of ∂X in the complement of $\cup_{j=1}^m \{x_j\}$. Let $X^+ = X \cup_{\partial X} \bar{c}(\partial X)$, and let $N^+ = N \cup_{\partial X} \bar{c}(\partial X)$. Notice that $N^+ \cong c(\partial X)$. We filter X^+ so that if v is the cone vertex of $c(\partial X)$, then $X^+ - \{v\}$ is filtered homeomorphic to $X - \partial X$, and we let $\{v\}$ be a zero-dimensional stratum. We also observe that the regular strata of X^+ are the regular strata of $X^+ - \{v\}$, and so are bijectively paired with (in fact, homeomorphic to) the regular strata of $X - \partial X$. In particular, the set $\{x_j\}_{j=1}^m$ contains one point in each regular stratum of X^+.

Let $\bar{p}^+$ be a perversity on X^+ that agrees with $\bar{p}$ on $X^+ - \{v\}$ and such that $\bar{p}^+(\{v\}) \geq n$. Then $I^{\bar{p}^+}H_*(c(\partial X); R) = 0$ by the cone formula (Theorem 6.2.13). We have a commutative diagram

$$\begin{array}{ccccccc}
I^{\bar{p}}H_n(X,\partial X;R) & \xrightarrow{\cong} & \begin{array}{c} I^{\bar{p}}H_n(X^+ - \{v\}, \\ N^+ - \{v\}; R) \end{array} & \xrightarrow{\cong} & \begin{array}{c} I^{\bar{p}}H_n(X^+, \\ N^+; R) \end{array} & \xleftarrow{\cong} & I^{\bar{p}}H_n(X^+;R) \\
\downarrow & & \downarrow & & \downarrow & & \downarrow \\
\begin{array}{c} \bigoplus_j I^{\bar{p}}H_n(X, \\ X - \{x_j\}; R) \end{array} & \xrightarrow{\cong} & \begin{array}{c} \bigoplus_j I^{\bar{p}}H_n(X^+ - \{v\}, \\ X^+ - (\{x_j\} \cup \{v\}); R) \end{array} & \xrightarrow{\cong} & \begin{array}{c} \bigoplus_j I^{\bar{p}^+}H_n(X^+, \\ X^+ - \{x_j\}; R) \end{array} & \xleftarrow{=} & \begin{array}{c} \bigoplus_j I^{\bar{p}^+}H_n(X^+, \\ X^+ - \{x_j\}; R). \end{array}
\end{array}$$

The isomorphisms in the top row are due, respectively, to stratified homotopy equivalence, excision, and the long exact sequence of the pair (using the relation $I^{\bar{p}^+}H_*(c(\partial X); R) = 0$). The isomorphisms in the bottom row are by stratified homotopy equivalence and excision. The commutativity comes from the commutativity of the space maps. The vertical map on the far right is an isomorphism by Theorem 8.1.18, so the map on the left is also an isomorphism, as desired. □

Remark 8.3.4 If X has no codimension-one strata and $\bar{p}$ is a GM perversity, then by employing Lemma 5.5.6 and Corollary 5.5.7, we need not even assume in the proof of part (2) of Theorem 8.3.3 that our collars were formed in the stratification X. They might just as well be collars from another stratification X', restratified to inherit the stratification from X. This observation will be handy below in proving the topological invariance of fundamental classes of ∂-stratified pseudomanifolds (Theorem 8.3.7).

By Lemma 8.3.2, if X is an R-orientable ∂-stratified pseudomanifold, then so is ∂X. We will next show that, just like for manifolds, the boundaries of fundamental classes of compact oriented ∂-stratified pseudomanifolds are fundamental classes on the boundaries.

Proposition 8.3.5 *Suppose X is a compact R-oriented n-dimensional ∂-stratified pseudomanifold, and let Γ_X be the fundamental class of X with respect to the given R-orientation.*

Then $\partial_* : I^{\bar{0}}H_n(X, \partial X; R) \to I^{\bar{0}}H_{n-1}(\partial X; R)$ *takes* Γ_X *to* $\Gamma_{\partial X}$, *a fundamental class of* ∂X. *We define the orientation of* ∂X *consistent with the fundamental class* $\Gamma_{\partial X}$ *to be the* induced R-orientation of ∂X *determined by the given orientation of* X.

The following lemma will be useful in then proving the proposition below. The unusual choice of generator e for $H_1([0,1], \{0,1\}; R)$ is to keep consistent with 0 being the collar coordinate for boundaries, which is the notation we will use in the proof of the proposition.

Lemma 8.3.6 *Let* X *be a CS set and* R *a Dedekind domain. Let* $e : [0,1] \to [0,1]$ *be the* 1*-simplex given by* $e(t) = 1 - t$ *so that* e *represents a generator of* $H_1([0,1], \{0,1\}; R)$. *Then the composition*

$$I^{\bar{p}}H_i([0,1] \times X, \{0,1\} \times X; R) \xrightarrow{\partial_*} I^{\bar{p}}H_{i-1}(\{0,1\} \times X, \{1\} \times X; R) \cong I^{\bar{p}}H_{i-1}(X; R)$$

is an isomorphism with inverse given by

$$I^{\bar{p}}H_{i-1}(X; R) \xrightarrow{e\times} I^{\bar{p}}H_i([0,1] \times X, \{0,1\} \times X; R).$$

Proof We first demonstrate that our inverse map $e\times$ is an isomorphism. In fact, this is just the composition

$$\begin{aligned} I^{\bar{p}}H_{i-1}(X; R) &\cong R \otimes I^{\bar{p}}H_{i-1}(X; R) \underset{\cong}{\xrightarrow{e\otimes \mathrm{id}}} H_1([0,1], \{0,1\}; R) \otimes I^{\bar{p}}H_{i-1}(X; R) \\ &\underset{\cong}{\xrightarrow{\varepsilon}} I^{\bar{p}}H_i([0,1] \times X, \{0,1\} \times X; R). \end{aligned}$$

Here the first line takes $\xi \in I^{\bar{p}}H_{i-1}(X; R)$ to $1 \otimes \xi$, and the second line takes $1 \otimes \xi$ to $e \otimes \xi$. The last map is the cross product, and so the full composition takes ξ to $e \times \xi$. Even though our unfiltered $[0,1]$ is not a CS set, this last map is an isomorphism by Proposition 7.3.69, using the pair $((-\epsilon, 1+\epsilon), (-\epsilon, \epsilon) \cup (1-\epsilon, 1+\epsilon))$ as a CS model for $([0,1], 0, 1)$. Of course we also use that $H_*([0,1], \{0,1\}; R)$ is trivial except in degree one, where it is isomorphic to R.

Now, suppose we start with an element of $I^{\bar{p}}H_{i-1}(X; R)$ represented by the cycle ξ. Then $\partial(e \times \xi) = (\partial e) \times \xi = \xi_0 - \xi_1$, where we let ξ_j represent the copy of ξ in $\{j\} \times X$. As the image of the connecting map ∂_* can be determined by taking the boundaries of representing chains, we see that $\partial_*(e \times \xi)$ is represented by $\xi_0 - \xi_1$ in $I^{\bar{p}}H_{i-1}(\{0,1\} \times X, \{1\} \times X; R)$. The image under the evident isomorphism

$$I^{\bar{p}}H_{i-1}(\{0,1\} \times X, \{1\} \times X; R) \cong I^{\bar{p}}H_{i-1}(\{0\} \times X; R) = I^{\bar{p}}H_{i-1}(X; R)$$

is then represented by ξ_0, corresponding to our original ξ. So, omitting basic isomorphisms from the notation, we see that the composition

$$\partial_* \circ (e \times \cdot) : I^{\bar{p}}H_{i-1}(X; R) \to I^{\bar{p}}H_{i-1}(X; R)$$

is the identity map. It follows that ∂_* is also an isomorphism, the inverse to $e\times$. □

The proof of Proposition 8.3.5 is now based on that of [220, Corollary 6.3.10].

Proof of Proposition 8.3.5 Let $N' \cong [0,2) \times \partial X$ be a filtered collar neighborhood of ∂X. Then let N be the image of $[0,1) \times \partial X$, which will be a smaller filtered collar neighborhood. We write $\bar{N} \cong [0,1] \times \partial X$ and $\mathring{N} \cong (0,1) \times \partial X$, and let N_0 and N_1 denote the images of $\{0\} \times \partial X$ and $\{1\} \times \partial X$, respectively, in X. Then we have the following commutative diagram with R coefficients omitted.

$$
\begin{array}{ccc}
I^{\bar{0}}H_n(X,\partial X) & \xrightarrow{\partial_*} & I^{\bar{0}}H_{n-1}(\partial X) \\
\downarrow & & \downarrow \cong \\
I^{\bar{0}}H_n(X, X - \mathring{N}) & \xrightarrow{\partial_*} & I^{\bar{0}}H_{n-1}(X - \mathring{N}, X - N) \\
\uparrow \cong & & \uparrow \cong \\
I^{\bar{0}}H_n(N', N' - \mathring{N}) & \xrightarrow{\partial_*} & I^{\bar{0}}H_{n-1}(N' - \mathring{N}, N' - N) \\
\uparrow \cong & & \uparrow \cong \\
I^{\bar{0}}H_n(\bar{N}, \bar{N} - \mathring{N}) & \xrightarrow{\partial_*} & I^{\bar{0}}H_{n-1}(\bar{N} - \mathring{N}, N_1) \\
\uparrow \cong & & \uparrow \cong \\
I^{\bar{0}}H_n([0,1] \times \partial X, \{0,1\} \times \partial X) & \xrightarrow{\partial_*} & I^{\bar{0}}H_{n-1}(\{0,1\} \times \partial X, \{1\} \times \partial X).
\end{array}
$$

All the vertical maps are induced by spatial inclusions or filtered isomorphisms, and so the diagram commutes by the naturality of the connecting morphisms ∂_*. The top right vertical map is evidently an isomorphism, as $X - \mathring{N}$ is the disjoint union of $X - N$ and ∂X. The other vertical maps along the right side are isomorphisms for analogous reasons. The maps on the left marked as isomorphisms are isomorphisms by excision, stratified homotopy equivalence, and filtered homeomorphism, respectively. The bottom horizontal map is an isomorphism by Lemma 8.3.6.

Now, let z be any point in a regular stratum of ∂X, let $t \in (0,1)$, and let $x \in X$ be the point of $N \subset X$ corresponding to the coordinates (t,z). By definition, the image of Γ_X in $I^{\bar{0}}H_n(X, X - \{x\}; R) \cong R$ is a generator, and it follows that if we let $\gamma \in I^{\bar{0}}H_n([0,1] \times \partial X, \{0,1\} \times \partial X; R)$ denote the image of Γ_X under the left-hand maps of the diagram, then γ further maps to a generator of $I^{\bar{0}}H_n([0,1] \times \partial X, [0,1] \times X - \{(t,z)\}; R)$ thanks to the diagram

$$\begin{array}{ccc}
I^{\bar{0}}H_n(X,\partial X;R) & & \\
\downarrow & \searrow & \\
I^{\bar{0}}H_n(X,X-\mathring{N};R) & \longrightarrow & I^{\bar{0}}H_n(X,X-\{x\};R) \\
\uparrow\cong & & \uparrow\cong \\
\begin{array}{c} I^{\bar{0}}H_n([0,1]\times\partial X, \\ \{0,1\}\times\partial X;R) \end{array} & \longrightarrow & \begin{array}{c} I^{\bar{0}}H_n([0,1]\times\partial X, \\ [0,1]\times\partial X-\{(t,z)\};R), \end{array}
\end{array}$$

in which the left-hand maps are those of the preceding diagram and the right-hand map is an isomorphism by excision, up to filtered homeomorphism.

By Lemma 8.3.6, the element $\gamma \in I^{\bar{0}}H_n([0,1]\times\partial X,\{0,1\}\times\partial X;R)$ is equal to $e\times\partial_*(\gamma)$, with e a generator of $H_1([0,1],\{0,1\};R)$. But from our first diagram and the simple nature of the maps on the right of that diagram, we see that

$$\partial_*(\gamma) = \partial_*(\Gamma_X) \in I^{\bar{0}}H_{n-1}(\{0,1\}\times\partial X,\{1\}\times\partial X;R) \cong I^{\bar{0}}H_{n-1}(\partial X;R).$$

So $\gamma = e\times\partial_*(\Gamma_X)$.

Finally, we consider the diagram (coefficients tacit)

$$\begin{array}{ccc}
\begin{array}{c} H_1([0,1],\{0,1\})\otimes \\ I^{\bar{0}}H_{n-1}(\partial X) \end{array} & \xrightarrow[\cong]{\varepsilon} & \begin{array}{c} I^{\bar{0}}H_n([0,1]\times\partial X, \\ \{0,1\}\times\partial X) \end{array} \\
\downarrow & & \downarrow \\
\begin{array}{c} H_1([0,1],[0,1]-\{t\})\otimes \\ I^{\bar{0}}H_{n-1}(\partial X,\partial X-\{z\}) \end{array} & \xrightarrow[\cong]{\varepsilon} & \begin{array}{c} I^{\bar{0}}H_n([0,1]\times\partial X, \\ [0,1]\times\partial X-\{(t,z)\}). \end{array}
\end{array}$$

The horizontal maps are isomorphisms by the Künneth Theorem (Theorem 6.4.7 and Proposition 7.3.69) and as $H_*([0,1],\{0,1\};R)$ is trivial except in degree one, where it is isomorphic to R. Since e is a generator of $H_*([0,1],\{0,1\};R)$, and so also of $H_*([0,1],[0,1]-\{t\};R)\cong R$, and since $e\times\partial_*(\Gamma_X)$ maps to a generator of $I^{\bar{0}}H_n([0,1]\times\partial X,[0,1]\times\partial X-\{(t,z)\};R)\cong R$, it follows that $\partial_*(\Gamma_X)$ must represent a generator of $I^{\bar{0}}H_{n-1}(\partial X,\partial X-\{z\};R)\cong R$.

As z was an arbitrary regular point of ∂X, we see that the section $\mathfrak{s}_{\partial_*(\Gamma_X)}$ of the orientation bundle of ∂X determined by $\partial_*(\Gamma_X)$ evaluates to a generator at each such regular point. Thus the restriction of $\mathfrak{s}_{\partial_*(\Gamma_X)}$ to $\partial X-\Sigma_{\partial X}$ is an orientation section. By Lemma 8.1.14, the global section $\mathfrak{s}_{\partial_*(\Gamma_X)}$ is the unique extension of this orientation section over $\partial X-\Sigma_{\partial X}$ to all of ∂X, and, in particular, it is thus itself an orientation section of ∂X. As the image of

$\partial_*(\Gamma_X)$ in $I^{\bar{0}}H_{n-1}(\partial X, \partial X - \{y\}; R)$ for each $y \in \partial X$ is $\mathfrak{s}_{\partial_*(\Gamma_X)}(y)$ by definition, $\partial_*(\Gamma_X)$ is therefore the fundamental class determined by this orientation section, by Theorem 8.1.18.

This completes the proof of Proposition 8.3.5. □

(a) Topological Invariance

As we did in Section 8.1.5 for pseudomanifolds, we can also discuss the invariance of the fundamental classes when working with pseudomanifolds with boundary. The treatment over varying perversities is essentially equivalent to our work in Proposition 8.1.25 and its corollary, so we will not run through all the details again. By contrast, invariance of stratification is a bit trickier because, as noted in Remark 2.10.24, we do not have intrinsic stratifications for pseudomanifolds with boundary. Thus we do not have an $\mathfrak{X}$ to use as an obvious intermediary as we did in Proposition 8.1.29. Hence, if we have two ∂-stratified pseudomanifolds with the same underlying space pairs $(|X_1|, |\partial X_1|) = (|X_2|, |\partial X_2|)$, we need to begin by constructing an isomorphism $I^{\bar{0}}H_n(X_1, \partial X_1; R) \cong I^{\bar{0}}H_n(X_2, \partial X_2; R)$, ideally in as canonical a way as possible. Then we need to show that this isomorphism takes the fundamental class of X_1 to that of X_2, assuming compatible orientations. We will carry out this program and then show that it produces a topological invariance of fundamental classes in the vein of Remark 8.1.30.

We begin with our construction of the isomorphism, assuming that X_1 and X_2 are compact n-dimensional ∂-stratified pseudomanifolds with the same underlying space pairs $(|X_1|, |\partial X_1|) = (|X_2|, |\partial X_2|)$ and without codimension-one strata, compatibly R-oriented for the Dedekind domain R in the sense of Corollary 8.1.11 (applied to $|X_i| - |\partial X_i|$). We will in fact construct the more general isomorphism $\phi : I^{\bar{p}}H_*(X_1, \partial X_1; R) \to I^{\bar{p}}H_*(X_2, \partial X_2; R)$ for any GM perversity $\bar{p}$. We will make two choices, but then we will show that the isomorphism does not depend on the choices and hence is canonical in this sense.

First, we let N_2 be a filtered open collar neighborhood of ∂X_2 in X_2. Then let $N_{2\to 1}$ denote $|N_2|$ but now filtered by the filtration it inherits as a subspace of X_1. As $|N_2| = |N_{2\to 1}|$ is an open set in $|X_2| = |X_1|$, the filtered space $N_{2\to 1}$ is also a ∂-stratified pseudomanifold by Lemma 2.7.8, using that $|\partial X_1| = |\partial X_2|$ is compact, and we have $\partial N_{2\to 1} = \partial X_1$. Therefore, ∂X_1 has an open filtered collar neighborhood N_1 in $N_{2\to 1}$. Note that N_1 is also a filtered collar of ∂X_1 in X_1. Finally, let $\mathfrak{X}$ be the intrinsic filtration of $|X_1 - \partial X_1| = |X_2 - \partial X_2|$, and let $\mathfrak{N}$ be the filtered subspace of $\mathfrak{X}$ with underlying set $|N_2 - \partial X_2| = |N_{2\to 1} - \partial X_1|$. By Lemma 2.10.10, this subspace filtration of $\mathfrak{N}$ is also its intrinsic filtration.

We now consider the following composition of isomorphisms, coefficients omitted:

$$\begin{aligned} I^{\bar{p}}H_*(X_1, \partial X_1) &\xrightarrow{\cong} I^{\bar{p}}H_*(X_1, N_1) \xrightarrow{\cong} I^{\bar{p}}H_*(X_1, N_{2\to 1}) \\ &\xleftarrow{\cong} I^{\bar{p}}H_*(X_1 - \partial X_1, N_{2\to 1} - \partial X_1) \xrightarrow{\cong} I^{\bar{p}}H_*(\mathfrak{X}, \mathfrak{N}) \\ &\xleftarrow{\cong} I^{\bar{p}}H_*(X_2 - \partial X_2, N_2 - \partial X_2) \xrightarrow{\cong} I^{\bar{p}}H_*(X_2, N_2) \xleftarrow{\cong} I^{\bar{p}}H_*(X_2, \partial X_2). \end{aligned}$$

The first and last maps are isomorphisms by stratified homotopy invariance as N_i has a stratified deformation retraction to ∂X_i. The second map is an isomorphism by Lemma 5.5.6 as $\bar{p}$ is a GM perversity. The third and sixth arrows are isomorphisms by excision. The fourth and fifth maps are isomorphisms by the topological invariance given by Theorem 5.5.1 (and Proposition 6.2.9, by which $I^{\bar{p}}H_*^{\mathrm{GM}}(X;R) \cong I^{\bar{p}}H_*(X;R)$, as we have no codimension-one strata and $\bar{p}$ is a GM perversity).

Altogether, this composition gives an isomorphism

$$\phi : I^{\bar{p}}H_*(X_1, \partial X_1) \cong I^{\bar{p}}H_*(X_2, \partial X_2)$$

that might seem to depend on our choices of N_2 and N_1. We next show that ϕ is actually independent of these choices. Suppose we had chosen instead a filtered collar N_2', constructed $N_{2\to 1}'$ by filtering $|N_2|$ as a subspace of X_1, and then chosen N_1' with $N_1' \subset N_{2\to 1}'$. Then we can also find filtered collars N_i'' with $N_i'' \subset N_i \cap N_i'$ and with $N_1'' \subset N_{2\to 1}''$, letting $N_{2\to 1}''$ be $|N_2''|$ with the filtration from X_1.

Now we have the big commutative diagram (coefficients omitted) shown in Figure 8.1. All the arrows so marked are isomorphisms by our previous discussion, and it follows that all arrows are isomorphisms. This shows that our isomorphism ϕ is independent of our choices.

It will be useful below in discussing Lefschetz duality to also formulate an isomorphism for absolute intersection homology $\phi : I^{\bar{p}}H_*(X_1;R) \xrightarrow{\cong} I^{\bar{p}}H_*(X_2;R)$. This can be found from the relative isomorphism by simply leaving out the boundary and its neighborhoods. We obtain

$$\begin{aligned} I^{\bar{p}}H_*(X_1;R) &\xleftarrow{\cong} I^{\bar{p}}H_*(X_1 - \partial X_1;R) \xrightarrow{\cong} I^{\bar{p}}H_*(\mathfrak{X};R) \xleftarrow{\cong} I^{\bar{p}}H_*(X_2 - \partial X_2;R) \\ &\xrightarrow{\cong} I^{\bar{p}}H_*(X_2;R). \end{aligned}$$

In this case, the first and last isomorphisms are by stratified homotopy invariance.

Now that we have somewhat canonical isomorphisms ϕ, we can formulate a ∂-pseudomanifold version of Proposition 8.1.29, which concerned invariance under restratification of fundamental classes for stratified pseudomanifolds without boundary:

Proposition 8.3.7 *Let R be a Dedekind domain, and let X_1 and X_2 be compact n-dimensional ∂-stratified pseudomanifolds with the same underlying space pairs $(|X_1|, |\partial X_1|) = (|X_2|, |\partial X_2|)$ and without codimension-one strata. Suppose X_1 and X_2 are compatibly R-oriented in the sense of Corollary 8.1.11 (applied to $|X| - |\partial X|$). Let $\Gamma_1 \in I^{\bar{0}}H_n(X_1, \partial X_1; R)$ and $\Gamma_2 \in I^{\bar{0}}H_n(X_2, \partial X_2; R)$ be the fundamental classes with respect to these R-orientations. Then the isomorphism $\phi : I^{\bar{0}}H_n(X_1, \partial X_1; R) \to I^{\bar{0}}H_n(X_2, \partial X_2; R)$ takes Γ_1 to Γ_2.*

Remark 8.3.8 Analogously to Remark 8.1.30, it follows from Proposition 8.3.7 that, if X is a compact n-dimensional R-oriented ∂-stratified pseudomanifold without codimension-one strata, then the fundamental class Γ_X is a topological invariant of the pair $(|X|, |\partial X|)$ in the following sense: Suppose that Y is another compact R-oriented ∂-stratified pseudomanifold without codimension-one strata and that $f : (|X|, |\partial X|) \to (|Y|, |\partial Y|)$ is a *topological homeomorphism*, i.e. that it is a homeomorphism of the underlying spaces without regard to the stratifications. Then f induces an image stratification, say Y', on $|Y|$, and an

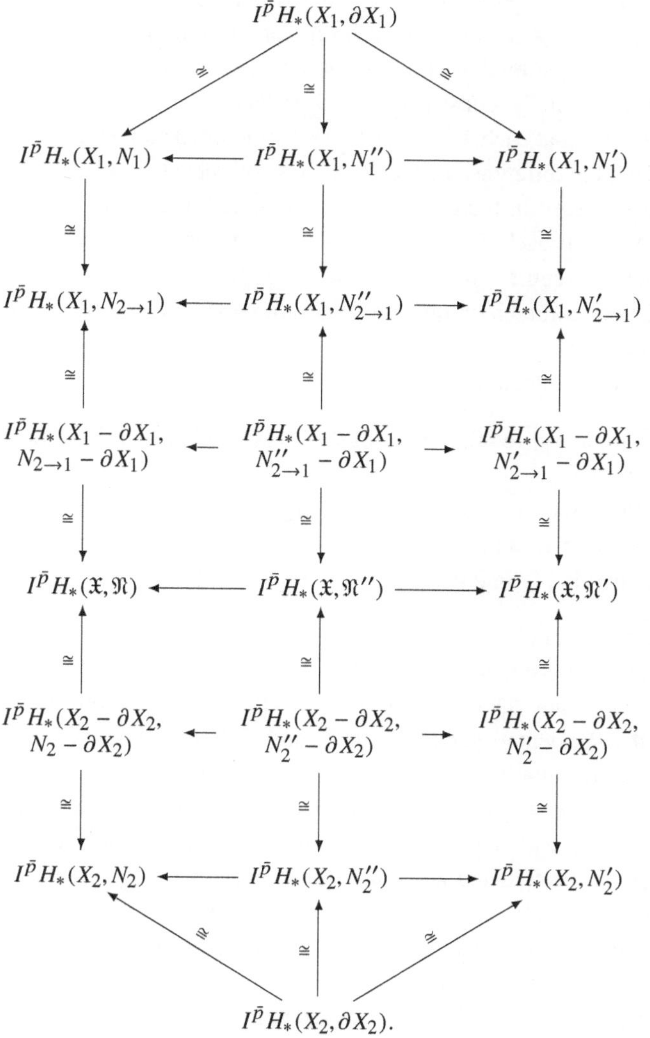

Figure 8.1

image R-orientation on Y' (via the pointwise isomorphisms $f : I^{\bar{0}}H_n(X, X - \{x\}; R) \to I^{\bar{0}}H_n(Y', Y' - \{f(x)\}; R)$ for $x \in |X| - |\partial X|$). Suppose that f is orientation-preserving, i.e. that the image R-orientation is compatible with the given R-orientation on Y in the sense of Corollary 8.1.11 applied to $|Y| - |\partial Y|$. Then it must also be the case, applying Proposition 8.3.7, that $f(\Gamma_X) \in I^{\bar{0}}H_n(Y', \partial Y'; R)$ corresponds to Γ_Y under the isomorphism $\phi : I^{\bar{0}}H_n(Y', \partial Y'; R) \xrightarrow{\cong} I^{\bar{0}}H_n(Y, \partial Y; R)$.

Proof of Proposition 8.3.7 We must show that ϕ takes Γ_1 to Γ_2. As neither X_i has any codimension-one strata by assumption, each regular stratum of $X_i - \partial X_i$ is contained as a dense subset of one of the regular strata of the intrinsic filtration of $|X_i - \partial X_i|$, which we continue to denote $\mathfrak{X}$ as in the construction of ϕ. In fact, it follows from the argument in the proof of Lemma 8.1.9 that each $X_i - (\Sigma_{X_i} \cup \partial X_i)$ is an open submanifold of $\mathfrak{X} - \Sigma_{\mathfrak{X}}$ that is dense and such that the difference of these sets has codimension at least two. Furthermore, by the argument provided in the proof of Lemma 8.1.10, the intersection of each regular stratum of $\mathfrak{X} - \Sigma_{\mathfrak{X}}$ with each $X_i - (\Sigma_{X_i} \cup \partial X_i)$ is path connected. Therefore, we have a bijection between the regular strata of $\mathfrak{X}$ and the regular strata of X_i, with each regular stratum of $\mathfrak{X} - \Sigma_{\mathfrak{X}}$ containing a unique regular stratum of each $X_i - (\Sigma_{X_i} \cup \partial X_i)$ as a dense, path-connected open set.

Thus, we can find a collection of points $\{x_j\}_{j=1}^m$ such that:

1. each x_j is contained in regular strata of $X_1 - \partial X_1$ and $X_2 - \partial X_2$, and therefore also a regular stratum of $\mathfrak{X}$,
2. each regular stratum of $X_1 - \partial X_1$ and $X_2 - \partial X_2$, and therefore each regular stratum of $\mathfrak{X}$, contains exactly one x_j, and
3. no x_j is contained in N_2, and hence also not in N_1 or $\mathfrak{N}$.

Let M be the manifold $(X_1 - (\partial X_1 \cup \Sigma_{X_1})) \cap (X_2 - (\partial X_2 \cup \Sigma_{X_2}))$. As in our discussion in Section 8.1.2, by the assumed compatibility of the orientations of X_1 and X_2 and the lack of codimension-one strata, the manifold M carries the orientation information for both X_i. In other words, the R-orientation bundles of X_1 and X_2 both restrict to the same orientation bundle over M (up to canonical isomorphisms induced locally by excision isomorphisms), and, conversely, any orientation section on M extends uniquely to the compatible orientation sections on $X_1 - \partial X_1$ and $X_2 - \partial X_2$ by Lemma 8.1.14. We then have a commutative diagram (coefficients omitted and where $\overline{X_1}$ etc. is short for $X_1 - \{x_j\}$) shown in Figure 8.2. The diagonal arrows are all excision isomorphisms, and the vertical arrows in the middle column are isomorphisms by excision and topological invariance. The top and bottom horizontal arrows are isomorphisms by Theorem 8.3.3, and so all the horizontal arrows are isomorphisms. By Theorem 8.3.3, the Γ_i are the unique elements of the $I^{\bar{0}}H_n(X_i, \partial X_i : R)$ whose images in the $\bigoplus I^{\bar{0}}H_n(X_i, X_i - \{x_j\}; R)$ are the direct sums of the local orientation classes at the x_j. The diagram thus demonstrates that ϕ, which is the composition along the left side of the diagram, must take Γ_1 to Γ_2, as $\phi(\Gamma_1)$ and Γ_2 both restrict to the common local orientation classes of X_1 and X_2 in each $I^{\bar{0}}H_n(X_2, X_2 - \{x_j\}; R)$. □

8.3.2 Lefschetz Duality

Now we can prove Lefschetz Duality Theorems for compact ∂-stratified pseudomanifolds. We will at first assume $\partial X = A \amalg B$ with each of A and B a union of components of ∂X, possibly empty. In particular, this implies that $A \cap B = \emptyset$. A more general theorem will follow as Corollary 8.3.10.

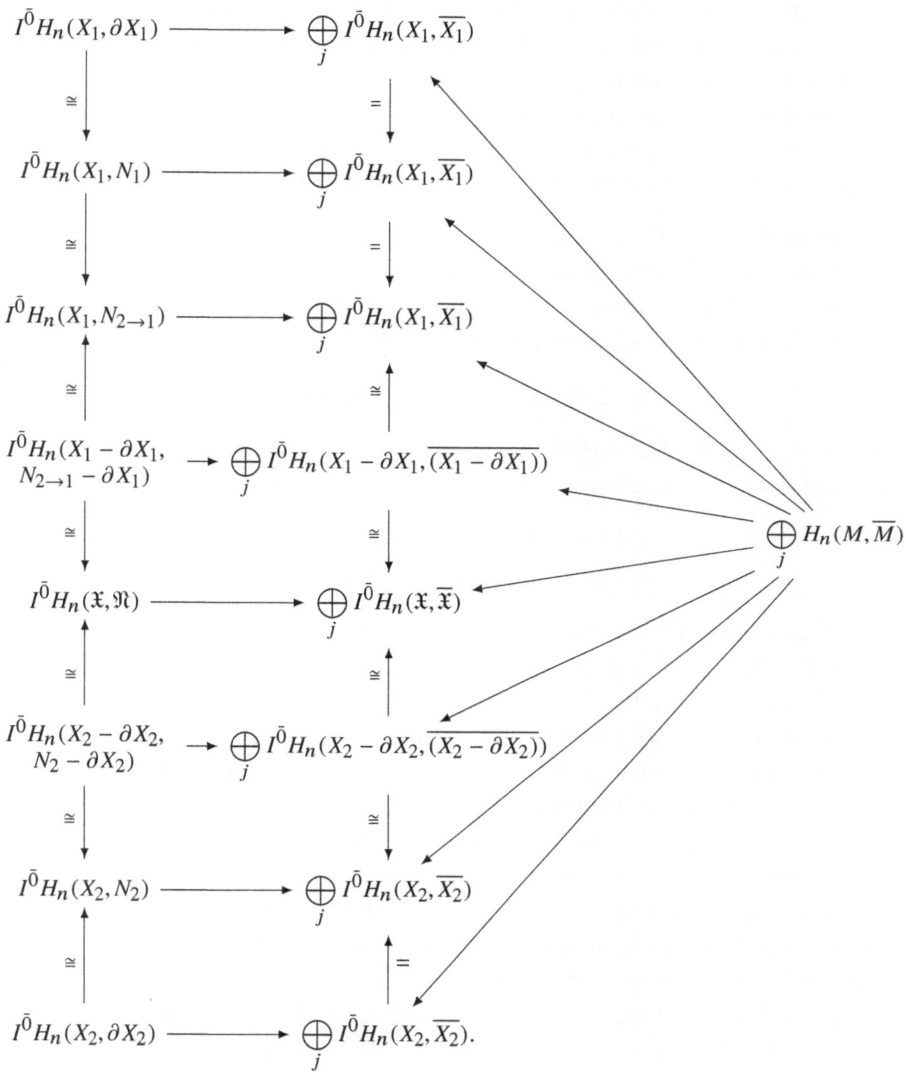

Figure 8.2 In this figure $\overline{X_1}$ etc. is short for $X_1 - \{x_j\}$.

If Γ_X is the fundamental class determined by an orientation on the compact n-dimensional ∂-stratified pseudomanifold X, then we have a duality map

$$\mathcal{D} : I_{\bar{p}}H^i(X, B; R) \to I^{D\bar{p}}H_{n-i}(X, A; R)$$

given by

$$\mathcal{D}(\alpha) = (-1)^{|\alpha| n}\alpha \frown \Gamma_X.$$

Even though X is not a CS set and A and B are not open subsets of X, this cap product is well defined by Theorem 7.3.72. We show that with an appropriate locally torsion-free hypothesis this duality map is an isomorphism.

We remark that the freedom to work with singularities allows us to provide a somewhat different proof from what is usually done for ∂-manifolds, e.g. [125, Theorem 3.43]; for a proof of the following theorem more akin to the one in [125], see [100, Theorem 7.10].

Theorem 8.3.9 (Lefschetz duality) *Suppose R is a Dedekind domain, and let X be a compact n-dimensional R-oriented locally $(\bar{p}; R)$-torsion-free*[18] *∂-stratified pseudomanifold. Let A and B be disjoint compact stratified pseudomanifolds with $A \cup B = \partial X$, i.e. each of A and B is a union of connected components of ∂X. Then the duality map*

$$\mathcal{D} : I_{\bar{p}}H^i(X, B; R) \to I^{D\bar{p}}H_{n-i}(X, A; R)$$

induced by the cap product with the fundamental class Γ_X is an isomorphism.

In particular, we have isomorphisms

$$\mathcal{D} : I_{\bar{p}}H^i(X; R) \to I^{D\bar{p}}H_{n-i}(X, \partial X; R),$$
$$\mathcal{D} : I_{\bar{p}}H^i(X, \partial X; R) \to I^{D\bar{p}}H_{n-i}(X; R).$$

Proof Using Lemma 8.3.2, the spaces A and B are compact orientable stratified pseudomanifolds. We can form a new stratified pseudomanifold[19] X^+ (without boundary) by coning off A and B. Specifically, let $X^+ = \bar{c}(A) \cup_A X \cup_B \bar{c}(B)$. Let v_A and v_B denote the cone vertices of the cones $\bar{c}A$ and $\bar{c}B$, respectively, and let $V = \{v_A, v_B\}$. Using the existence of a filtered collar of ∂X in X, it is easy to verify that X^+ is a stratified pseudomanifold and that $X^+ - V$ is filtered homeomorphic to $X - \partial X$. Therefore, there is a homeomorphism between the manifolds $X - (\partial X \cup \Sigma_X)$ and $X^+ - \Sigma_{X^+}$, and since X is R-orientable so is X^+. As a choice of global orientation is determined by a choice of local orientation at one point in each regular stratum (using that the orientability implies that the orientation sheaf $\mathcal{O}$ is isomorphic to the constant bundle with fiber R over each regular stratum), we can thus extend the given R-orientation of X to X^+.

Let N_A and N_B be disjoint open filtered collars of A and B, respectively, in X, let $N_A^+ = N_A \cup_A \bar{c}(A) \cong c(A)$, let $N_B^+ = N_B \cup_B \bar{c}(B) \cong c(B)$, and let $N^+ = N_A^+ \amalg N_B^+$. Let $\bar{p}^+$ be a perversity defined on X^+ whose value on each stratum of X^+ that is not in V agrees with the value of $\bar{p}$ on the corresponding stratum of X. Let $\bar{p}^+(\{v_A\}) = -2$ and $\bar{p}^+(\{v_B\}) = n$. Notice that the links of v_A and v_B in X^+ are A and B, respectively, and we have $\dim(A) - \bar{p}^+(\{v_A\}) - 1 = n$ and $\dim(B) - \bar{p}^+(\{v_B\}) - 1 = -2$. Clearly $I^{\bar{p}^+}H_{-2}(B; R) = 0$, while $I^{\bar{p}^+}H_n(A; R) = 0$ by item (1) of Theorem 8.1.18. Therefore, X^+ is locally $(\bar{p}^+; R)$-torsion-free at $\{v_A\}$ and $\{v_B\}$. But $\bar{p}^+ = \bar{p}$ for all other strata, and we have assumed that X is locally $(\bar{p}; R)$-torsion-free. Thus X^+ is locally $(\bar{p}^+; R)$-torsion-free.

[18] Or, equivalently, X can be locally $(D\bar{p}; R)$-torsion-free; see Corollary 8.2.5.

[19] Compare Example 6.3.15, and note that we use the notation a bit differently here, although with comparable computational results.

Consider now the following diagram:

$$
\begin{array}{ccccccc}
I_{\bar{p}}H^i(X,B;R) & \longleftarrow & \begin{array}{c} I_{\bar{p}^+}H^i(X^+-V, \\ N_B^+-\{v_B\};R) \end{array} & \longleftarrow & I_{\bar{p}^+}H^i(X^+,N_B^+;R) & \longrightarrow & I_{\bar{p}^+}H^i(X^+;R) \\
\downarrow{\scriptstyle \frown\Gamma_X} & & \downarrow{\scriptstyle \frown\Gamma_{X^+-V,N^+-V}} & & \downarrow{\scriptstyle \frown\Gamma_{X^+,N^+}} & & \downarrow{\scriptstyle \frown\Gamma_{X^+}} \\
I^{D\bar{p}}H_{n-i}(X,A;R) & \longrightarrow & \begin{array}{c} I^{D\bar{p}^+}H_{n-i}(X^+-V, \\ N_A^+-\{v_A\};R) \end{array} & \longrightarrow & \begin{array}{c} I^{D\bar{p}^+}H_{n-i}(X^+, \\ N_A^+;R) \end{array} & \longleftarrow & I^{D\bar{p}^+}H_{n-i}(X^+;R).
\end{array}
$$

We claim that, with appropriate definitions of the various Γ, this diagram commutes and that all the horizontal arrows, which are all induced by inclusions, are isomorphisms. The right-hand vertical map is an isomorphism by Poincaré duality (Theorem 8.2.4). It would follow that all of the vertical maps are isomorphisms. The left-hand vertical map is, up to sign, our Lefschetz duality map, so this would prove the theorem.

We will work right to left through the diagram. As $D\bar{p}^+(\{v_A\}) = n-2-\bar{p}^+(\{v_A\}) = n$, we have $I^{D\bar{p}^+}H_*(N_A^+;R) = 0$ by the cone formula (Theorem 6.2.13). Similarly, $\bar{p}^+(\{v_B\}) = n$, so $I_{\bar{p}^+}H^*(N_B^+;R) = 0$ by the cone formula and the Universal Coefficient Theorem (Theorem 7.1.4). Therefore, the horizontal maps in the right-hand square are isomorphisms by the long exact sequences of the pairs. We define Γ_{X^+,N^+} to be the image in $I^{\bar{0}}H_n(X^+,N^+;R)$ of $\Gamma_{X^+} \in I^{\bar{0}}H_n(X^+;R)$. The right-hand square then commutes by naturality of the cap product (Proposition 7.3.6) with respect to the inclusion map $(X^+;\emptyset,\emptyset) \to (X^+;N_A^+,N_B^+)$.

In the middle square, we consider the inclusions $(X^+-V;N_A^+-\{v_A\},N_B^+-\{v_B\}) \to (X^+;N_A^+,N_B^+)$. By Proposition 8.1.24, having $D\bar{p}^+(\{v_B\}) = -2 < 0$ implies that the inclusion-induced map $I^{D\bar{p}^+}H_*(X^+-\{v_B\};R) \to I^{D\bar{p}^+}H_*(X^+;R)$ is an isomorphism, so $I^{D\bar{p}^+}H_*(X^+-\{v_B\},N_A^+;R) \to I^{D\bar{p}^+}H_*(X^+,N_A^+;R)$ is an isomorphism from the long exact sequences and the Five Lemma. We also have that $I^{D\bar{p}^+}H_*(X^+-V,N_A^+-\{v_A\};R) \to I^{D\bar{p}^+}H_*(X^+-\{v_B\},N_A^+;R)$ is an isomorphism, by excision of $\{v_a\}$. Together, the composite isomorphism

$$
\begin{aligned}
I^{D\bar{p}^+}H_*(X^+-V,N_A^+-\{v_A\};R) &\to I^{D\bar{p}^+}H_*(X^+-\{v_B\},N_A^+;R) \\
&\to I^{D\bar{p}^+}H_*(X^+,N_A^+;R)
\end{aligned}
$$

is the bottom map of the square. Using the Universal Coefficient Theorem and the Five Lemma to dualize the isomorphisms to intersection cohomology, the argument that the top map in the square is an isomorphism is essentially the same, interchanging A with B and $D\bar{p}^+$ with $\bar{p}^+$.

We also observe that $(N_A^+-\{v_A\}) \cup (N_B^+-\{v_B\}) = N^+-V$; then the inclusion map $I^{\bar{0}}H_n(X^+-V,N^+-V;R) \to I^{\bar{0}}H_n(X^+,N^+;R)$ is also an excision isomorphism, so we can let $\Gamma_{X^+-V,N^+-V} \in I^{\bar{0}}H_n(X^+-V,N^+-V;R)$ be the image of Γ_{X^+,N^+} under the inverse isomorphism. Once again the square commutes by naturality (Proposition 7.3.6).

For the leftmost square, we have the inclusion $(X;A,B) \to (X^+-V;N_A^+-\{v_A\},N_B^+-\{v_B\})$. Note that the perversity $\bar{p}^+$ on X^+ restricts to the perversity $\bar{p}$ on X. All three inclusions are stratified homotopy equivalences (in fact, each image subset is a stratified deformation retract of its codomain), so the horizontal maps are isomorphisms using stratified homotopy invariance and the Five Lemma applied to the long exact sequence of the pair. Theorem 7.3.72 provides the cap product with Γ_X. We also claim that the image of $\Gamma_X \in I^{\bar{0}}H_n(X, \partial X; R)$ in $I^{\bar{0}}H_n(X^+ - V, N^+ - V; R)$ is Γ_{X^+-V,N^+-V}. Then we can once again apply naturality (Proposition 7.3.6 and Theorem 7.3.72), which will complete the argument.

To verify the claim, let the set $\{x_1, \dots, x_m\} \subset X^+ - N^+$ consist of one point from each regular stratum. Then this set also provides one point in each regular stratum of X. By items (2) and (3) of Theorem 8.1.18, the class $\Gamma_{X^+} \in I^{\bar{p}^+}H_n(X^+; R)$ is the unique class whose images in the $I^{\bar{0}}H_n(X^+, X^+ - \{x_i\}; R)$ agree with the orientations at x_i. As $x_i \notin N^+$ for each i, the images of Γ_{X^+,N^+} also give the local orientations at the x_i. But we have a commutative diagram

$$\begin{array}{ccccc} I^{\bar{0}}H_n(X, \partial X; R) & \xrightarrow{\cong} & I^{\bar{0}}H_n(X^+ - V, N^+ - V; R) & \xrightarrow{\cong} & I^{\bar{0}}H_n(X^+, N^+; R) \\ \downarrow \cong & & \downarrow & & \downarrow \\ \bigoplus_i I^{\bar{0}}H_n(X, X - \{x_i\}; R) & \xrightarrow{\cong} & \bigoplus_i I^{\bar{0}}H_n(X^+ - V, (X^+ - V) - \{x_i\}; R) & \xrightarrow{\cong} & \bigoplus_i I^{\bar{0}}H_n(X^+, X^+ - \{x_i\}; R), \end{array}$$

with the left-hand vertical isomorphism due to item (3) of Theorem 8.3.3 and the horizontal maps isomorphisms by stratified homotopy invariance and excision. It follows from the diagram that Γ_X must map across the composition in the top line to Γ_{X^+,N^+}, so the image of Γ_X in the middle term must be Γ_{X^+-V,N^+-V} by the definition of Γ_{X^+-V,N^+-V}. Thus Γ_X maps to Γ_{X^+-V,N^+-V} as claimed.

This concludes the demonstration that $\mathcal{D} : I_{\bar{p}}H^i(X, B; R) \to I^{D\bar{p}}H_{n-i}(X, A; R)$ is an isomorphism and completes the proof of Theorem 8.3.9. □

With Theorem 8.3.9 in hand, we can prove an even more general form of Lefschetz duality. The following version of Poincaré duality, as well as the broad strokes of the following proof, can be found for manifolds in [125, Theorem 3.43]. We include the extra details necessary to verify commutativity of the main diagram of the proof for the cap product as we have defined it here; this commutativity is more transparent in the manifold setting in [125] as the front face/back face construction of cap products is available in that context.

Corollary 8.3.10 *Suppose R is a Dedekind domain, and let X be a compact n-dimensional R-oriented locally $(\bar{p}; R)$-torsion-free*[20] *∂-stratified pseudomanifold. Let A and B be compact ∂-stratified pseudomanifolds with $A \cup B = \partial X$ and such that $A \cap B = \partial A = \partial B$. Then there is a duality isomorphism*

$$\mathcal{D} : I_{\bar{p}}H^i(X, B; R) \to I^{D\bar{p}}H_{n-i}(X, A; R)$$

induced by the cap product with the fundamental class Γ_X.

[20] Or, equivalently, X can be locally $(D\bar{p}; R)$-torsion-free; see Corollary 8.2.5.

$$
\begin{array}{ccccc}
\downarrow & & & & \downarrow \\
I_{\bar{p}}H^i(X,\partial X;R) & & \xrightarrow{\frown \Gamma_X} & & I^{D\bar{p}}H_{n-i}(X;R) \\
\downarrow & & & & \downarrow \\
I_{\bar{p}}H^i(X,A;R) & & \xrightarrow{\frown \Gamma_X} & & I^{D\bar{p}}H_{n-i}(X,B;R) \\
\downarrow & & & & \downarrow{\scriptstyle \partial_*} \\
I_{\bar{p}}H^i(\partial X,A;R) & \xrightarrow{\cong} & I_{\bar{p}}H^i(B,\partial B;R) & \xrightarrow{\frown \Gamma_B} & I^{D\bar{p}}H_{n-i-1}(B;R) \\
{\scriptstyle d^*}\downarrow & & & & \downarrow \\
I_{\bar{p}}H^{i+1}(X,\partial X;R) & & \xrightarrow{\frown \Gamma_X} & & I^{D\bar{p}}H_{n-i-1}(X;R) \\
\downarrow & & & & \downarrow
\end{array}
$$

Figure 8.3

Proof We will show that there is an up-to-sign commutative diagram, given in Figure 8.3, in which the columns are long exact. Then the corollary will follow from the Five Lemma, as the top displayed horizontal map and the map $\frown \Gamma_B$ are isomorphisms by Theorem 8.3.9 (using Lemma 8.3.2 to ensure that ∂X, and hence A and B, is R-orientable). The left column here is the long exact sequence of the triple $(X, \partial X, A)$, and the right column is the exact sequence of the pair (X, B). We will check that the rectangles commute up to sign.

The top square can be viewed as the composite of two squares (reoriented):

$$
\begin{array}{ccccc}
I_{\bar{p}}H^i(X,\partial X;R) & \xleftarrow{=} & I_{\bar{p}}H^i(X,\partial X;R) & \longrightarrow & I_{\bar{p}}H^i(X,A;R) \\
{\scriptstyle \frown \Gamma_X}\downarrow & & {\scriptstyle \frown \Gamma_X}\downarrow & & {\scriptstyle \frown \Gamma_X}\downarrow \\
I^{D\bar{p}}H_{n-i}(X;R) & \longrightarrow & I^{D\bar{p}}H_{n-i}(X,B;R) & \xleftarrow{=} & I^{D\bar{p}}H_{n-i}(X,B;R).
\end{array}
$$

These two squares each commute by the naturality of the cap product (Proposition 7.3.6 and Theorem 7.3.72) applied to the map $(X; \emptyset, \partial X) \to (X; B, \partial X)$ for the first square and the map $(X; B, A) \to (X; B, \partial X)$ for the second square. We have already seen above how to replace pairs of the form $(X, \partial X)$ with open pairs that stratified deformation retract to them by using collars. For the partial boundaries A and B, we can use the collars of $\partial A = \partial B$ in A and B to similarly create neighborhoods around A and B in ∂X that stratified deformation retract to A and B, respectively. These neighborhoods can then be extended to open subsets of ∂X in an extension of X to the union of X with an external collar on ∂X. Such neighborhoods can be used to ensure the existence of the necessary cap products and the naturality.

The middle rectangle of our diagram of exact sequences commutes up to sign by Proposition 7.3.38 via Theorem 7.3.72, Lemma 7.3.73, and Example 7.3.74. Notice that we are notationally reversed from the statement of Proposition 7.3.38, i.e. the A there corresponds to B here and vice versa. We also note that the map $\mathfrak{i}^*$ of the proposition corresponds to the composition of the maps right then down one arrow in our diagram here. We use that, up to signs, $\Gamma_B \in I^{\bar 0}H_{n-1}(B,\partial B;R)$ maps to $\partial_*(\Gamma_X) \in I^{\bar 0}H_{n-1}(\partial X,A;R)$ under the excision isomorphism guaranteed by the first statement of Lemma 7.3.73; in fact, we know that $\partial_*\Gamma_X = \Gamma_{\partial X}$ and then the chains representing Γ_B and $\Gamma_{\partial X}$ represent the same element of $I^{\bar 0}H_{n-1}(\partial X,A;R)$, as each of these must restrict to the local orientation classes at points in $B - \partial B$. Note also that by applying the Universal Coefficient Theorem and the Five Lemma, the first statement of Lemma 7.3.73 also provides the isomorphism $I_{\bar p}H^i(\partial X,A;R) \xrightarrow{\cong} I_{\bar p}H^i(B,\partial B;R)$.

For the commutativity up to sign of the final square, we reorient and expand it as

$$\begin{array}{ccccc}
I_{\bar p}H^i(\partial X,A;R) & \longrightarrow & I_{\bar p}H^i(\partial X;R) & \xrightarrow{d^*} & I_{\bar p}H^{i+1}(X,\partial X;R)\\
\downarrow{\cong} & \searrow{\frown\Gamma_{\partial X}} & \downarrow{\frown\Gamma_{\partial X}} & & \downarrow{\frown\Gamma_X}\\
I_{\bar p}H^i(B,\partial B;R) & & & & \\
\downarrow{\frown\Gamma_B} & & & & \\
I^{D\bar p}H_{n-i-1}(B;R) & \longrightarrow & I^{D\bar p}H_{n-i-1}(\partial X;R) & \longrightarrow & I^{D\bar p}H_{n-i-1}(X;R).
\end{array}$$

The composition along the bottom is equivalent to the intersection homology map induced by the inclusion $B \hookrightarrow X$. The composition along the top is equal to the connecting map in the long exact sequence of the triple; this can be seen by looking at the map of cohomology exact sequences of triples induced by $(X,\partial X,\emptyset) \hookrightarrow (X,\partial X,A)$. Then the left triangle commutes by naturality (Proposition 7.3.6) applied to the map of triples $(B;\emptyset,\partial B) \to (\partial X;\emptyset,A)$. This uses again that chains representing Γ_B in $I^{\bar 0}H_{n-1}(B,\partial B;R)$ and $\Gamma_{\partial X}$ in $I^{\bar 0}H_{n-1}(\partial X;R)$ both represent the same element in $I^{\bar 0}H_{n-1}(\partial X,A;R) \cong I^{\bar 0}H_{n-1}(B,\partial B;R)$ as they each represent the local orientation class at each point of $B - \partial B$. Similarly, the triangle on the right commutes by naturality (Proposition 7.3.6), applied to the map of triples $(\partial X;\emptyset,\emptyset) \to (\partial X;\emptyset,A)$. Finally, the square on the right commutes up to sign by Proposition 7.3.37 via Theorem 7.3.72. Note that the triple $(X;A,B)$ in the statement of

Proposition 7.3.37 becomes here $(X; \emptyset, \partial X)$, and so the necessary excision isomorphisms e and e' both become identity maps. □

Example 8.3.11 Our definition of ∂-stratified pseudomanifolds includes the assumption that the boundaries possess filtered collar neighborhoods. We have made regular use of this in our proofs, and in this example we show that the collar condition is also necessary, in general, for our results.

Let X be the two-dimensional closed disk D^2, and let z be a point in $\partial X = \partial D^1 = S^1$. If we filter X as $\{z\} \subset X$, then X satisfies all the conditions to be a ∂-stratified pseudomanifold except for the filtered collar condition. Let's do some computations of $I^{\bar{0}}H_*(X)$ and $I^{\bar{0}}H_*(X, \partial X)$. Letting z have its natural codimension of two, we have $\bar{0}(\{z\}) = 0 = \bar{t}(\{z\})$, and so by Proposition 6.2.9, we have $I^{\bar{0}}H_*(X) = I^{\bar{0}}H_*^{\mathrm{GM}}(X)$ and $I^{\bar{0}}H_2(X, \partial X) = I^{\bar{0}}H_2^{\mathrm{GM}}(X, \partial X)$, which will simplify our computations.

First we observe that $I^{\bar{0}}H_2^{\mathrm{GM}}(X) = 0$: Suppose that $\xi \in I^{\bar{0}}S_2^{\mathrm{GM}}(X)$ is a cycle. Let $\bar{c}\xi$ be the singular cone on ξ (see Example 3.4.7) with vertex at the center of the disk. If σ is a simplex of $\bar{c}\xi$, then σ has the form $\bar{c}\tau$ for some simplex τ of ξ, and $\sigma^{-1}(\{z\}) = \tau^{-1}(\{z\})$. But since τ must be allowable and σ has higher degree, the simplex σ will also be allowable. Thus $\bar{c}\xi$ will be allowable with $\partial(\bar{c}\xi) = \xi$. Therefore, all cycles in $I^{\bar{0}}S_2^{\mathrm{GM}}(X)$ bound.

Next, let's consider $I^{\bar{0}}H_1^{\mathrm{GM}}(\partial X)$. Recall from Section 4.3 that we let the filtration and perversity on ∂X be inherited from X so that $\{z\}$ continues to have codimension two as a stratum of ∂X, which has formal dimension two. If σ is a 1-simplex of ∂X, then for σ to be $\bar{0}$-allowable, we must have $\sigma^{-1}(\{z\})$ contained in the skeleton of Δ^1 of dimension $\dim(\sigma) - \operatorname{codim}(\{z\}) + \bar{0}(\{z\}) = 1 - 2 + 0 = -1$. So no allowable simplex in ∂X can intersect $\{z\}$, and it follows that $I^{\bar{0}}H_1^{\mathrm{GM}}(\partial X) = H_1(\partial X - \{z\}) = 0$.

Therefore, from the long exact sequence of the pair $(X, \partial X)$, we must have $I^{\bar{0}}H_2(X, \partial X) = I^{\bar{0}}H_2^{\mathrm{GM}}(X, \partial X) = 0$. This shows that X cannot have a fundamental class. Furthermore, we must have $I^{\bar{0}}H_0(X) = I^{\bar{0}}H_0^{\mathrm{GM}}(X) \cong \mathbb{Z}$ by Example 3.4.6, and so $I_{\bar{0}}H^0(X) \cong \mathbb{Z}$. As $D\bar{0} = \bar{t} = \bar{0}$ for this example, we see that Lefschetz duality also fails.

So, this example shows that filtered collars are necessary, in general, to have fundamental classes and Lefschetz duality for ∂-stratified pseudomanifolds as formulated here. However, a modified version of Lefschetz duality, not via cap products but omitting the collar condition, has been formulated in the PL setting by Valette [236]. △

(a) Topological Invariance

For Lefschetz duality, we have the following analogue of Theorem 8.2.6. The vertical isomorphisms labeled ϕ in the diagram are those constructed in the earlier topological invariance subsection, Section 8.3.1(a), while ϕ^* is obtained by taking the cohomological duals of the maps involved in the construction of ϕ. Applying the Universal Coefficient Theorem and the Five Lemma, it follows from the maps that are used to construct ϕ being isomorphisms that their duals are also isomorphisms, and so ϕ^* is an isomorphism.

Theorem 8.3.12 *Suppose R is a Dedekind domain and $\bar{p}$ is a GM perversity. Let X_1 and X_2 be two n-dimensional compact ∂-stratified pseudomanifolds with no codimension-one strata and with the same underlying space pairs $(|X_1|, |\partial X_1|) = (|X_2|, |\partial X_2|)$. Suppose X_1 and X_2 are compatibly R-oriented in the sense of Corollary 8.1.11 (applied to $|X_i| - |\partial X_i|$) and that X_1 and X_2 are locally $(\bar{p}; R)$-torsion-free*[21]*. Then there are diagrams of isomorphisms*

$$\begin{array}{ccc} I_{\bar{p}}H^i(X_1,\partial X_1;R) & \xrightarrow{\mathcal{D}} & I^{D\bar{p}}H_{n-i}(X_1;R) \\ \cong \uparrow \phi^* & & \cong \downarrow \phi \\ I_{\bar{p}}H^i(X_2,\partial X_2;R) & \xrightarrow{\mathcal{D}} & I^{D\bar{p}}H_{n-i}(X_2;R) \end{array} \quad \text{and} \quad \begin{array}{ccc} I_{\bar{p}}H^i(X_1;R) & \xrightarrow{\mathcal{D}} & I^{D\bar{p}}H_{n-i}(X_1,\partial X_1;R) \\ \cong \uparrow \phi^* & & \cong \downarrow \phi \\ I_{\bar{p}}H^i(X_2;R) & \xrightarrow{\mathcal{D}} & I^{D\bar{p}}H_{n-i}(X_2,\partial X_2;R) \end{array},$$

Remark 8.3.13 Analogously to Remark 8.2.7, it follows from such invariance results that Lefschetz duality is a topological invariant in the following broader sense: Suppose X and Y are compact n-dimensional R-oriented ∂-stratified pseudomanifolds without codimension-one strata, and suppose that $f : (|X|, |\partial X|) \to (|Y|, |\partial Y|)$ is a *topological homeomorphism*, i.e. that it is a homeomorphism of the underlying spaces without regard to the stratifications. Then X induces an image stratification, say Y', on Y, and an image R-orientation on Y'. Suppose that f is orientation-preserving in the sense the image R-orientation is compatible with the given R-orientation on Y in the sense of Corollary 8.1.11. Then employing Remark 8.3.8, Theorem 8.3.12, and naturality, we arrive at a diagram of isomorphisms of the following form, and analogously for the other duality diagram of Theorem 8.3.12:

$$\begin{array}{ccc} I_{\bar{p}}H^i(X,\partial X;R) & \xrightarrow{\mathcal{D}} & I^{D\bar{p}}H_{n-i}(X;R) \\ \cong \uparrow f^* & & \cong \downarrow f \\ I_{\bar{p}}H^i(Y',\partial Y';R) & \xrightarrow{\mathcal{D}} & I^{D\bar{p}}H_{n-i}(Y';R) \\ \cong \uparrow \phi^* & & \cong \downarrow \phi \\ I_{\bar{p}}H^i(Y,\partial Y;R) & \xrightarrow{\mathcal{D}} & I^{D\bar{p}}H_{n-i}(Y;R). \end{array}$$

Proof of Theorem 8.3.12 The proof runs through essentially the same sorts of isomorphisms as the proof of Proposition 8.3.7. We continue with the notation established there in the construction of the isomorphism ϕ and construct the diagram (coefficients tacit) in Figure 8.4. The composition up the left side is ϕ^*, while the composition down the right is ϕ. Recall that we have shown that these composites are independent of the precise choices of N_1 and N_2, and note that $D\bar{p}$ is a GM perversity as we have assumed that $\bar{p}$ is.

[21] By the argument of Proposition 5.5.9, both spaces are locally $(\bar{p}; R)$-torsion-free if either is.

$$
\begin{array}{ccc}
I_{\bar{p}}H^i(X_1,\partial X_1) & \xrightarrow[\cong]{\mathcal{D}} & I^{D\bar{p}}H_{n-i}(X_1) \\
\uparrow{\cong} & & \downarrow{=} \\
I_{\bar{p}}H^i(X_1,N_1) & \xrightarrow{\mathcal{D}} & I^{D\bar{p}}H_{n-i}(X_1) \\
\uparrow{\cong} & & \downarrow{=} \\
I_{\bar{p}}H^i(X_1,N_{2\to 1}) & \xrightarrow{\mathcal{D}} & I^{D\bar{p}}H_{n-i}(X_1) \\
\downarrow{\cong} & & \uparrow{\cong} \\
I_{\bar{p}}H^i(X_1-\partial X_1,N_{2\to 1}-\partial X_1) & \xrightarrow{\mathcal{D}} & I^{D\bar{p}}H_{n-i}(X_1-\partial X_1) \\
\uparrow{\cong} & & \downarrow{\cong} \\
I_{\bar{p}}H^i(\mathfrak{X},\mathfrak{N}) & \xrightarrow{\mathcal{D}} & I^{D\bar{p}}H_{n-i}(\mathfrak{X}) \\
\downarrow{\cong} & & \uparrow{\cong} \\
I_{\bar{p}}H^i(X_2-\partial X_2,N_2-\partial X_2) & \xrightarrow{\mathcal{D}} & I^{D\bar{p}}H_{n-i}(X_2-\partial X_2) \\
\uparrow{\cong} & & \downarrow{\cong} \\
I_{\bar{p}}H^i(X_2,N_2) & \xrightarrow{\mathcal{D}} & I^{D\bar{p}}H_{n-i}(X_2) \\
\downarrow{\cong} & & \uparrow{=} \\
I_{\bar{p}}H^i(X_2,\partial X_2) & \xrightarrow[\cong]{\mathcal{D}} & I^{D\bar{p}}H_{n-i}(X_2)
\end{array}
$$

Figure 8.4

The diagram commutes by the naturality of cap products (Proposition 7.3.6), which holds in each case due to Theorem 7.3.72, using Remark 7.3.70 to generalize for the pairs (X_1,N_1), $(X_1,N_{2\to 1})$, and (X_2,N_2). The vertical isomorphisms are those of ϕ and ϕ^*. All of the horizontal arrows are signed cap products with the fundamental classes or their images under the maps in the construction of ϕ for perversity $\bar{0}$. Proposition 8.3.7 shows that the image in $I^{\bar{0}}H_n(X_2,\partial X_2;R)$ of the fundamental class for X_1 is the fundamental class of X_2.

So the top and bottom horizontal maps are the Lefschetz duality isomorphisms (Theorem 8.3.9) induced by the compatible orientations. The compositions along the perimeter of the diagram provide our first claimed diagram of isomorphisms.

The argument for the second claimed diagram is analogous. □

Remark 8.3.14 We leave the reader to formulate and verify the analogous topological invariance property of Lefschetz duality of the form $\mathcal{D} : I_{\bar{p}}H^i(X, B; R) \to I^{D\bar{p}}H_{n-i}(X, A; R)$.

8.4 The Cup Product and Torsion Pairings

In this section, we discuss the nonsingular pairings that arise as a consequences of Poincaré and Lefschetz duality. The first, the cup product pairing on the torsion-free quotients of the cohomology groups, is well known from manifold theory and is a standard topic in introductory texts. The torsion product pairing is also classical for manifolds but is less often treated in textbooks. We also discuss, in Section 8.4.5, the "image pairings" that can be used to construct signatures on ∂-stratified pseudomanifolds.

As usual, throughout this section we continue to assume that our base ring R is a Dedekind domain.

8.4.1 Some Algebra

We begin by introducing some notation and recalling some algebraic background.

(a) Pairings

First, let us recall what we mean by a *nonsingular pairing*, starting with the general definition of a pairing.

Definition 8.4.1 Let A, B, C be R-modules. A homomorphism $P : A \otimes B \to C$ is called a *pairing*. If $a \in A$ and $b \in B$, then we typically write $P(a \otimes b) = P(a, b)$. Of course, not every element of $A \otimes B$ has the form $a \otimes b$, but knowing all the $P(a, b)$ (in fact, just knowing $P(a, b)$ as a and b run over sets of generators) is enough to determine P completely, as P is a homomorphism and so behaves bilinearly with respect to its two inputs.

We say that two pairings $P : A \otimes B \to C$ and $Q : D \otimes E \to C$ are isomorphic if there are isomorphisms $f : A \to D$ and $g : B \to E$ such that the following diagram commutes:

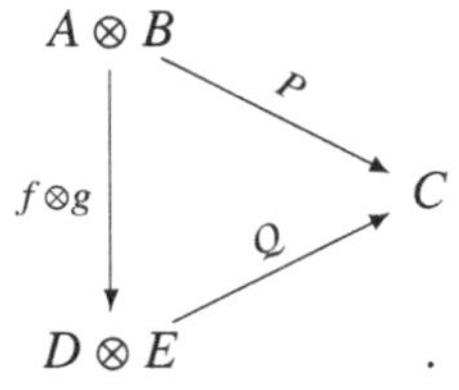

.

Recall that there is an adjunction isomorphism

$$\Lambda : \mathrm{Hom}(A \otimes B, C) \xrightarrow{\cong} \mathrm{Hom}(A, \mathrm{Hom}(B, C))$$

(see e.g. [238, Proposition 2.6.3]): Given $P \in \mathrm{Hom}(A \otimes B, C)$, then $\Lambda(P)$ is defined to take $a \in A$ to the homomorphism $P(a, \cdot) \in \mathrm{Hom}(B, C)$, which takes $b \in B$ to $P(a, b)$. That is, $((\Lambda(P))(a))(b) = P(a, b)$. Conversely, if $F \in \mathrm{Hom}(A, \mathrm{Hom}(B, C))$, then $\Lambda^{-1}(F) \in \mathrm{Hom}(A \otimes B, C)$ is the unique pairing that takes $a \otimes b$ to $(F(a))(b)$. Analogously, there is an adjunction isomorphism

$$\Lambda' : \mathrm{Hom}(A \otimes B, C) \xrightarrow{\cong} \mathrm{Hom}(B, \mathrm{Hom}(A, C))$$

such that $\Lambda'(P)(b) = P(\cdot, b)$.

Definition 8.4.2 The pairing $P : A \otimes B \to C$ is called *nonsingular* if the corresponding adjoint maps $\Lambda(P) : A \to \mathrm{Hom}(B, C)$ and $\Lambda'(P) : B \to \mathrm{Hom}(A, C)$ are both isomorphisms.

A slightly weaker notion that will concern us in Section 8.4.5 is that of a pairing being *nondegenerate*, which means that $\Lambda(P) : A \to \mathrm{Hom}(B, C)$ and $\Lambda'(P) : B \to \mathrm{Hom}(A, C)$ are injective. This is equivalent to saying that $P(a, b) = 0$ for all $b \in B$ if and only if $a = 0$ and that $P(a, b) = 0$ for all $a \in A$ if and only if $b = 0$. One is frequently concerned with pairings of finitely generated vector spaces with image in the ground field. In this case, a pairing is nondegenerate if and only if it is nonsingular, so the two expressions tend to be used interchangeably in that context.

(b) Torsion Submodules and Torsion-Free Quotients

Next we need some notation and background results about torsion submodules and their torsion-free quotients:

Definition 8.4.3 If A is an R-module, let $T(A)$ denote the *R-torsion submodule of A*,

$$T(A) = \{a \in A \mid \exists\, r \in R, r \neq 0, \text{ such that } ra = 0\}.$$

Let $F(A) = A/T(A)$ be the *torsion-free quotient* of A.

The module $F(A)$ is torsion-free: if $a \in A$ and $ra \in T(A)$ for some $r \in R$, $r \neq 0$, then there is some $s \in R$, $s \neq 0$, such that $s(ra) = (sr)a = 0$; as R is a domain, this means that $a \in T(A)$, so a represents 0 in $F(A)$. Recall also that torsion-free modules over Dedekind domains are flat; see Section A.4.2. Furthermore, if $F(A)$ is finitely generated (in particular, if A is finitely generated), then $F(A)$ is projective using that R is a Dedekind domain and hence Noetherian [30, Theorem VII.2.2.1] and that finitely-generated flat modules over Noetherian rings are projective [146, Theorem 4.38].

Example 8.4.4 As we saw in the proof of Corollary 8.2.5, the module $\mathrm{Hom}(A, R)$ is torsion-free for any R-module A, so $T(\mathrm{Hom}(A, R)) = 0$ and $F(\mathrm{Hom}(A, R)) = \mathrm{Hom}(A, R)$. △

Recall next that the cohomological dimension of a Dedekind domain is ≤ 1 (see [197, Proposition 8.1] and use that Dedekind domains are hereditary by definition [197, page 161]). In particular, this means that $\text{Ext}^n(A, B) = 0$ for $n > 1$ and for R-modules A, B. This justifies our writing $\text{Ext}(A, B)$ to mean $\text{Ext}^1(A, B)$ throughout the text. Therefore, the right derived homology exact sequence for the functor $\text{Hom}(\cdot, R)$ (see [197, Corollary 6.62]) applied to the short exact sequence

$$0 \to T(A) \to A \to F(A) \to 0$$

yields the exact sequence

$$\begin{aligned} 0 \to \text{Hom}(F(A), R) \to \text{Hom}(A, R) \to \text{Hom}(T(A), R) \\ \to \text{Ext}(F(A), R) \to \text{Ext}(A, R) \to \text{Ext}(T(A), R) \to 0. \end{aligned}$$

But $\text{Hom}(T(A), R)$ must be zero because if $x \in T(A)$ with $rx = 0$, $r \neq 0$, then for any $f \in \text{Hom}(T(A), R)$ we have $rf(x) = f(rx) = f(0) = 0$; this implies $f(x) = 0$, as R is a domain. Additionally, if $F(A)$ is finitely generated (for example if A is finitely generated), then $F(A)$ is projective, so $\text{Ext}(F(A), R) = 0$. Thus we have the following lemma, which also incorporates Example 8.4.4.

Lemma 8.4.5 *If R is a Dedekind domain and A is an R-module, then:*

1. $\text{Hom}(A, R)$ *is torsion-free,*
2. *the canonical map* $\text{Hom}(F(A), R) \to \text{Hom}(A, R)$ *is an isomorphism, and*
3. *if $F(A)$ is finitely generated (in particular, if A is finitely generated), then the canonical map* $\text{Ext}(A, R) \to \text{Ext}(T(A), R)$ *is an isomorphism.*

Next, let $Q(R)$ denote the field of fractions of R (see [147, Section II.4]); an important special case is $R = \mathbb{Z}$ with $Q(\mathbb{Z}) = \mathbb{Q}$. There is an exact sequence

$$0 \to R \to Q(R) \to Q(R)/R \to 0$$

and, again using that R has cohomological dimension ≤ 1, the right derived homology exact sequence of the functor $\text{Hom}(A, \cdot)$ (see [197, Corollary 6.46] or [126, Section IV.8]) yields the six-term exact sequence

$$\begin{aligned} 0 \to \text{Hom}(A, R) \to \text{Hom}(A, Q(R)) \to \text{Hom}(A, Q(R)/R) \\ \to \text{Ext}(A, R) \to \text{Ext}(A, Q(R)) \to \text{Ext}(A, Q(R)/R) \to 0. \end{aligned} \tag{8.9}$$

In this sequence, $\text{Ext}(A, Q(R))$ is trivial, because $Q(R)$ is a field. Hence $\text{Ext}(A, Q(R)/R) = 0$ also. We will be particularly interested in the case where A is replaced with its torsion submodule $T(A)$. In this case, we have seen just above that $\text{Hom}(T(A), R) = 0$ and $\text{Hom}(T(A), Q(R)) = 0$ by the same argument, as $Q(R)$ is also a domain. This yields the first part of the following lemma.

Lemma 8.4.6 *If R is a Dedekind domain and A is an R-module, then:*

1. *the connecting map* $\operatorname{Hom}(T(A), Q(R)/R) \to \operatorname{Ext}(T(A), R)$ *is an isomorphism, and*
2. *if A is finitely generated, then* $\operatorname{Hom}(T(A), Q(R)/R) \cong \operatorname{Ext}(T(A), R) \cong \operatorname{Ext}(A, R)$ *is a torsion module.*

Proof The first item has already been shown in the discussion just above. For the second, we first observe that if A is finitely generated, then so is $F(A)$, so the isomorphisms come from Lemma 8.4.5 and the first part of this lemma. Additionally, $T(A)$ will be finitely generated, as R is Noetherian (see [30, Theorem VII.2.2.1] and [147, Section X.1]). So, suppose $f \in \operatorname{Hom}(T(A), Q(R)/R)$, and let $\{x_i\}$ be a finite set of generators of $T(A)$. For each $f(x_i) \in Q(R)/R$, there is some $r_i \in R$, $r_i \neq 0$, such that $r_i f(x_i) = 0$; for example, we can let r_i be the denominator of any fraction in $Q(R)$ representing $f(x_i)$. If we let $r = \prod_i r_i$, then rf takes all generators of $T(A)$ to zero and so $rf = 0$, $r \neq 0$. □

8.4.2 The Cup Product Pairing

We now turn to demonstrating that Poincaré and Lefschetz duality imply that the cup product determines nonsingular pairings. For the sake of generality, we state the result for a ∂-stratified pseudomanifold, but of course ∂X may be empty.

Theorem 8.4.7 *Suppose R is a Dedekind domain, and let X be a compact n-dimensional R-oriented locally* $(\bar{p}; R)$*-torsion-free* ∂*-stratified pseudomanifold. Then the composition*[22]

$$F(I_{\bar{p}}H^i(X;R)) \otimes F(I_{D\bar{p}}H^{n-i}(X,\partial X;R)) \xrightarrow{\smile} F(I_{\bar{0}}H^n(X,\partial X;R)) \xrightarrow{\mathcal{D}} F(I^{\bar{t}}H_0(X;R)) \xrightarrow{\mathbf{a}} R \tag{8.10}$$

is a nonsingular pairing.

Definition 8.4.8 We will refer to the pairing of (8.10) as the *cup product pairing*. Evidently, there is a similar cup product pairing

$$F(I_{\bar{p}}H^i(X,\partial X;R)) \otimes F(I_{D\bar{p}}H^{n-i}(X;R)) \xrightarrow{\smile} F(I_{\bar{0}}H^n(X,\partial X;R)) \xrightarrow{\mathcal{D}} F(I^{\bar{t}}H_0(X;R)) \xrightarrow{\mathbf{a}} R.$$

Proof of Theorem 8.4.7 First, we verify that the given composition makes sense.

As X is locally $(\bar{p}; R)$-torsion-free, the triples $(\bar{p}, D\bar{p}; \bar{0})$ and $(\bar{t}, \bar{0}; \bar{0})$ are both agreeable by Corollary 7.2.10 and so the underlying cup and cap products are defined. In case $\partial X \neq \emptyset$, these exist by Corollary 7.3.71.

If $\alpha \in I_{\bar{p}}H^i(X;R)$ and $\beta \in I_{D\bar{p}}H^{n-i}(X,\partial X;R)$ and either α or β is a torsion element, then $\alpha \smile \beta$ must be a torsion element as the cup product is bilinear. For example, if $r\alpha = 0$ for $r \in R$, $r \neq 0$, then $r(\alpha \smile \beta) = (r\alpha) \smile \beta = 0$. Similarly, if $\gamma \in I_{\bar{0}}H^n(X,\partial X;R)$

[22] In fact, by an easy generalization of Example 3.4.6, the module $I^{\bar{t}}H_0(X;R)$ is free and so $F(I^{\bar{t}}H_0(X;R)) = I^{\bar{t}}H_0(X;R)$. Consequently, by Lefschetz duality (Theorem 8.3.9), we also have $F(I_{\bar{0}}H^n(X,\partial X;R)) = I_{\bar{0}}H^n(X,\partial X;R)$.

is a torsion element then $\mathcal{D}(\gamma)$ is a torsion element, and if $\xi \in I^{\bar{t}}H_0(X;R)$ is a torsion element then $\mathbf{a}(\xi) = 0$. So each map $\smile$, $\frown$, and $\mathbf{a}$ descends to a well-defined map of torsion-free quotient modules as indicated, and, in particular, the cup product pairing defined as the composition is well defined on $F(I_{\bar{p}}H^i(X;R)) \otimes F(I_{D\bar{p}}H^{n-i}(X,\partial X;R))$. It remains to show that the adjoint maps $F(I_{\bar{p}}H^i(X;R)) \to \mathrm{Hom}(F(I_{D\bar{p}}H^{n-i}(X,\partial X;R)),R)$ and $F(I_{D\bar{p}}H^{n-i}(X,\partial X;R)) \to \mathrm{Hom}(F(I_{\bar{p}}H^i(X;R)),R)$ determined by the pairing are isomorphisms.

Let us see how the adjoint to the pairing operates. Given $\alpha \in F(I_{\bar{p}}H^i(X;R))$, its image in $\mathrm{Hom}(F(I_{D\bar{p}}H^{n-i}(X,\partial X;R)),R)$ takes $\beta \in F(I_{D\bar{p}}H^{n-i}(X,\partial X;R))$ to[23] $\mathbf{a}(\mathcal{D}(\alpha \smile \beta)) = (-1)^n\mathbf{a}((\alpha \smile \beta) \frown \Gamma)$. We want to rewrite this formula using associativity of cup and cap products. Using our hypotheses, the torsion-free properties of $\bar{t}$ (Example 6.3.22), and Corollary 8.2.5, the space X is locally torsion-free with respect to all perversities involved, and so the associativity property of Proposition 7.3.35 holds by Remark 7.3.36 as $D\bar{0} = D\bar{t} + D\bar{p} + D(D\bar{p})$. When $\partial X \neq \emptyset$, we further invoke item (7) of Theorem 7.3.72, noting that our space pairs satisfy the necessary requirements by Example 7.3.68. So now we have

$$\mathbf{a}(\mathcal{D}(\alpha \smile \beta)) = (-1)^n\mathbf{a}((\alpha \smile \beta) \frown \Gamma) = (-1)^n\mathbf{a}(\alpha \frown (\beta \frown \Gamma)).$$

By Proposition 7.3.25 and item (6) of Theorem 7.3.72, this is further equivalent to the evaluation $(-1)^n\alpha(\beta \frown \Gamma)$. Once again, these formulas and identities descend in a well-defined manner to the torsion-free quotient modules.

Let $\kappa : I_{\bar{p}}H^i(X;R) \to \mathrm{Hom}(I^{\bar{p}}H_i(X;R),R)$ be the universal coefficient (Kronecker) evaluation map in the universal coefficient sequence of Theorem 7.1.4:

$$0 \leftarrow \mathrm{Hom}(I^{\bar{p}}H_i(X;R),R) \xleftarrow{\kappa} I_{\bar{p}}H^i(X;R) \leftarrow \mathrm{Ext}(I^{\bar{p}}H_{i-1}(X;R),R) \leftarrow 0.$$

Then $(-1)^n\alpha(\beta \frown \Gamma)$ can be written more pedantically as $(-1)^n(\kappa(\alpha))(\beta \frown \Gamma)$. Furthermore, as $I^{\bar{p}}H_i(X;R)$ is finitely generated by Corollary 6.3.40, the Ext term is a torsion module by Lemma 8.4.6, while $\mathrm{Hom}(I^{\bar{p}}H_i(X;R),R)$ is torsion-free. It follows that κ induces an isomorphism $F(I_{\bar{p}}H^i(X;R)) \cong \mathrm{Hom}(I^{\bar{p}}H_i(X;R),R) \cong \mathrm{Hom}(F(I^{\bar{p}}H_i(X;R)),R)$ and that $T(I_{\bar{p}}H^i(X;R)) \cong \mathrm{Ext}(I^{\bar{p}}H_{i-1}(X;R),R)$.

Next, we compute that

$$\begin{aligned}(-1)^n\kappa(\alpha)(\beta \frown \Gamma) &= (-1)^{n+n(n-i)}\kappa(\alpha)(\mathcal{D}(\beta))\\ &= (-1)^{n+n(n-i)+in}(\mathcal{D}^*(\kappa(\alpha)))(\beta)\\ &= \mathcal{D}^*(\kappa(\alpha))(\beta),\end{aligned}$$

where $\mathcal{D}^*$ is the $\mathrm{Hom}(\cdot,R)$ dual of $\mathcal{D}$ and we have used the Koszul convention for the interchange of the degree-n operator $\mathcal{D}$ with the degree-i operator $\kappa(\alpha)$.

[23] Our conventions for the Poincaré duality map are responsible for the sign $(-1)^n$, which is probably not commonly in use for the cup product pairing. However, the most important uses of the cup product pairing in the literature are, no doubt, those involving the symmetric and antisymmetric self-pairings on the middle-dimensional cohomology of even-dimensional manifolds, in which case the sign is 1.

So, in sum, our adjoint map $F(I_{\bar{p}}H^i(X;R)) \to \mathrm{Hom}(F(I_{D\bar{p}}H^{n-i}(X,\partial X;R)),R)$ takes α to $\mathcal{D}^*(\kappa(\alpha))$. But we have just observed that κ is an isomorphism,

$$F(I_{\bar{p}}H^i(X;R)) \to \mathrm{Hom}(F(I^{\bar{p}}H_i(X;R)),R),$$

and Theorem 8.3.9 shows that $\mathcal{D}^*$ is an isomorphism from $\mathrm{Hom}(F(I^{\bar{p}}H_i(X;R)),R)$ to $\mathrm{Hom}(F(I_{D\bar{p}}H^{n-i}(X,\partial X;R)),R)$. Therefore we have shown that the desired map $F(I_{\bar{p}}H^i(X;R)) \to \mathrm{Hom}(F(I_{D\bar{p}}H^{n-i}(X,\partial X;R)),R)$ is an isomorphism.

For the map $F(I_{D\bar{p}}H^{n-i}(X,\partial X;R)) \to \mathrm{Hom}(F(I_{\bar{p}}H^i(X;R)),R)$, we use that, by Proposition 7.3.15 and item (4) of Theorem 7.3.72, we have $\mathbf{a}((\alpha \smile \beta) \frown \Gamma) = \mathbf{a}((\beta \smile \alpha) \frown \Gamma)$, up to sign. From here, we can utilize an equivalent argument to that above, interchanging the roles of α and β and of $(X,\emptyset)$ and $(X,\partial X)$.

This completes the proof of Theorem 8.4.7. □

Example 8.4.9 Let $X = S(\mathbb{R}P^2)$ be the suspension of $\mathbb{R}P^2$, stratified in the natural way with just two singular points at the north and south poles. Let $\bar{0}$ and $\bar{1}$ be the perversities that take values, respectively, 0 or 1 at both singular points. These are dual Goresky–MacPherson perversities. Using Theorem 6.3.13, we compute

$$\begin{aligned}
I^{\bar{0}}H_3(X;\mathbb{Z}_2) &\cong \mathbb{Z}_2, & I^{\bar{1}}H_3(X;\mathbb{Z}_2) &\cong \mathbb{Z}_2,\\
I^{\bar{0}}H_2(X;\mathbb{Z}_2) &\cong 0, & I^{\bar{1}}H_2(X;\mathbb{Z}_2) &\cong \mathbb{Z}_2,\\
I^{\bar{0}}H_1(X;\mathbb{Z}_2) &\cong \mathbb{Z}_2, & I^{\bar{1}}H_1(X;\mathbb{Z}_2) &\cong 0,\\
I^{\bar{0}}H_0(X;\mathbb{Z}_2) &\cong \mathbb{Z}_2, & I^{\bar{1}}H_0(X;\mathbb{Z}_2) &\cong \mathbb{Z}_2.
\end{aligned}$$

Thus, taking $\mathbb{Z}_2$ as our ground field, the Universal Coefficient Theorem for cohomology implies

$$\begin{aligned}
I_{\bar{0}}H^3(X;\mathbb{Z}_2) &\cong \mathbb{Z}_2, & I_{\bar{1}}H^3(X;\mathbb{Z}_2) &\cong \mathbb{Z}_2,\\
I_{\bar{0}}H^2(X;\mathbb{Z}_2) &\cong 0, & I_{\bar{1}}H^2(X;\mathbb{Z}_2) &\cong \mathbb{Z}_2,\\
I_{\bar{0}}H^1(X;\mathbb{Z}_2) &\cong \mathbb{Z}_2, & I_{\bar{1}}H^1(X;\mathbb{Z}_2) &\cong 0,\\
I_{\bar{0}}H^0(X;\mathbb{Z}_2) &\cong \mathbb{Z}_2, & I_{\bar{1}}H^0(X;\mathbb{Z}_2) &\cong \mathbb{Z}_2.
\end{aligned}$$

Theorem 8.4.7 now provides nontrivial nonsingular pairings

$$\begin{aligned}
&I_{\bar{0}}H^3(X;\mathbb{Z}_2) \otimes I_{\bar{1}}H^0(X;\mathbb{Z}_2) \to \mathbb{Z}_2,\\
&I_{\bar{0}}H^1(X;\mathbb{Z}_2) \otimes I_{\bar{1}}H^2(X;\mathbb{Z}_2) \to \mathbb{Z}_2,\\
&I_{\bar{0}}H^0(X;\mathbb{Z}_2) \otimes I_{\bar{1}}H^3(X;\mathbb{Z}_2) \to \mathbb{Z}_2.
\end{aligned}$$

In particular, if $\alpha \in I_{\bar{0}}H^1(X;\mathbb{Z}_2)$ and $\beta \in I_{\bar{1}}H^2(X;\mathbb{Z}_2)$ are the non-zero elements, then $\alpha \smile \beta \neq 0$.

This is quite different from the situation for cup products in ordinary cohomology on suspensions; in that setting, the cup product on reduced cohomology is always trivial (see e.g. [231, Corollary 13.66]). △

8.4.3 The Torsion Pairing

We continue to assume that R is a Dedekind domain and X is a compact R-oriented ∂-stratified pseudomanifold. The proof of Theorem 8.4.7, which establishes the nonsingularity of the cup product pairing, shows that the adjoint of this pairing is the composition $\mathcal{D}^*\kappa$, where $\mathcal{D}^*$ is the $\operatorname{Hom}(\cdot, R)$ dual of the Poincaré/Lefschetz duality map $\mathcal{D}$ and κ is the Kronecker evaluation map $\kappa : I_{\bar{p}}H^i(X;R) \to \operatorname{Hom}(I^{\bar{p}}H_i(X;R),R)$. Relatedly, there is another pairing that can be obtained by a similar composition, this time using the torsion side of the universal coefficient splitting

$$I_{\bar{p}}H^i(X;R) \cong \operatorname{Hom}(I^{\bar{p}}H_i(X;R),R) \oplus \operatorname{Ext}(I^{\bar{p}}H_{i-1}(X;R),R).$$

Applying the Universal Coefficient Theorem together with Lemmas 8.4.5 and 8.4.6, we have a composite isomorphism

$$\begin{aligned}\lambda : T(I_{\bar{p}}H^i(X;R)) &\cong \operatorname{Ext}(I^{\bar{p}}H_{i-1}(X;R),R) \cong \operatorname{Ext}(T(I^{\bar{p}}H_{i-1}(X;R)),R)\\ &\cong \operatorname{Hom}(T(I^{\bar{p}}H_{i-1}(X;R)),Q(R)/R). \qquad (8.11)\end{aligned}$$

We will study the precise maps in this composition more carefully in the coming pages, but for now we note that if we compose λ with a different $\mathcal{D}^*$, the $\operatorname{Hom}(\cdot, Q(R)/R)$ dual of $\mathcal{D}$, restricted to be an isomorphism between torsion submodules, then we arrive at an isomorphism

$$\mathcal{D}^*\lambda : T(I_{\bar{p}}H^i(X;R)) \to \operatorname{Hom}(T(I_{D\bar{p}}H^{n-i+1}(X,\partial X;R)),Q(R)/R).$$

The adjoint is thus a pairing

$$L : T(I_{\bar{p}}H^i(X;R)) \otimes T(I_{D\bar{p}}H^{n-i+1}(X,\partial X;R)) \to Q(R)/R.$$

It will take some work to unravel this pairing in terms of cup and cap products and to show that the other adjoint

$$T(I_{D\bar{p}}H^{n-i+1}(X,\partial X;R)) \to \operatorname{Hom}(T(I_{\bar{p}}H^i(X;R)),Q(R)/R)$$

is also an isomorphism. However, we can state now the end result:

Theorem 8.4.10 *Suppose R is a Dedekind domain and that X is a compact n-dimensional R-oriented locally $(\bar{p};R)$-torsion-free ∂-stratified pseudomanifold. Then the composition of isomorphisms*

$$\begin{aligned}T(I_{\bar{p}}H^i(X;R)) &\xrightarrow{\lambda} \operatorname{Hom}(T(I^{\bar{p}}H_{i-1}(X;R)),Q(R)/R)\\ &\xrightarrow{\mathcal{D}^*} \operatorname{Hom}(T(I_{D\bar{p}}H^{n-i+1}(X,\partial X;R)),Q(R)/R)\end{aligned}$$

determines an adjoint nonsingular pairing

$$L_{\bar{p},D\bar{p}} : T(I_{\bar{p}}H^i(X;R)) \otimes T(I_{D\bar{p}}H^{n-i+1}(X,\partial X;R)) \to Q(R)/R.$$

Analogously, there is a nonsingular pairing

$$L'_{D\bar{p},\bar{p}} : T(I_{D\bar{p}}H^{n-i+1}(X,\partial X;R)) \otimes T(I_{\bar{p}}H^i(X;R)) \to Q(R)/R.$$

If α *and* β *are cochains representing elements in* $T(I_{\bar{p}}H^i(X;R))$*, respectively* $T(I_{D\bar{p}}H^{n-i+1}(X,\partial X;R))$*, and if* $t\beta = d\mathfrak{b}$ *for* $\mathfrak{b} \in I_{D\bar{p}}S^{n-i}(X,\partial X;R)$*,* $t \in R$*,* $t \neq 0$*, and* $r\alpha = d\mathfrak{a}$ *for* $\mathfrak{a} \in I_{\bar{p}}S^{i-1}(X;R)$*,* $r \in R$*,* $r \neq 0$*, then*

$$L_{\bar{p},D\bar{p}}(\alpha \otimes \beta) = (-1)^n \frac{\mathbf{a}((\alpha \smile \mathfrak{b}) \frown \Gamma)}{t},$$

$$L'_{D\bar{p},\bar{p}}(\beta \otimes \alpha) = (-1)^n \frac{\mathbf{a}((\beta \smile \mathfrak{a}) \frown \Gamma)}{r}.$$

Furthermore,

$$L_{\bar{p},D\bar{p}}(\alpha \otimes \beta) = (-1)^{1+n+in} L'_{D\bar{p},\bar{p}}(\beta \otimes \alpha).$$

The proof of the theorem will proceed over several steps in the subsections (a) to (d) below. First we study the components of the map λ in more detail in order ultimately to come up with an explicit expression for $\lambda(\alpha)(\beta)$ when $\alpha \in T(I_{\bar{p}}H^i(X;R))$ and $\beta \in T(I^{\bar{p}}H_{i-1}(X;R))$. We use this expression to write the torsion pairing explicitly in terms of cup and cap products. This then provides the symmetry formula and allows us to show that the torsion pairing is nonsingular. All this work will take us through most of the remainder of this section. After proving the theorem, we briefly discuss another common approach to torsion pairings.

(a) The Components of λ

To begin, we need to understand the map λ in more detail, starting with specifying the identification of the Ext summand of the Universal Coefficient Theorem as the torsion submodule of $I_{\bar{p}}H^i(X;R)$. For this, we bring back some of our convenient notation from Section 6.4.5 as well as adding some more: Let $C_i = I^{\bar{p}}S_i(X;R)$, let $Z_i = \ker(\partial : I^{\bar{p}}S_i(X;R) \to I^{\bar{p}}S_{i-1}(X;R))$, and let $B_i = \operatorname{im}(\partial : I^{\bar{p}}S_{i+1}(X;R) \to I^{\bar{p}}S_i(X;R))$. Further, let $W_i \subset C_i$ be the submodule of *weak boundaries*, i.e.

$$W_i = \{w \in C_i \mid \exists\, r_w \neq 0 \in R \text{ such that } r_w w \in B_i\}.$$

In other words, elements of W_i need not necessarily be boundaries, but they are the elements of C_i that have non-zero scalar multiples that are boundaries. Note that, for a given $w \in W_i$, there is not a unique associated r_w, but for the purposes of our current discussion we can assume that some particular r_w has been fixed for each w. Observe that $B_i \subset W_i \subset Z_i$, the first inclusion by using $r_w = 1$ and the second because if $w \in W_i$ then $0 = \partial(r_w w) = r_w(\partial w)$, so $\partial w = 0$ as C_i is projective and hence torsion-free.

Now, by definition $\operatorname{Ext}(\cdot, R)$ is the right derived functor of $\operatorname{Hom}(\cdot, R)$, so using the projective resolution

$$0 \to B_{i-1} \xrightarrow{\mathfrak{i}} Z_{i-1} \to H_{i-1}(C_*) \to 0$$

with $\mathfrak{i}$ the inclusion, we can realize $\operatorname{Ext}(I^{\bar{p}}H_{i-1}(X;R), R)$ concretely as the cokernel of the dual $\mathfrak{i}^* : \mathrm{Hom}(Z_{i-1}, R) \to \mathrm{Hom}(B_{i-1}, R)$. It follows that the embedding map $\mathrm{Ext}(I^{\bar{p}}H_{i-1}(X; R), R) \to I_{\bar{p}}H^i(X; R)$ can be written in terms of an appropriate map $\mathrm{Hom}(B_{i-1}, R) \to I_{\bar{p}}H^i(X; R)$, and this is essentially what is done in proofs of the Universal Coefficient Theorem with the map being essentially the adjoint of the boundary map $\partial : C_i \to B_{i-1}$. Indeed, if $f \in \mathrm{Hom}(B_{i-1}, R)$, then $\partial^* f$ acts on elements $x \in C_i$ by $f(\partial x)$ (up to sign), and so $d(\partial^* f)$ acts by $f(\partial\partial x) = 0$. Therefore, $\partial^* f$ is a cocycle and represents a class in $I_{\bar{p}}H^i(X; R)$. This is the basic idea of the following lemma, which indeed maps $\mathrm{Ext}(I^{\bar{p}}H_{i-1}(X; R), R)$ isomorphically to a submodule of $I_{\bar{p}}H^i(X; R)$ via ∂^*.

Remark 8.4.11 As the boundary map ∂ lowers degrees by one, the adjoint ∂^* should technically be treated as a map of cohomological degree one. However, in this section we will not really be concerned with chain maps so much as the torsion pairings in fixed degrees, and ultimately what we really want is the torsion pairing in the form stated in Theorem 8.4.10 – the λ maps we are developing are more of a means to an end. Hence, to simplify some of the coming formulas, we will abuse the Koszul convention for the definition of ∂^* used here and let $\partial^* f$ act on a chain x by $(\partial^* f)(x) = f(\partial x)$.

Lemma 8.4.12 *The torsion submodule of $I_{\bar{p}}H^i(X; R)$ is the image of the map*

$$\partial^* : \mathrm{Hom}(B_{i-1}, R) \to H^i(\mathrm{Hom}(C_i, R)) = I_{\bar{p}}H^i(X; R).$$

Furthermore, the kernel of ∂^* is the image of $\mathfrak{i}^* : \mathrm{Hom}(Z_{i-1}, R) \to \mathrm{Hom}(B_{i-1}, R)$, the dual of the inclusion $\mathfrak{i} : B_{i-1} \hookrightarrow Z_{i-1}$. Therefore, the map ∂^* induces an isomorphism $\partial^* : \mathrm{Ext}(I^{\bar{p}}H_{i-1}(X; R), R) \xrightarrow{\cong} T(I_{\bar{p}}H^i(X; R))$.

Proof We begin by recalling the proof of the Universal Coefficient Theorem (see [126], [182], or [238]; the discussion is also similar to our look at the algebraic Künneth Theorem in Section 6.4.5). Continuing the notation introduced just before the statement of the lemma, let Z_* be the chain complex with modules Z_i and all boundary maps zero, and let B'_* be the chain complex with modules $B'_i = B_{i-1}$ and all boundary maps zero. Note that, as R is a Dedekind domain, each Z_i and B'_i is projective, being submodules of the projective $I^{\bar{p}}S_i(X; R)$ and $I^{\bar{p}}S_{i-1}(X; R)$, respectively.

We have a short exact sequence of chain complexes[24]

$$0 \to Z_* \xrightarrow{\mathfrak{j}} C_* \xrightarrow{\delta} B'_* \to 0,$$

where $\mathfrak{j}$ is inclusion and δ is the boundary map in $I^{\bar{p}}S_*(X; R)$, treated as a degree-zero chain map. As each module is projective, this sequence dualizes to the short exact sequence of cochain complexes and degree-zero chain maps

$$0 \leftarrow \mathrm{Hom}(Z_*, R) \xleftarrow{\mathfrak{j}^*} \mathrm{Hom}(C_*, R) \xleftarrow{\delta^*} \mathrm{Hom}(B'_*, R) \leftarrow 0.$$

Taking (co)homology, setting $\mathrm{Hom}(B'_*, R) = \mathrm{Hom}(B_{*-1}, R)$, and recognizing δ^* now as

[24] We use δ here rather than the β of Section 6.4.5 to leave β free for cochains.

∂^* up to signs, the resulting long exact sequence includes sections of the form

$$\leftarrow \mathrm{Hom}(Z_*, R) \overset{\mathrm{i}^*}{\leftarrow} I_{\bar{p}}H^i(X;R) \overset{\partial^*}{\longleftarrow} \mathrm{Hom}(B_{*-1}, R) \leftarrow \mathrm{Hom}(Z_{*-1}, R) \leftarrow .$$

The cohomology symbols H^i do not need to appear in several terms because the differentials of the complexes $\mathrm{Hom}(Z_*, R)$ and $\mathrm{Hom}(B'_*, R)$ are trivial. A diagram chase with the zig-zag construction shows that the connecting morphisms are, up to sign, simply the restrictions of $\mathrm{Hom}(Z_*, R)$ to $\mathrm{Hom}(B_*, R)$.

Next, we use that

$$0 \to B_{i-1} \overset{\mathrm{i}}{\to} Z_{i-1} \to H_{i-1}(C_*) \to 0 \tag{8.12}$$

is a projective resolution of $H_{i-1}(C_*)$, so the cokernel of $\mathrm{i}^* : \mathrm{Hom}(B_{*-1}, R) \leftarrow \mathrm{Hom}(Z_{*-1}, R)$ is precisely $\mathrm{Ext}(H_{i-1}(C_*), R)$ by definition. But from our long exact sequence, this cokernel is isomorphic to the image of ∂^*. Furthermore, if each $H_j(C_*)$ is finitely generated, as we are assuming in the case at hand, then the discussion in the proof of Theorem 8.4.7 shows that $\mathrm{im}(\partial^*) \cong \mathrm{Ext}(I^{\bar{p}}H_{i-1}(X;R), R)$ is the torsion subgroup $T(I_{\bar{p}}H^i(X;R))$. So we have shown that

$$\begin{aligned} T(I_{\bar{p}}H^i(X;R)) &\cong \mathrm{Ext}(H_{i-1}(C_*), R) \cong \mathrm{im}(\partial^*) \\ &\cong \frac{\mathrm{Hom}(B_{i-1}, R)}{\mathrm{im}(\mathrm{Hom}(Z_{i-1}, R) \to \mathrm{Hom}(B_{i-1}, R))}. \end{aligned}$$

□

The next lemma will be useful in constructing an explicit isomorphism

$$\mathrm{Ext}(I^{\bar{p}}H_{i-1}(X;R), R) \cong \mathrm{Hom}(T(I^{\bar{p}}H_{i-1}(X;R)), Q(R)/R).$$

Lemma 8.4.13 *Let $B_i \subset W_i \subset Z_i$ be as defined above. Then* $\mathrm{im}(\mathrm{Hom}(Z_i, R) \to \mathrm{Hom}(B_i, R)) = \mathrm{im}(\mathrm{Hom}(W_i, R) \to \mathrm{Hom}(B_i, R))$, *where both maps are the* $\mathrm{Hom}(\cdot, R)$ *duals of the inclusions.*

Proof Since W_i consists precisely of those non-zero cycles whose scalar multiples are boundaries, $W_i/B_i \cong T(I^{\bar{p}}H_i(X;R))$. Then

$$Z_i/W_i \cong (Z_i/B_i)/(W_i/B_i) \cong I^{\bar{p}}H_i(X;R)/T(I^{\bar{p}}H_i(X;R)) = F(I^{\bar{p}}H_i(X;R)).$$

So

$$0 \to W_i \to Z_i \to F(I^{\bar{p}}H_i(X;R)) \to 0$$

is a projective resolution of $F(I^{\bar{p}}H_i(X;R))$, and

$$\frac{\mathrm{Hom}(W_i, R)}{\mathrm{im}(\mathrm{Hom}(Z_i, R) \to \mathrm{Hom}(W_i, R))} \cong \mathrm{Ext}(F(I^{\bar{p}}H_i(X;R)), R).$$

But, as $F(I^{\bar{p}}H_i(X;R))$ is torsion-free and finitely generated, it is projective, so

$$\mathrm{Ext}(F(I^{\bar{p}}H_i(X;R)), R) = 0,$$

and it follows that $\mathrm{Hom}(Z_i, R) \to \mathrm{Hom}(W_i, R)$ is surjective. Therefore,

$$\mathrm{im}(\mathrm{Hom}(Z_i, R) \to \mathrm{Hom}(B_i, R)) = \mathrm{im}(\mathrm{Hom}(W_i, R) \to \mathrm{Hom}(B_i, R)),$$

as desired. □

Remark 8.4.14 The preceding lemma provides another, perhaps more direct, proof that $\mathrm{Ext}(I^{\bar{p}}H_i(X;R), R) \cong \mathrm{Ext}(T(I^{\bar{p}}H_i(X;R)), R)$: We showed in the proof of Lemma 8.4.12 that we have

$$\frac{\mathrm{Hom}(B_i, R)}{\mathrm{im}(\mathrm{Hom}(Z_i, R) \to \mathrm{Hom}(B_i, R))} \cong \mathrm{Ext}(I^{\bar{p}}H_i(X;R), R),$$

using the definition of Ext and the projective resolution (8.12). Meanwhile,

$$0 \to B_i \to W_i \to T(I^{\bar{p}}H_i(X;R)) \to 0 \tag{8.13}$$

is a projective resolution of $T(I^{\bar{p}}H_i(X;R))$. So, applying $\mathrm{Hom}(\cdot, R)$, we see

$$\frac{\mathrm{Hom}(B_i, R)}{\mathrm{im}(\mathrm{Hom}(W_i, R) \to \mathrm{Hom}(B_i, R))} \cong \mathrm{Ext}(T(I^{\bar{p}}H_i(X;R)), R),$$

using the definition of Ext and the projective resolution (8.13). But Lemma 8.4.13 demonstrates that these are the same module, identically.

Next, we use Lemma 8.4.13 to look more closely at the isomorphism

$$\mathrm{Ext}(T(I^{\bar{p}}H_{i-1}(X;R)), R) \cong \mathrm{Hom}(T(I^{\bar{p}}H_{i-1}(X;R)), Q(R)/R)$$

of Lemma 8.4.6. From the exact sequence $0 \to R \to Q(R) \to Q(R)/R \to 0$, we can build a commutative diagram

$$\begin{array}{ccccccccc} 0 & \longrightarrow & \mathrm{Hom}(W_{i-1}, R) & \longrightarrow & \mathrm{Hom}(W_{i-1}, Q(R)) & \longrightarrow & \mathrm{Hom}(W_{i-1}, Q(R)/R) & \longrightarrow & 0 \\ & & \downarrow & & \downarrow & & \downarrow & & \\ 0 & \longrightarrow & \mathrm{Hom}(B_{i-1}, R) & \longrightarrow & \mathrm{Hom}(B_{i-1}, Q(R)) & \longrightarrow & \mathrm{Hom}(B_{i-1}, Q(R)/R) & \longrightarrow & 0. \end{array}$$

The rows are short exact, as W_{i-1} and B_{i-1} are projective. Thinking of the columns as representing the only nontrivial modules in complexes, the resulting long exact homology sequence (or, equivalently, the Snake Lemma) yields the six-term Hom–Ext sequence (8.9), as, for each R-module M and using the projective resolution (8.13), we have

$$\ker(\mathrm{Hom}(W_{i-1}, M) \to \mathrm{Hom}(B_{i-1}, M)) = \mathrm{Hom}(T(I^{\bar{p}}H_{i-1}(X;R)), M)$$

and

$$\mathrm{cok}(\mathrm{Hom}(W_{i-1}, M) \to \mathrm{Hom}(B_{i-1}, M)) = \mathrm{Ext}(T(I^{\bar{p}}H_{i-1}(X;R)), M).$$

We are interested in the connecting map that takes $\mathrm{Hom}(T(I^{\bar{p}}H_{i-1}(X;R)), Q(R)/R)$ to $\mathrm{Ext}(T(I^{\bar{p}}H_{i-1}(X;R)), R)$. We already know this map should be an isomorphism by Lemma 8.4.6 with our standing assumptions, and we would like to construct an explicit inverse.

Given $\bar{f} \in \operatorname{Ext}(T(I^{\bar{p}}H_{i-1}(X;R)),R)$, let $f \in \operatorname{Hom}(B_{i-1},R)$ represent $\bar{f}$. As we already know that the six-term exact sequence degenerates to yield an isomorphism from $\operatorname{Ext}(T(I^{\bar{p}}H_{i-1}(X;R)),R)$ to $\operatorname{Hom}(T(I^{\bar{p}}H_{i-1}(X;R)),Q(R)/R)$, there must be a unique $g \in \operatorname{Hom}(T(I^{\bar{p}}H_{i-1}(X;R)),Q(R)/R) \subset \operatorname{Hom}(W_{i-1},Q(R)/R)$ that maps to $\bar{f}$ via the connecting "zig-zag" map. Unwinding the zig-zag map, our $g \in \operatorname{Hom}(W_{i-1},Q(R)/R)$ must be the image of some $h \in \operatorname{Hom}(W_{i-1},Q(R))$, and h must restrict on B_{i-1} to a representative of $\bar{f}$. In particular, the restriction of h to an element of $\operatorname{Hom}(B_{i-1},R)$ must agree with f up to an element of $\operatorname{Hom}(W_{i-1},R)$. So if we can find an h whose restriction agrees with f exactly, then certainly its image $g \in \operatorname{Hom}(T(I^{\bar{p}}H_{i-1}(X;R)),Q(R)/R)$ must map to $\bar{f}$ by the Zig-Zag Lemma (see [182, Lemma 24.1]).

So let us deduce such an h from f. Suppose $w \in W_{i-1}$. Then for some $r_w \in R$ with $r_w \neq 0$, we have $r_w w \in B_{i-1}$, so $f(r_w w) \in R \subset Q(R)$ is defined, and we want $h(r_w w) = f(r_w w)$. But h is a $Q(R)$-module homomorphism, so we must have $h(r_w w) = r_w h(w)$, which determines $h(w) = f(r_w w)/r_w$. Let us show that this yields a well-defined element of $\operatorname{Hom}(W_{i-1},Q(R))$, in particular that it does not depend on our choice of r_w. Suppose that $r'_w \in R$ with $r'_w \neq 0$ and $r'_w w \in B_{i-1}$. Then we want to know that $f(r_w w)/r_w = f(r'_w w)/r'_w \in Q(R)$. But this is equivalent to asking that $r'_w f(r_w w) = r_w f(r'_w w) \in Q(R)$, which is certainly true as f is an R-module homomorphism so both sides are equal to $f(r_w r'_w w)$. This latter expression is well defined, as $r_w r'_w w \in B_{i-1}$. This also shows that h so defined is a homomorphism, as, if $w, v \in W_{i-1}$, then $r_v(r_w w), r_w(r_v v) \in B_{i-1}$, so

$$\begin{aligned} h(w) + h(v) &= \frac{f(r_w w)}{r_w} + \frac{f(r_v v)}{r_v} = \frac{r_v f(r_w w) + r_w f(r_v v)}{r_v r_w} = \frac{f(r_v r_w (w+v))}{r_v r_w} \\ &= h(w+v); \end{aligned}$$

and of course for any $r \in R$ we have $r r_w w \in B_{i-1}$, so

$$h(rw) = \frac{f(r_w r w)}{r_w} = \frac{r f(r_w w)}{r_w} = r h(w).$$

Finally, if $w \in B_{i-1}$, then we can take $r_w = 1$, and so $h(w) = f(w)$, as desired.

We infer that if $\bar{f} \in \operatorname{Ext}(T(I^{\bar{p}}H_{i-1}(X;R)),R)$ is the image of $f \in \operatorname{Hom}(B_{i-1},R)$ then the image of $\bar{f}$ in $\operatorname{Hom}(T(I^{\bar{p}}H_{i-1}(X;R)),Q(R)/R)$ under the connecting isomorphism acts by taking the class of the chain w to $f(r_w w)/r_w \in Q(R)/R$, where $r_w \in R$ is any non-zero element such that $r_w w$ is a boundary.

(b) Assembling λ

We can now put our discussion thus far together with Lemma 8.4.12 to find an explicit description of $\lambda : T(I_{\bar{p}}H^i(X;R)) \xrightarrow{\cong} \operatorname{Hom}(T(I^{\bar{p}}H_{i-1}(X;R)),Q(R)/R)$ as defined by the composition (8.11). Suppose $\alpha \in T(I_{\bar{p}}H^i(X;R))$, and let $w \in C_{i-1}$ represent an element of $T(I^{\bar{p}}H_{i-1}(X;R))$. We will write down an explicit formula for $(\lambda(\alpha))(w)$. By Lemma 8.4.12, there is an $f \in \operatorname{Hom}(B_{i-1},R)$ such that $\partial^*(f) = \alpha$. Let $z \in C_i$ be such that $\partial z = rw$

for some $r \neq 0$. By the preceding arguments, we must have[25]

$$(\lambda(\alpha))(w) = f(rw)/r = f(\partial z)/r = ((\partial^* f)(z))/r$$
$$= \alpha(z)/r.$$

So $(\lambda(\alpha))(w) = \alpha(z)/r$. While this formula should be well defined by our preceding discussion, it is reassuring to observe that this construction is independent of our choice of a cochain representative for α, as, for any coboundary $d\gamma$, we have

$$\frac{(d\gamma)(z)}{r} = \frac{\pm\gamma(\partial z)}{r} = \frac{\pm\gamma(rw)}{r} = \frac{\pm r\gamma(w)}{r} = \pm\gamma(w) = 0 \in Q(R)/R.$$

Similarly, if w is a boundary, so that $w = \partial z$, then $(\lambda(\alpha))(w) = \alpha(z) = 0 \in Q(R)/R$. So this provides a nice independent verification that our formula is indeed independent of our choices of cochain and chain representatives of our cohomology and homology classes.

Putting together our discussion thus far, we have obtained the following lemma:

Lemma 8.4.15 *Let R be a Dedekind domain and X a compact ∂-stratified pseudomanifold. Then the Universal Coefficient Theorem and the connecting isomorphism of the six-term* Hom–Ext *sequence result in an isomorphism*

$$\lambda : T(I_{\bar{p}}H^i(X;R)) \cong \operatorname{Hom}(T(I^{\bar{p}}H_{i-1}(X;R)), Q(R)/R).$$

If α is a cochain representing an element of $T(I_{\bar{p}}H^i(X;R))$ and w is a chain representing an element of $T(I^{\bar{p}}H_{i-1}(X;R))$, we have $(\lambda(\alpha))(w) = \alpha(z)/r$, where $z \in I^{\bar{p}}S_i(X;R)$ is a chain with $\partial z = rw$, $r \neq 0$.

(c) The Torsion Pairing Made Explicit

The map λ was the first ingredient in our torsion pairing

$$L : T(I_{\bar{p}}H^i(X;R)) \otimes T(I_{D\bar{p}}H^{n-i+1}(X,\partial X;R)) \to Q(R)/R.$$

The second ingredient is the $\operatorname{Hom}(\cdot, Q(R)/R)$ dual of the Poincaré duality isomorphism. So, altogether, our pairing is determined by the composition

$$T(I_{\bar{p}}H^i(X;R)) \xrightarrow{\lambda} \operatorname{Hom}(T(I^{\bar{p}}H_{i-1}(X;R)), Q(R)/R)$$
$$\xrightarrow{\mathcal{D}^*} \operatorname{Hom}(T(I_{D\bar{p}}H^{n-i+1}(X,\partial X;R)), Q(R)/R).$$

Now that we have studied λ in some detail, we are almost ready to write down a formula for $L(\alpha,\beta)$.

Let us assume that we have chosen a fixed algebraic diagonal on X so that it makes sense to discuss cup and cap products at the chain level. Ultimately, $L(\alpha,\beta)$ is independent of this choice because it is defined at the level of cohomology; we're using cup and cap products simply to find chain-level formulas for the torsion pairing.

25 Recall Remark 8.4.11.

Let $\alpha \in I_{\bar{p}}S^i(X;R)$ and $\beta \in I_{D\bar{p}}S^{n-i+1}(X,\partial X;R)$ represent elements of $T(I_{\bar{p}}H^i(X;R))$ and $T(I_{D\bar{p}}H^{n-i+1}(X,\partial X;R))$, respectively. Then there are $r,t \in R$, $r,t \neq 0$, and some $\mathfrak{a} \in I_{\bar{p}}S^{i-1}(X;R)$ and $\mathfrak{b} \in I_{D\bar{p}}S^{n-i}(X,\partial X;R)$ such that $d\mathfrak{a} = r\alpha$ and $d\mathfrak{b} = t\beta$. Then $\mathfrak{D}(\beta) = (-1)^{(n-i+1)n}\beta \frown \Gamma$ is a torsion element of $I^{p}H_{i-1}(X;R)$. In fact, as $t\beta = 0 \in I_{D\bar{p}}H^{n-i+1}(X,\partial X;R)$, we must have $t\mathfrak{D}(\beta) = 0 \in I^{p}H_{i-1}(X;R)$. So, the chain $t\mathfrak{D}(\beta)$ is a boundary. Let us see what it is the boundary of: By Lemma 7.2.19,

$$\partial(\gamma \frown \xi) = (d\gamma) \frown \xi + (-1)^{|\gamma|}\gamma \frown \partial\xi, \tag{8.14}$$

for a cochain γ and chain ξ (of appropriate perversities). In our setting, ξ will be Γ, which is a cycle in $I^{\bar{0}}S_n(X,\partial X;R)$. So, we have

$$t(\beta \frown \Gamma) = (t\beta) \frown \Gamma = (d\mathfrak{b}) \frown \Gamma = \partial(\mathfrak{b} \frown \Gamma).$$

Using this and Lemma 8.4.15, we now have[26]

$$\begin{aligned}(\mathfrak{D}^*\lambda(\alpha))(\beta) &= (-1)^{(|\alpha|-1)n}\lambda(\alpha)(\mathfrak{D}(\beta))\\ &= (-1)^{(|\alpha|-1)n+|\beta|n}\lambda(\alpha)(\beta \frown \Gamma)\\ &= (-1)^{n(|\alpha|+|\beta|-1)}\frac{\alpha(\mathfrak{b} \frown \Gamma)}{t}\\ &= (-1)^{n^2}\frac{\alpha(\mathfrak{b} \frown \Gamma)}{t}\\ &= (-1)^{n}\frac{\alpha(\mathfrak{b} \frown \Gamma)}{t}.\end{aligned}$$

So, the adjoint pairing to $\mathfrak{D}^*\lambda$ is $L : T(I_{\bar{p}}H^i(X;R)) \otimes T(I_{D\bar{p}}H^{n-i+1}(X,\partial X;R)) \to Q(R)/R$, defined by

$$L(\alpha,\beta) = (-1)^{n}\frac{\alpha(\mathfrak{b} \frown \Gamma)}{t}$$

when $d\mathfrak{b} = t\beta$, $t \neq 0$.

To study the other adjoint to this pairing, and so to verify that L is nonsingular, we would like to move toward a more symmetric expression for L, which means rewriting our formula for L in terms of the cup product. This is the purpose of our next proposition.

Proposition 8.4.16 *Let R be a Dedekind domain and X a compact n-dimensional R-oriented locally $(\bar{p};R)$-torsion-free ∂-stratified pseudomanifold. Let $\alpha \in I_{\bar{p}}S^i(X;R)$ and $\beta \in I_{D\bar{p}}S^{n-i+1}(X,\partial X;R)$ represent elements of $T(I_{\bar{p}}H^i(X;R))$ and $T(I_{D\bar{p}}H^{n-i+1}(X,\partial X;R))$, respectively, and suppose $d\mathfrak{b} = t\beta$ for some $\mathfrak{b} \in I_{D\bar{p}}S^{n-i}(X,\partial X;R)$ and $t \in R$, $t \neq 0$. Then*

$$L(\alpha,\beta) = (-1)^{n}\frac{\alpha(\mathfrak{b} \frown \Gamma)}{t} = (-1)^{n}\frac{\mathbf{a}((\alpha \smile \mathfrak{b}) \frown \Gamma)}{t}.$$

Proof We will employ now the somewhat mysterious formula from Remark 7.3.26, which

26 Note that $\lambda(\alpha)$ is an object of degree $|\alpha| - 1$, and we will employ the Koszul convention for our chain map $\mathfrak{D}$.

states that for α a cochain on X and ξ a chain on X, each with the same degree and perversity, we have

$$\alpha(\xi) = \mathbf{a}(\alpha \frown \xi) + \alpha(D\partial\xi + \partial D\xi); \tag{8.15}$$

here D is the chain homotopy given by Lemma 7.3.20, in this case from $I^{\bar{p}}S_*(X;R)$ to itself. The precise details of this chain homotopy will not be needed here, but see Remark 7.3.26 for more background if desired, noting that the discussion carries over to the context of ∂-stratified pseudomanifolds by our work in Section 7.3.10.

Taking $\xi = \mathfrak{b} \frown \Gamma$ in (8.15), we have

$$\alpha(\mathfrak{b} \frown \Gamma) = \mathbf{a}(\alpha \frown (\mathfrak{b} \frown \Gamma)) + \alpha(D\partial(\mathfrak{b} \frown \Gamma)) + \alpha(\partial D(\mathfrak{b} \frown \Gamma)).$$

Now $\alpha(\partial D(\mathfrak{b} \frown \Gamma)) = 0$ because our α is a cocycle. And by (8.14), using that Γ is a cycle in $I^{\bar{0}}S_*(X, \partial X; R)$,

$$\alpha(D\partial(\mathfrak{b} \frown \Gamma)) = \alpha(D((d\mathfrak{b}) \frown \Gamma)) = \alpha(D(t\beta \frown \Gamma)) = t\alpha(D(\beta \frown \Gamma)).$$

So

$$\begin{aligned}\frac{\alpha(\mathfrak{b} \frown \Gamma)}{t} &= \frac{\mathbf{a}(\alpha \frown (\mathfrak{b} \frown \Gamma))}{t} + \frac{t\alpha(D(\beta \frown \Gamma))}{t}\\ &= \frac{\mathbf{a}(\alpha \frown (\mathfrak{b} \frown \Gamma))}{t} + \alpha(D(\beta \frown \Gamma))\\ &= \frac{\mathbf{a}(\alpha \frown (\mathfrak{b} \frown \Gamma))}{t} \in Q(R)/R.\end{aligned}$$

The point at the end here is that, although we have no idea what $\alpha(D(\beta \frown \Gamma))$ might be, we do know it's in R, so it's zero in $Q(R)/R$. Therefore,

$$L(\alpha, \beta) = (-1)^n \frac{\mathbf{a}(\alpha \frown (\mathfrak{b} \frown \Gamma))}{t}.$$

Finally, we would like to be able to write

$$L(\alpha, \beta) = (-1)^n \frac{\mathbf{a}((\alpha \smile \mathfrak{b}) \frown \Gamma))}{t}.$$

For this last step, we must have a closer look at the proof of associativity of the cap product in Lemma 7.3.32. There, we saw at the chain level that $(\alpha \smile \beta) \frown x = \Phi(\mathrm{id} \otimes \Theta(\alpha \otimes \beta))(\mathrm{id} \otimes \bar{\mathbf{d}})\bar{\mathbf{d}}(x)$, while $\alpha \frown (\beta \frown x) = \Phi(\mathrm{id} \otimes \Theta(\alpha \otimes \beta))(\bar{\mathbf{d}} \otimes \mathrm{id})\bar{\mathbf{d}}(x)$. The identification of these in homology follows from $(\bar{\mathbf{d}} \otimes \mathrm{id})\bar{\mathbf{d}}$ and $(\mathrm{id} \otimes \bar{\mathbf{d}})\bar{\mathbf{d}}$ being chain homotopic, by Lemma 7.3.30. If D denotes the chain homotopy, we therefore have

$$(\alpha \smile \beta) \frown \Gamma - \alpha \frown (\beta \frown \Gamma) = \Phi(\mathrm{id} \otimes \Theta(\alpha \otimes \beta))((D\partial + \partial D)\Gamma).$$

Because Γ is a cycle in $I^{\bar{0}}S_*(X, \partial X; R)$, this difference reduces to

$$\Phi(\mathrm{id} \otimes \Theta(\alpha \otimes \beta))(\partial D\Gamma).$$

But $\partial D\Gamma$ is a cycle, while $\Theta(\alpha \otimes \beta)$ is a cocycle, as α and β are cocycles and Θ is a chain map by Lemma 7.2.1. Following the computation of Lemma 7.2.20, where we show that the cap product is well defined, $\Phi(\mathrm{id} \otimes \Theta(\alpha \otimes \beta))(\partial D\Gamma)$ must therefore be a boundary, so $\mathbf{a}(\Phi(\mathrm{id} \otimes \Theta(\alpha \otimes \beta))(\partial D\Gamma)) = 0$. Thus $\mathbf{a}((\alpha \smile \beta) \frown \Gamma) = \mathbf{a}(\alpha \frown (\beta \frown \Gamma))$. □

(d) Symmetry and Nonsingularity

Now that we have an expression for the pairing L in terms of the cup product, we can provide a symmetry formula, which will also allow us to show that L is nonsingular. More specifically, we have defined $L = L_{\bar{p},D\bar{p}}$ as the adjoint of $\mathcal{D}^*\lambda$, which is thus $\Lambda(L)$. As λ and $\mathcal{D}$ (and hence $\mathcal{D}^*$) are isomorphisms, the adjoint $\Lambda(L)$ is thus an isomorphism directly. We need to show that the other adjoint $\Lambda'(L) : T(I_{D\bar{p}}H^{n-i+1}(X, \partial X; R)) \to \mathrm{Hom}(T(I_{\bar{p}}H^i(X;R)), Q(R)/R)$ is also an isomorphism. We know that

$$(\Lambda'(L)(\beta))(\alpha) = L(\alpha, \beta) = (-1)^n \frac{\mathbf{a}((\alpha \smile \mathfrak{b}) \frown \Gamma)}{t},$$

and we will show that

$$\frac{\mathbf{a}((\alpha \smile \mathfrak{b}) \frown \Gamma)}{t} = \pm \frac{\mathbf{a}((\beta \smile \mathfrak{a}) \frown \Gamma)}{r}, \tag{8.16}$$

continuing to assume $d\mathfrak{a} = r\alpha$ and $d\mathfrak{b} = t\beta$ with $r, t \neq 0$ and with the sign depending only on the degrees of α and β. Up to sign, the expression on the right has again the exact form we have discovered for the pairing $L(\alpha, \beta)$ but with the roles of α and β reversed, and so also with the roles of the pairs $(X, \emptyset)$ and $(X, \partial X)$ reversed. But each of our preceding computations involving $I_{\bar{p}}H^*(X;R)$ in our discussion of λ hold just as well applied to $I_{D\bar{p}}H^*(X, \partial X; R)$. Therefore, this expression has the form of the adjoint evaluation $(\Lambda(L'_{D\bar{p},\bar{p}})(\beta))(\alpha)$, where we have defined L' to be the pairing that is adjoint to the composition of isomorphisms

$$\begin{aligned} T(I_{D\bar{p}}H^{n-i+1}(X, \partial X; R)) &\xrightarrow{\lambda} \mathrm{Hom}(T(I^{D\bar{p}}H_{n-i}(X, \partial X; R)), Q(R)/R) \\ &\xrightarrow{\mathcal{D}^*} \mathrm{Hom}(T(I_{\bar{p}}H^i(X;R)), Q(R)/R). \end{aligned}$$

In other words, the equation (8.16) will show that $\Lambda'(L) = \pm\Lambda(L')$, which is an isomorphism by construction and hence establishes the nonsingularity of L.

Lemma 8.4.17 *Suppose R is a Dedekind domain and that X is a compact n-dimensional R-oriented locally $(\bar{p}; R)$-torsion-free ∂-stratified pseudomanifold. Let*

$$L_{\bar{p},D\bar{p}} : T(I_{\bar{p}}H^i(X;R)) \otimes T(I_{D\bar{p}}H^{n-i+1}(X, \partial X; R)) \to Q(R)/R$$

and

$$L'_{D\bar{p},\bar{p}} : T(I_{D\bar{p}}H^{n-i+1}(X, \partial X; R)) \otimes T(I_{\bar{p}}H^i(X;R)) \to Q(R)/R$$

be the torsion pairings defined above. If

$$\alpha \in T(I_{\bar{p}}H^i(X;R)) \quad \text{and} \quad \beta \in T(I_{D\bar{p}}H^{n-i+1}(X, \partial X; R)),$$

then

$$L_{\bar{p},D\bar{p}}(\alpha \otimes \beta) = (-1)^{1+n+in} L'_{D\bar{p},\bar{p}}(\beta \otimes \alpha) \in Q(R)/R.$$

Remark 8.4.18 The sign $(-1)^{1+n+in}$ in the lemma looks somewhat asymmetric, but recall that we have $|\alpha| + |\beta| = n+1$. A quick computation then shows that $(-1)^{1+n+in} = (-1)^{1+(|\alpha|+1)(|\beta+1)}$, or, equivalently, $(-1)^{1+n+in} = (-1)^{1+|\mathfrak{a}||\mathfrak{b}|}$, if $d\mathfrak{a} = r\alpha$ and $d\mathfrak{b} = t\beta$ with $r,t \neq 0$. A quick comparison with, for example, [67, Exercise 56] demonstrates that this sign yields the correct symmetries for "middle-dimensional" self-pairings when $X = M$ is a manifold (without boundary) and $n = 4m+1$ or $4m+3$. In particular, in this case we get an antisymmetric pairing on $H^{2m+1}(M;R)$ when $n = 4m+1$ and a symmetric pairing on $H^{2m+2}(M;R)$ when $n = 4m+3$.

Proof of Lemma 8.4.17 We continue to assume our earlier notation, in particular that $d\mathfrak{a} = r\alpha$ and $d\mathfrak{b} = t\beta$ for some $r,t \in R$, $r,t \neq 0$, and that we have chosen a fixed algebraic diagonal with which to define cup and cap products.

We begin by observing that

$$d(\mathfrak{a} \smile \mathfrak{b}) = (d\mathfrak{a}) \smile \mathfrak{b} + (-1)^{|\mathfrak{a}|}\mathfrak{a} \smile d\mathfrak{b} = r\alpha \smile \mathfrak{b} + (-1)^{|\mathfrak{a}|}\mathfrak{a} \smile t\beta,$$

using that the cup product is defined via a composition of chain maps. So we have

$$\begin{aligned} L_{\bar{p},D\bar{p}}(\alpha \otimes \beta) &= (-1)^n \frac{\mathbf{a}((\alpha \smile \mathfrak{b}) \frown \Gamma)}{t} = (-1)^n \frac{r\mathbf{a}((\alpha \smile \mathfrak{b}) \frown \Gamma)}{rt} \\ &= (-1)^n \frac{\mathbf{a}((r\alpha \smile \mathfrak{b}) \frown \Gamma)}{rt} \\ &= (-1)^n \frac{\mathbf{a}((d(\mathfrak{a} \smile \mathfrak{b}) - (-1)^{|\mathfrak{a}|}\mathfrak{a} \smile t\beta) \frown \Gamma)}{rt} \\ &= (-1)^n \frac{\mathbf{a}((d(\mathfrak{a} \smile \mathfrak{b})) \frown \Gamma)}{rt} - (-1)^{n+|\mathfrak{a}|} \frac{\mathbf{a}((\mathfrak{a} \smile t\beta) \frown \Gamma)}{rt} \\ &= -(-1)^{n+|\mathfrak{a}|} \frac{\mathbf{a}((\mathfrak{a} \smile \beta) \frown \Gamma)}{r}. \end{aligned}$$

In the last line, we have used that

$$d(\mathfrak{a} \smile \mathfrak{b}) \frown \Gamma = \pm\partial((\mathfrak{a} \smile \mathfrak{b}) \frown \Gamma),$$

by Lemma 7.2.19, as Γ is a cycle $I^{\bar{0}}S_*(X, \partial X; R)$, and the augmentation of a boundary is zero.

Next, a close look at the proof of Proposition 7.3.15 shows that at the chain level we have

$$\mathfrak{a} \smile \beta - (-1)^{|\beta||\mathfrak{a}|}\beta \smile \mathfrak{a} = (Dd + dD)\Theta(\mathfrak{a} \otimes \beta),$$

where D is the chain homotopy between $\bar{\mathbf{d}}^*$ and $(\tau\bar{\mathbf{d}})^*$ guaranteed by Lemma 7.3.14 and using that the dual of a chain homotopy is a chain homotopy.

So

$$\begin{aligned}
&\frac{\mathbf{a}((\mathfrak{a} \smile \beta) \frown \Gamma)}{r} - (-1)^{|\beta||\mathfrak{a}|}\frac{\mathbf{a}((\beta \smile \mathfrak{a}) \frown \Gamma)}{r} \\
&= \frac{\mathbf{a}(((Dd + dD)\Theta(\mathfrak{a} \otimes \beta)) \frown \Gamma)}{r} \\
&= \frac{\mathbf{a}((Dd\Theta(\mathfrak{a} \otimes \beta)) \frown \Gamma)}{r} + \frac{\mathbf{a}((dD\Theta(\mathfrak{a} \otimes \beta)) \frown \Gamma)}{r}.
\end{aligned}$$

The second expression on the last line is zero, because $(dD\Theta(\mathfrak{a} \otimes \beta)) \frown \Gamma = \pm\partial((D\Theta(\mathfrak{a} \otimes \beta)) \frown \Gamma)$, using again Lemma 7.2.19, that Γ is a cycle, and that the augmentation of a cycle is zero. For the first expression, we use again that Θ is a chain map by Lemma 7.2.1. So, as β is a cocycle,

$$\begin{aligned}
\frac{\mathbf{a}((Dd\Theta(\mathfrak{a} \otimes \beta)) \frown \Gamma)}{r} &= \frac{\mathbf{a}((D\Theta((d\mathfrak{a}) \otimes \beta)) \frown \Gamma)}{r} \\
&= \frac{\mathbf{a}((D\Theta(r\alpha \otimes \beta)) \frown \Gamma)}{r} \\
&= \mathbf{a}((D\Theta(\alpha \otimes \beta)) \frown \Gamma),
\end{aligned}$$

which must be in R. So this term also vanishes in $Q(R)/R$.

Altogether, we have now shown that

$$\begin{aligned}
L_{\bar{p},D\bar{p}}(\alpha \otimes \beta) &= (-1)^n \frac{\mathbf{a}((\alpha \smile \mathfrak{b}) \frown \Gamma)}{t} \\
&= -(-1)^{n+|\mathfrak{a}|}\frac{\mathbf{a}((\mathfrak{a} \smile \beta) \frown \Gamma)}{r}. \\
&= (-1)^{1+n+|\mathfrak{a}|+|\beta||\mathfrak{a}|}\frac{\mathbf{a}((\beta \smile \mathfrak{a}) \frown \Gamma)}{r} \\
&= (-1)^{1+|\mathfrak{a}|+|\beta||\mathfrak{a}|} L'_{D\bar{p},\bar{p}}(\beta, \alpha).
\end{aligned}$$

To simplify the signs, let us recall that $|\alpha| = i$ and $|\beta| = n - i + 1$, so $|\mathfrak{a}| = i - 1$. Thus, mod 2, we have

$$\begin{aligned}
1 + |\mathfrak{a}| + |\beta||\mathfrak{a}| &\equiv 1 + i - 1 + (n - i + 1)(i - 1) \\
&\equiv 1 + i + 1 + ni + i + i + n + i + 1 \\
&\equiv 1 + ni + n.
\end{aligned}$$

This completes the proof of Lemma 8.4.17. □

As noted just before the preceding lemma, the symmetry demonstrated by the lemma implies the second condition of nonsingularity for the linking pairing $L_{\bar{p},D\bar{p}}$. The nonsingularity of $L'_{D\bar{p},\bar{p}}$ follows analogously. So, summarizing all the work in this section, we arrive at the statement of Theorem 8.4.10, presented at the beginning of this section, the proof of which is now complete.

(e) Another Approach to the Torsion Pairing

As a note for the interested reader, let us mention an alternative definition of the torsion pairing that is often employed for manifolds; see, for example, [67, Exercise 56]. For simplicity, we will assume that M is a compact oriented manifold and take $R = \mathbb{Z}$. The short exact sequence

$$0 \to \mathbb{Z} \to \mathbb{Q} \to \mathbb{Q}/\mathbb{Z} \to 0$$

yields a long exact sequence

$$\longrightarrow H^j(M;\mathbb{Q}) \longrightarrow H^j(M;\mathbb{Q}/\mathbb{Z}) \xrightarrow{\delta} H^{j+1}(M;\mathbb{Z}) \longrightarrow H^{j+1}(M;\mathbb{Q}) \longrightarrow .$$

As the image of $T(H^{j+1}(M;\mathbb{Z}))$ must vanish in $H^{j+1}(M;\mathbb{Q})$, the connecting morphism δ is surjective onto $T(H^{j+1}(M;\mathbb{Z}))$. A pairing

$$T(H^i(M;\mathbb{Z})) \otimes T(H^{n-i+1}(M;\mathbb{Z})) \to \mathbb{Q}/\mathbb{Z}$$

can then be defined by

$$\alpha \otimes \beta \to (\alpha \smile \delta^{-1}(\beta))(\Gamma),$$

where $\delta^{-1}(\beta)$ is any element of $H^{n-i}(M;\mathbb{Q}/\mathbb{Z})$ in the preimage of β. More precisely, the cup product being used here has the form

$$H^i(M;\mathbb{Z}) \otimes H^{n-i}(M;\mathbb{Q}/\mathbb{Z}) \to H^n(M;\mathbb{Q}/\mathbb{Z}),$$

and the Universal Coefficient Theorem provides a map

$$H^n(M;\mathbb{Q}/\mathbb{Z}) \to \operatorname{Hom}(H_n(M;\mathbb{Z}),\mathbb{Q}/\mathbb{Z}),$$

accounting for the evaluation. Among its other features, our development of the torsion pairing allowed us to avoid such cup products with varying coefficients. We will not attempt to develop this approach to the torsion pairing in detail for intersection cohomology, but let us at least demonstrate in the manifold case that the two definitions are compatible (up to signs). Thus the torsion pairing we have developed for intersection cohomology restricts to the perhaps more familiar definition for manifolds.

Once again we abuse notation and let β also denote a cocycle representing the cohomology class β. As we assume this is a torsion class, we can find again $\mathfrak{b} \in S^{n-i}(M;\mathbb{Z})$ such that $d\mathfrak{b} = t\beta$ for some $t \in \mathbb{Z}$, $t \neq 0$.

Let us define $f \in \operatorname{Hom}(S_{n-i}(M),\mathbb{Q}/\mathbb{Z}) = S^{n-i}(M;\mathbb{Q}/\mathbb{Z})$ by $f(x) = \frac{\mathfrak{b}(x)}{t} \in \mathbb{Q}/\mathbb{Z}$. Then for $x \in S_{n-i+1}(M;\mathbb{Z})$,

$$\begin{aligned}(df)(x) = (-1)^{n-i+1}f(\partial x) &= (-1)^{n-i+1}\frac{\mathfrak{b}(\partial x)}{t} = \frac{d\mathfrak{b}(x)}{t} = \frac{t\beta(x)}{t} = \beta(x)\\ &= 0 \in \mathbb{Q}/\mathbb{Z}.\end{aligned}$$

So f is a cocycle in $S^*(M;\mathbb{Q}/\mathbb{Z})$. Furthermore, by interpreting $\frac{\mathfrak{b}(x)}{t}$ in $\mathbb{Q}$, we see that f is really just the mod $\mathbb{Z}$ reduction of a function to $\mathbb{Q}$, and so by the zig-zag construction, $\delta(f)$

is represented by df, which we have just seen agrees with β as a cochain with values in $\mathbb{Z}$. In other words, the cohomology class of the cocycle $f \in S^{n-i}(M;\mathbb{Q}/\mathbb{Z})$ is a preimage of β under δ. So writing $f = \frac{\mathfrak{b}}{t}$, the torsion pairing described above takes the form

$$\alpha \otimes \beta \to (\alpha \smile \delta^{-1}(\beta))(\Gamma) = \left(\alpha \smile \frac{\mathfrak{b}}{t}\right)(\Gamma).$$

Using the basic properties of ordinary homology/cohomology products [220, Section V.6], this becomes

$$\mathbf{a}\left(\left(\alpha \smile \frac{\mathfrak{b}}{t}\right) \frown \Gamma\right).$$

Applying the standard Alexander–Whitney formulation of the cup product in terms of front and back faces, it is easy to compute that $\alpha \smile \frac{\mathfrak{b}}{t} = \frac{\alpha \smile \mathfrak{b}}{t}$, thinking of the left side as a cup product of chains in $S^*(M;\mathbb{Z})$ and $S^*(M;\mathbb{Q}/\mathbb{Z})$ and the right side as $\frac{1}{t}$ times the image of a cup product $S^*(M;\mathbb{Z}) \otimes S^*(M;\mathbb{Z}) \to S^*(M;\mathbb{Z})$. Similarly,

$$\frac{\alpha \smile \mathfrak{b}}{t} \frown \Gamma = \frac{1}{t}(\alpha \smile \mathfrak{b}) \frown \Gamma.$$

So, altogether,

$$(\alpha \smile \delta^{-1}(\beta))(\Gamma) = \frac{\mathbf{a}((\alpha \smile \mathfrak{b}) \frown \Gamma)}{t},$$

which agrees up to signs with our computation of $L(\alpha,\beta)$ in Theorem 8.4.10, as claimed.

8.4.4 Topological Invariance of Pairings

Let us show that the cup and torsion product pairings are topological invariants for appropriate perversities. Ultimately, we will use this to show that the signature is a topological invariant of Witt spaces in Theorem 9.3.16.

Theorem 8.4.19 *Suppose R is a Dedekind domain and $\bar{p}$ is a GM perversity. Let X_1 and X_2 be two n-dimensional compact ∂-stratified pseudomanifold stratifications with no codimension-one strata of the same underlying space pairs $(|X_1|,|\partial X_1|) = (|X_2|,|\partial X_2|)$. Suppose X_1 and X_2 are compatibly R-oriented in the sense of Corollary 8.1.11 (applied to $|X_i| - |\partial X_i|$) and that X_1 and X_2 are locally $(\bar{p};R)$-torsion-free*[27]*. Then the cup product and torsion pairings are independent of the stratification in the sense that there are canonical commutative diagrams*

$$\begin{array}{ccc} F(I_{\bar{p}}H^i(X_1;R)) \otimes F(I_{D\bar{p}}H^{n-i}(X_1,\partial X_1;R)) & & \\ \downarrow \cong & \searrow^{\mathbf{a}\mathfrak{D}\circ\smile} & \\ & & R \\ & \nearrow_{\mathbf{a}\mathfrak{D}\circ\smile} & \\ F(I_{\bar{p}}H^i(X_2;R)) \otimes F(I_{D\bar{p}}H^{n-i}(X_2,\partial X_2;R)) & & \end{array}$$

[27] By the argument of Proposition 5.5.9, both spaces are locally $(\bar{p};R)$-torsion-free if either is.

and

$$\begin{array}{ccc} T(I_{\bar{p}}H^i(X_1;R)) \otimes T(I_{D\bar{p}}H^{n-i+1}(X_1,\partial X_1;R)) & \xrightarrow{L_{\bar{p},D\bar{p}}} & \\ \downarrow \cong & & Q(R)/R \\ T(I_{\bar{p}}H^i(X_2;R)) \otimes T(I_{D\bar{p}}H^{n-i+1}(X_2,\partial X_2;R)) & \xrightarrow{L_{\bar{p},D\bar{p}}} & \end{array} \tag{8.17}$$

and similarly for the other torsion pairing L'.

Remark 8.4.20 Yet again, as in Remarks 8.2.7 and 8.3.13, this statement can be extended immediately to the observation that the pairings are topological invariants in the sense that, if there is an orientation-preserving topological homeomorphism $f : (|X|, |\partial X|) \to (|Y|, |\partial Y|)$ between the compact n-dimensional R-oriented ∂-stratified pseudomanifolds X and Y without codimension-one strata, then f, together with the isomorphisms of the theorem, induces isomorphisms between the pairings for X and the pairings for Y.

Proof of Theorem 8.4.19 For the cup product, we have already established the basic tools of the proof in our prior proofs of the topological invariance of products (Theorem 7.3.10), fundamental classes (Propositions 8.1.29 and 8.3.7), and duality (Theorems 8.2.6 and 8.3.12). We did not establish invariance of products for ∂-pseudomanifolds, but as in the proofs of Proposition 8.3.7 and Theorem 8.3.12, we can apply the techniques of Section 7.3.10 to ensure that the relevant products exist and satisfy the needed naturality properties; see Theorem 7.3.72. Such arguments are sufficient to demonstrate invariance of the cup product pairing.

For the torsion pairing, the proof is a little trickier, as our direct formula for computing the pairing involves a cup product of cochains, not just cohomology classes. At the chain level, maps such as $I_{\bar{p}}S^*(\mathfrak{X};R) \to I_{\bar{p}}S^*(X_i;R)$ are not, in general, isomorphisms, so we cannot invert them as we have when working at the cohomology level. Rather, let us revert to our original definition of the torsion product by its adjoint. Then we can consider the diagram in Figure 8.5.

As in the proofs of Proposition 8.3.7 and Theorem 8.3.12, we here let $\mathfrak{X}$ be the intrinsic stratification of $|X_1 - \partial X_1| = |X_2 - \partial X_2|$. The top row here is then a compressed version of the absolute cohomology analogue of the chain of cohomology isomorphisms along the left side of Figure 8.4 in the proof of Theorem 8.3.12, restricted to the torsion submodules. In other words, these are the maps that would appear on the left side of the analogous diagram one would use for proving the commutativity of the second diagram in the statement of Theorem 8.3.12, for which we did not provide explicit details. So, the composition left to right across the top of our diagram here is just the restriction to torsion submodules of the inverse of the isomorphism labeled ϕ^* in the second diagram in the statement of Theorem 8.3.12. Let us call this left-to-right composition here $\psi_{\bar{p}}$.

The top three pairs of squares commute by the naturality of the Universal Coefficient Theorem, by functoriality of Hom and Ext, and by the naturality of the six-term exact

$$
\begin{array}{ccccc}
T(I_{\bar{p}}H^{i}(X_1;R)) & \xleftarrow{\cong} & T(I_{\bar{p}}H^{i}(\mathfrak{X};R)) & \xrightarrow{\cong} & T(I_{\bar{p}}H^{i}(X_2;R)) \\
\uparrow{\scriptstyle\cong} & & \uparrow{\scriptstyle\cong} & & \uparrow{\scriptstyle\cong} \\
\operatorname{Ext}(I^{\bar{p}}H_{i-1}(X_1;R),R) & \longleftarrow & \operatorname{Ext}(I^{\bar{p}}H_{i-1}(\mathfrak{X};R),R) & \longrightarrow & \operatorname{Ext}(I^{\bar{p}}H_{i-1}(X_2;R),R) \\
\downarrow{\scriptstyle\cong} & & \downarrow{\scriptstyle\cong} & & \downarrow{\scriptstyle\cong} \\
\operatorname{Ext}(T(I^{\bar{p}}H_{i-1}(X_1;R)),R) & \longleftarrow & \operatorname{Ext}(T(I^{\bar{p}}H_{i-1}(\mathfrak{X};R)),R) & \longrightarrow & \operatorname{Ext}(T(I^{\bar{p}}H_{i-1}(X_2;R)),R) \\
\uparrow{\scriptstyle\cong} & & \uparrow{\scriptstyle\cong} & & \uparrow{\scriptstyle\cong} \\
\operatorname{Hom}(T(I^{\bar{p}}H_{i-1}(X_1;R)),Q(R)/R) & \longleftarrow & \operatorname{Hom}(T(I^{\bar{p}}H_{i-1}(\mathfrak{X};R)),Q(R)/R) & \longrightarrow & \operatorname{Hom}(T(I^{\bar{p}}H_{i-1}(X_2;R)),Q(R)/R) \\
{\scriptstyle\cong}\downarrow{\scriptstyle\mathcal{D}^*} & & {\scriptstyle\cong}\downarrow{\scriptstyle\mathcal{D}^*} & & {\scriptstyle\cong}\downarrow{\scriptstyle\mathcal{D}^*} \\
\begin{array}{c}\operatorname{Hom}(T(I_{D\bar{p}}H^{n-i+1}(X_1,\\ \partial X_1;R)),Q(R)/R)\end{array} & \longrightarrow & \operatorname{Hom}(T(I_{D\bar{p}}H^{n-i+1}(\mathfrak{X},\mathfrak{N};R)),Q(R)/R) & \longleftarrow & \operatorname{Hom}(T(I_{D\bar{p}}H^{n-i+1}(X_2,\partial X_2;R)),Q(R)/R).
\end{array}
$$

Figure 8.5

sequence [126, Section IV.8], applied to all the isomorphisms involved in the compositions in the top row. The bottom two squares of the diagram can be obtained by taking the diagram of Figure 8.4 in the proof of Theorem 8.3.12, interchanging $\bar{p}$ and $D\bar{p}$, restricting to torsion submodules, taking the $\operatorname{Hom}(\cdot, Q(R)/R)$ dual, and then compressing the diagram. In particular, the composition right to left along the bottom is the $\operatorname{Hom}(\cdot, Q(R)/R)$ dual of the relative version of our map ψ with perversity $D\bar{p}$, so let us call this composition $\psi^*_{D\bar{p}}$. The composition of all vertical maps on the left and right, except for the bottom map in each column, is the composition of isomorphisms we called λ in Section 8.4.3. The vertical maps in the middle are obtained in the same way and are isomorphisms for the same reasons, noting that $I_{\bar{p}}H^i(\mathfrak{X}; R)$ is finitely generated because it is isomorphic to $I_{\bar{p}}H^i(X_1; R)$. The left and right $\mathcal{D}^*$ are isomorphisms by Lefschetz duality, while the middle $\mathcal{D}^*$ is thus an isomorphism from Figure 8.4 again. It follows that all the horizontal maps here are also isomorphisms.

Now, let $L^j_{\bar{p},D\bar{p}}$ be the torsion pairing on X_j. Let λ_j and $\mathcal{D}^*_j$ be the appropriate maps on X_1 and X_2. Then, using the commutativity of the diagram and the definition of the $L^j_{\bar{p},D\bar{p}}$, we compute

$$\begin{aligned}
L^1_{\bar{p},D\bar{p}}(\alpha \otimes \beta) &= (\mathcal{D}^*_1\lambda_1(\alpha))(\beta) \\
&= (\psi^*_{D\bar{p}}\mathcal{D}^*_2\lambda_2\psi_{\bar{p}}(\alpha))(\beta) = (\mathcal{D}^*_2\lambda_2\psi_{\bar{p}}(\alpha))(\psi_{D\bar{p}}(\beta)) \\
&= L^2_{\bar{p},D\bar{p}}((\psi_{\bar{p}}(\alpha)) \otimes (\psi_{D\bar{p}}(\beta))).
\end{aligned}$$

Therefore, letting the map on the left side of diagram (8.17) be $\psi_{\bar{p}} \otimes \psi_{D\bar{p}}$, the diagram commutes as claimed. This choice is canonical as each of $\psi_{\bar{p}}$ and $\psi_{D\bar{p}}$ is derived from one of the canonical maps involved in Theorem 8.3.12.

The argument for the L' pairings is analogous. The proof of Theorem 8.4.19 is now complete. □

8.4.5 Image Pairings

When M^{2k} is a compact R-oriented even-dimensional manifold, the cup product pairing of Theorem 8.4.7 gives us a nonsingular pairing

$$F(H^k(M; R)) \otimes FH^k((M; R)) \to R.$$

Notice that the two input modules to the pairing are identical. In this setting, it is possible to tease out further invariants that have proven important in manifold theory. For example, taking $R = \mathbb{Q}$ and k even, the symmetric pairing $H^k(M; \mathbb{Q}) \otimes H^k(M; \mathbb{Q}) \to \mathbb{Q}$ yields the *signature* invariant by subtracting the number of negative eigenvalues of a matrix representing the pairing from the number of positive eigenvalues of the matrix. Details of the signature will be developed in Chapter 9.

When M is a compact R-oriented ∂-manifold with non-empty boundary, then the cup product pairing only has the form

$$F(H^k(M;R)) \otimes F(H^k(M,\partial M;R)) \to R.$$

Even though Poincaré duality can be used to show that $FH^k(M;R)$ and $FH^k(M,\partial M;R)$ will be abstractly isomorphic, they are not identical, and so we do not obtain the same sort of self-pairing that occurs when $\partial M = \emptyset$. Nonetheless, there is a way to recover a self-pairing in this setting that allows one to define a signature for ∂-manifolds of dimension 0 mod 4. It turns out that the cup product pairing between $FH^k(M;R)$ and $FH^k(M,\partial M;R)$ induces a nondegenerate pairing on $\text{im}(i^* : FH^k(M,\partial M;R) \to FH^k(M;R))$. If $\alpha,\beta \in \text{im}(i^*)$ and $\bar{\alpha},\bar{\beta}$ are their preimages in $FH^k(M,\partial M;R)$, then our new pairing takes $\alpha \otimes \beta$ to

$$\mathbf{a}(\mathcal{D}(\bar{\alpha} \smile \bar{\beta})) \in R.$$

We will see below that this is well defined, though it's worth noting now that, thanks to naturality with respect to the maps $(M;\partial M,\emptyset) \to (M;\partial M,\partial M)$ and $(M;\emptyset,\partial M) \to (M;\partial M,\partial M)$, we have $\bar{\alpha} \smile \bar{\beta} = \bar{\alpha} \smile \beta = \alpha \smile \bar{\beta} \in H^n(M,\partial M;R)$. We will call this pairing the *image pairing*.

In the world of intersection homology, there are two issues with constructing a self-pairing from the cup product pairing

$$F(I_{\bar{p}}H^i(X;R)) \otimes F(I_{D\bar{p}}H^{n-i}(X,\partial X;R)) \to R.$$

We still have the possibility of nontrivial boundaries to contend with, but there is also a lack of symmetry due to the difference between the two perversities. In Chapter 9, we will discuss conditions that can be imposed on spaces to ensure that $I_{\bar{p}}H^*(X;R) \cong I_{D\bar{p}}H^*(X;R)$. For now, we will look at an intersection homology version of the image pairing that arises if we make the assumption that $\bar{p} \leq D\bar{p}$.

(a) Nondegeneracy

Before moving on to intersection homology, we observe that, for image pairing, the best we can hope for is nondegeneracy, not, in general, nonsingularity.

First, recall from Section 8.4.1 that a pairing $P : A \otimes B \to C$ is called *nonsingular* if the adjoint homomorphisms $A \to \text{Hom}(B,C)$ and $B \to \text{Hom}(A,C)$ are both isomorphisms, while it is called *nondegenerate* if these homomorphisms are only assumed injective. Equivalently, P is nondegenerate if and only if

1. $P(a,b) = 0$ for all $b \in B$ if and only if $a = 0$, and
2. $P(a,b) = 0$ for all $a \in A$ if and only if $b = 0$.

If A and B are finitely generated vector spaces over a field F and if $C = F$, then nonsingularity and nondegeneracy are equivalent, as having injections $A \to \text{Hom}(B,F) \cong B$ and $B \to \text{Hom}(A,F) \cong A$ implies that A and B must have the same dimension and the injections must therefore be isomorphisms. However, when the ground ring is not a field, nondegeneracy and nonsingularity are not equivalent, as the following example shows.

Example 8.4.21 Let $P : \mathbb{Z} \otimes \mathbb{Z} \to \mathbb{Z}$ be the pairing such that $P(1,1) = r$ with $r \neq 0$. Then by the bilinearity of P, for any $a, b \in \mathbb{Z}$ we have $P(a,b) = abP(1,1) = rab$. It is easy to observe that this pairing is nondegenerate: clearly $P(a,b) = 0$ only if a or b is equal to zero. In fact, we can compute that each of the two adjoint maps $\mathbb{Z} \to \mathrm{Hom}(\mathbb{Z},\mathbb{Z})$ takes 1 to $\phi_1 \in \mathrm{Hom}(\mathbb{Z},\mathbb{Z})$ with $\phi_1(1) = r$. Thus the adjoints are injective. However, the map $\mathbb{Z} \to \mathrm{Hom}(\mathbb{Z},\mathbb{Z})$ determined by $1 \to \phi_1$ will be an isomorphism only if $r = \pm 1$, as otherwise ϕ_1 is not a generator of $\mathrm{Hom}(\mathbb{Z},\mathbb{Z})$. △

While we will show below that the image pairing is nondegenerate, the next example shows that it need not be nonsingular, even when working with ∂-manifolds, which of course constitute special cases of ∂-stratified pseudomanifolds.

Example 8.4.22 Let M be the disk bundle associated to the tangent bundle of the sphere S^2. Then ∂M is the sphere bundle associated to the tangent bundle of the sphere S^2, and in fact $\partial M \cong \mathbb{R}P^3$. To see this, let $x \in S^2$ and let v be a unit vector in $T_x S^2$, the tangent space to S^2 at x, which we can think of as embedded in $\mathbb{R}^3$ in the standard way as the tangent plane to the unit sphere. Then the triple $(x, v, x \times v)$ is an ordered, right-handed, orthonormal triple of vectors in $\mathbb{R}^3$, which prescribes the rotation in $SO(3)$ that takes the standard basis to the triple. In fact, this assignment describes a homeomorphism from ∂M to $SO(3)$ (exercise!), which is well known to be homeomorphic to $\mathbb{R}P^3$ (see, e.g. [125, Additional Topic 3.D]).

The space M itself is compact, four-dimensional, and orientable, as M is homotopy equivalent to S^2 and so simply connected. Let us compute some cohomology groups:

- M is homotopy equivalent to S^2, so $H^i(M) \cong \mathbb{Z}$ if $i = 0, 2$ and $H^i(M) = 0$ otherwise.
- By Lefschetz duality, $H_i(M, \partial M) \cong \mathbb{Z}$ if $i = 2, 4$ and $H_i(M, \partial M) = 0$ otherwise. So, by the Universal Coefficient Theorem, the only nontrivial $H^i(M, \partial M)$ are $H^2(M, \partial M) \cong H^4(M, \partial M) \cong \mathbb{Z}$.
- As $\partial M \cong \mathbb{R}P^3$, we have $H_0(\partial M) \cong H_3(\partial M) \cong \mathbb{Z}$, $H_1(\partial M) \cong \mathbb{Z}_2$, and $H_i(\partial M) = 0$ otherwise, by standard computations (e.g. [125, Example 2.42]). By Poincaré duality, we have $H^3(\partial M) \cong H^0(\partial M) \cong \mathbb{Z}$, $H^2(\partial M) \cong \mathbb{Z}_2$, and $H^i(\partial M) = 0$ otherwise.

The middle portion of the exact sequence of the pair $(M, \partial M)$ thus looks like

$$\begin{array}{ccccccccc}
H^1(\partial M) & \longrightarrow & H^2(M, \partial M) & \xrightarrow{\mathrm{i}^*} & H^2(M) & \longrightarrow & H^2(\partial M) & \longrightarrow & H^3(M, \partial M) \\
\| & & \| & & \| & & \| & & \| \\
0 & & \mathbb{Z} & & \mathbb{Z} & & \mathbb{Z}_2 & & 0,
\end{array}$$

and it follows that i^* must take a generator of $H^2(M, \partial M)$ to twice a generator of $H^2(M)$. We can assume we have chosen generators $\bar{\alpha} \in H^2(M, \partial M)$ and $\gamma \in H^2(M)$ so that $\mathrm{i}^*(\bar{\alpha}) = 2\gamma$, which is a generator of $\mathrm{im}(\mathrm{i}^*)$.

The nonsingular pairing $H^2(M, \partial M) \otimes H^2(M) \to \mathbb{Z}$ guarantees that if $\bar{\alpha}$ is a generator of $H^2(M,\partial M)$ and γ is a generator of $H^2(M)$, then $\mathbf{a}((\bar{\alpha} \smile \gamma) \frown \Gamma) = \pm 1 \in \mathbb{Z}$. So then, by definition, the image pairing acting on the generators 2γ of $\operatorname{im}(\mathfrak{i}^*)$ takes $2\gamma \otimes 2\gamma$ to $\mathbf{a}((\bar{\alpha} \smile \bar{\alpha}) \frown \Gamma)$. By naturality of the cup product with respect to the map $(M;\partial M,\emptyset) \to (M;\partial M,\partial M)$, we have $\bar{\alpha} \smile \bar{\alpha} = \bar{\alpha} \smile \mathfrak{i}^*(\bar{\alpha}) = \bar{\alpha} \smile 2\gamma \in H^4(M,\partial M)$, so $\mathbf{a}((\bar{\alpha} \smile \bar{\alpha}) \frown \Gamma) = \pm 2$. Thus the image pairing takes the tensor product of generators of $\operatorname{im}(\mathfrak{i}^*)$ to $\pm 2 \in \mathbb{Z}$.

So by Example 8.4.21, the image pairing is nondegenerate, but it is not nonsingular.

This example also illustrates why the intersection cohomology pairing over $\mathbb{Z}$ need not be nonsingular in general, as the ∂-manifold image pairing is a special case of the intersection cohomology image pairing. △

(b) The Intersection Cohomology Image Pairing

Suppose X is a ∂-stratified pseudomanifold with a perversity $\bar{p}$ such that $\bar{p} \le D\bar{p}$, i.e. $\bar{p}(S) \le D\bar{p}(S)$ for all strata S. This is certainly possible; for example if X is a classical ∂-stratified pseudomanifold then $\bar{0} \le D\bar{0} = \bar{t}$. We also have $\bar{m} \le \bar{n}$, where $\bar{m}$ and $\bar{n}$ are the lower and upper middle perversities of Definition 3.1.10. Given this assumption, the identity map $X \to X$ is $(\bar{p}, D\bar{p})$-stratified. If we let $\mathfrak{i} : (X,\emptyset) \to (X,\partial X)$ be the identity/inclusion, we can then consider

$$\mathfrak{i}^* : F(I_{D\bar{p}}H^*(X,\partial X;R)) \to F(I_{\bar{p}}H^*(X;R)).$$

This is well defined, as $\mathfrak{i}^*$ takes torsion elements of $I_{D\bar{p}}H^*(X,\partial X;R)$ to torsion elements of $I_{\bar{p}}H^*(X;R)$.

In the following proposition, we use the notation $\mathfrak{i}^*_i$ to specify the map $\mathfrak{i}^*$ in degree i.

Proposition 8.4.23 *Suppose R is a Dedekind domain, and X is a compact n-dimensional R-oriented locally $(\bar{p};R)$-torsion-free ∂-stratified pseudomanifold. Suppose that $\bar{p} \le D\bar{p}$, and let $\alpha \in \operatorname{im}(\mathfrak{i}^*_i : F(I_{D\bar{p}}H^i(X,\partial X;R)) \to F(I_{\bar{p}}H^i(X;R)))$ and $\beta \in \operatorname{im}(\mathfrak{i}^*_{n-i} : F(I_{D\bar{p}}H^{n-i}(X,\partial X;R)) \to F(I_{\bar{p}}H^{n-i}(X;R)))$. Let $\bar{\alpha}$ and $\bar{\beta}$ be preimages of α and β in $F(I_{D\bar{p}}H^*(X,\partial X;R))$. Then the pairing $\operatorname{im}(\mathfrak{i}^*_i) \otimes \operatorname{im}(\mathfrak{i}^*_{n-i}) \to R$ given by*

$$\alpha \otimes \beta \to \mathbf{a}\mathcal{D}(\alpha \smile \bar{\beta}) = \mathbf{a}\mathcal{D}(\bar{\alpha} \smile \bar{\beta}) = \mathbf{a}\mathcal{D}(\bar{\alpha} \smile \beta)$$

is well defined and nondegenerate.

Proof Notice that $\alpha \in F(I_{\bar{p}}H^i(X;R))$ and $\bar{\beta} \in F(I_{D\bar{p}}H^{n-i}(X,\partial X;R))$, and therefore $\mathbf{a}\mathcal{D}(\alpha \smile \bar{\beta})$ is the well-defined image of the cup product pairing of Theorem 8.4.7. Similarly for $\mathbf{a}\mathcal{D}(\bar{\alpha} \smile \beta)$. So to show that our image pairing is well defined, we must demonstrate independence of the choices of $\bar{\alpha}$ and $\bar{\beta}$ and show that the claimed equalities are valid.

Consider the following diagram:

$$\begin{array}{ccc} F(I_{D\bar{p}}H^i(X,\partial X;R)) \otimes F(I_{D\bar{p}}H^{n-i}(X,\partial X;R)) & \xrightarrow{\smile} & F(I_{\bar{0}}H^n(X,\partial X;R)) \\ \Big\downarrow {\scriptstyle \mathrm{i}^*\otimes \mathrm{id}} & & \Big\downarrow {\scriptstyle \mathrm{id}} \\ F(I_{\bar{p}}H^i(X;R)) \otimes F(I_{D\bar{p}}H^{n-i}(X,\partial X;R)) & \xrightarrow{\smile} & F(I_{\bar{0}}H^n(X,\partial X;R)). \end{array} \tag{8.13}$$

By Lemma 7.2.8, the triple of perversities $(D\bar{p}, D\bar{p}; \bar{0})$ is agreeable, as

$$D\bar{0} = \bar{t} = \bar{p} + D\bar{p} \geq \bar{p} + \bar{p} = DD\bar{p} + DD\bar{p}.$$

Here, the inequality is via our assumption that $\bar{p} \leq D\bar{p}$, and we use Corollary 8.2.5 to know that X is locally $(D\bar{p}; R)$-torsion-free. Thus the upper cup product is defined by Definition 7.2.16, and it restricts to a well-defined map on torsion-free quotients by the proof of Theorem 8.4.7. The vertical maps of the diagram are induced by the inclusion/identity maps $(X; \emptyset, \partial X) \to (X; \partial X, \partial X)$. So the diagram commutes by naturality of the cup product (Proposition 7.3.5 and Theorem 7.3.72), which induces commutativity on the torsion-free quotients. But this says that $\bar{\alpha} \smile \bar{\beta} = \mathrm{i}^*(\bar{\alpha}) \smile \bar{\beta} = \alpha \smile \bar{\beta}$. A similar diagram shows that $\bar{\alpha} \smile \beta = \bar{\alpha} \smile \bar{\beta}$. So

$$\alpha \smile \bar{\beta} = \bar{\alpha} \smile \bar{\beta} = \bar{\alpha} \smile \beta,$$

showing that our pairing formulas are equal. Additionally, as the left-hand side of the equation does not depend on the choice of preimage of α and the right-hand side does not depend on the choice of the preimage of β, our pairing is independent of these choices.

Next, we must show that the image pairing is nondegenerate. For this, first suppose that $\alpha \in \mathrm{im}(\mathrm{i}_i^*) \subset F(I_{\bar{p}}H^i(X;R))$ with $\alpha \neq 0$. By the nonsingularity of the cup product pairing demonstrated in Theorem 8.4.7, there exists some $\bar{\beta} \in F(I_{D\bar{p}}H^{n-i}(X,\partial X;R))$ such that $\mathbf{a}(\mathcal{D}(\alpha \smile \bar{\beta})) \neq 0$. But then if $\beta = \mathrm{i}^*(\bar{\beta})$, the image pairing is non-zero on $\alpha \otimes \beta$. Thus the image pairing takes $\alpha \otimes \beta$ to 0 for all β only if $\alpha = 0$. The equivalent argument holds interchanging the roles of α and β, so we see that the image pairing is nondegenerate. □

Similarly, we can consider the image torsion pairing:

Proposition 8.4.24 *Suppose R is a Dedekind domain, and let X be a compact n-dimensional R-oriented locally $(\bar{p}; R)$-torsion-free ∂-stratified pseudomanifold. Suppose that $\bar{p} \leq D\bar{p}$, and let $\alpha \in \mathrm{im}(\mathrm{i}_i^* : T(I_{D\bar{p}}H^i(X,\partial X;R)) \to T(I_{\bar{p}}H^i(X;R)))$ and $\beta \in \mathrm{im}(\mathrm{i}_{n-i+1}^* : T(I_{D\bar{p}}H^{n-i+1}(X,\partial X;R)) \to T(I_{\bar{p}}H^{n-i+1}(X;R)))$. Let $\bar{\alpha}$ and $\bar{\beta}$ be the preimages of α and β in $T(I_{D\bar{p}}H^*(X,\partial X;R))$. Then the pairing on the image modules given by*

$$\alpha \otimes \beta \to L_{\bar{p},D\bar{p}}(\alpha, \bar{\beta}) = L'_{\bar{p},D\bar{p}}(\bar{\alpha}, \beta)$$

is well defined and nondegenerate.

Proof The proof is analogous to that of Proposition 8.4.23. In particular, assuming the pairing is well defined, we obtain nondegeneracy as follows: Suppose that $\alpha \in \operatorname{im}(\mathfrak{i}_i^* : T(I_{D\bar{p}}H^i(X,\partial X;R)) \to T(I_{\bar{p}}H^i(X;R)))$ with $\alpha \neq 0$. By the nonsingularity of the torsion pairing demonstrated in Theorem 8.4.10 there exists some $\bar{\beta} \in T(I_{D\bar{p}}H^{n-i+1}(X,\partial X;R))$ such that $L_{\bar{p},D\bar{p}}(\alpha,\bar{\beta}) \neq 0$. But then if $\beta = \mathfrak{i}^*(\bar{\beta})$, the image pairing is non-zero on $\alpha \otimes \beta$ by definition. Thus the image pairing takes $\alpha \otimes \beta$ to zero for all β only if $\alpha = 0$. The equivalent argument holds interchanging the roles of α and β, so we see that the image pairing is nondegenerate.

To show that the pairing is well defined, we need to show that $L_{\bar{p},D\bar{p}}(\alpha,\bar{\beta}) = L'_{\bar{p},D\bar{p}}(\bar{\alpha},\beta)$. This will demonstrate the independence of the choices of $\bar{\alpha}$ and $\bar{\beta}$, as the left-hand term does not depend on the choice of $\bar{\alpha}$ and the right-hand term does not depend on the choice of $\bar{\beta}$.

Abusing notation, let $\bar{\beta}$ also denote a cochain representing $\bar{\beta}$ as a cohomology class. Suppose we let $\bar{\mathfrak{b}} \in I_{D\bar{p}}S^{n-i}(X,\partial X;R)$ such that $d\bar{\mathfrak{b}} = t\bar{\beta}$, $t \neq 0$. Then $d\mathfrak{i}^*(\bar{\mathfrak{b}}) = \mathfrak{i}^*(d\bar{\mathfrak{b}}) = \mathfrak{i}^*(t\bar{\beta}) = t\mathfrak{i}^*(\bar{\beta}) = t\beta$. So if we set $\mathfrak{b} = \mathfrak{i}^*(\bar{\mathfrak{b}})$, we have $d\mathfrak{b} = t\beta$. By Theorem 8.4.10, we thus have

$$L_{\bar{p},D\bar{p}}(\alpha,\bar{\beta}) = (-1)^n \frac{\mathbf{a}((\alpha \smile \bar{\mathfrak{b}}) \frown \Gamma)}{t}$$

and

$$L'_{\bar{p},D\bar{p}}(\bar{\alpha},\beta) = (-1)^n \frac{\mathbf{a}((\bar{\alpha} \smile \mathfrak{b}) \frown \Gamma)}{t}.$$

Therefore it suffices to show that both these expressions equal $(-1)^n \frac{\mathbf{a}((\bar{\alpha} \smile \bar{\mathfrak{b}}) \frown \Gamma)}{t}$. We will provide the argument for $\frac{\mathbf{a}((\alpha \smile \bar{\mathfrak{b}}) \frown \Gamma}{t} = \frac{\mathbf{a}((\bar{\alpha} \smile \bar{\mathfrak{b}}) \frown \Gamma)}{t}$, the other argument being equivalent.

We begin by observing that the cup product in these formulas is the chain-level cup product, as $\mathfrak{b}$ and $\bar{\mathfrak{b}}$ are not cocycles; therefore the naturality argument of Diagram (8.18) with respect to the map $(X;\emptyset,\partial X) \to (X;\partial X,\partial X)$ used in the proof of Proposition 8.4.23 to show that $\bar{\alpha} \smile \beta = \bar{\alpha} \smile \bar{\beta}$ does not quite apply. However, based on Lemma 7.3.4 and the proof of Proposition 7.3.5, there is still a cochain level version of Diagram (8.18) (without the F functor), but it only commutes up to chain homotopy. If we let D denote the chain homotopy, we therefore have that

$$\frac{\mathbf{a}((\alpha \smile \bar{\mathfrak{b}}) \frown \Gamma)}{t} - \frac{\mathbf{a}((\bar{\alpha} \smile \bar{\mathfrak{b}}) \frown \Gamma)}{t} = \frac{\mathbf{a}(((Dd + dD)(\bar{\alpha} \otimes \bar{\mathfrak{b}})) \frown \Gamma)}{t}.$$

To evaluate the right-hand side, we use that $\bar{\alpha}$ is a cocycle, that $d\bar{\mathfrak{b}} = t\bar{\beta}$, that the augmentation of a boundary must be zero, and the formula $\partial(\gamma \frown \xi) = (d\gamma) \frown \xi + (-1)^{|\gamma|}\gamma \frown \partial\xi$ from

Lemma 7.2.19, noting that Γ is a cycle in $I^{\bar{0}}S_n(X, \partial X; R)$:

$$\begin{aligned}
&\frac{\mathbf{a}(((Dd + dD)(\bar{\alpha} \otimes \bar{\mathfrak{b}})) \frown \Gamma)}{t} \\
&= \frac{\mathbf{a}((Dd(\bar{\alpha} \otimes \bar{\mathfrak{b}})) \frown \Gamma)}{t} + \frac{\mathbf{a}((d(D(\bar{\alpha} \otimes \bar{\mathfrak{b}}))) \frown \Gamma)}{t} \\
&= \frac{\mathbf{a}((D((d\bar{\alpha}) \otimes \bar{\mathfrak{b}} \pm \bar{\alpha} \otimes d\bar{\mathfrak{b}})) \frown \Gamma)}{t} \\
&\quad + \frac{\mathbf{a}(\partial(D(\bar{\alpha} \otimes \bar{\mathfrak{b}}) \frown \Gamma) \pm (D(\bar{\alpha} \otimes \bar{\mathfrak{b}})) \frown \partial\Gamma)}{t} \\
&= \pm\frac{\mathbf{a}((D(\bar{\alpha} \otimes t\bar{\beta})) \frown \Gamma)}{t} \\
&= \pm\mathbf{a}((D(\bar{\alpha} \otimes \bar{\beta})) \frown \Gamma).
\end{aligned}$$

This remaining expression is in R, and therefore $\frac{\mathbf{a}((\alpha \smile \bar{\mathfrak{b}}) \frown \Gamma)}{t} = \frac{\mathbf{a}((\bar{\alpha} \smile \bar{\mathfrak{b}}) \frown \Gamma)}{t}$ in $Q(R)/R$. □

8.5 The Goresky–MacPherson Intersection Pairing

At this point in our development of the subject, we would be remiss not to discuss the intersection pairing and the original Goresky–MacPherson approach to Poincaré duality for pseudomanifolds in [105]. This is the other big "intersection" in "intersection homology," the first of course being the description of allowability in terms of how chains may intersect strata. Unfortunately, the only route of which the author is aware for linking the intersection pairing to the duality via cup and cap products that we have been discussing runs through the derived category of sheaves [99] and so is beyond our current purview. Therefore, rather than develop a thorough presentation of intersection products, we will instead treat this section as a survey, referring to other sources for the details, particularly to [105] and [95], though see also [89, 168].

We will begin by discussing the intersection pairing on manifolds, and then we will move on to the pseudomanifold pairing.

8.5.1 The Intersection Pairing on Manifolds

Assume M is a compact oriented n-dimensional manifold. Simplifying to this setting and to integer coefficients, we showed in Section 8.4.2 that a consequence of Poincaré duality is the nonsingular cup product pairing

$$F(H^i(M)) \otimes F(H^{n-i}(M)) \xrightarrow{\smile} F(H^n(M)) \to \mathbb{Z},$$

where we recall that $F(H^*(M))$ is the quotient of $H^*(M)$ by its torsion subgroup. In fact, ignoring torsion, this pairing contains essentially the same information as Poincaré duality, as the nonsingularity and the finite generation due to compactness tell us that

$$F(H^i(M)) \cong \mathrm{Hom}(F(H^{n-i}(M)),\mathbb{Z}) \cong F(H_{n-i}(M)),$$

the first isomorphism from the nonsingularity of the cup product pairing and the second using the version of the Universal Coefficient Theorem that can be found, for example, as [182, Theorem 56.1]. One could then work backward from the definitions and what we did in Section 8.4.2 to relate this chain of isomorphisms to the cap product with the fundamental class. In fact, to those of a more analytic bent, Poincaré duality *is* the nonsingular pairing on de Rham cohomology $H^i_{dR}(M;\mathbb{R}) \otimes H^{n-i}_{dR}(M;\mathbb{R}) \to \mathbb{R}$ given by the wedge product and integration over M, and this is equivalent to the cup product pairing over $\mathbb{R}$; see [29].

Anyway, once we know about the Poincaré duality isomorphism $\mathcal{D}$ and the cup product pairing, we can observe that there is a *homology* pairing

$$F(H_{n-i}(M)) \otimes F(H_i(M)) \xrightarrow{(\mathcal{D}\otimes\mathcal{D})^{-1}} F(H^i(M)) \otimes F(H^{n-i}(M)) \xrightarrow{\smile} F(H^n(M)) \to \mathbb{Z} \tag{8.19}$$

given by inverting the duality isomorphisms and then applying the cup product pairing. Now, recall that the cohomology cup product is induced by the cochain-level cup product $S^*(M) \otimes S^*(M) \to S^*(M)$, and the cup product on cochains is defined on any space. But, as most algebraic topology textbooks will tell you (for example, Section 61 of [182]), there is no similar product on chains, in general, because the diagonal map $\mathbf{d} : X \to X \times X$, $x \to (x,x)$, goes the wrong way to induce a product together with the chain cross product. In other words, the cup product of cochains α and β is $\mathbf{d}^*(\alpha \times \beta)$, but if ξ and η are chains, there is no way to apply the diagonal map to $\xi \times \eta$ to obtain a chain in X. Yet we have just seen that there is a homology pairing on manifolds, which tells us that manifolds must be special in some way[28].

So what is the special property of manifolds that comes in here? At least in the smooth and piecewise linear categories, it is general position[29], which turns out to be strong enough to induce an intersection pairing on homology, arising from a partially defined pairing on chains; partially defined because we need chains to be in general position in order to define their intersection product[30]. The intersection pairing is so named because the image of two

[28] There are other spaces that are special in different ways. For example, if X is a topological group, or even if it just has a suitably defined multiplication, a chain pairing can be defined using the multiplication and this is called the Pontrjagin product; see [71, Section VII.3].

[29] There are also notions of general position for topological manifolds, but we will not discuss this here.

[30] There is a fascinating side plot here in that the fact that such an intersection pairing can only be partially defined is related to the fact that when it is defined it is graded commutative. The *commutative cochain problem*, which is more often formulated in cohomological language, says that it is not possible to construct a graded commutative differential graded algebra, functorial in the input space X, that is chain equivalent to the usual algebra $S^*(X)$ of singular cochains with cup product as algebra operation. Note that the cup product is graded commutative as an operation on *cohomology*, but it does not commute at the cochain level. The defect in commutativity is measured by something called cup-i products, which can be used to construct the Steenrod square operations on $\mathbb{Z}_2$ cohomology. So the existence of nontrivial Steenrod squares shows that we can't have a graded commutative cochain differential graded algebra. In the cochain world, what is sacrificed is commutativity. In the dual chain world, the intersection pairing we will discuss *is* commutative, but this

chains (in general position) under such a pairing is supported in their geometric intersection; consequently, the intersection product is much more geometrically accessible than the cup product, which is really pretty abstract. In fact, because it is more geometrically evident, the intersection pairing predates the cup product[31], whose arrival, in the words of McClure [168], made the intersection product "temporarily obsolete."

(a) What Should the Intersection Product Be?

To get an idea of how the intersection pairing should work, let's consider a simple example:

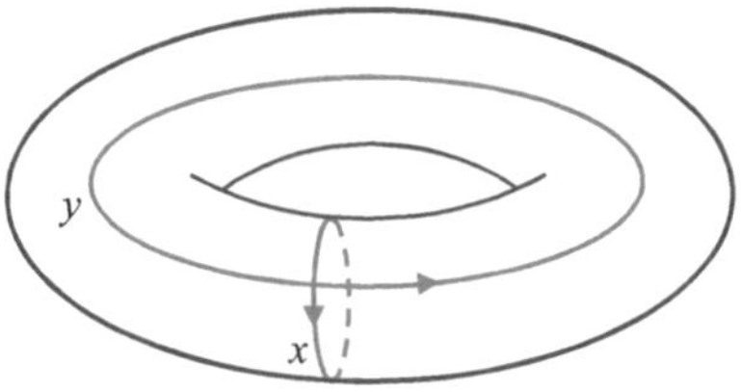

Figure 8.6. The torus with two intersecting chains.

Let T be the torus with some orientation, and let x and y be the standard generators of $H_1(T)$. For example, we can take x to be a meridian and y a longitude. See Figure 8.6. With the standard smooth embeddings, x and y intersect at a single point. We can assign that intersection point a number by looking at the ordered basis of its tangent plane given by the pair (tangent vector to x, tangent vector to y) and assign a 1 or -1 as this basis agrees or disagrees with our orientation of the torus. For now, let's assume agreement with the orientations, and then we can write the intersection product[32] $x \pitchfork y = 1$. And if we reverse the order, treating the tangent vector to y as coming first, the same argument says that $y \pitchfork x = -1$.

Now, an interesting thing happens if we start to manipulate x and y. Suppose we take two smooth deformations of x and y such that all of the intersection points are *transverse*, meaning that the tangent vectors to x and y at such points span the tangent plane to the torus. If we assign the numbers ± 1 at each intersection point as above, once again always treating x first, and then add up all of these *intersection numbers*, we still get 1. In other words, the intersection product does not depend on how we embed the curves realizing the generators. See Figure 8.7.

comes at the expense of not being fully defined! Incidentally, the commutative cochain problem does have a solution with $\mathbb{Q}$ coefficients, and this is (one of) the beginning(s) of rational homotopy theory; for example, see [116, Chapter 10].

31 According to Dieudonné [69, page 92], the cup product for simplicial complexes was introduced incorrectly by Alexander in 1935 [7], corrected by Alexander and Čech in 1936 [8, 50], and found independently by Whitney in 1938 [244]. The singular cochain cup product is due to Eilenberg in 1944 [74]. By contrast, intersection techniques seem to go back as far as Poincaré [190] and Lefschetz, with the intersection product of chains being first formalized by Lefschetz in 1926–27 [148, 149].

32 In other sources, the symbol $\pitchfork$ is often used just to denote transversality, and the intersection product is denoted by something like $\bullet$.

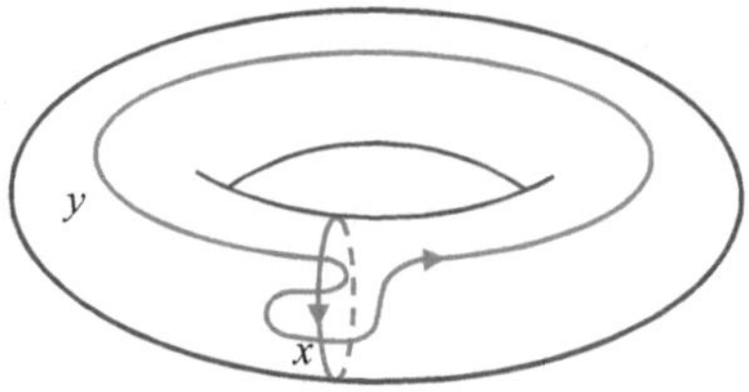

Figure 8.7 The chain y has been deformed but the intersection number (counted with signs) is unchanged.

Of course, this goes further. If we want to think about $(2x) \pitchfork y$, we can either extend our previous computation linearly and declare $(2x) \pitchfork y = 2(x \pitchfork y) = 2$, or we can represent $2x$ by, say, two parallel copies x' and x'' of the meridian. But these options are consistent, as the latter case yields two separate intersection points each with intersection number one, and adding up the intersection numbers gives us $x' \pitchfork y + x'' \pitchfork y = 2$. By now, the reader might have guessed that this process of counting intersection numbers among transverse curves, perhaps weighted with coefficients, depends only on the homology classes and not on the representatives. All in all, though we will not provide the detailed proofs, this procedure amounts to an intersection pairing $H_1(T) \otimes H_1(T) \to \mathbb{Z}$.

Our discussion so far has hinged on the fact that two smooth curves in a surface generically intersect transversely at points (and in fact, if we start with two curves that do not intersect transversely at points, we can smoothly deform them until they do, and the intersection product does not depend on the choice of deformation). But this is a special case of a more general phenomenon that is well known in the smooth category, which is that if N_1 and N_2 are two smooth submanifolds of a smooth n-manifold M, then we can deform N_1 and N_2 to be transverse, meaning that at each point $p \in N_1 \cap N_2$, the tangent spaces of N_1 and N_2 span the tangent space of M, i.e. $T_p(N_1) + T_p(N_1) = T_p(M)$ [38, Corollary II.15.4]. When this happens, the intersection is a submanifold of M of dimension $\dim(N_1) + \dim(N_2) - \dim(M)$ [38, Theorem II.7.7]. As (triangulated) closed submanifolds represent homology classes, we can therefore imagine a more general intersection product $H_i(M) \otimes H_j(M) \to H_{i+j-n}(M)$, perhaps even induced by a chain map that looks something like $S_i(M) \otimes S_j(M) \to S_{i+j-n}(M)$, though only partially defined when chains satisfy certain transversality relations.

This last thought is the basic idea for the "chain-level" intersection pairing, though in order to really make this work, we will abandon the smooth category and regroup in the (larger) PL category. One reason is that it is a well-known result of Thom's [235] that not every homology class in a smooth manifold can be represented by a smooth submanifold[33], so we would like to work with a broader class of subspaces to represent our homology classes. But working in the PL category has certain other benefits; for example, rather than

33 Relaxing the expectation to immersions doesn't work either [115].

requiring the fairly strict condition of transversality, we will be able to use chains that are only in *general position*:

Definition 8.5.1 Let K, L be two PL subsets of the n-dimensional PL manifold M. We say that K and L are in general position if $\dim(K \cap L) \leq \dim(K) + \dim(L) - n$.

Notice that the definition is satisfied if $K \cap L = \emptyset$, which is one reason for not writing the condition as an equality. Of course if $\dim(K) + \dim(L) < n$, then general position requires that $K \cap L = \emptyset$. So back in our torus example, any two PL curves that intersect only in a finite union of points are in general position. In particular, two PL curves that intersect at one point but that do not cross at that point are in general position but are not transverse.

Here is a useful theorem mirroring the transversality results for smooth manifolds. See Hudson [130, Lemma IV.4.6] for a proof of a stronger version of this statement.

Theorem 8.5.2 *If K, L are closed PL subsets of a PL manifold M, then there is a PL ambient isotopy*[34] $h : I \times M \to M$ *such that $h(1, K)$ and L are in general position. Furthermore, this isotopy can be made arbitrarily small in the sense that if $\varepsilon : M \to \mathbb{R}$ is any continuous positive function and d is a metric on M consistent with the topology, then h can be found such that $d(h(t,x), x) < \epsilon(x)$ for all x.*

(b) The PL Intersection Pairing

Having now developed some intuition, how do we officially define the intersection pairing of two PL chains in general position?

The first detailed work on intersection products was done by Lefschetz in [148, 149, 150]. Some historical discussion of Lefschetz's work can be found in [69, Section II.4.D] and the introduction to [168]. It seems to be agreed from a modern viewpoint that Lefschetz's work on intersection products was both complex and not completely satisfactory, explaining both its temporary obsolescence when the cup product arrived and the fact that we won't delve into details. Actually, the intersection product wasn't abandoned completely, but either it tended to rely on complicated processes such as replacing the chains with approximations living in sufficiently fine dual cellular subdivisions of M (see e.g. Seifert and Threlfall [213, Chapter 10]), or it was defined only at the level of homology via Poincaré duality as in (8.19) (see Dold[35] [71, Section VIII.13] or Bredon [38, Section VI.11]).

A modern revival of the intersection product, in particular as a product on *any two PL chains* (not homology classes) in general position, began with Goresky and MacPherson's work on intersection homology [105]. Their fundamental insight was to reverse history and use the cup product to define the intersection product. This technique identifies chains with certain homology classes, dualizes the homology classes using Poincaré duality, takes the cup product, and then dualizes back. This is reminiscent of (8.19) but incorporates a few

[34] This means that $\bar{h} : I \times M \to I \times M$ given by $\bar{h}(t,x) = (t, h(x,t))$ is a PL homeomorphism and that $h(0,\cdot)$ is the identity. The restriction $h(1,\cdot) : M \to M$ is then also a PL homeomorphism.

[35] Dold's intersection product looks at first a bit different from our (8.19), but it is noted that they are the same (up to sign) in Exercise 4 of Bredon [38, Section VI.11].

subtleties that let the whole thing work as a *chain* pairing. In fact, Goresky and MacPherson define their product in a PL stratified pseudomanifold, but for now we simplify their setting to manifolds to start with a more straightforward version of the details.

First recall our Useful Lemma (Lemma 3.3.10) which said that if X is a PL space and $B \subset A$ are closed PL subspaces of X such that $\dim(A) \le i$ and $\dim(B) \le i - 1$, then the group $\mathfrak{C}_i^{A,B} = \{\xi \in \mathfrak{C}_i(X) \mid |\xi| \subset A, |\partial\xi| \subset B\}$ is isomorphic to $\mathfrak{H}_i(A,B) \cong H_i(A,B)$. In other words, we can identify PL chains with certain homology classes.

Next, we need to reference some slightly more elaborate versions of Poincaré duality on manifolds:

Theorem 8.5.3 (Poincaré duality) *Let M be an oriented n-dimensional PL manifold, and let $L \subset K$ be a pair of compact PL subsets of M. Then there is a duality isomorphism $\mathcal{D} : H^i(K,L) \to H_{n-i}(M - L, M - K)$.*

This theorem follows, for example, from the version of Poincaré duality in Dold [71, Proposition VIII.7.2]. In fact, this can be strengthened to topological manifolds and arbitrary compact subsets if we replace $H^i(K,L)$ with the Čech cohomology $\check{H}^i(K,L)$. The isomorphism is given by a certain "cap product with the fundamental class" that is essentially the usual cap product with the fundamental class, though accompanied by some well-chosen inclusion maps and excisions. We refer to [71] or [95] for the details. What we really need is the following corollary:

Corollary 8.5.4 (Goresky–MacPherson duality) *Let M be a compact oriented n-dimensional PL manifold, and let $L \subset K$ be a pair of compact PL subsets of M. Then there is a duality isomorphism $\mathbb{D} : H^i(M - L, M - K) \to H_{n-i}(K,L)$.*

Notice that the difference between the theorem and the corollary lies entirely in whether the compact sets (respectively their open complements) appear in the cohomology groups or the homology groups. The corollary follows from Dold's duality theorem by some further homotopy equivalences and excisions. See [95] for a corrected version of the original construction in [105, Appendix].

With these tools, the Goresky–MacPherson intersection product on M, simplified to live on a PL manifold M and to the intersection of cycles, proceeds like this: Let $\xi \in \mathfrak{C}_i(M)$ and $\eta \in \mathfrak{C}_j(M)$ be two PL *cycles* in general position, and let $[\xi] \in H_i(|\xi|)$ and $[\eta] \in H_i(|\eta|)$ represent ξ and η under the isomorphisms of Lemma 3.3.10. Now apply to $[\xi] \otimes [\eta]$ the composite map

$$\begin{aligned} H_i(|\xi|) \otimes H_j([\eta]) &\xrightarrow{(\mathbb{D}\otimes\mathbb{D})^{-1}} H^{n-i}(M, M - |\xi|) \otimes H^{n-j}(M, M - |\eta|) \\ &\xrightarrow{\smile} H^{2n-i-j}(M - (|\xi| \cap |\eta|)) \\ &\xrightarrow{\mathbb{D}} H_{i+j-n}(|\xi| \cap |\eta|). \end{aligned} \tag{8.20}$$

We can then again apply Lemma 3.3.10 to identify the resulting homology class with a cycle $\xi \pitchfork \eta$ supported in $|\xi| \cap |\eta|$ using the general position assumption to assure $\dim(|\xi| \cap |\eta|) \leq i + j - n$; if $\dim(|\xi| \cap |\eta|) < i + j - n$ then the result will be trivial. Notice that the procedure is formally similar to that of (8.19), but the inputs and outputs are now *chains*.

While we have focused on cycles to clarify the exposition, it is not too difficult to generalize this construction to include chains with boundaries, provided we also assume that $\partial\xi$ and η are in general position and that ξ and $\partial\eta$ are in general position. Then the intersection $\xi \pitchfork \eta$ corresponds to an element of $H_{i+j-n}(|\xi| \cap |\eta|, (|\partial\xi| \cap |\eta|) \cup (|\xi| \cap |\partial\eta|))$; see [105, Section 2.1] or our construction below of the intersection pairing on pseudomanifolds, which we allow to include boundaries.

One can also show that, defined this way, the intersection product has versions of the nice properties one associates with the cup product, except that our insistence that everything be done with chain maps does lead to some odd signs compared with other choice conventions[36].

For example, one can check by hand using the definitions and the properties of the cup product that $(\xi \pitchfork \eta) \pitchfork \zeta = (-1)^{n+n|\xi|}\xi \pitchfork (\eta \pitchfork \zeta)$. One way to avoid such signs is by using shifts to redefine the intersection product as a degree 0 map; see [89, 99, 168, 169]. For more about signs when pairings are transferred using maps of non-zero degree, see [99, Appendix B].

Other properties of the intersection product include:

1. bilinearity,
2. signed commutativity: $\xi \pitchfork \eta = (-1)^{n+|\xi||\eta|}\eta \pitchfork \xi$, and a
3. boundary formula[37]: As $\pitchfork$ is induced by chain maps of total (homological) degree $-n$, we have $\partial \circ \pitchfork = (-1)^n \pitchfork \circ \partial$ so that the boundary formula becomes $\partial(\xi \pitchfork \eta) = (-1)^n(\partial\xi) \pitchfork \eta + (-1)^{n+|\xi|}\xi \pitchfork \partial\eta$.

Using these properties, we see that if η is a cycle and ξ is a homology between two cycles, say $\partial\xi = \zeta_1 - \zeta_2$, then $\zeta_1 \pitchfork \eta$ and $\zeta_2 \pitchfork \eta$ are homologous. Furthermore, by a slight strengthening of Theorem 8.5.2 (see [105, Section 2.2]), the following facts hold:

1. Given any two cycles ξ and η there is an ambient isotopy taking ξ to ξ' such that ξ' and η are in general position.
2. Given a chain ξ and a cycle η such that $\partial\xi$ and η are in general position, there is an

[36] We feel this is an acceptable price to pay for consistency with the Koszul conventions in general; see Section A.1. In particular, our conventions differ from both Goresky–MacPherson [105] and Dold [71, Section VIII.13]. There is already some confusion about signs in Dold owing to the omission of some signs in the definition of the umkehr map that would make it a chain map; see the comment on [71, page 314]. Dold also does not use our sign convention for the duality map. Consequently, some formulas in Dold are much nicer than those here, e.g. the associativity. But the Dold intersection product, as defined, does not act like a chain map of degree n, as we can see from the boundary formula [71, Section VIII.13.11]. See [99, Appendix B] for further discussion.

[37] Although not mentioned explicitly in [105], presumably one must also assume that $\partial\xi$ and $\partial\eta$ are in general position so that the terms on the right are well defined.

ambient isotopy taking ξ to ξ' and fixing $\partial\xi$ so that $\partial\xi = \partial\xi'$ and ξ and η are in general position.

Together, these observations imply that there is a well-defined product $H_i(M)\otimes H_j(M) \to H_{i+j-n}(M)$: By the first fact, if we are given any two cycles ξ and η, we can replace ξ by a homologous cycle ξ' such that ξ' and η are in general position, and we can use the isotopy to construct a homology by the usual prism argument, e.g. [125, proof of Theorem 2.10] or our proof here of Proposition 4.1.10. The second fact says that if we construct two such homologies from ξ to, say, ξ' and ξ'', each in general position with η, then we can put the composite homology from ξ' to ξ'' into general position without moving ξ' and ξ''. The boundary formula then tells us that $\xi' \pitchfork \eta$ and $\xi'' \pitchfork \eta$ are homologous. Of course the same arguments can be made in the second coordinate, showing that the chain-level pairing induces a pairing $H_i(M) \otimes H_j(M) \to H_{i+j-n}(M)$ independent of the choices made in picking representative cycles.

Furthermore, the intersection pairing conforms to our motivation:

Theorem 8.5.5 *Let M be a compact oriented PL manifold of dimension n. The following diagram commutes:*

$$\begin{array}{ccc} H^{n-i}(M) \otimes H^{n-j}(M) & \xrightarrow{\smile} & H^{2n-i-j}(M) \\ \downarrow{\scriptstyle \mathcal{D}\otimes\mathcal{D}} & & \downarrow{\scriptstyle \mathcal{D}} \\ H_i(M) \otimes H_j(M) & \xrightarrow{\pitchfork} & H_{i+j-n}(M). \end{array}$$

This theorem is not so hard to prove from the definitions using the preceding discussion together with the properties of the cup and cap products. See [95] for details.

Remark 8.5.6 Following on from our earlier observation about the trickiness of tracking signs in intersection products, we should explicitly note the following formula: If $\alpha \in H^{n-i}(M)$, $\beta \in H^{n-j}(M)$, $\mathcal{D}(\alpha) = x$, and $\mathcal{D}(\beta) = y$, then the diagram says that

$$x \pitchfork y = (-1)^{n(n-i)}\mathcal{D}(\alpha \smile \beta).$$

The sign comes from moving the $\mathcal{D}$ past α to apply the tensor product $\mathcal{D} \otimes \mathcal{D}$ to $\alpha \otimes \beta$. While this sign seems counterintuitive, this definition is consistent with that of McClure in [168, 169] (though without the chain complex shifts; see [95, Section 7] for details). As noted above, this convention differs from those in Dold [71, Section VIII.13].

The careful reader will notice that we have only treated here the intersection of two chains in suitable general position, though this is sufficient to arrive at our homology pairing $H_i(M) \otimes H_j(M) \to H_{i+j-n}(M)$. It is possible to generalize further so that this homology map is induced by appropriate chain maps, though they cannot be maps $\mathfrak{C}_*(M) \otimes \mathfrak{C}_*(M) \to \mathfrak{C}_*(M)$, as an arbitrary element of $\mathfrak{C}_*(M) \otimes \mathfrak{C}_*(M)$ does not satisfy enough general position requirements. Rather, the correct approach is to use chain maps

$$\mathfrak{C}_*(M) \otimes \mathfrak{C}_*(M) \hookleftarrow G_* \xrightarrow{\pitchfork} \mathfrak{C}_*(M),$$

where G_* is a *domain* subcomplex of $\mathfrak{C}_*(M) \otimes \mathfrak{C}_*(M)$ meeting sufficient general position requirements, the leftward inclusion map induces homology isomorphisms, and the map to the right is an appropriate version of the intersection pairing. Such a scenario then induces homology maps $H_*(M) \otimes H_*(M) \to H_*(M)$ more broadly. This approach to PL intersection pairings on manifolds is due to McClure [168]. While we will not discuss it here, an intersection chain analogue of this approach to intersection products on pseudomanifolds is developed in [89, 95].

8.5.2 The Intersection Pairing on PL Pseudomanifolds

Now we come to the intersection pairing of intersection chains on pseudomanifolds and to the original form of Poincaré duality on pseudomanifolds from [105], which says (from our current point of view) that if X is a compact oriented PL stratified pseudomanifold without codimension-one strata and $\bar{p}$ is a GM perversity then there is an intersection pairing

$$I^{\bar{p}}H_i^{\text{GM}}(X) \otimes H^{D\bar{p}}H_{n-i}^{\text{GM}}(X) \to I^{\bar{t}}H_0^{\text{GM}}(X) \xrightarrow{\mathfrak{a}} \mathbb{Z},$$

which becomes a nonsingular pairing upon tensoring with $\mathbb{Q}$. This is the original version of intersection homology duality due to Goresky and MacPherson in [105], with cup and cap products and notions of being locally torsion-free not being developed until later. So how did Goresky and MacPherson construct this intersection product?

The process is basically the same as that we have considered for manifolds, though we'll need to generalize our notion of general position and our Poincaré duality result. We will not need to change the third ingredient, representing chains by homology classes, because an intersection chain $\xi \in I^{\bar{p}}\mathfrak{C}_i^{\text{GM}}(X)$ is still a chain in $\mathfrak{C}_i(X)$ and so can still be represented by a class in the ordinary homology group $H_i(|\xi|, |\partial\xi|)$.

So first we need a new notion of general position because general position as we know it will no longer hold in a pseudomanifold.

Example 8.5.7 For example, for any compact manifold M, consider the space $X = (S^1 \times M)/(\{1\} \times M)$, and for any two distinct points $x, y \in M$, consider the images in X of $S^1 \times \{x\}$ and $S^1 \times \{y\}$. These two curves are disjoint except where they intersect at the "pinch point" corresponding to the image of $\{1\} \times M$, and it is evident that no deformations can separate the curves. When $n = \dim(M) > 1$, we have $1 + 1 - (n + 1) < 0$, so the failure to separate the curves is a violation of general position. See Figure 8.8. △

The appropriate generalization of general position is *stratified general position.*

Definition 8.5.8 Let K, L be two PL subsets of the n-dimensional PL stratified pseudomanifold X. We say that K and L are in *stratified general position* if for each stratum Z of

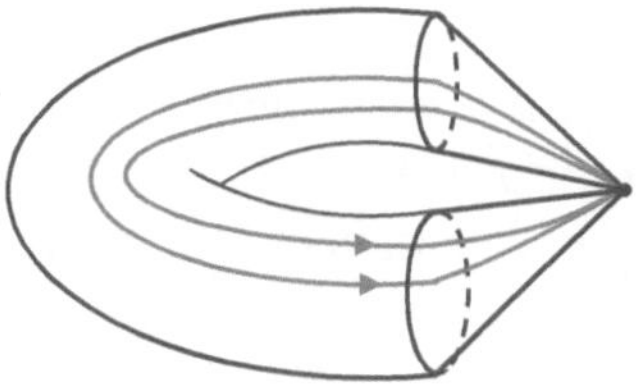

Figure 8.8 Schematic of general position failure at a singularity. Note that general position does not fail if we take the figure literally, as two 1-chains intersecting zero-dimensionally in a two-dimensional pseudomanifold does not violate general position.

X we have

$$\dim(K \cap L \cap Z) \leq \dim(K \cap Z) + \dim(L \cap Z) - \dim(Z).$$

In other words, K and L are in stratified general position if their intersections with each stratum are in general position within that stratum.

McCrory [170] showed that there is a stratified version of Theorem 8.5.2 for pseudomanifolds, i.e. that two PL subspaces can be put into stratified general position by an ambient isotopy that preserves the filtration.

So now suppose that $\xi \in I^{\bar{p}}\mathfrak{C}_i^{\mathrm{GM}}(X)$ and $\eta \in I^{\bar{q}}\mathfrak{C}_j^{\mathrm{GM}}(X)$ are in stratified general position in the n-dimensional stratified pseudomanifold X. Then the intersection of ξ and η in the regular strata will have the expected dimension $i+j-n$. Let's see what happens in a singular stratum. Suppose Z is a singular stratum of dimension $n-k$, and so codimension k. From the allowability conditions, $\dim(|\xi| \cap Z) \leq i - k + \bar{p}(Z)$ and $\dim(|\eta| \cap Z) \leq j - k + \bar{q}(Z)$. So, by stratified general position, we must have

$$\begin{aligned}\dim(|\xi| \cap |\eta| \cap Z) &\leq \dim(|\xi| \cap Z) + \dim(|\eta| \cap Z) - \dim(Z)\\ &\leq i - k + \bar{p}(Z) + j - k + \bar{q}(Z) - (n-k)\\ &= i + j - n - k + \bar{p}(Z) + \bar{q}(Z).\end{aligned} \tag{8.21}$$

But this is precisely the condition for an $(i+j-n)$-dimensional chain to be $(\bar{p}+\bar{q})$-allowable! Thus, if we can define $\xi \pitchfork \eta$ as an $(i+j-n)$-chain supported in $|\xi| \cap |\eta|$, it will furthermore be $(\bar{p}+\bar{q})$-allowable.

Remark 8.5.9 Technically, Goresky and MacPherson approach matters from the other direction by using this $(\bar{p}+\bar{q})$-allowability condition to define a notion of "dimensionally transverse chains," which is the hypothesis for their construction of the intersection pairing. McCrory's theorem is used in [105] to show that chains can be made dimensionally transverse.

Let us determine when the stratified general position of ξ and η implies that $\dim(|\xi| \cap |\eta \cap Z) \leq i + j - n$ for all Z and so consequently that $\dim(|\xi| \cap |\eta|) \leq i + j - n$. We have already observed that this must be the case on regular strata. If Z is a singular stratum, the above computation shows that it will be guaranteed if $\bar{p}(Z) + \bar{q}(Z) \leq k$. This is forced in [105] by the assumption that $\bar{p} + \bar{q} \leq \bar{r}$, with $\bar{r}$ being another GM perversity. By definition a GM perversity satisfies the stronger condition $\bar{r}(Z) \leq k - 2$; we will use this extra strength in a moment.

So, that takes care of general position, and, as noted, intersection chains can still be represented by *ordinary* homology classes as in the manifold case. What about the other ingredient, Poincaré duality? It turns out that Theorem 8.5.3 and Corollary 8.5.4 have a generalization for pseudomanifolds. The upshot is the following theorem; we refer to [95] for a corrected version of the original proof in [105]:

Theorem 8.5.10 *Let (X, S) be a compact PL space pair such that $X - S$ is an oriented n-dimensional manifold, and let $L \subset K \subset X$ be compact PL subspaces such that $S \subset L$. Then there is an isomorphism $\mathbb{D} : H^i(X - L, X - K) \to H_{n-i}(K, L)$ composed of excisions, isomorphisms induced by inclusions, and the duality isomorphism of Theorem 8.5.3.*

The basic idea of the proof is that, since S is contained within L, we can thicken K and L to open subsets by homotopy equivalences and then excise S, leaving a manifold as the ambient space. Notice that the manifold M in Theorem 8.5.3 is not required to be compact.

We can now define the chain-level intersection pairing on a PL stratified pseudomanifold X as follows. Since we will already need some extra steps to account for the singular set, it no longer adds much more clutter to include chains with nontrivial boundaries. So assume $\xi \in I^{\bar{p}}\mathfrak{C}_i^{\mathrm{GM}}(X)$ and $\eta \in I^{\bar{q}}\mathfrak{C}_j^{\mathrm{GM}}(X)$ are two chains in stratified general position and such that the pairs $(\xi, \partial\eta)$ and $(\partial\xi, \eta)$ are also in general position. Let $[\xi] \in H_i(|\xi|, |\partial\xi|)$ and $[\eta] \in H_i(|\eta|, |\partial\eta|)$ represent ξ and η under the isomorphisms of Lemma 3.3.10. Let Σ be the singular set of X, and let $J = \Sigma \cup |\partial\xi| \cup |\partial\eta|$.

Now we apply to $[\xi] \otimes [\eta]$ the following generalization of the composite map (8.20):

$$\begin{aligned}
&H_i(|\xi|, |\partial\xi|) \otimes H_j(|\eta|, |\partial\eta|)\\
&\quad\to H_i(|\xi| \cup J, J) \otimes H_j(|\eta| \cup J, J)\\
&\quad\xrightarrow{(\mathbb{D}\otimes\mathbb{D})^{-1}} H^{n-i}(X - J, X - (|\xi| \cup J)) \otimes H^{n-j}(X - J, X - (|\eta| \cup J))\\
&\quad\xrightarrow{\smile} H^{2n-i-j}(X - J, X - ((|\xi| \cap |\eta|) \cup J))\\
&\quad\xrightarrow{\mathbb{D}} H_{i+j-n}((|\xi| \cap |\eta|) \cup J, J)\\
&\quad\xleftarrow{\cong} H_{i+j-n}(|\xi| \cap |\eta|, (|\xi| \cap |\eta|) \cap J)\\
&\quad\xleftarrow{\cong} H_{i+j-n}(|\xi| \cap |\eta|, (|\partial\xi| \cap |\eta|) \cup (|\xi| \cap |\partial\eta|)).
\end{aligned} \tag{8.22}$$

The first leftward arrow is an isomorphism by excision; we can use simplicial excision (see [125, Corollary 2.24]) here as this is ordinary homology and we can find PL triangulations with respect to which all our subspaces are simplicial. To see that the second leftward arrow is an isomorphism, we consider the long exact sequence of the triple, the third term of which is

$$H_*((|\xi| \cap |\eta|) \cap J, (|\partial\xi| \cap |\eta|) \cup (|\xi| \cap |\partial\eta|)).$$

This, in turn, is isomorphic by excision to[38]

$$H_*((|\xi| \cap |\eta|) \cap \Sigma, ((|\partial\xi| \cap |\eta|) \cup (|\xi| \cap |\partial\eta|)) \cap \Sigma).$$

But we have seen in (8.21) and the discussion following the computation there that if $\bar{p} + \bar{q} \le \bar{r}$ for a GM perversity $\bar{r}$ (so that $\bar{r}(k) \le k - 2$) then the intersection of $|\xi| \cap |\eta|$ with Σ must have dimension smaller than $i + j - n - 1$. In particular, these last homology groups vanish for $* \ge i + j - n - 1$, providing the claimed isomorphism.

So we have $\dim(|\xi| \cap |\eta|) \le i + j - n$, and a similar argument holds to show $\dim((|\partial\xi| \cap |\eta|) \cup (|\xi| \cap |\partial\eta|)) \le i + j - n - 1$. Therefore, we can again apply Lemma 3.3.10 to identify the image of the composition in $H_{i+j-n}(|\xi| \cap |\eta|, (|\partial\xi| \cap |\eta|) \cup (|\xi| \cap |\partial\eta|))$ with a chain supported in $|\xi| \cap |\eta|$ and with boundary in $(|\partial\xi| \cap |\eta|) \cup (|\xi| \cap |\partial\eta|)$. This chain is our intersection product $\xi \pitchfork \eta$. Stratified general position arguments completely analogous to those discussed above for manifolds then imply that we obtain a well-defined map

$$I^{\bar{p}}H_i^{\mathrm{GM}}(X) \otimes I^{\bar{q}}H_j^{\mathrm{GM}}(X) \to I^{\bar{r}}H_{i+j-n}^{\mathrm{GM}}(X),$$

where $\bar{p}, \bar{q}, \bar{r}$ are GM perversities with $\bar{p} + \bar{q} \le \bar{r}$.

So that is the original Goresky–MacPherson intersection pairing, and if $i + j = n$ and $\bar{q} = D\bar{p}$, we get a pairing

$$I^{\bar{p}}H_i^{\mathrm{GM}}(X) \otimes I^{D\bar{p}}H_{n-i}^{\mathrm{GM}}(X) \to I^{\bar{t}}H_0^{\mathrm{GM}}(X) \xrightarrow{\mathbf{a}} \mathbb{Z}.$$

It is interesting to note that the $\bar{t}$-allowability conditions for a 0-cycle ensure that if $\xi \in I^{\bar{p}}\mathfrak{C}_i^{\mathrm{GM}}(X)$ and $\eta \in I^{D\bar{p}}\mathfrak{C}_{n-i}^{\mathrm{GM}}(X)$ are in stratified general position then $|\xi| \cap |\eta|$ lies in the union of regular strata of X. So, near these points, the intersection number looks just like what happens in a manifold.

The Poincaré Duality Theorem of Goresky and MacPherson [105, Theorem 3.3] says that this last pairing is nondegenerate when tensored with $\mathbb{Q}$. The proof of this in [105] is quite different from the proof of duality we have presented in this book and even quite different from the sheaf-theoretic proofs. There, starting with an arbitrary triangulation T of X, the authors define a collection of "basic sets" $\cdots \subset Q_i^{\bar{p}} \subset Q_{i+1}^{\bar{p}} \subset \cdots$ that are certain simplicial subcomplexes of X that, roughly speaking, carry the intersection homology information. More precisely, they show that $I^{\bar{p}}H_i(X) \cong \mathrm{im}(H_i(Q_i^{\bar{p}}) \to H_i(Q_{i+1}^{\bar{p}}))$,

[38] Again we may use simplicial excision, noting that

$$\begin{aligned}
&[(|\xi| \cap |\eta|) \cap \Sigma] \cap [(|\partial\xi| \cap |\eta|) \cup (|\xi| \cap |\partial\eta|)]\\
&= (|\xi| \cap |\eta|) \cap [(|\partial\xi| \cap |\eta|) \cup (|\xi| \cap |\partial\eta|)] \cap \Sigma\\
&= \{[(|\xi| \cap |\eta|) \cap (|\partial\xi| \cap |\eta|)] \cup [(|\xi| \cap |\eta|) \cap (|\xi| \cap |\partial\eta|)]\} \cap \Sigma\\
&= [(|\partial\xi| \cap |\eta|) \cup (|\xi| \cap |\partial\eta|)] \cap \Sigma.
\end{aligned}$$

but also that $I^{D\bar{p}}H_i(X) \cong \operatorname{im}(H^{n-i}(Q^{\bar{p}}_{n-i+1}) \to H^{n-i}(Q^{\bar{p}}_{n-i}))$. The proof that the rational intersection pairing is nonsingular then follows by observing that the Kronecker evaluation pairing between $\operatorname{im}(H_i(Q^{\bar{p}}_i) \to H_i(Q^{\bar{p}}_{i+1}))$ and $\operatorname{im}(H^{n-i}(Q^{\bar{p}}_{n-i+1}) \to H^{n-i}(Q^{\bar{p}}_{n-i}))$ is rationally nonsingular (this follows from some basic linear algebra) and corresponds to the intersection pairing under the given isomorphisms. See [105] for details.

(a) A Non-GM Intersection Product

So far we have outlined the construction of the Goresky–MacPherson intersection product of intersection chains from [105], which assumes GM perversities and GM intersection chains. It is possible, however, to extend this construction to non-GM intersection chains and arbitrary perversities satisfying $\bar{p} + \bar{q} \le \bar{r}$. The basic idea is as follows:

Assuming that $\xi \in I^{\bar{p}}\mathfrak{C}_i(X)$ and $\eta \in I^{\bar{q}}\mathfrak{C}_j(X)$ are in stratified general position then the above definition of the intersection product[39] (8.22) can be applied through to the point where we obtain an element in $H_{i+j-n}((|\xi| \cap |\eta|) \cup J, J)$. By stratified general position, it remains true that $\dim(|\xi| \cap |\eta| \cap Z) \le i + j - n$ for all regular strata Z, i.e. $\dim((|\xi| \cap |\eta|) - \Sigma) \le i + j - n$. Then, rather than applying the isomorphisms of the previous discussion, there is instead a generalization of Lemma 3.3.10 that holds in this context and lets us identify an element of $H_{i+j-n}((|\xi| \cap |\eta|) \cup J, J)$ with a degree-$(i + j - n)$ PL chain supported in the closure of $((|\xi| \cap |\eta|) \cup J) - J$ and with its boundary in J; see[40] [95, Lemma 4.3]. The same computation as above shows that stratified general position implies that this resulting chain $\xi \pitchfork \eta$ is $(\bar{p} + \bar{q})$-allowable, and similar arguments show that the boundary is also allowable. Thus we have $\xi \pitchfork \eta \in I^{\bar{p}+\bar{q}}\mathfrak{C}_{i+j-n}(X)$, and once again this induces a pairing of intersection homology groups.

For the more general development of intersection chain pairings based on chain maps (as opposed to pairings only of two chains in stratified general position), see [89] for GM perversities and GM intersection chains and [95] for the non-GM case.

(b) Almost Full Circle

We close this section by observing that, in some sense, the theory has nearly come full circle. Although duality for manifolds was originally conceived in terms of intersections, the intersection product was supplanted historically by the cup product, to which it is dual by Theorem 8.5.5. The cup product was then used by Goresky and MacPherson to construct a better intersection product that could be used to produce nonsingular pairings on pseudomanifolds. Now we have cup and cap products on pseudomanifolds, and in the setting where we have Poincaré duality for intersection homology, i.e. when X satisfies the

[39] Using the full boundaries for ξ and η in $H_i(|\xi|, |\partial\xi|)$ and $H_j(|\eta|, |\partial\eta|)$, not the boundaries $\hat{\partial}$ used in the definition of the non-GM intersection chain complex.

[40] In slightly more detail and specializing the situation of [95, Lemma 4.3] to the notation and details here, this generalization of Lemma 3.3.10 says that if X is a PL space with $C \subset B \subset A$ closed PL subspaces such that $\dim(A - B) = p$ and $\dim(B - C) < p$, then there is an isomorphism between $H_p(A, B)$ and $\{\xi \in \mathfrak{C}_p(X) \mid |\xi| \subset cl(A - B), |\partial\xi| \subset B\}$. For our use here, we take $(A, B, C) = ((|\xi| \cap |\eta|) \cup J, J, \Sigma)$.

appropriate torsion-free conditions or when substituting in appropriate field coefficients, we expect the intersection pairing

$$I^{\bar{p}}H_i(X) \otimes I^{D\bar{p}}H_{n-i}(X) \xrightarrow{\pitchfork} I^{\bar{t}}H_0^{\mathrm{GM}}(X) \xrightarrow{\mathbf{a}} \mathbb{Z}$$

to be Poincaré dual to the cup product pairing in the same manner that it is dual for manifolds. In fact, we have the following theorem generalizing Theorem 8.5.5, at least for field coefficients. Note that, due to the field coefficients, Lemma 7.2.8 shows that $(D\bar{p}, D\bar{q}; D\bar{r})$ being agreeable is equivalent to the original Goresky–MacPherson condition $\bar{p} + \bar{q} \leq \bar{r}$.

Theorem 8.5.11 *Let F be a field, and let X be a compact oriented n-dimensional PL stratified pseudomanifold with $(D\bar{p}, D\bar{q}; D\bar{r})$ an agreeable triple of perversities. Then there is a commutative diagram of nonsingular pairings and isomorphisms*

$$\begin{array}{ccc}
I_{D\bar{p}}H^{n-i}(X;F) \otimes I_{D\bar{q}}H^{n-j}(X;F) & \xrightarrow{\smile} & I_{D\bar{r}}H^{2n-i-j}(X;F) \\
\downarrow {\scriptstyle \mathcal{D}\otimes\mathcal{D}} & & \downarrow {\scriptstyle \mathcal{D}} \\
I^{\bar{p}}H_i(X;F) \otimes I^{\bar{q}}H_j(X;F) & \xrightarrow{\pitchfork} & I^{\bar{r}}H_{i+j-n}(X;F).
\end{array}$$

The author had originally hoped to include in this space a proof of Theorem 8.5.11 for Dedekind domain coefficients (with appropriate torsion-free conditions), but a proof that does not involve both sheaf theory and field coefficients remains elusive. This is somewhat surprising in light of a non-sheaf proof of the version for manifolds (Theorem 8.5.5), for which details are provided in [95], but the technical details do not quite go through for intersection homology. See the end of [95] for a discussion of the difficulty. For a sheaf-theoretic proof of Theorem 8.5.11, see [99].

Even assuming that the cup and intersection pairings are dual in full generality, we hope that this time around the cup product does not render the Goresky–MacPherson intersection product once again "obsolete" but rather that the intersection product has now proven its value, playing a fundamental role in precipitating the entire subject of this book!

8.5.3 An Intersection Pairing on Topological Pseudomanifolds and Some Relations of Goresky and MacPherson

In the preceding section we discussed the relationship between the intersection cohomology cup product pairing and Goresky–MacPherson intersection pairing for PL pseudomanifolds. Motivated by this relationship, we can *define* an intersection pairing for topological pseudomanifolds in terms of the cup product:

Definition 8.5.12 Suppose R is a Dedekind domain, and let X be a compact R-oriented n-dimensional stratified pseudomanifold. Suppose that X is locally $(\bar{p}; R)$-torsion-free, locally

$(\bar{q}; R)$-torsion-free, and locally $(\bar{p} + \bar{q}; R)$-torsion-free. Define the *intersection product* $\pitchfork: I^{\bar{p}}H_i(X) \otimes I^{\bar{q}}H_j(X) \to I^{\bar{p}+\bar{q}}H_{i+j-n}(X)$ so that the following diagram commutes:

$$\begin{array}{ccc} I_{D\bar{p}}H^{n-i}(X) \otimes I_{D\bar{q}}H^{n-j}(X) & \xrightarrow{\smile} & I_{D(\bar{p}+\bar{q})}H^{2n-i-j}(X) \\ \downarrow{\scriptstyle \mathcal{D}\otimes\mathcal{D}} & & \downarrow{\scriptstyle \mathcal{D}} \\ I^{\bar{p}}H_i(X) \otimes I^{\bar{q}}H_j(X) & \overset{\pitchfork}{\dashrightarrow} & I^{\bar{p}+\bar{q}}H_{i+j-n}(X). \end{array}$$

Of course this definition is somewhat unnecessary, as we already have the cup product and we are simply defining the intersection product in terms of it. However, we have seen that the dual intersection pairing can sometimes be useful, especially if we have a nice geometric interpretation of it to work with, such as when X is PL. Given the close relation to the cup product, we won't do much with the intersection product here except to point out some nice relations with ordinary homology observed by Goresky and MacPherson in [105, Section 2.4]. This continues our discussion from Section 8.1.6 of the factorization of the ordinary cap product with the fundamental class through intersection homology.

Recall from that section the following definitions. We assume X to be a compact oriented n-dimensional topological stratified pseudomanifold and $\bar{p}$ a perversity such that $\bar{0} \le \bar{p} \le \bar{t}$ and X is locally $(\bar{p}; R)$-torsion-free for a Dedekind domain R. Then we define $\omega_{\bar{p}}: I^{\bar{p}}H_*(X) \to H_*(X)$ to be the map induced by the inclusion $I^{\bar{p}}S_*^{\mathrm{GM}}(X) \hookrightarrow S_*(X)$, and we let $\alpha_{\bar{p}}$ be the composition

$$H^{n-*}(X; R) \xrightarrow{\omega^*_{D\bar{p}}} I_{D\bar{p}}H^{n-*}(X; R) \xrightarrow{\mathcal{D}} I^{\bar{p}}H_*(X; R).$$

Proposition 8.1.31 says the ordinary cap product $\frown \Gamma_X : H^{n-*}(X; R) \to H_*(X; R)$ factors as

$$H^{n-*}(X; R) \xrightarrow{(-1)^{n(n-*)}\alpha_{\bar{p}}} I^{\bar{p}}H_*(X; R) \xrightarrow{\omega_{\bar{p}}} H_*(X; R).$$

The sign comes from using $\mathcal{D}$ rather than simply the cap product with Γ in the definition of $\alpha_{\bar{p}}$.

Let us demonstrate a topological version of the following Goresky–MacPherson relations, observed for GM perversities, though for arbitrary PL pseudomanifolds, in [105]. Our signs differ from those in [105] due to our alternative conventions, especially in the definition of the duality map $\mathcal{D}$; see the discussion on page 602 and in footnote 36 thereat.

Proposition 8.5.13 *Suppose R is a Dedekind domain. Let X be a compact oriented n-dimensional topological stratified pseudomanifold and $\bar{p}, \bar{q}$ perversities such that $\bar{0} \le \bar{p} \le \bar{t}$, $\bar{0} \le \bar{q} \le \bar{t}$, $\bar{0} \le \bar{p} + \bar{q} \le \bar{t}$, and such that X is $(\bar{p}; R)$-torsion-free, $(\bar{q}; R)$-torsion-free, and*

$(\bar{p}+\bar{q};R)$-torsion-free. If $A \in H^i(X;R)$, $B \in H^j(X;R)$, and $\xi \in I^{\bar{p}}H_k(X;R)$, then

$$\alpha_{\bar{p}+\bar{q}}(A \smile B) = (-1)^{ni}\alpha_{\bar{p}}(A) \pitchfork \alpha_{\bar{q}}(B), \tag{8.23}$$
$$A \frown \omega_{\bar{p}}(\xi) = \omega_{\bar{p}+\bar{q}}(\alpha_{\bar{q}}(A) \pitchfork \xi), \tag{8.24}$$
$$A(\omega_{\bar{p}}(\xi)) = \mathbf{a}\omega_{\bar{p}+\bar{q}}(\alpha_{\bar{q}}(A) \pitchfork \xi). \tag{8.25}$$

Proof The verifications of these formulas utilize the basic properties of the cup and cap products. As observed at the end of Section 7.3.1, e.g. in Proposition 7.3.8, the naturality relations apply to the maps ω given our current assumptions. We will use these relations mostly without further comment.

To verify the first formula, we compute

$$\begin{aligned}
\alpha_{\bar{p}+\bar{q}}(A \smile B) &= \mathcal{D}\omega^*_{D(\bar{p}+\bar{q})}(A \smile B)\\
&= \mathcal{D}(\omega^*_{D\bar{p}}(A) \smile \omega^*_{D\bar{q}}(B))\\
&= \mathcal{D}(\mathcal{D}^{-1}\mathcal{D}(\omega^*_{D\bar{p}}(A)) \smile \mathcal{D}^{-1}\mathcal{D}(\omega^*_{D\bar{q}}(B)))\\
&= (-1)^{n(n-i)}\mathcal{D}\circ \smile \circ(\mathcal{D}^{-1} \otimes \mathcal{D}^{-1})(\mathcal{D}(\omega^*_{D\bar{p}}(A)) \otimes \mathcal{D}(\omega^*_{D\bar{q}}(B)))\\
&= (-1)^{n(n-i)+n}\mathcal{D}\circ \smile \circ(\mathcal{D} \otimes \mathcal{D})^{-1}(\alpha_{\bar{p}}(A) \otimes \alpha_{\bar{q}}(B))\\
&= (-1)^{ni}\alpha_{\bar{p}}(A) \pitchfork \alpha_{\bar{q}}(B).
\end{aligned}$$

The sign $n(n-i)$ comes from passing an operator of (cohomological) degree n, namely $\mathcal{D}^{-1}$, across $\mathcal{D}(\omega^*_{D\bar{p}}(A))$, which has (homological) degree $n-i$. The other signs come from the Koszul convention[41]. The second equality uses Proposition 7.3.9, with $(D\bar{p}, D\bar{q}; D(\bar{p}+\bar{q}))$ being an agreeable triple by Lemma 7.2.8 and Corollary 8.2.5.

For equation (8.24), let $E \in I_{D\bar{p}}H^{n-k}(X)$ be a class such that $\mathcal{D}(E) = \xi$; such an E

[41] In particular, note that

$$\begin{aligned}
\alpha \otimes \beta &= (\mathcal{D}^{-1}\mathcal{D}(\alpha)) \otimes (\mathcal{D}^{-1}\mathcal{D}(\beta))\\
&= (-1)^{n(|\alpha|-n)}(\mathcal{D}^{-1} \otimes \mathcal{D}^{-1})((\mathcal{D}\alpha) \otimes (\mathcal{D}\beta))\\
&= (-1)^{n(|\alpha|-n)+n|\alpha|}(\mathcal{D}^{-1} \otimes \mathcal{D}^{-1})(\mathcal{D} \otimes \mathcal{D})(\alpha \otimes \beta)\\
&= (-1)^{n}(\mathcal{D}^{-1} \otimes \mathcal{D}^{-1})(\mathcal{D} \otimes \mathcal{D})(\alpha \otimes \beta),
\end{aligned}$$

so $(\mathcal{D} \otimes \mathcal{D})^{-1} = (-1)^n(\mathcal{D}^{-1} \otimes \mathcal{D}^{-1})$.

exists by Poincaré duality. We then have

$$\begin{aligned}
\omega_{\bar{p}+\bar{q}}(\alpha_{\bar{q}}(A) \pitchfork \xi) &= \omega_{\bar{p}+\bar{q}}(\mathcal{D}(\omega^*_{D\bar{q}}(A)) \pitchfork \mathcal{D}(E)) \\
&= (-1)^{in}\omega_{\bar{p}+\bar{q}} \circ \pitchfork \circ (\mathcal{D} \otimes \mathcal{D})(\omega^*_{D\bar{q}}(A) \otimes E) \\
&= (-1)^{in}\omega_{\bar{p}+\bar{q}}\mathcal{D}(\omega^*_{D\bar{q}}(A) \smile E) \\
&= (-1)^{in+n(i+n-k)}\omega_{\bar{p}+\bar{q}}((\omega^*_{D\bar{q}}(A) \smile E) \frown \Gamma) \\
&= (-1)^{n(n-k)}\omega_{\bar{p}+\bar{q}}(\omega^*_{D\bar{q}}(A) \frown (E \frown \Gamma)) \\
&= (-1)^{n(n-k)+n(n-k)}\omega_{\bar{p}+\bar{q}}(\omega^*_{D\bar{q}}(A) \frown \mathcal{D}(E)) \\
&= \omega_{\bar{p}+\bar{q}}(\omega^*_{D\bar{q}}(A) \frown \xi) \\
&= A \frown \omega_{\bar{p}}(\xi).
\end{aligned}$$

Here we have used a variety of properties of cup and cap products from Section 7.3; we invite the reader to check that the perversities work out so that those lemmas apply. We have also used Proposition 7.3.8 for the last equation.

Finally, we deduce equation (8.25). This follows from the preceding formula and the standard fact that $\mathbf{a}(A \frown x) = A(x)$. We've given a proof here in the intersection context as Proposition 7.3.25. □

9
Witt Spaces and IP Spaces

If M is a compact oriented $4k$-dimensional manifold, then we have seen in Section 8.4.2 that we have a nonsingular[1] cup product pairing

$$H^{2k}(M;\mathbb{Q}) \otimes H^{2k}(M;\mathbb{Q}) \xrightarrow{\smile} H^{4k}(M;\mathbb{Q}) \xrightarrow{\mathcal{D}} H_0(M;\mathbb{Q}) \xrightarrow{\mathbf{a}} \mathbb{Q}. \tag{9.1}$$

By Proposition 7.3.15, this pairing is symmetric, i.e. $\alpha \otimes \beta$ and $\beta \otimes \alpha$ have the same image in $\mathbb{Q}$. This algebraic situation yields an integer invariant, the *signature* of the pairing, which turns out to be an important topological invariant of M. According to Gromov [118, Section $7\frac{1}{4}$], the signature "is not just 'an invariant' but *the invariant* which can be matched in beauty and power only by the Euler characteristic."

In this chapter, we will explore when signatures, and related invariants, can be extended to pseudomanifolds. It will turn out that some restrictions are needed on the space, but that on such spaces we obtain invariants with very nice properties.

In Section 9.1.1, we introduce the Witt spaces of Siegel [218], which possess self-duality analogous to (9.1) over fields, and the more restrictive IP spaces (intersection homology Poincaré spaces) of Pardon [187], which possess self-duality over Dedekind domains. We demonstrate the existence of these pairings in Section 9.2. In Section 9.3, we discuss signatures of rational Witt spaces after providing a review of the needed background from linear algebra. We also briefly discuss "perverse signatures" arising from image pairings in Section 9.3.4.

As an application of the existence of the signature for rational Witt spaces, we follow Goresky and MacPherson [105] by constructing characteristic homology L-classes for such spaces in Section 9.4. If the space is a smooth manifold, these are the Poincaré duals of the classical L-classes of the tangent bundle defined via the Pontrjagin classes. The construction involves an excursion into cohomotopy theory and transverse inverse images in the PL category. As a hint toward further applications of intersection homology theory, we end the chapter by providing a survey of pseudomanifold bordism theories in Section 9.5.

[1] Recall Definition 8.4.2.

9.1 Witt and IP Spaces

In this first section of the chapter we introduce Witt spaces and IP spaces and then prove some basic properties concerning products and stratification invariance.

9.1.1 Witt Spaces

When we turn to expanding the signature invariant to stratified spaces, there is an immediate problem: the signature of a closed oriented manifold M^{4k} is defined using the nonsingular symmetric cup product pairing

$$H^{2k}(M;\mathbb{Q}) \otimes H^{2k}(M;\mathbb{Q}) \to H^{4k}(M;\mathbb{Q}) \to \mathbb{Q}.$$

But for closed oriented pseudomanifolds, the Poincaré duality studied in Chapter 8 only provides nonsingular pairings

$$I_{\bar{p}}H^{2k}(X;\mathbb{Q}) \otimes I_{D\bar{p}}H^{2k}(X;\mathbb{Q}) \to \mathbb{Q}.$$

There is no reasonable way to interpret this as a symmetric pairing unless $I_{\bar{p}}H^{2k}(X;\mathbb{Q})$ and $I_{D\bar{p}}H^{2k}(X;\mathbb{Q})$ are isomorphic[2]. So at first glance, we want to find perversities $\bar{p}$ such that $\bar{p} = D\bar{p}$. However, this would require $\bar{p}(S) = D\bar{p}(S) = \operatorname{codim}(S) - 2 - \bar{p}(S)$, or $\bar{p}(S) = \frac{\operatorname{codim}(S)-2}{2}$, for all singular strata S. Clearly this is not possible if X has strata of odd codimension.

In fact, an early solution to this problem [105, Section 5] was to work with spaces with only even-codimension singularities. For example, every complex algebraic variety can be given such a stratification [109, Section I.1.7], so certainly this limitation was not completely unreasonable. However, one can do better.

To start, let us consider perversities $\bar{p}$ so that $\bar{p}(S) = \frac{\operatorname{codim}(S)-2}{2}$ when the codimension of S is even. For such a $\bar{p}$, we have $\bar{p} = D\bar{p}$ on spaces with only strata of even codimension. Now, what should we do when the codimension of S is odd? To stay as close to self-dual as possible, we should choose perversities that round $\frac{\operatorname{codim}(S)-2}{2}$ up or down to the nearest integer. If we round down, we obtain $\bar{m}$, the *lower-middle perversity*, defined by $\bar{m}(S) = \left\lfloor \frac{\operatorname{codim}(S)-2}{2} \right\rfloor$. This extends the Goresky–MacPherson lower middle perversity[3]

$$\bar{m} = [0, 0, 1, 1, 2, 2, 3, \dots].$$

[2] Actually, the image pairing studied in Section 8.4.5 does provide a self-pairing without further conditions, but its properties are not as nice or as well understood as those for the Witt and IP spaces we introduce in this section, so we'll mostly stick with the historical development and discuss Witt spaces first. The image pairing will return, however, as one of our ways to define *Witt signatures* for Witt spaces with boundary in Section 9.3 and in our brief discussion of *perverse signatures* for general perversities in Section 9.3.4.

[3] Recall from Definition 3.1.4 that Goresky–MacPherson perversities are functions of codimension $\{2, 3, \dots\} \to \mathbb{Z}$ and so can be described by the sequence of values $[\bar{p}(2), \bar{p}(3), \dots]$. Our definition here of $\bar{m}$ and $\bar{n}$ extend the Goresky–MacPherson definitions only in the sense that we include the possibility $\operatorname{codim}(S) = 1$.

When we round up, we get the dual perversity $D\bar{m} = \bar{n}$, the *upper middle perversity* with $\bar{n}(S) = \left\lceil \frac{\operatorname{codim}(S)-2}{2} \right\rceil$ that extends the Goresky–MacPherson upper middle perversity

$$\bar{n} = [0, 1, 1, 2, 2, 3, \ldots].$$

These two perversities are as close to each other as it is possible for two dual perversities to be, assuming we wish to make a consistent choice of which one is larger than the other on the odd-codimension strata. This is a useful assumption because it allows us to construct a "comparison map" via the inclusion $I^{\bar{m}}S_*(X; G) \to I^{\bar{n}}S_*(X; G)$ for any coefficient system; in fact, if $\bar{p}, \bar{q}$ are any perversities with $\bar{p}(S) \le \bar{q}(S)$ for all S, then $I^{\bar{p}}S_*(X; G) \subset I^{\bar{q}}S_*(X; G)$. One can then pose the natural question: for what spaces beyond those with only even-codimension strata does the inclusion $I^{\bar{m}}S_*(X; G) \hookrightarrow I^{\bar{n}}S_*(X; G)$ induce isomorphisms on homology? For $\mathbb{Q}$ coefficients, such spaces are candidates to possess signatures, and for other ring coefficients there are other invariants of self-duality that might be exploited.

Remark 9.1.1 Since the upper and lower middle perversities, and GM perversities in general, depend only on the codimensions of strata, in what follows we often write, for example, $\bar{m}(\ell)$ rather that $\bar{m}(S)$ when $\operatorname{codim}(S) = \ell$.

A natural class of spaces for which the inclusion $I^{\bar{m}}S_*(X; \mathbb{Q}) \to I^{\bar{n}}S_*(X; \mathbb{Q})$ induces an isomorphism was discovered by Paul Siegel [218] in his thesis. He named these spaces "Witt spaces" because he was able to prove that the $4n$-dimensional ($n > 0$) oriented bordism groups of PL Witt spaces are isomorphic to the Witt group[4] $W(\mathbb{Q})$ via the map that takes the self-dual cup product pairing[5] $I_{\bar{m}}H^{2k}(X; \mathbb{Q}) \otimes I_{\bar{m}}H^{2n}(X; \mathbb{Q}) \to \mathbb{Q}$, which we shall construct below, to its class as an element of $W(\mathbb{Q})$.

The defining condition for Witt spaces arises from wanting $I^{\bar{m}}H_*(X; \mathbb{Q}) \cong I^{\bar{n}}H_*(X; \mathbb{Q})$. As we have seen many times, intersection homology is in many ways controlled by what happens on cones, and if L is an $(n-1)$-dimensional link, then for n odd we have

$$I^{\bar{m}}H_i(cL; \mathbb{Q}) \cong \begin{cases} 0, & i \ge n - \left\lfloor \frac{n-2}{2} \right\rfloor - 1, \\ I^{\bar{m}}H_i(L; \mathbb{Q}), & i < n - \left\lfloor \frac{n-2}{2} \right\rfloor - 1, \end{cases}$$

$$I^{\bar{n}}H_i(cL; \mathbb{Q}) \cong \begin{cases} 0, & i \ge n - \left\lceil \frac{n-2}{2} \right\rceil - 1, \\ I^{\bar{n}}H_i(L; \mathbb{Q}), & i < n - \left\lceil \frac{n-2}{2} \right\rceil - 1. \end{cases}$$

Now, if we assume (by an induction on depth) that $I^{\bar{m}}H_i(L; \mathbb{Q}) \cong I^{\bar{n}}H_i(L; \mathbb{Q})$, then the only

4 By definition, for a field F, the Witt group $W(F)$ is the group generated by isomorphism classes of symmetric pairings on vector spaces, with the group operation being direct sum and with additional relations such that pairings with matrices of the form $\begin{pmatrix} 0 & I \\ I & A \end{pmatrix}$, with A arbitrary and I an identity matrix, are set to zero. See [176, Chapter I] for more details.

5 Technically, Siegel worked with the dual intersection pairing on homology, which is equivalent by Theorem 8.5.11.

difference between these two formulas is in dimension

$$n - \left\lceil \frac{n-2}{2} \right\rceil - 1 = n - \frac{n-1}{2} - 1 = \frac{2n-(n-1)-2}{2} = \frac{n-1}{2},$$

in which $I^{\bar{m}}H_{(n-1)/2}(cL;\mathbb{Q}) \cong I^{\bar{m}}H_{(n-1)/2}(L;\mathbb{Q})$, but $I^{\bar{n}}H_{(n-1)/2}(cL;\mathbb{Q}) = 0$. So to ensure that $I^{\bar{m}}H_*(cL;\mathbb{Q}) \cong I^{\bar{m}}H_*(cL;\mathbb{Q})$, we also need to have $I^{\bar{m}}H_{(n-1)/2}(cL;\mathbb{Q}) = 0$. This requirement is called the *Witt condition* for $\mathbb{Q}$.

Definition 9.1.2 Let G be an abelian group. Then a ∂-stratified pseudomanifold X is a *G-Witt space* if, for any point $x \in X$ contained in a stratum of odd codimension, there is a link L of x such that $I^{\bar{m}}H_{\dim(L)/2}(L;G) = 0$.

We provide some examples below after first making some remarks about the definition and then proving that Witt spaces do have isomorphic $\bar{m}$ and $\bar{n}$ intersection homology.

Remark 9.1.3 Since the point x in the definition is assumed to be in a stratum of odd codimension S, the link L must have even dimension because $\dim(X) = \dim(L)+\dim(S)+1$ and odd codimension means precisely that $\dim(X) - \dim(S)$ is odd. Also, notice that the condition on links is really a condition on the strata by Corollary 6.3.24, which extends easily to ∂-pseudomanifolds and which showed us that any two links of the same stratum have the same intersection homology. In particular, if $I^{\bar{m}}H_{\dim(L)/2}(L;G) = 0$ for one link L of x, then this property holds for any link of x.

Remark 9.1.4 Since the boundary of a ∂-stratified pseudomanifold is assumed to have a collar that is filtered homeomorphic to $[0,1) \times \partial X$, the stratified spaces X and $X - \partial X$ have the same links. Therefore, X will be G-Witt if and only if $X - \partial X$ is G-Witt. Furthermore, if X is G-Witt then ∂X is G-Witt.

Remark 9.1.5 It is not uncommon for other underlying conditions, such as orientability, compactness, empty boundary, or being PL (as in [218]), to be assumed as part of the definition for being G-Witt. We will not make such assumptions here in order to allow for greater flexibility, but the reader should pay careful attention to such assumptions when reading the literature. Furthermore, although we have formulated the definition of a G-Witt space for ∂-pseudomanifolds, the definition clearly extends to CS sets. To avoid confusion, we will refer to such spaces as CS G-Witt spaces when they arise.

Remark 9.1.6 As $I^{\bar{m}}H_0(L;G) = 0$ is impossible for a link with $\dim(L) = 0$, Witt spaces cannot possess codimension-one strata. Therefore, all Witt spaces are classical ∂-stratified pseudomanifolds.

Remark 9.1.7 If R is a commutative ring, then the notion of a ∂-stratified pseudomanifold X being R-Witt is the same whether we treat R as an abelian group or as the ground ring in its own right because intersection homology computations do not depend on ring structures. However, unless noted otherwise, when working with R-Witt spaces for a commutative ring R, we will also assume that R is the ground ring for the purposes of any homological algebra that may arise.

Proposition 9.1.8 *If X is a G-Witt space, then the inclusion*

$$I^{\bar{m}}S_*(X;G) \to I^{\bar{n}}S_*(X;G)$$

induces a homology isomorphism $I^{\bar{m}}H_*(X;G) \to I^{\bar{n}}H_*(X;G)$.

Proof For a stratified pseudomanifold X, we can use a Mayer–Vietoris argument (Theorem 5.1.4) with $F_*(U) = I^{\bar{m}}H_*(U;G)$, $G_*(U) = I^{\bar{n}}H_*(U;G)$, and Φ induced by the inclusion of chain groups. We know that we have Mayer–Vietoris sequences for intersection homology, and, using stratified homotopy invariance, our above computation for cones on links (generalized to use G coefficients) provides the needed isomorphism on distinguished neighborhoods. Since X is a pseudomanifold, the only open subsets of strata homeomorphic to Euclidean space are Euclidean subsets of the top stratum, on which both functors reduce to ordinary homology. We can also apply Lemma 5.1.6 together with Lemma 6.3.16 for the ascending chain condition. Thus $I^{\bar{m}}H_*(X;G) \to I^{\bar{n}}H_*(X;G)$ by Theorem 5.1.4.

If X is a ∂-stratified pseudomanifold with non-empty boundary, then we can use that X and $X-\partial X$ are stratified homotopy equivalent so that the inclusion map $I^{\bar{p}}H_*(X-\partial X;G) \to I^{\bar{p}}H_*(X;G)$ is an isomorphism. The proposition then follows from the empty-boundary case via the commutative diagram

$$\begin{array}{ccc} I^{\bar{m}}H_*(X-\partial X;G) & \xrightarrow{\cong} & I^{\bar{m}}H_*(X;G) \\ \cong\downarrow & & \downarrow \\ I^{\bar{n}}H_*(X-\partial X;G) & \xrightarrow{\cong} & I^{\bar{n}}H_*(X;G). \end{array}$$ □

Remark 9.1.9 The proof of Proposition 9.1.8 given in Siegel is more elaborate, utilizing spectral sequences (though, to be fair, it took us a certain amount of work with machinery to set up Theorem 5.1.4). Siegel also notes that this proposition is proven via the sheaf-theoretic formulation of intersection homology in [106]; that proof is also very straightforward, once again using only the local cone computation together with the sheaf-theoretic machinery developed in [106].

Corollary 9.1.10 *If R is a Dedekind domain and X is an R-Witt space, then the inclusion $I^{\bar{m}}S_*(X;R) \to I^{\bar{n}}S_*(X;R)$ induces a cohomology isomorphism $I_{\bar{n}}H^*(X;R) \to I_{\bar{m}}H^*(X;R)$.*

Proof This follows from Proposition 9.1.8, the Universal Coefficient Theorem (Theorem 7.1.4), and the Five Lemma. □

Example 9.1.11 Suppose X is a stratified pseudomanifold whose non-empty strata all have even codimension. Then X is automatically a G-Witt space for any G. Although this example appears somewhat trivial, all irreducible complex algebraic varieties can be given such stratifications! See [109, Section I.1.7]. △

Example 9.1.12 Let M be a (trivially stratified) compact ∂-manifold with $\partial M = \emptyset$, and let $M^+ = M \cup_{\partial M} \bar{c}(\partial M)$ as in Example 6.3.15. Then the cone vertex v is the only singularity. If $\dim(M)$ is even, then M^+ is automatically a G-Witt space for any G. If $\dim(M) = 2k+1$, then to determine whether or not M^+ is G-Witt we must check the homology of the link of v, which is ∂M. Since ∂M is a manifold, $I^{\bar{m}}H_k(\partial M; G) \cong H_k(\partial M; G)$, and so X is a G-Witt space if and only if $H_k(\partial M; G) = 0$. △

Example 9.1.13 As another easy example, let M be a compact $2k$-dimensional manifold with $H_k(M)$ finite but non-zero. Then $H_k(M; \mathbb{Q}) = 0$, but for some prime p, we have $H_k(M; \mathbb{Z}_p) \neq 0$. Thus the suspension SM will be a $\mathbb{Q}$-Witt space, but not a $\mathbb{Z}_p$-Witt space. Using the Universal Coefficient Theorem, SM will also be $\mathbb{Z}_{p'}$-Witt for any prime p' that does not divide the order of any element of $H_k(M)$ or $H_{k-1}(M)$.

On the other hand, if M is odd-dimensional, then the suspension points of SM have even codimension and SM is a G-Witt space for any G. △

Example 9.1.14 More generally, let X be a compact G-Witt space and consider the suspensions SX. First suppose $(t, x) \in (-1, 1) \times X = SX - \{\mathfrak{n}, \mathfrak{s}\}$, where $\mathfrak{n}$ and $\mathfrak{s}$ are the suspension points. Then if x has a distinguished neighborhood in X of the form $\mathbb{R}^k \times cL$, then (t, x) has a neighborhood in SX of the form $(-1, 1) \times \mathbb{R}^k \times cL \cong \mathbb{R}^{k+1} \times cL$. So L is also a link of (t, x), and L must satisfy the G-Witt condition by assumption. So whether or not SX is G-Witt depends entirely on the links of the suspension points, which are X itself. If $\dim(X)$ is odd, then the suspension points have even codimensions and SX is automatically G-Witt. If $\dim(X) = 2k$ is even, then SX is G-Witt if and only if $I^{\bar{m}}H_k(X; G) = 0$. △

These examples demonstrate that the coefficient choice matters in Definition 9.1.2.

(a) Dependence of Witt Spaces on Coefficient Choices

In Example 9.1.13, we saw how to construct spaces that are $\mathbb{Q}$-Witt but not $\mathbb{Z}_p$-Witt for some p. Here we will briefly explore related issues concerning how the Witt property depends on the coefficient choice. The reader eager to move on toward applications can safely skip forward to the next section.

In the next example, we show that there are spaces that are $\mathbb{Z}_p$-Witt for some $p > 1$ but not $\mathbb{Q}$-Witt, though the construction is a bit more elaborate.

Example 9.1.15 To find spaces that are $\mathbb{Z}_p$-Witt but not $\mathbb{Q}$-Witt, we need to take advantage of the failure of the Universal Coefficient Theorem for intersection homology. It will suffice for us to find a $2k$-dimensional stratified pseudomanifold X with only point singularities such that $I^{\bar{m}}H_k(X; \mathbb{Z}_p) = 0$ but $I^{\bar{m}}H_k(X; \mathbb{Q}) \neq 0$. Then the suspension SX will be a $\mathbb{Z}_p$-Witt space by Examples 9.1.11 and 9.1.14, but it will not be a $\mathbb{Q}$-Witt space. To construct X, let M be a compact connected oriented k-manifold, $k > 0$, equipped with a k-dimensional vector bundle V with Euler number p. For example, since the Euler number of the tangent bundle is the Euler characteristic [177, Corollary 11.12], we could use the complex projective space $\mathbb{C}P^{p-1}$ with its tangent bundle. Let X be the Thom space of V, which is the one-point compactification of the bundle. Then X is a stratified pseudomanifold with one singular stratum corresponding to the point at infinity. In fact, X can be identified as the disk bundle $D(V)$ of V with a cone adjoined on the boundary sphere bundle $S(V)$, so computations analogous to those of Example 4.4.22 apply. As V is homotopy equivalent to M and as $H_i(D(V), S(V)) \cong H_i(D(V)/S(V)) \cong H_i(X)$ for $i > 0$ by basic homology computations, the results of Examples 4.4.22 and 6.3.15 and Proposition 6.2.9 give us

$$I^{\bar{m}}H_i(X;G) \cong I^{\bar{m}}H_i^{\mathrm{GM}}(X;G) \cong \begin{cases} H_i(X;G), & i > k, \\ \mathrm{im}(H_i(M;G) \to H_i(X;G)), & i = k, \\ H_i(M;G), & i < k, \end{cases}$$

using $\bar{m}(2k) = k - 1$.

So the key term is $\mathrm{im}(H_k(M;G) \to H_k(X;G))$. We claim that

$$H_k(M;G) \cong H_k(X;G) \cong G$$

and that the image $\mathrm{im}(H_k(M;G) \to H_k(X;G))$ is isomorphic to pG. This claim is proven in Lemma 9.1.16, below. Assuming the lemma, we see that if $G = \mathbb{Z}_p$ then $I^{\bar{m}}H_k^{\mathrm{GM}}(X;G) = 0$, while if $G = \mathbb{Q}$ then $I^{\bar{m}}H_k^{\mathrm{GM}}(X;G) \cong \mathbb{Q}$. This is our desired result. △

Lemma 9.1.16 *If M is a compact connected oriented k-manifold, $k > 0$, and X is the Thom space of a k-dimensional oriented vector bundle over M with Euler number χ, then $H_k(M;G) \cong H_k(X;G) \cong G$, and the inclusion map $i : M \hookrightarrow X$ induces a homology homomorphism corresponding (up to sign) to multiplication by χ.*

Proof Since M is closed and oriented, $H_k(M) \cong \mathbb{Z}$ and $H_{k-1}(M)$ is torsion-free: By Poincaré duality and the Universal Coefficient Theorem

$$H_{k-1}(M) \cong H^1(M) \cong \mathrm{Hom}(H_1(M), \mathbb{Z}) \oplus \mathrm{Ext}(H_0(M), \mathbb{Z}),$$

but $\mathrm{Ext}(H_0(M), \mathbb{Z}) = 0$ due to $H_0(M)$ being free while $\mathrm{Hom}(H_1(M), \mathbb{Z})$ is torsion-free. So

$$H_k(M;G) \cong (H_k(M) \otimes G) \oplus (H_{k-1}(M) * G) \cong H_k(M) \otimes G \cong G,$$

by the homology Universal Coefficient Theorem.

Next, let ∞ be the point at infinity in the Thom space X, let V_0 be the vector bundle V with the zero section deleted, and let $D(V)$ and $S(V)$ be respectively the unit disk and unit sphere bundles associated with V. Then

$$\begin{aligned}
&H_k(X;G) \\
&\cong H_k(X,\infty;G) && \text{by long exact sequence since } k>0 \\
&\cong H_k(X,X-M;G) && \text{by homotopy equivalence} \\
&\cong H_k(D(V),D(V)-M;G) && \text{by excision} \\
&\cong H_k(D(V),S(V);G) && \text{by homotopy equivalence} \\
&\cong H^k(D(V);G) && \text{by Poincaré–Lefschetz duality} \\
&\cong H^k(M;G) && \text{by homotopy equivalence} \\
&\cong \mathrm{Hom}(H_k(M),G)\oplus \mathrm{Ext}(H_{k-1}(M),G) && \text{by Universal Coefficient Theorem} \\
&\cong \mathrm{Hom}(\mathbb{Z},G)\cong G,
\end{aligned}$$

where the last line follows from our preceding computations.

For the claim regarding the map, consider the map of universal coefficient sequences:

$$\begin{array}{ccccccccc}
0 & \longrightarrow & \mathrm{Ext}(H^{k+1}(M),G) & \longrightarrow & H_k(M;G) & \longrightarrow & \mathrm{Hom}(H^k(M);G) & \longrightarrow & 0 \\
 & & \downarrow & & \downarrow & & \downarrow & & \\
0 & \longrightarrow & \mathrm{Ext}(H^{k+1}(X),G) & \longrightarrow & H_k(X;G) & \longrightarrow & \mathrm{Hom}(H^k(X);G) & \longrightarrow & 0;
\end{array}$$

see [182, Theorems 56.1 and 56.2] and note that all the homology groups of M are finitely generated since M is a compact manifold, and all the homology groups of X are finitely generated since $\tilde{H}_*(X) \cong \tilde{H}_*(D(V),S(V))$ and $D(V)$ and $S(V)$ are compact ∂-manifolds. The group $H^{k+1}(M)$ vanishes, as M is k-dimensional, and by an argument similar to the above computation,

$$H^{k+1}(X) \cong H^{k+1}(D(V),S(V)) \cong H_{k-1}(D(V)) \cong H_{k-1}(M),$$

which is torsion-free. So the two left-hand terms vanish, and we see that the map $H_k(M;G) \to H_k(X;G)$ is isomorphic to the dual of the restriction map $i^* : H^k(X) \to H^k(M)$.

By homotopy equivalence and the Thom Isomorphism Theorem (see [177, Theorem 9.1]), there is an isomorphism

$$\mathbb{Z} \cong H^0(M) \cong H^0(D(V)) \xrightarrow{\smile u} H^k(D(V),S(V)) \cong H^k(X),$$

where $u \in H^k(D(V),S(V))$ is the Thom class of the bundle V. Since $1 \in H^0(D(V))$ is the generator, the generator of $H^k(D(V),S(V))$ is just the Thom class u. The restriction of the Thom class to $H^k(M)$ is precisely the Euler class $e \in H^k(M)$ by the definition on page 98 of [177]. Since $H^k(M) \cong \mathrm{Hom}(H_k(M),\mathbb{Z}) \cong \mathbb{Z}$, we can determine the class e by computing $e([M])$, where $[M]$ is the fundamental class of M. But $e([M])$ is precisely the Euler number χ by definition (if V is the tangent bundle of M, this is the Euler characteristic by [177, Corollary 11.2]). So we conclude that the image of $i^* : H^k(X) \to H^k(M)$ is $\chi\mathbb{Z}$. The lemma now follows from the universal coefficient diagram. □

To provide a flavor of more general possibilities, we quote the following theorem that is proven in [88].

Theorem 9.1.17 *Let F denote a field, and let $\mathbb{Z}_p$ denote the field of p elements, with p prime.*

1. *If F has characteristic $p > 0$, then X is F-Witt if and only if X is $\mathbb{Z}_p$-Witt; if F has characteristic zero, then X is F-Witt if and only if X is $\mathbb{Q}$-Witt.*
2. *If $n > 4$ and P is a finite set of primes, then there is a compact orientable n-dimensional stratified pseudomanifold that is $\mathbb{Z}_p$-Witt for any $p \in P$ but that is not $\mathbb{Q}$-Witt and not $\mathbb{Z}_p$-Witt for $p \notin P$.*
3. *If $n > 4$ and P is a finite set of primes, then there are $\mathbb{Q}$-Witt spaces that are not $\mathbb{Z}_p$-Witt for any $p \in P$ and are $\mathbb{Z}_p$-Witt for $p \notin P$.*
4. *If X is a 3- or 4-dimensional $\mathbb{Z}_p$-Witt space, then X is a $\mathbb{Q}$-Witt space.*
5. *If X is a 3- or 4-dimensional $\mathbb{Q}$-Witt space, then X is a $\mathbb{Z}_p$-Witt space for any $p \neq 2$. If X is also $\mathbb{Q}$-orientable, then it is also a $\mathbb{Z}_2$-Witt space. However, there are non-orientable 3- and 4-dimensional $\mathbb{Q}$-Witt spaces that are not $\mathbb{Z}_2$-Witt spaces.*
6. *All 0-, 1-, and 2-dimensional pseudomanifolds are F-Witt for all F.*

We refer the reader to [88] for the constructions of these examples, many of which are in the same vein as the Thom space example above. We will, however, provide a proof of the first fact of this theorem:

Proposition 9.1.18 *Let X be a ∂-stratified pseudomanifold and F a field of characteristic p, possibly with $p = 0$. Then X is F-Witt if and only if X is $\mathbb{Z}_p$-Witt (taking $\mathbb{Z}_0 = \mathbb{Q}$).*

Proof Let L be an even-dimensional link of X. As we work with field coefficients, every space is locally torsion-free, so we can apply Theorem 6.3.25 to compute that $I^{\bar m}H_i(L; F) \cong H_i(I^{\bar m}S_*(L; \mathbb{Z}_p) \otimes_{\mathbb{Z}_p} F)$. But now the algebraic Universal Coefficient Theorem [238, Theorem 3.6.1] shows that the latter is isomorphic to $I^{\bar m}H_i(L; \mathbb{Z}_p) \otimes_{\mathbb{Z}_p} F$. It follows immediately that if $I^{\bar m}H_{\dim(L)/2}(L; \mathbb{Z}_p) = 0$, then $I^{\bar m}H_{\dim(L)/2}(L; F) = 0$. But also if $I^{\bar m}H_{\dim(L)/2}(L; \mathbb{Z}_p)$ is not 0, it is a $\mathbb{Z}_p$-vector space of some dimension $m > 0$, and so $I^{\bar m}H_{\frac{\dim(L)}{2}}(L; \mathbb{Z}_p) \otimes_{\mathbb{Z}_p} F$ is a F-vector space of dimension m. Thus if $I^{\bar m}H_{\dim(L)/2}(L; F)$ vanishes, so does $I^{\bar m}H_{\dim(L)/2}(L; \mathbb{Z}_p)$. □

9.1.2 IP Spaces

When Siegel first introduced Witt spaces in [218], he was interested exclusively in rational coefficients. At the time, intersection homology Poincaré duality was only known for field coefficients, and so it was natural to focus on this case. However, the work of Goresky and Siegel in [111] made it apparent that duality could be extended to more general rings (at least to the integers) by introducing local torsion-free conditions. This led to Pardon's formulation in [187] of the definition for what he called *IP spaces*, short for *intersection homology Poincaré spaces*. In our language, this definition amounts to the following:

Definition 9.1.19 Let R be a Dedekind domain. Then a ∂-stratified pseudomanifold X is an IP space (with respect to R) if

1. X is R-Witt, and
2. X is locally $(\bar{m}, R)$-torsion-free[6].

If we need to specify the ring, we will refer to an *R-IP space*. If we simply say IP space with no ring specified by context, then we assume $R = \mathbb{Z}$.

Remark 9.1.20 As mentioned for Witt spaces in Remark 9.1.5, we have tried to keep the definition of IP spaces fairly general, though other authors often include other assumptions, such as orientability, compactness, empty boundary, or being PL.

Remark 9.1.21 In order that X is locally $(\bar{m}, R)$-torsion-free, each $I^{\bar{m}}H_{\dim(L)-\bar{m}(S)-1}(L; R)$ has to be flat for every singular stratum S with link L. Let us compute these dimensions more explicitly, recalling that $\bar{m}(S) = \left\lfloor \frac{\operatorname{codim}(S)-2}{2} \right\rfloor$ for each singular stratum S. Using that $\dim(L) + 1 = \operatorname{codim}(S)$, when $\operatorname{codim}(S) = 2k$ we have

$$\begin{aligned}\dim(L) - \bar{m}(S) - 1 &= (2k-1) - \left\lfloor \frac{2k-2}{2} \right\rfloor - 1 = (2k-1) - (k-1) - 1 = k-1 \\ &= \frac{\dim(L) - 1}{2}.\end{aligned}$$

When $\operatorname{codim}(S) = 2k + 1$, we have

$$\dim(L) - \bar{m}(S) - 1 = 2k - \left\lfloor \frac{2k-1}{2} \right\rfloor - 1 = 2k - (k-1) - 1 = k = \frac{\dim(L)}{2}.$$

In the latter case, we know that $I^{\bar{p}}H_{\dim(L)/2}(L; R)$ in fact vanishes if X is R-Witt, so there is some redundancy in Definition 9.1.19. Some authors therefore only state the torsion-free condition for IP spaces in terms of the links of the even-codimension strata. Our current formulation does, however, have the advantage of clearly enunciating the two nice things that the definition is doing: guaranteeing duality with the torsion-free condition and guaranteeing $I^{\bar{m}}H_* = I^{\bar{n}}H_*$ with the Witt condition.

Remark 9.1.22 By definition, every R-IP space is also R-Witt. Conversely, if F is a field, then any F-Witt space is automatically an F-IP space, as all spaces are automatically locally $(\bar{p}, F)$-torsion-free for all $\bar{p}$. Thus the terms "Witt space" and "IP space" as defined here are identical when working over field coefficients. Due to the historical development described above, it remains common in the literature to utilize the expression "Witt space" when

[6] If X is R-oriented then by Corollary 8.2.5 we could equivalently require that X be locally $(\bar{n}, R)$-torsion-free.

working with field coefficients and to reserve the expression "IP spaces" for work over more general rings.

As part of the definition, an R-IP space is also an R-Witt space. To finish this section, we show that R-IP spaces are also Witt with respect to the fraction field of R, among other fields, though the converse is not necessarily true.

Lemma 9.1.23 *Let R be a Dedekind domain and K a field that is also a flat R-module. Then if X is an R-IP space, it is also a K-Witt space. In particular, every R-IP space is a $Q(R)$-Witt space, where $Q(R)$ is the field of fractions of R, and so every $\mathbb{Z}$-IP space is a $\mathbb{Q}$-Witt space.*

Proof By definition, if X is an R-IP space then X is R-Witt and locally $(\bar{m}, R)$-torsion-free. Let L be an even-dimensional link of X. As every link of L is also a link of X by Remark 2.4.14, the link L is itself locally $(\bar{m}, R)$-torsion-free, and we can apply Theorem 6.3.25 to compute that $I^{\bar{m}}H_i(L; K) \cong H_i(I^{\bar{m}}S_*(L; R) \otimes_R K)$. But now the algebraic Universal Coefficient Theorem [238, Theorem 3.6.1] shows that the latter is isomorphic to $I^{\bar{m}}H_i(X; R) \otimes_R K$; to verify the hypotheses, we use that each $I^{\bar{m}}S_*(X; R)$ is projective over R and so each submodule of $I^{\bar{m}}S_*(X; R)$ is projective over R, as R is Dedekind. We also use that the torsion product $I^{\bar{m}}H_{i-1}(X; R) *_R K$ is zero as we have assumed that K is flat over R. It follows now that $I^{\bar{m}}H_{\dim(L)/2}(L; K) = I^{\bar{m}}H_{\dim(L)/2}(L; R) \otimes_R K = 0$, as $I^{\bar{m}}H_{\dim(L)/2}(L; R)$ vanishes by the assumption that X is R-Witt. This shows that X is K-Witt.

For the second statement of the theorem, we utilize that any localization of R is a flat R-module [197, Theorem 4.80]. □

The following example provides a $\mathbb{Q}$-Witt space that is not $\mathbb{Z}$-IP.

Example 9.1.24 Let $X = X^4$ be the suspension of $\mathbb{R}P^3$ with the natural filtration as $X^0 \subset X^4$ with X^0 equal to the set of suspension points. The only singular strata have even codimension, so X is a Witt space for any coefficients. However, the links of the singular points, which are homeomorphic to $\mathbb{R}P^3$, are 3-dimensional, and so, using Remark 9.1.21, the groups we need to check for the torsion-free condition are $I^{\bar{m}}H_{(3-1)/2}(\mathbb{R}P^3) \cong H_1(\mathbb{R}P^3)$. But this group is $\mathbb{Z}_2$, so X cannot be a $\mathbb{Z}$-IP space. △

9.1.3 Products and Stratification Independence

In this section we consider two basic properties of Witt and IP spaces: their behavior under taking products and the preservation of these classes of spaces with respect to change of stratification.

(a) Products of Witt and IP Spaces

One nice property of Witt spaces and IP spaces as classes is that they are preserved under products:

Proposition 9.1.25 *Let X, X' be R-Witt spaces for a Dedekind domain R. If X or X' is locally $(\bar{m}, R)$-torsion-free, then $X \times X'$ is an R-Witt space. In particular, if X, X' are F-Witt spaces for a field F, then $X \times X'$ is F-Witt.*

Proof The product $X \times X'$ is a ∂-stratified pseudomanifold by Lemma 2.11.7. For the Witt condition, we need to examine the links of odd-codimension strata. Since the product of a stratum of X of codimension k and a stratum of X' of codimension ℓ has codimension $k - \ell$ in $X \times X'$, in order for a stratum of $X \times X'$ to have odd codimension, it must be a product $S \times S'$ of strata of X and X' such that one of S and S' has odd codimension and the other has even codimension.

We first assume S and S' are singular strata. Let L be the[7] link of a point in S, and let L' be the link of a point of S'. Then the corresponding link in the product will be the join $L * L'$; see Section 2.11.

We have computed some of the intersection homology of joins in our Künneth Theorem computation, which required using the Künneth Theorem itself, inductively. Here we wish to use $\bar{m}$ for all perversities, so we need to check that this is consistent with the hypotheses of the Künneth Theorem. We compute in the following table, using codimensions k and ℓ.

k	ℓ	$\bar{m}(k)$	$\bar{m}(\ell)$	$\bar{m}(k+\ell)$
even	even	$\frac{k}{2} - 1$	$\frac{\ell}{2} - 1$	$\frac{k+\ell}{2} - 1$
odd	even	$\frac{k+1}{2} - 2$	$\frac{\ell}{2} - 1$	$\frac{k+\ell+1}{2} - 2$
even	odd	$\frac{k}{2} - 1$	$\frac{\ell+1}{2} - 2$	$\frac{k+\ell+1}{2} - 2$
odd	odd	$\frac{k+1}{2} - 2$	$\frac{\ell+1}{2} - 2$	$\frac{k+\ell}{2} - 1$

Thus $\bar{m}(k) + \bar{m}(\ell) \le \bar{m}(k+\ell) \le \bar{m}(k) + \bar{m}(\ell) + 2$ for all k, ℓ, and since we are working with Dedekind domain coefficients and a torsion-free assumption, this demonstrates that $\bar{m}$ is $(\bar{m}, \bar{m})$-compatible on $X \times X'$, which is sufficient for the Künneth Theorem (Theorem 6.4.7).

Now, supposing $\dim(L) = k - 1$, $\dim(L') = \ell - 1$, and let us first suppose that k is odd and ℓ is even. We need to compute $I^{\bar{m}}H_{(k+\ell-1)/2}(L * L'; F)$. We wish to use formula (6.10) on page 310, which, with the notation and assumptions of the current situation[8], holds in

[7] Recall again that links are not unique, but their intersection homology is, by Corollary 6.3.24, so we may blur this technical point with definite articles.

[8] Recall also that $cL \times cL' - \{v \times w\}$ is stratified homotopy equivalent to $L * L'$.

dimensions $< k + \ell - \bar{m}(k) - \bar{m}(\ell) - 2$. But we observe that

$$\begin{aligned} k + \ell - \bar{m}(k) - \bar{m}(\ell) - 3 &= k + \ell - \left(\frac{k+1}{2} - 2\right) - \left(\frac{\ell}{2} - 1\right) - 3 \\ &= k + \ell - \frac{k + \ell + 1}{2} \\ &= \frac{2k + 2\ell}{2} - \frac{k + \ell + 1}{2} \\ &= \frac{k + \ell - 1}{2}. \end{aligned}$$

So, to compute $I^{\bar{m}}H_{k+\ell-1/2}(L * L'; F)$, formula (6.10) applies to give us

$$\begin{aligned} I^{\bar{m}}H_{(k+\ell-1)/2}(L * L'; F) \cong & \bigoplus_{\substack{i+j=(k+\ell-1)/2 \\ i<k-\bar{m}(k)-1 \\ j<\ell-\bar{m}(\ell)-1}} I^{\bar{m}}H_i(L; R) \otimes I^{\bar{m}}H_j(L'; R) \\ & \oplus \bigoplus_{\substack{i+j=(k+\ell-1)/2-1 \\ i<k-\bar{m}(k)-1 \\ j<\ell-\bar{m}(\ell)-1}} I^{\bar{m}}H_i(L; R) * I^{\bar{m}}H_j(L'; R). \end{aligned}$$

But

$$k - \bar{m}(k) - 1 = k - \left(\frac{k+1}{2} - 2\right) - 1 = \frac{k-1}{2} + 1,$$

and

$$\ell - \bar{m}(\ell) - 1 = \ell - \left(\frac{\ell}{2} - 1\right) - 1 = \frac{\ell}{2}.$$

If

$$i < k - \bar{m}(k) - 1 = \frac{k-1}{2} + 1$$

and

$$j < \ell - \bar{m}(\ell) - 1 = \frac{\ell}{2},$$

then $i \leq (k-1)/2$ and $j \leq (\ell/2) - 1$, so

$$i + j \leq \frac{k-1}{2} + \frac{\ell}{2} - 1 = \frac{k + \ell - 1}{2} - 1.$$

So the only possible non-vanishing summand of the above expression is

$$I^{\bar{m}}H_{(k-1)/2}(L; R) * I^{\bar{m}}H_{(\ell/2)-1}(L'; R).$$

Since $\dim(L) - \bar{m}(k) - 1 = k - 1 - \bar{m}(k) - 1 = (k-1)/2$ and $\dim(L') - \bar{m}(\ell) - 1 = \ell - 1 - \bar{m}(\ell) - 1 = (\ell/2) - 1$, we see that the torsion-free assumption implies that this term vanishes as well. Furthermore, this vanishing is (remarkably!) independent of which space is locally $(\bar{m}, R)$-torsion-free, and so the same result holds if k is even and ℓ is odd.

If S' is a regular stratum and S has odd codimension k, then the link is L and we need $I^{\bar{m}}H_{(k-1)/2}(L; R) = 0$, which follows from X being R-Witt. Similarly, if S is regular and S' has odd-codimension ℓ, then the link is L' and $I^{\bar{m}}H_{(\ell-1)/2}(L'; R) = 0$ because X' is R-Witt. □

Remark 9.1.26 It is interesting to see in the proof of the proposition that the portions of the products over the regular strata are the only places where we need to utilize the assumption that X and X' are R-Witt; we did not need that for the portion of the argument over the products of singular strata.

Proposition 9.1.27 *Let X, X' be R-IP spaces for a Dedekind domain R. Then $X \times X'$ is an R-IP space.*

Proof By Proposition 9.1.25, the product $X \times X'$ is an R-Witt space, so it suffices to consider the locally torsion-free condition. As in the proof of Proposition 9.1.25, suppose $x \in X \times X'$ is contained in a product $S \times S'$ of singular strata with $\operatorname{codim}(S) = k$ and $\operatorname{codim}(S') = \ell$ and that L is the link of S with $\dim(L) = k - 1$ and L' is the link of S' with $\dim(L') = \ell - 1$. We need to consider

$$I^{\bar{m}}H_{\dim(L*L')-\bar{m}(k+\ell)-1}(L * L'; R) = I^{\bar{m}}H_{k+\ell-1-\lfloor(k+\ell-2)/2\rfloor-1}(L * L'; R).$$

If $k + \ell$ is odd, then

$$k + \ell - 1 - \left\lfloor \frac{k + \ell - 2}{2} \right\rfloor - 1 = k + \ell - 1 - \frac{k + \ell - 3}{2} - 1 = \frac{k + \ell - 1}{2},$$

and we have already seen in the proof of Proposition 9.1.25 that this module must vanish as X and X' are R-Witt.

So let $k + \ell$ be even. Then

$$k + \ell - 1 - \left\lfloor \frac{k + \ell - 2}{2} \right\rfloor - 1 = k + \ell - 1 - \frac{k + \ell - 2}{2} - 1 = \frac{k + \ell}{2} - 1.$$

Furthermore, using our computations in the table in Proposition 9.1.25, we have $\bar{m}(k) + \bar{m}(\ell) = \frac{k+\ell}{2} - C$ with $C \in \{2, 3\}$, the choice of C depending on whether k and ℓ are both even or both odd. So

$$\begin{aligned}\frac{k + \ell}{2} - 1 &= k + \ell - \frac{k + \ell}{2} - 1\\ &= k + \ell - (\bar{m}(k) + \bar{m}(\ell) + C) - 1\\ &= k + \ell - \bar{m}(k) - \bar{m}(\ell) - C - 1.\end{aligned}$$

So whether C is equal to 2 or 3, this expression is $< k + \ell - \bar{m}(k) - \bar{m}(\ell) - 2$, which means that again as in the proof of Proposition 9.1.25 we can use formula (6.10) on page 310 to get

$$I^{\bar{m}}H_{[(k+\ell)/2]-1}(L*L';F) \cong \bigoplus_{\substack{i+j=[(k+\ell)/2]-1\\ i<k-\bar{m}(k)-1\\ j<\ell-\bar{m}(\ell)-1}} I^{\bar{m}}H_i(L;R)\otimes I^{\bar{m}}H_j(L';R)$$
$$\oplus \bigoplus_{\substack{i+j=[(k+\ell)/2]-2\\ i<k-\bar{m}(k)-1\\ j<\ell-\bar{m}(\ell)-1}} I^{\bar{m}}H_i(L;R)*I^{\bar{m}}H_j(L';R). \tag{9.2}$$

If k and ℓ are both even, then $k-\bar{m}(k)-1 = k-(\frac{k}{2}-1)-1 = \frac{k}{2}$ and similarly for $\bar{m}(\ell)$, and so if $i<k-\bar{m}(k)-1$ and $j<\ell-\bar{m}(\ell)-1$ then $i+j\le \frac{k+\ell}{2}-2$, and the only nontrivial summand is $I^{\bar{m}}H_{(k/2)-1}(L;R)*I^{\bar{m}}H_{(\ell/2)-1}(L';R)$. But

$$\frac{k}{2}-1 = k-1-\frac{k-2}{2}-1 = \dim(L)-\bar{m}(k)-1,$$

so this summand is also zero as X is locally $(\bar{m},R)$-torsion-free.

Next, suppose k and ℓ are both odd. Then

$$k-\bar{m}(k)-1 = k-\frac{k+1}{2}-2-1 = \frac{k+1}{2}$$

and similarly for ℓ. So if $i<k-\bar{m}(k)-1$ and $j<\ell-\bar{m}(\ell)-1$ then

$$i+j\le \frac{k+1}{2}-1+\frac{\ell+1}{2}-1 = \frac{k+\ell}{2}-1.$$

So then for degree reasons the only terms in (9.2) that are possibly nontrivial are

$$I^{\bar{m}}H_{(k-1)/2}(L;R)\otimes I^{\bar{m}}H_{(\ell-1)/2}(L';R)$$
$$\oplus I^{\bar{m}}H_{[(k-1)/2]-1}(L;R)*I^{\bar{m}}H_{(\ell-1)/2}(L';R)$$
$$\oplus I^{\bar{m}}H_{(k-1)/2}(L;R)*I^{\bar{m}}H_{[(\ell-1)/2]-1}(L';R).$$

But $I^{\bar{m}}H_{(k-1)/2}(L;R)$ and $I^{\bar{m}}H_{(\ell-1)/2}(L';R)$ both vanish by the R-Witt condition because $\dim(L)=k-1$ and $\dim(L')=\ell-1$.

We have now seen that $I^{\bar{m}}H_{\dim(L*L')-\bar{m}(k+\ell)-1}(L*L';R)$ always vanishes when L and L' are links of singular strata S and S'. If S' is a regular stratum, then the link of $S\times S'$ is L, and if S is regular, then the link of $S\times S'$ is L'; both of these links satisfy the torsion-free condition by assumption. So, altogether, $X\times X'$ is locally $(\bar{m},R)$-torsion-free, and so $X\times X'$ is an R-IP space. □

(b) Independence of Stratification of the Witt and IP Conditions.

To end this section, we demonstrate that, suitably interpreted, the conditions of being G-Witt or R-IP are properties of a space and not of its stratification. First, notice that a space with codimension-one strata can never be G-Witt for any (nontrivial) G since the link L of a codimension-one stratum of a pseudomanifold is a disjoint union of points, trivially filtered and with formal dimension zero, and so $I^{\bar{m}}H_{\dim(L)/2}(L;G)\cong H_0(L;G)\neq 0$. On the other hand, the next proposition says that if X is a ∂-stratified pseudomanifold without codimension-one strata, then the property of being G-Witt turns out to depend only on the underlying space and not on the choice of stratification, assuming we rule out stratifications with codimension-one strata. The proof is quite analogous to that of Proposition 5.5.9:

Proposition 9.1.28 *If X and X' represent two different ∂-pseudomanifold stratifications, without codimension-one strata, of the same underlying space, then X is a G-Witt space if and only if X' is.*

Proof The property of being a G-Witt space is contingent only on the intersection homology of the links of the strata. Since the links of points in the boundary of a ∂-pseudomanifold are the same as the links of interior points, it suffices to prove the proposition for pseudomanifolds without boundaries. In that setting, we will show that X is G-Witt if and only if $\mathfrak{X}$ is a CS G-Witt space (see Remark 9.1.5), where $\mathfrak{X}$ is $|X|$ with its intrinsic filtration (see Section 2.10). As X and X' have the same intrinsic filtration, the result will follow. Recall that as $\bar{m}$ is a GM perversity it depends only on the codimensions of strata, so if $\operatorname{codim}(S) = \ell$ we can write $\bar{m}(\ell)$ for $\bar{m}(S)$.

First, assume that X is G-Witt. Recall that every stratum S of $\mathfrak{X}$ is a union of strata of X of dimension $\leq \dim(S)$ (see Section 2.10). So let S be a stratum of $\mathfrak{X}$ of odd codimension, so that the dimension of its link is even, and let x be a point of X contained in a stratum T of X with $T \subset S$ and $\dim(S) = \dim(T)$; such a stratum T must exist because the local distinguished neighborhood structure of a CS set implies that the union of strata of X of dimension $< \dim(S)$ must also have dimension $< \dim(S)$. Let L be a link of x in X and let $\mathscr{L}$ be a link of x in $\mathfrak{X}$. As $\dim(S) = \dim(T)$, we have $\dim(L) = \dim(\mathscr{L})$, and, using the computation preceding Definition 9.1.2, $I^{\bar{m}}H_{\dim(L)/2}(cL; G) \cong I^{\bar{m}}H_{\dim(L)/2}(L; G)$ and $I^{\bar{m}}H_{\dim(L)/2}(c\mathscr{L}; G) \cong I^{\bar{m}}H_{\dim(\mathscr{L})/2}(\mathscr{L}; G)$. It follows that if N and N^* are distinguished neighborhoods of x in X and $\mathfrak{X}$, respectively, we have

$$I^{\bar{m}}H_{\dim(L)/2}(L; G) \cong I^{\bar{m}}H_{\dim(L)/2}(N; G) \cong I^{\bar{m}}H_{\dim(L)/2}(N^*; G) \cong I^{\bar{m}}H_{\dim(L)/2}(\mathscr{L}; G),$$

using Corollary 5.5.4 for the middle isomorphism and stratified homotopy invariance for the others. As we have assumed that X is G-Witt, the link L of x in X satisfies the G-Witt condition $I^{\bar{m}}H_{\dim(L)/2}(L; G) = 0$, and so $\mathscr{L}$ also satisfies the G-Witt condition. Since the G-Witt condition is satisfied for a link at one point in S, it is satisfied at all points in S by Corollary 6.3.24.

Conversely, suppose $\mathfrak{X}$ is CS G-Witt. Let $x \in X$ be a point with distinguished neighborhood $N \cong \mathbb{R}^k \times cL$. Suppose $\dim(L) = \ell$ is even. As observed in the preceding paragraph, we have $I^{\bar{m}}H_{\ell/2}(L; G) \cong I^{\bar{m}}H_{\ell/2}(N; G)$. Now, let N^* be a distinguished neighborhood of x in $\mathfrak{X}$. By Corollary 5.5.4, we have $I^{\bar{m}}H_{\ell/2}(N; G) \cong I^{\bar{m}}H_{\ell/2}(N^*; G)$. But $N^* \cong \mathbb{R}^m \times c\mathscr{L}$ for some link $\mathscr{L}$ and some $\mathbb{R}^m$ with $m \geq k$, since the stratification of $\mathfrak{X}$ is coarser than that of X. If $m = k$, then $\dim(\mathscr{L}) = \ell$ as well, and, by the same argument as above, $I^{\bar{m}}H_{\ell/2}(N^*; G) \cong I^{\bar{m}}H_{\ell/2}(\mathscr{L}; G)$, which is zero by the assumption that $\mathfrak{X}$ is CS G-Witt. So suppose $m > k$, which implies that $d = \dim(\mathscr{L}) < \ell$. By stratified homotopy invariance,

$I^{\bar{m}}H_{\ell/2}(N^*;G) \cong I^{\bar{m}}H_{\ell/2}(c\mathscr{L};G)$, which, by the cone formula, is zero if $\ell/2 \geq d-\bar{m}(d+1)$. To see that this is indeed the case, we use that $\frac{\ell}{2} = \ell - \bar{m}(\ell+1) - 1$, which is easy to verify. We need to show that $\ell - \bar{m}(\ell+1) \geq d - \bar{m}(d+1) + 1$. Since $d < \ell$, let's see what happens to the quantity $i - \bar{m}(i+1)$ if we start with $i = \ell$ and step down from ℓ to d with step size 1. Obviously, the i summand will decrease by one with each step, while the term $\bar{m}(i+1)$ alternates between decreasing by one and not changing at all. So as we decrease from i to $i-1$, the expression $i - \bar{m}(i+1)$ either decreases by one or stays the same. Let's see what happens in the first step from ℓ to $\ell - 1$. Since we have assumed that ℓ is even,

$$\bar{m}(\ell+1) = \left\lfloor \frac{(\ell+1)-2}{2} \right\rfloor = \left\lfloor \frac{\ell-1}{2} \right\rfloor = \frac{\ell}{2} - 1,$$

while

$$\bar{m}((\ell-1)+1) = \bar{m}(\ell) = \left\lfloor \frac{\ell-2}{2} \right\rfloor = \frac{\ell}{2} - 1.$$

In other words, we have $\bar{m}(\ell) = \bar{m}(\ell+1)$, and so

$$\ell - \bar{m}(\ell+1) > \ell - 1 - \bar{m}(\ell).$$

It thus follows that $\ell - \bar{m}(\ell+1) \geq d - \bar{m}(d+1) + 1$ for any $d < \ell$, as desired. Therefore, $0 = I^{\bar{m}}H_{\ell/2}(N^*;G) \cong I^{\bar{m}}H_{\ell/2}(N;G) \cong I^{\bar{m}}H_{\ell/2}(L;G)$. Altogether, we see that, in all circumstances, the G-Witt condition is satisfied for all links of points in X, so X is G-Witt. □

Proposition 9.1.29 *If X and X' represent two different ∂-pseudomanifold stratifications, without codimension-one strata, of the same underlying space, then X is an R-IP space if and only if X' is.*

Proof Proposition 9.1.28 shows that X is R-Witt if and only if X' is, so it suffices to consider the local torsion property. But X is locally $(\bar{m}, R)$-torsion-free if and only if X' is by Proposition 5.5.9, using that we have no codimension-one strata so that the hypotheses apply to $\bar{m}$. □

Given the results of this section, we will sometimes say that X is a G-Witt space if it is a G-Witt space with respect to some filtration. It then follows that it is a G-Witt space with respect to any classical pseudomanifold stratification. IP spaces can be treated similarly.

9.2 Self-Pairings

The importance of Witt and IP spaces is that they possess self-pairings, meaning pairings of the form $P : A \otimes A \to R$. Unlike more general pairings, self-pairings allow for the possibility of symmetries:

Definition 9.2.1 A pairing $P : A \otimes A \to R$ of R-modules is $(-1)^\ell$*-symmetric* if $P(x, y) = (-1)^\ell P(y, x)$ for any $x, y \in A$. If ℓ is even, such pairings are also simply called *symmetric*; if ℓ is odd, such such pairings are sometimes called *skew symmetric* or *antisymmetric*.

We first discuss the cup product self-pairing for F-Witt spaces with F being a field; this constitutes the most important case and the one that provides signatures. We then state the corresponding result for IP spaces.

The following is the fundamental theorem for Witt spaces over a field. In Section 9.3, we use this theorem to obtain signatures on $\mathbb{Q}$-Witt spaces.

Proposition 9.2.2 *Suppose that the compact stratified pseudomanifold X is 2ℓ-dimensional, F-oriented, and F-Witt for some field F. Then the composition*

$$I_{\bar n}H^\ell(X;F) \otimes I_{\bar n}H^\ell(X;F) \xrightarrow{\smile} I_{\bar 0}H^{2\ell}(X;F) \xrightarrow{\mathcal{D}} I^{\bar t}H_0(X;F) \xrightarrow{\mathbf{a}} F$$

is a nonsingular $(-1)^\ell$-symmetric pairing.

Proof We consider the diagram

$$\begin{array}{ccc} I_{\bar n}H^\ell(X;F) \otimes I_{\bar n}H^\ell(X;F) & \xrightarrow{\mathrm{id}\otimes \mathrm{i}^*} & I_{\bar n}H^\ell(X;F) \otimes I_{\bar m}H^\ell(X;F) \\ & \searrow^{\smile} & \downarrow{\smile} \\ & & \mathbb{R}^k I_{\bar 0}H^{2\ell}(X;F), \end{array} \tag{9.3}$$

where $\mathrm{i}^* : I_{\bar n}H^*(X;F) \to I_{\bar m}H^*(X;F)$ is induced by the inclusion $\mathrm{i} : I^{\bar m}S_*(X;R) \to I^{\bar n}S_*(X;R)$. The vertical cup product is well defined as $\bar m = D\bar n$, so $(\bar n, \bar m; \bar 0)$ is agreeable by Corollary 7.2.10. Similarly, the triple $(\bar n, \bar n; \bar 0)$ is agreeable by Corollary 7.2.12. So the cup products in the diagram are well defined.

The diagram commutes by naturality of the cup product (Proposition 7.3.5), as the identity map $X \to X$ is $(\bar p, \bar q)$-stratified with respect to any pair of perversities with $\bar p \le \bar q$. So the pairing described in the proposition factors as

$$\begin{aligned} I_{\bar n}H^\ell(X;F) \otimes I_{\bar n}H^\ell(X;F) &\xrightarrow{\mathrm{id}\otimes \mathrm{i}^*} I_{\bar n}H^\ell(X;F) \otimes I_{\bar m}H^\ell(X;F) \xrightarrow{\smile} I_{\bar 0}H^{2\ell}(X;F) \\ &\xrightarrow{\mathcal{D}} I^{\bar t}H_0(X;F) \xrightarrow{\mathbf{a}} F. \end{aligned}$$

By Corollary 9.1.10, the first map is an isomorphism, and the composition of the remaining maps is a nonsingular pairing by Theorem 8.4.7. Therefore, the full composition is also a nonsingular pairing. The symmetry properties follow from the graded commutativity of the cup product (Proposition 7.3.15). □

The point of IP spaces is that, like F-Witt spaces for a field F, an IP space (over R) satisfies both Poincaré duality and $I_{\bar n}H^*(X;R) \cong I_{\bar m}H^*(X;R)$. Thus, one obtains for compact R-orientable IP spaces without boundary both a cup product self-pairing (now over R, which might provide more delicate information) and a torsion self-pairing. In particular, we have the following version of Proposition 9.2.2.

Proposition 9.2.3 *Suppose that R is a Dedekind domain and that the compact stratified pseudomanifold X is 2ℓ-dimensional, R-oriented, and an R-IP space. Then the composition*

$$F(I_{\bar n}H^{\ell}(X;R)) \otimes F(I_{\bar n}H^{\ell}(X;R)) \xrightarrow{\smile} I_{\bar 0}H^{2\ell}(X;R) \xrightarrow{\mathcal{D}} I^{\bar t}H_0(X;R) \xrightarrow{\mathbf{a}} R$$

is a nonsingular $(-1)^{\ell}$-symmetric pairing.

Similarly, if $\dim(X) = 2\ell - 1$, then we have a $(-1)^{\ell}$-symmetric pairing

$$T(I_{\bar n}H^{\ell}(X;R)) \otimes T(I_{\bar n}H^{\ell}(X;R)) \to Q(R)/R$$

that takes $\alpha \otimes \beta$ to $L_{\bar n,\bar m}(\alpha, \mathfrak{i}^*(\beta))$, where $\mathfrak{i} : X \to X$ is the identity map, thought of as an $(\bar m, \bar n)$-stratified map.

Proof The proof for the cup product pairing is analogous to that for Proposition 9.2.2.

For the linking pairing, the composition

$$\begin{aligned} T(I_{\bar n}H^{\ell}(X;R)) \otimes T(I_{\bar n}H^{\ell}(X;R)) &\xrightarrow{\mathrm{id}\otimes \mathfrak{i}^*} T(I_{\bar n}H^{\ell}(X;R)) \otimes T(I_{\bar m}H^{\ell}(X;R)) \\ &\xrightarrow{L_{\bar n,\bar m}} Q(R)/R \end{aligned}$$

is the composition of an isomorphism and a nonsingular pairing, so it is nonsingular. For the symmetry, we need to show that $L_{\bar n,\bar m}(\alpha, \mathfrak{i}^*(\beta)) = (-1)^{\ell} L_{\bar n,\bar m}(\beta, \mathfrak{i}^*(\alpha))$, but, by Theorem 8.4.10,

$$L_{\bar n,\bar m}(\beta, \mathfrak{i}^*(\alpha)) = (-1)^{1+(2\ell-1)+\ell(2\ell-1)} L'_{\bar m,\bar n}(\mathfrak{i}^*(\alpha), \beta) = (-1)^{\ell} L'_{\bar m,\bar n}(\mathfrak{i}^*(\alpha), \beta).$$

So it suffices to show $L_{\bar n,\bar m}(\alpha, \mathfrak{i}^*(\beta)) = L'_{\bar m,\bar n}(\mathfrak{i}^*(\alpha), \beta)$.

Suppose $d\mathfrak{b} = t\beta \in I_{\bar n}S^{\ell}(X;R)$, $t \in R$, $t \neq 0$, with β representing a cocycle as well as a cohomology class, as usual. Then $d\mathfrak{i}^*(\mathfrak{b}) = t\mathfrak{i}^*(\beta) \in I_{\bar m}S^{\ell}(X;R)$, and, by Theorem 8.4.10, we have

$$L_{\bar n,\bar m}(\alpha, \mathfrak{i}^*(\beta)) = (-1)^{2\ell-1}\frac{\mathbf{a}((\alpha \smile \mathfrak{i}^*(\mathfrak{b})) \frown \Gamma)}{t}$$

and

$$L'_{\bar m,\bar n}(\mathfrak{i}^*(\alpha), \beta) = (-1)^{2\ell-1}\frac{\mathbf{a}((\mathfrak{i}^*(\alpha) \smile \mathfrak{b}) \frown \Gamma)}{t}.$$

From here, we can proceed as in the proof of Proposition 8.4.24 by showing that each of these expressions is equal to $\frac{\mathbf{a}((\alpha\smile\mathfrak{b})\frown\Gamma)}{t}$ in $Q(R)/R$. In fact, by the arguments of Proposition 9.2.2, this cup product $\alpha \smile \mathfrak{b} \in I_{\bar 0}S^{2\ell-1}(X;R)$ based on the agreeable triple $(\bar n, \bar n; \bar 0)$ is well defined, and Diagram (9.3) commutes up to homotopy at the (co)chain level by the arguments of Lemma 7.3.4 and Proposition 7.3.5. But now the argument that our two expressions both equal $\frac{\mathbf{a}((\alpha\smile\mathfrak{b})\frown\Gamma)}{t}$ in $Q(R)/R$ is completely analogous to the argument in the last paragraph of the proof of Proposition 8.4.24. ◻

9.3 Witt Signatures

It is now possible to define signatures for compact oriented $4k$-dimensional $\mathbb{Q}$-Witt spaces, which includes $\mathbb{Z}$-IP spaces by Lemma 9.1.23, using the symmetric middle-dimensional middle-perversity self-pairing (Proposition 9.2.2). We first provide the definitions and basic properties before moving on to more sophisticated results.

9.3.1 Definitions and Basic Properties

In this section, we define Witt signatures, first for $\mathbb{Q}$-Witt spaces without boundaries and then for those that may possess boundaries, and then demonstrate that these are topological invariants in an appropriate sense[9]. We begin with some algebraic definitions.

(a) Signatures of Matrices and Pairings

More algebraic background regarding signature invariants can be found in Appendix A.5. Here we recall just the definitions needed to define the Witt signature, as well as some properties, leaving the proofs for the appendix.

Definition 9.3.1 If M is a symmetric matrix of rational numbers, then the *signature* $\sigma(M)$ is defined to be

$$\sigma(M) = \#\{\text{positive eigenvalues of } M\} - \#\{\text{negative eigenvalues of } M\}.$$

Note that the symmetry of the matrix M ensures that all eigenvalues will be real[10].

Definition 9.3.2 Suppose that $(V, (\cdot,\cdot))$ is a finite-dimensional rational vector space together with a symmetric bilinear pairing $(\cdot,\cdot) : V \times V \to \mathbb{Q}$. If $\{e_i\}$ is a chosen basis of V, then the *pairing matrix* of $(\cdot,\cdot)$ with respect to this basis is defined to be the matrix M with entries $M_{ij} = (e_i, e_j)$.

If M is a pairing matrix for $(V, (\cdot,\cdot))$, then we define the *signature of the pairing* $\sigma(V, (\cdot,\cdot))$ to be $\sigma(M)$.

It turns out that the signature of a pairing is independent of the choice of the pairing matrix, as changing the basis will have the effect of changing the pairing matrix from M to a congruent matrix $Q^t M Q$, with Q an invertible matrix. The signature is an invariant of such congruence classes. A proof can be found as Lemma A.5.4 in the appendix.

Here are some further useful properties of pairings and signatures. Proofs can be found in the appendix at the indicated locations.

[9] It is also not unusual to define signatures with $\mathbb{R}$ as the ground field, but by Theorem 9.1.17 a pseudomanifold is $\mathbb{R}$-Witt if and only if it is $\mathbb{Q}$-Witt. So working over $\mathbb{R}$ is equivalent to working over $\mathbb{Q}$ for the purpose of defining signatures. See Section A.5 for more details.

[10] Let $\langle \cdot, \cdot \rangle$ denote the standard complex inner product on $\mathbb{C}^n$, let M be a symmetric $n \times n$ matrix with real entries so that the conjugate transpose M^* is equal to M, and suppose $Mv = \lambda v$ for some unit vector v. Then $\lambda = \lambda\langle v, v\rangle = \langle v, \lambda v\rangle = \langle v, Mv\rangle = \langle M^* v, v\rangle = \langle Mv, v\rangle = \langle \lambda v, v\rangle = \bar{\lambda}\langle v, v\rangle = \bar{\lambda}$.

Lemma 9.3.3 (see Lemma A.5.4) *If $f : V \to W$ is an isomorphism of finite-dimensional rational vector spaces that induces an isomorphism (in the sense of Definition 8.4.1) between the pairings $(\cdot,\cdot)_V$ on V and $(\cdot,\cdot)_W$ on W (i.e. if f is an* isometry*), then $\sigma(V, (\cdot,\cdot)_V) = \sigma(W, (\cdot,\cdot)_W)$.*

Lemma 9.3.4 (see Lemma A.5.13) *If M has the block form*

$$\begin{pmatrix} A_1 & 0 & \cdots & 0 \\ 0 & A_2 & \cdots & 0 \\ \vdots & \vdots & \ddots & \vdots \\ 0 & 0 & \dots & A_m \end{pmatrix},$$

then $\sigma(M) = \sum_{i=1}^{m} \sigma(A_i)$. Consequently, if $(V, (\cdot,\cdot)_V)$ is a direct sum of orthogonal subspaces W_i, then

$$\sigma(V, (\cdot,\cdot)_V) = \sum_i \sigma((W_i, (\cdot,\cdot)_{W_i})),$$

where $(\cdot,\cdot)_{W_i}$ is the restriction of $(\cdot,\cdot)_V$ to W_i.

Lemma 9.3.5 (see Lemma A.5.8) *Let $(V, (\cdot,\cdot))$ be a symmetric pairing on the finite-dimensional rational vector space V, and let M be the matrix of the pairing with respect to some basis. Then the pairing is nonsingular*[11] *if and only if* $\det(M) \neq 0$.

Lemma 9.3.6 (see Lemma A.5.11) *Suppose $(V, (\cdot,\cdot))$ is a rational vector space together with a nonsingular symmetric bilinear pairing $(\cdot,\cdot) : V \times V \to \mathbb{Q}$. If there is a subspace $A \subset V$ of dimension $\dim(A) = \frac{1}{2}\dim(V)$ such that $(x, y) = 0$ for all $x, y \in A$, then $\sigma(V, (\cdot,\cdot)) = 0$.*

Lemma 9.3.7 (see Lemma A.5.5) *Let $(V, (\cdot,\cdot))$ be a finite-dimensional rational vector space with symmetric bilinear pairing. Then there is a basis of V with respect to which the pairing matrix is a diagonal matrix.*

Lemma 9.3.8 (see Corollary A.5.15) *If $(V, (\cdot,\cdot))$ is a finite-dimensional vector space with a nonsingular antisymmetric pairing, there is a vector space of half the dimension of V on which the pairing is trivial.*

(b) Witt Signatures

We first define the signatures of $\mathbb{Q}$-Witt spaces without boundary.

Definition 9.3.9 Suppose X is a closed (compact without boundary) oriented $\mathbb{Q}$-Witt space of dimension $4k$. Then the *Witt signature* $\sigma(X)$ is defined to be the signature of the symmetric cup product self-pairing

$$I_{\bar{n}}H^{2k}(X;\mathbb{Q}) \otimes I_{\bar{n}}H^{2k}(X;\mathbb{Q}) \xrightarrow{\smile} I_{\bar{0}}H^{4k}(X;\mathbb{Q}) \xrightarrow{\mathcal{D}} I^{\bar{t}}H_0(X;\mathbb{Q}) \xrightarrow{\mathbf{a}} \mathbb{Q}.$$

[11] See Definition 8.4.2.

If X has dimension $\not\equiv 0 \mod 4$, we set $\sigma(X) = 0$.

Remark 9.3.10 If M is a manifold, then all intersection homology/cohomology groups in the definition reduce to ordinary homology/cohomology, and we recover the classical manifold signature. It is also standard in manifold theory to define the signature to be zero when the dimension is not a multiple of 4; this is primarily a convenience for formulas that involve signatures of spaces of various dimensions, such as in Theorem 9.3.17, below.

Witt Signatures of Witt Spaces with Boundaries. For $\mathbb{Q}$-Witt spaces that may have non-empty boundary, we need a different approach to Witt signatures as the cup product pairs $I_{\bar{n}}H^i(X;\mathbb{Q})$ with $I_{\bar{n}}H^{n-i}(X,\partial X;\mathbb{Q})$, again breaking the symmetry. We will see two approaches. First we define the Witt signature in terms of the image pairing of Proposition 8.4.23, but then we will prove that this signature can also be computed as the signature of the $\mathbb{Q}$-Witt space without boundary obtained by coning off the boundary of X.

Definition 9.3.11 Let X be a compact $4k$-dimensional oriented $\mathbb{Q}$-Witt space. Then the *Witt signature* $\sigma(X)$ is defined to be the signature of the image pairing on $\operatorname{im}(I_{\bar{n}}H^{2k}(X,\partial X;\mathbb{Q}) \xrightarrow{i} I_{\bar{m}}H^{2k}(X;\mathbb{Q}))$ (Proposition 8.4.23).

Note that the image pairing already pairs $\operatorname{im}(I_{\bar{n}}H^{2k}(X,\partial X;\mathbb{Q}) \xrightarrow{i} I_{\bar{m}}H^{2k}(X;\mathbb{Q}))$ with itself without even needing to invoke the Witt property! And the pairing is symmetric using the formula $\alpha \otimes \beta \to \mathbf{a}\mathcal{D}(\bar{\alpha} \smile \bar{\beta})$ of Proposition 8.4.23 together with the symmetry of the cup product. So in fact this definition of a signature works for any ∂-stratified pseudomanifold and any perversities with $\bar{p} \leq D\bar{p}$. We take up this idea below in Section 9.3.4. For now, however, our focus will remain on Witt and IP spaces, for which this signature has a useful alternative formulation that we shall arrive at in Corollary 9.3.14. We also observe that when X is $\mathbb{Q}$-Witt then Definitions 9.3.9 and 9.3.11 are consistent when $\partial X = \emptyset$ because in this case if X is $\mathbb{Q}$-Witt then $I_{\bar{n}}H^{2k}(X,\partial X;\mathbb{Q}) = I_{\bar{n}}H^{2k}(X;\mathbb{Q})$ and the map $I_{\bar{n}}H^{2k}(X;\mathbb{Q}) \xrightarrow{i} I_{\bar{m}}H^{2k}(X;\mathbb{Q})$ is an isomorphism. Furthermore, the image pairing is defined by applying the cup product to elements in the preimage, in this case $I_{\bar{n}}H^{2k}(X;\mathbb{Q})$. So when $\partial X = \emptyset$, Definition 9.3.11 reduces to Definition 9.3.9.

It turns out that there is a useful alternative approach to Witt signatures for $\mathbb{Q}$-Witt spaces with non-empty boundary. To explain this, let X be a compact oriented $4k$-dimensional $\mathbb{Q}$-Witt space with $\partial X \neq \emptyset$ (so $k > 0$), and let $X^+ = X \cup_{\partial X} \bar{c}(\partial X)$. Let v be the cone vertex. As $X^+ - \{v\} \cong X - \partial X$, all of the links of X^+ are links of X except for the link of v. But v has even codimension, and so X^+ is also a $\mathbb{Q}$-Witt space. In particular, $I_{\bar{m}}H^i(X^+;\mathbb{Q}) \cong I_{\bar{n}}H^i(X^+;\mathbb{Q})$, and we can compute these groups analogously to the computations in Examples 4.4.22 and 6.3.15, using the intersection cohomology cone computation of Proposition 7.1.5. As

$$\dim(X) - \bar{m}(\{v\}) - 1 = 4k - \left\lfloor \frac{4k-2}{2} \right\rfloor - 1 = 2k,$$

we obtain

$$I_{\bar m}H^i(X^+;\mathbb{Q}) \cong I_{\bar n}H^i(X^+;\mathbb{Q}) \cong \begin{cases} I_{\bar n}H^i(X,\partial X;\mathbb{Q}), & i > 2k, \\ \operatorname{im}(I_{\bar n}H^i(X,\partial X;\mathbb{Q}) \to I_{\bar n}H^i(X;\mathbb{Q})), & i = 2k, \\ I_{\bar n}H^i(X;\mathbb{Q}), & i < 2k. \end{cases}$$

So, in this case, the nonsingular cup pairing $I_{\bar n}H^{2k}(X^+;\mathbb{Q}) \otimes I_{\bar n}H^{2k}(X^+;\mathbb{Q}) \to \mathbb{Q}$ can be thought of as a pairing on $\operatorname{im}(I_{\bar n}H^k(X,\partial X;\mathbb{Q}) \to I_{\bar n}H^k(X;\mathbb{Q}))$. In fact, as we will now show, this is precisely the image pairing, up to identifying $I_{\bar n}H^*$ with $I_{\bar m}H^*$.

While we are interested primarily in $\mathbb{Q}$-Witt spaces, we proceed more generally in order to deduce a nice corollary about torsion in the boundary.

Proposition 9.3.12 *Let R be a Dedekind domain, and let X be a compact $2k$-dimensional oriented R-IP space with $\partial X \neq \emptyset$ and $I_{\bar n}H^k(\partial X;R)$-torsion-free. Let $X^+ = X \cup_{\partial X} \bar c(\partial X)$. Then the cup product pairing on X^+ is isomorphic to the image pairing on $\operatorname{im}(F(I_{\bar n}H^k(X,\partial X;R)) \xrightarrow{\mathfrak{i}} F(I_{\bar m}H^k(X;R)))$.*

Proof Let us first establish clearly the isomorphism between $I_{\bar m}H^k(X^+;R)$ and $\operatorname{im}(I_{\bar n}H^k(X,\partial X;R) \xrightarrow{\mathfrak{i}} I_{\bar m}H^k(X;R))$. For simplicity, we write all our maps in the following argument without the dual * label, e.g. we write $\mathfrak{i}$ rather than $\mathfrak{i}^*$.

The long exact sequences of the pairs give us the following cohomological analogue of the second diagram in Example 4.4.22, coefficients tacit:

$$\begin{array}{ccccccccc}
\longleftarrow & I_{\bar n}H^k(\partial X) & \longleftarrow & I_{\bar n}H^k(X) & \xleftarrow{\mathfrak{h}} & I_{\bar n}H^k(X,\partial X) & \xleftarrow{d^*} & I_{\bar n}H^{k-1}(\partial X) & \longleftarrow \\
 & \uparrow & & \uparrow{\scriptstyle \mathfrak{j}} & & \cong\uparrow{\scriptstyle \mathfrak{f}} & & \uparrow\cong & \\
\longleftarrow & 0 & \longleftarrow & I_{\bar n}H^k(X^+) & \xleftarrow{\mathfrak{g}} & I_{\bar n}H^k(X^+, c(\partial X)) & \xleftarrow{d^*} & I_{\bar n}H^{k-1}(c(\partial X)) & \longleftarrow .
\end{array}$$

We here use the open cone instead of the closed cone. To have $\partial X \subset cX$, we can let N be a filtered collar neighborhood of ∂X in X; then as $N \cup_{\partial X} \bar c(\partial X) \cong c(\partial X)$, we can just relabel $N \cup_{\partial X} \bar c(\partial X)$ as $c(\partial X)$. The 0 term in the bottom left is $I_{\bar n}H^k(c(\partial X);R)$, which is zero by the cohomology cone formula (Proposition 7.1.5): As $\dim(\partial X) = 2k-1$, we have $2k - \bar n(2k) - 1 = 2k - (k-1) - 1 = k$, and the Ext vanishes by our torsion-free assumption on ∂X. So, from the exactness of the rows, we have

$$\begin{aligned} I_{\bar n}H^k(X^+;R) &\cong \operatorname{cok}(I_{\bar n}H^{k-1}(c(\partial X);R) \xrightarrow{d^*} I_{\bar n}H^k(X^+,c(\partial X);R)) \\ &\cong \operatorname{cok}(I_{\bar n}H^{k-1}(\partial X;R) \xrightarrow{d^*} I_{\bar n}H^k(X,\partial X;R)) \\ &\cong \operatorname{im}(\mathfrak{h}). \end{aligned}$$

Some diagram chasing shows that this isomorphism is induced by the restriction map $\mathfrak{j}$ and that $\mathrm{im}(\mathfrak{j}) = \mathrm{im}(\mathfrak{h})$. Furthermore, we can extend the central part of the diagram to

$$\begin{array}{ccc} I_{\bar m}H^k(X) & & \\ \uparrow{\scriptstyle \cong\;\mathfrak{i}} & \nwarrow{\scriptstyle i} & \\ I_{\bar n}H^k(X) & \xleftarrow{\ \mathfrak{h}\ } & I_{\bar n}H^k(X,\partial X) \\ \uparrow{\scriptstyle \mathfrak{j}} & & \uparrow{\scriptstyle \cong\;\mathfrak{f}} \\ I_{\bar n}H^k(X^+) & \xleftarrow{\ \mathfrak{g}\ } & I_{\bar n}H^k(X^+,c(\partial X)), \end{array} \tag{9.4}$$

showing that we have isomorphisms $I_{\bar n}H^k(X^+;R) \xrightarrow{\mathfrak{j}} \mathrm{im}(\mathfrak{h}) \xrightarrow{\mathfrak{i}} \mathrm{im}(i)$.

To relate the image pairing on X to the cup product pairing on X^+, we also need the diagram

$$\begin{array}{ccccccc} \begin{array}{c} I_{\bar n}H^k(X,\partial X)\otimes \\ I_{\bar n}H^k(X,\partial X)\end{array} & \xrightarrow{\ \smile\ } & I_{\bar 0}H^{2k}(X,\partial X) & \xrightarrow{\ \mathcal{D}\ } & I^{\bar t}H_0(X) & & \\ \uparrow{\scriptstyle \cong\;\mathfrak{f}\otimes\mathfrak{f}} & & \uparrow{\scriptstyle \cong} & & \downarrow{\scriptstyle \cong} & \searrow{\scriptstyle \mathbf{a}} & \\ \begin{array}{c} I_{\bar n}H^k(X^+,c(\partial X))\otimes \\ I_{\bar n}H^k(X^+,c(\partial X))\end{array} & \xrightarrow{\ \smile\ } & I_{\bar 0}H^{2k}(X^+,c(\partial X)) & \xrightarrow{\ \mathcal{D}\ } & I^{\bar t}H_0(X^+) & \xrightarrow{\ \mathbf{a}\ } & R \\ \downarrow{\scriptstyle \mathfrak{g}\otimes\mathfrak{g}} & & \downarrow & & \uparrow{\scriptstyle =} & \nearrow{\scriptstyle \mathbf{a}} & \\ \begin{array}{c} I_{\bar n}H^k(X^+)\otimes \\ I_{\bar n}H^k(X^+)\end{array} & \xrightarrow{\ \smile\ } & I_{\bar 0}H^{2k}(X^+) & \xrightarrow{\ \mathcal{D}\ } & I^{\bar t}H_0(X^+). & & \end{array}$$

The left side of the diagram commutes by the naturality of the cup product (Proposition 7.3.5 and Theorem 7.3.72). The right-hand squares commute by the naturality of the cap product (Proposition 7.3.6 and Theorem 7.3.72), using that the fundamental classes of $(X,\partial X)$ and X^+ each map to the fundamental class of $(X^+,c(\partial X))$ (properly labeled $\Gamma_{X^+-c(\partial X)}$), as can be seen by applying Theorems 8.1.18 and 8.3.3. The left two upper vertical maps are isomorphisms by excision and stratified homotopy invariance, while the upper vertical map on the right is an isomorphism via Example 3.4.6. The commutativity of the triangles is straightforward. Furthermore, this diagram induces the analogous diagram on torsion-free quotients as in the proof of Theorem 8.4.7.

Now, let P denote the image pairing on X. Recall from Proposition 8.4.23 that if $\alpha,\beta \in \mathrm{im}(F(I_{\bar n}H^k(X,\partial X;\mathbb{R})) \to F(I_{\bar m}H^k(X;R)))$, then the image pairing can be computed as $P(\alpha,\beta) = \mathbf{a}\mathcal{D}(\bar\alpha \smile \bar\beta)$, where $\bar\alpha,\bar\beta \in I_{\bar n}H^k(X,\partial X;R)$ are preimages of α and β. Our diagram shows that this is equal to $\mathbf{a}\mathcal{D}(\mathfrak{g}\mathfrak{f}^{-1}(\bar\alpha) \smile \mathfrak{g}\mathfrak{f}^{-1}(\bar\beta))$.

But now Diagram (9.4) shows that our isomorphism $\mathfrak{lj} : I_{\bar{n}}H^k(X^+; R) \to \operatorname{im}(\mathfrak{i})$ must take $\mathfrak{gf}^{-1}(\bar{\alpha})$ to α up to torsion elements. So the assignment $\alpha \to \mathfrak{gf}^{-1}(\bar{\alpha})$ is an inverse to $\mathfrak{lj}$ modulo torsion. Thus we obtain an isomorphism of pairings

$$\begin{array}{ccc} \operatorname{im}(\mathfrak{i}) \otimes \operatorname{im}(\mathfrak{i}) & & \\ \uparrow{\scriptstyle \mathfrak{lj}\otimes\mathfrak{lj}} & \searrow & \\ & & R, \\ & \nearrow & \\ F(I_{\bar{n}}H^k(X^+; R)) \otimes F(I_{\bar{n}}H^k(X^+; R)) & & \end{array}$$

now letting $\operatorname{im}(\mathfrak{i})$ denote the image of the torsion-free quotient mapping. □

As the cup product pairing on IP spaces without boundary is nonsingular by Theorem 8.4.7, Proposition 9.3.12 has the following interesting corollary, which is well known in manifold theory by other means.

Corollary 9.3.13 *Let R be a Dedekind domain, and let X be a compact $2k$-dimensional oriented R-IP space with $\partial X \neq \emptyset$. If $I_{\bar{n}}H^k(\partial X; R)$ is torsion-free, then the image pairing on* $\operatorname{im}(F(I_{\bar{n}}H^k(X, \partial X; R)) \xrightarrow{\mathfrak{i}} F(I_{\bar{m}}H^k(X; R)))$ *is nonsingular. Contrapositively, if the image pairing on* $\operatorname{im}(\mathfrak{i})$ *fails to be nonsingular, then $I_{\bar{n}}H^k(\partial X; R)$ must have torsion.*

Putting the proposition together with Lemma 9.3.3 leads to the following alternative characterization of the Witt signature for $\mathbb{Q}$-Witt spaces with boundary.

Corollary 9.3.14 *Let X be a compact $4k$-dimensional oriented $\mathbb{Q}$-Witt space with $\partial X \neq \emptyset$, and let $X^+ = X \cup_{\partial X} \bar{c}(\partial X)$. Then $\sigma(X) = \sigma(X^+)$.*

Remark 9.3.15 An interesting feature of Witt signatures in the presence of boundaries appears when we consider the case where M is an unfiltered ∂-manifold with $\partial M \neq \emptyset$. Classically, the only way to define the signature of M within manifold theory was as the signature of the image pairing because M^+ is not generally a manifold. By contrast, Corollary 9.3.14 shows that in some sense the theory of Witt signatures for ∂-stratified pseudomanifolds is contained entirely within the theory of Witt signatures for stratified pseudomanifolds without boundaries.

On the other hand, Corollary 9.3.13 shows that when working with more refined invariants that can be defined using integer coefficients (such as the element of the Witt group $W(\mathbb{Z})$ represented by the cup product self-pairing of an IP space – see Section 9.5.2) then the image pairing for non-empty boundaries may indeed access some invariants that cannot be obtained from pseudomanifolds without boundary, as we have seen in Example 8.4.22 that there are image pairings that fail to be nonsingular over $\mathbb{Z}$.

(c) Topological Invariance of Witt Signatures

As the signature of a $\mathbb{Q}$-Witt space is defined in terms of the cup product pairing, which is a topological invariant in the absence of codimension-one strata and when utilizing

GM perversities (such as $\bar{m}$ and $\bar{n}$), it follows that the signature is an oriented topological invariant:

Theorem 9.3.16 *If X and Y are two compatibly-oriented (in the sense of Corollary 8.1.11) $\mathbb{Q}$-Witt spaces with $|X| = |Y|$, then $\sigma(X) = \sigma(Y)$. So the Witt signature can be considered an invariant of the underlying oriented pseudomanifold $|X|$.*

More generally, if X and Y are oriented $\mathbb{Q}$-Witt spaces and $f : |X| \to |Y|$ is an orientation-preserving[12] *topological homeomorphism, then $\sigma(X) = \sigma(Y)$, i.e. the Witt signature is an "oriented topological invariant" of $\mathbb{Q}$-Witt spaces.*

Proof Just as in Remarks 8.1.30, 8.2.7, and 8.3.13, we need only show that $\sigma(X) = \sigma(Y)$ when $|X| = |Y|$. So suppose $|X| = |Y|$.

As $\mathbb{Q}$-Witt spaces cannot have codimension-one strata, Proposition 2.7.4 implies that also $|\partial X| = |\partial Y|$. Therefore if we form X^+ and Y^+, we'll also have $|X^+| = |Y^+|$. By Corollary 9.3.14, to show that $\sigma(X) = \sigma(Y)$, it thus suffices to show that $\sigma(X^+) = \sigma(Y^+)$. This now follows essentially from the topological invariance of the cup product pairing given by Theorem 8.4.19. The pairings there are in terms of dual perversities, but we are free to use the agreeable triple $(\bar{n}, \bar{n}; \bar{0})$ rather than the agreeable triple $(\bar{n}, \bar{m}; \bar{0})$ where that proof invokes Theorem 7.3.10. □

9.3.2 Properties of Witt Signatures

The Witt signature possesses some fundamental properties that are well known from manifold theory:

Theorem 9.3.17 *If X, X' are closed oriented $\mathbb{Q}$-Witt spaces and $\amalg$ denotes disjoint union, then*

1. *if $-X$ denotes X but with the opposite orientation, then $\sigma(-X) = -\sigma(X)$,*
2. $\sigma(X \amalg X') = \sigma(X) + \sigma(X')$,
3. $\sigma(X \times X') = \sigma(X)\sigma(X')$,
4. *if X is the boundary of a compact oriented $\mathbb{Q}$-Witt space, then $\sigma(X) = 0$.*

Before proving the theorem, we note the following important corollary:

Corollary 9.3.18 *If X, X' are closed oriented $\mathbb{Q}$-Witt spaces and W is a $\mathbb{Q}$-Witt space with $\partial W = X \amalg -X'$, then $\sigma(X) = \sigma(X')$. In other words, σ is a* $\mathbb{Q}$-Witt bordism invariant.

Proof From the properties of Theorem 9.3.17, we have

$$0 = \sigma(\partial W) = \sigma(X \amalg -X') = \sigma(X) + \sigma(-X') = \sigma(X) - \sigma(X').$$ □

12 Meaning that the orientation on $|Y|$ induced from that of X by f is compatible with the given orientation of Y.

The proof of Theorem 9.3.17 is essentially identical to that for manifolds, and in fact, as manifolds are $\mathbb{Q}$-Witt spaces, our proof will reduce to the classical one in this case. For completeness, we provide all the details of this beautiful theorem, in four stages.

Proof of Theorem 9.3.17(1) Let M be a matrix for the cup product pairing on X with respect to some basis. We have $I^{\bar{m}}H_*(X;\mathbb{Q}) \cong I^{\bar{m}}H_*(-X;\mathbb{Q})$, so choosing the same basis for $-X$, the only difference in computing the pairing is reversing the sign of the fundamental class. So the pairing matrix for $-X$ is $-M$. The eigenvalues of $-M$ are the negatives of the eigenvalues of M, so $\sigma(-M) = -\sigma(M)$, and the result follows. □

Proof of Theorem 9.3.17.(2) As the cup product between an element of $I_{\bar{n}}H^*(X;\mathbb{Q})$ and an element of $I_{\bar{n}}H^*(X';\mathbb{Q})$ must be trivial, the pairing matrix must have block form $M = \begin{pmatrix} A & 0 \\ 0 & B \end{pmatrix}$, and so $\sigma(M) = \sigma(A) + \sigma(B)$, with A and B corresponding to the pairing matrices on X and X' respectively; see Lemma 9.3.4. □

Proof of Theorem 9.3.17(3) We first recall that the signature of a space that is not of dimension $4k$ is zero by definition. So if $\dim(X \times X')$ is odd, then so must be the dimension of one of X or X' and the statement is true trivially. If $\dim(X \times X') \equiv 2 \mod 4$, then it can't be that both $\dim(X) \equiv 0 \mod 4$ and $\dim(X') \equiv 0 \mod 4$, so again the statement holds trivially. So the only nontrivial situation is when $\dim(X \times X') \equiv 0 \mod 4$. Here we must look at the pairing.

By Example 6.4.11, the perversity $\bar{n}$ is $(\bar{n}, \bar{n})$-compatible on $X \times X'$. Therefore, as X and X' are compact, we have an isomorphism $I_{\bar{n}}H^*(X;\mathbb{Q}) \otimes_{\mathbb{Q}} I_{\bar{n}}H^*(X';\mathbb{Q}) \xrightarrow{\times} I_{\bar{n}}H^*(X \times X';\mathbb{Q})$ by the intersection cohomology Künneth Theorem (Theorem 7.3.63), using Corollary 6.3.40 for the finite generation condition. Therefore, each group $I_{\bar{n}}H^i(X \times X';\mathbb{Q})$ is generated by elements of the form $\alpha \times \beta$ with $\alpha \in I_{\bar{n}}H^a(X;\mathbb{Q})$ for some a and $\beta \in I_{\bar{n}}H^{i-a}(X';\mathbb{Q})$.

Assume $\dim(X \times X') = 4K$. Then we are interested in

$$I_{\bar{n}}H^{2K}(X \times X';\mathbb{Q}) \cong \bigoplus_{i+j=2K} I_{\bar{n}}H^i(X;\mathbb{Q}) \otimes_{\mathbb{Q}} I_{\bar{n}}H^j(X';\mathbb{Q}).$$

An important fact to observe is that many of the cup products of elements in $I_{\bar{n}}H^{2K}(X \times X';\mathbb{Q})$ are automatically zero. In fact, suppose $\alpha_1 \times \beta_1$ and $\alpha_2 \times \beta_2$ are two elements of $I_{\bar{n}}H^{2K}(X \times X';\mathbb{Q})$ corresponding to $\alpha_1 \otimes \beta_1 \in I_{\bar{n}}H^{i_1}(X;\mathbb{Q}) \otimes_{\mathbb{Q}} I_{\bar{n}}H^{j_1}(X';\mathbb{Q})$ and $\alpha_2 \otimes \beta_2 \in I_{\bar{n}}H^{i_2}(X;\mathbb{Q}) \otimes_{\mathbb{Q}} I_{\bar{n}}H^{j_2}(X';\mathbb{Q})$. Then by the interchange rule for cross products and cup products (which holds in this setting by Example 7.3.58),

$$(\alpha_1 \times \beta_1) \smile (\alpha_2 \times \beta_2) = (-1)^{j_1 i_2}(\alpha_1 \smile \alpha_2) \times (\beta_1 \smile \beta_2).$$

By Theorem 8.1.18.1 and the Universal Coefficient Theorem, the intersection cohomology of X is trivial above dimension $n = \dim(X)$ and the intersection cohomology of X' is trivial above dimension $n' = \dim(X')$. So this product will be zero if either $i_1 + i_2 > n$ or $j_1 + j_2 > n'$. But since we must have $i_1 + j_1 + i_2 + j_2 = 4K = n + n'$, we see that we can only have non-zero cup products if in fact $i_1 + i_2 = n$ and $j_1 + j_2 = n'$. Thus each summand

$W_{i,j} = I_{\bar{n}}H^i(X;\mathbb{Q}) \otimes_{\mathbb{Q}} I_{\bar{n}}H^j(X';\mathbb{Q})$ of $I_{\bar{n}}H^*(X \times X';\mathbb{Q})$ can have nontrivial cup products only with the complementary summand $W_{n-i,n'-j} = I_{\bar{n}}H^{n-i}(X;\mathbb{Q}) \otimes_{\mathbb{Q}} I_{\bar{n}}H^{n'-j}(X';\mathbb{Q})$. Therefore, the pairing matrix on $X \times X'$ has a block sum decomposition of the form

$$\begin{pmatrix} A & 0 & 0 & \cdots \\ 0 & B & 0 & \cdots \\ 0 & 0 & C & \cdots \\ \vdots & \vdots & \vdots & \ddots \end{pmatrix},$$

where each matrix on the diagonal is the restriction of the pairing to a subspace of the form $W_{i,j} \oplus W_{n-i,n'-j}$, with the exception of restriction of the pairing to the subspace $W_{n/2,n'/2}$, which pairs with itself if n and n' are even so that it exists. As the determinant of this matrix is equal to the product $\det(A)\det(B)\cdots$, by Lemma 9.3.5, each of these restricted pairings must be nonsingular in order for the full matrix to be nonsingular. By Lemma 9.3.4, the signature of $X \times X'$ will be the sum of the signatures of these diagonal blocks.

First, let's consider the pairings on $W_{i,j} \oplus W_{n-i,n'-j}$, where it is not the case that both $i = n/2$ and $j = n'/2$. By the separate dualities on the $\mathbb{Q}$-Witt spaces X and X', we have $\dim(I_{\bar{n}}H^i(X;\mathbb{Q})) = \dim(I_{\bar{n}}H^{n-i}(X;\mathbb{Q}))$ and $\dim(I_{\bar{n}}H^j(X';\mathbb{Q})) = \dim(I_{\bar{n}}H^{n'-j}(X';\mathbb{Q}))$. So $\dim(W_{i,j}) = \dim(W_{n-i,n'-j})$. But since we are assuming that we do not have both $i = n/2$ and $j = n'/2$, one of the following must be true:

- $i + i > n$,
- $(n - i) + (n - i) > n$,
- $j + j > n'$,
- $(n' - j) + (n' - j) > n'$.

Without loss of generality, let's assume that it's the first situation that holds. But then, again by Theorem 8.1.18(1), the cup product $I_{\bar{n}}H^i(X;\mathbb{Q}) \otimes I_{\bar{n}}H^i(X;\mathbb{Q}) \to I_{\bar{n}}H^{2i}(X;\mathbb{Q})$ must be trivial, so the cup product on $X \times X'$ restricted to $W_{i,j}$ must be trivial. Since $W_{i,j}$ and $W_{n-i,n'-j}$ have equal dimensions and the pairing is nonsingular, it follows from Lemma 9.3.6 that the signature of the pairing on $W_{i,j} \oplus W_{n-i,n'-j}$ is zero.

So the only possibly non-zero signature in our block decomposition of the cup product on $I_{\bar{n}}H^{2K}(X \times X';\mathbb{Q})$ comes from the self-pairing on $W_{n/2,n'/2}$. This summand can only exist if both n and n' are even, so if one of n or n' is odd, we must have $\sigma(X \times X') = 0$, which then agrees with $\sigma(X)\sigma(X')$. So now suppose that n and n' are both even. Since $n + n' = 4K$, either $n \equiv n' \equiv 0 \mod 4$ or $n \equiv n' \equiv 2 \mod 4$. In the latter case, $n/2$ and $n'/2$ are both odd. But that implies that the cup product pairing on $I_{\bar{n}}H^{n/2}(X;\mathbb{Q})$ is antisymmetric. So by Lemma 9.3.8, there is a subspace of $I_{\bar{n}}H^{n/2}(X;\mathbb{Q})$ of half its dimension on which the cup product pairing is trivial. Let $\{a_1, \ldots, a_m\}$ be a basis for this subspace. But then if $\{c_j\}$ is a basis for $I_{\bar{n}}H^{n'/2}(X';\mathbb{Q})$, the collection $\{a_i \otimes c_j\}$ is a basis for a subspace of half the dimension of $I_{\bar{n}}H^{n/2}(X;\mathbb{Q}) \otimes_{\mathbb{Q}} I_{\bar{n}}H^{n'/2}(X';\mathbb{Q})$. For the corresponding $a_i \times c_j$ we then have $(a_i \times c_j) \smile (a_k \times c_\ell) = \pm(a_i \smile a_k) \times (c_j \smile c_\ell) = 0$. So the pairing is trivial on this subspace, and the signature of $X \times X'$ is thus zero, which equals $\sigma(X)\sigma(X')$, which is a product of zeros.

The last remaining case is that for which $\dim(X) \equiv \dim(X') \equiv 0 \bmod 4$, and the signature of $X \times X'$ reduces to that of the cup product pairing on

$$I_{\bar{n}}H^{n/2}(X;\mathbb{Q}) \otimes_{\mathbb{Q}} I_{\bar{n}}H^{n'/2}(X';\mathbb{Q}).$$

In this case, the separate pairings, say $(\cdot,\cdot)_X$ and $(\cdot,\cdot)_{X'}$ on $I_{\bar{n}}H^{n/2}(X;\mathbb{Q})$ and $I_{\bar{n}}H^{n'/2}(X';\mathbb{Q})$, are symmetric and nondegenerate, so by Lemma 9.3.7 we can find respective orthogonal bases $\{a_1,\dots,a_{r+s}\}$ and $\{b_1,\dots,b_{r'+s'}\}$ such that $(a_i,a_i)_X > 0$ for $i \le r$, $(b_i,b_i)_{X'} > 0$ for $i \le r'$, $(a_i,a_i)_X < 0$ for $i > r$, and $(b_i,b_i)_{X'} < 0$ for $i > r'$. We then note that the collection $\{a_i \otimes b_j\}$ is a basis for $I_{\bar{n}}H^{n/2}(X;\mathbb{Q}) \otimes_{\mathbb{Q}} I_{\bar{n}}H^{n'/2}(X';\mathbb{Q})$ that is orthogonal with respect to the pairing on this space. Hence the corresponding pairing matrix M is diagonal, and we compute the signature by counting the positive and negative elements on the diagonal. If we put the basis in the dictionary order, then we can decompose M into blocks corresponding to the subspaces obtained by fixing an a_i and considering the span of basis elements $\{a_i \otimes b_1,\dots,a_i \otimes b_{r'+s'}\}$. If $(a_i,a_i)_X = m_i$, then this matrix has the form m_iB, where B is the pairing matrix for X' in the basis $\{b_i\}$. So if $m_i > 0$, the signature of this block is just $\sigma(X')$, and if $m_i < 0$, the signature of the block is $-\sigma(X')$. But then the signature of all of M is $\sum_{i=1}^{r+s}\operatorname{sgn}(m_i)\sigma(X') = \left(\sum_{i=1}^{r+s}\operatorname{sgn}(m_i)\right)\sigma(X') = \sigma(X)\sigma(X')$.

This completes the proof of this part of the theorem. □

Proof of Theorem 9.3.17(4) We need to establish that if $\dim(X) = 4k$ and $X = \partial W$ then $\sigma(X) = 0$. As the cup product pairing is nonsingular by Theorem 8.4.7, we will use Lemma 9.3.6 by finding a self-annihilating subspace A of $I_{\bar{n}}H^{2k}(X;\mathbb{Q})$ of half the dimension. In fact, we let A be the image of the restriction map $j^* : I_{\bar{n}}H^{2k}(W;\mathbb{Q}) \to I_{\bar{n}}H^{2k}(X;\mathbb{Q})$.

Let us first verify that A is self-annihilating. Let $\alpha,\beta \in I_{\bar{n}}H^{2k}(W;\mathbb{Q})$. From the definitions, we must compute

$$\mathbf{a}_X(((j^*\alpha) \smile (j^*\beta)) \frown \Gamma),$$

where Γ is the fundamental class of X and $\mathbf{a}_X$ is the augmentation on X. Notice that $\mathbf{a}_X$ factors as $I^{\bar{t}}H_0(X;\mathbb{Q}) \xrightarrow{j} I^{\bar{t}}H_0(W;\mathbb{Q}) \xrightarrow{\mathbf{a}_W} \mathbb{Q}$ with $\mathbf{a}_W$ the augmentation on W. Via the naturality properties of cup and cap products (Propositions 7.3.5 and 7.3.6 and Theorem 7.3.72), we compute

$$\begin{aligned}\mathbf{a}_X(((j^*\alpha) \smile (j^*\beta)) \frown \Gamma) &= \mathbf{a}_W j(((j^*\alpha) \smile (j^*\beta)) \frown \Gamma)\\ &= \mathbf{a}_W j(j^*(\alpha \smile \beta) \frown \Gamma)\\ &= \mathbf{a}_W((\alpha \smile \beta) \frown j\Gamma).\end{aligned}$$

But since $X = \partial W$, it follows from Proposition 8.3.5 that $\Gamma_X = \partial_*(\Gamma_W)$, and so $j\Gamma = 0$ by the long exact sequence of the pair. Therefore, the above expression is zero.

Next we must show that $\dim(A) = \frac{1}{2}\dim(I_{\bar{n}}H^{2k}(X;\mathbb{Q}))$. For this, consider the long exact sequence of the pair (W, X). Leaving coefficients tacit, this has the following form:

$$\begin{aligned} 0 &\longrightarrow I_{\bar{n}}H^0(W,X) \longrightarrow I_{\bar{n}}H^0(W) \longrightarrow I_{\bar{n}}H^0(X) \longrightarrow I_{\bar{n}}H^1(W,X) \\ &\longrightarrow I_{\bar{n}}H^1(W) \longrightarrow \cdots \longrightarrow I_{\bar{n}}H^{4k}(W) \\ &\longrightarrow I_{\bar{n}}H^{4k}(X) \longrightarrow I_{\bar{n}}H^{4k+1}(W,X) \longrightarrow I_{\bar{n}}H^{4k+1}(W) \longrightarrow 0. \end{aligned}$$

Now, by duality (Theorem 8.3.9), and as all spaces are Witt spaces, $I_{\bar{n}}H^0(W,X;\mathbb{Q})$ is dual to $I_{\bar{n}}H^{4k+1}(W;\mathbb{Q})$, $I_{\bar{n}}H^0(W;\mathbb{Q})$ is dual to $I_{\bar{n}}H^{4k+1}(W,X;\mathbb{Q})$, $I_{\bar{n}}H^0(X;\mathbb{Q})$ is dual to $I_{\bar{n}}H^{4k}(X;\mathbb{Q})$, and so on symmetrically inward till we arrive at $I_{\bar{n}}H^{2k}(X;\mathbb{Q})$, which is dual to itself. Since each pair of dual spaces has the same dimension, we can complete the argument using the following linear algebra lemma, taking $C_0 = I_{\bar{n}}H^{2k}(X;\mathbb{Q})$. This completes the proof of Theorem 9.3.17. □

Lemma 9.3.19 *Let*

$$\cdots \xrightarrow{d_2} C_1 \xrightarrow{d_1} C_0 \xrightarrow{d_0} C_{-1} \xrightarrow{d_{-1}} \cdots$$

be an exact *sequence of finite-dimensional vector spaces such that* $C_i = 0$ *for* $|i| > m$ *and* $\dim(C_i) = \dim(C_{-i})$ *for all* i. *Then* $\dim(\mathrm{im}(d_i)) = \dim(\mathrm{im}(d_{-i+1}))$ *for* $i > 0$, *and* $\dim(\mathrm{im}(d_1)) = \frac{1}{2}\dim(C_0)$.

Proof This is clearly true for $i > m$. We will apply downward induction.

As the base case, since $C_m \xrightarrow{d_m} C_{m-1}$ is injective, $\dim(\mathrm{im}(d_m)) = \dim(C_m)$. Similarly, since $C_{-m+1} \xrightarrow{d_{-m+1}} C_{-m}$ is surjective, $\dim(\mathrm{im}(d_{-m+1})) = \dim(C_{-m})$. But $\dim(C_m) = \dim(C_{-m})$, so this case holds.

Now, assume that the claim has been verified for $i > n > 0$, and consider $d_n : C_n \to C_{n-1}$. By elementary algebra, for any i we have

$$\dim(\mathrm{im}(d_i)) = \dim(\mathrm{coim}(d_i)) = \dim(\mathrm{cok}(d_{i+1})) = \dim(C_i) - \dim(\mathrm{im}(d_{i+1})).$$

So

$$\dim(C_i) = \dim(\mathrm{im}(d_i)) + \dim(\mathrm{im}(d_{i+1})).$$

Since $\dim(C_n) = \dim(C_{-n})$, we obtain

$$\dim(\mathrm{im}(d_n)) + \dim(\mathrm{im}(d_{n+1})) = \dim(\mathrm{im}(d_{-n+1})) + \dim(\mathrm{im}(d_{-n})).$$

By the induction hypothesis, $\dim(\mathrm{im}(d_{n+1})) = \dim(\mathrm{im}(d_{-n}))$, and therefore $\dim(\mathrm{im}(d_n)) = \dim(\mathrm{im}(d_{-n+1}))$. This establishes the first claim of the lemma by induction.

For the last claim, we have $\dim(C_0) = \dim(\mathrm{im}(d_0)) + \dim(\mathrm{im}(d_1))$. But we have established that $\dim(\mathrm{im}(d_1)) = \dim(\mathrm{im}(d_0))$, hence the dimension of each of these is half the dimension of C_0. □

9.3.3 Novikov Additivity

Another nice property of signatures is how they behave under gluing. In particular, let M_1, M_2 be $4k$-dimensional compact oriented ∂-manifolds with $\partial M_1 = -\partial M_2$, where we recall that $-X$ denotes X but with its orientation reversed. Let us call the common boundary N, and let $M = M_1 \cup_N M_2$. The union M is also a closed (compact without boundary) oriented manifold with the orientation that restricts to the given orientations on M_1 and M_2. By Definition 9.3.11, the ∂-manifolds M_1 and M_2 possess signatures $\sigma(M_1)$ and $\sigma(M_2)$ defined in terms of the image pairing on $\operatorname{im}(H^{2k}(M, \partial M; \mathbb{Q}) \to H^{2k}(M; \mathbb{Q}))$ (see Proposition 8.4.23). The following formula is called Novikov additivity:

$$\sigma(M) = \sigma(M_1) + \sigma(M_2).$$

The classical proof involves some homological algebra, Lefschetz duality, and algebraic properties of signatures; a readable account can be found in [10, pages 587–590]. Novikov additivity generalizes to signatures of Witt spaces, where the proof becomes a very pleasant consequence of a geometric argument and the basic properties of signatures that we have already studied. This observation is due to Siegel [218, Proposition II.3.1].

To prove Novikov additivity for signatures of $\mathbb{Q}$-Witt spaces, we begin with a minor lemma and corollary that are sometimes useful in other contexts:

Lemma 9.3.20 *Let Y be a $(2k-1)$-dimensional compact filtered space, and let SY be its suspension. Then $I^{\bar{m}}H_k(SY; G) = 0$ for any coefficient group G.*

Proof We can compute $I^{\bar{m}}H_k(SY; G)$ using the suspension formula of Theorem 6.3.13. By that theorem, the intersection homology of a suspension is always trivial in degree $2k - p - 1$, where p is the value of $\bar{m}$ at either suspension point of SY. As the suspension points have codimension $2k$, we have

$$2k - p - 1 = 2k - \left\lfloor \frac{2k-2}{2} \right\rfloor - 1 = k. \qquad \square$$

Corollary 9.3.21 *Suppose X is a $(4k-1)$-dimensional stratified pseudomanifold and that SX is an orientable $\mathbb{Q}$-Witt space. Then $\sigma(SX) = 0$.*

Proof From Poincaré duality (Theorem 8.2.4), we obtain $I_{\bar{n}}H^{2k}(SX; \mathbb{Q}) \cong I^{\bar{m}}H_{2k}(SX; \mathbb{Q})$, which is zero by the lemma. $\square$

Now here is Siegel's argument for Novikov additivity of $\mathbb{Q}$-Witt spaces:

Theorem 9.3.22 (Novikov additivity) *Suppose $X = X_1 \cup X_2$ is a compact oriented $4k$-dimensional $\mathbb{Q}$-Witt space with $X_1, X_2 \subset X$ compact oriented ∂-stratified pseudomanifolds such that $X_1 \cap X_2 = \partial X_1 = -\partial X_2$. Then $\sigma(X) = \sigma(X_1) + \sigma(X_2)$.*

Proof As the links of X_1 and X_2 must also be links of X, the subspaces X_1 and X_2 are also $\mathbb{Q}$-Witt, and they inherit orientations from X. Let Y denote the common boundary of X_1 and X_2, ignoring orientation.

By Proposition 9.3.12, the signatures $\sigma(X_1)$ and $\sigma(X_2)$ are equal to the Witt signatures of X_1^+ and X_2^+, respectively, where $X_i^+ = X_i \cup_{\partial X_i} \bar{c}(\partial X_i)$. We will construct a $\mathbb{Q}$-Witt space W such that $\partial W = X_1^+ \amalg X_2^+ \amalg -X$. The theorem will then follow from Theorem 9.3.17.

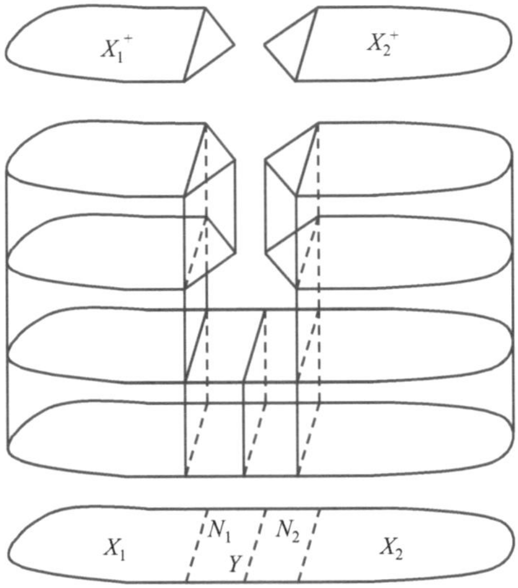

Figure 9.1 A schematic construction of W_0 with $X_1 \cup_Y X_2$ shown at the bottom and $X_1^+ \amalg X_2^+$ at the top. Note that X_1 and X_2 are intended to be glued along their entire boundaries, not just a piece of each boundary.

As X_1 and X_2 are ∂-stratified pseudomanifolds, there are closed filtered collar neighborhoods of their boundaries, say N_1 and N_2. Let $N = N_1 \cup_Y N_2$. Then $N \cong [-1,1] \times Y$; let $\mathring{N} = N - \partial N \cong (-1,1) \times Y$. Notice that $X - \mathring{N}$ can also be identified with a subset of $X_1^+ \amalg X_2^+$. We construct a preliminary space W_0 as follows (see Figure 9.1):

$$W_0 = ([0,1] \times X) \cup_{\{1\}\times(X-\mathring{N})} \left([1,2] \times (X - \mathring{N})\right) \cup_{\{2\}\times(X-\mathring{N})} ([2,3] \times (X_1^+ \amalg X_2^+)).$$

Alternatively, we can imagine beginning with $[0,3] \times (X - \mathring{N})$, but then over $[0,1]$ we fill back in the rest of $[0,1]\times X$ and over $[2,3]$ we fill back in the rest of $X_1^+ \amalg X_2^+$. Our new space W_0 is a ∂-stratified pseudomanifold with three boundary pieces: $\{0\} \times X$, $\{3\} \times (X_1^+ \amalg X_2^+)$, and a third piece that has the form

$$(\{1\} \times N) \cup_{\{1\}\times\partial N} ([1,2] \times \partial N) \cup_{\{2\}\times\partial N} \left(\{2\} \times ((N_1 \cup_Y \bar{c}(Y)) \amalg (N_2 \cup_Y \bar{c}(Y)))\right).$$

Note that this last piece is homeomorphic to the suspension SY, and we will identify it with SY notationally in what follows. The space W_0 is also $\mathbb{Q}$-Witt as all of the links in W_0 are all links of X, X_1^+, or X_2^+. To show that W_0 is really a ∂-stratified pseudomanifold, we should check that the boundary has a filtered collar. This is clear for the "top" and "bottom" pieces that are filtered homeomorphic to X and $X_1^+ \amalg X_2^+$. It is not too hard to see that SY also has a filtered collar, which can be obtained using the obvious collars on the pieces and then appropriately bending. The basic idea is that the "corners" have neighborhoods filtered homeomorphic to $Y \times ([(-1,1)\times(-1,1)] - [(0,1)\times(0,1)])$ (with the second factor unfiltered), which can be "unbent" to $Y \times (-1,1) \times (0,1]$; see Figure 9.2.

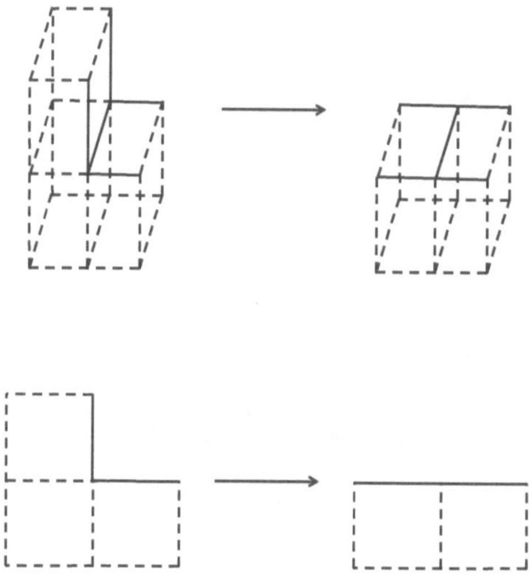

Figure 9.2 Schematic of unfolding at corners.

Now define $W = W_0 \cup_{SY} \bar{c}(SY)$; see Figure 9.3. Taking orientations into account, we then have $\partial W \cong X_1^+ \amalg X_2^+ \amalg -X$. Furthermore, W is $\mathbb{Q}$-Witt: the only point of W that has a link possibly different from the links in W_0 is the new cone vertex, whose link is SY, but $I^{\bar{m}}H_{2k}(SY) = 0$ by Lemma 9.3.20. □

Remark 9.3.23 In the statement of Novikov additivity, the boundary of M_1 must be glued completely to the boundary of M_2. Novikov additivity does not remain true if one glues boundaries only partially. In other words, suppose ∂M_1 and ∂M_2 are not necessarily equal any more but that ∂M_1 and ∂M_2 can be decomposed as $\partial M_1 = N_0 \cup_P N_1$ and $\partial M_2 = N_0 \cup_P N_2$, where N_0, N_1, N_2 are compact oriented $(4k-1)$-dimensional ∂-manifolds all with common boundary P (up to matching the orientations); see Figure 9.4. One could then form $M = M_1 \cup_{N_0} M_2$, which is a ∂-manifold with boundary $\partial M = N_1 \cup_P N_2$. In this case, it is *not* true in general that $\sigma(M) = \sigma(M_1) + \sigma(M_2)$. In fact, this is a good thing: if Novikov additivity held in this more general way, then every triangulable manifold would have signature zero: Every such manifold can be constructed by a sequence of such partial gluings, in this case the gluings that attach the simplices together, but in dimensions > 0 the signature of a simplex is trivial!

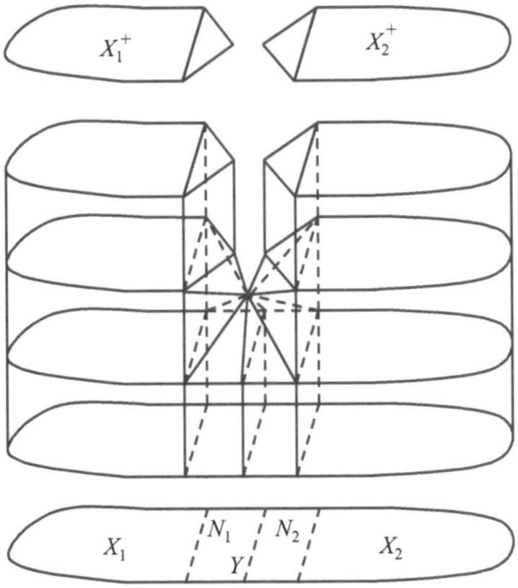

Figure 9.3 Coning off $\bar{c}(SY)$ to get the full cobordism W from $X_1 \cup_Y X_2$ to $X_1^+ \amalg X_2^+$.

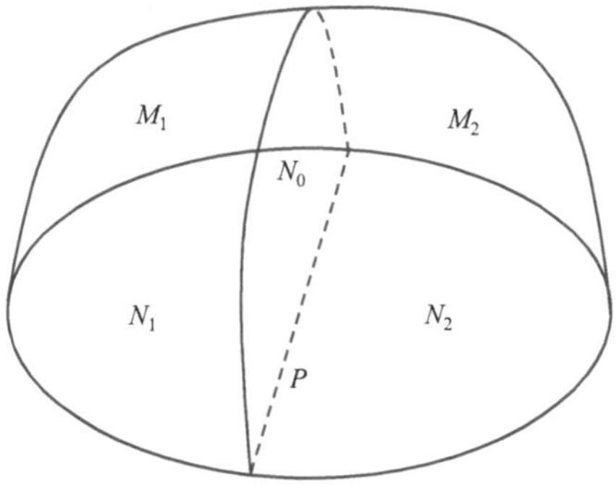

Figure 9.4 Gluing M_1 and M_2 along the shared partial boundary N_0.

It turns out that there is a formula for the signature of M due to Wall [237], but it involves a correction term

$$\sigma(M) = \sigma(M_1) + \sigma(M_2) + \mu(M; M_1, M_2).$$

This formula is sometimes called "Wall non-additivity." Of course there is always a correction formula; we could just define the correction term to be

$$\mu(M; M_1, M_2) = \sigma(M) - \sigma(M_1) - \sigma(M_2).$$

The interesting thing is that this particular correction term is computable as a *Maslov index*. The Maslov index is another linear algebra invariant a bit more complex than the signature. It takes as input a vector space V with a nonsingular antisymmetric pairing together with three Lagrangian subspaces – subspaces of half the dimension of V on which the pairing vanishes identically. In this case, converting the homological language that Wall uses to our cohomological language, we have $V = H^{2k-1}(P;\mathbb{Q})$, the pairing on V is the cup product pairing, and the Lagrangian subspaces are the subspaces $\operatorname{im}(H^{2k-1}(N_i;\mathbb{Q}) \to H^{2k-1}(P;\mathbb{Q}))$, $i = 0, 1, 2$. We will not get into the details further here, but rather recommend the very readable account in[13] [237]. There is also an intersection homology version of Wall non-additivity, at least for PL Witt spaces, which can be found in [96], and which also provides a review of the needed Maslov index construction. A very nice broader survey of Maslov indices and their applications can be found in [42].

9.3.4 Perverse Signatures

We have seen that $\mathbb{Q}$-Witt spaces carry signatures. What about more general spaces? In [134], Hunsicker noticed that it is possible to define a signature on *any* $4k$-dimensional compact oriented ∂-stratified pseudomanifold given to any perversity $\bar{p}$ such that $\bar{p} \le D\bar{p}$. In the notation of [96], if we have perversities $\bar{p}, \bar{q}$ with $\bar{p} \le \bar{q}$ and $\bar{q} = D\bar{p}$ then the *perverse signature* is denoted $\sigma_{\bar{p}\to\bar{q}}(X)$. With our notation here, this would be denoted $\sigma_{\bar{p}\to D\bar{p}}(X)$, and so perhaps it makes sense to shorten the notation to just $\sigma_{\bar{p}}(X)$, as the second perversity is determined by the first.

Remark 9.3.24 Such perversities with $\bar{p} \le D\bar{p}$ do exist. For example $\bar{m} \le D\bar{m} = \bar{n}$. We also have $\bar{0} \le D\bar{0} = \bar{t}$, so long as we are on a space X with no codimension-one strata. For such perversities, we observe that since $\bar{p} \le D\bar{p} = \bar{t} - \bar{p}$, we have $2\bar{p} \le \bar{t}$ and hence $D(2\bar{p}) \ge D\bar{t} = \bar{0}$.

The perverse signature is defined to be the signature of the nonsingular symmetric image pairing on $\operatorname{im}(i^* : I_{D\bar{p}}H^{2k}(X, \partial X;\mathbb{Q}) \to I_{\bar{p}}H^{2k}(X;\mathbb{Q}))$, where the map i^* is induced by the space pair inclusion $(X, \emptyset) \to (X, \partial X)$, which is also $(\bar{p}, D\bar{p})$-allowable using that $\bar{p} \le D\bar{p}$. In the special case where $X = M$ is an unfiltered manifold, this is just the classical image pairing on $\operatorname{im}(i^* : H^{2k}(M, \partial M;\mathbb{Q}) \to H^{2k}(M;\mathbb{Q}))$; if X is a $\mathbb{Q}$-Witt space, then $\sigma_{\bar{m}}(X)$ is just the Witt signature defined in the preceding section.

So the perverse signature is a legitimate generalization of the Witt signature. Unfortunately, its properties and applications mostly remain mysterious. Some basic properties are obvious. For example, by the arguments in the proof of Theorem 9.3.17, if $-X$ is the same pseudomanifold as X but with the opposite orientation then $\sigma_{\bar{p}}(-X) = -\sigma_{\bar{p}}(X)$, and also $\sigma_{\bar{p}}(X \amalg X') = \sigma_{\bar{p}}(X) + \sigma_{\bar{p}}(X')$. However, we also saw in Corollary 9.3.18 that the Witt signature is a bordism invariant. But while the perverse signatures are defined on all oriented stratified pseudomanifolds, they cannot be bordism invariants in this larger class as every compact orientable stratified pseudomanifold X of dimension > 0 is the boundary of

[13] Though be aware of some typos in [237] which are pointed out in [96].

the compact oriented ∂-stratified pseudomanifold $\bar{c}(X)$. So if the perverse signatures were bordism invariants, they would all have to be zero. But manifolds are (trivially stratified) pseudomanifolds, and there are certainly manifolds with non-zero signatures. For example, recalling that $\mathbb{C}P^m$ is a real manifold of dimension $2m$ and that $H^m(\mathbb{C}P^m) \cong \mathbb{Z}$, the non-singularity of the cup product pairing implies that $\sigma(\mathbb{C}P^{2n}) = \pm 1 \neq 0$; in fact, one can show that $\sigma(\mathbb{C}P^{2n}) = 1$ with its standard orientation as a complex variety [177, page 225]. So if the various $\sigma_{\bar{p}}$ are to be interesting bordism invariants, then they must be bordism invariants only on some restricted class of oriented pseudomanifolds, but just which class has not yet been determined.

It is also not true, in general, that the perverse signatures satisfy Novikov additivity, but they do have a Wall non-additivity formula, which can be found in [96].

9.4 *L*-Classes

A remarkable fact about signatures, which are numerical invariants, is that they provide the means to define homology invariants of spaces. To see why this should be so, let us perform a thought experiment with the warning that the details will be largely incorrect but with the enticement that they can be modified to obtain something useful.

Suppose X is a space, and let's imagine that all its homology classes can be represented by oriented manifolds. In other words, suppose that if $\xi \in H_k(X)$ then there exists a closed oriented k-dimensional manifold M^k and map $f : M^k \to X$ such that ξ is the image of the fundamental class of M under f. Furthermore, let's assume that all homologies in X can be realized by manifold bordism. In other words, suppose that if $f : M \to X$ and $f' : M' \to X$ represent the same homology class then there is an oriented ∂-manifold W^{k+1} and a map $W^{k+1} \to X$ that restricts to $f \amalg f'$ on $\partial W = M \amalg -M'$. (For more about homology theories that do behave this way, see Section 9.5, below.) Pretending that ordinary homology works this way, we could assign to ξ, represented by $f : M \to X$, the signature of M, and it would follow from our assumptions that this assignment $\xi \to \sigma(M)$ is well defined, since if we represent ξ by $f' : M' \to X$ instead, then M and M' cobound some W and hence have the same signature. It is not difficult to see that this would induce a homomorphism $H_k(X) \to \mathbb{Z}$, since reversing the orientation of a manifold changes the sign of its signature and signature is additive over disjoint unions. Tensoring with $\mathbb{Q}$ makes this a homomorphism $H_k(X; \mathbb{Q}) \to \mathbb{Q}$, i.e. an element of $\operatorname{Hom}(H_k(X; \mathbb{Q}), \mathbb{Q})$, which can then be identified as an element of $H^k(X; \mathbb{Q})$. So, starting with signatures of manifolds, we obtain an element of $H^k(X; \mathbb{Q})$.

One major problem with this argument is that our assumption that we can identify homology classes in terms of images of fundamental classes of manifolds is simply not true, even when X is itself a manifold. This was historically an important question, which was answered in the negative by Thom [235]. However, there turns out to be another interesting way to find manifolds and bordisms within a closed (i.e. compact without boundary) oriented

PL manifold M, but in terms of rational *co*homology rather than homology. The idea, roughly, is that there turns out to be a correspondence between elements of $H^m(M;\mathbb{Q})$ and homotopy classes of maps $M \to S^m$ in certain dimension ranges[14]. With certain further conditions, the inverse images of generic points of such maps are embedded manifolds in M; replacing such maps with homotopic maps replaces the embedded manifolds with bordant manifolds possessing the same signatures. So taking signatures of these embedded manifolds determines an element of $\mathrm{Hom}(H^m(M;\mathbb{Q});\mathbb{Q})$. Since M is compact, we have $\mathrm{Hom}(H^m(M;\mathbb{Q});\mathbb{Q}) \cong H_m(M;\mathbb{Q})$, and so we wind up with a class $\mathscr{L}_m(M) \in H_m(M;\mathbb{Q})$. If $\dim(M) = n$, the embedded manifolds have dimension $n-m$, and so $\mathscr{L}_m(M)$ will be nontrivial only when $n-m$ is a multiple of 4.

If we make the stronger assumption that M is a smooth manifold, the classes $\mathscr{L}_{n-4k}(M) \in H_{n-4k}(M;\mathbb{Q})$ turn out to be Poincaré dual to the classical Thom–Hirzebruch L-classes[15] $L^k(M) \in H^{4k}(M;\mathbb{Q})$. Recall that these classes arise as certain polynomials in the Pontrjagin characteristic classes of the tangent bundle of M; see [177, Chapter 19]. The cohomology L-classes possess the property that they evaluate on the fundamental class Γ_M to the signature of M, i.e.

$$(L^*(M))(\Gamma_M) = \sigma(M).$$

This is the Hirzebruch Signature Theorem, for which a proof can be found in[16] [177, Chapter 19]. The Signature Theorem motivates the construction of homology L-classes for PL manifolds just described, which is due to Thom [234]; an exposition is contained in Chapter 20 of Milnor and Stasheff [177].

The construction of homology L-classes using signatures that we have just outlined can further be extended to closed oriented PL $\mathbb{Q}$-Witt spaces[17]. In this setting, it is still possible to identify ordinary rational cohomology with maps to spheres (in fact, this part of the argument works for any space of the homotopy type of a CW complex; see [125]), but now the inverse images of generic points will be PL $\mathbb{Q}$-Witt spaces. As these have Witt signatures, and since these signatures are invariants of $\mathbb{Q}$-Witt bordism, we can enact our program with a Witt space X to get elements of $H_*(X;\mathbb{Q})$ that deserve to be called characteristic classes. This basic idea was described already by Goresky and MacPherson in [105] for stratified PL pseudomanifolds with only even-codimension strata, before the discovery of Witt spaces in [218]. Further generalizations to "twisted L-classes" and to more general spaces have been carried out by Cappell and Shaneson [47], Banagl, Cappell and Shaneson [20], and Banagl [14].

In Section 9.4.1, we will carefully formulate the details of the construction of L-classes

[14] Technically the group of such homotopy classes of maps to S^m also needs to be tensored with $\mathbb{Q}$; details will be provided below.

[15] Take note of the standard labeling with the class denoted L^k living in $H^{4k}(M;\mathbb{Q})$.

[16] See [128] for a very nice discussion by Hirzebruch of what led him to conjecture and prove this theorem.

[17] Note that although we will be assuming PL spaces throughout this section, we will not need to utilize PL intersection homology or cohomology. Rather, we are free to work with singular intersection (co)homology and so to utilize our results about duality, signatures, etc.

for PL $\mathbb{Q}$-Witt spaces, which include PL manifolds, and in the following sections we will provide the proofs. We will generally follow the exposition of [177, Chapter 20], though we adapt several results as necessary for our needs and fill in some additional details.

9.4.1 Outline of the Construction of L-Classes (without Proofs)

In this section, we describe in detail the construction of the L-classes $\mathscr{L}_m(X) \in H_m(X;\mathbb{Q})$ for a closed oriented n-dimensional PL $\mathbb{Q}$-Witt space X. These will be nontrivial only when m has the form $m = n - 4k$ for $k \in \mathbb{Z}$. The construction relies on a variety of results that will be stated here and then proven in subsequent sections.

(a) Maps to Spheres and Embedded Subspaces

We begin with the following proposition, which implies that, in suitable situations, we can assign Witt signature invariants to maps $f : X \to S^m$.

Proposition 9.4.1 *Let X be a closed oriented n-dimensional PL $\mathbb{Q}$-Witt space. Suppose that S^m, $m > 0$, has been given an orientation and that $f : X \to S^m$ is a PL map. Then for almost all $y \in S^m$ the inverse image $f^{-1}(y)$ can be filtered as a closed oriented $(n-m)$-dimensional PL $\mathbb{Q}$-Witt space (possibly empty) embedded in X. Furthermore, for almost all $y, y' \in S^m$ the Witt spaces $f^{-1}(y)$ and $f^{-1}(y')$ have the same signature; this common signature depends only on the PL homotopy class of f in $[X, S^m]_{\mathrm{PL}}$.*

Here, $[\cdot,\cdot]_{\mathrm{PL}}$ denotes the set of PL homotopy classes of PL maps, and, as in [177], "almost all y" means that the statement applies to all y in an open dense subset of S^m. In particular, we will see that for a fixed f we can choose y to be any point not belonging to the simplicial $(m-1)$-skeleton of a triangulation of S^m with respect to which $f : X \to S^m$ is simplicial. We also refer to these as "generic points of S^m."

The basic idea of the proof of the second statement of the proposition is that changing from y to y' or from f to a homotopic map will result in PL $\mathbb{Q}$-Witt space bordisms between the point inverses. Since signature is independent of filtration and $\mathbb{Q}$-Witt bordism class by Theorem 9.3.16 and Corollary 9.3.18, almost all point inverses will have the same signature.

Let us briefly describe how the orientation of $f^{-1}(y)$ is chosen. We will see in Lemma 9.4.19 that for almost all $y \in S^m$ there is a Euclidean neighborhood U of y and a homeomorphism $h : U \times f^{-1}(y) \to f^{-1}(U)$ such that fh is equal to the projection to the first factor. We will choose the orientation of $f^{-1}(y)$ so that h is orientation-preserving, giving U the orientation it inherits from the chosen orientation on S^m.

We can extend Proposition 9.4.1 to $m = 0$ by letting $S^0 = \{0, 1\}$ and declaring $y = 1$ to be the "generic point." Every map $f : X \to S^0$ takes some connected components of f to $0 \in S^0$ and some to $1 \in S^0$. So then $Z = f^{-1}(y)$ is a union of connected components of X, which of course is a closed oriented PL $\mathbb{Q}$-Witt space with the orientation it inherits from X, and we may consider its signature $\sigma(Z)$.

(b) Cohomotopy

If X is a $\mathbb{Q}$-Witt space, Proposition 9.4.1 shows how to assign an integer $\sigma(f)$ to each element of $[X, S^m]_{\text{PL}}$, the PL homotopy set of PL maps from X to S^m. If we have a map $g : X \to S^m$ that is not necessarily PL, then by the PL Approximation Theorem (see Theorem B.2.24), g is homotopic (by a small homotopy) to a PL map. Furthermore, by the same theorem, any continuous homotopy of PL maps is homotopic to a PL homotopy of the same PL maps. Therefore, $[X, S^m]_{\text{PL}} \cong [X, S^m]$, the full set of topological homotopy classes of maps $X \to S^m$. So we in fact obtain a well-defined function $[X, S^m] \to \mathbb{Z}$.

For obvious reasons, $[X, S^m]$ is also called the *cohomotopy* set $\pi^m(X)$. Since S^m is simply connected for $m > 1$ and since S^1 is an H-space, we have $[X, S^m] \cong [X, S^m]_0$, the basepoint-preserving homotopy set, by [125, Proposition 4A.2 and Example 4A.3]. So for $m > 0$ we do not need to take care with basepoints when discussing $\pi^m(X)$.

Putting together the discussion so far, we have functions

$$F : \pi^m(X) \to \mathbb{Z},$$

for $m > 0$, such that $[f] \in \pi^m(X)$ gets taken to the common value $\sigma(f^{-1}(y))$ for the generic y in S^m. In general we can only call this a function as $\pi^m(X)$ is not always a group. However, it turns out that $\pi^m(X)$ is a group when m is sufficiently large compared to $n = \dim(X)$:

Lemma 9.4.2 *If $m > (n+1)/2$ then $F : \pi^m(X) \to \mathbb{Z}$ is a homomorphism of abelian groups.*

So now we have constructed for any closed oriented PL $\mathbb{Q}$-Witt space X and any $m > (n+1)/2$ a homomorphism $F : \pi^m(X) \to \mathbb{Z}$, which, by tensoring with $\mathbb{Q}$ over $\mathbb{Z}$, can be made into a map $F \otimes \text{id}_{\mathbb{Q}} : \pi^m(X) \otimes \mathbb{Q} \to \mathbb{Q} \otimes \mathbb{Q} \cong \mathbb{Q}$. On the other hand, there is a homomorphism $\mathfrak{c} : \pi^m(X) \to H^m(X)$ that, in this degree range, becomes an isomorphism when tensored with $\mathbb{Q}$. This follows from deep work of Serre [214, Proposition 2′, page 289]:

Theorem 9.4.3 (Serre) *Let S^m be oriented, and let $u \in H^m(S^m)$ be the generator satisfying $u(\Gamma_{S^m}) = 1$. Suppose X^n is a compact CW complex and $m > (n+1)/2$. Define $\mathfrak{c} : \pi^m(X) \to H^m(X)$ by $\mathfrak{c}([f]) = f^*(u) \in H^m(X)$. Then $\mathfrak{c} \otimes \text{id}_{\mathbb{Q}} : \pi^m(X) \otimes \mathbb{Q} \to H^m(X) \otimes \mathbb{Q}$ is an isomorphism.*

(c) The L-Classes

We can now define the L-classes for $m > (n+1)/2$; the extension for all m will require a bit more work below.

Let X^n be a closed oriented PL $\mathbb{Q}$-Witt space, let $m > (n+1)/2$, and consider the composite homomorphism

$$H^m(X) \otimes \mathbb{Q} \xrightarrow{(\mathfrak{c} \otimes \text{id}_{\mathbb{Q}})^{-1}} \pi^m(X) \otimes \mathbb{Q} \xrightarrow{F \otimes \text{id}_{\mathbb{Q}}} \mathbb{Q},$$

which is an element of $\operatorname{Hom}_{\mathbb{Q}}(H^m(X) \otimes \mathbb{Q}, \mathbb{Q})$. A standard algebra identity says that $\operatorname{Hom}_{\mathbb{Q}}(H^m(X) \otimes_{\mathbb{Z}} \mathbb{Q}, \mathbb{Q}) \cong \operatorname{Hom}_{\mathbb{Z}}(H^m(X), \mathbb{Q})$ (see [182, Lemma 53.4]), and since X is compact, and thus $H^*(X)$ is finitely generated in each degree, the Universal Coefficient Theorem [182, Theorem 56.1] says that the evaluation homomorphism gives an isomorphism $\operatorname{Hom}_{\mathbb{Z}}(H^m(X), \mathbb{Q}) \cong H_m(X; \mathbb{Q})$. So $(F \otimes \mathrm{id}_{\mathbb{Q}})(\mathfrak{c} \otimes \mathrm{id}_{\mathbb{Q}})^{-1}$ corresponds to an element of $H_m(X; \mathbb{Q})$. This will be our L-class, up to a sign we introduce to better reconcile our sign conventions with the classical results about L-classes. It is convenient to apply [182, Theorem 56.1] once again to observe $H^m(X) \otimes \mathbb{Q} \cong H^m(X; \mathbb{Q})$ so that we can think of the evaluation map as running $H_m(X; \mathbb{Q}) \to \operatorname{Hom}_{\mathbb{Q}}(H^m(X; \mathbb{Q}), \mathbb{Q})$ in the following official definition.

Definition 9.4.4 If X^n is a closed oriented PL $\mathbb{Q}$-Witt space and $m > (n+1)/2$, let $\mathscr{L}_m(X) \in H_m(X; \mathbb{Q})$ be $(-1)^m$ times the element corresponding to $(F \otimes \mathrm{id}_{\mathbb{Q}})(\mathfrak{c} \otimes \mathrm{id}_{\mathbb{Q}})^{-1}$ under the universal coefficient evaluation isomorphism

$$\mathrm{ev} : H_m(X; \mathbb{Q}) \to \operatorname{Hom}_{\mathbb{Q}}(H^m(X; \mathbb{Q}), \mathbb{Q}).$$

In other words,

$$\mathscr{L}_m(X) = (-1)^m \mathrm{ev}^{-1}((F \otimes \mathrm{id}_{\mathbb{Q}}) \circ (\mathfrak{c} \otimes \mathrm{id}_{\mathbb{Q}})^{-1}) \in H_m(X; \mathbb{Q}).$$

The class $\mathscr{L}_m(X)$ is called the mth homology L-class of X.

Remark 9.4.5 If $n - m$ is not a multiple of 4, then the signature of any point inverse of any map $X \to S^m$ must be zero by definition. In this case the L-class is trivial. Therefore the L-classes are typically only defined in dimensions $m = n - 4k$, $k \geq 0$.

Remark 9.4.6 Taking into account Koszul sign conventions, the evaluation isomorphism $\mathrm{ev} : H_m(X; \mathbb{Q}) \to \operatorname{Hom}(H^m(X; \mathbb{Q}), \mathbb{Q})$ of the Universal Coefficient Theorem takes an element $\xi \in H_m(X; \mathbb{Q})$ to a homomorphism that acts on the class of a cocycle α by

$$\mathrm{ev}([\xi])[\alpha] = (-1)^m \alpha(\xi);$$

see Section A.1.5. This sign, which often does not appear in the literature, is essentially the reason for our extra sign in Definition 9.4.4. This sign yields a more familiar formula in Proposition 9.4.8 below, which relates these homology L-classes with the Thom–Hirzebruch L-classes in cohomology when X is a smooth manifold.

As our definition for $\mathscr{L}_m(X)$ looks pretty mysterious, let's see what this all actually means in terms of our constructions to this point. In particular, let's compute $\beta(\mathscr{L}_m(X))$ for an arbitrary $\beta \in H^m(X; \mathbb{Q})$. By Serre's Theorem (Theorem 9.4.3), we know that such a β is in the image of $\mathfrak{c} \otimes \mathrm{id}_{\mathbb{Q}}$. If G is any abelian group, then by finding common denominators and pulling numerators across the tensor product, any element of $G \otimes \mathbb{Q}$ can be written in the form $g \otimes r$ for some[18] $g \in G$ and $r \in \mathbb{Q}$. If we abbreviate such an element as rg, then, abusing

[18] For example, the expression $g_1 \otimes \frac{1}{2} + g_2 \otimes \frac{2}{3}$ can be rewritten as $g_1 \otimes \frac{3}{6} + g_2 \otimes \frac{4}{6} = (3g_1 + 4g_2) \otimes \frac{1}{6}$.

notation, we can write $\beta \in H^m(M;\mathbb{Q})$ as $\beta = (\mathfrak{c} \otimes \mathrm{id}_{\mathbb{Q}})([f] \otimes r) = \mathfrak{c}([f]) \otimes r = r\mathfrak{c}([f])$ for some PL $f : M \to S^m$ and $r \in \mathbb{Q}$. With these assumptions, we can compute

$$\begin{aligned}
\beta(\mathscr{L}_m(X)) &= (-1)^m \mathrm{ev}(\mathscr{L}_m)(\beta) && \text{see Remark 9.4.6}\\
&= (-1)^m(-1)^m(F \otimes \mathrm{id}_{\mathbb{Q}})(\mathfrak{c} \otimes \mathrm{id}_{\mathbb{Q}})^{-1}(\beta) && \text{by definition of } \mathscr{L}_m\\
&= (F \otimes \mathrm{id}_{\mathbb{Q}})(\mathfrak{c} \otimes \mathrm{id}_{\mathbb{Q}})^{-1}(\mathfrak{c}([f]) \otimes r)\\
&= (F \otimes \mathrm{id}_{\mathbb{Q}})([f] \otimes r)\\
&= rF([f])\\
&= r\sigma(f^{-1}(y)) && \text{by definition of } F.
\end{aligned}$$

Here, as usual, y is a suitable generic point of S^m.

We state this convenient and important formula as a proposition:

Proposition 9.4.7 *Let X^n be a closed oriented PL $\mathbb{Q}$-Witt space. Suppose that $m > (n+1)/2$ and that $\beta \in H^m(X;\mathbb{Q})$ is the image of $[f] \otimes r \in \pi^m(X) \otimes \mathbb{Q}$ under the isomorphism of Serre's Theorem. Let $f : X \to S^m$ be a PL representative of $[f]$, and let $y \in S^m$ be a generic point. Then*

$$\beta(\mathscr{L}_m(X)) = rF([f]) = r\sigma(f^{-1}(y)).$$

(d) L-Classes on Smooth Manifolds

Now let us state the promised relation to the cohomology L-classes on smooth manifolds, which justifies calling the homology classes we have constructed L-classes:

Proposition 9.4.8 *If M^n is a closed oriented smooth n-manifold and $m > (n+1)/2$, then for $m = n - 4k$ the class $\mathscr{L}_m(M)$ is the Poincaré dual of the rational Thom–Hirzebruch L-class $L^k(M) \in H^{4k}(M;\mathbb{Q})$. Here $L^k(M)$ is the degree k term of the multiplicative sequence associated to the power series of $\sqrt{t}/\tanh\sqrt{t}$ and taking as its variables the Pontrjagin classes of the tangent bundle of M (see [177, Chapter 19]).*

(e) L-Classes for Small Degrees

It remains to define the L-classes when $m \le (n+1)/2$. In this range, the cohomotopy sets $[X, S^m]$ are not groups, and an alternative procedure is necessary. When $k + m > (n + k + 1)/2$, then $\mathscr{L}_{k+m}(S^k \times X)$ can be defined, noting that if X is a $\mathbb{Q}$-Witt space then so is $S^k \times X$. For any fixed m and n, this condition will hold for any $k > n - 2m + 1$. We can now use the class $\mathscr{L}_{k+m}(S^k \times X)$ to define $\mathscr{L}_m(X)$ because it follows from the Künneth Theorem that if $k > n$ then the cross product with Γ_{S^k} gives an isomorphism[19] $\Gamma_{S^k} \times \cdot : H_m(X;\mathbb{Q}) \to H_m(S^k \times X;\mathbb{Q})$.

[19] If we want to think of $\Gamma_{S^k} \in H_k(S^k;\mathbb{Z})$ as being the fundamental class with respect to the integers, then either we can interpret this product in terms of the cross product $H_*(S^k;\mathbb{Z}) \otimes H_*(X;\mathbb{Q}) \to H_*(S^k \times X;\mathbb{Q})$ or we can first map $H_*(S^k;\mathbb{Z})$ to $H_*(S^k;\mathbb{Q})$ and then employ the cross product $H_*(S^k;\mathbb{Q}) \otimes H_*(X;\mathbb{Q}) \to H_*(S^k \times X;\mathbb{Q})$. These are equivalent thanks to the functoriality of the cross product; see for example [220, Theorem 5.3.3 and Corollary 5.3.4], noting that $\mathbb{Z} \otimes_{\mathbb{Z}} \mathbb{Q} \cong \mathbb{Q} \otimes_{\mathbb{Z}} \mathbb{Q} \cong \mathbb{Q}$. For this reason, we will treat Γ_{S^k} either as an element of $H_k(S^k;\mathbb{Z})$ or as also representing its image in $H_k(S^k;\mathbb{Q})$, as convenient for a given context and without further comment.

Definition 9.4.9 If X^n is a closed oriented PL $\mathbb{Q}$-Witt space and $m \le (n+1)/2$, we denote by $\mathcal{L}_m(X)$ the image of $\mathcal{L}_{k+m}(S^k \times X)$ under the isomorphism $(\Gamma_{S^k} \times \cdot)^{-1}$: $H_{k+m}(S^k \times X;\mathbb{Q}) \to H_m(X;\mathbb{Q})$ for $k > n+1$.

To see that $\mathcal{L}_m(X)$ is well defined, we will show that this construction is independent of k so long as $k > n+1$:

Proposition 9.4.10 *Let X^n be a closed oriented PL $\mathbb{Q}$-Witt space, and let $k, k' > n+1$. Consider for $0 \le m \le n$ the isomorphisms $H_m(X;\mathbb{Q}) \xrightarrow{\Gamma_k \times} H_{k+m}(S^k \times X;\mathbb{Q})$ and $H_m(X;\mathbb{Q}) \xrightarrow{\Gamma_{k'} \times} H_{k'+m}(S^{k'} \times X;\mathbb{Q})$. If $\mathcal{L}_{k+m}(S^k \times X) \in H_{k+m}(S^k \times X;\mathbb{Q})$ and $\mathcal{L}_{k'+m}(S^{k'} \times X) \in H_{k'+m}(S^{k'} \times X;\mathbb{Q})$ are the respective homology L-classes, then*

$$(\Gamma_{S^k}\times)^{-1}\mathcal{L}_{k+m}(S^k \times X) = (\Gamma_{S^{k'}}\times)^{-1}\mathcal{L}_{k'+m}(S^{k'} \times X) \in H_m(X;\mathbb{Q}).$$

As motivation for this definition, let us show that it is consistent with the behavior of the L-classes on smooth manifolds. So let M^n be a smooth manifold, and let $L^i(M) \in H^{4i}(M;\mathbb{Q})$ be its ith cohomology L-class [177, Chapter 19]. As M is a smooth manifold, these cohomology L-classes are defined for all i as characteristic classes of the tangent bundle TM of M. We already know that $\mathcal{L}_{n-4i}(M)$ is the Poincaré dual of $L^i(M)$ when $n - 4i > (n+1)/2$ by Proposition 9.4.8. In this smooth setting we will show that the following statements are equivalent for any $n - 4i \ge 0$:

1. $\mathcal{L}_{n-4i}(M) = (\Gamma_{S^k} \times \cdot)^{-1}(\mathcal{L}_{k+n-4i}(S^k \times M)$, for any $k > n+1$,
2. $\mathcal{L}_{n-4i}(M) = \mathcal{D}(L^i(M))$.

The cohomology L-classes are constructed from the Pontrjagin classes $p^i(TM) \in H^{4i}(M;\mathbb{Q})$ by a universal formula, so we can begin by working with these instead. Let $\pi_1 : S^k \times M \to S^k$ and $\pi_2 : S^k \times M \to M$ be the projections. By the naturality of characteristic classes we have $\pi_j^* p^i(TM) = p^i(\pi_j^* TM)$ for $j \in \{1,2\}$. Meanwhile, $T(S^k \times M) \cong \pi_1^* TS^k \oplus \pi_2^* TM$. But the tangent bundle of a sphere is stably trivial, so it follows from the product formula for Pontrjagin classes [177, Theorem 15.3] and the formula relating cup and cross products (Proposition 7.3.46) that

$$\begin{aligned} p^i(T(S^k \times M)) &= p^i(\pi_1^* TS^k \oplus \pi_2^* TM) \\ &= 1_{S^k \times M} \smile p^i(\pi_2^* TM) = \pi_1^*(1_{S^k}) \smile \pi_2^*(p^i(TM)) \\ &= 1_{S^k} \times p^i(TM), \end{aligned}$$

where we have written 1_{S^k} for the class $1 \in H^0(S^k;\mathbb{Q})$. As the L-classes are polynomials in the p^i, this implies via the formulas relating cup and cross products (Proposition 7.3.54 or [71, Section VII.8.16]) that $L^i(S^k \times M) = 1_{S^k} \times L^i(M)$. Therefore, if the $\mathcal{L}_{n-4i}(M)$ satisfy

$\Gamma_{S^k} \times \mathscr{L}_{n-4i}(M) = \mathscr{L}_{k+n-4i}(S^k \times M)$ for sufficiently large k, as proposed, then we have

$$\begin{aligned}\Gamma_{S^k} \times \mathscr{L}_{n-4i}(M) &= \mathscr{L}_{k+n-4i}(S^k \times M)\\ &= L^i(S^k \times M) \frown \Gamma_{S^k \times M} = (1_{S^k} \times L^i(M)) \frown (\Gamma_{S^k} \times \Gamma_M)\\ &= (-1)^{4ik}(1_{S^k} \frown \Gamma_{S^k}) \times (L^i(M) \frown \Gamma_M)\\ &= \Gamma_{S^k} \times (L^i(M) \frown \Gamma_M).\end{aligned}$$

The fourth equality uses the interchange property of cap and cross products (Proposition 7.3.55 or [71, Section VII.12.17]) while the fifth uses the unital property of cap products [71, Section VII.12.9]. But now we have assumed S^k large enough that $\Gamma_{S^k}\times$ is an isomorphism, and so

$$\mathscr{L}_{n-4i}(M) = L^i(M) \frown \Gamma_M = \mathcal{D}(L^i(M)),$$

using that $L^i(M)$ has even degree.

Conversely, if we instead begin with the requirement that $\mathscr{L}_{n-4i}(M) = L^i(M) \frown \Gamma_M$ then we can reorder the above computation to read

$$\begin{aligned}\mathscr{L}_{k+n-4i}(S^k \times M) &= L^i(S^k \times M) \frown \Gamma_{S^k \times M}\\ &= (1_{S^k} \times L^i(M)) \frown (\Gamma_{S^k} \times \Gamma_M)\\ &= (-1)^{4ik}(1_{S^k} \frown \Gamma_{S^k}) \times (L^i(M) \frown \Gamma_M)\\ &= \Gamma_{S^k} \times (L^i(M) \frown \Gamma_M)\\ &= \Gamma_{S^k} \times \mathscr{L}_{n-4i}(M).\end{aligned}$$

This argument demonstrates the following lemma:

Lemma 9.4.11 *Suppose M is a closed oriented smooth n-manifold and that $k > n + 1$. Then $\mathscr{L}_{n-4i}(M) = L^i(M) \frown \Gamma_M$ for all $n - 4i \geq 0$ if and only if $\Gamma_{S^k} \times \mathscr{L}_{n-4i}(M) = \mathscr{L}_{k+n-4i}(S^k \times M)$ for all $n - 4i \geq 0$.*

The lemma shows that, in the smooth setting, defining $\mathscr{L}_m(M) \in H_m(X; \mathbb{Q})$ for $m = n - 4i \leq (n + 1)/2$ to be the image of $\mathscr{L}_{k+m}(S^k \times M)$ under the isomorphism $(\Gamma_{S^k} \times \cdot)^{-1} : H_{k+m}(S^k \times M; \mathbb{Q}) \to H_m(M; \mathbb{Q})$ for $k > n+1$ is consistent with having $\mathscr{L}_m(M) = \mathscr{L}_{n-4i}(M)$ be the Poincaré dual to the cohomology L-class $L^i(M) \in H^{4i}(M; \mathbb{Q})$. In fact, it is the only possible consistent choice. This justifies the corresponding definition of $\mathscr{L}_m(X)$ as $(\Gamma_{S^k} \times \cdot)^{-1}(\mathscr{L}_{k+m}(S^k \times X))$ for $m \leq (n + 1)/2$ in the more general setting of PL $\mathbb{Q}$-Witt spaces, at least once we have shown in this setting that this formula is independent of the choice of $k > n + 1$.

Once we have shown that the L-classes for small m are well defined we will be able to make some further statements that apply for all m, including results that we have so far shown only for high or low degrees. In particular, we will be able to show the following:

Lemma 9.4.12 *If X^n is a closed oriented PL $\mathbb{Q}$-Witt space and $m \geq 0$, then $\Gamma_{S^k} \times \mathscr{L}_m(X) = \mathscr{L}_{k+m}(S^k \times X)$ for any $k > n + 1$.*

Proposition 9.4.13 *Let X be a closed oriented PL $\mathbb{Q}$-Witt space, and let $f : X \to S^m$ be a PL map for $m > 0$. If $y \in S^m$ is a generic point, $r \in \mathbb{Q}$, and $u \in H^m(S^m)$ is the cohomology class such that $u(\Gamma_{S^m}) = 1$, then*

$$(f^*(u) \otimes r)(\mathscr{L}_m(X)) = r\sigma(f^{-1}(y)).$$

If $m = 0$, the formula will hold if we take $y = 1 \in S^0 = \{0, 1\}$ and let $u = 1_1 \in H^0(S^0)$ be the cocycle that restricts to the augmentation class in $H^0(\{1\})$ and to 0 in $H^0(\{0\})$.

As a corollary, we will see that $\mathscr{L}_0$ takes a particularly nice form:

Proposition 9.4.14 *Let X be a closed oriented PL $\mathbb{Q}$-Witt space, and let $\{X_j\}$ be the connected components of X. Then*

$$\mathscr{L}_0(X) = \sum_j \sigma(X_j)\xi_j \in H_0(X; \mathbb{Q}),$$

where ξ_j is any 0-simplex in X_j.

(f) Characterizing the L-Classes

So far, our construction of the classes $\mathscr{L}_m(X)$ has been fairly explicit in terms of invariants on the space X itself. We will conclude by showing that there is an axiomatic characterization of the L-classes across all closed oriented $\mathbb{Q}$-Witt spaces simultaneously. There are two axioms: One is essentially the content of Proposition 9.4.14, which serves as something of a normalization condition by setting the L-class $\mathscr{L}_0(X)$ to be the signature of X for each connected X. The other axiom concerns the relation between the L-classes of X and the L-classes of certain nice subspaces $Z \subset X$ that are also closed orientable PL $\mathbb{Q}$-Witt spaces. We will see that the L-classes of X determine those of Z via a sort of homology pullback map. It turns out that these two properties are enough to completely determine the L-classes across all closed oriented PL $\mathbb{Q}$-Witt spaces. This will be the content of Theorem 9.4.18 below. First we have to set up some needed technical background.

PL Trivial Normally Nonsingular Subspaces. First we need to describe our "nice" subspaces. For this, recall from Definition 2.9.8 that a subspace $Z \subset X$ of a filtered space is called a normally nonsingular subspace of codimension m if the inclusion $i_Z : Z \hookrightarrow X$ extends to a filtered homeomorphism from some $\mathbb{R}^m$-vector bundle over Z (filtered by the inverse images of the skeleta of Z) onto some neighborhood W of Z. We will need a version of such subspaces here. On the one hand, we will only need to be concerned with the situation in which the $\mathbb{R}^m$-bundle is the trivial bundle, and we will want our homeomorphisms to be PL. On the other hand, we will not need to be so concerned about filtrations. The reason for this is that all of our L-class machinery has been built up from signatures of Witt spaces, which we know are independent of the precise filtration, and the rest of our considerations have been with respect to ordinary homology groups, which don't care about filtrations. Furthermore, suppose we have any PL homeomorphism from $\mathbb{R}^m \times Z$ onto a subspace of

X; then by Lemma 2.10.17 this must be a filtered homeomorphism if Z and X are both given their intrinsic filtrations[20]. We could just work with intrinsic filtrations throughout this section, but instead it is simpler to adopt the following definition:

Definition 9.4.15 Suppose that X is a PL Witt space, that $Z \subset X$ is a closed subspace, and that the inclusion $\mathfrak{i}_Z : Z \hookrightarrow X$ extends to a (not necessarily filtered) PL homeomorphism $\mathfrak{i}_W$ from $\mathbb{R}^m \times Z$ onto a neighborhood W of Z in X with $\mathfrak{i}_W|_{\{0\}\times Z} = \mathfrak{i}_Z$. In this case we will say that Z is a *PL trivial*[21] *normally nonsingular subspace of* X. To shorten this expression, we may say that Z is a *PL trivial nns*.

We will show below in Corollary 9.4.28 that if Z is a PL trivial nns of X then Z is itself a PL $\mathbb{Q}$-Witt space with respect to its intrinsic filtration, and so with respect to any classical pseudomanifold filtration by Proposition 9.1.28.

Our characterization of the L-classes involves certain "wrong-way" or *umkehr*[22] maps determined by the inclusions $\mathfrak{i}_W$. We next review such maps.

Umkehr Maps. Umkehr maps are so named because they run against the usual direction of functoriality. For example, given a map $f : Z \to X$, we might have umkehr maps of the form $f^! : H_*(X) \to H_*(Z)$ or $f_! : H^*(Z) \to H^*(X)$, though such maps do not typically preserve degree. For example, when Z^{n-m} and X^n are closed oriented manifolds, we can obtain such maps by employing duality as follows:

$$f^! = \mathcal{D}_Z f^* \mathcal{D}_X^{-1} : H_i(X) \to H_{i-m}(Z), \qquad f_! = \mathcal{D}_X^{-1} f \mathcal{D}_Z : H^i(Z) \to H^{i+m}(X).$$

However, even when the spaces are not manifolds, there are still some settings in which umkehr maps can be constructed, such as when Z is embedded as a closed subset in X and possesses a neighborhood W that is homeomorphic to an oriented $\mathbb{R}^m$-vector bundle, $m > 0$, with Z corresponding to the zero section. We will often find it useful to identify W itself as a bundle via the given homeomorphism, and so we employ bundle language for W. By the Thom Isomorphism Theorem [177, Theorem 10.4 and Corollary 10.7], there is a unique *Thom class* $\mu \in H^m(W, W - Z) \cong H^m((\mathbb{R}^m, \mathbb{R}^m - \{0\}) \times Z)$ such that

1. μ restricts on each fiber $H^m(\mathbb{R}^m, \mathbb{R}^m - \{0\})$ to a chosen generator consistent with the orientation,
2. $H^i(W) \xrightarrow{\smile\mu} H^{i+m}(W, W - Z)$ is an isomorphism for all i, and
3. $H_i(W, W - Z) \xrightarrow{u\frown} H_{i-m}(W)$ is an isomorphism for all i.

Using this Thom class, we define the map $\mathfrak{i}_W^!$ as follows:

[20] As bundles are locally products, this same statement can be extended to any homeomorphism of an $\mathbb{R}^m$-bundle over Z into X.

[21] It would probably be more correct to say such subspaces are "trivially normally nonsingular," but this would be even more awkward to say than "trivial normally nonsingular."

[22] The German word "Umkehr" can be translated as "reversal."

Definition 9.4.16 Suppose $Z \subset X$ is a closed subspace such that the embedding $\mathrm{i}_Z : Z \hookrightarrow X$ extends to an embedding i_W of an $\mathbb{R}^m$-bundle over Z, $m > 0$, onto a neighborhood W of X and such that i_W restricts to i_Z on the zero section of the bundle. Then we define $\mathrm{i}_W^! : H_i(X) \to H_{i-m}(Z)$ as the composition

$$H_i(X) \to H_i(X, X - Z) \overset{\cong}{\leftarrow} H_i(W, W - Z) \xrightarrow{\mu\frown} H_{i-m}(W) \overset{\cong}{\to} H_{i-m}(Z).$$

Here the left-facing arrow is an excision isomorphism, μ is the Thom class of the bundle W, and the last arrow is induced by the projection $W \to Z$, which is a homotopy equivalence[23]. Of course we can also construct $\mathrm{i}_W^!$ with respect to other coefficient groups, in particular $\mathbb{Q}$; we can even continue to assume $\mu \in H^m(W, W - Z; \mathbb{Z})$ and employ the cap product $H^m(W, W - Z; \mathbb{Z}) \otimes H_i(W, W - Z; \mathbb{Q}) \to H_{i-m}(W; \mathbb{Q})$ (see [220, Section 5.6]).

We can further extend this construction to the case $m = 0$. In this case, the space Z is a union of connected components of X. As homology is additive over disjoint unions, we have $H_i(X) \cong \bigoplus_j H_i(X_j)$, where the X_j are the connected components of X. We then let $\mathrm{i}^!$ be simply the projection

$$H_i(X) \cong \bigoplus_{X_j} H_i(X_j) \to \bigoplus_{X_j \subset Z} H_i(X_j) \cong H_i(Z).$$

Remark 9.4.17 In the special case in which we can identify W with the trivial bundle $W \cong \mathbb{R}^m \times Z$, $m > 0$, by an orientation-preserving homeomorphism then we can take the Thom class to be $\mu = a \times 1_Z$, where $1_Z \in H^0(Z)$ and a is the generator of $H^m(\mathbb{R}^m, \mathbb{R}^m - \{0\})$ that takes the fundamental class for the standard orientation to 1. Another sometimes useful way to describe μ in this case is as follows: Identifying $\mathbb{R}^m$ with $S^m - \{z_0\}$ for some $z_0 \in S^m$, let $f : X \to S^m$ be the map that takes each $(y, z) \in \mathbb{R}^m \times Z$ to $y \in \mathbb{R}^m \subset S^m$ and takes $X - W$ to z_0; note that then $f^{-1}(0) = Z$. Let $u \in H^m(S^m)$ be the generator such that $u(\Gamma_{S^m}) = 1$, and let a be the image of u under the isomorphisms $H^m(S^m) \cong H^m(S^m, S^m - \{0\}) \cong H^m(\mathbb{R}^m, \mathbb{R}^m - \{0\})$. Then as f_W is simply the projection of $W = \mathbb{R}^m \times Z$ onto $\mathbb{R}^m$, we have $\mu = f|_W^*(a) \in H^m(W, W - Z)$ by the singular homology version of Proposition 7.3.24 (see [71, Section VII.7.10]).

This special case is the only one that will be of importance to us in what follows.

Axiomatic Characterization. We can now state our characterizing result for the L-classes:

Theorem 9.4.18 *The L-classes $\mathscr{L}_*$ defined on closed oriented PL $\mathbb{Q}$-Witt spaces possess the following properties:*

1. $\mathbf{a}(\mathscr{L}_0(X)) = \sigma(X)$,
2. *if Z is a PL trivial normally nonsingular subset of X and $\mathrm{i}_W : \mathbb{R}^m \times Z \overset{\cong}{\to} W \subset X$ is the orientation-preserving PL homeomorphism of a trivial $\mathbb{R}^m$-bundle onto a neighborhood W of Z in X then $\mathrm{i}_W^!(\mathscr{L}_{j+m}(X)) = \mathscr{L}_j(Z)$ for all j.*

[23] This definition doesn't quite utilize the full power of the Thom Isomorphism Theorem as we don't use that $\mu\frown$ is an isomorphism. However, this will be the definition we need below.

Furthermore, the collection of classes $\{\mathscr{L}_*(X)\}$ as X ranges over all closed oriented PL $\mathbb{Q}$-Witt spaces is the unique collection with these properties.

(g) Some Notation

The following notation will be used throughout the remainder of our discussion of L-classes:

- If Z is a union of connected components of a space X, we let $1_Z \in H^0(X)$ denote the class that restricts to the augmentation class $1 \in H^0(Z)$ and to 0 on the complement of Z.
- For $m > 0$, we orient the sphere S^m consistently with viewing it as the one-point compactification of $\mathbb{R}^m$ with the standard orientation. It follows that the standard smash product identification
$$S^p \wedge S^q = \frac{S^p \times S^q}{(S^p \times \text{pt}) \cup (\text{pt} \times S^q)} \cong S^{p+q}$$
(see [125, Chapter 0]) is orientation-preserving, and so the quotient takes $\Gamma_{S^p} \times \Gamma_{S^q} \in H_{p+q}(S^p \times S^q)$ to $\Gamma_{S^{p+q}} \in H_{p+q}(S^{p+q})$.
- We let $u \in H^m(S^m)$ denote the class such that[24] $u(\Gamma_{S^m}) = 1$. If we wish to be particularly clear about the degree m under consideration, we may write u_m instead of u.
- When $m = 0$, we let $S^m = \{0, 1\}$, and we let $u_0 = u \in H^0(S^0)$ denote the class 1_1 that restricts to the augmentation class in $H^0(\{1\})$ and to 0 in $H^0(\{0\})$. If we take the fundamental class Γ_{S^0} to be the image of the standard generator of $H_1(D^1, S^0)$ under $\partial_* : H_1(D^1, S^0) \to H_0(S^0)$ then the property $u(\Gamma_{S^0}) = 1$ continues to hold when $m = 0$. Whenever we speak of a generic point $y \in S^m$, if $m = 0$ we take $y = 1$.
- We will also sometimes abuse notation and allow Γ_{S^m} and u to stand also for their images under the standard homomorphisms $H_m(S^m) \to H_m(S^m) \otimes \mathbb{Q} \cong H_m(S^m; \mathbb{Q})$ and $H^m(S^m) \to H^m(S^m) \otimes \mathbb{Q} \cong H^m(S^m; \mathbb{Q}) \cong \operatorname{Hom}_{\mathbb{Z}}(H_m(S^m); \mathbb{Q}) \cong \operatorname{Hom}_{\mathbb{Q}}(H_m(S^m; \mathbb{Q}), \mathbb{Q})$, as context requires. In general, if G is an abelian group and V is a $\mathbb{Q}$-vector space, then we have $\mathbb{Q}$-vector space isomorphisms $A \otimes_{\mathbb{Z}} V \cong (A \otimes_{\mathbb{Z}} \mathbb{Q}) \otimes_{\mathbb{Q}} V$ and $\operatorname{Hom}_{\mathbb{Z}}(A, V) \cong \operatorname{Hom}_{\mathbb{Q}}(A \otimes_{\mathbb{Z}} \mathbb{Q}, V)$. So, as $\mathscr{L}_m(X) \in H_m(X; \mathbb{Q})$ we can use these identifications to interpret expressions such as $\Gamma_{S^k} \otimes \mathscr{L}_m(X)$ as living over the ground rings $\mathbb{Z}$ or $\mathbb{Q}$ as convenient (cf. footnote 19 on page 653). In what follows, we will use this observation without further comment.

(h) The Proofs

We now turn to proving the various claims of this section. This will occupy the next several sections.

[24] This convention is not necessarily consistent with all other sources.

9.4.2 Maps to Spheres and Embedded Subspaces

We first turn to proving Proposition 9.4.1. Recall that this proposition says that if X is a closed oriented n-dimensional PL $\mathbb{Q}$-Witt space and $f : X \to S^m$ is a PL map to an oriented S^m then for almost all $y \in S^m$ the inverse image $f^{-1}(y)$ is a $\mathbb{Q}$-Witt space whose signature is independent both of the choice of y (within an open dense set depending on f) and of the PL homotopy class of f.

We will need the following lemma, which is Lemma 20.5 of [177]. We give a slightly more detailed proof.

Lemma 9.4.19 *If $f : K \to L$ is a simplicial map of simplicial complexes and if y is contained in the interior U of an m-simplex of L, then there is a homeomorphism $h : U \times f^{-1}(y) \to f^{-1}(U)$ such that $fh = \pi_1$, where $\pi_1 : U \times f^{-1}(y) \to U$ is the projection. Furthermore, the space $f^{-1}(y)$ is homeomorphic to $f^{-1}(y')$ for any other $y' \in U$.*

Proof The lemma is trivial when $m = 0$, so we can take $m > 0$.

Suppose that U is the interior of a simplex τ of L with $\tau = [w_0, \dots, w_m]$. Let $y = \sum_{i=0}^m t_i w_i$ represent y in barycentric coordinates. Let $\sigma = [v_0, \dots, v_N]$ be a simplex of K. If the mapping f is not surjective from σ onto τ, then $f(\sigma) \cap U = \emptyset$ and the lemma holds vacuously. So suppose instead that f maps σ onto τ.

Assuming f maps σ onto τ, then for each vertex w_k of τ there is some vertex v_i of σ that maps to w_k. Let us relabel the vertices of σ as $\{v_{ij}\}$ so that $f(v_{ij}) = w_i$ for all i. Then every element $x \in \sigma$ can be written $x = \sum_{i=0}^m \sum_{j=1}^{k(i)} a_{ij} v_{ij}$ in barycentric coordinates, where $k(i)$ is the number of vertices of σ that map to w_i. Then $f(x) = \sum_{i=0}^m \left(\sum_{j=1}^{k(i)} a_{ij}\right) w_i$. So $f^{-1}(y) \cap \sigma = \{x \in \sigma \mid \sum_{j=1}^{k(i)} a_{ij} = t_i \text{ for each } i\}$. This is a system of $m+1$ linear equations in $N+1 \geq m+1$ unknowns. Because we know that the linear map $f|_\sigma$ is onto τ, the solution set will be a linear subspace of σ of dimension $N - m$.

Next, suppose $x \in f^{-1}(U) \cap \sigma$ and let us rewrite $x = \sum_{i=0}^m \sum_{j=1}^{k(i)} a_{ij} v_{ij}$ as

$$x = \sum_{i=0}^{m} \left(\left(\sum_{\ell=1}^{k(i)} a_{i\ell} \right) \left(\sum_{j=1}^{k(i)} \frac{a_{ij}}{\sum_{\ell=1}^{k(i)} a_{i\ell}} v_{ij} \right) \right).$$

This is possible because if $\sum_{\ell=1}^{k(i)} a_{i\ell} = 0$ for any i then the coefficient of w_i in $f(x)$ is 0, which is impossible if $f(x) \in U$. The point of this rewrite is that

$$\sum_{j=1}^{k(i)} \frac{a_{ij}}{\sum_{\ell=1}^{k(i)} a_{i\ell}} = 1,$$

so for each i, the expression

$$u_i(x) = \sum_{j=1}^{k(i)} \frac{a_{ij}}{\sum_{\ell=1}^{k(i)} a_{i\ell}} v_{ij}$$

is a description in barycentric coordinates of a point on the face of σ spanned by $\{v_{ij}\}_{j=1}^{k(i)}$, i.e. the face $f^{-1}(w_i)$. Note that $u_i(x)$ depends continuously on the barycentric coordinates of x; see Figure 9.5.

Define $g : f^{-1}(U) \cap \sigma \to U \times (f^{-1}(y) \cap \sigma)$ by

$$g(x) = \left(f(x), \sum_{i=0}^{m} t_i u_i(x) \right);$$

see Figure 9.5. This is well defined because $f(x) \in U$ by assumption and

$$f\left(\sum_{i=0}^{m} t_i u_i(x) \right) = \sum_{i=0}^{m} t_i w_i = y.$$

We will show that g is a homeomorphism; it is clear that $fg^{-1} = \pi_1$, the projection to U.

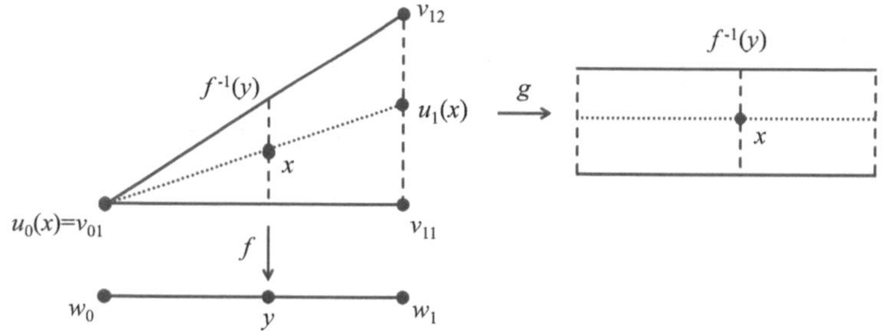

Figure 9.5 On the left, an example of a surjective map f from the simplex $\sigma = [v_{01}, v_{11}, v_{12}]$ onto the simplex $\tau = [w_0, w_1]$. Within σ we show a point $x \in f^{-1}(y)$ and the corresponding $u_0(x)$ and $u_1(x)$. The map g illustrates the homeomorphism from $f^{-1}(U)$ to $U \times f^{-1}(y)$, which for each $x \in f^{-1}(y)$ takes the open line segment from $u_0(x)$ to $u_1(x)$ to the line segment $U \times \{x\}$.

First we show that the map g is surjective. Choose any $x \in f^{-1}(y) \cap \sigma$ and a point $y' = \sum_{i=0}^{m} s_i w_i$ in U. Write x as $x = \sum a_i u_i(x)$. Since $f(x) = \sum_i a_i w_i = y$, by assumption, and since barycentric coordinates are unique, we must have $a_i = t_i$. Therefore we actually have $x = \sum t_i u_i(x)$. Now, let $z = \sum s_i u_i(x)$. Then $f(z) = \sum s_i f(u_i(x)) = \sum s_i w_i = y'$. We claim that $u_i(z) = u_i(x)$, in which case $g(z) = (y', \sum t_i u_i(z)) = (y', \sum t_i u_i(x)) = (y', x)$; as the choices of x and y' were arbitrary, surjectivity will follow. To prove the claim, by definition we have

$$u_i(x) = \sum_{j=1}^{k(i)} \frac{a_{ij}}{\sum_{\ell=1}^{k(i)} a_{i\ell}} v_{ij},$$

where the a_{ij} are the coefficients of v_{ij} in the barycentric coordinates for x. The corresponding coefficient of v_{ij} for z is $s_i\left(a_{ij}/\sum_{\ell=1}^{k(i)} a_{i\ell}\right)$. But then the coefficient for v_{ij} in $u_i(z)$

is

$$\frac{s_i\left(a_{ij}/\sum_{\ell=1}^{k(i)} a_{i\ell}\right)}{\sum_{r=1}^{k(i)}\left(s_i\left(a_{ir}/\sum_{\ell=1}^{k(i)} a_{i\ell}\right)\right)} = \frac{a_{ij}}{\sum_{r=1}^{k(i)} a_{ir}}.$$

So the coefficient of v_{ij} in $u_i(z)$ is identical to the corresponding coefficient in $u_i(x)$, which proves the claim.

Now we show that g is injective. Suppose $x, x' \in f^{-1}(U) \cap \sigma$, with $x = \sum b_i u_i(x)$ and $x' = \sum b_i' u_i(x')$, and that $g(x) = g(x')$. This implies $f(x) = \sum b_i w_i = \sum b_i' w_i = f(x')$, so we must have $b_i = b_i'$ for all i by uniqueness of barycentric coordinates. It also implies that $\sum_{i=0}^m t_i u_i(x) = \sum_{i=0}^m t_i u_i(x')$. Writing out u_i and u_i', this becomes

$$\sum_{i=0}^m t_i\left(\sum_{j=1}^{k(i)} \frac{a_{ij}}{\sum_{\ell=1}^{k(i)} a_{i\ell}} v_{ij}\right) = \sum_{i=0}^m t_i\left(\sum_{j=1}^{k(i)} \frac{a_{ij}'}{\sum_{\ell=1}^{k(i)} a_{i\ell}'} v_{ij}\right),$$

where a_{ij} and a_{ij}' are the barycentric coordinates of x and x' with respect to the v_{ij}. By the uniqueness of barycentric coordinates, we must have

$$\frac{t_i a_{ij}}{\sum_{\ell=1}^{k(i)} a_{i\ell}} = \frac{t_i a_{ij}'}{\sum_{\ell=1}^{k(i)} a_{i\ell}'}$$

for each i, j, so

$$\frac{a_{ij}}{\sum_{\ell=1}^{k(i)} a_{i\ell}} = \frac{a_{ij}'}{\sum_{\ell=1}^{k(i)} a_{i\ell}'}$$

for each i, j. But now comparing the definitions of $u_i(x)$ and $u_i(x')$, we see that we must actually have $u_i(x) = u_i(x')$ for all i. So, altogether,

$$x = \sum b_i u_i(x) = \sum b_i' u_i(x') = x'.$$

So, we have constructed homeomorphisms $f^{-1}(U) \cap \sigma$ to $U \times (f^{-1}(y) \cap \sigma)$ for any σ in K. Our above construction assumed that σ maps onto τ, but if σ does not map onto τ then both expressions are empty. Furthermore, if we rewrite the homeomorphism $f^{-1}(U) \cap \sigma \to U \times (f^{-1}(y) \cap \sigma)$ as g_σ and if η is a face of σ that also maps onto τ, then it is not difficult to verify that g_η agrees with the restriction of g_σ to η. In fact, if $x \in \eta \subset \sigma$ then the definition of $u_i(x)$ is the same whether we think of x as belonging to η or to σ: if we think in terms of σ then the coefficients of the vertices of σ not belonging to η must vanish in the expression for x and hence also in the expression for $u_i(x)$. It follows that we can piece the g_σ together over all the simplices of K that intersect $f^{-1}(y)$ in order to obtain a global homeomorphism $G : f^{-1}(U) \cong U \times f^{-1}(y)$. To complete the first statement of the lemma, we now take $h = G^{-1}$.

For the last statement of the proposition, that $f^{-1}(y)$ is homeomorphic to $f^{-1}(y')$ for any $y, y' \in U$, let $G : f^{-1}(U) \to U \times f^{-1}(y)$ be the homeomorphism constructed above. Within each simplex σ of K, the map G restricts to the homeomorphism $g_\sigma : f^{-1}(U) \cap \sigma \to$

$U \times (f^{-1}(y) \cap \sigma)$ defined by $g_\sigma(x) = (f(x), \sum_{i=0}^m t_i u_i(x))$, where $y = \sum t_i w_i$. So we see that f maps a point $x \in f^{-1}(U) \cap \sigma$ to a point $z \in U$ if and only if the first component of $g_\sigma(x)$ is z. Since σ was chosen arbitrarily, we see that, more generally, f maps a point $x \in f^{-1}(U)$ to a point $z \in U$ if and only if the first component of $G(x) \in U \times f^{-1}(y)$ is z. In particular then, we have $f^{-1}(y') = h(\{y'\} \times f^{-1}(y))$, which is a homeomorphic image of $f^{-1}(y)$.

This completes the proof of Lemma 9.4.19. □

Remark 9.4.20 Our proof of Lemma 9.4.19 only provides a topological homeomorphism $f^{-1}(U) \cong U \times f^{-1}(y)$, but it is possible to strengthen the argument to construct a homeomorphism that is PL. We will only need this stronger result later in Section 9.4.6 for the characterization theorem, but as the proof incorporates results that are far afield for us, we refer the reader to [249, Theorem 1.3.1].

Remark 9.4.21 Continuing the notation of the proof, suppose $x \in f^{-1}(y)$ is contained in the interior of the simplex σ of K. We have seen that any point in $f^{-1}(U) \cap \sigma$, and so in particular in $f^{-1}(y) \cap \sigma$, can be written as $x = \sum s_i u_i(x)$ for some coefficients s_i. But we also know that $f(x) = \sum s_i f(u_i(x)) = \sum s_i w_i$. If $f(x) = y$, then we must have $s_i = t_i$, so $x = \sum t_i u_i(x)$. Therefore $G(x) = (y, x)$ for $x \in f^{-1}(y)$. We can therefore think of our homeomorphism $f^{-1}(U) \to U \times f^{-1}(y)$ as being the identity on $f^{-1}(y)$ in this sense. We will implicitly utilize this identification in what follows.

Now we can return to Proposition 9.4.1, which we restate here:

Proposition (Proposition 9.4.1) *Let X be a closed oriented n-dimensional PL $\mathbb{Q}$-Witt space. Suppose that S^m, $m > 0$, has been given an orientation and that $f : X \to S^m$ is a PL map. Then for almost all $y \in S^m$ the inverse image $f^{-1}(y)$ can be filtered as a closed oriented*[25] *$(n-m)$-dimensional PL $\mathbb{Q}$-Witt space (possibly empty) embedded in X. Furthermore, for almost all $y, y' \in S^m$ the Witt spaces $f^{-1}(y)$ and $f^{-1}(y')$ have the same signature; this common signature depends only on the PL homotopy class of f in $[X, S^m]_{\mathrm{PL}}$. This result also holds for $m = 0$ allowing only $y = y' = 1 \in S^0$.*

Proof Once again the $m = 0$ case is trivial, so we assume $m > 0$.

Since X is compact, the map $f : X \to S^m$ is proper. Therefore, by Theorem B.2.19, there are triangulations K of X and L of S^m with respect to which f is simplicial. Let y be a point in the interior of an m-simplex τ of L, and let U be the interior of τ. By the preceding lemma, $f^{-1}(U) \cong U \times f^{-1}(y)$.

$f^{-1}(y)$ is a PL $\mathbb{Q}$-Witt Space. First we show that $f^{-1}(y)$ is an $(n-m)$-dimensional PL $\mathbb{Q}$-Witt space. As this is trivial if $f^{-1}(y)$ is empty, we will suppose $f^{-1}(y) \neq \emptyset$. It is clearly compact, being the inverse image of a closed point in a compact space X. Furthermore, the space $f^{-1}(y)$ can be triangulated. In fact, we can subdivide L to a triangulation L' so that

[25] Recall from page 650 that we will orient $f^{-1}(y)$ so that the homeomorphism $h : U \times f^{-1}(y) \to f^{-1}(U)$ of Lemma 9.4.19 is orientation-preserving.

y is a vertex by Example B.4.3. Then by Theorem B.2.19 we can subdivide K to K' and L' to L'' so that $f : K' \to L''$ is simplicial. Then $f^{-1}(y)$ is a subcomplex of K'.

Next, we will invoke Proposition 2.10.18 to show that the intrinsic filtration of $f^{-1}(y)$ has the structure of a classical PL stratified pseudomanifold; we do not claim that the proposition makes $f^{-1}(y)$ a stratified PL pseudomanifold with the filtration inherited from X (this won't be necessary). According to Proposition 2.10.18, it suffices to show that $f^{-1}(y)$ contains a dense $(n - m)$-dimensional PL manifold M such that $f^{-1}(y) - M$ has dimension $\leq n - m - 2$; then $f^{-1}(y)$ will be a classical pseudomanifold with respect to its intrinsic filtration. For this, let K and L continue to denote triangulations of X and S^m with respect to which f is simplicial, and let K^{n-2} be the $(n - 2)$-skeleton of K. Recall that by Corollary 2.5.21 the complex K is a union of n-simplices and every $(n - 1)$-simplex is a face of exactly two n-simplices. Identifying the spaces with their triangulations, we let $M = (f|_{K-K^{n-2}})^{-1}(y) = (K - K^{n-2}) \cap f^{-1}(y)$. We saw in the proof of Lemma 9.4.19 that if a simplicial map takes a k-simplex σ onto an m-simplex τ, then the inverse image of an interior point of τ will be a $(k - m)$-dimensional linear subspace of σ. So for any n-simplex σ of K, any point of $f^{-1}(y)$ in the interior of σ is contained in the interior of an $(n - m)$-dimensional linear subspace of σ; in particular such points have neighborhoods PL homeomorphic to $\mathbb{R}^{n-m}$. Any point of M not in the interior of an n-simplex is in the interior of some $(n - 1)$-simplex, δ. Once again by the proof of Lemma 9.4.19, the intersection of δ with $f^{-1}(y)$ is an $(n - m - 1)$-dimensional linear subspace of δ. If δ is a face of the n-simplices σ_1 and σ_2, then this linear subspace in δ is the intersection of the $(n - m)$-dimensional $f^{-1}(y) \cap \sigma_1$ and $f^{-1}(y) \cap \sigma_2$. In particular, any point in the intersection of $f^{-1}(y)$ with the interior of δ has a neighborhood in $f^{-1}(y)$ consisting of the union of two $(n - m)$-dimensional linear half-spaces along an $(n - m - 1)$-dimensional linear space. So such a point has a neighborhood in $f^{-1}(y)$ PL homeomorphic to $\mathbb{R}^{n-m}$. Thus every point of M has a neighborhood PL homeomorphic to $\mathbb{R}^{n-m}$, and M is an $(n - m)$-dimensional PL manifold.

Now we must show that M is dense in $f^{-1}(y)$ and that $f^{-1}(y) - M$ has dimension $\leq n - m - 2$. If $x \in K^{n-2}$, then x is contained in a simplex η of K^{n-2}. Let σ be an n-simplex of K with η as a face. Such a simplex must exist by Corollary 2.5.21. In order for $f^{-1}(y) \cap \sigma$ to be non-empty, σ must map onto the m-simplex of S^m containing y, so the intersection $\sigma \cap f^{-1}(y)$ is an $(n-m)$-plane through σ. This plane must intersect the interior of σ because f is continuous and simplicial, so if the interior of σ does not have any points that map to y, which lies in the interior of a face of L, then no point of σ can map to y, a contradiction. Thus every point of σ in $f^{-1}(y)$ must be in the closure of a plane that runs through the interior of σ, and every point in $f^{-1}(y) \cap \sigma$ is therefore in the closure of $f^{-1}(y) \cap \mathring{\sigma}$. This suffices to show that M is dense in $f^{-1}(y)$. To see that $f^{-1}(y) - M$ has dimension $\leq n - m - 2$, we need only note that $f^{-1}(y) - M \subset K^{n-2}$, and we know that the intersection $f^{-1}(y) \cap \eta$ for any $\eta \in K^{n-2}$ must have dimension $\leq n - m - 2$ by the proof of Lemma 9.4.19. This implies that $\dim((f|_{K^{n-2}})^{-1}(y)) = \dim(f^{-1}(y) - M) \leq n - m - 2$.

So, we have now shown that $f^{-1}(y)$ is a classical pseudomanifold with respect to its intrinsic filtration, and, momentarily ignoring filtrations, we know that $f^{-1}(y)$ has a neighborhood W in X homeomorphic to $U \times f^{-1}(y)$, where U is the interior of the m-simplex of S^m containing y. Pick a classical pseudomanifold filtration of $f^{-1}(y)$, and let $U \times f^{-1}(y)$ have the product stratification. Let the homeomorphic W have the filtration inherited from X. This W is a $\mathbb{Q}$-Witt space, since the $\mathbb{Q}$-Witt condition is local. Since the property of being a $\mathbb{Q}$-Witt space is independent of the filtration by Proposition 9.1.28, we see that $U \times f^{-1}(y)$ is also a $\mathbb{Q}$-Witt space. But the links of $f^{-1}(y)$ are all also links of $U \times f^{-1}(y)$, so $f^{-1}(y)$ is a PL $\mathbb{Q}$-Witt space.

Orientability. We have assumed that X is oriented, and as any filtration of a Witt space must be as a classical pseudomanifold (Remark 9.1.6), we have by Corollary 8.1.11 that X can be given compatible orientations for any such filtrations. So, continuing the notation just above, these orientations on X induce an orientation on the neighborhood W of $f^{-1}(y)$, which is homeomorphic to $U \times f^{-1}(y)$, which is thus also orientable. By Lemma 8.1.38, the product of stratified pseudomanifolds is orientable if and only if both factors are, so it follows that $f^{-1}(y)$ is orientable. Letting $h : U \times f^{-1}(y) \to f^{-1}(U)$ continue to denote the homeomorphism such that $\pi_1 h^{-1} = f$ and h is the "identity" from $(y, f^{-1}(y))$ to $f^{-1}(y)$, we choose to orient $f^{-1}(y)$ so that h is orientation-preserving, letting $U \times f^{-1}(y)$ have the product orientation and using the orientation on U inherited from S^m.

Homotopies of f. Next, we investigate the effect on $f^{-1}(y)$ of changing f by a homotopy. Suppose that f and g are PL homotopic PL maps $X \to S^m$ by a PL homotopy $H : I \times X \to S^m$. By adding closed collars to $I \times X$ if necessary, we can assume that there is an $\epsilon > 0$ such that $H(t,x) = f(x)$ for $t \in [0,\epsilon]$ and $H(t,x) = g(x)$ for $t \in [1-\epsilon, 1]$.

Since $I \times X$ is compact, we can find triangulations $\bar{K}$ and $\bar{L}$ of $I \times X$ and S^m with respect to which H is simplicial, again using Theorem B.2.19; we may also take $\bar{L}$ to be a subdivision of L. Without loss of generality, we can assume that $y \in S^m$ is contained in the interior of an m-simplex of $\bar{L}$ (if not, let y' be a point from the interior of the simplex of L containing y such that y' is in the interior of an m-simplex of $\bar{L}$; by Lemma 9.4.19 we have $f^{-1}(y) \cong f^{-1}(y')$ so we can relabel y' to y). Consider $Y = (H|_{(0,1)\times X})^{-1}(y)$. The above arguments demonstrate that the intrinsic filtration of Y gives it the structure of a stratified pseudomanifold, and, in fact, a $\mathbb{Q}$-Witt space. The spaces $(H|_{(0,\epsilon)\times X})^{-1}(y)$ and $(H|_{(1-\epsilon,1)\times X})^{-1}(y)$ are respectively homeomorphic to $(0,\epsilon)\times f^{-1}(y)$ and to $(1-\epsilon,1)\times g^{-1}(y)$. By Lemma 2.10.17, the intrinsic PL filtrations on these spaces (which are compatible by restriction with the intrinsic PL filtration on Y as intrinsic filtrations are defined by local properties) have the forms $(0,\epsilon) \times f^{-1}(y)^*$ and $(1-\epsilon, 1) \times g^{-1}(y)^*$, where $f^{-1}(y)^*$ and $g^{-1}(y)^*$ denote $f^{-1}(y)$ and $g^{-1}(y)$ with their intrinsic PL filtrations. This implies that if we filter $H^{-1}(y)$ with the intrinsic PL filtration on Y together with the collars $[0,1) \times f^{-1}(y)^*$ and $(1-\epsilon, 1] \times g^{-1}(y)^*$, then the filtrations will all be compatible so that $H^{-1}(y)$ can be filtered as a PL ∂-stratified pseudomanifold with boundary $g^{-1}(y)^* \amalg -f^{-1}(y)^*$.

Furthermore, as we have already seen that Y, $f^{-1}(y)^*$, and $g^{-1}(y)^*$ are PL $\mathbb{Q}$-Witt spaces, all of $H^{-1}(y)$ is a PL $\mathbb{Q}$-Witt space providing a $\mathbb{Q}$-Witt bordism between $f^{-1}(y)^*$ and

$g^{-1}(y)^*$. Thus $f^{-1}(y)^*$ and $g^{-1}(y)^*$ have the same Witt signature. By Theorem 9.3.16 these signatures are also independent of the choices of filtration. So $\sigma(f^{-1}(y)) = \sigma(g^{-1}(y))$.

Independence of y. To finish the proof of Proposition 9.4.1, we now need only see what happens if we change our choice of $y \in S^m$. Let y, y' be two points of S^m that are each in the interior of an m-simplex of the triangulation L (not necessarily the same simplex). We can find a PL homeomorphism h that takes y' to y and that is PL homotopic to the identity (this follows from the material in [130, Chapter VI], for example the Isotopy Extension Theorem [130, Theorem 6.12]). Therefore, f and hf are PL homotopic. Let $H : I \times X \to S^m$ be the homotopy, and let L' be a subdivision of L with respect to which H is simplicial. As y and y' may no longer be generic with respect to H, we work with nearby points. So choose a point $z' \in S^m$ that satisfies the following conditions:

1. z' and y' are contained in the interior of the same m-simplex of L,
2. z' is contained in the interior of an m-simplex of L',
3. the image $z = h(z')$ is contained in the interior of the m-simplex of L that contains y and in the interior of an m-simplex of L'.

Such choices are possible by the continuity of h and by the density in S^m of the complements of the $(m-1)$-skeleta of L and L'.

By our previous results, the space $f^{-1}(z)$ is $\mathbb{Q}$-Witt bordant to $(hf)^{-1}(z) = f^{-1}(h^{-1}(z)) = f^{-1}(z')$. But we also know from Lemma 9.4.19 that $f^{-1}(y) \cong f^{-1}(z)$ and $f^{-1}(z') \cong f^{-1}(y')$. Therefore, we conclude that for almost all y', the spaces $f^{-1}(y)$ and $f^{-1}(y')$ are $\mathbb{Q}$-Witt bordant with respect to their intrinsic filtrations; thus all these spaces have the same signature.

This completes the proof of Proposition 9.4.1. □

So, we have now shown that if X is a closed oriented PL $\mathbb{Q}$-Witt space then given any PL map $f : X \to S^m$ there is an open dense set of points $y \in S^m$ such that the signatures $\sigma(f^{-1}(y))$ are all well defined and identical. Furthermore, this common value does not change if we alter f by a PL homotopy. So we have a well-defined function $[X, S^m]_{\text{PL}} \to \mathbb{Z}$.

9.4.3 Cohomotopy

We now turn to a brief discussion of the cohomotopy sets $\pi^m(X) = [X, S^m]$. We could simply cite the needed results, which are a bit of a departure from the topic of the text, but since it is not too difficult to explain the basic ideas enough to at least sketch a proof of Lemma 9.4.2, we do so here. More thorough treatments of cohomotopy can be found in [219] and [129, Chapter VII].

Let us first recall Lemma 9.4.2. By the PL Approximation Theorem (Theorem B.2.24), we have $[X, S^m]_{\text{PL}} \cong [X, S^m]$, where $[X, S^m]_{\text{PL}}$ denotes the set of PL homotopy classes of PL maps $X \to S^m$ and $[X, S^m]$ denotes the set of homotopy classes of topological maps $X \to S^m$. In the preceding section, we constructed a function $[X, S^m]_{\text{PL}} \to \mathbb{Z}$. Composing with the isomorphism $[X, S^m]_{\mathrm{PL}} \cong [X, S^m]$, we obtain a function $F : \pi^m(X) \to \mathbb{Z}$, which takes the homotopy class $[f]$ to the signature of the inverse image of a generic point under a PL map in the homotopy class of f. Lemma 9.4.2 states the following:

Lemma (Lemma 9.4.2) *If $m > (n+1)/2$ then $F : \pi^m(X) \to \mathbb{Z}$ is a homomorphism of abelian groups.*

In particular, this lemma includes the statement that $\pi^m(X)$ can be given an abelian group structure when $m > (n+1)/2$, which we must explain. If $f : X \to S^m$ is a map, we let $[f]$ denote the homotopy class of f. The notation $f \sim g$ will mean that f and g are homotopic.

Proof

The Group Operation. The basic idea for turning the set of homotopy classes $\pi^m(X) = [X, S^m]$ into a group is the following: Let S^m have a basepoint s_0, and suppose $f, g : X \to S^m$ are two maps from an n-dimensional CW complex to S^m. We need to define a product $[f] + [g]$. For this we consider the map $(f, g) : X \to S^m \times S^m$ defined by $(f, g)(x) = (f(x), g(x))$. Using the structure of S^m as a CW complex with two cells, one in dimension zero and one in dimension m, the product $S^m \times S^m$ can be written as a product CW complex with four cells. If $n < 2m$, the Cellular Approximation Theorem [125, Section 4.1] allows us to deform (f, g) to a map to the $(2m-1)$-skeleton of $S^m \times S^m$, which in this case is just $(S^m \times \{s_0\}) \cup (\{s_0\} \times S^m) = S^m \vee_{s_0} S^m$. Let us call the deformed map $h : X \to S^m \vee S^m$, and let us denote by $H : I \times X \to S^m \times S^m$ the homotopy from (f, g) to h. Next, we employ the fold map $\Omega : S^m \vee S^m \to S^m$, which is the identity on each copy of S^m. Let $f +_h g$ denote the composition $X \xrightarrow{h} S^m \vee S^m \xrightarrow{\Omega} S^m$; then $f +_h g$ depends on h, but we'd like to show that the homotopy class of $f +_h g$ does not, so that we can then define the group operation on $\pi^m(X)$ by $[f] + [g] = [f +_h g]$. For this we need to consider what happens if we use an alternative homotopy to H, or, for that matter, alternative representatives of the homotopy classes of f and g. For this, it is necessary to strengthen the assumption on dimension to $n + 1 < 2m$. But now suppose that $f \sim f'$, $g \sim g'$, and that H' is a homotopy from (f', g') to h' with the image of h' in $S^m \vee S^m$. Then $h \sim (f, g) \sim (f', g') \sim h'$ as maps $X \to S^m \times S^m$, and so there is a homotopy $K : I \times X \to S^m \times S^m$ from h to h'. But now by another application of the Cellular Approximation Theorem, as $n + 1 < 2m$ there is a homotopy relative to $(\{0\} \times X) \amalg (\{1\} \times X)$ from K to a map $K' : I \times X \to S^m \vee S^m$, which then provides a homotopy $\Omega \circ K'$ from $f +_h g$ to $f' +_{h'} g'$.

Commutativity of the Operation. To see that our proposed group operation is commutative, let $T : S^m \times S^m$ interchange coordinates; we also let T denote the restriction of T to $S^m \vee S^m$. Then, with the notation above, $TH : I \times X \to S^m \times S^m$ is a homotopy from (g, f) to Th, the latter of which has image in $S^m \vee S^m$. But evidently $\Omega T = \Omega$. So $f +_h g = \Omega h = \Omega T h = g +_{Th} f$. So $[f] + [g] = [f +_h g] = [g +_{Th} f] = [g] + [f]$, using the independence of the homotopy to $S^m \vee S^m$ just demonstrated.

Associativity of the Operation. Associativity is technically more difficult, though the concepts are not more complicated; one needs to use suitable maps to $S^m \times S^m \times S^m$. We refer the reader to [219, Theorem 6.3] or [129, Theorem VII.5.2].

The Unit Element. The identity of $\pi^m(X)$ is the homotopy class of the map e that takes X to the point $s_0 \in S^m$. Then (f,e) and (e,f) already map into $S^m \vee S^m$ and clearly $[f] = [\Omega(e,f)] = [e] + [f]$.

Inverses. The inverse $-[f]$ is represented by the composition $X \xrightarrow{f} S^m \xrightarrow{\rho} S^m$, where ρ is a map of degree -1. To see this, let us fix ρ more precisely as a reflection map across the plane separating two hemispheres E_1 and E_2 of S^m. Let $s_0 \in E_1 \cap E_2$. Let $R : I \times S^m \to S^m$ be a homotopy that retracts E_1 into s_0, and let $r = R|_{\{1\}\times S^m}$ be the non-identity end of the homotopy. Then $rf \sim f$, and similarly $r\rho f \sim \rho f$. Notice that rf maps $M_1 = f^{-1}(E_1)$ into s_0, and $r\rho f$ maps $M_2 = f^{-1}(E_2)$ into s_0. Since every point of X is in either M_1 or M_2 (possibly in their intersection), it follows that $(rf, r\rho f) : X \to S^m \times S^m$, which is homotopic to $(f, \rho f)$, has image in $S^m \vee S^m$. Then $\Omega(rf, r\rho f)$ is well defined. Furthermore, $\Omega(rf, r\rho f)|_{M_1} = r\rho f|_{M_1} \sim \rho f|_{M_1}$, and $\Omega(rf, r\rho f)|_{M_2} = rf|_{M_2} \sim f|_{M_2}$. The two homotopies of the last sentence agree on $M_1 \cap M_2$ since in both cases we can use the same homotopy R and since points in $M_1 \cap M_2$ map under f to $E_1 \cap E_2$, which is fixed by ρ. So $\Omega(rf, r\rho f)$, which represents $[rf] + [r\rho f] = [f] + [\rho f]$, is homotopic to a map G that is f on M_2 and ρf on M_1. But, by definition, f takes M_2 into E_2 and M_1 into E_1, and so ρf takes M_1 into E_2. Together, then, G takes all of X to E_2, which implies that G is homotopically trivial and hence represents the identity.

Functoriality. We have now established that $\pi^m(X)$ is an abelian group whenever X is a CW complex of dimension $n < 2m - 1$. It is also worth observing that any map $\phi : X \to Y$ induces a function $\phi^* : \pi^m(Y) \to \pi^m(X)$ by taking the homotopy class $[g] \in \pi^m(Y) = [Y, S^m]$ to $[g\phi] \in [X, S^m] = \pi^m(X)$. If ϕ is a homotopy equivalence then ϕ^* is a bijection. If $\dim(X)$ and $\dim(Y)$ are in the dimension ranges such that $\pi^m(X)$ and $\pi^m(Y)$ are groups, then ϕ^* is a homomorphism (see [129, Proposition VII.5.4]): If $f, g : Y \to S^n$, let H denote the homotopy from (f,g) to a map $h : Y \to S^m \vee S^m$. Then Ωh represents $[f] + [g]$, and $\Omega h\phi$ represents $\phi^*([f] + [g])$. On the other hand, $f\phi$ and $g\phi$ represent $\phi^*[f]$ and $\phi^*[g]$, and $H \circ (\mathrm{id}_I \times \phi)$ is a homotopy from $(f,g)\phi = (f\phi, g\phi)$ to $h\phi$, so $\Omega h\phi$ also represents $\phi^*[f] + \phi^*[g]$.

F is a Homomorphism. To finish proving Lemma 9.4.2, we want to show that when X is a closed oriented PL $\mathbb{Q}$-Witt space and $m > (dim(X) + 1)/2$ then the assignment $F : \pi^m(X) \to \mathbb{Z}$ that takes $[f]$ to the signature of $f^{-1}(y)$ for a PL f and generic y is a homomorphism. By Proposition 9.4.1, we have already shown that F is well defined as a function on individual elements of $\pi^m(X)$. Now suppose $f, g : X \to S^m$ are PL maps, and let H be a homotopy from $(f,g) : X \to S^m \times S^m$ to $h : X \to S^m \vee S^m$. By the PL Approximation Theorem, we can assume that H and h are PL maps. Let $\pi_1, \pi_2 : S^m \times S^m \to S^m$ be the two projections, which are also PL maps. Then $\pi_1 H$ and $\pi_2 H$ are PL homotopies from f and g to respective maps $\tilde{f}, \tilde{g} : X \to S^m$. Notice that if we write

$S^m \vee S^m$ as the union of $S^m \vee \{s_0\}$ and $\{s_0\} \vee S^m$, then we can write $X = h^{-1}(S^m \vee S^m)$ as the union of $h^{-1}(S^m \vee \{s_0\})$ and $h^{-1}(\{s_0\} \vee S^m)$. If we let $\mathfrak{i}_1, \mathfrak{i}_2 : S^m \to S^m \vee S^m$ be the two inclusions, we observe that the restriction of h to $h^{-1}(S^m \vee \{s_0\})$ satisfies $h|_{h^{-1}(S^m \vee \{s_0\})} = \mathfrak{i}_1 \pi_1 h|_{h^{-1}(S^m \vee \{s_0\})} = \mathfrak{i}_1 \tilde{f}|_{h^{-1}(S^m \vee \{s_0\})}$, and similarly on $h^{-1}(\{s_0\} \vee S^m)$ we have $h|_{h^{-1}(\{s_0\} \vee S^m)} = \mathfrak{i}_2 \tilde{g}|_{h^{-1}(\{s_0\} \vee S^m)}$.

Now, as $[f] + [g]$ is represented by $\Omega h : X \to S^m$ and as Ω is also PL, the value of $F([f] + [g])$ is by definition the signature of $(\Omega h)^{-1}(y)$ for a generic $y \in S^m$. We may assume that $y \neq s_0$. We have $(\Omega h)^{-1}(y) = h^{-1}\Omega^{-1}(y)$, and $\Omega^{-1}(y) \in S^m \vee S^m$ consists of two copies of y, say $y_1 = \mathfrak{i}_1(y)$ and $y_2 = \mathfrak{i}_2(y)$, one in each of the spheres of $S^m \vee S^m$. And by the computation at the end of the last paragraph, we have $h^{-1}(y_1) = \tilde{f}^{-1}\mathfrak{i}_1^{-1}(y_1) = \tilde{f}^{-1}(y)$ and $h^{-1}(y_2) = \tilde{g}^{-1}\mathfrak{i}_2^{-1}(y_1) = \tilde{g}^{-1}(y)$. Therefore,

$$F([f] + [g]) = \sigma(\tilde{f}^{-1}(y) \amalg \tilde{g}^{-1}(y)) = \sigma(\tilde{f}^{-1}(y)) + \sigma(\tilde{g}^{-1}(y)).$$

But $\tilde{f}$ and $\tilde{g}$ are PL homotopic to f and g, and so up to possibly rechoosing our generic y, we have by Proposition 9.4.1 that

$$\sigma(\tilde{f}^{-1}(y)) + \sigma(\tilde{g}^{-1}(y)) = \sigma(f^{-1}(y)) + \sigma(g^{-1}(y)) = F([f]) + F([g]).$$

This establishes that F is a homomorphism.

We have now proven Lemma 9.4.2. □

To conclude this section, we provide the one further input we will need about cohomotopy by stating once again Serre's theorem (Theorem 9.4.3) relating cohomotopy and cohomology groups after rationalizing [214, Proposition 2′, page 289], though we do not attempt to indicate the proof[26]:

Theorem (Serre) *Let S^m be oriented, and let $u \in H^m(S^m)$ be the generator satisfying $u(\Gamma_{S^m}) = 1$. Suppose X^n is a compact CW complex and $m > (n+1)/2$. Define $\mathfrak{c} : \pi^m(X) \to H^m(X)$ by $\mathfrak{c}([f]) = f^*(u) \in H^m(X)$. Then $\mathfrak{c} \otimes \mathrm{id}_{\mathbb{Q}} : \pi^m(X) \otimes \mathbb{Q} \to H^m(X) \otimes \mathbb{Q}$ is an isomorphism.*

9.4.4 The L-Classes

Now that we have developed the necessary pieces, let us recall our definition of the L-classes $\mathscr{L}_m(X) \in H_m(X; \mathbb{Q})$ for X^n a closed oriented PL $\mathbb{Q}$-Witt space with $m > (n+1)/2$. We have defined the homomorphism $F : \pi^m(X) \to \mathbb{Z}$ that assigns to a homotopy class $[f] \in \pi^m(X)$ with a PL representative f the signature $\sigma(f^{-1}(y))$ for a generic point $y \in S^m$. We have also defined $\mathfrak{c} : \pi^m(X) \to H^m(X)$ so that $\mathfrak{c}([f]) = f^*(u)$, where $u \in H^m(S^m)$ is the class such that $u(\Gamma_{S^m}) = 1$. By Serre's Theorem, the map $\mathfrak{c} \otimes \mathrm{id}_{\mathbb{Q}} : \pi^m(X) \otimes \mathbb{Q} \to H^m(X) \otimes \mathbb{Q}$ is an isomorphism. Finally, we have the evaluation isomorphism

[26] In terms of spectra, this result appears in the form of the fact that the $\mathbb{Q}$-localization of the sphere spectrum is the rational Eilenberg–Mac Lane spectrum; see [199, Theorem II.7.1].

$\mathrm{ev} : H_m(X;\mathbb{Q}) \longrightarrow \mathrm{Hom}(H^m(X;\mathbb{Q});\mathbb{Q})$ such that $\mathrm{ev}(\xi)(\beta) = (-1)^m\beta(\xi)$. We have defined $\mathscr{L}_m(X)$ in Definition 9.4.4 to be the class determined by the composition

$$\mathscr{L}_m(X) = (-1)^m \mathrm{ev}^{-1}((F \otimes \mathrm{id}_{\mathbb{Q}}) \circ (\mathfrak{c} \otimes \mathrm{id}_{\mathbb{Q}})^{-1}),$$

which is zero if $n-m$ is not a multiple of 4. Note that we leave the isomorphism $H^*(X)\otimes\mathbb{Q} \cong H^*(X;\mathbb{Q})$ unlabeled in the composition. More computationally, we have also seen that if $\beta \in H^m(X;\mathbb{Q})$ is the image of $[f] \otimes r \in \pi^m(X) \otimes \mathbb{Q}$ under the isomorphism of Serre's Theorem, then $\beta(\mathscr{L}_m(X)) = r\sigma(f^{-1}(y))$ for a generic y; see Proposition 9.4.7.

To justify this definition, we will prove Proposition 9.4.8 in this section, which says that if M^n is a closed oriented smooth n-manifold, then for each $m = n - 4k$, $m > (n+1)/2$, the class $\mathscr{L}_m(M)$ is the Poincaré dual of the rational Thom–Hirzebruch L-class $L^k(X) \in H^{4k}(X;\mathbb{Q})$ [177, Chapter 19].

Both for proving Proposition 9.4.8 and for our characterization of L-classes in Theorem 9.4.18, the following lemma will be needed for the computations. We will state and prove a more general version than we need in this section in order to have the more general result for later. The statement involves the maps $\mathfrak{i}^!_W$ of Definition 9.4.16; we refer the reader back to Section 9.4.1 for the details of the definition.

Lemma 9.4.22 *Let $Z \subset X$ be space pairs such that Z has a neighborhood W that can be identified with a trivial $\mathbb{R}^m$-bundle $W = \mathbb{R}^m \times Z$, $m > 0$. Let $\mathfrak{i}_W : \mathbb{R}^m \times Z = W \hookrightarrow X$ be the inclusion, and let $\mathfrak{i}_W|_{\{0\}\times Z} = \mathfrak{i}_Z$ be the inclusion of Z. Identify S^m with the one point compactification of $\mathbb{R}^m$ with its standard orientation, letting z_0 denote the point at infinity. Let $u \in H^m(S^m)$ be the generator such that $u(\Gamma_{S^m}) = 1$. Finally, let $f : X \longrightarrow S^m$ be the map that takes each fiber $\mathbb{R}^m$ of the normal bundle W identically to $\mathbb{R}^m = S^m - \{z_0\}$ (i.e. $f(\mathfrak{i}_W(v,x)) = v$) and takes $X - W$ to z_0. Then for any homology class $\xi \in H_i(X)$ we have*

$$f^*(u) \frown \xi = \mathfrak{i}_Z\mathfrak{i}^!_W(\xi).$$

This statement also holds when $m = 0$, letting Z be a union of connected components of X and letting $u = 1_1$ as defined on page 659.

In particular, if X and Z are closed oriented classical pseudomanifolds and $\mathfrak{i}_W : \mathbb{R}^m\times Z = W \hookrightarrow X$ is orientation-preserving (giving $\mathbb{R}^m$ the standard orientation), then $f^*(u) \frown \Gamma_X = \mathfrak{i}_Z(\Gamma_Z)$.

This lemma also holds with $\mathbb{Q}$ coefficients.

Proof The proof is the same with $\mathbb{Z}$ or $\mathbb{Q}$ coefficients so we work over $\mathbb{Z}$ for simplicity. We will use a nice diagram. For $m > 0$, let us write $0 \in \mathbb{R}^m \subset S^m$. Then $Z = f^{-1}(0)$, and we have the following commutative diagram:

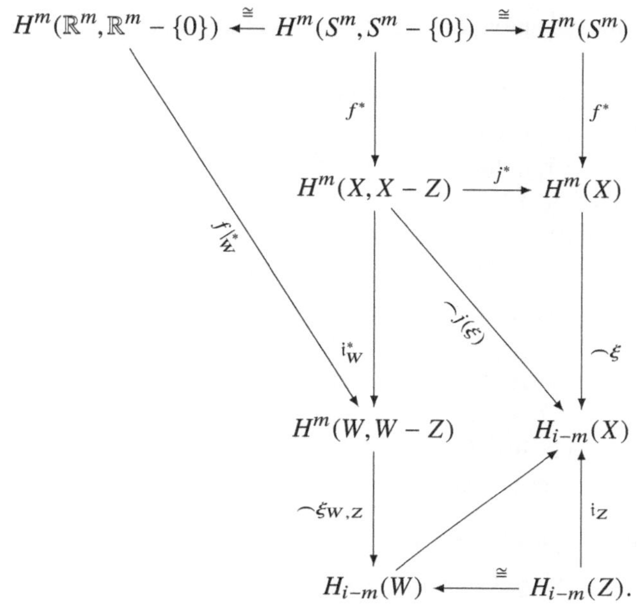

We will first explain the notation and commutativity of the diagram. Then we will show that the path counterclockwise around the outside of the diagram from the top right to the bottom right takes u to $\mathrm{i}_W^!(\xi) \in H_{i-m}(Z)$. It will then follow from the commutativity of the diagram that the image of $\mathrm{i}_W^!(\xi)$ in $H_{i-m}(X)$ under i_Z is equal to the composite $f^*(u) \frown \xi$, which proves the first claim.

The left portion of the diagram commutes by naturality of cohomology with $H^m(\mathbb{R}^m, \mathbb{R}^m - \{0\}) \cong H^m(S^m, S^m - \{0\})$ by excision.

The map j^* is induced by the inclusion $(X, \emptyset) \hookrightarrow (X, X - Z)$. The top square commutes by naturality of cohomology. The top right horizontal map is an isomorphism by the long exact sequence of the pair.

For the uppermost triangle on the right, we have commutativity from the naturality of the cap product (Proposition 7.3.6) with respect to $(X; \emptyset, \emptyset) \longrightarrow (X; \emptyset, X - Z)$.

Similarly, for the middle triangle on the right, we use the inclusion $(W; \emptyset, W - Z) \longrightarrow (X; \emptyset, X - Z)$. This inclusion induces an isomorphism by excision $H_i(W, W - Z) \longrightarrow H_i(X, X - Z)$. We let $\xi_{W,Z}$ be the image of $j(\xi)$ under the inverse isomorphism. Then we have commutativity again by naturality of the cap product.

Finally, the bottom triangle is induced by a triangle of inclusions of spaces; the bottom inclusion is a homotopy equivalence.

Next we show that the counterclockwise composition takes u to $\mathrm{i}_W^!(\xi)$. By Remark 9.4.17, we have that the Thom class μ is just $\mu = f|_W^*(a) = a \times 1_Z \in H^m(W, W - Z)$, where $a \in H^m(\mathbb{R}^m, \mathbb{R}^m - \{0\})$ is the image of $u \in H^m(S^m)$ under the isomorphisms $H^m(S^m) \cong H^m(S^m, S^m - \{0\}) \cong H^m(\mathbb{R}^m, \mathbb{R}^m - \{0\})$, the second isomorphism being an excision. So the image of u in the bottom left of the diagram is $\mu \frown \xi_{W,Z}$. As the projection $W = \mathbb{R}^m \times Z \to Z$ induces the inverse isomorphism to the inclusion $H_*(Z) \to H_*(W)$, it now follows from the definitions of $\mathfrak{i}^!_W$ and $\xi_{W,Z}$ that the image in the bottom right is $\mathfrak{i}^!_W(\xi)$ in $H_{i-m}(Z)$.

When $m = 0$, we have $f^*(u) = 1_Z$, the class that is 1 on the connected components of Z and 0 on the other components. Therefore, if we write the components of X as $\{X_j\}$ and $\xi = \sum \xi_j$ with $\xi_j \in H_i(X_j)$, then $f^*(u) \frown \xi = 1_Z \frown \xi = \sum_{X_j \subset Z} \xi_j$. But this is precisely $\mathfrak{i}_Z \mathfrak{i}^!(\xi)$ by Definition 9.4.16.

To complete the lemma, we need to show that when X and Z are closed oriented classical pseudomanifolds and $\xi = \Gamma_X$ then $\mathfrak{i}^!_W(\Gamma_X) = \Gamma_Z$. This is clear when $m = 0$. For $m > 0$, let $\Gamma_{W,W-Z} \in H_n(W, W-Z)$ denote the image of Γ_X under the maps $H_n(X) \to H_n(X, X-Z) \xleftarrow{\cong} H_n(W, W-Z)$, and let $\Gamma_{\mathbb{R}^m,\mathbb{R}^m-\{0\}}$ denote the fundamental class in $H_m(\mathbb{R}^m, \mathbb{R}^m - \{0\})$ consistent with the standard orientation. Since we have chosen orientations so that the product orientation on $\mathbb{R}^m \times Z$ is consistent with the orientation of X, we have $\Gamma_{W,W-Z} = \Gamma_{\mathbb{R}^m,\mathbb{R}^m-\{0\}} \times \Gamma_Z$ in $H_n(W, W-Z) = H_n((\mathbb{R}^m, \mathbb{R}^m - \{0\}) \times Z)$ by Proposition 8.1.39. Letting $\xi_0 \in H_0(\mathbb{R}^m)$ denote the canonical generator represented by any 0-simplex, we compute

$$\begin{aligned}
\mu \frown \Gamma_{W,W-Z} &= (a \times 1_Z) \frown (\Gamma_{\mathbb{R}^m,\mathbb{R}^m-\{0\}} \times \Gamma_Z) \\
&= (a \frown \Gamma_{\mathbb{R}^m,\mathbb{R}^m-\{0\}}) \times (1_Z \frown \Gamma_Z) \qquad \text{by Proposition 7.3.55} \\
&= \xi_0 \times \Gamma_Z.
\end{aligned}$$

The last equality employs Propositions 7.3.25 and 7.3.22. The latter tells us that $\mathbf{a}(a \frown \Gamma_{\mathbb{R}^m,\mathbb{R}^m-\{0\}}) = a(\Gamma_{\mathbb{R}^m,\mathbb{R}^m-\{0\}})$, but $a(\Gamma_{\mathbb{R}^m,\mathbb{R}^m-\{0\}}) = u(\Gamma_{S^m}) = 1$, and so $a \frown \Gamma_{\mathbb{R}^m,\mathbb{R}^m-\{0\}}$ must be the generator of $H_0(\mathbb{R}^m)$ that goes to 1 under the isomorphism $\mathbf{a} : H_0(\mathbb{R}^m) \to \mathbb{Z}$. Finally, the class $\xi_0 \times \Gamma_Z$ lives in $H_{n-m}(\mathbb{R}^m \times Z) \cong H_{n-m}(Z)$, and it is the image of Γ_Z under the inclusion $Z = \mathrm{pt} \times Z \hookrightarrow \mathbb{R}^m \times Z$; cf. Proposition 5.2.21. So the inverse isomorphism $H_{n-m}(\mathbb{R}^m \times Z) = H_{n-m}(W) \to H_{n-m}(Z)$ takes this class to Γ_Z.

This completes the proof of Lemma 9.4.22. □

Now we prove Proposition 9.4.8, which we restate here:

Proposition (Proposition 9.4.8) *If M^n is a closed oriented smooth n-manifold and $m > (n+1)/2$, then for $m = n - 4k$ the class $\mathscr{L}_m(M)$ is the Poincaré dual of the rational Thom–Hirzebruch L-class $L^k(M) \in H^{4k}(M;\mathbb{Q})$. Here $L^k(M)$ is the degree-k term of the multiplicative sequence associated to the power series of $\sqrt{t}/\tanh\sqrt{t}$ and taking as its variables the Pontrjagin classes of the tangent bundle of M (see [177, Chapter 19]).*

Proof For simplicity, we write $L^k = L^k(M)$ and $\mathscr{L}_m = \mathscr{L}_m(M)$. Under the assumption that $m > n + 1/2$ and $n - m \equiv 0 \bmod 4$, we must show that $L^{(n-m)/4} \frown \Gamma_M = \mathscr{L}_m$, where M is a closed oriented smooth manifold, $L^{(n-m)/4}$ is the rational Thom–Hirzebruch L-class in $H^{n-m}(M;\mathbb{Q})$ of the tangent bundle of M, and $\Gamma_M \in H_n(M;\mathbb{Q})$ is the rational fundamental class. Notice that since the degree of $L^{(n-m)/4}$ is even, the Poincaré duality map is simply the cap product with the fundamental class, with no sign. To show that

$L^{(n-m)/4} \frown \Gamma_M = \mathscr{L}_m$, we will show that for any $\beta \in H^m(M;\mathbb{Q}) \cong \operatorname{Hom}(H_m(M;\mathbb{Q}),\mathbb{Q})$ we have $\beta(L^{(n-m)/4} \frown \Gamma_M) = \beta(\mathscr{L}_m)$.

By Serre's Theorem, we know that any $\beta \in H^m(M;\mathbb{Q})$ can be written as $\beta = \mathfrak{c}([f]) \otimes r = r\mathfrak{c}([f])$ for some $f : M \to S^m$ and $r \in \mathbb{Q}$. Furthermore, recall that we have seen above in Proposition 9.4.7 that in this case

$$\beta(\mathscr{L}_m) = rF([f]) = r\sigma(f^{-1}(y)),$$

with y our generic point of S^m.

On the other hand, for $f : M \to S^m$, recall that $\mathfrak{c}([f]) = f^*(u)$ where $u \in H^m(S^m)$ satisfies $u(\Gamma_{S^m}) = 1$, and let $\mathbf{a} : H_0(M;\mathbb{Q}) \to \mathbb{Q}$ be the augmentation map. By a homotopy, we may assume f to be smooth [38, Theorem II.11.8], and then, for possibly a different generic $y \in S^m$, transversality theory says that the inverse image $N = f^{-1}(y)$ will be a smooth submanifold with a trivial normal bundle in M (see [38, Section II.16, particularly Theorem II.16.6]). Let $i : N \hookrightarrow M$ be the embedding. Then the restriction of the tangent bundle of M to N will be isomorphic to the direct sum of the tangent bundle TN with the trivial normal bundle of N in M. It follows from the basic properties of characteristic classes[27] that if L is the total L-class of M, then i^*L will be the total L-class of N. Thus $(i^*L)(\Gamma_N) = \sigma(N)$ by the Hirzebruch Signature Theorem [177, Theorem 19.4]. Using these observations and that $L^{(n-m)/4}$ has even degree, we compute

$$\begin{aligned}
&\beta(L^{(n-m)/4} \frown \Gamma_M)\\
&\quad= \mathbf{a}(\beta \frown (L^{(n-m)/4} \frown \Gamma_M)) && \text{by Proposition 7.3.25}\\
&\quad= \mathbf{a}((\beta \smile L^{(n-m)/4}) \frown \Gamma_M) && \text{by Proposition 7.3.35}\\
&\quad= \mathbf{a}((L^{(n-m)/4} \smile \beta) \frown \Gamma_M) && \text{by Proposition 7.3.15}\\
&\quad= \mathbf{a}((L^{(n-m)/4} \smile rf^*(u)) \frown \Gamma_M) &&\\
&\quad= r\mathbf{a}(L^{(n-m)/4} \frown (f^*(u) \frown \Gamma_M)) && \text{by Proposition 7.3.35}\\
&\quad= rL^{(n-m)/4}(f^*(u) \frown \Gamma_M) && \text{by Proposition 7.3.25}\\
&\quad= rL^{(n-m)/4}(i\Gamma_N) && \text{by Lemma 9.4.22}\\
&\quad= r(i^*L^{(n-m)/4})(\Gamma_N) &&\\
&\quad= r\sigma(N) = r\sigma(f^{-1}(y)) && \text{by the Hirzebruch Signature Theorem.}
\end{aligned}$$

This completes the proof of Proposition 9.4.8. □

9.4.5 L-Classes in Small Degrees

We now return to the situation $m \le (n+1)/2$, where $n = \dim(X)$. Recall that in Section 9.4.1 we defined $\mathscr{L}_m(X)$ for $m \le (n+1)/2$ as the image of $\mathscr{L}_{k+m}(S^k \times X)$ under the inverse

[27] Specifically, the L-classes are polynomials in the Pontrjagin classes p^j, so if we let ν denote the normal bundle of N in M and TM and TN the respective tangent bundles, then $i^*(p^j(TM)) = p^j(i^*(TM)) = p^j(\nu \oplus TN) = p^j(TN)$, as ν is trivial; see [177, Lemma 15.2]. As the polynomial defining L-classes is universal, we similarly have $i^*L(TM) = L(TN)$.

isomorphism $(\Gamma_{S^k}\times)^{-1} : H_{k+m}(S^k \times X; \mathbb{Q}) \to H_m(X; \mathbb{Q})$ for $k > n+1$. This range of k both makes $\Gamma_{S^k} \times \cdot$ an isomorphism and ensures that $\mathscr{L}_{k+m}(S^k \times X)$ is well defined by our first procedure. This definition was further motivated by Lemma 9.4.11, which demonstrated that in the category of smooth manifolds this is the only definition of $\mathscr{L}_m(X)$ for small $m = n - 4i$ that remains consistent with $\mathscr{L}_m(X) = \mathscr{L}_{n-4i}(X)$ being Poincaré dual to the Thom–Hirzebruch L-class $L^i(X) \in H^{4i}(X; \mathbb{Q})$.

It remains to show that the construction does not depend on the precise choice of k, which is the content of Proposition 9.4.10.

We first discuss an important tool in the proof of this proposition, namely a PL version of the quotient map $q : S^j \times S^k \to S^j \wedge S^k \cong S^{j+k}$. We will always assume that at least one of j and k is > 0. Recall that the smash product $S^j \wedge S^k$ is defined to be the quotient of $S^j \times S^k$ by the subspace $S^j \vee S^k = (S^j \times \text{pt}) \cup (\text{pt} \times S^k)$ and that the smash product of spheres is again a sphere; see [125, pages 10 or 223]. We will describe a simplicial map, also denoted $q : S^j \times S^k \to S^{j+k}$, that then gives us a PL map that will be sufficient for our purposes. To describe this map, consider any triangulation of $S^j \times S^k$ compatible with the product PL structure; for example, given triangulations of S^j and S^k we can use the construction of a product triangulation from Section B.6. Let s be any $(j+k)$-simplex of this triangulation. We can form the simplicial complex $J = s \cup \bar{c}(\partial s)$ by adjoining to s the cone on its boundary. We label the new vertex v and observe that J is the boundary of a $(j+k+1)$-simplex and so J is a simplicial S^{j+k}. Identifying the triangulations with the underlying spaces, we now define $q : S^j \times S^k \to S^{j+k}$ to be the simplicial map that is the identity on s and that takes all other vertices of $S^j \times S^k$ to v. In particular, this means that all simplices that are disjoint from s map to v. As s is top-dimensional, any simplex $\eta \neq s$ of $S^j \times S^k$ that is not disjoint from s must be spanned by some proper face τ of s together with some subset of vertices not in s. The map q takes all vertices of η to vertices of $\bar{c}\tau$, and so q is well defined on η. The map q takes the copy of s in $S^j \times S^k$ identically to the corresponding copy of s in S^{j+k}; we shall abuse notation below and refer to both of these simplices as s. Note that when $k = 0$ the space $S^j \times S^k$ is the disjoint union of $S^j \times \{0\}$ and $S^j \times \{1\}$. In this case, we shall assume that s is contained in $S^j \times \{1\}$. So then $q : S^j \times S^k \to S^{j+k} = S^j$ takes $S^j \times \{1\}$ to S^j by a degree-one map, while the restriction of q to $S^j \times \{0\}$ takes it to a point of S^j. If we like, we can instead let q be the identity map $S^j \times \{1\} \to S^j$ while still collapsing $S^j \times \{0\}$ to a point.

The PL map q we have just constructed is a surjection $S^j \times S^k \to S^{j+k}$ and a PL homeomorphism from s onto its identical image in S^{j+k}. If we assume s to be oriented based on an orientation of $S^j \times S^k$ then this carries over to determine an orientation of S^{j+k}, and we assume such compatible orientations in what follows. For any x in the interior of s, the map q induces an isomorphism $q : H_{j+k}(S^j \times S^k, S^j \times S^k - \{x\}) \to H_{j+k}(S^{j+k}, S^{j+k} - \{q(x)\})$. So by Theorem 8.1.18(3) we must then have $q(\Gamma_{S^j} \times \Gamma_{S^k}) = q(\Gamma_{S^j \times S^k}) = \Gamma_{S^{j+k}} \in H_{j+k}(S^{j+k})$.

We can now prove Proposition 9.4.10, which we restate:

Proposition (Proposition 9.4.10) *Let X^n be a closed oriented PL $\mathbb{Q}$-Witt space, and let*

$k, k' > n+1$. Consider for $0 \le m \le n$ the isomorphisms $H_m(X;\mathbb{Q}) \xrightarrow{\Gamma_k \times} H_{k+m}(S^k \times X;\mathbb{Q})$ and $H_m(X;\mathbb{Q}) \xrightarrow{\Gamma_{k'} \times} H_{k'+m}(S^{k'} \times X;\mathbb{Q})$.

If $\mathscr{L}_{k+m}(S^k \times X) \in H_{k+m}(S^k \times X;\mathbb{Q})$ and $\mathscr{L}_{k'+m}(S^{k'} \times X) \in H_{k'+m}(S^{k'} \times X;\mathbb{Q})$ are the respective homology L-classes, then

$$(\Gamma_{S^k}\times)^{-1}\mathscr{L}_{k+m}(S^k \times X) = (\Gamma_{S^{k'}}\times)^{-1}\mathscr{L}_{k'+m}(S^{k'} \times X) \in H_m(X;\mathbb{Q}).$$

Proof We first observe that because $k, k' > n+1$ the Künneth Theorem implies that the maps $\Gamma_{S^k}\times$ and $\Gamma_{S^{k'}}\times$ are isomorphisms. Furthermore, since $m \ge 0$ and $k, k' > n+1$ we have $2m + 2k > n + k + 1$, so $m + k > (n + k + 1)/2$ and similarly for k', meaning that $\mathscr{L}_{k+m}(S^k \times X)$ and $\mathscr{L}_{k+m'}(S^{k'} \times X)$ are defined by Definition 9.4.4. Now consider the following diagram

$$\begin{array}{ccc}
H_m(X;\mathbb{Q}) & \xrightarrow{\Gamma_{S^k}\times} & H_{k+m}(S^k \times X;\mathbb{Q}) \\
\Big\downarrow \Gamma_{S^{k'}}\times & \searrow^{\Gamma_{S^K}\times} & \Big\downarrow \\
H_{k'+m}(S^{k'} \times X;\mathbb{Q}) & \longrightarrow & H_{K+m}(S^K \times X;\mathbb{Q}).
\end{array}$$

We will show that for sufficiently large K we can define the maps on the right and bottom of the diagram so that the diagram commutes and so that these maps take $\mathscr{L}_{m+k}(S^k \times X)$ and $\mathscr{L}_{m+k'}(S^{k'} \times X)$ respectively to $\mathscr{L}_{m+K}(S^K \times X)$. Since the diagonal map will also be an isomorphism for any $K > n$, it will follow that

$$(\Gamma_{S^k}\times)^{-1}\mathscr{L}_{k+m}(S^k \times X) = (\Gamma_{S^K}\times)^{-1}\mathscr{L}_{K+m}(S^K \times X) = (\Gamma_{S^{k'}}\times)^{-1}\mathscr{L}_{m+k'}(S^{k'} \times X).$$

The constructions of the two commuting triangles in the diagram are equivalent, so we may focus on just the upper triangle, which will correspond to the following diagram:

$$\begin{array}{ccc}
H_m(X;\mathbb{Q}) & \xrightarrow{\Gamma_{S^k}\times} & H_{k+m}(S^k \times X;\mathbb{Q}) \\
 & & \Big\downarrow \Gamma_{S^{K-k}}\times \\
 & \searrow^{\Gamma_{S^K}\times} & H_{K+m}(S^{K-k} \times S^k \times X;\mathbb{Q}) \\
 & & \Big\downarrow q\times\mathrm{id} \\
 & & H_{K+m}(S^K \times X;\mathbb{Q}).
\end{array}$$

Here q is the PL quotient map $q : S^{K-k} \times S^k \to S^K$ constructed above.

By the naturality of the Künneth Theorem, the composition down the right side of the diagram takes a class of the form $\Gamma_{S^k} \times \xi$ to $q(\Gamma_{S^{K-k}} \times \Gamma_{S^k}) \times \xi$. By Proposition 8.1.39 we have $\Gamma_{S^{K-k}} \times \Gamma_{S^k} = \Gamma_{S^{K-k} \times S^k}$, but we have also seen that $q(\Gamma_{S^{K-k} \times S^k}) = \Gamma_{S^K}$. Thus the composition right then down in the above diagram takes ξ to $\Gamma_{S^K} \times \xi$, as desired. It remains to show that each of the vertical maps takes L-classes to L-classes. This is the content of the following two lemmas, which will complete the proof. In addition to having $k, k' > n+1$, which implies $m + k > (n + k + 1)/2$ as already observed, the degree condition of the first lemma translates here to the assumption that $K - k > k + n$ while the degree condition of the second requires $K - k \geq \max\{1, n - m - k + 3\}$. Both of these can be assured for this triangle and in the corresponding k' triangle by taking K to be sufficiently large.

This completes the proof of Proposition 9.4.10. □

Lemma 9.4.23 *If X^n is a closed oriented PL $\mathbb{Q}$-Witt space and $m > (n+1)/2$, then $\Gamma_{S^k} \times \mathscr{L}_m(X) = \mathscr{L}_{k+m}(S^k \times X)$ for any $k > n+1$.*

Proof We first consider the following diagram. We will show the diagram commutes for any $k > 0$ and $m \geq 0$; we do not need this generality here but will use it below in the proof of Proposition 9.4.13:

$$\begin{array}{ccccc} H^m(X) & \xleftarrow{\ c\ } & \pi^m(X) & & \\ \Big\downarrow{\scriptstyle (-1)^{km} u_k \times \cdot} & & \Big\downarrow{\scriptstyle q \circ (\mathrm{id}_{S^k} \times \cdot)} & \searrow^{F} & \\ & & & & \mathbb{Z}. \\ & & & \nearrow_{F} & \\ H^{k+m}(S^k \times X) & \xleftarrow{\ c\ } & \pi^{k+m}(S^k \times X) & & \end{array}$$

The diagonal maps are each the maps F that take a PL representative of a cohomotopy set to the signature of the inverse image of a generic point. The right vertical map takes $[f] \in \pi^m(X)$ to the class of the composition of $\mathrm{id}_{S^k} \times f : S^k \times X \to S^k \times S^m$ with the quotient map $q : S^k \times S^m \to S^{k+m}$ defined above.

To see that the right-hand triangle commutes, we may assume $f : X \to S^m$ is PL, and let us choose $(x, y) \in S^k \times S^m$ so that $(x, y) \subset s$ (recall the simplex s shared by $S^k \times S^m$ and S^{k-m} in our definition of q above), y is generic with respect to f (when $m = 0$ we take $y = 1 \in S^0$), and $z = q((x, y))$ is generic with respect to $q(\mathrm{id}_{S^k} \times f)$; this is possible using the density of generic image points of PL maps. Then $(q(\mathrm{id}_{S^k} \times f))^{-1}(z) = (q(\mathrm{id}_{S^k} \times f))^{-1}(q(x, y)) = (\mathrm{id}_{S^k} \times f)^{-1}(x, y) = \{x\} \times f^{-1}(y) \cong f^{-1}(y)$. As id_{S^k} preserves orientations, we see that the orientations of $f^{-1}(y)$ are consistent in the two constructions so that the triangle commutes. It follows that if y is any generic point of f and z is any generic point of $q(\mathrm{id}_{S^k} \times f)$ then $\sigma(f^{-1}(y)) = \sigma((q(\mathrm{id}_{S^k} \times f))^{-1}(z))$.

Turning to the square, let $u_i \in H^i(S^i)$ be the cohomology class such that $u_i(\Gamma_{S^i}) = 1$; we have arranged in our notation on page 659 that such a formula also holds when $i = 0$. We know that $q(\Gamma_{S^k} \times \Gamma_{S^m}) = q(\Gamma_{S^k \times S^m}) = \Gamma_{S^{k+m}}$, so we have

$$(q^*(u_{k+m}))(\Gamma_{S^k \times S^m}) = u_{k+m}(q(\Gamma_{S^k \times S^m})) = u_{k+m}(\Gamma_{S^{k+m}}) = 1.$$

On the other hand,

$$(u_k \times u_m)(\Gamma_{S^k \times S^m}) = (u_k \times u_m)(\Gamma_{S^k} \times \Gamma_{S^m}) = (-1)^{km} u_k(\Gamma_{S^k}) u_m(\Gamma_{S^m}) = (-1)^{km}.$$

So, as $H^{k+m}(S^k \times S^m) \cong \mathbb{Z}$ when $m > 0$, we must have in this case $u_k \times u_m = (-1)^{km} q^*(u_{k+m})$. When $m = 0$, the formula holds directly as $u_k \times u_m$ and $q^*(u_{k+m})$ both correspond to the cocycle that restricts to $u_k = u_k \times 1$ on $S^j \times \{1\}$ and that is trivial on $S^j \times \{0\}$.

Therefore, from the definitions, if $[f] \in \pi^m(X)$, then going left then down in the diagram gives us $(-1)^{km} u_k \times f^*(u_m)$. And going down then left gives us $(q(\mathrm{id}_{S^k} \times f))^*(u_{k+m}) = (\mathrm{id}_{S^k} \times f)^* q^*(u_{k+m}) = (-1)^{km} (\mathrm{id}_{S^k} \times f)^*(u_k \times u_m) = (-1)^{km} u_k \times f^*(u_m)$. So the diagram commutes.

Furthermore, when $m > (n+1)/2$, upon tensoring the diagram with $\mathbb{Q}$ the horizontal maps become group isomorphisms by Serre's Theorem, and the left-hand vertical map is an isomorphism already over $\mathbb{Z}$ by the Künneth Theorem as $k > n+1$. So the right side of the square also becomes an isomorphism when tensored with $\mathbb{Q}$, and this implies that every element of $\pi^{k+m}(S^k \times X) \otimes \mathbb{Q}$ has the form $[q(\mathrm{id}_{S^k} \times f)] \otimes r$ for some $r \in \mathbb{Q}$ and $f : X \to S^m$ (see footnote 18 on page 652). As

$$\mathfrak{c} \otimes \mathrm{id}_{\mathbb{Q}} : \pi^{k+m}(S^k \times X) \otimes \mathbb{Q} \to H^{k+m}(S^k \times X) \otimes \mathbb{Q} \cong H^{k+m}(S^k \times X; \mathbb{Q})$$

is an isomorphism, we conclude that every element of $H^{k+m}(S^k \times X; \mathbb{Q})$ has the form

$$\mathfrak{c}([q(\mathrm{id}_{S^k} \times f)]) \otimes r = r\mathfrak{c}([q(\mathrm{id}_{S^k} \times f)])$$

for some $r \in \mathbb{Q}$ and $f : X \to S^m$.

Now, let $\beta \in H^m(X; \mathbb{Q}) \cong H^m(X) \otimes \mathbb{Q}$, and suppose $\beta = \mathfrak{c}([f]) \otimes r = r\mathfrak{c}([f])$. From Proposition 9.4.7 we know that $\beta(\mathscr{L}_m(X)) = r\sigma(f^{-1}(y))$ for a generic $y \in S^m$. Similarly, we know by definition and the constructions above that if we let

$$\beta' = \mathfrak{c}([q(\mathrm{id}_{S^k} \times f)]) \otimes r = r\mathfrak{c}([q(\mathrm{id}_{S^k} \times f)]) \in H^{k+m}(S^k \times X; \mathbb{Q})$$

then $\beta'(\mathscr{L}_{k+m}(S^k \times X)) = r\sigma((q(\mathrm{id}_{S^k} \times f))^{-1}(z))$ for a generic $z \in S^{k+m}$, and we have already seen that this must equal $r\sigma(f^{-1}(y))$.

On the other hand, using the commutativity of the above diagram, we compute

$$\begin{aligned}
\beta'(\Gamma_{S^k} \times \mathscr{L}_m(X)) &= r\mathfrak{c}([q(\mathrm{id}_{S^k} \times f)])(\Gamma_{S^k} \times \mathscr{L}_m(X)) \\
&= (-1)^{km} r (u_k \times \mathfrak{c}([f]))(\Gamma_{S^k} \times \mathscr{L}_m(X)) \\
&= (-1)^{km}(-1)^{km} r u_k(\Gamma_{S^k}) \cdot \mathfrak{c}([f])(\mathscr{L}_m(X)) \\
&= \beta(\mathscr{L}_m(X)) \\
&= r\sigma(f^{-1}(y)).
\end{aligned}$$

So for all $\beta' \in H^{k+m}(S^k \times X; \mathbb{Q})$, we see that β' evaluates identically on $\Gamma_{S^k} \times \mathscr{L}_m(X)$ and $\mathscr{L}_{k+m}(S^k \times X)$. As $H^{k+m}(S^k \times X; \mathbb{Q}) \cong \mathrm{Hom}(H_{k+m}(S^k \times X; \mathbb{Q}), \mathbb{Q})$, this shows that $\mathscr{L}_{k+m}(S^k \times X) = \Gamma_{S^k} \times \mathscr{L}_m(X)$, completing the proof of Lemma 9.4.23. □

Lemma 9.4.24 *Suppose X^n is a closed oriented PL $\mathbb{Q}$-Witt space, $k > n+1$, and $j \geq \max\{1, n-m-k+3\}$. Then $q \times \mathrm{id} : H_{j+k+m}(S^j \times S^k \times X; \mathbb{Q}) \to H_{j+k+m}(S^{j+k} \times X; \mathbb{Q})$ takes $\mathscr{L}_{j+k+m}(S^j \times S^k \times X)$ to $\mathscr{L}_{j+k+m}(S^{j+k} \times X)$.*

Proof This time we begin with the diagram

$$\begin{array}{ccccc} H^{j+k+m}(S^{j+k} \times X) & \xleftarrow{\mathfrak{c}} & \pi^{j+k+m}(S^{j+k} \times X) & \searrow^{\mathscr{F}} & \\ \Big\downarrow (q \times \mathrm{id})^* & & \Big\downarrow (q \times \mathrm{id})^* & & \mathbb{Z}. \\ H^{j+k+m}(S^j \times S^k \times X) & \xleftarrow{\mathfrak{c}} & \pi^{j+k+m}(S^j \times S^k \times X) & \nearrow_{\mathfrak{F}} & \end{array}$$

The square commutes because, given $f : S^{j+k} \times X \to S^{j+k+m}$ and our usual generator, $u \in H^{j+k+m}(S^{j+k+m})$, then

$$\begin{aligned}(q \times \mathrm{id})^* \mathfrak{c}([f]) &= (q \times \mathrm{id})^* f^*(u) = (f(q \times \mathrm{id}))^*(u) = \mathfrak{c}([f(q \times \mathrm{id})]) \\ &= \mathfrak{c}(q \times \mathrm{id})^*([f]).\end{aligned}$$

For the triangle, let $f : S^{j+k} \times X \to S^{j+k+m}$ be a PL map representing an arbitrary class in $\pi^{j+k+m}(S^{j+k} \times X)$, and let $y \in S^{j+k+m}$ be generic. By Proposition 9.4.1, the inverse image $f^{-1}(y)$ is a PL subspace of dimension $n-m$. Consider the projection map $\pi : S^{j+k} \times X \to S^{j+k}$, which is PL and so can be made simplicial for some triangulations[28]. The image of an $(n-m)$-dimensional simplicial complex under a simplicial map must have dimension $\leq n-m$, and since $n-m \leq n < j+k$ there must be a point $z \in S^{j+k}$ that is disjoint from $\pi(f^{-1}(y))$. Let $g : S^{j+k} \to S^{j+k}$ be a PL homeomorphism that is PL isotopic to the identity and that shrinks[29] $\pi(f^{-1}(y))$ into the interior of the simplex s used in the construction of q (though note that s is here being used as a subspace, not as a simplex in any particular triangulation). Then $\tilde{f} = f(g^{-1} \times \mathrm{id})$ is PL homotopic to f and

$$\tilde{f}^{-1}(y) = (f(g^{-1} \times \mathrm{id}))^{-1}(y) = (g^{-1} \times \mathrm{id})^{-1}(f^{-1}(y)) = (g \times \mathrm{id})(f^{-1}(y))$$

is contained in $s \times X$.

So now consider $\tilde{f}(q \times \mathrm{id}) : S^j \times S^k \times X \to S^{j+k+m}$. We have just seen that $\tilde{f}^{-1}(y) \subset s \times X$, and as we think of q as restricting to the identity between our two copies of s, it follows that $(\tilde{f}(q \times \mathrm{id}))^{-1}(y) \cong \tilde{f}^{-1}(y)$. But using again $[\tilde{f}] = [f] \in \pi^{j+k+m}(S^{j+k} \times X)$, we thus have

$$\mathscr{F}([f]) = \sigma(\tilde{f}^{-1}(y)) = \sigma((\tilde{f}(q \times \mathrm{id}))^{-1}(y)) = \mathfrak{F}((q \times \mathrm{id})^*[f]),$$

and the triangle commutes. We have now shown that the whole diagram commutes.

28 In fact, using the construction of Section B.6 to triangulate a product $X \times Y$ of simplicial complexes, it is not difficult to write down simplicial projection maps explicitly.

29 To construct such a homeomorphism, using probably a much bigger hammer than necessary, we can apply engulfing techniques, for example [65, Theorem 3.1.3]. In that theorem we take $W = S^{j+k}$, U the interior of s, $K = \pi(f^{-1}(y))$, $L = \emptyset$. Then certainly the pair (S^{j+k}, U) is $n-m$ connected, as required, and the only other condition is the requirement that $n - m \leq j + k - 3$, which is the reason for our assumption on j. Then [65, Theorem 3.1.3] says that there is a PL ambient isotopy h_t of S^{j+k} such that h_0 is the identity and $h_1(U)$ contains K. We can then let our g be h_1^{-1}.

Now, as $H_*(S^{j+k} \times X; \mathbb{Q}) = \mathrm{Hom}(H^*(S^{j+k} \times X); \mathbb{Q})$, to show that

$$(q \times \mathrm{id})(\mathscr{L}_{j+k+m}(S^j \times S^k \times X)) = \mathscr{L}_{j+k+m}(S^{j+k} \times X)$$

it suffices to show that we have

$$\alpha((q \times \mathrm{id})(\mathscr{L}_{j+k+m}(S^j \times S^k \times X))) = \alpha(\mathscr{L}_{j+k+m}(S^{j+k} \times X))$$

for any $\alpha \in H^{j+k+m}(S^{j+k} \times X; \mathbb{Q})$. As $k > n+1$ and $j, m \geq 0$, by Serre's Theorem we can write $\alpha = \mathfrak{c}([f]) \otimes r = r\mathfrak{c}([f])$ for some PL map $f : S^{j+k} \times X \to S^{j+k+m}$, and by Proposition 9.4.7 we know that then $\alpha(\mathscr{L}_{j+k+m}(S^{j+k} \times X)) = rF([f])$.

For $(q \times \mathrm{id})(\mathscr{L}_{j+k+m}(S^j \times S^k \times X))$, we have

$$\begin{aligned}
\alpha((q \times \mathrm{id})(\mathscr{L}_{j+k+m}(S^j \times S^k \times X))) &= ((q \times \mathrm{id})^*\alpha)(\mathscr{L}_{j+k+m}(S^j \times S^k \times X)) \\
&= ((q \times \mathrm{id})^*(r\mathfrak{c}([f])))(\mathscr{L}_{j+k+m}(S^j \times S^k \times X)) \\
&= r(\mathfrak{c}((q \times \mathrm{id})^*([f])))(\mathscr{L}_{j+k+m}(S^j \times S^k \times X)) \\
&= rF((q \times \mathrm{id})^*([f])) \\
&= rF([f]).
\end{aligned}$$

For the third and fifth equalities we have used the commutativity of the above diagram, while for the fourth we use again Proposition 9.4.7 applied to the homology class $\mathscr{L}_{j+k+m}(S^j \times S^k \times X)$ and the cohomology class represented by $r[\mathfrak{c}((q \times \mathrm{id})^*([f]))] = \mathfrak{c}((q \times \mathrm{id})^*([f])) \otimes r$. As $\alpha \in H^{j+k+m}(S^{j+k} \times X; \mathbb{Q})$ was arbitrary, we see that $(q \times \mathrm{id})(\mathscr{L}_{j+k+m}(S^j \times S^k \times X)) = \mathscr{L}_{j+k+m}(S^{j+k} \times X)$. This completes the proof of Lemma 9.4.24. □

(a) Extending Properties to Small Degrees

We have now shown that $\mathscr{L}_m(X)$ is well defined for all m. Let us extend some of the properties we have already verified when $m > (n+1)/2$ to other values of m.

First, putting the definition of $\mathscr{L}_m(X)$ for $m \leq (n+1)/2$ together with Lemma 9.4.23 we immediately have the following general statement for all $m \geq 0$:

Lemma (Lemma 9.4.12) *If X^n is a closed oriented PL $\mathbb{Q}$-Witt space and $m \geq 0$, then $\Gamma_{S^k} \times \mathscr{L}_m(X) = \mathscr{L}_{k+m}(S^k \times X)$ for any $k > n+1$.*

Next we see that the L-classes for small m continue to satisfy what is essentially their defining property as stated in Proposition 9.4.7. This was our computation showing that when $\beta \in H^m(X; \mathbb{Q})$ is the image of $[f] \otimes r \in \pi^m(X) \otimes \mathbb{Q}$ under the isomorphism of Serre's Theorem, i.e. when $\beta = f^*(u) \otimes r$, then $\beta(\mathscr{L}_m(X)) = rF([f]) = r\sigma(f^{-1}(y))$. In small degrees we don't have Serre's Theorem available (in fact, we don't have cohomotopy groups), but we can still evaluate $\mathscr{L}_m(X)$ with respect to those cohomology classes that happen to have the form $f^*(u) \otimes r \in H^*(X) \otimes \mathbb{Q} \cong H^*(X; \mathbb{Q})$, and in this case we maintain our formula.

Proposition (Proposition 9.4.13) *Let X be a closed oriented PL $\mathbb{Q}$-Witt space, and let $f : X \to S^m$ be a PL map for $m > 0$. If $y \in S^m$ is a generic point, $r \in \mathbb{Q}$, and $u \in H^m(S^m)$ is the cohomology class such that $u(\Gamma_{S^m}) = 1$, then*

$$(f^*(u) \otimes r)(\mathscr{L}_m(X)) = r\sigma(f^{-1}(y)). \tag{9.5}$$

If $m = 0$, the formula will hold if we take $y = 1 \in S^0 = \{0, 1\}$ and let $u = 1_1 \in H^0(S^0)$ be the cocycle that restricts to the augmentation class in $H^0(\{1\})$ and to 0 in $H^0(\{0\})$.

Proof For $m > (n + 1)/2$, equation (9.5) is just the statement of Proposition 9.4.7, using that $c([f]) \otimes r = f^*(u) \otimes r$ by definition.

For $0 \le m \le (n + 1)/2$, we use the arguments from the proof of Lemma 9.4.23. From that proof, we know that $q(\mathrm{id}_{S^k} \times f)$ also has $f^{-1}(y)$ as a generic point inverse, and so for a large enough k we have

$$((q(\mathrm{id}_{S^k} \times f))^*(u_{k+m}) \otimes r)(\mathscr{L}_{k+m}(S^k \times X)) = r\sigma(f^{-1}(y))$$

by the preceding case, letting $u_i \in H^i(S^i)$ be our generating class in degree i (or $1_1 \in H^0(S^0)$). But by definition in this range we have $\mathscr{L}_{k+m}(S^k \times X) = \Gamma_{S^k} \times \mathscr{L}_m(X)$, and also from the proof of Lemma 9.4.23 we have

$$(q(\mathrm{id}_{S^k} \times f))^*(u_{k+m}) = (-1)^{km} u_k \times f^*(u_m)$$

for any $k > 0$ and $m \ge 0$. And so

$$\begin{aligned} r\sigma(f^{-1}(y)) &= ((q(\mathrm{id}_{S^k} \times f))^*(u_{k+m}) \otimes r)(\mathscr{L}_{k+m}(S^k \times X)) \\ &= (-1)^{km}((u_k \times f^*(u_m)) \otimes r)(\Gamma_{S^k} \times \mathscr{L}_m(X)) \\ &= u_k(\Gamma_{S^k}) \cdot (f^*(u_m) \otimes r)(\mathscr{L}_m(X)) \\ &= (f^*(u_m) \otimes r)(\mathscr{L}_m(X)). \end{aligned}$$

We have now demonstrated the desired formula for all m, and the proof of Proposition 9.4.13 is complete. □

As a consequence of this proposition, we can see that $\mathscr{L}_0(X)$ takes a particularly pleasant form. Suppose we let $\{X_j\}$ be the connected components of X. Then because homology is additive over disjoint unions we must have $\mathscr{L}_0(X) = \sum_j c_j \xi_j \in H_0(X; \mathbb{Q})$, where ξ_j is any 0-simplex in X_j. If Z is any connected component of X, then let $f_Z : X \to S^0$ be the map such that $f(Z) = 1 \in S^0$ and $f(X - Z) = 0 \in S^0$. From our definition of $u_0 \in H^0(S^0)$, we have $f_Z^*(u) = 1_Z$, the class represented by the cochain that evaluates to 1 on all 0-simplices of Z and is trivial otherwise. Then

$$\sigma(Z) = \sigma\left(f_Z^{-1}(1)\right) = (f^*(u))(\mathscr{L}_0(X)) = 1_Z\left(\sum_j c_j \xi_j\right) = c_j.$$

So we have proven the following:

Proposition (Proposition 9.4.14) *Let X be a closed oriented PL $\mathbb{Q}$-Witt space, and let $\{X_j\}$ be the connected components of X. Then*

$$\mathscr{L}_0(X) = \sum_j \sigma(X_j)\xi_j \in H_0(X;\mathbb{Q}),$$

where ξ_j is any 0-simplex in X_j.

9.4.6 Characterizing the L-Classes

Recall that our last goal is to provide a single characterization of all our L-classes on closed oriented PL $\mathbb{Q}$-Witt spaces in terms of their behavior with respect to "wrong-way" maps induced by PL trivial normally nonsingular inclusions together with a normalization condition imposed on the classes $\mathscr{L}_0$. In particular, we now turn to proving Theorem 9.4.18, which we restate here:

Theorem (Theorem 9.4.18) *The L-classes $\mathscr{L}_*$ defined on closed oriented PL $\mathbb{Q}$-Witt spaces possess the following properties:*

1. $\mathbf{a}(\mathscr{L}_0(X)) = \sigma(X)$,
2. *if Z is a PL trivial normally nonsingular subset of X and $\mathfrak{i}_W : \mathbb{R}^m \times Z \xrightarrow{\cong} W \subset X$ is the orientation-preserving PL homeomorphism of a trivial $\mathbb{R}^m$-bundle onto a neighborhood W of Z in X then $\mathfrak{i}_W^!(\mathscr{L}_{j+m}(X)) = \mathscr{L}_j(Z)$ for all j.*

Furthermore, the collection of classes $\{\mathscr{L}_*(X)\}$ as X ranges over all closed oriented PL $\mathbb{Q}$-Witt spaces is the unique collection with these properties.

Let us recall the definitions involved. As defined in Definition 9.4.15, the subspace $Z \subset X$ of the PL $\mathbb{Q}$-Witt space X is a PL trivial normally nonsingular subspace (PL trivial nns for short) if for some m the inclusion map $\mathfrak{i}_Z : Z \hookrightarrow X$ extends to a PL homeomorphism $\mathfrak{i}_W : \mathbb{R}^m \times Z \to W \subset X$ onto a neighborhood W of Z in X. Via this homeomorphism we will typically identify W with $\mathbb{R}^m \times Z$. When X and Z are oriented and $m > 0$, the map $\mathfrak{i}_W^! : H_i(X) \to H_{i-m}(Z)$ is then defined in Definition 9.4.16 as the composition

$$H_i(X) \to H_i(X, X - Z) \xleftarrow{\cong} H_i(W, W - Z) \xrightarrow{\mu\frown} H_{i-m}(W) \xrightarrow{\cong} H_{i-m}(Z),$$

where μ is the Thom class of the bundle. We will orient the bundle so that $\mathfrak{i}_W$ is orientation-preserving, giving $\mathbb{R}^m$ the standard orientation and $\mathbb{R}^m \times Z$ the product orientation. This can always be done by composing $\mathfrak{i}_W$ with a reflection of $\mathbb{R}^m$ if necessary. Identifying W with $\mathbb{R}^m \times Z$, we can then take $\mu = a \times 1_Z$, where $1_Z \in H^0(Z)$ and a is the generator of $H^m(\mathbb{R}^m, \mathbb{R}^m - \{0\})$ that takes the fundamental class with respect to the standard orientation to 1. When $m = 0$, then Z is a union of connected components of $X = \amalg X_j$, and we then let $\mathfrak{i}^!$ be the projection $H_i(X) \cong \bigoplus_{X_j} H_i(X_j) \to \bigoplus_{X_j \subset Z} H_i(X_j) \cong H_i(Z)$.

The connection between L-classes and normally nonsingular subspaces is provided by the following definition (see [249, Section 1.3]), lemma, and corollary:

Definition 9.4.25 Let $f : X \to Y$ be a PL map. Then $y \in Y$ is called a *regular point* if there is a PL neighborhood U of y in Y and a PL homeomorphism $h : U \times f^{-1}(y) \to f^{-1}(U)$ such that $fh = \pi_1$, where $\pi_1 : U \times f^{-1}(y) \to U$ is the projection.

Remark 9.4.26 If $h : U \times f^{-1}(y) \to f^{-1}(U)$ is any PL homeomorphism such that $fh = \pi_1$ then h restricts to a PL homeomorphism $\tilde{h}$ from $\{y\} \times f^{-1}(y)$ to $f^{-1}(y)$. By precomposing h with $\mathrm{id} \times \tilde{h}^{-1}$, we can always assume that h acts as an extension of the canonical inclusion $f^{-1}(y) \hookrightarrow f^{-1}(U) \subset X$. We will do so implicitly from here on.

Lemma 9.4.27 *Let X be a closed oriented PL $\mathbb{Q}$-Witt space. Then $Z \subset X$ is a PL trivial normally nonsingular subspace of codimension m, $m > 0$, if and only if there exists a PL map $f : X \to S^m$ such that $Z = f^{-1}(y)$ for a regular point $y \in S^m$. This result extends to $m = 0$ by taking $y = 1 \in S^0$.*

Proof First suppose $m = 0$. If Z is any nns, then Z is a union of connected components of X, so in this case we let f take Z to $1 \in S^0$ and its complement to $0 \in S^0$. Conversely, given any $f : X \to S^0$, the space $Z = f^{-1}(1)$ is a union of connected components and so normally nonsingular. For the rest of the proof we assume $m > 0$.

If y is a regular point of a PL map $f : X \to S^m$, then $f^{-1}(y)$ is a PL trivial nns directly from the definitions, as any neighborhood U of y contains a smaller neighborhood PL homeomorphic to $\mathbb{R}^m$ and the restriction of a PL homeomorphism to an open subspace continues to be a PL homeomorphism onto its image.

Conversely, suppose $Z \subset X$ is a PL trivial nns, and let us identify a neighborhood W of Z with $\mathbb{R}^m \times Z$, with Z being identified with the 0-section of the trivial bundle. Identifying S^m with the one-point compactification of $\mathbb{R}^m$ with z_0 denoting the point at infinity, we define a map $\bar{f} : X \to S^m$ that takes each point $(x, z) \in \mathbb{R}^m \times Z = W$ to $x \in \mathbb{R}^m = S^m - \{z_0\}$ and maps $X - W$ to z_0. It is not clear as defined that $\bar{f}$ is a PL map on all of X, but it is PL on W as a projection map. If we let $U \subset V \subset \mathbb{R}^m$ be neighborhoods of 0 with $\bar{U} \subset V$ and $\bar{V}$ compact in $\mathbb{R}^m$, then by the PL Approximation Theorem (Theorem B.2.24) we can find a PL map $f : X \to S^m$ that is homotopic to $\bar{f}$, that agrees with $\bar{f}$ on $\bar{V} \times Z$, and such that no point of $X - (V \times Z)$ maps under f into U. It follows that $f^{-1}(\{0\}) = Z$ with U a neighborhood of 0 in $\mathbb{R}^m \subset S^m$ such that $f^{-1}(U)$ is PL homeomorphic to $U \times Z$ with f acting as the projection onto U. So 0 is a regular point and $Z = f^{-1}(0)$. □

Corollary 9.4.28 *If Z is a PL trivial normally nonsingular subspace of a closed orientable PL $\mathbb{Q}$-Witt space X, then Z is also a closed orientable PL $\mathbb{Q}$-Witt space (with respect to some filtration). If Z is oriented so that the embedding $i_W : \mathbb{R}^m \times Z \hookrightarrow X$ is orientation-preserving and $f : X \to S^m$ is a PL map such that $Z = f^{-1}(y)$ for a regular point $y \in S^m$, then*[30] *$\sigma(Z) = F([f])$.*

Proof Again, this is trivial if $m = 0$, i.e. if Z is a union of connected components of X. So we assume $m > 0$.

Suppose Z is a PL trivial nns of X, and let $f : X \to S^m$ be a PL map such that

[30] Recall the definition of F on page 651.

$Z = f^{-1}(0)$ with $0 \in \mathbb{R}^m \subset S^m$ a regular point; we know such a map exists from the proof of the previous lemma. In particular, this means that 0 has a neighborhood U in S^m such that $f^{-1}(U)$ is PL homeomorphic to $U \times Z$, with the homeomorphism compatible with the projection. For any $y \in U$, we must have $Z \cong f^{-1}(y)$. It is also clear that if we orient Z and $f^{-1}(y)$ so that their product neighborhoods are compatibly oriented with X then the homeomorphism $Z \cong f^{-1}(y)$ also preserves orientation. By Theorem B.2.19, there are triangulations of X and S^m with respect to which f is simplicial. As U is a neighborhood of 0, it contains some y in the interior of an m-simplex of the triangulation of S^m, and $f^{-1}(y)$ is a closed PL Witt space by Proposition 9.4.1. Therefore, Z is also a closed PL Witt space, and $\sigma(Z) = \sigma(f^{-1}(y))$ by the invariance of signatures under orientation-preserving homeomorphisms (Theorem 9.3.16). Now we just apply that $\sigma(f^{-1}(y)) = F([f])$ by definition of the map $F : \pi^m(X) \to \mathbb{Z}$. □

Now we turn to proving Theorem 9.4.18. The proof will rely on two lemmas that will be proved after the proof of the theorem. For simplicity, we state and prove the lemmas with $\mathbb{Z}$ coefficients, but the arguments work just as well over $\mathbb{Q}$ (or other rings).

Lemma 9.4.29 *Let $Z \subset X$ be a PL trivial normally nonsingular subspace with PL homeomorphism $\mathfrak{i}_W : \mathbb{R}^m \times Z \to W \subset X$. Then $S^k \times Z$ is a PL trivial normally nonsingular subspace of $S^k \times X$ with neighborhood embedding $\mathfrak{i}_{S^k \times W} : \mathbb{R}^m \times S^k \times Z \xrightarrow{\cong} S^k \times W$. Letting $\Gamma_{S^k} \in H_k(S^k)$ be the standard orientation class, $k > 0$, then for $\xi \in H_{*+m}(X)$ we have*

$$\Gamma_{S^k} \times \mathfrak{i}_W^!(\xi) = \mathfrak{i}_{S^k \times W}^!(\Gamma_{S^k} \times \xi) \in H_{*+k}(S^k \times Z).$$

Lemma 9.4.30 *Let $Z \subset X$ be a PL trivial normally nonsingular subspace with PL homeomorphism $\mathfrak{i}_W : \mathbb{R}^m \times Z \to W \subset X$, and let $A \subset Z$ be a PL trivial normally nonsingular subspace with PL homeomorphism $\mathfrak{i}_V : \mathbb{R}^j \times A \to V \subset Z$. Let $\mathfrak{i}_U$ be the composition*

$$\mathbb{R}^{m+j} \times A = \mathbb{R}^m \times \mathbb{R}^j \times A \xrightarrow{\mathrm{id} \times \mathfrak{i}_V} \mathbb{R}^m \times Z \xrightarrow{\mathfrak{i}_W} X,$$

taking $\mathbb{R}^{m+j} \times A$ to a neighborhood U of A in X. Then

$$\mathfrak{i}_U^! = \mathfrak{i}_V^! \mathfrak{i}_W^! : H_{*+m+j}(X) \to H_*(A).$$

We will prove the lemmas after using them to prove Theorem 9.4.18.

Proof of Theorem 9.4.18 We first show that the L-classes have the claimed properties. Let X be a closed oriented PL $\mathbb{Q}$-Witt space. The fact that $\mathbf{a}(\mathscr{L}_0(X)) = \sigma(X)$ follows immediately from Proposition 9.4.14, according to which $\mathscr{L}_0(X) = \sum_j \sigma(X_j)\xi_j \in H_0(X; \mathbb{Q})$ with ξ_j being any 0-simplex in the connected component X_j, and the additivity of signatures over disjoint union (Theorem 9.3.17(2)).

Next suppose Z is a PL trivial nns in $X = X^n$ and $\mathfrak{i}_W : \mathbb{R}^m \times Z \xrightarrow{\cong} W \subset X$ is the PL homeomorphism onto a neighborhood W of Z in X. We want to show that $\mathfrak{i}_W^!(\mathscr{L}_{j+m}(X)) = \mathscr{L}_j(Z)$ for any $j \geq 0$, the case $j < 0$ being trivial. As $\Gamma_{S^k} \times$ is an isomorphism for large enough k, it suffices to show that $\Gamma_{S^k} \times \mathfrak{i}_W^!(\mathscr{L}_{j+m}(X)) = \Gamma_{S^k} \times \mathscr{L}_j(Z) \in H_{k+j}(S^k \times Z; \mathbb{Q})$ for large k. By Lemma 9.4.12, we have $\Gamma_{S^k} \times \mathcal{L}_j(Z) = \mathcal{L}_{k+j}(S^k \times Z)$, while by that corollary and Lemma 9.4.29 we have

$$\Gamma_{S^k} \times \mathfrak{i}_W^!(\mathcal{L}_{j+m}(X)) = \mathfrak{i}_{S^k\times W}^!(\Gamma_{S^k} \times \mathcal{L}_{j+m}(X)) = \mathfrak{i}_{S^k\times W}^!(\mathcal{L}_{k+j+m}(S^k \times X)),$$

so it suffices to show that $\mathcal{L}_{k+j}(S^k \times Z) = \mathfrak{i}_{S^k\times W}^!(\mathcal{L}_{k+j+m}(S^k \times X))$ for sufficiently large k. This is just a special case of what we are trying to show in the first place(!), but now with the benefit that by choosing k large we can assume that we are in the degree range where we can apply Serre's Theorem. So let us now reset the notation back to showing that $\mathfrak{i}_W^!(\mathcal{L}_{j+m}(X)) = \mathcal{L}_j(Z)$, but now with the added assumption that $j > (\dim(Z)+1)/2$. This also implies that $j + m > (\dim(Z) + m + 1)/2 = (\dim(X) + 1)/2$.

So consider again Z a PL trivial normally nonsingular subset of X, a PL homeomorphism $\mathfrak{i}_W : \mathbb{R}^m \times Z \xrightarrow{\cong} W \subset X$, and the desired $\mathfrak{i}_W^!(\mathcal{L}_{j+m}(X)) = \mathcal{L}_j(Z) \in H_j(Z;\mathbb{Q})$, but now with the assurance that $j > (\dim(Z) + 1)/2$. In this range, Serre's Theorem applies so that $\mathfrak{c} \otimes \mathrm{id}_{\mathbb{Q}} : \pi^j(Z) \otimes \mathbb{Q} \to H^j(Z;\mathbb{Q})$ is an isomorphism. As $H^j(Z;\mathbb{Q})$ is dual to $H_j(Z;\mathbb{Q})$, to test whether $\mathfrak{i}_W^!(\mathcal{L}_{j+m}(X)) = \mathcal{L}_j(Z)$ we need only check that $\alpha(\mathfrak{i}_W^!(\mathcal{L}_{j+m}(X))) = \alpha(\mathcal{L}_j(Z))$ for all $\alpha \in H^j(Z;\mathbb{Q})$. In fact, it suffices to demonstrate this equality for a basis of $H^j(Z;\mathbb{Q}) \cong H^j(Z) \otimes \mathbb{Q}$, and we can choose this basis to be represented by elements of $H^j(Z)$. In particular, we can find a basis of $H^j(Z;\mathbb{Q})$ consisting of elements of the form $\mathfrak{c}([f]) \otimes 1$, as every element of $H^j(Z) \otimes \mathbb{Q}$ has a rational multiple of this form. We know by Proposition 9.4.7 that if α corresponds to $\mathfrak{c}([f]) \otimes 1$ under Serre's isomorphism then $\alpha(\mathcal{L}_j(Z)) = F([f]) = \sigma(f^{-1}(y))$ for a generic point $y \in S^m$ and a PL representative f of $[f]$. So we need to see that the same is true of $\alpha(\mathfrak{i}_W^!(\mathcal{L}_{j+m}(X)))$ for such an α.

Continuing to assume that $\alpha = \mathfrak{c}([f]) \otimes 1$, recall that $\mathfrak{c}([f]) = f^*(u_j)$ by definition for $u_j \in H^j(S^j)$ with $u_j(\Gamma_{S^j}) = 1$. We can also assume we have chosen a specific PL map f representing $[f]$. Abusing notation, we also let $f^*(u_j)$ represent $f^*(u_j) \otimes 1 \in H^j(Z) \otimes \mathbb{Q} \cong H^j(Z;\mathbb{Q})$. We need to compute $f^*(u_j)(\mathfrak{i}_W^!(\mathcal{L}_{j+m}(X)))$.

Let y be a generic point of S^j with respect to the map $f : Z \to S^j$. By Remark 9.4.20, Lemma 9.4.19 can be strengthened to show that y is a regular point of f and so by Lemma 9.4.27 the space $A = f^{-1}(y)$ is a PL trivial nns of Z with neighborhood embedding $\mathfrak{i}_V : \mathbb{R}^j \times A \xrightarrow{\cong} V \subset Z$. By the composition

$$\mathbb{R}^{m+j} \times A = \mathbb{R}^m \times \mathbb{R}^j \times A \xrightarrow{\mathrm{id}\times\mathfrak{i}_V} \mathbb{R}^m \times Z \xrightarrow{\mathfrak{i}_W} X,$$

we see that A is a PL trivial nns of X, and we let U denote the image neighborhood of A in X. Identifying U with $\mathbb{R}^{m+j} \times A$, let $\mathfrak{f} : X \to S^{m+j}$ be the map that projects U to $\mathbb{R}^{m+j} \subset S^{m+j}$ and takes the complement of U to the point at infinity. As in the proof of Lemma 9.4.27, we can alter $\mathfrak{f}$ outside a small neighborhood of A to make it PL; we continue to call the map $\mathfrak{f}$ and note that we can then find a generic point y' in S^{j+m} with $\mathfrak{f}^{-1}(y') \cong A$.

Let $\mathfrak{i}_{A,Z} : A \hookrightarrow Z$ and $\mathfrak{i}_{A,X} : A \hookrightarrow X$ denote the inclusions. We compute:

$$\begin{aligned}
f^*(u_j)(\mathfrak{i}^!_W(\mathscr{L}_{j+m}(X))) &= \mathbf{a}(f^*(u_j) \frown \mathfrak{i}^!_W(\mathscr{L}_{j+m}(X))) \\
&= \mathbf{a}(\mathfrak{i}_{A,Z}\mathfrak{i}^!_V(\mathfrak{i}^!_W(\mathscr{L}_{j+m}(X)))) && \text{by Lemma 9.4.22}^{31} \\
&= \mathbf{a}(\mathfrak{i}_{A,Z}\mathfrak{i}^!_U(\mathscr{L}_{j+m}(X))) && \text{by Lemma 9.4.30} \\
&= \mathbf{a}(\mathfrak{i}_{A,X}\mathfrak{i}^!_U(\mathscr{L}_{j+m}(X))) \\
&= \mathbf{a}(\mathfrak{f}^*(u_{j+m}) \frown \mathscr{L}_{j+m}(X)) && \text{by Lemma 9.4.22} \\
&= \mathfrak{f}^*(u_{j+m})(\mathscr{L}_{j+m}(X)) \\
&= \sigma(\mathfrak{f}^{-1}(y')) && \text{by Proposition 9.4.7} \\
&= \sigma(A) && \text{by Corollary 9.4.28.}
\end{aligned}$$

The first and sixth equalities here are by the ordinary homology version of the evaluation property, Proposition 7.3.25 (see [71, Section VII.12.8]), and the fourth equality is because augmentation commutes with maps of spaces. We have also used that Lemmas 9.4.22 and 9.4.30 continue to hold with $\mathbb{Q}$ coefficients.

As we have already seen that $\alpha(\mathscr{L}_j(Z)) = \sigma(f^{-1}(y)) = \sigma(A)$ when $\alpha = f^*(u_j)$, we see that $\alpha(\mathscr{L}_j(Z))$ and $\alpha(\mathfrak{i}^!_W(\mathscr{L}_{j+m}(X)))$ do indeed agree. As we have chosen α arbitrarily within a basis for $H^j(Z;\mathbb{Q})$, we conclude that $\mathscr{L}_j(Z) = \mathfrak{i}^!_W(\mathscr{L}_{j+m}(X))$ as claimed.

This concludes the proof that the L-classes have the listed properties. Now we turn to showing that these properties completely characterize the $\mathscr{L}_*$.

Suppose there are two collections of classes $\mathscr{L}_*$ and $\mathscr{L}'_*$ that possess the stated properties, and first consider $\mathscr{L}_m(X)$ and $\mathscr{L}'_m(X)$ with $m > (\dim(X)+1)/2$. Once again, in this case we know by Serre's Theorem that $H^m(X;\mathbb{Q})$ is generated by classes of the form $f^*(u_m)$, again identifying $f^*(u_m)$ with $f^*(u_m) \otimes 1 \in H^m(X;\mathbb{Q})$. If Z is a generic point inverse of f such that W is a neighborhood of Z PL homeomorphic to $\mathbb{R}^m \times Z$ that is compatible with the projection, then by Lemma 9.4.22 we know that[32]

$$f^*(u_m)(\mathscr{L}_m(X)) = \mathbf{a}(f^*(u) \frown \mathscr{L}_m(X)) = \mathbf{a}(\mathfrak{i}_Z\mathfrak{i}^!_W\mathscr{L}_m(X)).$$

By hypothesis now $\mathfrak{i}^!_W\mathscr{L}_m(X) = \mathscr{L}_0(Z)$ and $\mathbf{a}(\mathfrak{i}_Z(\mathscr{L}_0(Z))) = \mathbf{a}(\mathscr{L}_0(Z)) = \sigma(Z)$. By the same computation, $f^*(u_m)(\mathscr{L}'_m(X)) = \sigma(Z)$. So, as we can vary f sufficiently that the $f^*(u_m)$ generate $H^m(X;\mathbb{Q})$, which is dual to $H_m(X;\mathbb{Q})$, it follows that $\mathscr{L}_m(X) = \mathscr{L}'_m(X)$. If $m \leq (\dim(X)+1)/2$, then we can observe that the embedding that takes X to $\mathrm{pt} \times X \subset S^k \times X$ can be extended to a PL homeomorphism from $\mathbb{R}^k \times X$ to a neighborhood of $\mathrm{pt} \times X$ in $S^k \times X$, and so $\mathrm{pt} \times X$ is a PL trivial nns of $S^k \times X$. By assumption, we have $\mathscr{L}_m(X) = \mathfrak{i}^!_W(\mathscr{L}_{m+k}(S^k \times X))$ and $\mathscr{L}'_m(X) = \mathfrak{i}^!_W(\mathscr{L}'_{m+k}(S^k \times X))$. But for sufficiently

31 Since we modified the map $\mathfrak{f}$, it does not necessarily have precisely the form of the map in the hypotheses of Lemma 9.4.22, but it is homotopic to such a map. As Lemma 9.4.22 is a statement about homology, it therefore remains true up to altering maps by homotopies.

32 Once again, the map f does not necessarily have precisely the form of the map in the hypotheses of Lemma 9.4.22, but this time it is homotopic to such a map by composing f with a homotopy equivalence that retracts the complement of $f(\mathbb{R}^m \times Z) \cong \mathbb{R}^m$ to the point at infinity of S^m.

large k, we have $m + k > (\dim(X) + k + 1)/2$, in which case we have just seen that $\mathscr{L}'_{m+k}(S^k \times X) = \mathscr{L}_{m+k}(S^k \times X)$. It follows that $\mathscr{L}_m(X) = \mathscr{L}'_m(X)$ for all m. As X has been arbitrary in this argument, we have that $\mathscr{L}_* = \mathscr{L}'_*$ in all cases.

Theorem 9.4.18 is now proved. □

Proof of Lemma 9.4.29 The proof is straightforward when $m = 0$, so we assume $m > 0$.

We will have to be careful about some orientation issues. In particular, recall that our definition for our $i^!_W$ maps assumes that the homeomorphisms $i_W : \mathbb{R}^m \times Z \to W \subset X$ are chosen to be orientation-preserving. So when identifying the neighborhood $S^k \times W \cong S^k \times \mathbb{R}^m \times Z \cong \mathbb{R}^m \times S^k \times Z$ of $S^k \times Z$ in $S^k \times X$ as a bundle, we need to utilize an orientation-preserving homeomorphism $\mathbb{R}^m \times S^k \times Z \to S^k \times W$. As the transposition map $t : \mathbb{R}^m \times S^k \times Z \to S^k \times \mathbb{R}^m \times Z$ defined by $t(x, y, z) = (y, x, z)$ is only orientation-preserving up to the sign $(-1)^{km}$, we must instead use a map we will label $\tilde{t}$ obtained by composing t with a reflection of one of the $\mathbb{R}^m$ coordinates if km is odd. For the sake of clarity, we include this homeomorphism explicitly in the following diagram.

$$
\begin{array}{ccccc}
H_k(S^k) \otimes H_{i+m}(X) & \xrightarrow{\times} & H_{k+i+m}(S^k \times X) & & \\
\downarrow & & \downarrow & & \\
\begin{array}{c} H_k(S^k) \otimes \\ H_{i+m}(X, X - Z) \end{array} & \xrightarrow{\times} & \begin{array}{c} H_{k+i+m}(S^k \times X, \\ S^k \times (X - Z)) \end{array} & & \\
\uparrow \cong & & \uparrow \cong & & \\
\begin{array}{c} H_k(S^k) \otimes \\ H_{i+m}(W, W - Z) \end{array} & \xrightarrow{\times} & \begin{array}{c} H_{k+i+m}(S^k \times W, \\ S^k \times (W - Z)) \end{array} & \xleftarrow[\cong]{\tilde{t}} & \begin{array}{c} H_{k+i+m}(\mathbb{R}^m \times S^k \times Z, \\ (\mathbb{R}^m - \{0\}) \times S^k \times Z) \end{array} \\
\downarrow {\scriptstyle (-1)^{km} \mathrm{id} \otimes (\mu \frown)} & & \downarrow {\scriptstyle (-1)^{km} (1_{S^k} \times \mu) \frown} & & \downarrow {\scriptstyle (a \times 1_{S^k} \times 1_Z) \frown} \\
H_k(S^k) \otimes H_i(W) & \xrightarrow{\times} & H_{k+i}(S^k \times W) & \xleftarrow[\cong]{\tilde{t}} & H_{k+i}(\mathbb{R}^m \times S^k \times Z) \\
\downarrow \cong & & \downarrow \cong & \swarrow \cong & \\
H_k(S^k) \otimes H_i(Z) & \xrightarrow{\times} & H_{k+i}(S^k \times Z). & &
\end{array}
$$

We next discuss the commutativity of the diagram. The top two and bottom squares commute by the naturality of the cross product (Proposition 5.2.17 with all filtrations trivial). The bottom right triangle commutes at the level of spaces with the downward maps projecting out the $\mathbb{R}^m$ coordinate.

For the left square involving cap products, if $\gamma \otimes \xi \in H_k(S^k) \otimes H_{i+m}(W, W - Z)$, the two ways around the square both yield

$$(-1)^{km+km}\gamma \times (\mu \frown \xi) = \gamma \times (\mu \frown \xi),$$

using the unfiltered version of Propositions 7.3.55 and 7.3.22. So this square commutes. We have chosen the signs here so that if $\gamma = \Gamma_{S^k}$ then the sign vanishes and the counterclockwise composition from the top left to the bottom right applied to $\Gamma_{S^k} \otimes \eta$ is precisely $\Gamma_{S^k} \times \mathrm{i}_W^!(\eta)$.

The square on the right will commute by the naturality of cap products (Proposition 7.3.6) once we show that $\tilde{t}^*((-1)^{km}(1_{S^k} \times \mu)) = a \times 1_{S^k} \times 1_Z$. For this, recall that a here stands for the class in $H^m(\mathbb{R}^m, \mathbb{R}^m - \{0\})$ that evaluates to 1 on the generator of $H^m(\mathbb{R}^m, \mathbb{R}^m - \{0\})$ with the standard orientation and that we can write $\mu = a \times 1_Z$ under the identification $W = \mathbb{R}^m \times Z$; see Remark 9.4.17. So if we write $1_{S^k} \times \mu = 1_{S^k} \times a \times 1_Z$, then $t^*(1_{S^k} \times a \times 1_Z) = a \times 1_{S^k} \times 1_Z$ by the commutativity properties of the cross product (Proposition 7.3.13). The difference between t and $\tilde{t}$ is that when km is odd $\tilde{t}$ also includes an orientation-reversing reflection, say r, of $\mathbb{R}^m$. Such a reflection takes a to $r^*(a) = (-1)^{km}a$. So in this case we have

$$\begin{aligned}\tilde{t}^*((-1)^{km}(1_{S^k} \times \mu)) &= (r \times \mathrm{id}_{S^k \times Z})^* t^*((-1)^{km}(1_{S^k} \times a \times 1_Z)) \\ &= (r \times \mathrm{id}_{S^k \times Z})^*((-1)^{km}(a \times 1_{S^k} \times 1_Z)) \\ &= (-1)^{km} r^*(a) \times 1_{S^k} \times 1_Z \\ &= a \times 1_{S^k} \times 1_Z.\end{aligned}$$

Now, in our definition of the $\mathrm{i}_W^!$ maps, the use of the cup product $\mu \frown$ has always involved implicitly identifying W with a bundle via an orientation-preserving homeomorphism. The path around the right of the diagram simply makes this explicit in the current case. So traveling from the top left around the diagram clockwise takes $\Gamma_{S^k} \otimes \eta \in H_k(S^k) \otimes H_{i+m}(X)$ to $\mathrm{i}_{S^k \times W}^!(\Gamma_{S^k} \times \eta)$.

This completes the proof of Lemma 9.4.29. □

Proof of Lemma 9.4.30 For ease of notation, we identify W with $\mathbb{R}^m \times Z$, V with $\mathbb{R}^j \times A$, and U with $\mathbb{R}^m \times \mathbb{R}^j \times A = \mathbb{R}^{m+j} \times A$. We also write $\mathbb{R}_0^k$ for the pair $(\mathbb{R}^k, \mathbb{R}^k - \{0\})$. We let $a_k \in H^k(\mathbb{R}_0^k)$ be the class that evaluates to 1 on the generator of $H_k(\mathbb{R}_0^k)$ consistent with the orientation. Then $a_k \times a_\ell = (-1)^{k\ell} a_{k+\ell}$, as can be seen by evaluating on the orientation class in $H_{k+\ell}(\mathbb{R}_0^{k+\ell})$, which is the product of those in $H_k(\mathbb{R}_0^k)$ and $H_\ell(\mathbb{R}_0^\ell)$.

Consider the diagram in Figure 9.6. The boxes not involving cap products commute at the space level, and the marked isomorphisms either are excision isomorphisms or are induced by space homeomorphisms or homotopy equivalences.

For the rightmost square involving cap products, we have $a_m \times 1 \in H^m(\mathbb{R}_0^m \times Z)$, and so the square commutes by naturality of the cap product with respect to the map

$$(\mathbb{R}^m \times Z; \emptyset, (\mathbb{R}^m - \{0\}) \times Z) \to (\mathbb{R}^m \times Z; \mathbb{R}^m \times (Z - A), (\mathbb{R}^m - \{0\}) \times Z).$$

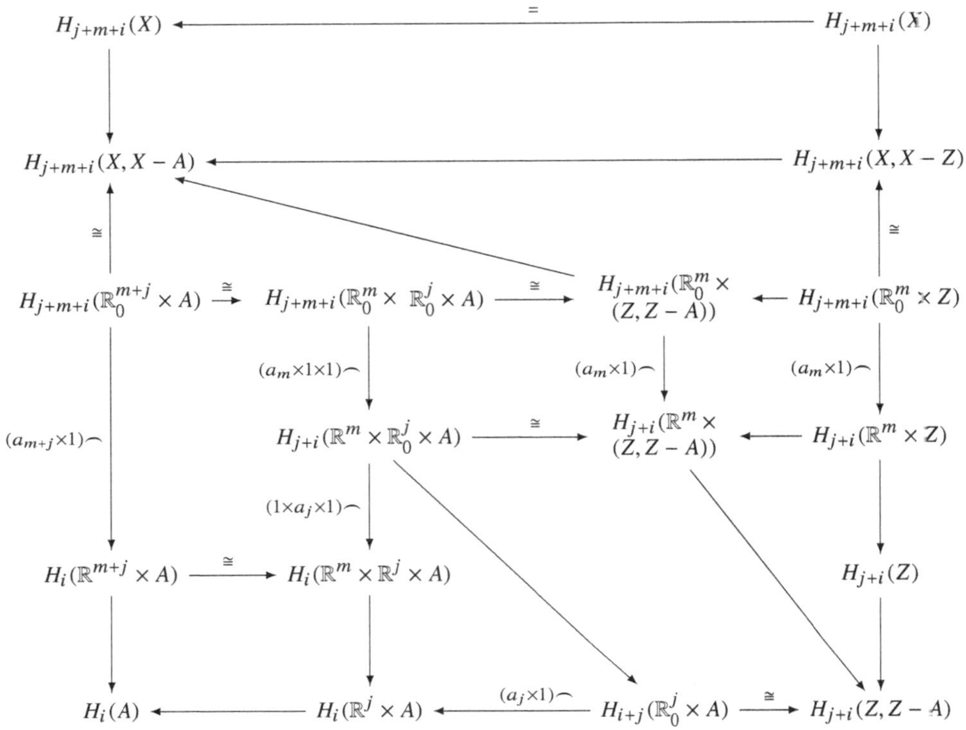

Figure 9.6

Similarly, the middle square commutes by naturality of the map

$$(\mathbb{R}^m \times \mathbb{R}^j \times A; \mathbb{R}^m \times (\mathbb{R}^j - \{0\}) \times A, (\mathbb{R}^m - \{0\}) \times \mathbb{R}^j \times A)$$
$$\to (\mathbb{R}^m \times Z; \mathbb{R}^m \times (Z - A), (\mathbb{R}^m - \{0\}) \times Z).$$

Here we note that the restriction of $a_m \times 1 \in H^m(\mathbb{R}_0^m \times Z)$ to $H^m(\mathbb{R}_0^m \times \mathbb{R}^j \times A)$ is indeed $a_m \times 1 \times 1$, as 1_Z restricts to $1_{\mathbb{R}^j \times A}$, which is equal to $1_{\mathbb{R}^j} \times 1_A$.

For the piece of the diagram involving maps down and diagonally down from $H_{j+1}(\mathbb{R}^m \times \mathbb{R}_0^j \times A)$, the naturality is with respect to the projection map

$$(\mathbb{R}^m \times \mathbb{R}^j \times A; \emptyset, \mathbb{R}^m \times (\mathbb{R}^j - \{0\}) \times A) \to (\mathbb{R}^j \times A; \emptyset, (\mathbb{R}^j - \{0\}) \times A).$$

For the region involving the maps down and to the right from $H_{j+m+i}(\mathbb{R}_0^{m+j} \times A)$, the two horizontal arrows can be treated as equalities, and then we have for any $\xi \in H_{j+m+i}(\mathbb{R}_0^{m+j} \times A)$

that

$$\begin{aligned}(1 \times a_j \times 1) \frown ((a_m \times 1 \times 1) \frown \xi) &= ((1 \times a_j \times 1) \smile (a_m \times 1 \times 1)) \frown \xi \\ &= (-1)^{jm}(a_m \times a_j \times 1) \frown \xi \\ &= (a_{m+j} \times 1) \frown \xi,\end{aligned}$$

using the associativity of cup and cap products (Proposition 7.3.35), the interchange property of cup and cross products (Proposition 7.3.54), the identity properties of 1 (Proposition 7.3.21), and the fact that $a_m \times a_j = (-1)^{jm} a_{m+j}$ as observed at the start of the proof.

Thanks to the maps that are isomorphisms, commutativity of the interior polygons implies commutativity around the outside of the diagram. Starting in the upper right, the path down then left to $H_i(A)$ is precisely the composition $i^!_V i^!_W$, while the path left then down is $i^!_U$.

This completes the proof of Lemma 9.4.30. □

9.5 A Survey of Pseudomanifold Bordism Theories

We briefly survey some of the further outgrowths of the material presented in this chapter, focusing particularly on bordism groups and the resulting generalized homology theories. However, we make no attempt to be comprehensive or to provide all details. Rather, we hope that we provide sufficient references for the interested reader to find the original sources or more thorough expositions.

Unfortunately, an analogous survey of characteristic classes on stratified spaces is beyond our scope. Many such generalizations of the classical characteristic classes now exist, some utilizing intersection homology and related tools but many not. Let us just mention a few expository sources and surveys the reader might consult: [9, 11, 32, 33, 35, 164, 188, 208, 209, 210, 229].

9.5.1 Bordism

We begin with a quick introduction to the idea of *bordism groups*[33] and *bordism homology theories*. A good first introduction can be found within the Milnor and Stasheff book on characteristic classes [177]. An introduction to the approach to bordism using spectra occurs within the text of Davis and Kirk [67]. For more thorough treatments, see [199, 224].

[33] Particularly in the older literature, these are sometimes called *cobordism* groups. The original language reflects the notion that two "cobordant" n-manifolds together bound some $(n + 1)$-manifold, i.e. they co-bound. But, as we will describe momentarily, this leads naturally to generalized homology theories which, if called cobordism homology theories, would lead to dual cohomology theories that would have to be called co-cobordism. To avoid this awkwardness, the original theory came to be called "bordism" theory, leaving "cobordism" free for the dual theory.

(a) Bordism Groups

To illustrate the basic idea, let's begin by considering the class of smooth manifolds, not necessarily oriented. We say that two closed smooth manifolds M_1, M_2 are *bordant* if there is a compact smooth ∂-manifold W such that $\partial W \cong M_1 \amalg M_2$. This is an equivalence relation: Symmetry is clear. Reflexivity comes from considering $[0,1] \times M$. And for transitivity, if we have W_1, W_2 with $\partial W_1 \cong M_1 \amalg M_2$ and $\partial W_2 \cong M_2 \amalg M_3$, we can form a W by gluing W_1 and W_2 along M_2, and then $\partial W \cong M_1 \amalg M_3$. Denote the equivalence class of M by $[M]$.

It turns out with a little more work that the equivalence classes in each dimension n constitute an abelian group Ω_n^O; the standard notation Ω_n^O is due to O_n being the structure group for the tangent bundles of closed smooth n-manifolds. More precisely, Ω_n^O consists of the bordism equivalence classes of n-dimensional closed manifolds, and the group operation is disjoint union: $[M_1] + [M_2] = [M_1 \amalg M_2]$. The identity for each n is the empty manifold, which we consider a manifold of every dimension (or, perhaps more accurately, we can think of having one empty manifold of each dimension): $[M_1] + [\emptyset] = [M_1 \amalg \emptyset] = [M_1]$. Notice that $[M] = [\emptyset]$ for any closed M that is a boundary of a compact manifold. The inverse of $[M]$ is $[M]$, as $M \amalg M \cong \partial([0,1] \times M)$, so $[M] + [M] = [M \amalg M] = [\emptyset]$.

Bordism groups of closed *oriented* manifolds are defined analogously, though now with M_1 and M_2 defined to be bordant if there is an oriented W with $\partial W = M_1 \amalg -M_2$, where $-M_2$ is M_2 with its orientation reversed. The resulting groups are denoted Ω_n^{SO}, as now the structure groups associated to the oriented tangent bundles are the SO_n. In these bordism groups, $-[M] = [-M]$, as $M \amalg -M \cong \partial([0,1] \times M)$, so $[M] + [-M] = [M \amalg -M] = [\emptyset]$. Further analogous groups Ω_n^G can be defined by imposing additional structure hypotheses on the involved manifolds and ∂-manifolds. Similarly, we could consider bordism in other categories, such as bordisms of closed PL manifolds or closed oriented PL manifolds, Ω_*^{PL} or Ω_*^{SPL}.

In fact, the groups Ω_*^O and Ω_*^{SO} can be given the structure of graded rings with the product being $[M_1] \times [M_2] = [M_1 \times M_2]$. A computation of Ω_*^O can be pieced together from [177, Section 4 and Exercises 4E and 16F] as an application of the study of Stiefel–Whitney numbers, which come from evaluating products of the Stiefel–Whitney characteristic classes on the $\mathbb{Z}_2$-fundamental classes of manifolds. A more thorough and quite readable treatment of $\Omega_*^{SO} \otimes \mathbb{Q}$ is given in [177, Section 17] via a similar study of Pontrjagin numbers, which come from evaluating products of the Pontrjagin characteristic classes on the $\mathbb{Z}$-fundamental classes of oriented manifolds. Complete computations of these bordism rings and many others can be found in [224].

(b) Bordism Homology Theories

Beyond the bordism groups of manifolds, each type of bordism yields a *generalized homology theory*. Recall [167] that these are functors that satisfy the Eilenberg–Steenrod axioms for a homology theory except perhaps for the Dimension Axiom. For ordinary homology with coefficients, the Dimension Axiom says that for the one-point space pt we have $H_0(\mathrm{pt}; G) \cong G$ and $H_i(\mathrm{pt}; G) \cong 0$ for $i \neq 0$.

We illustrate the idea of the bordism homology groups using oriented bordism as our example this time; the other bordism homology theories are defined similarly. Now instead of defining a group or graded ring Ω_*^{SO}, we need a functor $\Omega_*^{SO}(\cdot)$. On a space Z, the group $\Omega_n^{SO}(Z)$ is generated by continuous maps $f : M \to Z$, where M is a closed oriented smooth n-manifold. Two maps $f_i : M_i \to Z$, $i = 1, 2$, are equivalent if there are a compact oriented smooth ∂-manifold W with $\partial W \cong M_1 \amalg -M_2$ and a map $F : W \to Z$ that restricts to f_0 and f_1 on ∂W. Notice that this definition is quite analogous to how singular homology is defined except that, rather than working with chains of singular simplices, we work with entire manifolds at once. Addition is by disjoint union, and again we get a group. If $g : Z \to Y$ is a map, we have an induced homomorphism $\Omega_n^{SO}(Z) \to \Omega_n^{SO}(Y)$ via composition; i.e. a generator $[M \xrightarrow{f} Z]$ gets taken to the composition $[M \xrightarrow{f} Z \xrightarrow{g} Y]$. Notice also that if $Z =$ pt, the space with one point, then the maps contain no information and we have $\Omega_n^{SO}(\mathrm{pt}) = \Omega_n^{SO}$. The functor $\Omega_*^{SO}(\cdot)$ does not satisfy the Dimension Axiom, as there exist positive degrees in which Ω_n^{SO} is nontrivial.

Of course a proper homology theory should really take pairs of spaces as inputs. In this case, we generalize and let an element of $\Omega_n^{SO}(Z, A)$ consist of a compact oriented n-dimensional smooth ∂-manifold W together with a map $f : M \to Z$ such that $f(\partial M) \subset A$. In this case $[M_1 \xrightarrow{f_1} Z] = [M_2 \xrightarrow{f_2} Z]$ if there is a compact oriented smooth $(n+1)$-∂-manifold V such that $\partial V = M_1 \cup_{\partial M_1} M_0 \cup_{\partial M_2} -M_2$ and a map $F : V \to Z$ that restricts to f_i on M_i, $i = 1, 2$, and such that the restriction to M_0 is a bordism between ∂M_1 and ∂M_2 in A. One can also verify the other axioms of a generalized homology theory. We will not go further into detail here except to suggest that it is a good exercise to think through why such a bordism homology theory possesses the long exact sequence of the pair.

Bordism homology theories are quite interesting and have been put to various important uses. For one well-known example, see [63].

9.5.2 Pseudomanifold Bordism

Just as one can define bordism groups and bordism homology theories using classes of manifolds, one can consider bordism groups Ω_n^C generated by closed pseudomanifolds in a particular class C with relations given by $[X_1] = [X_2]$ if there is a compact W in the class with $\partial W = X_1 \amalg X_2$, or $\partial W = X_1 \amalg -X_2$ if our class allows for orientation information. For example, if we let C be the class of PL oriented G-Witt spaces, then we obtain bordism groups generated by the compact n-dimensional oriented PL G-Witt spaces without boundary and with $[X_1] = [X_2]$ if there is a compact $(n+1)$-dimensional oriented PL G-Witt space W with $\partial W \cong X_1 \amalg -X_2$. Similarly, one can define pseudomanifold bordism homology theories, at least when considering classes of PL pseudomanifolds, which allow one to verify the

Eilenberg–Steenrod axioms (minus the Dimension Axiom) without too much difficulty[34]; see [3, Proposition 7] and [94, Section 6].

One issue that was often not treated clearly in the early literature is whether or not to take the stratifications into account when defining bordism groups. For example, we saw in Proposition 9.1.28 that the property of being a G-Witt space does not depend upon the choice of (classical) stratification. So there are two possible definitions for the bordism groups of G-Witt spaces. We can conceive of them as being generated by G-Witt spaces thought of as stratified pseudomanifolds and with the bordism relation determined by $\partial W = X_1 \amalg -X_2$ assuming that X_1, X_2, and W are all compatibly stratified. Alternatively, we could define the generators to be pseudomanifolds $|X|$ that satisfy the G-Witt condition for some, and hence any, (classical) stratification, and with the relation $\partial W = X_1 \amalg -X_2$ as pseudomanifolds, again ignoring the stratifications. Happily, as shown in [94], if one forbids codimension-one strata (which are automatically disallowed for Witt spaces) and makes a reasonable choice of the class C, then we obtain the same bordism groups Ω_n^C and corresponding bordism homology theories $\Omega_*^C(\cdot)$ whether stratifications are taken into account or not. A detailed axiomatic study of permissible classes C and the construction of pseudomanifold bordism groups and homology theories can be found in [94].

Let us consider some examples. In each case, we assume all spaces to be PL and without codimension-one strata.

(a) All pseudomanifolds

A natural first example would be the class of all PL pseudomanifolds or of all oriented PL pseudomanifolds. Here the bordism groups turn out to be mostly trivial, as every closed pseudomanifold X of dimension > 0 is the boundary of the closed cone $\bar{c}X$. The only nontrivial bordism group is in dimension zero, and it is an easy exercise to check that the unoriented bordism group is $\mathbb{Z}_2$, while in the oriented case it is $\mathbb{Z}$. It follows from Eilenberg and Steenrod that the resulting homology theories are just $H_*(\cdot;\mathbb{Z}_2)$ and $H_*(\cdot;\mathbb{Z})$; see [75, Chapter III] and [167, Chapters 13–15]. While we get nothing new, this is still interesting: all homology classes in, say, $H_*(Z;\mathbb{Z}_2)$ or $H_*(Z;\mathbb{Z})$ can be represented by maps from pseudomanifolds $f : X \to Z$. An explicit map from the bordism theory to ordinary homology is given by taking f to the image under f of the $\mathbb{Z}_2$- or $\mathbb{Z}$-fundamental class of X to Z by f.

(b) Mod 2 Euler Spaces

Before the advent of intersection homology, Akin [3], following work of Sullivan [226], computed the bordism groups of what are essentially PL pseudomanifolds whose polyhedral links have vanishing Euler characteristic mod 2. He showed that the bordism groups[35] Ω_n

[34] In general, the study of topological bordism theories is much more complicated than that for spaces with more structure.

[35] Rather than make up notation for every class we shall discuss, we'll use simply Ω_* to denote whichever type of bordism is under discussion.

are isomorphic to $\mathbb{Z}_2$ for all $n \geq 0$, with $[X_1] = [X_2]$ if and only if X_1 and X_2 have the same mod 2 Euler characteristic.

(c) $\mathbb{Q}$-Witt Spaces

When Siegel introduced PL oriented $\mathbb{Q}$-Witt spaces in [218], he was also carrying out a program of Sullivan's, this time to find a geometric model for ko-homology theory at odd primes. KO-homology is the generalized homology theory dual to real topological K-theory, which, roughly speaking, is based on groups of stable isomorphism classes of real vector bundles over a space. The lower-case ko denotes the "connective" version of the theory[36]. The phrase "at odd primes" means that we ignore all 2-primary torsion.

Recall that we showed above in Corollary 9.3.18 that the signature of an oriented $\mathbb{Q}$-Witt space is a bordism invariant, and this remains true by the same argument if we limit ourselves to PL objects. But, in fact, something stronger holds, and the bordism classes preserve more information from the cup product pairing than just the signature. To explain the invariant, we need to introduce Witt groups in the next paragraph. See [176, Chapter I] for full details; the reader might also find it useful to review our treatment of signatures of symmetric pairings over $\mathbb{Q}$-vector spaces in Appendix A.5.

Let R be a commutative ring. We will consider nonsingular symmetric bilinear pairings on finitely generated projective R-modules. Such data is called a *symmetric inner product space*, and the pairing is called an *inner product*. Two inner product spaces $(V, (\cdot,\cdot)_V)$ and $(W, (\cdot,\cdot)_W)$ are isomorphic if there is an isomorphism of R-modules $f : V \to W$ such that $(v_1, v_2)_V = (f(v_1), f(v_2))_W$ for all $v_1, v_2 \in V$. There is an orthogonal sum operation on inner product spaces so that if $(V, (\cdot,\cdot)_V)$ and $(W, (\cdot,\cdot)_W)$ are two inner product spaces over R, then we can define their sum by defining an inner product on $V \oplus W$ by $(v_1 + w_1, v_2 + w_2)_{V \oplus W} = (v_1, v_2)_V + (w_1, w_2)_W$. If V and W are free modules and we represent their inner products by matrices with respect to some basis, then the orthogonal sum operation corresponds to taking a block sum of matrices. A symmetric inner product space $(S, (\cdot,\cdot)_S)$ is called *split* if there is a submodule $N \subset S$ so that $N = N^\perp$, i.e. if $N = \{s \in S \mid (s, n) = 0 \text{ for all } n \in N\}$. Finally, we define $(V, (\cdot,\cdot)_V)$ and $(W, (\cdot,\cdot)_W)$ to be in the same *Witt class* if there exist split inner product spaces S and S' such that $V \oplus S$ and $W \oplus S'$, with their inner products coming from the orthogonal sum, are isomorphic. Being in the same Witt class is an equivalence relation, and it turns out that the collection of Witt classes form an abelian group $W(R)$ under orthogonal sum. The identity is represented by any split inner product space; the additive inverse of $(V, (\cdot,\cdot)_V)$ is $(V, -(\cdot,\cdot)_V)$. In fact, $W(R)$ can be given the structure of a commutative ring with unity. The multiplication is given by taking tensor products; see [176].

In the case $R = \mathbb{Q}$, the split inner product spaces are those possessing a Lagrangian

[36] $KO^n(\text{pt})$ is periodic and so, in particular, can have non-zero values for negative n. The connective version is a modification, which can be done for any generalized homology theory that forces these "coefficient groups" to be zero for negative n. In the case of bordism theories, we clearly need such an assumption as there are no spaces of negative dimension and so no nontrivial bordism groups in negative degrees. See [1, Section III.6] or [199, Section II.4]

subspace, i.e. a subspace of half the dimension of the vector space on which the pairing is trivial; see Definition A.5.10. So the split inner product spaces all have signature equal to zero. In fact, there is a homomorphism $W(\mathbb{Q}) \to \mathbb{Z}$ given by the signature [176, Lemma III.2.6]. If we replace $\mathbb{Q}$ with $\mathbb{R}$, the signature homomorphism in fact yields an isomorphism $W(\mathbb{R}) \to \mathbb{Z}$ [176, Corollary III.2.7]. This is not the case for $\mathbb{Q}$. The Witt group $W(\mathbb{Q})$ is fairly complicated; a computation of the group structure can be found in [176, Section IV.2].

What does this all have to do with bordism theory? In [218], Siegel computed the bordism groups of oriented PL $\mathbb{Q}$-Witt spaces by showing that

$$\Omega_n^{\mathbb{Q}\text{-Witt}} \cong \begin{cases} \mathbb{Z}, & n = 0, \\ W(\mathbb{Q}), & n = 4k, k > 0, \\ 0, & \text{otherwise.} \end{cases}$$

When $k > 0$, the isomorphism $w : \Omega_{4k}^{\mathbb{Q}\text{-Witt}} \to W(\mathbb{Q})$ is given by taking a Witt space X to the nonsingular cup product pairing on[37] $I_{\bar{n}}H^{2k}(X;\mathbb{Q})$. The Witt signature $\sigma(X)$ is then the image of $w(X)$ under the signature map $W(\mathbb{Q}) \to \mathbb{Z}$.

This bordism computation plugs straight into some work of Sullivan's from [228]. It turns out that any class of spaces that has a theory of signatures that behaves like the Witt signatures[38] allows one to define on each such space X a ko-fundamental class $[X]_{ko} \in ko_*(X) \otimes \mathbb{Z}[1/2]$. Here $\mathbb{Z}[1/2]$ is the subring of $\mathbb{Q}$ consisting of fractions that can be written only with powers of 2 in their denominators; tensoring by $\mathbb{Z}[1/2]$ kills all 2-primary torsion. This ko-fundamental class is the $ko_* \otimes \mathbb{Z}[1/2]$ analogue of the more familiar fundamental classes for manifolds in ordinary homology.

So now suppose we are given a bordism class in $\Omega_*(Z)$ represented by $X \xrightarrow{f} Z$ for some closed oriented PL $\mathbb{Q}$-Witt space X. As $ko_*(\cdot) \otimes \mathbb{Z}[1/2]$ is a homology theory, we have an induced map $f : ko_*(X) \otimes \mathbb{Z}[1/2] \to ko_*(Z) \otimes \mathbb{Z}[1/2]$, and this takes $[X]_{ko}$ to a class $f([X]_{ko}) \in ko_*(Z) \otimes \mathbb{Z}[1/2]$. So, $[X \xrightarrow{f} Z] \in \Omega_*(Z)$ determines $f([X]_{ko}) \in ko_*(Z) \otimes \mathbb{Z}[1/2]$, and with a bit more work this construction yields a morphism of homology theories

$$\mu : \Omega_*^{\mathbb{Q}\text{-Witt}}(\cdot) \otimes \mathbb{Z}[1/2] \to ko_*(\cdot) \otimes \mathbb{Z}[1/2].$$

By Siegel's computation, the map μ on the space pt looks like the following in degree n:

$$\begin{aligned} W(\mathbb{Q}) \otimes \mathbb{Z}[1/2] &\to ko_n(\text{pt}) \otimes \mathbb{Z}[1/2], && n = 4k,\ k > 0, \\ \mathbb{Z} \otimes \mathbb{Z}[1/2] = \mathbb{Z}[1/2] &\to ko_0(\text{pt}) \otimes \mathbb{Z}[1/2], && n = 0, \\ 0 &\to ko_n(\text{pt}) \otimes \mathbb{Z}[1/2], && \text{otherwise.} \end{aligned}$$

But these "coefficient maps" all turn out to be isomorphisms! And it follows that μ is an

[37] Actually, Siegel worked with the dual intersection pairing on $I^{\bar{m}}H_{2k}(X;\mathbb{Q})$, though the result is the same.
[38] See [218] for precise details.

isomorphism of homology theories; see, for example, [199, Proposition II.3.19]. Thus Witt bordism turns out to provide a geometric description of the homology theory $ko_*(\cdot)\otimes\mathbb{Z}[1/2]$.

In fact, the group $W(\mathbb{Q}) \otimes \mathbb{Z}[1/2]$ actually turns out to be $\mathbb{Z}[1/2]$ because $W(\mathbb{Q})$ can be shown to be the direct sum of $\mathbb{Z}$ with a 2-group, i.e. a group whose elements all have order a power of 2. The signature map $W(\mathbb{Q}) \to \mathbb{Z}$ projects to the $\mathbb{Z}$ summand. Thus the same Sullivan argument can be applied to any bordism theory with an appropriately behaved signature that characterizes a $\mathbb{Z}$-summand of $\Omega_*(\mathrm{pt})$ in dimensions $4k$, $k \geq 0$, and so that all other bordism group summands are 2-groups. In particular, $\Omega_*(\cdot)\otimes\mathbb{Z}[1/2] \cong ko_*(\cdot)\otimes\mathbb{Z}[1/2]$ for any such bordism theory. We'll see other examples that fit this framework below.

Another property of $\mathbb{Q}$-Witt bordism is that there is a natural map of bordism theories $b : \Omega_*^{SO}(\cdot) \to \Omega^{\mathbb{Q}\text{-Witt}}(\cdot)$ because every smooth manifold is automatically a PL Witt space. Banagl, Cappell, and Shaneson showed in [20, Proposition 2] that the isomorphism $\mu : \Omega^{\mathbb{Q}\text{-Witt}}(\cdot) \otimes \mathbb{Z}[1/2] \to ko_*(\cdot) \otimes \mathbb{Z}[1/2]$ together with Sullivan's work [228] implies that $b \otimes \mathbb{Z}[1/2]$ gives a surjection

$$\Omega_*^{SO}(Z, A) \otimes \mathbb{Z}[1/2] \to \Omega^{\mathbb{Q}\text{-Witt}}(Z, A) \otimes \mathbb{Z}[1/2]$$

for any compact PL pair (Z, A). Consequently, one can "pull back" bordism invariant computations for Witt spaces to smooth manifolds, allowing one to invoke the myriad results of smooth manifold theory. This idea is exploited in [20] to study twisted L-classes and twisted signatures on Witt spaces.

Finally, let us mention another application of $\mathbb{Q}$-Witt bordism due to Jon Woolf in [250]. Woolf shows that if Z is a compact polyhedron and $i > \dim(Z)$ then $\Omega_i^{\mathbb{Q}\text{-Witt}}(Z)$ is isomorphic to $W_i^c(Z)$, where $W_i^c(\cdot)$ is another generalized homology theory defined in terms certain Witt groups of constructible sheaf complexes on Z over the ground field $\mathbb{Q}$. The theory $W_*^c(\cdot)$ is 4-periodic, meaning that $W_i^c(\cdot) \cong W_{i+4}^c(\cdot)$ for all i, and therefore $W_i^c(\cdot) \cong \varinjlim_k \Omega_{i+4k}^{\mathbb{Q}\text{-Witt}}(Z)$. This implies that every self-dual complex of $\mathbb{Q}$-sheaves on Z arises, up to Witt equivalence, as the pushforward of the intersection chain sheaf complex of some $\mathbb{Q}$-Witt space, i.e. that all such sheaf complexes are of "geometric origin"; see [250, Section 5.1] for a precise statement. As an application, it is shown that the L-classes determine homology operations $\mathscr{L}_* : \Omega_i^{\mathbb{Q}\text{-Witt}}(Z) \to H_i(Z; \mathbb{Q})$ obtained by pushing forward to Z the L-classes of the $\mathbb{Q}$-Witt space X by the map $f : X \to Z$ representing an element of $\Omega_i^{\mathbb{Q}\text{-Witt}}(Z)$.

(d) K-Witt Spaces

The bordism groups and homology theories for oriented PL K-Witt spaces for an arbitrary field K are computed in [88, 110]. In [88] it is shown that the bordism groups depend only on the characteristic of the field, reducing the computations to those for $\mathbb{Q}$ and the fields $\mathbb{Z}_p$. We have already seen the $\mathbb{Q}$-Witt bordism groups as computed by Siegel [218]. For a prime p, we denote the oriented $\mathbb{Z}_p$-Witt bordism groups by $\Omega_*^{\mathbb{Z}_p\text{-Witt}}$. It is shown in [88, Theorem

4.10] that, for $p \neq 2$,

$$\Omega_n^{\mathbb{Z}_p\text{-Witt}} \cong \begin{cases} \mathbb{Z}, & n = 0, \\ W(\mathbb{Z}_p), & n = 4k,\ k > 0, \\ 0, & \text{otherwise.} \end{cases}$$

Here $W(\mathbb{Z}_p)$ is the Witt group of symmetric inner product spaces over $\mathbb{Z}_p$. With some further work, it follows that as a bordism homology theory we have

$$\Omega_n^{\mathbb{Z}_p\text{-Witt}}(Z) \cong \bigoplus_{r+s=n} H_r(Z; \Omega_s^{\mathbb{Z}_p\text{-Witt}}), \tag{9.6}$$

so for a CW complex Z we do not wind up with quite as interesting a homology theory in this case. Incidentally, the same formula holds for $\mathbb{Q}$-Witt spaces if we localize[39] at 2:

$$\Omega_n^{\mathbb{Q}\text{-Witt}}(Z)_{(2)} \cong \bigoplus_{r+s=n} H_r(Z; (\Omega_s^{\mathbb{Q}\text{-Witt}})_{(2)});$$

see [16, Section 3] for details.

As observed in [92, 93], the computations given in [88] for the case $p = 2$ contain an error. However, it turns out that the $\mathbb{Z}$-oriented bordism groups of $\mathbb{Z}_2$-Witt spaces were computed already by Goresky and Pardon in [110, Section 10.5] to be

$$\Omega_n^{\mathbb{Z}_2\text{-Witt}} \cong \begin{cases} \mathbb{Z}, & n = 0, \\ \mathbb{Z}_2, & n = 4k,\ k > 0, \\ 0, & \text{otherwise.} \end{cases}$$

The invariant is the mod 2 Euler characteristic $\sum_i \dim(I^{\bar{m}}H_i(X; \mathbb{Z}_2)) \bmod 2$ in dimensions $4k$, $k > 0$. Note the similarity to Akin's computation of the bordism groups of mod 2 Euler spaces [3]. Equation (9.6) continues to hold with $p = 2$.

The bordism groups for not-necessarily-oriented $\mathbb{Z}_2$-Witt spaces were first computed by Goresky in [114, Section 5.1] to be 0 in odd degrees and $\mathbb{Z}_2$ in non-negative even degrees. In even degrees, the invariant is again the mod 2 Euler characteristic, and in [93] it is shown that once again the analogue of equation (9.6) holds.

The bordism computations for $\mathbb{Q}$- and $\mathbb{Z}_p$-Witt spaces, $p \neq 2$, all involve the nonsingular cup or intersection pairing on the Witt space X with coefficients in the corresponding field and so require the existence of a fundamental class over that field. As for manifolds, the existence of a $\mathbb{Q}$- or $\mathbb{Z}_p$-fundamental class, $p \neq 2$, is equivalent to the existence of a $\mathbb{Z}$-orientation. But, also as for manifolds, in the unoriented case we can only be sure to have $\mathbb{Z}_2$-fundamental classes. But there is no reason to suspect that a non-$\mathbb{Z}$-orientable $\mathbb{Q}$- or $\mathbb{Z}_p$-Witt space, $p \neq 2$, should possess a fundamental class or nonsingular cup product

[39] Recall that *localization at* 2 means taking the tensor product with $\mathbb{Z}_{(2)}$, the subring of $\mathbb{Q}$ consisting of fractions that can be written *without* powers of 2 in their denominators, i.e. $\mathbb{Z}_{(2)} = \mathbb{Z}\left[\frac{1}{3}, \frac{1}{5}, \ldots\right]$. So localization at 2 is something of a complementary process to our above study of $\mathbb{Q}$-Witt bordism *away* from 2, i.e. after tensoring with $\mathbb{Z}[1/2]$.

pairing with coefficients in the corresponding field $\mathbb{Q}$ or $\mathbb{Z}_p$, nor should we expect such a space to possess such pairings with $\mathbb{Z}_2$ coefficients. Therefore, it is not clear how to begin to study bordism of unoriented K-Witt spaces when $\text{char}(K) \neq 2$.

(e) IP bordism

Just as we have bordism of Witt spaces, one can define bordism of piecewise linear IP spaces, and this was done for $\mathbb{Z}$-IP spaces by Pardon in [187]. For the oriented bordism groups we have

$$\Omega_n^{\mathbb{Z}\text{-IP}} \cong \begin{cases} \mathbb{Z}, & n = 4k, \ k \geq 0, \\ \mathbb{Z}_2, & n = 4k+1, \ k > 0, \\ 0, & \text{otherwise.} \end{cases}$$

In non-negative degrees, these groups correspond to the Mishchenko–Ranicki symmetric L-groups $L^n(\mathbb{Z})$ for $n \neq 1$; see [194]. The invariant in dimensions 0 mod 4 is again the signature, but now in degrees $4k+1$, $k > 0$, we have a new invariant, the de Rham invariant. The de Rham invariant is defined on a $(4k+1)$-dimensional PL oriented IP space by counting mod 2 the number of $\mathbb{Z}_2$ summands in the torsion subgroup of $I^{\bar{m}}H_{2k}(X)$.

In [187], Pardon uses these computations together with technology of Sullivan [227] and Morgan (unpublished) to show that $\bar{\Omega}^*_{IP}(\cdot)_{per}$, a periodic version of the reduced cohomology theory dual to $\mathbb{Z}$-IP bordism, coincides with the cohomology theory coming from the infinite loop space G/TOP, which plays an important role in topological surgery theory. It follows that for a CW complex Z, the group $\bar{\Omega}^0_{IP}(Z)_{per}$ is isomorphic to the set of homotopy classes $[Z, G/TOP]$; Pardon calls this result the *Characteristic Variety Theorem*.

IP bordism yields another instance, as for $\mathbb{Q}$-Witt spaces, where the Sullivan–Siegel machinery holds, and so, as a homology theory,

$$\Omega_*^{\mathbb{Z}\text{-IP}}(\cdot) \otimes \mathbb{Z}[1/2] \cong ko_*(\cdot) \otimes \mathbb{Z}[1/2].$$

But in this case it turns out that we can say more even without inverting 2. The fact that the groups $\Omega_n^{\mathbb{Z}\text{-IP}}$ correspond to the symmetric L-groups $L^n(\mathbb{Z})$ except when $n = 1$ suggests that $\Omega_*^{\mathbb{Z}\text{-IP}}(\cdot)$ should be close to the homology theory of the connective symmetric L-spectrum[40] $\mathbb{L}^\bullet$, which plays an important role in surgery theory and thus in the classification of manifolds (see [195]). Analogously to the fundamental class $[X]_{ko} \in ko_*(X) \otimes \mathbb{Z}[1/2]$ that can be defined for a $\mathbb{Q}$-Witt space X, an IP space can be given a PL invariant fundamental class $[X]_{\mathbb{L}} \in \mathbb{L}_*^\bullet(X)$. A program for developing such fundamental classes was presented by Banagl[41] in [16] based on work of Eppelmann [77], and a detailed proof of all the necessary machinery has been more recently provided by Banagl, Laures, and McClure [23]. In particular, $[X]_{\mathbb{L}}$ is constructed as the image of $[X \xrightarrow{\text{id}} X] \in \Omega_*^{IP}(X)$ under a map of

[40] The connective version of the symmetric L-spectrum is also sometimes written $\mathbb{L}^\bullet\langle 0\rangle$ or $\mathbb{L}^\bullet\langle 0\rangle(\mathbb{Z})$ to distinguish it from other variants. Here $\mathbb{L}^\bullet$ will always mean $\mathbb{L}^\bullet\langle 0\rangle(\mathbb{Z})$.

[41] The survey [16] is also an excellent expository source in general for the connections between pseudomanifolds, signatures, and bordism theories.

homology theories $\Omega_*^{IP}(\cdot) \to \mathbb{L}_*^\bullet(\cdot)$ induced by a map of spectra (in the derived category). The class $[X]_{\mathbb{L}}$ maps to the total L-class of X in $\mathbb{L}_n^\bullet(X) \otimes \mathbb{Q} \cong \bigoplus_j H_{n-4j}(X;\mathbb{Q})$. At the same time, there is an *assembly map* $\mathbb{L}_n^\bullet(X) \to L^n(\mathbb{Z}[\pi_1(X)])$ that takes $[X]_{\mathbb{L}}$ to a stratified homotopy invariant called the *symmetric signature*. The symmetric signature for Witt spaces was constructed rationally in [98], integrally in [23]; it goes to the usual signature under the map $L^n(\mathbb{Z}[\pi_1(X)]) \to L^n(\mathbb{Z})$ induced by the natural coefficient map $\mathbb{Z}[\pi_1(X)] \to \mathbb{Z}$. This all generalizes known results for manifolds. See [23] for details and [194] for further background.

(f) Other "Witt-type" spaces

In a different direction, Goresky and Pardon [110] introduced and studied the bordism groups of several other classes of spaces, most of which are types of Witt spaces with additional conditions. Quickly summarizing, these are:

$\bar{s}$-duality spaces. Translated into our language, these are essentially $\mathbb{Z}_2$-IP spaces, though there is also an assumption of no singular strata of codimension ≤ 4. The oriented bordism groups are the symmetric L groups $L^n(\mathbb{Z}_{(2)})$ [110, Theorem 16.5].

Locally square-free (LSF) spaces. These are $\mathbb{Z}_2$-Witt spaces for which Goresky's Steenrod square operation[42] $Sq^1 : I^{\bar{m}}H_k(L;\mathbb{Z}_2) \to I^{\bar{m}}H_{k-1}(L;\mathbb{Z}_2)$ vanishes on links L of dimension $2k-1$. The oriented bordism groups are

$$\Omega_n \cong \begin{cases} \mathbb{Z}, & n = 0, \\ \mathbb{Z}_4, & n = 4k,\ k \geq 1, \\ \mathbb{Z}_2, & n = 4k+1,\ k > 0, \\ 0, & \text{otherwise.} \end{cases}$$

$\mathbb{Z}_2$-Witt spaces and locally orientable $\mathbb{Z}_2$-Witt spaces. We have already mentioned the computation of oriented $\mathbb{Z}_2$-bordism in [110], but as an intermediate between oriented and unoriented pseudomanifolds, one can also study locally orientable pseudomanifolds, i.e. those whose links are orientable. The unoriented bordism groups of locally orientable $\mathbb{Z}_2$-Witt spaces are all 2-torsion and can be found in [110, Section 10.5]. Goresky and Pardon also show that the bordism groups of arbitrary locally orientable pseudomanifolds are $\mathbb{Z}_2$ in degrees $2k$, $k \geq 0$, and 0 otherwise.

The computations in [110] are all carried out using characteristic numbers arising from intersection homology Wu classes, which are constructed using Goresky's intersection homology Steenrod operations. The bordism homology groups are not considered in [110], but the arguments of [88, Theorem 4.10 and Lemma 4.11] apply to each of these theories

42 The Steenrod operations on mod 2 intersection homology are defined using sheaves and so are beyond the purview of this book. See [114].

for which the bordism groups are all 2-primary in positive degrees (i.e. all of the examples just listed except the $\bar{s}$-duality spaces[43]), resulting in versions of equation (9.6).

(g) L-Spaces

Given the importance we have just seen of signature and pairing information, it is reasonable to ask what spaces more general than Witt spaces might possess self-dual pairings and hence signatures. In his thesis, published as [12] and following on earlier work of Cheeger and Morgan, Banagl showed that it is possible to define such invariants on a class generalizing the $\mathbb{R}$-Witt spaces. Recall from Proposition 9.1.17 that whether or not a space is K-Witt relies only on the characteristic of K; so there is no difference between $\mathbb{Q}$-Witt and $\mathbb{R}$-Witt spaces, but it is simpler for Banagl's machinery to work over $\mathbb{R}$. Banagl's spaces have been rechristened *L-spaces* (see [6]) due to their connection to L-classes and $\mathbb{L}^\bullet$-theory.

The L-spaces are oriented stratified pseudomanifolds possessing sheaf complexes that satisfy certain axioms, including self-duality under a sheaf-theoretic version of duality called Verdier duality (see [28, Section V] or Section 10.1.1 for further references). So now the data is no longer purely topological. To discuss this sheaf-theoretic machinery would take us too far afield, but the basic idea is as follows. One would like to define a kind of cohomology group $\mathfrak{B}^*(X;\mathbb{R})$ on a stratified pseudomanifold X that lives between the upper and lower middle perversity intersection cohomology groups such that the map

$$I_{\bar{n}}H^*(X;\mathbb{R}) \to I_{\bar{m}}H^*(X;\mathbb{R})$$

factors through $\mathfrak{B}^*(X;\mathbb{R})$. Furthermore, for a $2n$-dimensional L-space, $\mathfrak{B}^n(X;\mathbb{R})$ should possess a nonsingular $(-1)^n$-symmetric pairing. It turns out that in order to construct such a thing it is not completely necessary that $I^{\bar{m}}H_k(L;\mathbb{R})$ vanish for a link of dimension $2k$, as is required for a Witt space, but only that the self-pairing on $\mathfrak{B}^k(L;\mathbb{R})$ that has been built up inductively on L have a Lagrangian subspace, i.e. a space of half the dimension on which the pairing vanishes. In other words, $\mathfrak{B}^k(L;\mathbb{R})$ must be a split inner product space! Additionally, there is a monodromy condition that requires, roughly speaking, that we can choose the self-annihilating subspaces consistently for different links as we move around a stratum. For a given pseudomanifold X, it may or may not be possible to construct such *Lagrangian structures* on X by an inductive process over depth: note that to have $\mathfrak{B}^k(L;\mathbb{R})$ defined on L, we must have $\mathfrak{B}$ defined already on the links of L, meaning that we have already found Lagrangian structures on the lower-codimension strata. There is no obstruction to extending self-duality to even-codimension strata, including the regular strata, so the inductive process does have a natural starting point. The oriented pseudomanifolds that possess a compatible choice of Lagrangian structures over their odd-codimension strata are the L-spaces.

Analogously to Witt spaces, L-spaces possess bordism invariant signatures, where now

[43] According to [110, Theorem 16.5], these bordism groups in dimensions $4k$, $k > 1$, are isomorphic to $W(\mathbb{Z}_{(2)})$. By [176, Corollary IV.3.3] there is an exact sequence $0 \to W(\mathbb{Z}_{(2)}) \to W(\mathbb{Q}) \to W(\mathbb{Z}_2)$. But $W(\mathbb{Q})$ contains an infinite cyclic subgroup by [176, Theorem IV.2.1] and $W(\mathbb{Z}_2) \cong \mathbb{Z}_2$ by [176, Lemma IV.1.5]. So the positive-degree bordism groups of $\bar{s}$-duality spaces cannot all be 2-primary.

the bordisms also must carry self-dual sheaves[44]. One can even construct L-classes by a procedure analogous to that of Section 9.4. Banagl showed in [14] that these signatures and L-classes are independent of the choices of Lagrangian structures involved.

The bordism groups of L-spaces were computed in [12] to be

$$\Omega_n \cong \begin{cases} \mathbb{Z}, & n = 4k,\ k \geq 0, \\ 0, & \text{otherwise}. \end{cases}$$

Again the invariant in dimension $4k$ is the signature, and now we see it has become the only invariant! A version of this type of bordism as a homology theory, called *signature homology* $\mathrm{Sig}_*(\cdot)$, was introduced by Minatta [178] using Kreck's *stratifolds* (see [144]) in place of pseudomanifolds. Banagl showed in [13] that the homology theory can also be computed using oriented PL pseudomanifolds equipped with triangulations and their simplicial stratifications, i.e. using the filtration by simplicial skeleta omitting the codimension-one skeleton[45].

Minatta showed that signature homology at 2, i.e. when we invert all the other primes, is just

$$\mathrm{Sig}_*(\cdot) \otimes \mathbb{Z}_{(2)} \cong \bigoplus_{r+s=n} H_r(\cdot; \Omega_s \otimes \mathbb{Z}_{(2)}) \cong \bigoplus_j H_{n-4j}(\cdot; \mathbb{Z}_{(2)}).$$

Similarly,

$$\mathrm{Sig}_*(\cdot) \otimes \mathbb{Q} \cong \bigoplus_j H_{n-4j}(\cdot; \mathbb{Q}).$$

But, once again, Sullivan's work implies that when we invert just 2 in this sort of homology theory the result is an isomorphism

$$\mathrm{Sig}_*(\cdot) \otimes \mathbb{Z}[1/2] \cong ko_*(\cdot) \otimes \mathbb{Z}[1/2].$$

Using this isomorphism, one can construct a Sullivan orientation $[X]_{ko} \in ko_*(X) \otimes \mathbb{Z}[1/2]$ for an L-space X as the image of the signature fundamental class $[X]_{\mathrm{Sig}} \in \mathrm{Sig}_*(X)$. The class $[X]_{\mathrm{Sig}}$ is represented by $(X, \mathcal{S}) \to X$, where $\mathcal{S}$ is a self-dual sheaf complex over X and the map is the identity on X and discards the sheaf information. Furthermore, if $MSIG$ is the spectrum representing signature homology, there is a map of spectra $MSIG \to \mathbb{L}^\bullet$, and this map homotopy splits, implying that signature homology is a direct summand of $\mathbb{L}^\bullet$ homology. However, as shown by Banagl in [16], if M is a manifold (and so automatically carries a self-dual sheaf), the image of $[M]_{Sig}$ in $\mathbb{L}^\bullet_*(M)$ is not necessarily the $\mathbb{L}^\bullet_*$-fundamental class $[M]_{\mathbb{L}}$. Thus, unlike the case of IP bordism where such a map $\Omega^{IP}_*(\cdot) \to \mathbb{L}^\bullet_*(\cdot)$ can be used to define an L-fundamental class on an IP space by a procedure that yields the

[44] Note that the result of [94] stating that we achieve identical bordism theories whether or not we pay attention to stratification data does not consider bordisms involving sheaf complexes. In [12, 13, 178], the definition of the self-dual sheaf complexes utilizes the stratification, so stratification information seems unavoidable.

[45] The version of the theory in [13] uses a slightly more general type of self-dual sheaf complex that generalizes intersection homology with local coefficient systems. Interestingly, this does not alter the bordism computations.

"correct" classical result when applied to manifolds, such a procedure does not extend in a manifold-compatible way to L-spaces by way of signature homology.

Finally, we mention that the Banagl–Cappell–Shaneson argument [20], which we discussed above for $\mathbb{Q}$-Witt spaces, also generalizes to show there is a surjection $\Omega_*^{SO}(\cdot) \otimes \mathbb{Z}[1/2] \to \mathrm{Sig}_*(\cdot) \otimes \mathbb{Z}[1/2]$. This is used in [13] to construct twisted L-classes and twisted signatures on L-spaces.

(h) Novikov Conjectures

To conclude, we briefly discuss some relations between these circles of ideas and the Novikov Conjecture, one of the most famous open problems in topology. The story begins with the observation that the signature of a manifold is an oriented homotopy invariant. In other words, if $f : M_1 \to M_2$ is a homotopy equivalence of manifolds that is compatible with the orientations in the sense that f takes Γ_{M_1} to Γ_{M_2}, then $\sigma(M_1) = \sigma(M_2)$. This follows directly from the homotopy invariance of cohomology theory and the naturality of cup and cap products. We know that signatures are intimately related to L-classes, but the L-classes are *not* homotopy invariant. In fact, the L-classes, and the closely related Pontrjagin classes, are often used as the tool to show two homotopy equivalent manifolds are not equivalent in some stronger sense such as homeomorphism or diffeomorphism. For example, Pontrjagin classes are used in Milnor's proofs that there are manifolds that are homeomorphic but not diffeomorphic and that there are topological manifolds with no smooth structure [175].

The Novikov Conjecture concerns the homotopy invariance of *higher signatures*. Given a group G, let $BG = K(G,1)$ be its first Eilenberg–Mac Lane space (see [125, Additional Topic 1B]). Given a connected space X, there is a bijection between maps $\pi_1(X) \to G$ and basepoint-preserving homotopy classes of map $X \to BG$ [125, Proposition 1B.9]. Each such map $r : X \to BG$ determines a covering space of X by pulling back the universal cover of BG, so, for a manifold M, the data $(M, r : M \to B\pi_1(M))$ can be thought of as prescribing M together with one of its coverings.

Now, for each $x \in H^*(B\pi_1(M);\mathbb{Q})$, define the higher signature $\mathrm{sig}_x(M,r)$ to be the number $\mathrm{sig}_x(M) = \mathbf{a}((L^*(M) \smile r^*(x)) \frown \Gamma_M) \in \mathbb{Q}$, where $L^*(M)$ is the cohomological L-class of M. If r is the trivial map and $x = 1 \in H^0(B\pi_1(M);\mathbb{Q})$, we recover the usual signature of M. Using basic properties of (co)homology and that the cohomology L-classes are nontrivial only in dimensions a multiple of 4, we can also write

$$\begin{aligned}\mathrm{sig}_x(M) &= \mathbf{a}((L^*(M) \smile r^*(x)) \frown \Gamma_M) = \mathbf{a}((r^*(x) \smile L^*(M)) \frown \Gamma_M)\\ &= \mathbf{a}(r^*(x) \frown (L^*(M) \frown \Gamma_M)) = \mathbf{a}(r^*(x) \frown \mathscr{L}_*(M))\\ &= \mathbf{a}(x \frown r(\mathscr{L}_*(M))),\end{aligned}$$

where $\mathscr{L}_*(M)$ are the homology L-classes (see Proposition 9.4.8).

The Novikov Conjecture is that the higher signatures are all oriented homotopy invariants, i.e. that if $f : M_1 \to M_2$ is an orientation-preserving homotopy equivalence and $r : M_2 \to$

$B\pi_1(M_1) = B\pi_1(M_2)$, then $sig_x(M_2, r) = sig_x(M_1, rf)$. The Novikov Conjecture is related to several other important conjectures in manifold theory, including the Borel Conjecture, which states that a homotopy equivalence between closed aspherical manifolds, i.e. those for which $\pi_i(M) = 0$ for $i > 1$, must be homotopic to a homeomorphism. For more about the Novikov Conjecture, the reader can consult any number of surveys, including [66, 79, 145, 195, 196].

The oriented homotopy invariance of the higher signatures turns out to depend only on properties of the group $G = \pi_1(M)$, and in fact, for a given such G, it holds if and only if the assembly map $\mathbb{L}^\bullet_*(BG) \to L^*(\mathbb{Z}[G])$ is a split injection over $\mathbb{Q}$. This is known to hold for several classes of groups, for example free abelian groups, but the conjecture that it holds for all discrete groups has not been proven.

Returning to pseudomanifolds, we know that $\mathbb{Q}$-Witt spaces, and so IP-spaces, possess homology L-classes, and therefore the higher signatures $\text{sig}_x(X) = \mathbf{a}(x \frown r(\mathscr{L}_*(X)))$ are defined and one can formulate the Stratified Novikov Conjecture – that these higher signatures are oriented stratified homotopy-type invariants. It is shown in [23] that the Stratified Novikov Conjecture holds for groups for which the Novikov Conjecture holds. An analytic version of this result, using Dirac operators on the regular strata, had previously been proven in [6]. In fact, many of the topics of this section can be covered from an analytic viewpoint and connect with more analytic formulations of the Novikov and related conjectures; see [5, 6].

Signature homology also allows for an "integral refinement" of the Novikov Conjecture: Minatta shows in [178] that if we take the signature homology fundamental class $[M]_{\text{Sig}}$ of a closed oriented manifold M and tensor with $\mathbb{Q}$, then the resulting class in $\text{Sig}_*(M) \otimes \mathbb{Q} \cong \bigoplus_j H_{n-4j}(M; \mathbb{Q})$ is just the rational homology L-class. It follows that the Novikov Conjecture for $\pi_1(M)$ is equivalent to the oriented homotopy invariance of the image of $[M]_{\text{Sig}}$ in $\text{Sig}_*(B\pi_1(M)) \otimes \mathbb{Q}$ induced by $r : M \to B\pi_1(M)$. This leads to the following question due to Matthias Kreck, which can be viewed as an integral refinement of the Novikov Conjecture: When is $r([M]_{\text{Sig}}) \in \text{Sig}_*(B\pi_1(M))$ an oriented homotopy invariant? Unlike the Novikov Conjecture, there are groups for which such homotopy invariance is known to fail, but it remains to classify for precisely which groups invariance holds. The fact that we have maps $\text{Sig}_*(\cdot) \to \mathbb{L}^\bullet_*(\cdot)$ allows us to compose with the assembly map to obtain signature homology assembly maps $\text{Sig}_*(BG) \to L^*(\mathbb{Z}[G])$, which, as noted by Banagl [16], might be useful in attacking such questions. Of course as signature homology and $\mathbb{L}^\bullet$-homology agree rationally, another integral refinement, due to Ranicki, is to ask when the $\mathbb{L}^\bullet$-homology fundamental class is a homotopy invariant. As previously noted, it is well known that the L-classes are not generally homotopy invariants and so similarly the $\mathbb{L}^\bullet$-homology fundamental class is not generally a homotopy invariant. But as signature homology is a summand of $\mathbb{L}^\bullet$-homology, it is possible there may be groups for which signature homology fundamental classes are homotopy invariants while $\mathbb{L}^\bullet$-homology fundamental classes are not.

10
Suggestions for Further Reading

This chapter contains some suggestions for further reading, principally, though not exclusively, other expository sources. As noted in the Preface, there are by now hundreds, if not thousands, of works that involve intersection homology and related topics in some way, and so the discussion here will be limited to a small fraction of the possible routes the interested reader might explore. The suggestions that are provided necessarily reflect the author's own limited knowledge and particular interests, with his apologies for all that has been omitted due to ignorance or poor judgment. Of course, many further references can be found in the works cited below, especially [140].

10.1 Background, Foundations, and Next Texts

This section refers mostly to preparatory texts, especially those that will get the reader up to speed on the sheaf-theoretic version of intersection homology. This is the language in which many papers using intersection homology are formulated[1] and so is a prerequisite for many (though not all) of the other sources considered.

The sheaf-theoretic approach to intersection homology involves not just sheaf theory but also derived categories and Verdier duality. We provide below some suggestions for deeper background immersion, but the following textbook expositions are mostly self-contained,

[1] We should provide here one important terminological warning. In sheaf theory, one typically works with cohomologically graded complexes and computes what is called "sheaf cohomology", or, starting from a complex of sheaves, "hypercohomology". Thus sheaf-theoretic sources tend to speak of "intersection cohomology," which is typically isomorphic, at least on compact spaces, to our "intersection homology"; on non-compact spaces, sheaf-theoretic intersection cohomology (with closed supports) would correspond to intersection homology formed using locally finite chains, i.e. chains that may be formal sums of an infinite number of simplices (with coefficients) but such that every point in the space has a neighborhood intersecting the supports of only a finite number of simplices. In the sheaf-theoretic world, the dual theory is then also a version of (hyper)cohomology but applied to a sheaf complex obtained via the Verdier duality functor, which is the appropriate generalization of our Hom duals. The sheaf-theoretic version of Poincaré duality then says that the Verdier dual of the sheaf complex that computes perversity $\bar{p}$ intersection homology is the sheaf complex that computes the $D\bar{p}$ intersection homology. This is all explained in the cited references, but the reader is cautioned to exercise care in properly interpreting the word "cohomology."

including either overviews or more thorough treatments of all the material they require beyond basic algebraic topology:

- Markus Banagl, *Topological Invariants of Stratified Spaces* [11]. This text contains all the relevant background details on sheaf theory, derived categories, and Verdier duality, a thorough treatment of sheaf-theoretic intersection homology, and applications to characteristic classes, signatures, and their computations. The book also features introductions to related topics, including t-structures, perverse sheaves, characteristic classes on "non-Witt spaces," and L^2-cohomology.

- Frances Kirwan and Jonathan Woolf. *An Introduction to Intersection Homology Theory, Second Edition* [140]. The primary aim of this text is to provide an accessible overview of some of the most important applications of intersection homology, especially to fields beyond topology. As such, it does provide the necessary background material on sheaves and derived categories, but in a briefer treatment than is found in [11]. The bulk of the content is then dedicated to very readable accounts concerning the extension of the Kähler package to singular varieties, L^2-cohomology, perverse sheaves, intersection cohomology of fans and Stanley's conjectures on the combinatorics of polytopes, the Weil conjectures for singular varieties, $\mathcal{D}$-modules and the Riemann–Hilbert correspondence, and the Kazhdan–Lusztig conjecture. This is a great starting point for those interested in aspects of intersection homology in other areas than those considered below, and it contains many suggestions for further reading, some of which we repeat here.

- Jean-Paul Brasselet. *Introduction to Intersection Homology and Perverse Sheaves* [34]. Billed in its introduction as an "exploratory travel," this short book provides many hands-on examples, as well as introductions to sheaf theory, perverse sheaves, and the various approaches to intersection cohomology by differential forms.

- A. Borel et al., *Intersection Cohomology* [28]. This book is a collection of seminar notes by various authors, including Goresky and MacPherson, and it is not quite as self-contained as the others. The reader is assumed to already have some familiarity with sheaf theory and derived functors. The derived category is introduced in Section V.5 but some prior exposure wouldn't hurt. However, the reader who has acquired some of this background will find this to be both a readable exposition and an excellent resource. No prior knowledge of intersection homology is required, and the basic definitions and sheaf-theoretic approach are built up through the first few chapters. Many technical details concerning constructible sheaves that were not addressed directly in the original sheaf development of intersection cohomology by Goresky and MacPherson [106] can be found in Chapter V, in which Borel provides a thorough treatment of all the underlying machinery.

10.1.1 Deeper Background

(a) Sheaf Theory

As we discussed briefly in Section 8.1.3, sheaves can be thought of as generalized bundles of groups for which the group is allowed to vary from point to point on a space. Hence sheaf theory, which has many uses throughout mathematics, is particularly well suited to stratified phenomena, where the local structure of a space varies from stratum to stratum. Sheaf cohomology then provides a way to patch these local structures together to yield global invariants.

While most of the texts listed above provide overviews of the sheaf theory they need, the interested reader might desire a more thorough background concerning both sheaves and the related homological algebra. Here are some suggestions to take the novice to sheaf theory up through sheaf cohomology:

- B.R. Tennison, *Sheaf Theory* [233]. This is an elementary introduction that introduces all of the relevant homological algebra, including derived functors.
- Richard G. Swan, *The Theory of Sheaves* [230]. This is another nice introduction for beginners, though some of the notation has become somewhat outdated (for example, what are usually called "presheaves" are here called[2] "stacks".)
- Glen Bredon, *Sheaf Theory, Second Edition* [37]. This a very detailed account, perhaps better suited as a reference than as a first introduction (we have utilized this text as a reference at several points in the text). However, a good first course on sheaf theory could be assembled from Chapter I, Sections 1–10 of Chapter II, Chapter III, and Sections 1–8 of Chapter IV.
- Roger Godement, *Topologie Algébrique et Théorie des Faisceaux* [104]. This is the classic introduction to sheaf theory, only available in the original French and in Russian.

(b) Derived Categories and Verdier Duality

Sheaf cohomology is a derived functor, as are Ext and Tor: to compute any of these, one replaces the initial object (a sheaf, a group, a module, . . .) with an appropriate resolution (by a complex of sheaves, groups, modules, . . .), applies a functor to the resolution, and then takes (co)homology groups. It turns out that there is value in working in a category of complexes to begin with and in identifying objects (thought of as complexes that are trivial except in a single degree) with their resolutions. This leads to the notion of derived categories. Derived categories are constructed analogously to localizations in ring theory, but, rather than formally inverting elements, one formally inverts the morphisms that induce (co)homology isomorphisms, thus making all the resolutions of the same object isomorphic in the derived category itself (before taking (co)homology). As an intermediate between the original category and the derived category, one also has the homotopy category in which maps that are homotopic in an appropriate sense are identified (just as one can work in

[2] On page 25 of [230], Swan notes that "pre-sheaf" is the older term for "stacks," so presumably "presheaf" made a revival!

the homotopy category of topological spaces by treating homotopic maps as equivalent). Unfortunately, homotopy and derived categories are no longer *abelian categories*, meaning roughly that certain standard algebraic constructions such as forming kernels and cokernels no longer apply. Rather, these are examples of *triangulated categories*, which do have other useful formal properties. An important construction in the derived category of sheaves is the Verdier dualizing functor, a generalization of $\mathrm{Hom}(\cdot, R)$ that plays a critical role in sheaf-theoretic proofs of intersection homology Poincaré duality.

Here are some sources for this material:

- S.I. Gelfand and Yu I. Manin, *Methods of Homological Algebra, Second Edition* [102]. While not primarily focused on sheaf theory, this text contains a good account of derived and triangulated categories in Chapters III and IV. Section III.8 treats applications to sheaf theory, including Verdier duality (although it doesn't use that name).
- Alexandru Dimca, *Sheaves in Topology* [70]. Dimca's book is motivated by applications of sheaf theory to topology and contains material on derived categories and Verdier duality, as well as perverse sheaves and other topics related to intersection homology.
- Robin Hartshorne, *Residues and Duality* [124]. While these notes are ultimately concerned with a more specialized topic in algebraic geometry, the first chapter provides an introduction to triangulated categories and derived categories and functors.
- Masaki Kashiwara and Pierre Schapira, *Sheaves on Manifolds* [137]. This is a more advanced book primarily concerned with "microlocal analysis." However, it does include background on numerous topics including derived categories, sheaves, Verdier duality, constructible sheaves, and perverse sheaves. It is probably better consulted as a reference than as an introduction. We should also mention *Categories and Sheaves* [138] by the same authors, which treats categories and sheaves from a quite general abstract categorical point of view.
- B. Iversen, *Cohomology of Sheaves* [135]. This is another more advanced text that includes treatments of hypercohomology of sheaf complexes and Verdier duality on locally compact spaces.

10.2 Bordism

Various flavors of pseudomanifold bordism and signatures, as well as stratified versions of the Novikov conjecture, have already been discussed in Section 9.5. For the reader who wishes to pursue these topics further, we collect again here the references already provided, referring the reader back to that section for further discussion and context. With the exception of [16], none of these papers are primarily intended as expositions.

- Ethan Akin, *Stiefel–Whitney homology classes and bordism* [3]
- Pierre Albin, Éric Leichtnam, Rafe Mazzeo, and Paolo Piazza, *The signature package on Witt spaces* [5]

- Pierre Albin, Éric Leichtnam, Rafe Mazzeo, and Paolo Piazza, *The Novikov conjecture on Cheeger spaces* [6]
- Markus Banagl, *Extending intersection homology type invariants to non-Witt Spaces* [12]
- Markus Banagl, *Computing twisted signatures and L-classes of non-Witt spaces* [13]
- Markus Banagl, *The L-class of non-Witt spaces* [14]
- Markus Banagl, *The signature of singular spaces and its refinements to generalized homology theories* [16]
- Markus Banagl, Sylvain Cappell, and Julius Shaneson, *Computing twisted signatures and L-classes of stratified spaces* [20]
- Markus Banagl, Gerd Laures, and James E. McClure, *The L-homology fundamental class for IP-spaces and the stratified Novikov conjecture* [23]
- Greg Friedman, *Intersection homology with field coefficients: K-Witt spaces and K-Witt bordism* [88] (see also [92, 93])
- Greg Friedman, *Stratified and unstratified bordism of pseudomanifolds* [94]
- R. Mark Goresky, *Intersection homology operations* [114]
- Mark Goresky and William Pardon, *Wu numbers of singular spaces* [110]
- Augusto Minatta, *Signature homology* [178]
- William L. Pardon, *Intersection homology Poincaré spaces and the characteristic variety theorem* [187]
- P.H. Siegel, *Witt spaces: a geometric cycle theory for KO-homology at odd primes* [218]
- Jon Woolf, *Witt groups of sheaves on topological spaces* [250]

10.3 Characteristic Classes

As noted in the introduction to Section 9.5, there is an extensive theory of characteristic classes on stratified spaces, encompassing many different constructions and viewpoints. We provided there a sampling of expository references, which we list again here, excluding Banagl's [11], already listed above.

- Paolo Aluffi, *Characteristic classes of singular varieties* [9]
- Jean-Paul Brasselet, *Characteristic classes and singular varieties* [32]
- Jean-Paul Brasselet, *From Chern classes to Milnor classes – a history of characteristic classes for singular varieties* [33]
- Jean-Paul Brasselet, José Seade, and Tatsuo Suwa, *Vector fields on singular varieties* [35]
- Laurenţiu Maxim and Jörg Schürmann, *Characteristic classes of mixed Hodge modules and applications* [164]
- Adam Parusiński, *Characteristic classes of singular varieties* [188]
- Jörg Schürmann, *Lectures on characteristic classes of constructible functions* [208]
- Jörg Schürmann, *Nearby cycles and characteristic classes of singular spaces* [209]

- Jörg Schürmann and Shoji Yokura, *A survey of characteristic classes of singular spaces* [210]
- Tatsuo Suwa, *Characteristic classes of singular varieties* [229]

10.4 Intersection Spaces

Intersection homology theory recovers duality on stratified spaces by altering the homology theory depending on perversities. Banagl's theory of *intersection spaces* instead modifies the spaces in such a way that duality is recovered using ordinary homology. The resulting intersection space homology is different from intersection homology, though closely related to it, and so provides new invariants for stratified spaces.

- Markus Banagl, *Intersection spaces, spatial homology truncation, and string theory* [15], introduces intersection spaces as well as developing the homotopy-theoretic underpinnings of the *spatial homology truncation* used in their construction. Further work on intersection spaces, including extending the construction to more stratified spaces, developing connections to de Rham and L^2 cohomologies, and studying applications to algebraic varieties can be found in [17, 18, 19, 21, 22, 24, 25, 142, 143, 166].

10.5 Analytic Approaches to Intersection Cohomology

For those of a more analytic bent, there are several ways to define versions of intersection cohomology using differential forms. We provide some introductory references to each.

10.5.1 L^2-Cohomology

Originally developed by Cheeger independently of intersection homology [58, 59, 60], the idea of L^2-cohomology on stratified spaces is to consider differential forms on the regular strata that are square integrable (in an appropriate sense) with respect to an appropriately-chosen metric that reflects the stratified structure of the full space. On Witt spaces, Cheeger's original metrics turn out to give middle perversity intersection homology. Here are some further accounts:

- Jeff Cheeger, *On the Hodge theory of Riemannian pseudomanifolds* [59]. This is an early expository account by Cheeger of his work on L^2-cohomology and its relation to intersection homology.
- Xianzhe Dai, *An introduction to L^2-cohomology* [64]. This is a short introduction to L^2-cohomology and L^2 signatures.

- Leslie Saper and Steven Zucker, *An Introduction to L^2-cohomology* [203]. This is another introductory survey that discusses both the connection to intersection cohomology and L^2 Hodge decompositions for varieties endowed with Kähler metrics.
- Eugénie Hunsicker, *Hodge and signature theorems for a family of manifolds with fibre bundle boundary* [134]. The previous items concern only L^2-cohomology that corresponds to the middle perversity intersection homology groups. In this paper, the author demonstrates metrics that yield intersection homology with other perversities.
- Pierre Albin, *On the Hodge theory of stratified spaces* [4]. This is a survey of recent work on L^2 Hodge theory, including a discussion of the Novikov conjecture for stratified spaces.

10.5.2 Perverse Forms

Another approach to a de Rham theory of intersection cohomology does not use metrics to control differential forms defined on the top strata but rather places restrictions on the vanishing behavior of the forms in certain directions as they approach the singularities. This notion is very analogous to how the singular chain theory limits the intersection dimension of chains with strata. Such differential forms are sometimes called *perverse forms*.

- Jean-Luc Brylinski, *Equivariant intersection cohomology* [39]. This is the first implementation of the perverse forms construction, following a suggestion of Goresky and MacPherson.
- J.P. Brasselet, G. Hector, and M. Saralegi, *Théorème de de Rham pour les variétés stratifiées* [31]. Using a slight modification of the definition of perverse forms in [39], this paper provides a de Rham theorem relating cohomology of perverse forms to intersection homology by integration. This work is generalized to a broader range of perversities by Saralegi in [204, 205]; in particular [205] contains Saralegi's construction of non-GM intersection homology, which we discussed in Section 6.2.3.

(a) Chataur–Saralegi–Tanré Theory

The previous approaches to intersection homology using differential forms all assume smooth manifold regular strata on which differential forms satisfying various properties can be placed. Recent exciting work by David Chataur, Martin Saralegi, and Daniel Tanré takes these ideas a step further by constructing for any stratified space a complex of differential forms that live, roughly speaking, on the appropriately restricted singular simplices of the space. This construction is thus inspired by Dennis Sullivan's approach to rational homotopy theory.

- David Chataur, Martintxo Saralegi-Aranguren, and Daniel Tanré, *Intersection cohomology, simplicial blow-up and rational homotopy* [56]. This is the initial work containing all the basic details. Follow-up works, which are interesting in their own right and touch on many of the topics of this text, include [51, 52, 53, 54, 55, 57].

10.6 Stratified Morse Theory

Just as manifolds admit handle decompositions using Morse functions, sufficiently nice stratified spaces admit stratified analogues whose behavior near singularities of the Morse function is classified using intersection homology rather than ordinary homology. Stratified Morse theory is also deeply related to the theory of perverse sheaves[3], which is considered in the next section of references.

- Mark Goresky and Robert MacPherson, *Morse theory and intersection homology theory* [107]. This is a survey paper by the inventors of stratified Morse theory.
- Mark Goresky and Robert MacPherson, *Stratified Morse theory* [109]. This book provides the full account of the authors' stratified Morse theory.
- David B. Massey, *Stratified Morse theory: past and present* [159]. Another survey of stratified Morse theory that includes several applications.
- Jörg Schürmann, *Topology of Singular Spaces and Constructible Sheaves* [207]. This book contains several more advanced applications of the theory of constructible sheaves, including a development of stratified Morse theory for constructible sheaves.

10.7 Perverse Sheaves and the Decomposition Theorem

Perverse sheaves constitute a certain category of sheaf complexes[4] that generalize those whose cohomology gives intersection homology. In fact the intersection sheaves are the simple objects of this category. In the words of Kapranov and Schechtman, "The notion of a perverse sheaf . . . has come to play a central role in algebraic geometry and representation theory" [136]. As such, there are several good introductions, including [11, 70, 140], which have already been mentioned. Many more introductory notes can be found by searching online. We mention here only the following few examples:

- Alexander Beĭlinson, Joseph Bernstein, and Pierre Deligne, *Faisceaux pervers* [26]. This famous work, often known as "BBD," introduced perverse sheaves and used them to prove the equally famous Decomposition Theorem concerning the homological behavior of proper maps between algebraic varieties.
- Mark Andrea A. de Cataldo and Luca Migliorini, *The decomposition theorem, perverse sheaves and the topology of algebraic maps* [68]. This is an expository survey of perverse sheaves and the Decomposition Theorem, including several approaches and applications.

[3] On a complex analytic variety, the middle perversity perverse sheaves are exactly those whose normal Morse data vanishes except in a single degree; see [207, Remark 6.04]. Thanks to Jon Woolf for pointing this out to me.

[4] The famous quip, dating back to [26] itself, is that perverse sheaves are neither sheaves nor perverse; they are sheaf complexes and they behave rather nicely.

- Robert MacPherson, *Intersection homology and perverse sheaves* [156]. These unpublished[5] colloquium notes provide an approach to perverse sheaves through stratified Morse theory.
- Geordie Williamson, *Algebraic representations and constructible sheaves* [248]. This is a recent set of lecture notes in which one can see applications of intersection homology, perverse sheaves, and the Decomposition Theorem to the Kazhdan–Lusztig conjecture of representation theory. See also [76].
- David B. Massey, *Notes on perverse sheaves and vanishing cycles* [158]. Some notes on perverse sheaves with applications to the nearby and vanishing cycles associated to the Milnor fiber of an analytic singularity.

10.8 Hodge Theory

Hodge theory, in the form of Hodge decompositions, signature theorems, and further generalizations, extends to singular algebraic varieties using intersection homology. As such, Hodge theory already appears and has an important role in some of the references we have already cited, including [68, 76, 164, 203]. Here are some further references in this direction. Some additional background in algebraic geometry and Hodge theory would be useful, though most of these sources provide at least some review and, of course, further references.

- Morihiko Saito, *Introduction to mixed Hodge modules* [201] and *A young person's guide to mixed Hodge modules* [202]. These expository works provide an introduction to Saito's *Modules de Hodge polarisables*, [200], which first established intersection homology notions of Hodge theory.
- Eduard Looijenga, *Cohomology and intersection homology of algebraic varieties* [154]. This is a general introduction to the algebraic topology of complex algebraic varieties, including the extensions to singular varieties using intersection homology.
- Chris A.M. Peters and Joseph H.M. Steenbrink, *Mixed Hodge Structures* [189]. This is a full textbook account of mixed Hodge structures.
- Sylvain E. Cappell, Laurențiu G. Maxim, and Julius L. Shaneson, *Euler characteristics of algebraic varieties* [44] and *Hodge genera of algebraic varieties. I* [45]; Sylvain E. Cappell, Anatoly Libgober, Laurențiu G. Maxim, and Julius L. Shaneson, *Hodge genera of algebraic varieties. II* [43]. These papers apply intersection homology Hodge theory to obtain formulas for characteristic classes of singular varieties.

[5] Once hard to find, copies have begun to appear on the internet and can now usually be found with a bit of web searching.

10.9 Miscellaneous

- Shmuel Weinberger, *The Topological Classification of Stratified Spaces* [239]. This expanded version of seminar notes provides a blueprint for adapting surgery theory to the classification of stratified spaces.
- Dirk Schütz, *Intersection homology of linkage spaces* [211] and *Intersection homology of linkage spaces in odd-dimensional Euclidean space* [212]. These two recent papers provide an application of PL intersection homology to the study of linkage spaces.
- Jonathan Woolf, *The fundamental category of a stratified space* [251] and *Transversal homotopy theory* [252]; David A. Miller, *Strongly stratified homotopy theory* [174]. These papers explore various interesting aspects of the homotopy theory of stratified spaces.
- Richard P. Stanley, *Recent developments in algebraic combinatorics* [223]. This is a survey treatment of Stanley's work on the combinatorics of polytopes, including applications of intersection homology to the proof of the g-theorem.
- Martintxo Saralegi-Aranguren and Robert Wolak, *Poincaré duality of the basic intersection cohomology of a Killing foliation* [206]. This is one of the most recent papers concerning basic[6] intersection cohomology of singular foliations. It contains several references to previous work in this area.
- Sylvain E. Cappell and Julius L. Shaneson, *Singular spaces, characteristic classes, and intersection homology* [46]; Laurențiu Maxim, *Intersection homology and Alexander modules of hypersurface complements* [165]; Greg Friedman, *Intersection Alexander polynomials* [82]. These papers all explore applications of intersection homology to embeddings such as knots and hypersurfaces.
- Greg Friedman, Eugénie Hunsicker, Anatoly Libgober, and Laurențiu Maxim (editors), *Topology of Stratified Spaces* [97]. These conference proceedings contain a variety of survey articles, including [16, 64], already mentioned above.

[6] "Basic" is a technical term in foliation theory, essentially referring to objects whose information content is transverse to the leaves of the foliation.

Appendix A
Algebra

We here compile some useful facts from homological algebra, though many other bits are developed elsewhere in the main body of the text closer to where they are needed. This section can be thought to comprise "background results," including facts that are well known but not always easy to pinpoint in the literature or facts that are easy to pinpoint with certain restrictions, e.g. R being a principal ideal domain (PID), but for which we need slight generalizations, e.g. R being a Dedekind domain. We also recall some standard definitions for the reader's convenience. Some of this material overlaps with material treated in more detail elsewhere in the text.

We always assume that our rings R are commutative with unity.

A.1 Koszul Sign Conventions

A.1.1 Why Sign?

Signs are unavoidable in algebraic topology. As a first example, we know that to define the boundary map in a simplicial chain complex we need to take alternating sums in formulas that look something like $\partial[0,\ldots,n] = \sum_{i=1}^{n}(-1)^i[0,\ldots,\hat{i},\ldots,n]$. The signs are necessary to ensure that $\partial \circ \partial = 0$. But even once we've moved on from concrete geometric constructions to higher-level algebraic gizmos, there are still signs. For example, the reader likely knows that if $\alpha, \beta \in H^*(X)$ are cohomology classes, then $\alpha \smile \beta = (-1)^{|\alpha||\beta|}\beta \smile \alpha$. Here we use the notation $|\cdot|$ to take the degree of an element. So if $\alpha \in H^i(X)$, then $|\alpha| = i$.

Unfortunately, there are a variety of conventions for manipulating signs, and keeping all the signs consistent is certainly a nuisance. So why do we need them? Let's see that even when we move beyond topological constructions to pure homological algebra, we still can't do without them. Here's one example: Suppose C_* and D_* are chain complexes. We will always mean chain complexes of R-modules though we often omit the R from the explicit notation. Then one can form the tensor product chain complex $C_* \otimes D_*$. The module in degree n is sensibly defined to be $(C_* \otimes D_*)_n = \bigoplus_{i+j=n} C_i \otimes D_j$. But how should we define the boundary map? We can't let[1] $\partial(c \otimes d) = (\partial c) \otimes (\partial d)$, because such a map would lower degree by two. Another attempt, closer to the formula the reader might have seen, would be $\partial(c \otimes d) = (\partial c) \otimes d + c \otimes \partial d$. The problem then is that, after simplification and using that $\partial \circ \partial = 0$ in C_* and D_*, we would have $\partial \circ \partial(c \otimes d) = 2(\partial c) \otimes (\partial d)$. But we need $\partial \circ \partial = 0$ in $C_* \otimes D_*$ as well. One solution: set

$$\partial(c \otimes d) = (\partial c) \otimes d + (-1)^{|c|} c \otimes \partial d. \tag{A.1}$$

The reader can check that indeed $\partial \circ \partial = 0$. So signs are unavoidable even in pure homological algebra.

So, which signs? There are other options that would work to define the boundary map of $C_* \otimes D_*$ and get a chain complex. For example $\partial(c \otimes d) = (-1)^d \partial(c) \otimes d + c \otimes \partial d$. Ultimately, it is a matter of definition. There are some requirements; e.g. we want to make $C_* \otimes D_*$ into a chain complex. But how precisely to do that is often a matter of aesthetics and good compatibility with other sign conventions so that we wind up with "nice" formulas down the road. It also requires some care for sign conventions to be consistent in the sense that, for example, if some other construction is equivalent to forming a tensor product, then its sign convention should be compatible with that on the tensor product. Otherwise, we have some flexibility, and the Koszul convention that we will discuss has turned out to be favored in homological algebra. So how does it work?

One nice feature of the formula $\partial(c \otimes d) = (\partial c) \otimes d + (-1)^{|c|} c \otimes \partial d$ is that the sign only appears when the symbol ∂ jumps past the symbol c. If we assign ∂ the degree -1 (since it lowers degrees by one), then the sign $(-1)^{|c|} = (-1)^{|\partial||c|}$ is determined from the product of the degrees of the symbols being interchanged. This is what we see also in the cup product. This particular notion is called the *Koszul sign convention*. In this section, we review some important occurrences of the Koszul sign conventions, as well as an important place where we deviate from it in order for things to work out elsewhere[2]; some further references and details can be found in [71, Section VI.10] or [155, Section II.3].

A.1.2 Homological versus Cohomological Grading

Before going on, we should discuss the fact that there are two grading conventions in common usage: homological, with degree index written as a subscript, and cohomological, with degree index written as a superscript. For the purpose of formulas involving both types of objects, it is often useful to employ the convention $C^* = C_{-*}$. In other words, if C_* is a chain complex with homological indexing, then it can be made into a chain complex with cohomological indexing by defining $C^i = C_{-i}$ and letting $d : C^i \to C^{i+1}$ correspond to

[1] We will typically rely on context to make it clear which chain complexes our boundary maps are being computed in, in which case we simply write ∂ for all the boundary maps in a formula. If necessary for clarity, we will write things like ∂_{C_*} for the boundary map in C_*.

[2] As mentioned, this is all a nuisance.

$\partial : C_{-i} \to C_{-i-1}$. Of course this changes the degree of an element by a sign, but as we are typically most interested only in the parity of a degree, this usually causes no trouble.

Warning: When employing this convention, we need to be careful not to confuse this with the convention of letting $S_*(X)$ denote the singular chain complex on X and $S^*(X)$ denote the singular cochain complex whose groups are $\mathrm{Hom}(S_*(X), \mathbb{Z})$.

Throughout the book, we use ∂ to denote the boundary map in a homologically graded chain complex and d to denote the (co)boundary map in a cohomologically graded chain complex.

A.1.3 The Chain Complex of Maps of Chain Complexes

Many of our sign conventions involve not just elements with associated degrees but maps with degrees. Let us recall how this works.

For working with homomorphisms of chain complexes of R-modules, say $f : C_* \to D_*$, that might raise or lower the degree, it is useful to define the chain complex of R-modules $\mathrm{Hom}_*(C_*, D_*)$. These R-modules are defined so that $\mathrm{Hom}_n(C_*, D_*)$ consists of homomorphisms that raise the degree by n. In other words, $\mathrm{Hom}_n(C_*, D_*) = \prod_i \mathrm{Hom}(C_i, D_{i+n})$ so that an element $f \in \mathrm{Hom}_n(C_*, D_*)$ consists of a collection of R-module homomorphisms $f_i : C_i \to D_{i+n}$ for all $i \in \mathbb{Z}$. For example, the boundary map of C_* determines an element $\partial_{C_*} \in \mathrm{Hom}_{-1}(C_*, C_*)$.

Let us observe that this definition is consistent with the indexing typically used in algebraic topology to define the cochain complex $\mathrm{Hom}(S_*(X), \mathbb{Z})$. Here $\mathbb{Z}$ is treated as a homologically graded chain complex that is zero except in degree zero, where it is $\mathbb{Z}$. An element $f \in \mathrm{Hom}_n(S_*(X), \mathbb{Z})$ corresponds to an element of $\prod_i \mathrm{Hom}(S_i(X), (\mathbb{Z})_{i+n})$, and the components are trivial unless $i = -n$. So $\mathrm{Hom}_n(S_*(X), \mathbb{Z}) \cong \mathrm{Hom}(S_{-n}(X), \mathbb{Z})$. In cohomological indexing this translates to $\mathrm{Hom}^n(S_*(X), \mathbb{Z}) = \mathrm{Hom}_{-n}(S_*(X), \mathbb{Z}) \cong \mathrm{Hom}(S_{-n}(X), \mathbb{Z})$. So this is consistent with our notion of an n-cochain as something that acts on an n-chain.

So far we have defined the modules $\mathrm{Hom}_n(C_*, D_*)$. The boundary map of this complex is defined so that if $f \in \mathrm{Hom}_n(C_*, D_*)$, meaning $|f| = n$, then[3]

$$\partial_{\mathrm{Hom}_*(C_*, D_*)} f = \partial_{D_*} f - (-1)^{|f|} f \partial_{C_*}. \tag{A.2}$$

In other words, if $c \in C_*$ then

$$(\partial_{\mathrm{Hom}_*(C_*, D_*)} f)(c) = \partial_{D_*}(f(c)) - (-1)^{|f|} f(\partial_{C_*} c).$$

The reader can check that this makes $\mathrm{Hom}_*(C_*, D_*)$ into a chain complex because $\partial \circ \partial = 0$.

[3] An unfortunate notational ambiguity is that ∂f could mean the boundary of f as an element of $\mathrm{Hom}_*(C_*, D_*)$, or it could mean the composition of f with the boundary map of D_*. We can alleviate this ambiguity by writing $\partial_{\mathrm{Hom}_*(C_*, D_*)} f$ for the former and $\partial_{D_*} f$ for the latter, although these still look similar enough that the reader should exercise some care. The symbol $\partial_{D_*} \circ f$ might be better for the latter, but the notation is cluttered enough as it is, so we avoid this. Even worse, we do occasionally write just ∂f to further alleviate notational clutter if we feel that context alone should suffice. *Caveat lector.*

Warning: Here is the formula that violates the Koszul convention. We should expect that the piece of the formula in which ∂ passes through f picks up a sign $(-1)^{|f||\partial|} = (-1)^{|f|}$. But notice there is an extra minus sign in the formula coming from the subtraction. Alternatively, we could write $\partial_{\mathrm{Hom}_*(C_*,D_*)} f = \partial_{D_*} f + (-1)^{|f|+1} f \partial_{C_*}$. This extra minus sign is not critical – replacing it with a plus would still give a chain complex. So we should justify, as we will below, why this violation of the Koszul convention is useful.

The above definition of $\partial_{\mathrm{Hom}_*(C_*,D_*)} f$ has an important special case that also looks like it violates the Koszul convention: We have seen that if $C_* = S_*(X)$, the singular chain complex on X, and if $D_* = \mathbb{Z}$, the chain complex with just $\mathbb{Z}$ in degree zero, then $\mathrm{Hom}_{-n}(S_*(X), \mathbb{Z}) = \mathrm{Hom}^n(S_*(X), \mathbb{Z})$ is just the usual singular cochain complex. In this case, the boundary map on D_* is always zero, so the boundary formula becomes $\partial_{\mathrm{Hom}_*(S_*(X),\mathbb{Z})} f = (-1)^{|f|+1} f \partial_{C_*}$. If we change $\mathrm{Hom}_*(S_*(X), \mathbb{Z})$ over to cohomological indexing, in which case we switch our ∂ symbol to d, we get for $c \in C_*$,

$$(df)(c) = (-1)^{|f|+1} f(\partial c).$$

This is the standard formula for a cochain in books that are oriented toward the slightly fancier end of homological algebra. None of Munkres [182], Hatcher [125], or Spanier [220] utilize this convention, defining instead $(df)(c) = f(\partial c)$, demonstrating that one can do a lot without needing to impose a sign. By contrast, Dold first defines coboundaries without the sign but then notes in [71, Section VI.10.28] that it is useful to sometimes include the sign as it is "preferable from a systematic point of view." Furthermore, Mac Lane introduces the sign in [155, Section II.3] and provides some additional reasoning for it that we will see below.

A.1.4 Chain Maps and Chain Homotopies

An interesting thing happens if we consider the homology of the complex $\mathrm{Hom}_*(C_*, D_*)$. By (A.2), an element $f \in \mathrm{Hom}_*(C_*, D_*)$ is a cycle if $\partial_{D_*} f - (-1)^{|f|} f \partial_{C_*} = 0$, i.e. if $\partial_{D_*} f = (-1)^{|f|} f \partial_{C_*}$. When $|f| = 0$, this equation takes the form $\partial_{D*} f = f \partial_{C_*}$, which is the familiar formula for a (degree-zero) chain map of chain complexes. We hope the reader has seen enough topology that we do not need to emphasize how important this concept is. When $|f| \neq 0$, we *define* a *degree-$|f|$ chain map* to be one such that $\partial_{D*} f = (-1)^{|f|} f \partial_{C_*}$. Notice that here the Koszul convention is in force as $(-1)^{|f|} = (-1)^{|f||\partial|}$. So the "extra" sign in equation (A.2) is the price we pay for the cycles of $\mathrm{Hom}_*(C_*, D_*)$ to be chain maps by a definition that does obey the Koszul rule. Analogously to degree-zero chain maps, a chain map $f : C_* \to D_*$ of degree i takes cycles to cycles and boundaries to boundaries and so induces maps of homology groups $f : H_j(C_*) \to H_{j+i}(D_*)$ for all j.

When is $f \in \mathrm{Hom}_*(C_*, D_*)$ a boundary? This will be when there is an $F \in \mathrm{Hom}_*(C_*, D_*)$ such that $\partial_{\mathrm{Hom}_*(C_*,D_*)} F = f$, i.e. $f = \partial_{D_*} F - (-1)^{|f|+1} F \partial_{C_*}$. When $|f| = 0$, this becomes $f = \partial_{D_*} F + F \partial_{C_*}$, which the reader can recognize as the condition that f be chain homotopic to the zero map by the chain homotopy F. In general, we say that $f, g \in \mathrm{Hom}_*(C_*, D_*)$

are *chain homotopic* if there is an $E \in \mathrm{Hom}_*(C_*, D_*)$ such that $f - g = \partial_{\mathrm{Hom}_*(C_*,D_*)}E$. Putting this together, we see that $H_n(\mathrm{Hom}_*(C_*, D_*))$ is the module of degree-n chain maps $C_* \to D_*$ modulo chain homotopy, i.e. it is the module of chain homotopy equivalence classes of chain maps from C_* to D_*. As for degree-zero chain maps, chain homotopic chain maps of any degree determine the same map on homology, and so each element of $H_n(\mathrm{Hom}_*(C_*, D_*))$ determines maps $H_j(C_*) \to H_{j+n}(D_*)$. In fact, we get a homomorphism $H_n(\mathrm{Hom}_*(C_*, D_*)) \to \prod_j \mathrm{Hom}(H_j(C_*), H_{j+n}(D_*))$. In particular, $H_0(\mathrm{Hom}_*(C_*, D_*))$ is the module of degree-zero chain maps up to chain homotopy equivalence, and elements determine maps $H_j(C_*) \to H_j(D_*)$.

Key point: Chain maps are important, and we want them to satisfy the Koszul rule with respect to boundaries. One feature of this is that a composition of a degree-m chain map with a degree-n chain map is a degree-$(m + n)$ chain map, as is easy to verify, and so we will know immediately the sign properties, with respect to interchange with boundaries, of any map constructed by composing chain maps.

We make one final reassuring observation about this definition of the chain complex Hom_*: As Mac Lane notes [155, Proposition II.3.1], if the ring R is thought of as a chain complex with R in degree zero and trivial in other degrees, then $\mathrm{Hom}_*(R, D_*) \cong D_*$ via the map that takes $f \in \mathrm{Hom}_*(R, D_*)$ to $f(1)$, as the reader can easily check.

A.1.5 Consequences

So far, we have essentially made two choices of sign convention: equation (A.1) gives us a formula for the boundary map in $C_* \otimes D_*$ that makes it into a chain complex, and equation (A.2) gives us a formula for the boundary map in $\mathrm{Hom}_*(C_*, D_*)$ that makes it into a chain complex. The first satisfies the Koszul rule; the second does not but it leads to a definition of *chain* map that does, i.e. $\partial_{D_*} f = (-1)^{|f||\partial|} f \partial_{C_*} = (-1)^{|f|} f \partial_{C_*}$.

Let us see what other nice formulas follow from these decisions. The following hopefully justify our previous claims that we have made reasonable choices.

- The transposition map $\tau : C_* \otimes D_* \to D_* \otimes C_*$ determined by

$$\tau(c \otimes d) = (-1)^{|c||d|} d \otimes c$$

agrees with the Koszul convention and is a degree-zero chain map:

$$\begin{aligned}
\tau(\partial(c \otimes d)) &= \tau((\partial c) \otimes d + (-1)^{|c|} c \otimes \partial d) \\
&= (-1)^{(|c|-1)|d|} d \otimes \partial c + (-1)^{|c|+|c|(|d|-1)} \partial d \otimes c \\
&= (-1)^{|c||d|} \big((-1)^{|d|} d \otimes \partial c + \partial d \otimes c\big) \\
&= (-1)^{|c||d|} \partial(d \otimes c) \\
&= \partial \tau(c \otimes d).
\end{aligned}$$

- If $f : C_* \to E_*$ and $g : D_* \to F_*$ are chain maps, then $f \otimes g : C_* \otimes D_* \to E_* \otimes F_*$ defined by $(f \otimes g)(c \otimes d) = (-1)^{|c||g|} f(c) \otimes g(d)$ agrees with the Koszul convention and is a degree-$(|f| + |g|)$ chain map:

$$\begin{aligned}
&\partial(f \otimes g)(c \otimes d) \\
&= \partial((-1)^{|c||g|} f(c) \otimes g(d)) \\
&= (-1)^{|c||g|}((\partial f(c)) \otimes g(d) + (-1)^{|c|+|f|} f(c) \otimes \partial g(d)) \\
&= (-1)^{|c||g|}\big((-1)^{|f|} f(\partial c) \otimes g(d) + (-1)^{|c|+|f|+|g|} f(c) \otimes g(\partial d)\big) \\
&= (-1)^{|c||g|}\big((-1)^{|f|+|g|(|c|-1)}(f \otimes g)((\partial c) \otimes d) \\
&\qquad + (-1)^{|c|+|f|+|g|+|g||c|}(f \otimes g)(c \otimes \partial d)\big) \\
&= (-1)^{|f|+|g|}\big((f \otimes g)((\partial c) \otimes d) + (-1)^{|c|}(f \otimes g)(c \otimes \partial d)\big) \\
&= (-1)^{|f|+|g|}(f \otimes g)\partial(c \otimes d).
\end{aligned}$$

- The evaluation map $\mathrm{ev} : \mathrm{Hom}_*(C_*, D_*) \otimes C_* \to D_*$ defined by $\mathrm{ev}(f \otimes c) = f(c)$ agrees with the Koszul convention and is a degree-0 chain map:

$$\begin{aligned}
\mathrm{ev}(\partial(f \otimes c)) &= \mathrm{ev}(\partial_{\mathrm{Hom}_*(C_*,D_*)} f \otimes c + (-1)^{|f|} f \otimes \partial c) \\
&= \mathrm{ev}((\partial_{D_*} f - (-1)^{|f|} f \partial_{C*}) \otimes c + (-1)^{|f|} f \otimes \partial c) \\
&= \partial(f(c)) - (-1)^{|f|} f(\partial c) + (-1)^{|f|} f(\partial c) \\
&= \partial(f(c)) = \partial(\mathrm{ev}(f \otimes c)),
\end{aligned}$$

- Similarly, the evaluation map $\mathrm{ev} : C_* \otimes \mathrm{Hom}_*(C_*, D_*) \to D_*$ defined by $\mathrm{ev}(c \otimes f) = (-1)^{|f||c|} f(c)$ agrees with the Koszul convention and is a degree-0 chain map:

$$\begin{aligned}
&\mathrm{ev}(\partial(c \otimes f)) \\
&= \mathrm{ev}((\partial_{C_*} c) \otimes f + (-1)^{|c|} c \otimes \partial_{\mathrm{Hom}_*(C_*,D_*)} f) \\
&= \mathrm{ev}((\partial_{C_*} c) \otimes f + (-1)^{|c|} c \otimes (\partial_{D_*} f - (-1)^{|f|} f \partial_{C_*})) \\
&= (-1)^{|f|(|c|-1)} f(\partial_{C_*} c) + (-1)^{|c|+|c|(|f|-1)} \partial_{D_*} f(c) - (-1)^{|c|+|f|+|c|(|f|-1)} f \partial_{C_*} c \\
&= (-1)^{|f||c|-|f|} f(\partial_{C_*} c) + (-1)^{|c||f|} \partial_{D_*} f(c) - (-1)^{|f|+|c||f|} f \partial_{C_*} c \\
&= (-1)^{|c||f|} \partial_{D_*} f(c) = \partial_{D_*} \mathrm{ev}(c \otimes f).
\end{aligned}$$

- The composition map $c : \mathrm{Hom}_*(D_*, E_*) \otimes \mathrm{Hom}_*(C_*, D_*) \to \mathrm{Hom}_*(C_*, E_*)$ given by $c(g \otimes f) = gf$ follows the Koszul convention and is a degree-0 chain map:

$$
\begin{aligned}
\mathfrak{c}\partial(g\otimes f) &= \mathfrak{c}((\partial_{\mathrm{Hom}_*(D_*,E_*)}g)\otimes f + (-1)^{|g|}g\otimes \partial_{\mathrm{Hom}_*(C_*,D_*)}f)\\
&= \mathfrak{c}\big((\partial_{E_*}g - (-1)^{|g|}g\partial_{D_*})\otimes f + (-1)^{|g|}g\otimes(\partial_{D_*}f - (-1)^{|f|}f\partial_{C_*})\big)\\
&= \partial_{E_*}gf - (-1)^{|g|}g\partial_{D_*}f + (-1)^{|g|}g\partial_{D_*}f - (-1)^{|g|+|f|}gf\partial_{C_*}\\
&= \partial_{E_*}gf - (-1)^{|g|+|f|}gf\partial_{C_*} = \partial_{\mathrm{Hom}_*(C_*,E_*)}(gf)\\
&= \partial\mathfrak{c}(g\otimes f).
\end{aligned}
$$

- If $g : D_* \to E_*$ is a chain map, then $g_* : \mathrm{Hom}_*(C_*, D_*) \to \mathrm{Hom}_*(C_*, E_*)$ given by $g_*(f) = g\circ f$ is a degree $|g|$ chain map[4]:

$$
\begin{aligned}
\partial_{\mathrm{Hom}_*(C_*,E_*)}(g_*(f)) = \partial_{\mathrm{Hom}_*(C_*,E_*)}(gf) &= \partial_{E_*}gf - (-1)^{|g|+|f|}gf\partial_{C_*}\\
&= (-1)^{|g|}g\partial_{D_*}f - (-1)^{|g|+|f|}gf\partial_{C_*}\\
&= (-1)^{|g|}(g\partial_{D_*}f - (-1)^{|f|}gf\partial_{C_*})\\
&= (-1)^{|g|}g_*(\partial_{D_*}f - (-1)^{|f|}f\partial_{C_*})\\
&= (-1)^{|g|}g_*(\partial_{\mathrm{Hom}_*(C_*,D_*)}f).
\end{aligned}
$$

- If $g : B_* \to C_*$ is a chain map, then $g^* : \mathrm{Hom}_*(C_*, D_*) \to \mathrm{Hom}_*(B_*, D_*)$ given by $g^*(f) = (-1)^{|f||g|}f\circ g$ agrees with the Koszul convention and is a degree-$|g|$ chain map:

$$
\begin{aligned}
\partial_{\mathrm{Hom}_*(B_*,D_*)}(g^*(f)) &= (-1)^{|f||g|}\partial_{\mathrm{Hom}_*(B_*,D_*)}(fg)\\
&= (-1)^{|f||g|}(\partial_{D_*}fg - (-1)^{|g|+|f|}fg\partial_{B_*})\\
&= (-1)^{|f||g|}(\partial_{D_*}fg - (-1)^{|f|}f\partial_{C_*}g)\\
&= (-1)^{|g|}((-1)^{|g|(|f|-1)}\partial_{D_*}fg - (-1)^{|f|+|g|(|f|-1)}f\partial_{C_*}g)\\
&= (-1)^{|g|}g^*(\partial_{D_*}f - (-1)^{|f|}f\partial_{C_*})\\
&= (-1)^{|g|}g^*(\partial_{\mathrm{Hom}_*(C_*,D_*)}f).
\end{aligned}
$$

- As an interesting endnote to this section given our starting point, note that as elements of $\mathrm{Hom}_{-1}(C_*\otimes D_*, C_*\otimes D_*)$ we have $\partial_{C_*\otimes D_*} = \partial_{C_*}\otimes \mathrm{id}_{D_*} + \mathrm{id}_{C_*}\otimes \partial_{D_*}$, as the Koszul convention gives us the following computation for $x\in C_*$ and $y\in D_*$:

$$
\begin{aligned}
(\partial_{C_*}\otimes \mathrm{id}_{D_*} + \mathrm{id}_{C_*}\otimes\partial_{D_*})(x\otimes y) &= (-1)^{|x||\mathrm{id}_{D_*}|}((\partial_{C_*}x)\otimes y) + (-1)^{|x||\partial_{D_*}|}x\otimes\partial_{D_*}y\\
&= (\partial_{C_*}x)\otimes y + (-1)^{|x|}x\otimes\partial_{D_*}y.
\end{aligned}
$$

[4] In this computation and the next we'll use simply "∂" for the boundary maps in the Hom complexes and specify the boundary maps in the complexes C_*, D_*, E_*.

A.2 Some More Facts about Chain Homotopies

It is useful to know that the chain homotopy relation is preserved under various operations. We develop a few of these here.

Lemma A.2.1 *Let $f, g : C_* \to D_*$ and $h, k : D_* \to E_*$ be pairs of chain homotopic chain maps. Then hf is chain homotopic to kg.*

Proof Suppose $|f| = |g| = i$ and $|h| = |k| = j$ and that $\mathcal{D}_1 \in \mathrm{Hom}_*(C_*, D_*)$ and $\mathcal{D}_2 \in \mathrm{Hom}_*(D_*, E_*)$ are such that $\partial_{\mathrm{Hom}_*(C_*,D_*)}\mathcal{D}_1 = f - g$ and $\partial_{\mathrm{Hom}_*(D_*,E_*)}\mathcal{D}_2 = h - k$. Then

$$\begin{aligned} hf - kg &= hf - hg + hg - kg = h(f-g) + (h-k)g \\ &= h(\partial_{\mathrm{Hom}_*(C_*,D_*)}\mathcal{D}_1) + (\partial_{\mathrm{Hom}_*(D_*,E_*)}\mathcal{D}_2)g \\ &= h(\partial_{D_*}\mathcal{D}_1 - (-1)^{i+1}\mathcal{D}_1\partial_{C_*}) + (\partial_{E_*}\mathcal{D}_2 - (-1)^{j+1}\mathcal{D}_2\partial_{D_*})g \\ &= h\partial_{D_*}\mathcal{D}_1 - (-1)^{i+1}h\mathcal{D}_1\partial_{C_*} + \partial_{E_*}\mathcal{D}_2 g - (-1)^{j+1}\mathcal{D}_2\partial_{D_*}g \\ &= (-1)^j\partial_{E_*}h\mathcal{D}_1 - (-1)^{i+1}h\mathcal{D}_1\partial_{C_*} + \partial_{E_*}\mathcal{D}_2 g - (-1)^{j+1+i}\mathcal{D}_2 g\partial_{C_*} \\ &= \partial_{E_*}\mathcal{D}_2 g + (-1)^j\partial_{E_*}h\mathcal{D}_1 - (-1)^{j+1+i}\mathcal{D}_2 g\partial_{C_*} - (-1)^{i+1}h\mathcal{D}_1\partial_{C_*} \\ &= \partial_{E_*}(\mathcal{D}_2 g + (-1)^j h\mathcal{D}_1) - (-1)^{i+j+1}(\mathcal{D}_2 g + (-1)^j h\mathcal{D}_1)\partial_{C_*} \\ &= \partial_{\mathrm{Hom}_*(C_*,E_*)}(\mathcal{D}_2 g + (-1)^j h\mathcal{D}_1). \end{aligned}$$ □

Lemma A.2.2 *Suppose C_*, D_*, E_* are chain complexes and $f, g : C_* \to D_*$ are chain homotopic chain maps. Then $f^*, g^* : \mathrm{Hom}_*(D_*, E_*) \to \mathrm{Hom}_*(C_*, E_*)$ are chain homotopic.*

Proof Suppose $|f| = |g| = i$ and that $\mathcal{D} : C_* \to D_*$ is the chain homotopy so that $\partial_{\mathrm{Hom}_*(C_*,D_*)}\mathcal{D} = \partial_{D_*}\mathcal{D} - (-1)^{i+1}\mathcal{D}\partial_{C_*} = f - g$. We define $\mathfrak{D} : \mathrm{Hom}_*(D_*, E_*) \to \mathrm{Hom}_*(C_*, E_*)$ so that if $h \in \mathrm{Hom}_*(D_*, E_*)$ then $\mathfrak{D}(h) = (-1)^{|h|(i+1)} h \circ \mathcal{D}$. Note that $|f^*| = |g^*| = i$ and that $|\mathcal{D}| = |\mathfrak{D}| = i + 1$.

Now, suppose $h \in \mathrm{Hom}(D_*, E_*)$. Then we compute

$$
\begin{aligned}
&(\partial_{\mathrm{Hom}_*(C_*,E_*)}\mathfrak{D} - (-1)^{i+1}\mathfrak{D}\partial_{\mathrm{Hom}_*(D_*,E_*)})(h) \\
&= \partial_{\mathrm{Hom}_*(C_*,E_*)}\mathfrak{D}(h) - (-1)^{i+1}\mathfrak{D}\partial_{\mathrm{Hom}_*(D_*,E_*)}h \\
&= (-1)^{|h|(i+1)}\partial_{\mathrm{Hom}_*(C_*,E_*)}(h\mathcal{D}) - (-1)^{i+1}\mathfrak{D}(\partial_{E_*}h - (-1)^{|h|}h\partial_{D_*}) \\
&= (-1)^{|h|(i+1)}(\partial_{E_*}h\mathcal{D} - (-1)^{|h|+i+1}h\mathcal{D}\partial_{C_*}) \\
&\qquad - (-1)^{i+1}((-1)^{(i+1)(|h|-1)}\partial_{E_*}h\mathcal{D} - (-1)^{|h|+(i+1)(|h|-1)}h\partial_{D_*}\mathcal{D}) \\
&= (-1)^{|h|(i+1)}\partial_{E_*}h\mathcal{D} - (-1)^{|h|i+i+1}h\mathcal{D}\partial_{C_*} - (-1)^{|h|i+|h|}\partial_{E_*}h\mathcal{D} + (-1)^{|h|i}h\partial_{D_*}\mathcal{D} \\
&= -(-1)^{|h|i+i+1}h\mathcal{D}\partial_{C_*} + (-1)^{|h|i}h\partial_{D_*}\mathcal{D} \\
&= (-1)^{|h|i}h(\partial_{D_*}\mathcal{D} - (-1)^{i+1}\mathcal{D}\partial_{C_*}) \\
&= (-1)^{|h|i}h(f - g) \\
&= (f - g)^*(h) \\
&= (f^* - g^*)(h).
\end{aligned}
$$

So $\mathfrak{D}$ is a chain homotopy between f^* and g^*. □

Corollary A.2.3 *If the chain map $f : C_* \longrightarrow D_*$ is a chain homotopy equivalence, then so is $f^* : \mathrm{Hom}(D_*, E_*) \longrightarrow \mathrm{Hom}(C_*, E_*)$ for any chain complex E_*.*

Proof By definition, there is a $g : D_* \longrightarrow C_*$ such that gf and fg are chain homotopic to the respective identity maps id_C and id_D. But then, by the lemma, $(gf)^* = f^*g^*$ and $(fg)^* = g^*f^*$ are chain homotopic to the respective maps id_C^* and id_D^*. The dual of an identity map is an identity map. So f^* and g^* are chain homotopy inverses. □

Lemma A.2.4 *Suppose that $f, g : C_* \longrightarrow D_*$ are chain homotopic chain maps. Then $f \otimes \mathrm{id}, g \otimes \mathrm{id} : C_* \otimes E_* \longrightarrow D_* \otimes E_*$ are chain homotopic and $\mathrm{id} \otimes f, \mathrm{id} \otimes g : E_* \otimes C_* \longrightarrow E_* \otimes D_*$ are chain homotopic for any E_*.*

Proof Suppose $|f| = |g| = i$ and that $\partial_{\mathrm{Hom}_*(C_*,D_*)}\mathcal{D} = f - g$. Let $H_* = \mathrm{Hom}(C_* \otimes E_*, D_* \otimes$

E_*) and $H'_* = \mathrm{Hom}(E_* \otimes C_*, E_* \otimes D_*)$. Then

$$\begin{aligned}
&\partial_{H_*}(\mathcal{D} \otimes \mathrm{id}_{E_*}) \\
&= \partial_{D_* \otimes E_*}(\mathcal{D} \otimes \mathrm{id}_{E_*}) - (-1)^{i+1}(\mathcal{D} \otimes \mathrm{id}_{E_*})\partial_{C_* \otimes E_*} \\
&= (\partial_{D_*}\mathcal{D}) \otimes \mathrm{id}_{E_*} + (-1)^{i+1}\mathcal{D} \otimes \partial_{E_*} - (-1)^{i+1}(\mathcal{D} \otimes \mathrm{id}_{E_*})(\partial_{C_*} \otimes \mathrm{id}_{E_*} + \mathrm{id}_{C_*} \otimes \partial_{E_*}) \\
&= (\partial_{D_*}\mathcal{D}) \otimes \mathrm{id}_{E_*} + (-1)^{i+1}\mathcal{D} \otimes \partial_{E_*} - (-1)^{i+1}(\mathcal{D}\partial_{C*} \otimes \mathrm{id}_{E_*} + \mathcal{D} \otimes \partial_{E_*}) \\
&= (\partial_{D_*}\mathcal{D}) \otimes \mathrm{id}_{E_*} - (-1)^{i+1}\mathcal{D}\partial_{C*} \otimes \mathrm{id}_{E_*} \\
&= (\partial_{D_*}\mathcal{D} - (-1)^{i+1}\mathcal{D}\partial_{C*}) \otimes \mathrm{id}_{E_*} \\
&= (\partial_{\mathrm{Hom}_*(C_*,D_*)}\mathcal{D}) \otimes \mathrm{id}_{E*} \\
&= (f - g) \otimes \mathrm{id}_{E_*} \\
&= f \otimes \mathrm{id}_{E_*} - g \otimes \mathrm{id}_{E_*},
\end{aligned}$$

and

$$\begin{aligned}
&\partial_{H'_*}(\mathrm{id}_{E_*} \otimes \mathcal{D}) \\
&= \partial_{E_* \otimes D_*}(\mathrm{id}_{E_*} \otimes \mathcal{D}) - (-1)^{i+1}(\mathrm{id}_{E_*} \otimes \mathcal{D})\partial_{E_* \otimes C_*} \\
&= \partial_{E_*} \otimes \mathcal{D} + \mathrm{id}_{E_*} \otimes \partial_{D*}\mathcal{D} - (-1)^{i+1}(\mathrm{id}_{E_*} \otimes \mathcal{D})(\partial_{E_*} \otimes \mathrm{id}_{C_*} + \mathrm{id}_{E_*} \otimes \partial_{C_*}) \\
&= \partial_{E_*} \otimes \mathcal{D} + \mathrm{id}_{E_*} \otimes \partial_{D*}\mathcal{D} - (-1)^{i+1}((-1)^{i+1}\partial_{E_*} \otimes \mathcal{D} + \mathrm{id}_{E_*} \otimes \mathcal{D}\partial C_*) \\
&= \mathrm{id}_{E_*} \otimes \partial_{D*}\mathcal{D} - (-1)^{i+1}\mathrm{id}_{E_*} \otimes \mathcal{D}\partial C_* \\
&= \mathrm{id}_{E_*} \otimes (\partial_{D*}\mathcal{D} - (-1)^{i+1}\mathcal{D}\partial C) \\
&= \mathrm{id}_{E_*} \otimes (\partial_{\mathrm{Hom}_*(C_*,D_*)}\mathcal{D}) \\
&= \mathrm{id}_{E_*} \otimes (f - g) \\
&= \mathrm{id}_{E_*} \otimes f - \mathrm{id}_{E_*} \otimes g.
\end{aligned}$$ □

Corollary A.2.5 *If $f, g : C_* \longrightarrow D_*$ and $h, k : E_* \longrightarrow F_*$ are chain homotopic chain maps, then $f \otimes h, g \otimes k : C_* \otimes E_* \longrightarrow D_* \otimes F_*$ are chain homotopic chain maps.*

Proof We can write $f \otimes h$ as the composition $f \otimes h = (f \otimes \mathrm{id})(\mathrm{id} \otimes h)$. Applying Lemmas A.2.1 and A.2.4, we have that $(f \otimes \mathrm{id})(\mathrm{id} \otimes h)$ is chain homotopic to $(g \otimes \mathrm{id})(\mathrm{id} \otimes k) = g \otimes k$. □

Corollary A.2.6 *If $f : C_* \longrightarrow D_*$ and $h : E_* \longrightarrow F_*$ are chain homotopy equivalences, then so is $f \otimes h : C_* \otimes E_* \longrightarrow D_* \otimes F_*$.*

Proof Let $g : D_* \longrightarrow C_*$ and $k : F_* \longrightarrow E_*$ be chain homotopy inverses to f and h. Then, applying the preceding corollary, $(-1)^{|g||h|}(f \otimes h)(g \otimes k) = fg \otimes hk$ is chain homotopic to $\mathrm{id}_{D_*} \otimes \mathrm{id}_{F_*} = \mathrm{id}_{D_* \otimes F_*}$ and $(-1)^{|f||k|}(g \otimes k)(f \otimes h) = gf \otimes kh$ is chain homotopic to $\mathrm{id}_{C_*} \otimes \mathrm{id}_{E_*} = \mathrm{id}_{C_* \otimes E_*}$. As $|f| = -|g|$ and $|h| = -|k|$, we see that $(-1)^{|g||k|}g \otimes k$ is a chain homotopy inverse to $f \otimes h$. □

A.3 Shifts and Mapping Cones

Here, we briefly review some facts about shifting of complexes and algebraic mapping cones. The shift notation is useful for describing the mapping cones, which are used below in the proof of Lemma A.4.3 and for several proofs in Section 7.3.5. More about these objects is contained in the text proper in that section.

A.3.1 Shifts

It is useful to be able to reindex chain complexes. Most often, one sees this done for cohomologically indexed complexes, in which case, if D^* is such a chain complex (of R-modules) with (co)boundary map d_{D^*}, the shifted complex $D[k]^*$ is defined such that $D[k]^i = D^{k+i}$ and $d_{D[k]^*} = (-1)^k d_{D^*}$; see [102, Section III.3]. The boundary formula means that if $x \in D[k]^i = D^{k+i}$ then $d_{D[k]^*}(x) = (-1)^k d_{D^*}(x)$ treated as an element of $D[k]^{i+1}$. For homological indexing, using the standard bijection between cohomologically indexed complexes and homologically indexed complexes such that $C_i = C^{-i}$, we see that if C_* is a homologically indexed complex, then we should have

$$C[k]_i = C[k]^{-i} = C^{k-i} = C_{i-k}.$$

In other words, given C_*, we should let $C[k]_*$ be the chain complex with $C[k]_i = C_{i-k}$ and $\partial_{C[k]_*} = (-1)^k \partial_{C_*}$.

Taking $k = 1$ and C_* a chain complex, we obtain $C[1]_*$ with $C[1]_i = C_{i-1}$ and $\partial_{C[1]_*} = -\partial_{C_*}$. Let us define $\mathfrak{s} : C[1]_* \to C_*$ so that it takes $C[1]_i$ identically to the corresponding module C_{i-1}. Then from the definition of the boundary map on $C[1]_*$, we see that $\mathfrak{s}\partial_{C[1]_*} = -\partial_{C_*}\mathfrak{s}$, which is consistent with $\mathfrak{s}$ being a (homological) degree -1 chain map. Unfortunately, it is easy to get confused when attempting to consider C_{i-1} and $C[1]_i$ as two separate entities, especially when working with individual elements. Indeed, it is very tempting to write things like $\mathfrak{s}(x) = x$, which is right and wrong; right because C_{i-1} and $C[1]_i$ are identical modules, but wrong because they live in different chain complexes. In an attempt to mitigate the confusion, if x is an element of C_{i-1}, we will write $\bar{x}$ for the corresponding element of $C[1]_i$, i.e. $\mathfrak{s}(\bar{x}) = x$. Of course, we could also write $\mathfrak{s}^{-1}(x)$ instead of $\bar{x}$, but it is convenient to have both notations available.

A.3.2 Algebraic Mapping Cones

Suppose $f : C_* \to D_*$ is a degree-zero chain map of chain complexes. We let E^f_* (or simply E_* if there's no ambiguity) denote the algebraic mapping cone of $f : C_* \to D_*$ [102, Section III.3]. This means that $E_i = D_i \oplus C_{i-1}$ and $\partial(x, y) = (f(y) + \partial_{D_*}x, -\partial_{C_*}y)$. This is a chain complex, as

$$\partial(\partial(x, y)) = \partial(f(y) + \partial x, -\partial y) = (-f(\partial y) + \partial f(y) + \partial(\partial x), \partial(\partial y)) = 0.$$

This construction mimics algebraically the chain complex that one obtains from a topological mapping cone; the shift can be thought of as being due to taking the cone on the domain space, and so increasing the dimension by one. Lemma 7.3.39, in Section 7.3.5, should provide a more technically convincing version of this claim. We should also note that there are alternative conventions for the algebraic mapping cone construction; see, for example, [238, Section 1.5].

There is a short exact sequence of chain complexes

$$0 \longrightarrow D_* \xrightarrow{\mathfrak{e}} E_* \xrightarrow{\mathfrak{b}} C[1]_* \longrightarrow 0 \tag{A.3}$$

with $\mathfrak{e}(x) = (x, 0)$ and $\mathfrak{b}(x, y) = \bar{y}$, where $\bar{y}$ uses our notation for shifted elements from just above. It is immediate to verify that $\mathfrak{e}$ and $\mathfrak{b}$ are both chain maps of degree zero. Notice, however, that it is not true that $E_* = D_* \oplus C[1]_*$ as chain complexes, since the boundary map of E_* is not a direct sum of the boundary maps of the summands.

The following lemma shows that the connecting morphism in the long exact homology sequence associated to (A.3) is essentially just the map induced by f, up to shifts.

Lemma A.3.1 *Suppose $f : C_* \to D_*$ is a degree-zero chain map. Let ∂_* be the connecting morphism of the long exact homology sequence associated to the short exact sequence* (A.3)*. This map is the same as the map on homology induced by $f\mathfrak{s}$, where $\mathfrak{s} : C[1]_* \to C_*$ is the shift chain map.*

Proof Let $\bar{y} \in C[1]_*$ be a cycle and note that $(0, y) \in E_*$ is a preimage of $\bar{y}$ with respect to $\mathfrak{b}$, i.e. $\mathfrak{b}(0, y) = \bar{y}$. So, as $\partial(0, y) = (f(y), 0)$ in E_*, using that y is a cycle if $\bar{y}$ is, the zig-zag construction of ∂_* (see [182, Section 24]) shows that $\partial_*(\bar{y})$ is represented by $f(y) = f\mathfrak{s}(\bar{y})$. So ∂_* and $f\mathfrak{s}$ both take the homology class in $H_*(C[1]_*)$ represented by $\bar{y}$ to the same homology class in $H_*(D_*)$. □

A.4 Projective Modules and Dedekind Domains

The nice properties of projective modules in homological algebra are used regularly throughout the text, especially for projective modules over Dedekind domains. We mostly cite outside references, but we will need a few results for which such references were not easily available in our standard sources. So we provide a short treatment here, beginning with general properties of projectives.

A.4.1 Projective Modules

Recall that an R-module P is called *projective* if for every surjective map of R-modules $g : M \to N$ and every map $f : P \to N$ there exists a *lifting* $h : P \to M$ so that $gh = f$.

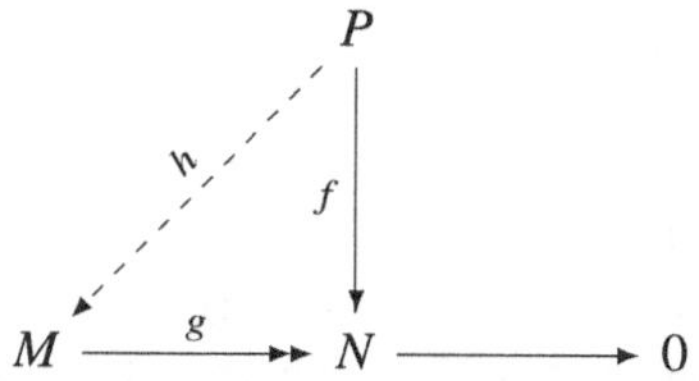

It is well known that being projective is equivalent to being a direct summand of a free module. The argument is standard, but we cannot resist providing it:

Lemma A.4.1 *The R-module P is projective if and only if P is a direct summand of a free R-module.*

Proof First suppose P is projective. Let F be the free module on the elements of P. Then there is a canonical surjection $p : F \twoheadrightarrow P$. If $\mathrm{id} : P \to P$ is the identity, the definition of projective gives a lift $s : P \to F$ so that $ps = \mathrm{id}$. So P is a direct summand of F.

Conversely, suppose P is a direct summand of a free module F. Then we can write $F = P \oplus Q$. Given a diagram as in the definition of projective, we can extend the map $f : P \to N$ to a map $\bar{f} : F \to N$ by taking $(x, y) \in P \oplus Q = F$ to $\bar{f}(x, y) = f(x)$. Now let $\{z_i\}$ be a basis for F, and for each z_i, let $h(z_i) \in M$ be an element such that $gh(z_i) = \bar{f}(z_i)$; such an element exists by the surjectivity of g. As $\{z_i\}$ is a basis for F, this determines a homomorphism $h : F \to M$ such that $gh = \bar{f} : F \to N$. The restriction of h to P is the desired lifting of f. □

The following lemma is also standard.

Lemma A.4.2 *Suppose* $0 \to A \to B \to C \to 0$ *is a short exact sequence of R-modules and that C is projective. Then the sequence splits and, in particular,* $B \cong A \oplus C$.

Proof We have a diagram

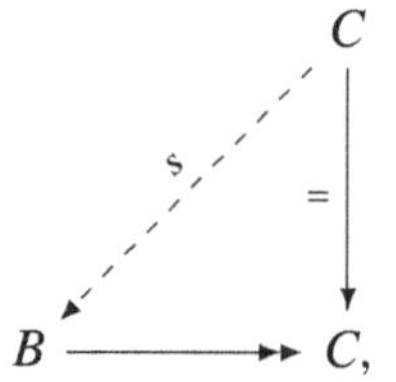

and, by the definition of projective, the map s exists, making the diagram commute. The map s provides a splitting of the exact sequence by standard homological algebra; see e.g. [182, Theorem 23.1] or [125, Section 2.2]. □

The next lemma is a basic fact of algebraic topology, but it is a bit hard to pin down a clean citation in our preferred sources. Munkres proves it in [182, Theorem 46.2] under the additional assumption that C and D are chain complexes of free modules. Hilton and Stammbach leave it as [126, Exercise IV.4.2]. The lemma also follows immediately from more elaborate theorems, such as the fact that if a category $\mathcal{A}$ has enough projectives, then the derived category of cochain complexes $D^-(\mathcal{A})$ is equivalent to the homotopy category $K^-(\mathcal{P})$, whose objects are bounded-above cochain complexes of projectives; see [238, Theorem 10.4.8] and note that the bounded-below condition of the lemma becomes a bounded-above condition when thinking of complexes as cochain complexes. This last argument is a somewhat big hammer that is not really necessary for this lemma. Really, all of the major pieces of the proof are provided between [182] and [126], but we will provide the details here for the convenience of the reader, beginning by assuming [126, Theorem IV.4.1], which is proven in [126] in detail. That theorem states that if A_* is a complex of projectives[5], B_* is acyclic, and $A_i = B_i = 0$ for $i < 0$, then for every homomorphism $\phi_0 : H_0(A_*) \to H_0(B_*)$, there is a chain map $\phi : A_* \to B_*$ inducing ϕ_0 and, furthermore, any two such chain maps are chain homotopic.

Lemma A.4.3 *Let $f : C_* \to D_*$ be a chain map of complexes of projective R-modules such that $C_i = D_i = 0$ if $i < 0$. If f induces isomorphisms in homology of all dimensions, then f is a chain homotopy equivalence.*

Proof Consider the algebraic mapping cone E_* of $f : C_* \to D_*$ as defined in Section A.3. As $E_i = D_i \oplus C_{i-1}$, each E_i is projective. The short exact sequence (A.3) generates a long exact homology sequence, and by Lemma A.3.1, the connecting morphism ∂_* is the same as the map induced by the composition $f\mathfrak{s}$, where $\mathfrak{s} : C[1]_* \to C_*$ is the shift map. As $\mathfrak{s}$ and f both induce isomorphisms on homology, it follows that ∂_* is an isomorphism, and, from the long exact sequence, E_* is acyclic. It therefore follows from [126, Theorem IV.4.1] that the maps $\mathrm{id} : E_* \to E_*$ and the zero map $0 : E_* \to E_*$ are chain homotopic, as they both induce the zero map $H_0(E) \to H_0(E)$. By definition, this means that there is a degree-one map $\mathcal{D} : E_* \to E_*$ such that $\partial\mathcal{D} + \mathcal{D}\partial = \mathrm{id}$.

From here, we follow the proof from Munkres [182, Theorem 46.2] and define $\theta, \psi, \lambda, \mu$ such that if $x \in D_i$ and $y \in C_{i-1}$, then

$$\mathcal{D}(x,0) = (\theta(x), \psi(x)) \in E_{i+1} = D_{i+1} \oplus C_i,$$
$$\mathcal{D}(0,y) = (\lambda(y), \mu(y)) \in E_{i+1} = D_{i+1} \oplus C_i.$$

Now, in the words of Munkres, "we compute like mad!":

$$\mathcal{D}\partial(x,0) = \mathcal{D}(\partial x, 0) = (\theta(\partial x), \psi(\partial x)),$$
$$\partial\mathcal{D}(x,0) = \partial(\theta(x), \psi(x)) = (f(\psi(x)) + \partial\theta(x), -\partial(\psi(x))),$$
$$\mathcal{D}\partial(0,y) = \mathcal{D}(0,-\partial y) + \mathcal{D}(f(y),0) = (-\lambda(\partial y), -\mu(\partial y)) + (\theta(f(y)), \psi(f(y))),$$
$$\partial\mathcal{D}(0,y) = \partial(\lambda(y), \mu(y)) = (f(\mu(y)) + \partial(\lambda(y)), -\partial\mu(y)).$$

[5] The terminology in [126] is actually "projective complex," by which is meant that each C_i is projective; see [126, page 126]. However, there is some danger of confusing "projective complex" with the requirement that C_* be projective as an object in the category of chain complexes, which is not the same thing. Thus we use the more precise terminology.

Since $\partial\mathcal{D} + \mathcal{D}\partial = \text{id}$, adding the first two equations implies that

$$\begin{aligned}(\theta(\partial x), \psi(\partial x)) &+ (f(\psi(x)) + \partial\theta(x), -\partial(\psi(x))) \\ &= (\theta(\partial x) + f(\psi(x)) + \partial\theta(x), \psi(\partial x) - \partial(\psi(x))) = (x, 0).\end{aligned}$$

Therefore, $\psi(\partial x) = \partial\psi(x)$, so ψ is a chain map, and $\theta(\partial x) + f(\psi(x)) + \partial\theta(x) = x$, which implies that θ is a chain homotopy between $f\psi$ and an identity. Adding the last two equations gives

$$\begin{aligned}(-\lambda(\partial y), -\mu(\partial y)) &+ (\theta(f(y)), \psi(f(y))) + (f(\mu(y)) + \partial(\lambda(y)), -\partial\mu(y)) \\ &= (-\lambda(\partial y) + \theta(f(y)) + f(\mu(y)) + \partial(\lambda(y)), -\mu(\partial y) + \psi(f(y)) - \partial\mu(y)) = (0, y).\end{aligned}$$

In the second coordinate, we obtain $y = -\mu(\partial y) + \psi(f(y)) - \partial\mu(y)$, which shows that μ provides a chain homotopy between the identity and ψf. The first coordinate tells us that $-\lambda(\partial y) + \theta(f(y)) + f(\mu(y)) + \partial(\lambda(y)) = 0$. I have no idea what this represents, but anyway, we have seen that ψf and $f\psi$ are each homotopic to the identity, and so f is a chain homotopy equivalence. □

A.4.2 Dedekind Domains

In several sections later in the book, it is necessary to work with Dedekind domains, a class of rings that includes all PIDs and fields. A *Dedekind domain* is an integral domain with the property that every submodule of a projective R-module is projective. This is essentially taken as the definition of a Dedekind domain in Cartan and Eilenberg [49, Section VII.5 and Theorem I.5.4]. Exercise 20 in Section 4 of Chapter VII of Bourbaki's *Commutative Algebra* [30] shows that this property can be derived from other, alternative, defining properties of Dedekind domains. A short literature search reveals that there are a very large number of equivalent definitions for Dedekind domains!

Another useful property of Dedekind domains is that any torsion-free module over a Dedekind domain is flat[6]. In fact, this is true more generally of Prüfer domains, which satisfy the weaker property that submodules of *finitely generated* projective modules are projective; a module over a Prüfer domain is torsion-free if and only if it is flat [146, Proposition 4.20].

Lemma A.4.4 *Let R be a Dedekind domain. Suppose D_* is a complex of R-modules with $D_i = 0$ for $i < 0$. Then there is a complex C_* of projective R-modules and a chain map $f : C_* \to D_*$ that induces isomorphisms $H_i(C_*) \to H_i(D_*)$ for all i. Furthermore, if $H_i(D_*)$ is finitely generated for all i, then we can choose C_i finitely generated for all i, and if D_* is a complex of projectives then f is a chain homotopy equivalence.*

Proof The construction of the chain complex C_* and a homotopy equivalence $f : C_* \to D_*$ that induces homology isomorphisms proceeds exactly as in the proof of [126, Proposition

[6] Recall that a module A is *flat* if the functor $A \otimes -$ is an exact functor, i.e. it preserves exact sequences.

V.2.4], replacing the free modules in that discussion with projective ones. As in [126], the proof is the consequence of two slightly more general lemmas that we will prove below.

Lemma A.4.5 *Let R be a Dedekind domain. Suppose D_* is a complex of R-modules. Then there is a complex C_* of projective R-modules such that $H_i(C_*) \cong H_i(D_*)$ for all i. Furthermore, if $H_i(D_*)$ is finitely generated for all i, then we can choose C_i finitely generated for all i, and if $H_i(D_*) = 0$ for all $i < 0$, then we can choose $C_i = 0$ for all $i < 0$.*

Lemma A.4.6 *Let R be a Dedekind domain. Suppose D_* is a complex of R-modules and that C_* is a complex of projective modules. Suppose $g_i : H_i(C_*) \to H_i(D_*)$ is any collection of homomorphisms. Then there is a chain map $f : C_* \to D_*$ that induces the g_i.*

It follows from Lemma A.4.5 that, given D_* as in the statement of Lemma A.4.4, there is a chain complex C_* with the desired characteristics (C_i projective, $C_i = 0$ for $i < 0$, C_i finitely generated if $H_i(D_*)$ is, and $H_i(C_*) \cong H_i(D_*)$), and it follows from Lemma A.4.6 that there is a chain map $f : C_* \to D_*$ inducing the isomorphisms $H_i(C_*) \to H_i(D_*)$. The map f is then a chain homotopy equivalence if the D_i are all projective by Lemma A.4.3. Thus we have proved Lemma A.4.4. □

Proof of Lemma A.4.5 Let F_p be a free R-module that surjects onto $H_p(D_*)$ by $q_p : F_p \to H_p(D_*)$. Then, as R is Dedekind, $K_p = \ker(q_p)$ is projective, so $0 \to K_p \to F_p \to H_p(D_*)$ is a projective resolution of $H_p(D_*)$. If $H_p(D_*)$ is finitely generated, we can choose F_p to be finitely generated. It follows that K_p will also be finally generated. This uses that Dedekind domains are Noetherian [30, Theorem VII.2.2.1], so submodules of finitely generated modules are finitely generated (see [147, Section X.1]).

Now, let $C_p = F_p \oplus K_{p-1}$, with $\partial(x, y) = (y, 0)$. If $H_p(D) = 0$ for $p < 0$, we can have $F_p = 0$ for $p < 0$ and so $C_p = 0$ for $p < 0$. This provides the desired chain complex C_* with $H_p(C_*) \cong H_p(D_*)$, as the cycle modules of C_* are the $F_p \oplus 0 \cong F_p$ and the boundary map corresponds to the natural embedding of K_p into F_p. □

Proof of Lemma A.4.6 Let Z_p, B_p be the cycle and boundary submodules of C_p, and let $\bar{Z}_p, \bar{B}_p$ be the corresponding submodules for D_*. As the C_p are projective and $B_p, Z_p \subset C_p$, the modules B_p and Z_p are projective. So by Lemma A.4.2 the short exact sequences

$$0 \longrightarrow Z_p \longrightarrow C_p \xrightarrow{\partial} B_{p-1} \longrightarrow 0$$

split, and we have $C_p \cong Z_p \oplus Y_p$, where Y_p maps isomorphically onto B_p by ∂. In fact, if $(z, y) \in Z_p \oplus Y_p$ is the general element, then $\partial(z, y) = \partial y \in B_{p-1} \subset Z_{p-1} \subset C_{p-1}$, corresponding to $(\partial y, 0)$ in the decomposition $Z_{p-1} \oplus Y_{p-1}$. Consider now the diagram

$$\begin{array}{ccccccccc} 0 & \longrightarrow & B_p & \longrightarrow & Z_p & \longrightarrow & H_p(C_*) & \longrightarrow & 0 \\ & & \Big\downarrow{\scriptstyle \phi} & & \Big\downarrow{\scriptstyle \theta} & & \Big\downarrow{\scriptstyle g_p} & & \\ 0 & \longrightarrow & \bar{B}_p & \longrightarrow & \bar{Z}_p & \longrightarrow & H_p(D_*) & \longrightarrow & 0. \end{array} \tag{A.4}$$

As Z_p is projective, the definition of projectivity yields the desired dashed map $\theta : Z_p \to \bar{Z}_p$, and this, in turn, induces the map of kernels $\phi : B_p \to \bar{B}_p$ by restriction. We also have a diagram

$$\begin{array}{ccc} Y_p & \xrightarrow[\cong]{\partial} & B_{p-1} \\ \Big\downarrow{\scriptstyle \psi} & & \Big\downarrow{\scriptstyle \phi} \\ D_p & \xrightarrow{\partial} & \bar{B}_{p-1}, \end{array}$$

and, once again, projectivity of $Y_p \cong B_{p-1}$ yields the map ψ.

We now define $f : C_p \cong Z_p \oplus Y_p \to D_p$ by $f(z, y) = \theta(z) + \psi(y)$. To see that f is a chain map, we check

$$\begin{aligned} f\partial(z,y) &= f(\partial y, 0) = \theta(\partial y) + \psi(0) = \theta(\partial y) = \phi(\partial y) \\ &= \partial\psi(y) = \partial\theta(z) + \partial\psi(y) = \partial(\theta(z) + \psi(y)) \\ &= \partial f(z, y). \end{aligned}$$

We have used here that $\theta(z)$ is a cycle in D_*. It follows from Diagram (A.4) that f induces the desired isomorphism on homology. □

A.5 Linear Algebra of Signatures

In this section, we collect some material from linear algebra regarding signatures of symmetric bilinear pairings. We will work primarily with the rational numbers as our ground field, but all results are equally valid for any ground field F with $\mathbb{Q} \subset F \subset \mathbb{R}$ unless noted otherwise. All vector spaces in this section are assumed to be *finite-dimensional.*

Definition A.5.1 If M is a symmetric matrix of rational numbers, then the *signature* $\sigma(M)$ is defined to be

$$\sigma(M) = \#\{\text{positive eigenvalues of } M\} - \#\{\text{negative eigenvalues of } M\}.$$

Notice that this makes sense because all eigenvalues of a real symmetric matrix will be real[7].

We will be interested in signatures that arise from symmetric bilinear pairings on vector spaces. So, let $(V, (\cdot,\cdot))$ be a finite-dimensional rational vector space together with a symmetric bilinear pairing $(\cdot,\cdot) : V \times V \to \mathbb{Q}$. In other words, for $u, v, w \in V$ and $c \in \mathbb{Q}$,

$$\begin{aligned}(u + v, w) &= (u, w) + (v, w),\\ (u, v + w) &= (u, v) + (u, w),\\ (cu, v) &= (u, cv) = c(u, v),\\ (u, v) &= (v, u).\end{aligned}$$

If we choose a basis $\{e_i\}$ for V, then we obtain the *pairing matrix* M with $M_{i,j} = (e_i, e_j)$. If $u, v \in V$, then $u = \sum_i a_i e_i$ and $v = \sum_i b_i e_i$ for some $a_i, b_i \in \mathbb{Q}$, so using the bilinearity of the pairing,

$$(u, v) = \left(\sum_i a_i e_i, \sum_j b_j e_j\right) = \sum_{i,j} a_i b_j (e_i, e_j) = \sum_{i,j} a_i M_{i,j} b_j = u^t M v,$$

where u^t is the transpose of u and we identify u and v with their vectors of coordinates in the given basis.

Definition A.5.2 If $(V, (\cdot,\cdot))$ is a symmetric bilinear pairing on a finite-dimensional rational vector space, we define the *signature of the pairing* $\sigma(V, (\cdot,\cdot))$ to be the signature $\sigma(M)$ of the pairing matrix M with respect to any basis of V.

Of course, we need to know that this is independent of the choice of basis. Suppose $\{f_i\}$ is another basis of V, that N is the pairing matrix with respect to this basis, i.e. $N_{i,j} = (f_i, f_j)$, and Q is the change-of-basis matrix such that $f_i = \sum_k Q_{k,i} e_k$. Then

$$\begin{aligned}N_{i,j} = (f_i, f_j) &= \left(\sum_k Q_{k,i} e_k, \sum_\ell Q_{\ell,j} e_\ell\right) = \sum_{k,\ell} Q_{k,i} Q_{\ell,j} (e_k, e_\ell) = \sum_{k,\ell} Q_{k,i} Q_{\ell,j} M_{k,\ell}\\ &= \sum_{k,\ell} (Q^t)_{i,k} M_{k,\ell} Q_{\ell,j},\end{aligned}$$

where Q^t is the transpose of Q. This computation shows that $N = Q^t M Q$. Thus a change of basis changes the pairing matrix to a congruent matrix. Conversely, the same computation shows that if Q is any nonsingular matrix and M represents a pairing on V with respect to a basis $\{e_i\}$, then $Q^t M Q$ represents the same pairing with respect to the basis $\{f_i\}$ given by $f_i = \sum_k Q_{k,i} e_k$.

So to see that the signature depends only on the pairing, we need to see that the number of eigenvalues of each sign of a symmetric matrix is independent of congruence by a

[7] Let $\langle \cdot, \cdot \rangle$ denote the standard complex inner product on $\mathbb{C}^n$, let M be a symmetric $n \times n$ matrix with real entries so that the conjugate transpose M^* is equal to M, and suppose $Mv = \lambda v$ for some unit vector v. Then $\lambda = \lambda\langle v, v\rangle = \langle v, \lambda v\rangle = \langle v, Mv\rangle = \langle M^* v, v\rangle = \langle Mv, v\rangle = \langle \lambda v, v\rangle = \bar{\lambda}\langle v, v\rangle = \bar{\lambda}$.

nonsingular matrix. This essentially follows from Sylvester's Law of Inertia [232], which classifies pairings over the reals, though we'll provide a proof of what we need here, relying only on the Spectral Theorem for real symmetric matrices.

The invariance of the signature with respect to matrix congruence is related to a useful alternative characterization of the signature in terms of positive definite and negative definite subspaces.

Definition A.5.3 Given a symmetric bilinear pairing $(V, (\cdot,\cdot))$, a subspace $W \subset V$ is called *positive definite* if for any $w \in W$, $w \neq 0$, we have $(w, w) > 0$. Similarly, a subspace $W \subset V$ is called *negative definite* if for any $w \in W$, $w \neq 0$, we have $(w, w) < 0$.

Even though we are primarily concerned with rational vector spaces, the proof we give requires working at a certain point with real matrices, so there is no extra work in stating our results for any subfield of $\mathbb{R}$.

Lemma A.5.4 *Let F be a field with $\mathbb{Q} \subset F \subset \mathbb{R}$, and let $(V, (\cdot,\cdot))$ be a finite-dimensional F-vector space with symmetric bilinear pairing. Let M be the pairing matrix with respect to some basis. Then the maximal dimension for a positive definite subspace is equal to the number of positive eigenvalues of M, and the maximal dimension for a negative definite subspace is equal to the number of negative eigenvalues of M. It follows that*

$$\sigma(M) = \max_{\{W^+ \subset V \text{ positive definite}\}} \dim(W^+) - \max_{\{W^- \subset V \text{ negative definite}\}} \dim(W^-),$$

which does not depend on the choice of basis.

Furthermore, two pairings $(V, (\cdot,\cdot))$ and $(V', (\cdot,\cdot)')$ have the same signature if they are isomorphic in the sense that there is an isomorphism $\phi : V \to V'$ and a commutative diagram of the form:

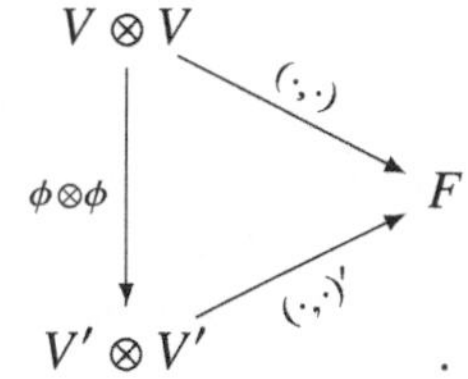

To prove this lemma, it helps to use another elementary lemma:

Lemma A.5.5 *Let F be a field with $\mathbb{Q} \subset F \subset \mathbb{R}$, and let $(V, (\cdot,\cdot))$ be a finite-dimensional F-vector space with symmetric bilinear pairing. Then there is a basis of V with respect to which the pairing matrix N is a diagonal matrix.*

Proof We will show that there is some basis $\{b_i\}$ of V that is orthogonal, in the sense that $(b_i, b_j) = 0$ if $i \neq j$. Then the pairing matrix N with respect to this basis will be diagonal.

Let $v \in V$ be such that $(v, v) \neq 0$; if there is no such v, it follows from the identity $(v + w, v + w) = (v, v) + 2(v, w) + (w, w)$ that $(v, w) = 0$ for all $v, w \in V$, and then any pairing matrix is the zero matrix, so we would be done. If we let $\langle v\rangle$ be the span of v, we will show that

$$V = \langle v\rangle \oplus \langle v\rangle^{\perp},$$

where for any subspace $W \subset V$, we let $W^{\perp} = \{u \in V \mid (u, w) = 0 \text{ for all } w \in W\}$. For this, let $u \in V$, and let $u_1 = \big((u,v)/(v,v)\big)v$ and $u_2 = u - \big((u,v)/(v,v)\big)v$. Note that all numerical expressions remain in the field F. In Euclidean space with the standard inner product, these would correspond to the projections of u to $\langle v\rangle$ and to its perpendicular subspace. Clearly $u = u_1 + u_2$, and

$$(u_2, v) = \left(u - \frac{(u,v)}{(v,v)}v, v\right) = (u,v) - \frac{(u,v)}{(v,v)}(v,v) = 0.$$

Thus any u is contained in $\langle v\rangle + \langle v\rangle^{\perp}$. Next, suppose $w \in \langle v\rangle \cap \langle v\rangle^{\perp}$. Then $w = \lambda v$ for some $\lambda \in F$. But then, since $w \in \langle v\rangle^{\perp}$, we have $0 = (v, w) = (v, \lambda v) = \lambda(v, v)$, which is impossible unless $\lambda = 0$. Hence $V = \langle v\rangle \oplus \langle v\rangle^{\perp}$.

Now we can complete the lemma using induction on $\dim(V)$. If $\dim(V) = 1$, there is nothing more to prove. So suppose we have proven the result whenever the dimension of the vector space is $< n$, and let $\dim(V) = n$. Again, we will also be done trivially if there is no v with $(v, v) \neq 0$. If there is a $v \in V$ with $(v, v) \neq 0$, then we have seen that $V = \langle v\rangle \oplus \langle v\rangle^{\perp}$. By induction there will be an orthogonal basis $\{b_1, \ldots, b_{n-1}\}$ of $\langle v\rangle^{\perp}$, and so $\{b_1, \ldots, b_{n-1}, v\}$ is the desired orthogonal basis for V. □

Proof of Lemma A.5.4 Throughout the proof we will adopt the following notation: If M is a matrix, we let $\sigma_+(M)$ denote the number of positive eigenvalues of M and $\sigma_-(M)$ denote the number of negative eigenvalues of M. If M is the matrix of a pairing with respect to some basis, then we let $d_+(M)$ denote the maximal dimension among positive definite subspaces of the pairing, and we let $d_-(M)$ denote the maximal dimension among the negative definite subspaces. We can write $d_\pm^F(M)$ if we wish to emphasize the field F.

First we consider the case where we have a diagonal pairing matrix N for the pairing with respect to some basis $\{b_i\}$. We are free to reorder the basis so that $(b_i, b_i) > 0$ for $1 \leq i \leq r$ while $(b_i, b_i) < 0$ for $r + 1 \leq i \leq r + s$ and $(b_i, b_i) = 0$ for $r + s + 1 \leq i \leq n$. In particular, this means that N is a diagonal matrix with r positive entries and s negative entries, and so $r = \sigma_+(N)$ and $s = \sigma_-(N)$.

It is then clear that $\{b_1, \ldots, b_r\}$ span a positive definite subspace W^+ of dimension r. On the other hand $\{b_{r+1}, \ldots, b_n\}$ span a subspace $W^{\leq 0}$ such that no $w \in W^{\leq 0}$ has the property that $(w, w) > 0$. Since $W^{\leq 0}$ has dimension $n-r$, every subspace of V of dimension $r+1$ must intersect $W^{\leq 0}$ in a subspace of dimension[8] ≥ 1 and hence possesses a w with $(w, w) \leq 0$.

[8] This property is well known when $F = \mathbb{R}$. Here's a proof that it continues to hold for any field $F \subset \mathbb{R}$: Suppose V is an F-vector space of dimension n and let U, W be subspaces of respective dimensions k, ℓ with $k + \ell > n$. Let $\{e_i\}_{i=1}^k$ and $\{f_j\}_{j=1}^\ell$ be bases of U and W. Together, the set of vectors $\{e_1, \ldots, e_k, f_1, \ldots, f_\ell\}$ must be linearly dependent as $\dim(V) = n$. This dependence can be written as $\sum_{i=1}^k a_i e_i = \sum_{j=1}^\ell b_j f_j$ for some $a_i, b_j \in F$ not all zero. As the $\{e_i\}$ and $\{f_j\}$ are bases, we can't have both sides equal to zero, and so neither side is zero and there is a non-zero vector in $U \cap W$.

So no positive definite subspace can have dimension $> r$. Thus $r = d_+(N)$, the maximal dimension for a positive definite subspace. A similar argument shows that $s = d_-(N)$.

For our next step, it is convenient to work over $\mathbb{R}$ as our field. Suppose that M is an $n \times n$ real symmetric matrix, which we can assume represents a pairing on $\mathbb{R}^n$ with respect to the standard basis $\{e_i\}$. Let W^+ and W^- be maximal positive and negative definite subspaces with respect to this pairing. By the Spectral Theorem [225, Section 5.5], there exists an invertible real matrix U with $U^t = U^{-1}$ such that $U^tMU = U^{-1}MU$ is diagonal. By elementary linear algebra, the matrix $U^{-1}MU$ has the same eigenvalues as M, so $\sigma_\pm(M) = \sigma_\pm(U^tMU)$. But U^tMU represents the same pairing on $\mathbb{R}^n$ as M but with respect to the basis $\{Ue_i\}$, and so $d^{\mathbb{R}}_\pm(M) = d^{\mathbb{R}}_\pm(U^tMU)$. But U^tMU is diagonal and so by our preceding results $d^{\mathbb{R}}_\pm(U^tMU) = \sigma_\pm(U^tMU)$. Thus, altogether, we have $d^{\mathbb{R}}_\pm(M) = \sigma_\pm(M)$.

Now, let's return to M being a symmetric matrix over F representing a pairing on the F-vector space V. Once again using Lemma A.5.5, there is a basis $\{b_i\}$ for V with respect to which the pairing is represented by a diagonal matrix of the form Q^tMQ for some invertible F-matrix Q. As we can interpret all our F-matrices as real matrices representing real pairings, the preceding argument shows that $\sigma_\pm(Q^tMQ) = \sigma_\pm(M)$. Meanwhile, as Q^tMQ is diagonal, the argument at the beginning of the proof shows that $\sigma_\pm(Q^tMQ) = d^F_\pm(Q^tMQ)$. But M and Q^tMQ represent the same F-vector space pairing (with respect to different bases), and so $d^F_\pm(Q^tMQ) = d^F_\pm(M)$. So $\sigma_\pm(M) = d^F_\pm(M)$.

The final statement of the lemma now follows from observing that ϕ must take positive-definite subspaces to isomorphic positive-definite subspaces and negative-definite subspaces to isomorphic negative-definite subspaces. □

A.5.1 Signatures of Nonsingular Pairings

Now that we have established that signatures of symmetric pairings are well defined, we turn to some important properties of nonsingular pairings.

Definition A.5.6 A symmetric pairing $(V, (\cdot,\cdot))$ on a rational vector space is called *nonsingular* or *nondegenerate* if $(v, w) = 0$ for all $w \in V$ implies $v = 0$.

Remark A.5.7 More generally, nondegeneracy of a symmetric pairing means that the adjoint homomorphism $V \to \mathrm{Hom}(V, \mathbb{Q})$ described by $v \to (v, \cdot)$ is injective, while being nonsingular means that it is an isomorphism. When V is finite-dimensional, these conditions are equivalent with field coefficients. However, it is possible to define symmetric bilinear pairings on free modules over other rings, such as $\mathbb{Z}$, in which case these become different conditions. For example, the pairing $\mathbb{Z} \otimes \mathbb{Z} \to \mathbb{Z}$ with pairing matrix $M = (2)$ is a nondegenerate pairing, but it is not nonsingular.

Note that the symmetry property we are assuming makes it unnecessary in this context to consider two different adjoints as in Definition 8.4.2 in the main body of the text.

Here is a useful way to tell if a pairing is nonsingular:

Lemma A.5.8 *Let $(V, (\cdot,\cdot))$ be a symmetric pairing on the finite-dimensional rational vector space V, and let M be the matrix of the pairing with respect to some basis. Then the pairing is nonsingular if and only if* $\det(M) \neq 0$.

Proof Let $\{b_i\}$ be the basis with respect to which M is defined, and define the dual basis $\{b_j^*\} \subset \mathrm{Hom}(V, \mathbb{Q})$ so that $b_j^*(b_i) = 0$ if $i \neq j$ and $b_i^*(b_i) = 1$ for all i. Let $\Lambda : V \to \mathrm{Hom}(V, \mathbb{Q})$ take $v \in V$ to $(v, \cdot)$. We compute the matrix of Λ with respect to the bases $\{b_i\}$ and $\{b_j^*\}$: By definition, we have $\Lambda(b_i) = \sum_j \Lambda_{ji} b_j^*$, so we have

$$M_{ik} = (b_i, b_k) = (\Lambda(b_i))(b_k) = \sum_j \Lambda_{ji} b_j^*(b_k) = \Lambda_{ki}.$$

So $\Lambda = M^t$. The map Λ is injective if and only if $\det(\Lambda) \neq 0$ by elementary linear algebra, and furthermore $\det(M^t) = \det(M)$, proving the lemma. □

Remark A.5.9 By the same arguments, we see that for a symmetric pairing of a free module over a commutative ring with unity we have that the pairing is nondegenerate if and only if $\det(M)$ is non-zero for the pairing matrix M, while it is nonsingular if and only if $\det(M)$ is a unit. Again, over a field these notions are equivalent.

Definition A.5.10 Given a nonsingular pairing $(V, (\cdot,\cdot))$, a subspace $A \subset V$ such that $\dim(A) = \frac{1}{2}\dim(V)$ and $(x, y) = 0$ for all $x, y \in A$ is called a *Lagrangian subspace*. Such subspaces are not generally unique.

Lemma A.5.11 *Suppose $(V, (\cdot,\cdot))$ is a rational vector space together with a nonsingular symmetric bilinear pairing $(\cdot,\cdot) : V \times V \to \mathbb{Q}$ with a Lagrangian subspace $A \subset V$. Then $\sigma(V, (\cdot,\cdot)) = 0$.*

Proof Let V^+ and V^- be respectively positive definite and negative definite subspaces of V of maximal dimensions. Let $\dim(V^+) = r$, $\dim(V^-) = s$, and $\dim(V) = n$. By Lemma A.5.8, if $(V, (\cdot,\cdot))$ is nonsingular then no pairing matrix can have a zero as an eigenvalue. Consequently, using Lemma A.5.4 to equate the number of positive eigenvalues with r and the number of negative eigenvalues with s, we must have $r + s = n$. Now, from the definitions, we must have $\dim(A \cap V^+) = \{0\}$ and $\dim(A \cap V^-) = \{0\}$. From the first equation, we must have that $\dim(A) \leq s$, and from the second, we must have $\dim(A) \leq r$; see the argument in footnote 8 on page 732. But since we have also assumed $\dim(A) = n/2$, this forces

$$n = 2\dim(A) \leq r + s = n.$$

So in fact all the inequalities of the discussion must be equalities, and $\dim(A) = r = s = \frac{n}{2}$. Thus $\sigma(V, (\cdot,\cdot)) = r - s = 0$. □

Remark A.5.12 The converse of Lemma A.5.11 is true if we work with ground field $\mathbb{R}$. To see this, suppose we have used Lemma A.5.5 to find an orthogonal basis for $(V, (\cdot,\cdot))$. Let us order and name the basis so that $\{a_1, \ldots, a_r\}$ are the orthogonal basis vectors with $(a_i, a_i) > 0$ and $\{b_1, \ldots, b_s\}$ are the orthogonal basis vectors with $(b_i, b_i) < 0$. We continue to assume that the pairing is nonsingular so that the $\{a_i\}$ and $\{b_i\}$ together constitute a full basis. We can now normalize the basis by setting

$$c_i = \frac{1}{\sqrt{(a_i, a_i)}} a_i \quad \text{and} \quad d_i = \frac{1}{\sqrt{|(b_i, b_i)|}} b_i.$$

With respect to this new basis consisting of the c_i and d_i, we have $(c_i, c_i) = 1$, $(d_i, d_i) = -1$, and all other pairings between basis elements yield zero.

Now suppose the signature is zero, so that $r = s = \frac{1}{2}\dim(V)$. Let $f_i = c_i + d_i$, $1 \leq i \leq \frac{1}{2}\dim(V)$. Then

$$\begin{aligned}(f_i, f_j) &= (c_i + d_i, c_j + d_j) = (c_i, c_j) + (c_i, d_j) + (d_i, c_j) + (d_i, d_j) \\ &= \delta_{i,j} + 0 + 0 - \delta_{i,j} = 0.\end{aligned}$$

It follows that the f_i span a Lagrangian subspace of V.

The converse of Lemma A.5.11 is not true over $\mathbb{Q}$. For example, consider the form with pairing matrix $\begin{pmatrix} 2 & 0 \\ 0 & -1 \end{pmatrix}$ with respect to some basis $\{a, b\}$. This pairing is nonsingular with index zero. In order to have a Lagrangian subspace, there would have to be rational numbers x, y, not both 0, such that $(xa + yb, xa + yb) = 2x^2 - y^2 = 0$. But since 2 is not the square of any rational number, this is impossible. This fact forms part of a rich theory of nonsingular symmetric forms over $\mathbb{Q}$, as can be found, for example, in [176], particularly in Section IV.2.

A.5.2 Signatures of Orthogonal Sums

Another useful situation that arises in practice is the one for which a pairing matrix for $(V, (\cdot, \cdot))$ has a block sum form, meaning that it has the form

$$M = \begin{pmatrix} A_1 & 0 & \cdots & 0 \\ 0 & A_2 & \cdots & 0 \\ \vdots & \vdots & \ddots & \vdots \\ 0 & 0 & \ldots & A_m \end{pmatrix}.$$

Here the A_i are square matrices of any size with their diagonals lying along the diagonal of M, and all other entries of M not in the A_i are zero. Such a form corresponds to a decomposition of V as a direct sum of subspaces $V = \bigoplus_i W_i$ such that the W_i are orthogonal to each other, i.e. $(w_i, w_j) = 0$ if $w_k \in W_k$ and $i \neq j$. In this case, each A_i represents the pairing restricted to W_i. Now, we can find an orthogonal basis spanning each W_i by Lemma A.5.5. If we do this for all subspaces simultaneously, we can find an orthogonal basis for V.

In this new basis we having a pairing matrix

$$M' = \begin{pmatrix} A_1' & 0 & \cdots & 0 \\ 0 & A_2' & \cdots & 0 \\ \vdots & \vdots & \ddots & \vdots \\ 0 & 0 & \ldots & A_m' \end{pmatrix}$$

in which all the A_i' are diagonal matrices. It now follows easily that $\sigma(M') = \Sigma_i\, \sigma(A_i')$. But since these invariants are independent of basis, in fact $\sigma(M) = \Sigma\, \sigma(A_i)$. We have shown:

Lemma A.5.13 *If M has the block form*

$$\begin{pmatrix} A_1 & 0 & \cdots & 0 \\ 0 & A_2 & \cdots & 0 \\ \vdots & \vdots & \ddots & \vdots \\ 0 & 0 & \ldots & A_m \end{pmatrix},$$

then $\sigma(M) = \Sigma_{i=1}^{m}\, \sigma(A_i)$. Consequently, if $(V, (\cdot,\cdot))$ is a direct sum of orthogonal subspaces W_i, then $\sigma(V, (\cdot,\cdot)) = \Sigma_i\, \sigma((W_i, (\cdot,\cdot)))$.

A.5.3 Antisymmetric Pairings

While the results so far in this section have been about symmetric bilinear pairings, there is one result we will need about antisymmetric pairings. An antisymmetric bilinear pairing over $\mathbb{Q}$ consists of a finite-dimensional rational vector space V and a map

$$(\cdot,\cdot) : V \times V \to \mathbb{Q}$$

such that for $u, v, w \in V$ and $c \in \mathbb{Q}$,

$$\begin{aligned} (u + v, w) &= (u, w) + (v, w), \\ (u, v + w) &= (u, v) + (u, w), \\ (cu, v) &= (u, cv) = c(u, v), \\ -(u, v) &= (v, u). \end{aligned}$$

As in the symmetric case, a choice of basis determines a pairing matrix, though now it will be antisymmetric, i.e. $M^t = -M$.

In analogy with the symmetric case, an antisymmetric pairing is called *nonsingular* if the assignment $v \to (v, \cdot)$ is an injection (and hence an isomorphism) $V \to \mathrm{Hom}(V, \mathbb{Q})$. Again, this corresponds to each pairing matrix having non-zero determinant by an easy modification of Lemma A.5.8.

In this setting, we have the following analogue of Lemma A.5.5.

Lemma A.5.14 *Given a nonsingular antisymmetric pairing $(V, (\cdot,\cdot))$ on a finite-dimensional rational vector space, there is a basis of V with respect to which the pairing matrix is a block sum of 2×2 matrices of the form $\begin{pmatrix} 0 & 1 \\ -1 & 0 \end{pmatrix}$. It follows as a consequence that V must be even-dimensional to have a nonsingular antisymmetric pairing.*

Proof We will construct a basis $\{a_1, b_1, a_2, b_2, \ldots, a_n, b_n\}$ such that $(a_i, b_i) = 1$, $(a_i, b_j) = 0$ for $i \neq j$, and $(a_i, a_j) = (b_i, b_j) = 0$ for all i, j. Then the pairing matrix N with respect to this basis will have the desired form.

Let b_1 be an arbitrary vector in V. Since being nonsingular implies $V \cong \mathrm{Hom}(V, \mathbb{Q})$, there is another vector in V, say a_1, such that $(a_1, b_1) = 1$. The pairing matrix restricted to the span of $\{a_1, b_1\}$ is now $\begin{pmatrix} 0 & 1 \\ -1 & 0 \end{pmatrix}$, using $\{a_1, b_1\}$ as a basis, noting that antisymmetry says that $(v, v) = -(v, v)$ and so $(v, v) = 0$ for any vector $v \in V$.

If V is two-dimensional, we are done. Otherwise, choose f_2 not in the span of $\{a_1, b_1\}$, and let $b_2 = f_2 + (b_1, f_2)a_1 - (a_1, f_2)b_1$. Then

$$\begin{aligned}
(a_1, b_2) &= (a_1, f_2 + (b_1, f_2)a_1 - (a_1, f_2)b_1) \\
&= (a_1, f_2) + (b_1, f_2)(a_1, a_1) - (a_1, f_2)(a_1, b_1) \\
&= (a_1, f_2) + 0 - (a_1, f_2) \\
&= 0.
\end{aligned}$$

Similarly

$$\begin{aligned}
(b_1, b_2) &= (b_1, f_2 + (b_1, f_2)a_1 - (a_1, f_2)b_1) \\
&= (b_1, f_2) + (b_1, f_2)(b_1, a_1) - (a_1, f_2)(b_1, b_1) \\
&= (b_1, f_2) - (b_1, f_2) - 0 \\
&= 0.
\end{aligned}$$

So b_2 is orthogonal to $\mathrm{span}(\{a_1, b_1\})$, but there must be some e_2 with $(e_2, b_2) = 1$ (note that e_1 is not in the span of $\{a_1, b_1, b_2\}$). Let $a_2 = e_2 + (b_1, e_2)a_1 - (a_1, e_2)b_1$. Then by the same calculations as for b_2, we see that a_2 is orthogonal to $\mathrm{span}(\{a_1, b_1\})$, while

$$\begin{aligned}
(a_2, b_2) &= (e_2 + (b_1, e_2)a_1 - (a_1, e_2)b_1, b_2) \\
&= (e_2, b_2) + (b_1, e_2)(a_1, b_2) - (a_1, e_2)(b_1, b_2) \\
&= 1 + 0 + 0 \\
&= 1.
\end{aligned}$$

We can continue in this manner: Once we have found $\{a_1, b_2, \ldots, a_k, b_k\}$, if these do not span V we can let f_{k+1} be any vector not in $\mathrm{span}(\{a_1, b_2, \ldots, a_k, b_k\})$ and then let

$b_{k+1} = f_{k+1} + \sum_{i=1}^{k}(b_i, f_{k+1})a_i - \sum_{i=1}^{k}(a_i, f_{k+1})b_i$. By analogous computations to those above, the vector b_{k+1} will be orthogonal to all the previous a_i and b_i. Then there must be some e_{k+1} not in the span of the established a_i and b_i such that $(e_{k+1}, b_{k+1}) = 1$. Let $a_{k+1} = e_{k+1} + \sum_{i=1}^{k}(b_i, e_{k+1})a_i - \sum_{i=1}^{k}(a_i, e_{k+1})b_i$. Again this vector will be orthogonal to all a_i and b_i, $1 \leq i \leq k$, but $(a_{k+1}, b_{k+1}) = 1$.

Eventually, the a_i and b_i span the space, and we are done. □

Corollary A.5.15 *If $(V, (\cdot, \cdot))$ is any finite-dimensional rational vector space with a non-singular antisymmetric pairing, there is a vector space of half the dimension of V on which the pairing is trivial.*

Proof Continuing to use the notation of the proof of the preceding lemma, the subspace spanned by $\{a_1, \ldots, a_n\}$ is such a subspace. □

Appendix B
An Introduction to Simplicial and PL Topology

In this appendix, we provide a brief survey introduction to PL spaces. Probably the most accessible introduction to PL spaces as subspaces of finite-dimensional Euclidean spaces is the book by Rourke and Sanderson [198]. However, for our purposes we will want to consider more general PL spaces, including those we wish to think about abstractly (not as concrete subspaces of Euclidean spaces), and PL spaces that have been given infinite (though locally finite) triangulations. To handle such generalities, we will refer primarily to Hudson [130]. Most of the following definitions and major results are taken from [130], though we add some additional arguments to tie the material together for our use. We assume that the reader is already familiar with simplicial complexes, e.g. from one of [125, 182, 220] (though they each take a somewhat different approach), but we also provide a very brief review of the main definitions.

In the abstract approach to PL spaces, one proceeds somewhat analogously to how one forms a smooth manifold by thinking of it as a collection of charts that have been glued together by appropriately smooth maps. Similarly, a PL space is defined by gluing together Euclidean polyhedra. So let us start with Euclidean polyhedra and a review of simplicial complexes.

B.1 Simplicial Complexes and Euclidean Polyhedra

This material is taken primarily from Chapters I and III of [130].

To begin, all of our constructions will at first live within some ambient Euclidean space $\mathbb{R}^n$. For infinite simplicial complexes, it is useful to allow $n = \infty$. We always think of each $\mathbb{R}^i$ as a subspace of $\mathbb{R}^{i+1}$ via the map given by the standard inclusion of the first i coordinates $(x_1, \dots, x_i) \to (x_1, \dots, x_i, 0)$. Then we let $\mathbb{R}^\infty$ be the union $\bigcup_{i=1}^\infty \mathbb{R}^i$ in the weak topology. So $\mathbb{R}^\infty$ consists of sequences of real numbers $(x_1, x_2, \dots)$ such that all but finitely many of the x_i are zero, and a subset of $\mathbb{R}^\infty$ is open (respectively, closed) if and only if its intersection with each $\mathbb{R}^i$, $i < \infty$, is open (respectively, closed).

B.1.1 Simplicial Complexes

Simplicial complexes are built up from simplices, which are a special case of the more general notion of convex cells.

Definition B.1.1 A *convex cell A in* $\mathbb{R}^n$, $n < \infty$, is a non-empty, compact subset of $\mathbb{R}^n$ that is the solution set to a finite number of linear equations $\{f_i(x) = 0\}$ and linear inequalities $\{g_i(x) \geq 0\}$.

Example B.1.2 Any convex polygon in the plane (regarded as a two-dimensional object including its interior) is a convex cell. △

Example B.1.3 The convex hull of a finite set of points in $\mathbb{R}^n$ is a convex cell; conversely, every convex cell is the convex hull of a finite set of points (the *vertices* of the cell) [130, page 2]. △

A simplex is a special kind of convex cell [130, Section 1.2]:

Definition B.1.4 The convex hull of $m + 1$ linearly independent points $\{v_i\}_{i=0}^{m}$ is called an *m-simplex* or *m-dimensional simplex*. Here linear independence means that the set of vectors $\{v_i - v_0\}_{i=1}^{m}$ is linearly independent. The v_i are called the *vertices* of the simplex. The convex hull of any subset of the $\{v_i\}$ is itself a simplex, called a *face* of the original simplex. If σ is a simplex and τ is a face of σ, we write $\tau < \sigma$. A face of σ that is not equal to σ is called a *proper face*.

Now that we have simplices, let us recall the definition of a simplicial complex; see Sections I.2 and III.2 of [130]:

Definition B.1.5 A locally finite *simplicial complex K* in $\mathbb{R}^n$, $n \leq \infty$, is a set of simplices in $\mathbb{R}^n$ such that

1. if $\sigma, \tau \in K$ and $\sigma \cap \tau \neq \emptyset$ then $\sigma \cap \tau$ is a face of both σ and τ,
2. if $\sigma \in K$ and τ is a face of σ, then $\tau \in K$,
3. (local finiteness) if x is contained in the union of the simplices of K, then x has a neighborhood in $\mathbb{R}^n$ that intersects only finitely many simplices of K.

The union of the i-simplices of K is called the i-skeleton of K and denoted K^i.

If K is a locally finite simplicial complex, we let $|K|$ denote the space consisting of the union of the simplices of K.

We say that m is the *dimension of K*, $\dim(K) = m$, if m is an integer such that every simplex of K has dimension $\leq m$ and K has at least one simplex of dimension m.

Remark B.1.6 It is possible to define more general simplicial complexes that are not necessarily locally finite. For example, see [182, Section 2]. We stick with the locally finite case, as this condition will be enforced for all spaces we will consider. Consequently, we will abbreviate "locally finite simplicial complex" to "simplicial complex" unless we want to particularly emphasize this property.

Example B.1.7 If σ is any simplex, then the set consisting of σ and all its faces is a simplicial complex. △

Remark B.1.8 If K contains only finitely many simplices, we call K a *finite simplicial complex*. In this case, $|K|$ must be a subset of some $\mathbb{R}^n$, $n < \infty$. It is also possible to have simplicial complexes with infinitely many simplices as a subspace of $\mathbb{R}^n$ for $n < \infty$. For example, we can construct the simplicial complex K in $\mathbb{R}^1$ whose vertices are the integer points and whose 1-simplices are the intervals $[i, i+1]$ for each integer i. Then $|K| = \mathbb{R}^1$.

Definition B.1.9 Suppose K, L are simplicial complexes and that every simplex of L is also a simplex of K. Then we say that *L is a subcomplex of K*. Equivalently, L is a subcomplex of K if it is a subset of K that is also itself a simplicial complex.

Example B.1.10 Let K be a simplicial complex, and let $\{\sigma_\alpha\}$ be any subset of the simplices of K. Let L consist of the simplices $\{\sigma_\alpha\}$ and all of their faces. Then L is a subcomplex. As a special case, suppose τ is any simplex of K and let $\{\sigma_\alpha\}$ be the set of simplices that have τ as a face. Then the union of the $\{\sigma_\alpha\}$ and all of their faces is called the *closed star of τ in K*, denoted $\overline{St}(\tau, K)$. △

Remark B.1.11 It is easy to verify from the definitions that if J, L are both subcomplexes of K, then $J \cap L$ is also a subcomplex of K.

It is also important to know about subdivisions of simplicial complexes:

Definition B.1.12 Let K be a simplicial complex. The simplicial complex K' is a *subdivision* of K if $|K'| = |K|$ and every simplex of K' is contained in some simplex of K.

Example B.1.13 Perhaps the most important subdivisions of a simplicial complex are the barycentric ones in which each simplex is replaced with its barycentric subdivision. These barycentric subdivisions of simplices are so important that it is unlikely that the reader familiar with algebraic topology has not seen them, utilized for example in a proof of the excision property for singular homology. See [125, Section 2.1], [182, Section 15], [220, Section 3.3], [71, Section III.6], or [130, Section I.2].

As a brief reminder, if $\{v_i\}_{i=0}^m$ are the vertices of an m-simplex $\sigma \subset \mathbb{R}^n$, then the *barycenter* of σ is

$$\hat{\sigma} = \frac{1}{m+1} \sum_{i=0}^{m} v_i.$$

The barycentric subdivision, say $\hat{K}$, of a simplicial complex K is then defined inductively over the skeleta of K: Let $\hat{K}^0 = K^0$. Now suppose that the skeleta $\hat{K}^i$ have been defined for $0 \le i \le n-1$ so that each simplex of $\hat{K}^i$ is contained in a simplex of K. Let σ be an n-simplex of K, and suppose that τ is a simplex of $\hat{K}^{n-1}$ contained in an $(n-1)$-dimensional face of σ. If $\{w_j\}_{j=0}^{n-1}$ are the vertices of τ, then there is an n-simplex whose vertices are $\hat{\sigma}$ and the w_j. The union of such n-simplices (and their faces) over all τ provides a subdivision of σ, and proceeding similarly for all n-simplices gives $\hat{K}^n$. The simplicial complex $\hat{K}$ is the union of all the $\hat{K}^i$; see Figure B.1. △

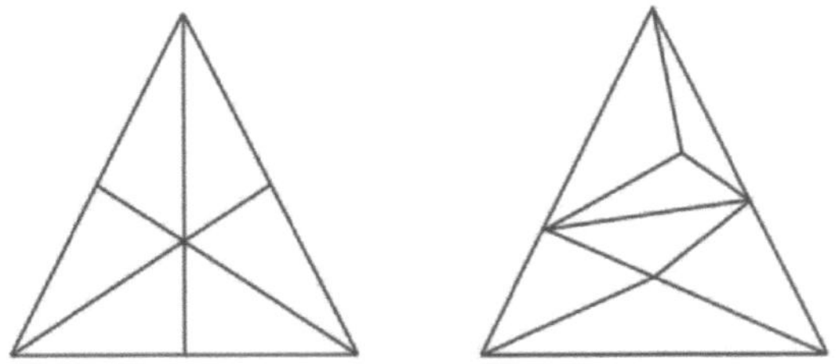

Figure B.1 Two subdivisions of a 2-simplex. The one on the left is barycentric, but that on the right is not.

Next we recall the definition of a simplicial map from [130, Section 1.4].

Definition B.1.14 A simplicial map $f : K \to L$ is a continuous function $f : |K| \to |L|$ that takes vertices of K to vertices of L and restricts to a linear map on each simplex of K, i.e. if the vertices $v_0, \ldots, v_n$ span a simplex of K and $x = \Sigma_{i=0}^n t_i v_i \in |K|$, then $f(x) = \Sigma_{i=0}^n t_i f(v_i)$.

Example B.1.15 If σ is any simplex and τ is a face, the inclusion map $\tau \hookrightarrow \sigma$ is simplicial, thinking of σ and τ as simplicial complexes as in Example B.1.7. △

Example B.1.16 If $V = \{v_0, \ldots, v_m\}$ are the vertices of a simplex σ and $W = \{w_0, \ldots, v_\ell\}$ are the vertices of a simplex τ, then any set map $V \to W$ determines a unique simplicial map of the simplicial complexes associated to σ and τ. △

The appropriate notion of equivalence among simplicial complexes is that of simplicial isomorphism; see [182, Lemma 2.8].

Definition B.1.17 A simplicial map $f : K \to L$ is a *simplicial isomorphism* if it induces a bijection between the vertices of K and L and is such that the vertices $v_0, \ldots, v_n$ of K span a simplex if and only if $f(v_0), \ldots, f(v_n)$ span a simplex of L. If $f : K \to L$ is a simplicial isomorphism, then so is $f^{-1} : L \to K$.

B.1.2 Euclidean Polyhedra

We can now define Euclidean polyhedra as in [130, Section I.1]. Essentially, these are the underlying sets of finite simplicial complexes.

Definition B.1.18 A *Euclidean polyhedron in* $\mathbb{R}^n$ is any finite union of convex cells in $\mathbb{R}^n$.

Remark B.1.19 As in Remark B.1.8, as a Euclidean polyhedron is a union of a finite number of cells, each contained in a finite-dimensional Euclidean space, every Euclidean polyhedron lives in some $\mathbb{R}^n$ with $n < \infty$.

Example B.1.20 There is no requirement in the definition concerning how the convex cells intersect. So, for example, the union of *any* two triangles in the plane is a Euclidean polyhedron. △

Example B.1.21 Any polygon in the plane, not necessarily convex, is a Euclidean polyhedron. △

Example B.1.22 If K is a finite simplicial complex, then $|K|$ is the union of a finite number of simplices, so it is a Euclidean polyhedron. The converse is also true, via the next proposition, which is [130, Corollary 1.7]. △

Proposition B.1.23 *Every Euclidean polyhedron is the underlying space of a finite simplicial complex.*

To build PL spaces, we will also need to know about piecewise linear maps of polyhedra:

Definition B.1.24 If $P \subset \mathbb{R}^n$ and $Q \subset \mathbb{R}^m$ are Euclidean polyhedra, then $f : P \to Q$ is *piecewise linear* if f is continuous and the graph

$$\Gamma = \{(x, f(x)) \mid x \in P\} \subset \mathbb{R}^n \times \mathbb{R}^m$$

is a Euclidean polyhedron.

Remark B.1.25 This definition seems a bit off-putting, but it has a more intuitive interpretation once we utilize the observation from Proposition B.1.23 that every Euclidean polyhedron is the underlying space of a simplicial complex. Suppose $P = |K|$ and $Q = |L|$ for some simplicial complexes K and L. If $f : |K| \to |L|$ is piecewise linear, then by [130, Lemma 1.10] there exist subdivisions K', L' of K, L such that $f : K' \to L'$ is simplicial. Conversely, every simplicial map of finite simplicial complexes is piecewise linear as a map of the underlying Euclidean polyhedra [130, Remark 2, page 15]. So a map $f : |K| \to |L|$ is piecewise linear if and only if there are subdivisions K', L' such that $f : K' \to L'$ is simplicial.

Example B.1.26 If K is a finite simplicial complex and L is a subcomplex, then the inclusion $|L| \hookrightarrow |K|$ is piecewise linear by Remark B.1.25. △

Example B.1.27 The composition of piecewise linear maps of Euclidean polyhedra is piecewise linear. See [130, Lemma 1.1]. △

The following lemmas will be needed below to prove Proposition B.5.3, which says that the product of PL spaces is a PL space.

Lemma B.1.28 *The product of two Euclidean polyhedra is a Euclidean polyhedron.*

Proof This is a consequence of the product of two convex cells being a convex cell [130, page 2]. □

Lemma B.1.29 *The product of two piecewise linear maps of Euclidean polyhedra is a piecewise linear map of Euclidean polyhedra.*

Proof Suppose that $f_i : P_i \to Q_i$ are piecewise linear maps of Euclidean polyhedra for $i = 1, 2$. Suppressing the ambient Euclidean spaces, the graph of $f_1 \times f_2$ is the Euclidean subset

$$\{((x_1, x_2), (f_1(x_1), f_2(x_2))) \mid x_i \in P_i\} \cong \{((x_1, f_1(x)), (x_2, f_2(x_2))) \mid x_i \in P_i\},$$

which is the product of the graphs of f_1 and f_2. As the graphs of f_1 and f_2 are Euclidean polyhedra by assumption, it follows from Lemma B.1.28 that the graph of the product is a Euclidean polyhedron. □

B.2 PL Spaces and PL Maps

We now turn to PL spaces, following Hudson [130, Chapter III]. Hudson's definition is not the one we have given in the main text in Section 2.5, although we will show that the category of spaces $\mathcal{PL}$ implicit in [130] (he does not use the language of categories and functors) is equivalent to our category $\mathcal{AT}$ of Definition 2.5.11. As we primarily follow Hudson's point of view through most of this appendix, we will use "PL map" and "PL space" here primarily in the sense of [130]; when we need to refer to the spaces of Section 2.5 in a way that distinguishes them from Hudson's PL spaces, we will simply call them "spaces with families of admissible triangulations," or "AT spaces" for short[1].

So far, our Euclidean polyhedra have been compact and our simplicial complexes have lived as subsets of Euclidean spaces in much the same way that one might first be introduced to smooth manifolds as submanifolds of Euclidean space. But just as one can abstract the definition of smooth manifold to one involving coordinate charts that needs no reference to an ambient space, so too can a PL space be defined as an abstract, not-necessarily-compact generalization of a Euclidean polyhedron that does not require an ambient space. Also as for manifolds, while this level of abstraction can be very powerful, it is in some sense not strictly necessary for reasonable spaces: by the Whitney embedding theorem, any finite-dimensional second-countable smooth manifold can be embedded as a closed subset of Euclidean space (see [38, Theorem II.10.8]), while Hudson shows that any PL space is homeomorphic to a simplicial complex in $\mathbb{R}^\infty$.

So let us proceed to the definitions, beginning with the structures that play the role of coordinate charts [130, Section III.2]:

[1] We do this with apologies, realizing it could cause some confusion, but since the spaces of the main body of this text *are* PL spaces, for all intents and purposes, and as this is already a recognized kind of space, we do not want to call them AT spaces throughout the book. But it would also be confusing to refer to the spaces in Hudson by any other name.

Definition B.2.1 Let X be a topological space. A *coordinate map* (f,P) is a topological embedding $f : P \to X$, where P is a Euclidean polyhedron (and so compact). The coordinate maps (f,P) and (g,Q) are deemed *compatible* if $f(P) \cap g(Q) = \emptyset$ or if $f(P) \cap g(Q) \neq \emptyset$ and there is a coordinate map (h,R) such that $h(R) = f(P) \cap g(Q)$ and $f^{-1}h$ and $g^{-1}h$ are piecewise linear maps in the sense of Definition B.1.24. Equivalently, (f,P) and (g,Q) are compatible if $f^{-1}(g(Q))$ is empty or if it is a subpolyhedron of P and $g^{-1}f : f^{-1}(g(Q)) \to Q$ is a piecewise linear map.

Example B.2.2 Suppose that (f,P) and (g,Q) are two compatible coordinate maps and that P' is a subpolyhedron of P, i.e. a Euclidean polyhedron that is also a subset of P. Then $(f|_{P'},P')$ is also compatible with (g,Q). Indeed, in this case $(f|_{P'})^{-1}(g(Q)) = P' \cap f^{-1}(g(Q))$, which is the intersection of two Euclidean polyhedra and hence a Euclidean polyhedron [130, page 2]. And $g^{-1}f|_{P'} : P' \cap f^{-1}(g(Q)) \to Q$ is piecewise linear as the restriction of a piecewise linear map of Euclidean polyhedra to a subpolyhedron [130, Lemma 1.1]. △

A compatible collection of coordinate maps yields a "PL structure" [130, Section III.2]:

Definition B.2.3 A *PL structure* $\mathcal{F}$ on a topological space X is a family of coordinate maps such that

1. any two coordinate maps of $\mathcal{F}$ are compatible in the sense of Definition B.2.1,
2. for each $x \in X$ there is some coordinate map (f,P) in $\mathcal{F}$ such that $f(P)$ is a topological neighborhood of x in X,
3. $\mathcal{F}$ is maximal in the sense that if the coordinate map (f,P) is compatible with every coordinate map in $\mathcal{F}$, then (f,P) is in $\mathcal{F}$.

Definition B.2.4 A *PL space* $(X,\mathcal{F})$ is a second-countable Hausdorff space X with a PL structure $\mathcal{F}$. We often speak of the "PL space X" leaving $\mathcal{F}$ tacit.

A PL space $(X,\mathcal{F})$ is called an *m-dimensional PL manifold* if for each $x \in X$ there is some coordinate map (h,Δ^m) in $\mathcal{F}$ such that Δ^m is an m-simplex and x is contained in the interior of the image $h(\Delta^m)$.

Remark B.2.5 Another similarity with smooth manifolds is that while one often includes having a maximal atlas as part of the definition of a smooth manifold, having enough charts to cover the manifold is really sufficient. Here, a collection of coordinate maps on X satisfying just the first two conditions of the definition is called a *base for a PL structure*, and every base can be completed to a unique PL structure [130, Lemma 3.1]. In particular, any two coordinate maps that are compatible with every coordinate map in a given base are compatible with each other.

Example B.2.6 If $|K|$ is the underlying space of a locally finite simplicial complex, we claim that $|K|$ can be given the structure of a PL space. The most obvious thing to try would be to take as coordinate maps the inclusions into $|K|$ of its simplices. While any two such inclusions are clearly compatible, this is not sufficient to give us a base because a point

$x \in |K|$ might not lie in the topological interior of any simplex of K (where here "topological interior" means the interior of the simplex as a topological subspace of $|K|$). For example, if K is the simplicial complex of Remark B.1.8 with $|K| = \mathbb{R}$, then no neighborhood of any integer point in $|K| = \mathbb{R}$ is contained in any one simplex.

However, there is a simple solution to this problem: We let the coordinate maps for X consist of *every* inclusion $|L| \hookrightarrow |K|$ where L is a finite subcomplex of K. These are compatible using Remark B.1.11 and Example B.1.26. To see that they satisfy the second condition to constitute a base, let x be point of K. By the local finiteness, x has a neighborhood in $\mathbb{R}^\infty$ that intersects only finitely many of the simplices of K. Let L be the union of these simplices and their faces, which is also a finite subcomplex of K. We see that $|L|$ must contain a neighborhood of x in $|K|$, and so the condition is met.

We will call this the *canonical PL structure on* $|K|$. This is perhaps an abuse of notation as the PL structure depends on K, not just $|K|$, so the "K" is critical to the meaning of the symbol. △

Example B.2.7 Let K be a simplicial complex and K' a subdivision of K. Let $\mathcal{B}$ be the base for the canonical PL structure on $|K|$ defined in the preceding example, and let $\mathcal{B}'$ be the analogous base for $|K'|$. We have that $|K| = |K'|$ and that $\mathcal{B}$ is a subset of $\mathcal{B}'$ because the underlying space of every subcomplex L of K is also the underlying space of L', where L' is the subdivision of L induced by the subdivision K' of K. By Remark B.2.5, both $\mathcal{B}$ and $\mathcal{B}'$ are bases for the same PL structure on $|K| = |K'|$ and so they determine the same PL space.

As a corollary, we note that if K and L are two simplicial complexes that have a common subdivision, then they both determine the same PL space. △

Example B.2.8 Building on Example B.2.6, let K be a simplicial complex and suppose there is a homeomorphism $h : |K| \to X$ for some topological space X. We can take as a base for a PL structure on X the collection $(h, |L|)$ as L runs over all finite subcomplexes of K. The verification that this is a base for a PL structure is essentially the same as the argument of Example B.2.6. Theorem B.2.9 now says that every PL space can be constructed in this manner; see [130, Section III.2]. △

Theorem B.2.9 *Let $(X, \mathcal{F})$ be a PL space. There exists a locally finite simplicial complex K and a homeomorphism $h : |K| \to X$ such that the restrictions of h to the finite subcomplexes of K are all coordinate maps of $\mathcal{F}$.*

Furthermore, if $(X, \mathcal{F})$ is an m-dimensional PL manifold then every point of $|K|$ lies in the interior of the image of a piecewise linear map of polyhedra $\Delta^m \to |K|$.

Definition B.2.10 If X is a topological space, K is a locally finite simplicial complex, and $h : |K| \to X$ is a homeomorphism, we say that the pair $T = (K, h)$ is a *triangulation* of X. In the case where $(X, \mathcal{F})$ is a PL space and the restrictions of h to the finite subcomplexes of K are all coordinate maps of $\mathcal{F}$, then we say that T is a *PL triangulation of* $(X, \mathcal{F})$. It is often common to abbreviate and speak of "a triangulation of X," or "a triangulation of X

by K," or "the triangulation $h : |K| \to X$," leaving the other elements of the definition tacit unless explicit reference to them is required.

Example B.2.8 says that if X is a topological space then any triangulation of X determines a PL structure $\mathcal{F}$ on X such that the triangulation is a PL triangulation of the PL space $(X, \mathcal{F})$.

Example B.2.11 Continuing Example B.2.6, we see that if K is a simplicial complex and if $|K|$ is given the canonical PL structure then the identity map $|K| \to |K|$ is a PL triangulation. △

Example B.2.12 More generally, suppose that $h : |K| \to X$ is a PL triangulation of $(X, \mathcal{F})$ and that K' is any subdivision of K. Then $h : |K'| \to X$ is a PL triangulation. To verify this, we need to know that the restriction of h to any finite subcomplex of K' is a coordinate map in $\mathcal{F}$. But every finite subcomplex L' of K' is contained in a finite subcomplex L of K; just let L be the union of all of the faces of all of the simplices of K that intersect simplices of L'. By assumption, the restriction of h to $|L|$ is a coordinate map, and so it is compatible with every other coordinate map in $\mathcal{F}$ by the first part of Definition B.2.3. Example B.2.2 then tells us that the restriction of h to $|L'|$ is also compatible with every coordinate map in $\mathcal{F}$, so by the third condition in Definition B.2.3, $(h, |L'|)$ is a coordinate map in $\mathcal{F}$. As L' was an arbitrary subcomplex of K', the map $h : |K'| \to X$ is a PL triangulation. △

A natural question to ask is when two triangulations of a space X determine the same PL structure. Equivalently, if X is already a PL space, we can ask when a triangulation is a PL triangulation. To discuss that question, we need PL maps [130, Section III.3]:

Definition B.2.13 Let $(X, \mathcal{F})$ and $(Y, \mathcal{G})$ be PL spaces. A topological[2] map $\phi : X \to Y$ is a *PL map* if for each pair of coordinate maps $(f, P) \in \mathcal{F}$ and $(g, Q) \in \mathcal{G}$ the set $f^{-1}(\phi^{-1}(g(Q)))$ is empty or a subpolyhedron of P and, if the latter case, the map $g^{-1}\phi f : f^{-1}(\phi^{-1}(g(Q))) \to Q$ is a piecewise linear map of Euclidean polyhedra.

Definition B.2.14 The PL spaces and maps form a category $\mathcal{PL}$. We leave verification of the category axioms as an exercise for the reader.

Following the definition of PL maps, Hudson goes on to give a second definition for PL maps in the case that $X = |K|$ and $Y = |L|$ for simplicial complexes K and L:

Definition B.2.15 If K, L are locally finite simplicial complexes and $\phi : |K| \to |L|$ is a topological map, we say that ϕ is a *PL map* if it maps each finite subcomplex of K piecewise linearly into some finite subcomplex of L.

Remark B.2.16 These two definitions of PL map are consistent: By [130, Remark, page 83], if $h : |K| \to X$ and $j : |L| \to Y$ are PL triangulations of the PL spaces X and Y and

[2] In fact, as noted in [130, note 1, page 83], the continuity follows from the other conditions, so we need not have mentioned it explicitly.

the diagram

$$\begin{array}{ccc} X & \xrightarrow{\phi} & Y \\ {\scriptstyle h}\uparrow & & \uparrow{\scriptstyle j} \\ |K| & \xrightarrow{\psi} & |L| \end{array}$$

commutes, then ϕ is a PL map in the sense of Definition B.2.13 if and only if ψ is a PL map in the sense of Definition B.2.15.

Notice that if $X = |K|$ and $Y = |L|$, i.e. if h and j are the identity maps so that X and Y are the canonical PL structures on $|K|$ and $|L|$, then this means that any PL map $|K| \to |L|$ in the second sense is also a PL map in the first sense.

Example B.2.17 We saw in Remark B.1.25 that any simplicial map of finite simplicial complexes is piecewise linear as a map of the underlying Euclidean polyhedra and so any simplicial map of locally finite simplicial complexes is a PL map. △

Example B.2.18 We saw in Theorem B.2.9 that if $(X, \mathcal{F})$ is a PL space then there exists a PL triangulation $h : |K| \to X$ such that the restrictions of h to the finite subcomplexes of K are all coordinate maps of $\mathcal{F}$. Let $\mathcal{K}$ denote $|K|$ as a PL space with the canonical PL structure coming from K, as in Example B.2.6. We claim that $h : \mathcal{K} \to X$ is a PL map. By Remark B.2.16, we may consider the diagram

$$\begin{array}{ccc} \mathcal{K} & \xrightarrow{h} & X \\ {\scriptstyle \mathrm{id}}\uparrow & & \uparrow{\scriptstyle h} \\ |K| & \xrightarrow{\mathrm{id}} & |K|, \end{array}$$

which clearly commutes. The map $h : |K| = \mathcal{K} \to X$ is a PL triangulation by hypothesis, and $\mathrm{id} : |K| \to \mathcal{K}$ is a PL triangulation by Example B.2.11. So by Remark B.2.16, the map $h : \mathcal{K} \to X$ is a PL map if and only if $\mathrm{id} : |K| \to |K|$ is a piecewise linear map, which is certainly true.

In fact, $h : \mathcal{K} \to X$ is a PL homeomorphism, as the same argument with the horizontal arrows reversed in the diagram shows that h^{-1} is also a PL map. △

We observed in Remark B.1.25 that piecewise linear maps of (compact) Euclidean polyhedra can be interpreted in simplicial terms: any piecewise linear map $|K| \to |L|$ of Euclidean polyhedra is a simplicial map with respect to some subdivisions K' and L' of K and L. Analogously, for PL maps, we have the following theorem, Theorems 3.6.B and 3.6.C of [130]:

Theorem B.2.19 *If K and L are locally finite simplicial complexes and $f : |K| \to |L|$ is* *a PL map then there is a subdivision K' of K such that f maps simplices of K' linearly into simplices of L. Furthermore, if f is proper*[3] *then there are subdivisions K' of K and L' of L such that the map $f : K' \to L'$ is a simplicial map.*

Remark B.2.20 The restriction to proper maps is necessary for the second conclusion of the theorem. Here is one of the standard counterexamples without that assumption: Let K be the simplicial complex with $|K| = \{x \in \mathbb{R} \mid x \geq 0\}$ and with vertices at the integers. Consider the map $f : \mathbb{N} \to \mathbb{R}$ defined by $f(2i) = 0$ and $f(2i+1) = 1/2^i$. Mapping each interval $[j, j+1]$ of K linearly to $\mathbb{R}$ based on where its endpoints are sent by f gives a piecewise linear map $f : |K| \to |K|$, but there are no locally finite subdivisions K_1' and K_2' of K such that $f : K_1' \to K_2'$ is simplicial.

Example B.2.21 Suppose X is a PL space. By Theorem B.2.9 there exist PL triangulations of X. Let $h : |K| \to X$ and $j : |L| \to X$ be two such triangulations. Then, by Remark B.2.16, $j^{-1}h : |K| \to |L|$ is a PL map, as the identity map $X \to X$ is certainly PL. As $j^{-1}h$ is a homeomorphism, it is proper. Therefore, by Theorem B.2.19, there are subdivisions K' and L' such that $j^{-1}h : K' \to L'$ is simplicial, and hence a simplicial isomorphism. In particular, if $|K| = |L|$ and h and j are both identity maps, this says that there are subdivisions K' and L' such that the identity map induces a simplicial isomorphism $K' \to L'$. This implies that $K' = L'$, i.e. K and L have a common subdivision. △

Here is a useful corollary to Theorem B.2.19:

Corollary B.2.22 *Suppose K and L are finite simplicial complexes and that $f : |K| \to |L|$ is a PL map that is also a bijection. Then f is a PL homeomorphism.*

Proof As K is finite, the space $|K|$ is compact. By definition of PL maps, f is continuous, and a continuous bijection of compact Hausdorff spaces is a topological homeomorphism [182, Theorem 26.6]. Furthermore, as any map with compact domain is proper, by Theorem B.2.19 there are subdivisions K' of K and L' of L such that $f : K' \to L'$ is simplicial. But since f is a homeomorphism, this implies that it restricts to a linear homeomorphism from each simplex of K' onto some simplex of L'. Thus $f : K' \to L'$ is a simplicial isomorphism, and hence so is f^{-1}, implying that f^{-1} is also a PL bijection. □

We can now answer our question concerning when two triangulations of a space X determine the same PL structure:

Theorem B.2.23 *Two triangulations (K, h) and (L, j) of the topological space X determine the same PL structure if and only if there are subdivisions (K', h) and (L', j) such that $j^{-1}h$ is a simplicial isomorphism. Therefore, if $(X, \mathcal{F})$ is a PL space, the triangulation (K, h) is a PL triangulation if and only if for any PL triangulation (L, j) there are subdivisions (K', h) and (L', j) such that $j^{-1}h$ is a simplicial isomorphism.*

[3] Recall that a map $f : X \to Y$ of topological spaces is called *proper* if for each compact set $K \subset Y$, the set $f^{-1}(K)$ is a compact subspace of X.

Proof First, suppose that (K,h) and (L,j) are both triangulations of X and that they determine the same PL structure $(X,\mathcal{F})$. Then they are both PL triangulations of $(X,\mathcal{F})$, and Example B.2.21 shows that there are subdivisions (K',h) and (L',j) such that $j^{-1}h$ is a simplicial isomorphism.

Conversely, suppose we have two triangulations (K,h) and (L,j) of the topological space X and that there are subdivisions (K',h) and (L',j) such that $j^{-1}h$ is a simplicial isomorphism. We know that (K,h) and (K',h) determine the same PL structure on X by Example B.2.12, and similarly (L,j) and (L',j) determine the same PL structure. So to show that (K,h) and (L,j) determine the same PL structure, it suffices to show that (K',h) and (L',j) determine the same PL structure. Let $\mathcal{B}_K$ and $\mathcal{B}_L$ be the bases of PL structures determined by the images of the finite subcomplexes of K' and L' under h and j, respectively; see Example B.2.8. Suppose $(h,P) \in \mathcal{B}_K$ and $(j,Q) \in \mathcal{B}_L$. Then $h^{-1}j(Q)$ is a subcomplex of K', as $h^{-1}j$ is a simplicial isomorphism, and, in particular, its intersection with P is a subcomplex of P. Furthermore, the restriction of $j^{-1}h$ to this subcomplex is simplicial and hence piecewise linear by Remark B.1.25. Therefore, by Definition B.2.1, these coordinate maps are compatible. As (h,P) and (j,Q) were arbitrary elements of $\mathcal{B}_K$ and $\mathcal{B}_L$, it follows from Remark B.2.5 that $\mathcal{B}_K$ and $\mathcal{B}_L$ are bases for the same PL structure.

For the second part of the theorem, if (L,j) is a PL triangulation of $(X,\mathcal{F})$, then the base $\mathcal{B}_L$ determines $\mathcal{F}$ by Definition B.2.10, and so (K,h) is a PL triangulation if and only if it also determines $\mathcal{F}$, which by the above argument is equivalent to (K,h) and (L,j) having subdivisions that are simplicial isomorphisms via $j^{-1}h$. □

One further important notion for this section is that of PL approximation, which says that arbitrary maps from PL spaces to PL manifolds can be approximated by nearby homotopic PL maps. Furthermore, if such a map is already PL on a closed subspace, then the homotopy can be taken as fixed on the subspace. The following technical statement is rephrased from Lemma 4.2 of [130]:

Theorem B.2.24 (PL Approximation Theorem) *Let $f : X \to M$ be a continuous map from the PL space X to the PL manifold M, and let $Z \subset X$ be a closed PL subspace such that $f|_Z : Z \to M$ is a PL map. Let $\epsilon : X \to \mathbb{R}$ be a continuous positive function. Then given a distance function d on X, there exists a PL map $f' : X \to M$ such that f' is homotopic to f relative to Z and $d(f(x), f'(x)) < \epsilon(x)$ for all $x \in X$.*

B.3 Comparing Our Two Notions of PL Spaces

In this section, we demonstrate that the category $\mathcal{AT}$ defined in Section 2.5 using admissible families of triangulations of a space is equivalent to the category $\mathcal{PL}$ of PL spaces as defined here in Definition B.2.14. As per our discussion at the beginning of Section B.2, in order to be very clear at all times about which definition we are using, we will in this section refer to the spaces defined in Section 2.5 as "AT spaces" and the maps as "AT maps," the "AT" referring to the families of admissible triangulations.

Let us briefly recall the definitions from Section 2.5, substituting in the "AT" language.

First recall our earlier definition of "triangulation," noting that there is no conflict with the definition here.

Definition B.3.1 (Definition 2.5.1) A *triangulation* T of a topological space X is a pair $T = (K, h)$, where K is a locally finite (possibly infinite) simplicial complex and $h : |K| \to X$ is a homeomorphism. A subdivision of $T = (K, h)$ is a pair $T' = (K', h)$, where K' is a *subdivision* of the simplicial complex K. If $T = (K, h)$ and $S = (L, j)$ are two triangulations of X, we say that T and S have a *common subdivision* if there are respective subdivisions $T' = (K', h)$ and $S' = (L', j)$ of T and S such that $j^{-1}h$ is a simplicial isomorphism from K' to L'. Of course in this case $h^{-1}j$ is also a simplicial isomorphism.

Definition B.3.2 (Definition 2.5.2) An *AT space* is a second-countable Hausdorff space X together with a family of triangulations $\mathcal{T}$ satisfying the following compatibility properties:

1. if $T \in \mathcal{T}$ and T' is any subdivision of T, then $T' \in \mathcal{T}$;
2. if $T, S \in \mathcal{T}$, then T and S have a common subdivision.

If $(X, \mathcal{T})$ is an AT space, we call the triangulations in $\mathcal{T}$ *admissible.*

Definition B.3.3 (Definition 2.5.4) If $(X, \mathcal{T})$ and $(Y, \mathcal{S})$ are two AT spaces, an *AT map* $(X, \mathcal{T}) \to (Y, \mathcal{S})$ is a (topological) map $f : X \to Y$ such that if given any admissible triangulations (K, h) of X and (L, j) of Y there is a subdivision K' of K such that $j^{-1}fh$ takes each simplex of K' linearly into a simplex of L.

Definition B.3.4 (Definition 2.5.11) The AT spaces and maps as we have defined them here form a category $\mathcal{AT}$.

Now let us construct functors $F : \mathcal{PL} \to \mathcal{AT}$ and $G : \mathcal{AT} \to \mathcal{PL}$, which we will then show yield an equivalence of categories. We begin by defining F and G on objects.

Suppose $(X, \mathcal{F})$ is a PL space. By Theorem B.2.9 and Definition B.2.10, there exists a PL triangulation (K, h) of X consisting of a homeomorphism $h : |K| \to X$ such that the restriction of h to the finite subcomplexes of K are all coordinate maps of $\mathcal{F}$. Let $\mathcal{T}$ be the collection of all such PL triangulations. We claim that $\mathcal{T}$ satisfies the conditions of Definition B.3.2.

Lemma B.3.5 *If $(X, \mathcal{F})$ is a PL space and $\mathcal{T}$ is the collection of all PL triangulations (K, h) of $(X, \mathcal{F})$ then the elements of $\mathcal{T}$ are compatible with each other in the sense of Definition B.3.2.*

Proof We saw in Example B.2.12 that if (K, h) is a PL triangulation then so is (K', h) for any subdivision K' of K.

So now suppose that (K, h) and (L, j) are two PL triangulations of K. By Example B.2.18, the maps h and j can be thought of as PL homeomorphisms $h : \mathcal{K} \to X$ and $j : \mathcal{L} \to X$, where $\mathcal{K}$ and $\mathcal{L}$ are $|K|$ and $|L|$ thought of as PL spaces with their canonical PL structures. By Theorem B.2.19, as $j^{-1}h$ is proper, there are subdivisions K' and L' of K and L such that $j^{-1}h : K' \to L'$ is a simplicial map, in fact a simplicial isomorphism as the map of underlying spaces is a homeomorphism. This provides the desired common subdivision. □

The lemma says that given any object $(X, \mathcal{F})$ in $\mathcal{PL}$ we obtain an object in $\mathcal{AT}$ with the same underlying space and with the admissible triangulations being all the PL triangulations of $(X, \mathcal{F})$ in the sense of Definition B.2.10. We define $F : \mathcal{PL} \to \mathcal{AT}$ to act on objects in this way.

Conversely, suppose we have an AT space $(X, \mathcal{T})$. If we let $(K, h) \in \mathcal{T}$ be any triangulation, then we can take as a base for a PL structure on X the collection $\{(h, |L|)\}$ as L ranges over all finite subcomplexes of K. This is a PL structure by Example B.2.8. We will call this the PL structure determined by the triangulation, and we let $G : \mathcal{AT} \to \mathcal{PL}$ be given on objects by this assignment. To show that this construction is well defined, we should check that it is independent of our choice of $(K, h) \in \mathcal{T}$.

Lemma B.3.6 *Let $(X, \mathcal{T})$ be an AT space. If $(K, h), (L, j) \in \mathcal{T}$ and $\mathcal{F}, \mathcal{F}'$ are the respective PL structures determined by these triangulations then $\mathcal{F} = \mathcal{F}'$.*

Proof Let (K, h) and (L, j) be any two triangulations of X in $\mathcal{T}$. By definition, there are subdivisions K' and L' such that $j^{-1}h : K' \to L'$ is a simplicial isomorphism. Theorem B.2.23 then tells us that (K, h) and (L, j) determine the same PL structures on X. □

So far we have seen that every AT space determines a PL space and vice versa. To elevate these assignments to functors, we must also consider what happens on maps.

Let $f : (X, \mathcal{F}) \to (Y, \mathcal{G})$ be a PL map of PL spaces. We need to define $F(f) : (X, \mathcal{T}) \to (Y, \mathcal{S})$, where $\mathcal{T}$ and $\mathcal{S}$ are the families of all PL triangulations of $(X, \mathcal{F})$, respectively $(Y, \mathcal{G})$. The map $F(f)$ will be the same map as f as a topological map $X \to Y$. We must show that f is also an AT map with respect to the given structures. For this, we let (K, h) and (L, j) be any PL triangulations of X and Y. We must show that there is a subdivision K' of K such that $j^{-1}fh$ takes each simplex of K' linearly into a simplex of L. As in the proof of Lemma B.3.5, we note that Example B.2.18 shows us that the triangulations $h : |K| \to X$ and $j : |L| \to Y$ are also PL maps, in fact PL homeomorphisms, using the canonical PL structures on $|K|$ and $|L|$. So the composition $j^{-1}fh$ is a PL map, and the existence of the desired K' is now guaranteed by Theorem B.2.19.

Conversely, let $g : (X, \mathcal{T}) \to (Y, \mathcal{S})$ be an AT map of AT spaces. Once again, we have the underlying map $g : X \to Y$, and we want to see that it's a PL map from $G(X) = (X, \mathcal{F})$ to $G(Y) = (Y, \mathcal{G})$, where $\mathcal{F}$ and $\mathcal{G}$ are the PL structures determined by the triangulations in $\mathcal{T}$ and $\mathcal{S}$. If (K, h) and (L, j) are such triangulations, we know from Remark B.2.16 that g is PL if and only if $j^{-1}gh$ is PL, meaning by Definition B.2.15 that $j^{-1}gh$ maps each finite subcomplex of K piecewise linearly into some finite subcomplex of L. We do know from the hypotheses that there is a subdivision K' of K such that $j^{-1}gh$ takes each simplex of K'

linearly into a simplex of L. The desired property that $j^{-1}gh$ takes any finite subcomplex of K' piecewise linearly into a finite subcomplex of L follows. This can be seen directly from Definition B.1.24: Let J be a finite subcomplex of $K \subset \mathbb{R}^k$, and let J' be the subdivision of J induced by the subdivision K' of K. As $j^{-1}gh$ takes each simplex of K' linearly into one simplex of L, there is a finite subcomplex I of L such that $f(|J|) \subset |I|$. If $|K| \subset \mathbb{R}^k$ and $|L| \subset \mathbb{R}^\ell$ then the graph of $f|_{|J|}$ in $\mathbb{R}^{k+\ell}$ is the union of the simplices that are the graphs of $j^{-1}gh$ restricted to the simplices of J'. These graphs are simplices because the restriction of $j^{-1}gh$ to each simplex of K' is linear by assumption. So the graph of $f|_{|J|}$ is a Euclidean polyhedron by Definition B.1.18. Thus $j^{-1}gh$ satisfies Definition B.2.15 to be a PL map.

So, the upshot of the preceding two paragraphs is that if f is a PL map then we can define $F(f)$ to be simply the same $f : X \to Y$ but interpreted as an AT map, and similarly if g is an AT map then we can define $G(g)$ to be simply the same $g : X \to Y$ but interpreted as a PL map. In particular, both functors F and G are identity functors at the level of topological spaces and topological maps; the only things they change are the additional PL or AT structures.

Theorem B.3.7 *The functors $F : \mathcal{PL} \to \mathcal{AT}$ and $G : \mathcal{AT} \to \mathcal{PL}$ determine an equivalence of categories.*

Proof We will show that there are natural isomorphisms of functors $\mathrm{id}_{\mathcal{AT}} \to FG$ and $GF \to \mathrm{id}_{\mathcal{PL}}$. The latter is the simpler: If $(X, \mathcal{F})$ is an object in $\mathcal{PL}$, then, from the definitions, $GF(X, \mathcal{F})$ is the topological space X with the PL structure induced by one of its PL triangulations, say (K, h). By definition, this is the structure determined by taking as a base the coordinate maps $(h, |J|)$, where J runs over the finite subcomplexes of K. This collection constitutes a base for a PL structure $\mathcal{F}'$ on X, but as (K, h) is a PL triangulation, each of these coordinate maps is compatible with the coordinate maps of $\mathcal{F}$ by definition. So as each base is compatible with a unique PL structure by [130, Lemma 3.1], we have $\mathcal{F} = \mathcal{F}'$. In other words, $GF(X, \mathcal{F}) = (X, \mathcal{F})$, so GF is the identity functor on spaces. We have already observed that each of G and F is the identity on maps at the level of topological spaces and topological maps, but knowing the topological map is enough to determine the PL or AT map. So GF is the identity functor.

By contrast, FG is not the identity functor. Rather, if $(X, \mathcal{T})$ is an AT space, $FG(X, \mathcal{T}) = (X, \hat{\mathcal{T}})$, where $\hat{\mathcal{T}}$ is the family of *all* PL triangulations of $(X, \mathcal{F})$, where $\mathcal{F}$ is the PL structure determined by the triangulations in $\mathcal{T}$ (recall that each triangulation determines the same PL structure by Lemma B.3.6). In general, we need not have $\mathcal{T} = \hat{\mathcal{T}}$. For example, if K is a simplicial complex and K' is some subdivision with $K \neq K'$, then we could let $\mathcal{T}$ be the family of triangulations of $|K| = |K'|$ consisting of $(K', \mathrm{id}_{|K|})$ and its subdivisions. In this case, $(K, \mathrm{id}_{|K|}) \notin \mathcal{T}$ even though we know K and K' determine the same PL structure on $|K|$ by Example B.2.7. However, we will have $\mathcal{T} \subset \hat{\mathcal{T}}$, as the elements of $\mathcal{T}$ will be *some* of the PL triangulations of $(X, \mathcal{F})$, and it follows from Proposition 2.5.7 in the main text that the identity map $X \to X$ induces an AT homeomorphism $(X, \mathcal{T}) \to (X, \hat{\mathcal{T}})$. If

$f : (X, \mathcal{T}) \to (Y, \mathcal{S})$ is an AT map, the diagram

$$\begin{array}{ccc} (X,\mathcal{T}) & \xrightarrow{\text{id}} & (X,\hat{\mathcal{T}}) = FG(X,\mathcal{T}) \\ \Big\downarrow f & & \Big\downarrow f \\ (Y,\mathcal{S}) & \xrightarrow{\text{id}} & (Y,\hat{\mathcal{S}}) = FG(Y,\mathcal{S}) \end{array}$$

certainly commutes at the space level, but this is enough to say that it commutes as a diagram of AT spaces because the horizontal maps are AT homeomorphisms. So we see that we have a natural isomorphism $\text{id}_{\mathcal{AT}} \to FG$, as desired. □

B.4 PL Subspaces

We will need just two facts about PL subspaces, contained in the following examples. See [130, Section III.4] for these and further results.

Definition B.4.1 If $(X, \mathcal{F})$ is a PL space and $(X_0, \mathcal{F}_0)$ is another PL space with $X_0 \subset X$, then $(X_0, \mathcal{F}_0)$ is a *PL subspace* of $(X, \mathcal{F})$ if X_0 has the subspace topology it inherits from X and the inclusion $X_0 \hookrightarrow X$ is a PL map.

In this case, $\mathcal{F}_0 = \{(f, P) \in \mathcal{F} \mid f(P) \subset X_0\}$.

Example B.4.2 If $(X, \mathcal{F})$ is a PL space, any open subset X_0 of X is a PL subspace with $\mathcal{F}_0 = \{(f, P) \in \mathcal{F} \mid f(P) \subset X_0\}$. It follows from Theorem B.2.19 that if $T = (K, h)$ is a triangulation of X and X_0 is an open subspace of X then there is a triangulation $S = (L, j)$ of X_0 that "subdivides" T in the sense that $h^{-1}j$ takes every simplex of L linearly and injectively into a simplex of K. △

Example B.4.3 Let $(X, \mathcal{F})$ be a PL space, and let $T = (K, h)$ be a triangulation of X. Suppose that L is a subcomplex of K. Then the restriction $h|_{|L|} : |L| \to X$ is PL by Remark B.2.16 if $|L|$ is given its canonical PL structure so that $h|_{|L|}$ is a PL triangulation. So $h(|L|)$, with its subspace topology and the PL structure it obtains from the triangulation $(L, h|_{|L|})$, is a PL subspace of X.

Conversely, by [130, Lemma 3.7] and its proof (which proves something a bit stronger than the statement of the lemma), if X is a PL space, $X_0 \subset X$ is a *closed* PL subspace, and $T = (K, h)$ is a triangulation of X, then there is a subdivision $T' = (K', h)$ of T and a subcomplex K_0' of K_0 with respect to which $X_0 = h(K_0')$.

So closed PL subspaces of a PL space X correspond to those subsets of X that can be triangulated as subcomplexes in PL triangulations of X. △

B.5 Cones, Joins, and Products of PL Spaces

In this section we demonstrate that basic operations with PL spaces yield new PL spaces.

Lemma B.5.1 *Let X, Y be compact PL spaces. Then the join $X * Y$ is a PL space.*

Proof Let $T_1 = (K, k)$ and $T_2 = (L, \ell)$ be PL triangulations of X and Y respectively. Then K and L are finite simplicial complexes, and the join $K * L$ is a simplicial complex whose simplices are the joins of the simplices of K and L [182, Section 62]. The join map $k * \ell : K * L \to X * Y$ is a triangulation of $X * Y$, and this gives a PL structure on $X * Y$ by Example B.2.6. To see that this structure does not depend on the choice of triangulation for X or Y, first let $T_1' = (K', k)$ be a subdivision of T_1. If the simplex τ of K' is contained in the simplex σ of K and if η is any simplex of L, then $\tau * \eta \subset \sigma * \eta$, so $K' * L$ is a subdivision of $K * L$. Hence we obtain a subdivision $(K' * L, k * \ell)$ of our first triangulation of $X * Y$. By Example B.2.7, both of these triangulations determine the same PL structure on $X * Y$. If (J, j) is another PL triangulation of X, then by Example B.2.21 there are subdivisions (J', j) of (J, j) and (K', k) of (K, k) so that $j^{-1}k$ and $k^{-1}j$ are simplicial isomorphisms. It follows that $(j * \ell)^{-1}(k * \ell) = (j^{-1} * \ell^{-1})(k * \ell) = j^{-1}k * \mathrm{id}_{|L|}$ is a simplicial isomorphism from $K' * L$ to $J' * L$. So by Theorem B.2.23, $(K * L, k * \ell)$ and $(J * L, j * \ell)$ determine the same PL structure on $X * Y$. Making a similar argument with Y, it follows that we obtain a PL structure of $X * Y$ that does not depend on our choices of triangulations for X and Y. □

Corollary B.5.2 *Let X be a compact PL space. Then $\bar{c}X$ and cX are PL spaces.*

Proof The space $\bar{c}X$ is a special case of the preceding lemma with Y being a single point. Then cX is an open subset of $\bar{c}X$ so we can invoke Example B.4.2. □

We next prove that the product of PL spaces is a PL space. In the next section, we will describe how to construct a specific triangulation of a product of simplicial complexes via the Eilenberg–Zilber shuffle product, which is used in Section B.6 in the main body of the text to construct the cross product of chains. The construction of the product triangulation culminates in Theorem B.6.6. Unfortunately, however, the resulting product triangulation does not behave well with respect to subdivisions; see Figure B.2, which will make more sense after reading the next section or consulting Section B.6. Thus we cannot directly emulate our proof of Lemma B.5.1 and must take a different tack, falling back to the abstract definition of PL spaces. However, once we have shown that the product of PL spaces is PL, Theorem B.6.6 shows that the Eilenberg–Zilber shuffle product triangulation is consistent with the product PL structure.

Proposition B.5.3 *Let X and Y be PL spaces. Then $X \times Y$ is a PL space.*

Proof Let $\mathcal{F}_X$ and $\mathcal{F}_Y$ be the respective PL structures on X and Y, and let $(f_X, P_X) \in \mathcal{F}_X$ and $(f_Y, P_Y) \in \mathcal{F}_Y$. Then $(f_X \times f_Y, P_X \times P_Y)$ is a coordinate map to $X \times Y$, using that the product of Euclidean polyhedra is a Euclidean polyhedron by Lemma B.1.28. Suppose $(g_X, Q_X) \in \mathcal{F}_X$ and $(g_Y, Q_Y) \in \mathcal{F}_Y$ are two other convex cells. We will check that $(f_X \times f_Y, P_X \times P_Y)$ is

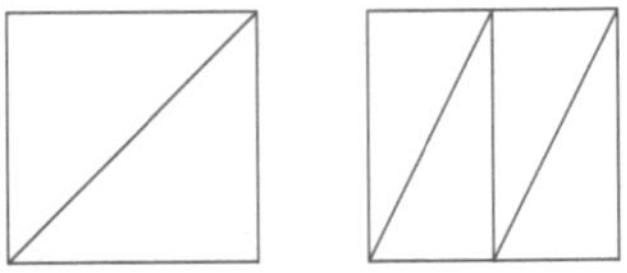

Figure B.2 Two different triangulations of $\Delta^1 \times \Delta^1$. On the left we have the Eilenberg–Zilber shuffle product triangulation of the product of two 1-simplices. On the right, the first 1-simplex has been subdivided into two 1-simplices. The triangulation on the right is not a subdivision of the one on the left.

compatible with $(g_X \times g_Y, Q_X \times Q_Y)$. It will then follow that the collection of all such products of coordinate maps in X and Y constitutes a base for a PL structure on $X \times Y$; it is clear that every point of $X \times Y$ must be in the interior of such a product coordinate map. Suppose that the intersection of

$$(f_X \times f_Y)(P_X \times P_Y) = f_X(P_X) \times f_Y(P_Y)$$

with

$$(g_X \times g_Y)(Q_X \times Q_Y) = g_X(Q_X) \times g_Y(Q_Y)$$

is non-empty. Then it is equal to

$$(f_X(P_X) \cap g_X(Q_X)) \times (f_Y(P_Y) \cap g_Y(Q_Y)).$$

By assumption, there exists a coordinate map (h_X, R_X) in $\mathcal{F}_X$ such that $h_X(R_X) = f_X(P_X) \cap g_X(Q_X)$ and $f_X^{-1} h_X$ and $g_X^{-1} h_X$ are both piecewise linear maps of Euclidean polyhedra, and similarly for Y. So then

$$(h_X \times h_Y)(R_X \times R_Y) = h_X(R_X) \times h_Y(R_Y) = (f_X(P_X) \cap g_X(Q_X)) \times (f_Y(P_Y) \cap g_Y(Q_Y)),$$

and it remains to observe that $(f_X \times f_Y)^{-1}(h_X \times h_Y) = (f_X^{-1} h_X) \times (f_Y^{-1} h_Y)$ is piecewise linear, and similarly replacing f with g. But the product of piecewise linear maps of polyhedra is piecewise linear by Lemma B.1.29. □

We leave the fact that the product of PL manifolds is a PL manifold as an exercise for the reader.

B.6 The Eilenberg–Zilber Shuffle Triangulation of Products

Our goal in this section is to construct an explicit standard triangulation of the product $|K| \times |L|$, where K and L are simplicial complexes. The construction assumes we have chosen partial orderings on the vertices of K and L that restrict to total orderings on each simplex, and the construction depends on this choice. These triangulations are critical in the main body of the text in order to define cross products of singular and PL chains explicitly in Section 5.2, based upon a simplicial cross product we define here in Section B.6.5.

B.6.1 The Definition of the Eilenberg–Zilber Triangulation

We begin by recalling the basics of *shuffles* presented in Section 5.2.

Definition B.6.1 Let p, q be non-negative integers. Then a *(p,q)-shuffle* is a partition of the ordered set $[1, 2, \dots, p+q]$ into two disjoint ordered sets $\mu = [\mu_1, \dots, \mu_p]$ and $\nu = [\nu_1, \dots, \nu_q]$ with $\mu_i < \mu_{i+1}$ for each i and similarly for the ν_j.

Such a partition (μ, ν) tells us how to shuffle together two ordered sets of respective cardinalities p and q to form a new ordered set of cardinality $p + q$: the elements of the first set occupy the spots labeled by the μs and the elements of the second set are placed in the spots corresponding to the νs. So, for example, if we have ordered sets $[A, B, C]$ and $[\alpha, \beta]$, and a $(3, 2)$-shuffle $([2, 3, 5], [1, 4])$, then we can shuffle our sets by this prescription to get the ordered set $[\alpha, A, B, \beta, C]$. Note that the elements of the first set go in order into spots 2, 3, and 5.

Using these shuffles, we construct embeddings $\Delta^{p+q} \to \Delta^p \times \Delta^q$, where each Δ^i is the i-simplex with the fixed standard ordering on its vertices. To fix notation, we let $\Delta^p = [u_0, \dots, u_p]$, $\Delta^q = [v_0, \dots, v_q]$, and $\Delta^{p+q} = [w_0, \dots, w_{p+q}]$. Let $\eta^\mu : \Delta^{p+q} \to \Delta^p$ take the vertex $w_i \in \Delta^{p+q}$ to the vertex $u_j \in \Delta^p$ if $\mu_j \le i < \mu_{j+1}$ (letting $\mu_0 = 0$ and $\mu_{p+1} = p+q+1$). Let $\eta^\nu : \Delta^{p+q} \to \Delta^q$ be defined similarly. We obtain a map $\eta_{\mu\nu} = (\eta^\mu, \eta^\nu) : \Delta^{p+q} \to \Delta^p \times \Delta^q$ by extending linearly from what this map must do to vertices, and it is a linear embedding. We denote the image of $\eta_{\mu\nu}$ by $\delta_{\mu\nu}$. We will show below that the collection of $\delta_{\mu\nu}$ and their faces gives a triangulation of $\Delta^p \times \Delta^q$. We will then use this to construct triangulations $K \times L$ of products of simplicial complexes K and L and to define a simplicial cross product chain map of the form $C_*(K) \otimes C_*(L) \to C_*(K \times L)$.

To better understand the local construction, let us see explicitly where the vertices $\{w_i\}$ of Δ^{p+q} get mapped by $\eta_{\mu\nu}$. Since $\nu_0 = \mu_0 = 0$ by definition, w_0 gets mapped to (u_0, v_0). Now, if $1 \in \mu$, then w_1 gets mapped to (u_1, v_0), and if $1 \in \nu$, then w_1 gets mapped to (u_0, v_1). In general, if $w_i = w_{j+k}$ goes to (u_j, v_k), then w_{i+1} will go to either (u_{j+1}, v_k) or (u_j, v_{k+1}) depending respectively on whether $i + 1$ is in μ or ν.

Another way to think of a (p, q)-shuffle is to imagine a walk on a $p \times q$ grid, where columns are labeled left to right by $\{0, \dots, p\}$ and the rows are labeled bottom to top by $\{0, \dots, q\}$. Then there is a bijection between (p, q)-shuffles and walks along the grid from $(0, 0)$ to (p, q) in which each step must move one unit either up or to the right: on the ith step, if $i \in \mu$ we move to the right and if $i \in \nu$ we move up; conversely, given such a path, if we move right on the ith step then we put $i \in \mu$: and if we move up on the ith step, we put $i \in \nu$. Then the sequence of labels (j, k) of the points of the grid along our path is the sequence of vertices (u_j, v_k) of $\delta_{\mu\nu}$.

To show that the construction so described really does provide a triangulation of $\Delta^p \times \Delta^q$, i.e. that $\Delta^p \times \Delta^q$ is homeomorphic to a simplicial complex whose $(p+q)$-simplices correspond bijectively to the $\delta_{\mu\nu}$ ranging over the index set of shuffles, we will utilize the techniques of Ramras [193]. This involves a specific realization procedure for finite abstract simplicial sets, particularly those associated to partially ordered sets. We review these in the next section.

B.6.2 Realization of Partially Ordered Sets

We first recall the definition of *abstract simplicial complex* [182, Section 3]. Such an object essentially contains all the combinatorial information of a simplicial complex without the specific geometric content.

Definition B.6.2 A *finite abstract simplicial complex* Λ consists of a finite vertex set $V(\Lambda)$ together with a collection of subsets of $V(\Lambda)$ such that (1) if $v \in V(\Lambda)$ then $\{v\} \in \Lambda$, and (2) if $B \subset A$ with $A \in \Lambda$ and $B \neq \emptyset$, then $B \in \Lambda$. The elements of Λ are called the *simplices* of Λ.

As noted, an abstract simplicial complex contains the same combinatorial information as a geometric simplicial complex. In fact, given a geometric simplicial complex K, it determines an abstract simplicial complex whose vertex set is the set of vertices of K, denoted $V(K)$, and whose simplices are the subsets of $V(K)$ spanned by simplices of K. Conversely, every abstract simplicial complex Λ determines a geometric simplicial complex $|\Lambda|$ obtained by taking a copy of Δ^i for each simplex of Λ containing $i+1$ vertices and gluing these together along appropriate faces via the combinatorial data (i.e. if $B \subset A$, $|B| = i$, $|A| = j$, then we glue the copy of Δ^i corresponding to B to the appropriate face of the Δ^j corresponding to A). Here we will describe a specific concrete realization of such a Λ as a subset of Euclidean space.

Let V be a set. We can let the elements $v_i \in V$ be generators of a real vector space isomorphic to $\mathbb{R}^{|V|}$. We identify the v_i with the standard unit basis vectors. Then we can identify $\Delta^{|V|-1}$ with the subset of $\mathbb{R}^{|V|}$ described as

$$\left\{ v \in \mathbb{R}^{|V|} \;\middle|\; v = \sum_{i=1}^{|V|} t_i v_i \text{ with } \sum_{i=1}^{|V|} t_i = 1, t_i \geq 0 \right\}.$$

This is just a description of the convex hull of the generators of $\mathbb{R}^{|V|}$ in barycentric coordinates; see [182, Section 1]. This is an alternative to the other standard embedding of Δ^i in $\mathbb{R}^i$ utilizing also the origin as a vertex; instead we use only the unit basis vectors as vertices.

Now, suppose Λ is any finite abstract simplicial complex with vertex set V. Then its realization $|\Lambda|$ can be realized as a subset of $\Delta^{|V|-1}$, as $\Delta^{|V|-1}$ contains faces corresponding to all possible subsets of V. In particular, if we define the support of $v = \sum_{i=1}^{|V|} t_i v_i \in \Delta^{|V|-1}$ to be $\text{supp}(v) = \{v_i \mid t_i \neq 0\}$, then

$$|\Lambda| = \left\{ v \in \mathbb{R}^{|V|} \;\middle|\; v = \sum_{i=1}^{|V|} t_i v_i \text{ with } \sum_{i=1}^{|V|} t_i = 1, t_i \geq 0, \text{supp}(v) \in \Lambda \right\}.$$

For example, the vectors in $\Delta^{|V|-1}$ with support $\{v_i, v_j\}$, $i \neq j$, are precisely those that lie on the open interval between the basis vectors corresponding to v_i and v_j.

We will be particularly interested in abstract simplicial complexes coming from finite partially ordered sets. If P is a finite partially ordered set, we let $\Lambda(P)$ be the abstract simplicial complex whose vertex set is P and whose k-simplices are the subsets consisting of totally ordered chains of the form $x_0 < x_1 < \cdots < x_k$ with each $x_i \in P$. As an example, if our partially ordered set is $N = \{0, 1, \ldots, n\}$ with its standard order coming from the integers, then every subset corresponds to a totally ordered chain. In this case, we write $\Lambda(N) = \underline{\Delta}^n$ and call it the abstract n-simplex[4]. Observe that $|\underline{\Delta}^n| = \Delta^n$.

We will need one other geometric construction associated to a partially ordered set P: Let

$$\lfloor P \rfloor = \left\{ z \in \mathbb{R}^{|P|} \;\middle|\; z = \sum_{i=1}^{n} \lambda_i x_i \text{ for some } n \text{ and some } x_1 < \cdots < x_n \in P, \lambda_i > 0, \sum_{i=1}^{n} \lambda_i \leq 1 \right\}.$$

Here the $x_i \in P$ are identified with standard basis vectors in $\mathbb{R}^{|P|}$, but we note that it is not necessary for all elements of P to occur in each sum. To understand this set, let $\lfloor P \rfloor_m$ denote the set of elements of $\lfloor P \rfloor$ for which $n = m$. Then $\lfloor P \rfloor_1$ is the union of the $|P|$ unit segments from the origin to the x_i, omitting the origin; $\lfloor P \rfloor_2$ is the union of the 2-simplices spanned by the origin and pairs of elements $x_i, x_j \in P$ with $x_i < x_j$, minus their intersections with the rays from the origin along the standard basis vectors (if x_i, x_j are not comparable in the partial order, they don't contribute to $\lfloor P \rfloor_2$); and so on. The $\lfloor P \rfloor_m$ are disjoint, so if $z \in \lfloor P \rfloor$, then $z \in \lfloor P \rfloor_m$ for some unique m that we denote $\nu(z)$. Additionally, each $\lfloor P \rfloor_m$ is partitioned according to the choice of chain $x_1 < \cdots < x_m$ of length m. So, since the elements of P represent linearly independent vectors in $\mathbb{R}^{|P|}$, each $z \in \lfloor P \rfloor$ determines uniquely a chain $x_1 < \cdots < x_{\nu(z)}$ and an ordered set of positive real numbers $[\lambda_1(z), \ldots, \lambda_{\nu(z)}(z)]$ so that $z = \sum_{i=1}^{\nu(z)} \lambda_i(z) x_i$. Let $\lambda(z) = \sum_{i=1}^{\nu(z)} \lambda_i(z)$. We observe that

$$|\Lambda(P)| = \{z \in \lfloor P \rfloor \mid \lambda(z) = 1\}.$$

[4] It is probably more common to use Δ^n for the abstract simplex and $|\Delta^n|$ for its realization, but since we have been using each of these notations for various other purposes throughout the book, we'll go with $\underline{\Delta}^n$ and Δ^n in this section.

B.6.3 Products of Partially Ordered Sets and Their Product Triangulations

Now, suppose P, Q are two finite partially ordered sets. Let the product $P \times Q$ have the partial ordering defined by $(x, y) \leq (u, w)$ if $x \leq u$ in P and $y \leq w$ in Q. We will show that $|\Lambda(P \times Q)|$ is PL homeomorphic to $|\Lambda(P)| \times |\Lambda(Q)|$. Note that we know that $|\Lambda(P)| \times |\Lambda(Q)|$ is a PL space by Proposition B.5.3.

In the case where $P = \{x_0, \ldots, x_p\}$ with $x_i < x_j$ when $i < j$, so that $\Lambda(P) = \underline{\Delta}^p$, and $Q = \{y_0, \ldots, y_q\}$ with $y_i < y_j$ when $i < j$, so that $\Lambda(Q) = \underline{\Delta}^q$, this homeomorphism therefore provides a triangulation of $\Delta^p \times \Delta^q$ by the simplicial complex realization $|\Lambda(P \times Q)|$. We will show in Corollary B.6.4 that this triangulation is the same as the one described in the preceding section using the Eilenberg–Zilber shuffles.

We prove the claimed homeomorphism following [193, Lemma 2.2.9]:

Lemma B.6.3 *For finite partially ordered sets P and Q, there is a PL homeomorphism $|\Lambda(P \times Q)| \cong |\Lambda(P)| \times |\Lambda(Q)|$.*

Proof The projections $P \times Q \to P$ and $P \times Q \to Q$ induce simplicial maps of the abstract simplicial complexes, and hence piecewise linear maps of the geometric realizations $|\Lambda(P \times Q)| \to |\Lambda(P)|$ and $|\Lambda(P \times Q)| \to |\Lambda(Q)|$. Together this gives a piecewise linear map $f : |\Lambda(P \times Q)| \to |\Lambda(P)| \times |\Lambda(Q)|$. In fact, if $w_i = (x_i, y_i)$ is a vertex of $\Lambda(P \times Q)$, then f takes w_i to the point of $|\Lambda(P)| \times |\Lambda(Q)|$ also represented by the pair (x_i, y_i), where now each coordinate is a vertex of $\Lambda(P)$ or $\Lambda(Q)$. So if $(x_1, y_1) < \cdots < (x_k, y_k)$ is a simplex of $|\Lambda(P \times Q)|$, then f just takes this simplex linearly into the product of the simplices of $|\Lambda(P)|$ and $|\Lambda(Q)|$ spanned respectively by the sets of vertices $\{x_i\}$ and $\{y_i\}$. Even more specifically, if $z \in |\Lambda(P \times Q)|$ has the form $z = \sum_{i=1}^n \lambda_i w_i \in \mathbb{R}^{|P \times Q|}$, then

$$f(z) = \left(\sum_{i=1}^n \lambda_i x_i, \sum_{i=1}^n \lambda_i y_i \right) \in \mathbb{R}^{|P|} \times \mathbb{R}^{|Q|}.$$

Note that we only have $x_i \leq x_j$ for $i < j$, not necessarily $x_i < x_j$, and similarly for the y_i.

By Corollary B.2.22, to show that f is a PL homeomorphism, we need only show that f is a bijection.

Proof that f is surjective Suppose it is true that every $(x, y) \in |\Lambda(P)| \times |\Lambda(Q)|$ can be written in the form

$$(x, y) = \left(\sum_{i=1}^n \lambda_i x_i, \sum_{i=1}^n \lambda_i y_i \right),$$

for some n, some $\lambda_i \in (0, 1]$ with $\sum_{i=1}^n \lambda_i = 1$, and some $x_1 \leq \cdots \leq x_n \in P$, and $y_1 \leq \cdots \leq y_n \in Q$, allowing that the x_i and y_i are not necessarily unique. Then we'll have $(x_i, y_i) \leq (x_{i+1}, y_{i+1})$ for all $1 \leq i < n$, so that the collection $\{(x_i, y_i)\}$ spans a simplex of $\Lambda(P \times Q)$, and hence $\sum_{i=1}^n \lambda_i (x_i, y_i) \in |\Lambda(P \times Q)|$. Furthermore, then $f\left(\sum_{i=1}^n \lambda_i (x_i, y_i)\right) = \left(\sum_{i=1}^n \lambda_i x_i, \sum_{i=1}^n \lambda_i y_i\right) = (x, y)$. So this would imply that f is surjective. We will show that any $(x, y) \in |\Lambda(P)| \times |\Lambda(Q)|$ can indeed be written in this form. Note that the catch is that we need to have the same sequence of λ_is in each factor.

In fact, we will show that if $(x, y) \in \lfloor P \rfloor \times \lfloor Q \rfloor$ and $\lambda(x) = \lambda(y)$, then there exist an integer n, real numbers $\lambda_1, \ldots, \lambda_n \in (0, 1]$, and elements $x_1 \leq \cdots \leq x_n \in P$ and $y_1 \leq \cdots \leq y_n \in Q$ such that

$$(x, y) = \left(\sum_{i=1}^{n} \lambda_i x_i, \sum_{i=1}^{n} \lambda_i y_i \right)$$

and such that $\sum_{i=1}^{n} \lambda_i = \lambda(x) = \lambda(y)$. As $|\Lambda(P)| = \{x \in \lfloor P \rfloor \mid \lambda(x) = 1\}$ and $|\Lambda(Q)| = \{y \in \lfloor Q \rfloor \mid \lambda(y) = 1\}$, this will imply the desired fact for elements of $|\Lambda(P)| \times |\Lambda(Q)|$.

Let $\nu(x, y) = \nu(x) + \nu(y)$, and let us induct on this number. The minimum value for $\nu(x, y)$ is 2, when $\nu(x) = \nu(y) = 1$. In this case we simply have $x \in \lfloor P \rfloor_1$ and $y \in \lfloor P \rfloor_1$. So, by definition, $x = \lambda(x) x_1$ for some $x_1 \in P$ and $y = \lambda(y) y_1$ for some $y_1 \in Q$. But then we just let $n = 1$ and $\lambda_1 = \lambda(x) = \lambda(y)$, and we have $(x, y) = (\lambda_1 x_1, \lambda_1 y_1)$, so we are done with the base case.

Now assume we have proven the claim for $\nu(x, y) \leq r$, and choose (x, y) with $\nu(x, y) = r + 1$ and $\lambda(x) = \lambda(y)$. Since $x \in \lfloor P \rfloor$ and $y \in \lfloor Q \rfloor$, we may write $x = \sum_{i=1}^{\nu(x)} t_i x_i$ for some $x_i \in P$ with $x_1 \leq \cdots \leq x_{\nu(x)}$ and all $t_i > 0$, and similarly $y = \sum_{i=1}^{\nu(y)} s_i y_i$ with $y_1 \leq \cdots \leq y_{\nu(y)}$ and all $s_i > 0$. By assumption, $\nu(x) + \nu(y) = r + 1$. Furthermore, as x is contained in the subspace of $\mathbb{R}^{|P|}$ spanned by $x_1, \ldots, x_{\nu(x)}$ in the usual linear algebra sense, we note that this implies that no other element of P can appear nontrivially in a linear combination representing x; similarly for y.

Now assume without loss of generality that $t_1 \leq s_1$, and write

$$(x', y') = (x, y) - (t_1 x_1, t_1 y_1) = \left(\sum_{i=2}^{\nu(x)} t_i x_i, (s_1 - t_1) y_1 + \sum_{i=2}^{\nu(y)} s_i y_i \right).$$

We now have $\nu(x', y') \leq r$ (it could be $r - 1$ if $s_1 = t_1$). Furthermore, $\lambda(x') = \sum_{i=2}^{\nu(x)} t_i = \lambda(x) - t_1$, and $\lambda(y') = (s_1 - t_1) + \sum_{i=2}^{\nu(y)} s_i = (\sum_{i=1}^{\nu(y)} s_i) - t_1 = \lambda(y) - t_1 = \lambda(x) - t_1$. So by induction hypothesis[5], there exist some integer n, real numbers $\lambda_1, \ldots, \lambda_n \in (0, 1]$, $x'_1 \leq \cdots \leq x'_n \in P$, and $y'_1 \leq \cdots \leq y'_n \in Q$ such that

$$(x', y') = \left(\sum_{i=1}^{n} \lambda_i x'_i, \sum_{i=1}^{n} \lambda_i y'_i \right).$$

So now,

$$(x, y) = (t_1 x_1 + x', t_1 y_1 + y') = \left(t_1 x_1 + \sum_{i=1}^{n} \lambda_i x'_i, t_1 y_1 + \sum_{i=1}^{n} \lambda_i y'_i \right)$$

has the desired form, provided $x_1 \leq x'_1$ and $y_1 \leq y'_1$. But since the λ_i are all non-zero, each

[5] In case the reader has been wondering, it is at this point that we use the possibility $\lambda(z) < 1$, which justifies our having introduced the sets $\lfloor P \rfloor$.

x'_i must be one of the $\{x_i\}$ (by the earlier observation that no element not in $\{x_i\}$ can appear nontrivially in a linear combination representing x), so as x_1 is the least of the $\{x_i\}$ in the partial order, we also have $x_1 \leq x'_1$, and similarly $y_1 \leq y'_1$. Also, we see that the sum of the coefficients on each side is again $\lambda(x) = \lambda(y)$, as required.

Proof that f is injective Let $z, z' \in |\Lambda(P \times Q)| \subset \lfloor P \times Q \rfloor$. Then $z = \sum_{i=1}^{\nu(z)} \lambda_i(x_i, y_i)$ and $z' = \sum_{i=1}^{\nu(z')} \lambda'_i(x'_i, y'_i)$ for some unique choices $\lambda_i, \lambda'_i > 0$, $\sum \lambda_i = \sum \lambda'_i = 1$, $(x_i, y_i) < (x_{i+1}, y_{i+1})$, and $(x'_i, y'_i) < (x'_{i+1}, y'_{i+1})$ for all relevant i; notice that we assume that vertices are not repeated within each of these representations of z and z'. We will show that if $f(z) = f(z')$, then $z = z'$.

If $f(z) = f(z')$, then $\sum_{i=1}^{\nu(z)} \lambda_i x_i = \sum_{i=1}^{\nu(z')} \lambda'_i x'_i$ and $\sum_{i=1}^{\nu(z)} \lambda_i y_i = \sum_{i=1}^{\nu(z')} \lambda'_i y'_i$. Since all the λ_i, λ'_i are > 0 and the elements of P and Q form respective bases of the Euclidean spaces containing $|\Lambda(P)|$ and $|\Lambda(Q)|$, it follows that we must have $\{x_1, \ldots, x_{\nu(z)}\} = \{x'_1, \ldots, x'_{\nu(z')}\}$ and $\{y_1, \ldots, y_{\nu(z)}\} = \{y'_1, \ldots, y'_{\nu(z')}\}$ as sets. These lists might each contain repeated elements, but we must have $x_1 = x'_1$ and $y_1 = y'_1$ since these are each the smallest elements in their respective sets under the order.

Our next goal is to begin to show that $\lambda_i = \lambda'_i$ for each i; we do not know yet that the λ_i and λ'_i have the same indexing sets, but this will follow from the argument. Since $(x_1, y_1) < (x_2, y_2)$, assume without loss of generality that $x_1 < x_2$ (otherwise $y_1 < y_2$ and we reverse the roles of P and Q in the following argument). Since $\sum_{i=1}^{\nu(z)} \lambda_i x_i = \sum_{i=1}^{\nu(z')} \lambda'_i x'_i$, we must have that $\lambda_1 = \sum_{\{i | x'_i = x_1\}} \lambda'_i \geq \lambda'_1$. If $x'_1 \neq x'_2$, then since $x'_1 \leq x'_2 \leq \cdots$, we would have only $x'_1 = x_1$, and so $\lambda_1 = \lambda'_1$. If $x'_1 = x'_2$, then $y'_1 < y'_2$, and by a symmetric argument to the above, $\lambda'_1 = \sum_{\{i | y_i = y'_1\}} \lambda_i \geq \lambda_1$. But we already know $\lambda_1 \geq \lambda'_1$, so $\lambda_1 = \lambda'_1$. This argument also implies that while $x_1 = x'_1$, no x_i or x'_i with $i > 1$ is equal to this element of P, and, symmetrically, while $y_1 = y'_1$, no y_i or y'_i with $i > 1$ is equal to this element of Q. It now follows that $x_2 = x'_2$ are the smallest terms in $\{x_2, \ldots, x_{\nu(z)}\}$ and $\{x'_2, \ldots, x'_{\nu(z')}\}$ and similarly for Q, and we can run the same argument inductively to eventually show that $\nu(z) = \nu(z')$, that $\lambda_i = \lambda'_i$ for all i, and that $x_i = x'_i$ and $y_i = y'_i$ for all i. So $z = z'$, as claimed.

□

Now we connect the preceding lemma to the Eilenberg–Zilber shuffle construction.

Corollary B.6.4 *If Δ^p and Δ^q are simplices with ordered vertex sets, then the $(p+q)$-simplices $\delta_{\mu\nu} \subset \Delta^p \times \Delta^q$ with vertices determined by the (p,q)-shuffles (μ, ν) are the $(p+q)$-simplices of a triangulation of $\Delta^p \times \Delta^q$ compatible with the product PL structure.*

Proof Let us take $P = [x_0, \ldots, x_p]$ with $x_i < x_j$ when $i < j$ and $Q = [y_0, \ldots, y_q]$ with $y_i < y_j$ when $i < j$. Then $\Lambda(P) = \underline{\Delta}^p$ and $\Lambda(Q) = \underline{\Delta}^q$. The homeomorphism of Lemma B.6.3 therefore provides a triangulation of $\Delta^p \times \Delta^q$ by the simplicial complex realization $|\Lambda(P \times Q)|$.

To see that this triangulation is the same as that given using the shuffles, we note by definition that the $(p+q)$-simplices of $P \times Q$ correspond to chains of elements $w_0 < w_1 < \cdots < w_{p+q} \in P \times Q$. Each $w_i \in P \times Q$ has the form $w_i = (x_{j_i}, y_{k_i})$ for $x_{j_i} \in P$ and $y_{k_i} \in Q$, and to have $w_i < w_{i+1}$, we must have $x_{j_i} \leq x_{j_{i+1}}$ and $y_{k_i} \leq y_{k_{i+1}}$, not both equalities. But there are only $p+1$ elements in P and $q+1$ elements in Q, so the only way to obtain a chain of length $p+q+1$ with these properties is to have $w_0 = (x_0, y_0)$, $w_{p+q} = (x_p, y_q)$, and, for each i, the vertex $(x_{j_{i+1}}, y_{k_{i+1}})$ is either (x_{j_i+1}, y_{k_i}) or (x_{j_i}, y_{k_i+1}). Considering our "walk in the plane" description of a shuffle, we see that we thus have a bijection between chains of length $p+q+1$ and (p,q)-shuffles! Furthermore, the homeomorphism of the lemma takes each w_i to the corresponding product vertex $(x_{j_i}, y_{k_i}) \in \Delta^p \times \Delta^q$, and it takes the $(p+q)$-simplex of $|\Lambda(P \times Q)|$ spanned by a $(p+q+1)$-chain of vertices linearly into $\Delta^p \times \Delta^q$ according to these vertex maps. But the image of such an embedding is the corresponding $\delta_{\mu\nu}$ by definition. So, the $\delta_{\mu\nu}$ are precisely the $(p+q)$-simplices of the triangulation described in the lemma. □

So we can triangulate products of simplices. In fact, as every finite simplicial complex is the realization of a finite partially ordered set, Lemma B.6.3 shows that every product of two finite simplicial complexes can be triangulated, employing the shuffle construction for each pair of simplices. However, as we are interested in possibly infinite simplicial complexes, we will take a slightly different route in the next section toward triangulating products of simplicial complexes by investigating the local situation a bit further and discussing how to piece things together.

B.6.4 Triangulations of Products of Simplicial Complexes and PL Spaces

We next notice the following easy corollary of Lemma B.6.3 and its proof.

Corollary B.6.5 *If P and Q are finite partially ordered sets and $A \subset P$ and $B \subset Q$ are subsets inheriting the partial ordering, then $|\Lambda(A \times B)| \subset |\Lambda(P \times Q)|$ is a subcomplex, and the triangulating homeomorphism $|\Lambda(P \times Q)| \to |\Lambda(P)| \times |\Lambda(Q)|$ restricts to the triangulating homeomorphism $|\Lambda(A \times B)| \to |\Lambda(A)| \times |\Lambda(B)|$.*

Proof By Lemma B.6.3, the subspace $|\Lambda(A)| \times |\Lambda(B)|$ is homeomorphic to the simplicial complex $|\Lambda(A \times B)|$. But, more than this, the construction of the proof demonstrates that this homeomorphism is compatible with the larger one $|\Lambda(P \times Q)| \cong |\Lambda(P)| \times |\Lambda(Q)|$: Clearly points in $|\Lambda(P \times Q)|$ that involve only the vertices in $A \times B$ have their image in $|\Lambda(A)| \times |\Lambda(B)|$ under the constructed homeomorphism $|\Lambda(P \times Q)| \to |\Lambda(P)| \times |\Lambda(Q)|$, and the restriction of the proof of Lemma B.6.3 shows that the induced map $|\Lambda(A \times B)| \to |\Lambda(A)| \times |\Lambda(B)|$ is surjective. □

Now suppose, in particular, that F is a face of $\underline{\Delta}^p$ and that G is a face of $\underline{\Delta}^q$. Then F and G correspond to respective subsets of the partially ordered sets $\{0, \ldots, p\}$ and $\{0, \ldots, q\}$. Let $F \times G$ be the abstract simplicial complex associated to the product of these corresponding partially ordered subsets. Then $F \times G$ is a subcomplex of $\underline{\Delta}^p \times \underline{\Delta}^q$ and, by the corollary, the PL homeomorphism $|\Lambda(\underline{\Delta}^p \times \underline{\Delta}^q)| \cong \Delta^p \times \Delta^q$ restricts to provide a triangulation $|F \times G| \cong |F| \times |G| \subset \Delta^p \times \Delta^q$.

In particular, let $F_0, \dots, F_p$ be the $(p-1)$-dimensional faces of $\underline{\Delta}^p$ and $G_0, \dots, G_q$ the $(q-1)$-dimensional faces of $\underline{\Delta}^q$. Then the restriction of $f : |\underline{\Delta}^p \times \underline{\Delta}^q| \cong \Delta^p \times \Delta^q$ to each $|F_i \times \underline{\Delta}^q|$ and $|\underline{\Delta}^p \times G_j|$ provides the triangulations of the $|F_i| \times \Delta^q$ and $\Delta^p \times |G_j|$. Collectively these give a triangulation of $((\partial\Delta^p) \times \Delta^q) \cup (\Delta^p \times \partial\Delta^q) \cong \partial(\Delta^p \times \Delta^q)$.

Now suppose we have simplicial complexes K and L with prescribed vertex partial orderings that restrict to total orderings on each simplex. If K comprises the simplices σ_i and L comprises the simplices τ_j, then $|K| \times |L|$ is the union of all the products $\sigma_i \times \tau_j$. For any fixed i, j, Corollary B.6.4 implies that we can use the Eilenberg–Zilber shuffle construction to triangulate $\sigma_i \times \tau_j$. Suppose then that σ_i' and τ_j' are two other simplices of K and L respectively. Then

$$(\sigma_i \times \tau_j) \cap (\sigma_i' \times \tau_j') = (\sigma_i \cap \sigma_i') \times (\tau_j \cap \tau_j').$$

But from the definition of a simplicial complex, $\sigma_i \cap \sigma_i'$ and $\tau_j \cap \tau_j'$ are each themselves simplices of K and L, say s and t respectively. From Corollary B.6.5, the set $s \times t$ is triangulated as a subcomplex of both $\sigma_i \times \tau_j$ and $\sigma_i' \times \tau_j'$. Now here's the key observation: these two subcomplex triangulations agree. To see this, we note that the Eilenberg–Zilber construction or, equivalently, the argument of Lemma B.6.3 only really depends on the vertex orderings. Once we know which vertices in which order determine the simplices of the product triangulation, we know everything. And the vertex order on s is the same whether it is restricted from the vertex order of σ_i or the vertex order of σ_i', and similarly for t.

It follows that the local triangulations of the $\sigma_i \times \tau_j$ are compatible at their intersections and so piece together to form a global simplicial complex that we denote by $K \times L$ with underlying space $|K \times L| = |K| \times |L|$. We claim that the PL structure on $|K \times L|$ coming from $K \times L$ is the same as the product PL structure on $|K| \times |L|$ given by Proposition B.5.3. In fact, we can state this more generally for arbitrary PL spaces as follows:

Theorem B.6.6 *Let X and Y be PL spaces with respective PL triangulations (K, h) and (L, j). Suppose we place orderings on the vertices of K and L (or partial orderings such that the vertices of each simplex are totally ordered). Then there is a simplicial complex with underlying space $|K| \times |L|$ and with simplicial complex structure given on each $\sigma \times \tau$, for each $\sigma \in K$ and $\tau \in L$, by the Eilenberg–Zilber shuffle construction. If we call this simplicial complex $K \times L$, then $(K \times L, h \times j)$ is a PL triangulation of $X \times Y$ with its product PL structure. In particular, taking h and j to be identity maps, we see that the product simplicial complex structure is compatible with the product PL structure on $|K| \times |L|$.*

Proof The existence of the simplicial complex $K \times L$ has been argued just above. To consider the PL structures, let $\mathcal{B}_K$ be the base for X determined by restricting h to each of the finite subcomplexes of K; see Example B.2.8. Define $\mathcal{B}_L$ similarly for Y. By the proof of Proposition B.5.3, if $(f, M) \in \mathcal{B}_K$ and $(g, N) \in \mathcal{B}_L$, then $(f \times g, M \times N)$ is a coordinate map for $X \times Y$, and in fact the collection of all such product coordinate maps is a base for the PL structure on $X \times Y$. Call this base $\mathcal{B}_K \times \mathcal{B}_L$, and let $\mathcal{B}_{K\times L}$ denote the set of coordinate maps obtained by restricting $h \times j$ to finite subcomplexes of $K \times L$.

Each simplex of $K \times L$ must be contained in the product of a simplex of K and a simplex of L, and, conversely, every product of simplices of K and L in $|K| \times |L|$ is a union of simplices of $K \times L$. So, in particular, for any M, N as above, $M \times N$ is a union of simplices of $K \times L$. Thus the coordinate map $(f \times g, M \times N) \in \mathcal{B}_K \times \mathcal{B}_L$ is also a coordinate map in $\mathcal{B}_{K\times L}$. In other words, $\mathcal{B}_K \times \mathcal{B}_L \subset \mathcal{B}_{K\times L}$. But these are each bases for a PL structure on $X \times Y$, so by Remark B.2.5 they determine the same PL structure. The theorem follows. □

B.6.5 The Simplicial Cross Product

Now that we have carefully studied the geometry of product triangulations, we can shift to algebraic topology and the cross product of simplicial chains. If K and L are simplicial complexes with chosen vertex partial orderings that restrict to total orderings on each simplex and if we again let $K \times L$ denote the simplicial complex given by the Eilenberg–Zilber shuffle product triangulation of $|K| \times |L|$, then we would like to have a chain map[6]

$$\bowtie\colon C_*(K) \otimes C_*(L) \to C_*(K \times L).$$

A review of the oriented simplicial chain complexes C_* can be found in Section 3.2.

To define $\bowtie$, we will define $\sigma \bowtie \tau$ for σ a p-simplex of K and τ a q-simplex of L and then extend bilinearly. We assume each simplex is given the orientation determined by its vertex ordering in K or L. To define $\sigma \bowtie \tau$, we again let $\delta_{\mu\nu}$ be the $(p+q)$-simplex corresponding to the (p, q)-shuffle (μ, ν) in the shuffle product triangulation of $\sigma \times \tau$, which, abusing notation, we also think of as an oriented simplex in $C_{p+q}(K \times L)$ oriented by its vertex ordering. Furthermore, let $\operatorname{sgn}(\mu, \nu)$ denote the sign of the permutation from $[1, 2, \dots, p+q]$ to $[\mu_1, \mu_2, \dots, \mu_p, \nu_1, \nu_2, \dots, \nu_q]$, i.e. 1 if the permutation is even and -1 if the permutation is odd. We define the *simplicial cross product* to be

$$\sigma \bowtie \tau = \sum \operatorname{sgn}(\mu, \nu)\delta_{\mu\nu},$$

where the sum is over all (p, q)-shuffles (μ, ν).

Our goal in the rest of this section is to prove the following proposition.

Proposition B.6.7 *The map* $\bowtie\colon C_*(K) \otimes C_*(L) \to C_*(K \times L)$ *is a chain map.*

Proof As $\bowtie$ is bilinear by definition, we only need to check the behavior with respect to boundaries. Let $\sigma = [x_0, \dots, x_p]$ be a simplex of K with vertices ordered as indicated, and similarly let $\tau = [y_0, \dots, y_q]$ be a simplex of L. We also use σ and τ to stand for their corresponding elements in $C_p(K)$ and $C_q(L)$. Throughout this proof, $\times$ will always denote

[6] We use the symbol $\bowtie$ here to avoid conflict with the many other related uses of $\times$ in close proximity and because it looks somewhat like $\times$ while also being a picture of a simplicial complex!

the product of spaces, given their Eilenberg–Zilber shuffle triangulation, and $\bowtie$ will denote the algebraic product.

We continue to write $\sigma \bowtie \tau = \sum \operatorname{sgn}(\mu, \nu)\delta_{\mu\nu}$. We must show that $\partial(\sigma \bowtie \tau) = (\partial\sigma) \bowtie \tau + (-1)^p \sigma \bowtie (\partial\tau)$.

Each $\delta_{\mu\nu}$ corresponds to an embedding of a $(p+q)$-simplex in $\sigma \times \tau$ with vertices of the form (x_i, y_j). When we look at a face $\mathcal{F}$ of $\delta_{\mu\nu}$ obtained by removing a vertex, there are two possibilities:

1. There is some x_i or y_j that no longer appears in any vertex of $\mathcal{F}$. Suppose for specificity that it is x_k. Then all vertices of $\mathcal{F}$ lie in the product $F_k \times \tau$, where $F_k = [x_0, \dots, \hat{x}_k, \dots, x_p]$ is the kth face of σ. Thus in the shuffle triangulation of $\sigma \times \tau$, we have $\mathcal{F} \subset F_k \times \tau \subset \partial(\sigma \times \tau)$, so $\mathcal{F}$ is a $(p+q-1)$-simplex lying on the boundary of $\sigma \times \tau$.
2. Each x_i and y_j appears in some vertex of $\mathcal{F}$. Suppose (x_k, y_l) is the vertex of $\delta_{\mu\nu}$ that is removed to create $\mathcal{F}$. Note that this can't be (x_0, y_0), since the next vertex must be either (x_0, y_1) or (x_1, y_0), and so if we remove (x_0, y_0), then either x_0 or y_0 will not occur in any vertex of $\mathcal{F}$; similarly, the removed vertex cannot be (x_p, y_q). So the removed vertex (x_k, y_l) must have a predecessor and successor among the vertices of $\delta_{\mu\nu}$. The next vertex in $\delta_{\mu\nu}$ is either (x_{k+1}, y_l) or (x_k, y_{l+1}), and the preceding vertex is either (x_{k-1}, y_l) or (x_k, y_{l-1}). But if the sequence is $(x_{k-1}, y_l), (x_k, y_l), (x_{k+1}, y_l)$, then $\mathcal{F}$ could not have x_k in a vertex, and similarly there would be a contradiction if the sequence were $(x_k, y_{l-1}), (x_k, y_l), (x_k, y_{l+1})$. So the only two possible subsequences in $\delta_{\mu\nu}$ are $(x_{k-1}, y_l), (x_k, y_l), (x_k, y_{l+1})$ or $(x_k, y_{l-1}), (x_k, y_l), (x_{k+1}, y_l)$. Suppose it is the former; we could use equivalent arguments to those below if it is the latter. Replacing this triple of vertices in $\delta_{\mu\nu}$ with $(x_{k-1}, y_l), (x_{k-1}, y_{l+1}), (x_k, y_{l+1})$ while leaving the other vertices unchanged gives us another $(p+q)$-simplex $\delta_{\mu'\nu'}$ with the same sequence of vertices except for the swap of (x_k, y_l) for (x_{k-1}, y_{l+1}). Thus $\delta_{\mu\nu}$ and $\delta_{\mu'\nu'}$ share the common face $\mathcal{F}$ obtained by removing the vertex (x_k, y_l) from $\delta_{\mu\nu}$ and (x_{k-1}, y_{l+1}) from $\delta_{\mu'\nu'}$. Furthermore, (x_k, y_l) and (x_{k-1}, y_{l+1}) are the only possible vertices of $\sigma \times \tau$ that are greater than (x_{k-1}, y_l) and less than (x_k, y_{l+1}) in the partial order, and so these are the only δs with $\mathcal{F}$ as a face.

Now let us compare $\operatorname{sgn}(\mu, \nu)$ with $\operatorname{sgn}(\mu', \nu')$. To do this, we continue to suppose that $\delta_{\mu\nu}$ contains the vertex sequence $(x_{k-1}, y_l), (x_k, y_l), (x_k, y_{l+1})$; again, if it contains $(x_k, y_{l-1}), (x_k, y_l), (x_{k+1}, y_l)$, there is an analogous argument to the following. Further, let us suppose that these correspond to the vertices w_{i-1}, w_i, and w_{i+1} of $\delta_{\mu\nu} = [w_0, \dots, w_{p+q}]$. As the x subscript increases going from w_{i-1} to w_i and the y subscript increases going from w_i to w_{i+1}, this means that $i \in \mu$ and $i+1 \in \nu$. On the other hand, $\delta_{\mu'\nu'}$ contains the sequence $(x_{k-1}, y_l), (x_{k-1}, y_{l+1}), (x_k, y_{l+1})$, and these are the vertices w'_{i-1}, w'_i and w'_{i+1}. As now the y subscript increases first, we have $i \in \nu'$ and $i+1 \in \mu'$. So if we write out the two sequences $(\mu_1, \dots, \mu_p, \nu_1, \dots, \nu_q)$ and $(\mu'_1, \dots, \mu'_p, \nu'_1, \dots, \nu'_q)$, with the μ, μ', ν, and ν' elements in order, the only difference between these two sequences is the interchange of i and $i+1$, with i moving from the μ

list to the ν' list and $i+1$ moving from the ν list to the μ' list. And because i and $i+1$ are adjacent, they simply move to each other's spots. Hence the sequences differ only by a transposition and it follows that $\text{sgn}(\mu, \nu) = -\text{sgn}(\mu', \nu')$. We see also that if $\mathcal{F}$ is the ith face of $\delta_{\mu\nu}$ it is also the ith face of $\delta_{\mu'\nu'}$, and so $\mathcal{F}$ occurs with the same sign in $\partial\delta_{\mu\nu}$ and $\partial\delta_{\mu'\nu'}$ and so opposite signs in $\partial(\text{sgn}(\mu, \nu)\delta_{\mu\nu})$ and $\partial(\text{sgn}(\mu', \nu')\delta_{\mu'\nu'})$. So altogether $\mathcal{F}$ appears with coefficient 0 in $\partial(\sum \text{sgn}(\mu, \nu)\delta_{\mu\nu})$. As mentioned already, the argument when we have a sequence $(x_k, y_{l-1}), (x_k, y_l), (x_{k+1}, y_l)$ in $\delta_{\mu\nu}$ is analogous.

So, our arguments so far show that $\partial(\sigma \bowtie \tau)$ is contained in $\partial(\sigma \times \tau)$. Of course $(\partial\sigma) \bowtie \tau + (-1)^p \sigma \bowtie (\partial\tau)$ is also supported in $\partial(\sigma \times \tau)$, so now we must show that each oriented $(p+q-1)$-simplex of $\partial(\sigma \times \tau)$ arises with the same coefficient in each expression.

For this, we will employ the following observation: As $\partial(\sigma \bowtie \tau)$ is supported in $\partial(\sigma\times\tau)$, the chain $\sigma \bowtie \tau$ represents an element of $H_{p+q}(\sigma \times \tau, \partial(\sigma \times \tau))$. In fact, as each $(p+q)$-simplex $\delta_{\mu\nu}$ has coefficient ± 1 in $\sigma \bowtie \tau$, it must be a generator of this homology group and so a fundamental class for some orientation of $\sigma \times \tau$, which is topologically a $(p+q)$-dimensional ball[7]. Therefore, $\partial(\sigma \bowtie \tau)$ is a generator for the degree $(p+q-1)$-homology of $\partial(\sigma \times \tau)$, which is topologically a $(p+q-1)$- sphere; see Proposition 8.3.5. So each coefficient of a simplex of $\partial(\sigma \bowtie \tau)$ is ± 1, and the sign on any $(p+q-1)$-simplex of this cycle must determine the signs of all the other simplices because they must all be consistent with some choice of orientation. Similarly, for each ordered/oriented $(p-1)$-simplex F of σ, the chain $F \bowtie \tau$ is a fundamental class for $F \times \tau$ with some orientation, and so the sign of one of its $(p+q-1)$-simplices determines the signs of all the others; analogously for each $\sigma \bowtie G$ with G a $(q-1)$-simplex of τ. So it will be sufficient to show that for each F, respectively G, there is one $(p+q-1)$-simplex of $F \times \tau$, respectively $\sigma \times G$, on which the coefficient signs in $\partial(\sigma \bowtie \tau)$ and $(\partial\sigma) \bowtie \tau + (-1)^p \sigma \bowtie (\partial\tau)$ agree.

First, let F_k be the kth face of σ, i.e. $F_k = [x_0, \dots, \hat{x}_k, \dots, x_p]$. Then F_k occurs with sign $(-1)^k$ in $\partial\sigma$. Let $F_k \bowtie \tau = \sum \text{sgn}(\xi, \zeta) d_{\xi\zeta}$, where the (ξ, ζ) run over the $(p-1, q)$-shuffles and the $d_{\xi\zeta}$ are the corresponding $(p+q-1)$-simplices of the shuffle product triangulation of $F_k \times \tau$. Since we have seen that it will be sufficient to know the signs on one such $(p+q-1)$-simplex, let's use the trivial shuffle (ξ_0, ζ_0) with $\xi_0 = [1, \dots, p-1]$ and $\zeta_0 = [p, \dots, p+q-1]$. Then d_{ξ_0, ζ_0} has vertices

$$[(x_0, y_0), \dots, \widehat{(x_k, y_0)}, \dots, (x_p, y_0), (x_p, y_1), \dots, (x_p, y_q)].$$

We have $\text{sgn}(\xi_0, \zeta_0) = 1$, so altogether the $(p+q-1)$-simplex $d_{\xi_0\zeta_0}$ has sign $(-1)^k$ in $(\partial\sigma) \bowtie \tau + (-1)^p \sigma \bowtie (\partial\tau)$. The computation for simplices in $\sigma \bowtie G_l$, where $G_l = [y_0, \dots, \hat{y}_l, \dots, y_q]$, is similar, except that there will be the additional sign $(-1)^p$, and the

[7] As $\sigma \times \tau$ is a topological $(p+q)$-ball, we know $H_{p+q}(\sigma \times \tau, \partial(\sigma \times \tau)) \cong \mathbb{Z}$. To see that $\sigma \bowtie \tau$ is a generator, we consider the associated singular chain determined by the vertex ordering; see Proposition 4.4.5. It suffices to show that this singular chain generates the singular homology group. We know by the theory of fundamental classes as provided in Section 8.1 or [125, Section 3.3] that there is an isomorphism $H_{p+q}(\sigma \times \tau, \partial(\sigma \times \tau)) \cong H_{p+q}(\sigma \times \tau, \sigma \times \tau - \{z\}) \cong \mathbb{Z}$ for any z in the interior of $\sigma \times \tau$. As the singular chain corresponding to $\pm\delta_{\mu\nu}$ generates $H_{p+q}(\sigma \times \tau, \sigma \times \tau - \{z\})$ for z in the interior of $\delta_{\mu\nu}$, it follows that the singular version of $\sigma \bowtie \tau$ must be a generator of $H_{p+q}(\sigma \times \tau, \partial(\sigma \times \tau))$.

simplices coming from the trivial $(p, q-1)$-shuffle will have the form

$$[(x_0, y_0), \dots, (x_p, y_0), (x_p, y_1), \dots, \widehat{(x_p, y_l)}, \dots, (x_p, y_q)].$$

So the overall sign of such a simplex in $(\partial\sigma) \bowtie \tau + (-1)^p \sigma \bowtie (\partial\tau)$ is $(-1)^{p+l}$.

Now let's turn to the corresponding terms in $\partial(\sigma \bowtie \tau) = \partial\left(\sum \operatorname{sgn}(\mu, \nu)\delta_{\mu\nu}\right)$. Letting (μ_0, ν_0) be the trivial (p, q)-shuffle, the $(p+q-1)$-simplex with the ordered vertex set

$$[(x_0, y_0), \dots, \widehat{(x_k, y_0)}, \dots, (x_p, y_0), (x_p, y_1), \dots, (x_p, y_q)]$$

is the face of $\delta_{\mu_0\nu_0}$ obtained by omitting the kth vertex (keeping in mind that there is also a 0th vertex). The trivial shuffle has sign 1, so the sign for this $(p+q-1)$-simplex in $\partial(\sigma \bowtie \tau)$ is $(-1)^k$. This corresponds with the computation in the preceding paragraph. Similarly, the $(p+q-1)$-simplex with vertex set

$$[(x_0, y_0), \dots, (x_p, y_0), (x_p, y_1), \dots, \widehat{(x_p, y_l)}, \dots, (x_p, y_q)]$$

is the face of $\delta_{\mu_0\nu_0}$ obtained by omitting the $(p+l)$th vertex (again accounting for a 0th vertex). Again, the trivial shuffle only contributes the sign 1, and since this is the $(p+l)$th face, the sign is $(-1)^{p+l}$ for this summand in $\partial(\sigma \bowtie \tau)$. This also corresponds with the computation in the preceding paragraph.

This completes the proof. □

There is one more useful fact about $\bowtie$ that we demonstrate here:

Proposition B.6.8 *Let $\sigma = [u_0, \dots, u_p]$ be a p-simplex and $\tau = [v_0, \dots, v_q]$ be a q-simplex. We also let σ and τ denote the generators of $H_p(\sigma, \partial\sigma) \cong \mathbb{Z}$ and $H_q(\tau, \partial\tau) \cong \mathbb{Z}$ oriented according to the given vertex orderings of σ and τ. Then $\sigma \bowtie \tau$ represents the generator of $H_{p+q}(\sigma \times \tau, \partial(\sigma \times \tau))$ oriented according to the product orientation of $\sigma \times \tau$, i.e. $\sigma \bowtie \tau$ is the fundamental class of the space $\sigma \times \tau$ consistent with the orientation.*

Proof That $\sigma \bowtie \tau = \sum_{\mu\nu} \operatorname{sgn}(\mu, \nu)\delta_{\mu\nu}$ is a generator of $H_{p+q}(\sigma \times \tau, \partial(\sigma \times \tau))$ is shown in the proof of Proposition B.6.7. In particular, all the $\operatorname{sgn}(\mu, \nu)\delta^{ij}_{\mu\nu}$ are oriented consistently, either with the same orientation as the space $\sigma \times \tau$ or its negative. Consider the trivial shuffle (μ_0, ν_0) with $\mu_0 = \{1, \dots, p\}$ and $\nu_0 = \{p+1, \dots, p+q\}$. It has sign 1, and $\delta_{\mu_0\nu_0}$ has vertices $[(u_0, v_0), \dots, (u_p, v_0), \dots, (u_p, v_q)]$. So it is oriented by the basis consisting of the vectors from (u_0, v_0) to the (u_c, v_0) in order followed by the vectors from (u_0, v_0) to the (u_p, v_d) in order. This agrees with the product orientation of $\sigma \times \tau$, so we see that each $\operatorname{sgn}(\mu, \nu)\delta_{\mu\nu}$ is oriented in agreement with the product orientation. □

References

[1] J.F. Adams, *Stable Homotopy and Generalised Homology*, Chicago Lectures in Mathematics. University of Chicago Press, Chicago, 1974.

[2] E. Akin, *Manifold phenomena in the theory of polyhedra*, Trans. Amer. Math. Soc. **143** (1969), 413–473.

[3] ———, *Stiefel–Whitney homology classes and bordism*, Trans. Amer. Math. Soc. **205** (1975), 341–359.

[4] P. Albin, *On the Hodge theory of stratified spaces*. In *Hodge Theory and L^2-Analysis*, Advanced Lectures in Mathematics, vol. 39, International Press, Somerville, MA, 2017, pp. 1–78.

[5] P. Albin, É. Leichtnam, R. Mazzeo, and P. Piazza, *The signature package on Witt spaces*, Ann. Sci. Éc. Norm. Supér. (4) **45** (2012), no. 2, 241–310.

[6] ———, *The Novikov conjecture on Cheeger spaces*, J. Noncommut. Geom. **11** (2017), no. 2, 451–506.

[7] J.W. Alexander, *On the ring of a compact metric space*, Proc. Natl. Acad. Sci. USA **21** (1935), 511–512.

[8] ———, *On the connectivity ring of an abstract space*, Ann. of Math. (2) **37** (1936), no. 3, 698–708.

[9] P. Aluffi, *Characteristic classes of singular varieties*. In *Topics in Cohomological Studies of Algebraic Varieties*, Trends in Mathematics, Birkhäuser, Basel, 2005, pp. 1–32.

[10] M.F. Atiyah and I.M. Singer, *The index of elliptic operators. III*, Ann. of Math. (2) **87** (1968), 546–604.

[11] M. Banagl, *Topological Invariants of Stratified Spaces*, Springer Monographs in Mathematics, Springer, Berlin, 2007.

[12] ———, *Extending intersection homology type invariants to non-Witt spaces*, Mem. Amer. Math. Soc. **160** (2002), no. 760, x+83.

[13] ———, *Computing twisted signatures and L-classes of non-Witt spaces*, Proc. London Math. Soc. (3) **92** (2006), no. 2, 428–470.

[14] ———, *The L-class of non-Witt spaces*, Ann. of Math. (2) **163** (2006), no. 3, 743–766.

[15] ———, *Intersection Spaces, Spatial Homology Truncation, and String Theory*, Lecture Notes in Mathematics, vol. 1997, Springer, Berlin, 2010.

[16] ———, *The signature of singular spaces and its refinements to generalized homology theories*. In *Topology of Stratified Spaces*, MSRI Publications, vol. 58, Cambridge University Press, Cambridge, 2011, pp. 223–249.

[17] ———, *First cases of intersection spaces in stratification depth 2*, J. Singul. **5** (2012), 57–84.

[18] ———, *Foliated stratified spaces and a de Rham complex describing intersection space cohomology*, J. Differential Geom. **104** (2016), no. 1, 1–58.

[19] M. Banagl, N. Budur, and L. Maxim, *Intersection spaces, perverse sheaves and type IIB string theory*, Adv. Theor. Math. Phys. **18** (2014), no. 2, 363–399.

[20] M. Banagl, S.E. Cappell, and J.L. Shaneson, *Computing twisted signatures and L-classes of stratified spaces*, Math. Ann. **326** (2003), no. 3, 589–623.

[21] M. Banagl and B. Chriestenson, *Intersection spaces, equivariant Moore approximation and the signature*, J. Singul. **16** (2017), 141–179.

[22] M. Banagl and E. Hunsicker, *Hodge theory for intersection space cohomology*, Geom. Topol. **23** (2019), no. 5, 2165–2225.

[23] M. Banagl, G. Laures, and J.E. McClure, *The L-homology fundamental class for IP-spaces and the stratified Novikov conjecture*, Selecta Math. (N.S.) **25** (2019), no. 1, art. 7, 104.

[24] M. Banagl and L. Maxim, *Deformation of singularities and the homology of intersection spaces*, J. Topol. Anal. **4** (2012), no. 4, 413–448.

[25] ———, *Intersection spaces and hypersurface singularities*, J. Singul. **5** (2012), 48–56.

[26] A.A. Beĭlinson, J. Bernstein, and P. Deligne, *Faisceaux pervers*. In *Analysis and Topology on Singular Spaces, I (Luminy, 1981)*, Astérisque, vol. 100, Soc. Math. France, Paris, 1982, pp. 5–171.

[27] A.C. Beshears, *G-isovariant Structure Sets and Stratified Structure Sets*. PhD Thesis, Vanderbilt University (1997). ProQuest LLC, Ann Arbor, MI.

[28] A. Borel et al., *Intersection Cohomology*. Notes on the seminar held at the University of Bern, Bern, 1983, Swiss Seminars, Progress in Mathematics, vol. 50, Birkhäuser, Boston, MA, 1984.

[29] R. Bott and L.W. Tu, *Differential Forms in Algebraic Topology*, Graduate Texts in Mathematics, vol. 82, Springer, New York, NY, 1982.

[30] N. Bourbaki, *Elements of Mathematics. Commutative Algebra*, Hermann, Paris, 1972, Translated from the French.

[31] J.-P. Brasselet, G. Hector, and M. Saralegi, *Théorème de de Rham pour les variétés stratifiées*, Ann. Global Anal. Geom. **9** (1991), no. 3, 211–243.

[32] J.-P. Brasselet, *Characteristic classes and singular varieties*, Vietnam J. Math. **33** (2005), Special Issue, 1–16.

[33] ———, *From Chern classes to Milnor classes – a history of characteristic classes for singular varieties*. In *Singularities – Sapporo 1998*, Advanced Studies in Pure Mathematics, vol. 29, Kinokuniya, Tokyo, 2000, pp. 31–52.

[34] ———, *Introduction to Intersection Homology and Perverse Sheaves*, Publicações Matemáticas do IMPA, 27th Brazilian Mathematics Colloquium, Instituto Nacional de Matemática Pura e Aplicada (IMPA), Rio de Janeiro, 2009.

[35] J.-P. Brasselet, J. Seade, and T. Suwa, *Vector Fields on Singular Varieties*, Lecture Notes in Mathematics, vol. 1987, Springer, Berlin, 2009.

[36] G.E. Bredon, *Introduction to Compact Transformation Groups*, Pure and Applied Mathematics, vol. 46, Academic Press, New York, 1972,

[37] ———, *Sheaf Theory*, 2nd edn, Graduate Texts in Mathematics, vol. 170, Springer, New York, NY, 1997.

[38] ———, *Topology and Geometry*, Graduate Texts in Mathematics, vol. 139, Springer, New York, NY, 1997, Corrected third printing of the 1993 original.

[39] J.-L. Brylinski, *Equivariant intersection cohomology*. In *Kazhdan–Lusztig theory and related topics (Chicago, IL, 1989)*, Contemporary Mathematics, vol. 139, American Mathematical Society, Providence, RI, 1992, pp. 5–32.

[40] S. Buoncristiano, C.P. Rourke, and B.J. Sanderson, *A Geometric Approach to Homology Theory*, London Mathematical Society Lecture Note Series, no. 18, Cambridge University Press, Cambridge, 1976.

[41] J.W. Cannon, *Shrinking cell-like decompositions of manifolds. Codimension three*, Ann. of Math. (2) **110** (1979), no. 1, 83–112.

[42] S. E. Cappell, R. Lee, and E.Y. Miller, *On the Maslov index*, Comm. Pure Appl. Math. **47** (1994), no. 2, 121–186.

[43] S.E. Cappell, A. Libgober, L.G. Maxim, and J.L. Shaneson, *Hodge genera of algebraic varieties. II*, Math. Ann. **345** (2009), no. 4, 925–972.

[44] S.E. Cappell, L.G. Maxim, and J.L. Shaneson, *Euler characteristics of algebraic varieties*, Comm. Pure Appl. Math. **61** (2008), no. 3, 409–421.

[45] ———, *Hodge genera of algebraic varieties. I*, Comm. Pure Appl. Math. **61** (2008), no. 3, 422–449.

[46] S.E. Cappell and J.L. Shaneson, *Singular spaces, characteristic classes, and intersection homology*, Ann. of Math. (2) **134** (1991), no. 2, 325–374.

[47] ———, *Stratifiable maps and topological invariants*, J. Amer. Math. Soc. **4** (1991), no. 3, 521–551.

[48] ———, *The mapping cone and cylinder of a stratified map*. In *Prospects in Topology (Princeton, NJ, 1994)*, Annals of Mathematics Studies, vol. 138, Princeton University Press, Princeton, NJ, 1995, pp. 58–66.

[49] H. Cartan and S. Eilenberg, *Homological Algebra*, Princeton University Press, Princeton, NJ, 1956.

[50] E. Čech, *Multiplications on a complex*, Ann. of Math. (2) **37** (1936), no. 3, 681–697.

[51] D. Chataur and J. Cirici, *Mixed Hodge structures on the intersection homotopy type of complex varieties with isolated singularities*, Preprint, `https://arxiv.org/abs/1603.09125`.

[52] D. Chataur, M. Saralegi-Aranguren, and D. Tanré, *Homologie d'intersection. Perversités générales et invariance topologique*, Preprint, `https://arxiv.org/abs/1602.03009`.

[53] ———, *Steenrod squares on intersection cohomology and a conjecture of M Goresky and W Pardon*, Algebr. Geom. Topol. **16** (2016), no. 4, 1851–1904.

[54] ———, *Singular decompositions of a cap product*, Proc. Amer. Math. Soc. **145** (2017), no. 8, 3645–3656.

[55] ———, *Blown-up intersection cohomology*. In *An Alpine Bouquet of Algebraic Topology*, Contemporary Mathematics, vol. 708, American Mathematical Society, Providence, RI, 2018, pp. 45–102.

[56] ———, *Intersection cohomology, simplicial blow-up and rational homotopy*, Mem. Amer. Math. Soc. **254** (2018), no. 1214, viii+108.

[57] ———, *Poincaré duality with cap products in intersection homology*, Adv. Math. **326** (2018), 314–351.

[58] J. Cheeger, *On the spectral geometry of spaces with cone-like singularities*, Proc. Natl. Acad. Sci. USA **76** (1979), no. 5, 2103–2106.

[59] ———, *On the Hodge theory of Riemannian pseudomanifolds*. In *Geometry of the Laplace Operator (Proc. Sympos. Pure Math., Univ. Hawaii, Honolulu, Hawaii, 1979)*, Proceedings of Symposia in Pure Mathematics, XXXVI, American Mathematical Society, Providence, RI, 1980, pp. 91–146.

[60] ———, *Spectral geometry of singular Riemannian spaces*, J. Differential Geom. **18** (1983), no. 4, 575–657.

[61] J. Cheeger, M. Goresky, and R. MacPherson, *L^2-cohomology and intersection homology of singular algebraic varieties*. In *Seminar on Differential Geometry*, Annals of Mathematics Studies, vol. 102, Princeton University Press, Princeton, NJ, 1982, pp. 303–340.

[62] D.C. Cohen, M. Goresky, and L. Ji, *On the Künneth formula for intersection cohomology*, Trans. Amer. Math. Soc. **333** (1992), no. 1, 63–69.

[63] P.E. Conner and E.E. Floyd, *Differentiable Periodic Maps*, Ergebnisse der Mathematik und ihrer Grenzgebiete, vol. 33, Springer, Berlin, 1964.

[64] X. Dai, *An introduction to L^2 cohomology*. In *Topology of Stratified Spaces*, MSRI Publications, vol. 58, Cambridge University Press, Cambridge, 2011, pp. 1–12.

[65] R.J. Daverman and G.A. Venema, *Embeddings in Manifolds*, Graduate Studies in Mathematics, vol. 106, American Mathematical Society, Providence, RI, 2009.

[66] J.F. Davis, *Manifold aspects of the Novikov conjecture*. In *Surveys on Surgery Theory, Vol. 1*, Annals of Mathematical Studies, vol. 145, Princeton University Press, Princeton, NJ, 2000, pp. 195–224.

[67] J.F. Davis and P. Kirk, *Lecture Notes in Algebraic Topology*, Graduate Studies in Mathematics, vol. 35, American Mathematical Society, Providence, RI, 2001.

[68] M.A.A. de Cataldo and L. Migliorini, *The decomposition theorem, perverse sheaves and the topology of algebraic maps*, Bull. Amer. Math. Soc. (N.S.) **46** (2009), no. 4, 535–633.

[69] J. Dieudonné, *A History of Algebraic and Differential Topology 1900–1960*, Modern Birkhäuser Classics, Birkhäuser , Boston, MA, 2009.

[70] A. Dimca, *Sheaves in Topology*, Universitext, Springer-Verlag, Berlin, 2004.

[71] A. Dold, *Lectures on Algebraic Topology*, Springer, Berlin, 1972.

[72] K.H. Dovermann and R. Schultz, *Equivariant Surgery Theories and their Periodicity Properties*, Lecture Notes in Mathematics, vol. 1443, Springer, Berlin, 1990.

[73] R.D. Edwards, *Suspensions of homology spheres*, Preprint, `http://arxiv.org/pdf/math/0610573v1.pdf`.

[74] S. Eilenberg, *Singular homology theory*, Ann. of Math. (2) **45** (1944), 407–447.

[75] S. Eilenberg and N. Steenrod, *Foundations of Algebraic Topology*, Princeton University Press, Princeton, NJ, 1952.

[76] B. Elias and G. Williamson, *Kazhdan–Lusztig conjectures and shadows of Hodge theory*. In *Arbeitstagung Bonn 2013*, Progress in Mathematics, vol. 319, Birkhäuser/Springer, Cham, 2016, pp. 105–126.

[77] T. Eppelmann, *Signature Homology and Symmetric L-Theory*, PhD Thesis, Rupprecht-Karls Universität Heidelberg, 2007, available at `http://deposit.ddb.de/cgi-bin/dokserv?idn=985798726`.

[78] B. Evans, *Proof that convex open sets in $\mathbb{R}^n$ are homeomorphic?*, Preprint, `http://math.stackexchange.com/questions/165629/proof-that-convex-open-sets-in-mathbbrn-are-homeomorphic`.

[79] S.C. Ferry, A. Ranicki, and J. Rosenberg, *A history and survey of the Novikov conjecture*. In *Novikov Conjectures, Index Theorems and Rigidity, Vol. 1 (Oberwolfach, 1993)*, London Mathematical Society Lecture Note Series, vol. 226, Cambridge University Press, Cambridge, 1995, pp. 7–66.

[80] D. Ferus, *Analysis III*, `http://page.math.tu-berlin.de/~ferus/ANA/Ana3.pdf`.

[81] G. Friedman, *Stratified fibrations and the intersection homology of the regular neighborhoods of bottom strata*, Topology Appl. **134** (2003), no. 2, 69–109.

[82] ______, *Intersection Alexander polynomials*, Topology **43** (2004), no. 1, 71–117.

[83] ______, *Intersection homology of regular and cylindrical neighborhoods*, Topology Appl. **149** (2005), no. 1-3, 97–148.

[84] ———, *Intersection homology of stratified fibrations and neighborhoods*, Adv. Math. **215** (2007), no. 1, 24–65.

[85] ———, *Singular chain intersection homology for traditional and super-perversities*, Trans. Amer. Math. Soc. **359** (2007), no. 5, 1977–2019.

[86] ———, *Intersection homology and Poincaré duality on homotopically stratified spaces*, Geom. Topol. **13** (2009), no. 4, 2163–2204.

[87] ———, *Intersection homology Künneth theorems*, Math. Ann. **343** (2009), no. 2, 371–395.

[88] ———, *Intersection homology with field coefficients: K-Witt spaces and K-Witt bordism*, Comm. Pure Appl. Math. **62** (2009), no. 9, 1265–1292.

[89] ———, *On the chain-level intersection pairing for PL pseudomanifolds*, Homology Homotopy Appl. **11** (2009), no. 1, 261–314.

[90] ———, *Intersection homology with general perversities*, Geom. Dedicata **148** (2010), 103–135.

[91] ———, *An introduction to intersection homology with general perversity functions*. In *Topology of Stratified Spaces*, MSRI Publications, vol. 58, Cambridge University Press, Cambridge, 2011, pp. 177–222.

[92] ———, *Corrigendum: Intersection homology with field coefficients: K-Witt spaces and K-Witt bordism*, Comm. Pure Appl. Math. **65** (2012), no. 11, 1639–1640.

[93] ———, *K-Witt bordism in characteristic 2*, Arch. Math. (Basel) **100** (2013), no. 4, 381–387.

[94] ———, *Stratified and unstratified bordism of pseudomanifolds*, Topology Appl. **194** (2015), 51–92.

[95] ———, *The chain-level intersection product for PL pseudomanifolds revisited*, J. Singul. **17** (2018), 330–367.

[96] G. Friedman and E. Hunsicker, *Additivity and non-additivity for perverse signatures*, J. Reine Angew. Math. **676** (2013), 51–95.

[97] G. Friedman, E. Hunsicker, A. Libgober, and L. Maxim (eds.), *Topology of Stratified Spaces*, MSRI Publications, vol. 58, Cambridge University Press, Cambridge, 2011.

[98] G. Friedman and J. McClure, *The symmetric signature of a Witt space*, J. Topol. Anal. **5** (2013), no. 2, 121–159.

[99] ———, *Intersection homology duality and pairings: singular, PL, and sheaf-theoretic*, Preprint.

[100] ———, *Cup and cap products in intersection (co)homology*, Adv. Math. **240** (2013), 383–426.

[101] W. Fulton and R. MacPherson, *Categorical framework for the study of singular spaces*, Mem. Amer. Math. Soc. **31** (1981), no. 243, vi+165.

[102] S.I. Gelfand and Yu.I. Manin, *Methods of Homological Algebra*, 2nd edn, Springer Monographs in Mathematics, Springer, Berlin, 2003.

[103] S. Geschke, *Convex open subsets of $\mathbb{R}^n$ are homeomorphic to n-dimensional open balls*, Preprint, `http://relaunch.hcm.uni-bonn.de/fileadmin/geschke/papers/ConvexOpen.pdf`.

[104] R. Godement, *Topologie Algébrique et Théorie des Faisceaux*, Troisième édition revue et corrigée, Publications de l'Institut de Mathématique de l'Université de Strasbourg, XIII, Actualités Scientifiques et Industrielles, no. 1252, Hermann, Paris, 1973.

[105] M. Goresky and R. MacPherson, *Intersection homology theory*, Topology **19** (1980), no. 2, 135–162.

[106] ———, *Intersection homology. II*, Invent. Math. **72** (1983), no. 1, 77–129.

[107] ———, *Morse theory and intersection homology theory*. In *Analysis and Topology on Singular Spaces, II, III (Luminy, 1981)*, Astérisque, vol. 101, Soc. Math. France, Paris, 1983, pp. 135–192.

[108] ______, *Lefschetz fixed point theorem for intersection homology*, Comment. Math. Helv. **60** (1985), no. 3, 366–391.

[109] ______, *Stratified Morse Theory*, Ergebnisse der Mathematik und ihrer Grenzgebiete (3), vol. 14, Springer, Berlin, 1988.

[110] M. Goresky and W. Pardon, *Wu numbers of singular spaces*, Topology **28** (1989), no. 3, 325–367.

[111] M. Goresky and P. Siegel, *Linking pairings on singular spaces*, Comment. Math. Helv. **58** (1983), no. 1, 96–110.

[112] R.M. Goresky, *Triangulation of stratified objects*, Proc. Amer. Math. Soc. **72** (1978), no. 1, 193–200.

[113] ______, *Whitney stratified chains and cochains*, Trans. Amer. Math. Soc. **267** (1981), no. 1, 175–196.

[114] ______, *Intersection homology operations*, Comment. Math. Helv. **59** (1984), no. 3, 485–505.

[115] M. Grant and A. Szücs, *On realizing homology classes by maps of restricted complexity*, Bull. Lond. Math. Soc. **45** (2013), no. 2, 329–340.

[116] P.A. Griffiths and J.W. Morgan, *Rational Homotopy theory and Differential Forms*, Progress in Mathematics, vol. 16, Birkhäuser, Boston, MA, 1981.

[117] M. Gromov, *Convex sets and Kähler manifolds*. In *Advances in Differential Geometry and Topology*, World Scientific, Teaneck, NJ, 1990, pp. 1–38.

[118] ______, *Positive curvature, macroscopic dimension, spectral gaps and higher signatures*. In *Functional Analysis on the Eve of the 21st Century, Vol. II (New Brunswick, NJ, 1993)*, Progress in Mathematics, vol. 132, Birkhäuser, Boston, MA, 1996, pp. 1–213.

[119] N. Habegger, *From PL to sheaf theory (rock to Bach)*. In *Intersection Cohomology (Bern, 1983)*, Progress in Mathematics, vol. 50, Birkhäuser, Boston, MA, 1984, pp. 23–34.

[120] N. Habegger and L. Saper, *Intersection cohomology of cs-spaces and Zeeman's filtration*, Invent. Math. **105** (1991), no. 2, 247–272.

[121] A. Haefliger, *Introduction to piecewise linear intersection homology*. In *Intersection Cohomology (Bern, 1983)*, Progress in Mathematics, vol. 50, Birkhäuser, Boston, MA, 1984, pp. 1–21.

[122] M. Handel, *A resolution of stratification conjectures concerning CS sets*, Topology **17** (1978), no. 2, 167–175.

[123] R.M. Hardt, *Triangulation of subanalytic sets and proper light subanalytic maps*, Invent. Math. **38** (1976/77), no. 3, 207–217.

[124] R. Hartshorne, *Residues and duality*, Lecture notes of a seminar on the work of A. Grothendieck, given at Harvard 1963/64. With an appendix by P. Deligne. Lecture Notes in Mathematics, No. 20, Springer-Verlag, Berlin, 1966.

[125] A. Hatcher, *Algebraic Topology*, Cambridge University Press, Cambridge, 2002.

[126] P.J. Hilton and U. Stammbach, *A Course in Homological Algebra*, Graduate Texts in Mathematics, Vol. 4, Springer, New York, 1971.

[127] H. Hironaka, *Triangulations of algebraic sets*. In *Algebraic Geometry, Humboldt State Univ., Arcata, Calif., 1974)*, Proceedings of Symposia in Pure Mathematics, Vol. 29, American Mathematical Society, Providence, RI, 1975, pp. 165–185.

[128] F. Hirzebruch, *The signature theorem: reminiscences and recreation*. In *Prospects in Mathematics (Proc. Sympos., Princeton Univ., Princeton, NJ, 1970)*, Annals of Mathematics Studies, vol. 70, Princeton University Press, Princeton, NJ, 1971, pp. 3–31.

[129] S.-T. Hu, *Homotopy Theory*, Pure and Applied Mathematics, vol. VIII, Academic Press, New York, 1959.

[130] J.F.P. Hudson, *Piecewise Linear Topology*, University of Chicago Lecture Notes prepared with the assistance of J.L. Shaneson and J. Lees, W.A. Benjamin, New York, NY, 1969.

[131] B. Hughes, *The approximate tubular neighborhood theorem*, Ann. of Math. (2) **156** (2002), no. 3, 867–889.

[132] B. Hughes, L.R. Taylor, S. Weinberger, and B. Williams, *Neighborhoods in stratified spaces with two strata*, Topology **39** (2000), no. 5, 873–919.

[133] B. Hughes and S. Weinberger, *Surgery and stratified spaces*. In *Surveys on Surgery Theory, Vol. 2*, Annals of Mathematical Studies, vol. 149, Princeton University Press, Princeton, NJ, 2001, pp. 319–352.

[134] E. Hunsicker, *Hodge and signature theorems for a family of manifolds with fibre bundle boundary*, Geom. Topol. **11** (2007), 1581–1622.

[135] B. Iversen, *Cohomology of Sheaves*, Universitext, Springer, Berlin, 1986.

[136] M. Kapranov and V. Schechtman, *Perverse schobers*, Preprint `https://arxiv.org/abs/1411.2772`.

[137] M. Kashiwara and P. Schapira, *Sheaves on Manifolds*, Grundlehren der Mathematischen Wissenschaften, vol. 292, Springer, Berlin, 1994.

[138] ———, *Categories and Sheaves*, Grundlehren der Mathematischen Wissenschaften, vol. 332, Springer, Berlin, 2006.

[139] H.C. King, *Topological invariance of intersection homology without sheaves*, Topology Appl. **20** (1985), no. 2, 149–160.

[140] F. Kirwan and J. Woolf, *An Introduction to Intersection Homology Theory*, 2nd edn, Chapman & Hall/CRC, Boca Raton, FL, 2006.

[141] S.L. Kleiman, *The development of intersection homology theory*, Pure Appl. Math. Q. **3** (2007), no. 1, Part 3, 225–282, Special issue in honor of Robert D. MacPherson.

[142] M. Klimczak, *Poincaré duality for spaces with isolated singularities*, Manuscripta Math. **153** (2017), no. 1-2, 231–262.

[143] ———, *Mixed Hodge structures on the rational models of intersection spaces*, J. Singul. **17** (2018), 428–482.

[144] M. Kreck, *Differential Algebraic Topology: From Stratifolds to Exotic Spheres*, Graduate Studies in Mathematics, vol. 110, American Mathematical Society, Providence, RI, 2010.

[145] M. Kreck and W. Lück, *The Novikov conjecture*, Geometry and algebra, Oberwolfach Seminars, vol. 33, Birkhäuser Verlag, Basel, 2005.

[146] T.Y. Lam, *Lectures on Modules and Rings*, Graduate Texts in Mathematics, vol. 189, Springer, New York, NY, 1999.

[147] S. Lang, *Algebra*, 3rd edn, Graduate Texts in Mathematics, vol. 211, Springer, New York, NY, 2002.

[148] S. Lefschetz, *Intersections and transformations of complexes and manifolds*, Trans. Amer. Math. Soc. **28** (1926), no. 1, 1–49.

[149] ———, *Manifolds with a boundary and their transformations*, Trans. Amer. Math. Soc. **29** (1927), no. 2, 429–462.

[150] ———, *Algebraic Topology*, American Mathematical Society Colloquium Publications, vol. 27, American Mathematical Society, New York, NY, 1942.

[151] J. Levine, *A characterization of knot polynomials*, Topology **4** (1965), 135–141.

[152] S. Łojasiewicz, *Triangulation of semi-analytic sets*, Ann. Scuola Norm. Sup. Pisa (3) **18** (1964), 449–474.

[153] ———, *Introduction to Complex Analytic Geometry*, Birkhäuser, Basel, 1991, Translated from the Polish by Maciej Klimek.

[154] E. Looijenga, *Cohomology and intersection homology of algebraic varieties*. In *Complex Algebraic Geometry (Park City, UT, 1993)*, IAS/Park City Mathematics Series, vol. 3, American Mathematical Society, Providence, RI, 1997, pp. 221–263.

[155] S. Mac Lane, *Homology*, Classics in Mathematics, Springer-Verlag, Berlin, 1995, Reprint of the 1975 edition.

[156] R. MacPherson, *Intersection Homology and Perverse Sheaves*, Unpublished Colloquium Lectures.

[157] R. MacPherson and K. Vilonen, *Elementary construction of perverse sheaves*, Invent. Math. **84** (1986), no. 2, 403–435.

[158] D.B. Massey, *Notes on perverse sheaves and vanishing cycles*, `https://arxiv.org/abs/math/9908107`.

[159] ______, *Stratified Morse theory: past and present*, Pure Appl. Math. Q. **2** (2006), no. 4, Part 2, 1053–1084, Special Issue in honor of Robert D. MacPherson.

[160] D.B. Massey and D.T. Lê, *Notes on real and complex analytic and semianalytic singularities*. In *Singularities in Geometry and Topology*, World Scientific, Hackensack, NJ, 2007, pp. 81–126.

[161] J. Mather, *Notes on topological stability*, Bull. Amer. Math. Soc. (N.S.) **49** (2012), no. 4, 475–506.

[162] K. Matthews, *Branched coverings of pseudomanifolds*, preprint.

[163] ______, *Universal Poincaré Duality for the Intersection Homology of Branched and Partial Coverings of a Pseudomanifold*, PhD Thesis, Texas Christian University, 2016.

[164] L. Maxim and J. Schürmann, *Characteristic classes of mixed Hodge modules and applications*. In *Schubert Varieties, Equivariant Cohomology and Characteristic Classes – IMPANGA* 15, EMS Series of Congress Reports, European Mathematical Society, Zürich, 2018, pp. 159–202.

[165] L. Maxim, *Intersection homology and Alexander modules of hypersurface complements*, Comment. Math. Helv. **81** (2006), no. 1, 123–155.

[166] ______, *Intersection spaces, perverse sheaves and string theory*, J. Singul. **15** (2016), 118–125. MR 3562857

[167] J.P. May, *A Concise Course in Algebraic Topology*, Chicago Lectures in Mathematics, University of Chicago Press, Chicago, IL, 1999.

[168] J.E. McClure, *On the chain-level intersection pairing for PL manifolds*, Geom. Topol. **10** (2006), 1391–1424.

[169] ______, *Erratum for the paper 'On the chain-level intersection pairing for PL manifolds'*, Geom. Topol. **13** (2009), no. 3, 1775–1777.

[170] C. McCrory, *Stratified general position*. In *Algebraic and Geometric Topology (Proc. Sympos., Univ. California, Santa Barbara, CA, 1977)*, Lecture Notes in Math., vol. 664, Springer, Berlin, 1978, pp. 142–146.

[171] C.G. McCrory, *Poincaré Duality in Spaces with Singularities*, PhD Thesis, Brandeis University, ProQuest LLC, Ann Arbor, MI, 1972.

[172] D.A. Miller, *Popaths and holinks*, J. Homotopy Relat. Struct. **4** (2009), no. 1, 265–273.

[173] ______, *Homotopy Theory of Stratified Spaces*, PhD Thesis, University of Aberdeen, 2010.

[174] ______, *Strongly stratified homotopy theory*, Trans. Amer. Math. Soc. **365** (2013), 4933–4962.

[175] J. Milnor, *On manifolds homeomorphic to the 7-sphere*, Ann. of Math. (2) **64** (1956), 399–405.

[176] J. Milnor and D. Husemoller, *Symmetric Bilinear Forms*, Ergebnisse der Mathematik und ihrer Grenzgebiete, vol. 73, Springer, Berlin, 1973,

[177] J.W. Milnor and J.D. Stasheff, *Characteristic Classes*, Annals of Mathematics Studies, vol. 76, Princeton University Press, Princeton, NJ, 1974,

[178] A. Minatta, *Signature homology*, J. Reine Angew. Math. **592** (2006), 79–122.

[179] H.R. Morton, *Joins of polyhedra*, Topology **9** (1970), 243–249.

[180] J.R. Munkres, *Elementary Differential Topology*, Lectures given at Massachusetts Institute of Technology, Fall, vol. 1961, Princeton University Press, Princeton, NJ, 1966.

[181] ______, *Topology: a First Course*, Prentice-Hall, Inc., Englewood Cliffs, NJ, 1975.

[182] ———, *Elements of Algebraic Topology*, Addison-Wesley, Menlo Park, CA, 1984.

[183] nLab, *Ball*, `http://ncatlab.org/nlab/show/ball`.

[184] L. Noirel, *Plongements Sous-Analytiques d'Espaces Stratifiés de Thom–Mather*, PhD Thesis, University of Provence, 1996.

[185] L. Noirel and D. Trotman, *Subanalytic and semialgebraic realisations of abstract stratified sets*. In *Real Algebraic Geometry and Ordered Structures (Baton Rouge, LA, 1996)*, Contemporary Mathemtatics, vol. 253, American Mathematical Society, Providence, RI, 2000, pp. 203–207.

[186] G. Padilla, *On normal stratified pseudomanifolds*, Extracta Math. **18** (2003), no. 2, 223–234.

[187] W.L. Pardon, *Intersection homology Poincaré spaces and the characteristic variety theorem*, Comment. Math. Helv. **65** (1990), no. 2, 198–233.

[188] A. Parusiński, *Characteristic classes of singular varieties*. In *Singularity Theory and its Applications*, Advanced Studies in Pure Mathematics, vol. 43, Mathematical Society of Japan, Tokyo, 2006, pp. 347–367.

[189] C.A.M. Peters and J.H.M. Steenbrink, *Mixed Hodge Structures*, Ergebnisse der Mathematik und ihrer Grenzgebiete (3), vol. 52, Springer, Berlin, 2008.

[190] H. Poincaré, *Analysis situs*, Journal de l'Ecole Polytechnique **1** (1895), 1–121.

[191] F. Quinn, *Intrinsic skeleta and intersection homology of weakly stratified sets*. In *Geometry and Topology (Athens, GA, 1985)*, Lecture Notes in Pure and Applied Mathematics, vol. 105, Dekker, New York, 1987, pp. 225–241.

[192] ———, *Homotopically stratified sets*, J. Amer. Math. Soc. **1** (1988), 441–499.

[193] D.A. Ramras, *Connectivity of the coset poset*, Undergraduate Thesis, `arXiv:math/0208176`.

[194] A.A. Ranicki, *Algebraic L-theory and Topological Manifolds*, Cambridge Tracts in Mathematics, vol. 102, Cambridge University Press, Cambridge, 1992.

[195] Andrew Ranicki, *On the Novikov conjecture*. In *Novikov Conjectures, Index Theorems and Rigidity, Vol. 1 (Oberwolfach, 1993)*, London Mathematical Society Lecture Note Series, vol. 226, Cambridge University Press, Cambridge, 1995, pp. 272–337.

[196] J. Rosenberg, *Novikov's conjecture*. In *Open Problems in Mathematics*, Springer, Cham, 2016, pp. 377–402.

[197] J.J. Rotman, *An Introduction to Homological Algebra*, 2nd ed., Universitext, Springer, New York, 2009.

[198] C.P. Rourke and B.J. Sanderson, *Introduction to Piecewise-Linear Topology*, Ergebnisse der Mathematik und ihrer Grenzgebiete, vol. 69, Springer, Berlin, 1972.

[199] Y.B. Rudyak, *On Thom Spectra, Orientability, and Cobordism*, with a foreword by Haynes Miller, Springer Monographs in Mathematics, Springer, Berlin, 1998.

[200] M. Saito, *Modules de Hodge polarisables*, Publ. Res. Inst. Math. Sci. **24** (1988), no. 6, 849–995.

[201] ———, *Introduction to mixed Hodge modules*, Actes du Colloque de Théorie de Hodge (Luminy, 1987). Astérisque (1989), vol. 179-180, pp. 145–162,

[202] ———, *A young person's guide to mixed Hodge modules*. In *Hodge theory and L^2-Analysis*, Advanced Lectures in Mathematics, vol. 39, International Press, Somerville, MA, 2017, pp. 517–553.

[203] L. Saper and S. Zucker, *An introduction to L^2-cohomology*. In *Several Complex Variables and Complex Geometry, Part 2 (Santa Cruz, CA, 1989)*, Proceedings of Symposia in Pure Mathematics, vol. 52, American Mathematical Society, Providence, RI, 1991, pp. 519–534.

[204] M. Saralegi, *Homological properties of stratified spaces*, Illinois J. Math. **38** (1994), no. 1, 47–70.

[205] M. Saralegi-Aranguren, *de Rham intersection cohomology for general perversities*, Illinois J. Math. **49** (2005), no. 3, 737–758 (electronic).

[206] M. Saralegi-Aranguren and R. Wolak, *Poincaré duality of the basic intersection cohomology of a Killing foliation*, Monatsh. Math. **180** (2016), no. 1, 145–166.

[207] J. Schürmann, *Topology of Singular Spaces and Constructible Sheaves*, Instytut Matematyczny Polskiej Akademii Nauk. Monografie Matematyczne (New Series), vol. 63, Birkhäuser, Basel, 2003.

[208] ______, *Lectures on characteristic classes of constructible functions*. In *Topics in Cohomological Studies of Algebraic Varieties*, Notes by Piotr Pragacz and Andrzej Weber, Trends in Mathematics, Birkhäuser, Basel, 2005, pp. 175–201.

[209] ______, *Nearby cycles and characteristic classes of singular spaces*. In *Singularities in Geometry and Topology*, IRMA Lectures in Mathematics and Theoretical Physics, vol. 20, European Mathematical Society, Zürich, 2012, pp. 181–205.

[210] J. Schürmann and S. Yokura, *A survey of characteristic classes of singular spaces*. In *Singularity Theory*, World Scientific, Hackensack, NJ, 2007, pp. 865–952.

[211] D. Schütz, *Intersection homology of linkage spaces*, J. Topol. Anal. **8** (2016), no. 1, 25–58.

[212] ______, *Intersection homology of linkage spaces in odd-dimensional Euclidean space*, Algebr. Geom. Topol. **16** (2016), no. 1, 483–508.

[213] H. Seifert and W. Threlfall, *Seifert and Threlfall: a Textbook of Topology*, Pure and Applied Mathematics, vol. 89, Academic Press, New York, NY, 1980, Translated from the German edition of 1934 by Michael A. Goldman, With a preface by Joan S. Birman, With "Topology of 3-dimensional fibered spaces" by Seifert, Translated from the German by Wolfgang Heil.

[214] J.-P. Serre, *Groupes d'homotopie et classes de groupes abéliens*, Ann. of Math. (2) **58** (1953), 258–294.

[215] M. Shiota, *Geometry of Subanalytic and Semialgebraic Sets*, Progress in Mathematics, vol. 150, Birkhäuser, Boston, MA, 1997.

[216] L. C. Siebenmann, *Pseudo-annuli and invertible cobordisms*, Arch. Math. (Basel) **19** (1968), 528–535.

[217] ______, *Deformation of homeomorphisms on stratified sets. I, II*, Comment. Math. Helv. **47** (1972), 123–136; 137–163.

[218] P.H. Siegel, *Witt spaces: a geometric cycle theory for KO-homology at odd primes*, Amer. J. Math. **105** (1983), no. 5, 1067–1105.

[219] E. Spanier, *Borsuk's cohomotopy groups*, Ann. of Math. (2) **50** (1949), 203–245.

[220] E.H. Spanier, *Algebraic Topology*, Springer, Berlin, 1981, Corrected reprint.

[221] J.R. Stallings, *On infinite processes leading to differentiability in the complement of a point*. In *Differential and Combinatorial Topology (A Symposium in Honor of Marston Morse)*, Princeton University Press, Princeton, NJ, 1965, pp. 245–254.

[222] ______, *Lectures on Polyhedral Topology*, Notes by G. Ananda Swarup. Tata Institute of Fundamental Research Lectures on Mathematics, no. 43, Tata Institute of Fundamental Research, Bombay, 1967.

[223] R.P. Stanley, *Recent developments in algebraic combinatorics*, Israel J. Math. **143** (2004), 317–339.

[224] R.E. Stong, *Notes on Cobordism Theory*, Mathematical Notes, Princeton University Press, Princeton, NJ; University of Tokyo Press, Tokyo, 1968.

[225] G. Strang, *Linear Algebra and its Applications*, 3rd edn, Harcourt Brace Jovanovich, San Diego, 1988.

[226] D. Sullivan, *Combinatorial invariants of analytic spaces*. In *Proceedings of Liverpool Singularities Symposium, I (1969/70)*, Springer, Berlin, 1971, pp. 165–168.

[227] ———, *Geometric periodicity and the invariants of manifolds*. In *Manifolds – Amsterdam 1970 (Proc. Nuffic Summer School)*, Lecture Notes in Mathematics, vol. 197, Springer, Berlin, 1971, pp. 44–75.

[228] ———, *Geometric Topology: Localization, Periodicity and Galois Symmetry*, The 1970 MIT notes, Edited and with a preface by Andrew Ranicki. K-Monographs in Mathematics, vol. 8, Springer, Dordrecht, 2005.

[229] T. Suwa, *Characteristic classes of singular varieties*, Sugaku Expositions **16** (2003), no. 2, 153–175. [Translation of Sūgaku **52** (2000), no. 4, 376–393; MR1802958]

[230] R.G. Swan, *The Theory of Sheaves*, University of Chicago Press, Chicago, 1964.

[231] R.M. Switzer, *Algebraic Topology – Homotopy and Homology*, Classics in Mathematics, Springer, Berlin, 2002, Reprint of the 1975 original.

[232] J.J. Sylvester, *A demonstration of the theorem that every homogeneous quadratic polynomial is reducible by real orthogonal substitutions to the form of a sum of positive and negative squares*, Phil. Mag. (Ser. 4) **4** (1852), no. 23, 138–142.

[233] B.R. Tennison, *Sheaf Theory*, London Mathematical Society Lecture Note Series, No. 20, Cambridge University Press, Cambridge, 1975.

[234] R. Thom, *Les classes caractéristiques de Pontrjagin des variétés triangulées*. In *Symposium Internacional de Topologia Algebraica*, Universidad Nacional Autónoma de México and UNESCO, Mexico City, 1958, pp. 54–67.

[235] ———, *Quelques propriétés globales des variétés différentiables*, Comment. Math. Helv. **28** (1954), 17–86.

[236] G. Valette, *A Lefschetz duality for intersection homology*, Geom. Dedicata **169** (2014), 283–299.

[237] C.T.C. Wall, *Non-additivity of the signature*, Invent. Math. **7** (1969), 269–274.

[238] C.A. Weibel, *An Introduction to Homological Algebra*, Cambridge Studies in Advanced Mathematics, Cambridge University Press, Cambridge, 1994.

[239] S. Weinberger, *The Topological Classification of Stratified Spaces*, Chicago Lectures in Mathematics, University of Chicago Press, Chicago, IL, 1994.

[240] S. Weinberger and M. Yan, *Equivariant periodicity for abelian group actions*, Adv. Geom. **1** (2001), no. 1, 49–70.

[241] ———, *Equivariant periodicity for compact group actions*, Adv. Geom. **5** (2005), no. 3, 363–376.

[242] G.W. Whitehead, *Elements of Homotopy Theory*, Graduate Texts in Mathematics, vol. 61, Springer, Berlin, 1978.

[243] J.H.C. Whitehead, *On C^1-complexes*, Ann. of Math. (2) **41** (1940), 809–824.

[244] H. Whitney, *On products in a complex*, Ann. of Math. (2) **39** (1938), no. 2, 397–432.

[245] ———, *Local properties of analytic varieties*. In *Differential and Combinatorial Topology (A Symposium in Honor of Marston Morse)*, Princeton University Press, Princeton, NJ, 1965, pp. 205–244.

[246] ———, *Tangents to an analytic variety*, Ann. of Math. (2) **81** (1965), 496–549.

[247] S. Willard, *General Topology*, Dover Publications, Inc., Mineola, NY, 2004, Reprint of the 1970 original (Addison-Wesley, Reading, MA).

[248] G. Williamson, *Algebraic representations and constructible sheaves*, Jpn. J. Math. **12** (2017), no. 2, 211–259.

[249] R.E. Williamson, Jr., *Cobordism of combinatorial manifolds*, Ann. of Math. (2) **83** (1966), 1–33.

[250] J. Woolf, *Witt groups of sheaves on topological spaces*, Comment. Math. Helv. **83** (2008), no. 2, 289–326.

[251] ———, *The fundamental category of a stratified space*, J. Homotopy Relat. Struct. **4** (2009), no. 1, 359–387.

[252] J. Woolf, *Transversal homotopy theory*, Theory Appl. Categ. **24** (2010), no. 7, 148–178.

[253] M. Yan, *The periodicity in stable equivariant surgery*, Comm. Pure Appl. Math. **46** (1993), no. 7, 1013–1040.

[254] E.C. Zeeman, *Seminar on combinatorial topology*, Institut des Hautes Etudes Scientifiques, 1963.

Glossary of Symbols

$\bar{0}$	zero perversity, xxi, 88
1_X	augmentation cocycle that evaluates to 1 on each 0-simplex of X, written 1 if X is clear, xxi
∂	boundary map of a chain complex, 91
$\hat{\partial}$	boundary map of the non-GM intersection chain complex $I^{\bar{p}}S_*(X)$; used only in Chapter 6, 267
d^*	connecting morphism of cohomologically indexed long exact sequence, xx
∂X	boundary of X, 52
$A * B$	torsion product $\mathrm{Tor}_R^1(A, B)$, xix, 220
$X * Y$	join of X and Y, 75
$[X, Y]$ or $[X, Y]_{\mathrm{PL}}$	set of (PL) homotopy classes of (PL) maps $X \to Y$, 650
$\prec$	partial ordering for strata of a stratified space, 25
$\subset$	subobject, including the possibility of equality, xix
$\times$	cross product or cohomology cross product, 201, 374
$\bowtie$	simplicial cross product, 200, 765
$\amalg$	disjoint union, xviii
$\smile$	cup product, xxi, 375
$\frown$	cap product, xxi, 376
$\pitchfork$	intersection product, xxii, 598
$\wedge$	smash product, 659
a	augmentation map, xx
$\alpha_{\bar{p}}$	map $H^{n-i}(X) \to I^{\bar{p}}H_i^{GM}(X)$, 530, 532
$A^{\bar{p}}S_i(X)$	chain group generated by $\bar{p}$-allowable i-chains in X, 269

$\mathcal{AT}$	category of PL spaces and maps defined in terms of admissible triangulations (compare $\mathcal{PL}$), 42, 751
$\mathfrak{b}$	projection map $E_* \xrightarrow{\mathfrak{b}} C[1]_*$ when E_* is the algebraic mapping cone of $f : C_* \to D_*$, 421, 724
cX	open cone on X, xviii, 23
$\bar{c}X$	closed cone on X, xviii, 23
c_rX	open cone on X of radius r, xviii, 23
$\bar{c}_rX$	closed cone on X of radius r, xviii, 23
$\bar{c}\sigma$	singular cone on the singular simplex σ, 131
$C_*(X)$	simplicial chain complex of X, xx, 39, 91, 166
$C_*^T(X)$	simplicial chain complex of X with respect to the triangulation T, 108
$\mathfrak{C}_*(X)$	PL chain complex of X, xx, 110
$\mathfrak{C}_i^{A,B}$	PL i-chains with support in A and boundary in B, 115
$C[k]_*$	shifted chain complex with $C[k]_i = C_{i-k}$, 421, 723
$\text{codim}_X(Z)$	codimension of Z in X, written $\text{codim}(Z)$ if X is clear, xviii, 21
$\mathcal{D}$	Poincaré duality map given by signed cap product with a fundamental class, xxi, 537, 558
d	coboundary map of a cochain complex, xix
∂_*	connecting morphism of homologically indexed long exact sequence, xx
$\mathbf{d}$	diagonal map, xviii
$\bar{\mathbf{d}}$	algebraic diagonal map, 366, 375
Δ^i	standard geometric i-simplex, xx
$(X, \mathcal{T})$	simplicial subdivision of Δ^i, 161
$\underline{\Delta}^i$	abstract i-simplex, 759
$D\bar{p}$	dual perversity of $\bar{p}$, xxii, 89
$\mathfrak{e}$	inclusion map $D_* \xrightarrow{\mathfrak{e}} E_*$ when E_* is the algebraic mapping cone of $f : C_* \to D_*$, 421, 724
E_*^f	algebraic mapping cone of $f : C_* \to D_*$, 421, 723
ε	cross product map of chain complexes, xxi, 198, 201
ev	evaluation map, 652, 718
$\text{Ext}(A, B)$	Ext group $\text{Ext}_R^1(A, B)$, xix
$F(A)$	torsion-free quotient module of A, 569

$f^!$	umkehr map $f^! : H_*(Y) \to H_*(X)$ associated to an inclusion $f : X \hookrightarrow Y$, 657
$f[k]$	shift of a chain map f by k degrees, 421
G	abelian group, xix
$\mathcal{G}$	bundle of groups, 299
Γ_K	fundamental class over the compact subset $K \subset X$, agrees with Γ_X when $K = X$ if X is compact, 508
Γ_X	fundamental class of X, written Γ if X is clear, xxi, 501, 509
$H_*(X)$	simplicial or singular homology of X, xx, 91
$\mathfrak{H}_*(X)$	PL homology of X, xx, 112
i	often used to denote a topological or algebraic inclusion map, xviii
IAW	intersection Alexander–Whitney map, 368, 371
$I^{\bar{p}}C_*(X)$	simplicial non-GM intersection chain complex of X, xxii, 267
$I^{\bar{p}}C_*^{GM}(X)$	simplicial GM-intersection chain complex of X, xxii, 92
$I^{\bar{p}}C_*^{GM,T}(X)$	simplicial GM-intersection chain complex of X with respect to the triangulation T, 120
$I^{\bar{p}}\mathfrak{C}_*(X)$	PL non-GM intersection chain complex of X, xxii, 267
$I_{\bar{p}}\mathfrak{C}^*(X)$	PL intersection cochain complex of X, 355
$I^{\bar{p}}\mathfrak{C}_*^{\mathcal{V}}(X,A)$	for an open cover $\mathcal{V}$ of X the complex $\sum_{V\in\mathcal{V}} I^{\bar{p}}\mathfrak{C}_*(V, A\cap V)$, 338
$I^{\bar{p}}\mathfrak{C}_*^{GM}(X)$	PL GM-intersection chain complex of X, xxii, 120
id	identity map, xxii, xxiii
$I^{\bar{p}}H_*(X)$	simplicial or singular non-GM intersection homology of X, xxii, 267
$I_{\bar{p}}H^*(X)$	singular intersection cohomology of X, 355
$I_{\bar{p}}H_c^*(X;R)$	intersection cohomology with compact support, 480
$I^{\bar{p}}H_*^{GM}(X)$	simplicial or singular GM-intersection homology of X, xxii, 92, 120, 129
$I^{\bar{p}}\mathfrak{H}_*(X)$	PL non-GM intersection homology of X, xxii, 267
$I_{\bar{p}}\mathfrak{H}^*(X)$	PL intersection cohomology of X, 355
$I^{\bar{p}}\mathfrak{H}_*^{GM}(X)$	PL GM-intersection homology of X, xxii
$I^{\bar{p}}S_*(X)$	singular non-GM intersection chain complex of X, xxii, 267
$I_{\bar{p}}S^*(X)$	singular intersection cochain complex of X, 355
$I^{\bar{p}}S_*'(X)$	first alternative definition of the singular non-GM intersection chain complex of X, 269

$I^{\bar{p}}S''_*(X)$	second alternative definition of the singular non-GM intersection chain complex of X, 271
$I^{\bar{p}}S^{GM}_*(X)$	singular GM-intersection chain complex of X, xxii, 129
$I^{\bar{p}}S^{GM}_*(X,Y)$	relative complex $I^{\bar{p}}S^{GM}_*(X)/I^{\bar{p}_Y}S^{GM}_*(Y)$, 146
$I^{\bar{p}}S^{GM}_*(Y \subset X)$	relative complex $I^{\bar{p}}S^{GM}_*(X) \cap S_*(Y) \subset S_*(X)$, 147
$I^{\bar{p}}S^{\mathcal{V}}_*(X,A)$	for an open cover $\mathcal{V}$ of X the complex $\sum_{V\in\mathcal{V}} I^{\bar{p}}S_*(V, A\cap V)$, 338
$\lvert K\rvert$	underlying space of the simplicial complex K, 38, 740
(K,h)	triangulation of a space X given by a homeomorphism $h: \lvert K\rvert \to X$ from the simplicial complex K, 39
L	link, xix, 28
$\mathbb{L}^\bullet$	connective symmetric L-spectrum, 697
$Lk(x)$	polyhedral link of the point x in a PL space, xix
$\mathscr{L}_m(X)$	mth L-class of X, 652
$L_{\bar{p},D\bar{p}}$ and $L'_{D\bar{p},\bar{p}}$	torsion pairings, 574
$\bar{m}$	lower middle perversity, xxi, 89
μ	Thom class of a bundle, 657
$\bar{n}$	upper middle perversity, xxi, 89
$\mathcal{O}$	orientation sheaf, 499
$\mathcal{O}^{\bar{p}}$	perversity $\bar{p}$ orientation sheaf, 506
$\mathfrak{o}$	orientation section, 500
$\mathfrak{o}^i$	generator of $C_i(\Delta^i)$ consistent with the given orientation, 167
Ω^C_*	bordism groups in the category C, 690
$\omega_{\bar{p}}$	inclusion map $I^{\bar{p}}S_*(X) \to S_*(X)$, 388
$\bar{p}$	an arbitrary perversity, xxi, 87
$\bar{p}_Y$	subspace perversity on Y inherited from perversity $\bar{p}$ on the space containing Y, 147
Φ	the canonical map of the form $A \otimes R \to A$ for a ring R and R-module A, 368
$\mathcal{PL}$	category of PL spaces and maps defined in terms of coordinate maps (compare $\mathcal{AT}$), 42, 747
$\mathfrak{q}$	quasi-isomorphism $H_*(E_*) \to H_*(D_*/C_*)$ when E_* is the algebraic mapping cone of an inclusion $C_* \hookrightarrow D_*$, 422

Q	typical notation for a perversity on a product space, especially a $(\bar{p}, \bar{q})$-compatible perversity, 313
$Q_{\bar{p},\bar{q}}$	maximal $(\bar{p}, \bar{q})$-compatible perversity on a product space, 371
$\hat{Q}^a_{\bar{p},\bar{q}}$	the $(\bar{p}, \bar{q})$-compatible perversity assigned the value $\hat{Q}^a_{\bar{p},\bar{q}}(S \times T) = \bar{p}(S) + \bar{q}(T) + a$ when S, T are singular strata, 444
R	commutative ring with unity, xix
$\mathfrak{s}$	chain complex degree-1 shift map $C[1]_* \to C_*$, 421, 723
SX	(unreduced) suspension of X, xviii
$S_*(X)$	singular chain complex of X, xx, 128
$S_i^{\bar{p}}(X)$	chain group generated by $\bar{p}$-allowable i-simplices of X not contained in Σ_X, 266
Σ_X	singular locus (or singular set) of X, written Σ if X is clear, xviii, 23
$\hat{\sigma}$	singular subdivision of the singular simplex σ, 163
$\bar{t}$	top perversity, xxi, 88
τ	transposition (interchange) map, 212, 392, 717
$\tau_{\bar{p},\bar{q}}$	map $I^{\bar{p}}S_*(X) \to I^{\bar{q}}S_*(X)$ when $\bar{p} \le \bar{q}$, 523
T	triangulation of a PL space, 39
T'	subdivision of the triangulation T, 39
$\mathcal{T}$	family of admissible triangulations of a PL space, 39
$T(A)$	torsion submodule of A, 569
Θ	algebraic map of the form $\mathrm{Hom}(A, R) \otimes \mathrm{Hom}(B, R) \to \mathrm{Hom}(A \otimes B, R)$, 365, 369
$[v_0, \dots, v_i]$	oriented simplex with vertices v_j, 91
$\lceil x \rceil$	round x up to the nearest integer, 89
$\lfloor x \rfloor$	round x down to the nearest integer, 89
$\bar{x}$	element of $C[1]_i$ corresponding to $x \in C_{i-1}$ under the inverse of the shift map $\mathfrak{s}$, i.e. $\mathfrak{s}(\bar{x}) = x$, 421, 723
$\mathfrak{X}$	space X with its intrinsic filtration, xix, 65, 66, 69
$\lvert X \rvert$	underlying topological space of X (typically used to neglect other structures such as filtrations), xviii
$\tilde{X}$	normalization of the pseudomanifold X, 50
X^+	∂-pseudomanifold X with its boundary coned off (i.e. $X^+ = X \cup_{\partial X} \bar{c}(\partial X)$), 183
$X^\bullet$	homogenization of X, 291

X^i	i-skeleton of X, xviii, 20
X_i	the space $X^i - X^{i-1}$, xviii, 21
$(X, \mathcal{T})$	a PL space and its family of admissible triangulations, 39
$[\xi]$	equivalence class of ξ; the equivalence relation varies by context, xix
$\lvert\xi\rvert$	degree of $\lvert\xi\rvert$ or support of $\lvert\xi\rvert$, depending on context, xx, 91
$\hat{\xi}$	singular subdivision of the singular chain ξ, 168

Index

KEY: Numbers in **bold** indicate where terms are defined